Software Defined Radio using MATLAB® & Simulink® and the RTL-SDR

Software Defined Radio using MATLAB® & Simulink® and the RTL-SDR

Robert W. Stewart

Kenneth W. Barlee

Dale S. W. Atkinson

Louise H. Crockett

Department of Electronic and Electrical Engineering

University of Strathclyde

Glasgow, Scotland, UK

1st Edition *(revised)*

First published September 2015 by Strathclyde Academic Media. This revised edition published October 2017. Version 1.17.1023.

Copyright © Robert W. Stewart, Kenneth W. Barlee, Dale S. W. Atkinson, and Louise H. Crockett.

Book PDF and Paper Materials Licence to Use and Reproduce for Teaching, Learning and Academic Purposes

This book is available for free as an electronic book (PDF format). (A printed version is available for purchase from Amazon and other retailers and from www.desktopSDR.com.) Text and diagrams from this book may be reproduced, partially or in their entirety and used in a manner consistent with applicable law. A clear reference to the original source must be made in all documents using text or diagrams from the book. The reference should be of the following form:

R. W. Stewart, K. W. Barlee, D. S. W. Atkinson, and L. H. Crockett, *Software Defined Radio using MATLAB & Simulink and the RTL-SDR*, Published by Strathclyde Academic Media, 2015 (Softback ISBN 9780992978716, Hardback ISBN 9780992978723).

Requests to use content from this book for purposes other than non-profit academic use should be directed to: info@desktopSDR.com. This book may *not* be reproduced in paper, PDF or other readable form and may not be sold or re-sold by any unauthorised third party.

Example Files and Software

Simulation examples, design exercises and files and associated software MATLAB and Simulink resources that accompany this book are available as a download from the book's companion website: www.desktopSDR.com or may be variously acquired by other means such as via USB storage, cloud storage, disk or any other electronic or optical or magnetic storage mechanism. These files and associated software may be used subject to the terms of the software license agreement which is reproduced on page ii, and referenced in each MATLAB and Simulink file.

For information, the files and associated software may be updated from time to time in order to address bugs, provide updates, or for version compatibility reasons. These updates will be made available from the book's companion website: www.desktopSDR.com.

Music, Vocal, Audio and Modulated Audio Files

The music and vocal files used within the Examples files and software within the book were variously written, arranged, performed, recorded and produced by Garrey Rice, Adam Struth, Jamie Struth, Iain Thistlethwaite and also Marshall Craigmyle who collectively, and individually where appropriate, assert and retain all of their copyright, performance and artistic rights. Permission to use and reproduce this music is granted for all purposes associated with MATLAB and Simulink software and the simulation examples and design exercises files that accompany this book. Requests to use the music for any other purpose should be directed to: info@desktopSDR.com. For information on music track names, full credits, and links to the musicians please refer to www.desktopSDR.com/more/audio.

Proper Use of the Radio Frequency (RF) Spectrum

Some simulation examples and design exercises in this book describe the use of hardware capable of radio frequency (RF) signal transmission and reception. Use of the RF spectrum is subject to regulations and restrictions that vary by country and geographical region. In all countries persons may only transmit on bands for which they have the appropriate licence or permission, or bands which are designated for access on an unlicensed or perhaps lightly licensed basis. RF transmissions must also meet all conditions of the use of that frequency band. Please refer to your local Government or other appropriate spectrum regulator for further information on RF reception and transmission in your country. In addition to restrictions on transmission, receiving and /or recording and/or attempting to decode certain types of RF signal may be restricted or illegal in some countries. It is the responsibility of the individual to comply with all spectrum access regulations and information laws that apply in their region or country and every effort should be made to confirm legal transmission and reception in any RF band.

Warning and Disclaimer

The material included is provided on an 'as is' basis in the best of faith, and with the best of educational intentions. Neither the authors nor publishers make any warranty of any kind, expressed or implied, with regard to the documentation or other content contained in this book. The authors and publisher shall not be held liable for any loss or damage resulting directly or indirectly from any information contained herein; neither shall the authors and publisher be held responsible for the non-compliance of any readers with their local RF spectrum access regulations or information laws.

Trademarks

MATLAB and Simulink are registered trademarks of MathWorks, Inc. in the United States and other countries.

Microsoft and Windows are registered trademarks of Microsoft Corporation in the United States and other countries.

Apple and OS X are registered trademarks of Apple Inc. in the United States and other countries.

Linux is the registered trademark of Linus Torvalds in the United States. and other countries.

Raspberry Pi is a trademark of the Raspberry Pi Foundation in the United Kingdom and other countries.

USRP, USRP2, UHD, and Ettus Research are trademarks of National Instruments Corp.

All other trademarks used in this book are acknowledged as belonging to their respective companies. The use of trademarks in this book does not imply any affiliation with, or endorsement of, this book by trademark owners.

Book and Support File Information

Obtaining a Printed Copy of this Book

If you are reading the PDF version of this book and wish to obtain a printed copy, you can buy one from amazon.com. Details about other distributors are provided at:

www.desktopSDR.com/print_version

Support File Download Information

A set of files accompany this book. These files provide the Simulink models, MATLAB scripts and data files that you require to work through the examples in this book. They have been compressed into a ZIP folder, which is around 1.5GB in size. It can be downloaded either from the book's accompanying website or directly from the MathWorks site, along with a PDF of this book. Links for the two are as follows:

www.desktopSDR.com
www.mathworks.com/SDR

You can also sign up to a mailing list on our website to keep up to date with the book and tutorials.

Errata

Any items of errata arising will be published at:

www.desktopSDR.com

Software, Simulation Examples and Design Exercises Licence Agreement

This licence agreement refers to the simulation examples, design exercises and files, and associated software MATLAB and Simulink resources that accompany the book:

Software Defined Radio using MATLAB & Simulink and the RTL-SDR
First published by Strathclyde Academic Media, 2015. This revised edition published 2017
Authored by Robert W. Stewart, Kenneth W. Barlee, Dale S.W. Atkinson, and Louise H. Crockett

and made available as a download from, www.desktopSDR.com or variously acquired by other means such as via USB storage, cloud storage, disk or any other electronic or optical or magnetic storage mechanism. These files and associated software may be used subject to the terms of agreement of the conditions below:

Copyright © 2015-2017 Robert W. Stewart, Kenneth W. Barlee, Dale S.W. Atkinson, and Louise H. Crockett
All rights reserved.

Redistribution and use in source and binary forms, with or without modification, are permitted provided that the following conditions are met:

1. Redistributions of source code must retain the above copyright notice, this list of conditions and the following disclaimer.
2. Redistributions in binary form must reproduce the above copyright notice, this list of conditions and the following disclaimer in the documentation and/or other materials provided with the distribution.
3. Neither the name of the copyright holder nor the names of its contributors may be used to endorse or promote products derived from this software without specific prior written permission.
4. In all cases, the software is, and all modifications and derivatives of the software shall be, licensed to you solely for use in conjunction with The MathWorks, Inc. products and service offerings.

THIS SOFTWARE IS PROVIDED BY THE COPYRIGHT HOLDERS AND CONTRIBUTORS "AS IS" AND ANY EXPRESS OR IMPLIED WARRANTIES, INCLUDING, BUT NOT LIMITED TO, THE IMPLIED WARRANTIES OF MERCHANTABILITY AND FITNESS FOR A PARTICULAR PURPOSE ARE DISCLAIMED. IN NO EVENT SHALL THE COPYRIGHT HOLDER OR CONTRIBUTORS BE LIABLE FOR ANY DIRECT, INDIRECT, INCIDENTAL, SPECIAL, EXEMPLARY, OR CONSEQUENTIAL DAMAGES (INCLUDING, BUT NOT LIMITED TO, PROCUREMENT OF SUBSTITUTE GOODS OR SERVICES; LOSS OF USE, DATA, OR PROFITS; OR BUSINESS INTERRUPTION) HOWEVER CAUSED AND ON ANY THEORY OF LIABILITY, WHETHER IN CONTRACT, STRICT LIABILITY, OR TORT (INCLUDING NEGLIGENCE OR OTHERWISE) ARISING IN ANY WAY OUT OF THE USE OF THIS SOFTWARE, EVEN IF ADVISED OF THE POSSIBILITY OF SUCH DAMAGE.

Table of Contents

Book and Support File Information i

Foreword xi

A Few SDR Thoughts xiii

Preface xv

Acknowledgements xix

1 Introduction 1
    1.1 Real Time Desktop Software Defined Radio 2
    1.2 What is the RTL-SDR? 3
    1.3 What Do I Need to Get Started? 5
    1.4 The Aim and Objectives of this Book 7
    1.5 The Evolution of the Software Defined Radio Architecture 8
    1.6 RTL-SDR Hardware 10
    1.7 Interfacing with the RTL-SDR from MATLAB and Simulink 17
    1.8 Practicalities and Some Challenges of (Low Cost) Desktop SDR . 20
    1.9 Working with Discrete and Continuous Time Signals and Equations 22
    1.10 The Structure of the Book and Format of the Exercises 22

2 Open the Box! First SDR with MATLAB and Simulink 25
    2.1 Getting Started: Hardware and Software Checklist 25
    2.2 Getting Started: Installing the RTL-SDR Hardware Support Package 27
    2.3 Getting Started: Book Support Files and the MATLAB Environment 31
    2.4 Running the First Desktop RTL-SDR Receiver Designs 34
    2.5 Summary 42

3 Radio Frequency Spectrum Viewing 43
    3.1 Different Signals, Different Frequencies 44
    3.2 Spectrum Usage and Allocations Around the Globe 45
    3.3 Working with a Suitable Antenna 46
    3.4 Go Forth and Explore the Spectrum! 49
    3.5 Spectral Viewing — Spectrum Analyser and Waterfall Plots 52
    3.6 Spectral Viewing — RTL-SDR Tuner GUI Controls 53
    3.7 Engineering Requirements — Eyeball Radio Tuning & More 54
    3.8 FM Radio Stations 60
    3.9 Mobile (Cell) Phone Signals—2G, 3G and 4G 64
    3.10 433MHz: Key Fobs and Wireless Sensors 75
    3.11 Digital Video & Audio Signals 79
    3.12 Using Multiple RTL-SDRs 83
    3.13 Sweeping the Spectrum: Receiving from 25MHz to 1.75GHz 89
    3.14 Summary 98

4 Getting Started with MATLAB and Simulink 99
    4.1 Introducing MATLAB 100

	4.2	MATLAB Functions	107
	4.3	Plotting in MATLAB	111
	4.4	MATLAB Arrays, Matrices, and Structures	115
	4.5	MATLAB System Objects	121
	4.6	Introducing Simulink	125
	4.7	Creating Simulink Models	127
	4.8	Variables and Parameters	142
	4.9	Generating Frequency Domain Plots	150
	4.10	Sampling Rates, Samples and Frames	152
	4.11	Data Types	159
	4.12	Working with Input and Output Files	161
	4.13	Saving and Re-importing RTL-SDR Data	165
	4.14	Summary	170
5	**Complex Signals, Spectra and Quadrature Modulation**		**171**
	5.1	Real and Complex Signals — it's all Sines and Cosines	172
	5.2	Viewing Real Signals in the Frequency Domain via Complex Spectra	173
	5.3	Standard Amplitude Modulation	182
	5.4	Quadrature Modulation and Demodulation (QAM)	187
	5.5	Quadrature Amplitude Modulation using Complex Notation	191
	5.6	Quadrature Amplitude Demodulation using Complex Notation	192
	5.7	Spectral Representation for Complex Demodulation	195
	5.8	Frequency Offset Error and Correction at the Receiver	199
	5.9	Frequency Correction using a Complex Exponential	199
	5.10	RTL-SDR Quadrature / Complex Architecture	201
	5.11	Summary	201
6	**Amplitude Modulation (AM) Theory and Simulation**		**203**
	6.1	Amplitude Modulation — An Introduction	203
	6.2	AM-DSB-SC: Double Sideband Suppressed Carrier AM	203
	6.3	AM-DSB-TC: Double Sideband Transmitted Carrier AM	210
	6.4	AM-SSB: Single Sideband AM	217
	6.5	AM-VSB: Vestigial Sideband AM	225
	6.6	Theoretical AM Demodulation	227
	6.7	Receiving and Downconverting AM-DSB-TC Signals to Complex Baseband	227
	6.8	Non-Coherent AM Demodulation: The Envelope Detector	231
	6.9	Summary	234
7	**Frequency Tuning and Simple Synchronisation**		**235**
	7.1	Selecting a Frequency Band: Tuning	235
	7.2	The Synchronisation Problem	238
	7.3	Demodulation of AM Signals	241
	7.4	Coherent Demodulation and Carrier Synchrony	242
	7.5	Introduction to Phase Locked Loops	245
	7.6	Discrete Time PLL Model	252
	7.7	PLL Behaviours, Parameters and Characteristics	256

Table of Contents v

    7.8   PLL Design 264
    7.9   PLL Performance in Noise 272
    7.10 Carrier Synchronisation 273
    7.11 Summary 277

8 Desktop AM Transmission and Reception 279
    8.1   Transmitting AM Signals with a USRP® Radio 280
    8.2   Implementing Non-Coherent AM Receivers with the RTL-SDR......... 290
    8.3   Implementing Coherent AM Receivers with the RTL-SDR 310
    8.4   Audio Multiplexing with the USRP® and RTL-SDR Hardware 316
    8.5   Alternative Hardware for Generating Desktop AM Signals 326
    8.6   Summary 328

9 Frequency Modulation (FM) Theory and Simulation 329
    9.1   The History of the FM Standard 329
    9.2   The Mathematics of FM & the Modulation Index 331
    9.3   FM Signal Bandwidth 335
    9.4   FM Demodulation Using Differentiation........................... 347
    9.5   Receiving and Downconverting FM Signals to Complex Baseband....... 348
    9.6   Non-Coherent FM Demodulation: The Complex Differentiation
          Discriminator 350
    9.7   Non-Coherent FM Demodulation: The Complex Delay Line
          Discriminator 355
    9.8   Coherent FM Demodulation: The Phase Locked Loop 358
    9.9   Demodulating Signals from Commercial FM Radio Stations............ 361
    9.10 Summary 366

10 Desktop FM Transmission and Reception............................. 367
    10.1 Transmitting Mono WFM Signals with the USRP® Hardware 368
    10.2 Implementing Mono FM Receivers with RTL-SDR and Simulink........ 376
    10.3 Transmitting Stereo WFM Signals with the USRP® Hardware........... 390
    10.4 Implementing Stereo FM Receivers with RTL-SDR and Simulink 396
    10.5 Manipulating the MPX: Transmitting AM Signals with FM Transmitters. 409
    10.6 Manipulating the MPX: Audio Multiplexing with FM Transmitters...... 415
    10.7 Alternative Hardware for Generating Desktop FM Signals................ 423
    10.8 Summary 425

11 Digital Communications Theory and Simulation 427
    11.1 Digital Modulation Schemes 427
    11.2 Pulse Shaping 437
    11.3 Digital Up and Downconversion................................. 442
    11.4 Carrier Synchronisation 448
    11.5 Timing Errors and Symbol Recovery 458
    11.6 Symbol Timing Synchronisation................................. 464
    11.7 Digital Receiver Design: Joint Carrier and Timing Synchronisation...... 477
    11.8 Coarse Frequency Synchronisation............................... 480
    11.9 Phase Ambiguity 488
    11.10 Differential Encoding and Decoding 489

11.11 Synchronisation with a Unique Word 499
11.12 Summary.. 502

12 Desktop Digital Communications: QPSK Transmission and Reception 503
12.1 Pulse Shaping with Real Time QPSK Transmitter and Receiver Designs. . 504
12.2 Coarse Frequency Synchronisation in a Real-time System.............. 512
12.3 Carrier and Timing Synchronisation with the RTL-SDR 516
12.4 Developing a Simple Communications Protocol 523
12.5 ASCII Encoding and Decoding.. 525
12.6 Data and Frame Synchronisation 528
12.7 ASCII Message Transmission and Reception 544
12.8 Transmitting Images Across the Desktop 551
12.9 Transmitting Data Using FM Transmitters 556
12.10 Summary.. 566

Appendix A: Hardware Setup 569
A.1 The RTL-SDR Hardware Support Package 569
A.2 The USRP® Hardware Support Package................................ 571
A.3 RTL-SDR Frequency Error Correction 577

Appendix B: Common Equations 581

Appendix C: Digital Filtering and Multirate................................ 583
C.1 Filter Classes and Characteristics 583
C.2 Filter Specification and Design 584
C.3 FIR Filter Processing Architecture 586
C.4 Computation and Trade-offs... 586
C.5 Multirate Filtering: The Motivation 587
C.6 Decimation... 588
C.7 Interpolation 590

Appendix D: PLL Design.. 593
D.1 Digital Type 2 PLL Linear Model and Z-Domain Transfer Function 593
D.2 Analogue Type 2 PLL Linear Model and S-Domain Transfer Function... 596
D.3 Extraction of Digital PLL Parameters Based on Analogue PLL Equivalence.. 597
D.4 Phase Detector Gain 602
D.5 Oscillator Gain.. 605

Appendix E: AM and FM Transmitters 607
E.1 Upconverting AM Radio Signals with the Ham It Up.................. 607
E.2 Building an 'RT4' 433.9MHz AM Transmitter.......................... 612
E.3 Using the Raspberry Pi as an FM Transmitter 619

References 629

List of Acronyms 635

Index 641

Software Defined Radio Simulation Examples & Design Exercises

This section lists the titles of the various simulation exercises and/or examples and their page locations in this workbook.

Please note that the licence given on page ii applies to all MATLAB & Simulink Simulation Examples and Design Exercises presented in this book, and listed below.

1   Introduction 1
2   Open the Box! First SDR with MATLAB and Simulink 25
    - Verify Software Setup: MATLAB and Simulink 26
    - Verify Software Setup: RTL-SDR Hardware Support Package 28
    - Verify Hardware Setup: RTL-SDR Hardware Support Package 29
    - MATLAB (and Simulink) Working Environment Setup 32
    - First Use of the RTL-SDR: Simulink 34
    - First Use of the RTL-SDR: MATLAB 37
3   Radio Frequency Spectrum Viewing 43
    - Opening the Spectrum Viewing Receiver Model 49
    - An Introduction to Tuning 54
    - Changing the Tuner Gain of the RTL-SDR 59
    - Searching for FM Radio Stations 61
    - Exploring the Mobile Spectrum: 2G GSM (800–1000MHz) 64
    - Exploring the Mobile Spectrum: 3G UMTS (800–900MHz) 66
    - Exploring the Mobile Spectrum: 4G LTE (700–900MHz) 68
    - Exploring the Mobile Spectrum: Challenges! 70
    - Searching for Key Fob and Wireless Sensor Signals 75
    - Searching for DVB-T Digital TV Signals 80
    - Searching for DAB Digital Audio Signals 82
    - Exploring the Spectrum with Multiple RTL-SDRs 84
    - Sweeping the Radio Frequency Spectrum: 25MHz to 1.75GHz 90
4   Getting Started with MATLAB and Simulink 99
    - MATLAB Orientation and Using the Command Window 100
    - MATLAB Scripts 105
    - Functions in MATLAB 108
    - Writing Your Own Functions 109
    - MATLAB Figures 111
    - MATLAB Arrays 116
    - Matrices in MATLAB 118
    - MATLAB Structures 120
    - An Introduction to System Objects 121
    - Simulink Orientation and the Simulink Library 125
    - Building and Simulating a First System 127
    - Manipulating Blocks and Wires 137

- Commenting Blocks Out, and Commenting Through Blocks 141
- Working with Variables and Parameters 142
- Setting up Variables as a Model Loads 145
- Writing Simulation Results to the Workspace 147
- Frequency Domain Plots: Spectrum Analyzer 151
- Upsampling and Downsampling 153
- Conversion Between Samples and Frames 158
- Real and Complex Data Types 159
- Write and Read .mat Data Files (General Case) 161
- Read and Write Audio Files 164
- Saving and Importing RTL-SDR Data in MATLAB and Simulink 166

5 Complex Signals, Spectra and Quadrature Modulation 171
- Generating Frequency Domain Plots with MATLAB Code 175
- Generating Frequency Domain Plots with Simulink 176
- Plotting Complex Spectra 177
- Complex Spectra for Three Real Sine Waves 181
- Modulation, Demodulation and Frequency Correction 194
- Complex Demodulation of a Signal 198
- Frequency Correction using Complex Exponential 199

6 Amplitude Modulation (AM) Theory and Simulation 203
- AM-DSB-SC Simulation 207
- AM-DSB-TC Simulation 214
- AM-SSB Simulation 220

7 Frequency Tuning and Simple Synchronisation 235
- 'Perfect' Modulation and Demodulation 243
- Modulation and Demodulation (out of synchrony) 244
- Phase Detector 253
- Loop Filters 254
- Phase Detector and Loop Filter 254
- Type 2 PLL 255
- PLL Linear Model: Steady State Error 263
- PLL Linear Model: Effect of Damping Ratio 263
- Challenge: Design of a Type 2 PLL 270
- Type 2 PLL: Performance in Noise 272
- Coherent Receiver for AM-DSB-TC 274
- Coherent Receiver for AM-DSB-SC (Costas Loop) 276

8 Desktop AM Transmission and Reception 279
- USRP® Radio: AM-DSB-SC Modulator and Transmitter 281
- USRP® Radio: AM-DSB-TC Modulator and Transmitter 285
- USRP® Radio: AM-SSB Modulator and Transmitter 288
- RTL-SDR: Envelope Detector for AM-DSB-TC Signals 292
- RTL-SDR: MATLAB Envelope Detector for AM-DSB-TC Signals 299
- RTL-SDR: Envelope Detector for AM-DSB-SC Signals 304
- RTL-SDR: Complex Demodulator for USRP® AM-SSB Signals 307

- RTL-SDR: PLL Demodulator for AM-DSB-TC Signals 311
- RTL-SDR: Costas Demodulator for AM-DSB-SC Signals 314
- FDM AM: FDM MPX'er, AM Modulator and USRP® Transmitter 319
- FDM AM: RTL-SDR AM Receiver and Demultiplexer 322

9 Frequency Modulation (FM) Theory and Simulation 329
- Narrowband FM (NFM) Simulation 336
- Wideband FM (WFM) Simulation 343
- Stereo FM Encoder and Multiplexer Simulation 365

10 Desktop FM Transmission and Reception 367
- USRP® Radio: Mono FM Modulator and Transmitter 370
- RTL-SDR: Mono FM Radio Receiver (Discriminator) 376
- RTL-SDR: MATLAB Mono FM Radio Receiver (Discriminator) 384
- RTL-SDR: Mono FM Radio Receiver (Complex Differentiation) 388
- USRP® Radio: Stereo FM Modulator and Transmitter 391
- RTL-SDR: Stereo FM Radio Receiver and Decoder (Discrim) 397
- RTL-SDR: Stereo FM Radio Receiver and Decoder (PLL) 405
- RTL-SDR: Stereo FM Radio Receiver (Slope Detector) 407
- AM in FM: Multiplexer, Modulator and USRP® Transmitter 410
- AM in FM: RTL-SDR FM Receiver and AM Demodulator 412
- FDM FM: FDM & FM MPXer, Mod and USRP® Transmitter 417
- FDM FM: RTL-SDR FM Receiver and Demultiplexer 420

11 Digital Communications Theory and Simulation 427
- Bit to Symbol Mapping and Demapping (QPSK) 432
- QPSK Symbol Mapping and Demapping: Separate I and Q 433
- Bit to Symbol Mapping and Demapping in Noise (QPSK) 435
- 16-QAM Symbol Mapping and Demapping 436
- Higher Order Constellations 437
- Pulse Shaping and Transmission Bandwidth 439
- Raised Cosine Inter-Symbol-Interference Properties 440
- RRC Inter-Symbol-Interference Properties 441
- Matched Filtering of QPSK Modulated Data 442
- Digital Upconverter (DUC): Filter Cascade & Modulation 446
- Digital Downconverter (DDC): Demodulation & Filter Cascade 447
- Carrier Synchronisation for QPSK (Demodulation) 453
- Carrier Synchronisation for QPSK (Baseband) 457
- Matched Filtering and Maximum Effect Points 458
- Symbol Timing Imperfections: Sampling Phase Error 461
- Symbol Timing Imperfections: Sampling Frequency Error 461
- Symbol Decisions: Sampling Phase Error 462
- Symbol Decisions: Sampling Frequency Error 463
- Numerically Controlled Clock (NCC) 469
- Early Late Timing Synchronisation (Raised Cosine Pulses) 475
- Early Late TED: Gain Coefficient 476
- Early Late Timing Synchronisation (Design Task) 476

- Joint Carrier and Timing Synchronisation for QPSK 479
- Coarse Frequency Correction 485
- QPSK Synchronisation (with Coarse Frequency Correction) 486
- Implementation of a BPSK Differential Encoder & Decoder 491
- Implementing a QPSK Differential Encoder & Decoder 497

12 Desktop Digital Communications: QPSK Transmission and Reception 503
- RRC Transmit Pulse Shaping with the USRP® Radio 505
- RRC Matched Filtering in an RTL-SDR Receiver Model 509
- Coarse Frequency Correction: Inspecting the Transmitter 513
- Coarse Frequency Correction: Investigation with the Receiver 513
- Carrier & Timing Synchronisation: Inspecting the Transmitter 516
- RTL-SDR and Theory Synchronisation Comparison 517
- Further Investigation of Real-time RTL-SDR Synchronisation 520
- ASCII Encoding using MATLAB Code 525
- ASCII Decoding using MATLAB Code 527
- Numbered ASCII Frame Generator for Transmitter Designs 537
- Frame Synchronisation using a Matched Filter 541
- ASCII Message Tx Rx: USRP® Transmitter 544
- ASCII Message Tx Rx: RTL-SDR Receiver 545
- Image Tx Rx: USRP® Transmitter 551
- Image Tx Rx: RTL-SDR Receiver 553
- Data in FM: ASCII/ Audio Signal Generator 558
- Data in FM: RTL-SDR FM Demod & ASCII/ Audio Receiver 560
- Data in FM: Image/ Audio Signal Generator 563
- Data in FM: RTL-SDR FM Demod & Image/ Audio Receiver 564

Appendix A: Hardware Setup 569
- Verify Software Setup: USRP® Hardware Support Package 571
- Verify Hardware Setup: USRP® Hardware Support Package 573
- Finding the PPM Error of your RTL-SDR 577

Appendix B: Common Equations 581

Appendix C: Digital Filtering and Multirate 583

Appendix D: PLL Design 593

Appendix E: AM and FM Transmitters 607
- Ham It Up: Hardware Setup 607
- RTL-SDR: Envelope Detector for HIU AM-DSB-TC Signals 610
- Build the RT4 AM Transmitter 613
- Test the RT4 AM Transmitter 617
- Refine your RT4 AM Transmitter 617
- PiFM: Backing Up A Blank SD Card 619
- PiFM: Setting up a Raspberry Pi FM Radio Station 621
- PiFM: Restoring the SD Card 627

Foreword

I have a DSP book on my book shelf. I have a lot of DSP books on my book shelf. More than a 100, probably every one that's ever been printed! What's that in your hand? Oh, you are holding and examining a new DSP book. A copy of this one will soon find its way to my book shelf. Many of the DSP books tell us that they are going to teach us how to design something; something like a software defined radio. I want to share with you that there is something different about this one. First let me tell you what's in all the others. They all contain an introduction chapter telling you where and how DSP is used. Then a chapter on Z-Transforms, followed by one on region of convergence of Z-Transforms, then one on linear systems, and one on spectral analysis, one of Finite Duration Impulse Response filters, one on Infinite Duration Impulse Response filters, Oh, and Fourier transforms, oops, one on phase lock loops, ah I almost missed one, RF propagation and channel modeling, and on and on and on. By now we are starting to lose sight of the path through the woods. When we reach the end of the book we are a bit startled to realize we never actually touch a radio. We learned how to simulate one, a very useful skill; but the fact is we never actually touch a radio. We might ask, "Is *there a radio in here somewhere? Must we caress every mathematical skill before we get to see the forest?*" Till now, it seems we had to!

This book is different, very different. It starts with an introduction to an inexpensive radio; a radio designed to be a television tuner, but thanks to some creative hackers and programmers, it has been directed to a new calling.... an SDR receiver. Right up front we are presented to a simple flexible radio plugged into your computer's USB port. "*Hello reader, let me introduce you to a feely touchy hardware radio, the RTL-SDR. Pleased to meet you Ms. Radio!*" Then we are presented with a description of what's in a typical radio and what is in this particular radio. We then run through a list of software and the tools required to operate it. We run through step by step preparation instructions required to have MATLAB and Simulink engage the radio. Amazingly we then operate the radio. All by the end of chapter 2. Not a bad way to start a book on DSP for software defined radios. Pretty neat in fact! I have no excuses any more. I have to acquire some of the RTL-SDR radios and then devour this book to learn how to play with these radios while folding them into my modem design class! Why don't you join me and do the same? Remember my advice: "If you are not having fun, you are probably not doing it right!"

Best regards and have fun with this innovative book and its message.

fred harris,   Professor, San Diego State University, California

Bernie Sklar,   Communications Engineering Services, Tarzana, California   May 2015

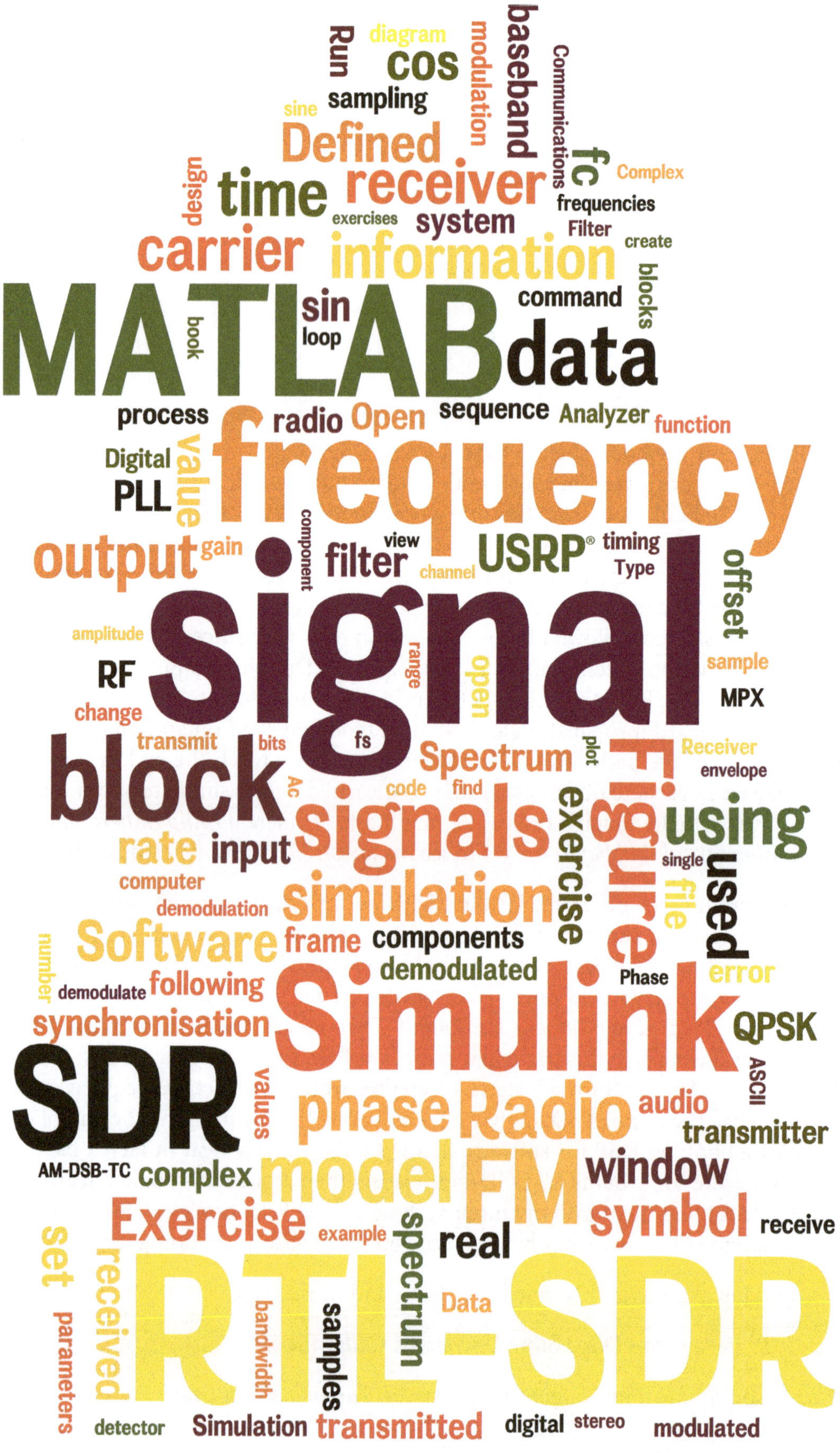
Run
diagram
cos
modulation
baseband
Communications
sine
sampling
Defined
design
time
receiver
fc
Complex
frequencies
exercises
system
Filter
carrier
information
create
MATLAB
sin
command
blocks
book
loop
data
process
radio
Open
sequence
Analyzer
function
Digital
value
frequency
PLL
component
output
gain
view
filter
USRP®
timing
Type
offset
channel
signal
amplitude
range
open
sample
RF
MPX
change
transmit
bits
fs
plot
Receiver
envelope
block
Spectrum
Ac
code
find
Figure
exercise
using
signals
rate
input
single
computer
used
simulation
file
demodulation
Software
frame
components
number
demodulated
Phase
error
demodulate
following
Simulink
QPSK
synchronisation
SDR
ASCII
values
phase
Radio
audio
transmitter
AM-DSB-TC
complex
model
FM
window
Exercise
example
symbol
receive
set
spectrum
real
received
Data
RTL-SDR
parameters
bandwidth
samples
detector
Simulation
transmitted
digital
stereo
modulated

A Few SDR *Thoughts*

Many of today's leading engineers in the communications industry may have begun their communications careers back in the 1960s and 1970s by building a crystal radio set, and perhaps listened to radio broadcasts while tucked under their pillows late at night, using their very own engineered radio. The ability to make a radio receiver that was portable, cheap and really worked was an amazing motivator for the aspiring engineer. The key element for the popularisation of the crystal radio was perhaps that it was simple and rewarding.

For the last 20 years, Software Defined Radio (SDR) has been a topic of great interest and extensive research and development. Developing hardware solutions has been the preserve of those who could afford turnkey systems and custom designs costing thousands of dollars or pounds. But not now; with the advent of devices such as the RTL-SDR, we just might be seeing a new motivation for the leading communications engineers of 2040 (who will perhaps be designing 10G radio by that time!). At the time of writing this book, the RTL-SDR costs around $15—and that includes an antenna; it only costs a few dollars to get started with a digital radio. With the ability to tune over the frequency range of 25MHz to 1.75GHz, an aspiring engineer can use this to receive a variety of radio signals. Using this simple device and the MathWorks softwares tools, one can implement FM receivers, spectral viewing designs and even construct systems to decode the ID of a nearby cellular base station.

Perhaps mimicking the excitement and discovery of the crystal radio hobbyist of 40 years ago, today you can wave your RTL-SDR antenna in the air, and get a simple receiver working in just a few lines of MATLAB code or a few blocks in Simulink, on your laptop. Casual experimentation can also teach you surprising things! For example in a recent class, students were designing FM receivers and were able to receive signals transmitted from local radio stations with good quality. One experimental student tried to demodulate an FM signal using a design intended for AM signal demodulation. And it worked! The audio was not high quality but was intelligible and the listener could interpret speech and recognise music. The teacher, who initially thought the student was plainly not thinking, did a 180-degree reversal when this was observed and realised this student really was thinking and had in fact implemented a form of SDR FM slope detector! (We added this example to the book, see page 407.)

We hope this book will allow you to quickly get started with software-defined radio. The early chapters ensure you learn what software is required — this can be the Professional, Student or Home Version of MATLAB & Simulink, with the Signal Processing, Communications and DSP System Toolboxes and Hardware Support Packages for the RTL-SDR. These chapters guide you toward getting started quickly for a very rewarding 'out of the box' experience. Where possible, examples work with off-the-air RF signals, but we also give a few examples of using inexpensive FM transmitters that plug into your phone, and some other simple transmitter circuits you can build. More involved designs in later chapters, such as a full QPSK digital communications system, require you to transmit from more sophisticated SDR hardware but should be readily available at universities and on some hobbyist benchtops, such as the Ettus Research USRP® hardware, ZedBoard (featuring the Xilinx *Zynq* System-on-Chip) with an Analog Devices FMCOMMS RF card, all of which are supported by MATLAB & Simulink and allow you to design and implement complete wireless systems.

The authors of the book and their engineering colleagues at MathWorks are confident that once you plug the RTL-SDR into your computer and start searching the local RF spectrum, you will be hooked. You might even become a *spectrum detective*—you find a signal, you start thinking about what kind of signal it is, and then begin to take a closer look with spectrum analysers, algorithms and receivers that you design! Or you may begin building simple receivers and start manipulating parameters, filter characteristics, synchronisers and so on.

The 1960s Crystal Radio Ham

The 2015 Software Defined Radio Ham

Unlike the crystal radios of 40 or more years ago, it may be hard with laptop in hand to listen to a station with radio tucked under your pillow, but perhaps the next-generation communications engineer can at least lean on his or her pillow, and code late into the night to receive radio transmissions from broadcast stations, weather satellites, time clocks, beacons and so on. So an update to this SDR text seems unavoidable in the near term as the momentum of change from the Internet of Things (IoT) and various related applications drives the market forward.

Stay tuned ...the wireless revolution is just beginning!

Don Orofino, Director of Signal Processing, MathWorks, USA

Bob Stewart,   MathWorks Professor of Signal Processing, University of Strathclyde, Scotland, UK

Preface

The availability of the RTL-SDR device for less than $20 brings Software Defined Radio (SDR) to the home and work desktops of EE students, professional engineers and the maker community. The RTL-SDR can be used to acquire and sample RF (radio frequency) signals transmitted in the frequency range 25MHz to 1.75GHz, and the MATLAB and Simulink environment can be used to develop receivers using first principles DSP (digital signal processing) algorithms. Signals that the RTL-SDR hardware can receive include: FM radio, UHF band signals, ISM signals, GSM, 3G and LTE mobile radio, GPS and satellite signals, and any that the reader can (legally) transmit of course! In this book we introduce readers to SDR methods by viewing and analysing downconverted RF signals in the time and frequency domains, and then provide extensive DSP enabled SDR design exercises which the reader can learn from. The hands-on SDR design examples begin with simple AM and FM receivers, and move on to the more challenging aspects of PHY layer DSP, where receive filter chains, real-time channelisers, and advanced concepts such as carrier synchronisers, digital PLL designs and QPSK timing and phase synchronisers are implemented. In the book we will also show how the RTL-SDR can be used with SDR transmitters to develop complete communication systems, capable of transmitting payloads such as simple text strings, images and audio across the lab desktop.

Accessible SDR

Designing real world SDR receivers for analogue and digital communications systems based on advanced DSP and digital communications theory is far from trivial (and of great educational value [6], [11]!), but gaining practical experience was, unfortunately, all the more difficult given the high cost of FPGA based SDR cards and systems. However with low cost and programmable RTL-SDR hardware that is easy to use and integrate with technical programming environments such as MATLAB and Simulink, then anyone who wants to implement and learn SDR by experiencing and doing, now can.

In this book we are also addressing another challenge and aiming to make real time SDR design accessible to readers from many different backgrounds and experience levels. We do of course realise that not everyone has extensive DSP and communications theory experience; nor that everyone has the intention of becoming a DSP/ SDR system designer. Many student and professional engineer readers perhaps just want to use the RTL-SDR at a system level, to learn and just receive RF signals (i.e. to use SDR implementations as parameterisable black boxes or as simple spectrum analysers). Whereas others would like to design and implement complete physical layers for digital communications receivers.

In order to make more rapid progress with this book, the fundamental theory and background prerequisites for readers are mainly in DSP. Therefore a working knowledge of the following will be useful, and are things we do *not* cover from first principles in the book, but found in many standard texts:

1. The ***Nyquist sampling theorem*** and time domain signal representation;
2. The ***frequency domain***, or spectral representation of signals using DFTs and FFTs;
3. Basic design and implementation parameters of filtering, ***digital filters*** and ***multirate***;
4. The generics of the ***radio frequency (RF) spectrum*** and using an antenna to receive RF signals.

Experience and Background of Different User Groups

We envisage a few of the groups of readers with different backgrounds using this book and running the real world SDR receiver exercises herein to be:

1. **Beginner Level;**
   No background in analogue/ digital communications and no MATLAB or Simulink experience:
   For readers new to analogue and digital communications, and perhaps to the concept of RF signals in general, we anticipate you will find this book a good way to start your learning. In Chapter 2 we will get you up and running with some preliminary spectrum viewing exercises using the RTL-SDR. Do not worry if you do not know how to use MATLAB or Simulink—in this chapter we tell you just enough to get you up and running with using the RTL-SDR, and then in Chapter 4 we provide you a fuller, more formal introduction to the software. This will focus on how it can be used for DSP and developing communications systems. In later chapters, we will present the theory behind a number of analogue and digital communications techniques, and show you how to implement SDR receivers that demodulate the associated signals.

2. **Introductory Level;**
   Some DSP/ communications experience and looking to learn by running & observing exercises:
   Readers who have some experience of working with MATLAB and Simulink (and having installed the RTL-SDR Hardware Support Package) should be able to open and run most of the examples and make appropriate observations at a system level. Seeing and hearing is believing; for instance, in exercises such as the SDR FM receivers in Chapter 10, being able to listen to a received and demodulated radio station *and* being able to visualise the RF spectral energy fluctuating with a Simulink *Spectral Analyzer*, is very informative. Many of the other exercises, such as those focused on viewing the spectra of mobile (cell) phone signals (Chapter 3), or watching QPSK constellations lock their phase (Chapter 11), should act as a great learning reinforcement for readers at this stage.

3. **Intermediate Level;**
   Those looking to understand the DSP and communication design of the exercises:
   For readers who have a good grounding in DSP and general linear systems theory, along with core trigonometric skills, who are looking to understand the mathematical and DSP functional implementations of many of the more advanced design exercises. Having a working knowledge of DSP theory, filtering methods, complex modulation and demodulation methods, and basic multirate systems design should be enough to get you going, and a number of these areas will be presented and reviewed in this book. For example, again with the FM receiver, seeing and hearing

Preface

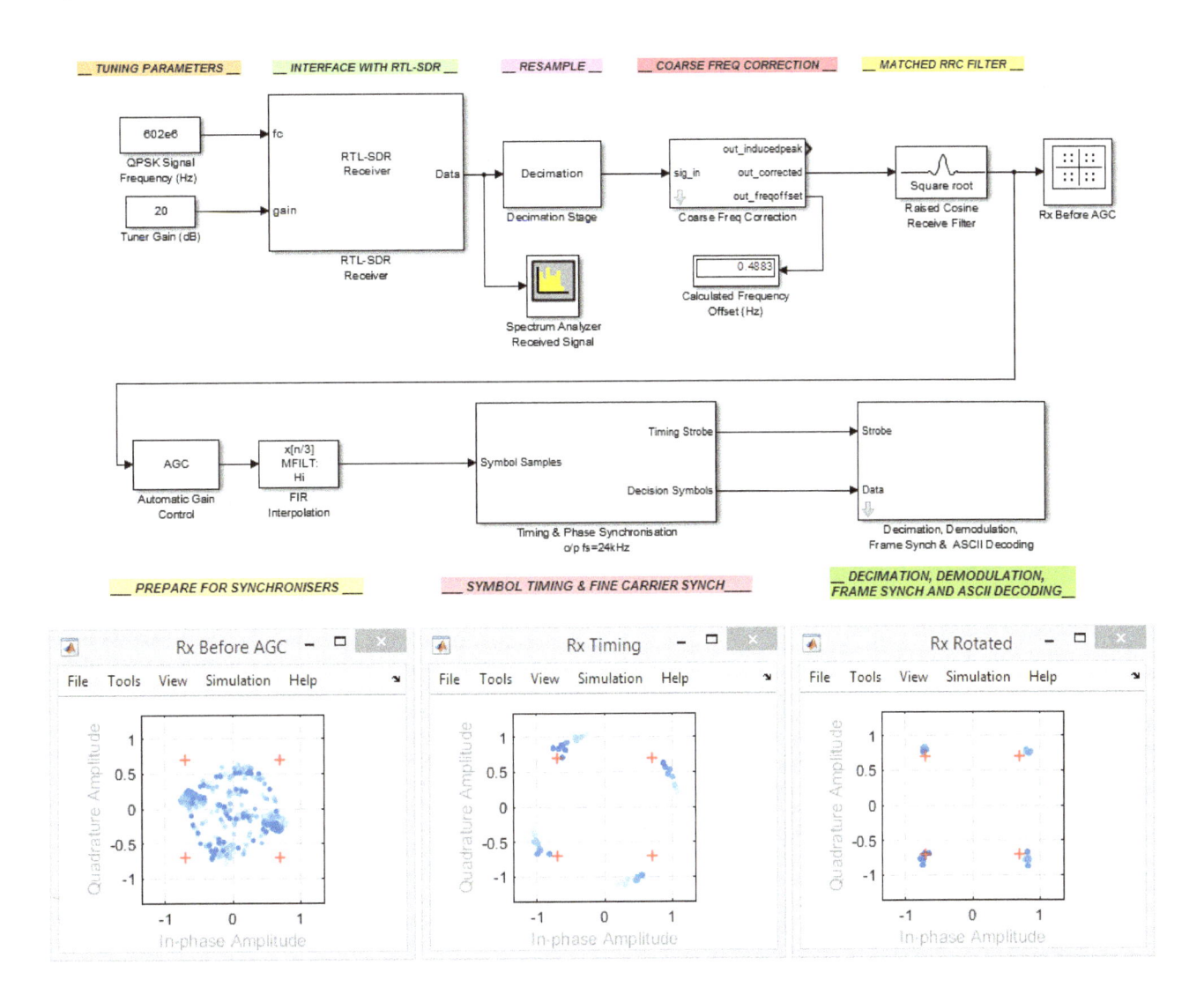

The block diagram of a Simulink receiver for a pulse shaped QPSK signal, that has been transmitted on a 602MHz carrier and received by the RTL-SDR

it working is informative at an introductory level, but moving to an intermediate level requires readers to understand how received complex FM signals are demodulated (i.e. understanding the mathematical analysis), as well as appreciating the spectral structure of the FM multiplex and the various filter design stages. Similarly in Chapter 11, the exercises allow you to observe synchronisation methods working, but we will also aim to explain the basic theory and core mathematics of digital synchronisation from first principles for the intermediate level reader.

4. **Advanced Level;**
   Those seeking the knowledge to design DSP and Communications SDRs from first principles:
   This will require a more advanced knowledge of DSP, such as filter design methods, Direct Digital Downconverter (DDC) implementations, and perhaps the most involved aspect of all; how the frequency tuning, synchronisation and timing circuits work. One of the main challenges in any radio system is tuning and synchronising to the carrier, as well as the symbol synchronisation for digital communications systems. For readers with a good working knowledge in these areas, after the first few introductory exercises in this book, you may well head off and restructure some of

the examples and designs to produce new receivers from first principles! This could be for GPS, or ISM band signals, satellite downlink signals, or even aircraft radio beacons. Any RF signal transmitted near you that operates inside the RTL-SDR's operating band (25MHz to 1.75GHz) has the potential to be received and examined in the time and frequency domains, and if it is legal to do so(!), you might just have the software defined radio expertise and knowledge to decode them.

Consider the example system shown above on page xvii, which is designed to receive an off-the-air, pulse-shaped QPSK signal from a 602MHz carrier using the RTL-SDR, and transmitted using a SDR transmitter (602MHz is a TV white space frequency where the authors had a test and development licence for the frequency band). Received signals pass through stages of decimation, matched filter implementation, time and phase synchronisation, frame and symbol synchronisation, and finally, ASCII message decoding. For the *beginner-level*, perhaps they might just add in a simple spectral analyser block at the data output port of the RTL-SDR block and then observe the frequency spectra of the input and output signals. We might then expect *introductory-level* readers just beginning their digital communications journey to open this example and learn by tuning the RTL-SDR, and observing the constellation diagrams with a frequency offset. *Intermediate-level* users may work to optimise the performance of the receiver's filters at a minimum cost, and the *advanced-level* user may look at designing and implementing their own timing and synchronisation systems. So in summary we anticipate readers with many different objectives and levels of experience, but feel that all can expand or consolidate their SDR knowledge from the exercises herein and have great fun while doing so.

The very fact you are reading this book means you wish to learn more about SDR in general, MATLAB and Simulink, or specifically about the RTL-SDR. I think we might already have said this, but; *stay tuned ...the wireless revolution is just beginning!*

Enjoy your SDR journey.

Acknowledgements

A number of people have contributed some way to the completion of this book and the many RTL-SDR and MATLAB and Simulink exercises, and we would like to extend our sincere thanks to them.

We would also acknowledge our close colleagues at the University of Strathclyde, who have been generous with their time, in terms of reading and reviewing chapters and giving their input and advice on topics related to the book. We are especially grateful to Douglas Allan, Dani Anderson, Fraser Robinson, Sarunas Kalade, Damien Muir, Shruthi Arenu, Valentin Dodonov, Iain Chalmers, David Crawford, and Martin Enderwitz. Our working environment has been the forum for much SDR-related discussion over the past few months, and we additionally appreciate the contributions of Ross Elliot, David Northcote, Ian Brown, Lakshmy Vazhayil Sasikumar, and Yousif Awad. Further, we would like to thank some of our taught programme students, who tried out early versions of our material in class and gave their constructive feedback. All of this has helped us to improve the final version that you are reading now!

At MathWorks, we acknowledge the support and prowess of engineering colleagues Ethem Sozer, Mike McLernon, Darel Linebarger, Ken Karnofsky and Don Orofino in producing an excellent DSP and Communications software environment for this RTL-SDR project. From the DSP and Communications Toolboxes, to the Hardware Support Packages for the RTL-SDR and USRP® devices, everything you need to build a real and parameterisable DSP enabled SDR is all there in MATLAB and Simulink. We also acknowledge the support of engineering colleagues in the MathWorks office in Glasgow, including Neil MacEwen, Garrey Rice and Daniel Garcia-Alis.

We must also thank those whose work we have referred to, and drawn inspiration from, in the writing of this book. Of particular note are the excellent *Digital Communications: Fundamentals and Applications* by our friend Bernard Sklar, and the fantastic *Digital Communications: A Discrete-Time Approach*, by Michael Rice. Both are highly recommended for anyone working in this area! It has been great fun to take some of these concepts from the printed page, and integrate them into our RTL-SDR MATLAB and Simulink designs.

Finally we acknowledge the efforts and support of NooElec, and everyone in the RTL-SDR and amateur radio community who have developed solutions, code and ideas to bring SDR to the desktop of students, scientists, professional engineers and home users.

Bob, Kenny, Dale, and Louise.   *August 2015, Glasgow, Scotland.*

1 Introduction

It's scarcely twenty years since the first second generation (2G) digital mobile phones appeared on the market. By the mid–1990s texting was widely adopted, and further, GSM and GPRS based modems became available—albeit only providing slow connectivity no higher than a few kbps. This was the start of the wireless digital data revolution. By the end of the 1990s and into the 2000s, third generation (3G) mobile connections became available at speeds of a few 100kbps and WiFi emerged as a means of connecting devices to wireless access points over distances below 20 metres or so. From early WiFi speeds of just a few Mbps, by 2005 we were running at 54Mbps in most implementations and soon to speeds of 300Mbps and higher when MIMO methods emerged in the early 2010s. Smartphones also just keep getting smarter. WiFi and Bluetooth are now considered standard wireless connectivity requirements on smartphones alongside 2G, 3G, and likely fourth generation (4G) LTE-Advanced connections. Homes and offices are now well served by superfast WiFi, and there are mobile basestations throughout cities, towns and the countryside, all around the world. In many ways though, the 'wireless revolution' is still just beginning. With more Short Range Devices (SRDs) being produced, and the so-called Internet of Things (IoT) continuing to evolve, we should expect in the very near future that there will be more than one wireless phone, laptop or tablet per person; and perhaps up to 10 other devices, ranging from keyfobs to sensors, GPS tracked objects, and so on. It will be wireless everything very soon.

Software Defined Radio (SDR) is a generic term which refers to radio systems in which almost all of the functionality associated with the Physical Layer (PHY) is implemented in software using Digital Signal Processing (DSP) algorithms [32], [33], [114]. An ideal SDR receiver would have a very small hardware front-end; only an antenna and a high speed GHz sampler that was capable of capturing and digitising a wide band of radio frequencies. Any demodulation, synchronisation, decoding or decryption required to recover information contained within a received signal would be performed in software that is executed on a superfast, dedicated processing device [42].

Many smartphones and similar devices currently have up to around 8 different radios optimised for receiving various signals from different frequency bands, such as those for WiFi (2.4GHz), LTE (Long Term Evolution, 800MHz), GSM (Global System for Mobile Communications, 900MHz), UMTS (Universal Mobile Telecommunications System, 2.1GHz) GPS (Global Positioning System, 1.5GHz),

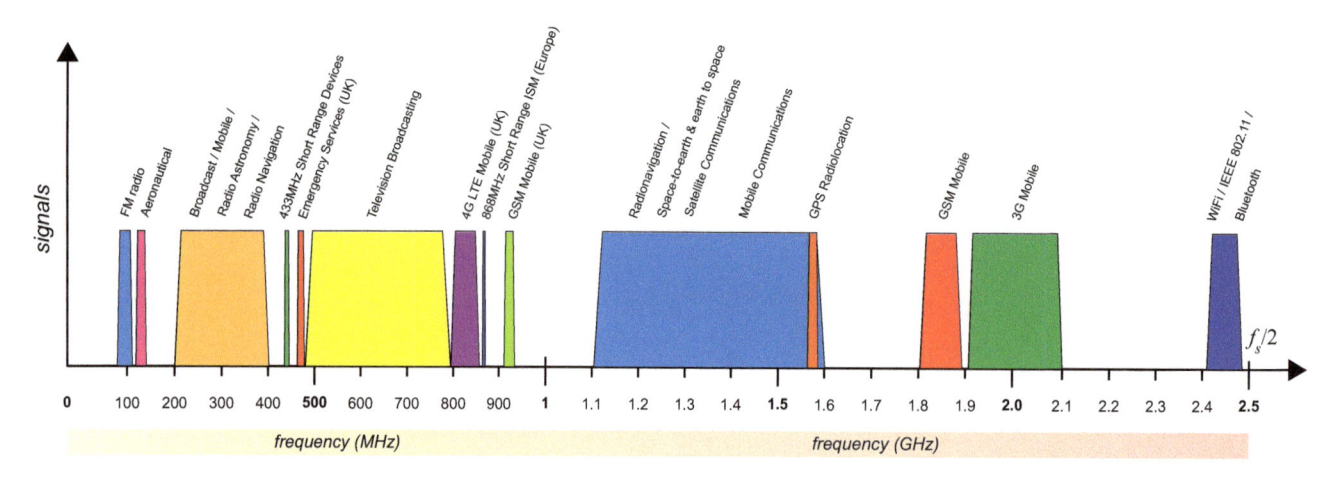

Figure 1.1: With the availability of devices that can sample up to a rate of say f_s = 5GHz, the sampling theorem dictates we can digitise all signals from baseband up to 2.5GHz. In practice its more likely that different bands would be filtered, and digitised at a lower sampling frequency, but in theory we could work with the entire band.

Bluetooth (2.4GHz), NFC (Near Field Communications, 13.56MHz), and FM Radio (100MHz). Soon they may even include radios to receive IoT and TV White Space UHF (Ultra High Frequency at 400MHz+) signals too. The ultimate solution here would be to utilise a single SDR that samples the spectrum at GHz rates to digitise and capture all signals from baseband to 2.5 or even 3GHz, and to implement all of these receivers in software code. Figure 1.1 illustrates the approximate frequency band positions for many of the broadcast radio signals that most of us receive and use on a daily basis.

In this book we will work with RTL-SDR USB SDR hardware, and MATLAB & Simulink software to design and implement real world desktop SDR systems. We will acquire signals from the frequency range 25MHz to 1.75GHz and digitise them with the hardware, then perform processing in software to demodulate and extract the signal's information. We will design Amplitude Modulation (AM) and Frequency Modulation (FM) receivers, Quadrature Amplitude Modulation (QAM) digital receivers, spectral viewers and frequency channeliser systems; all using the low cost *RTL-SDR receiver*, and SDR algorithms/ DSP algorithms implemented in *MATLAB* and *Simulink*. As we will discuss in Section 1.6, the RTL-SDR is what is termed an Intermediate Frequency (IF) sampling radio, rather than a Radio Frequency (RF) sampling radio, however we will demonstrate how this form of SDR can be used to receive signals from 25MHz all the way to 1.75GHz.

1.1 Real Time Desktop Software Defined Radio

The book introduces the technological breakthrough that is the RTL-SDR, and how it can be used in conjunction with MATLAB and Simulink to implement 'real world' desktop SDR systems, capable of receiving RF signals broadcast in the air around you. We cover the theory and implementations of a variety of transmitters and receivers for both analogue and digital communication systems, and have a large number of hands-on practical exercises throughout where you will have the opportunity to apply these techniques; both in simulation and in real world situations. You will have the chance to receive, demodulate and decode RF signals live off the air, and even generate your own to build complete SDR based communications systems. In later chapters of this book we will focus on the principles of real digital transmitter and receiver design, where DSP components such as digital filters, encoders and decoders, and phase/ timing synchronisers will be constructed in MATLAB and Simulink to building systems capable of sending text strings and images across a room from one computer to another.

Chapter 1: Introduction

The focus of this book is all about real-time desktop SDR.

Of course, we want you to get moving as soon as possible, so in the opening chapters we aim to provide you with the necessary background knowledge on SDR (and the RTL-SDR in particular), to get you up and running and acquiring RF signals within just a few minutes of '*opening the box*' (or unwrapping the shrink-wrap of the RTL-SDR more likely). If you know what you are doing or you can't wait to open the box and get started, then jump ahead to Chapter 2 (page 25).

1.2 What is the RTL-SDR?

The RTL-SDR is a *very* low cost (sub-$20), easy to use USB device that receives RF radio signals. Originally these devices were designed to be used as DVB-T (Digital Video Broadcast—Terrestrial) receivers, but it was discovered that they could be used as generic (receive only) SDRs by simply putting them into a different mode. In this mode, they are capable of receiving any signal in the range their tuner operates over; not just the Digital Television (DTV) signals they were designed to receive. This range varies from device to device depending on what components have been used, but is most commonly from 25MHz to 1.75GHz. The front end of the RTL-SDR receives RF signals live off the air, downconverts them to baseband, digitises them, and the device outputs samples of the baseband signal across its USB interface. Figure 1.2 shows three variants of the RTL-SDR made by NooElec.

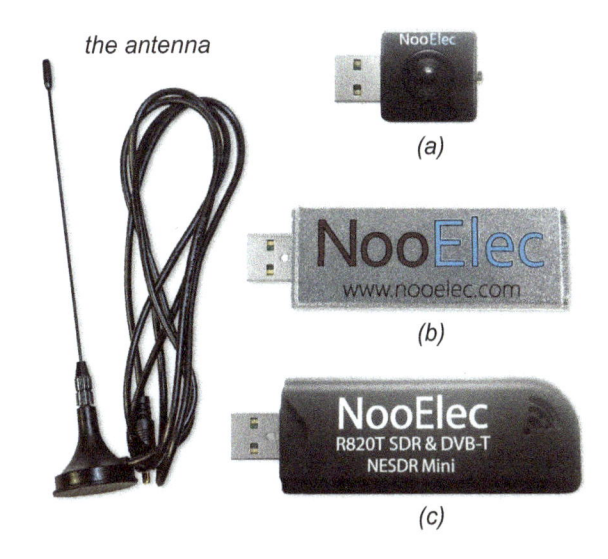

Figure 1.2: Three types of NooElec RTL-SDR: (a) the nano, (b) the RF shielded, and (c) the mini and their antenna

MathWorks released a Hardware Support Package for the RTL-SDR in early 2014 which enables both MATLAB and Simulink to interface with, and control the RTL-SDR. With this add-on, samples output from the device can be captured and brought into the software, enabling users to implement any kind of DSP receiver or spectrum sensing system they desire in either a Simulink model or MATLAB code. To give an example, connecting the RTL-SDR to your laptop and the antenna to the RTL-SDR's RF input port, you could write some MATLAB code that tuned the device to the centre frequency of an FM radio station, demodulated and decoded the received samples, and output the resulting audio signal to the laptop's speakers. As the entire demodulation process is carried out in software, this is a *Software Defined Radio* implementation.

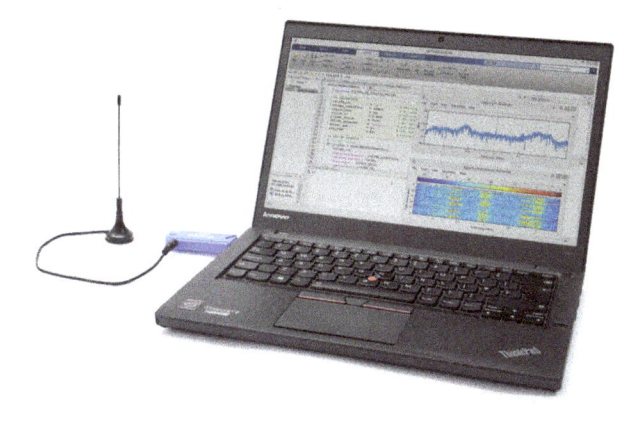

Figure 1.3: RTL-SDR with an omni-directional antenna connected to a laptop via a USB port. The computer is running MATLAB & Simulink to enable implementation of the DSP and SDR algorithms presented in this book. MathWorks tools can be used to monitor the signal in the frequency and time domains as it is demodulated [59]

The RTL-SDR set up of Figure 1.3 is represented in a block diagram form in Figure 1.4. RF signals are received at the antenna, quadrature downconverted by the RTL-SDR, and In Phase/ Quadrature Phase (IQ) samples are presented to the computer running MATLAB. The receiver design is implemented using the appropriate DSP algorithms to demodulate the signal to baseband and extract the information signal. This might be audio, video, images, or data.

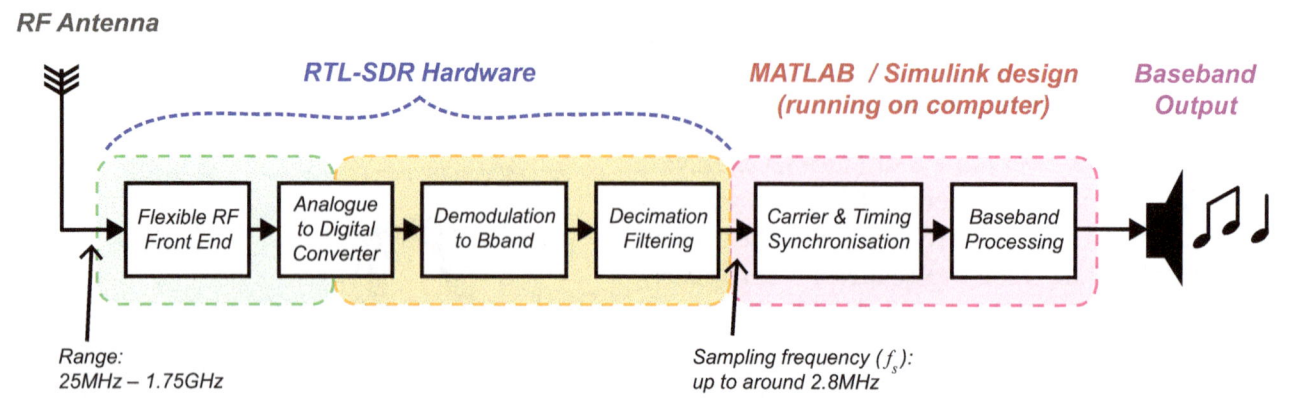

Figure 1.4: Block diagram of the RTL-SDR receiver chain.

Subject to appropriate hardware configuration (i.e. using the right antenna!) and signals being broadcast in your vicinity, the RTL-SDR allows you to receive not only FM radio signals, but also UHF/ DTV signals, Digital Audio Broadcast (DAB) radio, GPS signals, 2G, 3G and 4G cellular signals, transmissions in the Industrial, Scientific and Medical (ISM) bands, etc. — in fact any signal that is transmitted within the operating range of the tuner. Figure 1.5 shows some of the RF signals from the electromagnetic spectrum that can be received by the device.

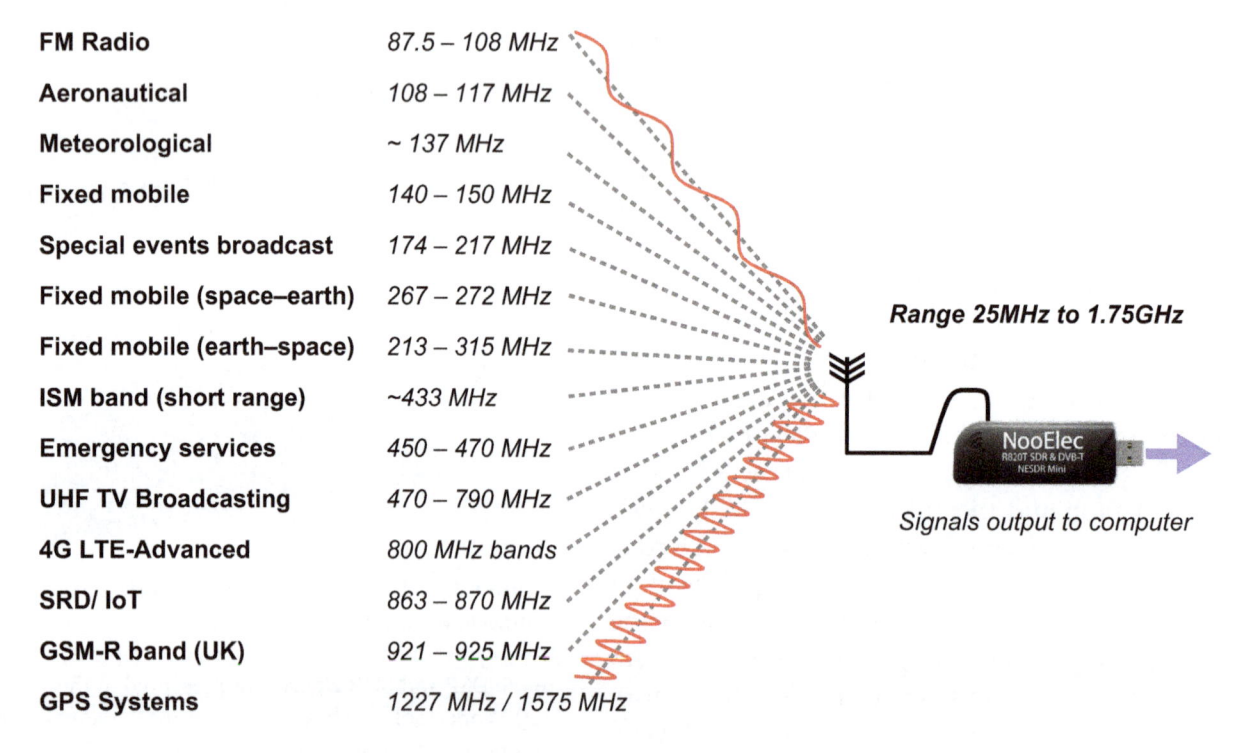

FM Radio	87.5 – 108 MHz
Aeronautical	108 – 117 MHz
Meteorological	~ 137 MHz
Fixed mobile	140 – 150 MHz
Special events broadcast	174 – 217 MHz
Fixed mobile (space–earth)	267 – 272 MHz
Fixed mobile (earth–space)	213 – 315 MHz
ISM band (short range)	~433 MHz
Emergency services	450 – 470 MHz
UHF TV Broadcasting	470 – 790 MHz
4G LTE-Advanced	800 MHz bands
SRD/ IoT	863 – 870 MHz
GSM-R band (UK)	921 – 925 MHz
GPS Systems	1227 MHz / 1575 MHz

Figure 1.5: A few of the signals available to the RTL-SDR (location dependent!)

Chapter 1: Introduction

1.3 What Do I Need to Get Started?

In order for you to be able to implement SDR systems, there are a few hardware and software requirements that must be taken into consideration.

1.3.1 SDR Hardware: NooElec Receiver

Firstly, and most importantly, you require a **NooElec NESDR RTL-SDR receiver.** You can source these directly from nooelec.com or from their store on amazon.com, along with numerous other outlets; and the price you should be expecting to pay for it is less than £15/ $20. There are a number different form factors and sizes of these devices (note again the Mini, Nano and RF Shielded versions shown in Figure 1.2), but all essentially have the same functionality and can be used in the same way. There are also some RTL-SDRs that use different RF tuners (notably the Elonics E4000 rather than the more common Rafael Micro R820T [113]); again all are functional, and you will be able to use any to get you started on real world SDR designs.

The NooElec RTL-SDR devices feature a Micro Coaxial (MCX) antenna port, and ship with an omni-directional MCX antenna. Although this antenna is tuned to perform best in the UHF band, it will be suitable for receiving a range of signals for preliminary 'out of the box' spectrum viewing and baseband DSP processing in MATLAB and Simulink.

1.3.2 SDR Hardware: Computer System Requirements

This workbook has been developed based on Windows 8.1, 64-bit edition. The necessary software is available for other Operating Systems (OSs) (Windows 7, 8 and 10 32- and 64-bit, OS X 10.7.4+ and numerous Linux distributions), and it should be possible to follow the exercises in this book using any of them with minor deviations.

It is not appropriate to list detailed system requirements (as so many variations are possible nowadays!), but we recommend you have a modern computer with a 'good' processor (such as a recent Intel i5, i7, AMD FX-8k series or similar core), at least 8GB of RAM, and 30GB of free hard drive space. You will also need at least one free USB2.0 (or higher) port, and we recommend a computer with a soundcard so that you can listen to any audio signals you demodulate. As long as your computer is fast enough, then for the most part, the SDR implementations with your RTL-SDR should be able to run in real-time (i.e. the processor will be able to process data as fast as it arrives). If your PC is not fast enough to run real-time, you can of course just record the signals to disk or cloud locations, and then process them offline.

1.3.3 SDR Software: MathWorks MATLAB & Simulink

Before you start building your own SDR systems, you will also need to have MATLAB and Simulink installed on your computer. This book is based in **MATLAB R2014b**, and although the *RTL-SDR Hardware Support Package* is compatible with versions R2013b and later [59], you may find some file compatibility issues if using an earlier version. We therefore recommend working with R2014b or later where possible. This software can be purchased from mathworks.com.

The RTL-SDR Hardware Support Package can be downloaded and installed by all users of MATLAB, whether you have a Professional, Home [56], or Student [57] licence. To complete the exercises in this book, you will also require the following MathWorks toolboxes to be installed on your computer. These are official software add-ons that provide extra features (such as filter designing tools and frequency domain scopes) in the development environment:

- DSP System Toolbox [55]
- Communications System Toolbox [54]
- Signal Processing Toolbox [63]

MATLAB and Simulink provide you an environment where you can conveniently code and build the receivers, and these toolboxes[1] will provide you the necessary means to implement any SDR receiver algorithm you desire. Not only do they give you the facilities to design things like the digital filters, decimators and synchonisers your system requires, but also provide tools that allow you to visualise signals in the time and frequency domains as they undergo the demodulation process.

1.3.4 SDR Software: MathWorks RTL-SDR Hardware Support Package

MathWorks Hardware Support Packages can be considered 'add-ons' that provide specific MATLAB and Simulink interfacing support for third-party hardware, including devices such as the Raspberry Pi, the Arduino, various Field Programmable Gate Arrays (FPGAs) and the Universal Software Radio Peripheral (USRP®). Most relevant to this book is the RTL-SDR Hardware Support Package, which enables interfacing with your RTL-SDR, although you may well use some of the others later on. We will talk you through downloading and installing the Hardware Support Package and USB drivers in Chapter 2, and we provide some troubleshooting information in Appendix A.1. Note that you will only be able to obtain the Hardware Support Package once you have MATLAB and Simulink installed.

1.3.5 Optional Extras

To complete some of the exercises later in this book you will require some low cost radio transmitters, which will allow you to generate your own low power RF signals from the desktop. This means that you can transmit (Tx) signals to your RTL-SDR, which will receive (Rx) them. Transmitters can either be purchased; built using instructions that we provide in Appendix E; or you may find some wireless sensors around the home (for example 433MHz or 868 MHz RF low temperature sensors) that transmit signals suitable for the tasks too.

For classroom/ lab environments, users may wish to procure a professional transmitting device such as the USRP® software-defined radio [86]. There is a MathWorks Hardware Support Package for this device, meaning it is easy to interface with via MATLAB and Simulink, and is capable of generating many different types of signal (it does however cost more at a few thousand $'s).

1.3.6 A Brief Recap

So in summary, to design the real time SDR systems that are presented in this book you need:

- *An RTL-SDR USB device*
- *A powerful computer running a recent operating system*

---

1. All in, MathWorks have produced over 80 toolboxes for MATLAB & Simulink supporting many other aspects of DSP and digital communications too; such as HDL coding, image processing, fixed point, phased arrays and RF tools. We will only use the minimum set of toolboxes required for this book.

- *MATLAB & Simulink Version 2014b or later*
- *MathWorks DSP System Toolbox*
- *MathWorks Communications System Toolbox*
- *MathWorks Signal Processing Toolbox*
- *MathWorks RTL-SDR Hardware Support Package*
- *(optional) Radio Transmitters*

1.4 The Aim and Objectives of this Book

For readers of this book we have a core aim:

> *To allow engineers to **learn** about, and **implement SDR designs** at very **low cost** using the **RTL-SDR** receiver, with communications algorithms and receivers implemented in **real-time** using the **MATLAB & Simulink** software tools.*

We can also present the following key objectives and learning opportunities from this book:

1. To understand the fundamental design and implementation of SDR systems;
2. Gain an appreciation of the many suitable applications of SDR in radio, mobile, wireless, instrumentation, and general RF receiving and 'listening';
3. To learn about the core components of an SDR system, and how DSP and PHY layer algorithms allow signals to be extracted and received information presented;
4. (If you are new to MATLAB and Simulink) to be able to competently design with, and run, MATLAB code and Simulink models, and work with associated toolboxes and support packages;
5. Be able to tune across the spectrum to capture and view different signals, and plot live RF spectra on screen;
6. Gain an appreciation of the different communications systems and standards in use, and the bands of RF frequencies they use;
7. Learn the fundamentals of DSP receiver design (filters, demodulators, decimators) and be able to build and run systems to bring quadrature-based signals to baseband;
8. Learn the fundamentals of the AM and FM analogue modulation schemes, and be able to understand and implement digital receivers for both AM and FM signals;
9. Learn how to use other hardware to generate radio signals from the desktop, using simple component level designs, and off-the-shelf environment sensors (temperature etc.);
10. Learn how to multiplex and codify audio signals, then transmit them via low cost FM radio transmitters, receive them with the RTL-SDR, demodulate, demultiplex, decode, and output each channel again;
11. Learn the requirements for tuning, synchronisation, and timing, and be able to understand and implement these components as part of an SDR receiver;
12. Learn how to implement simple SDR transmitters from the desktop using the USRP® transceiver, and generate QAM signals to transmit (via RF on the desktop) to RTL-SDR receivers.
13. Experiment with more complex SDR applications (working up the stack), capable of transmitting text strings and images across the classroom, using the RTL-SDR, USRP® and FM Transmitters.

1.4.1 Undergraduate and Postgraduate Class Teaching

The book is also intended for undergraduate, or postgraduate courses in DSP and SDR. Some additional teaching support materials information, syllabus and lab class suggestions and related information will be collated on the book's companion website, desktopSDR.com in the teaching section.

1.5 The Evolution of the *Software Defined Radio* Architecture

Over recent years, and probably since the late 1990s, SDR has often been presented — even heralded — as the future solution and design for all RF receivers [17]. It is probably true to say that SDR has been promising solutions for quite a few years, but only now (from 2014) has this become *very* low cost and widely available at the desktop level. In the past, SDR was more commonly associated with military and research applications, due to its (historically) high cost of implementation [19].

It is useful to review the 'ultimate' SDR architecture, along with more common practical implementations, and to understand the technology factors involved.

1.5.1 The Ultimate SDR Architecture

At its very simplest conceptual level, SDR comprises of an RF section (antenna, amplifiers and filters) and a very high speed Analogue-to-Digital Converter (ADC) and Digital-to-Analogue Converter (DAC) pair, interfaced with a powerful DSP processor and/ or computing system, as illustrated in Figure 1.6. Samples are passed into and out of the DSP section via the ADC and DAC, respectively. Recalling the Nyquist theorem (i.e. that you need to sample a signal at greater than twice the signal bandwidth to retain all information), sampling at, say, f_s=4GHz, would produce a baseband spectrum from 0 to 2GHz, or $f_s/2$.[1] Thus, the SDR would be capable of transmitting and receiving all frequencies up to 2GHz, with the modulation and demodulation undertaken in the digital domain.

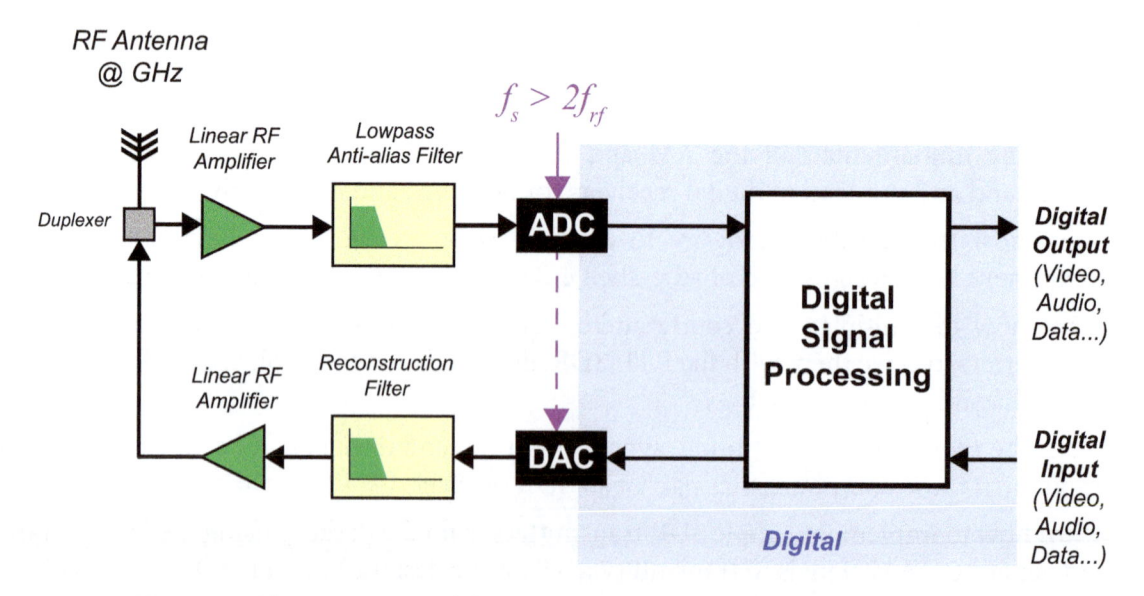

Figure 1.6: The components of the simple, conceptual, 'ultimate' Software Defined Radio which in theory will digitise the spectrum from 0Hz to 2GHz (or $f_s/2$)

Chapter 1: Introduction

To develop a complete radio receiver system, we do need to rely on our RF and broadband antenna colleagues to educate us on some of the engineering issues associated with operating over such a wide frequency range. Ensuring that we use appropriate analogue components are therefore also important considerations. Assuming this is all in order, then from a DSP theory point of view, if we can digitise the RF signal at these sampling frequencies, the rest of the receiver can be implemented purely in software [18]. The software would of course be required to run on very high speed processors, or perhaps even on powerful FPGAs (e.g. for extensive, complex standards-based signals such as LTE). These devices perform all of the DSP algorithmic arithmetic required, which are primarily multiplications and additions, as per usual for DSP.

Sometime soon, the super-highspeed ADCs and DACs needed to interface at multi-GHz sampling rates will become available, and thus the system indicated in Figure 1.6 will be readily realisable. To remind ourselves of technology timelines and evolutions, back in the late 1990s, a DAC/ADC running at f_s=100kHz with 16-bit resolution was the latest (and rather expensive) technology. Now, in 2015, devices running at f_s=100MHz with almost 14-bit resolution are widely available at reasonable cost from companies such as Analog Devices Inc.. Samplers operating at GHz frequencies are available now with 8-bit (or so) resolution, but they are expensive and not yet consumer grade. Despite these current limitations, one thing we can be sure of with technology is that it will advance — so get ready, GHz samplers in our consumer devices are just around the corner!

1.5.2 SDR Architectures: The Advance of Digital Processing

Next, we consider the evolution of a quadrature radio receiver, and the increasing role of digital processing (which supports SDR).

First generation 'digital radios' appeared back in the mid–1990s. As illustrated in Figure 1.7(a), the analogue part of this radio architecture downconverted signals from their RFs to an IF using a Local Oscillator (LO), and then, using a second LO, further downconverted the IF signal to baseband. The baseband signal was then sampled and digitised using an ADC (at a rate of no more than a few 10's of kHz), and then DSP was used to perform the final processing stages to recover the transmitted information. Second generation mobile phones of the 1990s — those that received GSM signals — were likely to have used this architecture. It is worth remembering the first wave of mobile phones in the early 1990s were in fact all analogue, using the old AMPs and TACs standards [15].

In the next generation of digital radios, which emerged in the 2000s, the sampling and digitisation processes started to be performed in some devices at IFs. IFs of around 40MHz (for example), could be supported by an ADC that sampled at, say, 125MHz. The first DSP stage of this architecture involved using a Direct Digital Downconverter (DDC) to shift IF signals to baseband using demodulation and decimation filtering, as shown in Figure 1.7(b). Further DSP processing was then performed once the signal was at baseband. In this architecture, more functionality was implemented in the digital domain, giving greater flexibility for SDR.

---

1. Because so many applications use baseband signals (rather than single tone sinusoidal signals), and the signal's frequency components start down near 0Hz, the Nyquist sampling theorem (often stated as sampling at greater than twice the maximum frequency of interest) should more correctly be defined as *twice the signal bandwidth*. Clearly for a baseband signal with highest frequency f_{max}, the frequency range from 0Hz to f_{max} is in fact the bandwidth.

Ultimately (and just about the state of the art at the time of writing of this book — check the date on the front cover!), the move has been made to sample RF signals directly as illustrated in Figure 1.7(c), and downconvert them from RF frequencies to baseband in a single stage, using DSP. This is possible today (again—check the date!) because we are now able to sample in the order of GHz. This is a shift towards the 'ultimate SDR' architecture presented in Section 1.5.1. [1]

1.5.3 *The RTL-SDR Architecture and Workflow*

In the context of these receivers, the architecture of the RTL-SDR corresponds to Figure 1.7(b). It has a two-stage demodulation process: RF to IF in analogue hardware; and IF to baseband implemented digitally. Importantly, control over the demodulation process can be exerted through software, which allows the desired band of RF frequencies to be selected.

Referring again to Figure 1.7(b), the output of the RTL-SDR is equivalent to the two inputs to the 'Baseband DSP' block seen at the right hand side (i.e. baseband samples from the I and Q branches). Once output by the RTL-SDR, these IQ samples can be brought into MATLAB and Simulink via the RTL-SDR Hardware Support Package — where designs for the 'Baseband DSP' section may be created to implement the final stages of the SDR receiver. The samples output by the RTL-SDR are in an 8-bit fixed point format, but systems designed in MATLAB or Simulink can be implemented using floating point arithmetic. More detailed information about the architecture and internal components of the RTL-SDR (and its variations) will be provided in Section 1.6.

Noting that the RTL-SDR architecture involves implementing the 'Baseband DSP' functionality in MATLAB or Simulink, it is clear that we have a great deal of flexibility in developing the 'software' side of the SDR receiver. It is possible to design baseband SDR software receivers that will run in real time on a standard, high speed computer processor. Using the RTL-SDR with MATLAB and Simulink will allow you to create a wide variety of systems, capable of receiving signals using many different radio standards; both from off-the-air broadcasts and locally generated and transmitted 'educational' RF signals.

1.6 RTL-SDR Hardware

The RTL-SDR USB hardware originated from consumer grade DVB-T receivers, designed to enable users to watch DTV on a computer. These receivers were *not* originally designed or conceived to be used as generic programmable SDRs, and we owe our thanks to a number of independent engineers and developers in the SDR community, who discovered their potential for use as SDRs. Specifically, they found that the devices could be placed in a 'test mode', which essentially bypassed the DVB decoding stage, thus allowing the device to tune over the range 25MHz to 1.75GHz, producing raw, 8-bit IQ data samples at a programmable baseband sampling rate!

1.6.1 *Device Operation*

The two main components used in the RTL-SDR DVB receivers are a DTV tuner (the Rafael Micro *R820T* [95] is most commonly used, although some RTL-SDRs utilise the Elonics *E4000* [69]) and the Realtek *RTL2832U* DVB-T Coded Orthogonal Frequency Division Multiplex (COFDM) demodulator

---

1. Logically, where is this all going? So as sampling rates increase, taking this to the mathematical limit well past GHz rates, we get to an "infinite" sampling rate. Then its the mind-bending realisation that we are back to analogue with an "infinite" sampling rate. One to think on.

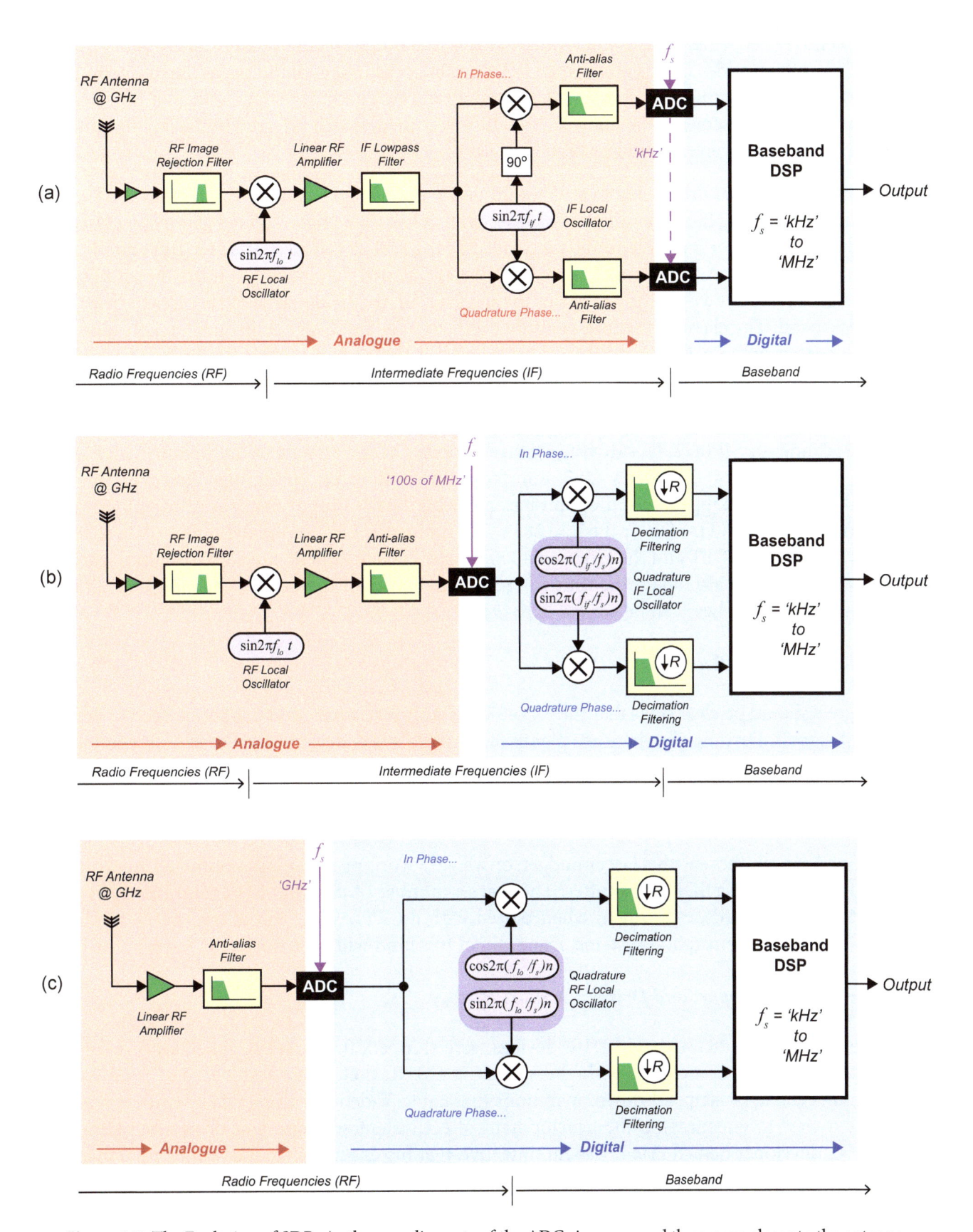

Figure 1.7: The Evolution of SDR: As the sampling rate of the ADCs increase and they move closer to the antenna. (a) the baseband Digital Radio evolves to (b) the IF Digital Radio, and then to (c) the RF (zero-IF) Digital Radio

[109]. A few interested, enterprising and knowledgeable individuals realised that this demodulator could be placed into a 'test mode'. In this mode, the DVB devices act as IF digital radios (as illustrated in Figure 1.7(b)), essentially becoming high speed ADCs, and output raw quadrature data with an 8-bit resolution [116]. Soon after this discovery, the name RTL-SDR was coined, which referred to the fact that the RTL (Realtek) based DVB receivers could be used as SDRs.

Figure 1.8 shows a signal processing flow diagram of the main stages that are carried out on the RTL-SDR. RF signals entering the tuner are downconverted to a low-IF using a Voltage Controlled Oscillator (VCO). The VCO is programmable, and is controlled by the RTL2832U over an Inter-Integrated Circuit (I^2C) interface. After an Active Gain Control (AGC) stage, which dynamically adjusts the amplitude of the input signal to suit the operating range of the device [39], the IF signal then requires to be brought down to baseband. The classical method of doing this is to pass the IF signal through an anti-alias filter, sample the output with an ADC, and then downconvert it to baseband using quadrature Numerically Controlled Oscillators (NCOs) (i.e. a sine and a cosine oscillating at the IF frequency).

When the RTL2832U is operating as it is designed to, this baseband IQ data would be DVB-T demodulated, and an MPEG2-TS (Moving Picture Experts Group Transmission Stream) video stream would be output over the device's USB interface. When it is in 'test mode' however, the final demodulation stage is skipped, and the 8-bit baseband IQ data is output instead. Although no formal documentation for the RTL-SDR or the RTL2832U demodulator has been made public, data sheets for the tuners (the Rafael R820T and the Elonics E4000 [111] can be obtained with a quick online search). A blogger from Japan has also taken the time to create schematics of the device, showing how all of the components interface. These can be viewed here [91].

1.6.2   *RTL-SDR—A Receive Only Device*

Something we should be clear on is that the RTL-SDR is ***only a receiver***, and cannot be used to generate or transmit signals. Fortunately, there are plenty of RF signals broadcast in the air around you (such as the ones shown in Figure 1.5), that you can receive with your RTL-SDR with little effort. You may also find that there are plenty of RF devices transmitting around your home and workplace, in the form of temperature and alarm sensors, car key fobs, and even wireless doorbells!

As mentioned earlier, those with a larger budget, or who are working in the classroom/ lab environment, may consider acquiring a SDR transmitter. There are a number of different SDR transmitters available (such as the USRP® hardware) which, when paired with an RTL-SDR, will allow you to implement complete Tx/ Rx communication systems. Examples of this kind will be featured throughout the book.

1.6.3   *RTL-SDR Support and Drivers: A Short History*

When the original RTL2832U based DVB devices were shipped, they were supplied with drivers and software that was only compatible with the Windows OS. Realtek, who designed and produced the demodulator, released a stripped-down operations manual to Video-4-Linux (V4L) developers in early 2010, with a view to encouraging the development of equivalent software and drivers for Linux OSs [116]. A V4L developer named Eric Fry spent time investigating exactly what data was being transmitted over the device's USB interface, and he created some low level drivers that facilitated communication between Linux OSs and the device.

Chapter 1: Introduction

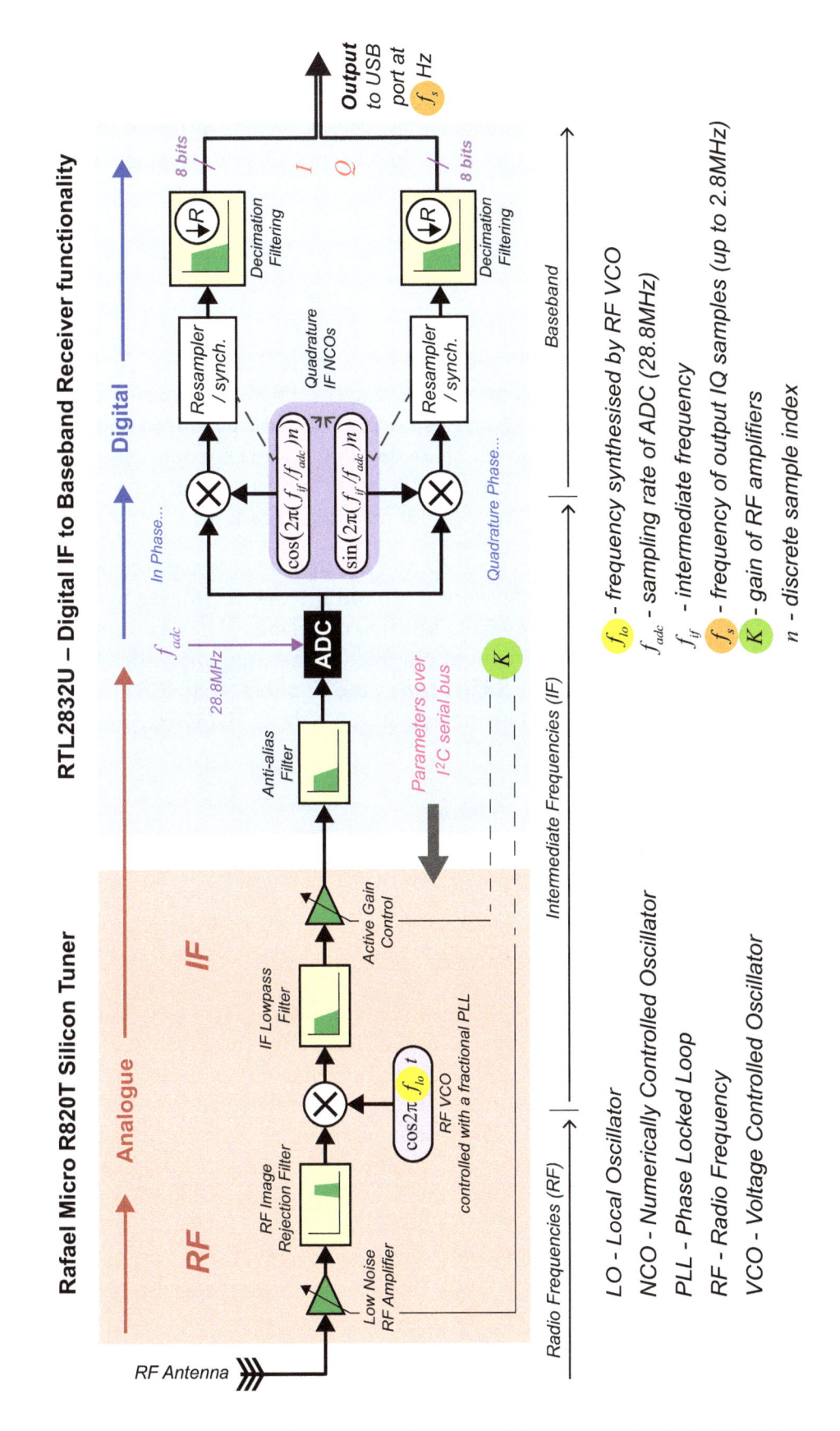

Figure 1.8: Block diagram showing the internal architecture of the R820T/RTL2832U RTL-SDR

In early 2012, Finnish engineering student and Linux developer Antti Palosaari posted on the V4L GMANE developer forum that he had taken *"radio sniffs"* using an RTL-based DVB device. He discovered that when the device was tuned to receive FM and DAB radio stations, it was programmed into a different mode, where raw, modulated data samples were transferred to the computer, and that demodulation was performed in software [46]. He captured 17 seconds-worth of data originating from a Finnish radio station, and posted it online asking if anyone could work out how to demodulate it manually. This was accomplished only 36 hours later, after some collaborative effort. He is quoted as saying in the original post, *"I smell a very cheap poor man's software defined radio here ☺"* [104]!

This discovery led to further investigation of the RTL-SDR's USB protocol. The commands transmitted when tuning to a radio station were captured, and used to force the device to stay in this special mode continuously. It turned out to be a test mode, and when the RTL2832U was in this mode, it output 8-bit unsigned samples of baseband IQ data, rather than decoded DVB signals in MPEG2-TS format (as it was designed to operate). Some developers from Osmocom, who had produced an independent SDR device called '*OSMO-SDR*', then became involved. These developers had experience in writing software that was able to program the DTV tuners used with the RTL2832U, and after spending some time examining the Windows drivers provided by Realtek, figured out how to program the tuner via the demodulator [116].

RTL-SDR exploded onto the scene in early 2013, and multiple devices and software kits became available, produced by various companies and developers around the world. Judging by the communities on the web, the RTL-based DVB-T devices appear to be more popular as SDR receivers than they were for their original intended purpose of DTV reception! NooElec is one company with worldwide distribution of these devices. Based on their use of the R820T tuner, the NooElec NESDR RTL-SDR devices are capable of receiving signals from the RF frequency range 25MHz to 1.75GHz, and reliably sampling the frequency spectrum at a rate of up to 2.8MHz [71].

1.6.4 RTL-SDR Tuning and Demodulation

The RTL-SDR is not quite the 'ultimate SDR' we illustrated in Figures 1.6 and 1.7(c) given that its ADC is not wideband and does not sample at GHz rates. There are actually two different architectures of the RTL-SDR, as summarised below. We will generally refer to the first of these in our discussions, which can be considered similar to the IF Digital SDR of Figure 1.7(b).

- **Rafael Micro R820T / Realtek RTL2832U Combination** [71] — The Rafael Micro R820T tuner (and its upgrade, the R820T2) use a *low-IF* of 3.57MHz, and downconvert a band of RF signals (roughly 6MHz wide) to this IF [95]. The resulting signal is input to an 'IF sampling' connection on the RTL2832U, which tunes to the IF centre frequency and downconverts the IF signal to baseband [98]. After this, the RTL2832U samples the signal with its ADC at a rate of 28.8MHz, and performs quadrature demodulation to produce IQ samples [105][1]. A decimation process is performed to reduce the sampling rate to a lower value, e.g. to 2.8MHz, and the samples are then output over its USB interface. Looking forward a few pages, Figure 1.11 shows the processing involved in this architecture, along with frequency spectra at significant stages.

---

1. In Figure 1.8 we represent the IF to baseband RealTek RTL2832U chip as a classical downconverter with a single ADC followed by a quadrature demodulator implemented with NCOs. The implementation described here with two analogue oscillators (a cosine and sine at the IF frequency), where the signal is digitised at baseband by two ADCs (one each for the I-channel and Q-channel) has an equivalent functionality.

- **Elonics E4000 / Realtek RTL2832U Combination** [69] — The Elonics E4000 tuner works differently to the Rafael R820T, as it is a direct downconverter and operates with *zero-IF* [46] [98]. This means that it downconverts a wide band of RF signals (roughly 10MHz wide) directly to baseband with *no IF stage*, performing quadrature demodulation as it does so [81]. The baseband I and Q signals are then input to the RTL2832U, sampled, decimated, and output over its USB interface as with the other architecture. Elonics ceased manufacture of the E4000 in 2012, and as a result, the stock of E4000 based devices is dwindling. For this reason, it is most likely that the RTL-SDR you have (or intend to get!) will use the R820T/ RTL2832U combination.

To give an example of the RTL-SDR in action, let's assume that you wanted it to receive a standard FM radio station centred at 102.5MHz. To do this, you would set the centre frequency of the RTL-SDR, f_c, to 102.5MHz, and apply an appropriate tuner gain. The signal would be downconverted by the tuner from RF to an IF (or zero-IF) using either the R820T or the E4000, sampled at 28.8MHz by the RTL2832U, and then output as a baseband IQ signal to your computer. At this point, the signal would still be FM modulated (at baseband), and would need to be demodulated in order to recover the audio signal. As you will see (and do!) later in this book, the demodulation process can be performed either in MATLAB or Simulink, both of which will demodulate in *real-time* and output the audio signal to your PC speakers.

The rate at which samples are output by the RTL-SDR can also be configured, and reducing the rate often improves the performance of baseband processing on less powerful computers. Setting the 'f_s' value to 2.4MHz, for instance, will configure the RTL2832U to decimate by a factor of 12 (noting the ADC samples at 28.8MHz), meaning that only 1 in every 12 IQ samples is output to the computer. With this reduced sampling rate, the signal the computer receives has a bandwidth of f_s, i.e. 2.4MHz (noting that it is a complex signal), and spans from

$$\left(-\frac{f_s}{2} \text{ to } \frac{f_s}{2}\right) \text{ Hz, i.e. } (-1.2 \text{ to } 1.2)\text{MHz},$$

as demonstrated in Figure 1.11 (on page 19). The 2.4MHz bandwidth means that, if you were tuning to a particular FM radio station, you would almost certainly receive several stations (FM radio stations in the UK are spaced 200kHz apart, so we could have up to 12 stations present, i.e. 12 x 200kHz = 2.4MHz). It would therefore be necessary to digitally filter the signal to remove the unwanted stations, before performing FM demodulation of the desired 200kHz wide station.

As mentioned previously, the nominal range of the R820T tuner is from 25MHz to 1.75GHz, and f_c can be set to any value within this range. The same is true with the E4000 tuner, which operates from 53MHz to 2.2GHz with a gap between 1.1GHz and 1.25GHz [113]. If you set f_c to one of these 'gap' frequencies when using the E4000 tuner, it would be unable to downconvert the signal to baseband without destroying it.

1.6.5 NooElec NESDR RTL-SDR Receivers

The first generation NooElec NESDR RTL-SDR itself is about the size of an ordinary USB memory drive, and comes packaged with an antenna and remote (which is included for anyone that wants to use the device for its intended purpose; see Figure 1.3), but with no software or any form of documentation. This is because the user is expected to source their own, depending on what they want to use the device for. More recently (in 2014), NooElec released some new RTL-SDRs that have a Printed Circuit Board (PCB) footprint smaller than 2cm² (the *Nano*), and also versions with higher quality components to improve tolerance issues (the *RF Shielded*—see Figure 1.2). The latter is more thoroughly shielded to reduce noise arising from its surroundings. All of these and more can be sourced from nooelec.com.

Figure 1.9 highlights some of the main components used on two of the more popular variations of the NooElec RTL-SDRs: the *NESDR Mini* and the *NESDR Nano*. These are:

- **MCX connector** — for attaching an antenna to the device (an omni-directional antenna is supplied with it);
- **Electrostatic Discharge (ESD) Diode** — protects the tuner from electrostatic discharge originating from the MCX antenna;
- **R820T** — the tuner chip, which selects a portion of the RF spectrum and downconverts to an IF;
- **RTL2832U** — the demodulator chip, which downconverts from IF to baseband, digitises the signal and reduces the sampling rate;

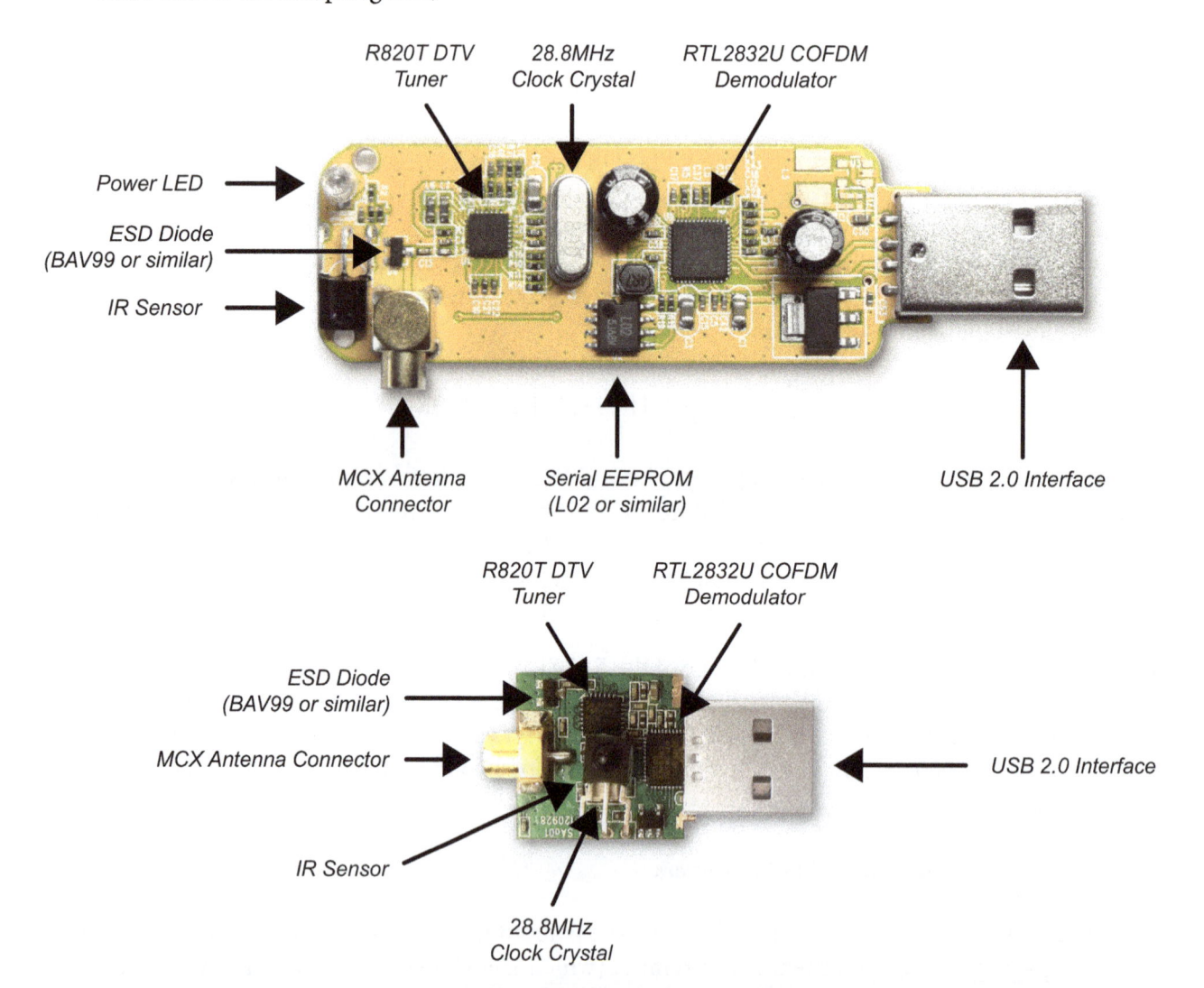

Figure 1.9: Key components of the NooElec NESDR Mini RTL-SDR (top) and the NooElec NESDR Nano RTL-SDR (bottom)

Chapter 1: Introduction

- **28.8MHz clock crystal** — provides a reference for frequency synthesis, and used for the generation of the local oscillator and clock (this component is common to both the R820T and RTL2382U);
- **USB 2.0 Interface** — part of the RTL2832U, used to transfer the baseband IQ data to the host PC;
- **The Infra Red (IR) Sensor** — used to interface with the remote control that is supplied with the device (neither of these are supported when the stick is working as an IF SDR);
- **Serial EEPROM (Electronically Erasable Programmable Read Only Memory)** — holds USB configuration information for the device, and is connected to the RTL2832U via an I²C bus [105].

1.7 Interfacing with the RTL-SDR from MATLAB and Simulink

After installing the RTL-SDR Hardware Support Package, the *RTL-SDR Receiver* Simulink block illustrated in Figure 1.10 will become available for use in your Simulink Library Browser. Similarly, the Hardware Support Package adds support to MATLAB in the form of the `comm.SDRRTLReceiver` System Object, also shown in the same figure. Both of these facilitate direct communication with any RTL-SDRs attached to your computer, and allow key parameters such as the RF centre frequency f_c, the sampling rate f_s, and the tuner gain K to be set. Additionally, a frequency correction value can be entered through both interfaces to compensate for any hardware tolerance issues associated with the R820T. We explain how to calculate the frequency correction value in Appendix A.3 (page 577), but you do not need to worry about this for now.

Looking at Figure 1.11, which shows the internal structure of the RTL-SDR, you may notice how these parameters relate to physical values in the hardware. What may not be obvious however, is that the VCO used by the R820T is not set with the value of parameter f_c, but is actually set to a value f_{lo}, where $f_{lo} = f_c - f_{if}$ and f_{if} is the tuner's IF frequency. As an example, lets consider that we would like to receive a signal that is being transmitted on an RF carrier with a frequency of 400MHz. This signal must be demodulated down to baseband by the RTL-SDR, but because the R820T has an IF value of $f_{if} = 3.57$MHz [95], its VCO is configured to oscillate at a value of:

$$f_{lo} = 400 - 3.57 = 396.43 \; MHz \, .$$

When the RF signal (centred at f_c) is mixed with the VCO's sinusoid (oscillating at f_{lo}), we get:

$$\left[\cos 2\pi(400e6)t\right]\cdot\left[\cos 2\pi(396.43e6)t\right] = \frac{1}{2}\left[\left(\cos 2\pi(3.57e6)t + \cos 2\pi(796.43e6)t\right)\right],$$

recalling the trigonometric identity $\cos(A)\cos(B) = 0.5[\cos(A-B) + \cos(A+B)]$. The component at 796.43MHz is attenuated by the R820T's IF lowpass filter and only the IF component (the energy around f_{if}) is propagated forward to the next stage, as shown in Figure 1.11. Fortunately when we set the centre frequency in MATLAB and Simulink we can simply enter the f_c centre frequency of interest (400MHz in this example), and the software will automatically perform this calculation and configure the hardware appropriately.

Software Defined Radio Using MATLAB & Simulink and the RTL-SDR

This is how the RTL-SDR Receiver block will appear in Simulink models

tunable parameters

Double clicking on the block opens its parameters window.

In the parameters window, you can configure the properties of your RTL-SDR

This is how an RTL-SDR System Object can be initialised in MATLAB code

Parameters can be set in a similar manner as with the Simulink block

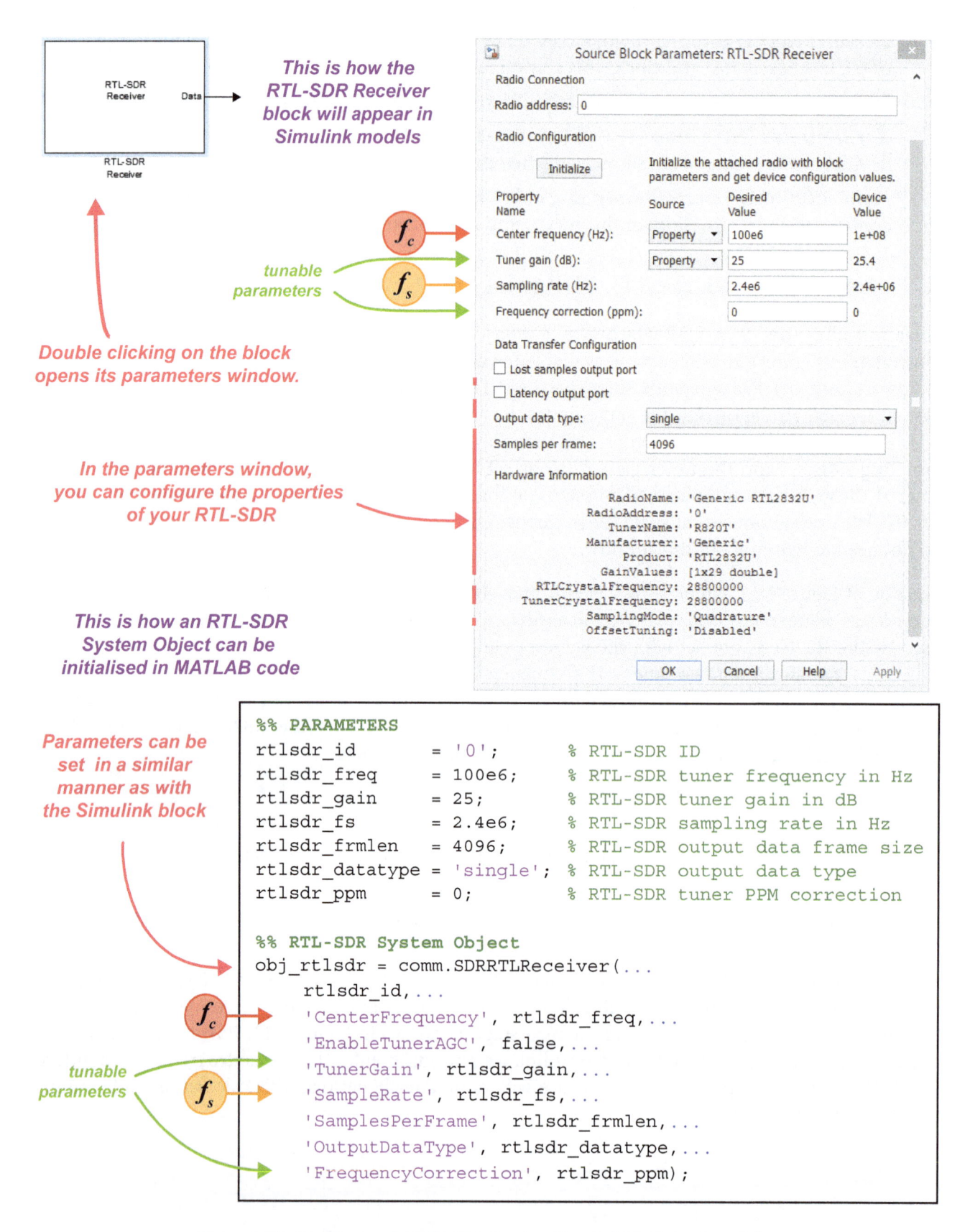

```
%% PARAMETERS
rtlsdr_id       = '0';       % RTL-SDR ID
rtlsdr_freq     = 100e6;     % RTL-SDR tuner frequency in Hz
rtlsdr_gain     = 25;        % RTL-SDR tuner gain in dB
rtlsdr_fs       = 2.4e6;     % RTL-SDR sampling rate in Hz
rtlsdr_frmlen   = 4096;      % RTL-SDR output data frame size
rtlsdr_datatype = 'single';  % RTL-SDR output data type
rtlsdr_ppm      = 0;         % RTL-SDR tuner PPM correction

%% RTL-SDR System Object
obj_rtlsdr = comm.SDRRTLReceiver(...
    rtlsdr_id,...
    'CenterFrequency', rtlsdr_freq,...
    'EnableTunerAGC', false,...
    'TunerGain', rtlsdr_gain,...
    'SampleRate', rtlsdr_fs,...
    'SamplesPerFrame', rtlsdr_frmlen,...
    'OutputDataType', rtlsdr_datatype,...
    'FrequencyCorrection', rtlsdr_ppm);
```

tunable parameters

Figure 1.10: Configuring the *RTL-SDR Receiver* Simulink block (top), and the RTL-SDR MATLAB System Object (bottom) with identical parameters

Chapter 1: Introduction

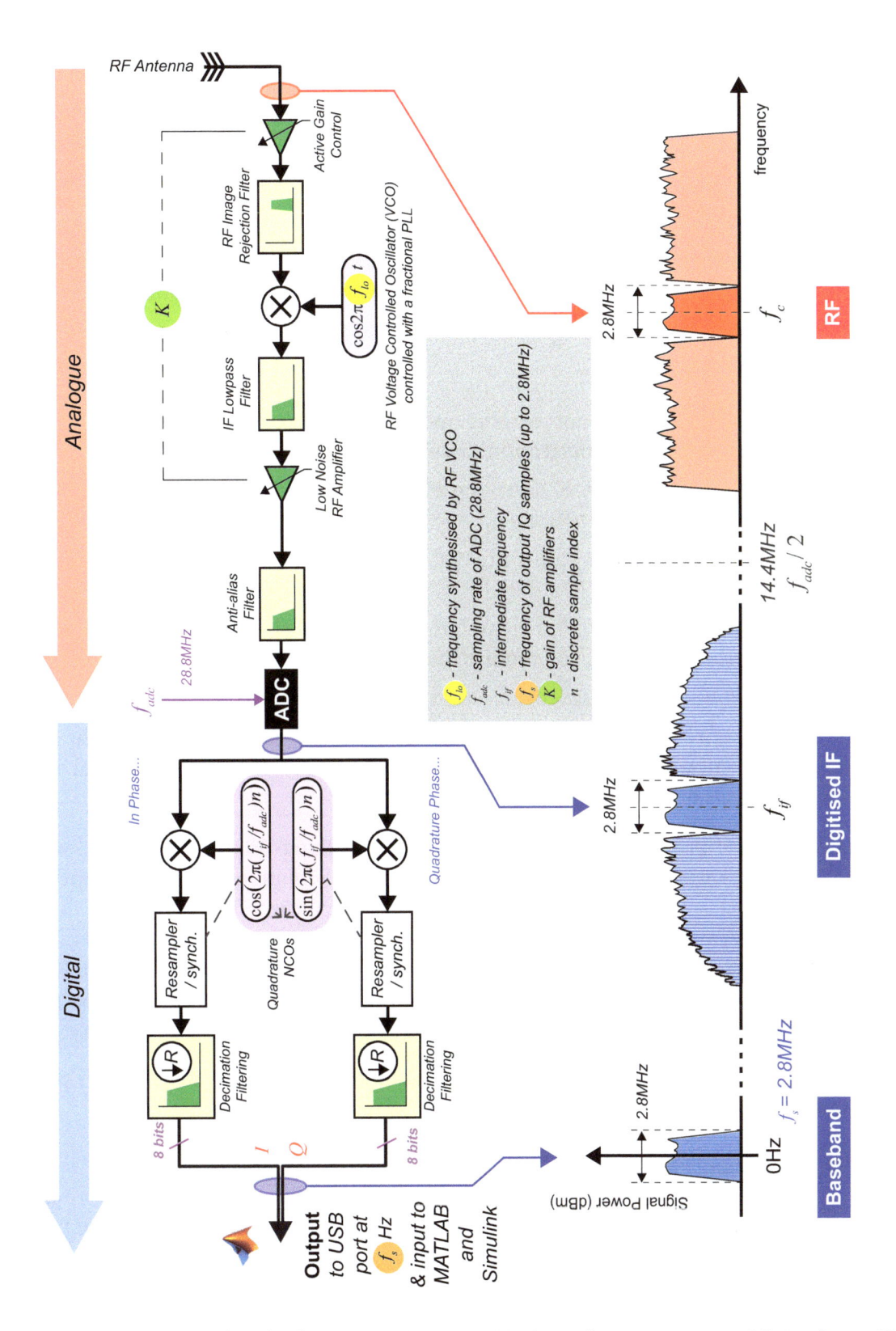

Figure 1.11: The process carried out by the R820T/RTL2832U RTL-SDRs: downconverting an RF signal to an IF, then digitising to baseband. (Note compared to Figure 1.8 this block diagram runs right to left, rather than left to right). The parameters of baseband sampling rate, f_s, the tunable gain, K, and the RF centre frequency, f_c are set in the Simulink *RTL-SDR Receiver* block or MATLAB `comm.SDRRTLReceiver` system object

1.8 Practicalities and Some Challenges of (Low Cost) Desktop SDR

Based on the authors' experiences, we are confident that you will be up and running quickly, and implementing SDR designs with the RTL-SDR, MATLAB and Simulink. Even so, there are a few potential challenges and pitfalls to be aware of. Some of these we will outline now, and others will be addressed where appropriate at various points throughout the book. The challenges you may encounter could vary from RF and antenna problems, hardware or software issues, not having the right components and tools, or software installation or run-time problems.... but of course this is engineering, and solving and addressing any or all of the preceding problems is what engineering is all about!

Examples throughout this book will challenge you to implement real SDR systems, and engineering issues you might encounter are:

1. The omnidirectional antenna not working optimally in the particular band you are using; i.e. not enough gain, or a sensitivity/ polarisation issue that needs to be addressed.
2. The tuner hardware in the RTL-SDR having a frequency offset or error, and further, that the error drifts with time, or as the device heats up.
3. Challenges with the front end gain and AGC of the RTL-SDR, leading to saturation or similar problems.
4. Your computer not being powerful enough to run the DSP algorithms in MATLAB or Simulink in real-time, perhaps meaning you require the use of a faster machine.
5. Complexity of implementing synchronisation and timing circuits.
6. Issues arising from being in a very busy RF environment, or, at the other end of the scale, perhaps in an environment where there are very few RF signals to receive (such as in rural areas or inside shielded buildings or rooms).

Hence to maximise the progress that can be made from working through this book, we have made available SDR exercises and examples that can optionally run *without* the RTL-SDR hardware, instead using pre-recorded signals (which we provide) that you can learn with! And even if you don't (yet) have MATLAB or Simulink, we also provide you with some videos to show you the simulations as screen recordings. Of course, the best way to learn about SDR is to ensure you have an RTL-SDR, the appropriate installation of MATLAB and Simulink (complete with the required toolboxes and hardware support packages), and to get hands-on!

In the following subsections, we provide a suggestions guide on what you can achieve if you are having problems, and/or don't (yet) have all the components.

1.8.1 "I do not actually have a working RTL-SDR device installed yet..."

We would of course recommend that you obtain one, given the sub-$20 cost. However if you can't get it running (or you are waiting on your delivery perhaps!), then **you can still learn about SDR and work through workbook examples without the RTL-SDR**. This is possible because, in many exercises and examples, we have provided a recording of a typical received RF signal that you can use as an input to the various SDR design examples. Importing a recorded signal is as simple as connecting our *Import RTL-SDR Data* block in place of the *RTL-SDR Receiver* block, as shown in Figure 1.12. Importing signals is outlined later in Section 4.13 of Chapter 4 (page 165).

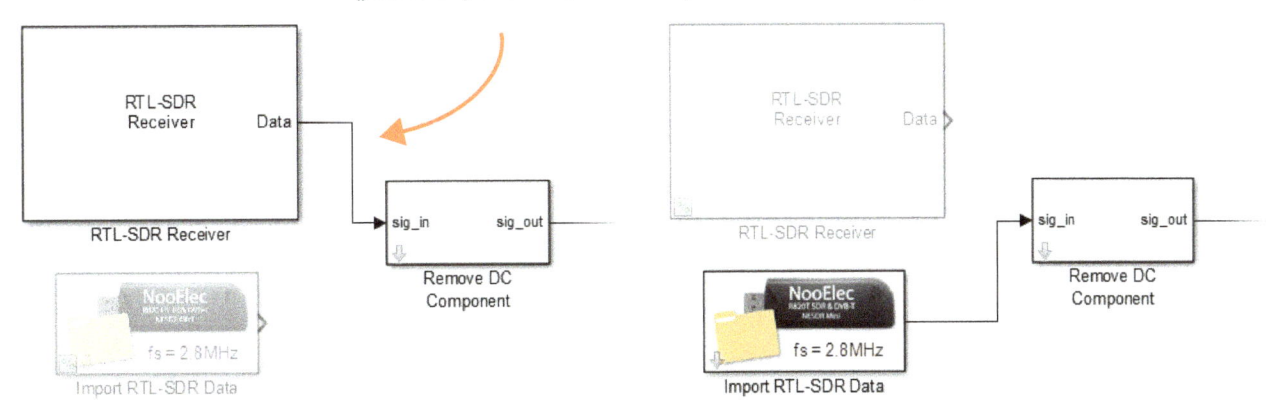

Figure 1.12: The *RTL-SDR Receiver* block is used to configure the RTL-SDR, and is also the source through which received RF signals enter the Simulink model. The *Import RTL-SDR Data* block has been developed by the authors to act in place of the *RTL-SDR Receiver*. It imports saved RTL-SDR data from a file, outputting it at the rate it was originally recorded at. This can be very useful for situations in which you do not have an RTL-SDR to hand, but still want to run and complete exercises to learn from them.

1.8.2 "I am having installation or setup problems with the RTL-SDR hardware..."

You can visit mathworks.com and look for support on any aspect of MATLAB or Simulink and the RTL-SDR Hardware Support Package. If you are having hardware problems or difficulties with the USB drivers, visit desktopSDR.com and look in the *Troubleshooting* section. You can also search for your issue online, and it may be that the extensive RTL-SDR community can provide an answer!

1.8.3 "I do not have MATLAB or Simulink installed yet..."

Once again, we strongly recommend that you procure a suitable installation (student, home or professional) of MATLAB and Simulink (2014b or later), the Communications and DSP System, and Signal Processing Toolboxes, and the RTL-SDR Hardware Support Package. If you are not able to install these in the short term however, you can still learn from this book. Many exercises provide a 'screen simulation' version of the example as a video file. Therefore, you can watch the video to observe what you would typically see on-screen if you were running the real example using the MATLAB/ Simulink software and an RTL-SDR device. We will discuss how these videos will be presented in Section 1.10.

1.8.4 "I do not really know what is meant by SDR and I am new to the software and hardware..."

In this case, you perhaps need some early conceptual support as to what radio transmitters and receivers are, and what SDR implementations will do. Similar to the screen simulation videos, we have a number of system setup videos, where we will show a video the complete setup, outside of the computer. To give an example, one video you will see later shows music being played from the audio output of a smartphone, modulated onto a carrier and transmitted at RF across the desk to the RTL-SDR. In the video you can also see and hear the signal being received (the MATLAB/ Simulink software receives baseband samples from the RTL-SDR device, demodulates to baseband, and then outputs to the audio from the computer speakers). More will be discussed about these videos in the following section.

1.9 Working with Discrete and Continuous Time Signals and Equations

In the book we are working with both continuous time (analogue) and discrete time (digital) signals. All signals received by the RTL-SDR and input to MATLAB and Simulink will of course be in discrete time, and we represent these sampled signals as $s[n]$ (or similar), using the time index n or sometimes k and square brackets. Signals from the 'real' world will of course be in continuous time, and we use the conventional notation $s(t)$ for a time varying signal. In cases where we require to go from continuous to discrete time (or visa versa) in a signal flow graph or set of equations, we will assume that suitable ADCs or DACs are available (deployable!) to allow this happen, as shown in Figure 1.13. As such, where it suits our notation and purpose, $s(t)$ and $s[n]$ can considered equivalent across the frequency range 0 to $f_s/2$.

$s(t)$ → [Anti-Alias Filter] → [ADC] → $s[n]$ $s[n]$ → [DAC] → [Reconstruction Filter] → $s(t)$

continuous time   **discrete time**   **continuous time**

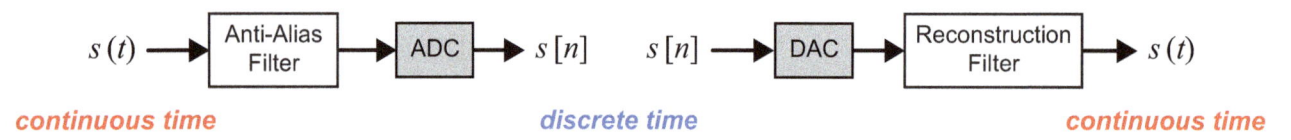

Figure 1.13: The equivalence of $s[n]$ and $s(t)$ when using DAC and ADC pairs

1.10 The Structure of the Book and Format of the Exercises

The remainder of this book is comprised of tutorial style review material *for reading*, and exercises *for doing*. All of this material is structured such that users can interact with it on different levels; and hence the early chapters are introductory. The main objectives of this book are to tutor unfamiliar readers in working with MATLAB, Simulink and the associated toolboxes, and thus to gain hands-on experience of building and designing DSP enabled SDR systems from first principles.

1.10.1 Exercise Layout

Each exercise has a title and an exercise number (you can find a complete list of all exercise titles and numbers on page vii). The start of exercises are clearly marked with a black bar, and they are distinguishable from the main 'body text' because they are presented in a different font (Arial and *Arial Italic*). The features of typical exercise presentation are demonstrated in Exercise 1.0.

Exercise 1.0   This is the start of an exercise, and the title would go here

When reading through the exercises, you will encounter a number of brightly coloured icons. These are used to represent the different types of activity presented. The activity icons are:

| Step-by-step tutorial | Hands-on model building, coding, (and also circuit building) | Reference design | Run/ simulate code or model | Watch video of on-screen design or simulation | Watch video of system setup or live operation | Question or challenge |

Chapter 1: Introduction 23

Exercises usually comprise a number of steps, each labelled with a letter, e.g. (a), (b), (c). Activity icons will normally appear to the left of the page, especially where a particular step moves onto a new type of activity, or extends across pages. The meanings of these icons are explained in more detail below.

For readers with no experience of using MATLAB or Simulink, some exercises show rudimentary steps such as how to set parameters, or fundamentals such as finding the 'run simulation' button in Simulink or just viewing and interpreting key information. When MATLAB or Simulink functionality is being explained we use the pink 'footstep' icon.

Some exercises will involve designing and building SDR systems from first principles, where readers will essentially start with a 'blank' design and under step by step instruction build the SDR implementations from scratch. In others, we will ask you to build physical circuits to construct transmitters. Where there are design activities in exercises, they will be denoted by the red 'tools' icon.

To bypass the design and build stage, or to troubleshoot your model, reference designs are provided within most exercises. The reference design will be denoted by the purple 'reference book' icon and the reader will be directed to the specific file via a pathname from the root folder.

Once a system is available (either from the design and build process, or from opening a reference design), it can then be run or executed, and parameters can be changed and their effects evaluated. Execute activities are denoted by the 'green running person' icon.

We also include screen recordings of some MATLAB and Simulink exercises. Watching these will allow users to confirm expected sequences, results, etc.. Also, perhaps, if a reader does not have an RTL-SDR nor a copy of MATLAB or Simulink installed yet, this allows them to at least get started. The orange 'signal on a screen' icon denotes a screen recording is available.

Several examples use external hardware (such as an audio output from a tablet or smartphone, or an additional transmitter) to generate RF signals from the desktop. For some of these, we provide a full video of the implementation showing both the on-screen SDR systems and video of the other devices on the desktop. An orange 'film strip' icon shows if there is a video available.

Finally, while we can provide many examples and designs, the best learning is often when working to produce your own designs. Therefore in some exercises, we will suggest some extensions to the exercise content that you might try. Challenge activities will be denoted by the blue 'question mark light bulb' icon.

When working with the MATLAB and Simulink software, certain mouse clicks are used. The examples in this book were developed with a Windows 8.1 OS running MATLAB and Simulink. As such, both left and right mouse buttons are used, with various combinations of clicks, double clicks and so on. As a shorthand for the appropriate mouse clicks in different scenarios, we will adopt the graphical notation shown in Figure 1.14.

If you are working through this book using a Linux or Mac computer with MATLAB and Simulink installed, you will need to use the equivalent mouse and keyboard combinations. Instructions on how to enable the 'right click' in OS X Yosemite can be found here [75].

1	Left mouse button: *click once*	1	Right mouse button: *click once*
H	Left mouse button: *hold down*	H	Right mouse button: *hold down*
2	Left mouse button: *double click*	2	Right mouse button: *double click*

Figure 1.14: Mouse click notation used throughout this book

1.10.2 Chapter Outline

A brief outline of the remaining chapters is given below.

- **Chapter 2** begins with first steps in using the RTL-SDR with MATLAB and Simulink. We will take you through the hardware and software setup, and get you up and running, receiving 'live' RF signals as soon as possible with the RTL-SDR!

- **Chapter 3** introduces the RF spectrum in more detail, and encourages you to explore the spectrum to find signals that are present around you (acknowledging that the usage of the spectrum varies geographically). MATLAB and Simulink features that help you to view the spectrum are reviewed.

- **Chapter 4** provides a tutorial based introduction to the MATLAB and Simulink software development tools, for those new to them. We will also introduce the use of the Communication, Signal and DSP toolboxes required for the exercises, and also how to import recorded RTL-SDR signals.

- **Chapter 5** gives an introduction to the concept of complex signals, and reviews how these relate to 'real' signals in quadrature modulated systems.

- **Chapter 6** covers analogue AM from a theory and simulation perspective, and introduces mathematics relevant to the modulation and demodulation process. This includes generating and analysing the different variants of AM signals.

- **Chapter 7** is about frequency tuning and synchronisation. These two concepts are key to almost all types of radio receiver. The major building block for synchronisation circuits, the Phase Locked Loop (PLL), is reviewed and simulation exercises are provided.

- **Chapter 8** is concerned with 'live' modulation and demodulation of AM signals, using the RTL-SDR; and some ideas for low-cost AM transmitters are also presented.

- **Chapter 9** considers analogue FM from a system and simulation perspective, and introduces mathematics relevant to the modulation and demodulation process.

- **Chapter 10** follows on from the theory in Chapter 9 with some 'live' FM modulation and demodulation examples using ambient radio signals, desktop transmitters, and the RTL-SDR.

- **Chapter 11** moves the focus from analogue to digital communications, and covers fundamentals on modulation schemes and pulse shaping, as well as more advanced topics like carrier synch, symbol timing, and phase ambiguity correction.

- **Chapter 12** continues with 'desktop digital comms' experiments, using the USRP® hardware to generate test signals we can receive with the RTL-SDR. We present some examples of real wireless transmission and reception of data over the air.

- A set of **Appendices** provides detailed information on installation and configuration of the key components, lists common equations, and expands some theoretical background, etc..

Next, we move on to our first exercises and begin to receive radio signals! It's open the box time.

2 Open the Box! First SDR with MATLAB and Simulink

This book is intended to accelerate your progress with real time desktop SDR, so in this chapter we move straight on to practical work with some 'getting started' exercises. We will help you check that you have the correct versions of MATLAB and Simulink installed (along with all of the required toolboxes), then move on to installing the RTL-SDR Hardware Support Package. After you confirm that MATLAB can communicate with your RTL-SDR, we will help you set up the MATLAB environment as required to complete the rest of the exercises in this book. Once all of these prerequisites are complete, we will move on to help you run your first real-time desktop SDR MATLAB and Simulink receivers. If you would like to become more familiar with MATLAB and/ or Simulink before completing the steps in this chapter, you should explore our *Getting Started with MATLAB and Simulink* guide, which starts on page 99.

2.1 Getting Started: Hardware and Software Checklist

Since you have read this far, you are obviously keen to begin working with the RTL-SDR! In order to be properly prepared to follow the exercises in this book, we just need to confirm that you have all of the necessary hardware and software, as outlined in Section 1.3 (page 5). To recap, these included

- *an RTL-SDR*
- *a powerful computer running a recent operating system (this book uses Windows 8.1)*
- *MATLAB & Simulink Version R2014b or later*
- *MathWorks DSP System Toolbox*
- *MathWorks Communications System Toolbox*
- *and the MathWorks Signal Processing Toolbox*

To confirm that you have the correct version of MATLAB and Simulink, and the required toolboxes, please complete Exercise 2.1.

Exercise 2.1 Verify Software Setup: MATLAB and Simulink

This exercise will be used to confirm that your installation of MATLAB and Simulink contains all of the necessary toolboxes required to proceed with the exercises in this book.

(a) **Start MATLAB.** MATLAB can be opened in a few different ways. You can 2️⃣ on the desktop icon, 1️⃣ on the Start menu, Start screen, Dock or Launchpad shortcuts, or run the MATLAB start script through Terminal, depending on your computer environment.

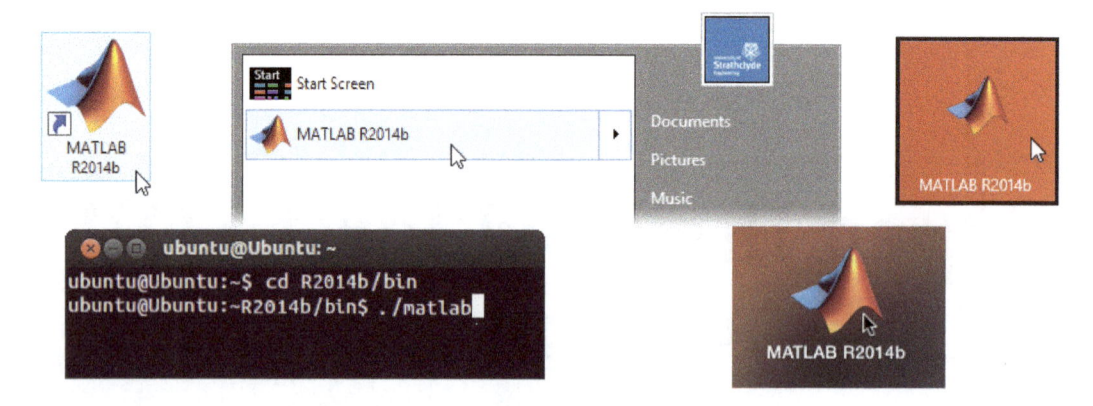

When MATLAB loads, you should see an environment layout similar to what is shown below.

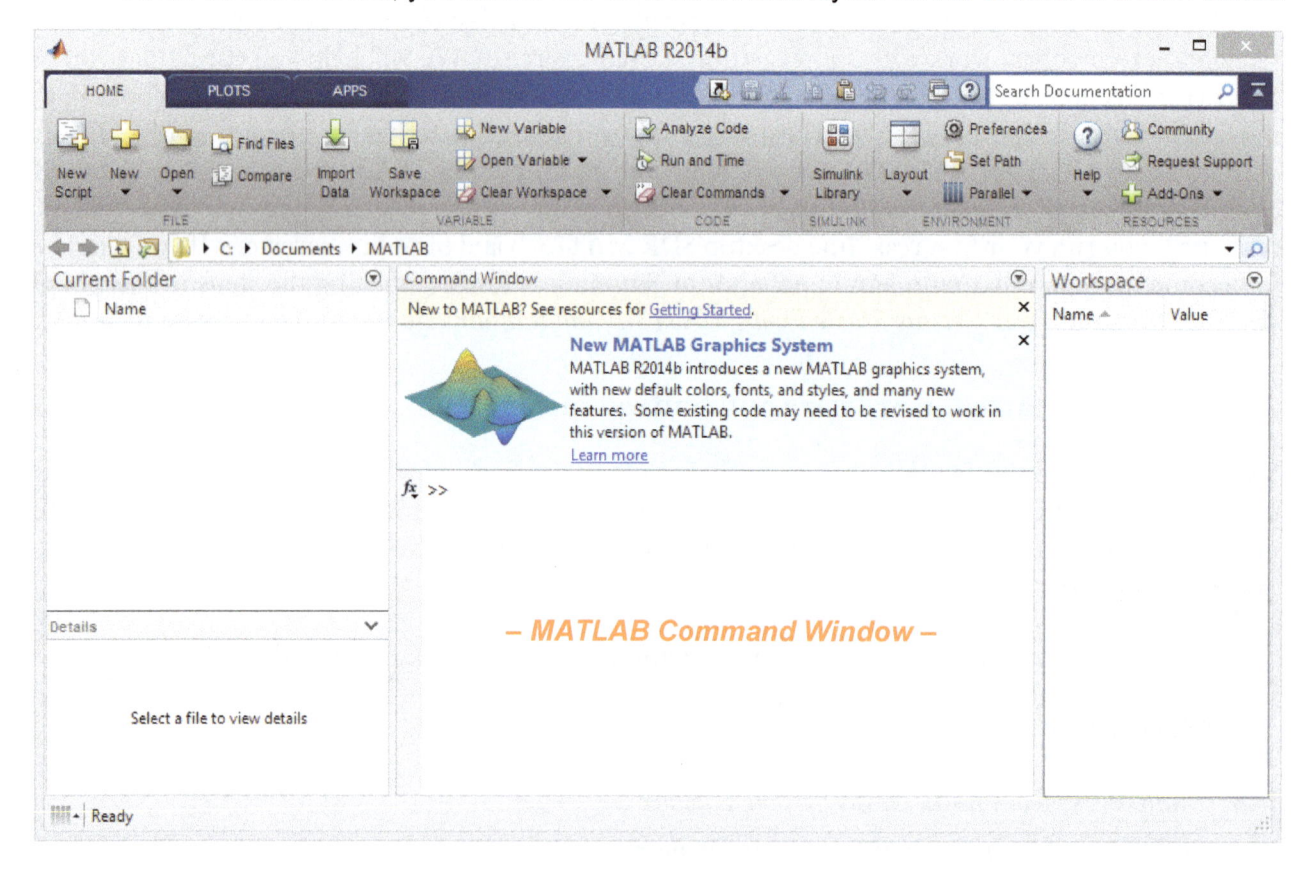

Chapter 2: Open the Box! First SDR with MATLAB and Simulink 27

(b) **Display software version information.** Type `ver` into the MATLAB command window, and then press the Enter key on your keyboard to print details about the software version and toolboxes that you currently have installed (the command window is the area highlighted in the screenshot opposite). This should return details similar to the following:

```
>> ver

-----------------------------------------------------------------
MATLAB Version: 8.4.0.150421 (R2014b)
MATLAB License Number: ######
Operating System: Microsoft Windows 8.1 Pro Version 6.3 (Build 9600)
Java Version: Java 1.7.0_11-b21 with Oracle Corp Java HotSpot(TM)
-----------------------------------------------------------------
MATLAB                                    Version 8.4    (R2014b)
Simulink                                  Version 8.4    (R2014b)
Communications System Toolbox             Version 5.7    (R2014b)
DSP System Toolbox                        Version 8.7    (R2014b)
Signal Processing Toolbox                 Version 6.22   (R2014b)

...details of other toolboxes may be displayed here too
```

If you are running a version of MATLAB lower that R2014b, we advise you to upgrade before proceeding. If you are missing any of the required toolboxes, please purchase copies of them from the MathWorks website. You can leave MATLAB open as you will need it in the following section.

2.2 Getting Started: Installing the RTL-SDR Hardware Support Package

It is now time to add the RTL-SDR Hardware Support Package to your installation of MATLAB and Simulink. The information given in reference [61] (the hyperlinked URL below)

mathworks.com/help/supportpkg/rtlsdrradio/ug/support-package-hardware-setup.html

provide step by step installation instructions for users of Windows, Linux and Mac operating systems. It covers:

- Launching the *Get Hardware Support Packages* wizard,
- Selecting and installing the RTL-SDR Hardware Support Package software (which requires a free MathWorks account),
- Installing the RTL-SDR USB driver,

and also contains some basic troubleshooting information. Make sure that you choose the 'Install from Internet' option, as shown in our install video:

desktopSDR.com/videos/#rtlsdr_hw_supportpkg

Once you have finished the installation, please try to complete Exercises 2.2 and 2.3.

| Exercise 2.2 | Verify Software Setup: RTL-SDR Hardware Support Package |

This exercise will be used to confirm that the RTL-SDR Hardware Support Package has been installed successfully.

(a) **Open MATLAB.** If you do not have MATLAB open, return to Exercise 2.1 (page 26), and repeat Part (a).

(b) **Start Simulink.** 1 on the *Simulink Library* button in the MATLAB Home ribbon at the top of the MATLAB window, as circled below:

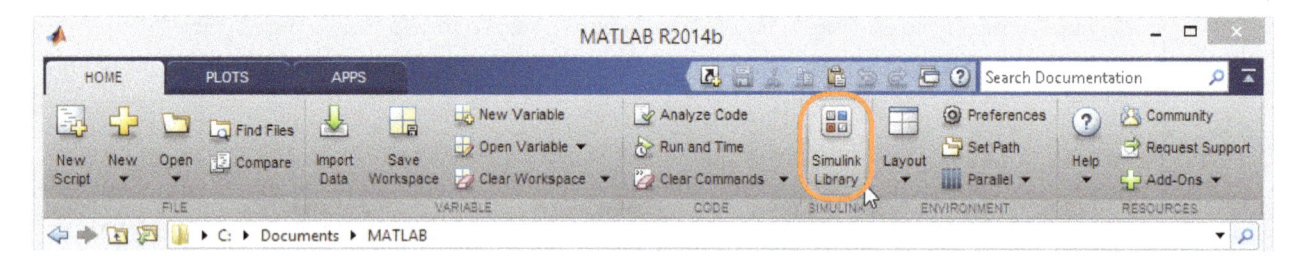

This should cause the Simulink Library Browser window to open, and it will display something similar to the screenshot shown below (your view may differ subject to your window size, and the libraries and toolboxes you have installed on your computer). Notice that the Signal Processing toolbox is not shown—it does not contain any Simulink blocks and hence is not listed.

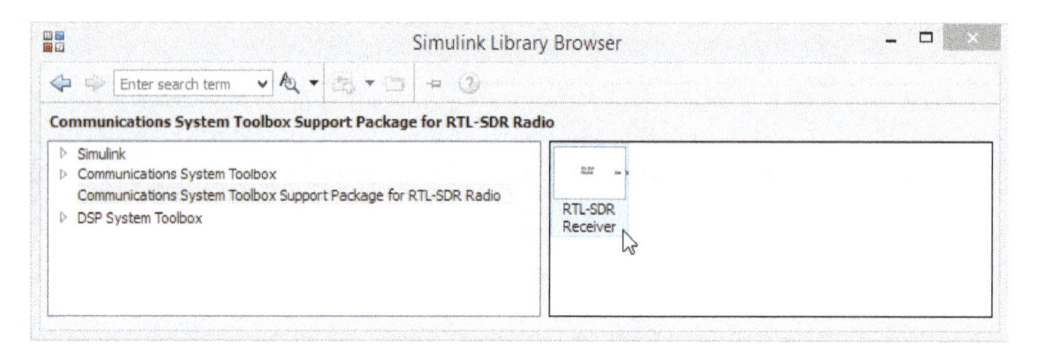

(c) **Confirm the RTL-SDR Hardware Support is present in Simulink.** You should see that one of the libraries listed is the ▦ > *Communications System Toolbox Support Package for RTL-SDR Radio*. This should have been installed automatically when the RTL-SDR Hardware Support Package was added to your computer. If you 1 on this library, you should notice that it contains a single block, titled *RTL-SDR Receiver*. This is a parameterisable RTL-SDR interface which is able to bring samples from the RTL-SDR into Simulink in real time.

If you do *not* see the RTL-SDR library, try typing `setupsdrr` into the MATLAB command line. If this command is not known, you will need to re-install the Hardware Support Package following the instructions detailed in [61].

(d) **Review the RTL-SDR Hardware Support Package documentation.** Close the Simulink Library browser, and type `sdrrdoc` into the MATLAB command window. If the Hardware

Chapter 2: Open the Box! First SDR with MATLAB and Simulink

Support Package has installed successfully, a help window should appear titled *Communications System Toolbox Support Package for RTL-SDR Radio*. Perhaps take some time now to read through this documentation.

The RTL-SDR Hardware Support Package documentation can be viewed using the `sdrrdoc` *MATLAB command*

→ Command Window
`>> sdrrdoc`

If the help window did not appear, the Hardware Support Package has not installed correctly. Repeat the installation process detailed in [61].

Exercise 2.3 Verify Hardware Setup: RTL-SDR Hardware Support Package

This exercise will be used to check that the RTL-SDR drivers have been installed on your computer. To complete this exercise you will require an RTL-SDR. If you do not currently have an RTL-SDR, return to this exercise once you have sourced one!

(a) **Open MATLAB.** If you do not have MATLAB open, return to Exercise 2.1 (page 26), and repeat Part (a).

(b) **Connect the RTL-SDR.** If you have not yet connected the RTL-SDR to your computer, plug it into a free USB2.0 (or USB3.0) port. Connecting the antenna is not necessary at this stage, but having it plugged in will do no harm.

(c) **Check that the RTL-SDR Hardware is Recognised.** Enter 'my_rtlsdr = sdrinfo' into the MATLAB command window. This should display the following information:

```
>> my_rtlsdr = sdrinfo

my_rtlsdr =

              RadioName: 'Generic RTL2832U OEM'
           RadioAddress: '0'
            RadioIsOpen: 0
              TunerName: 'R820T'
           Manufacturer: 'Realtek'
                Product: 'RTL2841UHIDIR'
             GainValues: [29x1 double]
     RTLCrystalFrequency: 28800000
   TunerCrystalFrequency: 28800000
           SamplingMode: 'Quadrature'
            OffsetTuning: 'Disabled'
```

For the procedure and result expected when you have more than one RTL-SDR device connected to your computer, see Appendix A.1 on page 569, where we will also explain a little more about the 'Radio address' of your RTL-SDR. (There are some exercises later in the book where more than one RTL-SDR is used at once.)

If your device was *not* recognised or was *not* plugged in, then you will get a null return as shown below (you can unplug your working RTL-SDR if you wish to see this result):

```
>> my_rtlsdr = sdrinfo

my_rtlsdr =

    {}
```

If MATLAB is unable to recognise your RTL-SDR (but knows what the `sdrinfo` command is), it is likely that the driver has not been successfully installed. Type `targetupdater` into the command window. This will open the Support Package Installer wizard you should have encountered previously. Repeat the steps detailed in [61] to reinstall the drivers.

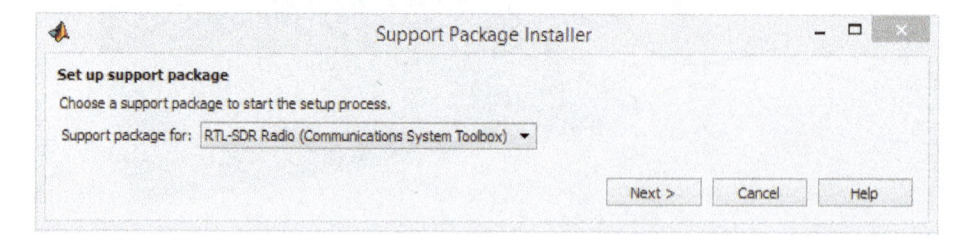

If after completing this, MATLAB is still unable to recognise your RTL-SDR, please take a look at our troubleshooting advice at desktopSDR.com, and/or in the RTL-SDR Hardware Support Package documentation from MathWorks [60] for more information.

Chapter 2: Open the Box! First SDR with MATLAB and Simulink

2.3 Getting Started: Book Support Files and the MATLAB Environment

A set of files accompany this book. These files provide the reference designs and other resources necessary to work through the exercises. As you will need regular access to this folder, it is worth positioning this at the top level of a drive, e.g.

`c:/rtlsdr_book/`   *or similar*

Irrespective of your chosen location, do ensure that (i) there are no spaces in the file path, and (ii) that you have full write control over the folder and its contents. We also advise against working from a network drive, as this may extend compilation times and affect your experience of working with the models.

From now on, the file path of the top level directory will be referred to graphically with a folder icon:

📁

During the remainder of the book, we will refer to examples within the folder by supplying the relative filepath, for instance,

📁`/intro/rtlsdr_rx_startup_simulink.slx`

If you are reading this in electronic form on your PC, it is also worth noting that clicking on the blue link text will open the referenced file (subject to appropriate settings on your computer). For example, clicking on the above link will open the `rtlsdr_rx_startup_simulink.slx` model in Simulink.

A custom RTL-SDR Simulink library called *RTL-SDR Book Library* has been prepared to accompany the exercises in this book. It provides a set of blocks which will be used as you work through the various Simulink based exercises. The library resides in the root directory of the support files provided with this book, and this folder must be added to your MATLAB path in order for the library to be listed in the Simulink Library Browser. Because some of the exercises in this book also use "Callback" routines to run functions located in the same folder, it is **essential** that this folder be added to the path. If the callback routines fail because of missing files, neither MATLAB or Simulink receivers will run and error messages will appear.

The following exercise will show you how to add the custom library to the path, and set the MATLAB *Current Folder*. Once this is done you will be ready to begin your first real-time desktop SDR exercises, where you will run both a simple RTL-SDR receiver Simulink model, and an RTL-SDR receiver MATLAB function. These will allow you to gain familiarity with the receiver and software tools before you begin hunting for RF signals in the next chapter!

As with the Hardware Support Package install, a video has been produced which shows the steps required to set up the MATLAB environment for the exercises in this book.

desktopSDR.com`/videos/#matlab_environment_setup`

| Exercise | 2.4 | **MATLAB (and Simulink) Working Environment Setup** |

We will use this exercise to demonstrate how to include the custom Simulink library in the MATLAB path, and how to set the MATLAB 'Current Folder'.

(a) **Open MATLAB.** If you do not have MATLAB open, return to Exercise 2.1 (page 26), and repeat Part (a).

(b) **Open the 'Set Path' dialogue.** on the 'Set Path' button in the 'Home' ribbon at the top of the MATLAB window. This will open the dialogue shown below. Note that the paths already set in your installation of MATLAB might be different to what is shown below.

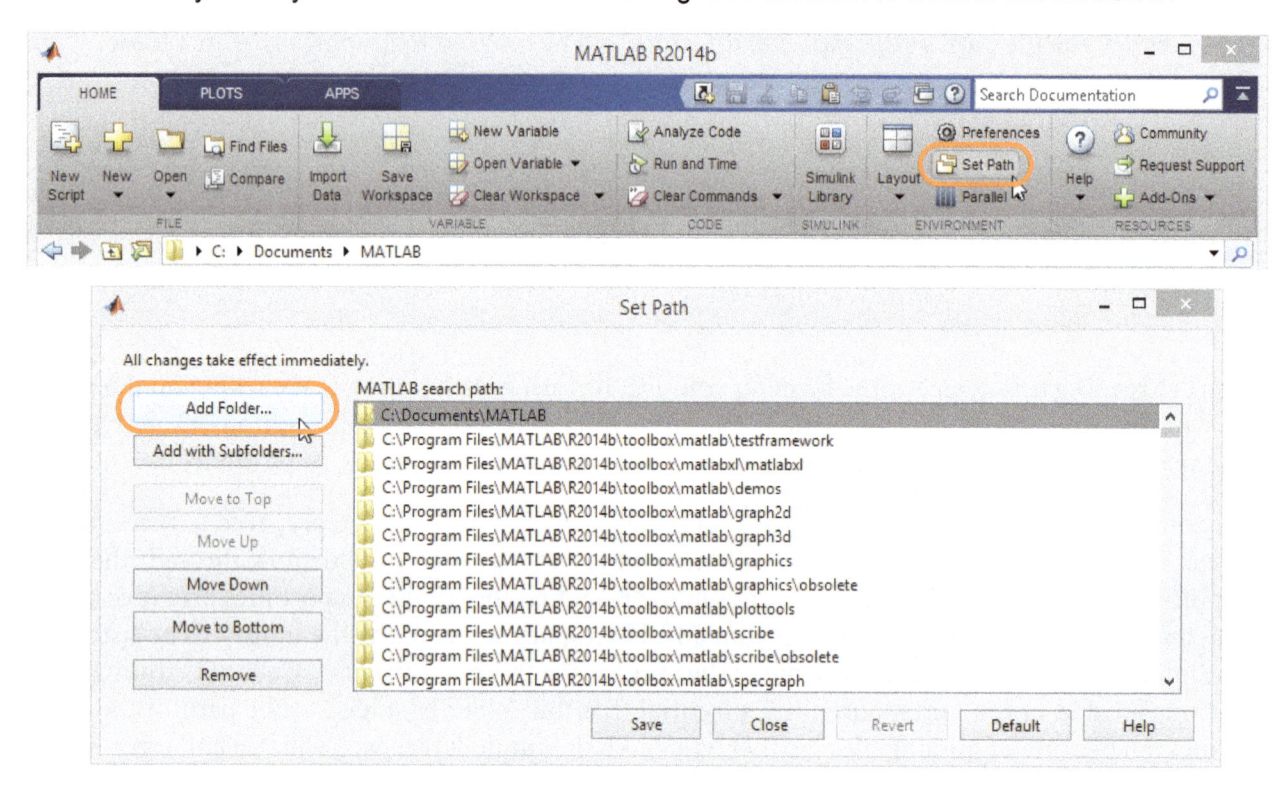

(c) the *Add Folder* button and navigate to:

/rtlsdr_book_library

Select Folder, then *Save* the changes and close the window. *Note that if you are not running MATLAB in an Administrator or Super User account, you may not be able to save these changes permanently. This means that you will need to repeat this process each time you open MATLAB.*

When you next open the Simulink Library Browser, you should find that a new library called **> RTL-SDR Book Library** is listed. If it does not immediately appear, in the left hand panel, and select *Refresh Library Browser*, or press the F5 key on your keyboard. There are a number of categories inside the top level library, and these contain blocks you will use throughout this book for a variety of different purposes. on each of these in turn and have a look at the various blocks provided.

Chapter 2: Open the Box! First SDR with MATLAB and Simulink

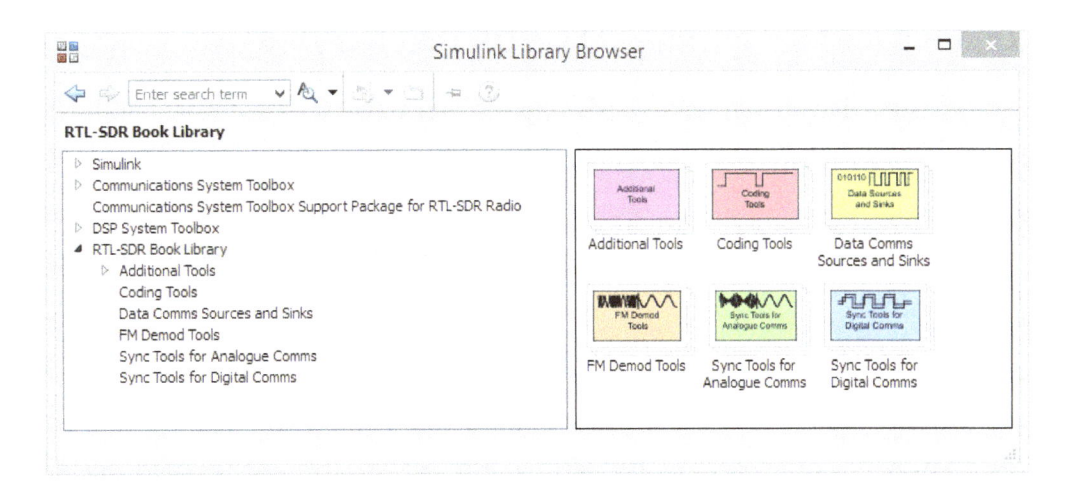

When you are finished exploring the ▦ > *RTL-SDR Book Library*, close the Simulink Library Browser and return to the MATLAB command window.

(d) To check that the other necessary files can be accessed, type

```
help import_rtlsdr_data
```

into the MATLAB command window. This will display help documentation for a System Object that has been created to accompany the book, called `import_rtlsdr_data` (to be reviewed in Section 4.13 of Chapter 4, page 165). If the documentation is displayed, then you have set the path correctly. If not, and it reports it not found, please repeat Part (b) to try adding the folder to the path again.

(e) **Set the MATLAB Current Folder.** In many of the exercises within the book, you will be asked to set the MATLAB *Current Folder*, such that you can easily access the various RTL-SDR MATLAB functions, scripts, data files and Simulink models as required.

For example, we might ask you to:

" *...set the working directory to:*

📁 `/intro/` "

To do this, simply enter the file path in the address bar at the top of the MATLAB window.

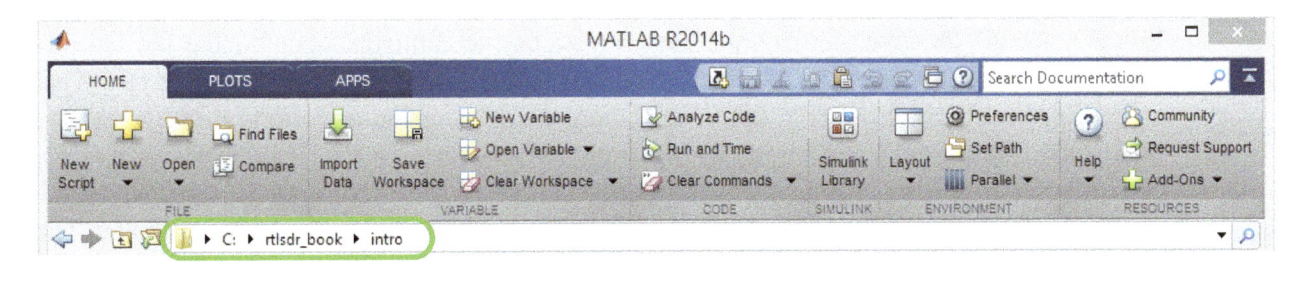

2.4 Running the First Desktop RTL-SDR Receiver Designs

Now that you have checked the Hardware Support Package is installed, that the RTL-SDR drivers are working, and have set the MATLAB path to include our custom Simulink library, it is time to move on and get started with your first desktop SDR receivers! This section contains two exercises, one where you will get the chance to run a simple Simulink based 'spectral analyser' receiver, and the other an equivalent receiver written in MATLAB code. In both cases, RF signals will be received by the RTL-SDR hardware, downconverted, digitised, and displayed in *Spectrum Analyzers* (frequency domain scopes) by the MathWorks software. A video of us completing these two exercises can be found here:

desktopSDR.com/videos/#rtlsdr_first_use

If you have no previous experience of using MATLAB or Simulink, and do not feel confident in starting with the RTL-SDR straight away, you might wish to work through the *Getting Started with MATLAB and Simulink* tutorial in Chapter 4 first, and then return to Exercises 2.5 and 2.6.

Exercise 2.5 First Use of the RTL-SDR: Simulink

This exercise gives you the chance to run your first RTL-SDR receiver Simulink design. The system will acquire and display the spectra of some of the RF signals that are broadcast in the air around you (the results of course will depend on where you are!). We will demonstrate one method of tuning the RTL-SDR centre frequency from Simulink, and also how to exert some control over its front end gain. Completion of this exercise also further ensures that the Simulink part of the Hardware Support Package has been installed successfully, and that your RTL-SDR hardware is functioning correctly.

(a) **Open MATLAB.** If you do not have MATLAB open, return to Exercise 2.1 (page 26), and repeat Part (a). After MATLAB has initialised, change the current folder to:

/intro/

(b) The file you need to open is called `rtlsdr_rx_startup_simulink.slx`, which will be displayed in the *Current Folder* pane of the MATLAB window. You can either open the model by 🖱² on it in this pane, or, if reading this document as a PDF on your computer (with the default and supplied file structure), by 🖱¹ on the link below to open it directly.

/intro/rtlsdr_rx_startup_simulink.slx

(c) **Browse the model.** When the model loads, you should see a Simulink window similar to that shown opposite. This model will allow you to receive your first signals with the RTL-SDR.

The *RTL-SDR Receiver* block (which was introduced to Simulink by installing the Hardware Support Package) acts as the interface to your RTL-SDR. The RTL-SDR is tuned to the value of the *Centre Frequency (MHz)* block, and tuner gain is applied according to the value of *RF Gain (dB)*. Figure 1.11 (on page 19) highlights which parts of the RTL-SDR hardware are configured with these parameters. As discussed in Section 1.6 (on page 10), signals received by the RTL-SDR are output from the device as a stream of complex 8-bit data samples. These IQ samples are output from the 'Data' port of the *RTL-SDR Receiver* block as a complex signal. For more detail on complex notation, please see Chapter 5 (page 171).

Chapter 2: Open the Box! First SDR with MATLAB and Simulink 35

[Screenshot of Simulink model `rtlsdr_rx_startup_simulink` showing the "First Use of the RTL-SDR: Simulink" title block (www.desktopSDR.com), with TUNING PARAMETERS (Centre Frequency 100 MHz, Gain 1e6, RF Gain 25 dB), INTERFACE WITH RTL-SDR (RTL-SDR Receiver block with fc, gain inputs and Data output), and PROCESS AND DISPLAY SIGNAL (Spectrum Analyzer FFT and Spectrum Analyzer Waterfall).]

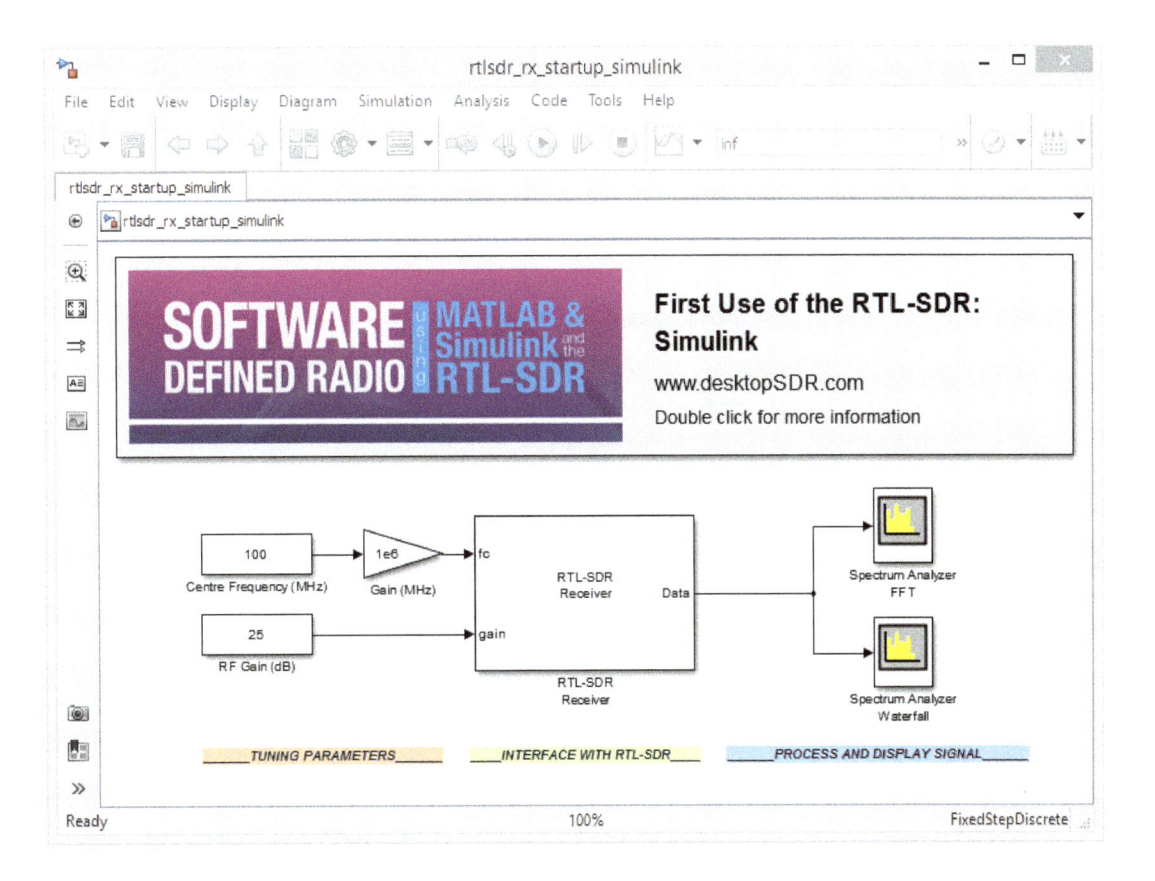

The signal is input to two *Spectrum Analyzer* blocks. The first of these is configured to show a Fast Fourier Transform (FFT) of the received signal, which will have a bandwidth of f_s MHz where f_s is the sampling frequency of the RTL-SDR. The second *Spectrum Analyzer* is configured to show a spectrogram (a waterfall), and will have the same bandwidth as the first one.

(d) **Open the receiver documentation.** The title bar displayed above the blocks in the Simulink model contains information about the file, the exercise it is related to, and also links to the .pdf copy of this book. 2 on the title bar and explore this tool.

(e) **Check the parameters of the *RTL-SDR Receiver* block.** When you are finished examining the model information, 2 on the *RTL-SDR Receiver* block to open its parameters window. You should notice that the 'Source' of the 'Center Frequency' and 'Gain' parameters have been configured to be 'Input ports'. Passing these values in like this allows them to be changed (or tuned) while the model runs. Other parameters that can be set in this dialogue window include the:

- RTL-SDR Radio address
- RTL-SDR Sampling rate
- RTL-SDR Tuner frequency ppm correction
- Block output data type
- The number of samples in an output frame
- Enabling of debugging outputs

... as shown in the dialogue box below.

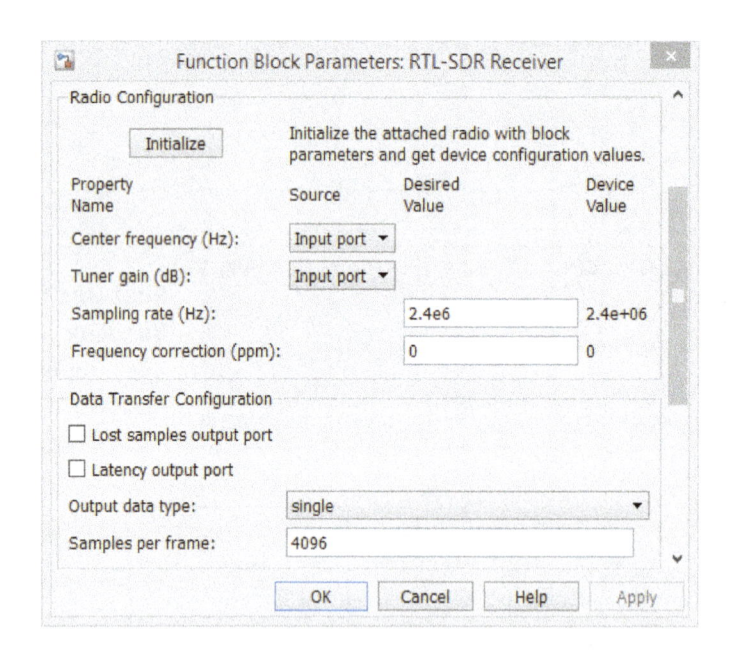

[Screenshot: Function Block Parameters: RTL-SDR Receiver dialog showing Radio Configuration with Center frequency (Hz): Input port, Tuner gain (dB): Input port, Sampling rate (Hz): 2.4e6 / 2.4e+06, Frequency correction (ppm): 0 / 0; Data Transfer Configuration with Lost samples output port and Latency output port unchecked, Output data type: single, Samples per frame: 4096.]

Refer back to Section 1.7 of the *Introduction* (page 17) and confirm that you know which of these parameters are used to configure the RTL-SDR hardware. When you have done this, click 'Cancel' to close the window and return to the Simulink model.

(f) **Prepare to run the simulation.** Now that you are familiar with the model, it is time to try it out, and view some signals! Ensure that the RTL-SDR is connected to your computer, and that it has an antenna attached as shown in Exercise 2.3. Check that MATLAB is able to communicate with the RTL-SDR by running the `sdrinfo` command in the command window, as in Part (c) of Exercise 2.3 (page 29).

Examine the top toolbar of the Simulink window, and notice that the *Simulation Stop Time* has been configured to be 'inf'. This means that the simulation will run continuously (i.e. for an *infinite* time), from when you start until you *manually* stop it, allowing you as much time as you need to view and analyse the signals. Take a note of the 'Run' ▶ and 'Stop' ■ buttons highlighted below.

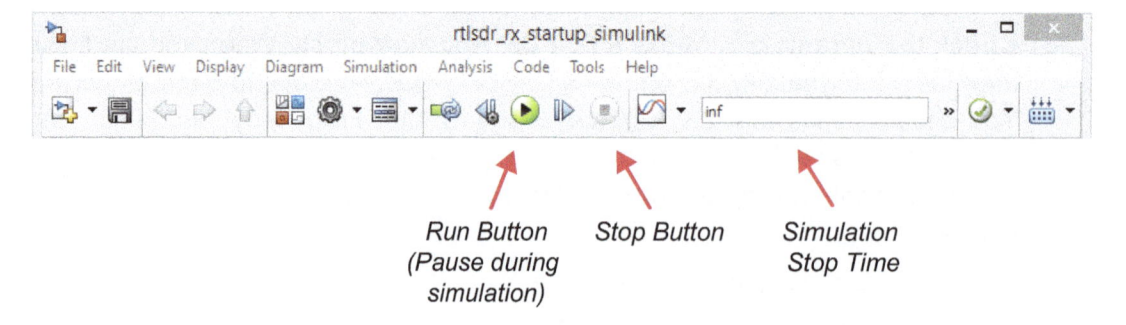

[Screenshot of Simulink toolbar with arrows pointing to: Run Button (Pause during simulation), Stop Button, and Simulation Stop Time.]

(g) **Run the simulation.** Begin the simulation by 🖱 on the 'Run' ▶ button at the top of the Simulink window. After a few seconds of displaying 'Initialising' in the bottom left of the status bar, the simulation will begin, and the two *Spectrum Analyzer* windows will open. If necessary, drag these two windows to arrange them as shown below. Hopefully your *Spectrum Analyzer* windows will show something similar, with one or more signals present! This example defaults to open at a centre frequency of 100MHz, which is in the FM radio band, so it is likely that any signals you see are FM radio stations. Don't worry if you can't see any obvious signals at this stage; count any sort of activity as a success!

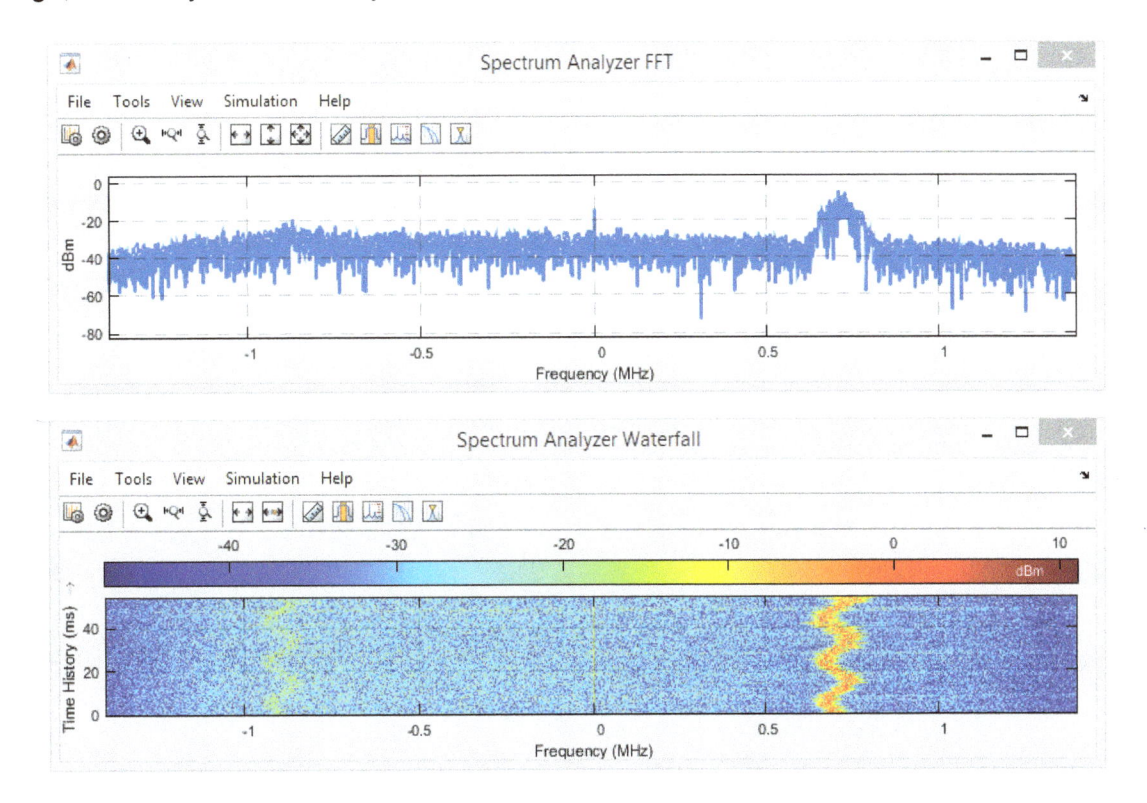

(h) If you can see any sort of signals in the *Spectrum Analyzers*, then congratulations! You have set everything up correctly! If not, stop the simulation by pressing the 'Stop' button and check again that the antenna is correctly attached to your RTL-SDR and that MATLAB can interact with it by running `sdrinfo` in the MATLAB command window. If spectrum viewing is still unsuccessful, please take a look at our troubleshooting advice at desktopSDR.com, and/ or in the RTL-SDR Hardware Support Package documentation from MathWorks [60] for more information.

Do not be too concerned if the information displayed in the *Spectrum Analyzer* windows is unfamiliar; you will find out more about what these actually show in the next chapter, which covers spectral viewing in detail. Once you're happy to move on, you can on the 'Stop' button to stop the simulation.

Exercise 2.6 First Use of the RTL-SDR: MATLAB

The SDR receiver implemented in this exercise is essentially the same as that implemented in Simulink in Exercise 2.5, in the sense that it acquires and displays the spectra of RF signals received by the RTL-SDR—but this time it is implemented in MATLAB m-code. In this example, we will show you how to tune the RTL-SDR from MATLAB, and also how to control hardware parameters such as the front end gain, centre frequency and so on. Completion of this exercise also ensures that the MATLAB part of the Hardware Support Package has been installed successfully, and that your RTL-SDR is functioning correctly. If you would prefer to know more about the rudiments of working with MATLAB m-code and functions before proceeding, then you can review this in Chapter 4 (page 99).

(a) **Open MATLAB.** If you do not have MATLAB open, return to Exercise 2.1 (page 26), and repeat Part (a). After MATLAB has initialised, change the current folder to:

> /intro/

(b) The file you need to open is called `rtlsdr_rx_startup_matlab.m`, which will be displayed in the *Current Folder* pane of the MATLAB window. You can either open the file by ② on it in this pane, or, if reading this document as a PDF on your computer (with the default and supplied file structure), by ① on the link below to open it directly.

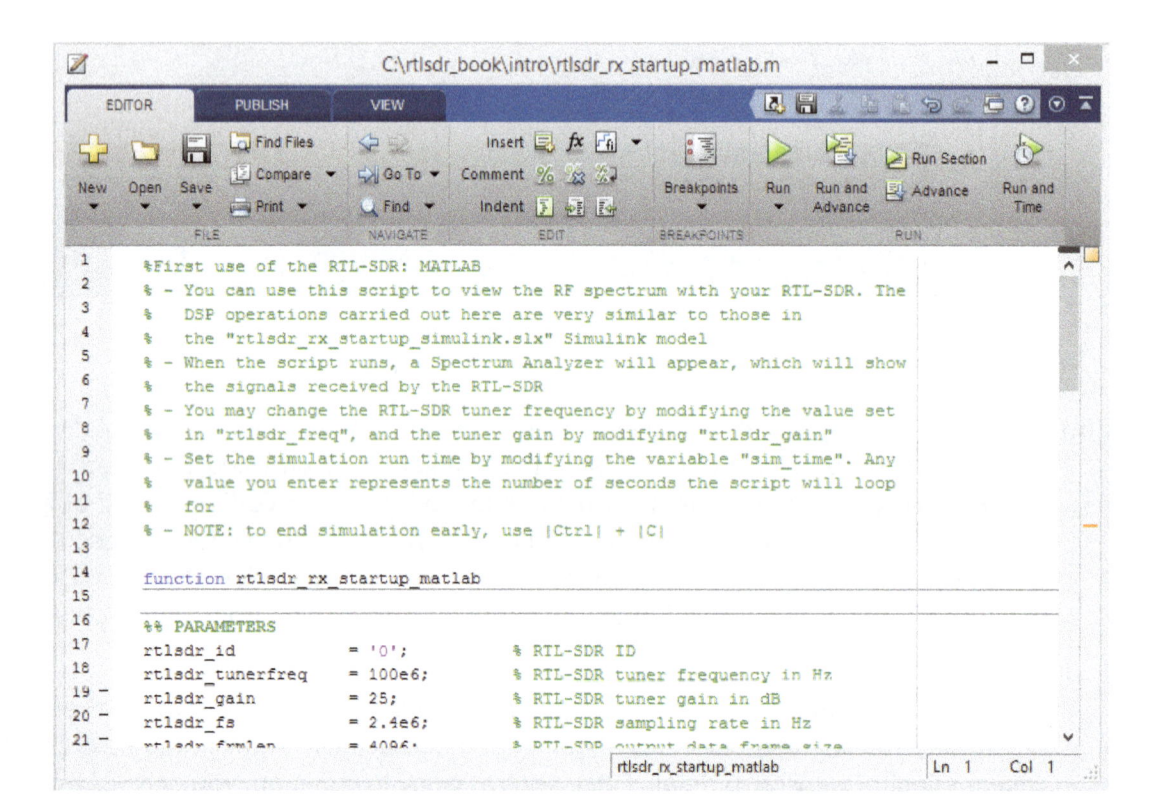

> /intro/rtlsdr_rx_startup_matlab.m

(c) **Examine the code.** There is a brief description of how to use this function in the first few lines, as you should see from the screenshot. It has been separated into a number of sections (indicated by the horizontal section breaks) to help document the code. The sections are titled as follows:

- Parameters
- System Objects
- Calculations
- Simulation

Chapter 2: Open the Box! First SDR with MATLAB and Simulink

(d) **Inspect the parameters.** A number of variables are initialised in the 'parameters' section, with values matching those set in the *RTL-SDR Receiver* Simulink block in Exercise 2.5. In other words, the centre frequency of the RTL-SDR is configured to be 100MHz (`100e6`), and the sampling frequency is set as 2.4MHz (`2.4e6`); the same default values as before. Other values set here include the output data frame length, type, and the simulation run time. Note that this script will by default run for 60 seconds. If desired, the duration can be increased from 60 seconds just by changing the value of the `sim_time` parameter.

```matlab
%% PARAMETERS
rtlsdr_id        = '0';       % RTL-SDR ID
rtlsdr_freq      = 100e6;     % RTL-SDR tuner frequency in Hz
rtlsdr_gain      = 25;        % RTL-SDR tuner gain in dB
rtlsdr_fs        = 2.4e6;     % RTL-SDR sampling rate in Hz
rtlsdr_frmlen    = 4096;      % RTL-SDR output data frame size
rtlsdr_datatype  = 'single';  % RTL-SDR output data type
rtlsdr_ppm       = 0;         % RTL-SDR tuner PPM correction
sim_time         = 60;        % simulation time in seconds
```

(e) **Review the system objects.** As this is purely a 'getting started' model, and there are very few components to this coded receiver, only three system objects are initialised. The first is for the RTL-SDR, and simply uses the values created earlier when setting each of the parameters of the object handle. The second is for a *Spectrum Analyser*, the same type of scope that was

```matlab
%% SYSTEM OBJECTS
% rtl-sdr object
obj_rtlsdr = comm.SDRRTLReceiver(...
    rtlsdr_id,...
    'CenterFrequency', rtlsdr_freq,...
    'EnableTunerAGC', false,...
    'TunerGain', rtlsdr_gain,...
    'SampleRate', rtlsdr_fs,...
    'SamplesPerFrame', rtlsdr_frmlen,...
    'OutputDataType', rtlsdr_datatype,...
    'FrequencyCorrection', rtlsdr_ppm);

% spectrum analyzer objects
obj_specfft = dsp.SpectrumAnalyzer(...
    'Name', 'Spectrum Analyzer FFT',...
    'Title', 'Spectrum Analyzer FFT',...
    'SpectrumType', 'Power density',...
    'FrequencySpan', 'Full',...
    'SampleRate', rtlsdr_fs);
obj_specwaterfall = dsp.SpectrumAnalyzer(...
    'Name', 'Spectrum Analyzer Waterfall',...
    'Title', 'Spectrum Analyzer Waterfall',...
    'SpectrumType', 'Spectrogram',...
    'FrequencySpan', 'Full',...
    'SampleRate', rtlsdr_fs);
```

used in the Simulink receiver. It is configured to act as *Spectrum Analyzer FFT* from Exercise 2.5. Thirdly, another *Spectrum Analyzer* is initialised, also configured to act as *Spectrum Analyzer FFT*, but in the waterfall style (also used in Exercise 2.5). If you feel that you do not fully understand how system objects work, we advise you to have a look at Section 4.5 (page 121) of *Getting Started with MATLAB and Simulink* before continuing.

(f) **Inspect the calculations section.** The value of `rtlsdr_frmtime` is found by calculating the length of time it takes to process one frame of data, and this is used later to control the duration of the simulation.

```matlab
%% CALCULATIONS
rtlsdr_frmtime = rtlsdr_frmlen/rtlsdr_fs;
```

(g) **Review the simulation section.** The final section is where the 'simulation' is carried out and the receiver is implemented. After double checking that MATLAB can communicate with your RTL-SDR, a `while` loop is instantiated that loops for the number of seconds specified by the `sim_time` variable. Using the `step` function, a frame of data is fetched from the RTL-SDR and then stored in the `rtlsdr_data` variable. The frame is a matrix of length `rtlsdr_frmlen` and by default it contains `4096` samples. Using the `step` function once again, this frame of data is passed to the *Spectrum Analyzers*, which prompts them to perform FFTs on the new batch of data and update their displays. Finally, the `run_time` counter that controls the loop is updated to incorporate the processing time associated with demodulating another frame from the RTL-SDR.

```matlab
%% SIMULATION
if isempty(sdrinfo(obj_rtlsdr.RadioAddress))
    error(['RTL-SDR failure. Please check connection to ',...
           'MATLAB using the "sdrinfo" command.']);
end

% reset run_time to 0 (secs)
run_time = 0;

% run while run_time is less than sim_time
while run_time < sim_time
    
    % fetch a frame from the rtlsdr
    rtlsdr_data = step(obj_rtlsdr);
    
    % update spectrum analyzer windows with new data
    step(obj_specfft, rtlsdr_data);
    step(obj_specwaterfall, rtlsdr_data);
    
    % update run_time after processing another frame
    run_time = run_time + rtlsdr_frmtime;
    
end
```

Chapter 2: Open the Box! First SDR with MATLAB and Simulink 41

(h) **Open the receiver documentation.** As this receiver has been developed in MATLAB, it has not been possible to include a double-clickable title bar which can be used to display the receiver documentation. What we have done instead however is create a function that carries out a similar operation. Enter

```
mfileinfo rtlsdr_rx_startup_matlab.m
```

into the MATLAB command window to view the file information, the exercise it is related to, and access the link to the `.pdf` copy of this book.

(i) **Prepare to run the script.** Now that you are familiar with the script, it is almost time to try it out and view some signals! Ensure that the RTL-SDR is connected to your computer, and that it has an antenna attached as shown in Exercise 2.3. Check that MATLAB is able to communicate with the RTL-SDR by running the `sdrinfo` command in the command window, as in Part (c) of Exercise 2.3 (page 29).

(j) **Run the script.** If you look at the 'Editor' ribbon at the top of the MATLAB window, you will see a 'Run' button as highlighted below.

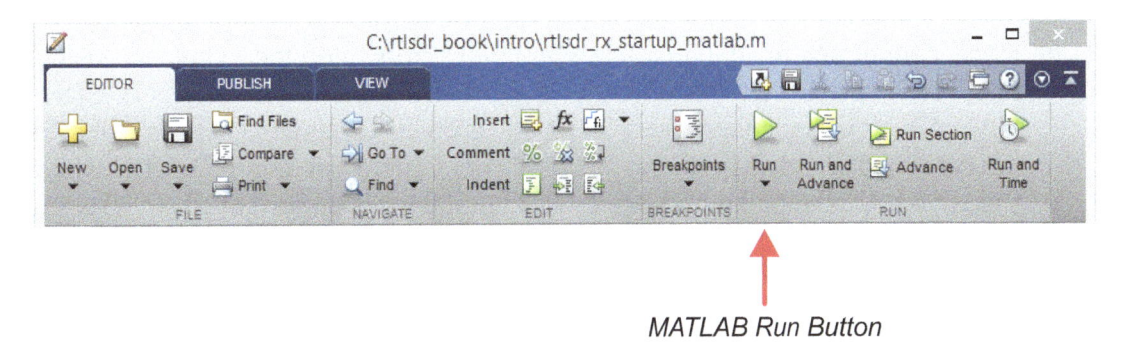

MATLAB Run Button

Execute the function by 🖱 on the 'Run' ▷ button. After a few seconds, MATLAB will establish a connection with your RTL-SDR and begin running, with the *Spectrum Analyzers* opening and displaying frequency domain views of the data. The function should remain in the `while` loop for `sim_time` seconds, regularly pulling frames of data from your RTL-SDR into MATLAB, and updating the *Spectrum Analyzer* as the new frames arrive.

If you wish to cancel the script execution before `sim_time` seconds have elapsed, navigate to the command window and use the classic key combination Control-C to break out of the loop. *Note: This will cause a red error message to be printed to the MATLAB command window! You have not done anything wrong, it is simply the case that ending the function during execution throws a MATLAB exception.*

(k) The *Spectrum Analyzer FFT* and *Spectrum Analyzer Waterfall* windows should appear as the function begins to run. Hopefully *Spectrum Analyzer FFT* will show something similar to the example below, with a recognisable signal in it! As with Exercise 2.5, the RTL-SDR is tuned to a centre frequency of 100MHz, which is in the FM radio band, so it is likely that any signals you see are FM radio stations. Don't worry if you can't see any obvious signals at this stage; count any sort of activity as a success!

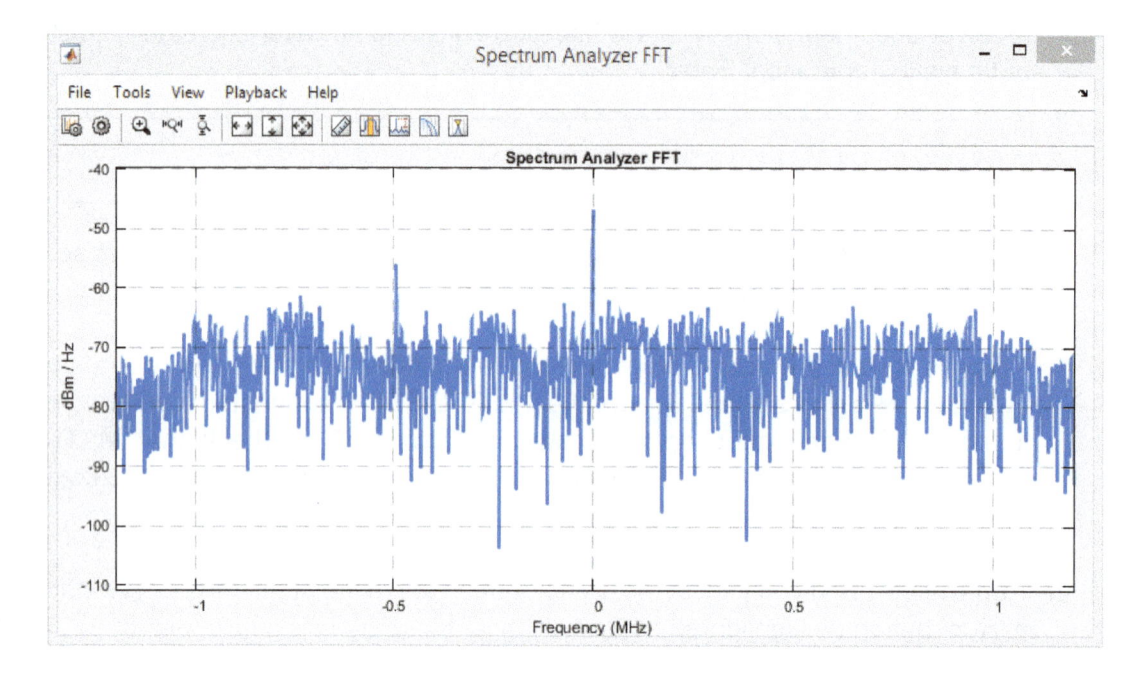

Spectrum Analyzer FFT showing signal power (dBm/Hz) from approximately -110 to -40 dBm/Hz across a frequency range of -1 to 1 MHz.

(l) If you cannot see anything or you encounter errors, stop the simulation using the Control-C command, check that the antenna is correctly attached to your RTL-SDR and also that MATLAB can communicate with it. If still unsuccessful, please take a look at our troubleshooting advice at desktopSDR.com, and/or in the RTL-SDR Hardware Support Package documentation from MathWorks [60] for more information.

2.5 Summary

In this chapter we have reviewed the fundamental steps required to get up and running with the RTL-SDR in MATLAB and Simulink. This includes some general housekeeping requirements such as ensuring installation of the appropriate MathWorks toolboxes, the setting of paths, installation of the book exercise files, and the style and nomenclature used in the remainder of the book. The chapter also aimed to walk you through a first few examples where you had the opportunity to plug in the device, confirm the software was all in place, and then receive and view the spectra of some off-the-air RF signals, in real-time. In the next chapter we will take spectrum viewing a little further, and provide some guidance and examples on viewing various signal types ranging from FM radio, to mobile GSM, to LTE and other wireless signals that are present in the air around you.

3 Radio Frequency Spectrum Viewing

The RF spectrum is a portion of the electromagnetic spectrum that ranges between 3kHz and 300GHz. We use the RF spectrum extensively for communications services in applications such as broadcasting signals for television, radio, mobile and WiFi, as well as navigation and detection systems such as radar, GPS, radio beacons, transponders and so on. The actual radio frequencies used in different applications depend on many different physical, economic and legal constraints including:

- Propagation characteristics of electromagnetic waves *(waves at different frequencies have distinct properties which affect the distance they can travel, their building penetration potential, and their ability to diffract around objects)*
- Antenna size and practicalities *(e.g. small antennas are desired for small devices)*
- Sharing the same bands of the spectrum with other users *(is there something already being transmitted in this band?)*
- Government (or other) licensing bodies *(am I allowed to transmit at this frequency, or even transmit at all?)*

Figure 3.1 illustrates a breakdown of the electromagnetic spectrum (top) and outlines some of the more well known communications services that operate within the RF spectrum (bottom). It is most likely that the RTL-SDR you have (or intend to get!) will use the Rafael Micro R820T tuner[1], meaning that it will be capable of receiving any signal transmitted in the frequency range **25MHz to 1.75GHz** [71]. The intention of this chapter is to help you explore this part of the spectrum, find, and identify some of the RF signals broadcast in this range at your locale.

---

1. A few RTL-SDRs are available with the Elonics E4000 tuner (instead of the R820T), which exhibits a wider tuning range from around 53MHz to 2.2GHz, with a gap between 1.1GHz and 1.25GHz [69]. The E4000 tuner was designed and made by a company called Elonics, from Livingston in Scotland, however they ceased manufactured in 2012. As a result, these particular RTL-SDRs are now in very short supply!

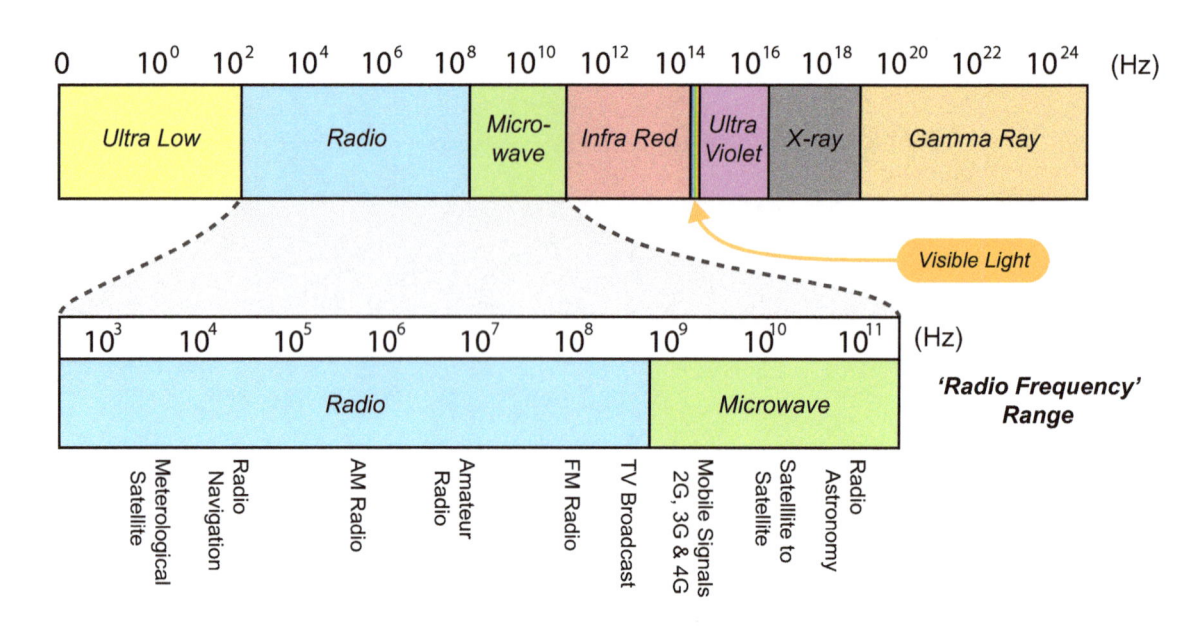

Figure 3.1: A diagram of the electromagnetic spectrum

3.1 Different Signals, Different Frequencies

Wireless communication via RF is now an essential part of life for many people. Whether it is wireless network access in offices, mobile basestation backhaul, consumers texting, using voice and video communications or social media, it is likely at some point that they will be communicating over various frequency bands using standards such as Bluetooth, WiFi, GSM, and LTE. Planning and licensing is required in order to set rules about power levels, and ensure that interference does not occur between different broadcasters. In the UK, these frequency planning and licensing tasks fall upon a government-approved body called the Office of Communications (Ofcom – ofcom.org.uk). The USA has the Federal Communications Commission (FCC – fcc.gov); and other countries have their own national agencies.

In all countries the RF spectrum has been split up into a number of defined frequency bands which are allocated to different users and for different applications. Generally no two countries are exactly the same in their frequency planning, and there can be considerable differences. To give an example, in the UK there are currently no mobile phone service frequencies below 800MHz, whereas in some countries in eastern Europe, they operate down as low as the 400MHz UHF band. In recent years there have been efforts via organisations such as the International Telecommunications Union (ITU – itu.int) to harmonise these bands where possible (particularly in areas where countries have borders). An example of successful harmonisation is WiFi (the IEEE 802.11 standard [94]) which operates in the 2.4GHz band. FM radio is also pretty standard across most countries, in the 88MHz to 108MHz band, and nearly all radio stations are spaced at least 200kHz apart. When frequencies are harmonised, equipment designed to work in one country will likely work in others; which helps bring costs down. Communications techniques where harmonisation is not as prevalent include mobile standards, with, for example, the USA and Europe using quite distinct frequencies for GSM, UMTS and LTE services. Modern phones are now able to address issues like this adequately however, by being tri- (or even quad-) band, and having the ability to receive and transmit at various frequencies using different front end RF hardware, which can be switched in and out depending on which country it is being used in.

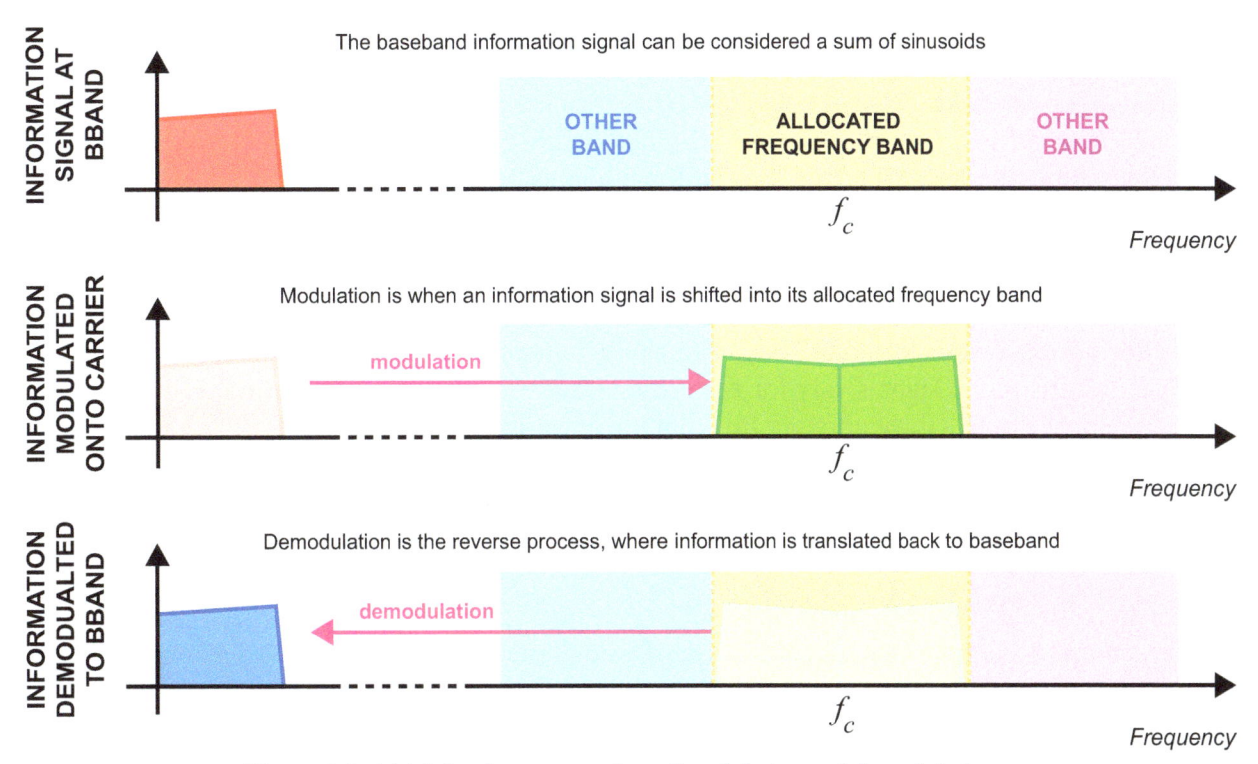

Figure 3.2: A high level representation of modulation and demodulation being performed on a baseband information signal

Virtually all modern radios and communications systems operate in a broadly similar way. Baseband information signals (which can be music, voice, or data) go through a process called *modulation* in RF transmitters to translate the baseband information into the allocated frequency band. The reverse process is carried out by RF receivers, and is referred to as *demodulation*. A simple visual representation of modulation and demodulation is illustrated in Figure 3.2.

3.2 Spectrum Usage and Allocations Around the Globe

Over the years, the spectrum management process has developed in an evolutionary fashion, as technology has brought new frequencies into common usage or as new services have been developed. Because of this, the usage of the spectrum is not always optimal, and further, as noted earlier, the frequency allocations and licensing methods and bodies are not the same around the world; meaning different countries have developed and used parts of the RF spectrum in different ways. Therefore at this point it is important to mention that the frequency bands we associate with some communications services in this chapter are primarily those used in the UK. Bands used for these services in your country may be the same or different from the ones mentioned in the following exercises, but you should easily be able to find a definitive breakdown of the spectrum for your country from a quick online search. We have catalogued a number of these at desktopsdr.com/more/worldwide-frequency-allocation-tables.

Regulatory authorities around the world play a large part in the development of the RF spectrum. They oversee, control and license the broadcasting and communications services that use it, and also police the airwaves to ensure there are no signals being transmitted in it illegally. Competition to transmit in parts of the spectrum used for services like radio, DTV and mobile communications is fierce, and these bodies aim to make sure that only authorised signals are broadcast. As well as strict regulation, Ofcom, the FCC, and other regulators are also heavily involved with the development of new standards; for instance, although 4G mobile was only rolled out in late 2013, they are already working on *"laying the foundations for the next generation of wireless communications"*, fifth generation (5G) mobile [102]. Requesting support from industry, academia, and international bodies, regulators aim to develop a framework that allows for innovation and worldwide uniformity for any new communications technology.

Different information signals have specific requirements when it comes to how they must be transmitted, and this means various modulation and multiplexing techniques are used when they are broadcast, to ensure information is transferred reliably. The characteristics of the bandwidth, amplitude, frequency and phase of transmitted signals vary for each of the modulation and multiplexing techniques. As you will see from the exercises throughout in this chapter, every distinct signal interacts with the spectrum in a different way.

3.3 Working with a Suitable Antenna

When you purchased your RTL-SDR, it probably came packaged with an omni-directional antenna which takes the form of a small metal rod (see Figure 3.3). This is only one of the many different types of antenna that can be used with the device however. An antenna is defined by the IEEE Standards Definitions of Terms for Antennas as the *"part of a transmitting or receiving system that is designed to radiate or to receive electromagnetic waves"* [24]. Simply put, it is the interface between the transmitter and receiver hardware, and the electromagnetic waves in the RF spectrum. It's an important and integral part of any wireless communication system, meaning that the choice of antenna (which often depends upon the application) can be vital to it's success. A brief overview of some of the properties of antennas will be covered in this section to aid the practical work in this book. More extensive coverage of the subject theory can be found in [2] and [25].

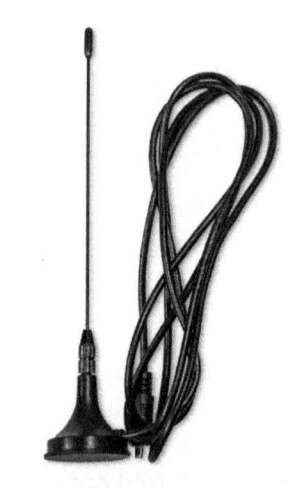

Figure 3.3: The omni-directional antenna supplied with the NooElec NESDR RTL-SDRs [71]

In general terms, antennas can be designed either to provide directional gain for specific frequencies in the spectrum (a narrowband antenna), or for non-directional low gain reception of a large range of frequencies (a broadband antenna). They can vary widely in their properties of gain, intended operating frequency, and polarisation; all of which affects how best to position the antenna to receive the desired signal. Because the tuners used in the RTL-SDR have such a wide operating range, ideally you should use a broadband antenna when exploring the spectrum later in this chapter (i.e. an antenna that works well across a broad band of frequencies, not one for fast internet!).

3.3.1 *Antenna Gain and Directivity*

Technically speaking, passive antennas do not apply any physical gain to transmitted or received RF signals. Antenna gain is a term used to describe how much more power can be transmitted or received with a particular antenna than could be with the *lossless isotropic antenna* [7]. It is measured in dBi, which refers to decibels over the isotropic. The theoretical lossless isotropic antenna is one which has no sensitivity to any particular direction, and offers the metric for comparing the 'gain' that any antenna provides.

Antennas emit electromagnetic waves with different radiation patterns. The isotropic antenna radiates waves uniformly in all directions, so has a *directivity* of zero [48]. Directional antennas on the other hand radiate mainly in one direction, and as such have a larger directivity and a higher gain. Often the most effective antennas (in terms of gain) are those which have high directivity, and they work best when both the transmitter and receiver antennas are aligned with each other.

Ideally, a transmitter would use a high gain directional antenna in order to transmit as strong a signal as possible towards the receiver. Using a similar antenna at the receiving end would mean that less amplification would be required in the receiver hardware, and the received signal would be of a higher quality.

3.3.2 *Tuned Antennas*

As well as being designed for directivity, antennas can also be constructed to operate optimally at specific frequencies. *Tuned* antennas are designed to resonate and receive particular bands of electromagnetic waves, meaning they have a high antenna gain for these intended frequencies, and low gain for all others. These antennas are used when transmitting and receiving mobile, WiFi, and DTV signals (amongst others) because they only need to operate over a finite frequency range. Although the stock omni-directional antenna supplied with the RTL-SDR is tuned to operate in the DTV part of the UHF band, it will be adequate for the purpose of exploring the spectrum later in this chapter.

Tuned antennas vary in size because of the fact that different frequencies have different wavelengths. Low frequencies have long wavelengths, and inversely, high frequencies have very short wavelengths. For best performance, the length of an antenna needs to match the wavelength of the wave.

3.3.3 *Antenna Polarisation and Positioning*

The final antenna characteristic to discuss is polarisation. Signals can either be transmitted as linearly polarised (meaning they are transmitted in one plane), or circularly polarised (meaning they are transmitted in two planes as a helix); and antennas are polarised to match. The two most common forms of linear polarisation are horizontal and vertical polarisation, however, signals can also be transmitted at any angle between these planes. If a signal was transmitted in the horizontal plane, a matched receiving antenna should be placed in the horizontal plane in order to maximise the power of the received signal. If the receiving antenna was in the vertical plane the signal could still be received, but its power would be greatly reduced. Note: the omni-directional antenna supplied with the RTL-SDR is vertically polarised.

Examples of some different types of polarisation are shown in Figure 3.4, where we illustrate the planes the waves are transmitted in as they propagate through the air, the recommended orientation of the receiving antenna, and give examples of the type of antenna that could be used to receive the signal.

Signals can take multiple paths with varying distances between transmitter and receiver antennas, due to the electromagnetic waves reflecting off walls and objects. This is known as multipath, and when these multipath signals arrive at a particular point out of time with each other, a destructive interference process called multipath fading occurs. It should be noted that this usually only affects signals with a large transmission distance; such as FM radio or mobile signals, and less so on the short distance desktop transmission you will be carrying out in later chapters. It is however worth bearing in mind that minimal movements of an antenna may mitigate the effects of the multipath fading, and greatly increase the quality of the received signal.

Horizontal Linear Polarisation:

Vertical Linear Polarisation:

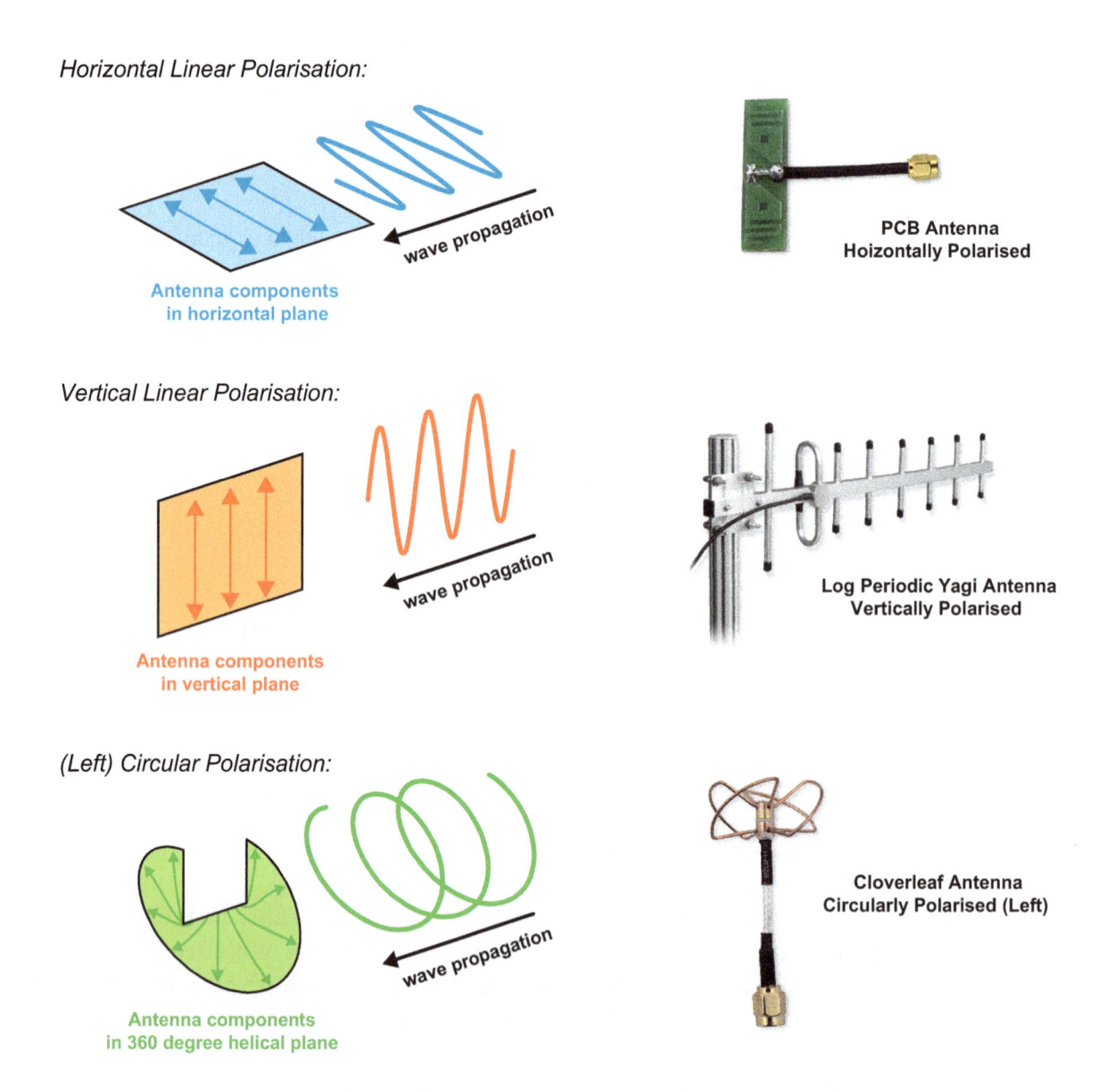

(Left) Circular Polarisation:

Figure 3.4: Diagram showing how to orientate antennas when receiving polarised signals, and examples of polarised antennas

3.3.4   Antenna Adapters

Although not an antenna characteristic, the last antenna related topic we must consider is the adapters used between antennas and receiver hardware. While tuned directional antennas can offer dB's of gain to received signals, incorrect adapter usage can negate this with attenuation. Adapters and connectors act like filters to signals, which is why it is important to keep the number of them to a minimum, and to use high quality ones. Having long lengths of cable and large numbers of adapters between the antenna and receiver is not recommended, as this reduces the signal power (and therefore the signal quality) due to voltage drop. You must attempt to keep the connection between the antenna and receiver as short (and simple) as possible to get the best results.

Chapter 3: Radio Frequency Spectrum Viewing

The NooElec RTL-SDRs feature an MCX socket, to which the supplied omni-directional antenna will easily connect. If you wish to use a different antenna, it's likely that you will require an adapter. Figure 3.5 shows the form of a standard MCX plug. Whatever adapter you use must have this on one end! A number of different adapters can be purchased from the 'SDR Adapters' page of the NooElec website [72].

Figure 3.5: MCX Plug

3.4 Go Forth and Explore the Spectrum!

It's now time to move on to some practical exercises where you will have the chance to receive some *real* RF signals *live* from the RF spectrum. Most of the exercises in the remainder of this chapter will use a Simulink model called `exploring_the_spectrum.slx`. This model contains a Graphical User Interface (GUI) which will assist you with viewing some of the different signals shown in Figure 3.1. Initially you will be asked to open the model and become familiar with the layout of the blocks in Simulink, and the '*Control_Panel*' GUI. After this brief introduction you can get on with the exploring!

For those readers who do not yet have an RTL-SDR or MATLAB/ Simulink: Videos are included for you throughout the following exercises to show you what you could be seeing!

Exercise 3.1 **Opening the Spectrum Viewing Receiver Model**

This exercise acts as a brief introduction to the model that will be used for many of the exercises in this chapter.

(a) **Open MATLAB.** If you cannot remember how to do this, repeat the first few steps of Exercise 2.4 (page 32). Set the working directory to the exercise folder,

   /spectrum

(b) Next, open the following model. You can either do this by 🖱️2 on the model name in the 'Current Folder' pane in MATLAB, or, if reading this document as a PDF on your computer, by 🖱️1 on the link below.

   .../exploring_the_spectrum.slx

(c) **Simulink window layout.** When the model opens, a Simulink window should appear at the top of your screen, along with two other windows below. These are the windows for the two spectrum analyzers—*Spectrum Analyzer FFT* and *Spectrum Analyzer Waterfall*—which will show spectral activity in the portion of the RF spectrum that the RTL-SDR has been tuned to receive.

There is also another window associated with this model, a GUI called *Control_Panel*, which provides interactive controls for the model shown above. This final window (shown in Figure 3.8) will be further explained in Section 3.6.

[Screenshot of Simulink model "exploring_the_spectrum" with Spectrum Analyzer FFT and Spectrum Analyzer Waterfall windows]

Simulink model that receives and displays signals from the RF spectrum

Spectrum Analyzer configured to perform a FFT and plot the result

Spectrum Analyzer configured to show spectral activity using a spectrogram

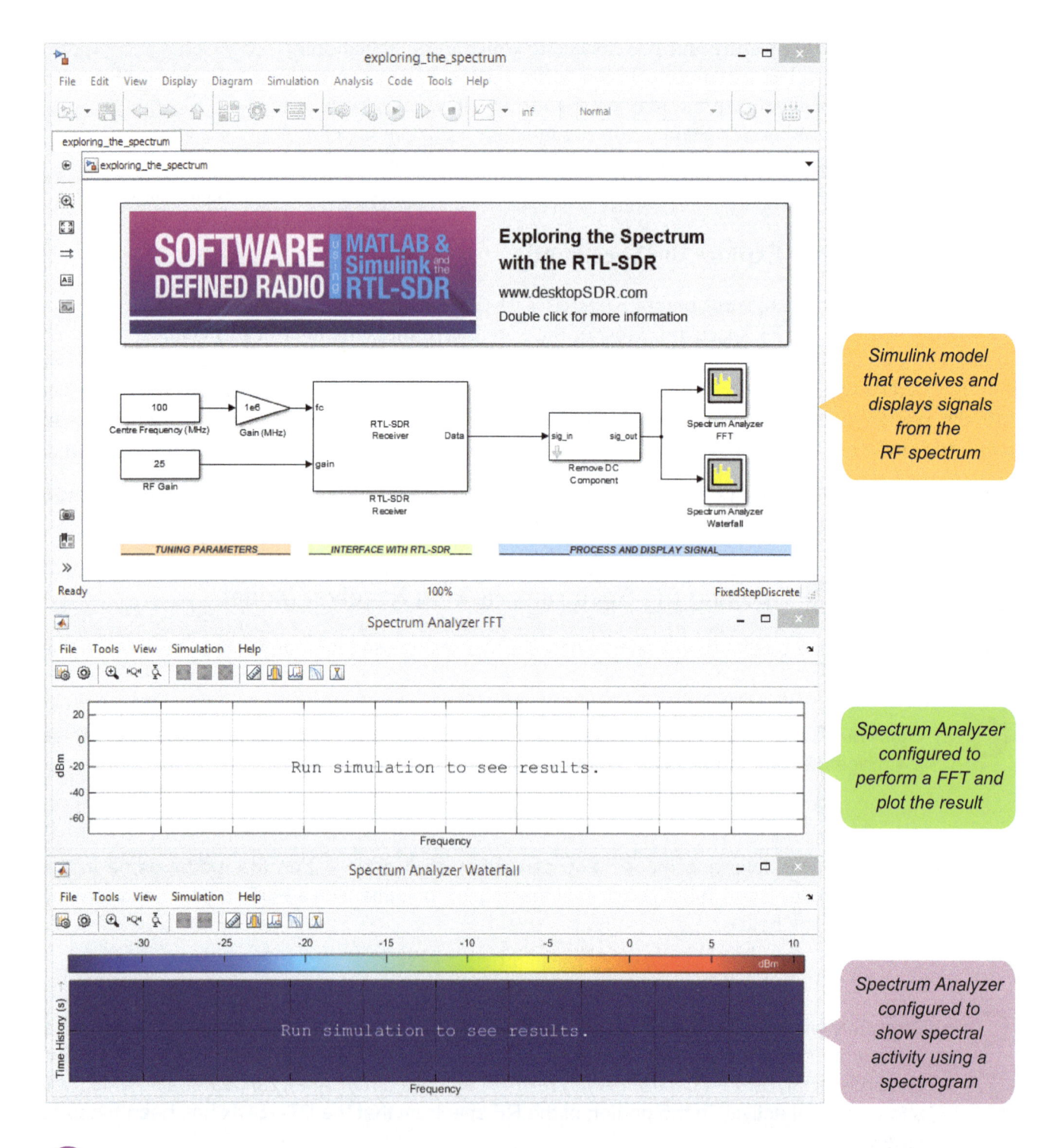

(d) **Browse the model.** Functionally, this model is essentially the same as the one you ran in Exercise 2.5 (page 34). The main difference is that it uses a GUI to control the centre frequency and gain of the tuner on your RTL-SDR.

The *RTL-SDR Receiver* block (which is introduced to Simulink by installing the Hardware Support Package) acts as the connection to your RTL-SDR. The RTL-SDR is tuned to the value of the *Centre Frequency (MHz)* block, and a tuner gain is applied according to the value

Chapter 3: Radio Frequency Spectrum Viewing 51

of *RF Gain*. 8-bit IQ samples are output by the *RTL-SDR Receiver* block, and these are passed to a *Remove DC Component* block.

This block finds the average value of a frame of data and subtracts this to remove the DC spike in the spectrum. Next, the signals are input to two *Spectrum Analyzer* blocks. The first of these is configured to show an FFT of the received signal, which will have a bandwidth of f_s MHz where f_s is the sampling frequency that has been set for the RTL-SDR. The second *Spectrum Analyzer* is configured to show a spectrogram (a waterfall), and will have the same bandwidth as the first one.

(e) **Run the simulation.** Ensure your chosen antenna is connected to your RTL-SDR, that the RTL-SDR is plugged into a USB port on your computer, and that MATLAB can communicate with it by running the `sdrinfo` command. Begin the simulation by 1 the 'Run' ▶ button at the top of the Simulink window. The status bar will indicate the model is 'initilaising' for a few seconds, then the simulation will begin.

If a Simulink diagnostics window like the one below appears instead of signals in the spectrum analyzers, you have encountered an error. The screenshot below shows the errors that result from attempting to run the model without having the RTL-SDR connected to the computer. A short description of the error explains why the model has been unable to run. 1 on the blue 'Message' hyperlink will give a more thorough explanation of how to resolve any problems that occur.

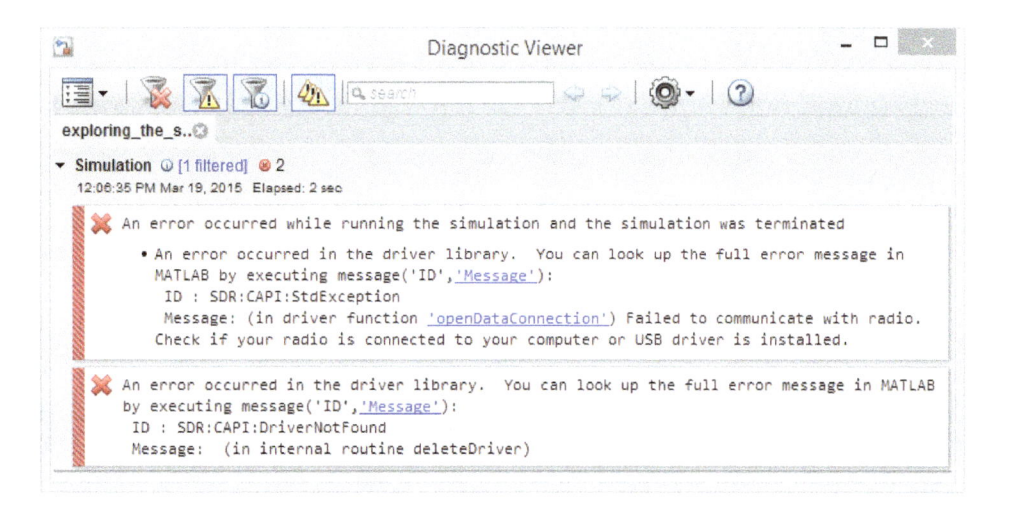

(f) You may be wondering what exactly the spectrum analyzer windows are showing you, and how to use the GUI to control your RTL-SDR. Explanations of these are given in the following two sections.

(g) You can either leave the model running (which will allow it to continually receive and display signals in real time), pause it by 1 on the 'Pause' ⏸ button, or stop it by 1 on the 'Stop' ■ button in the Simulink toolbar.

3.5 Spectral Viewing — Spectrum Analyser and Waterfall Plots

Two spectrum analysers blocks are used to display the RF signals received by the RTL-SDR in `exploring_the_spectrum.slx`. *Spectrum Analyzer FFT* shows the spectrum of the received complex signal, and has a bandwidth of f_s MHz. This means that it shows spectral activity in the range:

$$\left(f_c - \frac{f_s}{2}\right) \text{ to } \left(f_c + \frac{f_s}{2}\right) \text{ Hz,}$$

centred around f_c, the centre frequency of the RTL-SDR. To give an example, if we tuned the RTL-SDR to 801.4MHz and configured it's sampling rate to 2.8MHz, RF signals in the range 800MHz to 802.8MHz would be captured, downconverted and complex demodulated by the device. Plotting the complex baseband samples that enter Simulink with the Spectrum Analyzer (illustrated in Figure 3.6), you would note it had frequency components in the range -1.4MHz to 1.4MHz:

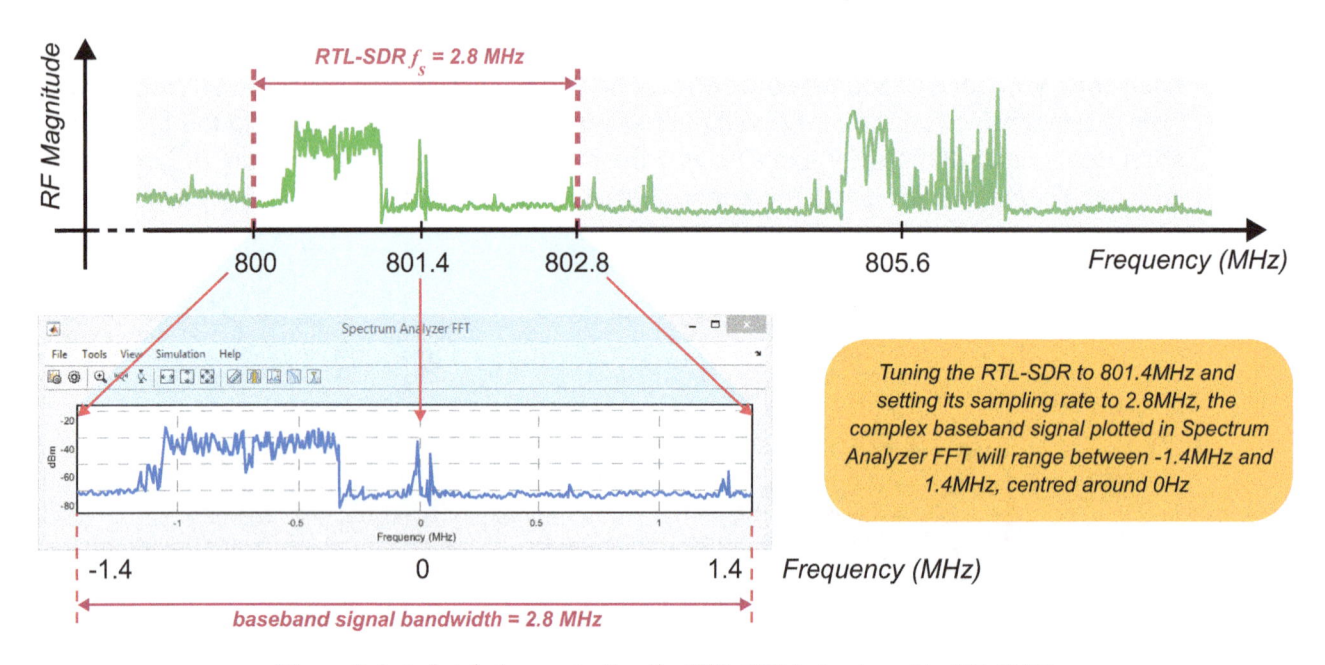

Figure 3.6: A sketch demonstrating the RTL-SDR being tuned to 801.4MHz, and downconverting RF signals from this frequency to complex baseband (0HZ)

This range is depicted by the green line in Figure 3.7. The power residing in each bin of the FFT is plotted in dBm (decibels per milliwatt), as illustrated by the vertical orange line. This scope shows you the instantaneous relative power levels of spectral components within the received band, and makes it easy to see signals that are transmitted above the noise floor. It is also very useful for manually tuning the RTL-SDR to any RF carrier frequency you are interested in receiving.

Spectrum Analyzer Waterfall shows exactly the same range of frequencies as *Spectrum Analyzer FFT*, although instead of presenting the activity as an instantaneous power, it shows activity through time (the last 50ms) as an intensity plot. The colour scale represents the power level (again in dBm), which shows the relative power levels of the frequency components in received signals. It ranges from dark blue (the noise floor) to dark red (high powered signals). This type of scope is known as a *spectrogram* or *waterfall* because it allows you to track bursts of activity over time, as illustrated by the vertical purple line.

Chapter 3: Radio Frequency Spectrum Viewing

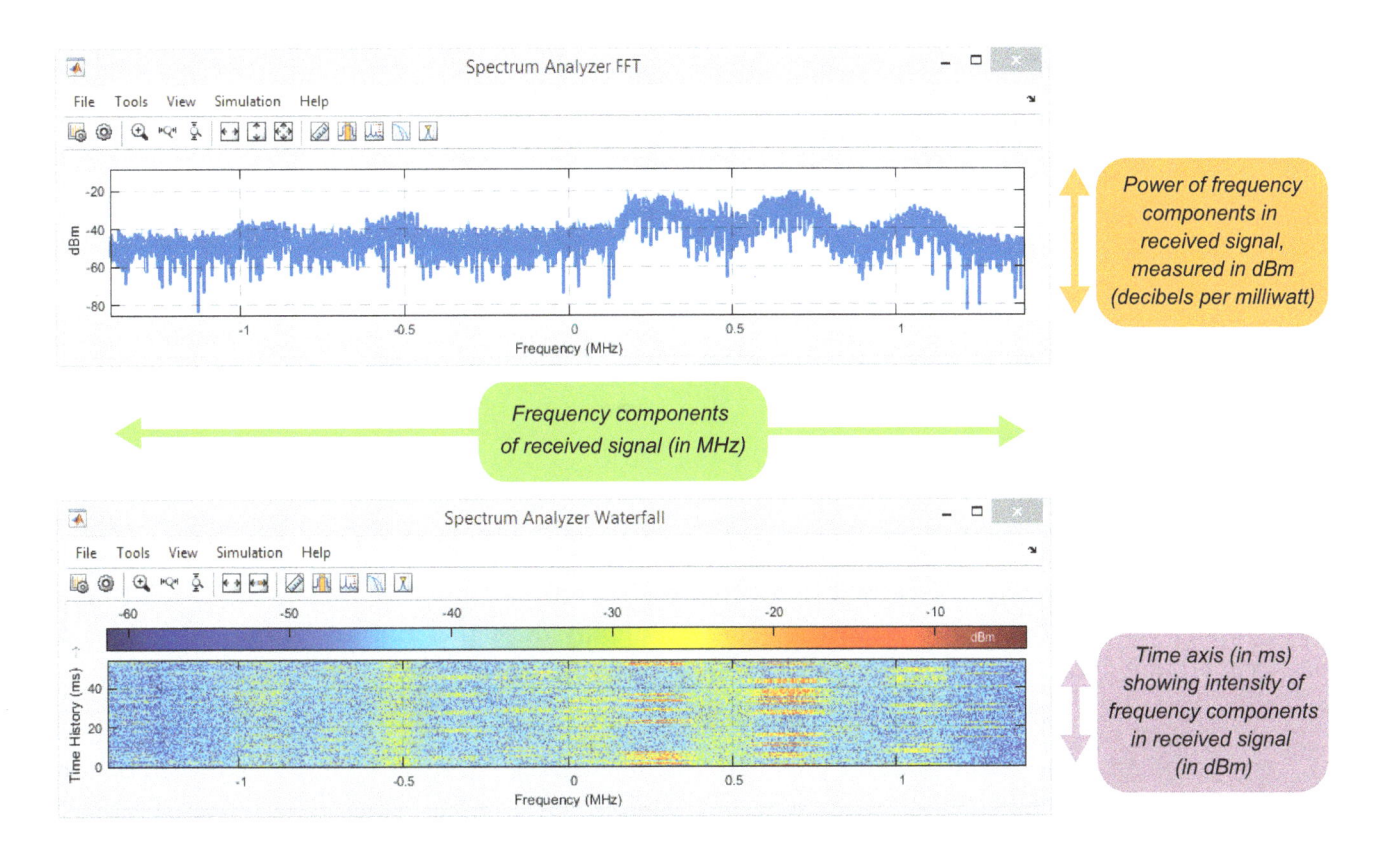

Figure 3.7: Annotated *Spectrum Analyzer FFT* (top) and *Spectrum Analyzer Waterfall* (bottom) scopes

Although we are calling it a waterfall, it should be noted that information plotted in the scope actually displays (and moves) from the bottom to the top, which correlates to the time axes shown on the left hand side of the window. Simulink allows the characteristic settings of both these scopes to be changed and modified to suit the user's requirements, but for now we shall leave them as they are. More information on these spectrum analysers will be given during Exercise 3.2.

3.6 Spectral Viewing — RTL-SDR Tuner GUI Controls

A MATLAB GUI called '*Control_Panel*' is used in `exploring_the_spectrum.slx` to aid tuning the RTL-SDR when navigating around the RF spectrum. Shown in Figure 3.8, this GUI will allow you to easily change the centre frequency and tuner gain parameters, and control the model execution (using the '*Start Simulation*' and '*Stop Simulation*' buttons).

Any changes made to the centre frequency or tuner gain reflect immediately in the *Centre Frequency (MHz)* and *Tuner Gain* blocks in the model. There are a number of ways to change these parameters: you can either enter specific values into the top text boxes, adjust the slider bars until you have the value you require, or for the centre frequency—by clicking on one of the radial buttons to select general regions.

If using the text boxes, both values should be entered as numbers—e.g. "25". Note that the centre frequency should be entered as a MHz value, for example "102.5" for 102.5MHz. The gain slider has the absolute range 0–50, because this matches the accepted gain values for the NooElec receiver (0 to 50dB). During the exercises you will have to alter the gain where appropriate; for example if using a transmitter

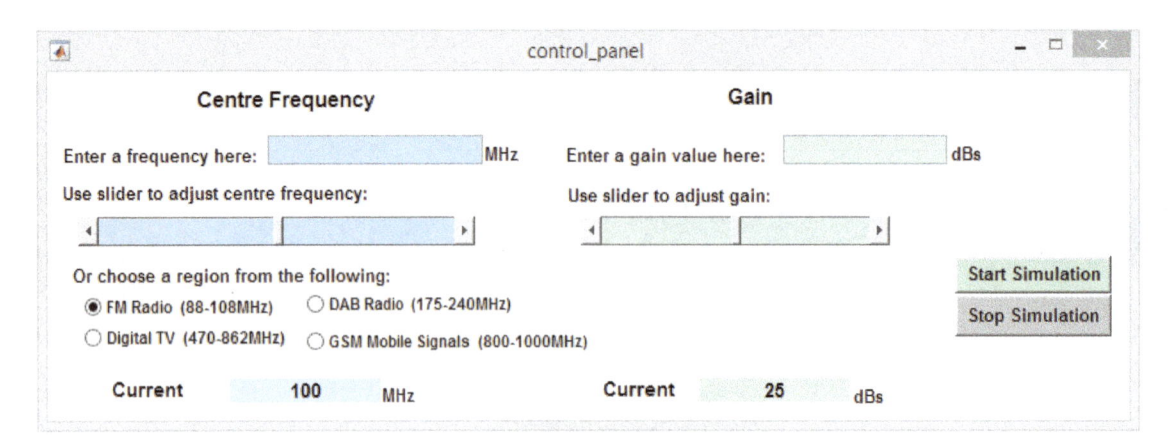

Figure 3.8: 'Control Panel' GUI for `exploring_the_spectrum.slx`

in close vicinity to the RTL-SDR's antenna, you should lower the gain so that the receiver does not saturate. If you were trying to receive a signal that has been transmitted from afar, such as FM radio signal, you would likely need to increase the gain to amplify the signal and help differentiate it from background noise. The centre frequency slider will automatically centre upon the chosen frequency, and will allow you to slide to ±500MHz of this value, moving in steps of 0.1MHz.

If at any point you accidentally close the GUI window, you can open it again by stopping the simulation and starting it again.

3.7 Engineering Requirements — Eyeball Radio Tuning & More

In order to make the best use of the RTL-SDR with MATLAB and Simulink, there are some basic RF engineering skills that are required of you. Firstly, you will need to be able to make cognitive changes to the centre frequency and tuner gain of the RTL-SDR, based on what you see in the spectrum analysers. Throughout the exercises in this book we will introduce various methods of changing these parameters, and these will be presented along with many other hints and tips that can help you use the tools more efficiently.

Another key requirement is *'Eyeball Radio Tuning'*, which can be thought of as an iterative feedback process where YOU will adjust the centre frequency of the RTL-SDR to tune it to an RF signal, using a spectrum analyser and your eyes! Getting the receiver as centred as possible on an RF signal is advantageous as the more accurate the initial tuning is, the easier the carrier synchronisation and phase locking processes are later on.

Exercise 3.2 **An Introduction to Tuning**

This short exercise will show readers how to change the sampling frequency of the RTL-SDR, and how to tune it to a different frequency using the 'Control Panel' GUI.

(a) **Open MATLAB.** Set the working directory to the exercise folder,

📁 `/spectrum`

Next, open the following model:

Chapter 3: Radio Frequency Spectrum Viewing

`.../exploring_the_spectrum.slx`

(b) **Organise the windows.** If you closed the model, rearrange the windows to match the layout shown in Exercise 3.1.

(c) **Run the simulation.** Ensure your chosen antenna is connected to your RTL-SDR, that the RTL-SDR is connected to your computer and that MATLAB can communicate with it via the `sdrinfo` command. Begin the simulation by [1] the 'Start Simulation' button in the GUI or by [1] on the 'Run' ▶ button in the Simulink toolbar.

The first thing to notice is that, as mentioned before, although the two scopes display different information, the x-axes scales are the same, and relate to the frequency components in the signal received by the RTL-SDR. If you think back to the RTL-SDR parameters window from earlier, there was an option to change the 'Sample rate'. Changing the sample rate would have the effect of increasing or decreasing the bandwidth observed in the spectrum analysers, as it displays from $-f_s/2$ to $f_s/2$.

By default, the sampling rate in this model is set to 2.8MHz, which means the spectrum analyzers will display a 2.8MHz band, centred around the chosen tuner frequency. This tuner frequency will be located at 0Hz in the *Spectrum Analyzer* windows (i.e. the centre frequency is downconverted to baseband), which is an important point to note when viewing and navigating the spectrum.

What you should see when the model runs is shown on the next page.

(d) **Change the Sampling Rate.** Stop the simulation by either [1] the 'Stop Simulation' button in the GUI or the 'Stop' ■ button in the Simulink toolbar, and open the parameters window of the *RTL-SDR Receiver* block by [2] on it. The 'Sampling rate' is set to a default value of '2.8e6', which means the device is set to output 2.8 million samples per second (a 2.8MHz sampling rate). The RTL-SDR can be set to multiple different sampling rates, and can even be set as high as '3.2e6'. It is recommended not to exceed 2.8MHz however as this rate ensures accurate and reliable sampling of the spectrum without any data loss.

To change the value, all you must do is to enter a new number into the text box that currently displays '2.8e6', and then [1] the 'Apply' button to set the new value. If the value entered is not suitable, the model will not run. The sampling rate can be set to any of the values in these ranges listed in Table 3.1:

Table 3.1: Accepted sampling rates on the RTL-SDR

RTL-SDR Sampling Rate Ranges [92]					
225,001 Hz	to	300,001 Hz	(225.001 kHz	to	300.001 kHz)
900,001 Hz	to	3,200,000 Hz	(900.001 kHz	to	3.2 MHz)

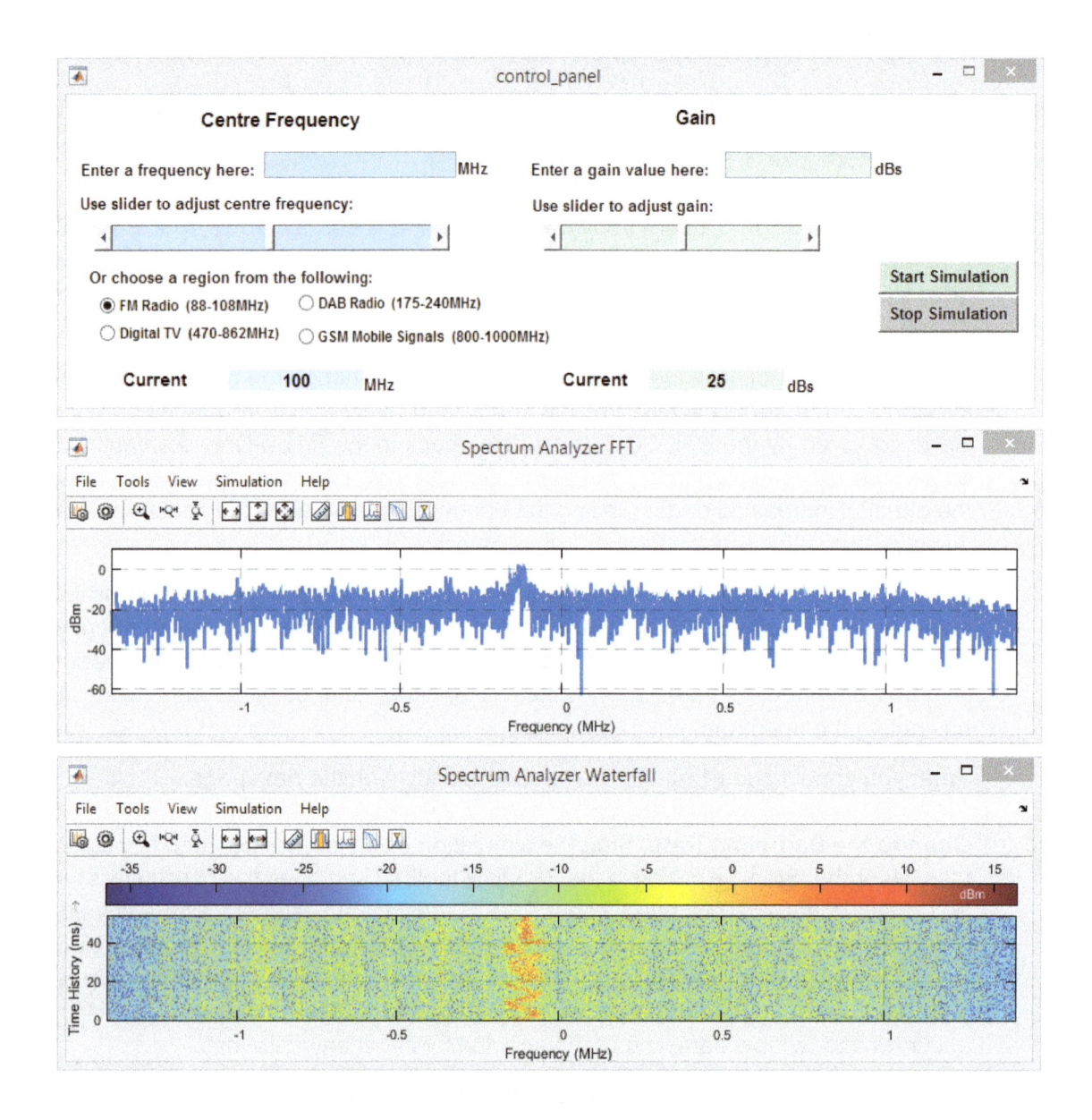

(e) Try changing the sampling rate to 2.4MHz ('2.4e6') using the method above. Once the change has been applied, close the properties window and 🖱 the 'Start Simulation' button in the GUI or the 'Run' ▶ button in the Simulink toolbar.

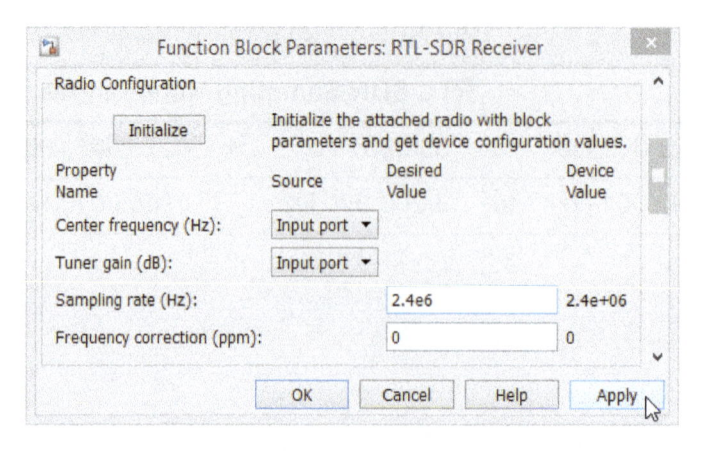

Chapter 3: Radio Frequency Spectrum Viewing

(f) Observe the change in the spectrum analysers. The x-axes of both scopes should update to match the new rate. As you can see, this affects the bandwidth of the spectrum, and this limits what you can view at one time. The higher the sampling rate, the wider the bandwidth. One of the skills you will require when working through this book is to be able to use the information presented in the spectrum analyzers to make decisions about what parameters need to be changed, to optimise the quality of the received signal.

(g) Now try doing the reverse task. Using the same process, set the rate back to 2.8MHz.

(h) **Eyeball tuning: getting centred.** Whenever you change the centre frequency parameter in the GUI, the RTL-SDR will be retuned to this new value. This means that it will downconvert a different band of frequencies from RF, to IF, to baseband. This new baseband signal will enter Simulink and be analysed in the frequency domain, then displayed in the spectrum analyzers.

(i) Ensure the model is running, then use the frequency slider in the GUI to change the centre frequency value. Search until you find a strong signal like the one shown in Figure 3.9 (a). Here, the desired signal has a frequency offset of approximately 285kHz, as shown by the annotations. If this was the RF signal you wanted to receive and demodulate, you would need to retune the RTL-SDR so that this signal was instead centred at 0Hz. While there is only a visual display to aid you with this process in this particular Simulink model, later in the book you will encounter things like FM radio receivers, which will have audio outputs too. In these models you will be able to use both your eyes and your ears when performing tuning!

Tuning the RTL-SDR to an RF signal offset from the current centre frequency can either be accomplished by using the frequency slider in the GUI, or by entering the required centre frequency in the GUIs text box.

(j) Adjust the centre frequency until the *Spectrum Analyzers* look like Figure 3.9 (b); until the signal you want to receive is centred around 0Hz, with no frequency offset.

In other exercises that do not include the GUI, this method can still be implemented by using the other tuning interfaces provided. When you are satisfied that you understand eyeball frequency tuning, move onto Exercise 3.3.

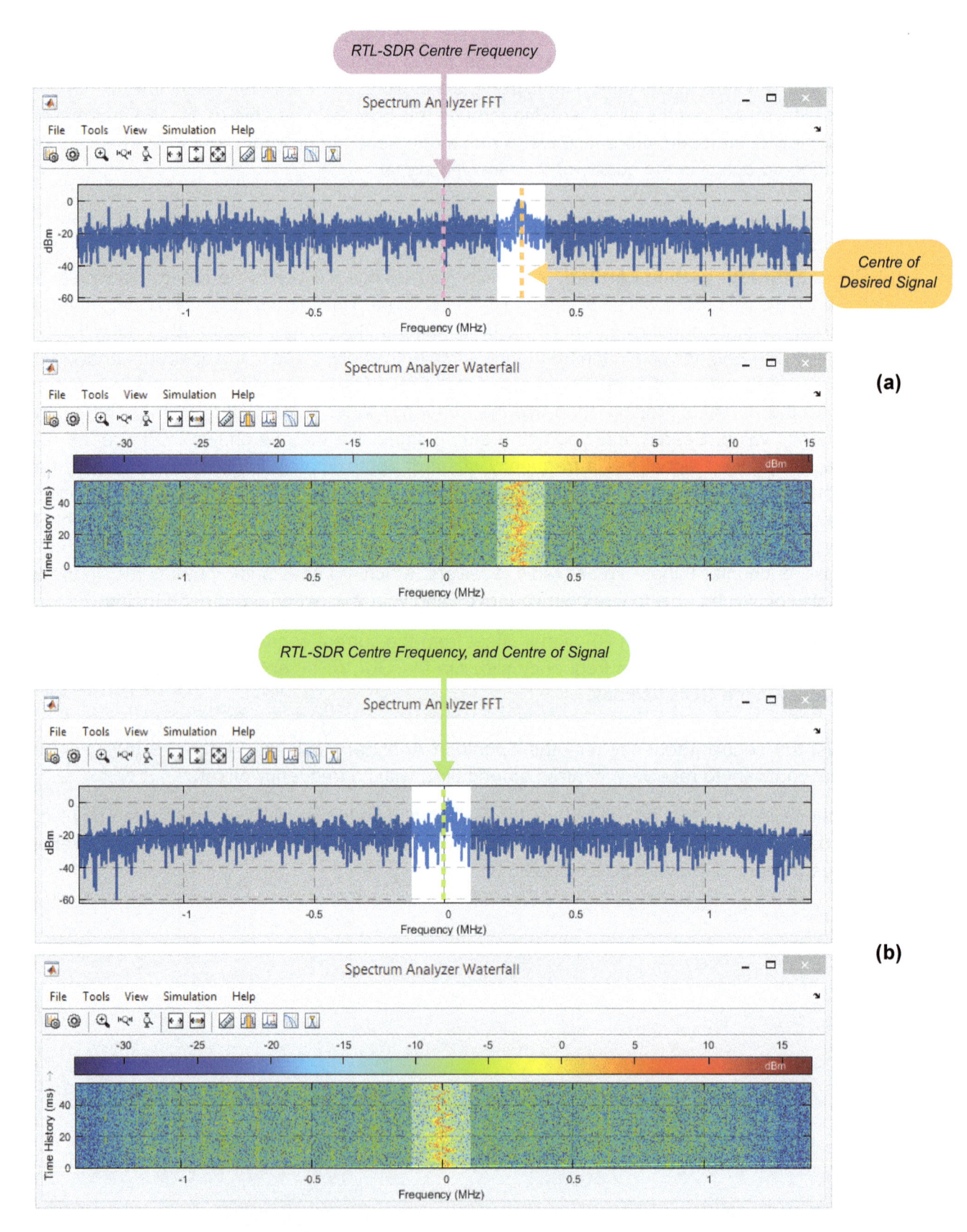

Figure 3.9: *Spectrum Analyzer FFT* and *Spectrum Analyzer Waterfall* showing (a) the received signal with a centre frequency offset, and (b) the received signal correctly centred around 0Hz

Chapter 3: Radio Frequency Spectrum Viewing

Exercise 3.3 Changing the Tuner Gain of the RTL-SDR

This short exercise will explain the relationship of the signal strength shown on the analysers, to the changes of the gain parameter in the *RTL-SDR Receiver* Simulink block.

(a) **Open MATLAB.** Set the working directory to the exercise folder,

>   /spectrum

Next, open the following model:

>   .../exploring_the_spectrum.slx

(b) **Organise the windows and run the simulation.** Hopefully by this stage you should be confident in laying out the windows and beginning the simulation. If not, look back to some of the previous exercises in this chapter. Ensure your chosen antenna is connected to your RTL-SDR, that the RTL-SDR is connected to your computer and that MATLAB can communicate with it via the `sdrinfo` command. Begin the simulation by 🖱 the 'Start Simulation' button in the GUI or by 🖱 on the 'Run' ▶ button in the Simulink toolbar.

(c) Changing the tuner gain in the GUI directly changes the gain that is applied by the tuner in your RTL-SDR. This can be one of the most important parameters when trying to configure the hardware to receive signals in high quality. The images in Figure 3.10 illustrate the difference between a very low gain and a very high gain when receiving a signal from the RF spectrum. As the gain increases you should notice that the noise floor—the unwanted spectral components—increase as well as the desired signal, however, this is still preferred as the signal will be received at a higher quality.

(d) **Change the tuner gain.** Use the centre frequency controls in the GUI to tune to a signal as shown in Figure 3.9, and then alter the gain to see what effect it has. The gain can be changed either using the gain slider, or by typing a different number into the text box in the GUI window. Even if there aren't any particularly strong signals present, you should still observe the difference that changing the gain has upon the signals shown in the spectrum analysers. Figure 3.10 shows a radio signal centred around 0Hz, (a) with a low gain, and (b) with a high gain.

(e) A point to note: increasing the gain of a signal too much can cause the tuner on the RTL-SDR to saturate, which will actually reduce the quality of the signal rather than increasing it. Figure 3.10 (b) illustrates a high tuner gain that is considered adequate for this signal, where a value higher than this would likely affect its reception quality. The key is to find the best signal quality without overdriving the amplifier in the tuner, and this will be much easier to do when you listen to demodulated signals in later exercises.

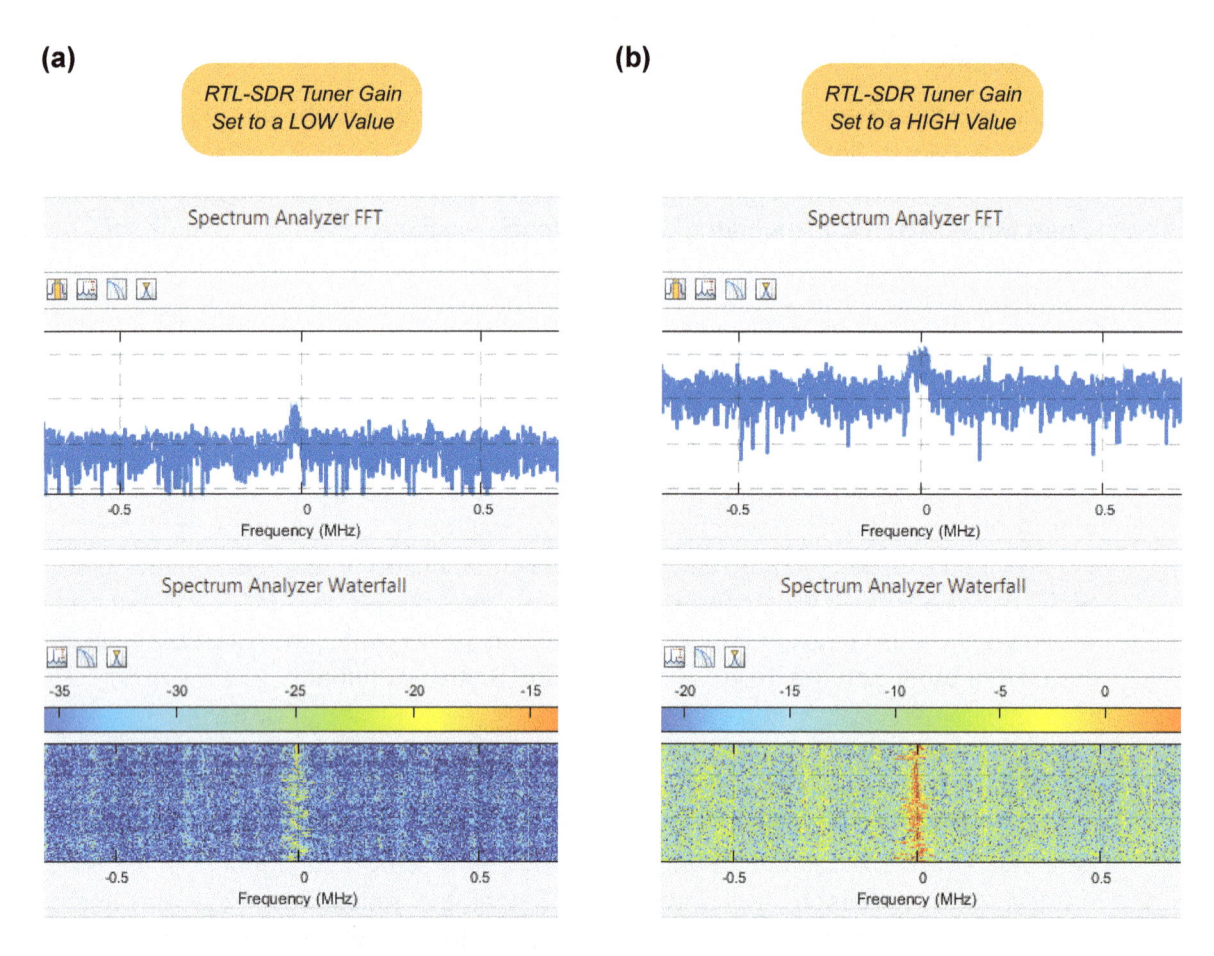

Figure 3.10: *Spectrum Analyzer FFT* and *Spectrum Analyzer Waterfall* showing (a) the received signal when the tuner gain is set to a low value and (b) the received signal when the tuner gain is set to a high value

3.8 FM Radio Stations

FM is the modulation process used for commercial 'FM' radio stations. FM signals are produced by modulating the phase of a carrier with an information signal. In the UK (and most other countries) FM radio signals are transmitted between 88 and 108MHz as shown in Figure 3.11, which allows them to propagate miles across the country. The exact frequencies of each radio station can change from location to location, and if you are in a different country, the region of the FM band may even be different. If this is the case then you should navigate to the region of the spectrum most appropriate to your location.

In the later chapters of this book, you will be taught the theory behind FM, and be shown how to construct a working FM receiver in Simulink that will allow you to listen to the radio stations you find in this next exercise. Here, you will use the `exploring_the_spectrum.slx` model to scan the FM frequency region to see how many stations you can find!

Chapter 3: Radio Frequency Spectrum Viewing

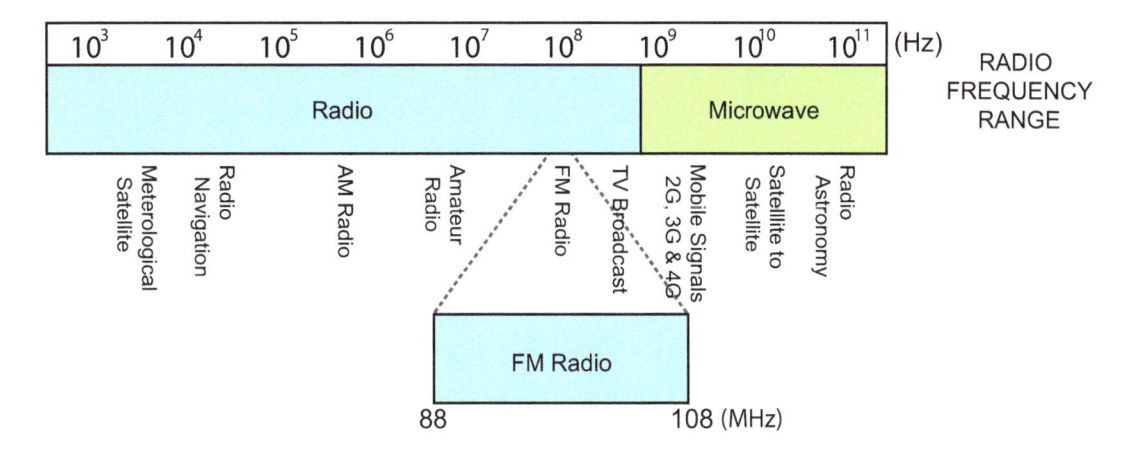

Figure 3.11: Illustration of the FM Radio band in the RF spectrum

Exercise 3.4 Searching for FM Radio Stations

This exercise is focused around finding FM radio signals. These reside in the frequency range 88MHz to 108MHz, and (in the UK anyway), the carriers are spaced 200kHz apart.

(a) **Open MATLAB.** Set the working directory to the exercise folder,

   `/spectrum`

Next, open the following model:

   `.../exploring_the_spectrum.slx`

(b) **Run the simulation.** Ensure your chosen antenna is connected to your RTL-SDR, that the RTL-SDR is connected to your computer and that MATLAB can communicate with it via the `sdrinfo` command. Begin the simulation by [1] the 'Start Simulation' button in the GUI or by [1] on the 'Run' ▶ button in the Simulink toolbar.

(c) **The FM radio region.** Navigate to the FM band by [1] the 'FM radio' button in the GUI, which will set the carrier frequency to a default of 100MHz. You can then move around using the frequency slider or tune to specific radio stations by entering their frequency into the frequency text box and pressing the enter key.

Depending upon your particular geographical location and the location of your chosen antenna, the strength of the radio stations can vary widely from one to another. This can be due to a number of factors, such as your distance from the transmitter, as well as its transmit power. In order to see some broadcast radio stations, you may have to increase the gain to a high value to help visually differentiate them from the surrounding noise floor.

(d) **Exploring the FM radio band.** A good way to quickly understand how many stations are in the FM band is by scanning all of the way through it. Begin by setting the centre frequency of the RTL-SDR to 88MHz. 🖱 the 'up' arrow button on the frequency slider to scan in a continuous motion to the other end of the FM band.

Scanning the band in this way should allow you to see spectral activity of the radio stations very briefly as you pass them by. Repeating this tuning process (between 88 and 108MHz) a few more times, should allow you to get a feel for where some of the strongest signals are.

(e) The screenshot below shows three individual FM radio stations present around a centre frequency of 100.5MHz. We have greyed out areas of the spectrum where there are no FM signals to help emphasise where they are. Stations contain far more power than the noise floor, hence they are easy to see both in *Spectrum Analyzer FFT* and *Spectrum Analyzer Waterfall*.

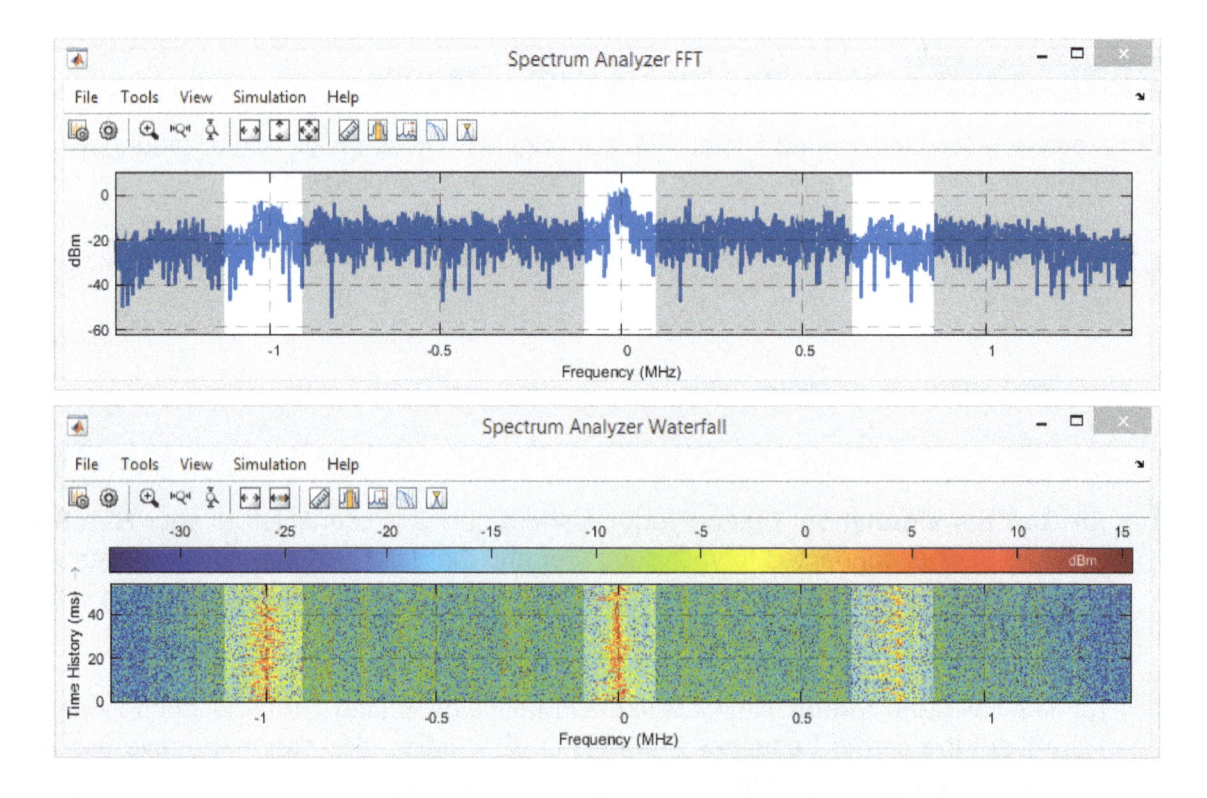

(f) Try finding an area of the FM band where you can see these differences for yourself with real signals.

(g) **Signal Analysis.** Tune to the strongest signal you have found in the FM band, and focus on this signal that is displayed in the two *Spectrum Analyzers*.

As mentioned briefly at the start of this section, FM radio signals are produced by modulating the phase of a carrier with an information signal, and this has the effect of changing the frequency of the carrier. Looking closely at the radio station you have chosen, you should hopefully be able to see the raised peaks moving from side to side around a central frequency. This movement may be more significant in *Spectrum Analyzer Waterfall* as it will illustrate the

Chapter 3: Radio Frequency Spectrum Viewing 63

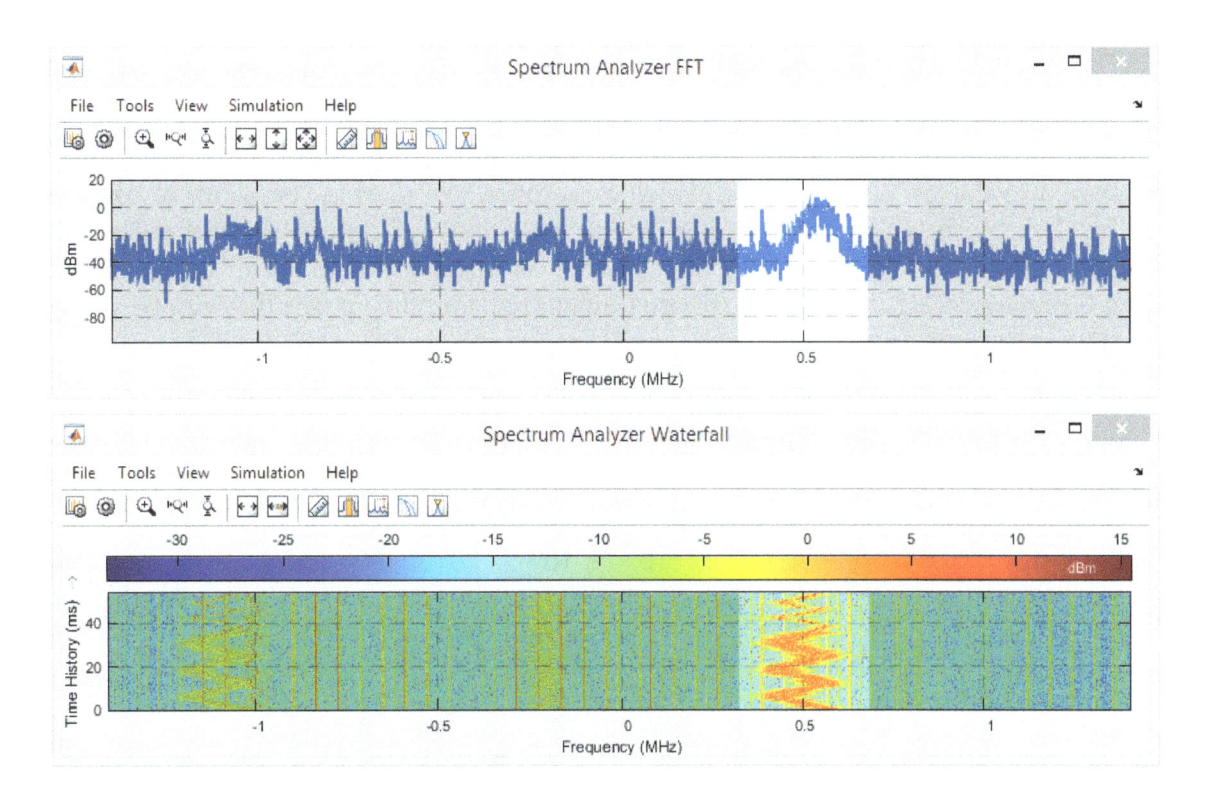

intensity of the signal through time, and should be shown in a colour between yellow and red on the dBm scale.

The waterfall scope above shows this movement occurring as the FM signal propagates through the air. You should hopefully be able to see this type of activity for the strongest radio stations in your area.

(h) **How many stations...?** Using the methods we have previously discussed, scan the FM band again to find out how many radio stations are operating near you. Note down their frequencies by tuning the RTL-SDR to each of them in turn, taking the value displayed in the 'Current Frequency' field of the GUI. Noting these frequencies will be helpful later in the book when you construct an FM radio receiver and will need some stations to tune to!

(i) **Hunting for FM radio stations.** We have recorded a video showing the spectral activity seen from a number of FM radio stations transmitted in Glasgow. Take a look at it and compare how the signals you are able to receive compare to what we saw. Is the station bandwidth the same? Are the stations spaced the same distance apart?

desktopSDR.com/videos/#fm_radio

(j) Once you are finished looking at the FM band, you can move on to learn about viewing mobile signals! As before, you can either leave the model running (which will allow it to continually receive and display signals in real time), pause it by 🖱 on the 'Pause' ⏸ button, or stop it by 🖱 on the 'Stop' ⏹ button in the Simulink toolbar. Instructions will be given in the next exercise if you need help re-opening MATLAB.

3.9 Mobile (Cell) Phone Signals—2G, 3G and 4G

Most of us use mobile (or cell) phones on a daily basis, but few actually have knowledge about how they operate with such flexibility and connectivity. The standards for mobile communications have been developed over the last 30 or so years; evolving continually to meet the demands of consumers. Signals for First Generation (1G) mobile phones (which were designed only to support voice calls) were first broadcast in the UK in the mid 1980s, and alas are now long gone from the airwaves. Second, Third and Fourth Generation (2G, 3G, 4G) signals are transmitted, and this means they can be received with the RTL-SDR!

Due to the manner in which regulating bodies have licenced mobile operators throughout the years, different countries transmit mobile signals at different frequencies. Figure 3.12 shows a rough breakdown of where in the spectrum 2G, 3G and 4G signals are transmitted in the UK.

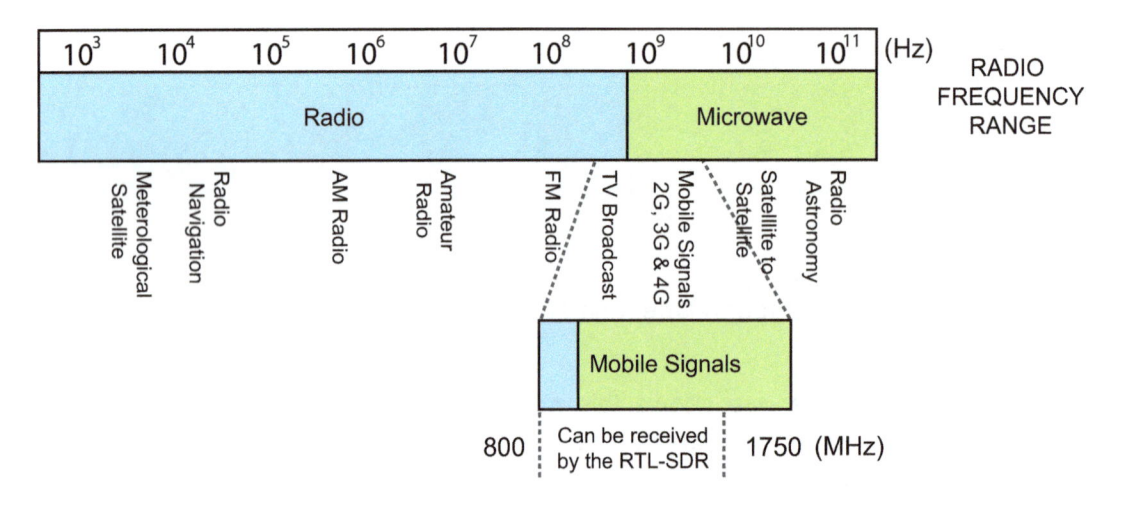

Figure 3.12: Illustration of mobile bands in the RF spectrum, highlighting the region that can be received by the RTL-SDR

3.9.1 2G Mobile Signals

In the UK, the 2G standard is the backbone of the network, and operates between 890 and 960MHz with carriers spaced 200kHz apart. The GSM standard provides support for voice calls, text and picture messages (SMS, MMS) and a slow internet connection (WAP) [82]. GSM signals transmitted from mobile base stations to your phone are transmitted on what is known as a '*downlink channel*', and in the UK, downlink channels lie in the region 935 to 960MHz. Signals sent from your phone to the base station are transmitted on an '*uplink channel*', and again in the UK, these lie between 890 and 915MHz. GSM signalling uses a technique called Time Division Multiple Access (TDMA), which allows multiple mobile phone users to have access and connectivity through a single channel by dividing the access time slots of the channel (and the carrier signal) between them. More detailed information on this standard can be found here [16].

Exercise 3.5 Exploring the Mobile Spectrum: 2G GSM (800–1000MHz)

This exercise is focused around finding GSM mobile signals. In the UK these lie between 890 and 960MHz, but are at various other frequencies in other countries. If you stay outside of the UK, you

Chapter 3: Radio Frequency Spectrum Viewing

should be able to find which frequencies the GSM uplink and downlink channels are transmitted on online.

(a) **Open MATLAB.** Set the working directory to the exercise folder,

   /spectrum

Next, open the following model:

   .../exploring_the_spectrum.slx

(b) **Run the simulation.** Ensure your chosen antenna is connected to your RTL-SDR, that the RTL-SDR is connected to your computer and that MATLAB can communicate with it via the `sdrinfo` command. Begin the simulation by clicking the 'Start Simulation' button in the GUI or by clicking on the 'Run' button in the Simulink toolbar.

(c) **Find some GSM mobile signals.** If GSM is transmitted between 800 and 1000MHz in your country, click the 'GSM Mobile Signals' button in the GUI, which will set the carrier frequency to a default frequency of 900MHz. If GSM is outwith this range (but still within the 25MHz to 1.75GHz range of the RTL-SDR), manually enter a frequency value to the centre frequency text box. As when you were searching for FM radio, you can adjust the frequency slider to navigate back and forth ±500MHz of the chosen centre frequency.

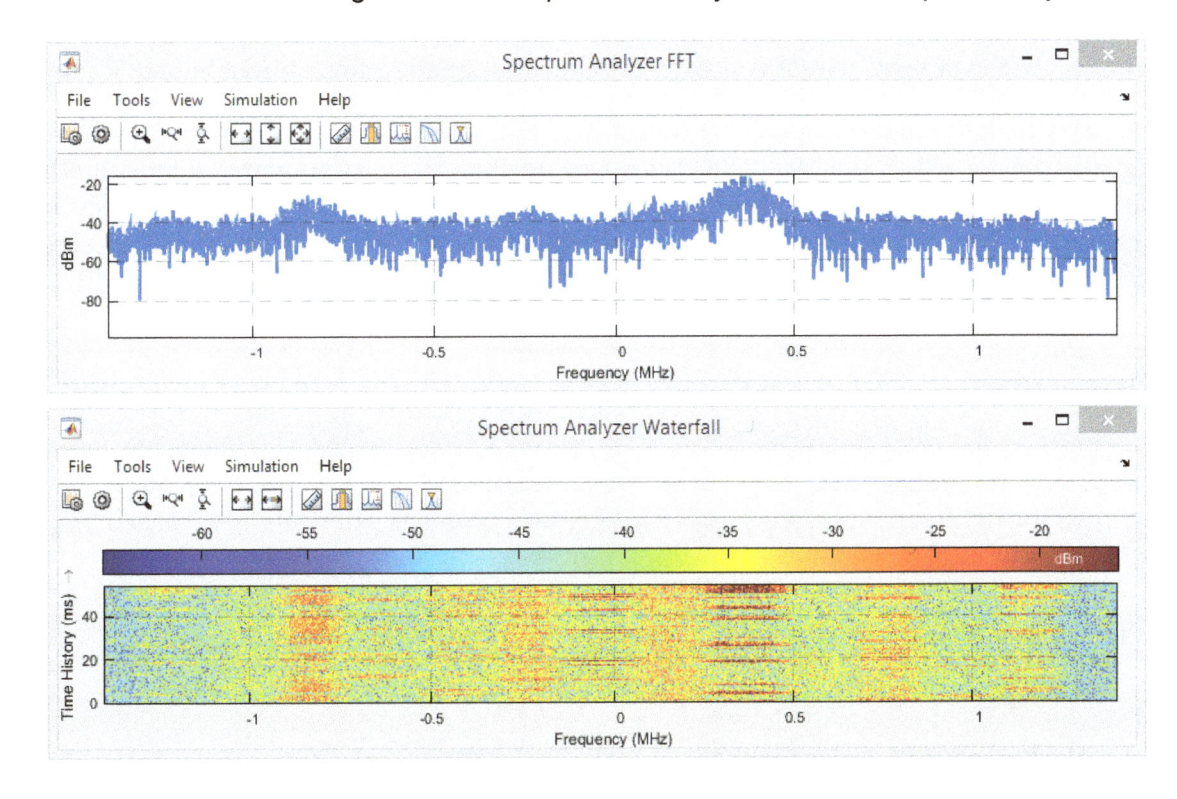

(d) As you scan this region you should hopefully come across some GSM signals. These will be clustered in small groups of 200kHz wide channels, all of which should operate independently of each other in a 'bursty' fashion. The screenshot below shows some GSM uplink channels we observed when tuning to 945MHz. *Spectrum Analyzer Waterfall* is particularly useful here as it highlights bursts of activity in the channels through time. These relate to the transmission of frames that form the TDMA link between users and the mobile basestation.

(e) Comparing the GSM signal to the FM signal you saw in Exercise 3.4, you should be able to see that they have very different characteristics! If you wish to confirm the comparison yourself you can easily navigate back and forth between regions by either using the radio buttons, or the frequency input box in the GUI.

(f) Try to see how many GSM channels you can find in the spectrum, and then search online to compare this value to the total number of channels that actually exist. Did you find all of them?

(g) **Watch GSM signals being transmitted from a nearby basestation.** We have recorded a video showing spectral activity from a GSM basestation around 200m from our location. Take a look and compare how the signals you have received compare to what we saw.

desktopSDR.com/videos/#gsm_basestation

3.9.2 3G Mobile Signals

The UMTS 3G standard was a further evolution to mobile communications that was orientated for higher amounts of packet data. It uses a connectivity method called Wideband Code Division Multiple Access (WCDMA). While GSM channels are only 200kHz wide, UMTS channels are 5MHz wide, and signals are sent across the full band as a *spread spectrum* signal. The wider bandwidth associated with this standard allowed an increase in data rates while keeping the transmission efficiency high by affording multiple users access at the same time [85]. Instead of relying on timeslots, each user's mobile phone can determine which broadcast signals are meant for it by matching its individual code with them. More information on the original standard and its subsequent upgrades can be found here [83] [16].

Although the majority of these signals are transmitted at 1700MHz and above, there are still a number of 3G signals transmitted in the 800–900MHz region in the UK. Unfortunately due to the bandwidth limitation of the RTL-SDR, you will not be able to view the full 5MHz band simultaneously when using a single device, but you will be able to see enough of the spread spectrum signal to recognise the differences between it and the previously discussed GSM signals.

Exercise 3.6 Exploring the Mobile Spectrum: 3G UMTS (800–900MHz)

This exercise is focused around finding UMTS mobile signals. In the UK these are transmitted on numerous different frequencies, some of which lie between 800 and 900MHz. A quick search online should reveal what frequency bands UMTS uplink and downlink channels are transmitted in at your location.

(a) **Open MATLAB.** Set the working directory to the exercise folder,

/spectrum

Next, open the following model:

Chapter 3: Radio Frequency Spectrum Viewing

.../exploring_the_spectrum.slx

(b) **Run the simulation.** Ensure your chosen antenna is connected to your RTL-SDR, that the RTL-SDR is connected to your computer and that MATLAB can communicate with it via the `sdrinfo` command. Begin the simulation by [1] the 'Start Simulation' button in the GUI or by [1] on the 'Run' ▶ button in the Simulink toolbar.

(c) **Find some UMTS mobile signals.** Use the GUI buttons to navigate to the mobile region if you aren't already there, and then use the slider to move to the frequency that UMTS is transmitted on in your region. In the UK, this is often between 915 and 935MHz. Pan around until you see some wideband UMTS signals like the one shown in the screenshot on the following page.

As explained in the earlier part of this section, the UMTS channels are 5MHz wide, and hence, the RTL-SDR can only show a fraction of one at any point in time. In the screenshot, we have captured the start of a channel, as this helps to show the difference between it and the noise floor. To view a full channel with one RTL-SDR you will have to use the frequency slider to scroll across the spectrum. The shaded area on the left-hand side of both spectrum analysers shows a UMTS guard band, which contains no signals and aims to separate the active channels.

(d) Scrolling through a full channel you should notice that there is a constant intensity (in terms of signal power). Unlike GSM signals, there are no noticeable frame breaks in UMTS channels, and this is due to users being distinguished by codes rather than time slots, meaning there is constant activity and intensity in the spectrum because many users are connecting at the same time.

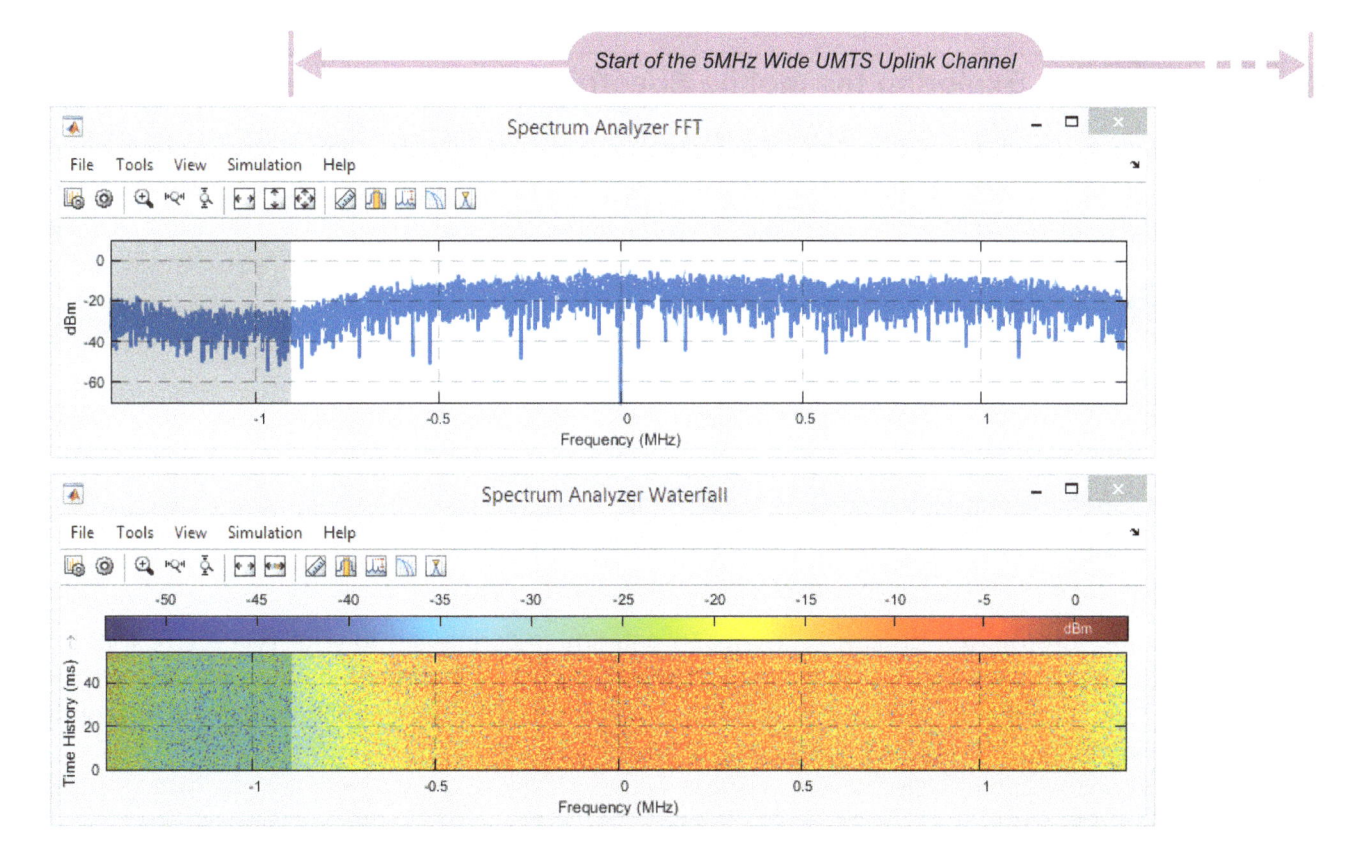

(e) **Watch UMTS signals being transmitted from a nearby basestation.** We have recorded a video showing UMTS downlink channels that we can receive with the RTL-SDR in our location. Take a look and compare how the signals you have received compare to what we saw.

desktopSDR.com/videos/#umts_basestation

3.9.3 4G Mobile Signals

LTE (Long Term Evolution) is one of the most recent mobile standards (4G) to be deployed around the world. It builds on the previous infrastructure, focusing once again on increasing data rates, due to the ever increasing societal requirement for connectivity. The headline 4G standard (into which LTE does not actually fall [15]) specifies data download speeds of up to 1Gbps to stationary devices, which put into context, is around x10 faster than the speed of most ethernet networks. An 'upgraded' edition of the LTE standard called LTE-Advanced is currently under trial in various cities around the world, and this aims to get a little closer to the mark [74].

LTE aims to offer long term benefits by its progression to a statistical multiplexing technique called Orthogonal Frequency Division Multiple Access (OFDMA). This provides multiple access connectivity by using both time and frequency division to assign resources to individual users, in order to provide the highest and most efficient throughput. Tx and Rx channels for LTE signals can be found at 800, 900, 1800 and 2600MHz in Europe, and at other frequencies around the globe. They can occupy various different bandwidths from 1.4MHz to 20MHz in the spectrum, as well as having the ability to aggregate together to achieve even larger bandwidths, as explained in [73].

Exercise 3.7 Exploring the Mobile Spectrum: 4G LTE (700–900MHz)

This exercise is focused around finding LTE mobile signals. Worldwide, these are transmitted on numerous different frequencies, most of which are outwith the range of the RTL-SDR. Some countries (including the UK) do transmit these 4G signals at lower frequencies, between 700 and 900MHz. A quick search online should reveal if you are able to receive any of these signals. If you can't, you can still watch the video to see some being transmitted in Glasgow.

(a) **Open MATLAB.** Set the working directory to the exercise folder,

/spectrum

Next, open the following model:

.../exploring_the_spectrum.slx

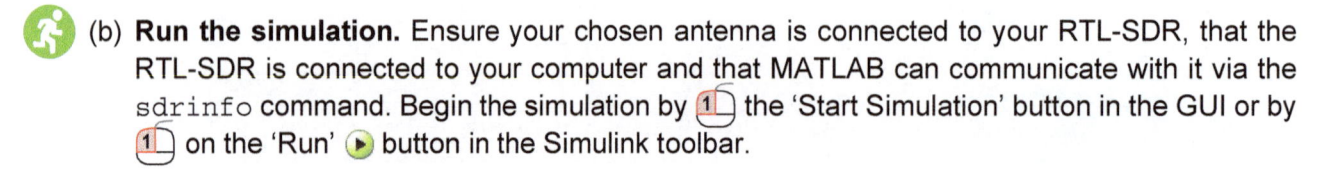

(b) **Run the simulation.** Ensure your chosen antenna is connected to your RTL-SDR, that the RTL-SDR is connected to your computer and that MATLAB can communicate with it via the `sdrinfo` command. Begin the simulation by 🖱 the 'Start Simulation' button in the GUI or by 🖱 on the 'Run' ▶ button in the Simulink toolbar.

Chapter 3: Radio Frequency Spectrum Viewing

(c) **Find some LTE mobile signals.** The last mobile signals we will try to view are in LTE 4G channels. Navigate to 800MHz (or another frequency if you know there to be LTE signals present) by entering the number into the centre frequency text box, then use the frequency slider to pan across the spectrum until you find a signal that looks similar to that shown in the screenshot below. Here it is clear to see with the spectrum analysers that the LTE signals have numerous high powered OFDM carriers that are equally spaced across the bandwidth of the channel. Due to the flexibility that the LTE standard allows, the bandwidth of these signals can theoretically be as small as 1.4MHz, and as large as 20MHz. For this reason, it's best to recognise an LTE channel by the presence of the OFDM carriers rather than by its bandwidth.

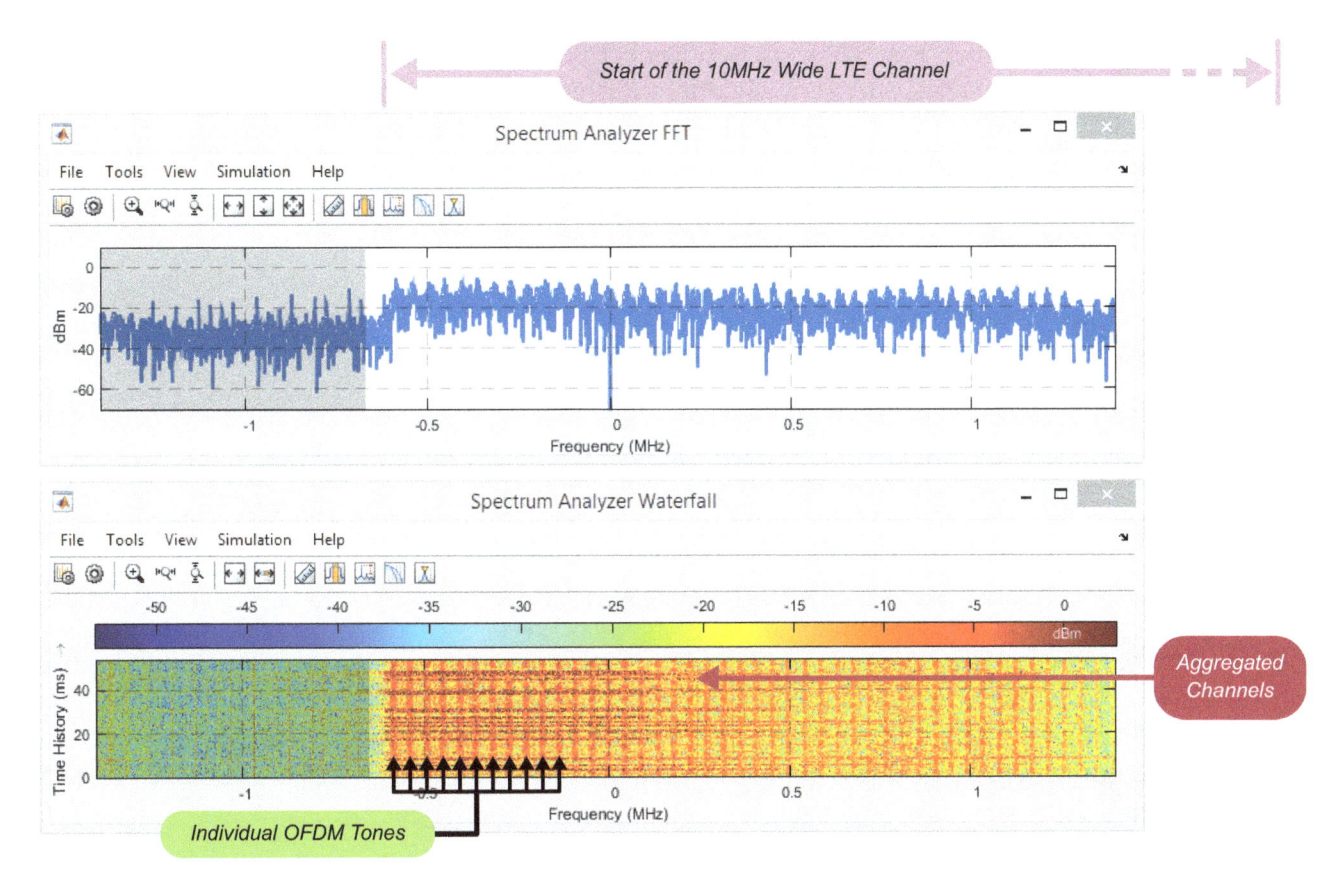

(d) As introduced earlier, the resources for LTE connectivity are assigned using both time and frequency division, as can be seen in the screenshot. The individual OFDM carriers spaced across the channel are clear to see from the vertical lines in *Spectrum Analyzer Waterfall*, and the spikes in the *Spectrum Analyzer FFT*. Each of the OFDM carriers are then further divided into time slots, which are illustrated by the breaks in intensity.

(e) **Watch LTE signals being transmitted from a nearby basestation.** We have recorded a video showing spectral activity from a mobile basestation outputting an LTE signal that is near our location. Have a look at the signal we saw, and see how it compares to any you have seen with your RTL-SDR. Do the bandwidths match?

desktopSDR.com/videos/#lte_basestation

3.9.4 Finding your phones current frequency

Viewing the spectrum in search of mobile signals being broadcast around you is extremely interesting, but a more exciting mission is trying to view signals being transmitted to and from your own mobile! To do this, you will need to find your mobile phone's current operating frequency. Mobiles are dynamic in their connectivity, and you may have noticed that your phone will change standard depending upon the area you are in and the network coverage. Your phone can even change frequency when using the same standard, as a result of the number of users trying to connect at your location, and your connection requirements.

In order to find the frequency that your phone is using at any point in time, you must access some hidden settings called '*field test mode*' that will show you in-depth connectivity information. These settings are present on most mobile phones, and normally offer information such as the current signal strength (in dBm), and most importantly the current uplink and downlink frequencies. These frequencies are often displayed as an Absolute Radio Frequency Channel Number (ARFCN) instead of an actual frequency value in MHz, but this number can easily be converted back with tools online, such as this one [77]. For readers who like a challenge, you can also use Eq. (3.1) to calculate the centre frequency of the GSM channel your phone has been allocated. You will need to do a little research as to what the other f values are, as these vary from country to country.

$$f_{gsm} = ARFCN \times (f_c + f_b + f_o) \text{ MHz} \tag{3.1}$$

where: f_{gsm} = GSM Channel Centre Frequency (MHz)
f_c = Channel Spacing (MHz)
f_b = Base Frequency (MHz)
f_o = Offset Frequency (MHz)

This document [119] provides information about how to access field test mode on a number of the most popular mobile phone handsets. If your phone is listed here, you will be able to follow the instructions given to access field test mode, and find all of the information required to carry on with the next exercise. If your phone doesn't have a field test mode, or you don't have it around, you can still move on to the exercise and search for generic mobile signals.

Exercise 3.8 Exploring the Mobile Spectrum: Challenges!

This exercise contains a series of challenges, where we will ask you to try and view mobile signals being transmitted to and from your phone using the various generational standards. To complete these challenges, you will require access to field test mode on your phone. If your phone does not support this and you cannot borrow one that does, it will not matter as videos are provided to show what you could be seeing.

(a) **Open MATLAB.** Set the working directory to the exercise folder,

/spectrum

Next, open the following model:

Chapter 3: Radio Frequency Spectrum Viewing

```
.../exploring_the_spectrum.slx
```

(b) **Run the simulation.** Ensure your chosen antenna is connected to your RTL-SDR, that the RTL-SDR is connected to your computer and that MATLAB can communicate with it via the `sdrinfo` command. Begin the simulation by [1] the 'Start Simulation' button in the GUI or by [1] on the 'Run' ▶ button in the Simulink toolbar.

(c) **Enter Field Test Mode using the steps in [119].** Menus for GSM, UMTS and possibly even LTE-Advanced should appear depending your phone's capabilities. Browsing these menus should indicate which standard you are currently using, as information such as the current channel power should only be given for an active connection. If it is GSM, and the value is given as an ARFCN, you can easily convert it to a frequency value using an online converter.

Unfortunately on most phones you cannot create network activity (for any of these standards) while the phone is in field test mode, which means you will almost certainly have to move in and out of it as you progress through this exercise. If you think that the activity on the spectrum does not fit with the operations you are performing on the phone, then it's worth checking the connection information again, as a number of factors could have caused the phone to change frequency.

Another point to note is that as the more recent standards (3G, 4G) are intended for data connectivity, it's often the case that the amount of traffic on the uplink and downlink channels varies greatly, as the channels are asymmetric. If you were, for example, streaming a video, the load on the downlink channel would be far higher than that of the uplink channel, because data is being downloaded to your device. If you were uploading a photo to a social network however, the reverse would be true.

(d) As your phone changes between the different standards, you will have to alter the type of communication operation you perform, to achieve the best visual results. For 2G, sending a text or making a phone call near your antenna should show activity on the uplink frequency, whereas, when trying to view 3G and 4G signals, you should perform an online search or stream video content to force data packets across the downlink connection. Many modern phones often provide the user the ability to restrict the connection type to the older standards, by turning 3G and 4G off. Finding out if this is possible on your phone could be helpful when trying to force your phone to use a 2G connection, as by default, phones tend to opt for the highest generation signal available. It may not always be possible to control the standard that the phone uses, but try to be patient with it, as it is only trying to give you the best possible connection!

(e) **Making a GSM phonecall on a mobile (cell) phone.** The following video shows field test mode being accessed on a mobile, the calculation required to convert an ARFCN number back to a frequency value, and spectral activity seen with the RTL-SDR/ Simulink receiver when making a call.

```
desktopSDR.com/videos/#gsm_fieldtest_phonecall
```

Frames of voice data transmitted from the phone will appear as 200kHz wide bursts.

(f) **GSM challenge.** Have you watched the video? It is your turn to try this now! Force your phone to use a GSM signal by setting it in '2G mode' (or turning off data) and then find the centre frequency of the GSM uplink channel. Tune the RTL-SDR to this frequency and then make a call. Can you see a higher intensity GSM signal? The screenshot shows what we saw when we tuned to our mobile's uplink frequency. Do you see something similar? Try sending a text message instead. Is there any difference?

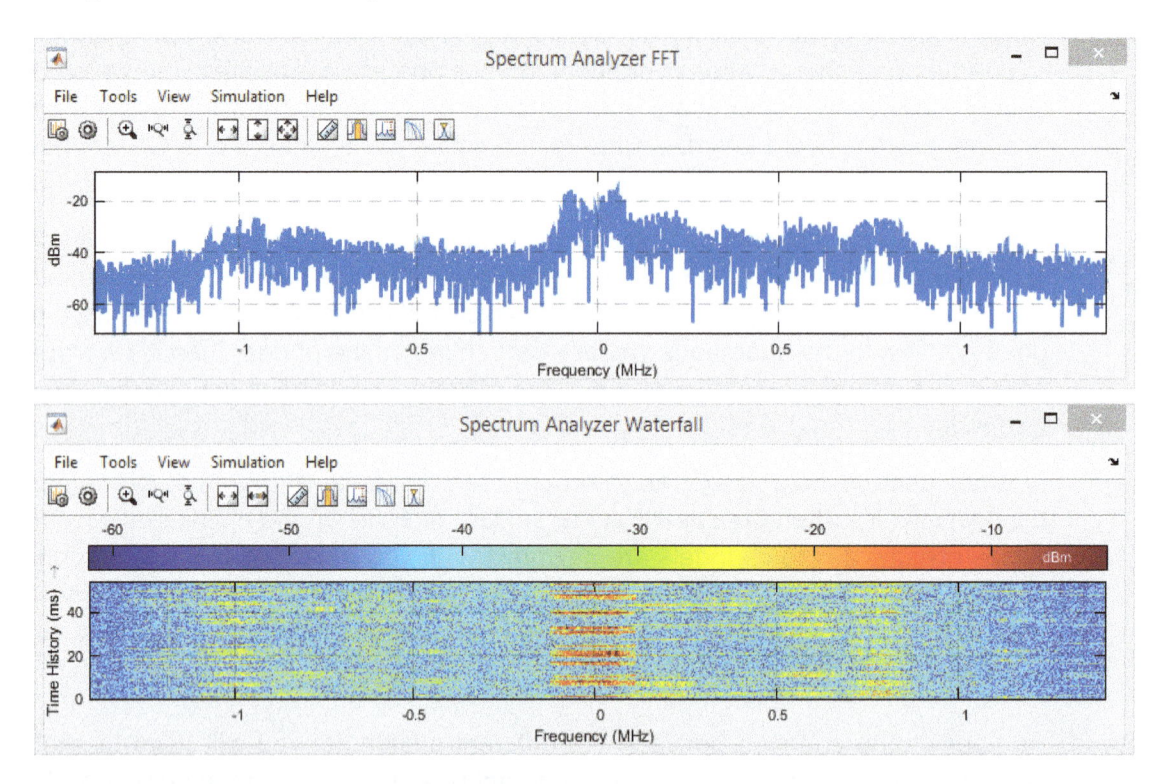

(g) **Streaming a video to a mobile (cell) phone over UMTS.** In the following video, we access field test mode on a mobile to find the UMTS uplink and downlink channel frequencies, then tune the RTL-SDR to the uplink frequency and stream a video from YouTube to the phone.

desktopSDR.com/videos/#umts_fieldtest_videostream

Signals are transmitted from the phone requesting a connection to the internet, and then to the video from YouTube. As the video is downloaded to the phone's buffer, requests and acknowledgements are transmitted from the phone to keep the data flowing.

(h) **UMTS challenge.** Have you watched the video? It is your turn to try this now! Make sure WiFi is disabled and force your phone to use a UMTS signal by setting it to '3G mode'. Find the centre frequency of the UMTS uplink and downlink channels in field test mode. Tune the RTL-SDR to the uplink frequency and then access the internet. The screenshot below shows what we would expect you to see.

(i) Do you see 5MHz wide bursts of data being transmitted from your phone? What happens if you make a call or send a text when in this mode? It should be noted that because the UMTS channels are shared between multiple users, the results may be clearer when carrying out this test in an uncrowded location. Being a singular user of this channel—obviously the ideal case—would allow you to notice the vast difference in intensity caused by your own phone's activity. If you are unsure whether the activity on the spectrum relates to your own phone connection, try switching cellular coverage off by placing the phone into 'Flight Mode' (or 'Airplane Mode'). Flight Mode guarantees to stop the phone's current connection and removes it's presence from the RF spectrum.

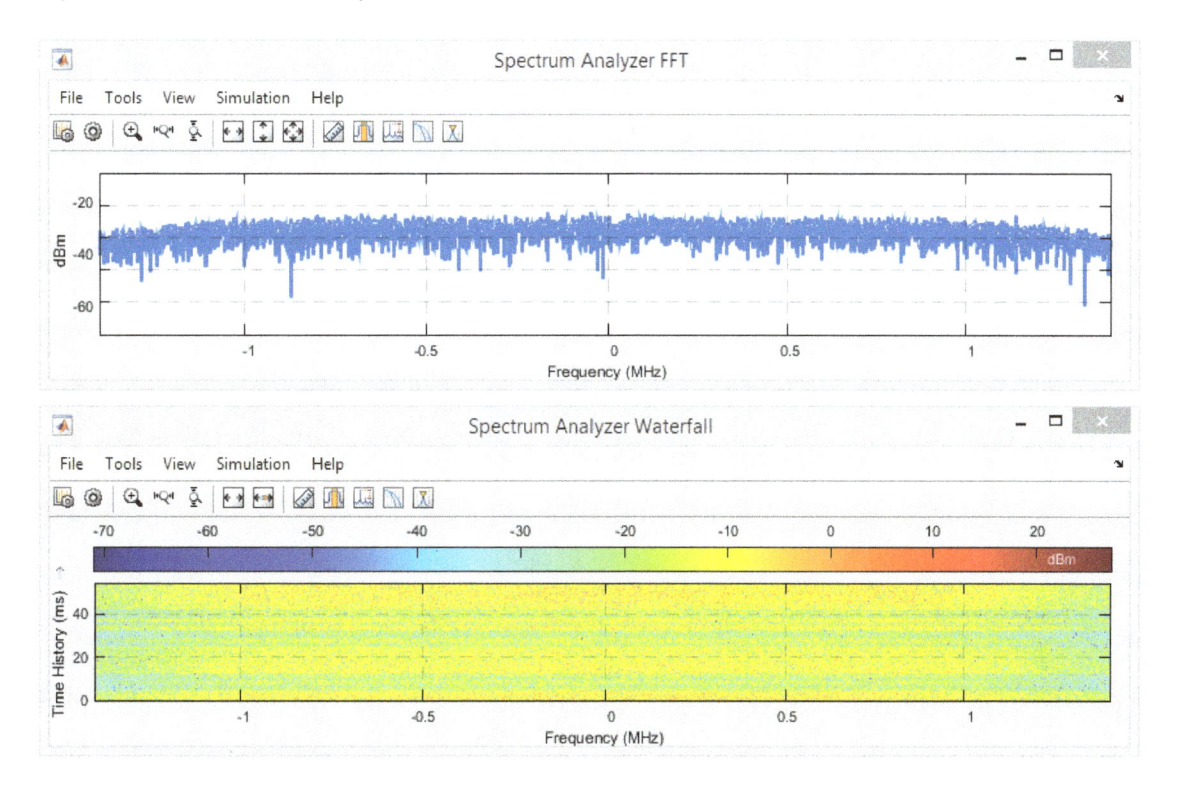

(j) **Asymmetric UMTS channels.** Following on from the previous challenge, the exercise now changes focus to viewing the difference between the uplink and downlink UMTS channels. Using the method discussed previously, obtain both the uplink and downlink frequencies for your mobile phone when it is set in '3G mode'.

(k) While streaming appropriately long video content, navigate to the uplink frequency of your mobile by entering the value into the centre frequency text box in the GUI. Even though you are receiving lots of data when streaming a video, your phone also sends large amounts of control data across the uplink channel back to the base station. This is because of the data transfer protocols being used by your phone, and may make the uplink channel look as heavily loaded or congested as the downlink channel. In reality, if you turn the gain down very low, similar to the 5dB gain shown in the screenshot on the following page, you should hopefully be able to see the 'bursts' of intensity that appear on the *Spectrum Analyzer* displays. During a burst, the FFT will jump to a higher power level, and then return back again. The waterfall scope will show these short-term increases in intensity with darker orange and red lines.

Retune to the downlink frequency, which needs much higher gain (in the region of 30dB or more), and compare and contrast this with what you observed previously. You should notice

UMTS Uplink Channel Tuner Gain = 5dB

UMTS Downlink Channel Tuner Gain = 50dB

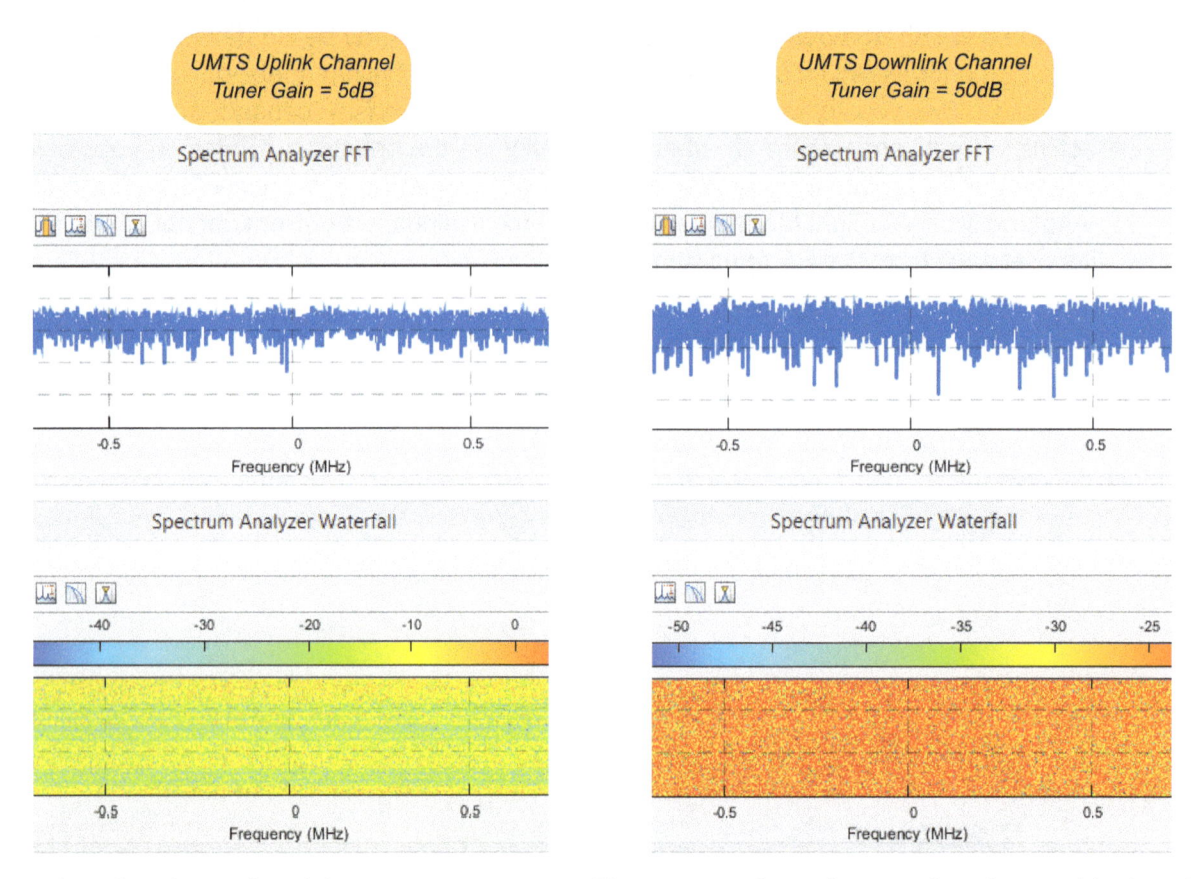

that the channel activity appears constant. The reason the gain must be changed is that the phone is transmitting from a very short distance away, while the base station may be 100s of metres away. The closer the antenna is to either the phone transmitter or the base station transmitter, the higher the initial received signal strength will be.

(l) When you are finished creating and viewing spectrum activity, you can either continue on to the final challenge, or move on to the next section.

(m) **Final mobile network challenge.** The aim of this challenge is to use a database such as the UK Ofcom 'Mobile Phone Base Station Database' known as 'Sitefinder', to locate the mobile base stations providing coverage to your area. If you stay outwith the UK, you shall need to research to find if such a database exists for your country first.

(n) For those in the UK, Sitefinder is an online resource that allows you to see the location of mobile base stations throughout the country, and displays information about which standards they support as well as the operator that owns them. Using Sitefinder, try to determine which local transmitter is supporting your current mobile phone connection! We will give you the following hints to do this:
- Discover the operator network that your phone is currently using
- Find what standard the phone is using, such as 2G, 3G or 4G
- Find your current uplink and downlink frequency
- Match these details to the nearest base station that can support these requirements

3.10 433MHz: Key Fobs and Wireless Sensors

Examples in Sections 3.8 and 3.9 have encouraged you to try and view spectral activity for some of the most well known RF signals that most of us use on a daily basis. But what about the other everyday signals that use the RF spectrum, that most of us don't even think about? Car key fobs that are used for central locking commonly operate on 433.9MHz to transmit the locking, unlocking and boot (trunk) opening signals. Note that the frequency used varies from manufacturer to manufacturer, and will also depend on localised frequency allocation rules. Every time a button is pressed on the fob, an encoded bit stream is sent via a 433.9MHz transmitter to a receiver inside the car. Other signals at 433MHz include those radiating from wireless sensors, remote controlled multi-socket extensions, doorbells and many other wireless controls for various devices.

Fortunately, the RTL-SDR receiver can be tuned to receive these signals as they are transmitted through the air! Figure 3.13 shows the region where these 433MHz wireless SRDs feature on the spectrum. If you happen to have a set of car keys, or any products that transmit at 433MHz (or indeed any other frequency within the RTL-SDRs operating range), retrieve them for the next section of practical work. More information about SRDs can be found here [84].

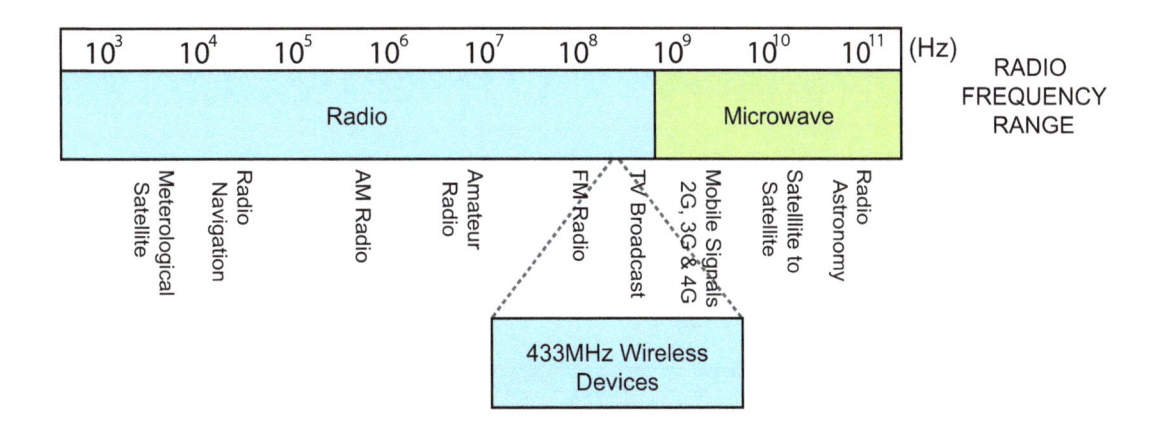

Figure 3.13: Illustration of the SRD bands in the RF spectrum

The following exercise aims to show a few examples of the activity in the spectrum that is caused by SRD sensors and transmitters. There are many different products that incorporate these transmitters, but they all have the same aim; to transmit an information signal to a receiver for an intended purpose. These information signals are often carried through the air using AM. Here, the amplitude of an RF carrier is modified in sympathy with the amplitude of the information signal.

Exercise 3.9 Searching for Key Fob and Wireless Sensor Signals

In this exercise, we aim to show you a couple of examples of spectral activity from SRD RF devices you use on a daily basis.

(a) **Open MATLAB.** Set the working directory to the exercise folder,

📁 `/spectrum`

Next, open the following model:

```
.../exploring_the_spectrum.slx
```

(b) **Run the simulation.** If you have any portable RF devices, locate them now. Normally these will state what frequencies they use to transmit information in inconspicuous locations, for example on the base, or where a key flips out from. A quick online search for any devices you have located should also provide this information. Make a note of these frequencies, as you will need to tune to them in the next step.

(c) **Find your wireless transmitter signals.** Now use these noted frequencies to tune the RTL-SDR. Do this by entering the value into the centre frequency text box and pressing the enter key.

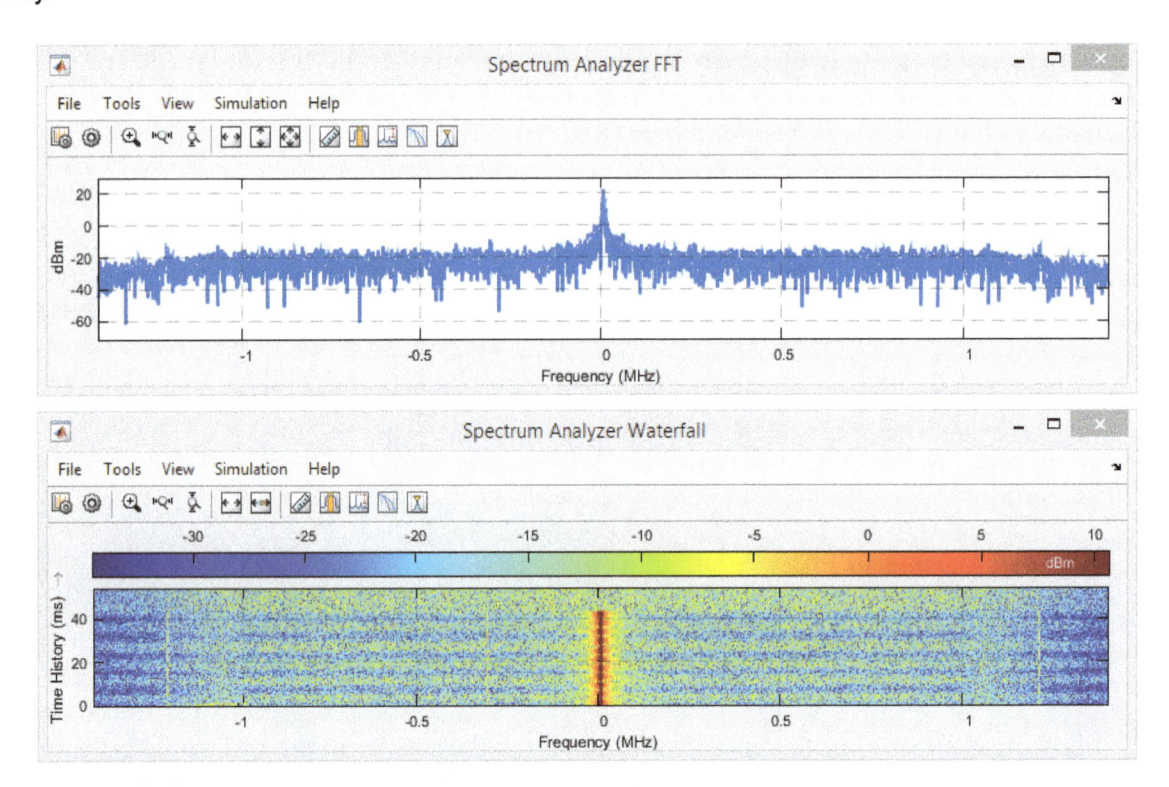

The wireless sensors and controllers can operate in different ways. Some require user interaction to transmit a signal, such as pressing the button on a remote control to open a garage door, and some simply transmit signals of their own accord at regular intervals, say every 30 minutes. The activity seen above is from a sensor that transmits temperature and humidity information every 30 minutes to a main display unit.

(d) Look at the screenshot, and compare it to any of the previous signals we have presented you with in this chapter. What do you notice? It is likely that the signals transmitted from your SRDs take a totally different form to any of the mobile signals, as most are narrowband. This can be seen here as the peak in the centre of *Spectrum Analyzer FFT* is very sharp and thin when compared to all of the previous signals, and the intensity in the waterfall is equally thin.

Chapter 3: Radio Frequency Spectrum Viewing

(e) If you have your own sensor or transmitter at this point, look for their signals in the spectrum analysers to see what form they take. This of course may require some user interaction to produce the signal. How does it compare to what we saw?

As the beginning of this chapter briefly explained, each service that uses the spectrum has regulations and specifications for the characteristics associated with them, such as the bandwidth and transmission power. Due to the short distance and low amount of data being transferred by the SRD transmitters, the bandwidth required is much less than that of other communications services. You may also notice that the receive power for the signal is very high. This is because the transmitter is close to the antenna, much like the results you saw when watching signals being transmitted from your phone in Exercise 3.8.

(f) **Receiving SRD signals.** In the following video, we show the signal received from a remote control for a garage door. This is a simple device that has only two functions: up and down.

desktopSDR.com/videos/#srd_garagedoor

Watch the video and see how this compares to any SRD signals you have been able to receive.

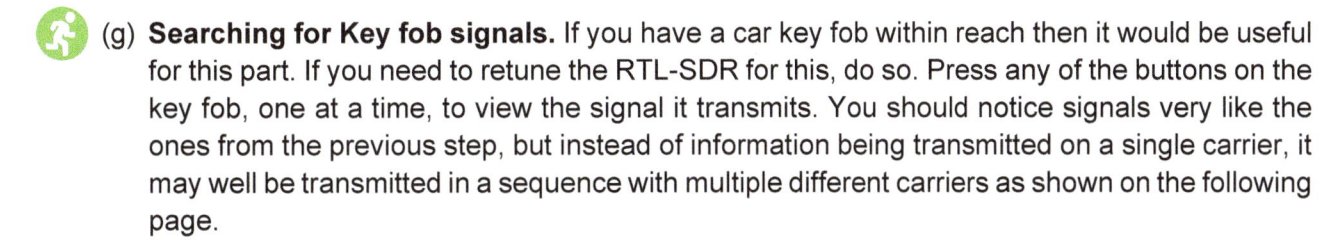

(g) **Searching for Key fob signals.** If you have a car key fob within reach then it would be useful for this part. If you need to retune the RTL-SDR for this, do so. Press any of the buttons on the key fob, one at a time, to view the signal it transmits. You should notice signals very like the ones from the previous step, but instead of information being transmitted on a single carrier, it may well be transmitted in a sequence with multiple different carriers as shown on the following page.

The particular key fob used in this example transmits a first 'burst' of digital data at a frequency lower than its 433.9MHz centre, then at the centre and finally at a frequency higher than the centre. This is then closely followed by a shorter 'burst' of data at the three frequencies in the same order. When viewing your own signals, you may even discover that the signals transmitted from your own key are different for each individual button.

(h) In the SRD example we gave, the hardware uses a 'crystal oscillator' to produce a fixed carrier of 433.9MHz, which will remain the same for the life of the device. In comparison, you should notice that in the screenshots the carrier frequency of the key actually changes between three predetermined frequencies. If your own car key shows something similar, then it's actually quite a sophisticated device!

This should help you understand that even services that operate in the same part of the RF spectrum can be designed to produce different transmission characteristics depending on their application. The main priority of the temperature sensor is to simply transfer the data reliably, whereas, the priority of the key fob is to transfer the digital sequence securely, without being easily copied.

(i) **Receiving car key fob signals.** We have included this video to demonstrate just how different the signals transmitted from car key fobs can be. Here we show example locking sequences from three different car manufacturers.

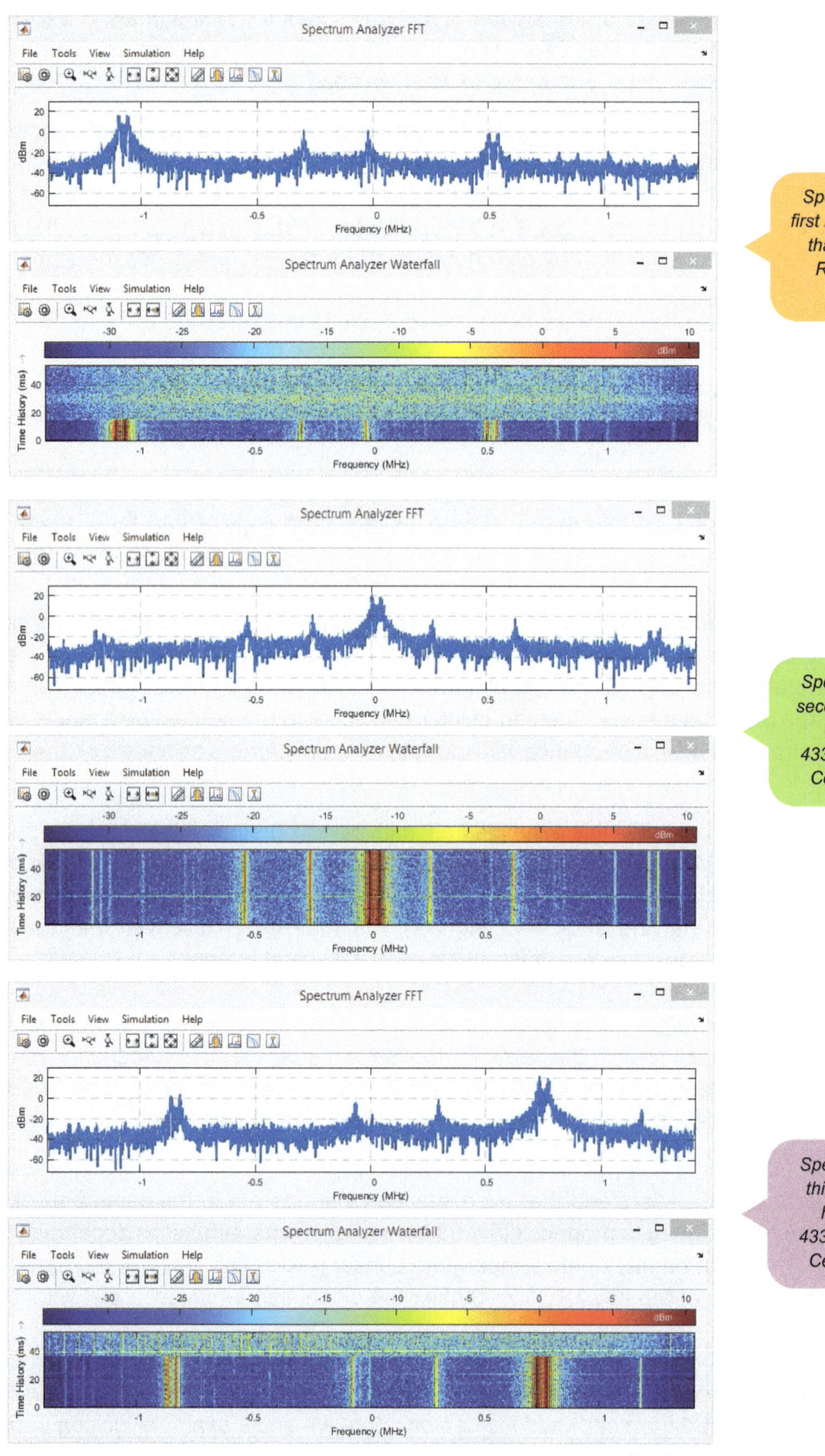

Spectra showing the first burst of data, lower than the 433.9MHz RTL-SDR Centre Frequency

Spectra showing the second burst of data, centred on the 433.9MHz RTL-SDR Centre Frequency

Spectra showing the third burst of data, higher than the 433.9MHz RTL-SDR Centre Frequency

Chapter 3: Radio Frequency Spectrum Viewing

desktopSDR.com/videos/#srd_carkey

(j) **SRD challenge.** The previous steps concentrated on two specific devices, the car key fob, and the wireless sensor, where the characteristics of these signals were shown to differ. The aim of this challenge is to find as many wireless SRDs as possible, and search for their signals in the RF spectrum, to learn about their characteristics.

(k) Although this exercise is primarily concerned with the 433MHz region, remember that these devices may transmit in other areas of the RF spectrum as well. When searching for these signals, pay close attention to their form as displayed in the spectrum analyzers. Are they similar to the ones shown previously, or are they different yet again?

3.11 Digital Video & Audio Signals

Digital TV and radio signals are broadcast in the frequency ranges 470–862MHz and 175–240MHz respectively, in the UK, and are known as DVB-T and DAB signals.

3.11.1 DVB-T DTV Signals

TV channels (frequency bands, not stations!) around the world have historically had a number of different bandwidths, so when the DVB-T standard was being created, it was designed to support 5, 6, 7, and 8MHz channels. This variable bandwidth means that the DVB-T signals you receive may look different to the ones we will show you in the next exercise. The band of the RF spectrum used to transmit DVB signals in the UK is highlighted in Figure 3.14.

Over the years, the increase in video quality has forced the creation of new modulation, multiplexing and coding techniques to meet data transfer requirements. To reduce error rates, a large amount of redundancy is added to every frame of information through a number of stages of coding, and although this process increases the amount of data that must be transmitted, it adds Forward Error Correction (FEC) to the signals. The information is statistically multiplexed and modulated using a technique called Coded Orthogonal Frequency Division Multiplexing (COFDM). COFDM transmits the coded

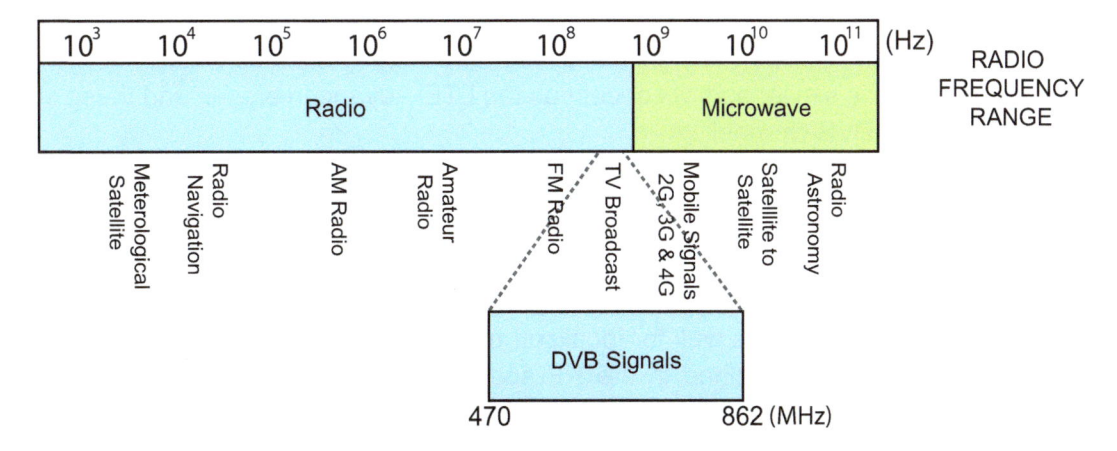

Figure 3.14: Illustration of the DVB band in the RF spectrum

information on 1000 or more orthogonal RF carriers, which further reduces errors resulting from interference and noise, while being able to transmit vast amounts of digital information successfully [16].

In the *Introduction* of this book, we discussed what the RTL-SDR had been born from. RTL2832U based DTV receivers were originally designed to receive, demodulate and decode DVB-T signals. Because you are using the RTL-SDR as an IF radio, all of the on-chip demodulation and decoding is bypassed, but there is no reason you cannot still receive and view these signals!

Exercise 3.10 Searching for DVB-T Digital TV Signals

The following exercise will again use the `exploring_the_spectrum.slx` model to find DVB-T signals. As they are often transmitted from a mast far away, you may have to move towards a window to observe the best results in this exercise.

(a) **Open MATLAB.** Set the working directory to the exercise folder,

   `/spectrum`

   Next, open the following model:

   `.../exploring_the_spectrum.slx`

(b) **Run the simulation.** Navigate to the DTV region by the 'Digital TV' button in the GUI.

(c) **Spotting DVB-T signals.** There are a large number of channels in which DTV signals are transmitted. If you have some sort of database available that tells you which of the 40-odd channels are used, it will give you a head start! The 'UK Free TV' website has this information for the UK [115].

Pan around the spectrum until you come across any strong signals like the one shown in the following screenshot. Although the RTL-SDR is not capable of showing the full 8MHz DVB channel, it should provide a good illustration of its characteristics.

(d) **Signal Analysis.** When tuning to the carrier frequency of a certain DVB channel, you should notice that some very uniformly intense signals will be present, spanning 4MHz above and below the central carrier. DVB channels have many subcarriers—the orthogonal frequencies mentioned earlier—that permit significantly higher throughput. Each subcarrier transmits part of the video data; a similar sort of concept as the LTE-Advanced signal, and these subcarriers fill the full 8MHz DVB channel.

DVB signals are primarily focused on reliably transmitting vast amounts of data for high quality DTV signals over great distances, and this is why the COFDM modulation scheme has been chosen for this standard. The WCDMA scheme used to transmit UMTS signals would not be well suited for this application, as it is designed to send varying amounts of data to different users, and really only works well for localised transmissions. Again this is an example of the multitude of signals transmitted in the RF spectrum that have been designed to meet the requirements of various different applications.

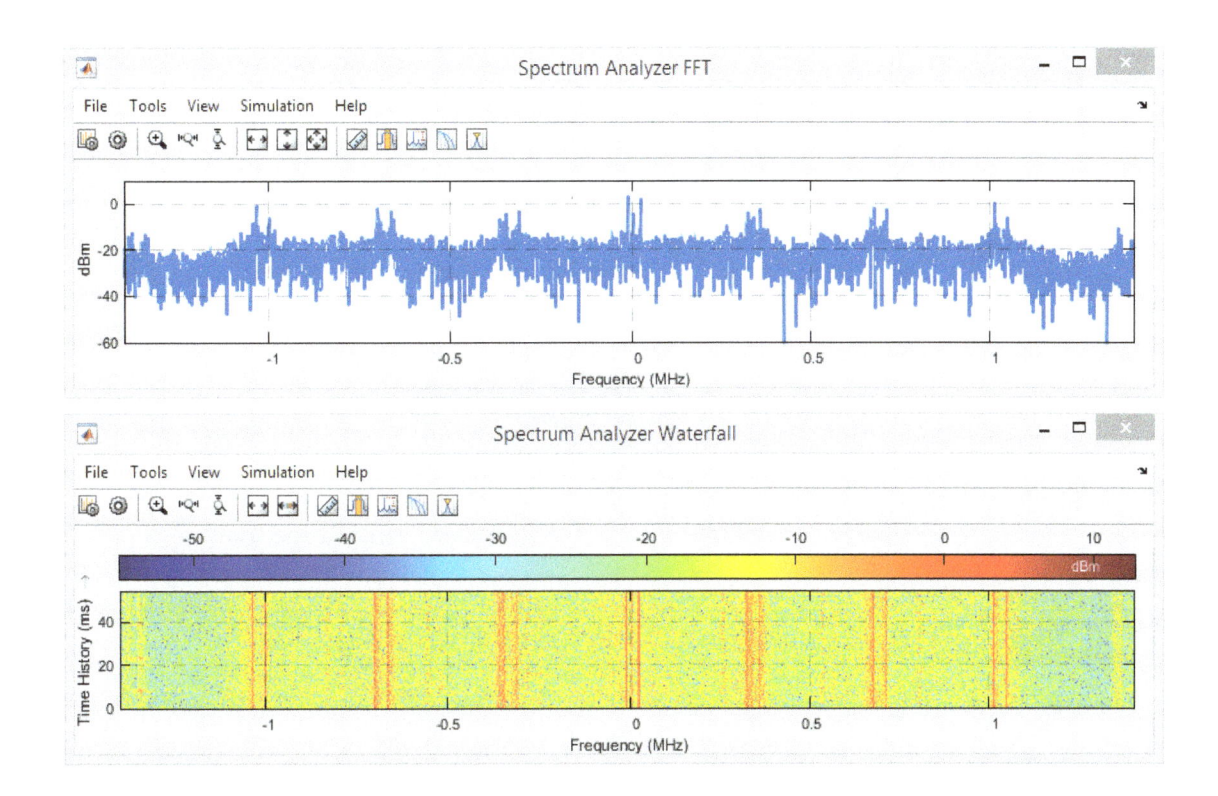

A point to note: If you tune the RTL-SDR to every potential DVB channel frequency, you will find that there are empty channels not populated with DVB signals all throughout the DVB region. This will be due to a number of factors, such as your location and the analogue to digital switchover.

(e) **How many populated channels can you find?** Do they all display in the spectrum analysers with the same power?

(f) **Watch us receive DAB radio signals.** We have recorded a video showing spectral activity in a DVB-T DTV channel. Have a look at it and see how the DVB-T signals we receive compare to what is broadcast near you.

desktopSDR.com/videos/#dvbt_tv

3.11.2 DAB Digital Radio Signals

DAB channels normally have a bandwidth of approximately 1.5MHz, which is significantly lower than that of DVB due to the lower data transmission requirement. DAB radio (or digital radio) is a far more robust form of transmitting radio signals than FM radio, and produces a higher quality signal at the receiver. Due to the way the digital data is coded and transmitted (it uses COFDM too), it also uses the spectrum more efficiently, as multiplexing can be used to transmit many stations in a single channel. Figure 3.15 shows where the DAB band is situated in the UK; somewhere between the FM radio and DVB bands.

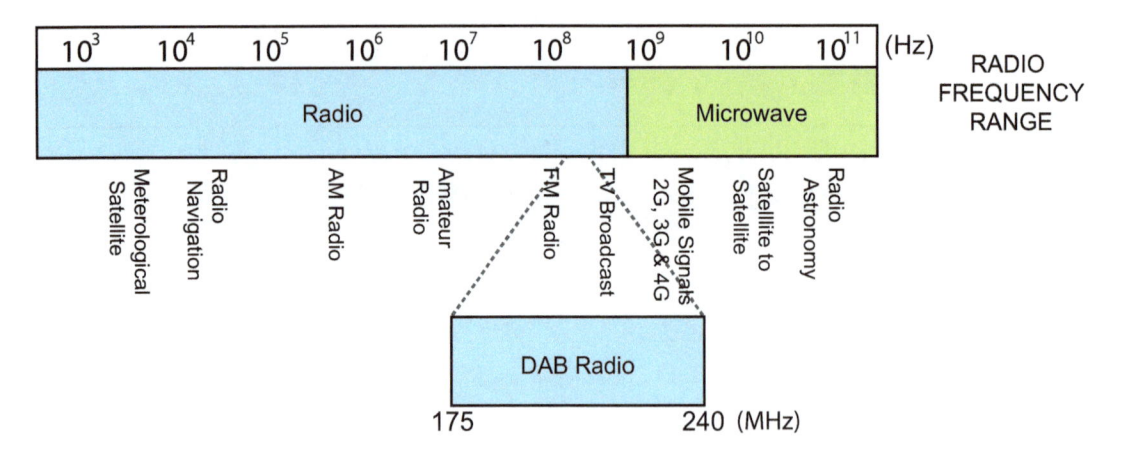

Figure 3.15: Illustration of the DAB Radio band in the RF spectrum

Exercise 3.11 Searching for DAB Digital Audio Signals

The following exercise will again use the `exploring_the_spectrum.slx` model, this time to find DAB signals. As they are often transmitted from a mast far away, you may have to move towards a window to observe the best results in this exercise.

(a) **Open MATLAB.** Set the working directory to the exercise folder,

   `/spectrum`

   Next, open the following model:

   `.../exploring_the_spectrum.slx`

(b) **Run the simulation.** Navigate to the DAB region by 🖱 the 'Digital Radio' button in the GUI.

(c) **Spotting DAB signals.** As discussed in the previous exercise, having a rough idea of where in the spectrum DAB radio signals lie will give you a bit of a head start here. Pan around the spectrum until you come across any strong signals like the ones shown on the following page.

Although the channels are organised in a continuous and sequential order, much like the DVB region there will likely be empty channels that are not used to transmit any signals, largely due to the number of DAB radio station broadcasts being location dependent. In the UK the DAB radio switchover is still in progress (at the time of writing), and this means that some people will be unable to receive any high quality signals.

(d) **Signal Analysis.** DAB signals use the COFDM modulation to transmit information, for the same reasons that DVB uses this standard. This is why the spectrum analysers depict very similar looking signals to those shown in the previous exercise. Probably the most significant difference between DVB and DAB signals is the data rate; DAB has a much lower rate as it is only required to transmit audio signals, as opposed to both audio and video.

Chapter 3: Radio Frequency Spectrum Viewing 83

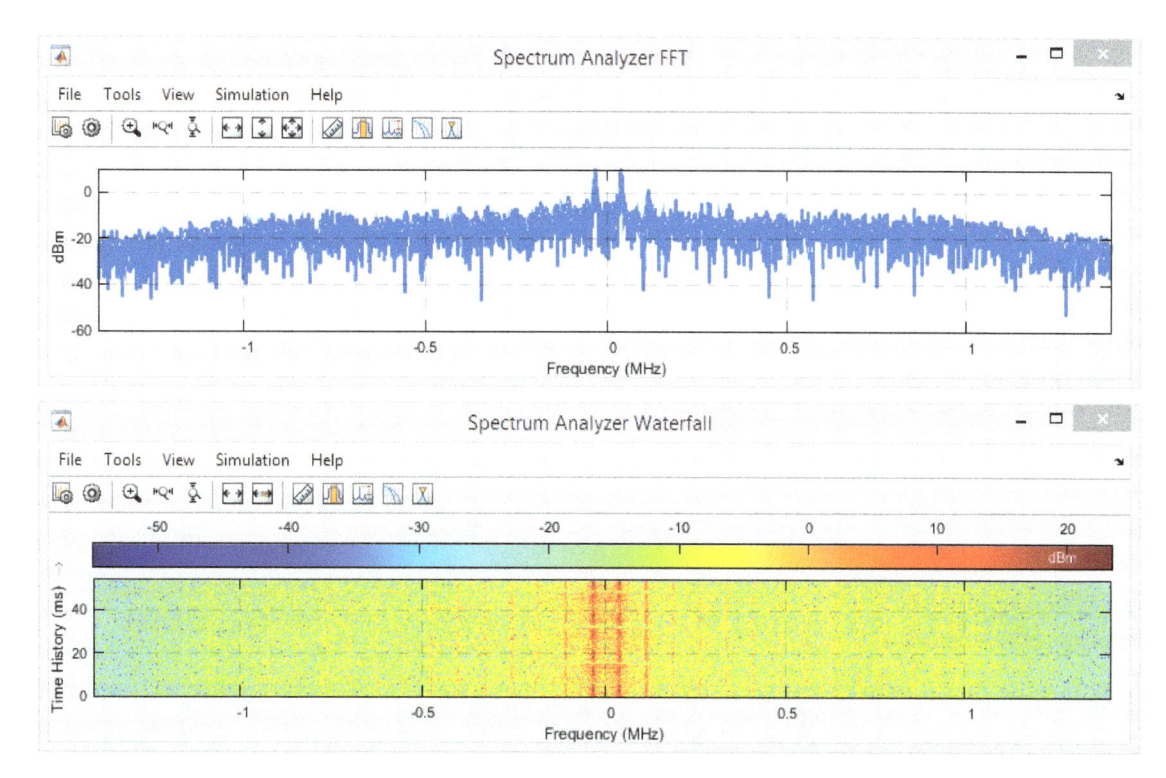

With DAB, many radio stations can be transmitted within one channel, and this significantly increases the transmission efficiency. Multiplexing information signals saves bandwidth in the already crowded spectrum, and also reduces power consumption through a more robust modulation technique.

(e) **Watch us receive DAB radio signals.** We have recorded a video showing spectral activity in a DAB digital radio channel. Have a look at it and see how the DAB signals we receive compare to what is broadcast near you.

desktopSDR.com/videos/#dab_radio

3.12 Using Multiple RTL-SDRs

Throughout the previous sections and examples, the focus has been primarily on using a single RTL-SDR, which has limited the bandwidth of the receiver to the sampling rate of the device, e.g. 2.8MHz. This is not to say that you can only use one at once however. Connecting multiple RTL-SDRs to your machine does however require the use of 'Radio Address' identification numbers, as explained in Appendix A.1 (page 569). This exercise introduces the simple commands used to obtain the ID of an RTL-SDR, and it may be useful for you to look at this before the next exercise.

The following exercise provides a new Simulink model which contains two *RTL-SDR Receiver* blocks. Unfortunately if you do not have multiple RTL-SDR devices you will not be able to complete this exercise. Instead move on to the final section in this chapter, where a MATLAB script will be used to scan the entire range of the RTL-SDR's tuner, and plot the whole of the receivable spectrum in a single figure window!

For those fortunate enough to have multiple devices, this exercise will effectively allow you to repeat Exercises 3.4 to 3.11 while having a receiver with double the bandwidth. While this does not add much to, for example an FM radio receiver, sampling a 5.6MHz band of the spectrum will allow you to view *full* UMTS channels along with larger portions of LTE-Advanced and DVB signals, as shown in Figure 3.16.

When using one RTL-SDR, receiver bandwidth is only f_s Hz

When using two RTL-SDRs, receiver bandwidth is $2f_s$ Hz

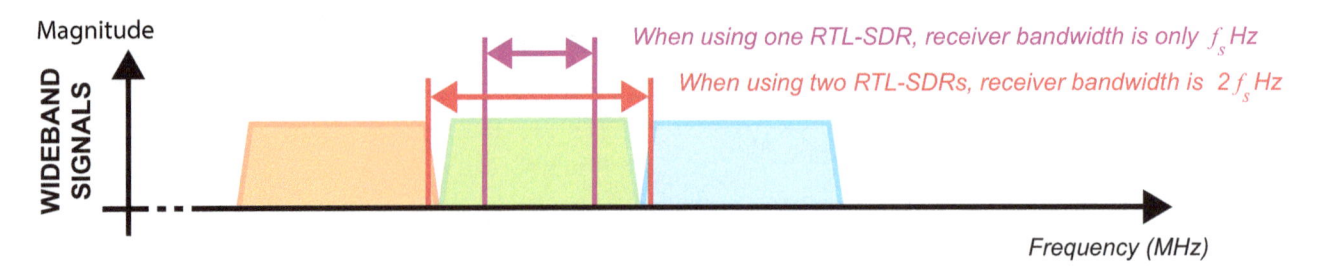

Figure 3.16: Illustration showing the bandwidth advantage when using more than one RTL-SDR

Exercise 3.12 Exploring the Spectrum with Multiple RTL-SDRs

The model used in this exercise requires that you have two RTL-SDR receivers at hand. If working in a classroom environment, working in pairs may permit this. The main aim of this exercise is to see the benefit that multiple RTL-SDRs bring to viewing the RF spectrum.

(a) **Open MATLAB.** Set the working directory to the exercise folder,

   /spectrum

(b) **Connect two RTL-SDRs to your computer.** Using the process described in Appendix A.1, check that MATLAB can communicate with the devices by entering the `sdrinfo` command. If both devices are connected and recognised, the information returned in the command window will display something similar to the following:

```
>> my_rtlsdr = sdrinfo

my_rtlsdr =

  Column 1

    [1x1 sdrr.internal.RTLSDRInfoContainer]

  Column 2

    [1x1 sdrr.internal.RTLSDRInfoContainer]
```

(c) If the command does not return a multi-dimension container, it is probably because the drivers for one or both of the RTL-SDRs need updated. Enter the command `targetupdater` into the MATLAB command window and press enter.

Chapter 3: Radio Frequency Spectrum Viewing

This will initialise the 'Set up Support Package' wizard you saw when you installed the RTL-SDR Hardware Support Package:

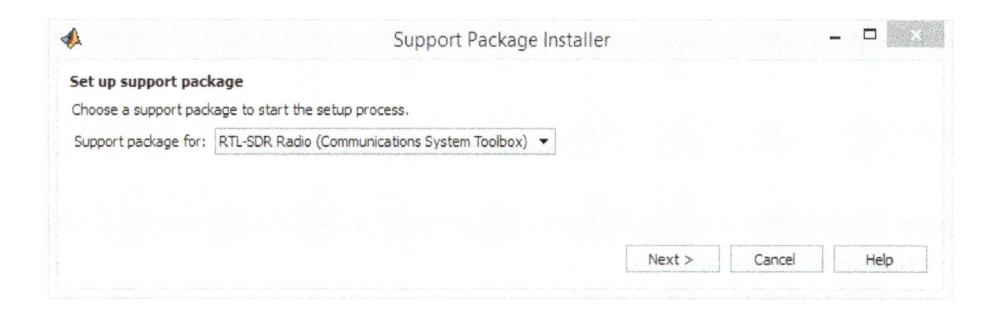

Follow the instructions in this wizard, allowing it to open the Zadig software that will replace the drivers for the RTL-SDRs. Once this process is complete you can return to the MATLAB command window and check if it recognises both devices yet. Hopefully this will resolve any issues and you can move on.

(d) Open the Simulink model:

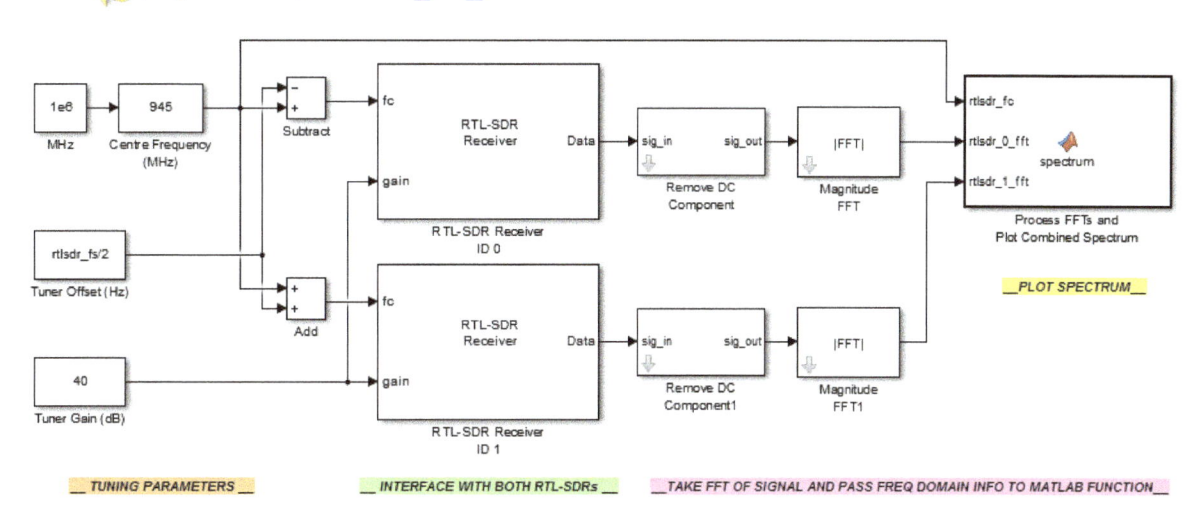

/spectrum/rtlsdr_rx_combinetwo.slx

Two *RTL-SDR Receiver* blocks have been used in this model to allow data to be received from the two RTL-SDR devices. They are both configured to a sample rate of 2.8MHz, but the top block connects to radio '0' and the bottom connects to radio '1'. The centre frequency value used to tune each RTL-SDR is calculated by taking the desired centre of the spectrum and adding or subtracting half the bandwidth of the signal. This is performed with the *Tuner Offset (Hz)* constant block and the arithmetic blocks. To give an example of how this would work, if the 'centre frequency' of the model was set to 100MHz, the top RTL-SDR would be tuned to 98.6MHz, and the bottom to 101.4MHz. Doing this means that you will effectively receive a 5.6MHz wide band of information.

The two received signals enter the Simulink model, where their DC components are firstly removed before FFTs are performed. The two FFT matrices are then passed into a MATLAB function block, which begins by rearranging the matrices to create a large combined matrix, and then plots the data in a figure window we have configured to look like a spectrum analyser. This process repeats every time new frames of data are received by the RTL-SDRs, and the figure appears to plot continuously. The downside of this is that the model will not run in real time; but it is fast enough for the purposes of this exercise!

(e) 2 on both the *RTL-SDR Receiver* blocks to open their parameter windows, then 1 on the 'Initialise' button. This has the same functionality as entering the `sdrinfo` command, and hardware information about the two devices should be returned as follows. You should find that both RTL-SDR radio addresses, '0' and '1', can be initialised during this process. 1 'OK' to close the parameter windows.

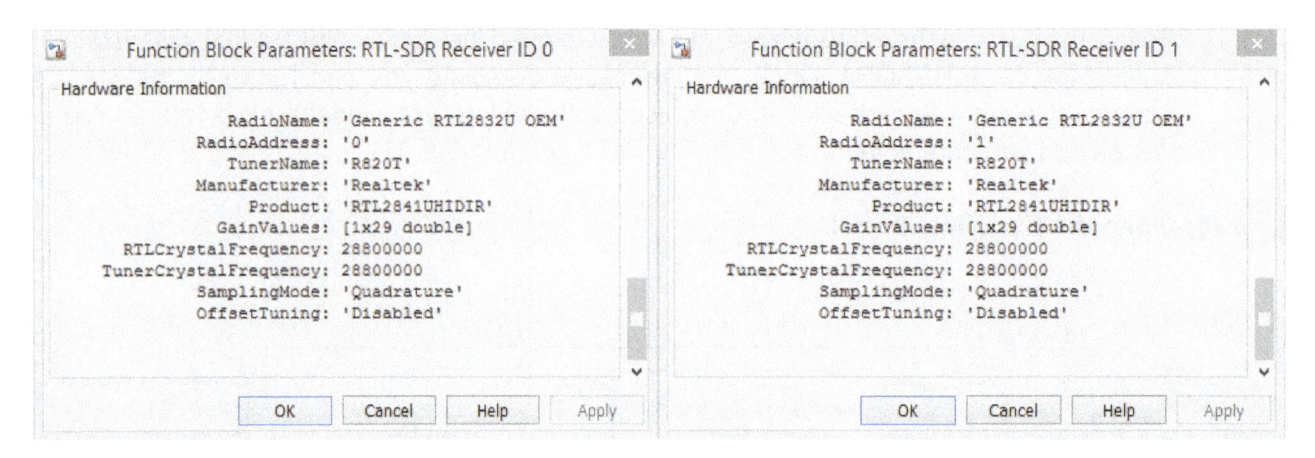

```
Function Block Parameters: RTL-SDR Receiver ID 0
Hardware Information
              RadioName: 'Generic RTL2832U OEM'
           RadioAddress: '0'
              TunerName: 'R820T'
           Manufacturer: 'Realtek'
                Product: 'RTL2841UHIDIR'
             GainValues: [1x29 double]
    RTLCrystalFrequency: 28800000
  TunerCrystalFrequency: 28800000
           SamplingMode: 'Quadrature'
            OffsetTuning: 'Disabled'
```

```
Function Block Parameters: RTL-SDR Receiver ID 1
Hardware Information
              RadioName: 'Generic RTL2832U OEM'
           RadioAddress: '1'
              TunerName: 'R820T'
           Manufacturer: 'Realtek'
                Product: 'RTL2841UHIDIR'
             GainValues: [1x29 double]
    RTLCrystalFrequency: 28800000
  TunerCrystalFrequency: 28800000
           SamplingMode: 'Quadrature'
            OffsetTuning: 'Disabled'
```

(f) **Run the simulation.** Begin the simulation by 1 the 'Run' ▶ button in the Simulink toolbar. As the simulation begins, a figure window will appear and when the first frame of data is received and processed, the combined FFT will be plotted.

To change the gain or centre frequency of the multi-dongle receiver, 2 on the 'Tuner Gain (dB)' constant or 'Centre Frequency (MHz)' slider gain blocks and change their values.

(g) **Signal Analysis.** Although you may not have noticed in the previous examples, it is likely that your RTL-SDR is non-linear across the full bandwidth of the signal it receives, as shown above. Another quirk you may come across is that the average power levels of the two RTL-SDRs may not match. This is normally caused by grounding problems.

As the two RTL-SDRs are not synchronised together; i.e. do not use a common crystal oscillator, it is highly unlikely that samples will be taken in time with each other. This, combined with the other two issues means that you would not be able to use this style of setup if you actually wanted to receive and demodulate information from a wideband signal. It is purely designed for viewing larger portions of the spectrum; and viewing a larger portion of the spectrum is exactly what we want to do in this exercise!

Chapter 3: Radio Frequency Spectrum Viewing

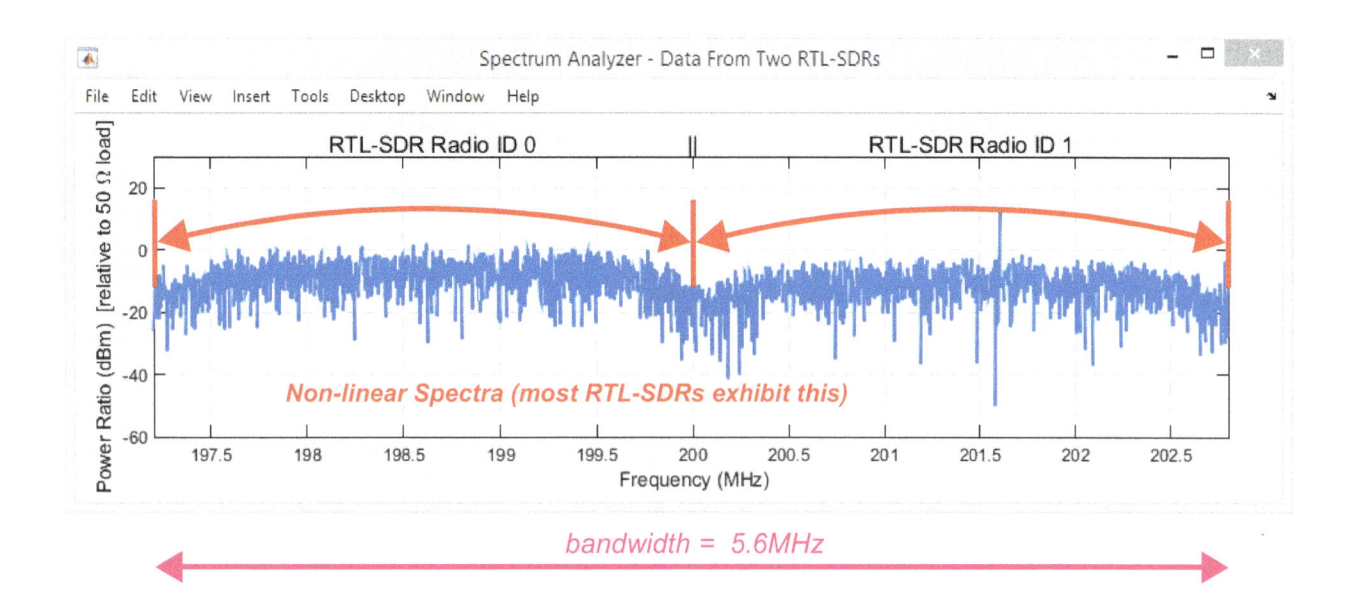

[Figure: Spectrum Analyzer - Data From Two RTL-SDRs, showing RTL-SDR Radio ID 0 and RTL-SDR Radio ID 1 across frequency range 197.5 to 202.5 MHz, with annotation "Non-linear Spectra (most RTL-SDRs exhibit this)" and bandwidth = 5.6MHz]

(h) **Multi RTL-SDR receiver challenge.** It is now time for you to go and do some exploring! Repeat some (or all) of the previous exercises and return to the different frequencies used to transmit wideband signals. To make exploring the spectrum easier, it can be useful to set the 'Low', 'Middle' and 'High' values of the 'Centre Frequency (MHz)' block to the rough ends of the bands. If looking for DVB signals, we would set them as follows:

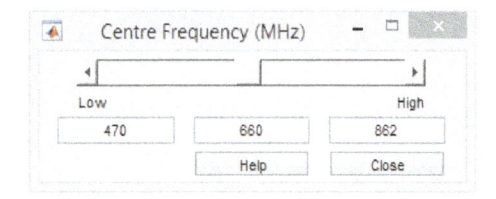

[Figure: Centre Frequency (MHz) dialog with Low = 470, Middle = 660, High = 862]

(i) How much of a difference does it make being able to see much more of the wideband signals? Can you clearly see the edges of channels? The following annotated figures show some of the spectra we saw.

FM Radio Stations

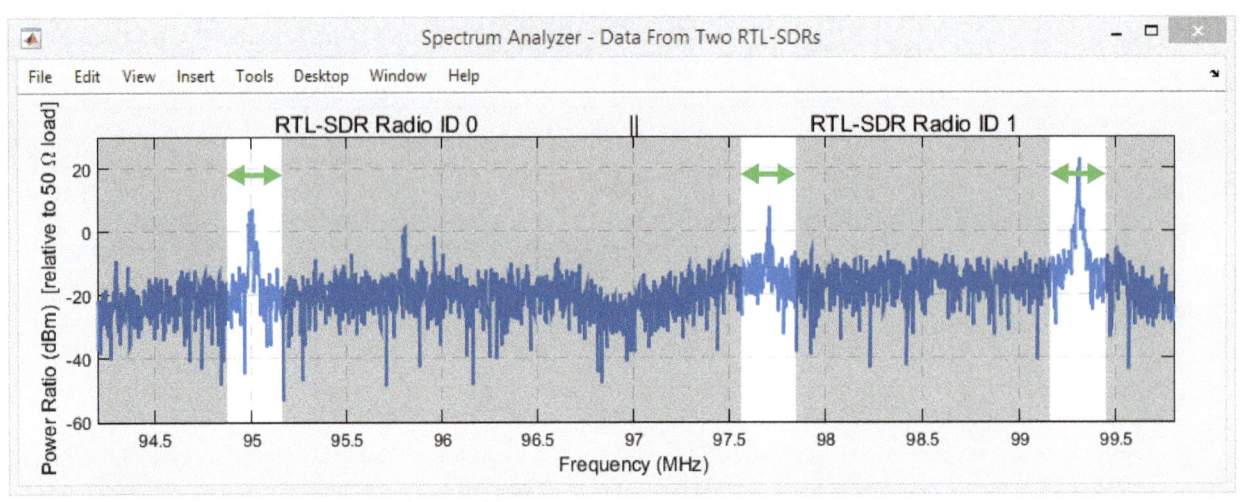

5MHz Wide UMTS (3G) Downlink Channel

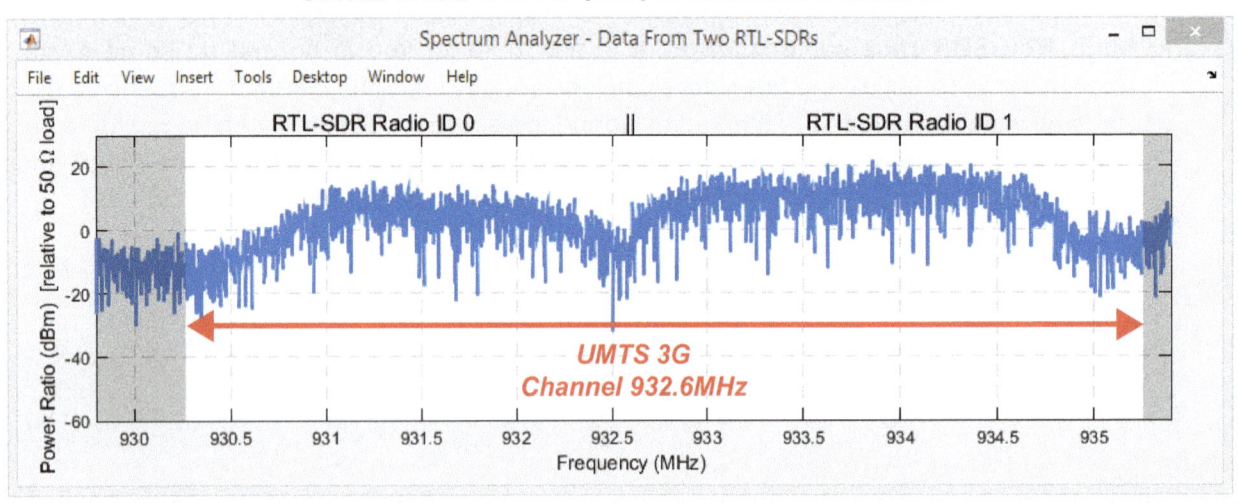

Parts of Two 10MHz Wide LTE-Advanced (4G) Channels and a Guard Band

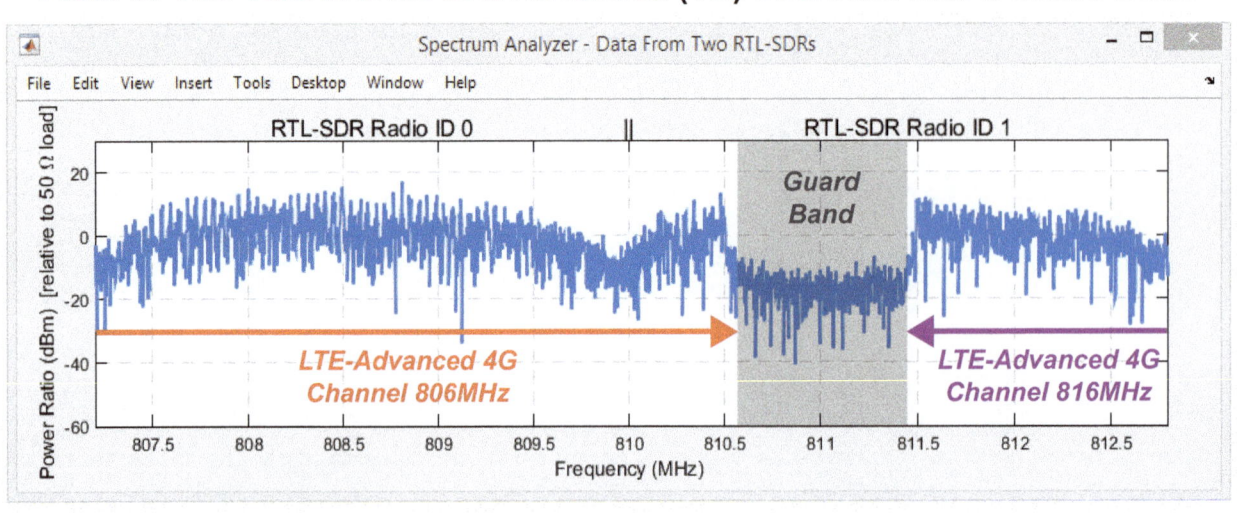

Chapter 3: Radio Frequency Spectrum Viewing

3.13 Sweeping the Spectrum: Receiving from 25MHz to 1.75GHz

In this section, we will move on to the final exercise in this chapter—performing a frequency sweep with a single RTL-SDR to receive all of the signals transmitted in the operating range of its tuner. For the most common RTL-SDR devices (those based on the R820T and as featured and reviewed in Section 1.6 (page 10)), this range will be from 25MHz to 1.75GHz; a 1.725GHz portion of the RF spectrum. Yes... your $20 SDR is capable of this! *(Note that for the less common, and less available RTL-SDR devices based on the Elonics E4000 tuner, the frequency range is slightly wider from 53MHz to 2.2GHz, but has a dead zone from 1.1GHz to 1.25GHz. We will show a couple of scans using the E4000 based devices for comparison with the R820T based devices.)*

We define the term 'sweep' here as a repetitive process of tuning and retuning the RTL-SDR to different centre frequencies in order to obtain spectral information for the full range of the sweep. Figure 3.17 gives an example of how this process can be used to build up spectral information.

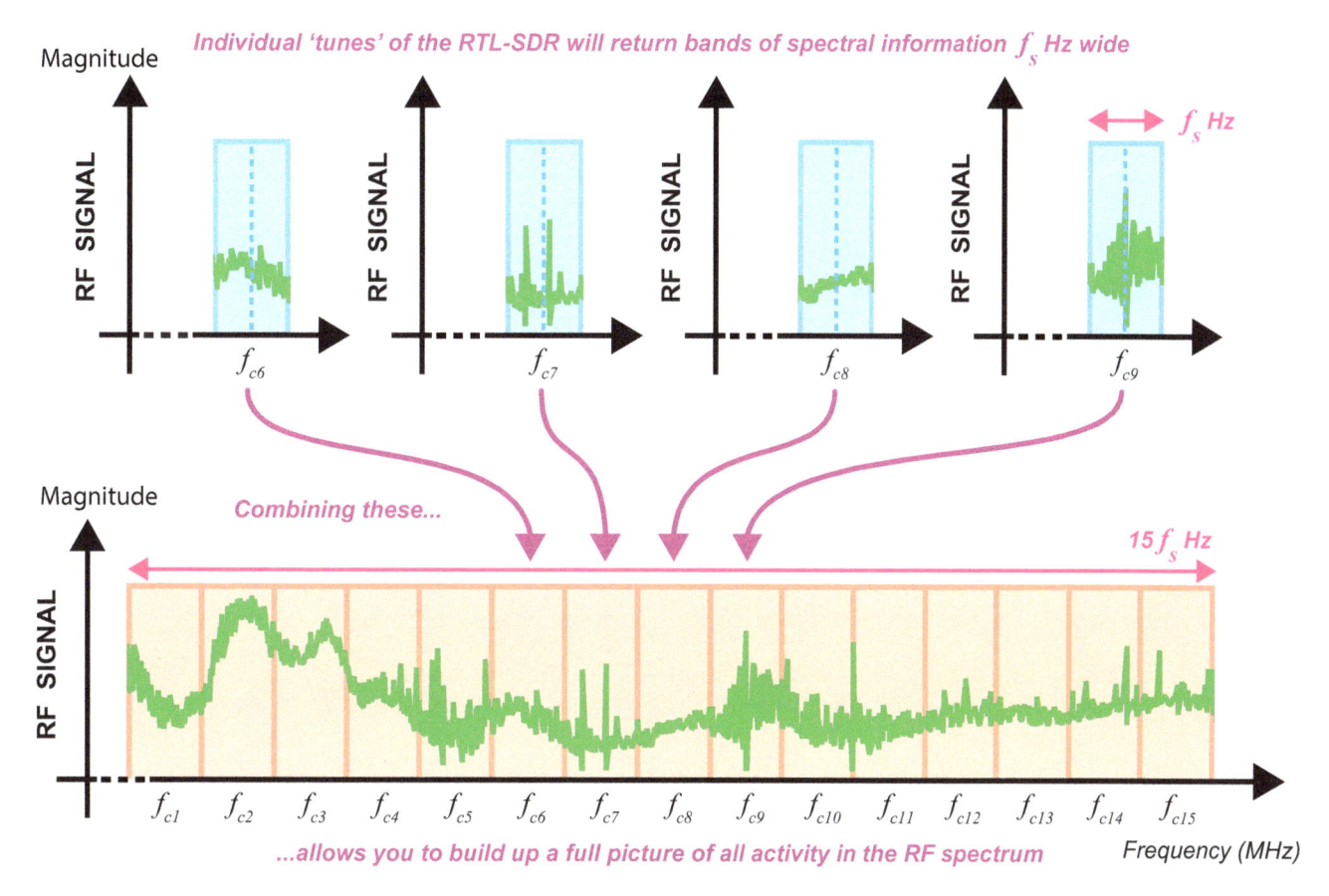

Figure 3.17: Diagram showing how data from multiple retunes of an RTL-SDR can be used to build a big picture of the RF spectrum

In an ideal situation we would set the sampling frequency of our SDR to a high value (e.g. 100MHz) in order to optimise this process, but unfortunately this is not possible with the RTL-SDR! If the sampling frequency in the above situation was 2.8MHz, each retune of the RTL-SDR would provide another 2.8MHz wide band of spectral information. Each centre frequency, f_c, selected during the retuning process would be 2.8MHz higher than the previous centre frequency, so in Figure 3.17,

$f_{c(i+1)} = f_{c(i)} + 2.8\,\text{MHz}$, and the fifteen retunes would result in a 42MHz wide band of information with none of the data captures frequency bands overlapping.

We have written a MATLAB script which will carry out the sweep process for you, and this is what you will have the opportunity to run in this final exercise.

Exercise 3.13 Sweeping the Radio Frequency Spectrum: 25MHz to 1.75GHz

In this exercise you will have the opportunity to use your RTL-SDR to sweep across its tunable range of the RF spectrum, and create a MATLAB figure showing all of the spectral activity it detects. We provide the MATLAB script so you will not need to write any code yourself—this is a 'click run' exercise. (If you prefer to know more about the rudiments of working with MATLAB scripts before proceeding, then you can review this in *Getting Started with MATLAB and Simulink* Exercise 4.2 (on page 105).)

(a) **Open MATLAB.** Set the working directory to the exercise folder,

```
/spectrum/sweep
```

Next, open the following MATLAB script:

```
.../rtlsdr_rx_specsweep.m
```

(b) **Examine the parameters.** There are a number of different parameters listed at the top of the `rtlsdr_rx_specsweep` function. These are used to configure features such as the start and stop frequencies of the sweep, and also to configure the RTL-SDR.

```
% PARAMETERS (can change)
location            = 'Glasgow';    % location used for figure name
start_freq          = 25e6;         % sweep start frequency
stop_freq           = 1750e6;       % sweep stop frequency
rtlsdr_id           = '0';          % RTL-SDR stick ID
rtlsdr_fs           = 2.8e6;        % RTL-SDR sampling rate in Hz
rtlsdr_gain         = 40;           % RTL-SDR tuner gain in dB
rtlsdr_frmlen       = 4096;         % RTL-SDR output data frame size
rtlsdr_datatype     = 'single';     % RTL-SDR output data type
rtlsdr_ppm          = 0;            % RTL-SDR tuner PPM correction
```

(c) Change the `location` string to match your location. By default, this will be set to our location, `Glasgow`. This affects the title and filename of the spectrum produced after the sweep is complete. Set the `start_freq` and `stop_freq` to set the range that you wish to sweep across. You will need to make sure that the tuner in your RTL-SDR can actually centre on these frequencies. The limits for both the R820T and E4000 tuners are given in Table 3.2.

Chapter 3: Radio Frequency Spectrum Viewing

Table 3.2: Possible centre frequencies for the two RTL-SDR architectures

R820T Tuner	E4000 Tuner
25 MHz to 1.75 GHz	53 MHz to 1.1 GHz
	1.25 GHz to 2.2 GHz

(Note: If using an E4000 RTL-SDR, you will need to sweep each of the ranges separately as the PLL inside the tuner will not lock to any frequency value in its 'dead zone'.)

As required, you can change the parameters relating to the RTL-SDR too—for example, the `rtlsdr_id` or `rtlsdr_gain`. You can also change `rtlsdr_fs`, however if you do the description we just gave you will become invalid!

By default, the RTL-SDR is configured to sample at 2.8MHz on both the I and Q channels. This means that the signal received in the MATLAB code will have an information bandwidth of 2.8MHz, and contains spectral information from f_c-1.4MHz to f_c+1.4MHz. Because of the non-flat frequency response of the tuner at the low and high ends of the band (as reviewed in Exercise 3.12, page 86), it is more practical to increment through the frequency spectrum in steps that are smaller than 2.8MHz. The MATLAB script has therefore been structured to keep information from f_c-1.2MHz to f_c+1.2MHz, as across this range the RTL-SDR response is *'reasonably'* flat. The RTL-SDR is retuned in increments of 1.4MHz, and the same processing is carried out to keep only half of the spectral information. Repeating this process, a fuller, flatter spectrum can be obtained.

Other parameters listed at the top of the function include:

```
% PARAMETERS (can change, but may break code)
nfrmhold      = 20;         % number of frames to receive
fft_hold      = 'avg';      % hold function "max" or "avg"
nfft          = 4096;       % number of points in FFT
dec_factor    = 16;         % output plot downsample
overlap       = 0.5;        % FFT overlap to counter rolloff
nfrmdump      = 100;        % number of frames to dump after
                                retuning (to clear buffer)
```

`nfrmhold` is used to set the number of frames to receive (and average over) every time the RTL-SDR is retuned. Ideally this number would be as big as possible, but increasing it exponentially increases the amount of time the code takes to execute. `fft_hold` controls the method in which the averaging is carried out. If it is set to `avg`, the mean received power will be calculated and returned for every spectral component, however, when it is set to `max`, a max-order-hold operation will be carried out and the largest power for each spectral component will be returned.

`nfft` sets the number of points in the FFT. `dec_factor` is used to control the level of decimation performed before plotting the results of the full sweep. We have set this to 16, as this value results in a plot that is not too crowded. `overlap` is used to set the percentage overlap of the retunes. As we discussed previously, a default overlap value of '0.5' means that the RTL-SDR will be retuned in increments of f_s/2Hz. Finally, `nfrmdump` is used to set the number of frames to be dumped after each retune takes place. When the RTL-SDR is retuned, it takes roughly 100 frames of IQ samples until data related to the new centre frequency can be accessed. In this script we simply dump these frames so that they do not interfere with the frame averaging process.

As we outline in the comments, changing these parameters may have some undesired effects on the code's execution! Leave them as they are for now, you can modify them later.

(d) **Examine the rest of the code.** Unlike the simple RTL-SDR receiver script you ran in Exercise 2.6 (page 37), this code is somewhat more complicated as it collects significantly more data before giving a graphical output. The general flow of this code is shown in Figure 3.18. You

```
Function: rtlsdr_rx_specsweep
    % user can set 'simulation' parameters and sweep range
    % calculations are performed to find all of the centre
      frequencies required to receive the sweep range
    % runs create_spectrum and then capture_and_plot
    % resulting figure is saved to MATLAB current folder

    Function: create_spectrum
        % creates and customises a figure window to make
          it look like the a spectrum analyzer
        % creates two axes inside this, runs axes_position

    Function: axes_position
        % uses current figure size (in pixels) to reposition
          the two axes inside figure window

    Function: resize_spectrum
        % callback that runs whenever a user resizes the
          figure window
        % runs axes_position to reposition axes inside window

    Function: capture_and_plot
        % creates RTL-SDR object and various other receiver
          components
        % runs 'simulation' to capture frames of data from
          the RTL-SDR
        % analyses data in the frequency domain
        % retunes RTL-SDR to next required centre frequency
          and repeats this process
        % after all data is collected, it is processed and
          plotted in the figure window (populates both axes)
```

Figure 3.18: Main components of the frequency sweep m-code, `rtlsdr_rx_specsweep.m`

Chapter 3: Radio Frequency Spectrum Viewing

should be able to match up the different sections described here to the code in you can see in MATLAB.

(e) **Signal Analysis: Provided Spectrum Sweeps.** Before you run the script to sweep the spectrum in your area, you can have a look at some sweeps made in other locations. Open the 'R820T Glasgow' spectrum sweep (shown in Figure 3.19) or click on the file links below to open the plots in MATLAB.

`.../R820T_25MHz_1750MHz_Glasgow.fig`

This image shows the generated MATLAB figure window (produced by the script to look like a *Spectrum Analyzer*), which depicts the information about the signal power of spectral components for the entire frequency range of the sweep. You may need to maximise the window to allow the axes and figure titles to display correctly. The top axes with the blue line shows the power levels of the spectral components in dBm, and the bottom with the orange line shows them as a relative power in Watts.

(f) Have a look at our spectrum and notice how many frequency peaks there are throughout the sweep—we can observe quite a large number of different signals being transmitted in Glasgow! Using the zoom and pan tools in the figure toolbar,

navigate around the spectrum to view the signals present. Try to correlate the activity with the frequency allocations listed for Glasgow in Table 3.3.

Table 3.3: Services that operate in Glasgow that are visible in Figure 3.19

Frequency	Communications Service
100MHz	FM Radio Stations
138MHz	Meteorological Satellites
208MHz	Mobile Broadcasting
299MHz, 355MHz	DAB Radio Channel
332MHz	UK Civil Aviation Authority/ Ministry of Defence
432MHz	SRD Activity
474MHz, 592MHz, 711MHz	DVB-T TV Channels
806MHz, 816MHz	LTE-Advanced 4G Mobile Channels
927MHz, 934MHz	UMTS 3G Mobile Downlink Channels
935MHz to 960MHz	GSM 2G Mobile Downlink Channels
1561MHz	Aero Radionav: Space to Earth & Earth to Space Satellites

Zooming into each of the centre frequencies, you should easily be able to see that all of these signals have different characteristics; in terms of amplitude, bandwidth, and modulation type.

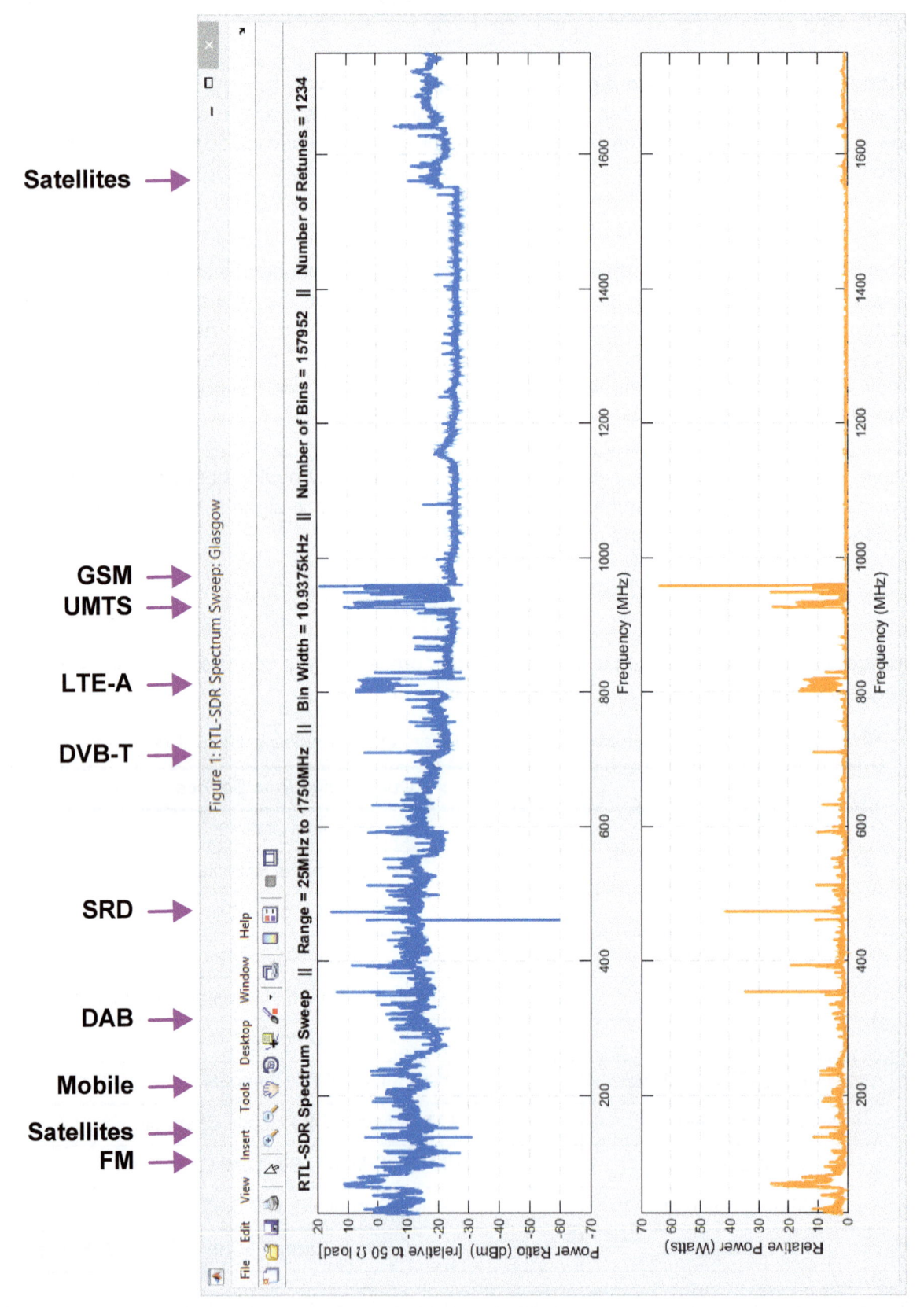

Figure 3.19: Frequency sweep in 1.4MHz steps from 25MHz to 1750MHz in the centre of Glasgow, Scotland

Chapter 3: Radio Frequency Spectrum Viewing

Some signals appear to be very high powered and narrowband, while others spread across a much larger range of frequencies and have multiple low powered carriers.

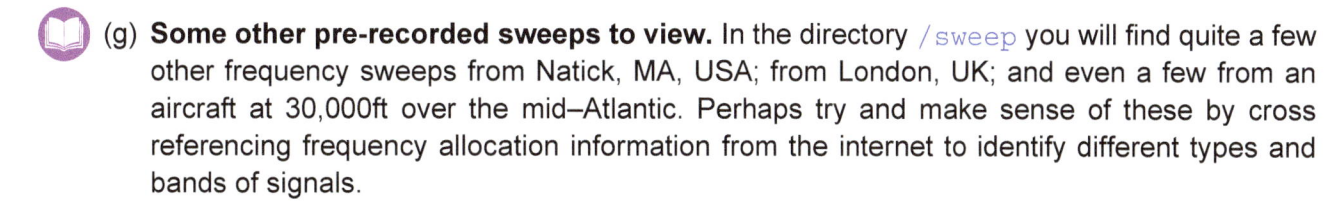

(g) **Some other pre-recorded sweeps to view.** In the directory `/sweep` you will find quite a few other frequency sweeps from Natick, MA, USA; from London, UK; and even a few from an aircraft at 30,000ft over the mid–Atlantic. Perhaps try and make sense of these by cross referencing frequency allocation information from the internet to identify different types and bands of signals.

NOTE: There is a known issue with MATLAB R2015b, whereby figures will not open with the axes scaled to their 'saved' positions. If you open any of the provided sweeps and the x axes is scaled from 0–1, you will need to rescale it. For a sweep between 25MHz and 1.75GHz, enter the following into the command window to fix this:

```
>> xlim([25,1750])     % to scale from 25MHz to 1.75GHz
```

(h) **Prepare to run the script.** Now you are familiar with the script and have an expectation of what it should return, it is now time for you to run it yourself. Make sure that the RTL-SDR is connected to your computer, and that MATLAB is able to communicate with it by running the `sdrinfo` command. Close any unnecessary programs to free up RAM.

(i) **Run the script.** Run the script by 1️⃣ on the 'Run' ▷ button in the MATLAB toolbar (refer back to page 41 if you need a reminder on how to do this). After a few seconds or so, MATLAB will establish a connection with your RTL-SDR, and it will start receiving frames of data. As this happens, information about the current centre frequency of the RTL-SDR will print out in the MATLAB command window:

```
>> rtlsdr_rx_specsweep

        fc = 25MHz
        fc = 26.4MHz
        fc = 27.8MHz
        fc = 29.2MHz
        ...
```

The script will continue executing until it has received enough information to show you all spectral activity in the range you specified. This can take a long time (up to 10 minutes), and may take significantly longer if your computer does not have a large amount of RAM. If you wish to cancel the script execution before it has finished, use the key combination:

CTRL + C

If you do cancel the script before it is finished, the figure will not be populated with progress 'to date', i.e. you will only see the results when you let it run its course and complete the sweep. When it finishes executing, the 'spectrum analyzer' figure window should appear, and graphical representations of all the spectral activity in your area should be plotted on the two axes. You may need to maximise the window to allow the axes and figure titles to display correctly.

(j) Unless you run the script in our location at the University of Strathclyde, Glasgow, what you see will look different, because the spectrum all around the world (and even from location to location in a particular city) is different, and that is what makes this exercise so interesting!

(k) **Explore your spectrum.** Have a look around the spectrum using the zoom tools. Do you notice large bands of activity that look similar to the ones we pointed out? Can you take a guess at what any of these signals are?

(l) **Match the signals up!** Take a note of the centre frequencies of anything that looks like a signal and compare these against any frequency allocation information you can obtain for your country, and try to identify what the different signals are. UK readers can use the Ofcom Frequency Allocation Tables [101] to help.

(m) **Results from the E4000 RTL-SDR:** The directory /sweep also contains plots from RTL-SDRs with the E4000 sweeping the same Glasgow spectrum. Do you notice anything different in the figures?

.../E4000_53MHz_1100MHz_Glasgow.fig

.../E4000_1250MHz_2200MHz_Glasgow.fig

Because the E4000 is a *zero-IF* tuner, it is often the case that the baseband signals output from it contain large DC spikes. This is not the case for the *IF* R820T, as no information resides around the DC component. The DC spike resulting from the E4000 is so large that it sometimes cannot be removed in the manner discussed in Exercise 3.1, and this means that it will remain in the spectrum, which poses a problem in this situation.

Zooming into the 2GHz region of the E4000 sweep you will find seven *2100MHz UMTS channels*, each 5MHz wide. Notice how spiky the spectrum appears to be here. These spikes are the DC components from each of the retunes, and every one marks a different centre frequency. This kind of distortion is not seen when using an R820T based RTL-SDR, however the R820T cannot tune past 1.75GHz!

Chapter 3: Radio Frequency Spectrum Viewing 97

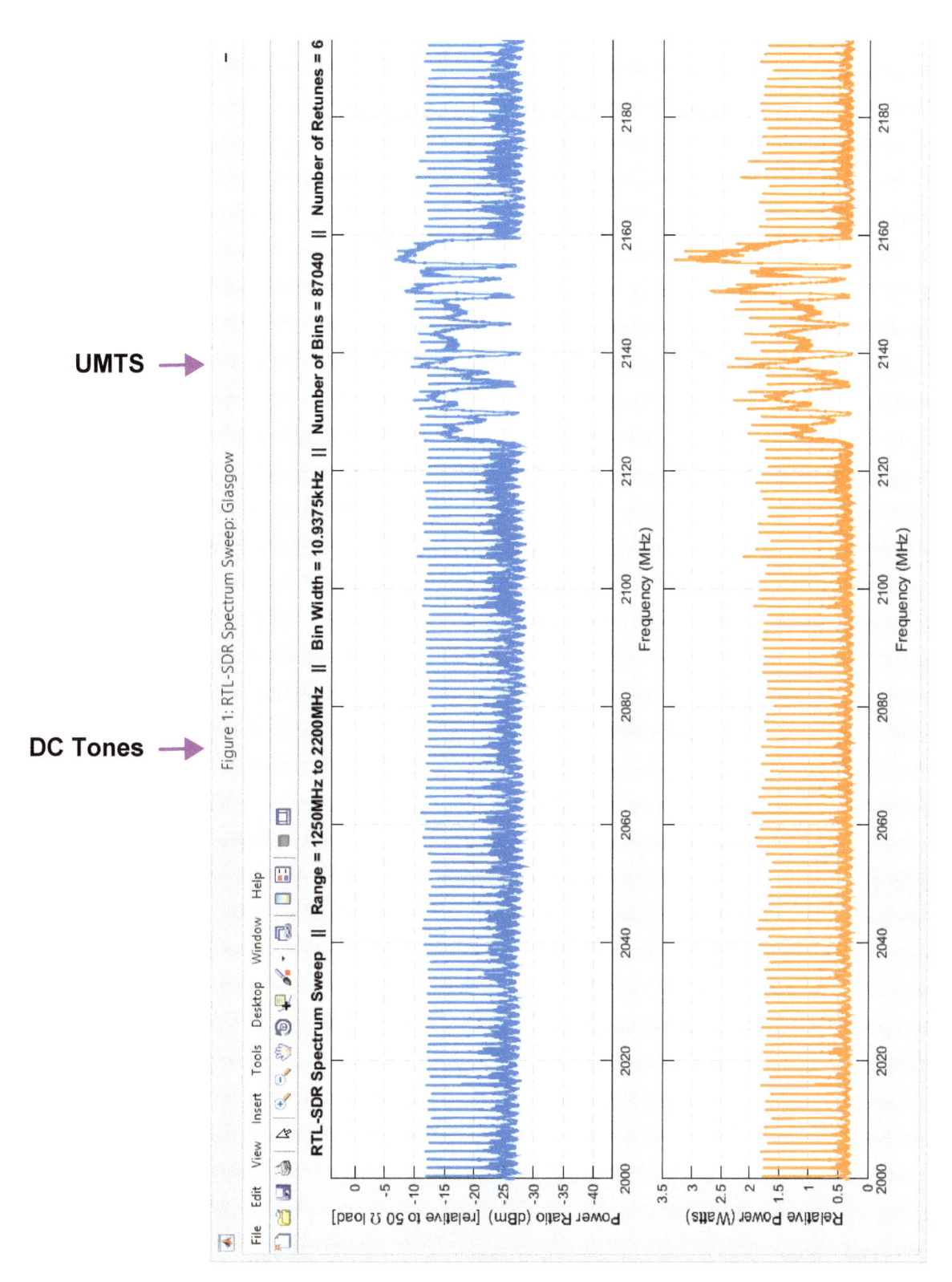

Figure 3.20: Frequency sweep in 1.4MHz steps from 2GHz to 2.2GHz in the centre of Glasgow, Scotland, obtained using the less common E4000 based RTL-SDR

3.14 Summary

The final exercise, along with all of the exercises throughout this chapter have hopefully helped you develop your understanding of how the RF spectrum is utilised by the communications services we use on a daily basis. You should have seen that numerous different modulation schemes are used for these services, and be able to distinguish between them simply from their spectral characteristics. We introduced the concept of 'eyeball radio tuning', and showed you how to adjust the gain of the RTL-SDR. Everything you have learned here will be crucial for you to complete the tasks in the remainder of this book.

4 Getting Started with MATLAB and Simulink

This chapter introduces the MATLAB and Simulink environments, and highlights some of the particular features relevant for the RTL-SDR examples that follow later in the book. It is intended as a step-by-step guide for those new to working with MATLAB and Simulink, and a recap for those with some prior experience. Those who are already familiar with the tools may wish to proceed directly to later chapters.

The chapter comprises a set of exercises, each of which ask you to follow a series of steps. Once you have worked through all of the exercises, you should have an understanding of the following:

- How to use the MATLAB command window and Workspace
- How to create, save, and run MATLAB scripts and MATLAB functions
- How to generate basic time domain plots
- How to work with System Objects
- How to create a new Simulink system, and save it
- How to navigate the Simulink library browser and search for blocks
- How to drag blocks into a system, connect and configure them
- How to add simulation sources and sinks
- How to run simulations and view the outputs
- How to work with different data types in MATLAB and Simulink
- How to work with sampling rates

MATLAB and Simulink have many more features than can be covered here, so we focus on the core aspects—additional features will be introduced later as needed.

4.1 Introducing MATLAB

The aim of this first exercise is simply to demonstrate the MATLAB interface, and highlight some important features. It is assumed that you have already opened MATLAB and followed the initial procedure to configure the environment described in Exercise 2.4 (page 32).

Exercise 4.1 MATLAB Orientation and Using the Command Window

In this first exercise, we will take a quick tour of the MATLAB interface, and point out important features. Note that the screenshots provided show MATLAB 2014b. If using a subsequent version, the appearance may differ slightly.

(a) **Open and configure MATLAB.** If you have not already done so, run through Exercise 2.4 to ensure that MATLAB is open with the correct libraries installed and paths set.

(b) **Explore the MATLAB interface.** When MATLAB is open, you should see an environment and layout similar to that shown in Figure 4.1.

The MATLAB interface comprises a number of panes which are highlighted in the diagram. In this workbook, we will work with two main file types: Simulink models and MATLAB scripts. These are opened in separate windows that provide further specific facilities, as we will see a little later.

(c) **Execute code via the command window.** MATLAB code can be entered and run directly in the command window. Let's try this now, by defining a variable and assigning it with a value (both occur at once). Type

```
x = 100;
```

at the window, followed by *Enter*. You should notice that the variable x now appears in the Workspace panel.

(d) **Create a second variable.** Write a similar line of code to create the variable y, and assign it with the value 3. Check that the variable appears in the Workspace panel.

(e) **Perform arithmetic operations.** Next, try finding the sum of x and y, by typing the following:

```
s = x + y;
```

Likewise, try to find the difference of x and y, and assign the value to the variable d.

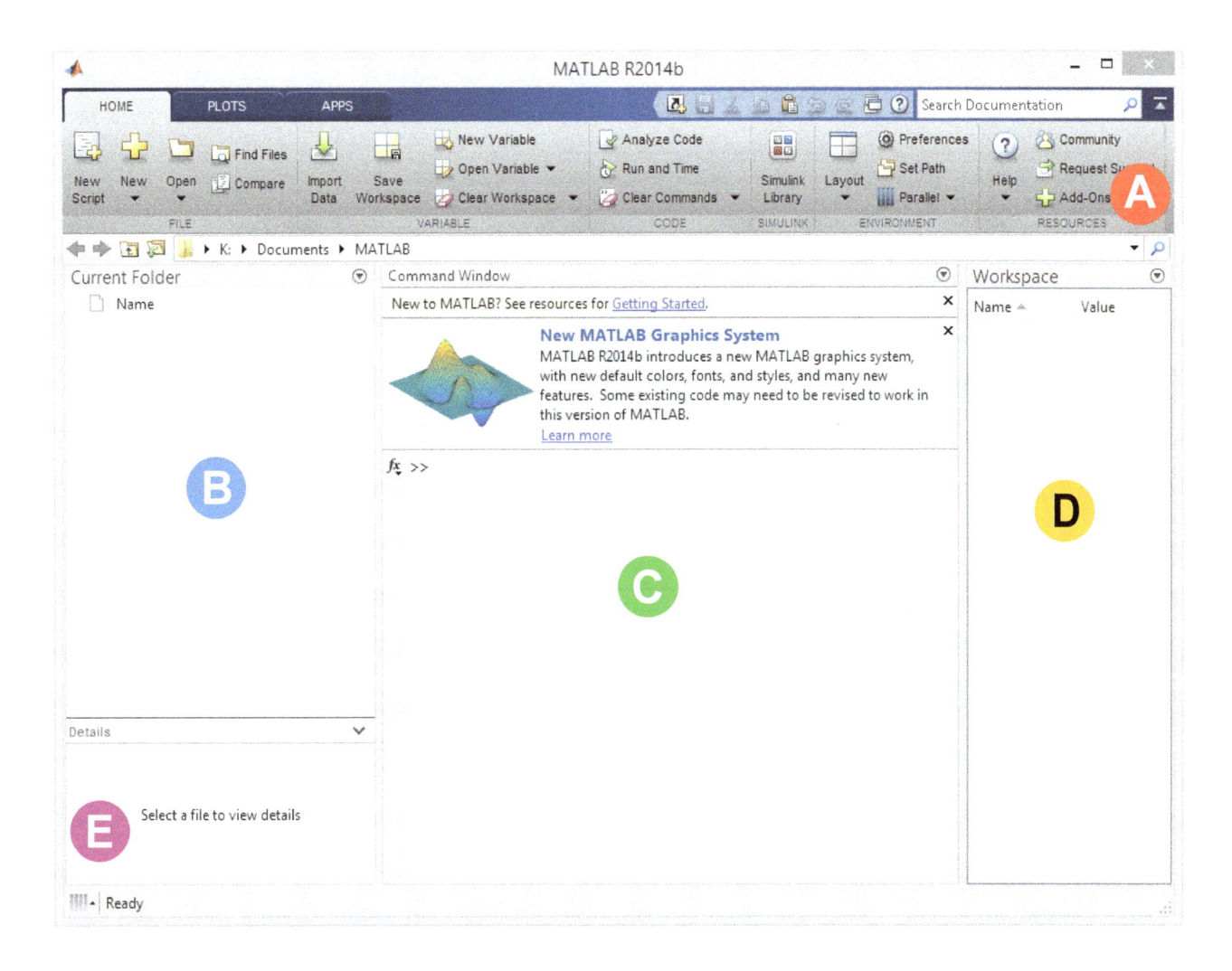

Chapter 4: Getting Started with MATLAB and Simulink 101

A The **Menu Ribbon**, shown here with the main 'Home' tab open. This allows you to work with files and variables, open Simulink, setup the environment, and access help and support.

B The **Current Folder** pane shows the files within the present folder. This facility acts like Windows Explorer — you can open files, as well as move, rename and delete them.

C The **Command Window** is the place for directly entering commands, including single statements or short sets of statements, running scripts and calling functions.

D The **Workspace** shows all of the variables currently held in memory, of various types (scalars, arrays, matrices, strings, structures, etc.). These can be inspected / edited here.

E The **File Details** pane simply displays a summary of the file currently selected in the *Current Folder* pane. This includes revision history details and a description / preview.

Figure 4.1: The MATLAB environment, shown with default layout

(f) **Use a function.** Next, enter the following code to find the remainder when x is divided by y. Notice that this code calls the function `rem`, which is an inbuilt function of MATLAB. Omitting a semi-colon from the end of the line causes the result to appear in the command window.

```
r = rem(x,y)
```

(g) Try assigning x and y with different values, and recalculating the value of r.

(h) **Obtain help on a function.** MATLAB and its toolboxes provide a large number of functions that users can leverage in their code. It is easy to obtain information about a particular function. To find out more about the `rem` function, type

```
help rem
```

You should now see details of the `rem` function appear within the command window:

```
>> help rem
 rem    Remainder after division.
    rem(x,y) is x - n.*y where n = fix(x./y) if y ~= 0.  If y is not
    an integer and the quotient x./y is within roundoff error of an
    integer, then n is that integer. The inputs x and y must be real
    arrays of the same size, or real scalars.

    By convention:
       rem(x,0) is NaN.
       rem(x,x), for x~=0, is 0.
       rem(x,y), for x~=y and y~=0, has the same sign as x.

    Note: MOD(x,y), for x~=y and y~=0, has the same sign as y.
    rem(x,y) and MOD(x,y) are equal if x and y have the same sign,
    but differ by y if x and y have different signs.

    See also mod.

    Overloaded methods:
       codistributed/rem
       gpuArray/rem
       sym/rem

    Reference page in Help browser
       doc rem
```

Chapter 4: Getting Started with MATLAB and Simulink 103

As hinted in the last few lines of the help text, you can enter

```
doc rem
```

at the command window, to view the same information in the *Help Browser*. Try this now and confirm that a window opens showing the desired help information. You should also notice that some examples are provided to demonstrate how to use the function. As shown in Figure 4.2, the Contents menu on the left hand side, and the Search facility at the top of the window, can be used to find information about other functions if desired.

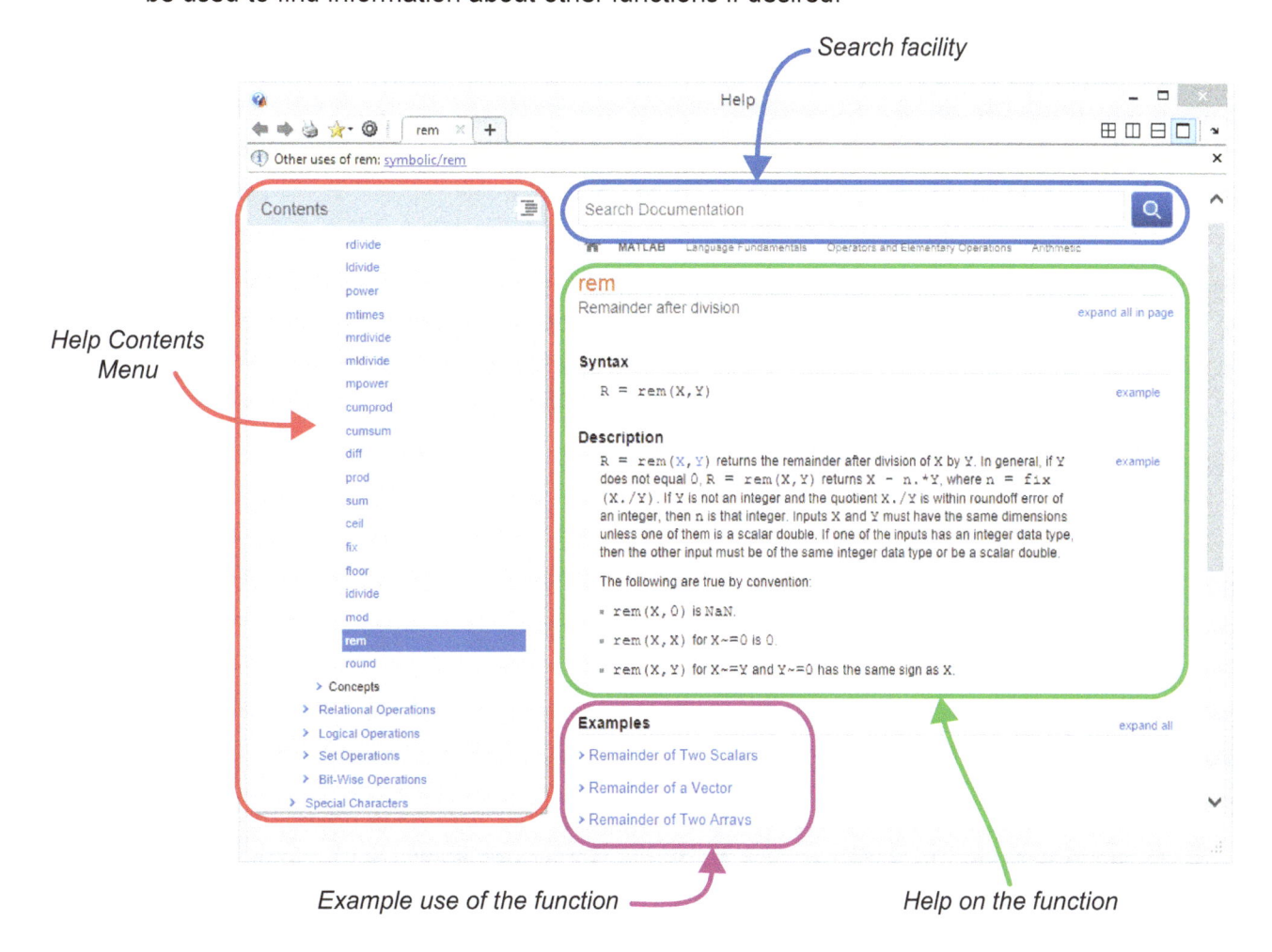

Figure 4.2: Layout of the Help Browser (typical view)

Try finding out about some other functions, such as `mod`, `power`, and `ceil`.

(i) **Using the command history.** MATLAB keeps a memory of commands executed in the command window. This can be used to view recently executed commands, and it also provides a quick and easy way to repeat a previous command (either directly, or with modifications).

Place the cursor at the command window and press the *UP* arrow key on your keyboard. This will cause a list of previous commands to be displayed (the command history). You can now use the **UP** and **DOWN** arrow keys to highlight previous commands from a list. As you change line, you should notice that the highlighted code appears at the command window.

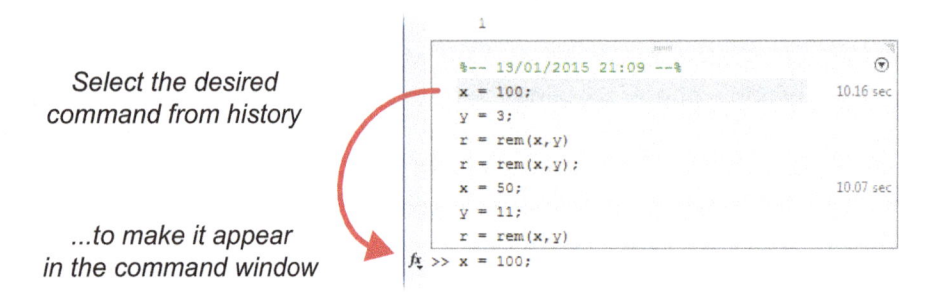

Select the desired command from history

...to make it appear in the command window

If you press the *Enter* key then MATLAB will execute the code currently shown at the command window. Alternatively, you can [1] on the command window, and edit the code before pressing *Enter* to execute.

(j) **Another command history tip!** If you would like the command history to form a subset of commands that begin in a certain way, type the first character (or few characters) before pressing the UP key. For instance, if you type x before pressing UP, the command history will allow you to select only those commands that begin with x. This can make it quicker to retrieve previously used commands.

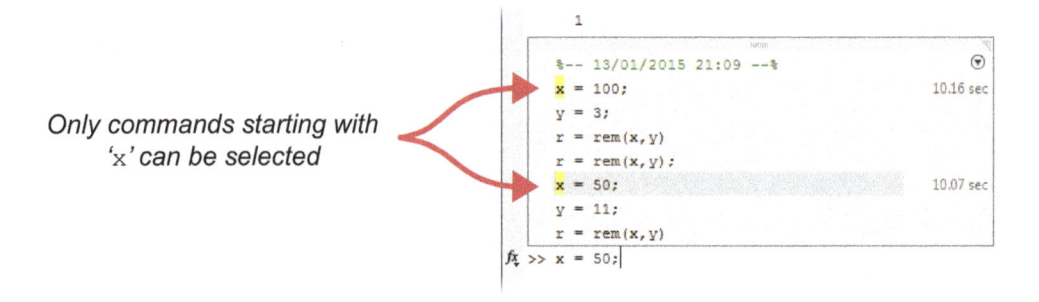

Only commands starting with 'x' can be selected

(k) **Inspecting and editing Workspace variables.** You may have noticed that the values you have assigned to x and y, and calculated for r, are shown as variables in the MATLAB Workspace, as shown below (your values may differ). The Workspace is useful for inspecting variables and values.

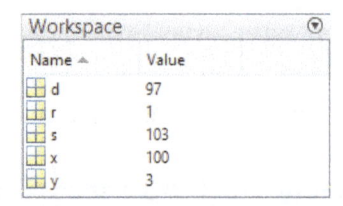

Name	Value
d	97
r	1
s	103
x	100
y	3

(l) We can also change the value of a variable directly in the Workspace pane (i.e. without writing code). To do this, [2] on the variable (name or value) in the Workspace. This will open a new pane showing details of the variable, similar to below (in this case there is just a single value,

Chapter 4: Getting Started with MATLAB and Simulink

but in other circumstances there may be an array or matrix of values). Next, [2] in the desired cell until a cursor appears, edit the value, and press *Enter* to finish. You should then see the value change in the Workspace pane as well.

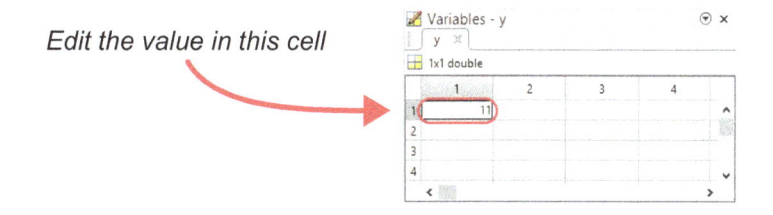

Edit the value in this cell

(m) **Confirming variable values at the command line.** If you would simply like to the confirm the current value of a variable, one easy way to do this is to type the variable name at the command window, without a semi colon afterwards. Try this now!

Exercise 4.2 MATLAB Scripts

When writing more involved MATLAB code, and particularly when there is a desire to reuse the code in the future, creating a MATLAB script is more appropriate than typing line-by-line at the command window. A script is a file that can be edited, commented, executed, and saved for future use, and it has the file extension .m. This exercise will demonstrate how to write a simple script.

(a) **Create a MATLAB script.** There are a few ways to create a new MATLAB script within the MATLAB environment. The most straightforward way is to [1] on the *New Script* button on the MATLAB *Home* tab.

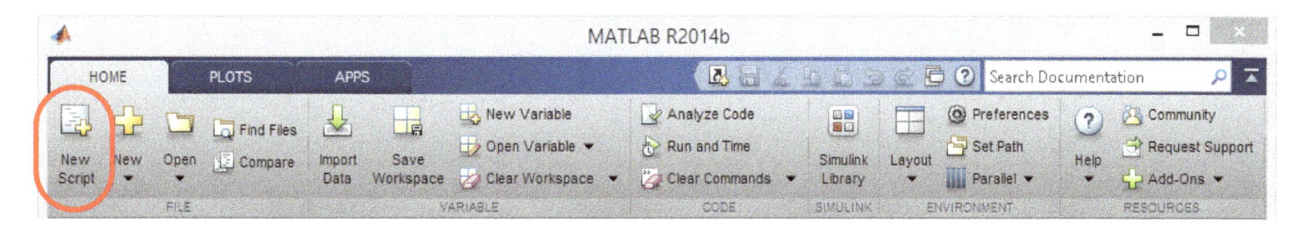

(b) This will open a script window, which will initially have the name 'Untitled'. It can be saved by [1] on the 'Save' button in the *Editor* tab:

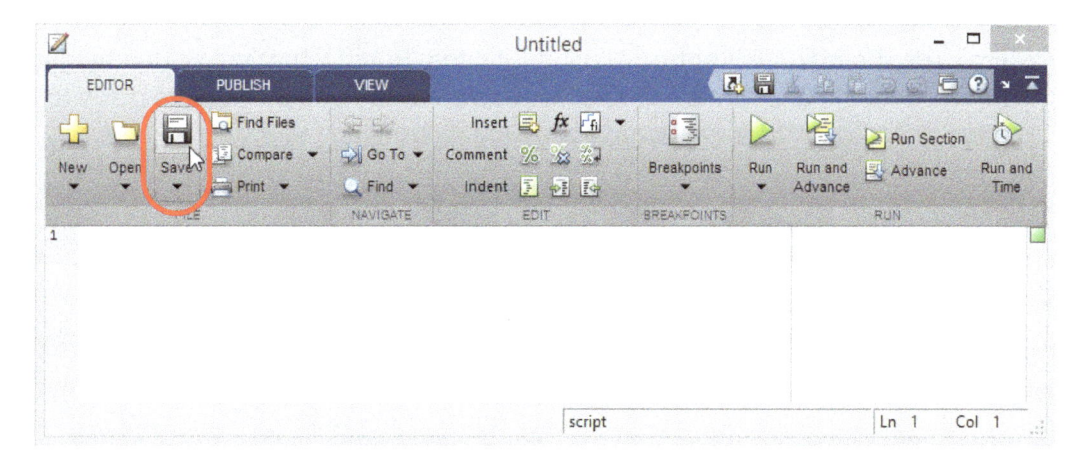

and then choosing the desired directory and file name. Save the file as:

📁 `/matlab_simulink/my_script.m`

(c) **Change the working directory.** Ensure that your current MATLAB working directory is set to the above folder. The working directory is shown in the bar immediately below the menu. You can change the working directory by 🖱 on this bar and typing into it, or by 🖱 on the *Browse Folder* icon (highlighted below) and then navigating to the desired location.

Browse for folder — *Current MATLAB folder*

(d) Next, enter some comments and code into the file, as shown below. Notice that comments are preceded with a `%` symbol, which turns the rest of the current line green. MATLAB will automatically recolour certain features of the code, in particular keywords and strings, while you type.

```matlab
% an introductory MATLAB script

a_array = [1 5 -4 9 7 -6 0];        % an array of numbers to test
thresh = 3;                          % threshold to test against

a_mean = mean(a_array);

if a_mean > thresh
    disp('The average value exceeds the threshold.');
else
    disp('The average value is less than or equal to the threshold.');
end
```

Save the file once you have finished entering the code. This simple script finds the average value of the set of 7 numbers defined in `a_array`, and then compares it to a threshold. One of two messages is printed to the MATLAB command window using the `disp` function, depending on whether the average of the array is greater than or less than the threshold.

(e) **Execute the MATLAB script.** Next, 🖱 on the 'Run' ▷ button to execute the script. View the MATLAB command window to see the results—there should be a message to tell you how the average value of the array compares to the threshold.

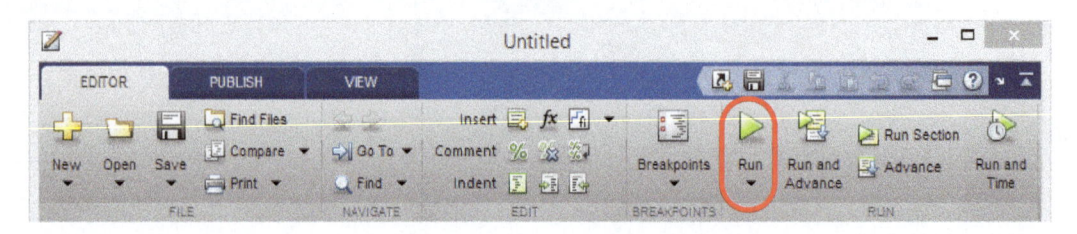

Chapter 4: Getting Started with MATLAB and Simulink

(f) **What is the average value?** Although the message states whether the average is above or below the threshold, it does not give the numerical value of the average. See if you can find this out!

(g) **Printing results to the command window.** It may be convenient to write results at the command window using a string format. This can be done by adding another line of code to display the average and threshold values. At the end of the file, add the following:

```
disp(['The threshold is ', num2str(thresh), ' and the average', ...
    ' value is ', num2str(a_mean),'.']);
```

This command builds a formatted string from different sections of text and numbers, using the `num2str` function to convert numbers to strings. Square brackets are placed around the different sections that are to be concatenated into a string, and commas separate the different sections. The use of dots (...) at the end of the first line allows the command to be split over two lines. (This method of splitting across lines can be used for other code too, not just this particular function.)

(h) **Execute the modified MATLAB script.** Next, save and run the modified script, and ensure that it prints the expected results to the command window.

(i) If you wish, try changing the numerical values in the array, or the value of the threshold, to verify that the correct results continue to be generated.

4.2 MATLAB Functions

There are many functions available in MATLAB, including simple arithmetic calculations like `rem`, and string formatting functions such as `num2str`, which was demonstrated in Exercise 4.2. Functions allow frequently used pieces of code to be packaged for easy reuse. For instance, we often want to find the average of a set of numbers—one method would be to add all the numbers up, and then divide by the number of numbers(!), but actually it is much easier to simply use the `mean` function.

There are also a wide variety of other functions available in MATLAB and its associated toolboxes. The particular selection of functions available to you will depend on the products you have installed on your computer, e.g. if you have the Signal Processing Toolbox installed, you will be able to use filter design functions such as `fir1`.

In the next couple of exercises, we will further investigate built-in functions of MATLAB and its toolboxes, and then go on to consider the creation of custom functions.

Exercise 4.3 Functions in MATLAB

In this exercise, we demonstrate how to investigate the functions available in MATLAB and its toolboxes. As you will see, there are too many functions to introduce them all, but that is exactly what the Help documentation is for!

(a) **Obtain information about functions.** At the MATLAB command window, type

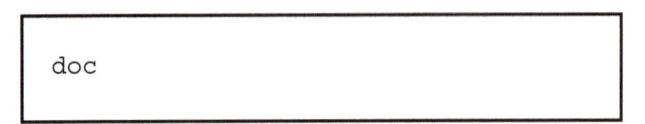

```
doc
```

and then *Enter* to open the Help browser. You should now see a window listing all of your installed components.

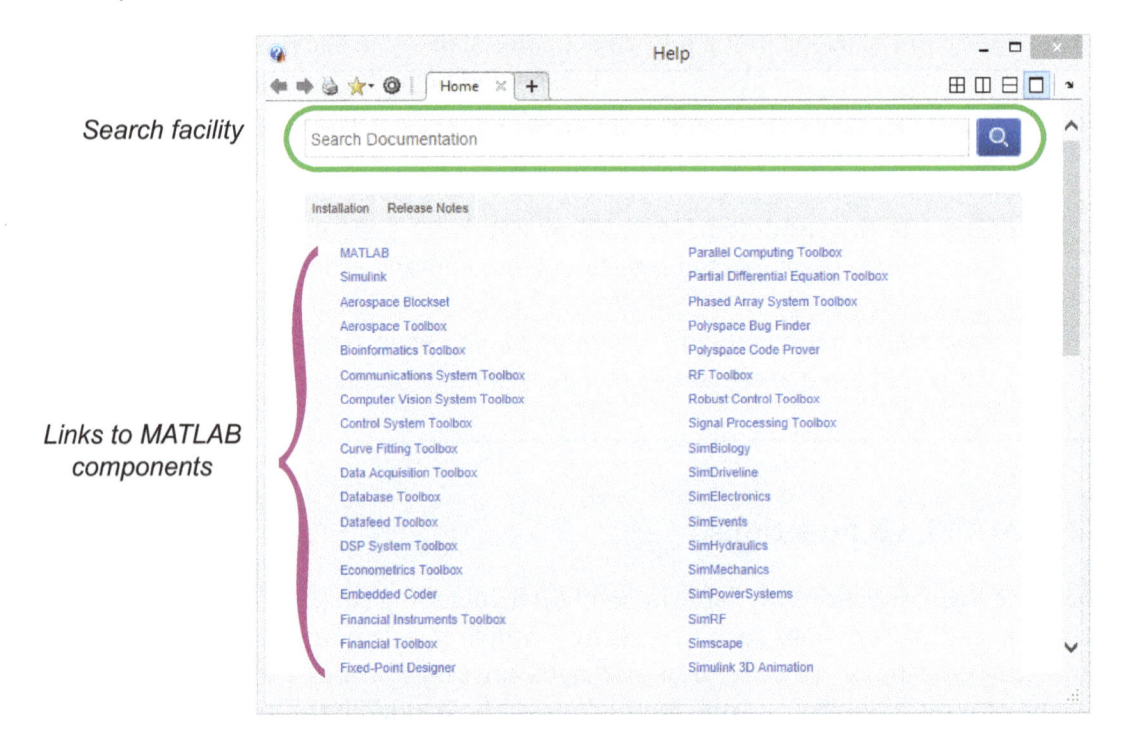

Search facility

Links to MATLAB components

If you are looking for a function to perform a certain task (e.g. for a DSP or communications application), then 🖱 on the link to the *DSP System Toolbox*, *Signal Processing Toolbox* or *Communications System Toolbox*, respectively. Alternatively, you can search using the box at the top of the window.

(b) Let's assume that we would like to find out about the `fir1` function mentioned earlier. This relates to digital filtering, so we will look in the *Signal Processing Toolbox* for more information. 🖱 on the link to the *Signal Processing Toolbox*.

(c) At this point, you should see the contents of the Help on the *Signal Processing Toolbox*. It is arranged based on different types of signal processing operations (and you may wish to

Chapter 4: Getting Started with MATLAB and Simulink

explore some of these!). You should also notice that there is a *Functions* link at the bottom left hand corner — ① on it.

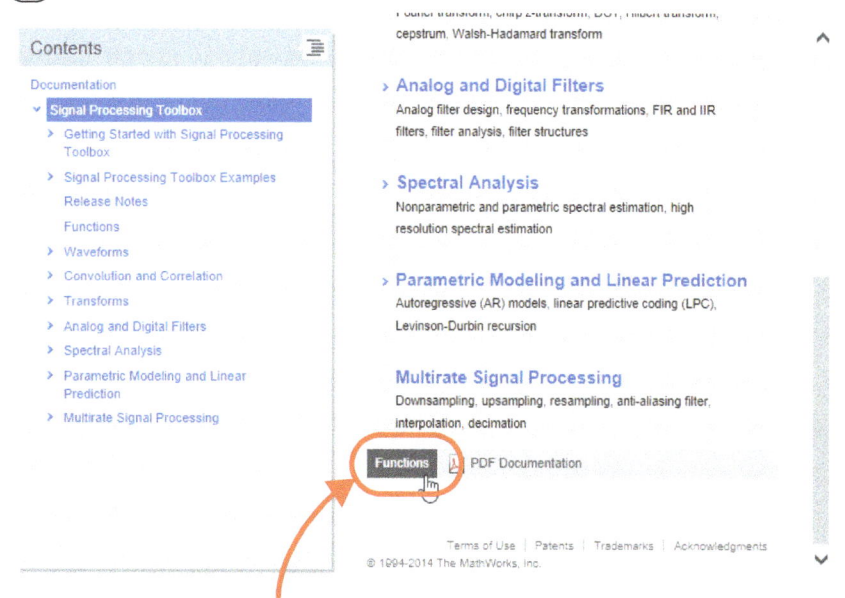

Link to Signal Processing Toolbox functions page

(d) The view should now change, to show the selection of signal processing functions available. Again, these are listed according to category. You may wish to take a few moments to review the available functions.

(e) Locate the `fir1` function in the list, and select it, to see more information about this function.

(f) Read through the information, noting that `fir1` can be called using different sets of arguments (inputs to the function). The *Description* section provides detailed information about the operation of the function, and towards the bottom of the page, some *Examples* are provided to demonstrate its use.

(g) **Replicate an example.** MATLAB Help often provides segments of example code. These can be cut and pasted into the MATLAB command window, and run, to replicate the results shown in the documentation. Try this now for Examples 1 and 2, and check that the results correspond with the Help file.

(h) What other functions are used within these example code segments? Can you use the documentation to find out more about them?

Exercise 4.4 Writing Your Own Functions

Although MATLAB contains many functions, in some cases you may wish to write your own, to undertake a custom operation. Next, we will demonstrate how to write a new function based on the averaging-and-thresholding script from Exercise 4.2.

Bear in mind that our example is a simple one, and functions can be written using many combinations of input and output types, options, etc. They can also be 'overloaded', such that the function behaves differently depending on the supplied set of input arguments. We do not have time to cover these more complex examples here, but it is useful to be aware that you can create your own, sophisticated functions in MATLAB—see the Help documentation for more information.

(a) **Create a new function.** In the main MATLAB window, choose the *New Function* option as shown:

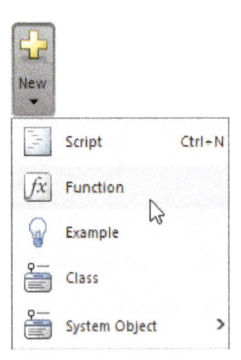

A new editor window will now open, containing a function template. The initial name of the file is 'Untitled' (or similar), and the name of the function is also `Untitled`. *It is important to note that the name of a function, and its file name, must be identical.*

(b) Save the function with the name `compare_mean_to_thresh.m`.

(c) Notice that two parts of the function are highlighted: the function name (which is still set to `Untitled`), and the list of input arguments.

```
function [ output_args ] = Untitled( input_args )
%UNTITLED Summary of this function goes here
%   Detailed explanation goes here


end
```

(d) **Customise the function interface.** Change the function name from `Untitled` to `compare_mean_to_thresh`, and replace the `input_args` placeholder with the argument names `num_array` and `thresh`, separated by a comma. Similarly, change `output_args` to `diff`.

(e) When called, the function will be passed an array of numbers, `num_array`, and a threshold value, `thresh`, as the input arguments. It will then calculate the difference between the average of the array values, and the threshold, and return the result as the output argument, `diff`. The value of `diff` should be positive if the mean is higher than the threshold.

Write some comments to explain the operation of the function, replacing the placeholder comments in the template.

Chapter 4: Getting Started with MATLAB and Simulink 111

(f) **Write the function!** Next, write your own MATLAB code to implement the operation of the function described above, and then save the file.

(g) **Call the function.** We can now call the function from other MATLAB code, provided that the file `compare_mean_to_thresh.m` is in the current working directory. The function can be called from a MATLAB script, or directly from the command window.

Type the following commands (either into a script, or individually at the command window), and execute them. You do not need to enter comments if typing at the command window.

```matlab
% code to demonstrate a function call

clear all;                              % clears Workspace of existing vars
nums = [3 10 7 -8 -2 5 -6 -1 4];        % an array of numbers to test
thresh = 2;                             % threshold to test against

% CALL THE FUNCTION!
% (omit semi-colon at end of line to print result in command window)
difference = compare_mean_to_thresh(nums,thresh)
```

Check that the function executes successfully and produces the correct result!

(h) **Further function investigation!** Here, we have taken the example of a function that returns a single result. It should be noted, however, that functions can recall more than one result, or indeed they can return no results! Find out more about this topic using the Help documentation, by typing

```matlab
doc function basics
```

at the command window.

4.3 Plotting in MATLAB

Our next example in this brief introduction to MATLAB demonstrates how to create basic figures. As with many other aspects of MATLAB, there are far more possibilities than can be covered here, but it is useful to provide a few simple examples to help new users get started.

Exercise 4.5 MATLAB Figures

This exercise will demonstrate how to plot sine and cosine waves, add axes labels, legends and titles, and how to customise different aspects of their appearance.

(a) **Create a new script.** Open a new MATLAB script and save it in your working directory, with the name `plot_sin_cos.m`.

(b) **Defining header information and parameters.** First, add a line or two of comment to explain that this file will be used to generate and plot sine and cosine waves. Then, enter the following lines of code to specify the sampling frequency to be used, the frequency of the sine and cosine waves, and the duration of the simulation.

```
% script to generate and plot sine and cosine waves

fs = 1000;        % sampling frequency in Hz
f1 = 100;         % frequency of sine and cosine waves in Hz
Tmax = 0.08;      % duration of simulation (stop time in seconds)
```

(c) **What is the sample period?** Add a line of code to calculate the sample period, and assign it to the variable `Ts`.

(d) **Generate the data arrays.** The next step is to generate an array of time samples extending from 0 to `Tmax = 0.08` seconds, which we will call `t`. The corresponding sine and cosine values can be generated according to equations:

$$A_s = \sin(2\pi f_1 t) \tag{4.1}$$

$$A_c = \cos(2\pi f_1 t) \tag{4.2}$$

... which will result in another two arrays for sine and cosine respectively. Add the following code to your MATLAB script to create the time, sine and cosine arrays.

```
t = [0:Ts:Tmax];        % create an array of time values from 0 to Tmax
                        % (in steps of Ts, the sample period)
As = sin(2*pi*f1*t);    % create a corresponding array of sine samples
Ac = cos(2*pi*f1*t);    % create a corresponding array of cosine samples
```

(e) **Run a preliminary simulation.** Often it is useful to run a script from time to time while you are writing it, so that you can detect any problems at an early stage. Run the script now and check that there are no errors (which would be shown in red in the command window), and that the variables have been created successfully. If any problems have cropped up, tend to these now before moving on! Note that the script will automatically be saved when you execute it.

(f) **Create a figure and plot data.** Next, we must add some lines of code to generate a new figure (it will be numbered as Figure 1), and plot the sine and cosine data within it. Copy the lines below into your script (the comments are optional!).

Chapter 4: Getting Started with MATLAB and Simulink

```
figure(1)                 % create a new figure
hold off                  % do not retain any previous data in plot
plot(t,As,'r-o');         % plot the sine wave data in red ('r') with
                          % a continuous line and a round marker
hold on                   % retain the sine while we add the cosine...
plot(t,Ac,'b-o');         % plot the cosine wave data in blue ('b')
                          % with a continuous line and a round marker
```

(g) **Run another preliminary simulation.** Re-simulate to inspect the results! You should see a figure window appear, something like this:

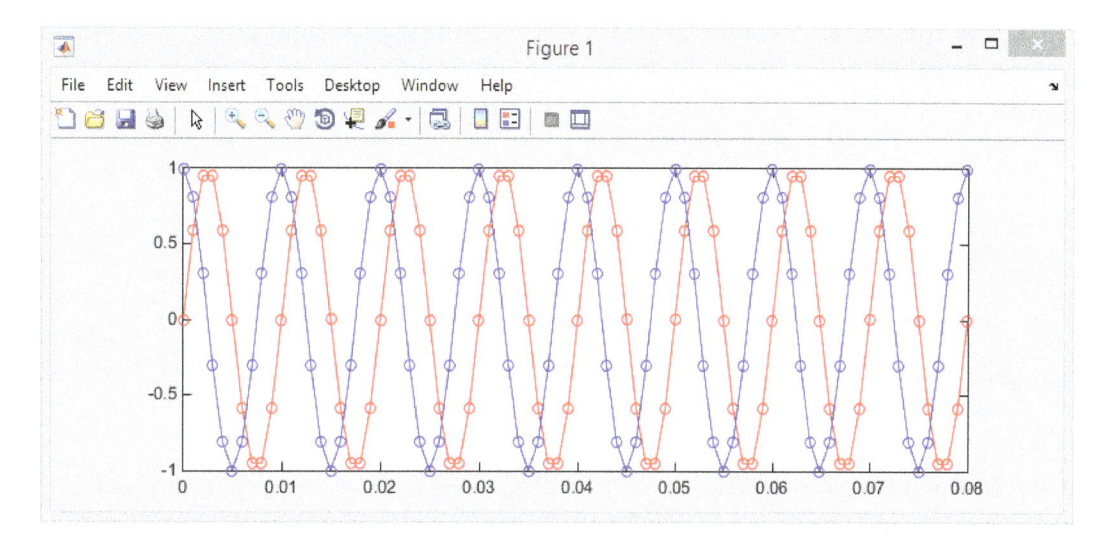

(h) Compare the appearance of the figure against the code written so far. It should be possible to see that the cosine appears in blue and the sine in red, as desired. Both are continuous lines with round markers, as a result of the `'b-o'` and `'r-o'` strings supplied as part of the `plot` function calls.

(i) **What does 'hold on' do?** Try commenting out `hold on` by adding a % symbol to the start of the line, and then resimulating. You should notice that only the cosine wave is shown now — the sine wave has not been retained when the next `plot` function is called. The purpose of the hold is to enable multiple sets of data to be plotted on the same figure.

(j) **Experiment with line and marker styles.** Try changing the specification of the two lines and markers to alter their appearance. For instance, you could test the effect of changing the specifier string to `'m:s'` or `'k--x'`.

Further possibilities include thickening the line, changing the size of the marker, or filling the marker if it is hollow—a square or circle, for instance. Other, non-fillable marker styles include crosses and asterisks. The fill colour of the marker can also be specified, and it need not match that of the line.

Try changing the code so that the two lines starting with `plot(...)` are as follows, then re-simulate and see the effect of your changes.

```
plot(t,As,'b--s','MarkerFaceColor','m','LineWidth',1,'MarkerSize',5);

plot(t,Ac,'k-o','MarkerFaceColor','y','LineWidth',2,'MarkerSize',10);
```

You might also wish to consult the MATLAB Help documentation to find out more about the possibilities of the `plot` function!

Once finished, return the line and marker styles to their original specifications.

(k) **Label the axes, and title the plot.** Although data has been drawn on the figure, it does not yet have any axes labels, or a title. You can add these now by appending the following lines of code to your script.

```
xlabel('Time (seconds)');      % add a label to the x (time) axis
ylabel('Amplitude');            % add a label to the y axis
title('Sine and Cosine');       % add a title for the plot
```

(l) Once incorporated, re-run the simulation to observe the changes. You should now see that the labels and title specified in the code have been added to the figure.

(m) **Add a legend and customise formatting.** The last step we will take is to add a legend to the figure, to clearly indicate which set of data represents the sine wave, and which represents the cosine wave. It would also be nice to improve the appearance by changing the axis limits, and showing a grid in the background. These three features can be introduced by the code shown below. Add this into your MATLAB script.

```
legend('sin(2\pif_1t)','cos(2\pif_1t)','location','NorthEast');
                                % add a legend to the graph
grid on;                        % show the grid
axis([0 max(t) -1.2 1.2]);      % configure the axis scaling
```

It is useful to explain a couple of points relating to the legend text, taking the first legend entry as an example.

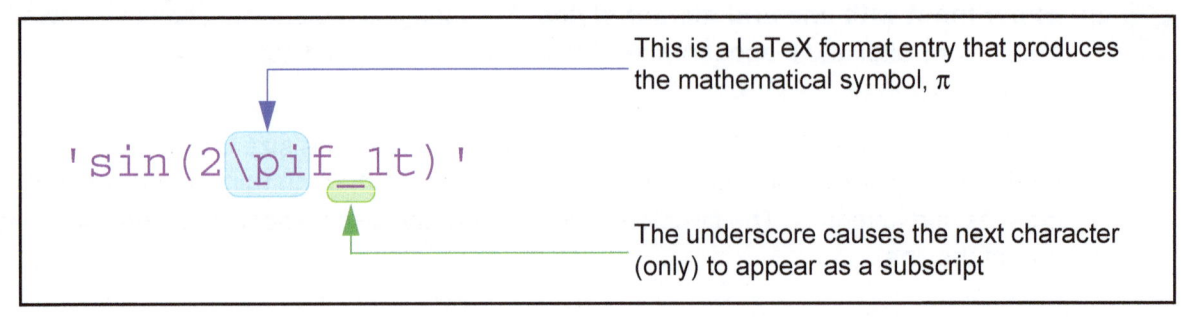

`'sin(2\pif_1t)'`

`\pi` — This is a LaTeX format entry that produces the mathematical symbol, π

`_` — The underscore causes the next character (only) to appear as a subscript

Chapter 4: Getting Started with MATLAB and Simulink

If you wish, you can also experiment with changing the position of the legend (acceptable entries include all of the major and minor compass points, e.g. `'North'`, `'South'`, `'SouthWest'`, etc., and also outside the axes, e.g. `'EastOutside'`, `'SouthOutside'`).

(n) Once you have added the extra code, re-run the simulation to see the effect. The legend should now be visible, along with the grid. Notice also that the axis limits have been changed to provide a little extra space above and below the sine and cosine waves. Your final version of the figure should look similar to this:

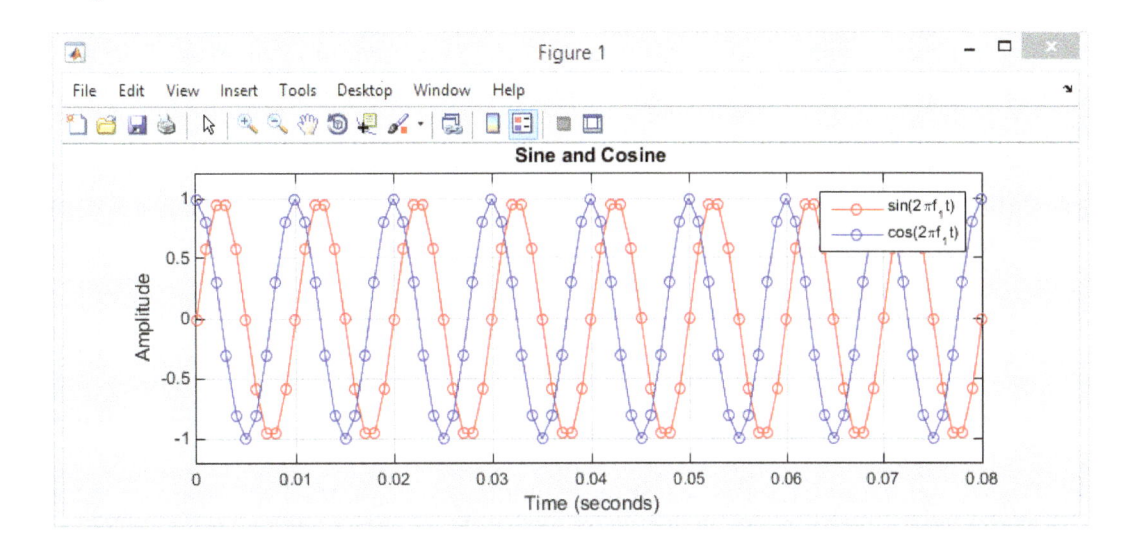

(o) **Challenges!** Now that you have followed through this example to produce the desired graph, consider the following points, and see if you can alter and simulate the script accordingly.

Can you change the duration of the simulation to 0.6 seconds, ensuring that the time axis is altered appropriately?

See if you can alter the sampling frequency such that the signal is sampled at 2kHz.

Can you change the amplitude of the sine wave to 0.7, and the amplitude of the cosine wave to 1.3? Make sure that the figure shows the full range of both.

Try adding a third data series to the figure. Add the sine and cosine together, and plot this in magenta with square markers. Make sure that your extra plot is also labelled.

What happens if you supply only the sine and cosine wave samples to the `plot` function, without the time information?

Check out the `stem` function... Replace `plot` with `stem` and see what happens!

4.4 MATLAB Arrays, Matrices, and Structures

MATLAB is a tool for technical computing, and it has extensive support for working with data in the form of arrays, matrices, and structures. Arrays and matrices are crucial for many types of mathematical analysis, while structures are convenient for keeping sets of related data together.

The next few exercises will introduce these data types and demonstrate some aspects of their use. There are many more functions relating to arrays and matrices (in particular) than can be covered here; however, following the examples in the coming exercises will provide a good start!

Exercise 4.6 MATLAB Arrays

Earlier exercises have introduced MATLAB variables, including scalars and arrays. In this exercise, we will work a little more with arrays, and highlight some useful functions for working with arrays.

The following steps can either be executed at the MATLAB command window, or in a MATLAB script, as you prefer. Remember that you can type `doc` followed by the name of the function, to open the help documentation on that topic.

(a) **Define arrays.** An array of values can be specified easily in MATLAB. Enter the following code to create two new arrays.

```
my_array_r = [1  -2  -5  6  9  -6  -4 0]    % define a row vector
my_array_c = [8; -9; -1; -4; 3;  6;   5]    % define a column vector
```

(b) **Create an array of zeros or ones.** Sometimes it is useful to create an array of a certain dimension, where all of the elements are zero (0), or all elements are one (1). This can be easily achieved via the following code:

```
my_ones = ones(1,10)      % create an array of ones, with 1 row
                          % and 10 columns
my_zeros = zeros(1,8)     % create an array of zeros, with 1 row
                          % and 8 columns
```

(c) **Find the length of an array.** To find the length of a previously declared array, type:

```
num_ones = length(my_ones)      % find the length of array my_ones
num_zeros = length(my_zeros)    % find the length of array my_zeros
```

It can be useful to find the size of an array when defining the limit of a `for` loop, for example (the `for` loop might be required to iterate once for every element in the array).

(d) **Find the maximum or minimum value within an array.** Try this code to find the maximum and minimum values of the arrays defined earlier.

```
max_r = max(my_array_r)      % find the largest value in my_array_r
min_c = min(my_array_c)      % find the smallest value in my_array_c
```

Chapter 4: Getting Started with MATLAB and Simulink

You can also combine the above with the `abs` function, if you would like to find the maximum and minimum absolute values.

(e) **Generate an array of random values.** To create an array of random numbers, type:

```
my_rands = rand(1,10)          % create an array of random values
```

This will create 10 random values from a uniform distribution, in the interval 0 to 1. If you would like to generate random values across a different range, then you can do this by scaling and/or offsetting the generated array. For instance, if we wanted to create random values in the range -6 to + 6, it could be done like this:

```
my_rands_6 = (12*rand(1,10))-6    % create randoms in range -6 to +6
```

(f) **Obtaining integer values.** If the desired random array is intended to contain integer numbers only, the `round` function can be incorporated to achieve this:

```
my_rand_ints = round((12*rand(1,10))-6)   % create random integers
```

(g) **Transpose an array.** Arrays can be transposed in a couple of different ways. Try these out, to transpose the `my_rands` and `my_rands_6` arrays (in this case, transposing will change both arrays from a single row to a single column).

```
my_rands_t = transpose(my_rands)    % transpose the first array
my_rands_6_t = my_rands_6'          % transpose the second array
```

(h) **Extract a specific element from an array.** If you would like to find the value of a particular element from an array, this can be done simply by giving the desired index in brackets. It is important to note that MATLAB indexing begins at 1 (rather than 0, as in some other programming languages).

```
first_elem = my_rands(1)     % pulls out first element
zero_elem = my_rands(0)      % try this, and confirm it fails!
```

(i) **Find particular elements in an array.** Sometimes it is useful to extract specific elements from an array, and the first step in doing so is to identify the indices of these elements. We might do so by testing which elements of an array meet a desired condition. For instance, suppose that we would like to find those elements of the array `my_rands_6` that are less than zero.

```
sub_zero_i = find(my_rands_6 < 0)        % find indices of elements < 0
my_rands_6_sz = my_rands_6(sub_zero_i)   % extract sub_zero elements
```

Note that `sub_zero_i` is an array of the *indices* of `my_rands_6` that meet the condition, NOT the values themselves. The second operation generates a new array of values that contains only those elements of `my_rands_6` (the original array) that meet the condition.

(j) How would you concatenate the arrays `my_ones` and `my_zeros`, to form a single, longer array? Use the Help documentation to find out.

(k) There are many more possibilities for working with arrays than we can cover here—to investigate further, please refer to the Help documentation:

```
doc matrices and arrays
```

Exercise 4.7 Matrices in MATLAB

Matrices are a natural extension of arrays, comprising two dimensions of data (or higher dimensions in some cases), and we can define and operate on matrices easily in MATLAB. This exercise introduces a few common operations, based on two dimensional matrices.

(a) **Define a matrix.** Matrices can be defined in a very similar manner to arrays. For instance, to define two matrices of values with 3 rows and 4 columns, we can simply write:

```
my_matrix_A = [1 2 3 4; 5 6 7 8; 9 10 11 12]
my_matrix_B = [3 3 3 3; 1 2 1 2; 4 4 5 5]
```

(b) **Find the dimensions of a matrix.** Similar to the `length` function which we applied to arrays, the `size` returns the dimensions of a matrix, giving first the number of rows, and then the number of columns. Try checking the dimensions of `my_matrix_B` by typing:

```
size(my_matrix_B)                % find the dimensions of the matrix
```

(c) **Create matrices containing zeros, ones, or random values.** These three types of matrices can be created using the same syntax as shown in Exercise 4.6 for arrays—the difference is now that both arguments supplied to each function are greater than 1, because there is no single dimension in a matrix, as there is in an array.

Chapter 4: Getting Started with MATLAB and Simulink

```
my_oneM = ones(4,2)        % create a matrix of ones, with 4 rows
                           % and 2 columns
my_zeroM = zeros(3,5)      % create a matrix of zeros, with 3 rows
                           % and 5 columns
my_randM = rand(5,5)       % create a matrix of random values, with
                           % 5 rows and 5 columns
```

(d) **Matrix addition and subtraction.** Matrices can be added and subtracted, meaning that these operations are performed on individual elements, provided that the dimensions of the two matrices are equal. For instance, try typing:

```
my_matrix_C = my_matrix_A + my_matrix_B    % add matrices
my_matrix_D = my_matrix_A - my_matrix_B    % subtract matrices
```

(e) **Element-wise matrix multiplication.** Following on from last point, matrices can also be multiplied on an element-wise basis, using the .* operator, again provided that they have the same dimensions. Try this now, using my_matrix_A and my_matrix_B as the multiplicands.

(f) **Matrix product.** Matrices may be multiplied, provided that the dimensions agree. This means that the number of columns in the first matrix, and the number of rows in the second matrix, must be the same. It would not be possible to multiply the matrices my_matrix_A and my_matrix_B defined above, for instance, because their dimensions are not compatible (if you wish, you can try this and see what happens!).

(g) Try defining matrices that *will* be compatible for multiplication, and find the matrix product (using the * operator).

(h) **Extracting a row or column from a matrix.** It is useful to review how to extract a row or column from a matrix, to form an array. This can be useful, for instance, when a matrix represents several channels of data, and we require to separate them. To illustrate retrieval of a row and column, we can extract the first row, and then the second column, of the matrix my_matrix_A using a :.

```
row_1 = my_matrix_A(1,:)     % extract first row
col_2 = my_matrix_A(:,2)     % extract second column
```

(i) **Transpose a matrix.** The transpose of a matrix can be found using the same method as the transpose of an array. Refer back to Exercise 4.6, and adapt this code to find the transpose of my_matrix_A.

(j) **Squeezing a matrix.** Occasionally, variables can be created with more dimensions than are necessary. For instance, the *To Workspace* Simulink block (to be covered in Exercise 4.16) can create a variable of dimensions 1 x 1 x *n*, depending on the simulation sources present in the system. There is effectively a superfluous dimension. This extra dimension can be removed, and the data converted to a 1 x *n* array, by applying the `squeeze` function.

First, let's create a variable with an 'extra' dimension...

```
my_var(1,1,:) = [1 2 3 4 5 6 7 8 9]      % create 1 x 1 x n data
```

You can confirm these dimensions by looking in the MATLAB Workspace.

(k) Next, we can 'squeeze' the variable to remove the unnecessary dimension:

```
my_var_2 = squeeze(my_var)      % reduce to 1 x n data
```

Confirm this has worked by examining `my_var_2`.

(l) As noted previously, there are many more possibilities for working with matrices than we can cover here—to investigate further, please refer to the Help documentation, by typing

```
doc matrices and arrays
```

Exercise 4.8 MATLAB Structures

MATLAB structures permit related variables to be grouped together, which can help to keep code organised and readable. This exercise briefly introduces the use of structures.

(a) Open the MATLAB script:

　　　　`/matlab_simulink/structures.m`

(b) **Inspect the code.** Take a few moments to read through the script. Notice that a structure called `weather` has been created in this script. There are numerous variables associated with `weather`, which are defined using the dot notation. For instance, we can reference the temperature as `weather.temperature`.

(c) **Run the script.** A figure should be created, which will plot two variables from the `weather` structure against each other—`weather.days` is plotted on the x axis, while `weather.temperature` corresponds to the y axis.

Chapter 4: Getting Started with MATLAB and Simulink

(d) **View the structure in the Workspace.** Check the MATLAB Workspace and confirm that there is a single item present—the `weather` structure, `weather`. The internal contents of this structure can be explored by 🖱 on it. Later, we will see that Simulink models can output simulation data in the format 'structure with time'. This allows data samples and their corresponding sample times to be conveniently stored together in a structure.

4.5 MATLAB System Objects

Many different programming languages are object orientated, which means that they support *data objects* as well as being able to perform computations. MATLAB (which is based on Java) is one of these languages, as you should have seen in the previous exercise, where structures were introduced. Many of the MathWorks Toolboxes (which contain the Simulink blocks and MATLAB functions) use a specific type of object called a *System Object* that allows dynamic systems to be initialised both in MATLAB and Simulink. These dynamic systems are systems that output different signals depending on the values of their inputs, which can change over time [65]. They contain memory which holds their parameters, current state and past behaviour, all of which are used during the next computational step. The easiest way to demonstrate how a system object works is to give an example, so we shall do this now.

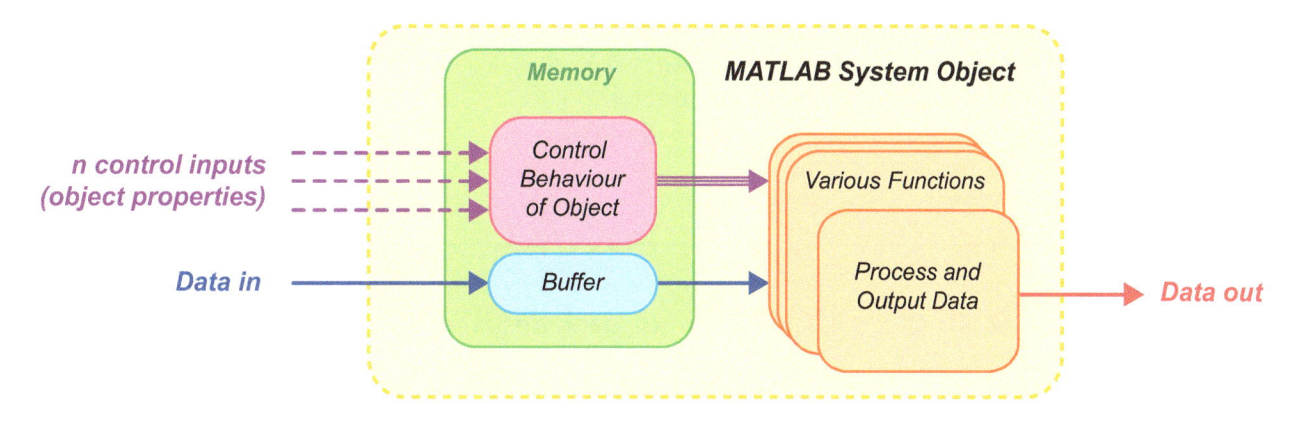

Figure 4.3: A block diagram showing the architecture of a basic System Object

Exercise 4.9 An Introduction to System Objects

This exercise will guide you through the process of creating a lowpass FIR filter system object, viewing its properties, filtering a signal with it, and finally, redesigning it. If you need to remind yourself about what an FIR filter is, have a look at Appendix C (page 583).

(a) **Create a lowpass filter.** Enter the following code into the MATLAB command window to create a new FIR filter system object; using the `fir1` filter design tool:

```
% create a lowpass FIR filter, and assign the handle obj_filter
obj_filter = dsp.FIRFilter('Numerator', fir1(50,0.25,'low'));
```

The `fir1` function designs a set of `50` filter coefficients that will implement a lowpass filter that cuts off at `0.25` (normalised frequency, i.e. 0.25* f_s /2 Hz). These coefficients will be used as the numerator of the filter, which is assigned the name `obj_filter`.

(b) The filter will remain in the MATLAB workspace until it is deleted, and it can be accessed at any time by calling its name. Information about the filter can be displayed either by 2️⃣ on 📄 obj_filter in the workspace panel, or by typing

```
obj_filter
```

into the command window. Try this now. You should see something similar to the following. Note that if you enter `obj_filter.Numerator` at the command line, you will print the array of filter coefficients, designed by `fir1`.

```
>> obj_filter

obj_filter =

  System: dsp.FIRFilter
  Properties:
              Structure: 'Direct form'
        NumeratorSource: 'Property'
              Numerator: [1x51 double]
       InitialConditions: 0
    FrameBasedProcessing: true
```

(c) **Use the fvtool.** To check that the filter has been designed correctly, you can use the Filter Visualisation Tool. Enter

```
fvtool(obj_filter);
```

into the command window. The magnitude response of the filter should be displayed as shown below.

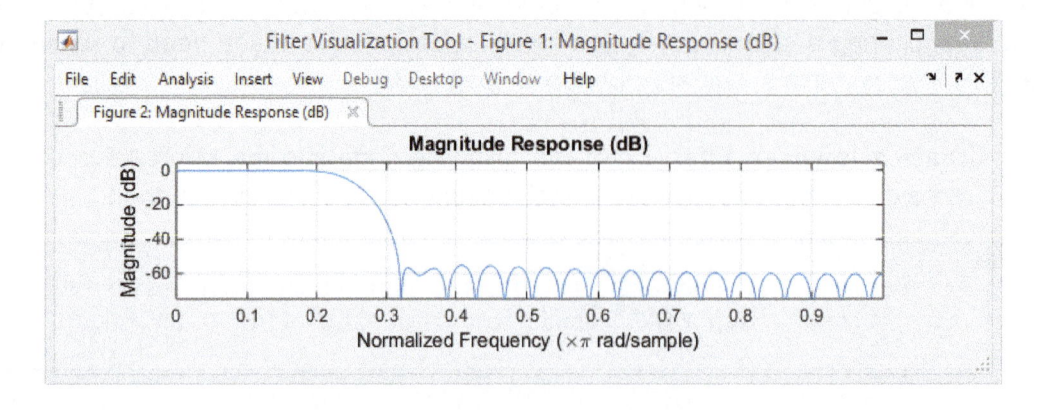

Chapter 4: Getting Started with MATLAB and Simulink

(d) **Using the filter.** Before you can use the filter, you will need to generate a signal to pass into it! A simple way to do this is to create a Gaussian White Noise array using the `randn` function. Creating signal `x`, the `step` function can be used to pass `x` to the filter, and the returned data can be stored in `y`. Enter the following code.

```
x = randn(125,1);              % generate a Gaussian White Noise signal
y = step(obj_filter,x);        % filter the signal with the lowpass filter
```

(e) To plot the original and filtered signals, type the following lines of code:

```
% create a figure
figure
grid on
hold on
xaxis = 1:100;

% plot the signals
plot(xaxis,x(1:100),'r');
plot(xaxis,y(26:125),'b');     % note the 25 sample delay
```

As the filter contains `50` coefficients, a 25 sample delay is required in order to keep the two signals synchronised in the figure. Notice that the blue, filtered signal is far smoother. This is because the high frequency components in the red signal have been filtered out.

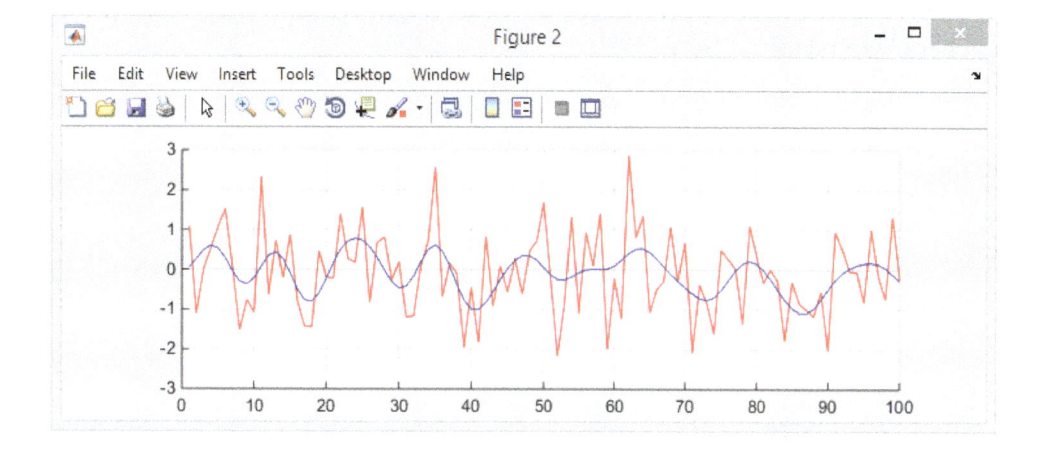

(f) **Redesign the filter.** If it was decided that the filter must be redesigned because it needs to become a highpass filter instead, the numerator of the filter would need to be updated with a new array of coefficients. Enter the following code to do this.

```
% redesign obj_filter to make it a highpass filter instead
obj_filter.Numerator = fir1(50,0.25,'high');
```

(g) Use the `fvtool` once again to check that the filter is now a highpass filter, with cutoff point `0.25`. Finally, repeat Part (d) and (e) to test the filter out.

Many of the components in the DSP and Communications System Toolboxes use system objects, and you can use the MATLAB autocomplete facility to list all of the ones available. Try typing '`dsp.`' or '`comm.`' in the MATLAB command window and pressing the *TAB* key on your keyboard.

Figure 4.4: Accessing the list of system objects using the autocomplete function

The ones you are most likely to use when developing MATLAB SDR receivers are the following:

- RTL-SDR Receiver   `comm.SDRRTLReceiver()`
- FIR Filter   `dsp.FIRFilter()`
- FIR Decimation   `dsp.FIRDecimator()`
- Spectrum Analyzer   `dsp.SpectrumAnalyzer()`
- Audio Player   `dsp.AudioPlayer()`

You can find information about each of these using the `help` function. More information on system objects in general can be found here [65].

4.6 Introducing Simulink

Simulink is a graphical, block-based tool that integrates with and complements MATLAB. There are some types of design that lend themselves more to a block-based than a code-based description, and designers may have an individual preference for one or the other.

It is worth being aware that, as Simulink and MATLAB are integrated, several mechanisms exist for combining block-based and code-based design to best effect. For instance, MATLAB code can be 'placed' inside a block in a Simulink system, or MATLAB code can be executed automatically as a model loads, defining variables necessary to parameterise the model (possible as Simulink shares the MATLAB workspace). There is great scope to jointly leverage the facilities of MATLAB and Simulink.

The exercises that follow in this section introduce the Simulink interface, and the library of blocks from which Simulink systems are constructed. We will then move on to build a first system.

Exercise 4.10 Simulink Orientation and the Simulink Library

This exercise provides an introduction to Simulink. We create a new Simulink model, and open and explore the Simulink library.

(a) **Create a new Simulink model.** Open a new model from the MATLAB *Home* tab, by [1] on *New* and then *Simulink Model*, as shown below.

This will open a new window—an empty Simulink model with the name 'untitled'.

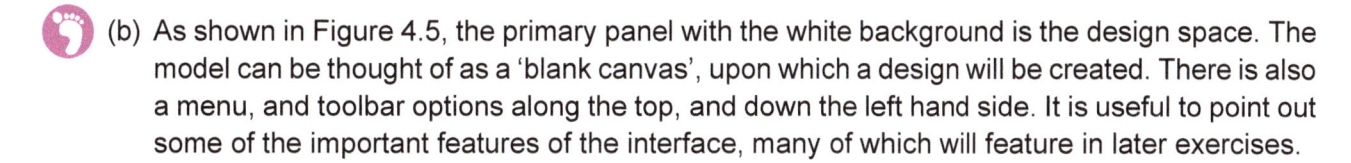

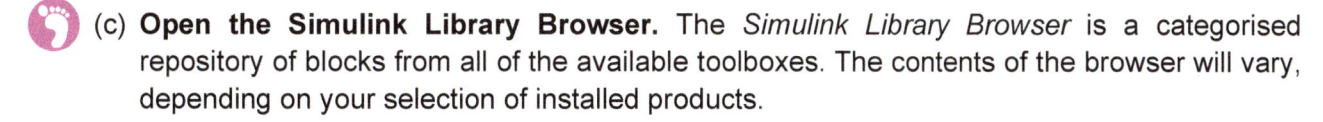

(b) As shown in Figure 4.5, the primary panel with the white background is the design space. The model can be thought of as a 'blank canvas', upon which a design will be created. There is also a menu, and toolbar options along the top, and down the left hand side. It is useful to point out some of the important features of the interface, many of which will feature in later exercises.

(c) **Open the Simulink Library Browser.** The *Simulink Library Browser* is a categorised repository of blocks from all of the available toolboxes. The contents of the browser will vary, depending on your selection of installed products.

The *Simulink Library Browser* can be opened with the button in the main MATLAB menu, or simply by typing

```
simulink
```

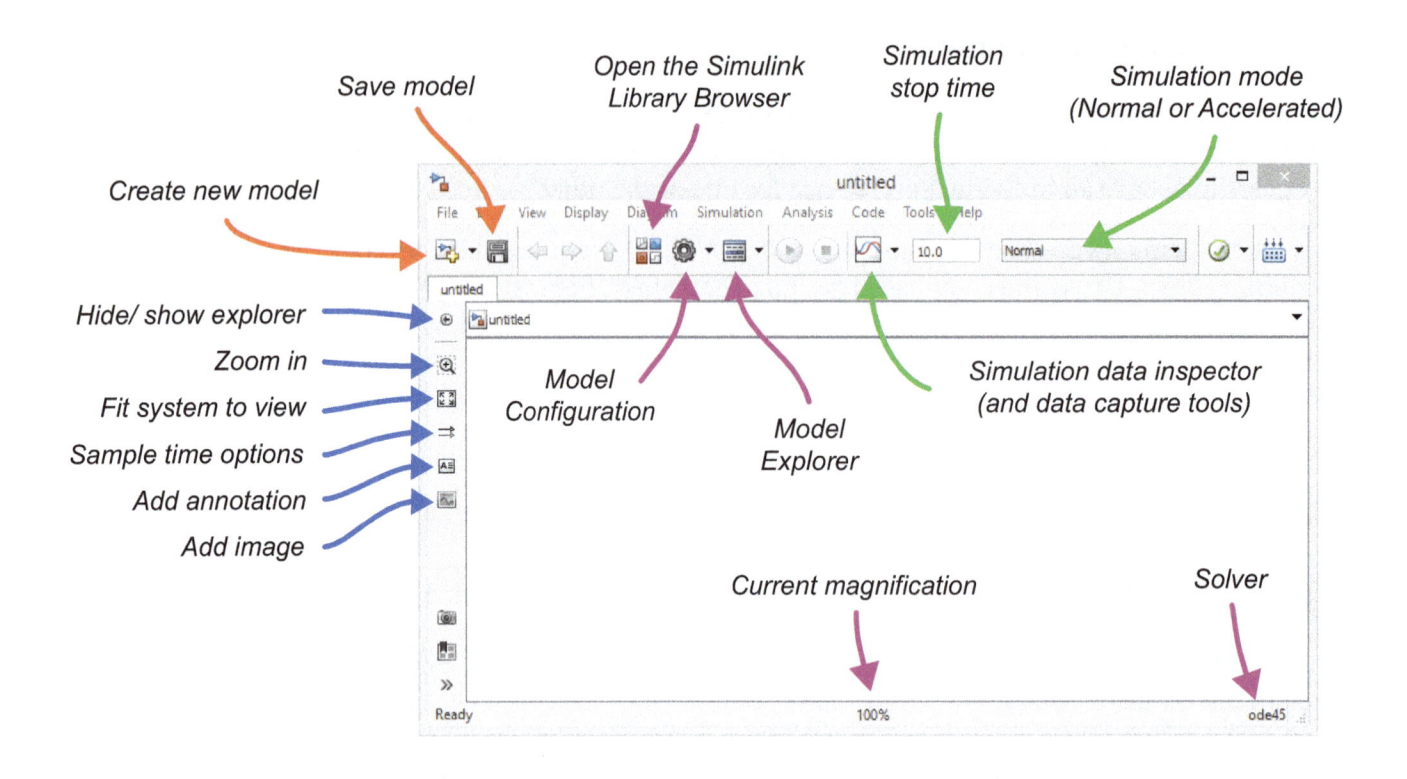

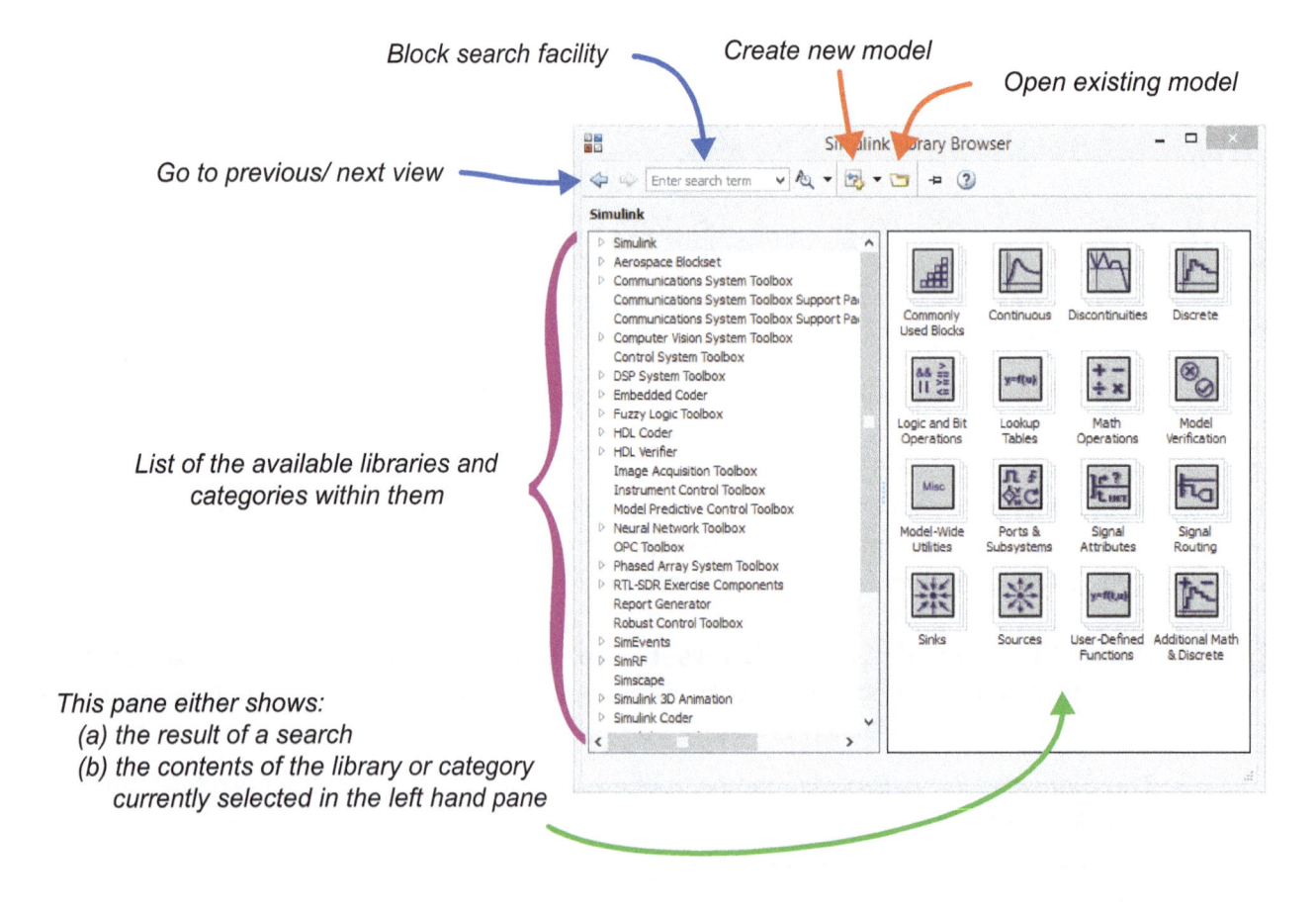

Figure 4.5: The Simulink design environment (top) and the Simulink Library Browser (bottom)

Chapter 4: Getting Started with MATLAB and Simulink

into the MATLAB command window, and then pressing *Enter* on your keyboard. Either of these actions will prompt the Simulink Library Browser to appear. Some of its important features are highlighted in Figure 4.5.

(d) **Saving the Simulink model.** Now let's return to the empty Simulink model that has been created, and save it. You can do this simply by 🔘 on the 'Save' 💾 button. You can name the model in your working directory as `first_system.slx`.

(e) **Accessing Help documentation for Simulink.** Next, type

```
doc simulink
```

into the MATLAB command window. This will open the *Help* documentation for Simulink in a browser window. From here, you can access details of the various aspects of Simulink's functionality (we only have scope for a brief introduction here), and information about functions and blocks. There is also 'getting started' guide, and a number of examples that you may wish to investigate.

This completes our basic introduction to the Simulink environment. In the next exercise, we will move on to design a first model in the blank Simulink file just created.

4.7 Creating Simulink Models

Having introduced the MATLAB and Simulink interfaces, and the libraries of blocks associated with Simulink, this section now demonstrates how to build a basic model and simulate it. At this stage, the focus is on system design and simulation in general, and the processes necessary to create and test a simple design.

Exercise 4.11 Building and Simulating a First System

This exercise walks through the construction of a very simple first system in Simulink—a single, 1 sample delay. We also review how to set the sampling rate and run a simulation.

(a) **Open and arrange the Simulink windows.** If you still have the Simulink model created during the last exercise (`first_system.slx`) open, then bring it to the front of any other windows you may have open. Otherwise, reopen the file

📁 `/matlab_simulink/first_system.slx`

Likewise bring the *Simulink Library Browser* to the front (or reopen it if necessary—refer back to Exercise 4.10 to recap), and arrange the two windows side-by-side on the screen:

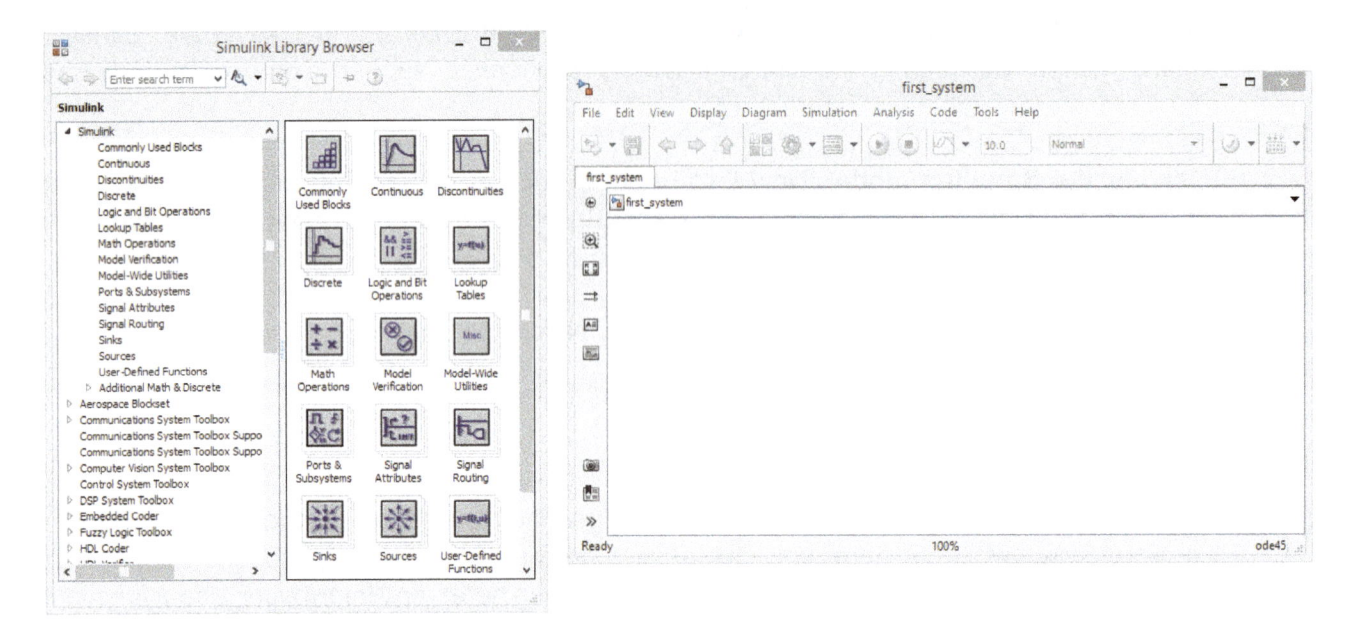

Simulink Library Browser

Simulink model (empty so far!)

🔧 **(b) Add a *Step* source block into the design.** The system to be constructed will comprise: (i) a *source*, to provide a simulation input; (ii) a *delay* element; and (iii) a *sink*, to view the simulation output. Each of these can be created by introducing a block from the Simulink library. Navigate to the 🔲 **> Simulink > Sources** category in the browser. You should note that a number of different types of source are available. Here we will use a *Step* input.

To introduce it into the design, 🖱 on the *Step* block, and drag it into the left hand side of the 'first_system' Simulink model, then release the mouse.

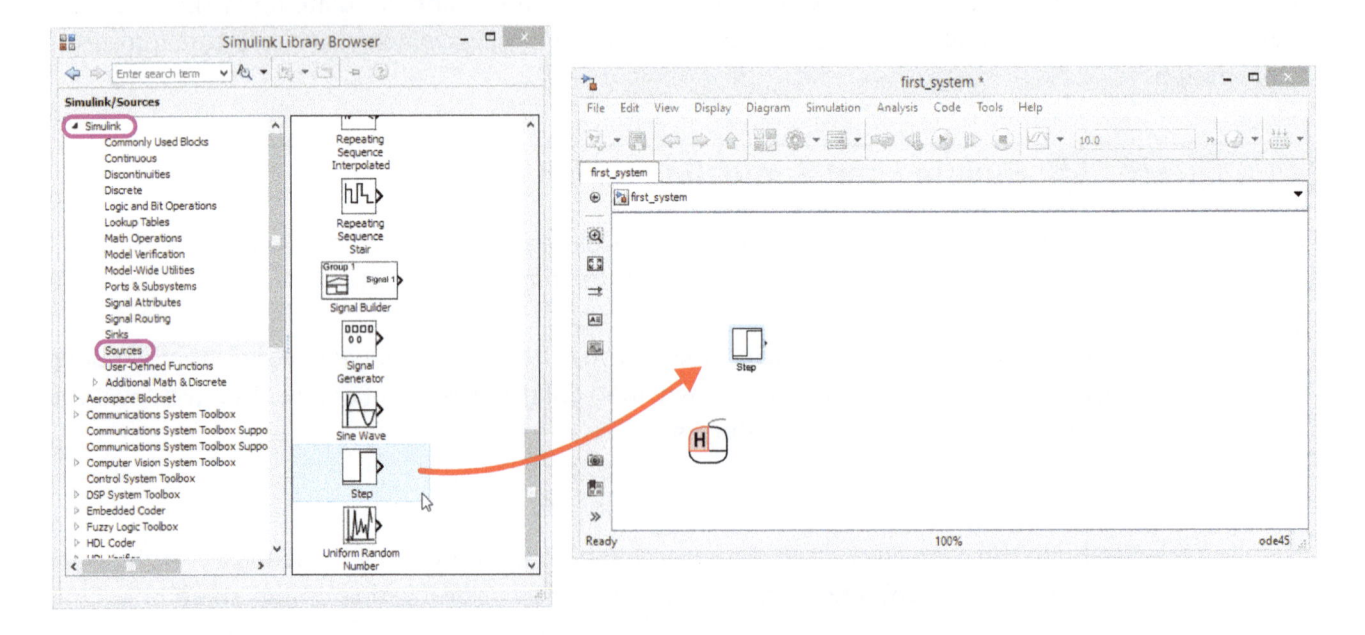

🔧 **(c) Add a *Time Scope* block.** Next, change to 🔲 **> DSP System Toolbox > Sinks**, and introduce a *Time Scope* block into the right hand side of the system, using the same method.

Chapter 4: Getting Started with MATLAB and Simulink 129

(d) **Add a *Delay* block.** Finally, we need to add in the "functional" part of the design, which in this case is simply a delay element. Navigate to ▦ > ***Simulink*** > ***Discrete***, and then drag a *Delay* block into the centre of the Simulink model, between the *Step* (source) and *Time Scope* (sink). At this point, your design should look like this:

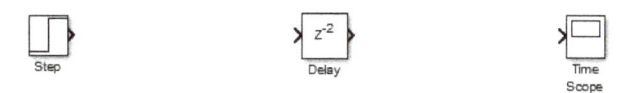

If you want to move any of the blocks around, simply [H] on the block to be moved, drag it to the desired location, then release the mouse.

(e) **Connect the blocks together.** The next step is simply to connect the three blocks together with wires. Firstly, hover the mouse over the output port of the *Step*, until you see a cross-hair appear. Then [H] and drag the mouse across to the input port of the *Delay* block. While you are making the connection, the wire will be shown as a red dashed line. Once the mouse has reached the input port of the *Delay* block, the wire will turn to a solid black arrow, indicating a valid connection. At this point the mouse button can be released and the wire will be complete.

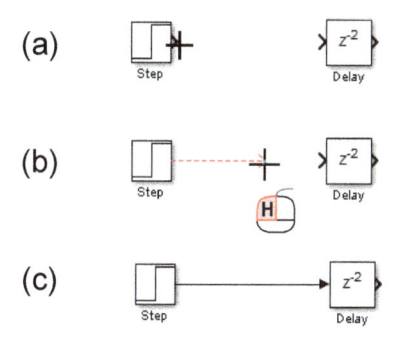

(a)

(b)

(c)

Secondly, move the mouse over the output port of the *Delay* block, and form a wire connection to the *Time Scope* in the same way. Now, the system will appear like this:

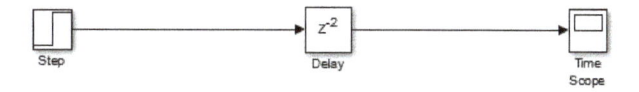

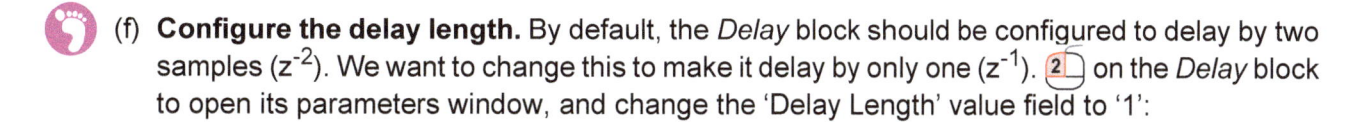

(f) **Configure the delay length.** By default, the *Delay* block should be configured to delay by two samples (z^{-2}). We want to change this to make it delay by only one (z^{-1}). [2] on the *Delay* block to open its parameters window, and change the 'Delay Length' value field to '1':

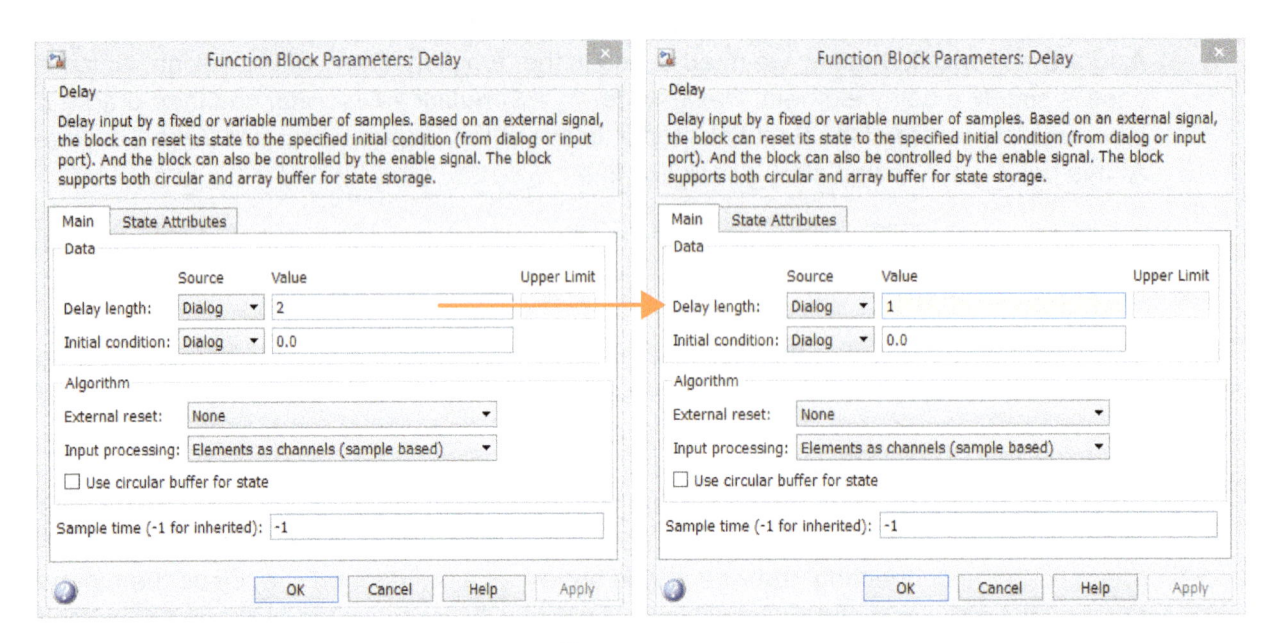

Once you have made the change, 🖱1 on *OK* to apply the changes and close the window.

(g) **Configure the sampling rate.** The designed system contains a discrete time delay (z^{-1}), and therefore it is a sampled system, and we must set the sampling rate.

Open the *Step* block by 🖱2 on it. This will open a dialogue box showing the parameter values for the step input. The initial settings are for a continuous time system (Sample time = 0), and this must be changed to discrete time. We desire a sampling rate of 1000Hz, which requires a 'Sample time' of 1/1000, or equivalently 0.001 (note that the units are seconds). The 'Step time' is the time at which the transition from the initial value to the final value occurs. It is currently set to 1, i.e. 1 second. This means that the transition would happen after 1000 samples. Change this value so that the transition takes place after 3 samples. This requires 'Step time' to be set to 3/1000, or 0.003. All of the other parameters can be left as their default values, and once the changes are made, 🖱1 on *OK* to finish.

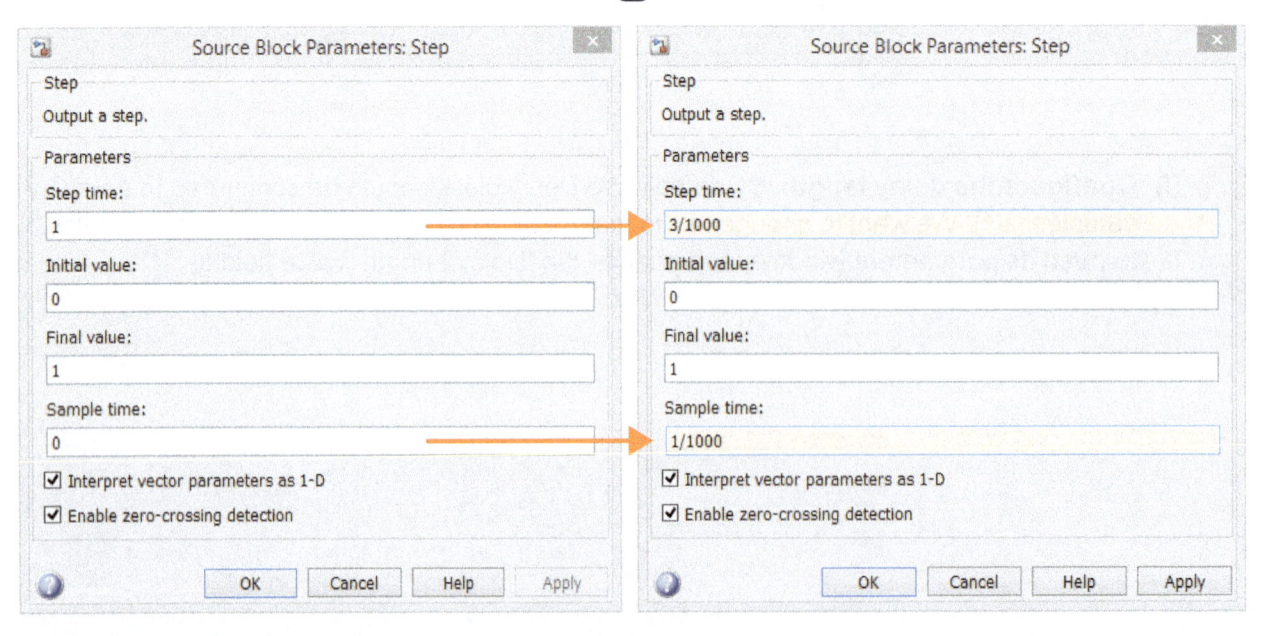

Chapter 4: Getting Started with MATLAB and Simulink

(h) **Set the simulation time.** Another aspect of timing that needs to be configured is the duration of the simulation. This is set in a box at the top of the Simulink model window. If you hover the mouse over the box, the annotation *Simulation Stop Time* appears, as shown.

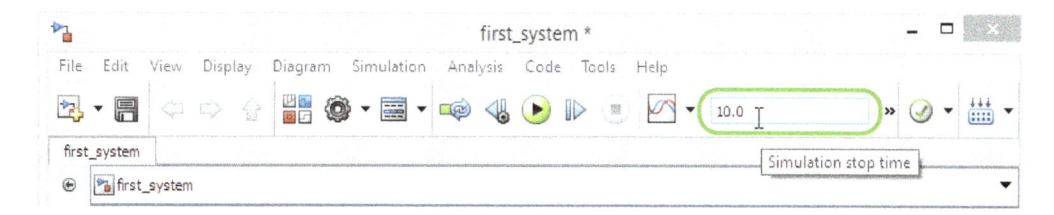

Currently the *Simulation Stop Time* is set to 10.0. The units are seconds, so this means that the duration of the simulation is 10s. At the sampling rate of 1000Hz, and with the only expected event in the simulation occurring after 0.003s, this is too long!

(i) Change the *Simulation Stop Time* to 10/1000, or 0.01. This means that the simulation will start at 0s, and finish at 0.01s.

(j) **Run a simulation.** With the parameters set, a simulation can be executed. To do this, 1 on the *Run* button ▶ at the top of the Simulink model.

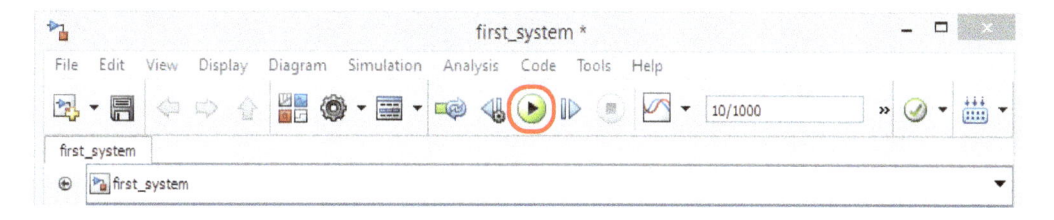

You should see the window change appearance momentarily, and then return to the original view. The simulation results have now been collected by the *Time Scope* block.

(k) **View the simulation results.** Next, 2 on the *Time Scope* block to view the simulation results. This will open the *Time Scope* waveform viewer (note that the colours may be different to what is shown here):

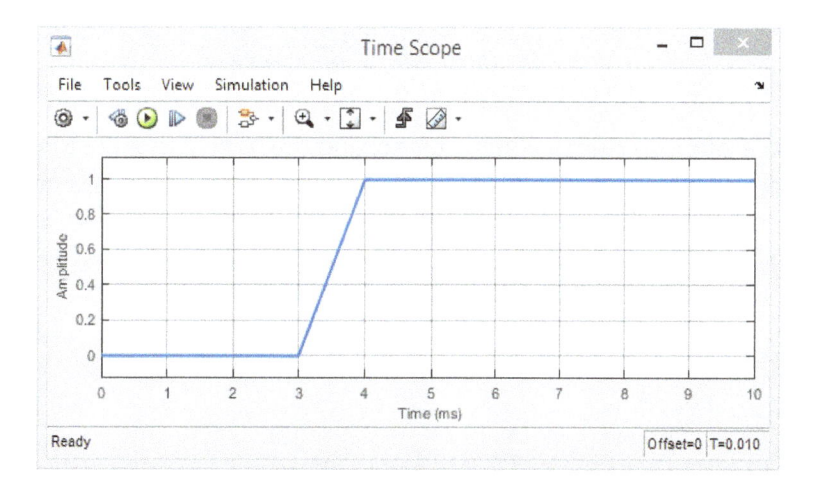

(l) **Zooming and scaling.** Here we will try out some of the features for changing the appearance of the plot. First, try zooming into the area around the transition using the zoom-X facility. 🖱 on the arrow of the 🔍 button, choose the 'Zoom-X' option,

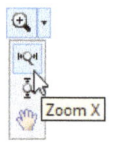

then 🖱 the mouse and drag a line along the X-axis, around the transition from 0 to 1, and release. The plot will then be scaled to a region around 3.5ms.

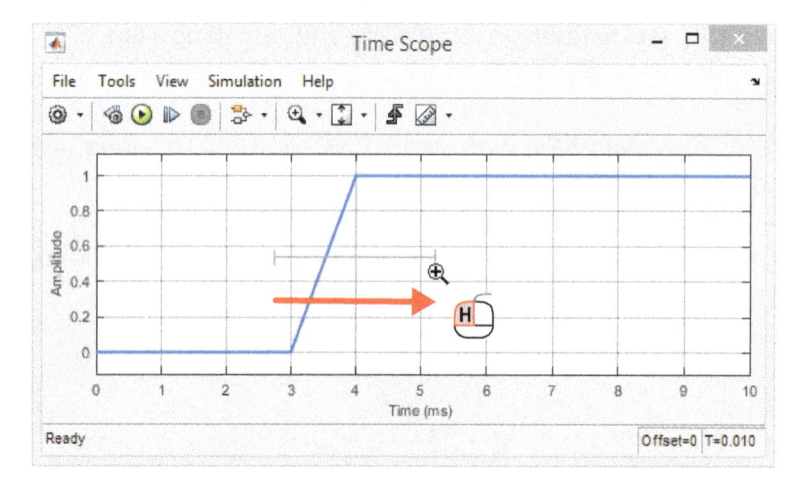

You can also zoom into the Y axis in a similar way, by selecting the zoom-Y 🔍 option, and then 🖱 and dragging the mouse around the desired range.

(m) **Rescale the *Time Scope* window.** Try 🖱 on the Scale X & Y Axes Limits button 🔳 to rescale the plot. This is found within the group of axes scaling options shown below. Note that the other two options scale the X and Y axes independently.

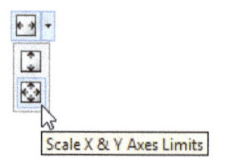

(n) **Scaling to a custom range.** Lastly, try scaling the Y axis to a custom range. You can do this by selecting *View* from the top menu, then *Configuration Properties*, then *Display*. Change the Y axis limits to a custom range of -0.5 to 1.5. You can also give the *Time Scope* window a title, if you wish. After making the changes, 🖱 on OK to finish.

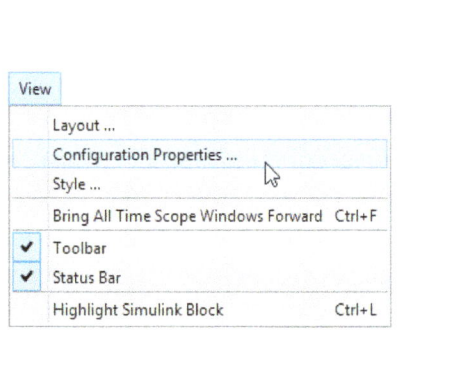

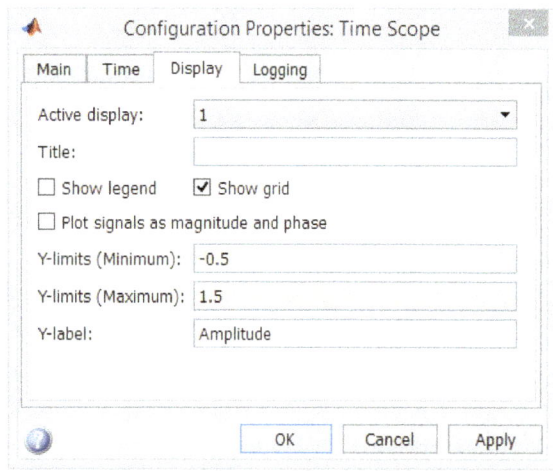

Chapter 4: Getting Started with MATLAB and Simulink

The *Time Scope* window should now look like this:

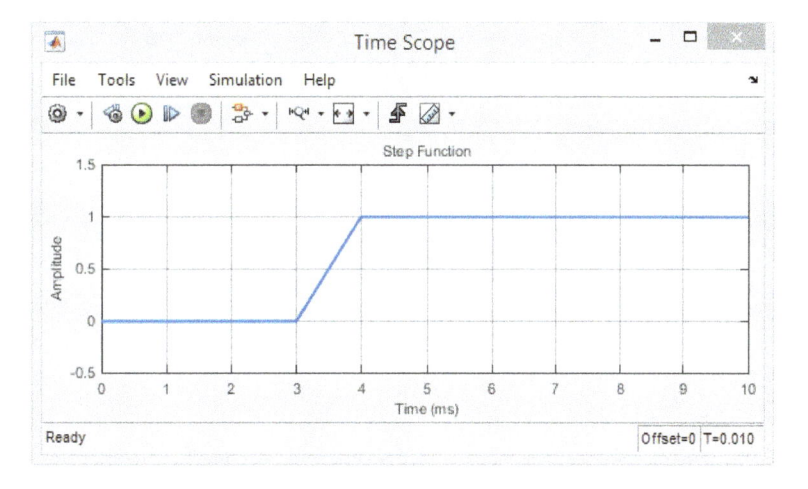

(o) **Viewing an additional signal.** At present, we can see only one signal in the *Time Scope*, i.e. the output of the *Delay* block. It would be useful to see the input to the *Delay* block too. To do this, 1 on the ⚙ icon on the toolbar of the *Time Scope* window, or alternatively, *View* from the top menu, then *Configuration Properties* (as demonstrated earlier). In the *Main* tab, change the number of input ports to 2, and then 1 on OK.

You should now notice that a second input has appeared on the *Time Scope* block in the design space:

new input

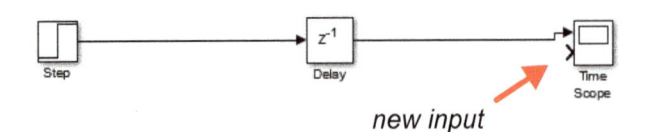

👣 (p) **View sample points.** It would also be helpful to see the individual samples on the waveform. To display sample points, choose *View* from the top menu, and then *Style*. This will open a dialogue that enables the appearance of a *Time Scope* plot to be customised. The colours used for the figure background, plot axes and labels can be specified. The line colour, pattern, and width can also be chosen, along with the shape of the markers (which indicate sample points). By default, markers are set to 'none', which explains why we could not see markers in the original plot! Set the *Marker* to the circle style, then 🖱 on *OK*.

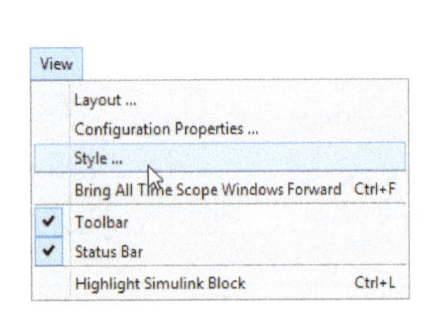

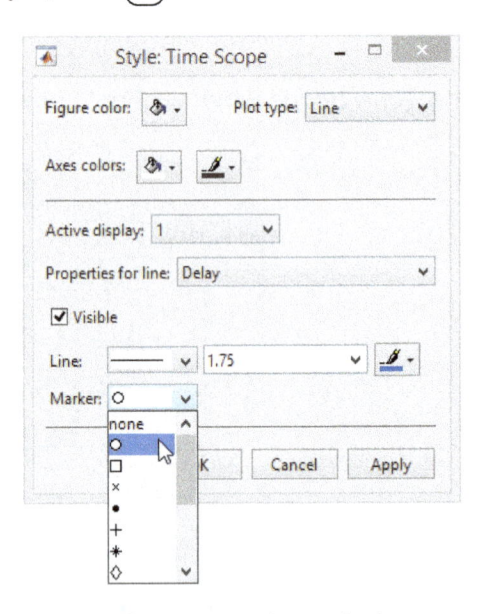

The samples should now be clearly visible in the *Time Scope*, as shown below.

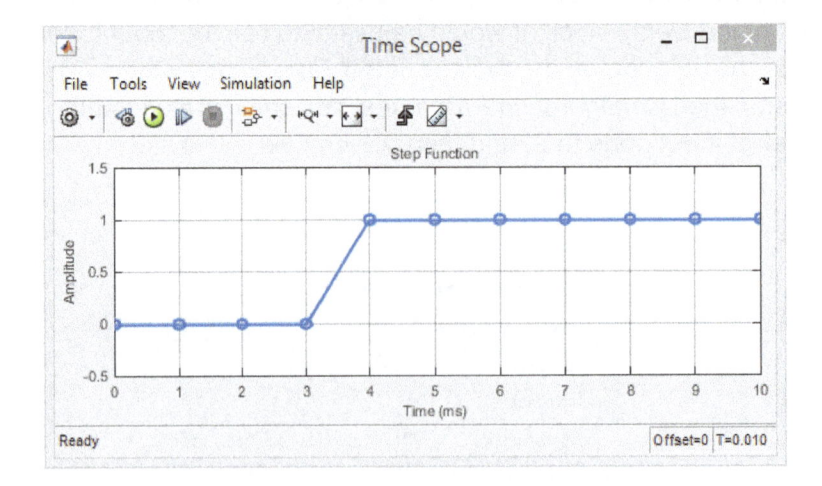

🔧 (q) **Connecting the second input to the *Time Scope*.** The last few steps have changed the configuration of the *Time Scope*, but it still only has one input connected to it! Check the design space to confirm this. Next, we will connect the output of the *Step* function to the second input of the *Time Scope*. There are two easy ways to add the new wire.

Chapter 4: Getting Started with MATLAB and Simulink

Method (1) — Destination Back to Source: and drag the mouse from the input of the *Time Scope* block, to a point on the wire between the *Step* and *Delay*. Release the mouse when the wire becomes a solid black line.

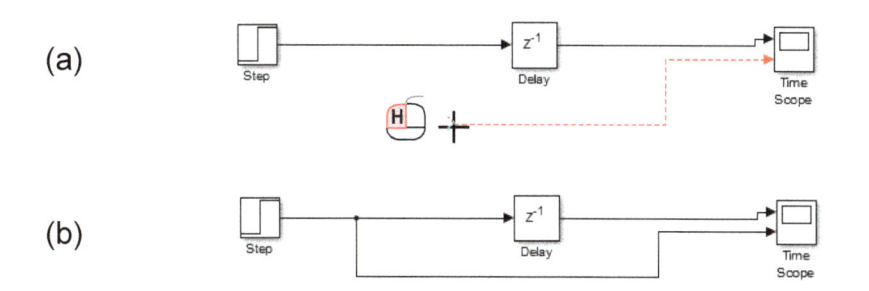

(a)

(b)

Method (2) — Source to Destination: Using the right hand mouse button, and drag from the wire connecting the *Step* and *Delay*, to the input of the *Time Scope*. Release the mouse when the wire becomes a solid black line.

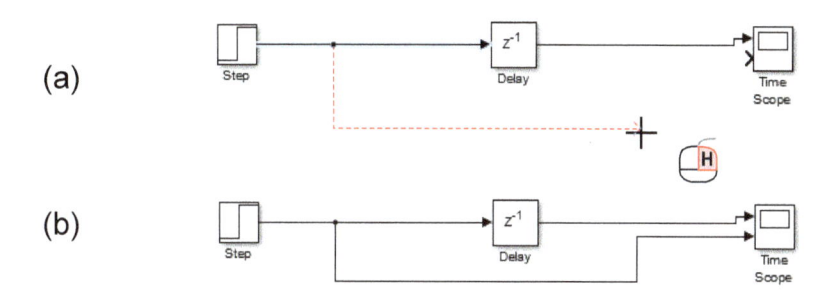

(a)

(b)

(r) **Re-run the simulation.** If you run a simulation, you should notice that a second waveform is visible in the *Time Scope*.

(s) **Add a legend.** As there are now two signals shown in the *Time Scope*, it would be useful to add a legend. You can do this via the *Configuration Properties* dialogue introduced earlier — go to the *Display* tab, and select the *Show legend* check box.

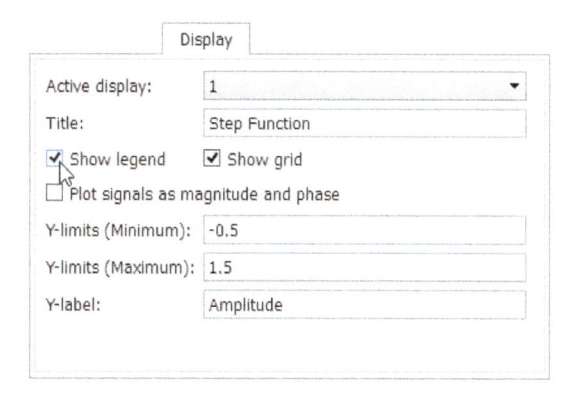

(t) **Change the appearance of the new signal.** Initially, the 'new' waveform does not have markers (samples), so repeat the steps from earlier to ensure that the markers are shown for *both* signals. It would be a good idea to choose a different shape of marker for the new signal (i.e. not a circle this time), so that it is easier to differentiate between the two signals. This will require you to select the 'Step' signal in the *Style* window, as shown below, before choosing your preferred type of marker.

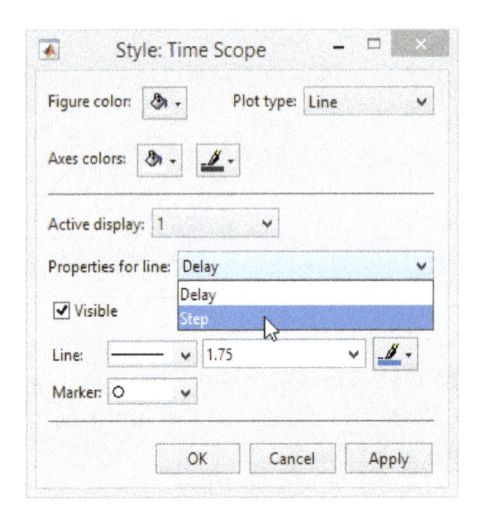

(u) **Rerun again!** Run the simulation to view the effects of your changes. Confirm that the *Delay* block has indeed introduced a 1 sample delay to the output of the *Step* source. Note that the colours you see may be different!

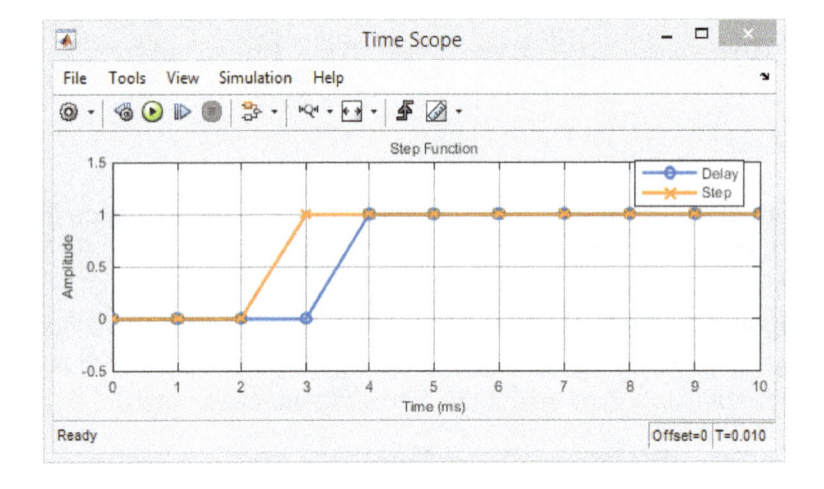

(v) **Configure the *Time Scope* with 2 Axes.** The last change we will make to the *Time Scope* is to view the signals on two separate axes. This can be achieved by going to the *View* Menu, then *Layout*. Hover your mouse over the second from top box on the left hand side, such that the top two boxes on the left are highlighted (as shown below), then 🖱.

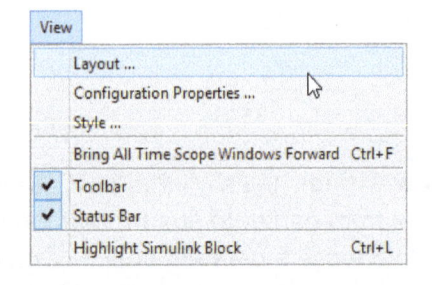

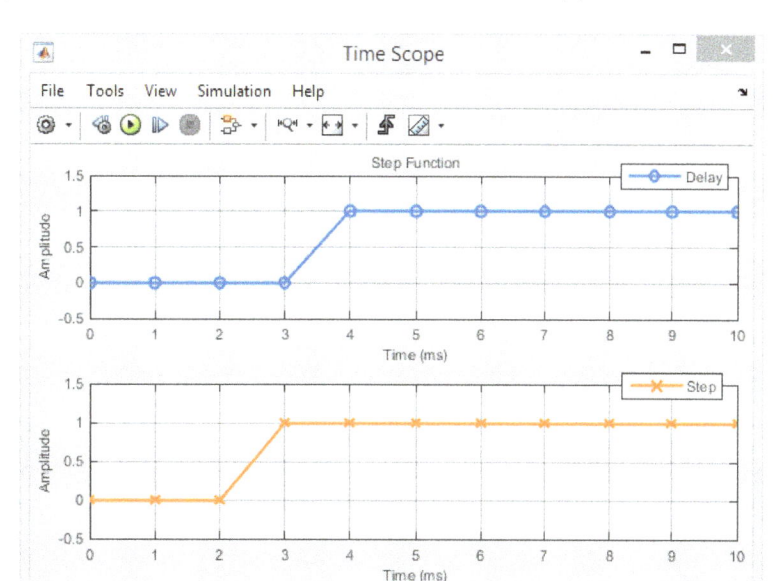

Chapter 4: Getting Started with MATLAB and Simulink

The *Time Scope* should now show the two waveforms on upper and lower axes, similar to this:

You can switch off the legend or rescale the axes if you wish.

(w) **Changing the *Delay* Parameter.** The simulations so far have been of a system with a 1 sample delay, while the input signal was a step with a transition after 3 samples.

Try changing the parameters of the *Step* and *Delay* blocks, such that the *Step* occurs after 4 sample periods, and the *Delay* block applies a 3 sample delay to the step input.

(x) **Re-run!** Remember that you will have to resimulate to view the results!

(y) **Save!** Save your work once finished. If you are carrying on to Exercise 4.12 now, keep the model open as it will be required again!

This exercise has demonstrated a number of simple techniques for building models in Simulink, and simulating them. These methods will be used extensively (and built upon) later in the book, so it may be worth bookmarking this exercise for reference later.

Exercise 4.12 Manipulating Blocks and Wires

Having built our first model, the next aim is to briefly introduce methods for copying and pasting blocks, positioning and deleting them, and for deleting wires. It is also shown how to name wires.

(a) This exercise follows on directly from the end of Exercise 4.11, using the same file. Therefore, if it is not already open, reopen the file:

   `/matlab_simulink/first_system.slx`

(b) **Create a second *Delay* block.** We start by making a copy of the *Delay* block from the original system. There are two simple ways to do this.

Method (1): 🖱 on the original *Delay* block, and choose 📋 *Copy* from the menu that appears.

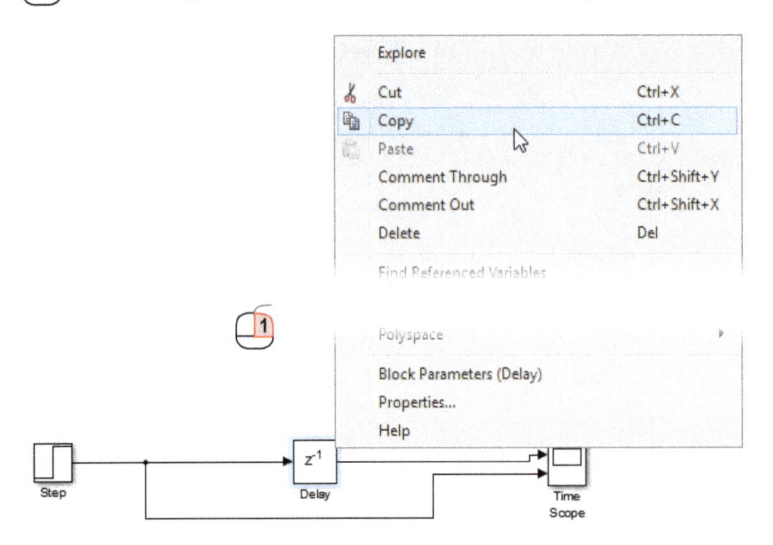

Then 🖱 again in the desired location, and this time choose 📋 *Paste* from the menu. A copy of the *Delay* block will appear with the name *Delay1*. (Note that keyboard shortcuts [*Ctrl + C* for copy; *Ctrl + V* for paste] could be used to achieve the same outcome.)

Method (2): An alternative technique uses a single mouse action. Simply 🖱 on the *Delay* block, and move the mouse away, releasing the button where you would like the new *Delay* block to be positioned.

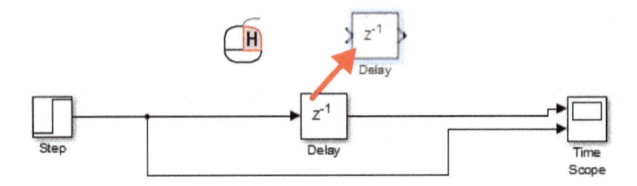

👣 (c) **Deleting a block.** If you tried out both of the methods from the last step, you will now have two new, unconnected *Delay* blocks in the system. We will now mention two methods of deleting a block. Simply return to the previous step to create more blocks if you need to, in order to practice deleting them!

Method (1): 🖱 on the block to be deleted, which will select it. At this point, the block should be highlighted with a light blue border. Then press *Delete* or *Backspace* on your keyboard to remove the block.

Method (2): 🖱 on the block to be deleted, and then choose *Delete* from the menu that appears.

👣 (d) **Dropping a block onto a wire.** In this step we will add a *Delay* into the existing system by directly 'dropping' it onto a section of wire. First, ensure that you have a spare *Delay* block

Chapter 4: Getting Started with MATLAB and Simulink 139

available (return to the earlier, block-copying step if need be). Next, [H] on the *Delay* block and drag it onto the wire in the position shown below.

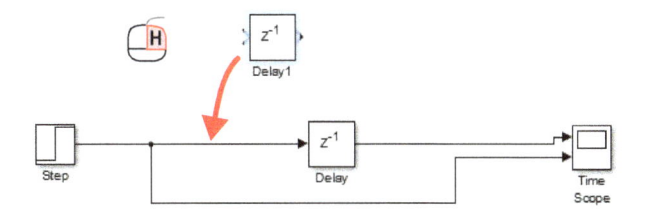

When the *Delay* block is positioned directly over the wire, it will be shown connected (highlighted in blue). At this point, the mouse button can be released, and the connection will be made.

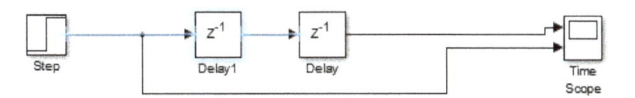

If necessary, blocks can be repositioned by simply [H] and dragging to the desired location.

(e) **Selecting and deleting wires.** Wires can be selected just by [1] on them. When selected, wires turns blue as indicated below.

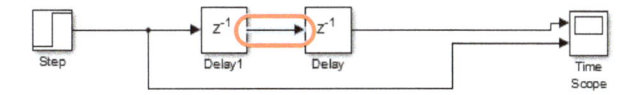

Once selected, a wire can then be deleted by pressing the *Delete* or *Backspace* key on your keyboard. You can also delete it by [1] on it and then choosing *Delete* from the menu. In other words, wires can be deleted using the same two methods as blocks. You can also delete wires and blocks at the same time, by [H] and dragging around all of the items to be removed.

(f) **Naming wires and blocks.** Sometimes it is useful to name a wire. This can be done by [2] on the wire and then typing in the box that appears adjacent to it. For example, below we name the signal as 'delayed'.

Names of wires propagate into sinks etc., which assists with the viewing of simulation results. You can also name (or rename) blocks, by [1] on the existing name and then editing the text.

(g) **Adding annotations.** Labels and notes can be added to the design space easily. To do so, simply [2] on the background of the design space, at the position where you would like to add a note. This should be away from wires and blocks. An edit box will then appear, into which you can type your desired text.

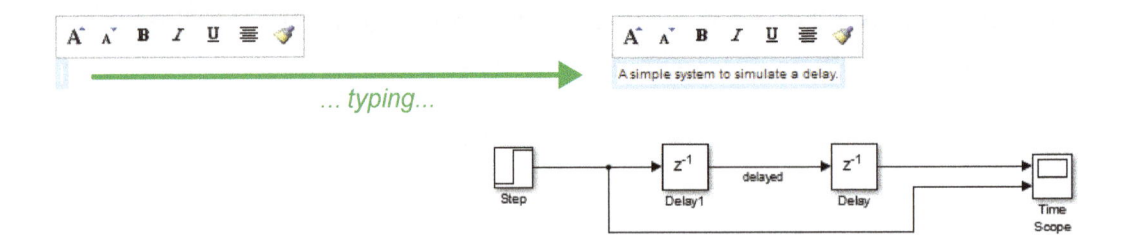

The formatting of the text can be changed (size, format, justification, etc.) using the icons visible above the text box. Further options are available by ④① on the text box and choosing *Format* from the menu, followed by *Font Style*. This will allow you to change the font, for instance.

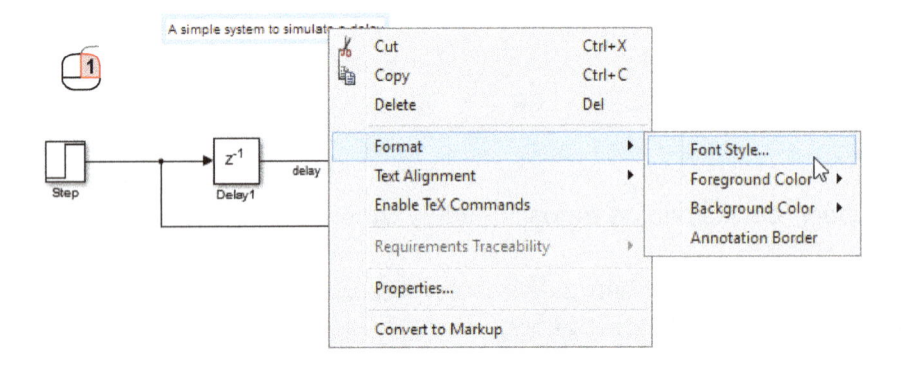

(h) **Customising colours.** It is easy to change the colours of blocks and annotations to suit user preferences. Colours can be chosen for two aspects:

Foreground colour — the text in a text box;
the border of a block, and graphical annotations within it.

Background colour — the fill colour of a text box;
the fill colour of a block.

Both of these options can be accessed by ④① on the item to be re-coloured, and choosing the *Format* option from the menu, followed by *Foreground Colour* or *Background Colour*. You can then either choose from the default colour selection, or define your own custom colour using the palette.

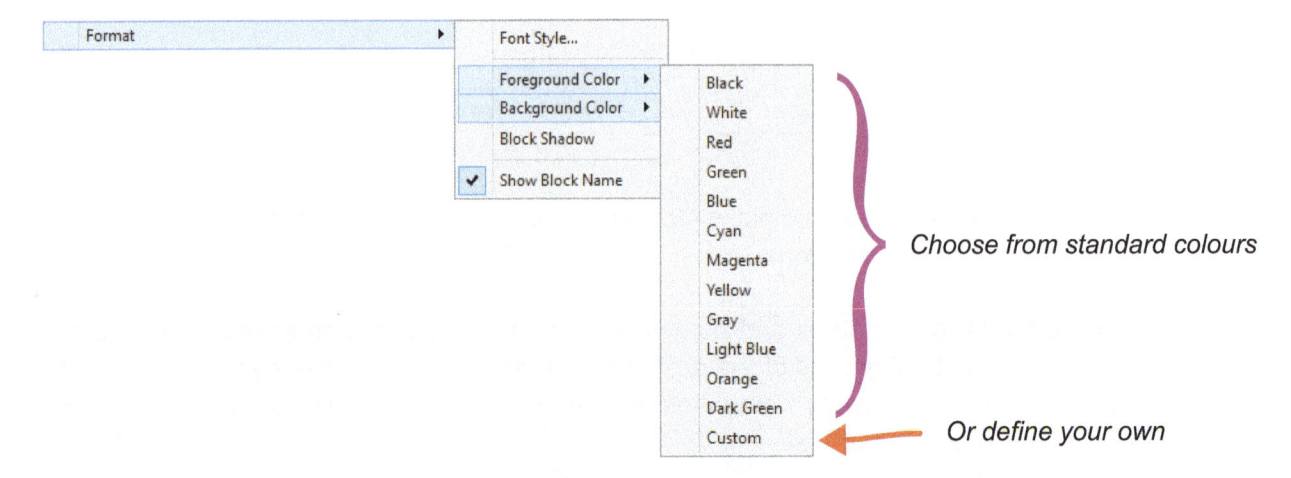

Choose from standard colours

Or define your own

Chapter 4: Getting Started with MATLAB and Simulink

(i) **Colour and annotate your model.** Spend a little time trying out the methods demonstrated in this exercise, by annotating your model and customising the colour scheme.

(j) **Save!** Once you are finished, save the modified model as:

   `/matlab_simulink/first_system_delay2.slx`

Exercise 4.13 — Commenting Blocks Out, and Commenting Through Blocks

Simulink includes a feature that allows Simulink blocks to be functionally removed from a model, without physically deleting them. This can be useful during design iterations when developing a model. This exercise briefly reviews the use of these 'commenting' features—so-called because they are analogous to 'commenting out' lines or sections of code in a script.

(a) Open the file:

   `/matlab_simulink/simulink_commenting.slx`

(b) **Inspect the system.** View the system and confirm that it implements some simple arithmetic, with the results displayed using *Time Scopes*.

(c) Run a simulation and view the results.

(d) **Commenting.** Next, [1] on the lower *Time Scope* block, and choose the option to *Comment Out*.

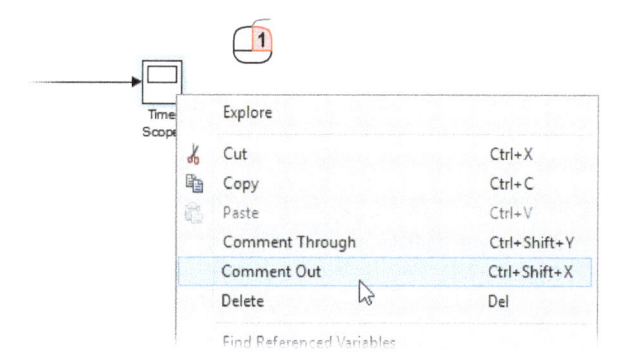

(e) **Re-run the simulation.** Open both *Time Scopes* to view the results. What happens?

(f) **Uncommenting.** Try commenting the *Time Scope* back in again (i.e. [1] and choose the option to *Uncomment*), then re-simulate.

(g) **Commenting through.** Next, [1] and [H] the mouse, dragging around and selecting the two *Gain* blocks, then [1] and choose *Comment Through*.

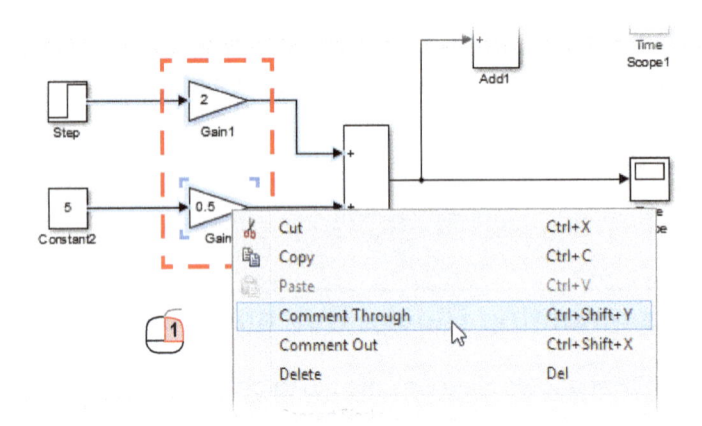

(h) Run a simulation and view the results, and hence confirm the effect of your 'commenting'!

4.8 Variables and Parameters

Parameterisation is an important aspect of achieving flexible and reusable models. Variables can be created in the MATLAB Workspace, and used to set the values within Simulink blocks (variable names can be entered directly into the parameter boxes, in place of numerical values), and to define other aspects of the model, such as the *Simulation Stop Time*.

Another useful technique is to write output data from Simulink simulations to the MATLAB Workspace, which creates a new variable (or set of variables). These new variables can subsequently be used to generate plots of the simulation outputs, or to perform other analysis on simulation results.

Exercise 4.14 Working with Variables and Parameters

In this exercise, we build upon the original *Step-Delay-Time Scope* system by introducing the use of variables to parameterise the system. This is a very simple design, but in more complex cases, parameterisation becomes increasingly useful—it means that changes can be propagated throughout a model quickly and reliably.

(a) **Reopen the Step-Delay-Scope model.** For this exercise, we will return to the single delay model from Exercise 4.12. Open the file:

/matlab_simulink/first_system.slx

(b) **Introducing a variable for sampling rate.** The initial design was based on a sampling rate of 1000Hz. This value was used to set the parameters in the *Step* source block, and also to set the *Simulation Stop Time*.

It would be helpful if a variable could be created in the MATLAB Workspace to represent the sample rate. This would allow values of time to be set with respect to the sample rate. As demonstrated in Exercise 4.1, such a variable can be created by typing

Chapter 4: Getting Started with MATLAB and Simulink

```
fs = 1000;
```

at the MATLAB command window, and then pressing *Enter*. If you inspect the main MATLAB window, you should now see the variable fs listed in the Workspace pane.

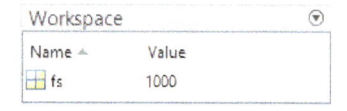

Name	Value
fs	1000

Note: there may also be one or more other variables present, resulting from previous simulations.

(c) **Parameterising the model using the variable `fs`.** It is now possible to set the sampling rate and step time of the *Step* source block using the variable `fs`. Open the *Step* block, and set the 'Step time' and 'Sample time' parameters to the following:

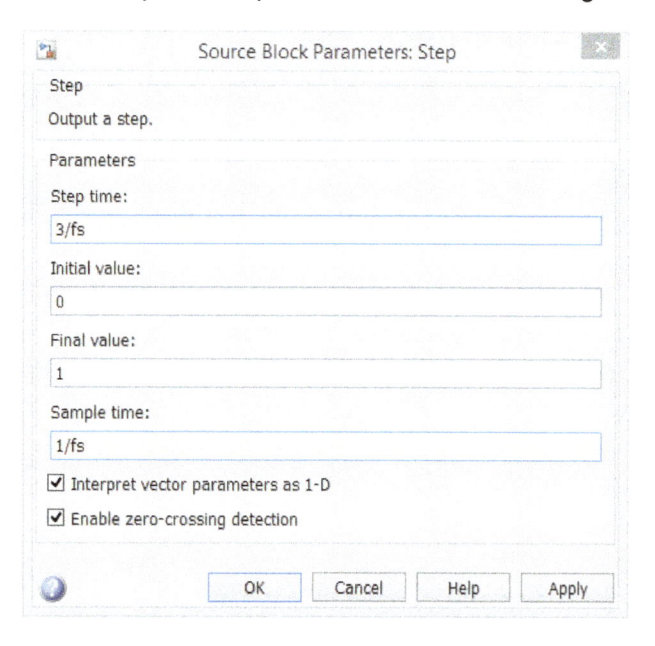

Source Block Parameters: Step

Step
Output a step.

Parameters

Step time:
3/fs

Initial value:
0

Final value:
1

Sample time:
1/fs

☑ Interpret vector parameters as 1-D
☑ Enable zero-crossing detection

then click on *OK* to complete.

(d) **Simulate!** Try simulating the modified system, to ensure it operates correctly (the observed behaviour should be the same as before).

(e) **Set the simulation stop time as a parameter.** Now change the *Simulation time* (at the top of the model) so that it is defined in terms of `fs` too:

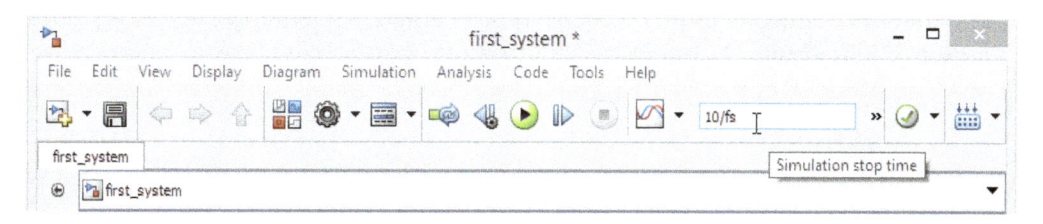

first_system *

10/fs   Simulation stop time

143

(f) Resimulate again, and ensure that the correct behaviour is maintained. Note from the *Time Scope* waveform that the simulation ends after 0.01s, as before. In other words, the parameter is understood, and defines the functionality of the model as intended.

(g) Next, try setting the value of `fs` to 500Hz by typing

```
fs = 500;
```

into the MATLAB command window.

(h) Resimulate and see what happens!

(i) **Introducing a parameter for *Delay*.** Having set and used a parameter for the sampling rate, there is no reason why we can't parameterise other aspects of the system as well.

Try setting a parameter for the delay. Type:

```
delay = 3;
```

at the MATLAB command window, and verify that this new variable is stored in the MATLAB workspace.

(j) Open the *Delay* block, and replace the numerical value with the name of the variable just set:

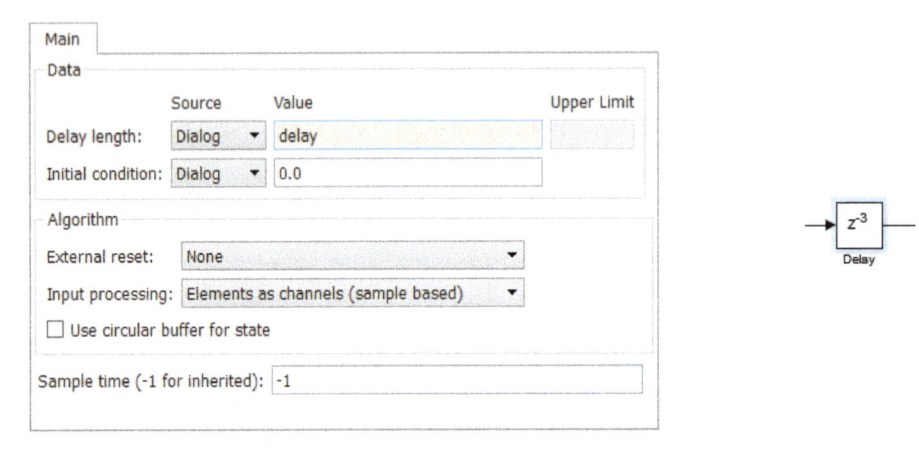

Note that the value of delay shown on the *Delay* block changes to z^{-3}.

(k) Simulate the design, and check the simulation results shown in the *Time Scope*, to ensure that a 3 sample delay has been correctly implemented. Try setting the `delay` parameter to another value, if you wish.

Chapter 4: Getting Started with MATLAB and Simulink

Exercise 4.15 Setting up Variables as a Model Loads

Variables were successfully used as parameters in Exercise 4.11, but these had to be created beforehand by manually setting them in the MATLAB command window. This form of parameterisation is not particularly amenable in terms of ease-of-use, or reusability of Simulink models. In this exercise, we first demonstrate what the issue is, and review a convenient method of creating and setting variable values as the model loads.

(a) **Clarifying the problem...** Here we simply note the disadvantages of setting up variables manually via the command window.

First, save and close the `first_system.slx` model that you have been working with. Then, type:

```
clear all;
```

at the MATLAB command window. You should notice that this command removes all variables from the Workspace. This means that the previously set variables, `fs` and `delay`, have now been deleted and are unavailable for use.

(b) Now reopen the `first_system.slx` model. Note that the *Delay* block shows a different annotation, indicating a delay of `d` clock cycles...

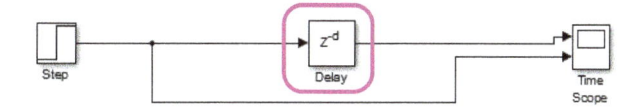

This means that the delay cannot be expressed as a numerical value. In fact, because the variable `delay` has been deleted and not reinstated, the block is unable to determine what the actual number of sample delays should be.

(c) **Simulate!** Now, attempt to simulate the system. You should encounter an error message like this one:

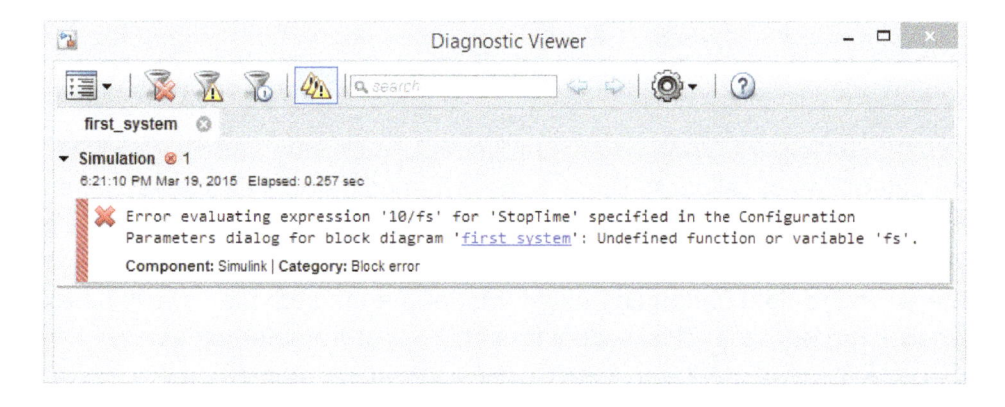

The error occurs because the variable `fs` is no longer available. The simulator is unable to execute a simulation, because the duration of the simulation cannot be resolved.

A possible resolution to these problems would be to repeat the process used in Exercise 4.14, i.e. to enter values for `fs` and `delay` at the MATLAB command window (again). However, this would be a little tedious. It would be even more tedious in a large model with many variables to configure! Fortunately there is a better way...

(d) **Setting up Pre-Load commands.** In this step, we incorporate commands into the Simulink model, to set up the required variables at the point of loading the model.

Open the *File* menu, and choose the *Model Properties* option, followed by *Model Properties* again.

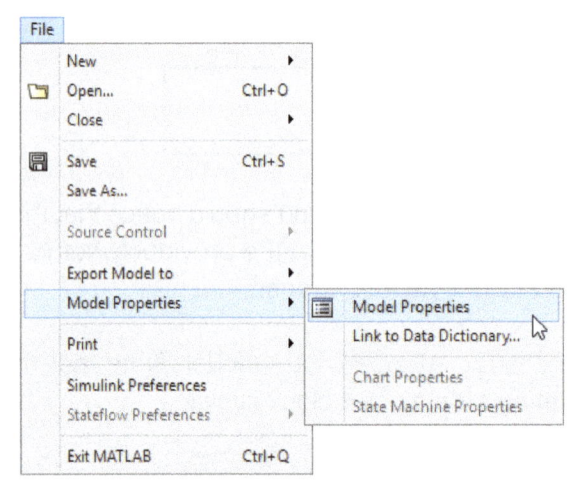

This will open a new window. Move to the *Callbacks* tab and select the *PreLoadFcn* callback:

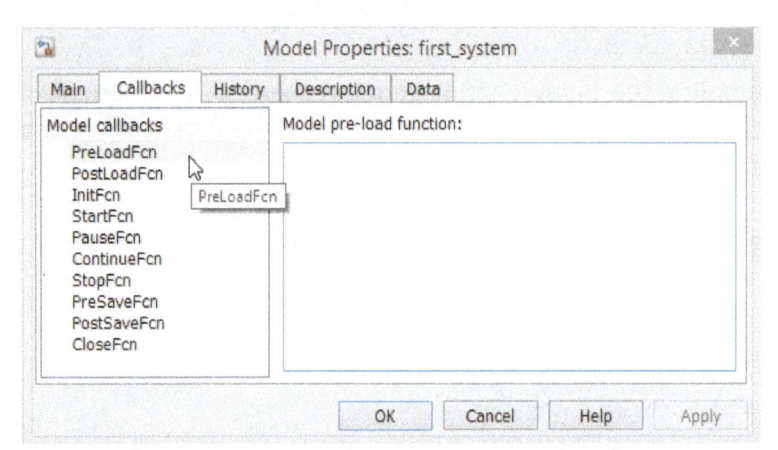

The categories of callbacks listed in the left hand, 'Model callbacks' pane relate to particular phases in the use of a Simulink model. This is reflected by their names. We have chosen the *PreLoadFcn* because it relates to the phase before the model loads. Any commands that are entered in the right hand pane (titled 'Model pre-load function:'), will be executed before the Simulink model loads. This is ideal for setting up parameter variables.

(e) **Enter PreLoadFcn code.** Type the same commands as used previously into the box, and then 1 on *OK*. (Note that the background is yellow while there are unapplied changes.) Now,

Chapter 4: Getting Started with MATLAB and Simulink

whenever the `first_system.slx` model loads, it will execute these commands first, and the sampling rate and delay will be setup correctly, *without* the need to type any further instructions.

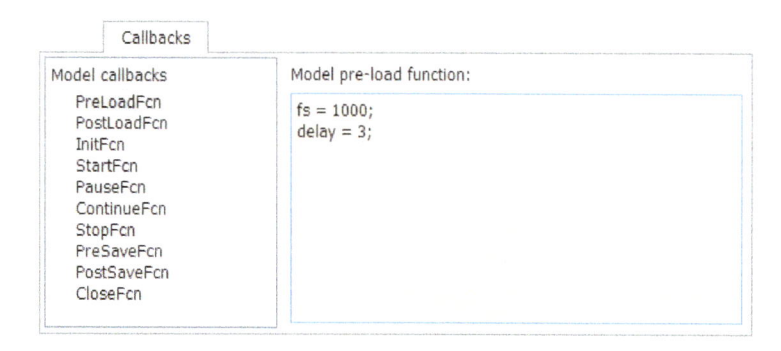

```
Callbacks
Model callbacks          Model pre-load function:
   PreLoadFcn
   PostLoadFcn              fs = 1000;
   InitFcn                  delay = 3;
   StartFcn
   PauseFcn
   ContinueFcn
   StopFcn
   PreSaveFcn
   PostSaveFcn
   CloseFcn
```

(f) **Test the PreLoadFcn.** We will now test out the *PreLoadFcn* functionality. By its very nature, the *PreLoadFcn* is only executed as the model loads, so we must close and reopen the model to properly test it.

Close the `first_system.slx` model, and then type:

```
clear all;
```

at the MATLAB command window, to make sure that no variables are retained in the MATLAB Workspace.

(g) Open the model again, and confirm that the `fs` and `delay` variables are initialised in the MATLAB Workspace. The *Delay* block will show a delay of (z^{-d}). Update the model diagram by 🖱 on *Update Diagram* in the *Simulation* toolbar.

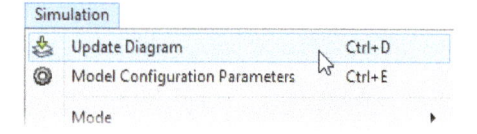

```
Simulation
   Update Diagram                  Ctrl+D
   Model Configuration Parameters  Ctrl+E
   Mode                            ▶
```

Note that the *Delay* block is now properly configured with the desired delay. Lastly, a simulation of the system should be executed successfully, without any error messages!

Exercise 4.16   Writing Simulation Results to the Workspace

The Simulink Library contains a number of different Sinks, and the DSP System Toolbox and Communications System Toolbox provide further, more specialist options. We do not have the opportunity to review all of them here, but one worth highlighting in particular is the *To Workspace* block, which is part of the main Simulink library. This provides the facility to write simulation results into the MATLAB Workspace, from where they can be further manipulated, plotted, saved, etc.

(a) **Open the model.** If it is not already open, reopen the model

📁 `/matlab_simulink/first_system.slx`

(b) **Introduce and connect a *To Workspace* block.** Open ▦ > **Simulink** > **Sinks**, then 🖱 and drag a copy of the *To Workspace* block into the system.

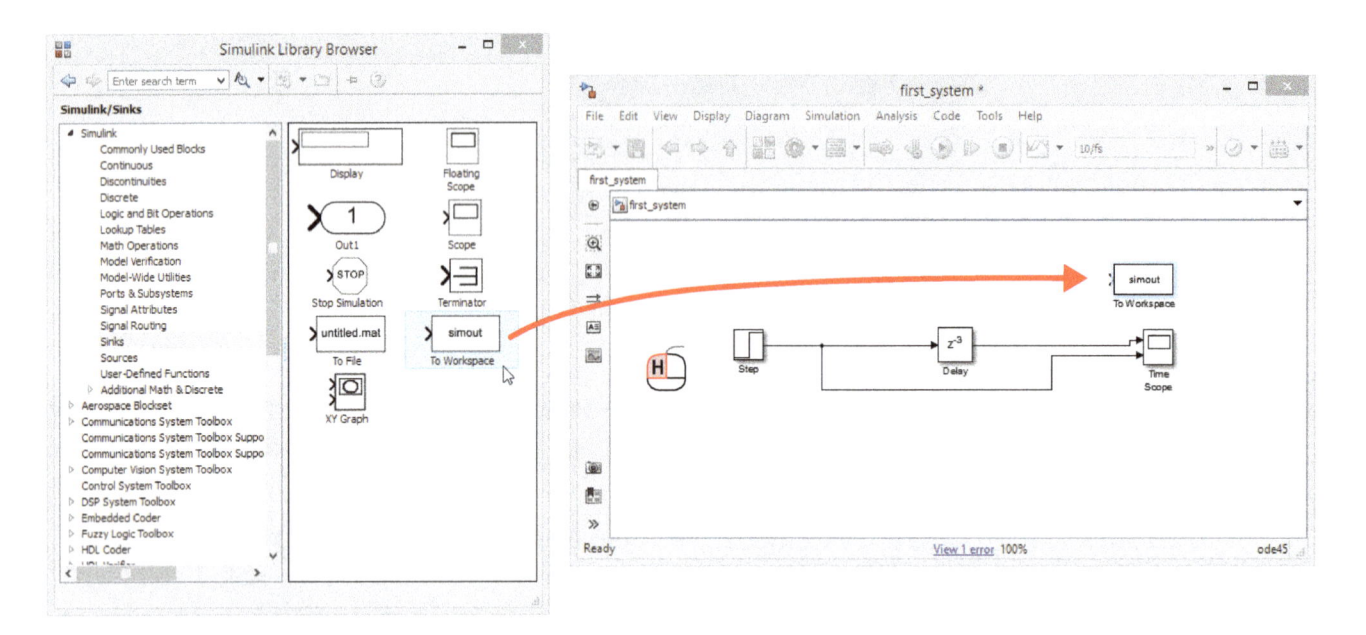

Connect the *To Workspace* block to the wire linking the *Delay* and *Time Scope* blocks — we will use the *To Workspace* block to collect the data output by the *Delay*. The system should now look as follows:

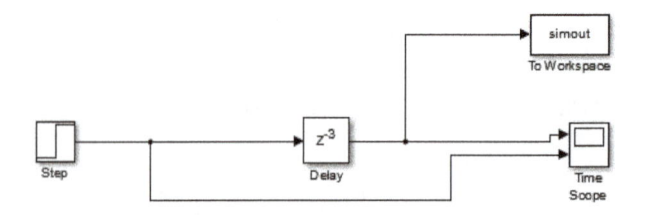

(c) **Configure the *To Workspace* block.** The *To Workspace* block creates a variable with a specified name in the MATLAB Workspace. The default variable name is 'simout', as shown in the screenshot, but it is easy to change the name to something more meaningful. The format for saving the data can also be chosen from a small set of alternatives.

First, 🖱 on the *To Workspace* block to open it. Change the name to 'del_samples', and the save format to 'Structure With Time', as shown on the following page.

The 'Structure with Time' format means that the data and corresponding sample times are captured and stored in a structure (recall that the general concepts of structures were introduced in Exercise 4.8).

The structure contains three fields:
        (1) the name of the *To Workspace* block;
        (2) the time series, i.e. an array of the simulation time steps at which data is captured;
        (3) a sub-structure containing the captured simulation data.

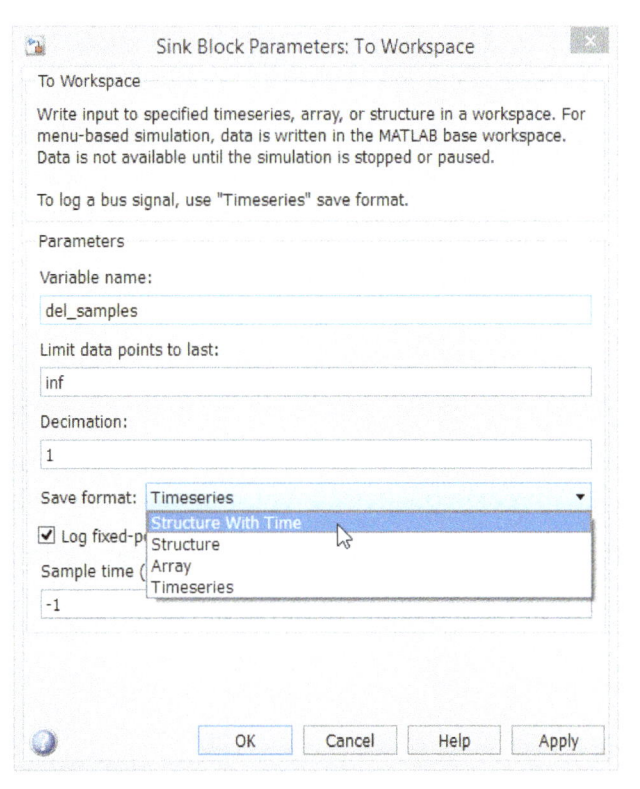

Chapter 4: Getting Started with MATLAB and Simulink

[1] on OK once finished.

(d) **Execute a simulation and capture the delayed signal.** Run a simulation, and confirm that a structure called del_samples has been created in the Workspace.

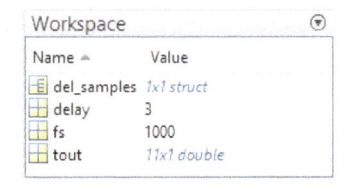

You can open and explore this structure by [2] on it. The structure will open in a 'Variables' pane, where you can further inspect its individual elements—for instance, [2] on time to view the array of simulation time steps.

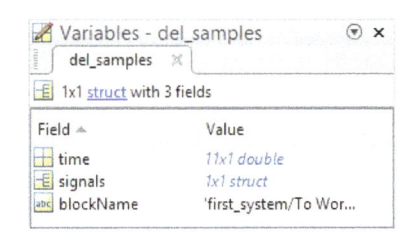

If you [2] on the signals structure, signals, then likewise you will see the content of this structure in the 'Variables' pane. Once the `signals` structure has been opened, the array of sample values can then be inspected by [2] on the values array.

(e) **Plot a graph using the `del_samples` Workspace data.** Now that the simulation data is present in the Workspace, it can be plotted in a figure.

Type the following MATLAB code, either at the command window, or into a MATLAB `.m` script, and execute it:

```
figure(1)
stem(del_samples.time, del_samples.signals.values);
xlabel('Time');
ylabel('Amplitude');
```

Note that you can adapt the code in terms of line and marker styles, etc., if you wish. Refer back to Exercise 4.5 for some guidance on how to customise the appearance of figures.

The result should be a stem plot of the captured data, similar to this:

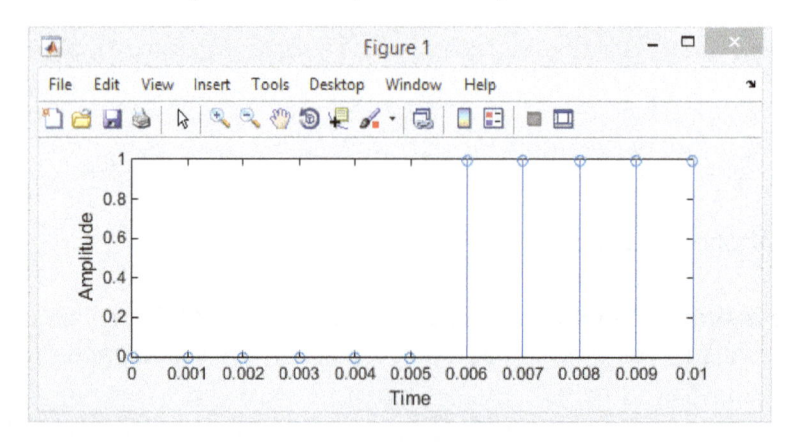

Although this is a very simple example, it should be clear that using the *To Workspace* block brings data into MATLAB, and thus opens up the full plotting capabilities of MATLAB. The possibilities are extensive!

Note that you could also add these commands to the *StopFcn* Callback (refer back to Exercise 4.15, where we entered commands under the *PreLoadFcn* — the same procedure is required, just using *StopFcn* rather than *PreLoadFcn*). Adding commands to the *StopFcn* means that they will be executed every time a simulation ends. This is particularly useful, as results can be captured using the *To Workspace* block, and then immediately plotted using the *StopFcn*, without any need to subsequently write or execute additional code.

4.9 Generating Frequency Domain Plots

The *Spectrum Analyzer* Simulink block is used extensively throughout this book, to enable the frequency content of signals to be viewed and analysed. The next exercise briefly introduces the *Spectrum Analyzer*. Further discussion of this block will follow later in the book, with particular features being highlighted as they are needed.

Exercise 4.17 Frequency Domain Plots: Spectrum Analyzer

Connecting a *Spectrum Analyzer* is a simple and intuitive method of generating a frequency domain plot. This exercise introduces the basic operation of the block.

(a) **Open the model:**

    `/matlab_simulink/spectrum_analyzer.slx`

(b) **View the system.** Confirm that it comprises the sum of three sine waves. Notice also that a *Spectrum Analyzer* block is connected at the output. Investigate the model and establish the system sampling frequency, and the frequencies of the three *Sine Waves*.

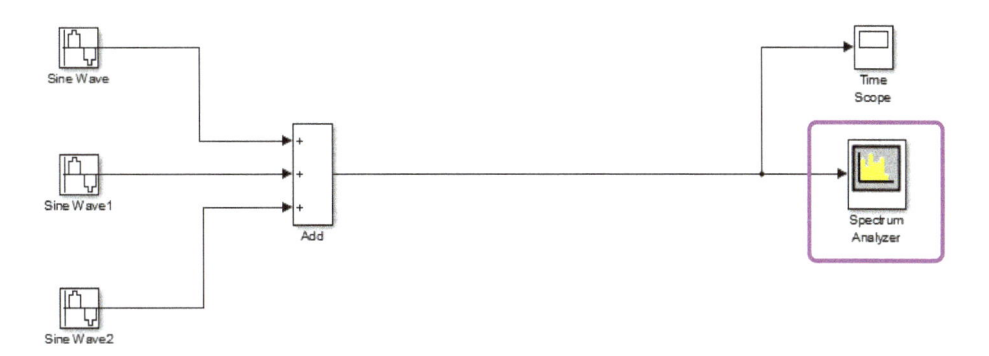

(c) **Simulate!** Run a simulation and view the outputs. Confirm that the time domain waveform looks as expected, and then open the *Spectrum Analyzer*. Make sure that you can identify the sine wave components on the plot. The zoom tools are available to help you inspect the peaks in more detail.

(d) **Open the configuration pane.** on the *Spectrum Settings* button to confirm how the FFT has been configured. This allows various aspects of the plot to be specified, such as whether a one or two sided plot is shown, the number of points in the FFT, and the window type used.

(e) **Investigate the peak finder feature.** Next, on the *Peak Finder* button , which will open a panel showing information about the peak frequency components. This will show you the frequencies and respective amplitudes of the most significant frequency components. Try on the individual check boxes, which will annotate the peaks onto the frequency plot.

(f) **Re-simulate with different parameters.** Try altering the frequencies of the sine waves, then re-simulate, and view and inspect the results. Remember from DSP theory that frequency accuracy is limited by the number of bins in the FFT, which is set within the *Spectrum Settings* pane.

4.10 Sampling Rates, Samples and Frames

So far, we have dealt with simple systems with a single sampling rate, defined mathematically as f_s and represented by the MATLAB variable `fs`. This parameter has been used to set the sample time associated with source blocks (i.e. `1/fs`), and also to define the simulation stop time of Simulink models (for instance, `(n-1)/fs`, where `n` is the total number of samples in the simulation.

In this section, we look at some more sophisticated examples, where Simulink models may have several different sample rates. The use of frames in Simulink is also explored, and conversion between samples and frames is demonstrated.

4.10.1 Sample Rate Changes: Upsampling and Downsampling

In some systems, there is a requirement to either increase or decrease the sampling rate within a model. In DSP systems, this is often done in conjunction with lowpass filtering, in order to appropriately modify the frequency content of the signal (see Appendix C (page 583) for further details on this). Here, we focus purely on the operations of *upsampling* (increasing the sampling rate by the factor L) and *downsampling* (decreasing the sampling rate by the factor M), where both L and M are assumed to be integers.

Upsampling by the factor L involves inserting $L-1$ zeros between the original samples. This is illustrated in Figure 4.6, where the upsampling ratio is specified as $L = 4$. The Simulink *Upsample* block also provides the option to specify a phase, which introduces a delay and thus adjusts the timing of the non-zero samples.

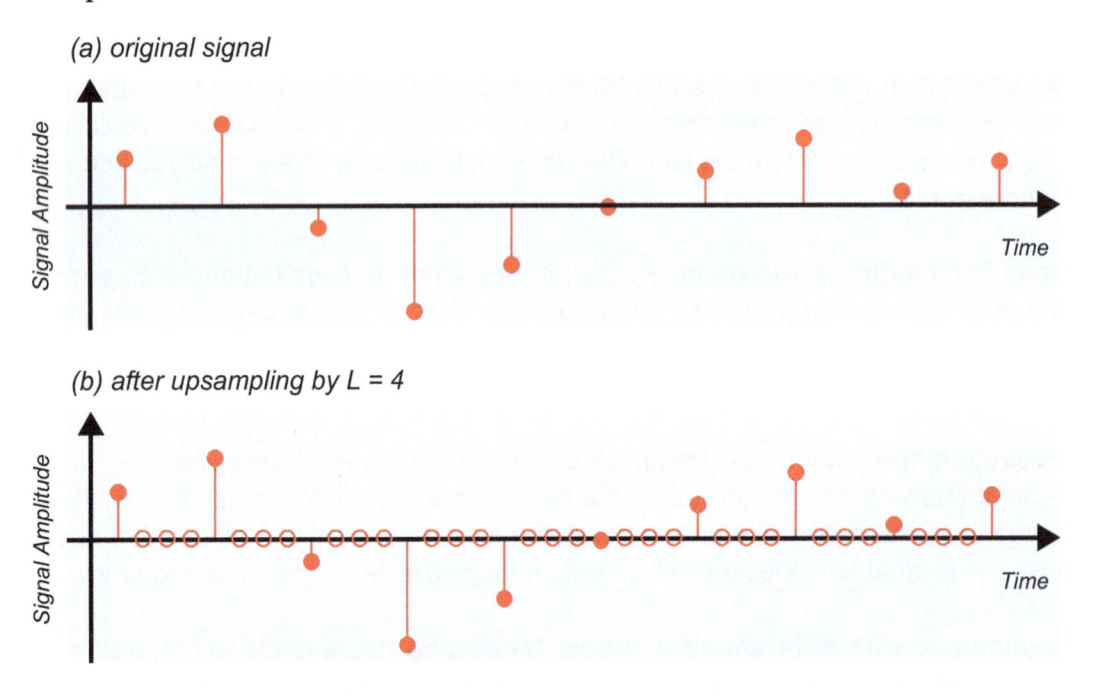

Figure 4.6: Upsampling operation with ratio $L = 4$

Downsampling by the factor M effectively means that only every M^{th} sample from the original set is retained, while the intermediate samples are discarded. The phase of downsampling is significant,

Chapter 4: Getting Started with MATLAB and Simulink

because there are M possible phases of samples that could be retained. For instance, in the case where $M = 4$, we would decide to retain one of the below sets of samples, with indices:

$$n = 0, 4, 8, 12...$$
$$n = 1, 5, 9, 13...$$
$$n = 2, 6, 10, 14...$$
$$n = 3, 7, 11, 15...$$

These four phases are indicated by the differently coloured sample points in Figure 4.7. Notice that, in this case, the **purple** phase has been retained by the downsample operation ($n = 2, 6, 10, 14...$).

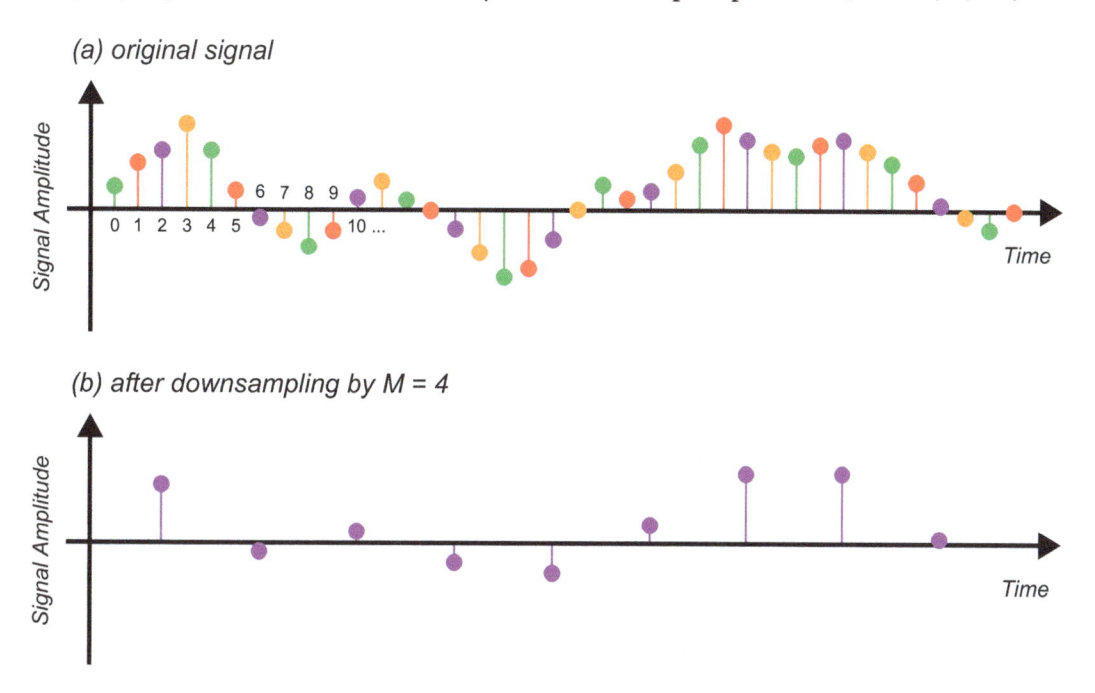

Figure 4.7: Downsampling operation with $M = 4$

In the next exercise, upsampling and downsampling operations are both demonstrated in Simulink. Note that **sample based processing** is adopted in this example. This means that the outputs resulting from upsampling and downsampling continue to be treated as individual *samples*, and passed through the Simulink model one-by-one, rather than being grouped into *frames* and processed several-at-a-time. The issues of samples versus frames will be discussed further in Section 4.10.2.

Upampling and downsampling are involved when performing the multirate DSP tasks of *interpolation* and *decimation*, respectively. We will not cover these operations here, as our current focus is on familiarisation with MATLAB and Simulink; further information on these can be found in Appendix C.

Exercise 4.18 Upsampling and Downsampling

In this simple example, *Upsample* and *Downsample* blocks are introduced into a Simulink model and their operation is confirmed. The Sample Time Colours and Annotations feature of Simulink is also demonstrated.

(a) **Open the model:**

`/matlab_simulink/upsampling_downsampling.slx`

(b) **Confirm operation.** Briefly inspect and run the model, and confirm that a *Sine Wave* is generated, and then displayed in the *Time Scope*.

(c) **Add an *Upsample* block.** Locate the *Upsample* block in the Simulink Library Browser, and introduce it between the source and sink. You can find this block in ▦ **> DSP System Toolbox > Signal Operations**. Configure it with an 'Upsample factor' of '3', and a 'Sample offset' of '0'. Ensure that the Input Processing type is set to *Elements as Channels (sample based)*:

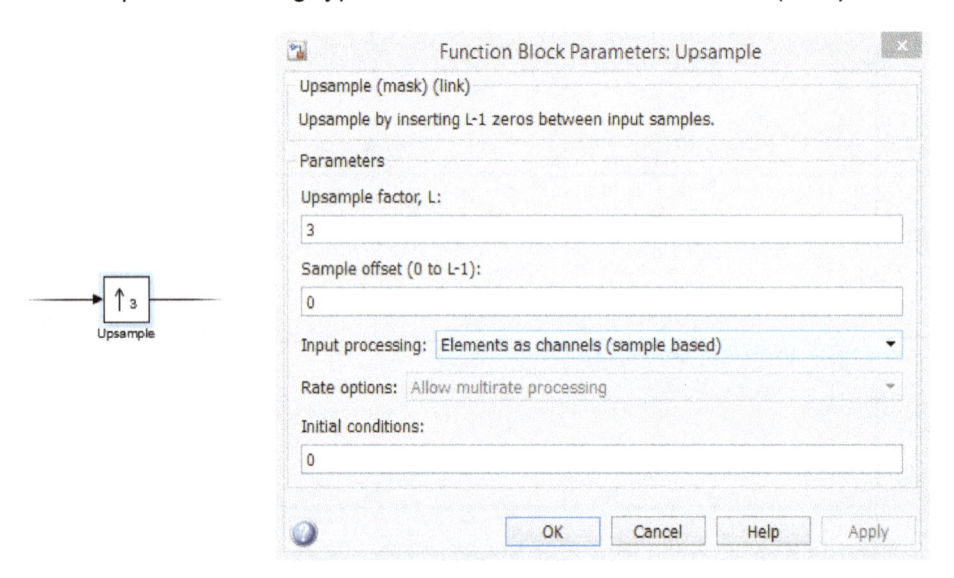

Notice that the other option here is *Columns as channels (frame based)*. We will discuss frames a little more shortly.

(d) **Simulate!** Run a simulation and view the output. You should be able to confirm that the Upsample operation has resulted in sets of two zero-valued samples being inserted between each of the original samples. (Here, the upsample ratio is $L = 3$, and so $L - 1 = 2$ zero samples are inserted.)

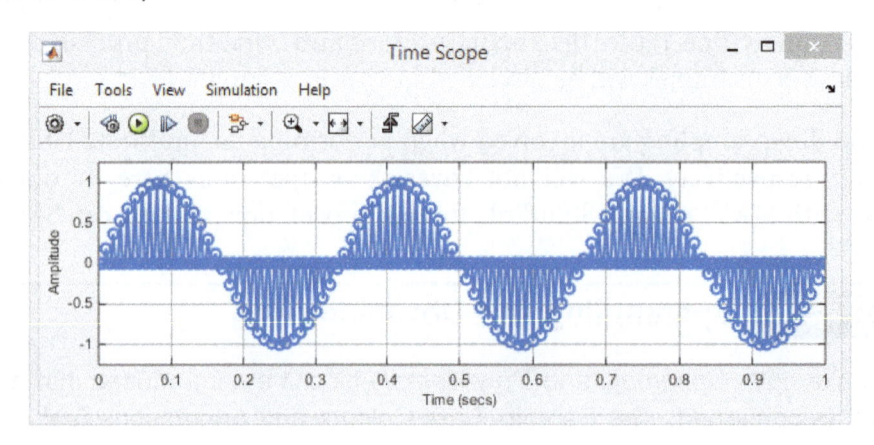

Chapter 4: Getting Started with MATLAB and Simulink 155

(e) **Change to downsampling.** Next, 'Comment Through' the *Upsample* block (refer back to Exercise 4.13 if you need a reminder of how to do this), and then add a *Downsample* block immediately after it. Configure the *Downsample* block as shown below.

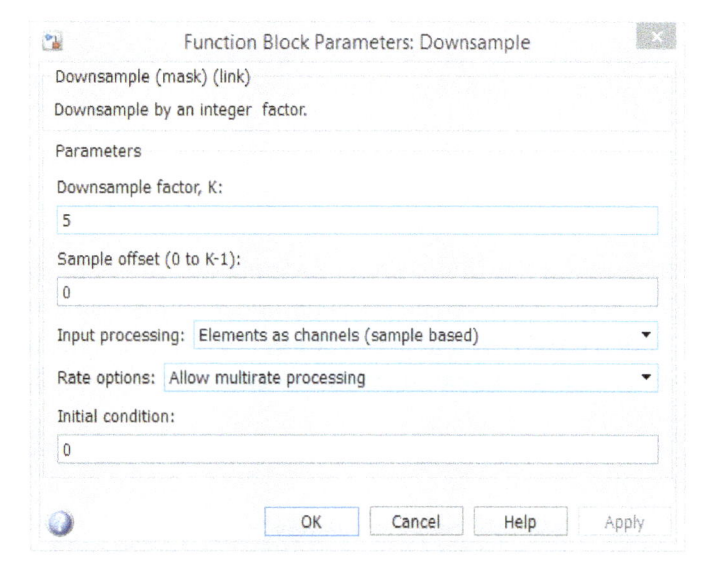

Function Block Parameters: Downsample

Downsample (mask) (link)
Downsample by an integer factor.

Parameters
Downsample factor, K:
5
Sample offset (0 to K-1):
0
Input processing: Elements as channels (sample based)
Rate options: Allow multirate processing
Initial condition:
0

OK   Cancel   Help   Apply

(f) **Simulate again!** Run a simulation and view the results. Does the waveform appear as you expect?

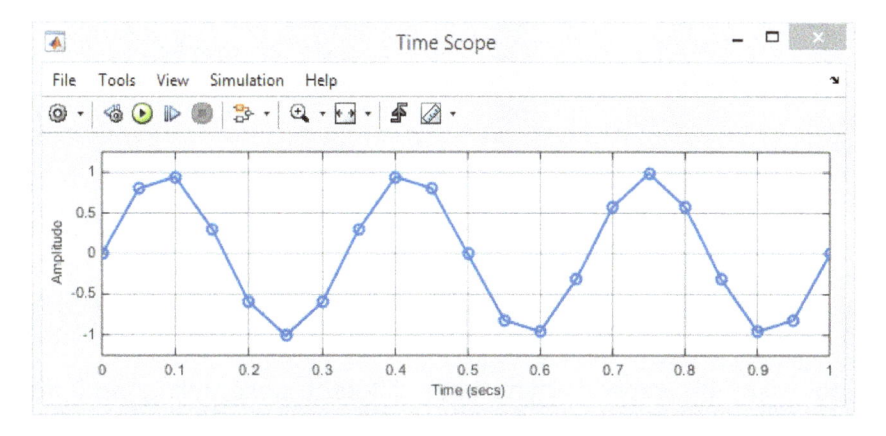

(g) Uncomment the *Upsample* block, and simulate again. This time, you should notice that most of the samples are zero—can you think why?

(h) **Introducing Sample Time Information.** Lastly, it is useful to highlight the Sample Time feature that can be used in Simulink. [1] in an empty area of the design space, to view the menu. Choose the *Sample Time Display* option, and then *All*.

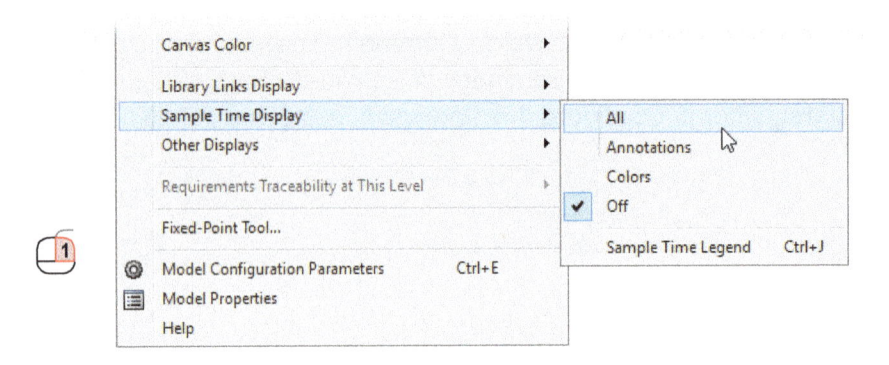

(i) The effect of this will be that the model is annotated with colours, and the wires are labelled D1, D2, and D3. There will also be a legend showing the sample times associated with each of these colours and labels, similar to the below. Take a little time to confirm the various sample rates involved, and that these correspond with the upsample and downsample factors specified.

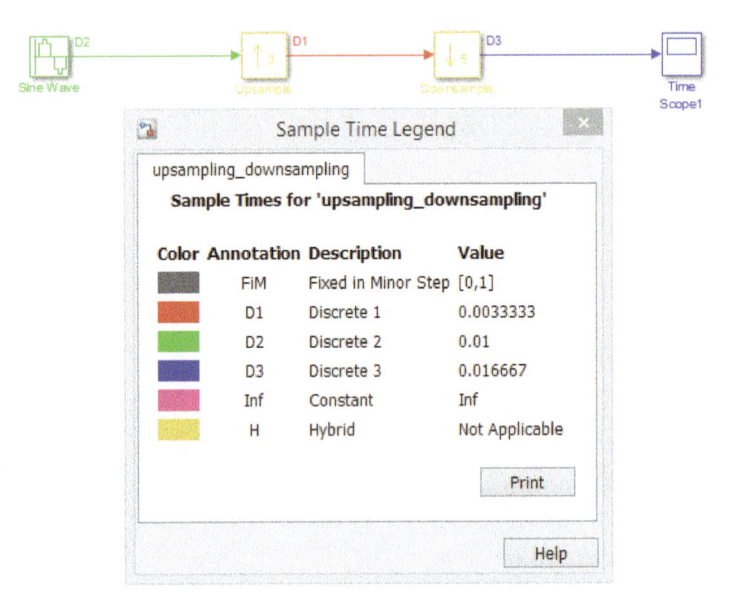

(j) If you wish, try experimenting with different upsampling and downsampling factors, or swapping the order of the *Upsample* and *Downsample* blocks.

4.10.2 Working with Samples and Frames

In the last section, we learned how to change the sampling rate in the system, while maintaining the use of *sample based processing*. We would now like to discuss the alternative, *frame based processing*.

Samples are very intuitive, particularly in DSP. Each sample corresponds to a particular sample index, or time instant, and these occur at regular intervals. It is easy to think about a Simulink model processing samples sequentially, one at a time. The impact of downsampling is to reduce the frequency of the samples, while upsampling has the opposite effect, and thus it follows that Simulink models can include several different sampling rates.

Chapter 4: Getting Started with MATLAB and Simulink

The alternative is frame based processing, which involves grouping sets of consecutive samples into frames. The frame rate is therefore lower than if samples were used, with the exact relationship between the two rates depending on the size of the frames. To give a simple example, if we took an input with a sampling rate of 1kHz and upsampled by a factor of 3, the output sample rate would be 3kHz. If frames were used, the frame size would be 3, and the output rate would be 1kframe/s, as sketched in Figure 4.8.

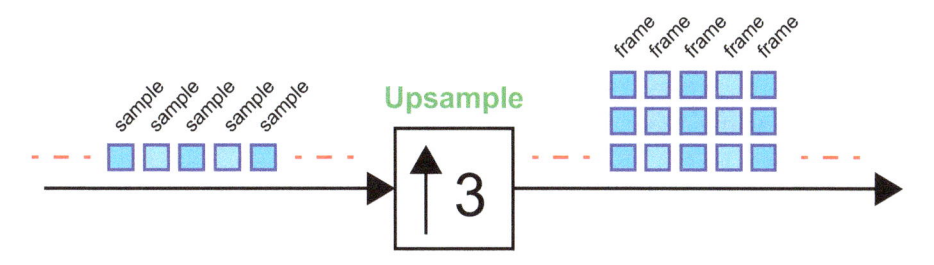

Figure 4.8: Upsampling with frame based processing at the output

If the relationship between sampling rates in a system is convenient, then single rate processing can be used throughout, meaning that the various sample rates are achieved simply by varying the sizes of the frames. In other words, all required rates result in frames with an integer number of elements. This is not always possible, however, and often multiple rates are needed. For instance, you might try returning to the model from Exercise 4.18, and setting both the *Upsample* (by 3) and *Downsample* (by 5) blocks to use frames, while enforcing single rate processing—you'll find that error messages occur!

One of the main reasons for using frames is that it allows the computer to process frames of data in batches, which allows simulations to run faster. With several samples available at each time step, computations can be undertaken in parallel and this allows more efficient processing. File accesses (reading from, or writing to, a file) can also be undertaken more quickly if frames are used, as larger sets of data can be read or written simultaneously. To give an example, if we converted a sample-based signal to frames of size 3 samples, then frames would be generated at one third of the sample rate. This scenario is depicted in Figure 4.9.

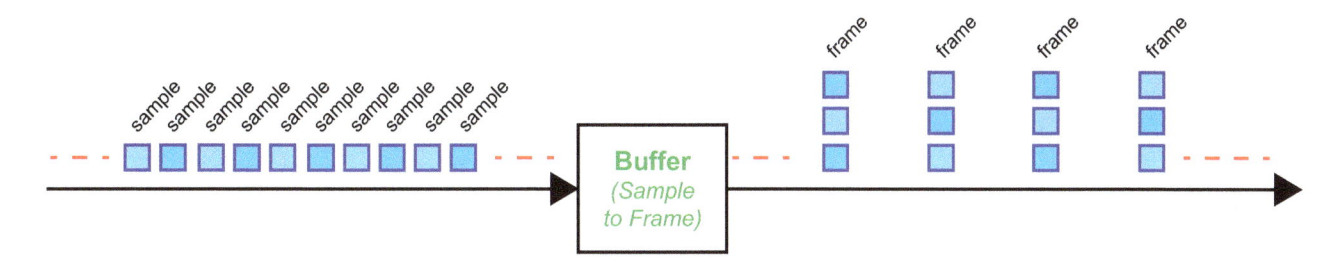

Figure 4.9: Sample to frame conversion (frame size of 3 samples)

It is also useful to compare Figure 4.8 with Figure 4.9, and confirm that there is no increase to the *sample rate* in the latter case. Upsampling creates more samples, whereas converting from samples to frames does not.

Frames are not suited to all situations, though. A particularly important case is that of feedback loops, where any given output sample is a function of previous input samples. For samples to be available on a consecutive basis, as required, sample based processing is a must.

In the next exercise, we consider another method of converting between samples and frames, and vice versa. This method is used when there is no requirement to change the sampling rate, and we simply wish to achieve the benefits of using frames in terms of simulation performance. That is, the data passing through the system is processed in parallel, at a lower rate, and the model can therefore be simulated more quickly.

Exercise 4.19 Conversion Between Samples and Frames

In this simple reference design, the conversions from samples to frames, and then frames back to samples, are demonstrated.

(a) **Open the model:**

   `/matlab_simulink/sample_frame_sample.slx`

(b) **Confirm operation.** Investigate the model, and confirm that the output of the source block is sample-based. Establish the sample rate at the output of the source block.

   Check also the parameter of the *Buffer* block. Samples arriving at the buffer are stored up until it reaches the required size, at which time the set of samples is converted to a frame. The buffer is then emptied, and it begins collecting samples with which to create the next frame. The parameter specifying the size of the buffer therefore defines the size of the frames that are created. Likewise, conversion from frames to samples requires an *Unbuffer* operation. The frames comprise a number of consecutive samples, which must then be pushed out one-by-one at the appropriate sampling rate.

(c) **View signal dimensions.** When working with frames, it is often convenient to view signal dimension annotations on the model. You can show these annotations by [1] in an empty area of the design space, then choosing *Other Displays*, *Signals & Ports*, and finally *Signal Dimensions*, as shown.

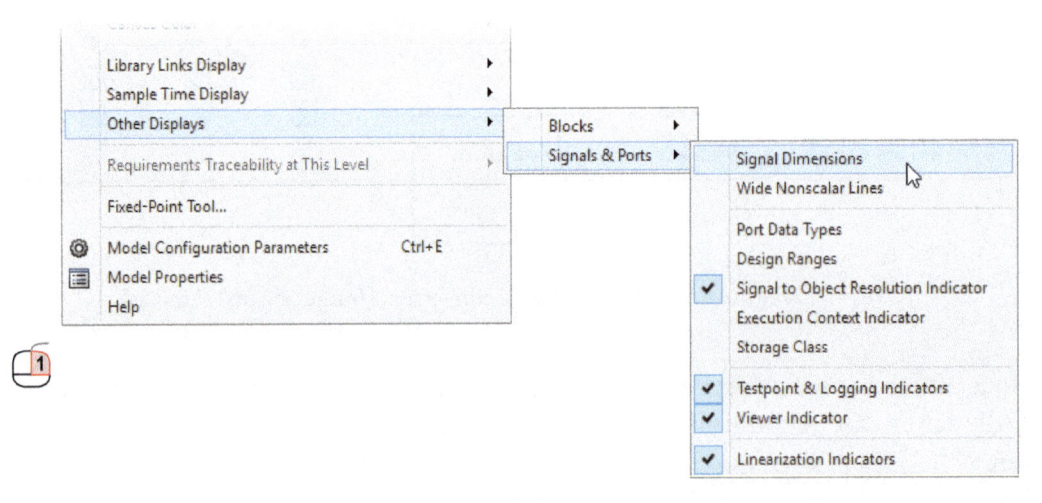

This will allow you to see the frame size annotated next to some of the wires in your model. Check that the dimensions match those specified in the *Buffer* block. Wires which do not have an annotated dimension are simply sample-based signals.

Chapter 4: Getting Started with MATLAB and Simulink

(d) **Simulate!** Run the model, and confirm that a sine wave is generated, and then displayed in a *Time Scope*. You may notice that there is a delay before meaningful (non-zero) data is shown in the *Time Scope*. This is due to the latency implied by the buffering operation.

(e) **Experiment...** Try changing the value in the Buffer block, and viewing the results in the *Time Scope*. What do you notice?

4.11 Data Types

A number of different data types can be used in MATLAB and Simulink, with the default in most cases being 'double', i.e. double precision floating point. Most Simulink source blocks produce outputs of the 'double' type, which is suitable for most of the systems featured in this book.

It is worth being aware of other data types that may be specified. To provide a brief summary (consult the Help documentation for more information):

- `Double`   double precision floating point
- `Single`   single precision floating point
- `Boolean`   1 or 0 (representing true or false)
- `uint8, uint16, uint32`   unsigned integer (8, 16, and 32 bits)
- `int8, int16, int32`   signed integer (8, 16, and 32 bits)

Fixed point types can also be defined using the general format:

- `fixdt(s,n,f)`   (e.g. `fixdt(1,16,14)`)

Where `s` is set to 0 for unsigned and 1 for signed; `n` specifies the total number of bits; and `f` gives the number of fractional bits.

These types can also be used to represent complex data. In this case, the data type is appended with a `(c)`; for instance `double (c)` denotes a complex, double precision floating point number.

Aspects of working with data types will generally be introduced on an as-needed basis during the rest of the book, but it is worthwhile briefly highlighting the issues of converting between real and complex signals, as complex representations will be used extensively in later chapters.

Exercise 4.20 Real and Complex Data Types

The majority of source blocks in Simulink generate real data types, such as single, double, and boolean. Some, such as the *Sine Wave* source block in the DSP System Toolbox, additionally provide the option to generate complex data. Often we will actually create complex data types by creating the real and imaginary parts separately (each of them using a real data type such as double), before combining them together to form a complex type.

(a) Open the model:

📁 `/matlab_simulink/real_complex.slx`

📖 (b) Briefly inspect the model and notice that it comprises three sinusoidal sources. Investigate the parameters of the source blocks to establish whether they are real or complex sources.

👣 (c) **View Signal Data Types.** Switch on 'Port Data Types', which will allow you to check on the data types associated with each wire in the system. You can do this by 🖱① in an empty area of the design space, then choosing *Other Displays*, *Signals & Ports*, and then *Port Data Types*, as shown below.

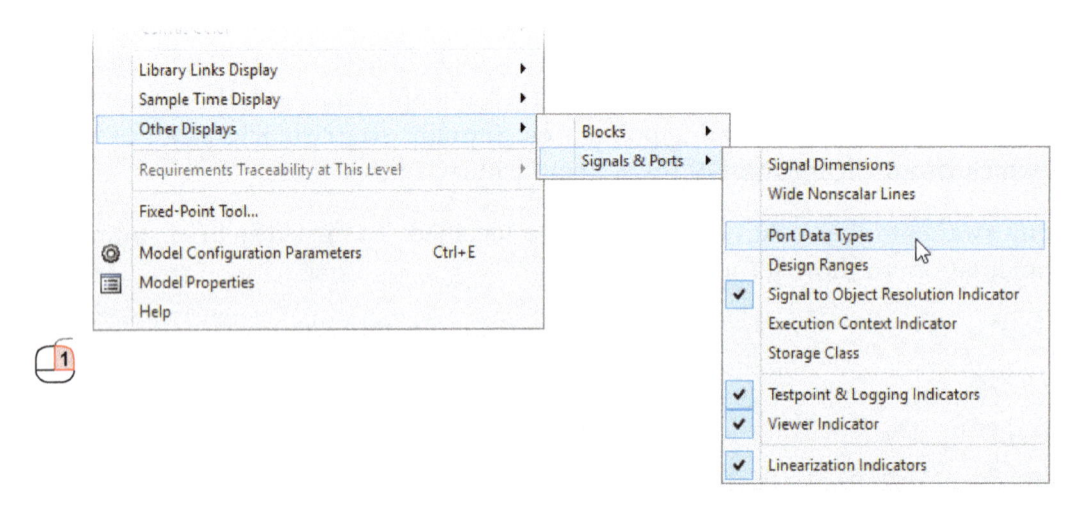

🔧 (d) **Convert Real to Complex.** We will now add a block to convert from two real signals, to a single complex data format. Open 🗂 > *Simulink > Math Operations*. Locate the *Real-Imag to Complex* block, and introduce it into the model, then connect as shown below, and add a *Time Scope* to inspect the output. Notice that the output wire from the *Real-Imag to Complex* block is annotated with 'double (c)', indicating that it is a complex signal.

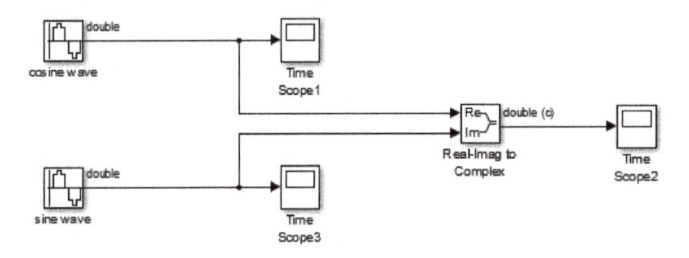

🏃 (e) **Simulate!** Run a simulation and compare the outputs of the system. In particular, check whether the output of the *Real-Imag to Complex* block looks similar to the output of the *DSP Sine Wave* block in the lower part of the system.

Chapter 4: Getting Started with MATLAB and Simulink 161

(f) **Convert Complex to Real.** Next, the conversion will be reversed. Introduce a *Complex to Real-Imag* block from the library, and then connect it to the output of the *Real-Imag to Complex* block. The the two outputs can be connected to *Time Scopes*.

(g) **Simulate again!** Run another simulation, and check that the complex to real and imaginary conversion has been successful in both cases.

4.12 Working with Input and Output Files

When working with Simulink and MATLAB designs, it is often useful to save simulation data to a file, such that it can be stored for later analysis. Later in the book, some exercises will involve reading data captured by the RTL-SDR from provided files, which has the obvious advantage that realistic simulations can be performed even in the absence of an RTL-SDR or a suitable reception environment. We will investigate these later in Section 4.13.

MATLAB supports several different file types, but the format adopted here for capturing and storing simulation data is the MAT file (*.mat* file extension). Custom Simulink library blocks have been created to ensure that the correct formatting is applied to data stored at the output of the *RTL-SDR Receiver* block in Simulink. The standard *To File* and *From File* blocks should be adopted for more general use. Reading and writing of multimedia files are further, commonly used operations, and audio files will be used extensively later in the book. These features allow audio samples to be used as the inputs for analogue modulation schemes, for instance.

The next few exercises demonstrate how to read and write the above file types in Simulink.

Exercise 4.21 **Write and Read *.mat* Data Files (General Case)**

In this exercise, we start with a model that generates data using standard Simulink blocks, which we desire to write to a *.mat* file for later use. We then switch to a second model that reads in data from the newly created *.mat* file.

(a) Open the model:

> /matlab_simulink/write_mat_file.slx

(b) Briefly inspect the model, and notice that it comprises three *Sine Waves*, which have been added together to form an output.

(c) **Simulate!** Run a simulation and view the results in the *Time Scope*.

(d) **Add a *To File* block.** Next, open 🗔 > *Simulink* > *Sinks*. Introduce a *To File* block into the model, and connect it such that the model looks like this:

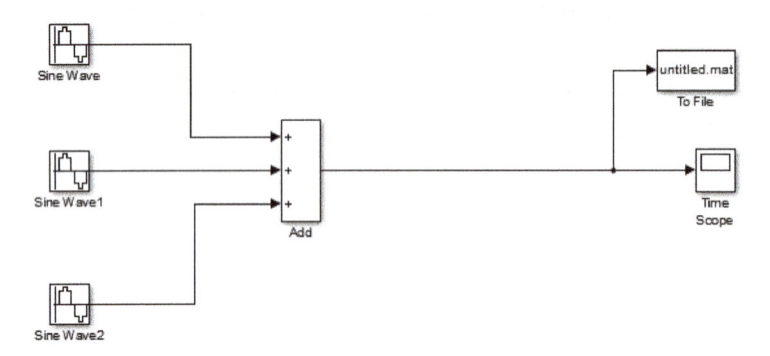

[Simulink model diagram showing three Sine Wave blocks connected to an Add block, whose output connects to both a To File block (untitled.mat) and a Time Scope.]

(e) **Set the parameters of the *To File* block.** Note that the block will, by default, write a file named `Untitled.mat`. Normally it is desirable to change this to something more meaningful. Open the *To File* block by 🖱 on it, and set the parameters as follows:

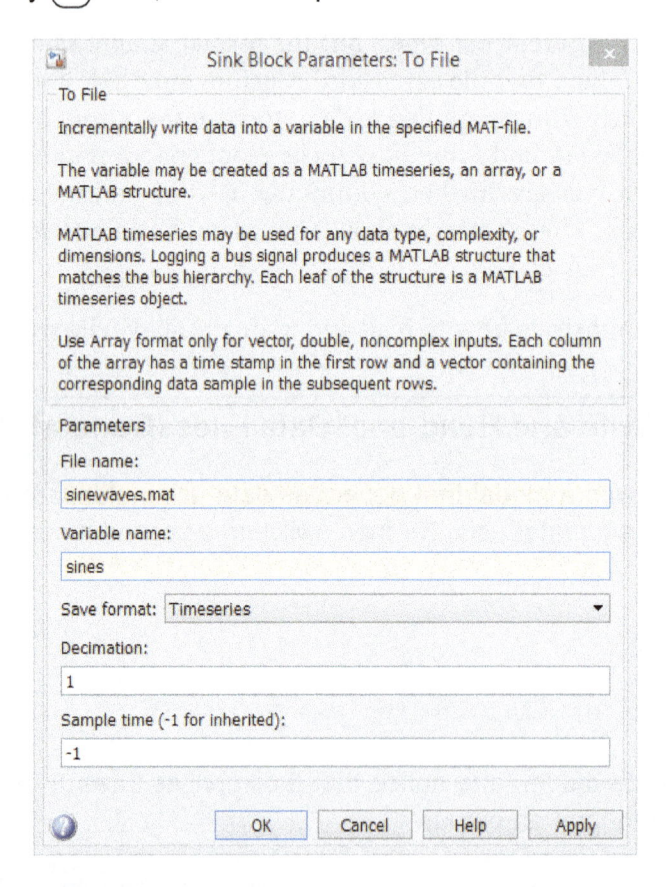

[Sink Block Parameters: To File dialog window. File name: sinewaves.mat; Variable name: sines; Save format: Timeseries; Decimation: 1; Sample time (-1 for inherited): -1]

This will create a file named `sinewaves.mat`, and the file will store a variable called `sines`, which will take the form of a Timeseries.

(f) **Simulate!** Run a simulation and then open the main MATLAB window to inspect the contents of your current working directory. You should notice that a new file has been created:

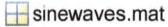

sinewaves.mat

Chapter 4: Getting Started with MATLAB and Simulink

The *Time Scope* can still be viewed as before, to check that the correct results have been produced.

(g) **Save and close the model.**

(h) **Open a model to read the .mat file.** Next, we will move to a new model that will read in the `.mat` file just created. Open:

   📁 `/matlab_simulink/read_mat_file.slx`

(i) **Introduce a *From File* source block.** Open ▦ > *Simulink* > *Sources*, and locate the *From File* block. Add this block to the model and connect it to the *Time Scope* block. 🖱 on the block to open its parameters window. We must set its parameters to match those of the `sinewaves.mat` file just created. Configure the block as shown below.

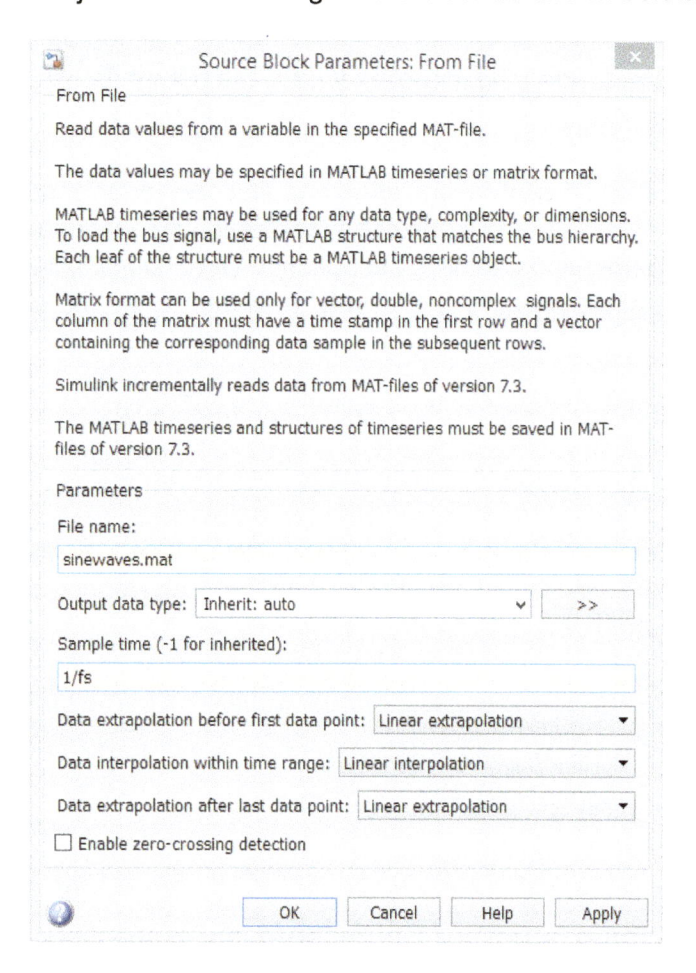

This ensures that the source block is specified, along with the sampling rate associated with the data.

(j) **Simulate!** Run a simulation and view the results in the *Time Scope*. Check that these correspond with the previous system, in which the sum-of-sines data was created. You should be able to confirm that the waveform is the same, which shows that the `.mat` file has successfully stored simulation data and allowed it to be ported between different Simulink files.

Exercise 4.22 Read and Write Audio Files

Reading and writing audio files is a commonly used feature of Simulink, and one which will be utilised frequently in later examples. In this exercise, we briefly demonstrate the methods necessary to export audio data from, and import audio data into, Simulink models.

(a) Open the model:

   `/matlab_simulink/write_audio_file.slx`

(b) **View the reference design.** Briefly inspect the model and notice that a (mono) audio signal is created by summing three tones together. The signal is buffered before being written to a `.wav` file. The use of frames means that multiple samples can be written to file at a time. Confirm the size of frames that are created.

(c) **Run a simulation.** When the model executes, the output data will be written to the output file, which has been specified in the *To Multimedia File* block as `tones.wav`.

(d) **Listen to the file.** Once the simulation has finished, open the main MATLAB window, and check that the `tones.wav` file has been created in your working directory. You can open it by on the file name, and choosing the option to *Open Outside MATLAB*. This will open the file in your default audio player. Alternatively, navigate to the file using Windows Explorer, and open it.

(e) Next, let's import a music file to Simulink. Open the model:

   `/matlab_simulink/read_audio_file.slx`

(f) **View the reference design.** Inspect the model, and notice that it refers to the file `music1_48kHz.wav`. Check that the correct file is referenced by opening the block and the 'Browse' button.

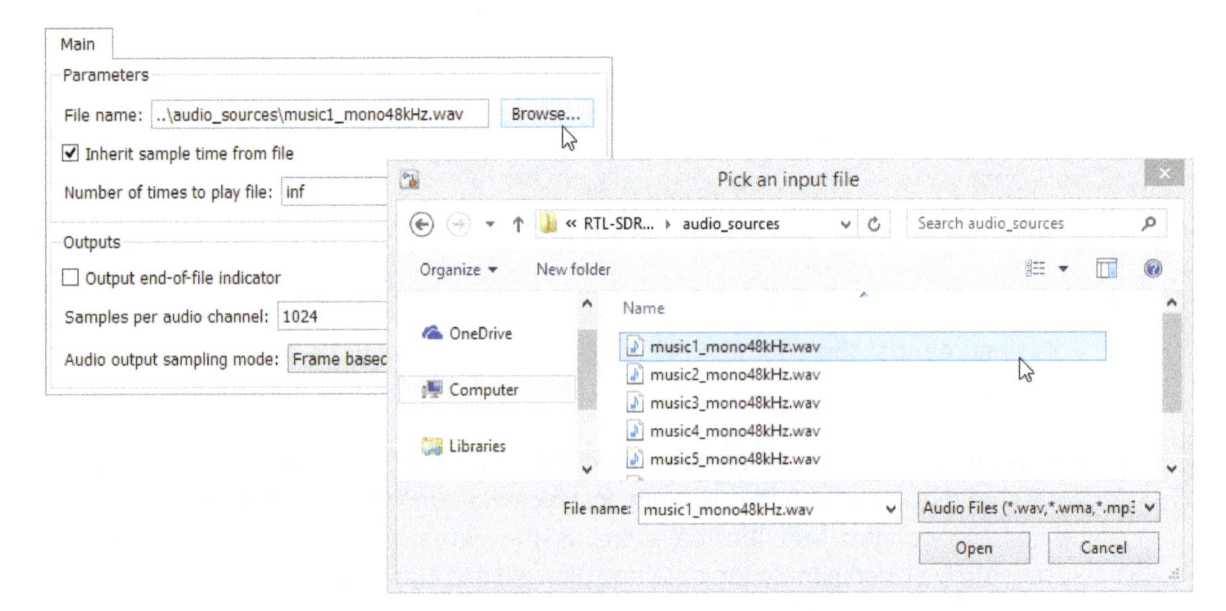

Chapter 4: Getting Started with MATLAB and Simulink 165

(g) **Simulate!** Run a simulation and confirm that audio data is successfully read from the file.

(h) **Add an audio output.** Locate the *To Audio Device* block from ▦ **> DSP System Toolbox > Sinks**, and connect it as an output. Try resimulating—you should now be able to hear the music while the model is running.

(i) Experiment with any other audio files you have on your computer, such as `.mp3` files. These can be read into Simulink too, and subsequently written to a `.wav` format if desired.

If using a stereo file, there are two channels (left and right), which can be isolated with the help of the *Submatrix* block. This can be found in ▦ **> DSP System Toolbox > Signal Management > Indexing**. Setting two such blocks to 'First' and 'Last' column respectively, enables the two individual audio channels to be extracted. An example is shown below:

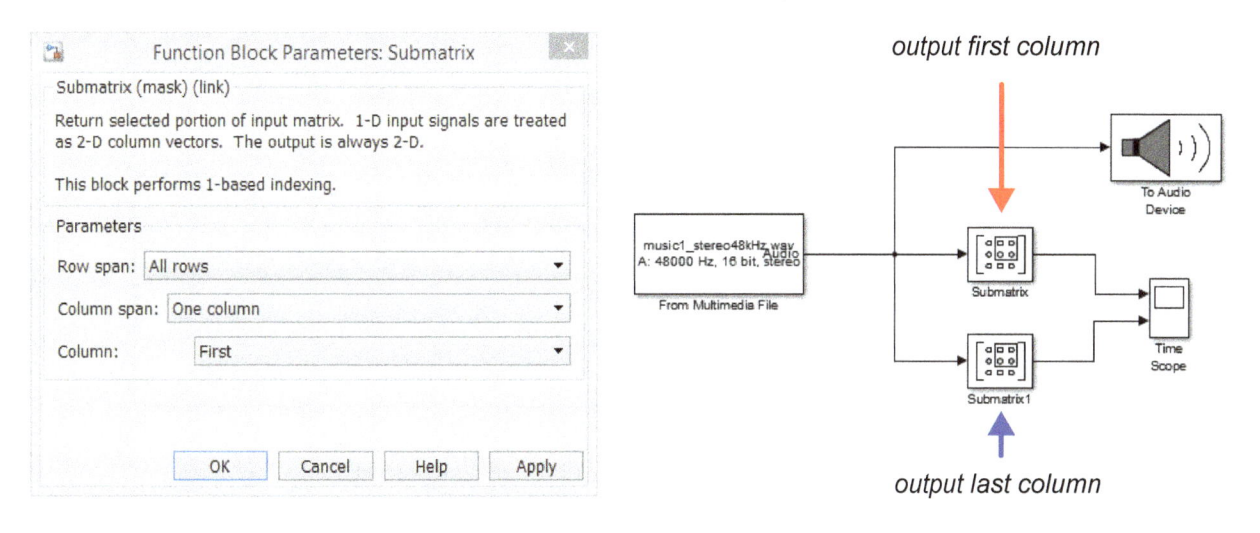

4.13 Saving and Re-importing RTL-SDR Data

We are aware that some of you may not have an RTL-SDR yet, and that others may not have a means to transmit RF signals. We want to cater for everyone in this book, so we are providing data files with many of the exercises throughout that contain recordings of the RF signals you should be aiming to receive. This allows you to test the receivers you build with 'real' signals, without actually having to transmit and receive them yourself.

We have developed a pair of blocks for this purpose that are packaged in our custom Simulink library. The first of these can be used to record signals received from the RTL-SDR, and this is what we used when recording the signals for the exercises in this book. The second block can be used to re-import the data to a Simulink receiver. They can be found in ▦ **> RTL-SDR Book Library > Additional Tools**.

We have also developed a system object called `import_rtlsdr_data` that allows you to import recorded RTL-SDR data to MATLAB receivers. This has exactly the same functionality and parameters as the Simulink block, and can be used in place of `comm.SDRRTLReceiver` objects in situations where you cannot use an RTL-SDR. Figure 4.10 compares the parameters that must be set in the Simulink block with the parameters set in the system object. Note that they are essentially the same.

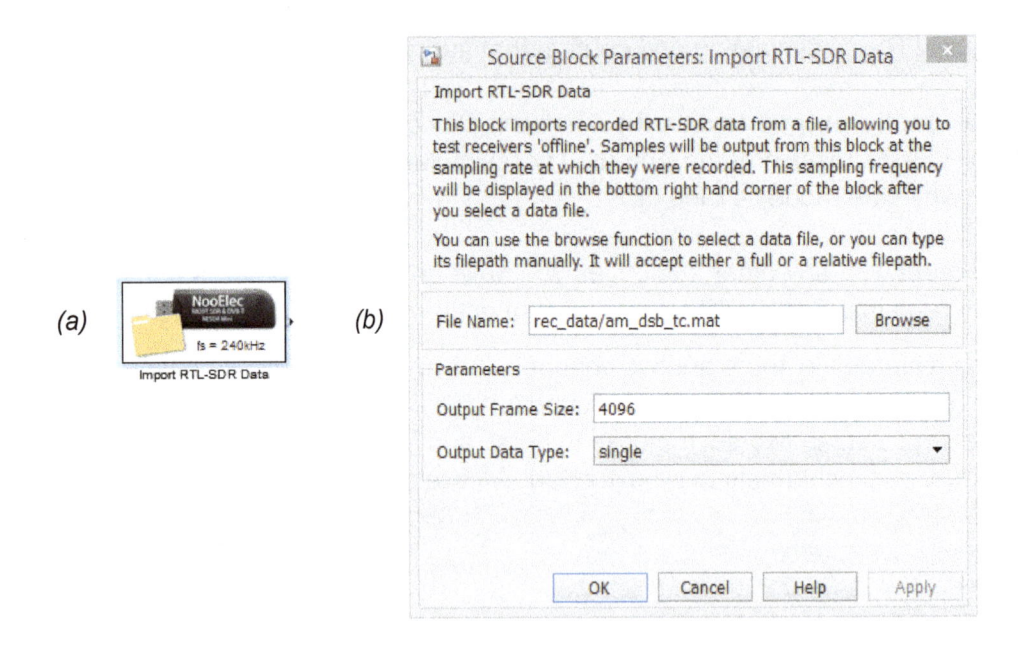

```
% create an 'import RTL-SDR data' object
obj_rtlsdr = import_rtlsdr_data(...
    'filepath', 'rec_data/am_dsb_tc.mat',...
    'frm_size', 4096,...
    'data_type', 'single');
```

Figure 4.10: Comparison between the two methods of importing RTL-SDR data to your receiver:
(a) the *Import RTL-SDR Data* Simulink block
(b) the `import_rtlsdr_data` system object

In the final exercise in this chapter, we will demonstrate how these Simulink blocks and the system object can be used to permit your receivers to run 'offline'.

Exercise 4.23 Saving and Importing RTL-SDR Data in MATLAB and Simulink

This exercise is split into three parts, to demonstrate:

(1) How to use the *Save RTL-SDR Data* block,
(2) How to use the *Import RTL-SDR Data* block, and
(3) How to use the `import_rtlsdr_data` system object.

Note: If you do not have an RTL-SDR, you will be unable to complete Parts (a) to (e), as this deals with saving received data. Simply skip forward to Part (f) where we will show you how to import data instead.

(a) Open the model:

`/matlab_simulink/rtlsdr_save_data.slx`

Chapter 4: Getting Started with MATLAB and Simulink 167

(b) This model contains an *RTL-SDR Receiver* block, which is configured to tune the RTL-SDR to 100MHz, and sample at a rate of 240kHz. It does not matter if there is not actually a signal transmitted at this frequency for this exercise; receiving noise will be fine.

(c) **Add a *Save RTL-SDR Data* block.** Open ▦ *> RTL-SDR Book Library > Additional Tools* and find the *Save RTL-SDR Data* block. Place one in your model, and connect it to the output of the *RTL-SDR Receiver*. ② on the block to open its parameters window, and make a note of the file name that is specified. The file will be saved by default into the MATLAB current folder, overwriting the `rtlsdr_data.mat` file that is already there. If you wanted to save it elsewhere, you could ① on the 'Browse' button and navigate to a new location. Apply any changes, and close the parameter window.

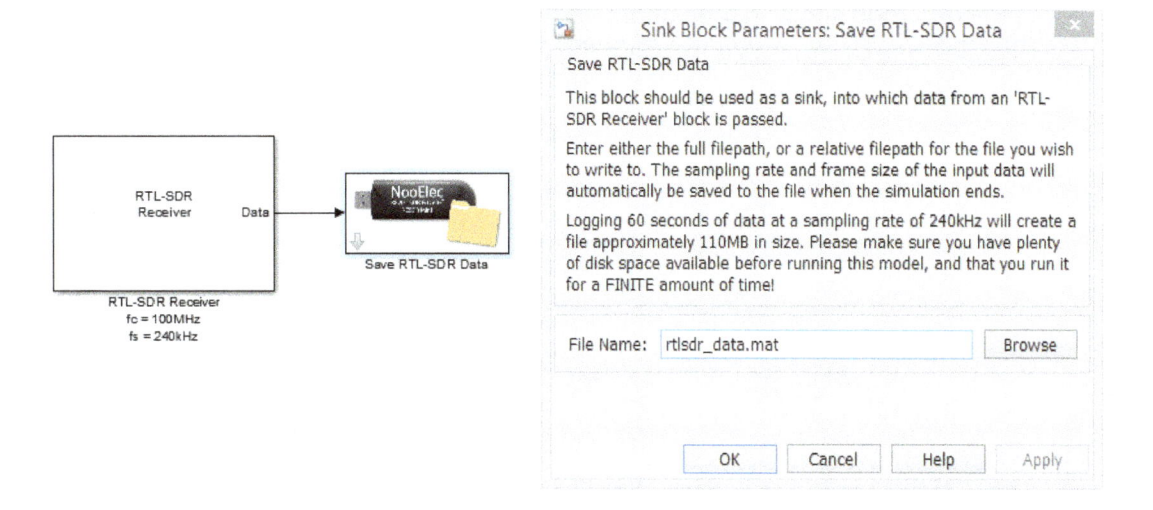

(d) **Prepare to run the simulation.** As this block can be used to save vast amounts of data, we advise that you do not have the *Simulation Stop Time* set greater than 120 seconds in any model using it, as the files it generates will be massive! As this exercise is purely being used as a test, set the *Simulation Stop Time* to 10 seconds by typing '10' into the Simulink toolbar.

Check that MATLAB can communicate with your RTL-SDR by typing `sdrinfo` into the MATLAB command window.

(e) **Run the simulation.** Begin the simulation by ① on the 'Run' ▶ button in the Simulink toolbar. After the simulation is complete, you should be able to see the file that has been overwritten if you look in Windows Explorer or MATLAB.

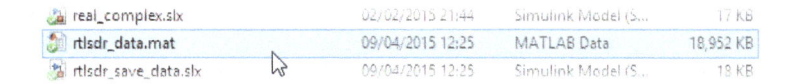

Save and close the Simulink model when this is complete.

(f) **Create a new Simulink model.** Place an *Import RTL-SDR Data* block from 🔲 > **RTL-SDR Book Library > Additional Tools** and a *Spectrum Analyzer* from 🔲 > **DSP System Toolbox > Sinks** into the model. Connect these together, and then 2️⃣ on the import block to open its parameters window.

Make sure that the block is set to import the correct file by 1️⃣ on the 'Browse' button and navigating to:

📁 /matlab_simulink/rtlsdr_data.mat

(or wherever you saved the file if you saved it elsewhere). Keep the 'Output Frame Size' set to '4096' and the 'Output Data Type' as 'single'. Apply any changes, and close the parameter window. The block should re-render as it locates the file, and display the sampling frequency of the recorded signal in the bottom right hand corner, as shown below. If this does not happen, Simulink has been unable to locate the file; and your should ensure that the correct file path is set in the block.

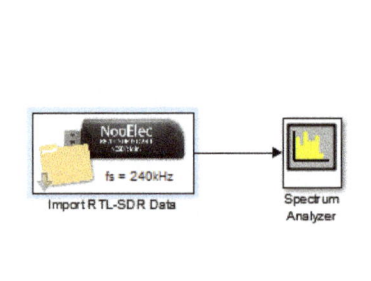

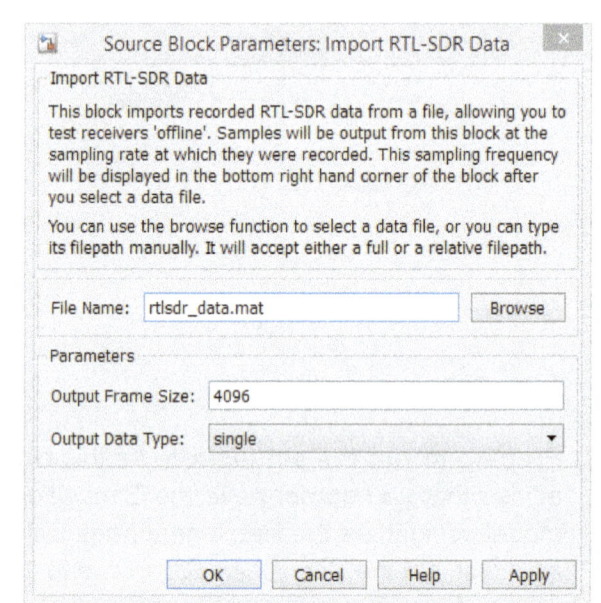

(g) **Prepare to run the simulation.** As the file only contains 10 seconds worth of data, set the *Simulation Stop Time* to 10 seconds.

(h) **Run the simulation.** The *Spectrum Analyzer* window should appear as the simulation starts, and display something similar to what is shown below.

Chapter 4: Getting Started with MATLAB and Simulink

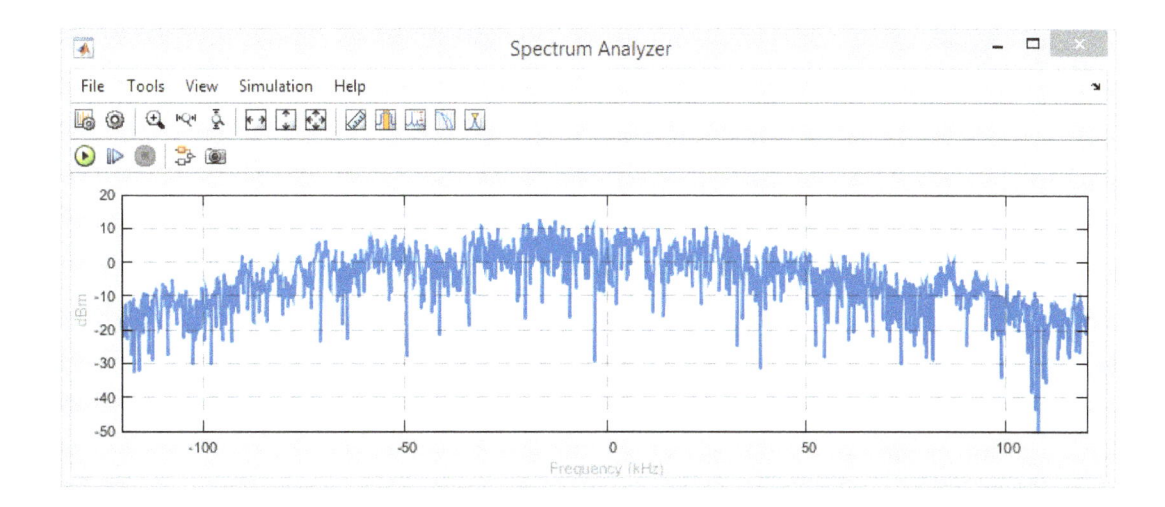

As there are no hardware interfacing blocks used in this model, it is likely that the simulation will process all of the data in 1 or 2 seconds, completing very quickly, but you should see activity for a short amount of time before the simulation ends. Save the model with the filename `rtlsdr_load_data_simulink.slx`, and then close it.

(i) **Create a new MATLAB script.** Enter the following code into it, and save the script with the filename `rtlsdr_load_data_matlab.m`.

(j) This code performs an equivalent operation to the simple two block Simulink model you implemented in Part (f). System objects are initialised in place of the import and scope blocks, and a simple while loop is used to pass data from one to another. The script is limited to run for only `10` seconds (the value of `sim_time`), and will complete when this amount of time has passed.

(k) **Run the function.** Run the function by on the 'Run' button in the MATLAB toolbar. The *Spectrum Analyzer* window should appear as the script executes, and display the same data that was shown previously. Although the function is designed to loop for `sim_time` seconds, you can cancel its execution early using the following key combination:

CTRL + C

(l) Once you are satisfied that the system object is the same as the *Import RTL-SDR Data* block, close the script. If you wish to view the help documentation for the `import_rtlsdr_data` object, type

```
help import_rtlsdr_data
```

into the MATLAB command window.

```matlab
% create an 'import RTL-SDR data' object
obj_rtlsdr = import_rtlsdr_data(...
    'filepath', 'rtlsdr_data.mat',...
    'frm_size', 4096,...
    'data_type', 'single');

% initialise obj_rtlsdr and aquire the sampling rate
rtlsdr_data = step(obj_rtlsdr);
rtlsdr_fs = obj_rtlsdr.fs;

% spectrum analyzer object
obj_spectrum = dsp.SpectrumAnalyzer(...
    'SpectrumType', 'Power density',...
    'FrequencySpan', 'Full',...
    'SampleRate', rtlsdr_fs);

% parameters
rtlsdr_frmtime = 4096/rtlsdr_fs;   % calc time for 1 frame of data
run_time = 0;                      % reset run_time to 0 (secs)
sim_time = 10;                     % simulation time in seconds

% loop while run_time is less than sim_time
while run_time < sim_time

    rtlsdr_data = step(obj_rtlsdr);      % fetch a frame from obj_rtlsdr
    step(obj_spectrum,rtlsdr_data);      % update the spectrum analyzer

    run_time = run_time + rtlsdr_frmtime;  % update run_time

end
```

4.14 Summary

We hope that this chapter has successfully helped you familiarise yourself with the MATLAB and Simulink software, and that you now feel confident about proceeding on with the practical work in this book. If you want to spend some more time learning how to use the software (and how to use some of the more advanced features), check out the *Getting Started* guide in the MATLAB help documentation. Additional information can be found about the features of the software and the programming language on the MathWorks website, mathworks.com/help.

5 Complex Signals, Spectra and Quadrature Modulation

Samples that arrive in MATLAB and Simulink from the RTL-SDR are in a form that is generally termed 'complex baseband'. In this chapter, we will review and explain what this means in the context of the overall treatment of radio signals, and specifically in terms of the RTL-SDR. Simulation examples will also be presented to demonstrate the processes of quadrature modulation and demodulation. Further, to ensure clarity, we will be using both trigonometric (sines and cosines) *and* complex (exponential) mathematics, to define and represent our quadrature systems — remembering that both representations are of course equivalent, and just alternative mathematical ways of presenting the same processing.

The World is Real, and not Imaginary!

It is worth re-stating here that we live in the real world, and therefore only 'real' signals actually exist. The imaginary component of a complex signal is as the name suggests; it is imaginary, and does not exist in the real world. However, we often use a complex signal representation, as it turns out to be a convenient and tractable mathematical notation to describe the modulation and demodulation of two independent, quadrature modulated real signals (where one is depicted as being *real* and the other as *imaginary*). This representation allows us to use the simpler mathematics of complex exponentials when manipulating these trigonometric expressions. We can motivate the use of *simpler* by asking you to write down the trigonometric identity of $\cos(A)\cos(B)$, expressed as a sum of individual sines and cosines[1]. Perhaps, (like the authors of this book!) you struggle to remember the *precise* answer, and deriving it from first principles is not trivial. If we were dealing with complex exponentials, however, and we asked you the question — to express the product $e^{jA}e^{jB}$ as a single exponential — then the mathematics of raising to the power gives the simple answer of $e^{j(A+B)}$.

---

[1]. If you cannot quite remember the trigonometric identities, you can find them listed in Appendix B. As a quick reminder, $\cos(A)\cos(B) = 0.5\cos(A+B) + 0.5\cos(A-B)$.

This is precisely the point — it is far easier mathematically to work with the multiplication of complex exponentials than it is with the multiplication of trigonometric functions. Recalling Euler's Formula that $e^{jA} = \cos(A) + j\sin(A)$, then it might be advantageous to find a way to express our cosines and sines as complex exponentials. If we can represent signals used during the modulation and demodulation processes in complex exponential form, rather than sine/cosine form, then they might just be easier to work with in a mathematical sense.

Before we move on, remember that, if you really do not want to spend any time in the *imaginary* mathematical world, then you don't need to. The systems we are building and using are all processing real world signals — the complex mathematics and representation of imaginary signals just makes this easier for us. Stay real if you prefer!

5.1 Real and Complex Signals — it's all Sines and Cosines

In the real world, signals are *real*. In signal processing we work with analogue voltages to represent real world quantities such as air pressure, electromagnetic fields or temperature, which vary over time and may take on both positive and negative values with respect to some reference point. For instance, a voltage is induced within the antenna connected to your RTL-SDR in response to the changing electromagnetic field around it. This is a 'real' signal. A microphone picking up sound waves from small variations in air pressure produces a real analogue voltage signal, too. There is no such thing as, for example, an *imaginary* audio signal (or is there — if you imagine it, does it make it real to you ?!).

In analogue and digital communications, we often choose to represent signals in a form that appears to yield a *complex* signal. As outlined above, complex signals are *analytic* representations that can make the mathematics more tractable, but which do not exist in the real world. So, simply, an analytic signal is one that is essentially for analysis purposes only.

As every engineer and mathematician should know, *Euler's Formula* relates to the trigonometric world such that:

$$e^{j\omega t} = \cos(\omega t) + j\sin(\omega t), \quad (5.1)$$

where e is the base of the natural logarithm (a constant approximately equal to 2.71828...), ω in this case is an angular frequency $\omega = 2\pi f$, $j = \sqrt{-1}$, and t is time. This means that when we have a negative exponential, e.g. $e^{-j\omega t}$, we can say:

$$e^{-j\omega t} = \cos(\omega t) - j\sin(\omega t) \quad (5.2)$$

as $\cos(-\omega t) = \cos(\omega t)$ and $\sin(-\omega t) = -\sin(\omega t)$.

If we add Eqs. (5.1) and (5.2) together, we note that:

$$2\cos(\omega t) = e^{j\omega t} + e^{-j\omega t}, \quad (5.3)$$

and if we subtract Eq. (5.2) from Eq. (5.1), we get

$$j2\sin(\omega t) = e^{j\omega t} - e^{-j\omega t}. \quad (5.4)$$

Chapter 5: Complex Signals, Spectra and Quadrature Modulation

With some re-organisation of the above two equations, we can express our cosines and sines in terms of positive and negative powered complex notation:

$$\cos(\omega t) = \frac{e^{j\omega t} + e^{-j\omega t}}{2} \quad \text{and} \quad \sin(\omega t) = \frac{e^{j\omega t} - e^{-j\omega t}}{2j}. \tag{5.5}$$

Just as a refresher (with some High School and Bachelor mathematics), we can try to derive the trigonometric identity Eq. (B.4) from Appendix B,

$$\cos(A)\cos(B) = \frac{1}{2}\cos(A+B) + \frac{1}{2}\cos(A-B)$$

from the complex exponential starting point:

$$\begin{aligned}
\cos(A)\cos(B) &= \left[\frac{e^{jA} + e^{-jA}}{2}\right]\left[\frac{e^{jB} + e^{-jB}}{2}\right] \\
&= \frac{e^{jA}}{2}\left[\frac{e^{jB} + e^{-jB}}{2}\right] + \frac{e^{-jA}}{2}\left[\frac{e^{jB} + e^{-jB}}{2}\right] \\
&= \frac{1}{2}\left\{\left[\frac{e^{jA+jB} + e^{jA-jB}}{2}\right] + \left[\frac{e^{-jA+jB} + e^{-jA-jB}}{2}\right]\right\} \\
&= \frac{1}{2}\left\{\frac{e^{j(A+B)}}{2} + \frac{e^{j(A-B)}}{2} + \frac{e^{-j(A-B)}}{2} + \frac{e^{-j(A+B)}}{2}\right\} \\
&= \frac{1}{2}\left\{\frac{e^{j(A+B)}}{2} + \frac{e^{-j(A+B)}}{2}\right\} + \frac{1}{2}\left\{\frac{e^{j(A-B)}}{2} + \frac{e^{-j(A-B)}}{2}\right\} \\
&= \frac{1}{2}\cos(A+B) + \frac{1}{2}\cos(A-B)
\end{aligned} \tag{5.6}$$

So that was in fact not so difficult, and it is a relief that after nearly 250 years, Euler's incredible body of work still holds true (!). Euler's Formula remains probably the most significant equation in mathematics. If you need more convincing and proof by action, try to repeat the above for the similar identity $\sin(A)\sin(B)$. If we didn't have the trigonometric identity tables available or memorised (note they can be found in Appendix B), how would you have worked out this identity by any other first principles method? The answer is probably with great difficulty!

5.2 Viewing Real Signals in the Frequency Domain via Complex Spectra

Signals can be viewed and analysed in the frequency domain as well as the time domain, and doing so makes it much easier to understand their frequency content. Up to this point, we have been viewing signals extensively in the frequency domain, particularly in Chapter 3 where we explored the radio frequency spectrum. The RTL-SDR is however a quadrature receiver device, as was shown in Figure 1.8 (page 13), meaning that the signal received has both I (In Phase) and Q (Quadrature Phase) components.

This all relates to complex signal notation and representation. We will now aim to make matters as clear as possible, regarding the *how* and the *why* of using quadrature receivers, and how we might represent these signals purely in the real domain, or by utilising complex notation.

In this section, we present the important concepts of the frequency domain at a high level, and provide a brief overview of the associated mathematics (for more details, please refer to [29]). Additionally, a short introduction is provided to generating MATLAB and Simulink plots that represent signals in the frequency domain using real notation only, and also using complex notation.

5.2.1 Simple 'sum of three tones' signal in the frequency domain

First of all, it is worth re-establishing from basic first principles the merit that the frequency domain representation offers when viewing the form of a time varying signal.

Consider a simple signal composed of three sine waves, with frequencies 100Hz, 200Hz and 300Hz, and respective amplitudes, 10, 1, and 4.

$$s_1(t) = 10\cos(2\pi 100t) + \cos(2\pi 200t) + 4\cos(2\pi 300t). \tag{5.7}$$

When a few milliseconds of this signal are plotted in the time domain, we see the signal as shown in the time domain plot in Figure 5.1(a). Even for a very simple three-tone signal, it is always difficult to deduce the precise sinusoidal/frequency components a signal is composed of, purely from viewing it in the time domain. In this plot, it is possible to see that the signal is periodic — and by inspection of the time domain plot we can probably interpret a fundamental period of 0.01s (i.e. 100Hz), which we know to be true of course (frequency spectra interpretation is much easier if you know what frequencies are present!).

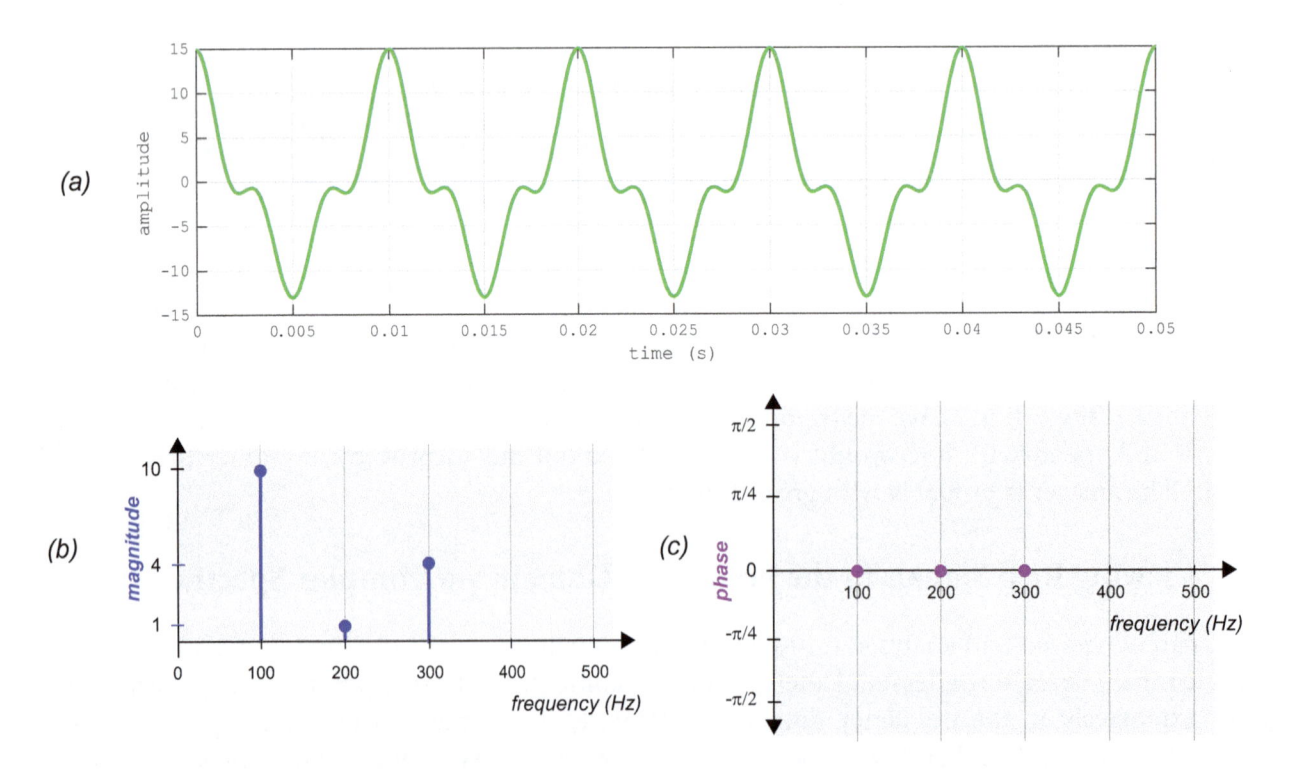

Figure 5.1: (a) Time domain representation of $s_1(t)$ from Eq. (5.7) versus time, t; (b) Magnitude spectrum of the $s_1(t)$ signal. (c) Phase spectrum of the $s_1(t)$ signal.

Chapter 5: Complex Signals, Spectra and Quadrature Modulation

If we were to represent the signal with magnitude-frequency and phase-frequency spectra, you would be able to more precisely determine its composition (in terms of sinusoidal frequency components) compared to simply inspecting its time domain plot. Figure 5.1(b) and (c) shows the magnitude and phase frequency plots of this signal which have been created by plotting the amplitudes of the composite tones at frequencies 100Hz, 200Hz, and 300Hz.

When signals are expressed by their equations, the process for plotting or representing in the frequency domain is a rather easy one, and we do not need to perform any frequency domain transforms or other analysis to do this — it can simply be done by inspection. Clearly when we do not have the equation of a signal however (usually the case!), we will take a computational approach and use the FFT functions within MATLAB and Simulink to calculate the spectra [29].

Exercise 5.1 Generating Frequency Domain Plots with MATLAB Code

In this exercise we will run a MATLAB script to plot the signal from Eq. (5.7) in the time and frequency domains.

(a) **Open MATLAB.** Set the working directory to the exercise folder,

   `/complex`

   This folder will be used for all of the examples in this section. Next, view the code at:

   `.../time_to_frequency_domain_cosines1.m`

   which contains the three sines waves of Eq. (5.7) sampled at $f_s = 10240$ Hz.

(b) **Inspect the m-code.** Note the generation of the three sine waves, the FFT function to calculate the frequency spectra, and code to generate the magnitude versus frequency plot.

(c) **Run the script.** Inspect the magnitude spectrum that is plotted. Note that the y-axis may have some scaling, however the relative amplitudes in the ratio of 10, 1, and 4 are correct for the three sine waves.

(d) **Challenge — spectral leakage.** In producing the frequency plots in this example we have *carefully* chosen the sampling frequency and the number of points in the FFT calculation to set a particular frequency *bin width*. The bin width determines the frequency resolution and discrete frequency spacing in the spectral plots. Bin width is given by $f_{bin} = f_s/\text{FFT Length}$, and for 1024 samples in the FFT, and $f_s = 10240$ Hz, the frequency resolution or bin width is exactly 10Hz. This is convenient as our frequencies of interest are all multiples of 10Hz and therefore sit exactly on FFT bins, and are easy to interpret when we view these plots.

Try changing the frequencies of the three sine waves to similar values of 97Hz, 203Hz and 302Hz, and then repeating the above steps. In the resulting spectral plots, note that the frequency energy has spread somewhat around the actual signal frequencies. The reason is that there are no frequency bins at precisely these new frequency values of 97Hz, 203Hz and 302Hz. The amplitudes are also no longer exactly in the correct ratios, albeit we can get an informative view of the energy/amplitude 'around' these frequencies. Viewing the phase plot is even more confusing!

These changes in the spectral plots are a result of **discontinuities** in the time domain signal, leading to **spectral leakage** in the frequency domain. It is the everyday problem of *windowing* and *FFTs*. If you have some time, being aware and knowledgeable of *FFT frequency resolution*, and *spectral leakage* and *windowing* is a very good thing. More on spectral leakage can be found in any good DSP textbook, for example [29].

Exercise 5.2 Generating Frequency Domain Plots with Simulink

In this exercise we will run a Simulink model that will plot the signal from Eq. (5.7) in the time and frequency domains.

(a) **Open MATLAB.** In this example we will repeat the same simulation as Exercise 5.1, but this time using Simulink. In the working directory set in Exercise 5.1, open the file:

   `/complex/time_to_frequency_domain_cosines2.slx`

which generates the three sines waves of Eq. (5.7) again sampled at $f_s = 10240$ Hz.

(b) **Run the simulation.** Note the spectrum plotted. This time the y-axis displays power (in dBm) on a logarithmic scale, but otherwise the spectrum shows the same information as was obtained by inspection in Figure 5.1(b). Therefore we can establish (as we already know), that we can extract the frequency content from a signal using the Fourier transform, calculated via the FFT routine.

5.2.2 'Sum of three tones' signal in the Complex Frequency Domain

For our three tone signal in Eq. (5.7), we can use the complex representation for the cosine function, as in Eq. (5.5), i.e. $\cos(\omega t) = (e^{j\omega t} + e^{-j\omega t})/2$, to represent the signal as a sum of complex exponentials:

$$\begin{aligned} s_1(t) &= 10\cos(2\pi 100 t) + \cos(2\pi 200 t) + 4\cos(2\pi 300 t) \\ &= 10\left(\frac{e^{j2\pi 100 t} + e^{-j2\pi 100 t}}{2}\right) + 1\left(\frac{e^{j2\pi 200 t} + e^{-j2\pi 200 t}}{2}\right) + 4\left(\frac{e^{j2\pi 300 t} + e^{-j2\pi 300 t}}{2}\right) \\ &= 5e^{j2\pi 100 t} + 5e^{-j2\pi 100 t} + \frac{1}{2}e^{j2\pi 200 t} + \frac{1}{2}e^{-j2\pi 200 t} + 2e^{j2\pi 300 t} + 2e^{-j2\pi 300 t} \end{aligned} \tag{5.8}$$

Grouping the positive (green) and negative (yellow) exponentials gives:

$$s_1(t) = 5e^{j2\pi 100 t} + \frac{1}{2}e^{j2\pi 200 t} + 2e^{j2\pi 300 t} \; + \; 5e^{-j2\pi 100 t} + \frac{1}{2}e^{-j2\pi 200 t} + 2e^{-j2\pi 300 t} . \tag{5.9}$$

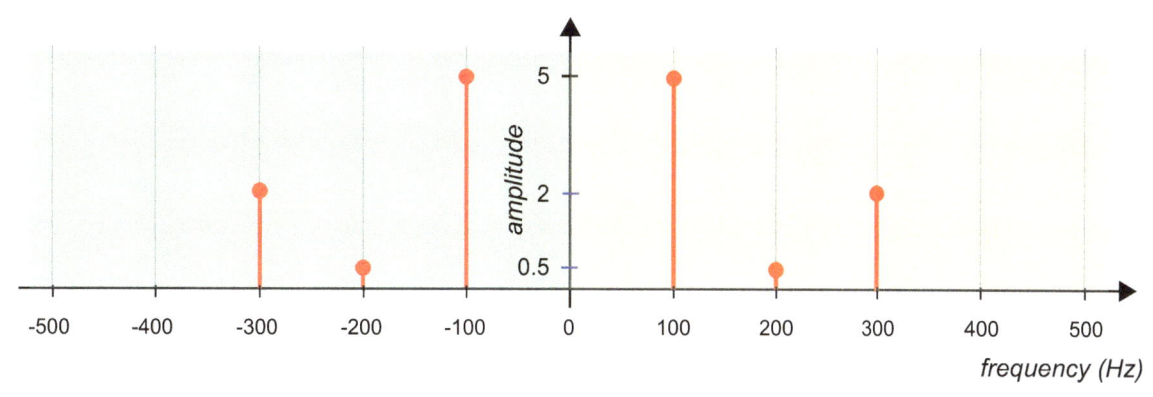

Chapter 5: Complex Signals, Spectra and Quadrature Modulation 177

Figure 5.2: Complex frequency spectrum of the signal $s_1(t)$ as defined in Eq. (5.7).

In Figure 5.1(b), we plotted the amplitudes of the cosine frequencies, whereas now as in Figure 5.2 we can display the same information by plotting the amplitudes of the *complex exponential terms*. This gives the positive exponential terms highlighted with yellow ☐, or sometimes called '*positive*' frequencies, and negative exponential terms highlighted with green ☐, sometimes called '*negative*' frequencies. The term 'negative' frequency is a confusing and perhaps misleading one. Preferably we should use the term negative complex exponential when making reference to a negative power exponential such as $e^{-j\omega t}$, but usually we do not; it's common practice to talk about positive and negative frequencies when referring to positive and negative power complex exponentials.

Looking at the right hand side of Figure 5.2, we can see that it is analogous to that of Figure 5.1(b). Noting again that $\cos(\omega t) = (e^{j\omega t} + e^{-j\omega t})/2$, the symmetric nature of the complex frequency plot means that a component at, for example 100Hz and -100Hz, add together to give a (real) cosine at 100Hz. The phase of the signal previously shown in Figure 5.1(c) is now implicitly presented in the complex spectra plots, although in this example there is no imaginary spectra. In the next example we will use MATLAB to generate the complex valued spectra for a real signal.

Exercise 5.3 Plotting Complex Spectra

This exercise is very similar to Exercise 5.1. Here you will run a MATLAB script to plot the signal from Eq. (5.7) in the time and frequency domains, but this time we plot the *complex frequency spectra*.

(a) **Open MATLAB.** Set the working directory an appropriate folder to open the following file

   /complex/time_to_complex_frequency_domain.m

   which contains the three sines waves of Eq. (5.7) sampled at $f_s = 10240\,\text{Hz}$.

(b) **Inspect the m-code.** Read through the code and note the generation of the three cosine waves, and the statements for computing and plotting the FFT.

(c) **Run the simulation.** The complex spectrum is plotted, showing the relative amplitudes of the positive and negative exponential terms. Aside from a different scaling on the y-axis, the magnitude spectrum is as expected, and equivalent to the sketch in Figure 5.2.

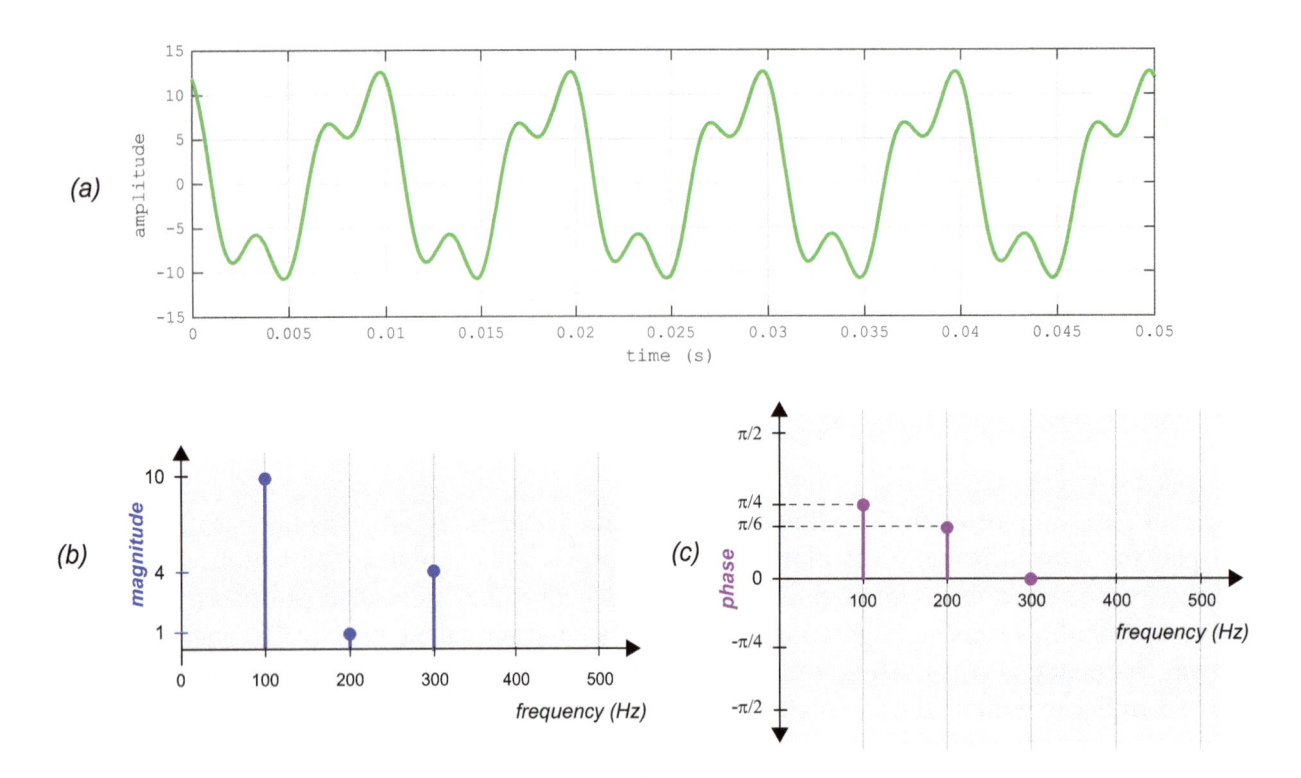

Figure 5.3: Phase Magnitude spectrum of the signal, $s_2(t)$ Eq. (5.10) versus time, t; (b) Magnitude spectrum of the $s_2(t)$ signal. (c) Phase spectrum of the $s_2(t)$ signal.

5.2.3   'Sum of three tones' signal, but now with phase shifts

For the signal described by Eq. (5.7) and shown in Figure 5.1, the phases of all components were zero (i.e. no deviation from pure cosine waves). Consider if the signal changed slightly and the 100Hz and 200Hz signals now had phase shifts:

$$s_2(t) = 10\cos\left(2\pi 100t + \frac{\pi}{4}\right) + \cos\left(2\pi 200t + \frac{\pi}{6}\right) + 4\cos(2\pi 300t) \tag{5.10}$$

We can see by comparing Figure 5.1(a) for Eq. (5.7) and Figure 5.3(a) for Eq. (5.10) that these signals have a different time domain form, but the magnitude-frequency plot of this signal as shown in Figure 5.3(b) is identical to Figure 5.1(b) (it has the same '*cosine*' frequency components with the same amplitude). To completely represent the signal in the frequency domain however, we also need to show the phase-frequency spectra which is the phase shift of a frequency from a standard cosine. While the phase spectra for signal $s_1(t)$ was zeroes for all frequency components as in Figure 5.1(c), for the signal $s_2(t)$ in Eq. (5.10) the phase spectrum $s_2(t)$ is as shown in Figure 5.3(c), and clearly highlights the phase shift components.

We can also plot the complex frequency spectra of the signal, $s_2(t)$. The phase shift components will make it slightly more complicated to derive than in Eq. (5.8) which produced the complex frequency spectra in Figure 5.2. In Figure 5.5 we however derive the complex exponential form from first principles, starting with the sum of cosines Eq. (5.10). Again we can see that the presence of the phase shifts makes the algebra more involved than in Eq. (5.8). Therefore in the next equation we can write Eq. (5.10) entirely in complex exponential form:

Chapter 5: Complex Signals, Spectra and Quadrature Modulation

$$s_2(t) = 10\cos(2\pi 100t + \pi/4) + \cos(2\pi 200t + \pi/6) + 4\cos(2\pi 300t)$$

$$= \left[3.53e^{j2\pi 100t} + 0.43e^{j2\pi 200t} + 2e^{j2\pi 300t}\right] + \left[3.53e^{-j2\pi 100t} + 0.43e^{-j2\pi 200t} + 2e^{-j2\pi 300t}\right] \quad (5.11)$$
$$+ j\left[3.53e^{j2\pi 100t} + 0.25e^{j2\pi 200t}\right] + j\left[-3.53e^{-j2\pi 100t} - 0.25e^{-j2\pi 200t}\right]$$

Note that compared to Eq. (5.9) and Figure 5.2 we now also have imaginary amplitude components. We can plot the spectra now (using four illustrative coloured components) as shown in Figure 5.4:

- ☐ real amplitude positive complex exponentials ('positive' frequencies),
- ☐ real amplitude negative complex exponentials ('negative' frequencies),
- ☐ imaginary amplitude positive complex exponentials ('positive' frequencies) and
- ☐ imaginary amplitude negative complex exponentials ('negative' frequencies).

There is no requirement to draw a 'phase' plot, since the phase is implicitly present with the real/imaginary representation. We can extract the phase information for our real signals in cosine form, as required, from the complex spectra using the $\tan^{-1}(.)$ function — see Figure 5.6. It is also a key point that, if we know a signal is real valued, then we also know that its real valued spectra is always even-symmetric (formed from any cosine terms), and that the imaginary valued spectra is always odd-symmetric (formed from any sine terms) — recall Eq. (5.5).

It is often more useful to plot the *magnitude spectra*, which gives the magnitude of the component at each complex exponential value, and is illustrated in Figure 5.6. This essentially takes us back to the magnitude/phase spectrum form of Figure 5.3, when just viewing the right hand side of Figure 5.6. Once again for a real valued signal only, this magnitude spectrum will always be even-symmetric, and hence we only plot the positive frequency values.

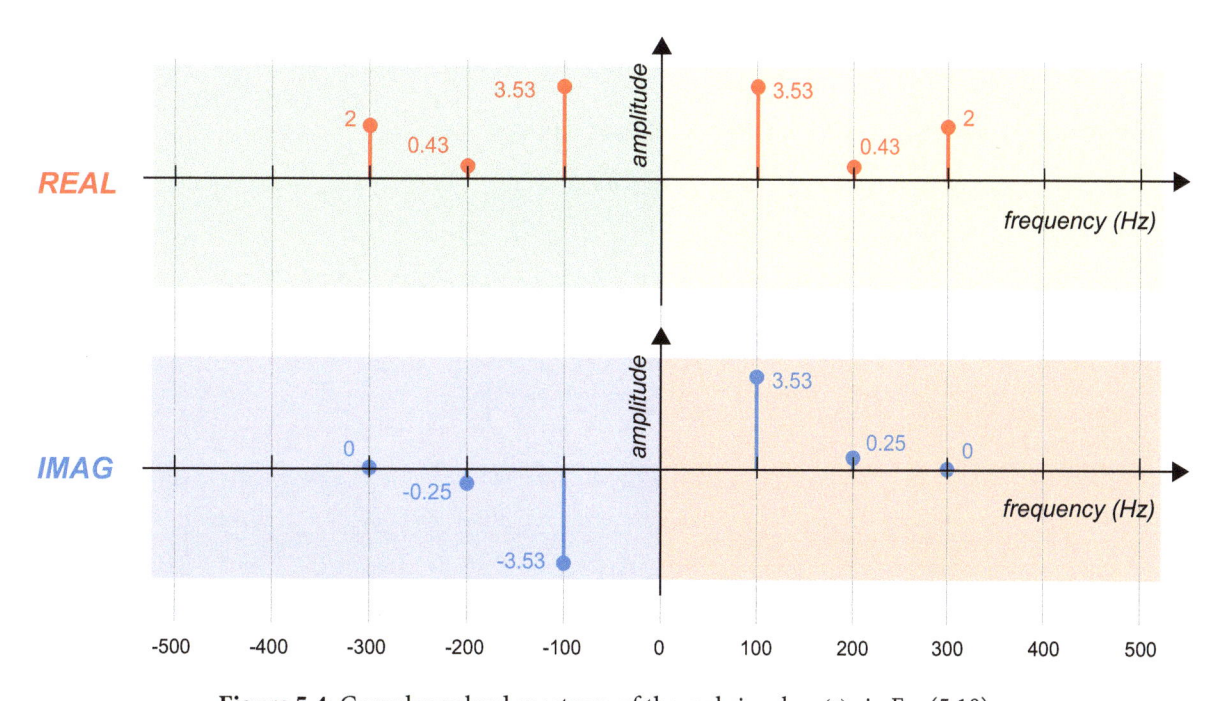

Figure 5.4: Complex valued spectrum of the real signal, $s_2(t)$, in Eq. (5.10)

$$s_2(t) = 10\cos(2\pi 100t + \pi/4) + \cos(2\pi 200t + \pi/6) + 4\cos(2\pi 300t)$$

First, simplifying the notation, denoting $A = 2\pi 100t$, $B = 2\pi 200t$ and $C = 2\pi 300t$:

$$= 10\cos(A + \pi/4) + \cos(B + \pi/6) + 4\cos(C)$$

Expressing the cosines as complex exponentials yields:

$$= 10\left[\frac{e^{j(A+\pi/4)} + e^{-j(A+\pi/4)}}{2}\right] + 1\left[\frac{e^{j(B+\pi/6)} + e^{-j(B+\pi/6)}}{2}\right] + 4\left[\frac{e^{jC} + e^{-jC}}{2}\right]$$

$$= 5\left[e^{j(A+\pi/4)} + e^{-j(A+\pi/4)}\right] + 0.5\left[e^{j(B+\pi/6)} + e^{-j(B+\pi/6)}\right] + 2\left[e^{jC} + e^{-jC}\right]$$

Noting that $e^{j(u+v)} = e^{ju}e^{jv}$, we can re-write as:

$$= 5e^{j(\pi/4)}e^{jA} + 5e^{-j(\pi/4)}e^{-jA} + 0.5e^{j(\pi/6)}e^{jB} + 0.5e^{-j(\pi/6)}e^{-jB} + 2e^{jC} + 2e^{-jC}$$

And using Euler's formula; $e^{j(\pi/4)} = \cos(\pi/4) + j\sin(\pi/4)$ and $e^{j(\pi/6)} = \cos(\pi/6) + j\sin(\pi/6)$

$$= 5(\cos(\pi/4) + j\sin(\pi/4))e^{jA} + 0.5(\cos(\pi/6) + j\sin(\pi/6))e^{jB} + 2e^{jC} + 2e^{-jC}$$
$$+ 5(\cos(\pi/4) - j\sin(\pi/4))e^{-jA} + 0.5(\cos(\pi/6) - j\sin(\pi/6))e^{-jB}$$

Noting $\cos(\pi/4) = 1/\sqrt{2}$, $\sin(\pi/4) = 1/\sqrt{2}$, $\cos(\pi/6) = \sqrt{3}/2$, & $\sin(\pi/6) = 1/2$

$$= 5\left(\frac{1}{\sqrt{2}} + j\frac{1}{\sqrt{2}}\right)e^{jA} + 5\left(\frac{1}{\sqrt{2}} - j\frac{1}{\sqrt{2}}\right)e^{-jA} + 0.5\left(\frac{\sqrt{3}}{2} + j\frac{1}{2}\right)e^{jB} + 0.5\left(\frac{\sqrt{3}}{2} - j\frac{1}{2}\right)e^{-jB} + 2e^{jC} + 2e^{-jC}$$

$$= \frac{5}{\sqrt{2}}e^{jA} + j\frac{5}{\sqrt{2}}e^{jA} + \frac{5}{\sqrt{2}}e^{-jA} - j\frac{5}{\sqrt{2}}e^{-jA} + \frac{\sqrt{3}}{4}e^{jB} + j\frac{1}{4}e^{jB} + \frac{\sqrt{3}}{4}e^{-jB} - j\frac{1}{4}e^{-jB} + 2e^{jC} + 2e^{-jC}$$

Grouping all of the positive exponentials, e^{jX}, and negative exponentials, e^{-jX} together:

$$= \left[\frac{5}{\sqrt{2}}e^{jA} + j\frac{5}{\sqrt{2}}e^{jA} + \frac{\sqrt{3}}{4}e^{jB} + j\frac{1}{4}e^{jB} + 2e^{jC}\right] + \left[\frac{5}{\sqrt{2}}e^{-jA} - j\frac{5}{\sqrt{2}}e^{-jA} + \frac{\sqrt{3}}{4}e^{-jB} - j\frac{1}{4}e^{-jB} + 2e^{-jC}\right]$$

Now grouping the real and imaginary scaled terms in each of the positive and negative exponentials:

$$= \left[\frac{5}{\sqrt{2}}e^{jA} + \frac{\sqrt{3}}{4}e^{jB} + 2e^{jC}\right] + j\left[\frac{5}{\sqrt{2}}e^{jA} + \frac{1}{4}e^{jB}\right] + \left[\frac{5}{\sqrt{2}}e^{-jA} + \frac{\sqrt{3}}{4}e^{-jB} + 2e^{-jC}\right] + j\left[-\frac{5}{\sqrt{2}}e^{-jA} - \frac{1}{4}e^{-jB}\right]$$

Replacing the scaling terms with numbers to two decimal places:

$$= \left[3.53e^{jA} + 0.43e^{jB} + 2e^{jC}\right] + j\left[3.53e^{jA} + 0.25e^{jB}\right]$$
$$+ \left[3.53e^{-jA} + 0.43e^{-jB} + 2e^{-jC}\right] + j\left[-3.53e^{-jA} - 0.25e^{-jB}\right]$$

$$= \left[3.53e^{jA} + 0.43e^{jB} + 2e^{jC}\right] + \left[3.53e^{-jA} + 0.43e^{-jB} + 2e^{-jC}\right]$$
$$+ j\left[3.53e^{jA} + 0.25e^{jB}\right] + j\left[-3.53e^{-jA} - 0.25e^{-jB}\right]$$

Substituting back for $A = 2\pi 100t$, $B = 2\pi 200t$ and $C = 2\pi 300t$,

$$s_2(t) = \left[3.53e^{j2\pi 100t} + 0.43e^{j2\pi 200t} + 2e^{j2\pi 300t}\right] + \left[3.53e^{-j2\pi 100t} + 0.43e^{-j2\pi 200t} + 2e^{-j2\pi 300t}\right]$$
$$+ j\left[3.53e^{j2\pi 100t} + 0.25e^{j2\pi 200t}\right] + j\left[-3.53e^{-j2\pi 100t} - 0.25e^{-j2\pi 200t}\right]$$

Figure 5.5: First principles calculation of the complex spectrum of Eq. (5.10).

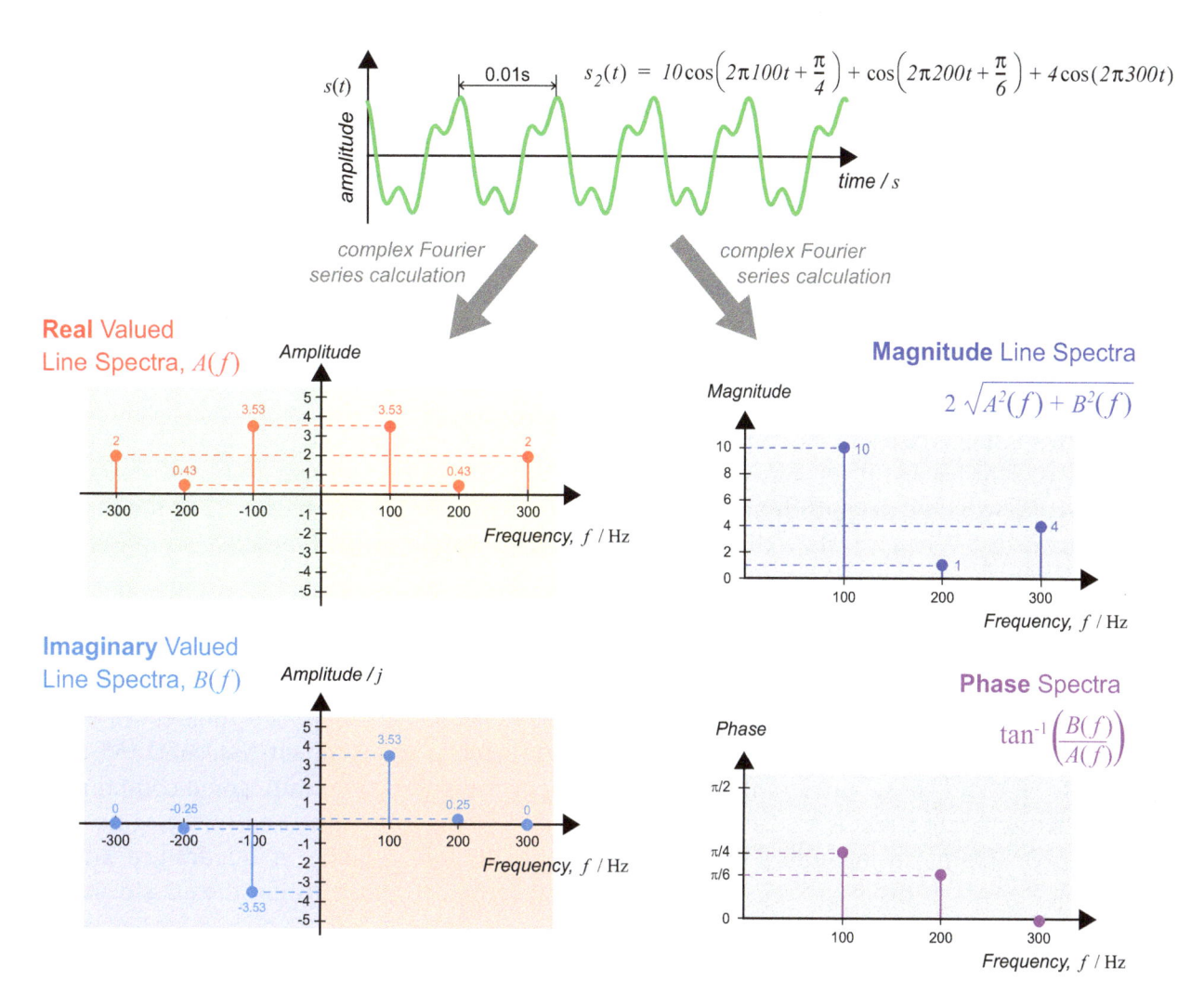

Chapter 5: Complex Signals, Spectra and Quadrature Modulation 181

$$s_2(t) = 10\cos\left(2\pi 100t + \frac{\pi}{4}\right) + \cos\left(2\pi 200t + \frac{\pi}{6}\right) + 4\cos(2\pi 300t)$$

Figure 5.6: Representing a real signal as a complex spectra (left hand side), and then realising the more intuitive magnitude and phase spectra (right hand side), where we can easily correlate with the signal equation, Eq. (5.10) and (top).

Exercise 5.4 Complex Spectra for Three Real Sine Waves

This exercise involves running a script to plot the complex magnitude and phase spectra of the three real cosine waves, where two of these cosines have phase shifts as in Eq. (5.10) and Figure 5.6.

(a) **Open MATLAB.** Set the working directory to the `complex` exercise folder and open,

   `/complex/three_cosines_complex_spectra_plot.m`

(b) **Inspect the m-code.** Inspect the m-code and note the generation of the same three cosine waves as in the previous exercise, and sampled again at $f_s = 10240\,\text{Hz}$.

(c) **Run the simulation.** View the spectra that are calculated by MATLAB and plotted. Aside from a different scaling on the y-axis, the spectra should be the same as derived from inspection of the last lines of Eq. (5.11) and plotted in Figure 5.4 and Figure 5.6.

> (d) **Challenge — phase plot 'decluttering'**. You might notice the very small impulse that is also added to this signal in the code. This is to ensure that finite arithmetic precision effects (floating point is accurate, but it's still finite!), do not cause random phase values (between 0 and 360°) to appear for those frequencies that have infinitesimally small amplitudes. The adding of the impulse generates a very small, zero phase cosine at all frequency points in the FFT. To learn more you could remove the impulse, and rerun — note the difference in the phase plot: it's still correct at the frequencies of interest, but you will now see that extensive clutter appears in the phase spectrum plot.

Based on this last exercise and looking again at Figure 5.4, does viewing the real signal in the complex world help you to visualise it? Probably not! But that's the way we often present and analyse signals. For a real signal, viewing the magnitude and phase spectra as in Figure 5.1 and Figure 5.3 is the easiest and most intuitive way. When we take a closer look at quadrature signals, we will find that complex notation and complex spectra become very useful.

5.3 Standard Amplitude Modulation

Often we work with quadrature signals, where two independent or separate 'phases' of data are transmitted. Quadrature signals are created by modulating two independent baseband information signals onto sine and cosine carriers at the same frequency, generating two orthogonal components. In this way, two separate streams of data can be transmitted at the same time and in the same band of frequencies, without causing interference to each other. Before looking at quadrature amplitude modulation (QAM) in detail in Section 5.4, we will first recap on, and consider the classic amplitude modulation process.

5.3.1 Double Sideband Suppressed Carrier Amplitude Modulation

The amplitude modulation of a low frequency baseband signal, $g(t)$, with a high frequency carrier, $c(t)$, is shown in Figure 5.7 and Figure 5.8. If $g(t)$ was just a single cosine wave, i.e. $g(t) = A\cos(2\pi f_b t)$, then simple trigonometry confirms that the modulated signal will be:

$$s(t) = A\cos(2\pi f_b t)\cos(2\pi f_c t) = \frac{A}{2}\left[\cos(2\pi(f_c - f_b)t) + \cos(2\pi(f_c + f_b)t)\right] \quad (5.12)$$

where A is the amplitude of the baseband signal, and f_b and f_c are the frequencies of the baseband signal, and carrier, respectively.

The result of plotting $s(t)$ with simple magnitude line spectra is illustrated in Figure 5.8. If we now consider the transmission of a more general and wider frequency baseband signal, $g(t)$, with frequency components from 0 to f_b Hz, and denote frequency representation of $g(t)$ as $G(f)$, we can illustrate the baseband frequencies being modulated as a double sideband suppressed carrier amplitude modulation (AM-DSB-SC). This produces an upper and a lower sideband as shown in Figure 5.9.

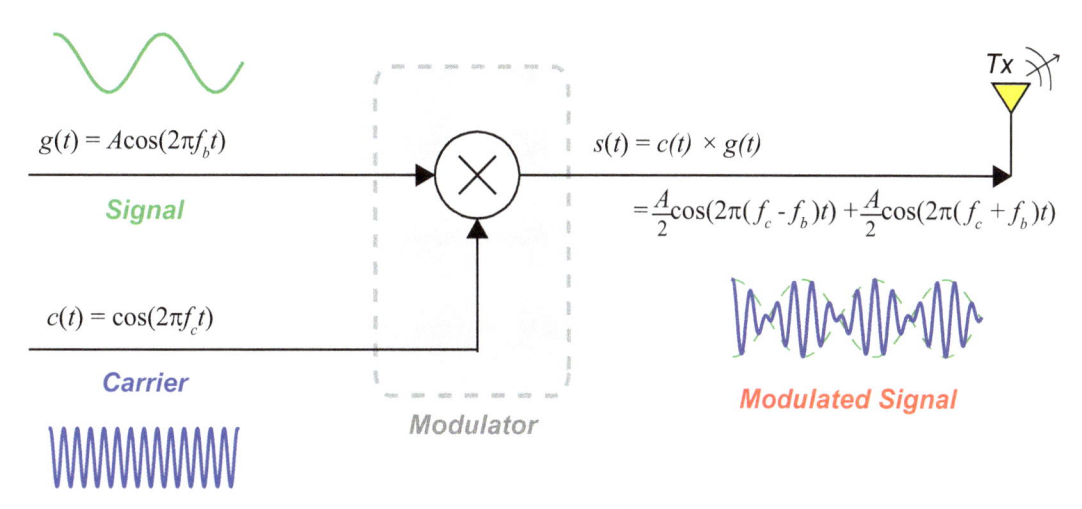

Chapter 5: Complex Signals, Spectra and Quadrature Modulation

Figure 5.7: Modulating a low frequency cosine signal onto a high frequency carrier.

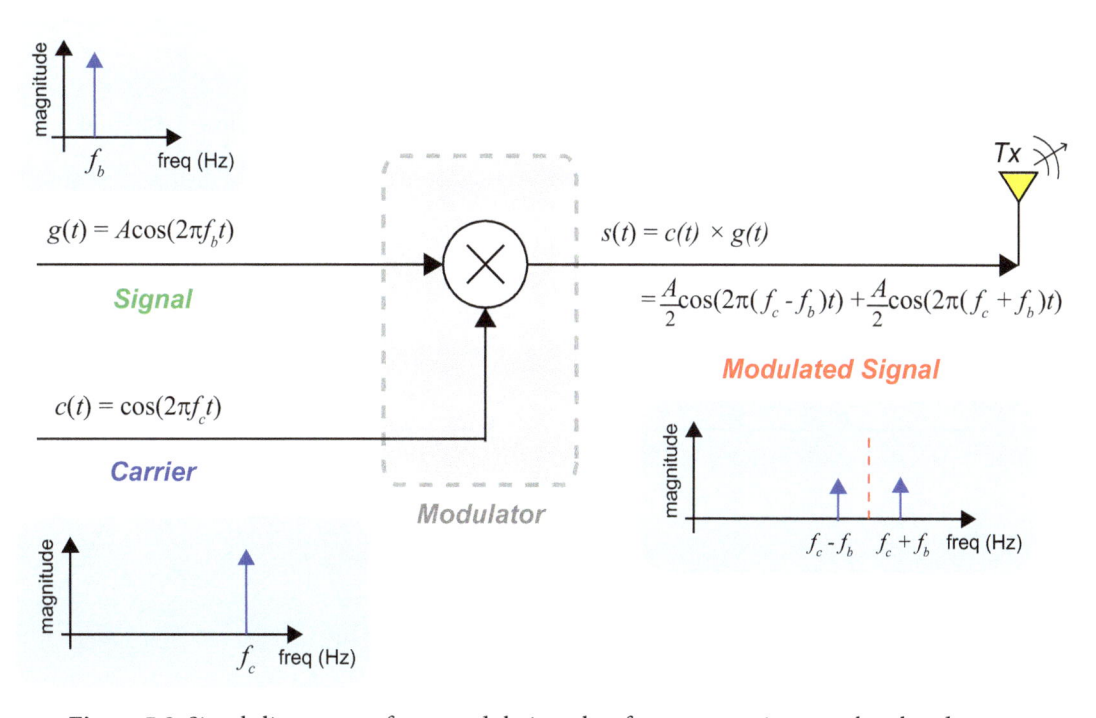

Figure 5.8: Simple line spectra from modulating a low frequency cosine wave baseband signal onto a high frequency carrier.

5.3.2 Amplitude Demodulation

Demodulation of a simple AM signal is relatively straightforward to represent mathematically, as long as the local oscillator in the receiver generates *exactly* the same frequency and phase as the carrier tone (cosine wave) and a lowpass filter is included, as shown in Figure 5.10. (The chances of the local oscillator at the receiver having *exactly* the same frequency and phase as the carrier is unlikely, and we will leave the practicalities of this challenge to a first discussion in Chapter 7, where we will consider phase locking and carrier synchronisation.)

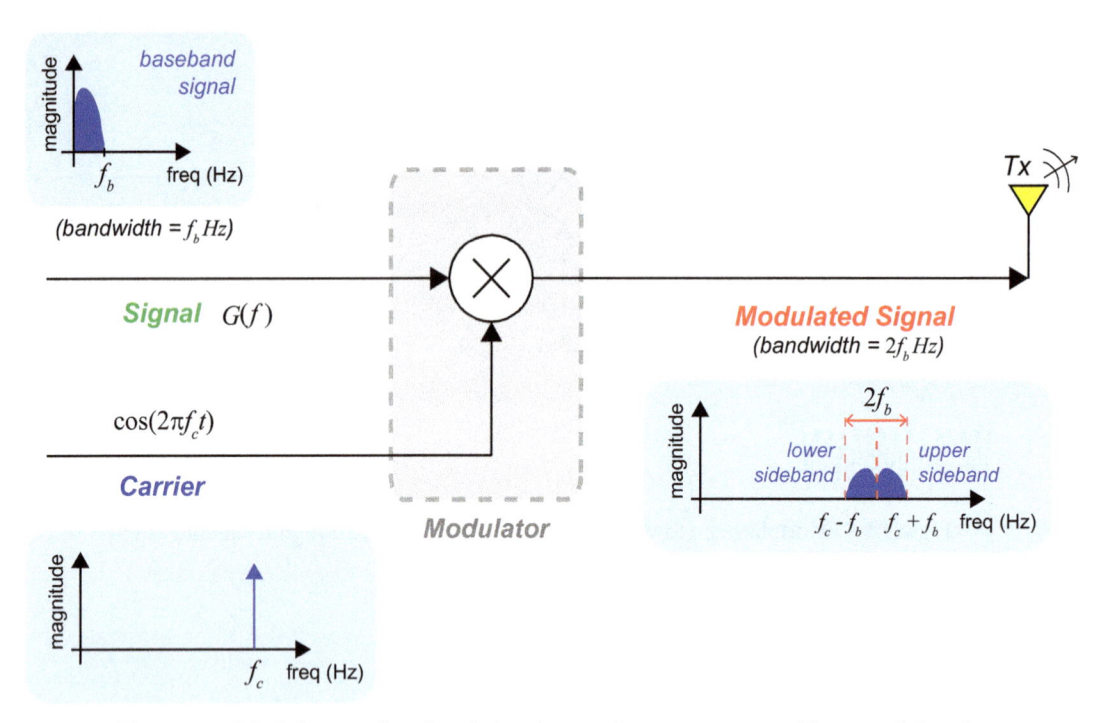

Figure 5.9: Modulating a baseband signal to produce an upper and lower sideband.

$$s(t) = \frac{A}{2}\cos(2\pi(f_c - f_b)t) + \frac{A}{2}\cos(2\pi(f_c + f_b)t)$$

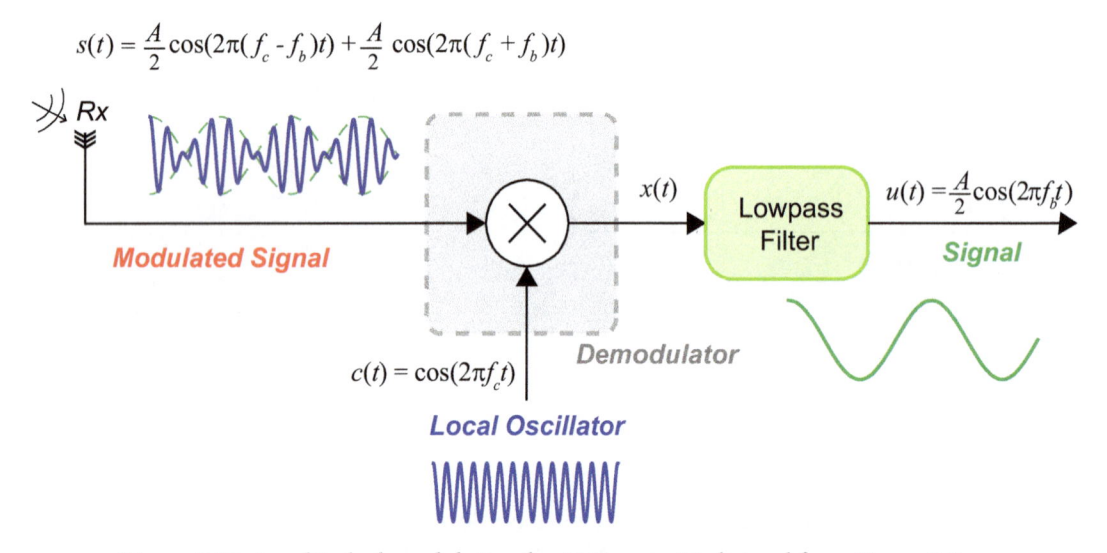

Figure 5.10: Amplitude demodulating the AM transmitted signal from Figure 5.7.

The received modulated signal from the antenna in Figure 5.10,

$$s(t) = \frac{A}{2}\cos\left(2\pi(f_c - f_b)t\right) + \frac{A}{2}\cos\left(2\pi(f_c + f_b)t\right) \tag{5.13}$$

is *'perfectly'* demodulated by multiplying with a local carrier signal, $c(t) = \cos(2\pi f_c t)$:

Chapter 5: Complex Signals, Spectra and Quadrature Modulation 185

$$x(t) = c(t) \times s(t)$$
$$= \frac{A}{2}\cos(2\pi f_c t)\left[\cos(2\pi(f_c - f_b)t) + \cos(2\pi(f_c + f_b)t)\right] \quad (5.14)$$
$$= \frac{A}{2}\cos(2\pi f_c t)\cos(2\pi(f_c - f_b)t) + \frac{A}{2}\cos(2\pi f_c t)\cos(2\pi(f_c + f_b)t)$$
$$= \frac{A}{4}\cos(2\pi(2f_c - f_b)t) + \frac{A}{4}\cos(2\pi f_b t) + \frac{A}{4}\cos(2\pi(2f_c + f_b)t) + \frac{A}{4}\cos(2\pi f_b t)$$
$$= \frac{A}{2}\cos(2\pi f_b t) + \left[\frac{A}{4}\cos(2\pi(2f_c - f_b)t) + \frac{A}{4}\cos(2\pi(2f_c + f_b)t)\right]$$

where we used the trigonometric identity $\cos(A)\cos(B) = 0.5\cos(A+B) + 0.5\cos(A-B)$ in lines 3 to 4 of Eq. (5.14). Applying the lowpass filter, the two terms at the high frequencies (sitting around $2f_c$, twice the carrier) are attenuated, leaving a scaled version of the signal that was transmitted in Figure 5.7:

$$\Rightarrow u(t) = \frac{A}{2}\cos(2\pi f_b t) + \underbrace{\left[\frac{A}{4}\cos 2\pi(2f_c - f_b) + \frac{A}{4}\cos 2\pi(2f_c + f_b)\right]}_{\text{lowpass filtered terms}}$$
$$= \frac{A}{2}\cos(2\pi f_b t) \quad (5.15)$$
$$= \frac{g(t)}{2}$$

When the baseband signal illustrated in Figure 5.9 (with frequencies over the range 0Hz to f_b Hz) is received, its magnitude spectrum has the form shown in Figure 5.11. The lowpass filter in the receiver attenuates frequencies above the maximum frequency of the baseband signal, i.e. cut-off is around f_b Hz.

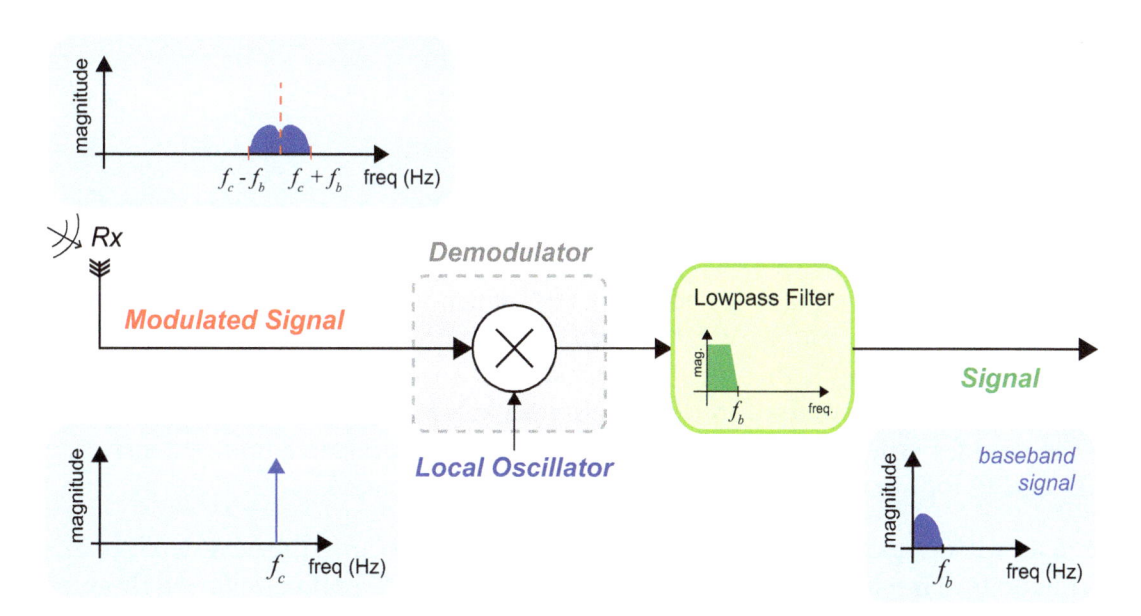

Figure 5.11: Amplitude demodulating the AM transmitted baseband signal from Figure 5.9.

5.3.3 Amplitude Demodulation Phase Error

Note that, if the local oscillator is not exactly in phase with the received signal, the resulting phase error may cause a variable gain in the output. Of course a phase error (or even a small frequency error) is almost definitely the case, as the local oscillator is likely to have a frequency/phase error associated with it, hence the reason why we need carrier synchronisation to be performed by the receiver.

To simply illustrate the effect of a phase error, we can add a phase shift of θ to the local oscillator, $c(t)$ in Figure 5.10. The demodulation presented previously in Eq. (5.14) now becomes:

$$
\begin{aligned}
x(t) &= c(t) \times s(t) \\
&= \cos(2\pi f_c t + \theta)\left[\frac{A}{2}\cos(2\pi(f_c - f_b)t) + \frac{A}{2}\cos(2\pi(f_c + f_b)t)\right] \\
&= \frac{A}{2}\left[\cos(2\pi f_c t + \theta)\cos(2\pi(f_c - f_b)t) + \cos(2\pi f_c t + \theta)\cos(2\pi(f_c + f_b)t)\right] \quad (5.16) \\
&= \frac{A}{4}\cos(2\pi(2f_c - f_b)t + \theta) + \frac{A}{4}\cos(2\pi f_b t + \theta) + \frac{A}{4}\cos(2\pi(2f_c + f_b)t + \theta) + \frac{A}{4}\cos(2\pi f_b t - \theta) \\
&= \frac{A}{4}\cos(2\pi f_b t + \theta) + \frac{A}{4}\cos(2\pi f_b t - \theta) + \left[\frac{A}{4}\cos(2\pi(2f_c - f_b)t + \theta) + \frac{A}{4}\cos(2\pi(2f_c + f_b)t + \theta)\right]
\end{aligned}
$$

After lowpass filtering of the high frequency components, we can use one of the trigonometric identities (see Appendix B on page 581) to obtain the following:

$$
\begin{aligned}
u(t) &= \frac{A}{4}\cos(2\pi f_b t + \theta) + \frac{A}{4}\cos(2\pi f_b t - \theta) + \left[\frac{A}{4}\cos(2\pi(2f_c - f_b + \theta)t) + \frac{A}{4}\cos(2\pi(2f_c + f_b + \theta)t)\right] \\
&\quad \text{\textit{lowpass filtered terms}} \\
&= \frac{A}{4}\cos(2\pi f_b t + \theta) + \frac{A}{4}\cos(2\pi f_b t - \theta) \quad (5.17) \\
&= \frac{A}{2}\cos(2\pi f_b t)\cos(\theta) \\
&= \frac{A}{2}g(t)\cos(\theta)
\end{aligned}
$$

Hence the amplitude of the output is scaled by $\cos(\theta)$, which will of course be a value between 1 and -1. In the extreme case of $\theta = \pm\pi/2$, or $\pm 90°$, then the output is zero! If we recognise any time varying phase error as denoted by $\theta(t)$, then this is equivalent to a frequency error and will manifest as the received signal being frequency offset, and the amplitude likely fading in and out. (Once again this simply illustrates the need for synchronisation and **phase locking** at the receiver, a topic extensively covered in Chapters 11 and 12 for our digital communications RTL-SDR designs.)

In Chapter 6 we will discuss a number of variants of simple AM, however the next step is to motivate and derive the quadrature modulation methods for transmission and reception, followed by an explanation of how we then analyse and define these processes using *complex baseband signals* and *complex exponentials*.

5.4 Quadrature Modulation and Demodulation (QAM)

We now get to the '*complex*' representation point where we can introduce quadrature amplitude demodulation (QAM) as performed by the RTL-SDR. First, let's motivate *why* we use QAM.

Referring back to Figure 5.9, note that to transmit a baseband signal of bandwidth f_b Hz, using simple AM modulation, we require $2f_b$ Hz. In an attempt to achieve more bandwidth-efficient signalling, we can present quadrature modulation, where we transmit two signals of bandwidth f_b Hz, both on the same carrier frequency, but we separate the carrier phases by 90° (i.e. quadrature carrier), one being a sine wave and the other a cosine wave. This is illustrated in Figure 5.12. Note that the sine wave has a negative amplitude (i.e. $-\sin 2\pi f_c t$). This is by convention in order that the complex representation derived in Section 5.5 matches more exactly. The QAM transmission and reception model can easily be shown to work with the quadrature pair of $\{\cos 2\pi f_c t, \sin 2\pi f_c t\}$ or $\{\cos 2\pi f_c t, -\sin 2\pi f_c t\}$.

To illustrate quadrature modulation, we will use a simple example of two independent baseband signals denoted as

$$g_1(t) \text{ and } g_2(t). \tag{5.18}$$

By convention the cosine modulating channel is referred to as the *In-Phase* or *I*-channel or '*real*' channel, and the sine modulating channel is referred to as the *Quadrature Phase* or *Q*-channel or '*imaginary*' channel. Note that '-sine' is one quadrant or 90° or $\pi/2$ radians away from the 'cosine', hence the name *quadrature*; $-\sin(2\pi f_c t) = \cos(2\pi f_c t + \pi/2)$.

If we have now quadrature modulated our baseband signals the transmitter will produce the following:

$$y(t) = g_1(t)\cos(2\pi f_c t) - g_2(t)\sin(2\pi f_c t) \tag{5.19}$$

Can we quadrature demodulate the received signals and get the baseband signals back? The answer is yes (of course!) and to illustrate this, we will quadrature demodulate Eq. (5.19) using the scheme shown in

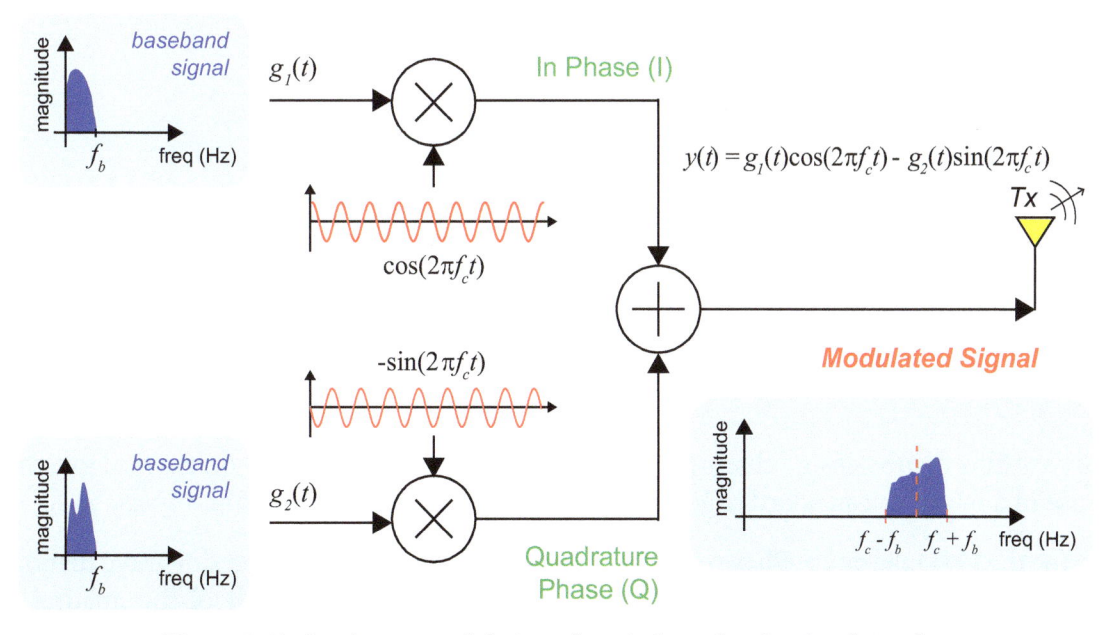

Figure 5.12: Quadrature modulation of two independent baseband signals.

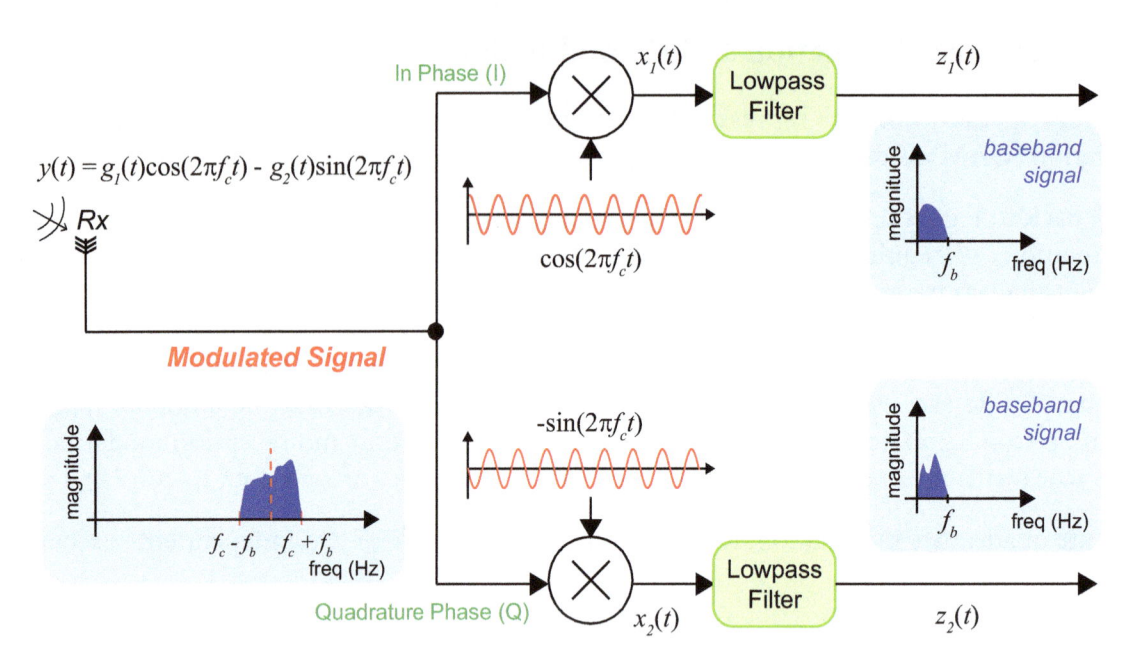

Figure 5.13: Quadrature demodulation of quadrature modulated signals

Figure 5.13. Consider the received signal on the left hand side of Figure 5.13, assuming the ideal case of no channel degradation (i.e. the perfect channel) for the signal transmitted in Figure 5.12 / Eq. (5.19).

We can now demonstrate with some simple trigonometry that the quadrature ampltitude modulated (QAM) receiver will work, and allow the $g_1(t)$ and $g_2(t)$ transmitted baseband signals to be recovered from the received $y(t)$ signal. Although we have a very simple block diagram of the quadrature receiver, if you compare Figure 5.13 to the RTL-SDR block diagram in Figure 1.8 (page 13), then you can see the same quadrature receiver components in the RTL2832U device on the right hand side of the figure.

For the I (In Phase, or cosine) channel, the output after the cosine demodulator is:

$$\begin{aligned} x_1(t) &= y(t)\cos(2\pi f_c t) \\ &= \left[g_1(t)\cos(2\pi f_c t) - g_2(t)\sin(2\pi f_c t)\right]\cos(2\pi f_c t) \\ &= g_1(t)\cos^2(2\pi f_c t) - g_2(t)\sin(2\pi f_c t)\cos(2\pi f_c t) \\ &= \frac{1}{2}g_1(t)[1 + \cos(4\pi f_c t)] - \frac{1}{2}g_2(t)\sin(4\pi f_c t) \\ &= \frac{1}{2}g_1(t) + \left[\frac{1}{2}g_1(t)\cos(4\pi f_c t) - \frac{1}{2}g_2(t)\sin(4\pi f_c t)\right] \text{ lowpass filtered terms} \end{aligned} \qquad (5.20)$$

After lowpass filtering the cosine channel output is: $z_1(t) = \text{LPF}\{x_1(t)\} = 0.5g_1(t)$. Note that we are making use of a few of the standard trigonometric identities listed in Appendix B (page 581).

Similarly for the Q (Quadrature Phase or sine) channel, the input signal to the sine wave demodulator is the same as that supplied to the cosine channel, Eq. (5.19), and the output of the quadrature (sine) demodulator and lowpass filter is:

Chapter 5: Complex Signals, Spectra and Quadrature Modulation

$$x_2(t) = y(t)(-\sin(2\pi f_c t))$$

$$= \left[g_1(t)\cos(2\pi f_c t) - g_2(t)\sin(2\pi f_c t)\right](-\sin(2\pi f_c t))$$

$$= [-g_1(t)\cos(2\pi f_c t)\sin(2\pi f_c t)] + g_2(t)\sin^2(2\pi f_c t) \quad (5.21)$$

$$= -\frac{1}{2}g_1(t)\sin(4\pi f_c t) + \frac{1}{2}g_2(t)[1 - \cos(4\pi f_c t)]$$

$$= \frac{1}{2}g_2(t) - \left[\frac{1}{2}g_1(t)\sin(4\pi f_c t) + \frac{1}{2}g_2(t)\cos(4\pi f_c t)\right] \text{ lowpass filtered terms}$$

Therefore, after lowpass filtering and scaling: $z_2(t) = \text{LPF}\{x_2(t)\} = 0.5g_2(t)$.

5.4.1 Quadrature Receiver Phase Shift

If there is a phase shift of θ on the local oscillator, then the quadrature output signal will be mixed with the In Phase component. This means that instead of the signal from Eq. (5.20), the I channel contains:

$$x_1(t) = y(t)\cos(2\pi f_c t + \theta)$$

$$= \left[g_1(t)\cos(2\pi f_c t) - g_2(t)\sin(2\pi f_c t)\right]\cos(2\pi f_c t + \theta)$$

$$= g_1(t)\cos(2\pi f_c t)\cos(2\pi f_c t + \theta) - g_2(t)\sin(2\pi f_c t)\cos(2\pi f_c t + \theta) \quad (5.22)$$

$$= \frac{1}{2}g_1(t)\left[\cos(-\theta) + \cos(4\pi f_c t + \theta)\right] - \frac{1}{2}g_2(t)\left[\sin(-\theta) + \sin(4\pi f_c t + \theta)\right]$$

Noting that $\cos(-x) = \cos(x)$ and $\sin(-x) = -\sin(x)$,

$$= \frac{1}{2}g_1(t)\left[\cos(\theta) + \cos(4\pi f_c t + \theta)\right] - \frac{1}{2}g_2(t)[-\sin(\theta) + \sin(4\pi f_c t + \theta)]$$

$$= \frac{1}{2}\left[g_1(t)\cos(\theta) + g_2(t)\sin(\theta)\right] + \left[\frac{1}{2}g_1(t)\cos(4\pi f_c t + \theta) - \frac{1}{2}g_2(t)\sin(4\pi f_c t + \theta)\right] \text{ lowpass filtered terms}$$

Similarly, the signal on the Q channel takes the following form:

$$x_2(t) = y(t)(-\sin(2\pi f_c t + \theta))$$

$$= \left[g_1(t)\cos(2\pi f_c t) - g_2(t)\sin(2\pi f_c t)\right](-\sin(2\pi f_c t + \theta))$$

$$= -g_1(t)\cos(2\pi f_c t)\sin(2\pi f_c t + \theta) + g_2(t)\sin(2\pi f_c t)\sin(2\pi f_c t + \theta) \quad (5.23)$$

$$= -\frac{1}{2}g_1(t)[-\sin(-\theta) + \sin(4\pi f_c t + \theta)] + \frac{1}{2}g_2(t)[\cos(-\theta) - \cos(4\pi f_c t + \theta)]$$

Noting that $\cos(-x) = \cos(x)$ and $\sin(-x) = -\sin(x)$,

$$= -\frac{1}{2}g_1(t)[\sin(\theta) + \sin(4\pi f_c t + \theta)] + \frac{1}{2}g_2(t)[\cos(\theta) - \cos(4\pi f_c t + \theta)]$$

$$= \frac{1}{2}\Big[-g_1(t)\sin(\theta) + g_2(t)\cos(\theta)\Big] - \Big[\frac{1}{2}g_1(t)\sin(4\pi f_c t + \theta) + \frac{1}{2}g_2(t)\cos(4\pi f_c t + \theta)\Big] \quad \text{lowpass filtered terms}$$

Therefore when the local oscillator has a shift of θ degrees, the demodulated signals are mixed versions of the two transmitted signals, $g_1(t)$ and $g_2(t)$, scaled by 0.5, i.e.

$$\begin{aligned} z_1(t) &= 0.5[g_1(t)\cos(\theta) + g_2(t)\sin(\theta)] \\ z_2(t) &= 0.5[-g_1(t)\sin(\theta) + g_2(t)\cos(\theta)] \end{aligned} \quad (5.24)$$

Figure 5.14 shows a simple quadrature modulator and demodulator with (lowpass) bandlimiting transmit filters, and matched filters included (a typical filter is a root raised cosine that will be used in the exercises in Chapter 11, Section 11.2, page 437).

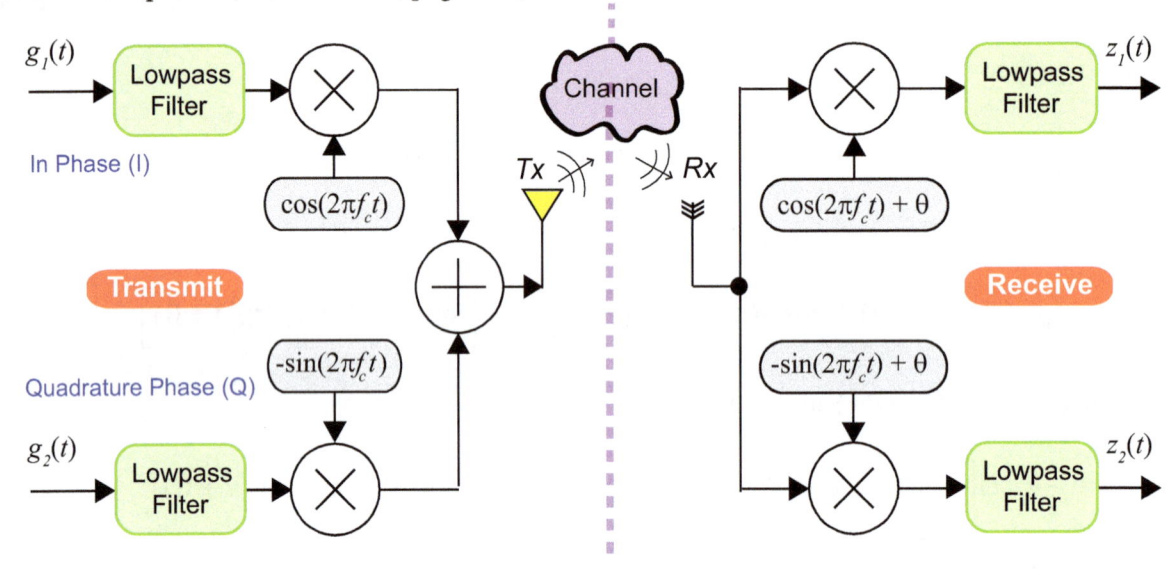

Figure 5.14: Quadrature modulation and demodulation with a phase shift.

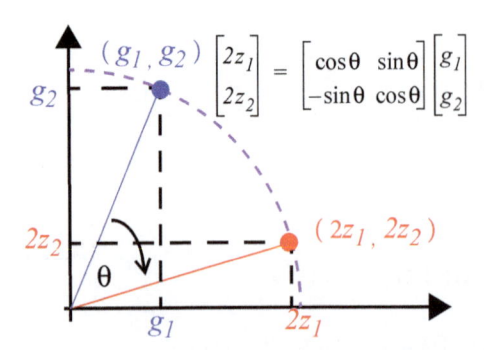

$$\begin{bmatrix} 2z_1 \\ 2z_2 \end{bmatrix} = \begin{bmatrix} \cos\theta & \sin\theta \\ -\sin\theta & \cos\theta \end{bmatrix} \begin{bmatrix} g_1 \\ g_2 \end{bmatrix}$$

Figure 5.15: Cartesian plane rotation.

Take a closer look at Eq. (5.24), if g_1 and g_2 were interpreted as points (at a given sample time) in the Cartesian (x-y) plane, the resulting points $2z_1$ and $2z_2$ are just the $(x, y) = (g_1, g_2)$ points rotated about the origin by θ degrees as illustrated in Figure 5.15. (Note we multiply by 2 to account for the 0.5 scaling in Eq. (5.24)). This is one of the classic problems of digital communications, where received signals are sampled and processed to recover the transmitted constellation, however a phase error means that the received points are 'rotated'. Or if the carrier has a small frequency error, then the phase error, $\theta(t)$, is constantly changing and we see a spinning constellation.

Chapter 5: Complex Signals, Spectra and Quadrature Modulation

In Chapter 11 and Chapter 12 we will observe exactly this problem, and will design DSP receivers to calculate the phase errors, and effectively *de-rotate* the constellation. If you look at the constellation plots in later chapters, such as in Chapter 11, Figure 11.16 (page 450) or in Chapter 12 in Exercise 12.7 (page 520) we see precisely the effect of this angle rotation.

5.5 Quadrature Amplitude Modulation using Complex Notation

It is easy to confirm that the QAM system of Figure 5.14 is entirely composed of real signals. There are no *imaginary* signals therein. So it's 100% *not* a complex system and does not use any complex arithmetic.

But, if we introduce some complex notation, one outcome will be that we can make the associated mathematics easier and more tractable. As above we can describe and analyse this QAM system entirely with real arithmetic and trigonometric identities, however we will now introduce complex notation to redefine the equations to describe the QAM modulation.

Consider the block diagram of Figure 5.16 which shows a complex baseband signal, $g(t)$ composed of real component, $g_1(t)$ and imaginary component, $g_2(t)$. Therefore, we have defined $g(t)$ as:

$$g(t) = g_1(t) + jg_2(t) \tag{5.25}$$

and have a complex exponential carrier frequency at f_c Hz, i.e. $e^{j2\pi f_c t} = \cos(2\pi f_c t) + j\sin(2\pi f_c t)$.

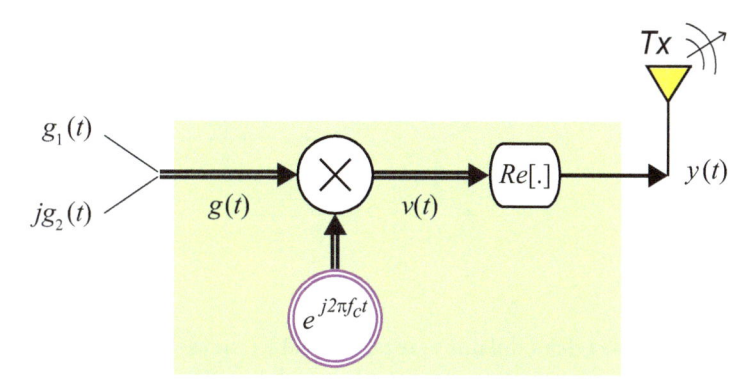

Figure 5.16: Real and Imaginary baseband signals modulated by a complex carrier and transmitting the real component only

This mixer or modulator creates the following signal, simply achieved by multiplying $g(t)$ with $e^{j2\pi f_c t}$:

$$\begin{aligned} v(t) &= g(t)e^{j2\pi f_c t} = \Big[g_1(t) + jg_2(t)\Big]e^{j2\pi f_c t} \\ &= \Big[g_1(t) + jg_2(t)\Big]\Big[\cos(2\pi f_c t) + j\sin(2\pi f_c t)\Big] \\ &= g_1(t)\cos(2\pi f_c t) + jg_2(t)\cos(2\pi f_c t) + jg_1(t)\sin(2\pi f_c t) - g_2(t)\sin(2\pi f_c t) \\ &= \underbrace{\Big[g_1(t)\cos(2\pi f_c t) - g_2(t)\sin(2\pi f_c t)\Big]}_{\textbf{Real}} + j\underbrace{\Big[g_1(t)\sin(2\pi f_c t) + g_2(t)\cos(2\pi f_c t)\Big]}_{\textbf{Imaginary}} \end{aligned} \tag{5.26}$$

The output $v(t)$ consists of a real and imaginary component and is hence a complex signal. We now just drop the imaginary part by taking the real part of $v(t)$ only, and as illustrated in Figure 5.16, we note that the resulting real signal to be transmitted is:

$$\Re\{v(t)\} = \Re\left\{\left[g_1(t)\cos(2\pi f_c t) - g_2(t)\sin(2\pi f_c t)\right] + j\left[g_1(t)\sin(2\pi f_c t) + g_2(t)\cos(2\pi f_c t)\right]\right\}$$
$$= g_1(t)\cos(2\pi f_c t) - g_2(t)\sin(2\pi f_c t)$$
$$= y(t) \tag{5.27}$$

This $y(t)$ is the same signal output that we achieved from the simple QAM modulation of Figure 5.12. We now have the required complex notation to create a real signal for transmission. What is the benefit of this complex notation? As we stated earlier, it's simpler, more tractable mathematics, with less need to recall trigonometric identities!

5.6 Quadrature Amplitude Demodulation using Complex Notation

The demodulation process can also be conveniently expressed using complex notation. Consider the proposed complex demodulator of Figure 5.17.

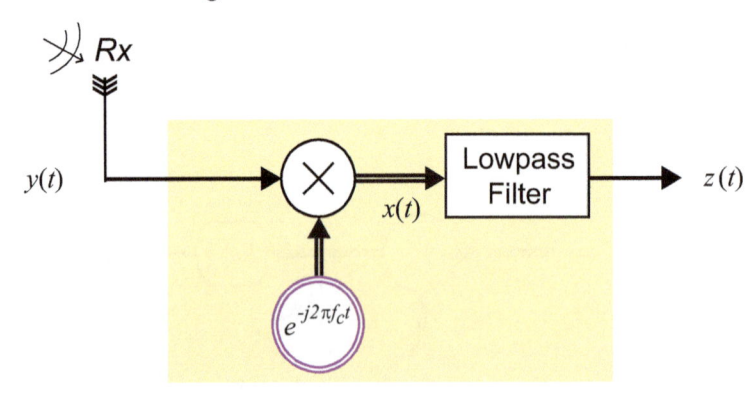

Figure 5.17: Using a complex demodulator to receive a real RF signal and demodulate to baseband to receive the signal sent by complex modulator in Figure 5.16.

The input to this complex demodulator is the *real* signal from the *real* world, and it is the same $y(t)$ as was generated in Figure 5.12. This we are assuming a perfect radio transmission channel — i.e. we sent $y(t)$ on the transmit antenna and received $y(t)$ on the receive antenna. After the multiplication of the received $y(t)$ with the complex exponential, the signal $x(t)$ is:

$$x(t) = y(t)e^{-j2\pi f_c t}$$
$$= \left[g_1(t)\cos(2\pi f_c t) - g_2(t)\sin(2\pi f_c t)\right] e^{-j2\pi f_c t} \tag{5.28}$$
$$= \left[g_1(t)\cos(2\pi f_c t) - g_2(t)\sin(2\pi f_c t)\right]\left[\cos(2\pi f_c t) - j\sin(2\pi f_c t)\right]$$

Chapter 5: Complex Signals, Spectra and Quadrature Modulation

As this is just trigonometry, we can change the notation to make the simple trigonometric manipulation and use of identities easier to read. Therefore, if we denote $A = g_1(t)$ and $B = g_2(t)$, and set $\phi = 2\pi f_c t$, the notation is a little clearer:

$$x(t) = \Big[g_1(t)\cos(2\pi f_c t) - g_2(t)\sin(2\pi f_c t)\Big]\Big[\cos(2\pi f_c t) - j\sin(2\pi f_c t)\Big]$$

$$= \Big[A\cos(\phi) - B\sin(\phi)\Big]\Big[\cos(\phi) - j\sin(\phi)\Big]$$

$$= A\cos(\phi)\Big[\cos(\phi) - j\sin(\phi)\Big] - B\sin(\phi)\Big[\cos(\phi) - j\sin(\phi)\Big]$$

Multiplying out the brackets and re-ordering gives:

$$= A\cos^2(\phi) - jA\cos(\phi)\sin(\phi) - B\sin(\phi)\cos(\phi) + jB\sin^2(\phi)$$

$$= A\cos^2(\phi) + jB\sin^2(\phi) - jA\cos(\phi)\sin(\phi) - B\sin(\phi)\cos(\phi)$$

Using trigonometric identities from Appendix B (page 581) gives:

$$= \frac{A}{2}\Big[1 + \cos(2\phi)\Big] + j\frac{B}{2}\Big[1 - \cos(2\phi)\Big] - j\frac{A}{2}\sin(2\phi) - \frac{B}{2}\sin(2\phi)$$

$$= \frac{A}{2} + \frac{A}{2}\cos(2\phi) + j\frac{B}{2} - j\frac{B}{2}\cos(2\phi) - j\frac{A}{2}\sin(2\phi) - \frac{B}{2}\sin(2\phi)$$

$$= \frac{A}{2} + j\frac{B}{2} + \frac{A}{2}\cos(2\phi) - j\frac{B}{2}\cos(2\phi) - j\frac{A}{2}\sin(2\phi) - \frac{B}{2}\sin(2\phi)$$

Substituting back for $g_1(t)$, $g_2(t)$ and $2\pi f_c t$, the signal $x(t)$ in Figure 5.17 is:

$$x(t) = \frac{1}{2}\Big[g_1(t) + jg_2(t)\Big] + \frac{1}{2}g_1(t)\cos(4\pi f_c t) - j\frac{1}{2}g_2(t)\cos(4\pi f_c t) \quad \text{lowpass filtered terms} \quad (5.29)$$

$$- j\frac{1}{2}g_1(t)\sin(4\pi f_c t) - \frac{1}{2}g_2(t)\sin(4\pi f_c t)$$

Hence after the lowpass filter in Figure 5.17, the output $z(t)$ is given by:

$$z(t) = \frac{1}{2}\Big[g_1(t) + jg_2(t)\Big] \qquad (5.30)$$

The mathematical equivalence of the standard quadrature demodulator of Figure 5.13 and the complex demodulator of Figure 5.17 is therefore confirmed, and we illustrate this equivalence in Figure 5.18.

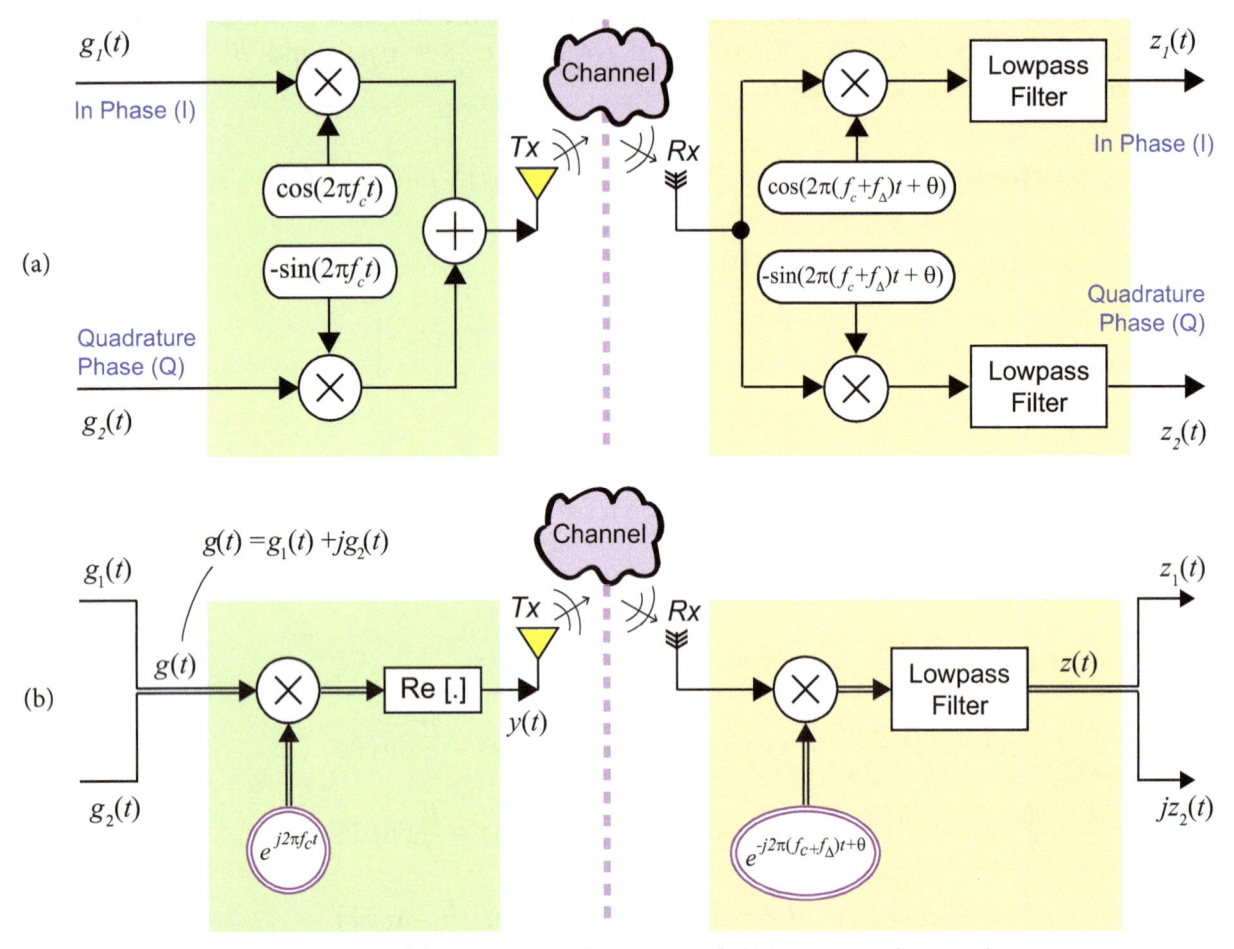

Figure 5.18: Equivalence of the (a) QAM architecture, and (b) QAM complex signal representation.

Exercise 5.5 Modulation, Demodulation and Frequency Correction

In this exercise we set up and run the two equivalent QAM modulators shown in Figure 5.18.

(a) **Open MATLAB** and within the `complex` working directory, open the example:

   `/complex/qam_mod_demod.slx`

   which contains the two QAM modulator/demodulator designs of Figure 5.18.

(b) **Inspect the Systems.** Confirm the various components, including sine and cosine signal generators, complex exponential signal generators, and multipliers and lowpass filters. The system transmits two frequency sweeps (or chirps) from the left hand side of the model, via a QAM modulation scheme, and receives them at the right side of the model.

(c) **Run the simulation.** Run the system, and thereafter by viewing the *Time Scopes* and spectra at the transmit and receive sides and the RF-transmit signal, confirm that both designs produce identical signals. This simulation confirms the equivalence and the mathematics for the design of Figure 5.18.

(d) Try changing one of the transmit signals to a different source (e.g. a single tone sine wave at 1250Hz... but you choose!) and run again to observe that two signals are being transmitted.

(e) Now change just the phase of the carriers on both receiver designs and see if the effect is the same (remember to change both sine and cosine by the same angle!). This is the θ at the receive side in Figure 5.18, and the effect will be to "mix" the in-phase, $g_1(t)$, and quadrature phase, $g_2(t)$, signals at the receiver as shown back in Eq. (5.24).

5.7 Spectral Representation for Complex Demodulation

If we sketch the complex spectra for the '*complex*' modulation and demodulation processes, we obtain the magnitude spectra shown in Figure 5.19. Noting that the two transmitted real signals are separate and independent, we choose to assign one as real, $g_1(t)$, and one as imaginary, $g_2(t)$. Then, if we represent them as

$$g(t) = g_1(t) + jg_2(t), \tag{5.31}$$

$g(t)$ is just a complex signal which can be shown in the complex frequency domain by taking the Fourier transform. As this signal is not real (it's complex!), the spectra will be non-symmetric. This is illustrated in Figure 5.19 where we show the spectra at point **A** in the signal flow graph. The spectra at **B** then shows the complex baseband signal modulated by the complex carrier,

$$v(t) = g(t)e^{j2\pi f_c t} \tag{5.32}$$

where the baseband signal spectrum is simply frequency shifted to be centred around f_c. When we then take the real part only, with the $\Re e[.]$ operator, we again see that, as in Eq. (5.26) and Eq. (5.27):

$$\Re e\left\{g(t)e^{j2\pi f_c t}\right\} = g_1(t)\cos(2\pi f_c t) - g_2(t)\sin(2\pi f_c t) \ /$$

Therefore, we obtain $y(t)$ at **C** which is the real signal for transmission, and of course this signal is now composed of the real part only, and is therefore symmetric in the frequency domain.

After applying the complex exponential multiplier, $e^{-j2\pi f_c t}$ at the receive side, the positive and negative frequencies of $y(t)$ are both shifted by f_c Hz to realise the analytic complex spectra shown at **D**. Finally both the real and imaginary signals are filtered by a suitable lowpass filter (i.e. one filter on each of the real and imaginary channels, or just input the complex signal to the real valued lowpass filter) and the complex baseband received signal, $z(t)$, is obtained at **E**. And if we have perfect synchronisation of carrier frequencies and perfect filters then we can see that

$$z(t) = 0.5g(t) = 0.5g_1(t) + j0.5g_2(t). \tag{5.33}$$

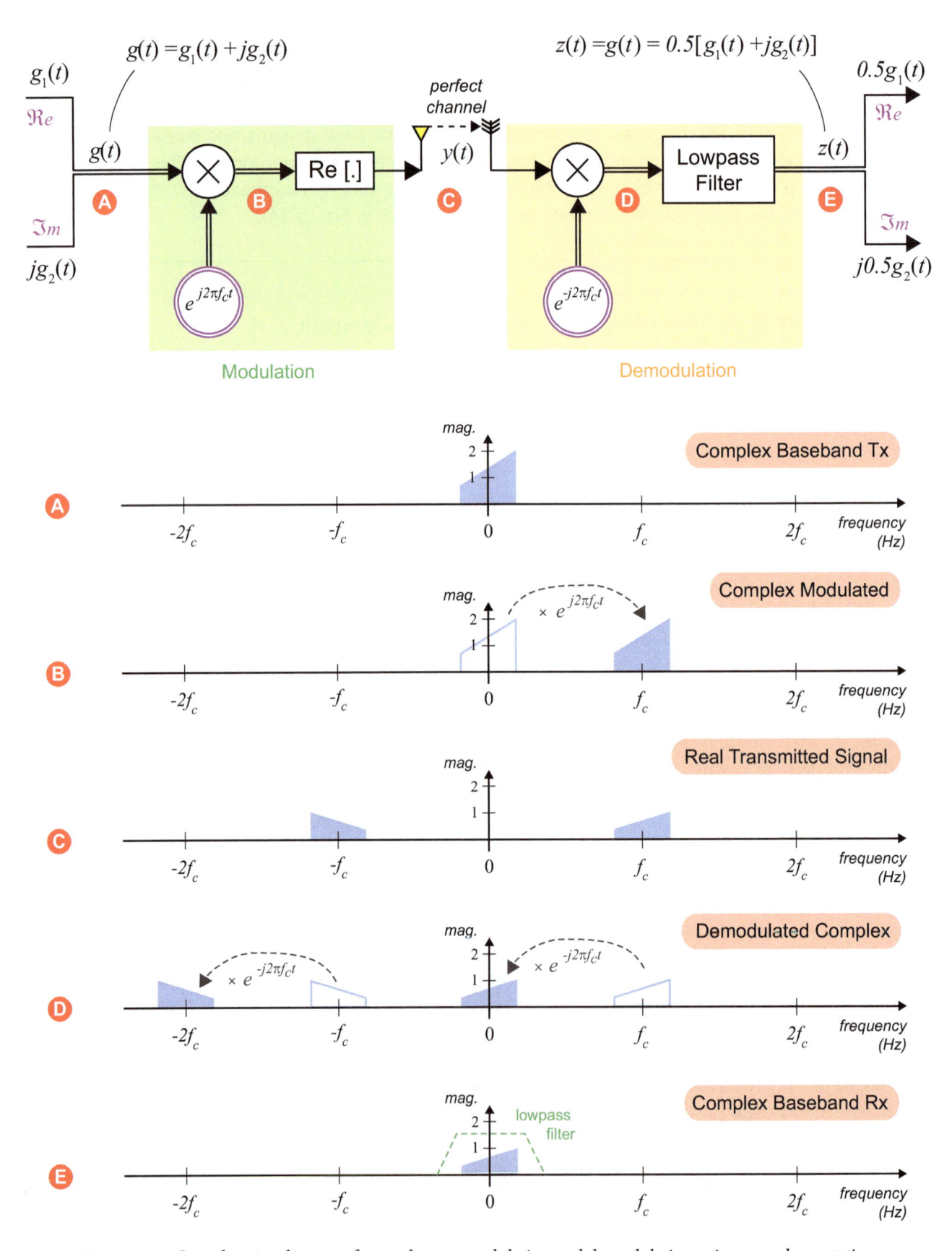

Figure 5.19: Complex signal spectra for quadrature modulation and demodulation using complex notation.

Chapter 5: Complex Signals, Spectra and Quadrature Modulation

5.7.1 Simple Mathematical Example of Passband to Complex Baseband

In order to present an example of demodulation using the mathematics of complex exponentials, consider the signal in Figure 5.20(a) which shows a bandpass signal in the frequency range of 800Hz to 1220Hz, and consists of four tones (cosine waves) at 800Hz, 900Hz, 1080Hz and 1220Hz. We can assume that a baseband signal has been generated and modulated onto a carrier at f_c = 1000Hz (this is a rather low frequency of course, and RF carriers are usually MHz rather than kHz — however we are just using this for illustration purposes and to keep the figures simple). The signal is of course real, and consists of four cosine signals at different frequencies:

$$y(t) = 8\cos(2\pi 800t) + 6\cos(2\pi 900t) + 4\cos(2\pi 1080t) + 2\cos(2\pi 1220t) \quad (5.34)$$

Figure 5.20(a) shows the complex spectra (again recalling that $\cos(\omega) = \frac{1}{2}(e^{j\omega} + e^{-j\omega})$):

$$\begin{aligned}
y(t) &= \frac{8}{2}(e^{j2\pi 800t} + e^{-j2\pi 800t}) + \frac{6}{2}(e^{j2\pi 900t} + e^{-j2\pi 900t}) \\
&\quad + \frac{4}{2}(e^{j2\pi 1080t} + e^{-j2\pi 1080t}) + \frac{2}{2}(e^{j2\pi 1220t} + e^{-j2\pi 1220t}) \\
&= 4e^{j2\pi 800t} + 3e^{j2\pi 900t} + 2e^{j2\pi 1080t} + e^{j2\pi 1220t} \\
&\quad + 4e^{-j2\pi 800t} + 3e^{-j2\pi 900t} + 2e^{-j2\pi 1080t} + e^{-j2\pi 1220t}
\end{aligned} \quad (5.35)$$

Looking again at Figure 5.19, consider that Eq. (5.35) as the signal available at point **C**.

Because the signal is a real cosine, then the complex frequency spectra in Figure 5.20(a) is symmetric, and as there are no sine wave components then the imaginary amplitude spectra has no non-zero components and is not drawn here. (Recall Exercises 5.3 and 5.4 to review on this point.)

If we now demodulate or multiply Eq. (5.35) with a complex exponential, then we get:

$$\begin{aligned}
y(t)e^{-j2\pi 1000t} &= 4e^{j2\pi(800-1000)t} + 3e^{j2\pi(900-1000)t} + 2e^{j2\pi(1080-1000)t} + e^{j2\pi(1220-1000)t} \\
&\quad + 4e^{-j2\pi(800+1000)t} + 3e^{-j2\pi(900+1000)t} + 2e^{-j2\pi(1080+1000)t} + e^{-j2\pi(1220+1000)t} \\
&= 4e^{-j2\pi 200t} + 3e^{-j2\pi 100t} + 2e^{j2\pi 80t} + e^{j2\pi 220t} \\
&\quad + 4e^{-j2\pi 1800t} + 3e^{-j2\pi 1900t} + 2e^{-j2\pi 2080t} + e^{-j2\pi 2220t}
\end{aligned} \quad (5.36)$$

Eq. (5.36) is equivalent to the spectra at **D** in Figure 5.19. Finally, after passing both the real and imaginary through a lowpass filter, we get just the complex baseband signal:

$$\begin{aligned}
z(t) = \text{LPF}\{y(t)(e^{-j2\pi 1000t})\} &= 4e^{-j2\pi 200t} + 3e^{-j2\pi 100t} + 2e^{j2\pi 80t} + e^{j2\pi 220t} \\
&\quad + 4e^{-j2\pi 1800t} + 3e^{-j2\pi 1900t} + 2e^{-j2\pi 2080t} + e^{-j2\pi 2220t}
\end{aligned} \quad (5.37)$$

Eq. (5.37) is equivalent to the spectra at **E** in Figure 5.19, and Figure 5.20(c).

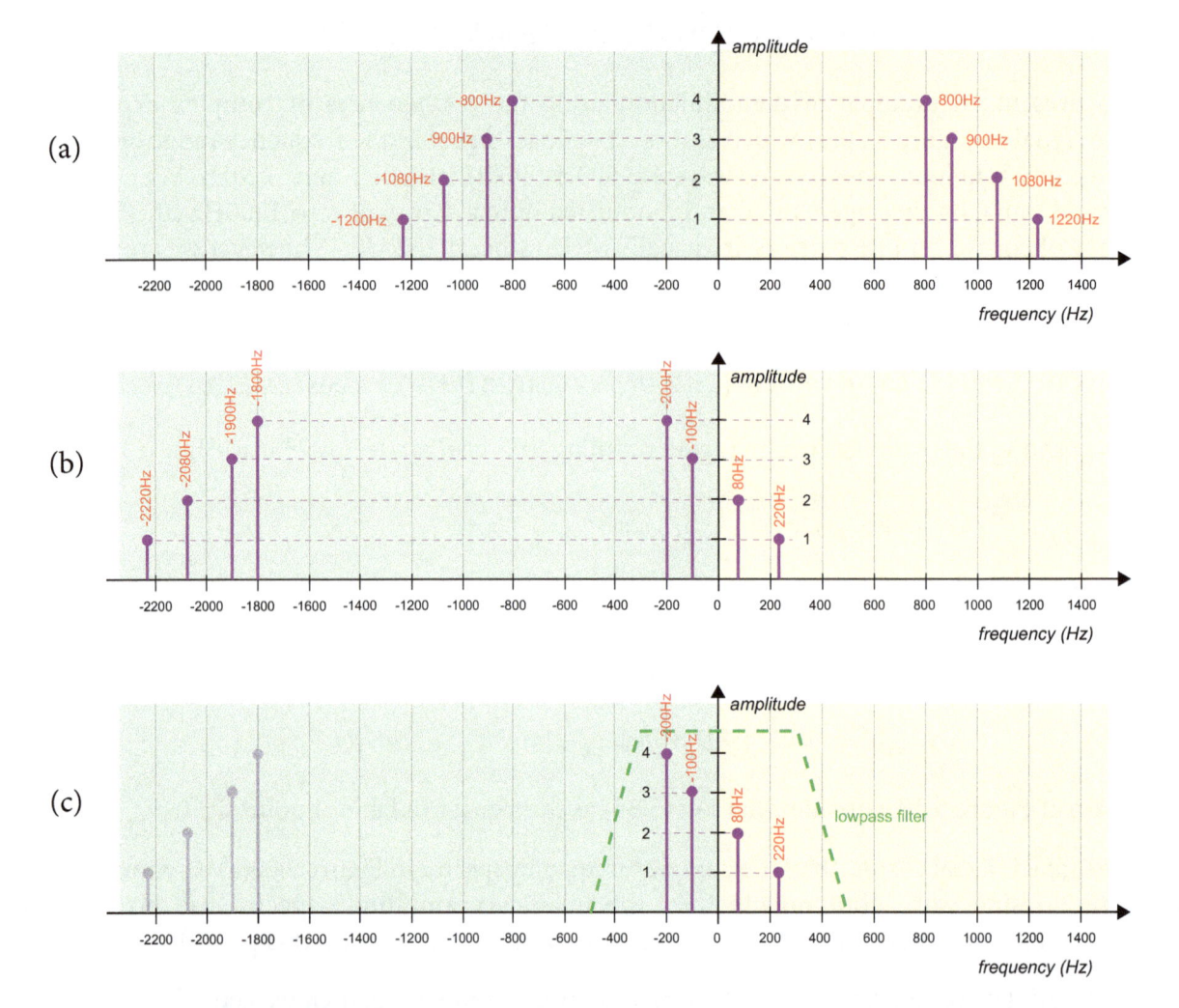

Figure 5.20: (a) A bandpass signal centred around 1000Hz is (b) Complex demodulated by a negative 1000Hz complex exponential creating (c) baseband components and components centred around $-2 \times 1000 \text{Hz} = -2000 Hz$ and lowpass filtered to leave just the baseband components.

Exercise 5.6 Complex Demodulation of a Signal

In this exercise we will perform the complex demodulation from Figure 5.20.

(a) **Open MATLAB** and within the `complex` working directory, open the example:

   `/complex/complex_demodulation.slx`

(b) **Inspect the systems.** Confirm the various components, including the sine and cosine signal generators, complex exponential signal generators, and multipliers and lowpass filters.

(c) **Run the simulation.** Run the system and note that the spectra obtained before and after multiplication by the complex exponential and lowpass filtering are exactly as represented in Figure 5.20.

5.8 Frequency Offset Error and Correction at the Receiver

Returning to Figure 5.18(b), a potential errors may be that the receiver carrier frequency f_c has a slight offset, and rather than being exactly at f_c Hz, it will be at $f_c + f_\Delta$.

At the receiver, we need knowledge of the transmit carrier frequency in most cases. However we will have some level of error in the receiver carrier frequency. If we refer to this error as Δf then in actual fact the receiver demodulation Eq. (5.28) is now:

$$\begin{aligned} x(t) &= y(t) e^{-j2\pi(f_c + \Delta f)t} \\ &= \left[g_1(t)\cos(2\pi f_c t) - g_2(t)\sin(2\pi f_c t) \right] e^{-j2\pi f_c t} \cdot e^{-j2\pi f_\Delta t} \end{aligned} \quad (5.38)$$

and the effect is to simply shift the spectrum by f_Δ Hz from 0 Hz, as illustrated in Figure 5.21.

This frequency offset error is one that we encounter with the RTL-SDR. Whereas we might set the RTL-SDR receive carrier frequency to 90MHz, it may be "out" and in error by a few 100Hz or even kHz, due to component tolerances in the receiver. This offset error is precisely the f_Δ Hz above, and that we consider in Chapter 7 (see for example Figure 7.2, page 237).

5.9 Frequency Correction using a Complex Exponential

After the complex signal is received, the first processing stage in the receiver may be a *frequency correction* which will multiply the incoming complex spectrum by:

$$e^{j2\pi \Delta f t} \quad (5.39)$$

as illustrated in Figure 5.21. Thereafter the final part of the DSP enabled SDR receiver is variously the carrier phase locking, synchronisation and/or timing sections followed by the signal decoding sections (depending on whether the signal is FM, AM, QPSK etc). This is the part that takes place in the MATLAB and Simulink code and is the core of the DSP enabled SDR which we will review, design and implement in the chapters following this one.

It is worth bearing in mind that this is an idealised model, which assumes that each of the operations are undertaken in a mathematically perfect fashion, and neglects the 'real world' effects of transmitting a signal across a communications channel. Nevertheless, it does provide a useful reference with which to consider the reception of real signals.

Exercise 5.7 Frequency Correction using Complex Exponential

In this exercise we will perform the simple frequency correction operation of Figure 5.21.

(a) **Open MATLAB** and within the `complex` working directory, open the example:

   `/complex/complex_frequency_correction.slx`

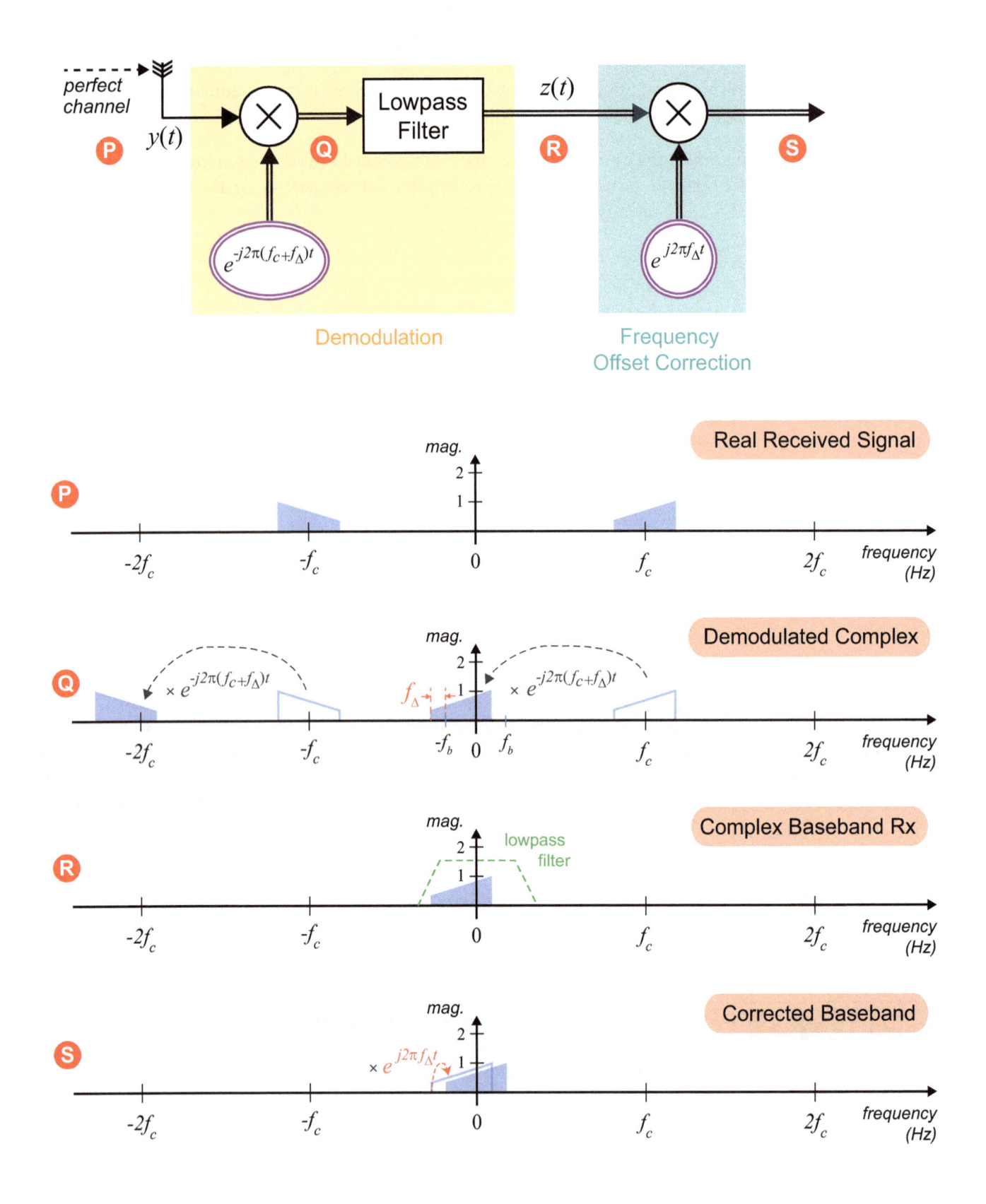

Figure 5.21: Effect of a frequency error in the complex receiver — spectrum shifts by Δf at baseband — and correction.

Chapter 5: Complex Signals, Spectra and Quadrature Modulation

(b) **Inspect the systems.** This example is the same as the example in Exercise 5.6 except there is now a (large!) frequency offset on the receive carrier which is at 19,927Hz. Note that the correction complex exponential frequency is set to 0 (zero), i.e. it multiplies by 1, as $e^{j0} = 1$.

(c) **Run the simulation.** Run the system and note that the spectra obtained (recall Figure 5.21).

(d) **Set the frequency correction parameter.** Calculate the frequency correction parameter and set the frequency of the complex exponential to this value. Does this correct the frequency spectra to the desired spectrum? If not try again (hint: 20,000 and 19,927!).

5.10 RTL-SDR Quadrature / Complex Architecture

The last stage in our review of complex signals and spectra is to confirm how the presented models relate to the RTL-SDR, and the types of models that you will work with via the exercises in this book (as well as your own personal projects!).

The RF signals that you receive have first been modulated, then transmitted through a wireless communications channel. At the receive side (i.e. your desktop RTL-SDR setup!), the signal has been received and demodulated by your RTL-SDR, and then passed into MATLAB or Simulink via the MathWorks RTL-SDR Hardware Support Package. At this point, the samples read into MATLAB or Simulink correspond to the output signal from the system model presented in Figure 5.22 (and the preceding equations) — these samples are of the demodulated signal. An initial, 'rough' stage of frequency offset correction is then one of the first operations performed in MATLAB or Simulink.

5.11 Summary

In this chapter we have reviewed the fundamentals of AM for both transmission and reception. We derived the QAM transmitter and receiver, and showed how they can be represented by complex exponential notation. In a more general sense, it was also demonstrated that complex mathematical representation is a tractable, descriptive way of working with communications signals and systems. We will go on to make use of this notation in later chapters.

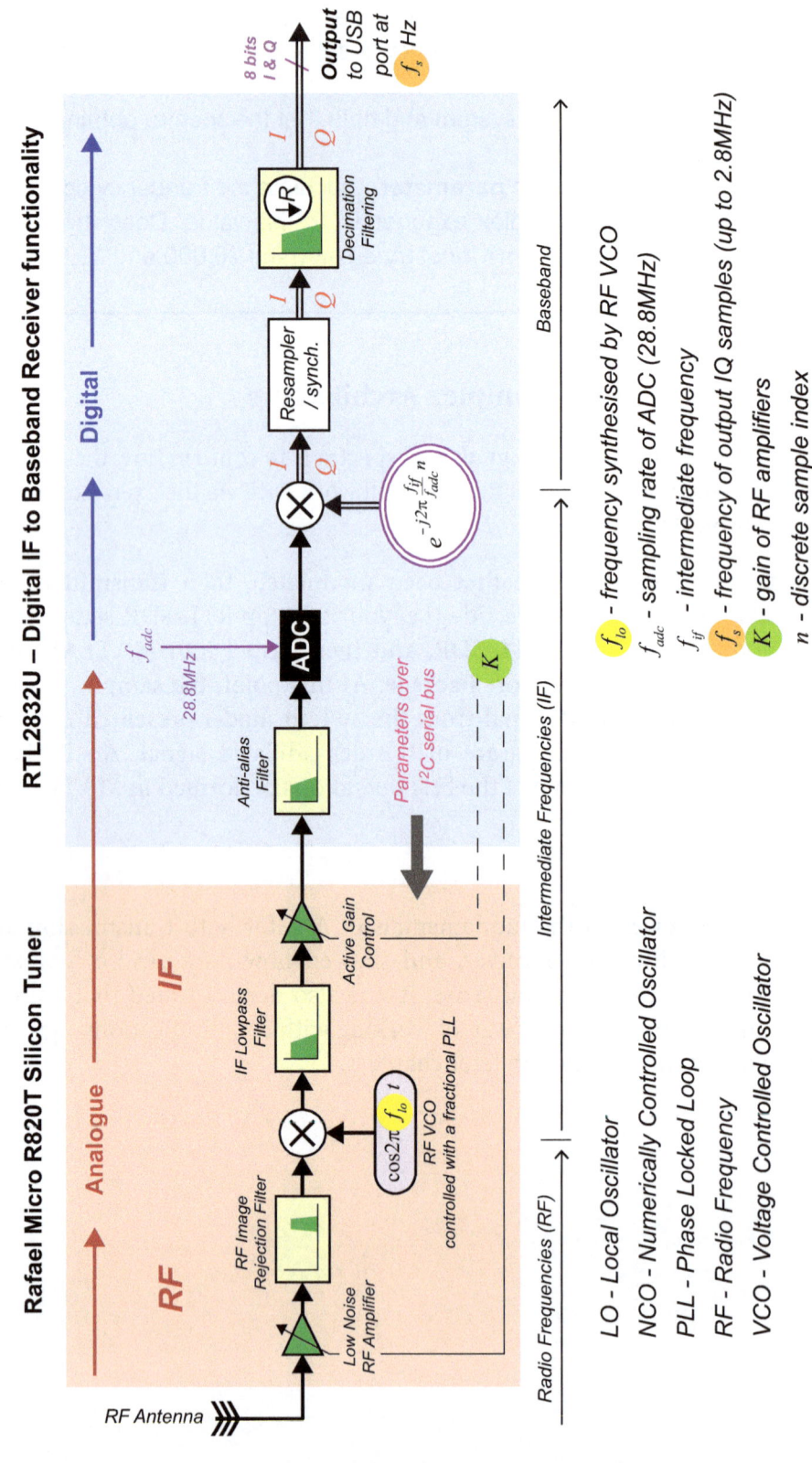

Figure 5.22: The RTL-SDR represented in complex form.

6 Amplitude Modulation (AM) Theory and Simulation

In this chapter, we will review extensively about classic AM — Amplitude Modulation. The theory behind conventional forms of AM will be presented, and there will be an opportunity to run simulations that demonstrate some of the different types of AM modulator. After this we will move on to review the various demodulators that will later be constructed in Chapter 8 and RTL-SDR AM receivers.

6.1 Amplitude Modulation — An Introduction

AM is intuitively the simplest modulation method and, as will be shown later in Chapter 11, it also forms the basis of many digital modulation schemes. In its most basic form, it is simply the process of mixing (*heterodyning*) an information signal with a carrier wave to produce a signal that has two sideband components in the frequency domain. This process is called Double Sideband (AM-DSB) modulation, and it results in the information signal being 'shifted up' from baseband to a carrier frequency. There are two variations of AM-DSB: Transmitted Carrier (AM-DSB-TC) and Suppressed Carrier (AM-DSB-SC). Further types of AM include Single Sideband (AM-SSB) and Vestigial Sideband (AM-VSB).

In the following sections we will discuss and give examples of each of these in turn, before we go on to discuss AM demodulators at the receive side.

6.2 AM-DSB-SC: Double Sideband Suppressed Carrier AM

Mathematically, the least complicated of the AM variants is AM-DSB-SC. Here, an information signal $s_i(t)$ is mixed with a carrier wave $s_c(t)$ to produce the AM signal $s_{am-dsb-sc}(t)$. A block diagram of this modulator is shown in Figure 6.1 (AM-DSC-SC was also reviewed earlier in Section 5.3 (page 182):

$$s_{am-dsb-sc}(t) = s_i(t) \times s_c(t) \qquad (6.1)$$

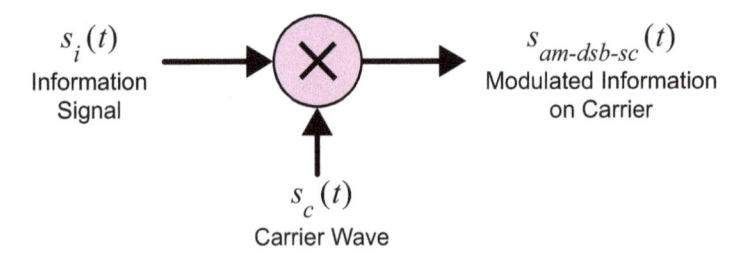

Figure 6.1: AM-DSB-SC Modulator block diagram

6.2.1 AM-DSB-SC Modulation: Modulating a Sine Wave

Presenting the maths associated with the modulation of a sinusoidal information signal is the most logical place to start. The information signal is defined as having amplitude A_i and the frequency f_i,

$$s_i(t) = A_i \cos(2\pi f_i t) = A_i \cos(\omega_i t) \tag{6.2}$$

where $\omega_i = 2\pi f_i$. The carrier has amplitude A_c and the (higher) frequency f_c,

$$s_c(t) = A_c \cos(2\pi f_c t) = A_c \cos(\omega_c t) \tag{6.3}$$

and mixing (multiplying) the two signals as in Eq. (6.1) yields:

$$s_{am-dsb-sc}(t) = A_i \cos(\omega_i t) A_c \cos(\omega_c t) \ . \tag{6.4}$$

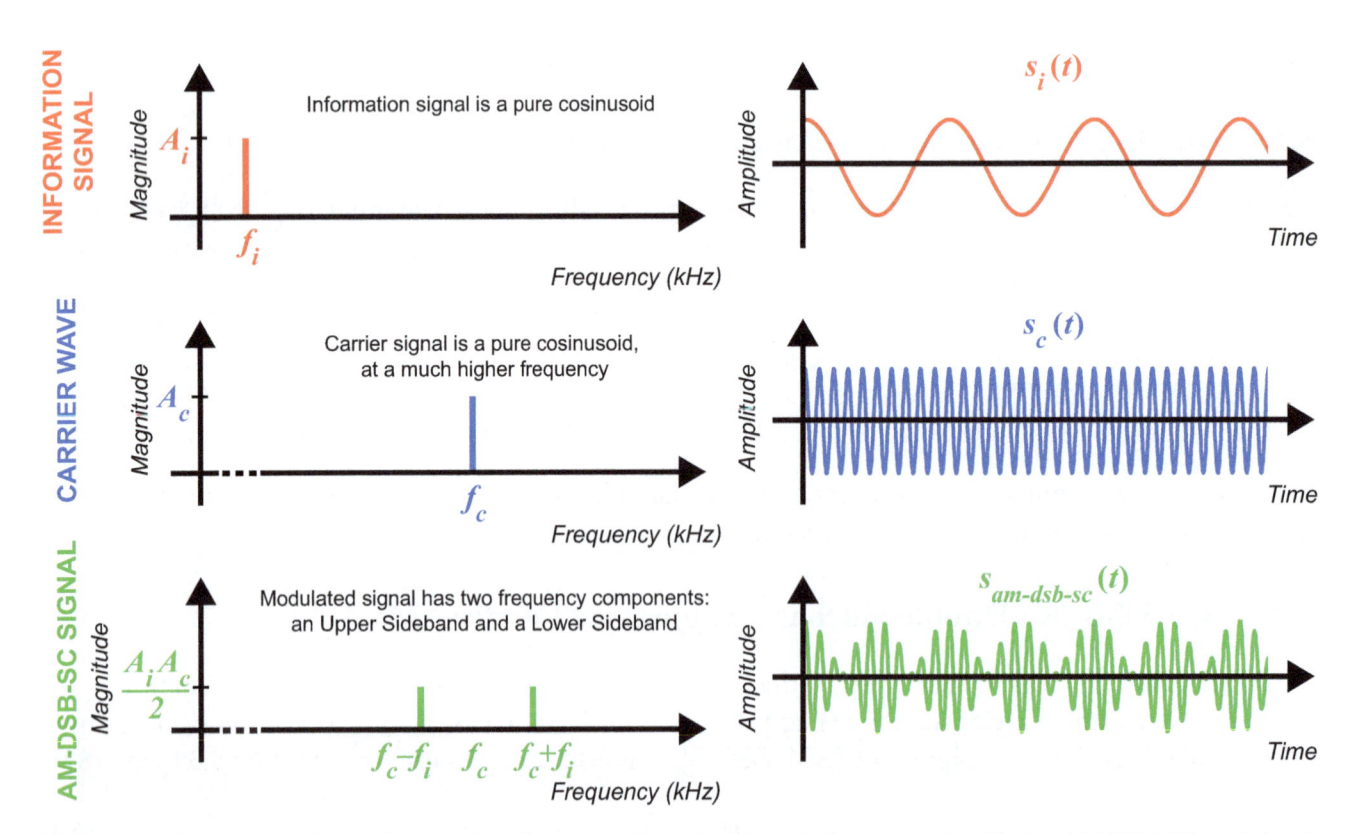

Figure 6.2: Frequency and time domain plots demonstrating a single tone information signal being AM-DSB-SC modulated

Solving this with the product-to-sum trigonometry rule from Eq. (B.4) in Appendix B (page 581) gives:

$$s_{am-dsb-sc}(t) = \frac{A_i A_c}{2} \Big(\cos(\omega_c - \omega_i)t + \cos(\omega_c + \omega_i)t \Big). \tag{6.5}$$

This modulation process results in two sinusoidal terms; one at frequency $f_c - f_i$, and the other at frequency $f_c + f_i$, as illustrated in Figure 6.2 [50]. The amplitude of the modulated signal fluctuates in sympathy with the amplitude of the information signal; hence the term amplitude modulation.

6.2.2 AM-DSB-SC Modulation: Modulating Baseband Signals

Usually information signals are far more complex than a single sinusoidal wave, and they are composed of a set of frequency components. If a baseband information signal $s_i(t)$ had a bandwidth of f_h Hz (where f_h was the highest frequency component contained within the signal), AM-DSB-SC modulating it would result in a signal with a bandwidth of $2f_h$ Hz. The modulated signal, which is centred around f_c, is shown in the time and frequency domains in Figure 6.3.

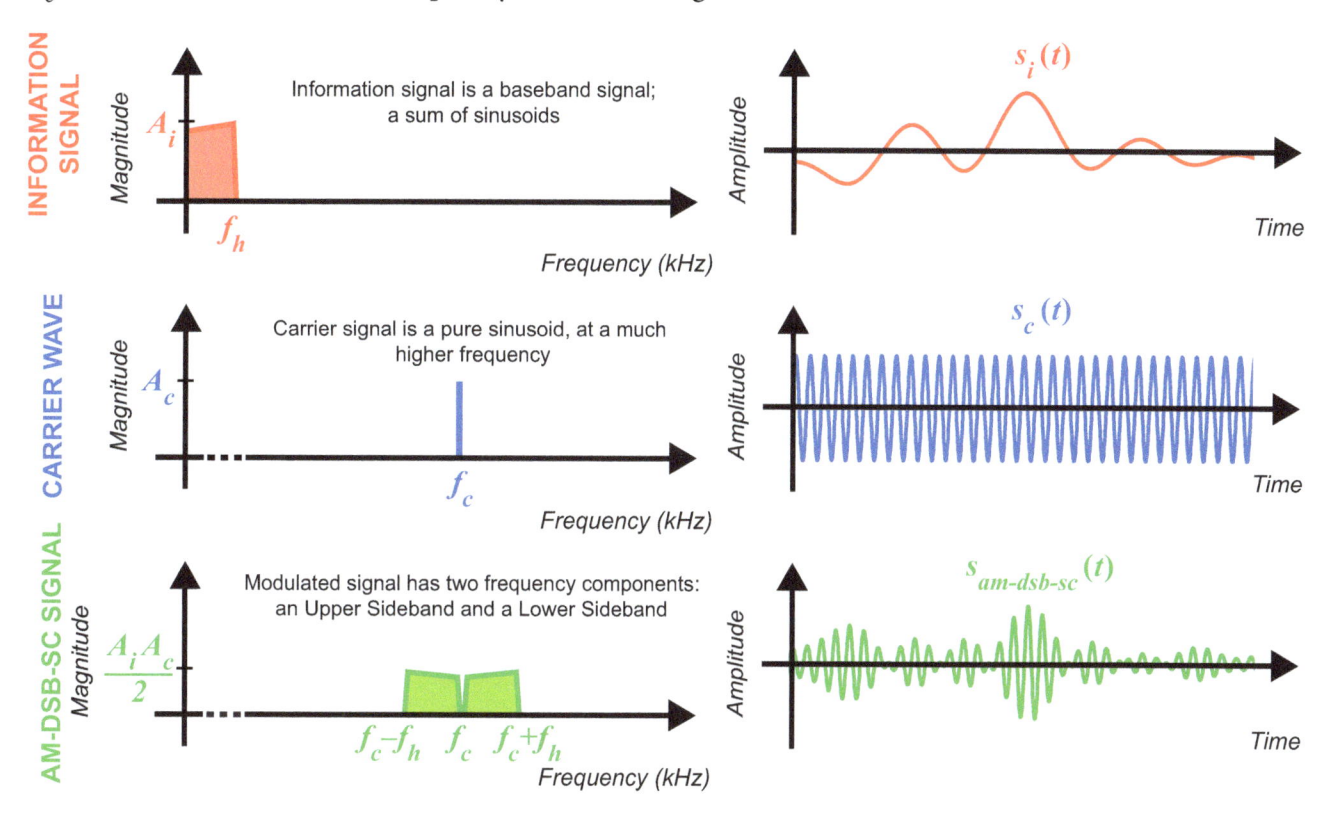

Figure 6.3: Frequency and time domain plots demonstrating a baseband information signal being AM-DSB-SC modulated

Mathematically, any baseband signal can be represented as a sum of weighted sinusoidal components using Fourier decomposition. If we define an information signal that is composed of the following three components:

$$s_i(t) = A_{i1}\cos(2\pi f_{i1} t) + A_{i2}\cos(2\pi f_{i2} t) + A_{i3}\cos(2\pi f_{i3} t) \tag{6.6}$$

$$= A_{i1}\cos(\omega_{i1} t) + A_{i2}\cos(\omega_{i2} t) + A_{i3}\cos(\omega_{i3} t),$$

then we would see a view similar to Figure 6.4 if we plotted it in the frequency domain:

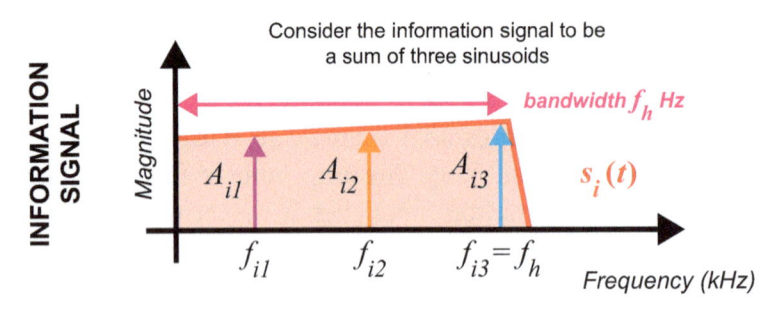

Figure 6.4: Baseband information signal composed of three weighted cosines

Mixing this baseband signal with the carrier from Eq. (6.3), we get:

$$s_{am-dsb-sc}(t) = \left[A_{i1}\cos(\omega_{i1} t) + A_{i2}\cos(\omega_{i2} t) + A_{i3}\cos(\omega_{i3} t) \right] A_c \cos(\omega_c t) . \quad (6.7)$$

Multiplying this out and applying the product-to-sum trigonometry rule from Eq. (B.4) gives:

$$s_{am-dsb-sc}(t) = \frac{A_{i1} A_c}{2} \Big(\cos(\omega_c - \omega_{i1})t + \cos(\omega_c + \omega_{i1})t \Big) \quad (6.8)$$

$$+ \frac{A_{i2} A_c}{2} \Big(\cos(\omega_c - \omega_{i2})t + \cos(\omega_c + \omega_{i2})t \Big)$$

$$+ \frac{A_{i3} A_c}{2} \Big(\cos(\omega_c - \omega_{i3})t + \cos(\omega_c + \omega_{i3})t \Big) .$$

This signal contains three pairs of tones, or six frequency components, all centred around the carrier f_c. The modulated signal can be shown in the frequency domain as in Figure 6.5; note from the sketch that the sidebands are symmetrical around f_c.

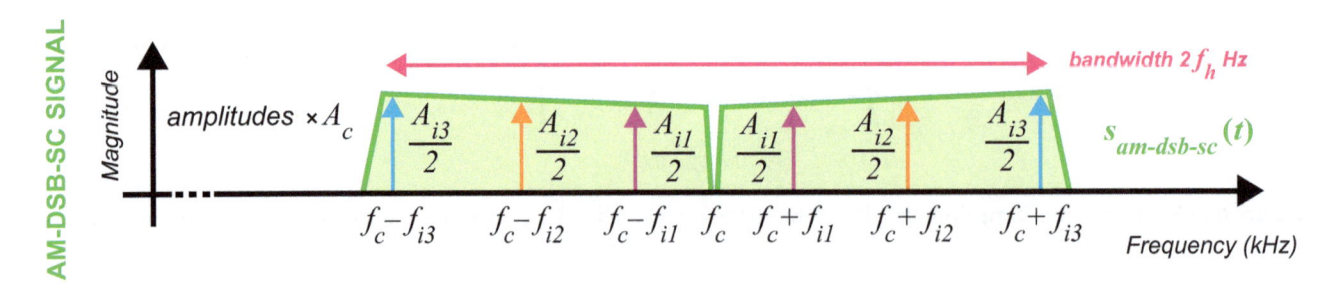

Figure 6.5: Modulated baseband information signal composed of three weighted cosines

Next, in Exercise 6.1, a simple Simulink simulation will confirm via simulation what happens to sinusoidal signals (in the time and frequency domains) as they are modulated onto a carrier to produce an AM-DSB-SC signal.

Chapter 6: Amplitude Modulation (AM) Theory and Simulation 207

Exercise 6.1 AM-DSB-SC Simulation

In this exercise we will test out the AM-DSB-SC modulator. Two different information signals are included in this model, with the option to switch between them to see the differences in the output. A *Spectrum Analyzer* and a *Time Scope* are connected to various points throughout the model which will allow the signals to be monitored in the time and frequency domains as they are modulated.

(a) **Open MATLAB.** Set the working directory to an appropriate folder so you can open this model:

/am/simulation/am_dsb_sc.slx

The block diagram should look like this:

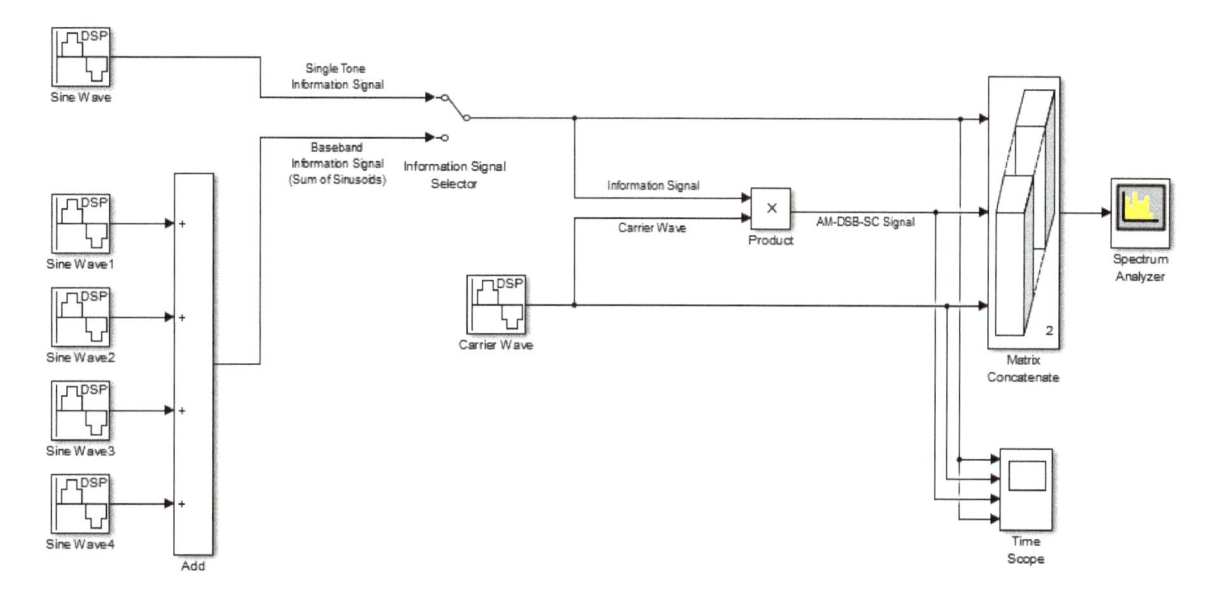

(b) The first information signal (top) is a single tone. The other is a sum of tones. They are input to the *Product* block, where they are mixed with the *Carrier Wave* to create the AM-DSB-SC signal. The *Information Signal Selector* switch allows you to switch between them. To change the selection, simply 2 on it. Leave this set on 'Single Tone' for now.

All of these signals are input to scopes that will allow you to visualise them in both the time and frequency domains.

(c) **Run the simulation.** Begin the simulation by 1 on the 'Run' ▶ button in the Simulink toolbar. The two scope windows should appear — position these so that you can see both of them. The information signal is blue, the carrier is orange, and the AM-DSB-SC signal is green, as shown in the legends.

(d) **Signal Analysis #1.** With the switch set to 'Single Tone', this is what you should see in the *Spectrum Analyzer* and *Time Scope* windows:

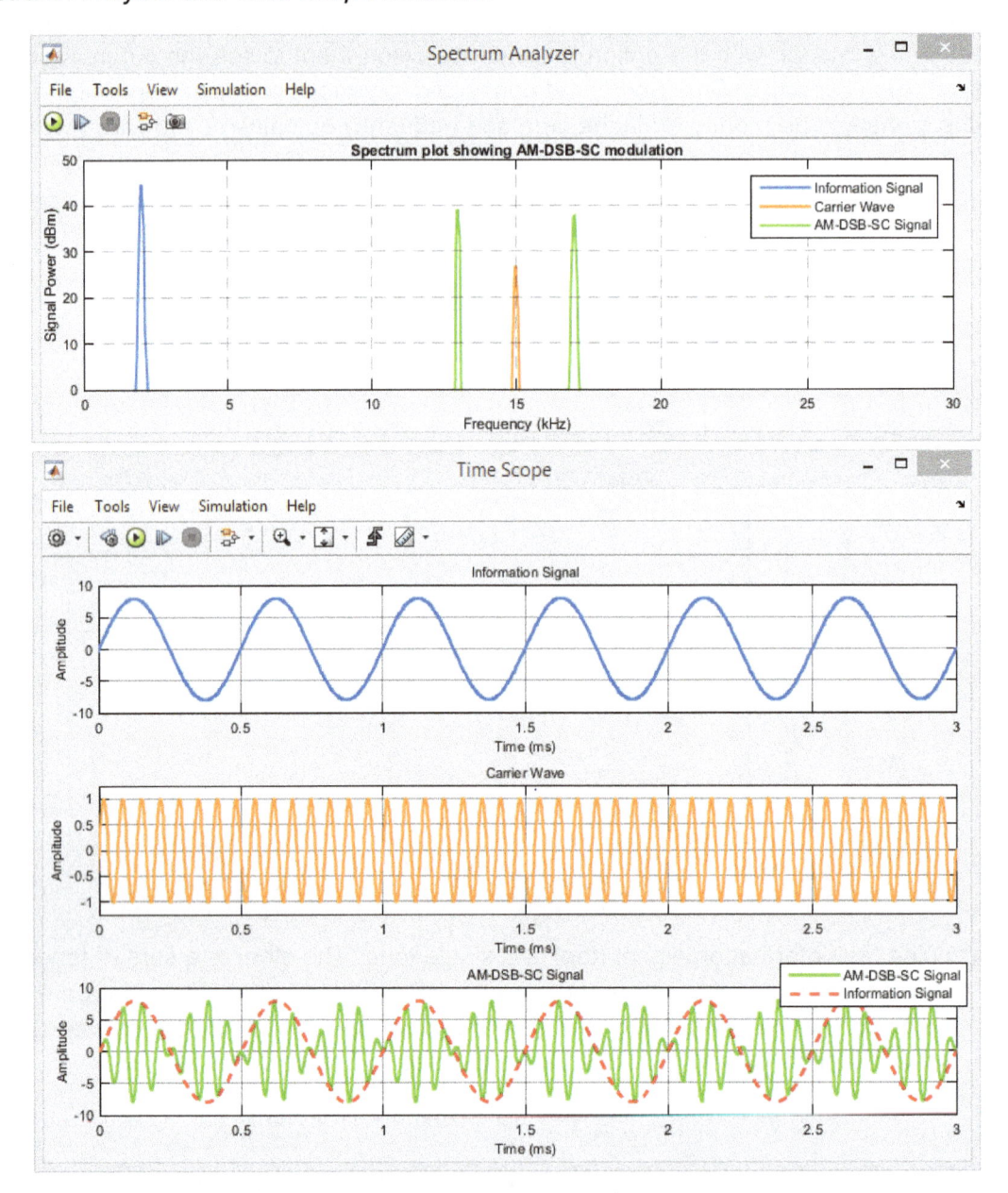

The information signal is a single tone with a frequency of 2kHz. The carrier (also a single tone), has a frequency of 15kHz. When these are mixed together the AM-DSB-SC signal is generated. You should see from the *Spectrum Analyzer* that the modulated signal only contains two frequency components, located at 13kHz and 17kHz. It is clear that there is no carrier present in this signal, as there is no green component at f_c.

(e) The red dashed line overlaid on the time domain plot of the AM-DSB-SC signal shows how the information signal has affected the amplitude of the carrier. As suppressed carrier modulation has been performed, the *modulation index* of the signal is >100%, which means that no information envelope is present. We will discuss this further in Section 6.3.

(f) **Signal Analysis #2.** Set the *Information Signal Selector* switch to 'Baseband Information Signal', and re-run the simulation. This is what the scopes should now show:

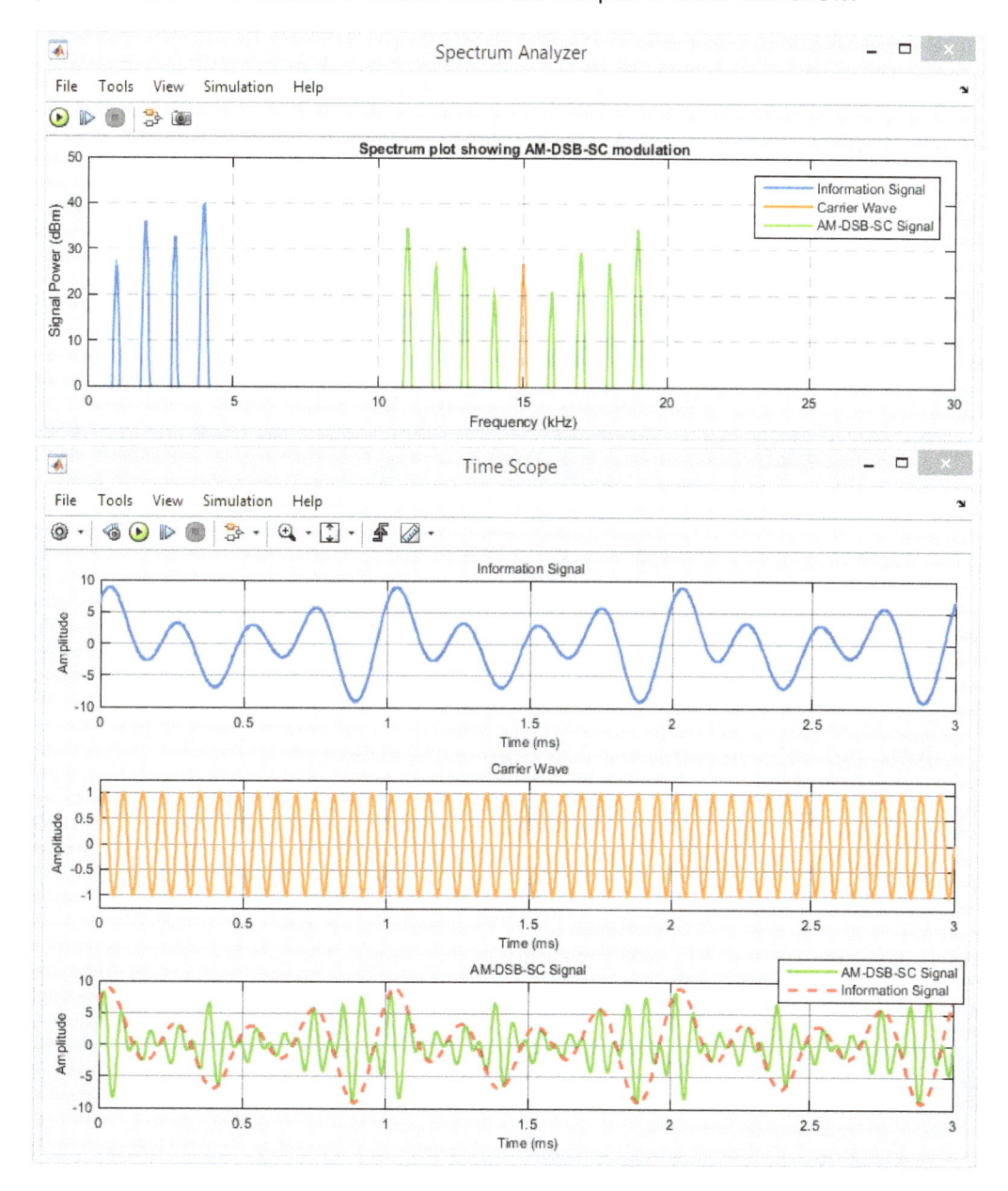

The information signal in this case contains four frequency components, located at 1kHz, 2kHz, 3kHz and 4kHz. When this is modulated with the same carrier signal, four pairs of tones are created, positioned at 14kHz & 16kHz, 13kHz & 17kHz, 12kHz & 18kHz, and finally, 11kHz & 19kHz. You should note that the amplitudes of these pairs of tones match, and thus that the sidebands are mirror images of each other.

6.3 AM-DSB-TC: Double Sideband Transmitted Carrier AM

While AM-DSB-SC signals are simple to generate and understand mathematically, radio receivers require relatively sophisticated coherent demodulators in order to extract any information from them, due to their lack of an information envelope. AM-DSB-TC is an alternative AM modulation scheme that enables the use of non-coherent demodulators. Here, an information signal $s_i(t)$ with a DC offset A_o is mixed with a carrier wave $s_c(t)$ to produce the AM signal $s_{am-dsb-tc}(t)$ [49].

$$s_{am-dsb-tc}(t) = \left[A_o + s_i(t) \right] \times s_c(t) \tag{6.9}$$

The block diagram for this modulator is shown in Figure 6.6:

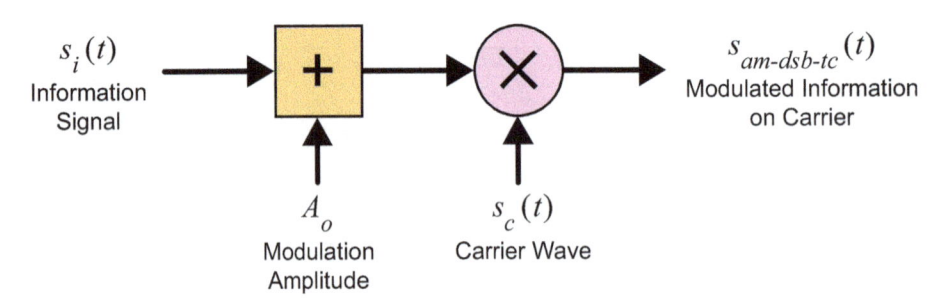

Figure 6.6: AM-DSB-TC Modulator block diagram

6.3.1 AM-DSB-TC Modulation: Modulating a Sine Wave

If we once again consider a sinusoidal 'information' signal with amplitude A_i and frequency f_i,

$$s_i(t) = A_i \cos(2\pi f_i t) = A_i \cos(\omega_i t) \tag{6.10}$$

and a carrier with amplitude A_c and the (higher) frequency f_c,

$$s_c(t) = A_c \cos(2\pi f_c t) = A_c \cos(\omega_c t) \tag{6.11}$$

then modulating as in Eq. (6.9) yields:

$$s_{am-dsb-tc}(t) = \left[A_o + A_i \cos(\omega_i t) \right] A_c \cos(\omega_c t) . \tag{6.12}$$

To solve this, we can use the product-to-sum trigonometry rule Eq. (B.4) from Appendix B (page 581):

$$s_{am-dsb-tc}(t) = A_o A_c \cos(\omega_c)t + \frac{A_i A_c}{2} \cos(\omega_c - \omega_i)t + \frac{A_i A_c}{2} \cos(\omega_c + \omega_i)t \tag{6.13}$$

$$= A_o A_c \cos(\omega_c)t + \frac{A_i A_c}{2} \Big(\cos(\omega_c - \omega_i)t + \cos(\omega_c + \omega_i)t \Big) .$$

Chapter 6: Amplitude Modulation (AM) Theory and Simulation   211

Sometimes this is represented a little differently, and instead of solving as in Eqs. (6.12) and (6.13) — where the modulated signal is expressed in terms of the information signal's amplitude A_i — it is expressed in terms of the *AM modulation index*, 'm':

$$s_{am-dsb-tc}(t) = A_o \left[1 + m\cos(\omega_i t) \right] A_c \cos(\omega_c t) \qquad (6.14)$$

$$= A_o A_c \cos(\omega_c t) + \frac{A_o A_c m}{2} \left(\cos(\omega_c - \omega_i)t + \cos(\omega_c + \omega_i)t \right)$$

$$\text{where: } m = \frac{A_i}{A_o}.$$

We will come back to consider the modulation index in a moment. Unlike AM-DSB-SC modulation, the AM-DSB-TC modulation process results in *three* sinusoidal terms, as illustrated in Figure 6.7: a lower sideband at $f_c - f_i$; an upper sideband at $f_c + f_i$; and the carrier component at f_c. In the diagram, the information envelope is also highlighted. This is a smooth line that connects between the upper peaks of the modulated signal, and (in this case) the envelope's amplitude fluctuations exactly match the fluctuations in the information signal.

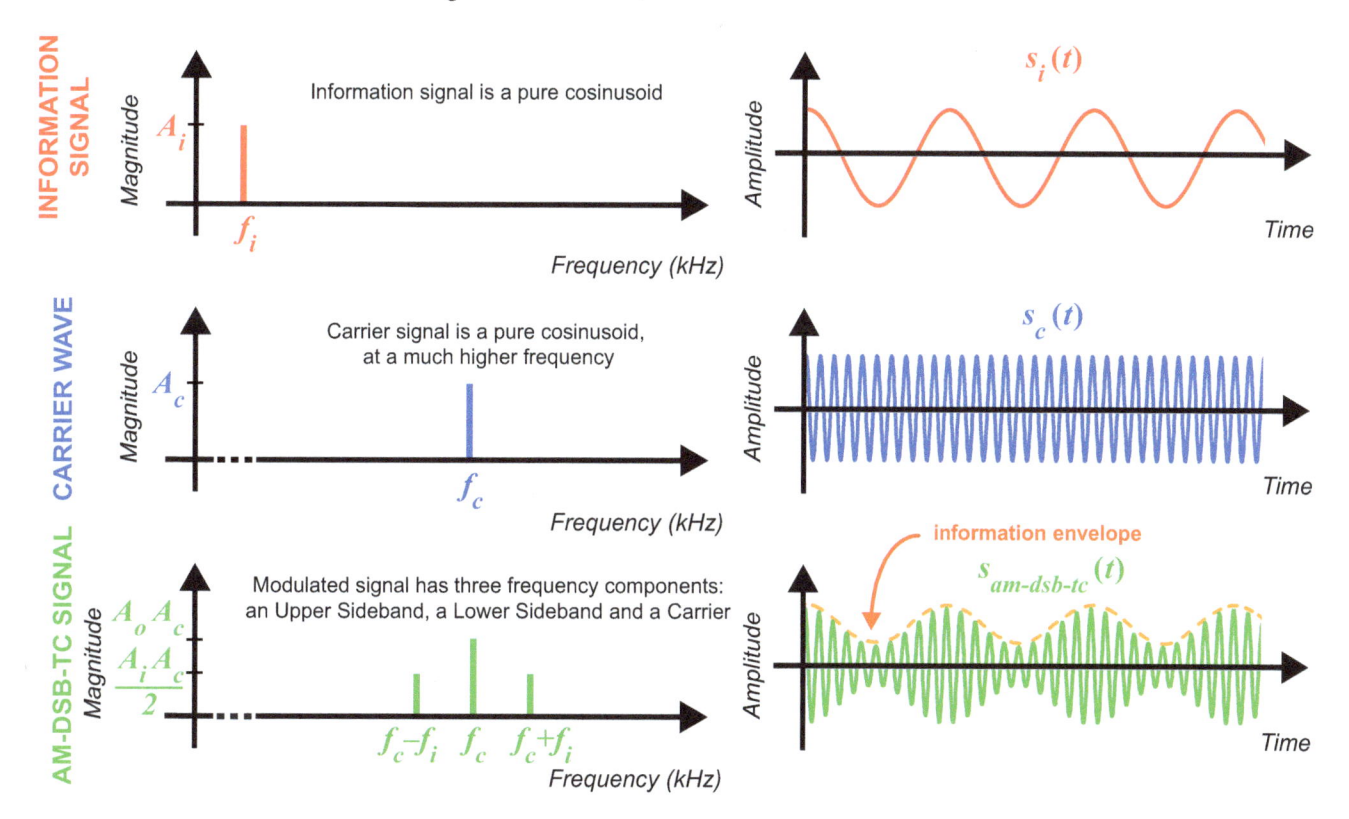

Figure 6.7: Frequency and time domain plots demonstrating a single tone information signal being AM-DSB-TC modulated

The modulation index expresses the level of modulation in an AM-DSB-TC signal, and it is often expressed as a percentage. The envelope of the modulated signal will fluctuate by $\pm m(A_o A_c)$, meaning that, if a signal had a modulation index of 0.5 (or 50%), the peak-to-peak amplitude of the envelope would be $A_o A_c$, and it would fluctuate in the range $A_o A_c + 0.5(A_o A_c)$ to $A_o A_c - 0.5(A_o A_c)$. The maximum value the modulation index can take whilst keeping the envelope intact is 1 (or 100%), as shown in Figure 6.8.

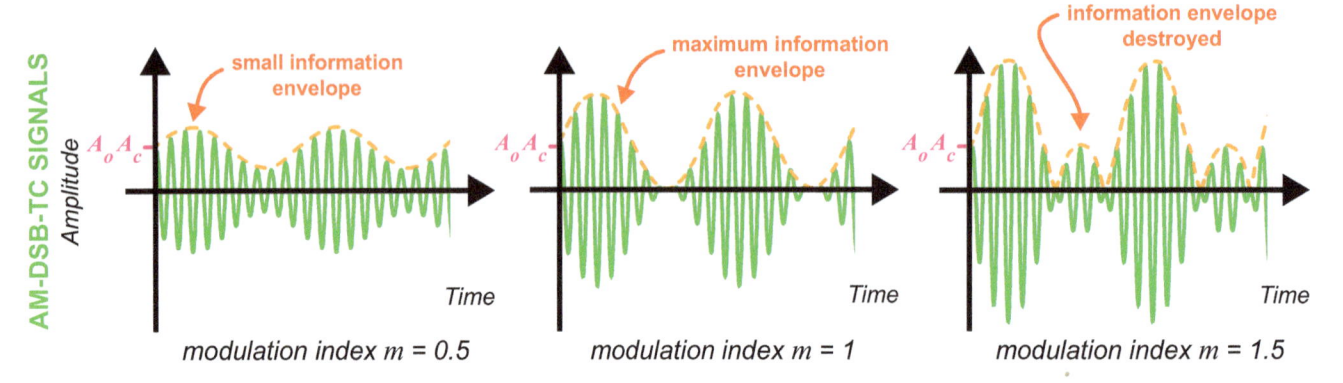

Figure 6.8: Time domain plots showing AM-DSB-TC signals with modulation indexes of 0.5, 1 and 1.5

6.3.2 AM-DSB-TC Modulation: Modulating Baseband Signals

Consider once again a baseband information signal, with a bandwidth of f_h Hz, where f_h is the highest frequency component contained within the signal. AM modulating it with a carrier would result in the modulated signal shown in Figure 6.9. Note that, when modulated, the signal occupies a bandwidth of $2f_h$ Hz, and is centred around f_c.

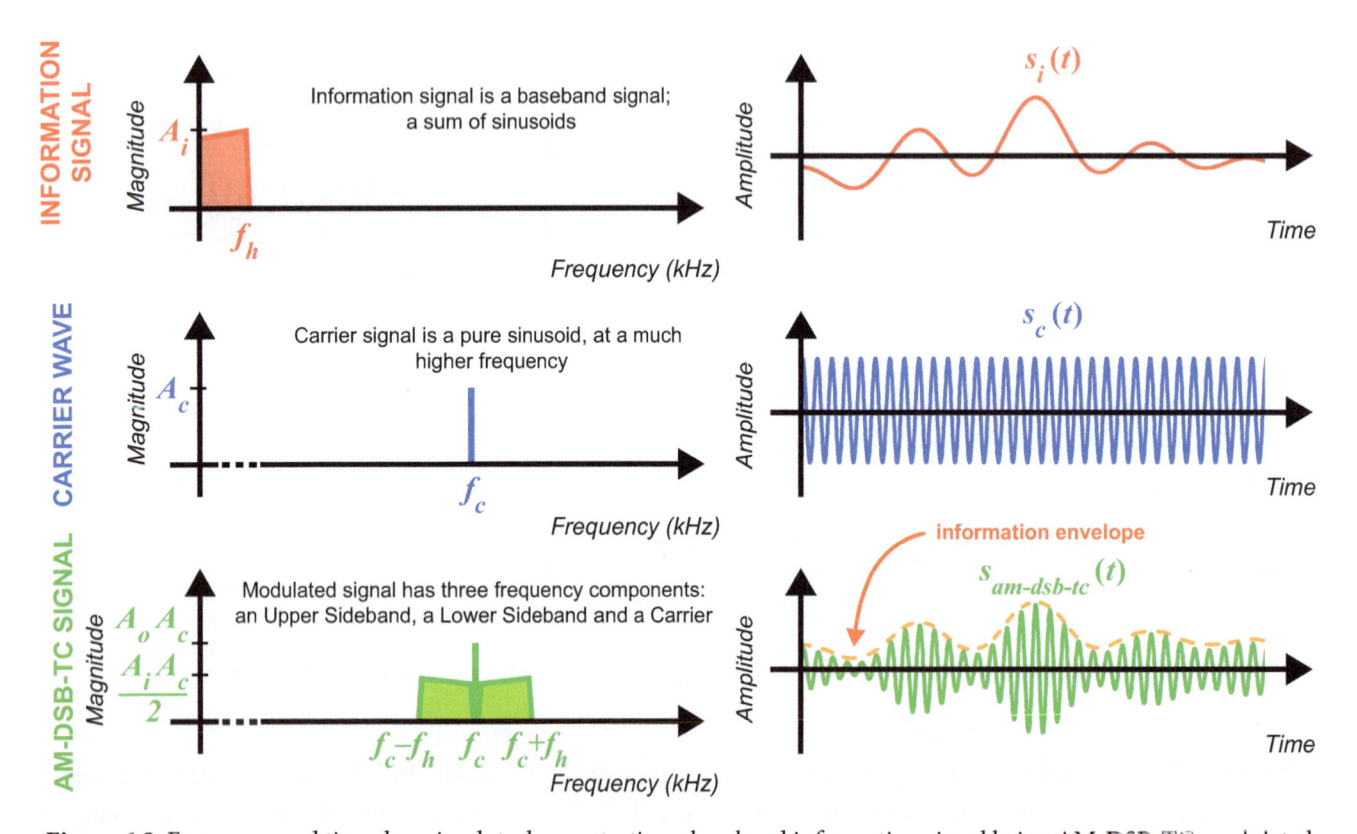

Figure 6.9: Frequency and time domain plots demonstrating a baseband information signal being AM-DSB-TC modulated

Chapter 6: Amplitude Modulation (AM) Theory and Simulation

As discussed previously in Section 6.2, any baseband information signal can be represented as a sum of weighted sinusoidal components using Fourier decomposition. If we define an information signal as being composed of the same three components used before:

$$s_i(t) = A_{i1}\cos(2\pi f_{i1} t) + A_{i2}\cos(2\pi f_{i2} t) + A_{i3}\cos(2\pi f_{i3} t) \qquad (6.15)$$

$$= A_{i1}\cos(\omega_{i1} t) + A_{i2}\cos(\omega_{i2} t) + A_{i3}\cos(\omega_{i3} t) \; ,$$

then it would appear as Figure 6.10 in the frequency domain.

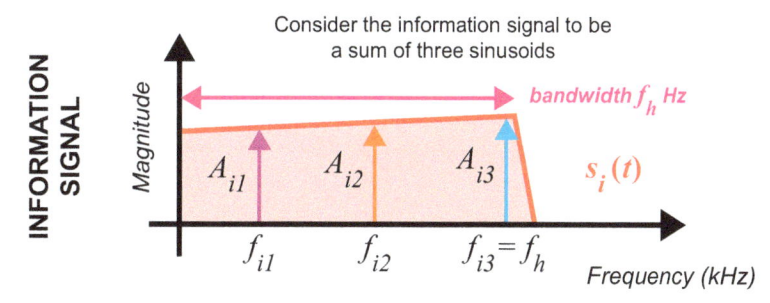

Figure 6.10: Baseband information signal composed of three weighted cosines

Modulating this baseband signal with the carrier from Eq. (6.11), we get:

$$s_{am-dsb-tc}(t) = \left[A_o + A_{i1}\cos(\omega_{i1} t) + A_{i2}\cos(\omega_{i2} t) + A_{i3}\cos(\omega_{i3} t) \right] A_c \cos(\omega_c t) \qquad (6.16)$$

Multiplying this out and applying the product-to-sum trigonometry rule from Eq. (B.4) gives:

$$s_{am-dsb-tc}(t) = A_o A_c \cos(\omega_c)t + \frac{A_{i1} A_c}{2}\left(\cos(\omega_c - \omega_{i1})t + \cos(\omega_c + \omega_{i1})t \right) \qquad (6.17)$$

$$+ \frac{A_{i2} A_c}{2}\left(\cos(\omega_c - \omega_{i2})t + \cos(\omega_c + \omega_{i2})t \right)$$

$$+ \frac{A_{i3} A_c}{2}\left(\cos(\omega_c - \omega_{i3})t + \cos(\omega_c + \omega_{i3})t \right) .$$

This signal contains three pairs of tones, or six frequency components, all centred around the carrier f_c. The frequency domain sketch is shown in Figure 6.11; note that the sidebands are symmetrical around f_c. It is also worth comparing this to Figure 6.5, i.e. the equivalent spectrum sketch for AD-DSB-SC.

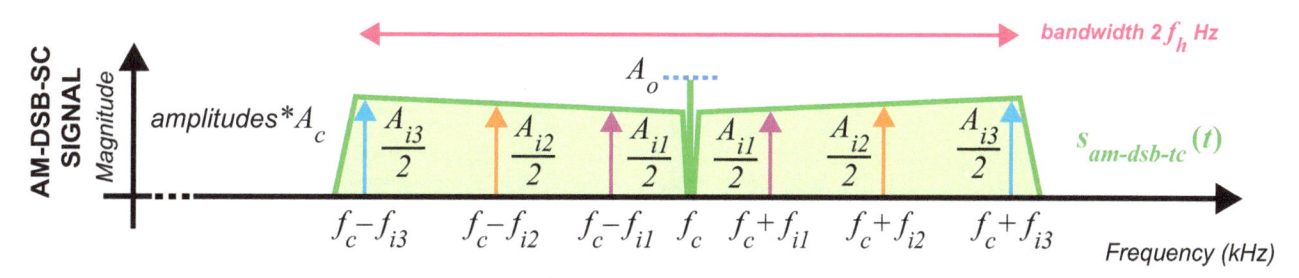

Figure 6.11: Modulated baseband information signal composed of three weighted cosines

Next, we will move on to Exercise 6.2, which involves running a simple Simulink simulation illustrating happens (in the time and frequency domains) to sinusoidal waves as they are modulated onto a carrier to produce an AM-DSB-TC signal.

Exercise 6.2 AM-DSB-TC Simulation

In this exercise we will test out the AM-DSB-TC modulator. Two different information signals are included in this model, and you will be able to switch between them to see the differences in the output. A *Spectrum Analyzer* and a *Time Scope* are connected to various points throughout the model which will permit inspection of the signals in the time and frequency domains during the modulation process.

(a) **Open MATLAB.** Set the working directory to an appropriate folder so you can open this model:

> /am/simulation/am_dsb_tc.slx

The block diagram should look like this:

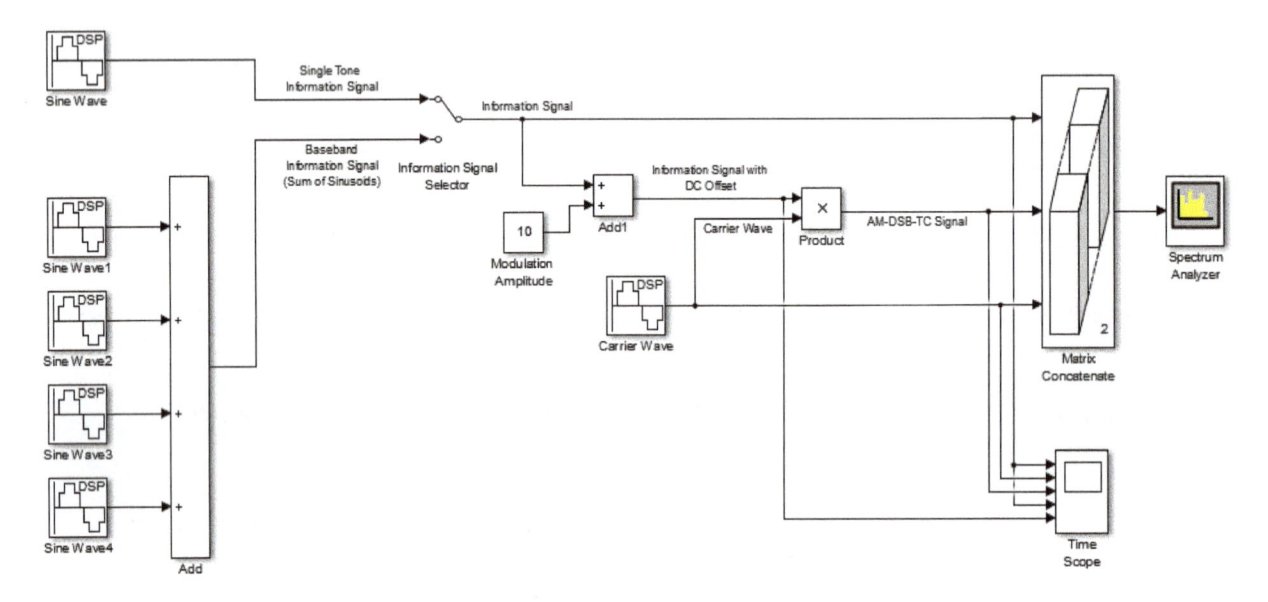

(b) Essentially, this model is the same as the one from Exercise 6.1, however a DC Offset has been added to the information signal prior to modulation. The first information signal (top) is a single tone, and the other is a sum of tones. The offset information signals are input to the *Product* block, where they are mixed with the *Carrier Wave* to create the AM-DSB-TC signal. The *Information Signal Selector* switch allows you to switch between them. To change the selection, simply ② on it. Leave this set on 'Single Tone' for now.

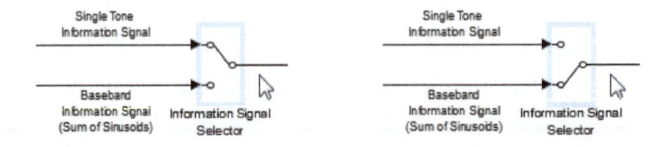

All of these signals are input to scopes to enable visualisation in both the time and frequency domains.

Chapter 6: Amplitude Modulation (AM) Theory and Simulation

(c) **Run the simulation.** Begin the simulation by 🖱 on the 'Run' ▶ button in the Simulink toolbar. The two scope windows will appear, and you should position them so they can both be seen at once. The information signal is blue, the carrier is orange, and the AM-DSB-TC signal is green, as indicated by the legends.

(d) **Signal Analysis #1.** If you left the switch set to 'Single Tone', this is what you should see in the *Spectrum Analyzer* and *Time Scope* windows:

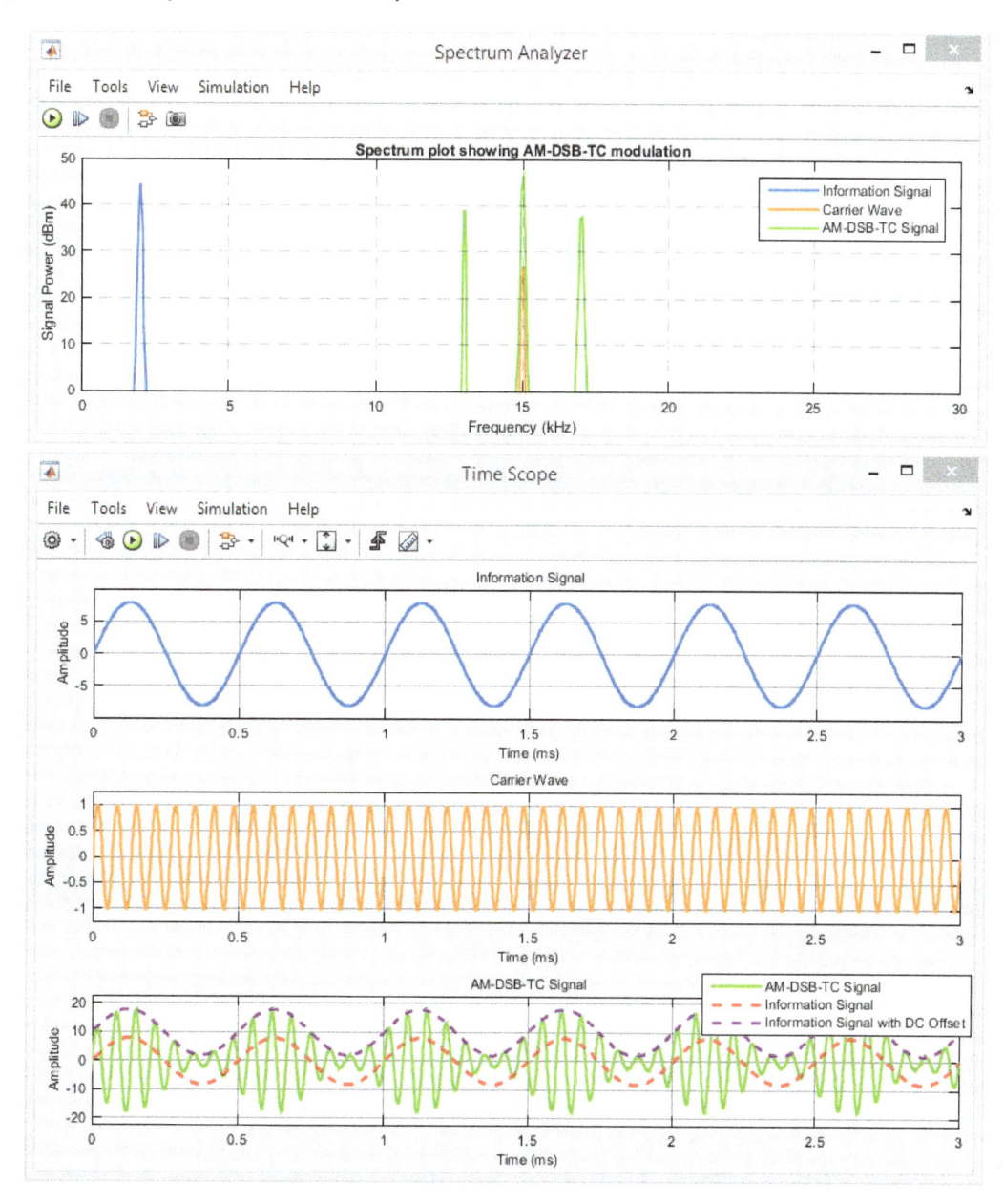

As before, both the information signal and carrier wave are single tones with frequencies of 2kHz and 15kHz, and these are mixed together to create the AM-DSB-TC signal. The modulated signal contains three frequency components, located at 13kHz, 15kHz and 17kHz. Note in particular that a carrier is present in this signal, thus there is an information envelope. The envelope is highlighted in the *Time Scope* by the dashed purple line.

(e) **Signal Analysis #2.** Set the *Information Signal Selector* switch to 'Baseband Information Signal' and re-run the simulation. This is what you should expect to see:

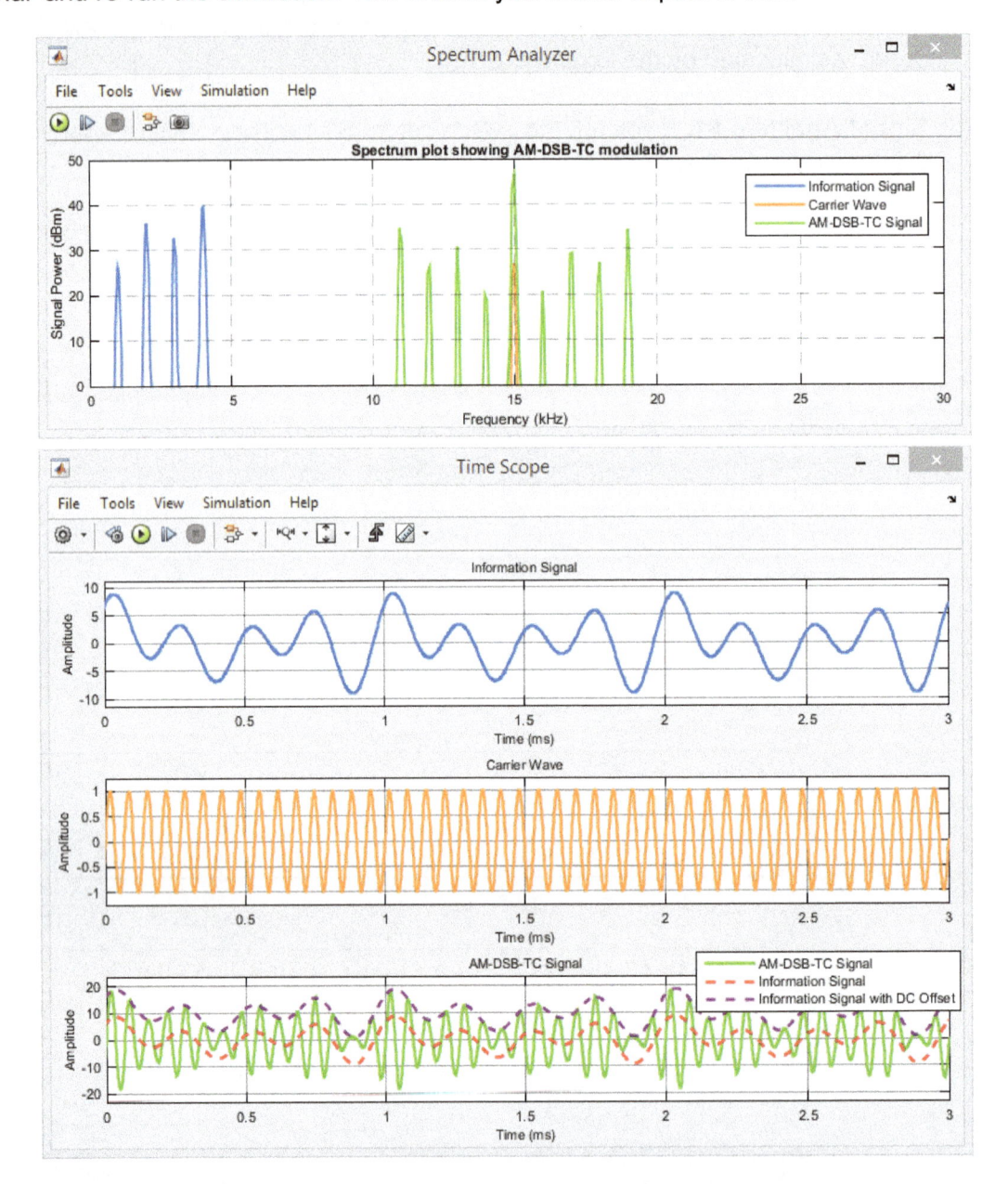

The information signal in this case contains four frequency components, located at 1kHz, 2kHz, 3kHz and 4kHz. When modulated with the same carrier signal, four pairs of tones are created, positioned at 14kHz & 16kHz, 13kHz & 17kHz, 12kHz & 18kHz, and finally, 11kHz & 19kHz. The carrier is also present in this modulated signal. You should note that the amplitudes of the tones in these pairs match, that the sidebands are mirror images of each other, and that there is an information envelope present.

(f) Can you estimate the value of the modulation index by looking at the *Time Scope*?

6.4 AM-SSB: Single Sideband AM

While mathematically simple, AM-DSB modulation is spectrally inefficient because the process creates a pair of identical sidebands. This means these signals have a wider bandwidth than is strictly necessary to carry the information. Therefore, if only a finite amount of bandwidth is available in a communications channel, this would not be the analogue modulation standard of choice. The main advantage of AM-SSB modulation is that the bandwidth occupied by the modulated signal is exactly the same as the bandwidth of the baseband signal. This is highlighted in Figure 6.12, where we also show the two variations of AM-SSB: Single Upper Sideband (AM-SUSB) and Single Lower Sideband (AM-SLSB).

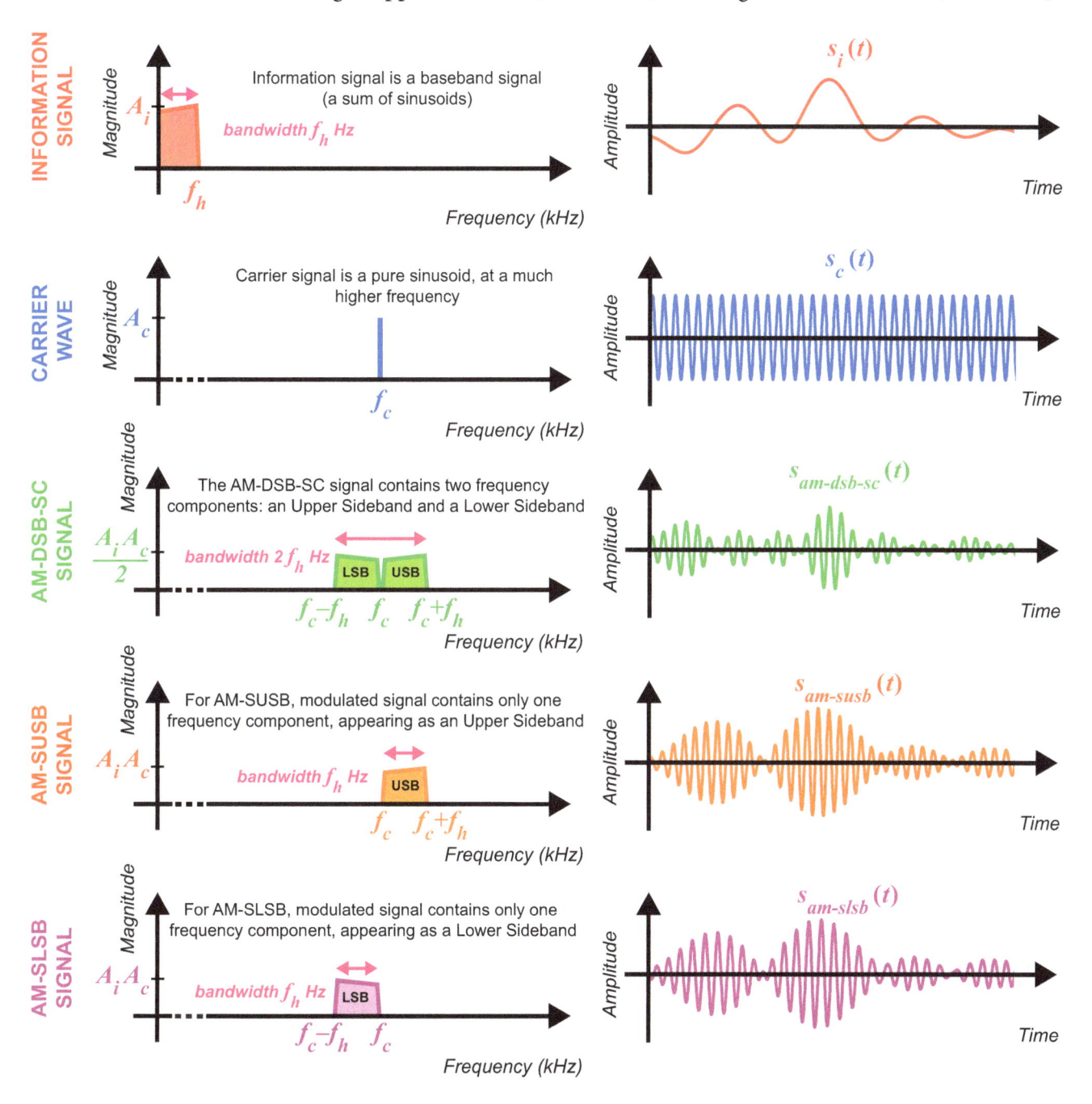

Figure 6.12: Comparison of AM-DSB-SC, AM-SUSB and AM-SLSB signals in the frequency and time domains

If a baseband information signal had a bandwidth of f_h Hz (where f_h is the highest frequency component contained within the signal), AM-DSB-SC modulation would result in a signal with a bandwidth of $2f_h$ Hz. On the other hand, if it were AM-SSB modulated, the modulated signal would only have a bandwidth of f_h Hz.

In the time domain, the upper and lower sideband AM-SSB signals appear to take a totally different form to the AM-DSB-SC signal, and they both have very similar envelopes. Neither of these AM-SSB signals contain the modulation carrier though, so neither have information envelopes.

6.4.1 Generating AM-SSB Signals

There are a number of ways you can create AM-SSB signals. One option is to simply bandpass filter an AM-DSB-SC signal to remove one of the sidebands; the issue with this though is that the filter would require a steep rolloff and a large attenuation to ensure that the unwanted sideband could be completely removed whilst keeping the required sideband intact. If you were going to implement this with a digital filter, you would require a very large number of filter weights. Unless there was dedicated hardware available such as an FPGA that would allow the filter's multiply-accumulate operations to be performed in parallel, this method would not normally be used for real time applications.

The alternative method is to use the AM-SSB modulator. This design is a little different to the AM-DSB modulator, in that it involves a Hilbert transform and a quadrature carrier. The output of the AM-SSB modulator is the sum of the in-phase (I) and quadrature-phase components (see Section 5.4 (page 187) for a review of quadrature modulator), as shown in Figure 6.13 [51].

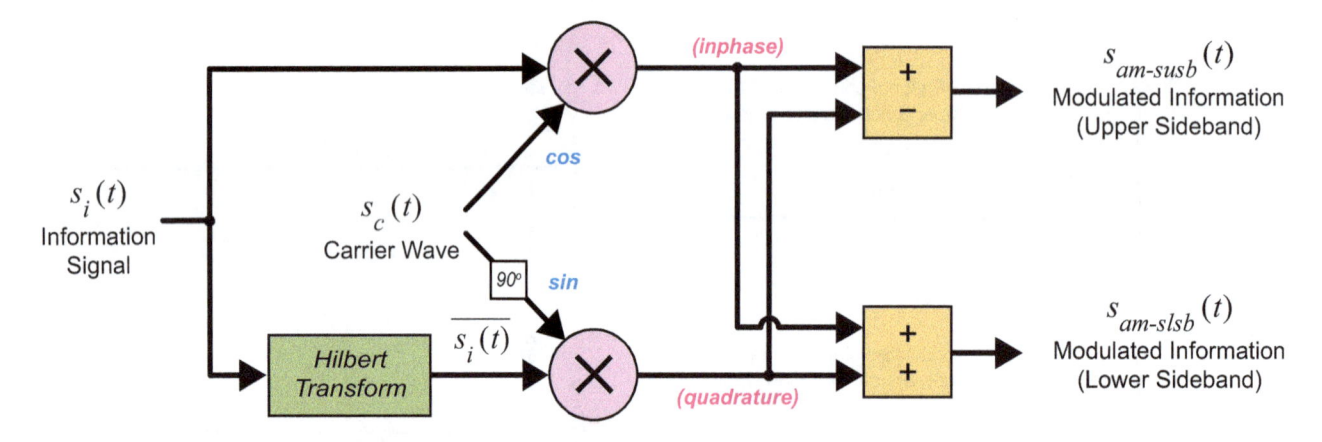

Figure 6.13: Block diagram of the AM-SSB modulator, showing configurations for both AM-SUSB and AM-SLSB

The general equation for this modulator is as follows:

$$s_{am-ssb}(t) = s_i(t)\,\Re e\!\left[s_c(t)\right] \mp \overline{s_i(t)}\,\Im m\!\left[s_c(t)\right] \tag{6.18}$$

where: $\overline{s_i(t)}$ is the Hilbert transform $H(s_i)(t)$ of the information signal $s_i(t)$,

$\Re e\!\left[s_c(t)\right]$ and $\Im m\!\left[s_c(t)\right]$ are the real and imaginary components of the quadrature carrier,

and the $\mp$ relates to whether the modulator is configured in AM-SUSB or AM-SLSB mode.

Chapter 6: Amplitude Modulation (AM) Theory and Simulation

6.4.2 AM-SSB Modulation: Modulating a Sine Wave

If we define an information signal as having amplitude A_i and frequency f_i,

$$s_i(t) = A_i \cos(2\pi f_i t) = A_i \cos(\omega_i t) \tag{6.19}$$

then the Hilbert Transform (the integral) of this signal would be:

$$s_i(t) \rightarrow \overline{s_i(t)}$$

$$A_i \cos(\omega_i t) \rightarrow A_i \sin(\omega_i t) \ . \tag{6.20}$$

If the quadrature carriers wave had amplitude A_c and frequency f_c, then it would take the form

$$s_c(t) = A_c \cos(2\pi f_c t) + A_c \sin(2\pi f_c t) = A_c \cos(\omega_c t) + A_c \sin(\omega_c t) \ . \tag{6.21}$$

Substituting Eqs. (6.19), (6.20) and (6.21) into Eq. (6.18) yields:

$$s_{am-ssb}(t) = A_i \cos(\omega_i t) A_c \cos(\omega_c t) \mp A_i \sin(\omega_i t) A_c \sin(\omega_c t) \ . \tag{6.22}$$

Solving this with the product-to-sum trigonometry rule from Eq. (B.4) gives:

$$s_{am-ssb}(t) = \frac{A_i A_c}{2} \Big(\cos(\omega_c - \omega_i)t + \cos(\omega_c + \omega_i)t \Big) \tag{6.23}$$

$$\mp \frac{A_i A_c}{2} \Big(\cos(\omega_c - \omega_i)t - \cos(\omega_c + \omega_i)t \Big) \ .$$

If we were to consider the configuration required to generate an AM-SUSB signal, we would solve this as follows:

$$s_{am-susb}(t) = \frac{A_i A_c}{2} \Big(\cos(\omega_c - \omega_i)t + \cos(\omega_c + \omega_i)t \Big) \tag{6.24}$$

$$- \frac{A_i A_c}{2} \Big(\cos(\omega_c - \omega_i)t - \cos(\omega_c + \omega_i)t \Big)$$

$$= A_i A_c \cos(\omega_c + \omega_i)t$$

which leaves us with only the upper sideband component. Solving Eq. (6.23) for the AM-SLSB configuration, we find the lower sideband component:

$$s_{am-slsb}(t) = \frac{A_i A_c}{2} \Big(\cos(\omega_c - \omega_i)t + \cos(\omega_c + \omega_i)t \Big) \tag{6.25}$$

$$+ \frac{A_i A_c}{2} \Big(\cos(\omega_c - \omega_i)t - \cos(\omega_c + \omega_i)t \Big)$$

$$= A_i A_c \cos(\omega_c - \omega_i)t \ .$$

When a baseband signal is AM-SSB modulated, the mathematical analysis for this process becomes rather complicated, so we will not go into it here. You will however be able to see it in action by completing Exercise 6.3, when we will run a simple Simulink simulation that shows what happens to sinusoidal waves as they are modulated with an AM-SSB modulator. It is configured in both AM-SUSB and AM-SLSB modes, enabling examination of both outputs.

Exercise 6.3 AM-SSB Simulation

In this exercise we will investigate the AM-SSB modulator. Two different information signals are included in this model, along with a switch to choose between them. A *Spectrum Analyzer* and a *Time Scope* are connected to various points throughout the model, which will allow the signals to be viewed in the time and frequency domains as they are modulated.

(a) **Open MATLAB.** Set the working directory to an appropriate folder so you can open this model:

/am/simulation/am_ssb.slx

The block diagram should look like this:

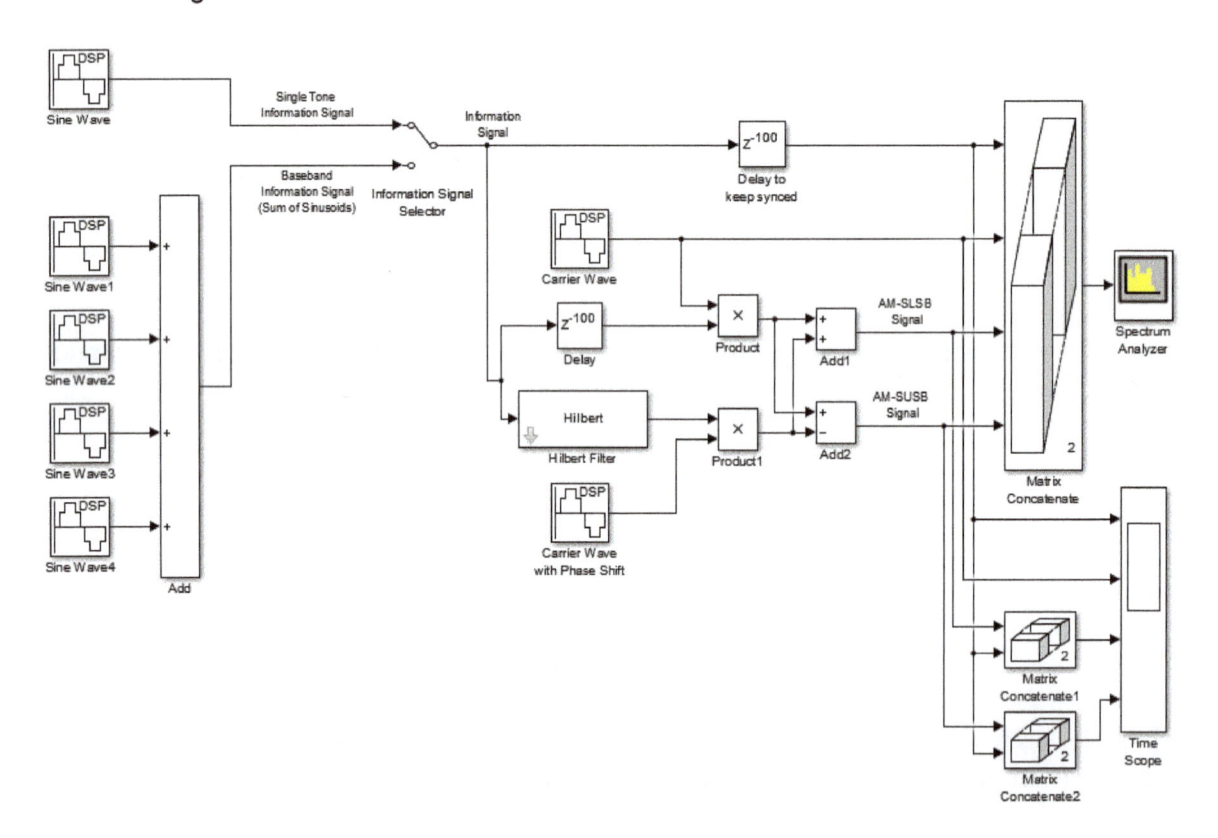

(b) In this model, the first information signal (top) is a single tone, while the second is a sum of tones. The *Information Signal Selector* switch allows you to switch between them. To change the selection, simply 2 on the switch. Leave this set on 'Single Tone' for now.

The selected information signal is split in two, and the first branch is mixed with a carrier wave to produce the $s_i(t) \Re[s_c(t)]$ signal. The second branch of the information signal is input to

Chapter 6: Amplitude Modulation (AM) Theory and Simulation 221

a *Hilbert Filter*, which behaves like a digital integrator. This filtered signal is then mixed with the imaginary part of the carrier to produce the $s_i(t)\, \Im[s_c(t)]$ signal. The two *Add* blocks are used to implement the two configurations of the AM-SSB modulator. These, along with the carrier and original information signal, are input to scopes to allow you to visualise the signals in the time and frequency domains.

(c) **Run the simulation.** Begin the simulation by clicking on the 'Run' ▶ button in the Simulink toolbar. The two scope windows should appear, showing the information signal (blue), the carrier (orange), the AM-SLSB signal (green), and the AM-SUSB signal (red).

(d) **Signal Analysis #1.** If the single tone information signal was selected, you should see the following in the scope windows:

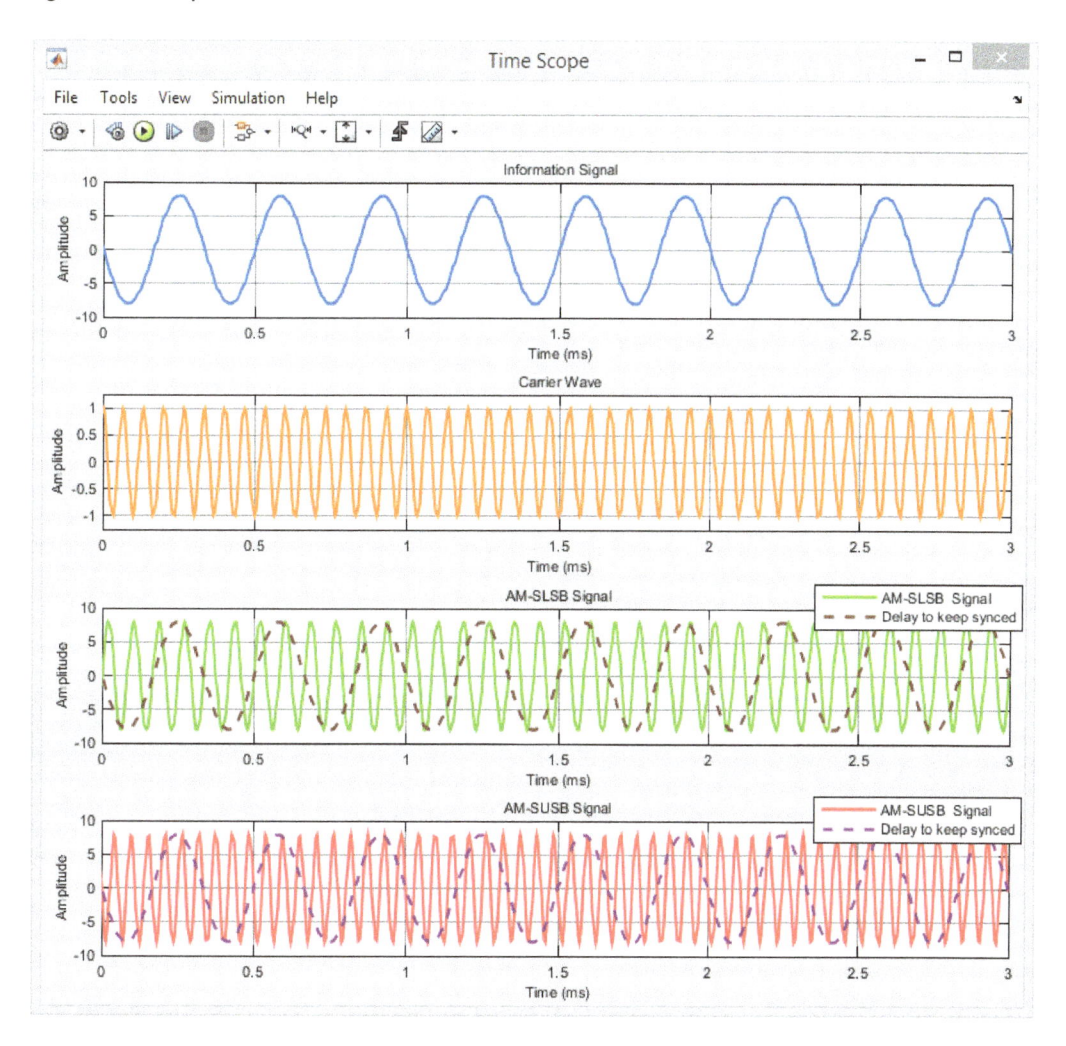

Once again, the information signal is a single tone with a frequency of 2kHz. The carrier (also a single tone) has a frequency of 15kHz. The AM-SLSB modulated signal is sinusoidal, and it has a lower frequency than the carrier wave. Conversely, you should see that the AM-SUSB modulated signal, also sinusoidal, has a higher frequency than the carrier wave.

The *Spectrum Analyzer* shows the signals in the frequency domain. The green AM-SLSB signal has a frequency component at 13kHz, i.e. $f_c - f_i$, and the component of the red AM-

SUSB signal is positioned at 17kHz, i.e. $f_c + f_i$. As the modulated signals are themselves only single tones, there are no information envelopes present within them.

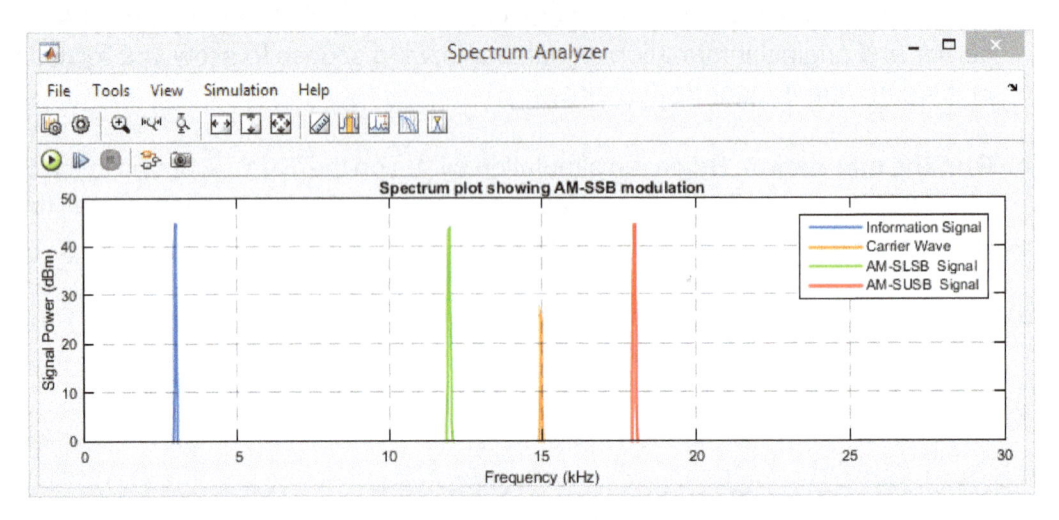

(e) 🖱 the 'Scale y axis' 🔘 button to zoom out. This will allow you to see how well attenuated the unwanted sidebands are:

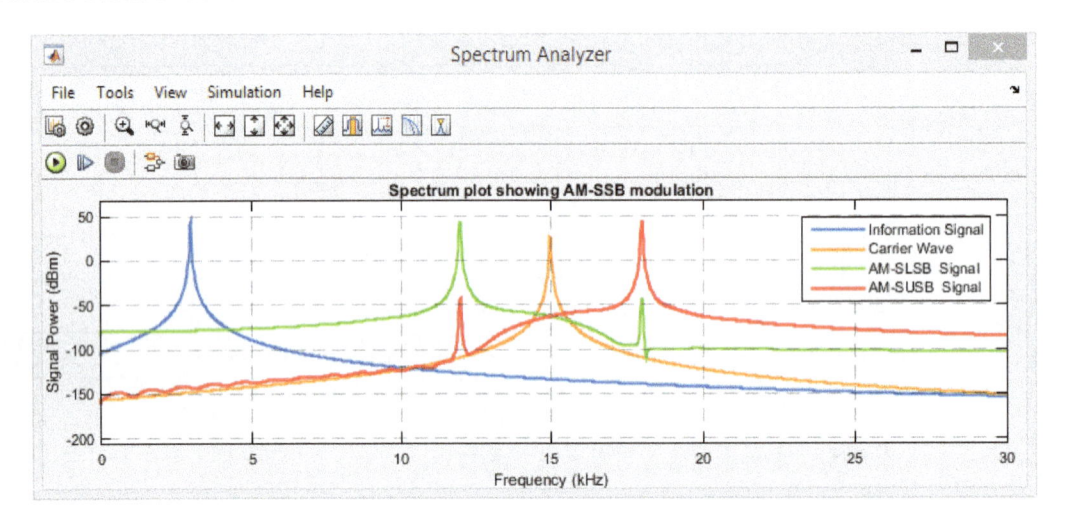

The AM-SSB modulator has attenuated the unwanted sideband in each of the signals to around -40dBm, whilst keeping the power of the primary sideband at about 40dBm. This means that the 'removed' sidebands do not totally disappear but, as they only contain a fraction of the power, these components are insignificant.

(f) 🖱 on the plot and select 'Configuration Properties'. Change the limits of the y axis to the following, then apply the changes to zoom back in:
- Y-limits (Minimum): 0
- Y-limits (Maximum): 50

Chapter 6: Amplitude Modulation (AM) Theory and Simulation 223

🏃 **(g) Signal Analysis #2.** With single tone AM-SSB modulation, the resulting signal is simply a sinusoidal tone at a higher frequency. This is not the case if the information signal is a baseband signal. Set the *Information Signal Selector* switch to 'Baseband Information Signal', and re-run the simulation. This is what you should expect to see:

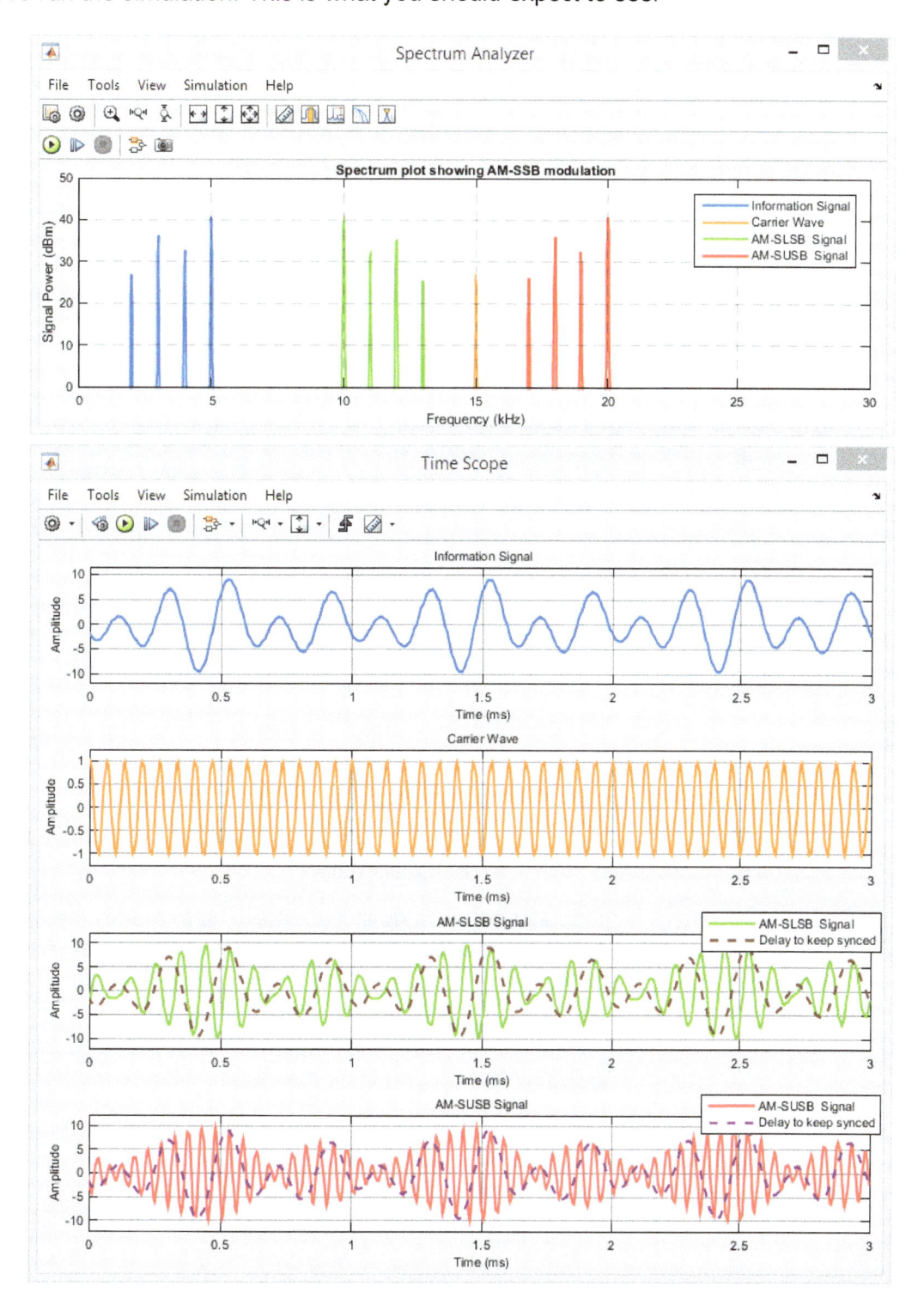

(h) This information signal contains four frequency components, located at 2kHz, 3kHz, 4kHz and 5kHz. The highest component in this signal, f_h, is 5kHz. When it is AM-SLSB modulated with the 15kHz carrier, a lower sideband is created, and this appears as a mirror image of the baseband information signal. It has components at 13kHz, 12kHz, 11kHz and 10kHz, relating to the 2kHz, 3kHz, 4kHz and 5kHz components of the baseband signal. When you examine the modulated signal in the time domain, notice that the signal *looks* like it has a carrier with a frequency of around 11.5kHz — this is simply a function of the AM-SSB modulation process.

(i) When the baseband signal is AM-SUSB modulated, an upper sideband is created, which exactly matches the baseband information signal, shifted up to higher frequencies. It has components at 17kHz, 18kHz, 19kHz and 20kHz, which relate to the 2kHz, 3kHz, 4kHz and 5kHz components of the baseband signal, respectively.

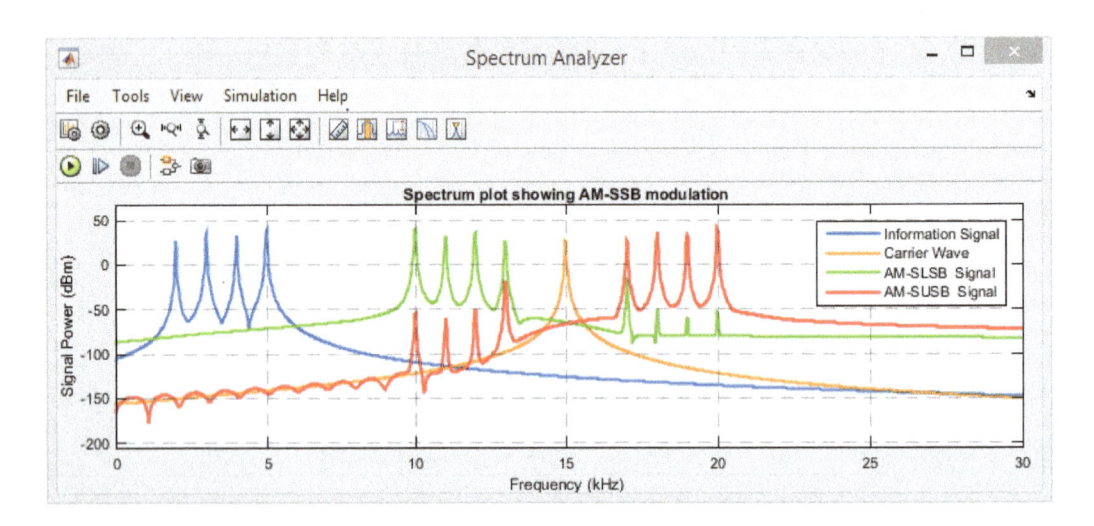

(j) the 'scale y axis' button to zoom out. Once again, this will allow you to see how well attenuated the unwanted sidebands are:

Components of the unwanted sidebands are present in each of the signals, but again, these are of insignificant power to affect the signal quality.

One way in which this modulator could be optimised would be to modify the parameters of the *Hilbert Filter* to ensure that a greater attenuation is applied to frequency components in the unwanted sidebands. You may want to investigate this as a side project!

(k) Can you think of any applications where this modulation scheme would be preferable to either AM-DSB-SC or AM-DSB-TC? What do you think the disadvantages are in terms of receiver complexity?

Chapter 6: Amplitude Modulation (AM) Theory and Simulation 225

6.5 AM-VSB: Vestigial Sideband AM

AM-VSB was created in order to reduce the spectral requirements of analogue TV signals. Baseband PAL and SECAM analogue multiplex signals had a bandwidth of around 6.75MHz, and contained information about the luminance and chroma of the picture, along with an audio signal. Critical information was stored within the signal's DC component, which meant that a carrier had to be transmitted with the signal when it was broadcast. The AM-DSB-TC modulation process resulted in a signal with a bandwidth of 13.5MHz, and regulating bodies were not happy with TV broadcasters using such a large portion of spectrum for every single TV channel. TV broadcasters were not overly keen on this either, as they were having to pay huge electricity bills to transmit a signal with two identical sidebands, and a high powered carrier.

What they decided to do was apply a Bandpass Filter (BPF) to the AM-DSB-TC signal to suppress most of one of the sidebands, in an effort to reduce its bandwidth. The partial sideband is called a Vestigial Sideband, and it remains because the filter roll-off and attenuation required to completely remove it cannot be achieved whilst keeping the carrier intact[53]. Figure 6.14 shows a baseband PAL TV signal, and highlights the reduced bandwidth associated with AM-VSB modulation.

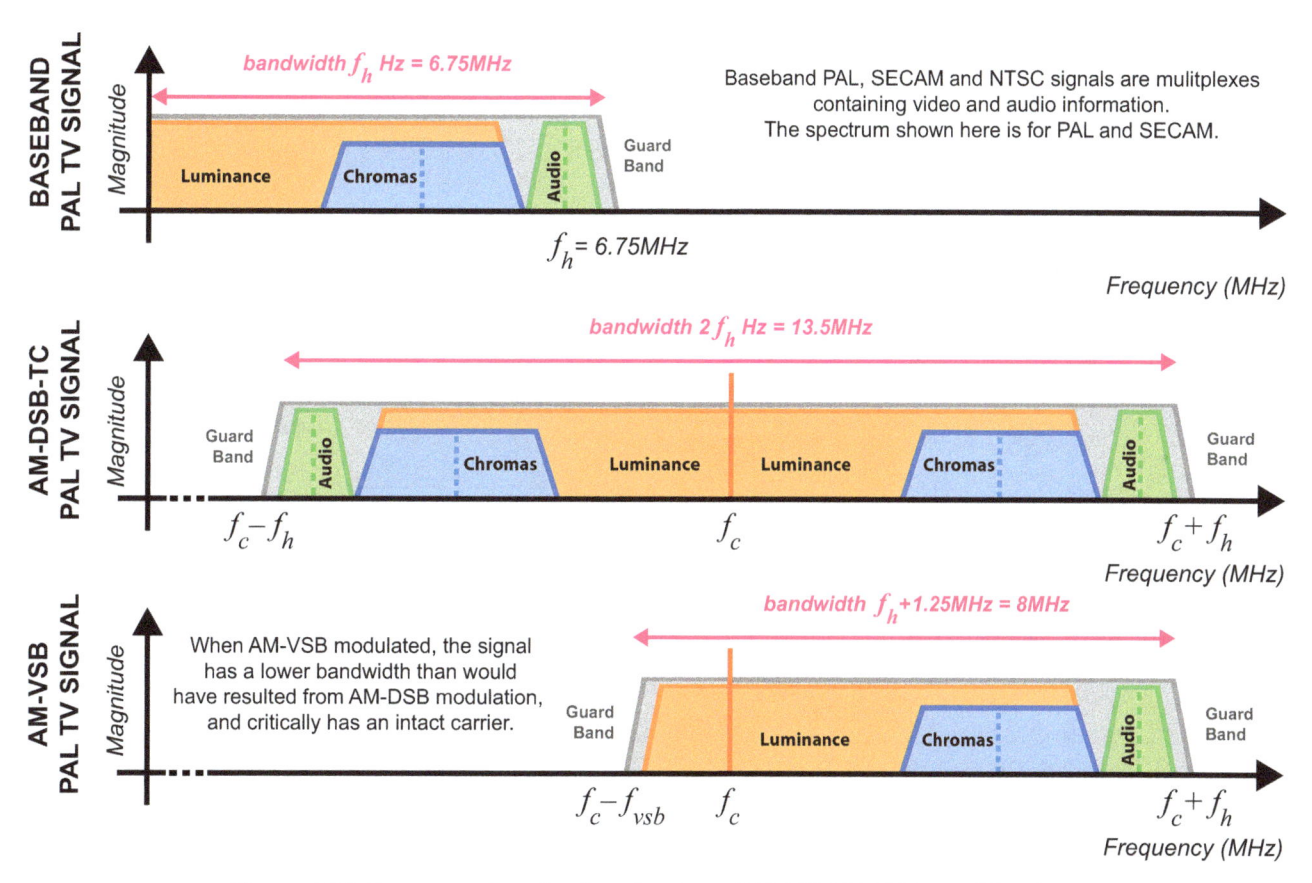

Figure 6.14: Frequency domain plots showing a PAL/ SECAM TV signal first being AM-DSB-TC modulated, then AM-VSB modulated

AM-VSB TV signals have not been transmitted in the UK since the *digital switchover* was completed in 2012. DVB, the new DTV standard, uses a statistical multiplexing and modulation process called COFDM, and as such, the signals appear totally different in the frequency domain — see Section 3.11 (page 79) for more information.

There are numerous countries around the world still using analogue TV that are yet to undergo a *digital switchover* and, if you live in one, it is likely that AM-VSB signals can be received with your RTL-SDR; so it is still worth briefly discussing. For the general case AM-VSB signals shown in Figure 6.15, the AM-VSB modulator takes the form shown in Figure 6.16.

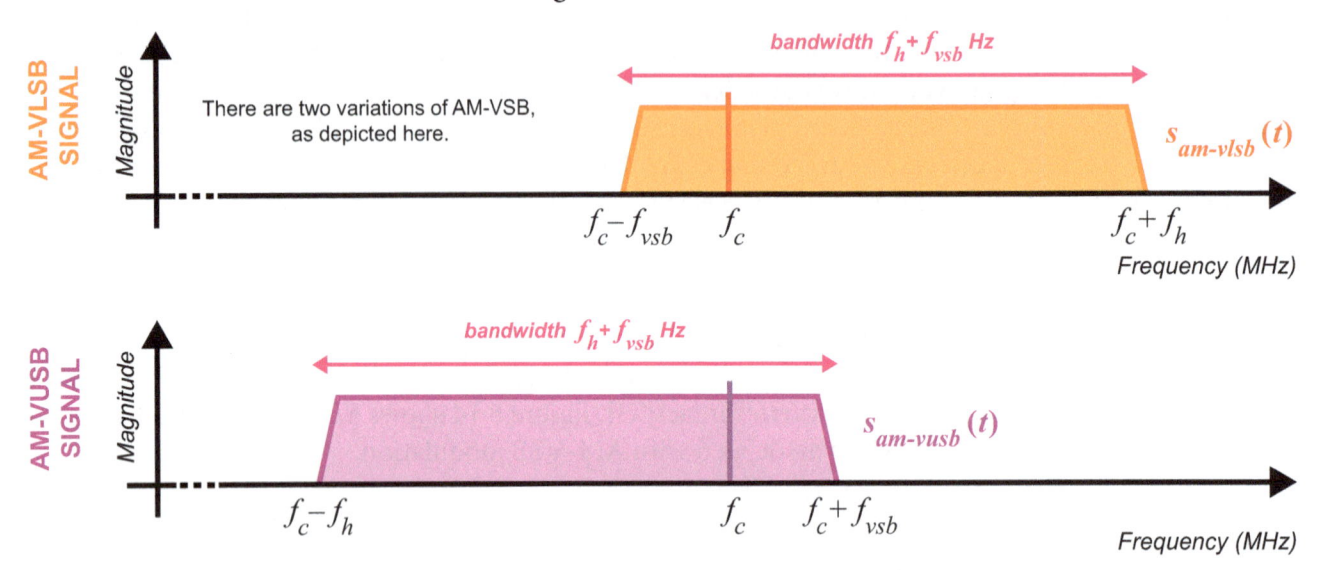

Figure 6.15: Frequency domain plots of AM-VSB signals showing the two variations: AM-VLSB and AM-VUSB

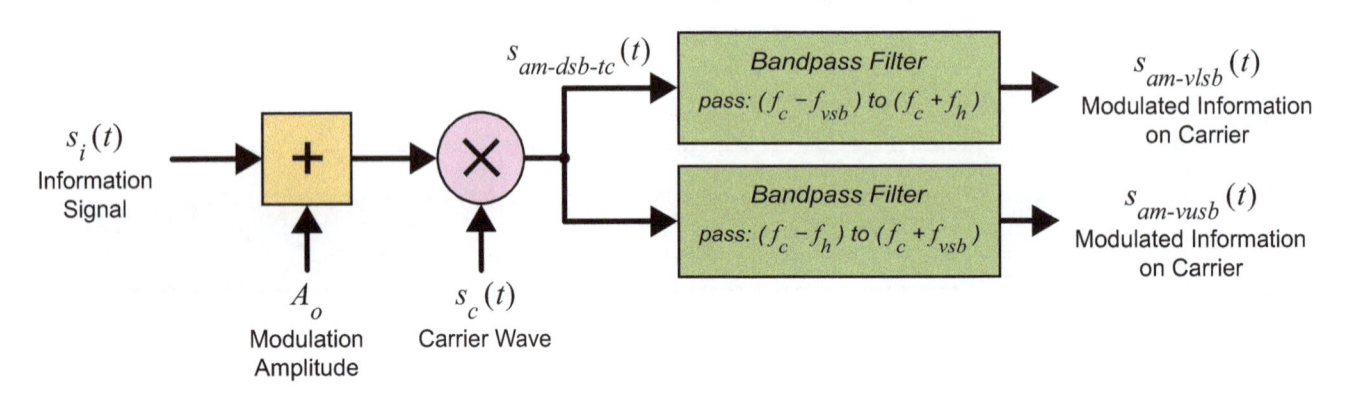

Figure 6.16: Block diagram of the AM-VSB modulator, showing configurations for both AM-VLSB and AM-VUSB

The general equation of an AM-VSB signal is as shown in Eq. (6.26), and depending upon how the bandpass filter is configured, it will either produce a Vestigial Upper Sideband (VUSB) or Vestigial Lower Sideband (VLSB) signal.

$$s_{am-vsb}(t) = BPF\left\{\left[A_o + s_i(t)\right] \times s_c(t)\right\} \quad (6.26)$$

where: $s_i(t)$ is the information signal as in Eq. (6.10), and $s_c(t)$ the carrier as in Eq. (6.11).

6.6 Theoretical AM Demodulation

Mathematically, to demodulate an AM signal, a receiver must simply mix (or multiply) a received signal with a sine wave that has *exactly* the same frequency and phase as the carrier embedded within it (the carrier originally used to modulate the signal). The mixing operation shifts the modulated information from being centred around the carrier frequency, back to baseband, and simultaneously creates a spectral image of the demodulated signal around $2f_c$ Hz. This approach can be considered a *synchronous* demodulation method, and it is often associated with *carrier synchronisation*. To give an example of how this method works, we shall present the demodulation of a single tone modulated AM-DSB-SC signal.

A single tone signal (Eq. (6.2)) modulated by a carrier (Eq. (6.3)) produces an AM-DSB-SC signal:

$$s_{am-dsb-sc}(t) = \frac{A_i A_c}{2}\Big(\cos(\omega_c - \omega_i)t + \cos(\omega_c + \omega_i)t\Big). \tag{6.27}$$

Mixing this with $\cos(\omega_c t)$ demodulates the signal, yielding $s_d(t)$:

$$s_d(t) = \frac{A_i A_c}{2}\Big(\cos(\omega_c - \omega_i)t + \cos(\omega_c + \omega_i)t\Big)\cos(\omega_c t) \tag{6.28}$$

$$= \frac{A_i A_c}{2}\Big[\cos\big((\omega_c - \omega_i)t\big)\cos(\omega_c t) + \cos\big((\omega_c + \omega_i)t\big)\cos(\omega_c t)\Big]$$

$$= \frac{A_i A_c}{2}\begin{bmatrix} \frac{1}{2}\cos(-\omega_i t) + \frac{1}{2}\cos\big((2\omega_c - \omega_i)t\big) \\ + \frac{1}{2}\cos(\omega_i t) + \frac{1}{2}\cos\big((2\omega_c + \omega_i)t\big) \end{bmatrix} \quad \text{(lowpass filtered)}$$

Lowpass filtering removes the high frequency components, resulting in the (scaled) information signal:

$$s_d(t) = \frac{A_i A_c}{4}\cos(-\omega_i t) + \frac{A_i A_c}{4}\cos(\omega_i t) = \frac{A_i A_c}{2}\cos(\omega_i t). \tag{6.29}$$

Unfortunately, it is very unlikely that the frequency (and phase) of the modulating and demodulating sine waves exactly match, and when there is any offset, this demodulation process is unsuccessful. Recall that in Section 5.3 (page 182) we reviewed the simple AM transmitter and receiver, and considered the mathematical theory and some of the reception issues.

6.7 Receiving and Downconverting AM-DSB-TC Signals to Complex Baseband

As reviewed in Chapter 5, real RF signals received by the RTL-SDR are *quadrature demodulated* to baseband before they are sampled — this means that the baseband samples entering our MATLAB and Simulink RTL-SDR receiver designs have both I and Q components, i.e. a complex signal.

When an RF AM-DSB-TC signal, denoted by $s_{amRF}(t)$, is received by the RTL-SDR, it is mixed with a complex exponential at frequency f_{lo} (representing the local oscillator frequency in the RTL-SDR) to demodulate the signal to complex baseband, as illustrated in Figure 6.17.

Assuming that the RTL-SDR receives the transmitted signal $s_{am-dsb-tc}$ from Eq. (6.13) on page 210,

$$s_{amRF}(t) = s_{am-dsb-tc}(t) = A_o A_c \cos(\omega_c)t + \frac{A_i A_c}{2}\cos(\omega_c - \omega_i)t + \frac{A_i A_c}{2}\cos(\omega_c + \omega_i)t \quad (6.30)$$

(as we assume a perfect radio channel), the signal output from the RTL-SDR can be modelled as:

$$s_{\text{RTL-SDR}}(t) = LPF\left[s_{bband}(t)\right] = LPF\left[s_{amRF}(t)e^{-j\omega_{lo}t}\right] \quad (6.31)$$

where $e^{-j\omega_{lo}t}$ represents the complex oscillator inside the RTL-SDR.

We can express the downconverted signal using Euler's Formula ($e^{j\omega t} = \cos(\omega t) + j\sin(\omega t)$):

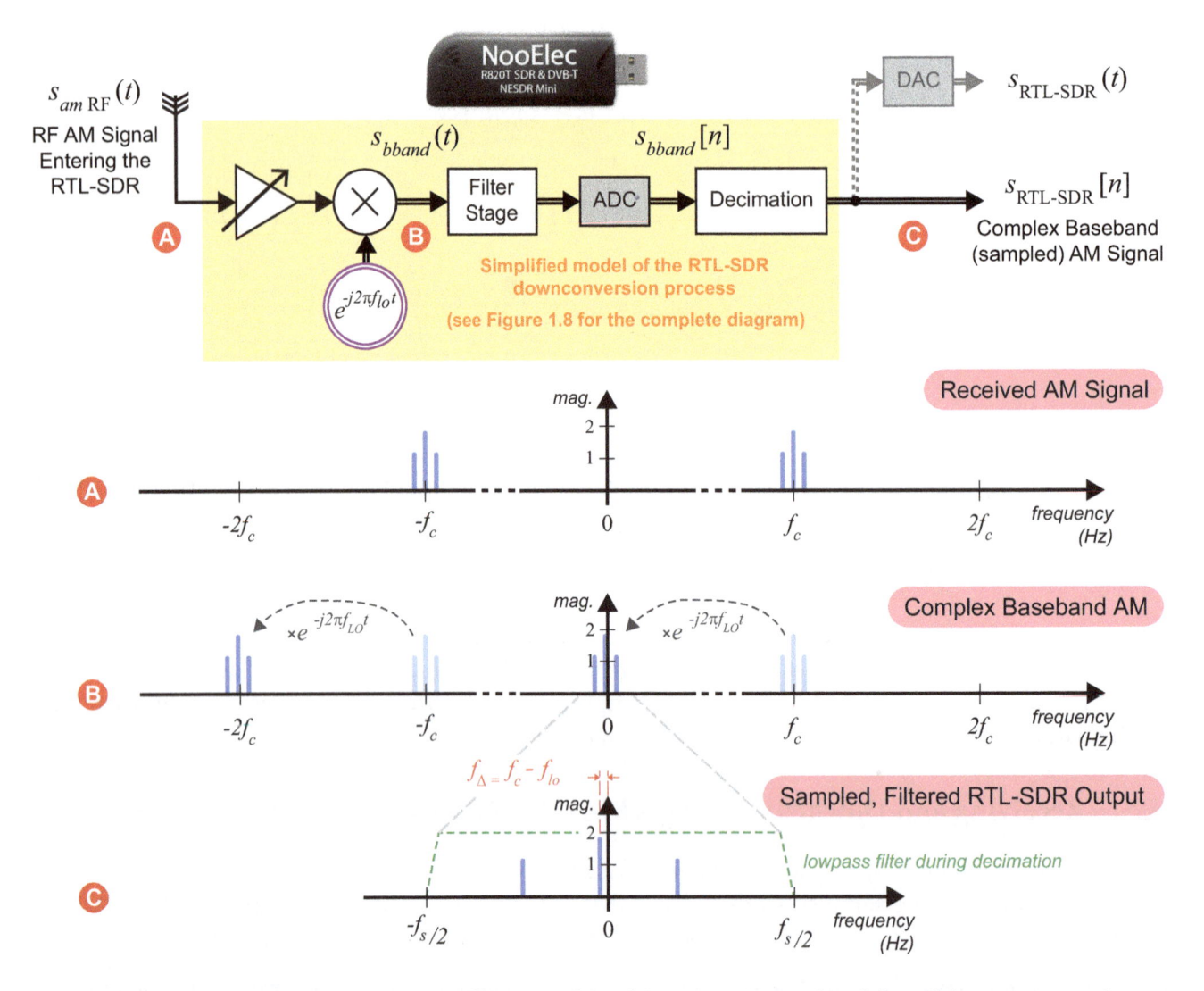

Figure 6.17: The RTL-SDR receiving a single tone modulated AM-DSB-TC signal, and demodulating it to complex baseband. Note that when a frequency offset exists between the original modulating carrier f_c and the local oscillator used during the demodulation process f_{lo}, a frequency offset of f_Δ occurs in the complex baseband output

Chapter 6: Amplitude Modulation (AM) Theory and Simulation 229

$$s_{bband}(t) = s_{amRF}(t)e^{-j\omega_{lo}t} \quad (6.32)$$

$$= s_{amRF}(t) \times \left(\cos(\omega_{lo}t) - j\sin(\omega_{lo}t)\right)$$

$$= A_o A_c \cos(\omega_c t) \times \left(\cos(\omega_{lo}t) - j\sin(\omega_{lo}t)\right)$$

$$+ \frac{A_i A_c}{2}\left(\cos(\omega_c t - \omega_i t) + \cos(\omega_c t + \omega_i t)\right) \times \left(\cos(\omega_{lo}t) - j\sin(\omega_{lo}t)\right).$$

Using the Eqs. (B.4) and (B.6) from Appendix B (page 581), Eq. (6.32) becomes:

$$s_{bband}(t) = \frac{A_o A_c}{2}\Big[\cos(\omega_c t - \omega_{lo}t) + \cos(\omega_c t + \omega_{lo}t)$$

$$- j\sin(\omega_c t - \omega_{lo}t) - j\sin(\omega_c t + \omega_{lo}t)\Big] \quad (6.33)$$

$$+ \frac{A_i A_c}{4}\Big[\cos(\omega_c t - \omega_i t - \omega_{lo}t) + \cos(\omega_c t - \omega_i t + \omega_{lo}t)$$

$$+ \cos(\omega_c t + \omega_i t - \omega_{lo}t) + \cos(\omega_c t + \omega_i t + \omega_{lo}t)$$

$$- j\sin(\omega_c t - \omega_i t - \omega_{lo}t) - j\sin(\omega_c t - \omega_i t + \omega_{lo}t)$$

$$- j\sin(\omega_c t + \omega_i t - \omega_{lo}t) - j\sin(\omega_c t + \omega_i t + \omega_{lo}t)\Big].$$

baseband components
high freq components

The high frequency components are filtered by the lowpass filters within the RTL-SDR, leaving only the complex baseband signal (which we express here in the continuous time domain for simplicity—note the signal $s_{\text{RTL-SDR}}(t)$ can be created with a DAC as shown in Figure 6.17):

$$s_{\text{RTL-SDR}}(t) = \frac{A_o A_c}{2}\Big[\cos(\omega_c t - \omega_{lo}t) - j\sin(\omega_c t - \omega_{lo}t)\Big]$$

$$+ \frac{A_i A_c}{4}\Big[\cos(\omega_c t - \omega_{lo}t - \omega_i t) + \cos(\omega_c t - \omega_{lo}t + \omega_i t) \quad (6.34)$$

$$- j\sin(\omega_c t - \omega_{lo}t - \omega_i t) - j\sin(\omega_c t - \omega_{lo}t + \omega_i t)\Big].$$

6.7.1 Perfect Demodulation

If the frequency of the modulating carrier (ω_c where $\omega_c = 2\pi f_c$) has exactly the same frequency as the local oscillator used during the demodulation process, i.e. $\omega_c = \omega_{lo}$, and if we can assume that no phase difference exists, then Eq. (6.34) becomes:

$$s_{\text{RTL-SDR}}(t) = \frac{A_o A_c}{2}\Big[\cos(0) - j\sin(0)\Big]$$

$$+ \frac{A_i A_c}{4}\Big[\cos(-\omega_i t) + \cos(\omega_i t) - j\sin(-\omega_i t) - j\sin(\omega_i t)\Big], \quad (6.35)$$

meaning that the real (In Phase) component can be expressed as:

$$s_{ip}(t) = \Re e\left[s_{\text{RTL-SDR}}(t)\right] = \frac{A_o A_c}{2} + \frac{A_i A_c}{4}\left[\cos(-\omega_i t) + \cos(\omega_i t)\right], \quad (6.36)$$

and the imaginary (Quadrature Phase) component as:

$$s_{qp}(t) = \Im m\left[s_{\text{RTL-SDR}}(t)\right] = -j\frac{A_i A_c}{4}\left[\sin(-\omega_i t) + \sin(\omega_i t)\right]. \quad (6.37)$$

Notice that when demodulation is performed perfectly, the real component contains a DC offset (i.e. the carrier's DC offset), however the imaginary component does not. This is because, as exploited in the transition from Eq. (6.35) to Eq. (6.36), $\cos(0) = 1$, and $\sin(0) = 0$.

6.7.2 Demodulation with a Frequency Offset

When a frequency offset exists between the modulating carrier and the local oscillator used during the demodulation process (i.e. $\omega_\Delta t = \omega_c t - \omega_{lo} t$), Eq. (6.34) is solved differently:

$$s_{\text{RTL-SDR}}(t) = \frac{A_o A_c}{2}\left[\cos(\omega_\Delta t) - j\sin(\omega_\Delta t)\right] \quad (6.38)$$

$$+ \frac{A_i A_c}{4}\left[\cos(\omega_\Delta t - \omega_i t) + \cos(\omega_\Delta t + \omega_i t) - j\sin(\omega_\Delta t - \omega_i t) - j\sin(\omega_\Delta t + \omega_i t)\right],$$

meaning that the real (In Phase) component can be expressed as

$$s_{ip}(t) = \Re e\left[s_{\text{RTL-SDR}}(t)\right] = \frac{A_o A_c}{2}\cos(\omega_\Delta t) + \frac{A_i A_c}{4}\left[\cos(\omega_\Delta t - \omega_i t) + \cos(\omega_\Delta t + \omega_i t)\right], \quad (6.39)$$

and the imaginary (Quadrature Phase) component as

$$s_{qp}(t) = \Im m\left[s_{\text{RTL-SDR}}(t)\right] = -j\frac{A_o A_c}{2}\sin(\omega_\Delta t) - j\frac{A_i A_c}{4}\left[\cos(\omega_\Delta t - \omega_i t) + \cos(\omega_\Delta t + \omega_i t)\right]. \quad (6.40)$$

It is likely that a complex AM signal of this form will be presented by the RTL-SDR to MATLAB and Simulink as baseband samples, as it would be very challenging to tune the RTL-SDR to the exact frequency of the RF carrier! It is also usually true that the signal being received contains a baseband information signal, rather than a single tone or set of discrete tones. Thus, it would be more realistic to draw Figure 6.18 (as compared to Figure 6.17) to represent the reception of signals with the RTL-SDR. This is therefore what we must aim to demodulate in order to recover the information signal. Back in Section 5.8 (page 199) we reviewed on frequency offset issues and methods of addressing and correcting this.

Coherent AM demodulators use carrier synchronisation techniques to generate local sine waves that do have exactly the same frequency and phase as the modulating carrier, and these will be discussed in depth in Section 7.10 (starting on page 273). For now though, we will focus on non-coherent demodulation

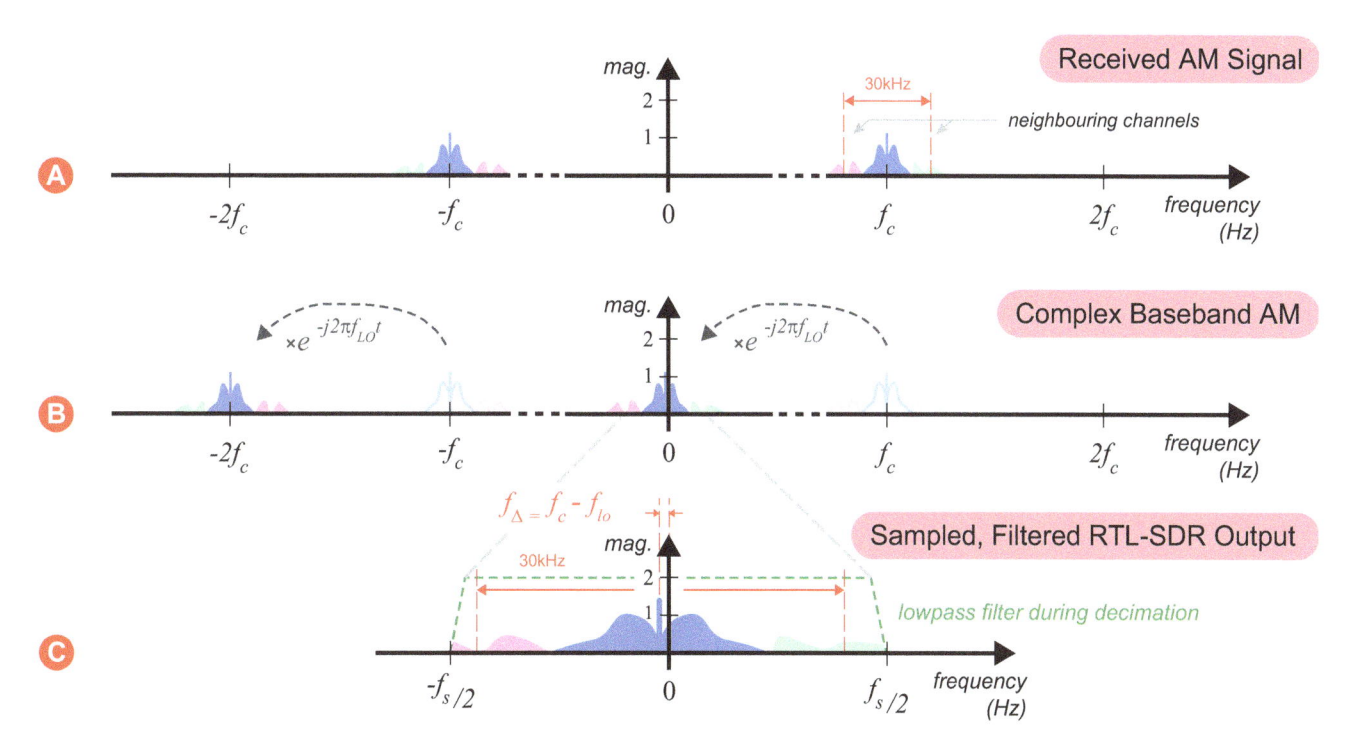

Chapter 6: Amplitude Modulation (AM) Theory and Simulation

Figure 6.18: The RTL-SDR receiving an information signal that has been AM-DSB-TC modulated, and demodulating it to complex baseband. Note, as with Figure 6.17, that when a frequency offset exists between the original modulating carrier, f_c, and the local oscillator used during the demodulation process, f_{lo}, a frequency shift of f_Δ occurs in the complex baseband output

methods. Non-coherent AM demodulators are conceptually simple and easy to implement, and we will discuss three different types in the next section.

6.8 Non-Coherent AM Demodulation: The Envelope Detector

In the analogue world, one of the simplest forms of non-coherent AM demodulator is the envelope detector. This is a physical circuit, comprising only three components: a diode, a resistor and a capacitor. The diode is used to block negative current flow in the circuit, and the remaining two components form a lowpass filter which is used to smooth the gaps between the peaks of the carrier wave. The envelope detector can only be used to demodulate AM-DSB-TC signals, as these are the only AM signals with information envelopes.

We cannot implement the standard analogue envelope detector to demodulate AM signals received by the RTL-SDR, because the complex baseband signals input to the computer are sampled, and are in the digital (discrete time) domain. This means that we must consider discrete time implementations of the demodulator, which operate on a sample-by-sample basis.

6.8.1 The Traditional Envelope Detector

The functionality of the envelope detector can easily be replicated in the digital domain. A saturating operation can be used to perform the same task as the diode, and an FIR filter can be used to implement the lowpass filter. The block diagram for the *traditional envelope detector* is shown in Figure 6.19, and Figure 6.20 shows how the AM-DSB-TC signal would be processed as it passes through the system.

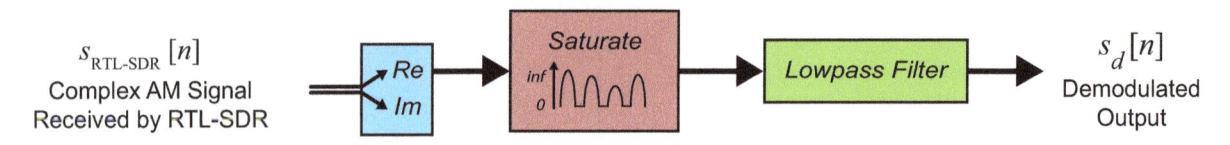

Figure 6.19: Discrete time implementation of the Traditional Envelope Detector

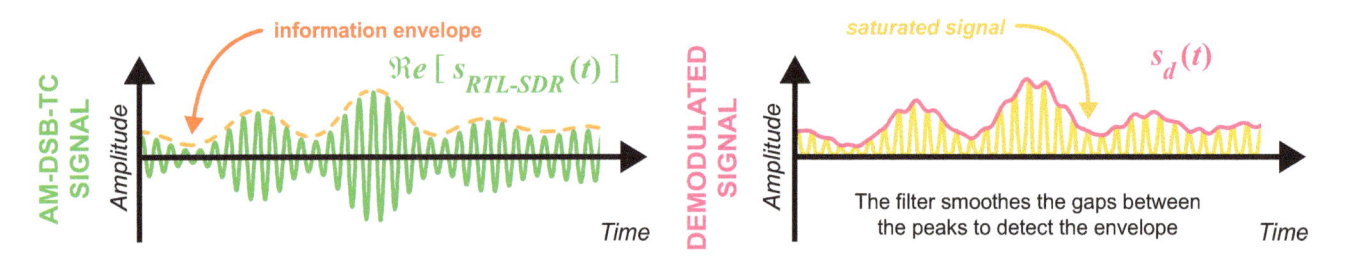

Figure 6.20: Sketch of an AM-DSB-TC signal in the time domain before (left) and after (right) Traditional Envelope Detection

As the envelope detector is a non-linear device, it is non-trivial to represent its demodulation of a signal mathematically. We can form an output equation for the demodulator however, as the following convolution operation:

$$s_d[n] = h_{lpf}[n] \otimes \left\{ \text{Sat}_0^{inf}\left(\Re e\left[s_{\text{RTL-SDR}}(t)\right]\right) \right\}. \tag{6.41}$$

In order for the lowpass filter (h_{lpf}) to successfully smooth the peaks and detect the information envelope, the frequency of the carrier in the AM-DSB-TC signal must be as high as possible, relative to the frequency of the information signal. If the carrier frequency is low (as it is above), the filter will struggle to attenuate it, resulting in a noisy output signal.

6.8.2 The Optimised Envelope Detector

We can improve the performance of the envelope detector by reducing the time gap between the peaks of the carrier. The traditional envelope detector only uses the positive part of the AM-DSB-TC signal, as the negative part is removed with the saturation operation. If we instead took the *magnitude* of the signal, the negative part would flip and become positive too, doubling the number of positive peaks. This effectively doubles the frequency of the carrier, meaning that the lowpass filter will have an easier task to attenuate it and detect the information envelope.

To avoid confusion, we shall call this the *optimised envelope detector*. The block diagram for this demodulator is shown in Figure 6.21, and Figure 6.22 shows what would happen to an AM-DSB-TC signal as it is processed. It is clear to see that replacing the saturate operation with a magnitude operation makes a huge difference to the quality of the output signal.

The output equation for the optimised envelope detector is as follows:

$$s_d[n] = h_{lpf}[n] \otimes \left| \Re e\left[s_{\text{RTL-SDR}}(t)\right] \right|. \tag{6.42}$$

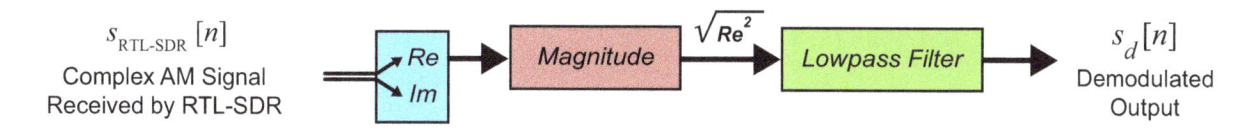

Chapter 6: Amplitude Modulation (AM) Theory and Simulation

Figure 6.21: Discrete time implementation of the Optimised Envelope Detector

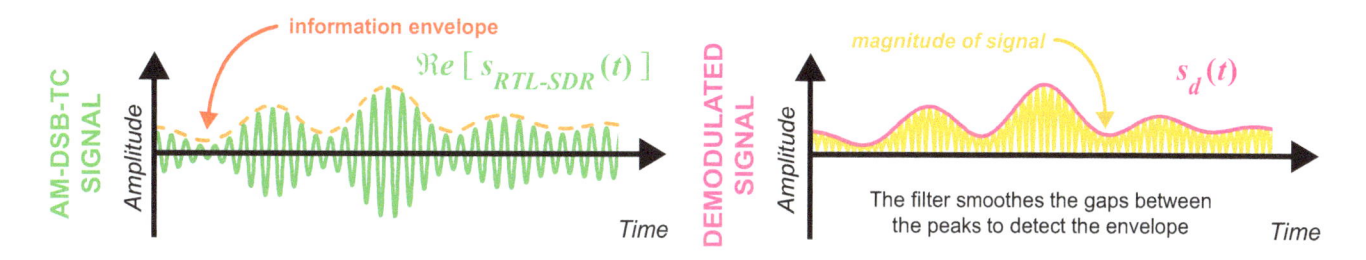

Figure 6.22: Sketch of an AM-DSB-TC signal in the time domain before (left) and after (right) Optimised Envelope Detection

It has been made clear in Figures 6.19 and 6.21 that both the traditional and the optimised envelope detectors only work with real signals. There is a final form of envelope detector that can be used with complex AM signals, like the ones received by the RTL-SDR.

6.8.3 The Complex Envelope Detector

The *complex envelope detector* is similar to the optimised envelope detector, in the sense that it is based around finding the magnitude of the complex baseband AM-DSB-TC signal. Both the In Phase and Quadrature Phase components (Eqs. (6.39) and (6.40)) contain envelopes of the information signal. Figure 6.23 shows what they would look like when plotted individually in the time domain.

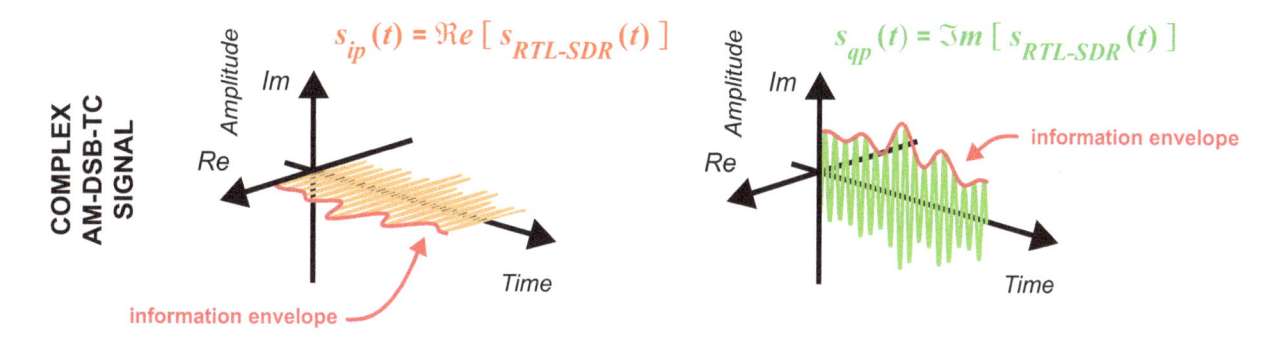

Figure 6.23: Time domain sketch of the components in a complex AM-DSB-TC signal

Plotting them against each other, however, produces what is shown in blue in Figure 6.24. Although this may be a little hard to picture, perhaps imagine it as a spiral travelling through time with a constantly fluctuating radius. As the radius changes in sympathy with the information envelope, you can actually demodulate the signal simply by finding this radius, which is equivalent to taking the magnitude of the complex signal.

$$s_{mag}(t) = \sqrt{s_{ip}(t)^2 + s_{qp}(t)^2} \tag{6.43}$$

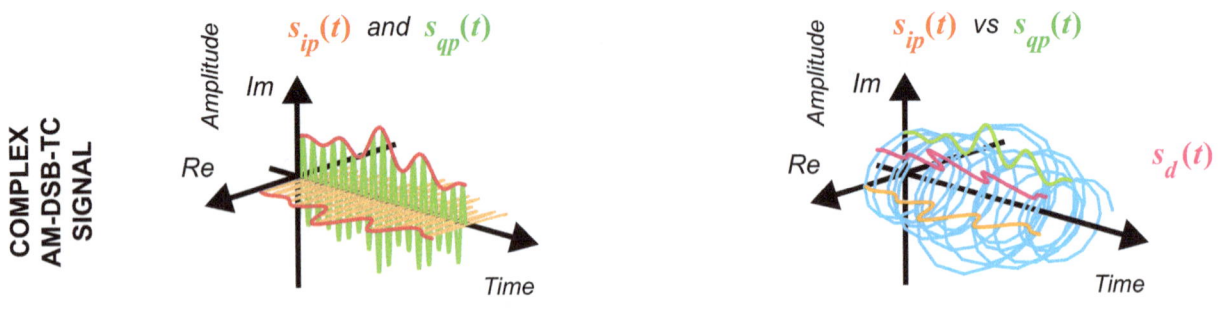

Figure 6.24: Time domain sketch of a complex AM-DSB-TC signal

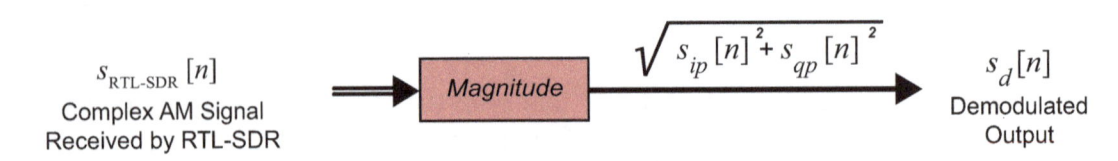

Figure 6.25: Discrete time implementation of the Complex Envelope Detector

A block diagram for the discrete time complex envelope detector is shown in Figure 6.25.

The main advantage of this demodulator is that a lowpass filter is not required — finding the magnitude of the complex signal perfectly demodulates it. The output equation of the complex envelope detector is as follows:

$$s_d[n] = \left| s_{ip}[n] + js_{qp}[n] \right|. \tag{6.44}$$

If you would like to non-coherently demodulate any AM-DSB-TC signals received by your RTL-SDR, this final envelope detector would be your best choice. We will test it out when working through the *Desktop AM Transmission and Reception* exercises in Chapter 8.

6.9 Summary

In this chapter we have reviewed the various AM modulation schemes in use around the world, presented the maths behind AM modulation and demodulation, and run some simulation-based exercises involving AM signals. Looking forward, Chapter 7 will present tuning and carrier synchronisation, leading to the development of some coherent demodulators. We will use these, along with the non-coherent demodulators presented here, to construct AM receivers in Chapter 8.

7 Frequency Tuning and Simple Synchronisation

An important aspect of all radios is their ability to select a frequency band of interest (or to 'tune'), and thereafter to capture and receive a desired signal from within that band. First, we consider tuning, which is independent of the modulation format and applies to both digital and analogue communications. Once tuning has taken place, the majority of modulation formats require some form of carrier synchronisation to be undertaken in the receiver. We discuss what synchronisation 'means' and why it is needed, before going on to look at the PLL as a basis for carrier synchronisation circuits.

7.1 Selecting a Frequency Band: Tuning

With the RTL-SDR, as with many other radio architectures, the first task undertaken by the receiver is to select a band of frequencies from the RF spectrum to demodulate to baseband for further processing. This stage is referred to as *tuning*, and is shown in Figure 7.1. The two main parameters of the tuning process can be defined as:

- The **position of the captured frequency window in the RF spectrum**, commonly defined based on its centre frequency, which is denoted as f_c in Figure 7.1. This can be defined by appropriately setting the LO frequency in the receiver.
- The **bandwidth of the frequency window**, i.e. the range of frequencies around f_c that will be demodulated (shown by the green bar in Figure 7.1). The bandwidth can be controlled by appropriate filtering and setting of sample rates.

We will now go onto consider the RTL-SDR tuning process more specifically. It is important to highlight that tuning takes place *inside the RTL-SDR hardware*.

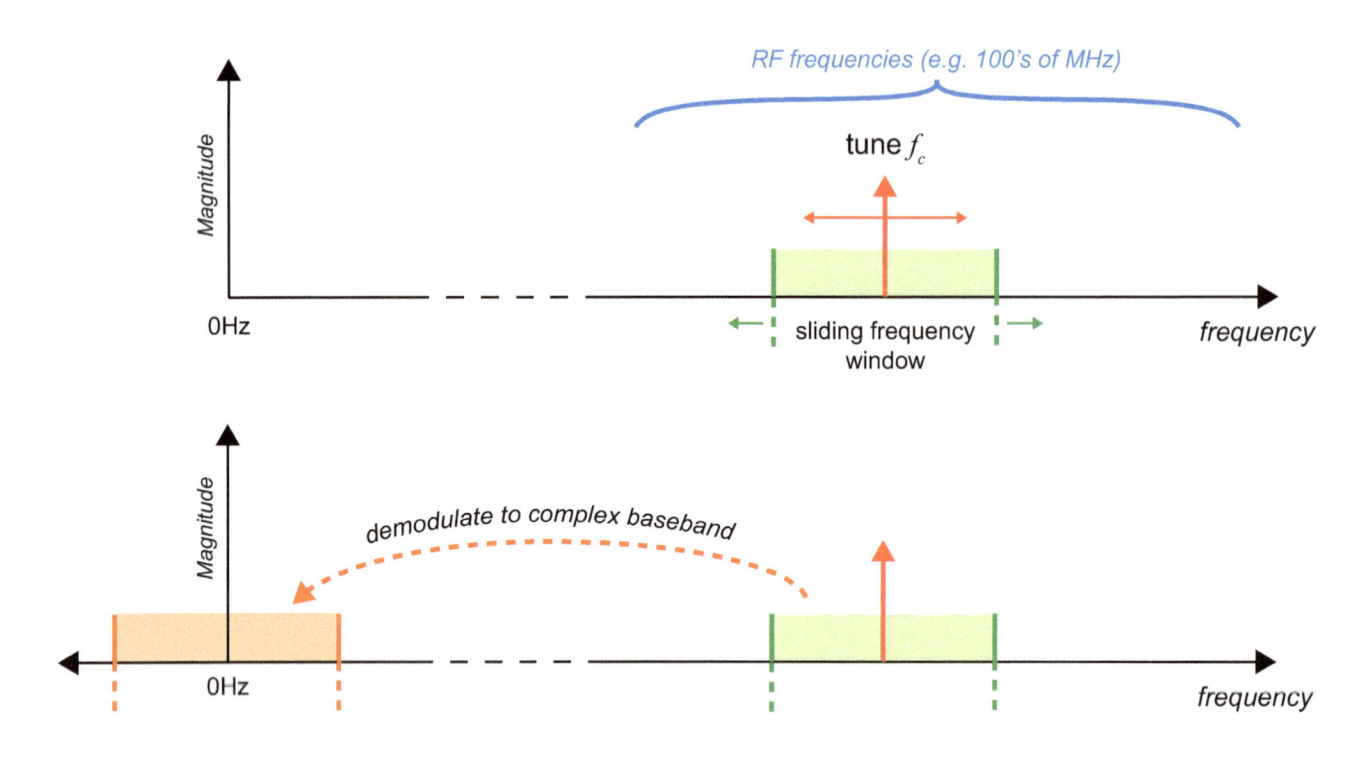

Figure 7.1: Tuning to select a band of frequencies

The user controls the RTL-SDR tuning from within MATLAB or Simulink, by specifying (i) a desired centre frequency and (ii) a sampling rate to define the signal bandwidth. These parameters are then passed into the RTL-SDR to control the behaviour of the demodulation process. The device digitises the signal and provides complex, 8-bit sampled data as the input to MATLAB or Simulink for further processing.

The most popular RTL-SDR hardware dongles are based on the *Rafael Micro R820T* tuner, although some use the *Elonics E4000* or the *Rafael Micro R820T2*, which can tune over the frequency ranges shown in Table 7.1. Sampling takes place in the RTL2832U, for which a formal data sheet in not available at the time of writing, but internet sources suggest that the sampling frequency (which defines the signal bandwidth) can be varied between 225kHz to 300kHz, and between 900kHz and 3.2MHz [92].

Table 7.1: Popular RTL-SDR tuners and frequency ranges [111]

Tuner	Approximate Frequency Range
Elonics E4000	52MHz to 2200MHz, with a gap from 1100MHz to 1250MHz (varies)
Rafael Micro R820T	25 to 1750MHz
Rafael Micro R820T2	25 to 1750MHz

The RTL-SDR interface in MATLAB and Simulink permits the user to define both the carrier frequency, which forms the centre of the band to be demodulated, *and* the bandwidth of that band (via the sampling rate parameter of the RTL-SDR). Figure 7.2 confirms the relevant settings in the Simulink *RTL-SDR Receiver* block. The user may view the spectrum and observe, for instance, that the signal of interest is centred at 102.5MHz with an approximate bandwidth of 250kHz, and configure the RTL-SDR block

Chapter 7: Frequency Tuning and Simple Synchronisation 237

accordingly. We have called this process 'eyeball tuning', as the user adjusts parameters based on visual observation of the spectrum. Equivalent RTL-SDR interfacing and control can be undertaken in MATLAB using a `comm.SDRRTLReceiver` system object.

The demodulation shown in Figure 7.1 takes place inside the RTL-SDR hardware, with the relevant parameters being supplied from Simulink. The sine wave used to demodulate the signal is generated within the RTL-SDR using a LO. In practical terms, the oscillator and demodulation circuitry introduces a degree of error (arising from component tolerances and temperature effects), and therefore readers should be aware that this method of 'eyeball tuning' may not necessarily result in a demodulated spectrum that is exactly centred at 0Hz.

To mitigate the imperfections of the hardware, steps can be taken to measure, and then compensate for, the actual frequency deviation of any individual RTL-SDR. The deviation is measured in ppm and can be supplied as a parameter to the RTL-SDR interface block in Simulink (or if using MATLAB, this can be supplied via the `comm.SDRRTLReceiver` system object). More information on how to measure and compensate for the oscillator error can be found in Appendix A.3 (page 577).

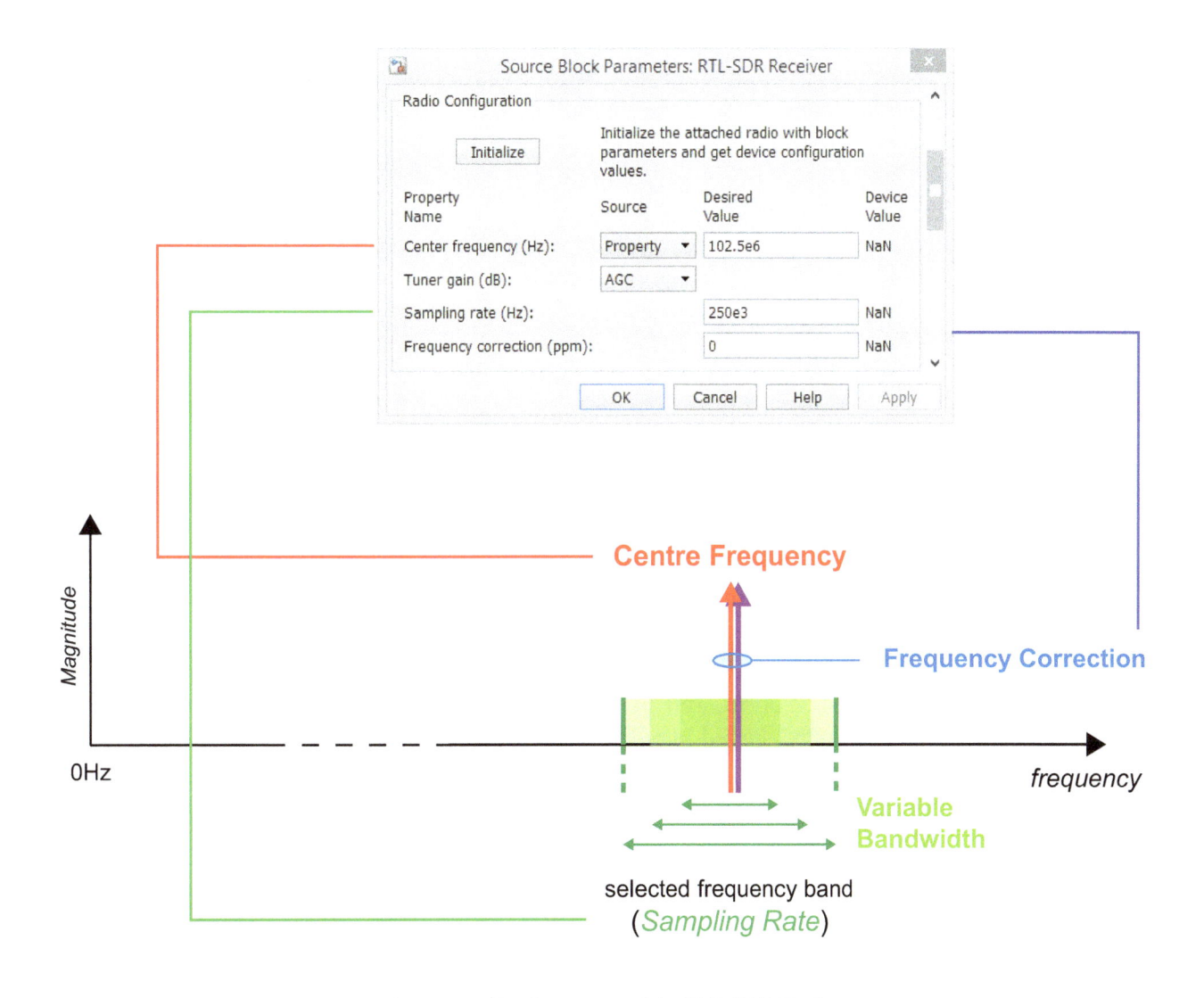

Figure 7.2: Configuration of the RTL-SDR block

7.2 The Synchronisation Problem

In communications systems, information is passed between a transmitter and receiver through a communication medium, or channel. In the context of this book, we consider a wireless radio channel, where there is no physical connection between the two terminals. They will be located some distance apart, ranging from a few centimetres to several kilometres, or even further in the case of satellite communications. In the case of the Earth to Mars Curiosity rover link, that distance is 225 million kilometres [99]! As entirely separate, physically distant devices, they do not share a common time or frequency reference. This is the basis of the synchronisation problem.

During the remainder of this section, we will review the basic principles of modulation and demodulation, and consider the impact of an imperfect radio channel and device characteristics on the synchronisation problem.

7.2.1 Modulation and Demodulation Recap

It is useful to briefly review the communications architecture being considered. Here, we amplitude modulate a signal onto an RF carrier, and the modulated signal is then transmitted across a radio channel and demodulated in the receiver. Modulation and demodulation are undertaken by multiplying the signal with a locally generated sine wave at the defined carrier frequency. This simple architecture is shown in Figure 7.3.

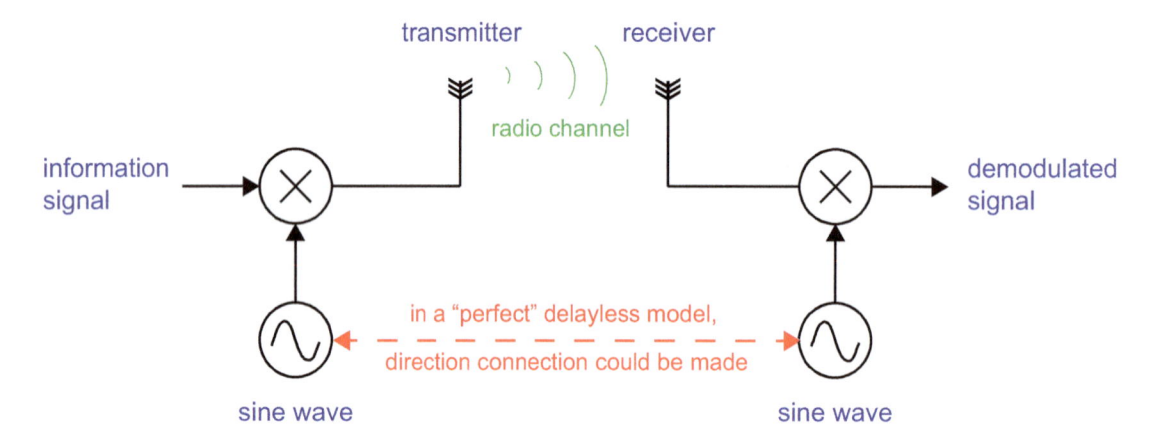

Figure 7.3: Transmitter and receiver with a 'virtual' connection

The information signal can be perfectly recovered (after filtering), provided that the channel does not introduce any delay or frequency offset, and provided that the LOs in the transmitter and receiver generate sine waves that are exactly aligned in both frequency and phase.

Of course, that's the trouble—in the real world, the channel is not trivial, and the transmit and receive local oscillators *do not* produce exactly the same sine waves. Receivers are designed to compensate, as far as possible, for these effects.

7.2.2 Channel Effects

Passing a signal through a radio channel causes a variety of unwanted effects to be introduced, including attenuation, noise, interference from other users, a fixed or time-varying propagation delay, and

Chapter 7: Frequency Tuning and Simple Synchronisation

multipath propagation. These channel effects occur in addition to device effects, which will be covered in Section 7.2.3.

Here, we highlight the two most significant channel effects in terms of synchronisation, using some familiar real-world examples to illustrate the points.

1. ***There will be a non-zero propagation delay through the channel.***
   It will take some finite amount of time for the signal leaving the transmitter to arrive at the receiver. The propagation delay varies depending on the distance covered, and if we assume that the transmitter and receiver are static, this will translate into a fixed carrier phase shift of θ degrees. An example of this scenario is depicted in Figure 7.4 (imagine that you are sitting at home using your tablet, which is connected to your wireless router).

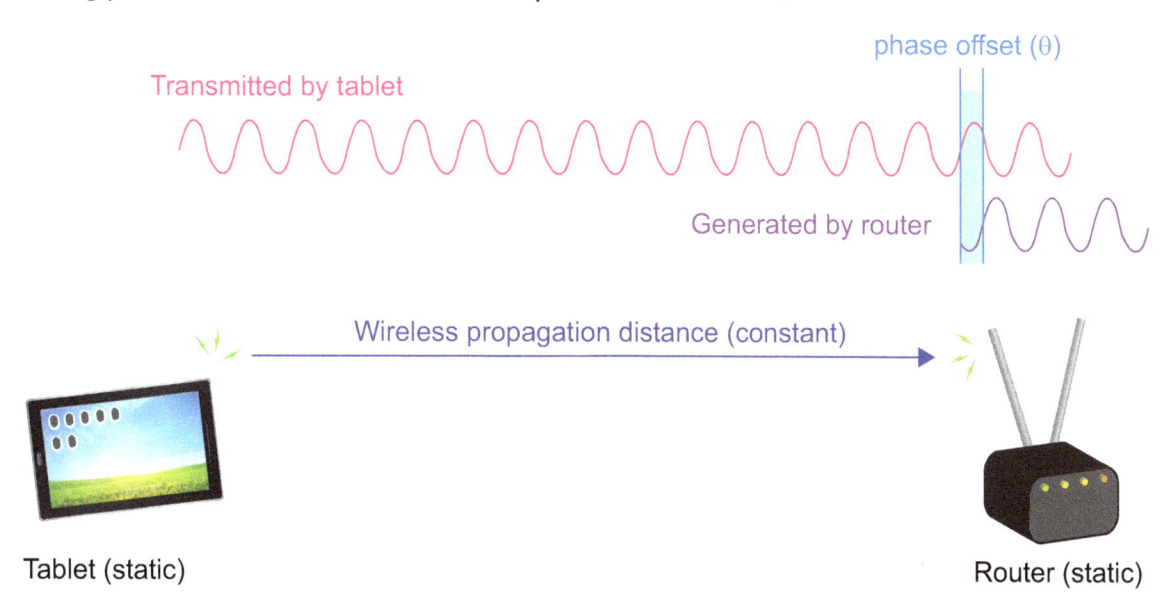

Figure 7.4: Communication between two static terminals, resulting in carrier phase shift

2. ***The propagation delay may change over time.***
   If the propagation delay depends on the distance between transmitter and receiver, it follows that the delay will change over time if the transmitter and receiver are moving relative to each other. This results in a time-varying phase shift, which is equivalent to a frequency shift (often referred to as Doppler shift). An example of this scenario would be making a phone call while travelling in a car or train—your mobile phone communicates with a fixed basestation, while your position changes. In Figure 7.5, the phone is moving towards the basestation, and this results in a higher perceived frequency at the receiver.

The frequency shift may itself change over time, if the relative motion is dynamic. A good example of this is GPS reception, wherein your GPS device receives signals from non-geostationary satellites. These satellites travel at high speed relative to the surface of the earth, hence the relative motion of the transmitter and receiver changes as they move across the sky.

transmitted at X MHz   received at $X+\delta$ MHz

wavefronts arrive more frequently at receiver... seen as higher frequency

mobile phone
(moving towards basestation)

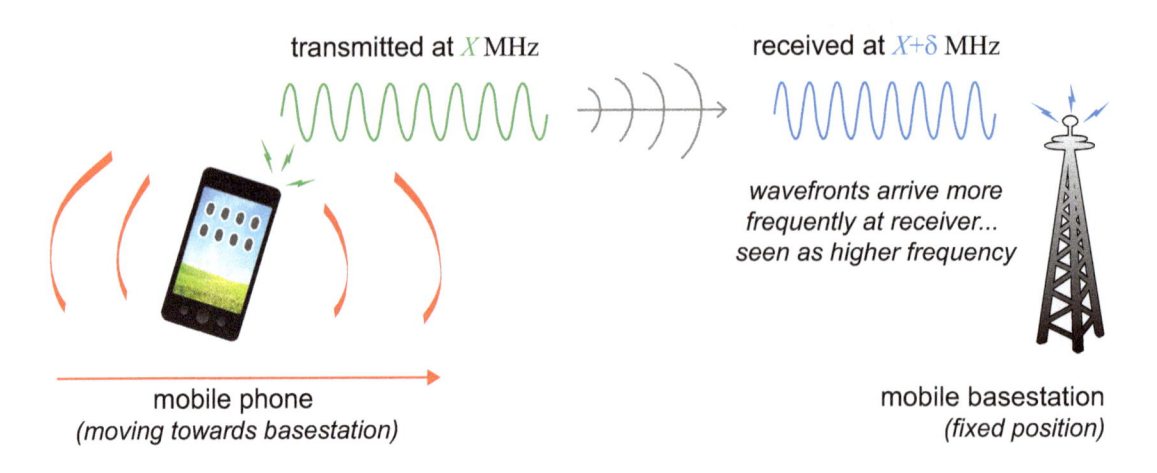

mobile basestation
(fixed position)

Figure 7.5: Communication between terminals that are moving closer together

7.2.3 Device Considerations

Transmitters and receivers rely on sine waves for modulation and demodulation. Sine waves are generated by oscillator components, which are subject to manufacturing variability and temperature-related effects, and therefore the frequencies they generate will vary within some defined limits. Higher quality oscillators will have tighter tolerances (perhaps a few ppm for a "good" oscillator).

Oscillator behaviour can change over time as a result of varying temperature, and also, in the longer term, due to ageing of the component. Taking all of these effects into account (manufacturing tolerances, temperature effects, ageing), it is clear that the oscillators in the transmitter and receiver cannot be said to produce exactly the same frequency.

Therefore, it must be assumed that the transmitter and receiver carriers are not synchronised—even if they were to start at the same phase, they would soon drift out of alignment. Figure 7.6 illustrates the effect using an exaggerated example to permit easy visualisation. In this example, the tolerance is ±5,000ppm, and taking frequencies at the limits, the blue sine wave has a frequency of 100.5MHz, and the red, 99.5MHz. It is clear that the two sine waves are aligned in phase at the beginning of the simulation, but soon begin to deviate.

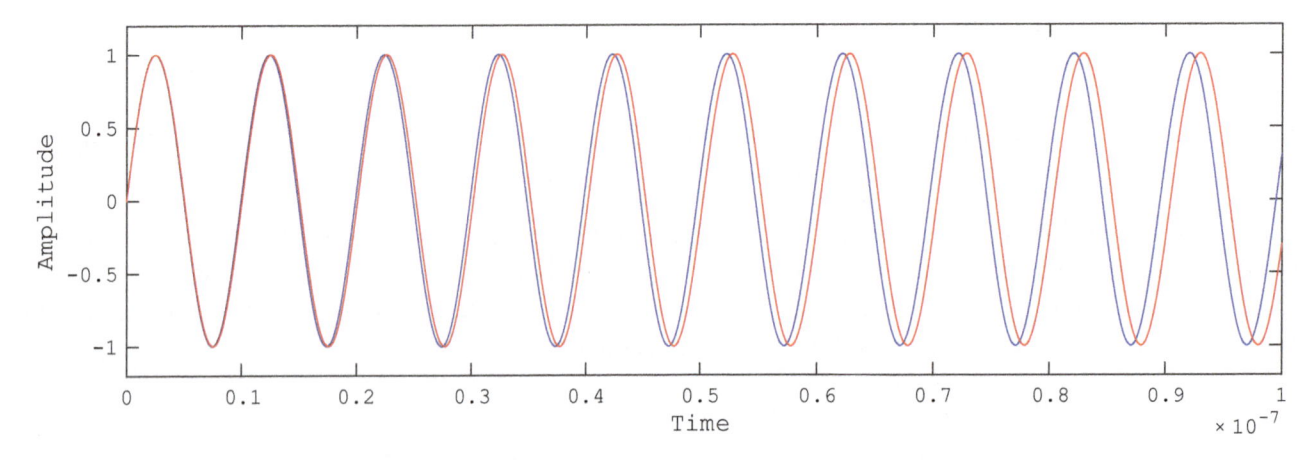

Figure 7.6: Sine waves with slightly different frequencies

Chapter 7: Frequency Tuning and Simple Synchronisation

7.2.4   Timing Synchronisation

So far, discussion has been restricted to the oscillators that modulate and demodulate the information signal. In digital communications, the clock oscillator from which bit and symbol operations are derived is an additional factor. Similarly to the carrier oscillator, the timing reference in the transmitter and receiver differs, and channel effects such as Doppler can introduce a further offset.

Symbol timing synchronisation will be the subject of more focused discussion later, in the *Digital Communications Theory and Simulation* Chapter 11 (page 427).

7.3   Demodulation of AM Signals

As discussed earlier, perfect demodulation of a modulated signal requires a locally generated sinusoid that exactly matches the frequency and phase of the received carrier. Figure 7.3 showed a conceptual diagram with a connection between the two carrier oscillators, but practical scenarios do not have this artificial link, and call for a different approach.

Demodulating with a LO that *does not* correspond with the carrier results in a baseband signal that is not perfectly centred at 0Hz. In practice, the two carrier frequencies are unlikely to be exactly matched due to device considerations and channel effects.

There are two possible approaches to this problem:

1. Transmit the carrier signal so that it can be used to demodulate the signal at the receiver;
2. Extract the carrier from the received signal using a synchronisation algorithm, and regenerate it in the receiver.

These options have advantages and disadvantages:

1. Transmitting the carrier requires extra power to be expended. This is wasteful because the carrier signal does not contain any information. Demodulation of the signal is less complex, however, meaning that it is possible to design a simpler and cheaper receiver.
2. If the carrier is not transmitted, this has the advantage that power is not wasted in transmitting a component that carries no information. It is possible to synchronise to the carrier embedded in the received signal, however a more sophisticated and costly receiver is needed.

Both of these approaches imply **coherent demodulation,** which involves synthesising a sine wave with the same frequency and phase as the carrier of the received signal, and using it to demodulate the received signal. Coherent demodulation can be used whether the carrier is explicitly transmitted or not. The performance of coherent demodulation typically exceeds that of non-coherent demodulation (covered in Section 6.8, (page 231), and therefore is usually the preferred method. This chapter focuses on coherent demodulation and associated synchronisation problems.

In digital communications, it is possible to demodulate using a carrier that does not match the frequency and phase of the received carrier, and to compensate for the offset in a later processing stage. This method will be explained in Section 11.4.3 (page 455).

7.4 Coherent Demodulation and Carrier Synchrony

In this section we consider coherent demodulation, and briefly confirm the effect of being synchronised (or not) on the successful demodulation of a transmitted signal. To keep things simple, in the first instance we assume that the 'information' signal of interest is a sine wave.

The architecture we will consider is as shown in Figure 7.7. This implements AM-DSB-SC modulation, where the carrier is NOT explicitly transmitted, and demodulation is undertaken using a carrier with a frequency offset of Δf.

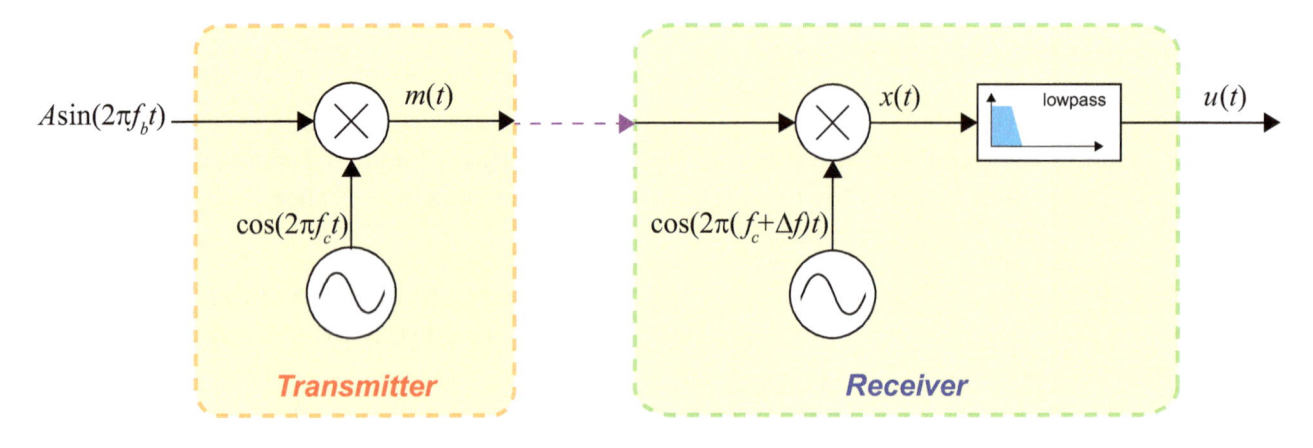

Figure 7.7: Demodulation with a frequency offset

A sine wave with amplitude A at frequency f_b, modulated onto a carrier at frequency f_c, is expressed as

$$m(t) = \frac{A}{2}\left[\sin(2\pi(f_c+f_b)t) - \sin(2\pi(f_c-f_b)t)\right] \tag{7.1}$$

In Chapter 5 (page 171), the mathematics for 'perfect' demodulation was presented, wherein the sine wave used to demodulate the signal had frequency and phase exactly matching those of the receiver carrier. In practice this is almost certainly not the case, because the carrier generators in the transmitter and receiver are physically disconnected and do not have a common frequency reference. Their carrier oscillators are subject to component tolerances, and channel propagation effects contribute further to phase and frequency offsets. Therefore, we must consider the case where the sine waves used to modulate and demodulate differ.

In order to model the unequal frequencies, it is assumed that demodulation is by multiplication with a cosine wave at frequency $f_c + f_\Delta$, where f_Δ is a frequency offset, resulting in

$$x(t) = \cos(2\pi(f_c+f_\Delta)t) \times \frac{A}{2}\left[\sin(2\pi(f_c+f_b)t) - \sin\left(2\pi(f_c-f_b)t\right)\right] \tag{7.2}$$

$$= \frac{A}{2}\left[\cos(2\pi(f_c+f_\Delta)t)\sin\left(2\pi(f_c+f_b)t\right) - \cos\left(2\pi(f_c+f_\Delta)t\right)\sin\left(2\pi(f_c-f_b)t\right)\right]. \tag{7.3}$$

Chapter 7: Frequency Tuning and Simple Synchronisation 243

Applying the product-to-sum trigonometric identity, Eq. (B.4) from Appendix B (page 581), it can be shown that

$$x(t) = \frac{A}{2}\left[\frac{1}{2}\sin\left(2\pi\left((f_c+f_\Delta)+(f_c+f_b)\right)t\right) - \frac{1}{2}\sin\left(2\pi\left((f_c+f_\Delta)-(f_c+f_b)\right)t\right)\right]$$
$$-\frac{A}{2}\left[\frac{1}{2}\sin\left(2\pi\left((f_c+f_\Delta)+(f_c-f_b)\right)t\right) - \frac{1}{2}\sin\left(2\pi\left((f_c+f_\Delta)-(f_c-f_b)\right)t\right)\right]$$
(7.4)

and Eq. (7.4) may subsequently be rearranged to give:

$$x(t) = \frac{A}{4}\sin\left(2\pi(2f_c+f_\Delta+f_b)t\right) - \frac{A}{4}\sin\left(2\pi(f_\Delta-f_b)t\right)$$
$$-\frac{A}{4}\sin\left(2\pi(2f_c+f_\Delta-f_b)t\right) + \frac{A}{4}\sin\left(2\pi(f_\Delta+f_b)t\right)$$
(7.5)

Although this expression includes two high frequency components, it is not equivalent to the transmitted signal even after lowpass filtering to remove the high frequency components. This is due to the presence of the f_Δ terms.

$$u(t) = \frac{A}{4}\sin\left(2\pi(2f_c+f_\Delta+f_b)t\right) - \frac{A}{4}\sin\left(2\pi(f_\Delta-f_b)t\right)$$
$$-\frac{A}{4}\sin\left(2\pi(2f_c+f_\Delta-f_b)t\right) + \frac{A}{4}\sin\left(2\pi(f_\Delta+f_b)t\right)$$
(7.6)

lowpass filtered term

Lastly, making use of the general property that $\sin(-x) = -\sin(x)$, we can write

$$u(t) = \frac{A}{4}\left[\sin(2\pi(f_b+f_\Delta)t) + \sin(2\pi(f_b-f_\Delta)t)\right].$$
(7.7)

Likewise, it can also be confirmed that via simulation that, in the presence of the frequency error, f_Δ, the result of demodulation is not equivalent to the original signal. This will be seen in the coming exercises.

Exercise 7.1 'Perfect' Modulation and Demodulation

This first model shows modulation and demodulation of a 1kHz sine wave onto a 10kHz carrier signal, where the local oscillators in the transmitter and receiver generate identical cosine waves. This is easy to do in a simulation model, but almost impossible in the real world!

(a) **Open MATLAB.** Set the working directory to an appropriate folder so you can open this model:

/synch/am_coherent/perfect_mod_demod.slx

(b) **Confirm the properties of the modulating and demodulating Cosine Waves.** Inspect the model, and confirm that the *Cosine Wave* blocks used to the generate the local oscillators in the transmitter and receiver have exactly the same frequency and phase. To do this, 2 on each of the blocks to open their parameter windows. If you wish, you could even delete the second *Cosine Wave*, and connect the first to both multiplier inputs, to make doubly sure that they are exactly the same!

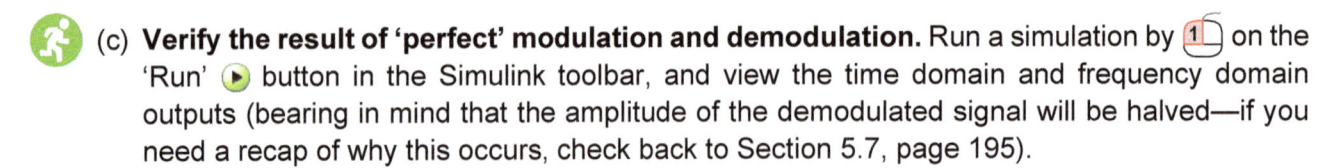

(c) **Verify the result of 'perfect' modulation and demodulation.** Run a simulation by clicking on the 'Run' button in the Simulink toolbar, and view the time domain and frequency domain outputs (bearing in mind that the amplitude of the demodulated signal will be halved—if you need a recap of why this occurs, check back to Section 5.7, page 195).

Confirm that the received signal is the same as the transmitted one, i.e. a sine wave at 1kHz. (You may observe an apparent phase difference, which is due to the group delay of the filter.)

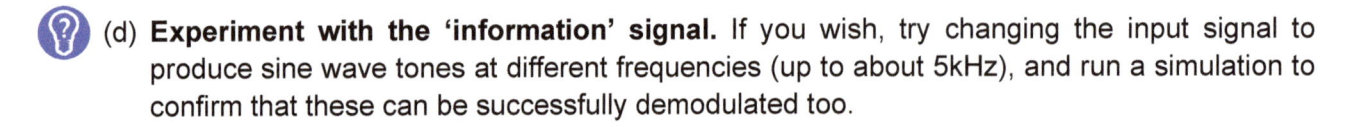

(d) **Experiment with the 'information' signal.** If you wish, try changing the input signal to produce sine wave tones at different frequencies (up to about 5kHz), and run a simulation to confirm that these can be successfully demodulated too.

Exercise 7.2 Modulation and Demodulation (out of synchrony)

Following on from the previous exercise, let's see what happens when the oscillator in the receiver does not have the same frequency as the oscillator in the transmitter.

(a) **Open the model:**

/synch/am_coherent/frequency_mismatch_mod_demod.slx

(b) **Confirm the properties of the modulating and demodulating *Sine Wave* blocks.** Check the frequencies of the transmit and receive oscillators, and confirm that they differ (in fact there is quite a big mismatch, to permit easy visualisation).

(c) **Simulate!** Run a simulation by clicking on the 'Run' button in the Simulink toolbar and view the results. What do you observe in the time and frequency domain plots? Is the transmitted signal recovered correctly?

(d) **Experiment with the 'information' signal.** Try changing the frequency of the input tone, and repeating the simulation—you should still observe a difference in the transmitted and received signals.

(e) **Experiment with the transmit and receive oscillator frequencies.** Next, try altering the frequencies of the transmit and receive oscillators, and resimulate. You should be able to confirm that the signal is not perfectly demodulated when the frequencies of the two oscillators differ.

The examples in this section have demonstrated that generating a sine wave in the receiver with the correct frequency and phase is very important. If the sine waves used to modulate and demodulate an AM signal do not match, then the input signal cannot be properly recovered in the receiver.

Chapter 7: Frequency Tuning and Simple Synchronisation

Later, in Section 7.10, we will consider synchronisation methods for AM signals in more detail. First, we turn to the problem of generating a sine wave with the same frequency and phase as a reference signal. The circuit required to do this—the PLL—is a fundamental component of carrier synchronisation systems.

7.5 Introduction to Phase Locked Loops

PLLs are essential building blocks for synchronisation, because they are able to recreate an input sinusoid, track deviations in its frequency, and reject noise. These simple circuits, which comprise of only three components, can be constructed in either the analogue or digital domains, or even a mixture of the two. Despite being apparently simple, PLLs have been extensively studied, and a number of papers, tutorials, and books have been written on the topic. Perhaps the most widely known book is *Phaselock Techniques* by Floyd M. Gardner [13], but there are several others that may be useful if you are interested in reading further on this topic, e.g. [9], [26], [43]. Our treatment of PLLs most closely follows that of the excellent *Digital Communications: A Discrete Time Approach*, by Michael Rice [38].

In this section we start by defining the structure of a PLL, and then go on to consider PLL behaviours, parameters, and detailed implementation. Discrete time PLLs are of primary interest in our SDR application, because sampled Simulink models will be used to implement receivers.

A simple block diagram of a PLL is shown in Figure 7.8. It contains the following three components:

- A **Phase Detector** — the Phase Detector generates a signal that varies in proportion to the difference in phase between the incoming signal, and the locally generated sine wave.
- A **Controllable Oscillator** — the oscillator is a VCO in analogue PLLs or an NCO in digital PLLs. The VCO or NCO has the task of generating a sinusoidal output (usually both sine and cosine, or else one is derived from the other) whose phase and frequency is controlled by a time-varying input signal.
- A **Loop Filter** — the Loop Filter acts upon the output of the Phase Detector to remove unwanted high frequency terms, and produce the signal that drives the VCO or NCO. The design of the Loop Filter has a fundamental effect on the behaviour of the PLL, as will be explained shortly.

We will go on to describe each of these components in more detail over the coming pages.

$x(t) = \cos\left(2\pi f_i t + \theta_i(t)\right)$ — Input

$\theta_e(t) = K_p\left(\theta_i(t) - \theta_o(t)\right)$

$s(t) = -\sin\left(2\pi f_o t + \theta_o(t)\right)$ — synthesised sine wave

$c(t) = \cos\left(2\pi f_o t + \theta_o(t)\right)$ — Output

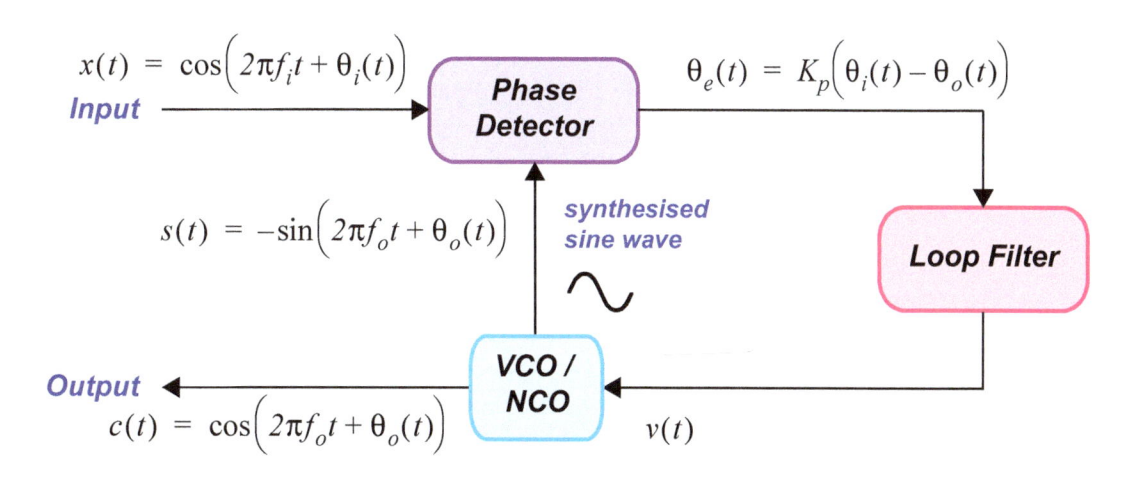

$v(t)$

Figure 7.8: Block diagram of a PLL

7.5.1 Phase Detector

The role of the phase detector is to generate a signal that varies in proportion to the difference in phase between the reference input signal, and the locally generated sine wave (i.e. the output of the VCO or NCO). This is known as the phase error, and for an ideal phase detector, the error is given by

$$\theta_e(t) = K_p\big(\theta_i(t) - \theta_o(t)\big) \tag{7.8}$$

where $\theta_i(t)$ and $\theta_o(t)$ are the phases of the input reference and local oscillator signals at time t, respectively, and K_p denotes the gain of the phase detector.

Phase detectors can generate the phase error in different ways, and often this component is simply implemented as a multiplier, as shown in Figure 7.9. This type of phase detector is easy to implement, and it produces a low frequency term in proportion to the phase difference, and higher frequency terms that are removed by filtering.

The behaviours of the ideal and multiplier-based phase detectors are described by their characteristic 's-curves' (which are explained in Appendix D, page 593). We generally assume that the value of K_p has a fixed value depending on the type of phase detector used.

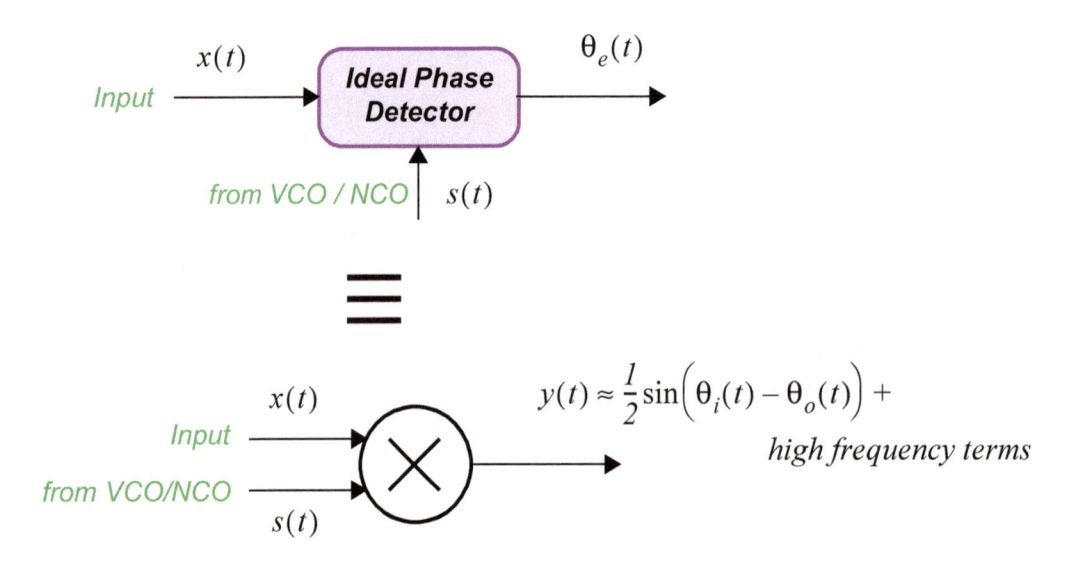

Figure 7.9: The PLL phase detector implemented as a multiplier

It can be established through trigonometry that the 'multiplier' phase detector produces two frequency components at the output:

- A low frequency term that varies with the phase difference; and
- A high frequency term at approximately double the input frequency.

This can be shown by working through the mathematics for the multiplier phase detector, where its inputs are defined as

$$x(t) = \cos\big(2\pi f_i t + \theta_i(t)\big) \tag{7.9}$$

Chapter 7: Frequency Tuning and Simple Synchronisation

and

$$s(t) = -\sin\left(2\pi f_o t + \theta_o(t)\right), \qquad (7.10)$$

as shown in Figure 7.8.

The output of the multiplier phase detector is:

$$y(t) = x(t) \times s(t) \qquad (7.11)$$

$$= \cos\left(2\pi f_i t + \theta_i(t)\right) \times -\sin\left(2\pi f_o t + \theta_o(t)\right)$$

$$= \frac{1}{2}\left[\sin\left((2\pi f_i t + \theta_i(t)) - (2\pi f_o t + \theta_o(t))\right) - \sin\left((2\pi f_i t + \theta_i(t)) + (2\pi f_o t + \theta_o(t))\right)\right]$$

$$= \frac{1}{2}\left[\sin\left(2\pi(f_i - f_o)t + (\theta_i(t) - \theta_o(t))\right) - \sin\left(2\pi(f_i + f_o)t + (\theta_i(t) + \theta_o(t))\right)\right].$$

Where the frequencies f_i and f_o are equal, the output reduces to:

$$y(t) = \underbrace{\frac{1}{2}\sin\left(\theta_i(t) - \theta_o(t)\right)}_{\text{low frequency term}} - \underbrace{\frac{1}{2}\sin\left(4\pi f_i(t) + \theta_i(t) + \theta_o(t)\right)}_{\text{high frequency term}}. \qquad (7.12)$$

Notice that, of the two frequency components, only the lower frequency component depends directly on the phase difference. The loop filter attenuates the higher (double frequency) term, along with any other unwanted noise components, which leaves only the phase-related term.

If the phase difference can be assumed to be small (i.e. as the PLL approaches phase lock), the approximation $\sin(x) \approx x$ can be applied, and thus the multiplier-based phase detector provides the output

$$y(t) \approx \frac{1}{2}\left(\theta_i(t) - \theta_o(t)\right), \qquad (7.13)$$

or equivalently,

$$y(t) \approx \theta_e(t) \qquad (7.14)$$

where $K_p = \frac{1}{2}$ (more generally, $K_p = \frac{A}{2}$, where A is the amplitude of the input reference signal, $x(t)$).

Further details regarding phase detectors, both multiplier-based and ideal, are provided in Appendix D (page 593).

7.5.2 The Loop Filter (and Type 1, Type 2 and Type 3 PLLs)

The Loop Filter has the task of filtering the error signal produced by the phase detector. In most cases, the loop filter is a simple lowpass filter, composed of a proportional path, and one or more integral paths. The design of the loop filter is vital in defining the overall characteristics and behaviour of the PLL.

The PLL *Type* corresponds to the number of integrators in the loop, with PLLs normally being Type 1 (one integrator), Type 2 (two integrators) or Type 3 (three integrators). The total number of integrators is the sum of integrators from the Loop Filter, *plus* the VCO or NCO (which contributes one integrator, because it integrates the feedback signal supplied to the VCO / NCO to control its current phase).

Table 7.2 summarises the composition of the above three PLL variations, in terms of integrators. We will discuss loop filters further in Section 7.6, when a discrete-time model of the PLL will be presented, along with signal flow graph examples.

Table 7.2: Type 1, 2, and 3 PLLs

PLL Type	VCO/NCO	Loop Filter	Total
Type 1	1	0	1
Type 2	1	1	2
Type 3	1	2	3

7.5.3 The Controllable Oscillator (VCO / NCO)

A PLL requires a VCO or NCO, and both have the same basic behaviour. Although VCOs are analogue components, a sampled version of the VCO can be used for modelling in Simulink, whereas NCOs are inherently digital and operate in discrete time only. In many of the models that follow in this section, either type of oscillator is applicable.

The VCO or NCO is an oscillator with a standard or *quiescent* frequency, f_o, and a control input that can adjust the output frequency upwards or downwards from the quiescent value. We will now consider VCOs and then NCOs in a little more detail.

Voltage Controlled Oscillator (VCO)

In the PLL model, the filtered phase difference acts as the control signal, $v(t)$, that is input to the VCO: the larger the phase difference, the greater the adjustment applied. The outputs $c(t)$ and $s(t)$ generated by the VCO are cosine and sine waves at the adjusted frequency.

The interface of the VCO is shown in Figure 7.10. Note that a generic VCO may generate both sine and cosine outputs, or have a single output only.

Taking the cosine output (and similarly for sine), the signal generated by the VCO can be expressed as

$$c(t) = \cos\left(2\pi f_o t + \hat{\theta}(t)\right) \tag{7.15}$$

Chapter 7: Frequency Tuning and Simple Synchronisation

where $\hat{\theta}(t)$ is the estimated phase at time t, generated by integrating the control input to the VCO over all time from *time = 0*, i.e.

$$\hat{\theta}(t) = k_o \int_0^t v(t) \, dt, \tag{7.16}$$

and k_o is the *gain* (or *sensitivity*) associated with the VCO, normally expressed in radians/V.

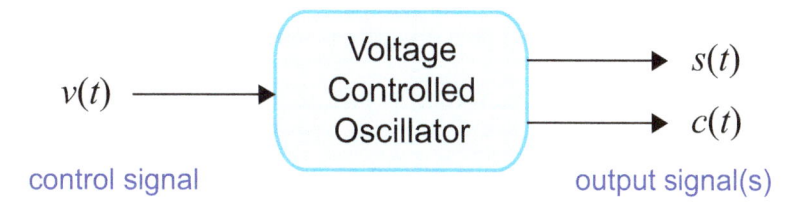

Figure 7.10: Interface of a VCO

Numerically Controlled Oscillator (NCO)

The NCO is a discrete time version of the VCO and has an equivalent interface, as shown in Figure 7.11, where m is the sample index.

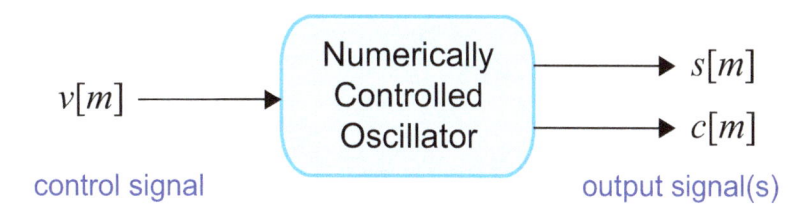

Figure 7.11: Interface of an NCO

Similar to the VCO, the NCO produces an output cosine wave at a specified quiescent frequency, adjusted by an input control signal. The output at sample m is given by

$$c[m] = \cos\left(2\pi f_o mT + \hat{\theta}[m]\right), \tag{7.17}$$

where T is the sample period, and $\hat{\theta}[m]$ is the estimated phase at sample m, generated by summing the control input to the NCO, $v[m]$, over all samples, i.e.

$$\hat{\theta}[m] = K_o \sum_{m=0}^{M} v[m]. \tag{7.18}$$

The parameter K_o is equivalent to k_o from the continuous time model for the VCO, and is likewise given in radians/V.

In order to understand the behaviour of the NCO, it is useful to consider how it operates on a sample-by-sample basis. The internal behaviour of an NCO is often described in terms of step size, μ. If we assume that the step size is fixed, then the phase of the NCO increments by the given step size on each sample, i.e.

$$\hat{\theta}[m] = \hat{\theta}[m-1] + \mu[m] \tag{7.19}$$

and hence the NCO progresses through a full cycle (2π radians) over some fixed number of sample periods. After the phase is generated, a *phase-to-amplitude* conversion is then undertaken to generate the amplitude of the sine and/or cosine wave at that specific phase.

The step size controls the speed of phase increments, and therefore it also controls the frequency of the generated cosine wave, with a larger step size producing a higher frequency. For instance, if the step size was equivalent to 30°, or $\pi/6$ radians, then it would require 12 samples to progress through one full cycle of a cosine wave. If the step size was 15°, or $\pi/12$ radians, then it would take 24 samples (twice as long) to complete a full cycle. Thus it is seen that the frequency of the generated cosine wave is numerically controllable. These two examples (for $\mu_1 = \pi/6$ and $\mu_2 = \pi/12$) are shown in Figure 7.12. Notice that the cosine waves complete a full cycle after 12 and 24 cycles respectively, in each case corresponding to a phase of 2π radians.

When operating in radians, the step size required to generate a sine or cosine output at a particular desired frequency, f_d, is given by:

$$\mu = \frac{2\pi f_d}{f_s}. \tag{7.20}$$

For instance, if $f_s = 100\,\text{Hz}$ and $f_d = 10\,\text{Hz}$, then $\mu = \frac{2\pi}{10}$.

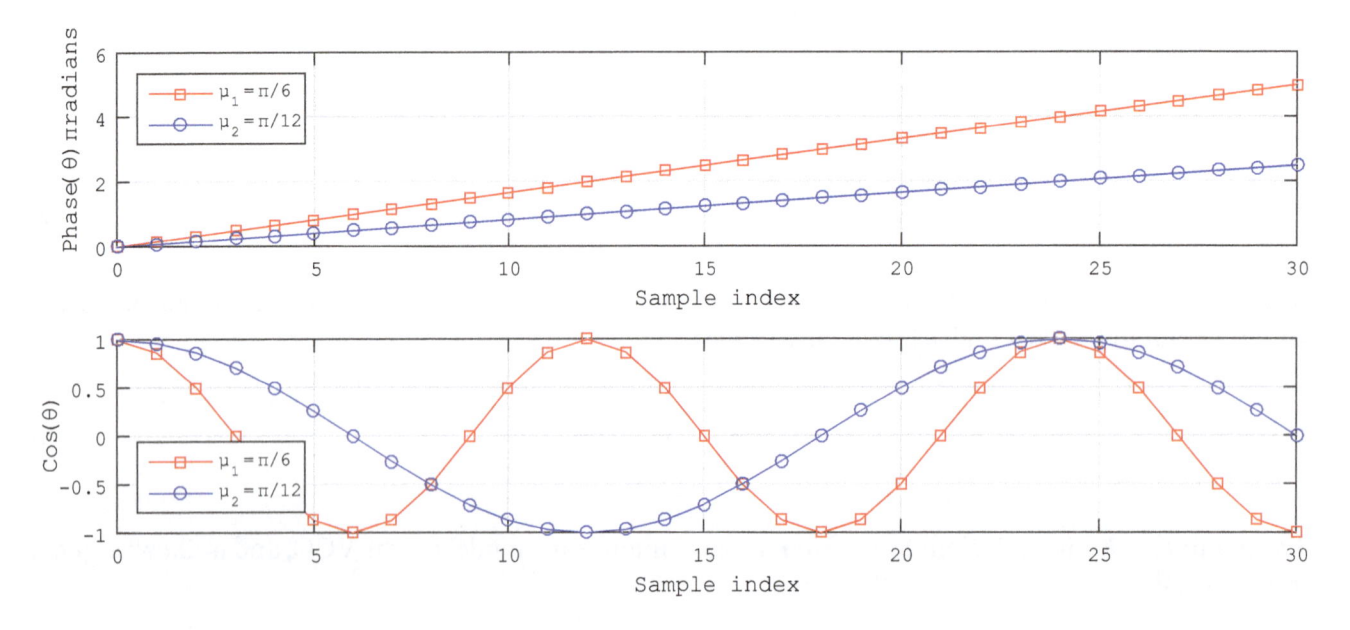

Figure 7.12: Example behaviour of an NCO for two different, fixed step sizes

Chapter 7: Frequency Tuning and Simple Synchronisation

The discussion so far has assumed that the step size, μ, is a constant value, producing a fixed frequency cosinusoid. The requirement in a PLL is to have a *dynamically controllable* cosine wave generator, and indeed this can be achieved using an NCO. In this scenario, the step size at sample m is formed by summing two components of the step size: μ_q, a constant value corresponding to the quiescent frequency; and $\mu_a[m]$, representing the adjustment term which may vary over time.

$$\mu[m] = \mu_q + \mu_a[m] \tag{7.21}$$

The adjustment term is supplied by the PLL feedback signal, i.e. $v[m]$, scaled by the oscillator gain, K_o, which has units of radians/V. Thus, the step size

$$\mu_a[m] = K_o v[m] \tag{7.22}$$

represents a phase angle. A larger value of oscillator gain therefore makes the NCO more sensitive to changes in the feedback signal.

The processing undertaken by the NCO can be summarised by Figure 7.13. Notice that the direction of data flow is from right to left in this diagram, reflecting the position of the NCO within the feedback loop. It should be noted that, if the gain constant K_o is not stated in the analysis, then implicitly $K_o = 1$ (and indeed this is a common choice).

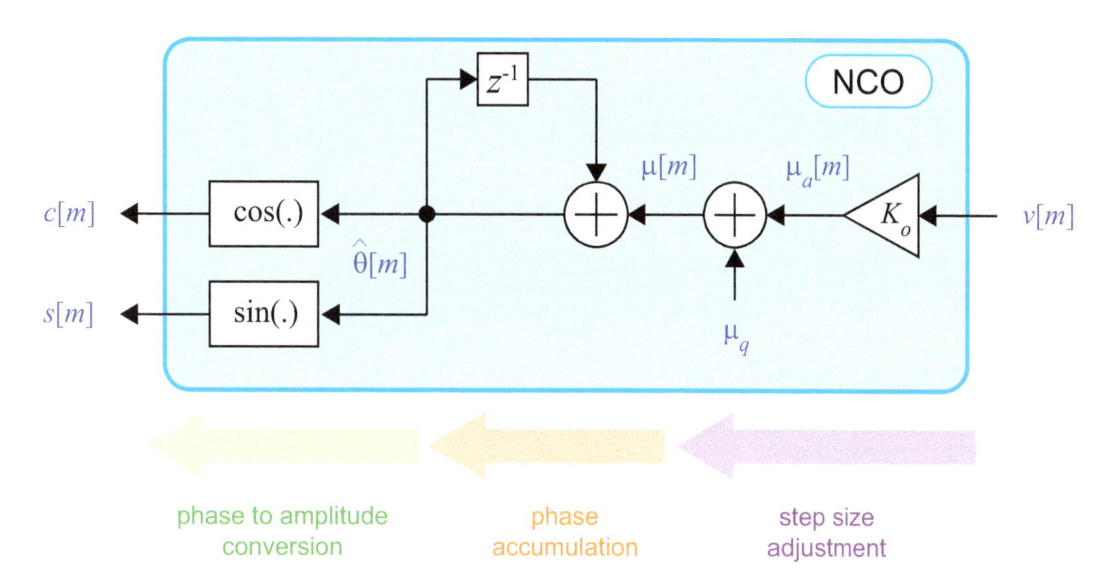

Figure 7.13: Overview of NCO processing in the PLL feedback loop

The error produced by the PLL phase detector diminishes to zero once the synthesised cosine wave has converged to the input reference signal (i.e. once the two cosine waves are in phase). As a result, the output of the Loop Filter adjusts such that it produces a constant output (or, in practice, it may fluctuate slightly around the desired value) and this provides the adjustment signal to the NCO. After lock has been attained, the NCO step size is not further adjusted by the feedback loop, aside from these small fluctuations.

It is worth noting that, while the phase shown in Figure 7.12 is permitted to increment without bounds, it is equally valid to express the phase '*modulo* 2π' (meaning that the phase wraps at 2π). Having a complete cycle of a sine wave correspond to 2π is convenient from a conceptual perspective. Those working with lookup-table-based hardware implementations may prefer a range of $0 \rightarrow 2^n - 1$, where n is some positive integer, to correspond with the number of memory locations used to store the sine wave samples. In this case, a further scaling factor must be incorporated into the gain block alongside K_o (shown in Figure 7.13), to reflect that a full cycle involves counting up to 2^{n-1} rather than 2π.

As we have seen, the PLL is simple in its construction, but even so, it can be tuned via a set of parameters to achieve a variety of behaviours. We go on to discuss these aspects of PLL design next.

7.6 Discrete Time PLL Model

We will now start to focus on digital PLLs, which can be expressed in terms of Z-domain notation. First, the loop filter component will be examined.

Figure 7.14 shows a general model for the construction of Type 1, Type 2, and Type 3 digital PLL loop filters. Note that the *Type* can be set by setting the appropriate multiplier coefficients.

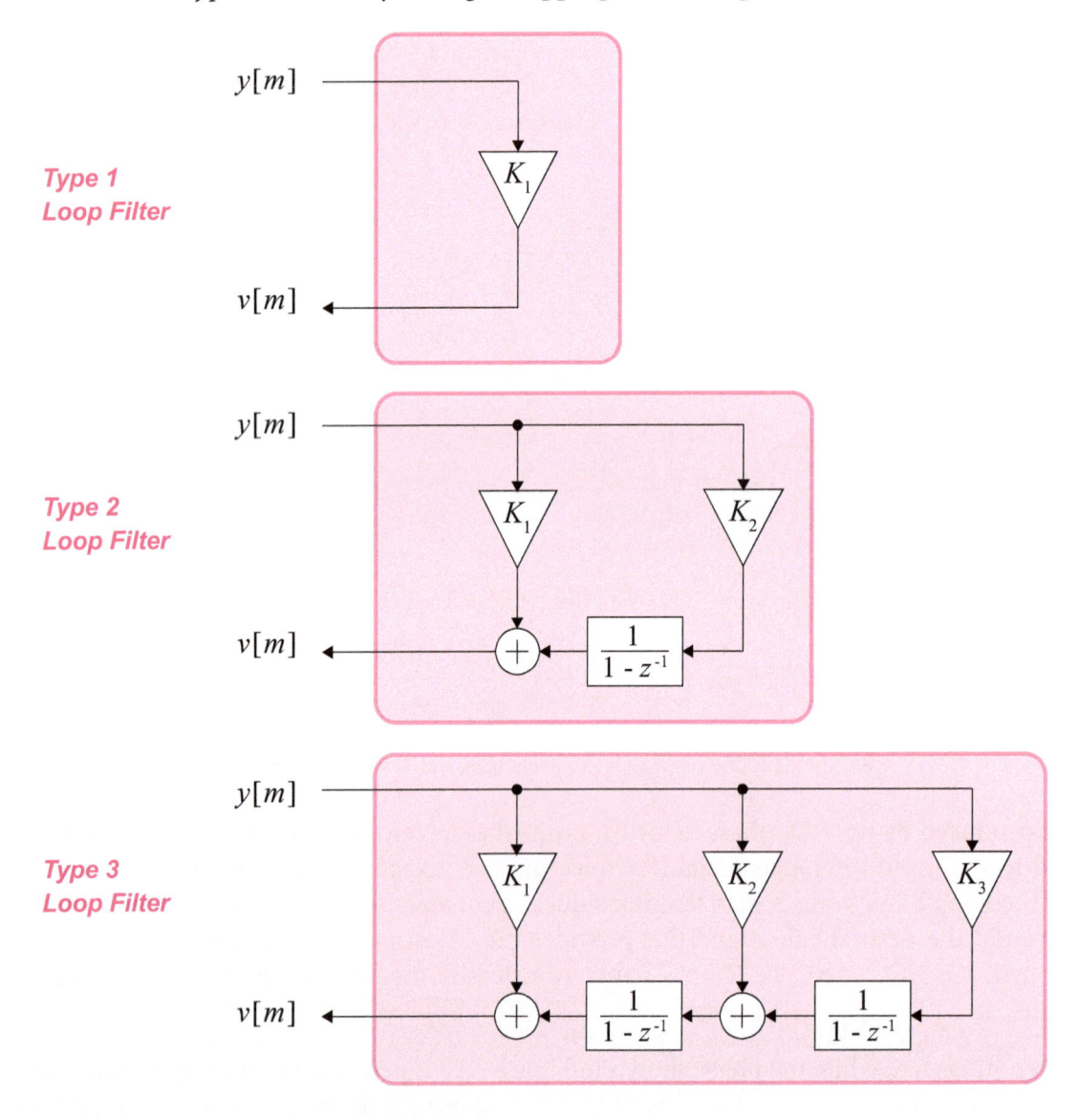

Figure 7.14: PLL loop filters for Types 1, 2 and 3

Chapter 7: Frequency Tuning and Simple Synchronisation

To summarise:

- **Type 1** — K_1 has a significant value; both K_2 and K_3 are set to zero (or branches are omitted);
- **Type 2** — K_1 and K_2 have significant values; K_3 is set to zero (or branch is omitted);
- **Type 3** — all coefficients have significant values.

Combining the loop filter with the other parts of the PLL results in the signal flow graph of Figure 7.15. Notice that the gains associated with the phase detector and NCO, K_p and K_o have been incorporated along with the coefficients within the loop filter.

Before going on to consider PLL design in detail, we will start with some exercises that illustrate the general construction and behaviour of PLLs, focusing on the Type 2 version.

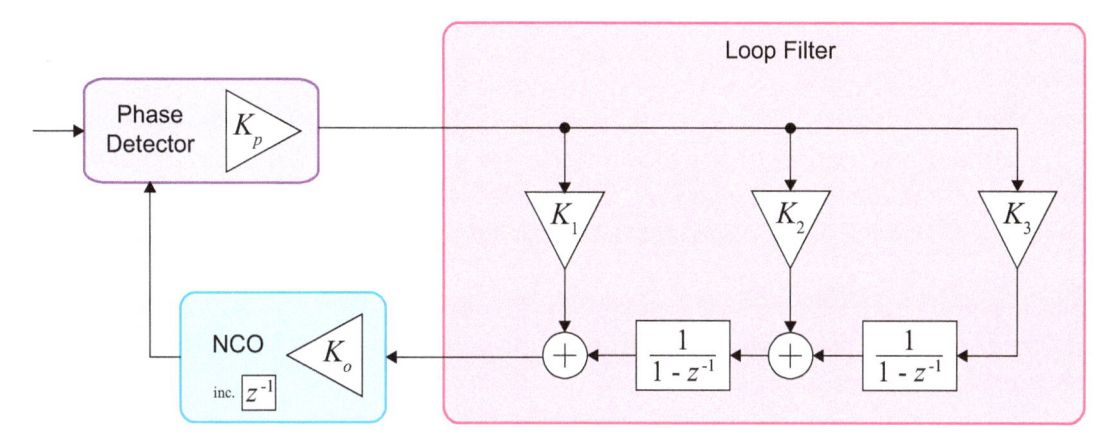

Figure 7.15: Model of a generic digital PLL

Exercise 7.3 Phase Detector

First of all we will look at the phase detector component, implemented as a multiplier. The phase detector multiplies two input sine waves, A and B, in order to find the phase difference between them.

(a) **Open MATLAB.** Set the working directory to an appropriate folder so you can open this model:

/synch/PLL_intro/phase_detector.slx

(b) **Inspect the system.** Note the frequencies and phases of the inputs, in MHz and radians.

(c) **Simulate!** Run the simulation by clicking on the 'Run' button in the Simulink toolbar and observe the signal in the *Time Scope*. What form does the output take? Can you account for the DC offset?

(d) Try changing the phase of *Sine Wave B*, then resimulate and note any changes in the detector output.

(e) **Spectral Analysis.** Inspect the spectrum of the phase detector output, using the *Spectrum Analyzer*. You should see two significant frequency components (zoom in around 0Hz if it initially appears that there is only one major component).

(f) **Confirm the maths...** Using trigonometry, can you relate the frequencies of these components to the frequencies of the inputs? Referring back to Section 7.5.1 on page 246 may help.

(g) Try repeating the process with different input frequencies.

Exercise 7.4 Loop Filters

The loop filter processes the phase detector output signal. This exercise introduces the properties of Type 1 and Type 2 loop filters, and we apply an impulse to find the frequency responses of two example designs.

(a) **Open the model:**

   /synch/PLL_intro/loop_filters.slx

(b) **Inspect the construction of the loop filters.** The parameters are set in the block mask (i.e. the dialogue that opens when you [2] on the block), and the underlying components can be viewed by [1] on the small arrow at the bottom left hand corner of the mask icon.

(c) **Run a simulation.** Simulate the model by [1] on the 'Run' ▶ button and examine the *Spectrum Analyzer* plots, which measure the powers of the loop filter outputs. Recall that an impulse excites all frequencies.

(d) **Consider the simulation results...** Zoom in to the frequency domain plots using the button, and you should see that the Type 2 filter has a larger gain at DC. What is the power of the outputs from the Type 1 and Type 2 loop filters at 0Hz?

(e) Recall the 10MHz component from the phase detector (i.e. the 'double frequency' term, which is not useful for correcting the phase). What are outputs powers at this frequency?

(f) You should be able to conclude that both Type 1 and Type 2 loop filters will greatly attenuate the 'double frequency' term. The Type 2 loop filter will apply higher gain to the 'phase' term close to DC than the Type 1.

Exercise 7.5 Phase Detector and Loop Filter

Now we will put the phase detector and Type 2 loop filter together, as we start to build up a Type 2 PLL.

(a) **Open the model:**

   /synch/PLL_intro/phase_detector_loop_filter.slx

(b) **Inspect the properties of the *Sine Wave* sources.** What are their frequencies and initial phases? With this in mind, at what frequencies do you expect to see the two spectral components output by the phase detector?

Chapter 7: Frequency Tuning and Simple Synchronisation

(c) **Simulate!** Run the simulation and compare the frequency domain results, by opening the two *Spectrum Analyzers* and placing them side by side. You should see something like the following. Look in particular at the circled components.

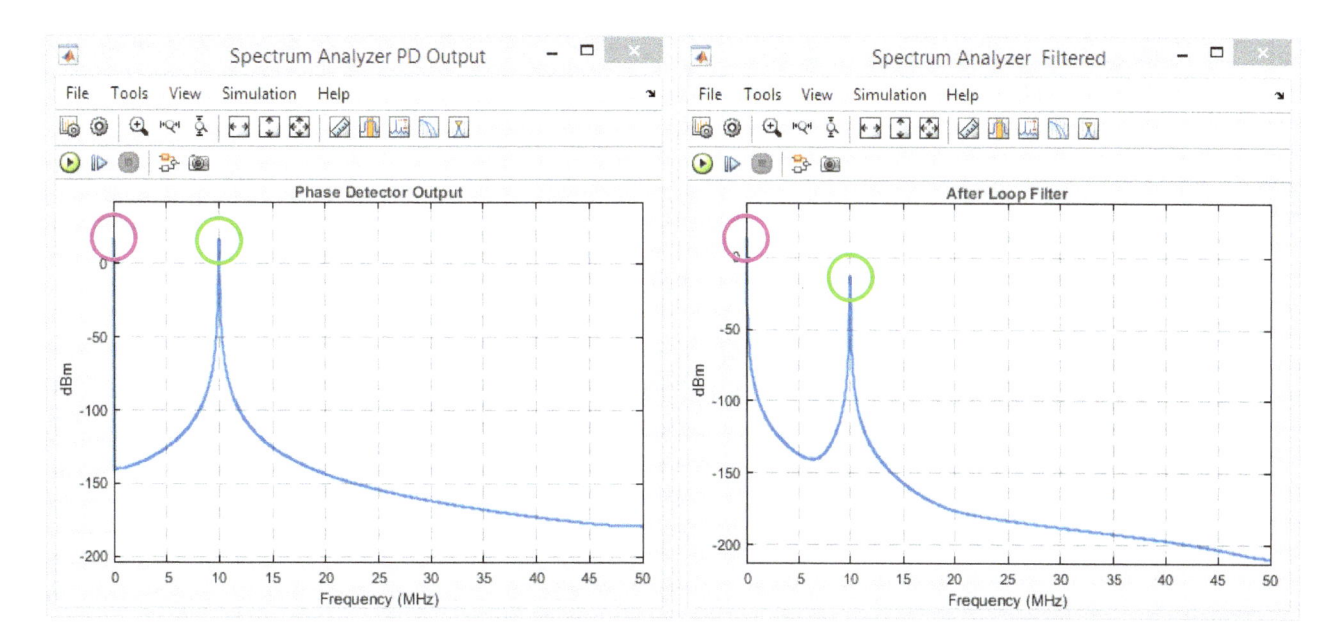

[Spectrum Analyzer PD Output — Phase Detector Output; Spectrum Analyzer Filtered — After Loop Filter. Both plots show dBm vs Frequency (MHz) from 0 to 50, with circled low-frequency and ~10 MHz peaks.]

(d) **Inspect the plots.** Notice that the loop filter has significantly attenuated the high frequency component. Zoom in to the two spectra, and quantify the amount of attenuation introduced by the loop filter.

(e) Now zoom in to view the low frequency component and find out how much the loop filter has attenuated it. Why might this be?

(f) Also observe the *Time Scope* plots for the two signals. Bearing in mind the high and low frequency components output by the phase detector, it should also be possible to identify them in the waveforms. The high frequency component appears as a sine wave (of reduced amplitude after the filter), and there is also a DC offset resulting from the low frequency term.

Exercise 7.6 Type 2 PLL

In this example, a VCO has been added, making a complete Type 2 PLL. When a Type 2 loop filter is used, the PLL should be able to track a step or ramp change in phase (deviations in both phase and frequency), with zero phase error once adapted. Later, we will compare its performance with that of a Type 1 PLL.

(a) **Open the model:**

/synch/PLL_intro/PLL_type2.slx

(b) **Investigate the model.** Note the parameters of the VCO block (in particular the quiescent frequency).

(c) **Simulate and inspect the various plot windows that appear.** Inspect the spectrum of the adapted output using the FFT Plot. Zoom in and find out whether the synthesised sine wave has attained the same frequency as the input sine wave.

(d) Finally, zoom into the initial section of the time domain plot of the sine waves (i.e. during the period of adaption, when the error signal is transient). Do the input and synthesised sine waves match? Now repeat the process at the end of the simulation run time.

(e) **Experiment with the sine wave frequencies.** Change the frequency of the input sine wave (i.e. the *Sine Wave* source block) to 3.72MHz and rerun the simulation. Does the PLL take more or less time to converge? Why do you think this is?

(f) **Experiment with a frequency sweep.** Now switch to the chirp input (i.e. a frequency ramp) by 2 on the *Manual Switch*, and rerun the simulation. View the time and frequency domain outputs. Has the PLL adapted to the frequency of the input, and if so, has it done so with zero phase error (zoom in very closely)?

(g) Feel free to experiment with other parameter combinations.

7.7 PLL Behaviours, Parameters and Characteristics

According to Gardner's definition, a PLL attains '*phaselock*' when the sine wave generated by the PLL converges to the same frequency as the input reference signal [13]. In other words, when there is one cycle of locally generated sine wave for every period of the input reference signal. This does not necessarily mean that the two signals converge to the same phase (although they may do), but rather that they have attained the same frequency. Other authors use the term 'phase lock' and 'locked' when the synthesised sine wave achieves both the same frequency *and* the same phase as the input reference signal. We will generally use the Gardner approach in this book.

The design of the PLL is based on the choice of the gains in the system: K_1, K_2, K_3 (in the loop filter), K_p (the phase detector gain), and K_o (the NCO gain). Together, these define the behaviour of the PLL as it achieves lock, and the time it takes to do so. If the parameters are not suitably chosen, then it may not converge or attain lock at all.

A few key characteristics of PLLs can be defined, and will be discussed over the next few pages. The designer may seek to optimise one or more of these factors when designing a particular PLL.

- Time to achieve lock
- Steady state error
- Transient behaviour and tracking capability
- Bandwidth

These choices depend on the application, and the expected operating circumstances. For instance, it would be preferable to minimise the bandwidth of the PLL if it is to work in a noisy environment, so as to minimise the amount of noise that enters the loop. On the other hand, the bandwidth needs to be large

Chapter 7: Frequency Tuning and Simple Synchronisation 257

enough to accommodate any likely deviation between the initial frequency of the PLL sine wave generator, and the input reference signal.

7.7.1 Time to Achieve Lock

The time a PLL takes to become locked with a reference signal is an important factor, particularly in communications systems. The longer the time to achieve lock, the greater the required synchronisation overhead, and the less efficient the system or protocol is. For instance, if it takes 10 symbol periods to achieve lock, this places a minimum length on the *preamble* (a signal that can be used to 'wake up' and help receivers synchronise to incoming information) that precedes information transmissions.

The time to achieve lock is affected by the design of the PLL, and also by the difference in frequency between the input reference signal, and the quiescent frequency of the NCO. If the PLL has to lock to a frequency far away from the 'expected' value, it will take longer.

Figure 7.16 (a) and (b) show two examples of Type 2 PLL behaviour as it adapts. In both cases, the design of the PLL is the same (we are not specifically interested in the details of the PLL here). Comparing the two plots, there is a clear difference in the time to achieve lock, and this is due to the deviation of the input reference signal from the quiescent frequency of the NCO. In the first example, the frequency deviation is larger and the PLL takes significantly longer to lock.

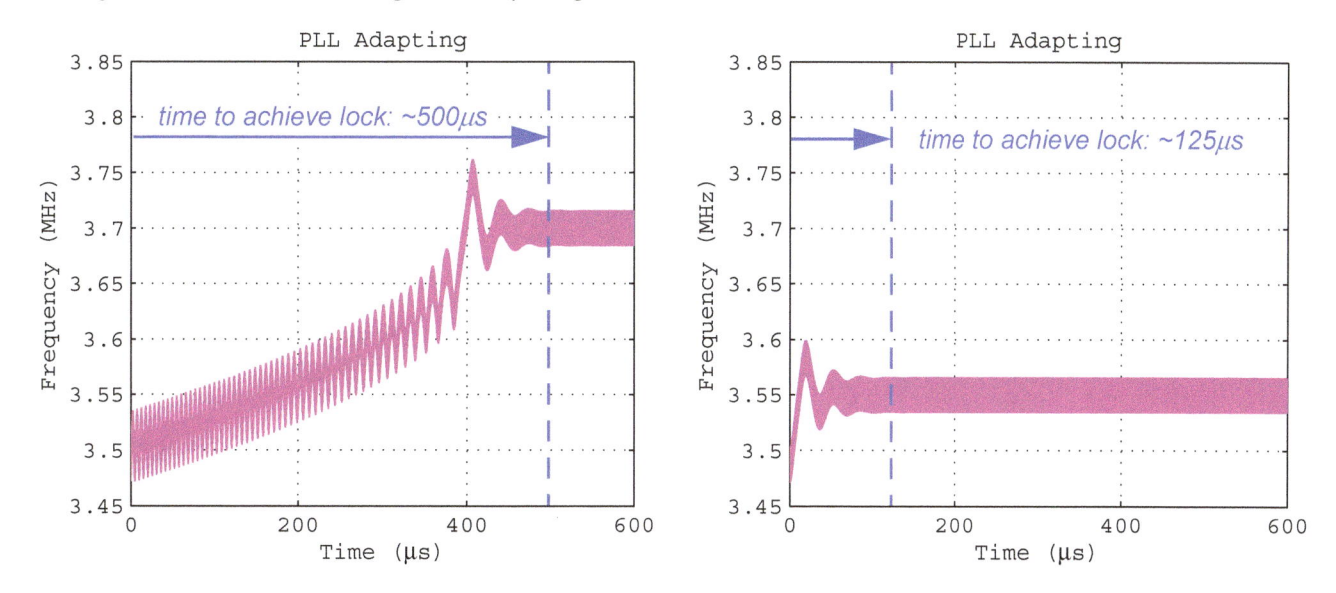

Figure 7.16: PLL adaption behaviour for a Type 2 PLL with quiescent frequency 3.5MHz:
(a) left: input frequency = 3.7MHz;
(b) right: input frequency = 3.55MHz

There are also other factors that influence the time to achieve lock, in particular the damping ratio and bandwidth (to be discussed in Section 7.7.3).

7.7.2 Steady State Error

Even once phaselock has been achieved, the PLL may have a steady state phase error. The steady state error is influenced by the choice of loop filter (which defines the PLL Type), and the nature of the synchronisation task. For instance, if a Type 1 PLL attempts to synchronise to an input reference signal with a different frequency than its local NCO, it will attain the correct frequency, but will not be able to synchronise to the exact phase of the input—in this case there will be a residual phase error. By comparison, applying a Type 2 PLL to the same scenario would result in zero phase error.

Table 7.3 summarises the steady state errors of Type 1, 2, and 3 PLLs, when different kinds of inputs are applied. It is clear that the *Type* of the PLL (which is defined by the loop filter design) dictates the nature of its steady state behaviour.

In most circumstances, it is desirable to achieve a zero phase error in the steady state, and this usually motivates the choice of the Type 2 PLL, particularly where the dominant effect is a frequency offset (i.e. where any dynamically changing frequency offsets are not significant). Type 3 loops are usually used in more complex scenarios, for instance GPS receivers, where large Doppler shifts are common [43].

The steady state error in each of the scenarios shown in Table 7.3 can be confirmed via the linear model of the PLL shown in Figure 7.17. The same model can also be used to evaluate transient behaviour.

Table 7.3: Summary of PLL Steady State Errors

PLL Type	Initial Phase or Frequency Difference compared to Reference		
	Phase Offset	Frequency Offset	Dynamic Frequency Offset
Type 1	zero phase error	residual error	does not track
Type 2	zero phase error	zero phase error	residual error
Type 3	zero phase error	zero phase error	zero phase error

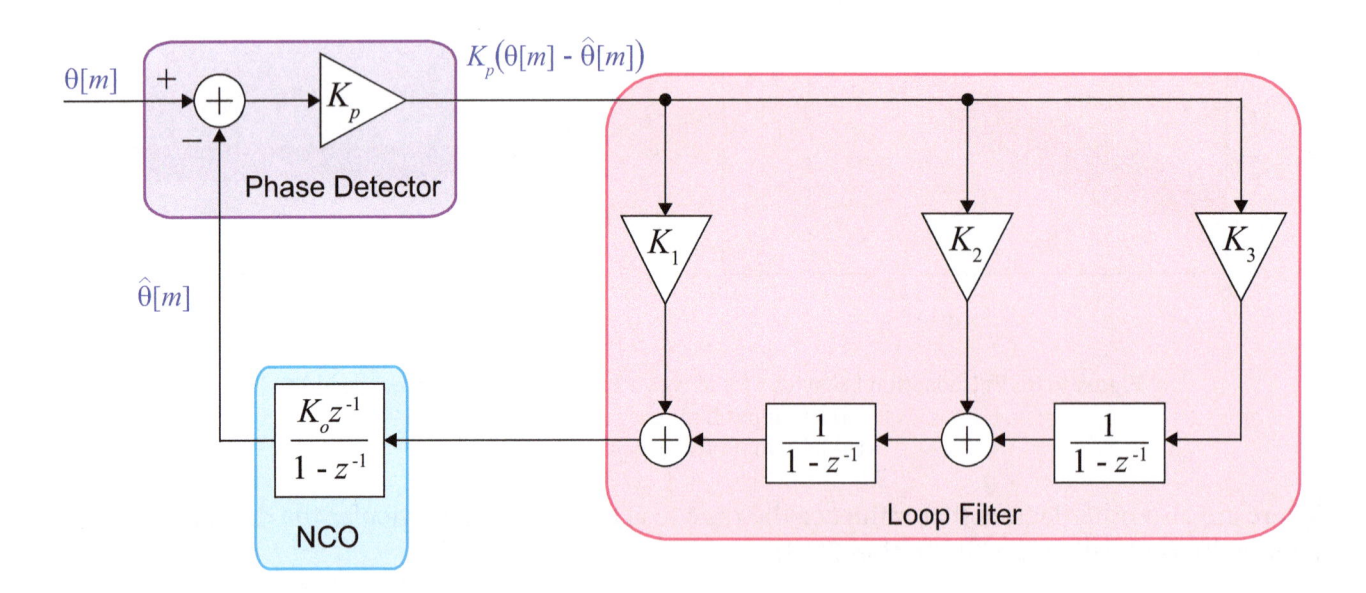

Figure 7.17: Generalised linear PLL model

The input to this Z-domain model of the PLL is the current phase at sample index m, i.e. $\theta[m]$.

Chapter 7: Frequency Tuning and Simple Synchronisation

The output generated by the NCO is the current phase estimate, denoted by $\hat{\theta}[m]$, and the output of the phase detector is the difference between the two. The phase error, given by

$$\theta_e[m] = K_p\Big(\theta[m] - \hat{\theta}[m]\Big) \qquad (7.23)$$

is subsequently processed by the loop filter and input to the NCO, which can be modelled as an integrator. There will be one or more samples of delay within the loop (shown here between the loop filter and NCO).

The response of the PLL can be tested by applying different inputs to the linear model, bearing in mind that the input signal represents *phase* (rather than a sine wave as in conventional operation).

- A *step* — representing a step change in phase offset
- A *ramp* — representing a constant rate of change of phase offset (i.e. a frequency offset)
- A *hyperbola* — representing an increasing rate of change of phase offset (i.e. an acceleration)

These three input types are shown in the upper part of Figure 7.18. The lower part of Figure 7.18 shows the response of the linear model to these three inputs, in each case where the Type 2 PLL has a damping ratio of 1. We will consider the impact of changing the damping ratio shortly.

Notice that Type 1, Type 2, and Type 3 PLLs can all adapt to a *step* input with zero steady state error. If the input is a *ramp*, the Type 2 and 3 PLLs can adapt with zero phase error, while the Type 1 adapts but has a residual phase error. For the *hyperbola* input, only the Type 3 PLL can adapt fully: the Type 2 does so but with a residual error, while the Type 1 cannot successfully adapt at all (the error keeps changing). This confirms the summary given in Table 7.3.

Relating this information back to the operation of 'real' PLLs, the following statements can be made about the three different Types of PLLs:

Type 1 PLLs (red waveforms in Figure 7.18)

- Can adapt in the presence of an initial *phase* difference with zero residual phase error (i.e. if the input reference signal had exactly the same *frequency* as the NCO quiescent frequency, but a different initial *phase*, it could adapt leaving zero phase error).
- Can adapt to the correct frequency in the presence of an initial *frequency* difference, but leaving a residual phase error.
- Cannot adapt in the presence of a *dynamic frequency* offset.

Type 2 PLLs (blue waveforms in Figure 7.18)

- Can adapt to both the correct frequency and the correct phase in the presence of an initial *frequency* difference, as well as an initial *phase* difference.
- Can adapt to the correct frequency in the presence of a *dynamic frequency* offset, but leaving a residual phase error.

Type 3 PLLs (green waveforms in Figure 7.18)

- Can adapt in the presence of an initial *phase* difference, an initial *frequency* difference, and a *dynamic frequency* offset, in each case with zero residual phase error.

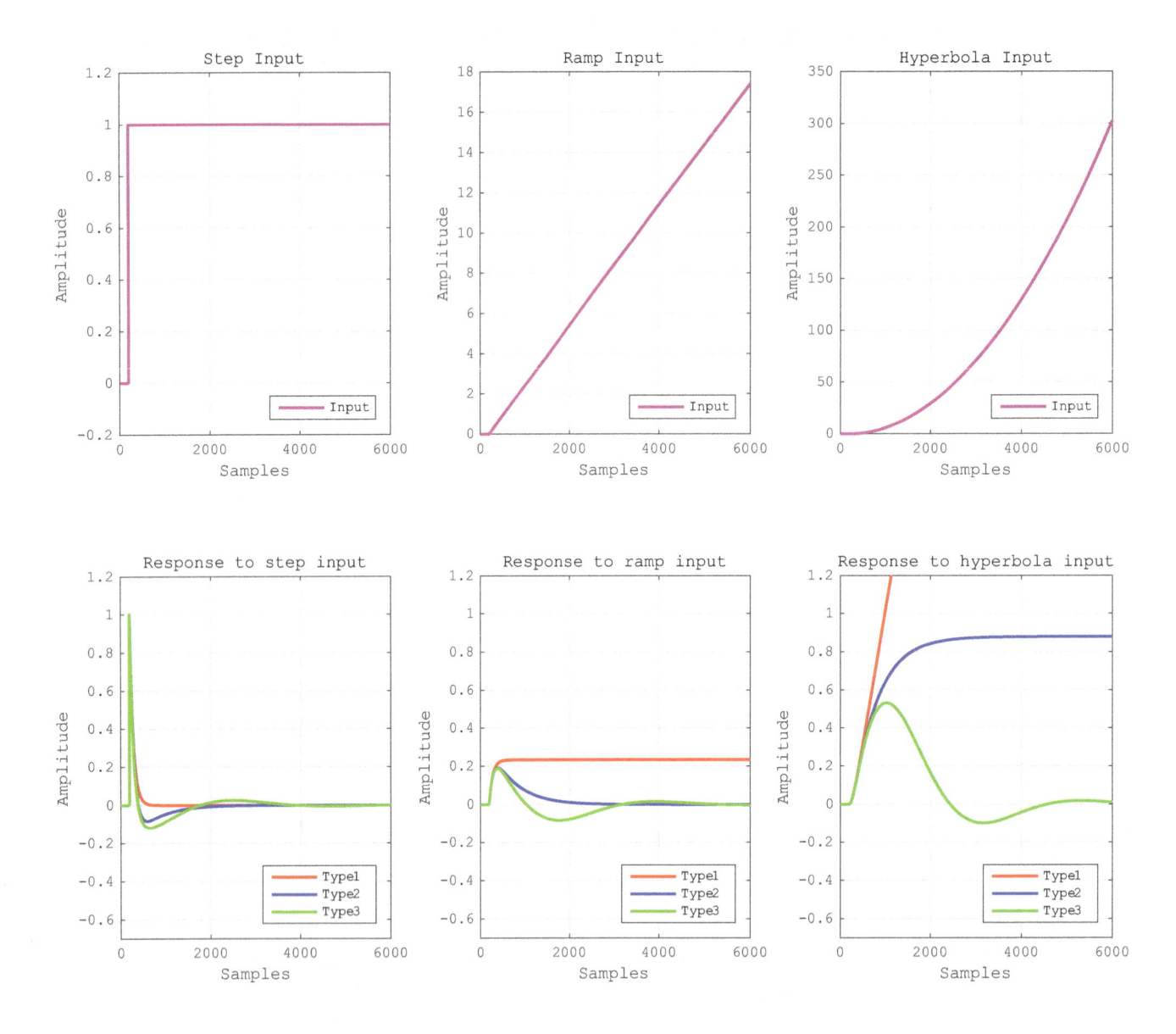

Figure 7.18: Response of Type 1, Type 2, and Type 3 PLLs (linear model)

Type 2 PLLs are commonly used, because they provide sufficient performance for most situations, without the additional complexity of a second integrator in the loop filter. Further, second order differential equations are very popular for modelling control systems, and the analysis and design of Type 2 PLLs is well understood and treated in the literature. Pragmatically, this makes it easier to design and implement Type 2 PLLs. For these various reasons, the rest of this section on will focus on Type 2 PLLs.

7.7.3 Transient Behaviour and Tracking Capability

A PLL will display some transient behaviour while adapting, and the nature of this transient behaviour depends on three factors in particular:

- The *Type* of the PLL (as defined by the loop filter and discussed in Section 7.7.2) determines whether it can adapt to the point of zero phase error, or will have a residual steady state error.

- There is an interaction between the **bandwidth** of the loop, and the initial **deviation** between the input and reference frequencies. For any given deviation in frequency, a PLL with a narrower bandwidth will take longer to adapt.
- The **damping ratio** of the PLL affects the pattern of the adaption behaviour, including its speed and the extent of overshoots.

These resulting behaviours can be evaluated with the aid of the linear model introduced in Figure 7.17.

The damping factor, or damping *ratio*, denoted by ζ, relates to the transient behaviour of the PLL as it achieves phaselock. This derives from control theory, with values of $\zeta < 1$ corresponding to an *underdamped* system (in which the transient displays overshoots and fluctuations around the steady state value before settling down), whereas when $\zeta > 1$, the system is *overdamped* and converges gradually to the steady state. If $\zeta = 1$, then the system is said to be *critically damped*, meaning that it has the shortest possible rise time without having overshoots [14].

Large damping ratios cause the PLL to converge quickly, while smaller damping ratios result in a longer period of transient behaviour with 'overshoots' and ripples before the loop finally converges. On the other hand, smaller damping ratios normally have a better ability to track changes, and therefore a trade-off is encountered. Realistic damping ratios are normally in the range $0.5 \leq \zeta \leq 2$, which achieves a good balance of properties [13].

Figure 7.19 shows the response of a Type 2 PLL to a frequency step, ramp, and hyperbola, for different damping ratios. Notice that overdamped systems can converge to zero phase error more quickly, in the case of the step input, whereas they take longer when the input is a ramp or hyperbola. This indicates that overdamped PLLs have a lesser ability to track time-varying inputs. On the other hand, very underdamped PLLs have significant oscillatory transients, which is normally undesirable.

In communications applications, it is reasonable to assume that the input reference signal will change over time, and therefore it is useful to have a PLL with good tracking capability. Values of $\zeta = 1$ and $\zeta = 1/\sqrt{2} = 0.707$ are commonly used to achieve this.

7.7.4 Bandwidth

The bandwidth refers to the range of frequencies over which the PLL operates, and also to the range of frequencies that can enter the PLL as noise.

There are several possible definitions of bandwidth, but in general terms, a PLL with a broader bandwidth can cope with larger deviations between the quiescent frequency of the NCO, and the input reference signal. In other words, the PLL can successfully adapt over a larger range of frequencies. On the other hand, the broad bandwidth also allows more noise to enter the PLL than a narrow-bandwidth PLL, thus degrading the purity of the synthesised sine wave.

To make a qualitative comparison, Figure 7.20 shows two examples of PLLs adapting (here using a 'real' PLL, rather than the linear model). Figure 7.20(a) represents a PLL with a 'narrow' bandwidth, while Figure 7.20(b) has a 'broad' bandwidth. Notice that the narrow bandwidth PLL takes longer to adapt. Once it has done so, however, its frequency fluctuates more closely around the desired frequency than the broad bandwidth PLL. This means that it produces a sine wave with superior spectral purity.

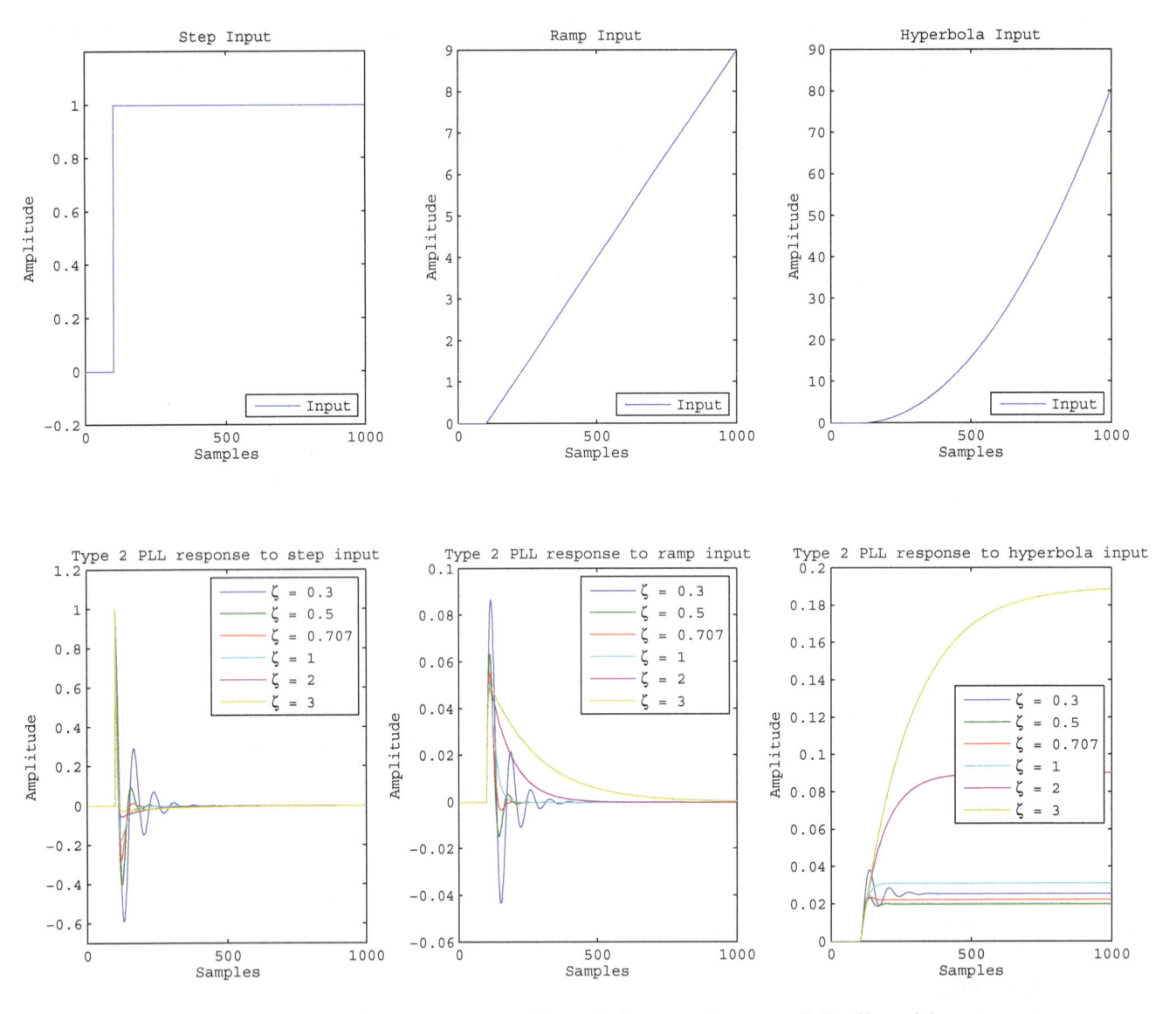

Figure 7.19: Response to phase step, ramp and hyperbola inputs (linear model): effect of damping ratio

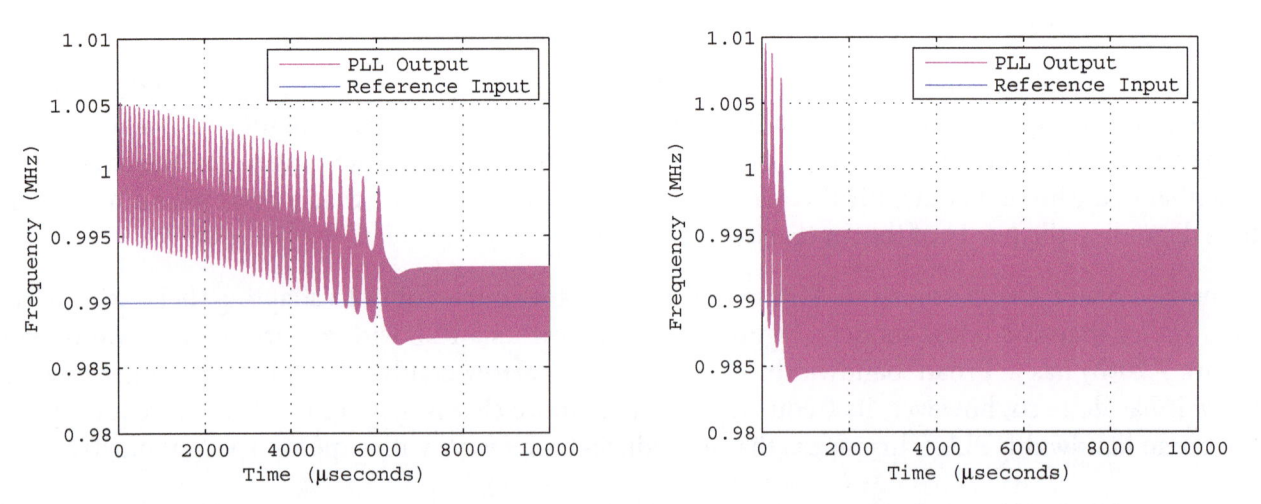

Figure 7.20: Adaptation behaviour for a Type 2 PLL: (a) narrow bandwidth, (b) broad bandwidth

Chapter 7: Frequency Tuning and Simple Synchronisation

Another advantage of the narrower bandwidth is that less noise is accepted into the PLL. The plots shown in Figure 7.20 show the response of two PLL designs to a perfect, noiseless sine wave input.

7.7.5 Experimenting with the Linear Model

Having spent some time reviewing various aspects of PLL performance with the aid of the linear model, it is useful to provide this as a Simulink file so that you can experiment with it! The next set of exercises will guide you through some simulations based on the linear model.

Exercise 7.7 PLL Linear Model: Steady State Error

In this exercise, we look at the linear model for a digital PLL, and confirm the residual phase errors in response to different types of inputs: a step, a ramp, and a hyperbola. As you work through this exercise, complete the relevant cells within the table below with the **steady state error**.

PLL Type	Input Signal		
	Step	Ramp	Hyperbola
Type 1			
Type 2			
Type 3			

Note that the damping ratio in set to $\zeta = 1$ for all cases considered in this model. We will consider the impact of changing the damping ratio later, in Exercise 7.8.

(a) **Open MATLAB.** Set the working directory to an appropriate folder so you can open this model:

    `/synch/PLL/PLL_linear_model_1.slx`

(b) **Inspect the system and note the inputs applied.** In this model, the gain coefficients for the phase detector, oscillator, and loop filter have all been set to numerical values. The model initially implements a Type 3 PLL. You can achieve a Type 2 loop by setting k_3 to zero, and a Type 1 loop by setting both k_2 and k_3 to zero.

(c) **Simulate!** Run a simulation by clicking on the 'Run' button in the Simulink toolbar for each of the three inputs, and in each case, note the residual error after the PLL has adapted, and enter these values in the corresponding cells of the table.

(d) **Compare.** How do your simulation results compare to Figure 7.18?

Exercise 7.8 PLL Linear Model: Effect of Damping Ratio

In this exercise, we again consider the linear model of a PLL, and this time look at the impact of the damping ratio. Here, two files are needed: a Simulink model of the linear model, and a MATLAB script to configure the coefficients within the PLL, to achieve the desired damping ratio.

(a) **Open the MATLAB script:**

　　/synch/PLL/configure_PLL_linear_model.m

(b) Inspect the contents of the MATLAB file. This is a simple script which sets the values of the loop gain coefficients k_p, k_o, and k_1, and then calculates the required value of k_2 based on the chosen damping ratio, d (ζ). The damping ratio is initially set to a value of 1, i.e. the system is said to be 'critically damped'.

(c) **Run the MATLAB script.** This will create the variables necessary to parameterise the Simulink model, in advance of opening it.

(d) **Open the Simulink model:**

　　/synch/PLL/PLL_linear_model_2.slx

Notice that this system represents a PLL, with a Type 3 loop filter. However, the last loop filter coefficient, k_3, is initially set to zero, meaning that it effectively forms a Type 2 PLL.

(e) **Check out the Step response.** Run the Simulink model by 🖱 on the 'Run' ▶ button, with the input selector set to the *Step* input, and view the results. You should see that the PLL is able to adapt with zero steady state phase error. It will follow a particular shape as it adapts.

(f) **Investigate the damping ratio.** Try changing to a damping ratio of 0.3, by making an appropriate change to the MATLAB script. After you have altered the code, first rerun the MATLAB script, and then secondly the Simulink model. What do you notice about the transient behaviour? You could also try other values, including a much higher damping ratio (say ζ = 3). You might also wish to experiment with different input types (ramp and hyperbola).

(g) **Try a Type 3!** Lastly, change the PLL to a Type 3 and see what happens. Set the value of the k_3 gain coefficient to be 2e-5 (you can make this alteration directly in the Simulink model), and switch the input to the *Hyperbola*, by changing the positions of the manual switches. Ensure that the MATLAB script is run for the case where ζ = 1 (you will need to change this back to its original value, then rerun).

(h) **Resimulate!** Can this Type 3 PLL adapt effectively in the case of the Hyperbola input?

7.8 PLL Design

Although some of the characteristics and parameters of PLLs have been introduced, we have not so far discussed how to design a PLL to achieve a certain set of requirements. This section presents a practical, step-by-step guide to setting the parameters of a Type 2 PLL, in order to achieve the desired characteristics. More detailed review and derivation may be found in Appendix D (page 593), which follows the analysis presented in [38].

Chapter 7: Frequency Tuning and Simple Synchronisation

7.8.1 Step 1: Determine the sample period

When designing a digital PLL, the sampling rate f_s is likely to be a pre-determined aspect of the system. The sample period can therefore be calculated simply as

$$T = \frac{1}{f_s} \; . \tag{7.24}$$

Note that the parameter T will be required in later calculations.

If the sampling rate is not already set, it should be specified as a value at least 5 times higher than the expected frequency of the input reference sine wave. This assumes the use of a multiplier-based phase detector, which will generate components at approximately double the input frequency. (It may be possible to run successfully with a lower sample rate, but using the above rule will keep matters simple and clear!)

7.8.2 Step 2: Establish the gain of the phase detector

In this book, we consider two alternative methods of implementing the phase detector: (i) an 'ideal' phase detector component that accurately measures the phase difference between two input signals; or (ii) as a multiplier. The two alternatives result in different values for the phase detector gain coefficient, K_p, which is due to their different characteristic 's-curves' (more on this in Appendix D, page 593).

For our current purposes, we simply note the following:

- For an 'ideal' phase detector: $K_p = 1$
- For a 'multiplier' phase detector: $K_p = A/2$

where A is the amplitude of the input reference signal applied to the multiplier phase detector.

The above indicates that, in the multiplier-based design, the phase detector gain varies with the amplitude of the input signal. This dependency is undesirable, because a change in the input signal amplitude could fundamentally alter the behaviour of the PLL. As a result, it is common to precede the PLL with a stage of AGC [39], which dynamically adjusts the gain applied to the input signal, such that it is scaled to the normalised range from -1 to +1.

It should be noted that the gain, K_p, is an inherent property of the phase detector, and does not refer to a separate gain coefficient set within the phase detector section of the PLL circuit.

7.8.3 Step 3: Establish the gain of the NCO

The structure and behaviour of the NCO component were reviewed in Section 7.5.3. The NCO is capable of generating sine and cosine waves of a dynamically variable frequency, with the adjustment made via a feedback signal supplied from the output of the loop filter. A gain of K_o is applied to this feedback signal, as shown in Figure 7.13. The gain coefficient can be freely chosen, but a common choice is to set $K_o = 1$.

7.8.4 Step 4: Choose the Damping Ratio

The damping ratio parameter ζ was introduced in Section 7.7.3. It was shown that the value of ζ affects the transient behaviour of the PLL as it adapts, and also its tracking capabilities. An appropriate value of ζ should be selected based on the nature of the synchronisation task; as stated earlier, values in the range 0.5 to 2 are common because they achieve a good balance of stability and tracking properties.

The damping ratio is not set explicitly within the PLL. Rather, the parameter is incorporated into the calculation of the loop filter coefficient values (as will be shown in Step 6 of our design process). The desired damping ratio is achieved by appropriately setting the various parameters of the PLL.

7.8.5 Step 5: Specify the Noise Bandwidth

It was earlier established that specifying the bandwidth of a PLL is an important aspect of its design. A narrow bandwidth lets very little noise into the PLL, but can result in the PLL taking a long time to achieve lock; while a broad bandwidth enables quick adaptation, but with poorer noise performance. Here, we will specify the noise bandwidth, B_n, of a PLL in Hz.

The maximum expected frequency offset (i.e. the difference between an incoming sine wave, and the locally generated sine wave) must be taken into account when defining B_n, in order to ensure that the PLL can successfully lock. The range of frequencies deviations over which the PLL can lock is referred to the *pull-in* range, and it is an interaction between B_n and damping ratio, ζ. Useful approximations for pull-in range, along with approximate times to attain lock, are given in [38]. Suitable operation over the range of expected offsets can also be confirmed through simulation.

Like damping ratio, PLL noise bandwidth is an implicit parameter of the PLL, rather than being set directly within the loop implementation. Instead, the chosen value of noise bandwidth, B_n, is incorporated into the equations from which the loop filter coefficient values are determined.

7.8.6 Step 6: Calculate the Loop Filter Gain Coefficients

Two gain coefficients are needed in the Type 2 PLL: K_1 and K_2, for the proportional and integral branches, respectively. This was previously shown in Figure 7.14. Values for the coefficients can be calculated, provided that the following parameters are known:

- Sampling period, T
- NCO gain, K_o
- Phase detector gain, K_p
- Damping ratio, ζ
- Noise bandwidth, B_n

To calculate the required values, Eqs. (D.50) and (D.51) from Appendix D (page 593) can be rearranged for K_1 and K_2, respectively, to give:

Chapter 7: Frequency Tuning and Simple Synchronisation

$$K_1 = \frac{4\zeta\left(\dfrac{B_n T}{\zeta + \dfrac{1}{4\zeta}}\right)}{K_o K_p\left(1 + 2\zeta\left(\dfrac{B_n T}{\zeta + \dfrac{1}{4\zeta}}\right) + \left(\dfrac{B_n T}{\zeta + \dfrac{1}{4\zeta}}\right)^2\right)} \tag{7.25}$$

$$K_2 = \frac{4\left(\dfrac{B_n T}{\zeta + \dfrac{1}{4\zeta}}\right)^2}{K_o K_p\left(1 + 2\zeta\left(\dfrac{B_n T}{\zeta + \dfrac{1}{4\zeta}}\right) + \left(\dfrac{B_n T}{\zeta + \dfrac{1}{4\zeta}}\right)^2\right)}. \tag{7.26}$$

Noting the repeated term, it is convenient to make the substitution

$$\eta = \frac{B_n T}{\zeta + \dfrac{1}{4\zeta}}, \tag{7.27}$$

which results in

$$K_1 = \frac{4\zeta\eta}{K_o K_p(1 + 2\zeta\eta + \eta^2)} \tag{7.28}$$

and

$$K_2 = \frac{4\eta^2}{K_o K_p(1 + 2\zeta\eta + \eta^2)}. \tag{7.29}$$

According to [38], a pair of simpler approximations can be used, where the bandwidth of the loop is defined as small in relation to the sample rate (i.e. where $B_n T \ll 1$). These simplified expressions are

$$K_1 \approx \frac{4\zeta B_n T}{K_o K_p\left(\zeta + \dfrac{1}{4\zeta}\right)} \tag{7.30}$$

and

$$K_2 \approx \frac{4(B_n T)^2}{K_o K_p \left(\zeta + \frac{1}{4\zeta}\right)^2} \tag{7.31}$$

respectively.

7.8.7 Step 7: Confirm the Complete PLL Model

We can now confirm the construction of a Type 2 PLL from low level blocks, and associate with it the various parameters reviewed over the preceding sections. A diagram is shown in Figure 7.21. Note that the signal supplied as the input reference signal is sampled at a rate of f_s Hz, which results in a sampling period of T seconds.

It should also be remembered that the value of K_p depends on the amplitude of the input signal, if using a multiplier phase detector (with $K_p = A/2$). As mentioned previously, it is usual to precede a PLL with an AGC stage where the input signal amplitude is variable, thus ensuring that the amplitude of the signal input to the PLL is approximately equal to 1, and therefore that the value of K_p is 0.5.

Finally, note that a 'modulo 2π' operation could also be inserted prior to the cos( ) function within the NCO. This may be preferable for visualisation purposes, and it would also prevent the possibility of numerical overflow occurring.

7.8.8 Step 8: Simulate the PLL and Verify Behaviour

At this stage, the parameters T, K_p, K_o, B_n and ζ have been established, and from these, the required values of K_1 and K_2 have been derived. The value of μ_q, the quiescent step size for the NCO, has been set based on the quiescent frequency, using Eq. (7.20).

The model can now be realised in MATLAB or Simulink, and simulated. The simulated outputs should be assessed in terms of transient behaviour, time to achieve lock, performance in noise, tracking capability, steady state behaviour, and any other desired characteristics. Adjustments can then be made, if necessary, by re-parameterising the model and repeating the simulation process.

7.8.9 Design Example

We will now consider an example with the following parameters:

- Sampling rate: $f_s = 10kHz$
- Sample period: $T = 1/f_s = 100e^{-6}s$
- Quiescent frequency: $f_q = 800Hz$
- Damping ratio: $\zeta = 0.707$
- Phase detector gain: $K_p = 0.5$
- Oscillator gain: $K_o = 1$
- Noise bandwidth: $B_n = 5Hz$

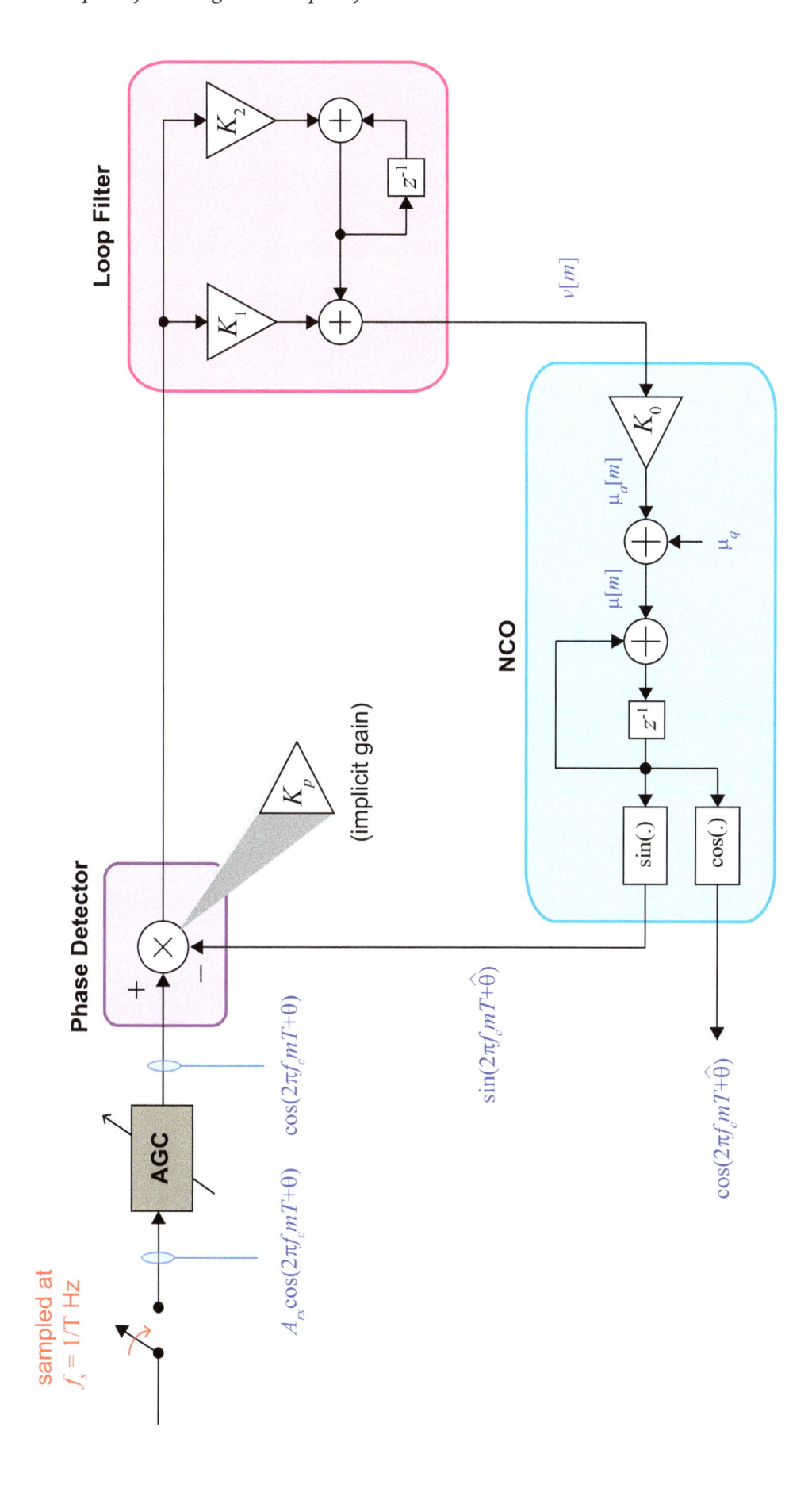

Chapter 7: Frequency Tuning and Simple Synchronisation

Figure 7.21: PLL model showing low level implementation

It can be calculated that the required NCO quiescent step size is

$$\mu_q = \frac{2\pi f_q}{f_s} = \frac{2\pi \times 800}{10,000} = 0.5027. \tag{7.32}$$

The value of loop filter coefficient K_1 can be obtained from Eq. (7.28), and K_2 from Eq. (7.29). We first determine the value of η using Eq. (7.27), i.e.

$$\eta = \frac{B_n T}{\zeta + \frac{1}{4\zeta}} = \frac{5 \times 100e^{-6}}{0.707 + \left(\frac{1}{4 \times 0.707}\right)} = 4.714e^{-4}. \tag{7.33}$$

The value for η can then be substituted into Eq. (7.28), i.e.

$$K_1 = \frac{4\zeta\eta}{K_o K_p \left(1 + 2\zeta\eta + \eta^2\right)} \tag{7.34}$$

$$= \frac{4 \times 0.707 \times 4.714e^{-4}}{1 \times 0.5 \times (1 + 2 \times 0.707 \times 4.714e^{-4} + 4.714e^{-4^2})} = 0.0027.$$

Likewise, K_2 can be obtained from Eq. (7.29) using the value for η calculated in Eq. (7.33).

$$K_2 = \frac{4\eta^2}{K_o K_p \left(1 + 2\zeta\eta + \eta^2\right)} \tag{7.35}$$

$$= \frac{4 \times 4.714e^{-4^2}}{1 \times 0.5 \times (1 + 2 \times 0.707 \times 4.714e^{-4} + 4.714e^{-4^2})} = 1.777e-6.$$

All of the necessary parameters are now available, and can be used to implement a PLL model and simulate it.

Exercise 7.9 Challenge: Design of a Type 2 PLL

This example follows on from the design example in Section 7.8.9. This is quite a challenging exercise that will require you to generate your own scripts and models (or if you wish, you could try to reuse and modify existing ones from earlier exercises!).

(a) **Open the MATLAB.** Create a new MATLAB script, and save it with the filename:

/my_models/my_PLL_parameters.m

Chapter 7: Frequency Tuning and Simple Synchronisation 271

(b) **Write code to define the PLL parameters given in Section 7.8.9.** Follow this with code that implements equations (7.32) to (7.35), and calculates the remaining parameters. If you need to refresh your memory on how to create MATLAB variables, please look back at the *Getting Started with MATLAB and Simulink* chapter.

(c) **Execute and check the results.** Run the script you have created, and ensure that it generates the expected numerical results.

(d) **Create a new Simulink model.** Save this with the filename:

   `/my_models/my_PLL.slx`

(e) **Build a PLL.** Build a Type 2 PLL, following the diagram given in Figures 7.21 and 7.22. You can leave out the AGC section if you wish, by ensuring that the input reference signal has an amplitude of 1.

Remember that, having run the script and created the variables in the MATLAB Workspace, you can now use the parameters directly in the model. For instance, you could use the variable names `K1` and `K2` to set the gain of the Constant Multiplier blocks from the Loop Filter.

All of the components you will require to build the PLL can be found inside the Simulink Library Browser. Blocks you are likely to need include:

Product	*> Simulink > Math Operations*
Gain	*> Simulink > Math Operations*
Add	*> Simulink > Math Operations*
Trigonometric Function	*> Simulink > Math Operations*
Delay	*> Simulink > Discrete*

(f) Make sure to place some input sources from *> Simulink > Sources* and a *Time Scope* from *> DSP System Toolbox > Sinks*, to allow you to test your PLL design out!

(g) **Run a simulation.** Run the model by clicking on the 'Run' button, and check that the PLL is working as expected.

(h) **Design PLL parameters #2.** Use your script to design PLL parameters for the following scenario, and hence update the configuration of your PLL.
  - Sampling rate of 48kHz
  - Expected tone frequency of 256Hz
  - Maximum expected frequency deviation 0.5Hz
  - Multiplier phase detector
  - Critically damped PLL.

(i) **Run a simulation!** Execute a simulation and ensure that the model behaves correctly.

7.9 PLL Performance in Noise

In this section, we briefly look at the behaviour of PLLs in noisy conditions, and the effect of noise bandwidth on performance. These simulations incorporate an Additive White Gaussian Noise (AWGN) source, which is added to the reference sine wave at the input to the PLL.

Exercise 7.10 Type 2 PLL: Performance in Noise

In this example, we will consider the performance of a Type 2 PLL in noise, and note the impact of altering the noise bandwidth.

(a) **Open MATLAB.** Set the working directory to an appropriate folder so you can open this script:

/synch/PLL/configure_noise_model.m

(b) **Run the script to generate the PLL parameters.** You may also wish to read through this script to see which parameters have been created.

(c) **Open the Simulink model:**

/synch/PLL/PLL_type2_noise.slx

(d) **Simulate the model.** First, run the simulation in zero noise conditions (check the position of the *Manual Switch*) and confirm that the PLL converges to the frequency of the input. How long does it take?

(e) **Re-simulate with added noise.** Now flick the manual switch so that the input is corrupted by noise, and rerun the simulation. Hence, ensure that the PLL converges in the presence of noise. (You should notice from the time domain output that the PLL is able to synthesise a sine wave of the correct frequency and phase, even in noisy conditions.)

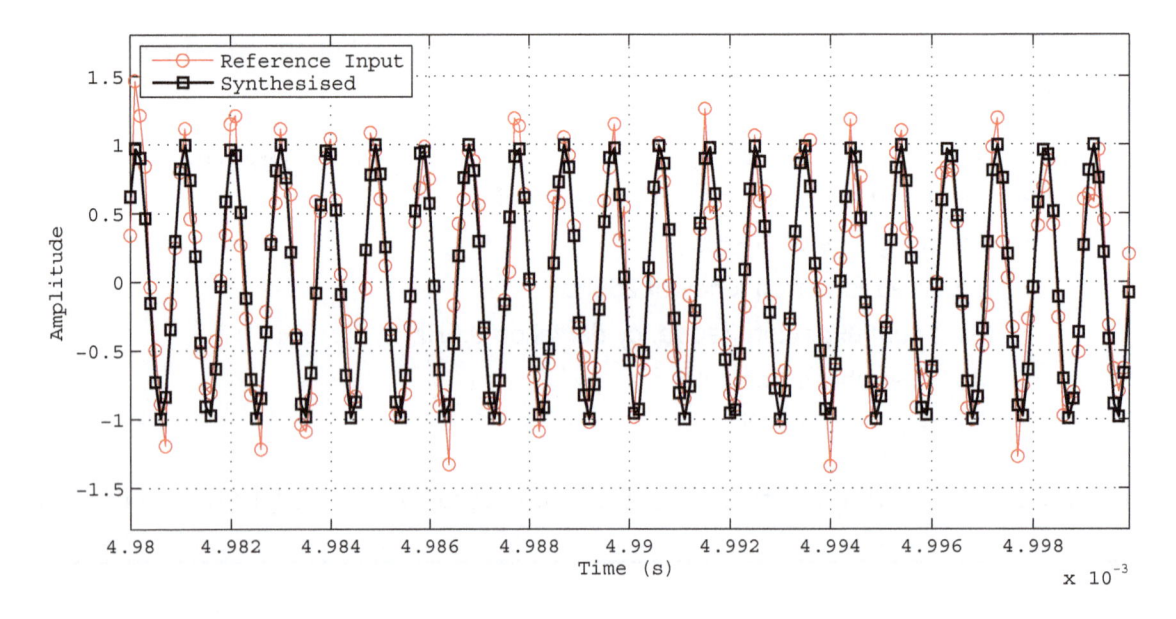

Chapter 7: Frequency Tuning and Simple Synchronisation 273

(f) **Inspect the spectra.** Spectrally, how well has the PLL 'cleaned up' the input sine wave? Try estimating the distance between the spectral peak and the noise floor, in both cases.

(g) **Experiment with the noise bandwidth.** Next, decrease the noise bandwidth of the PLL by altering the parameters in the `configure_noise_model.m` script (try dividing the noise bandwidth by a factor of 2).

(h) **Rerun the simulation.** What do you observe this time, particularly in terms of adaption time and noise rejection?

Try again, this time doubling the noise bandwidth. Do the results match your expectations?

(i) What general statement can you make linking PLL bandwidth, noise rejection, and adaption time?

7.10 Carrier Synchronisation

Coherent receivers employ an aspect of synchronisation with the received carrier signal. The effect of NOT being carrier synchronised was shown in Section 7.4, in the context of analogue AM communications. It was apparent that the demodulated signal was not, after appropriate lowpass filtering, equivalent to the original. Similar issues occur in digital communications, and this will be discussed later, in Section 11.4 (page 448).

As reviewed in Section 6 (page 203), there are different variants of AM modulation. From the perspective of carrier synchronisation, it is significant whether the carrier is transmitted or not. If the carrier is transmitted as a separate component (AM-DSB-TC), then the synchronisation task is relatively easy, because the carrier is explicitly present and can provide the reference input to a PLL synchronisation circuit. On the other hand, an AM-DSB-SC signal does not contain a distinct carrier component, and the receiver must instead synchronise using the *modulated* version of the carrier.

These two variations call for different synchroniser architectures, as explained over the next two sections.

7.10.1 *Carrier Synchronisation for Transmitted Carrier AM (AM-DSB-TC)*

In DSB-TC AM, a coherent receiver extracts the transmitted carrier component, and uses it to demodulate the received signal to baseband. This can be achieved using a PLL, with the carrier acting as the input reference signal. The PLL operates in the manner discussed in Section 7.5, and can be designed as a Type 1, 2, or 3, with an appropriately chosen damping ratio and bandwidth.

The PLL regenerates the carrier sine wave from the received signal, and then demodulation is performed. The architecture for this synchronisation circuit is shown in Figure 7.22.

We can now confirm the operation of this receiver for DSB-TC AM with a Simulink exercise.

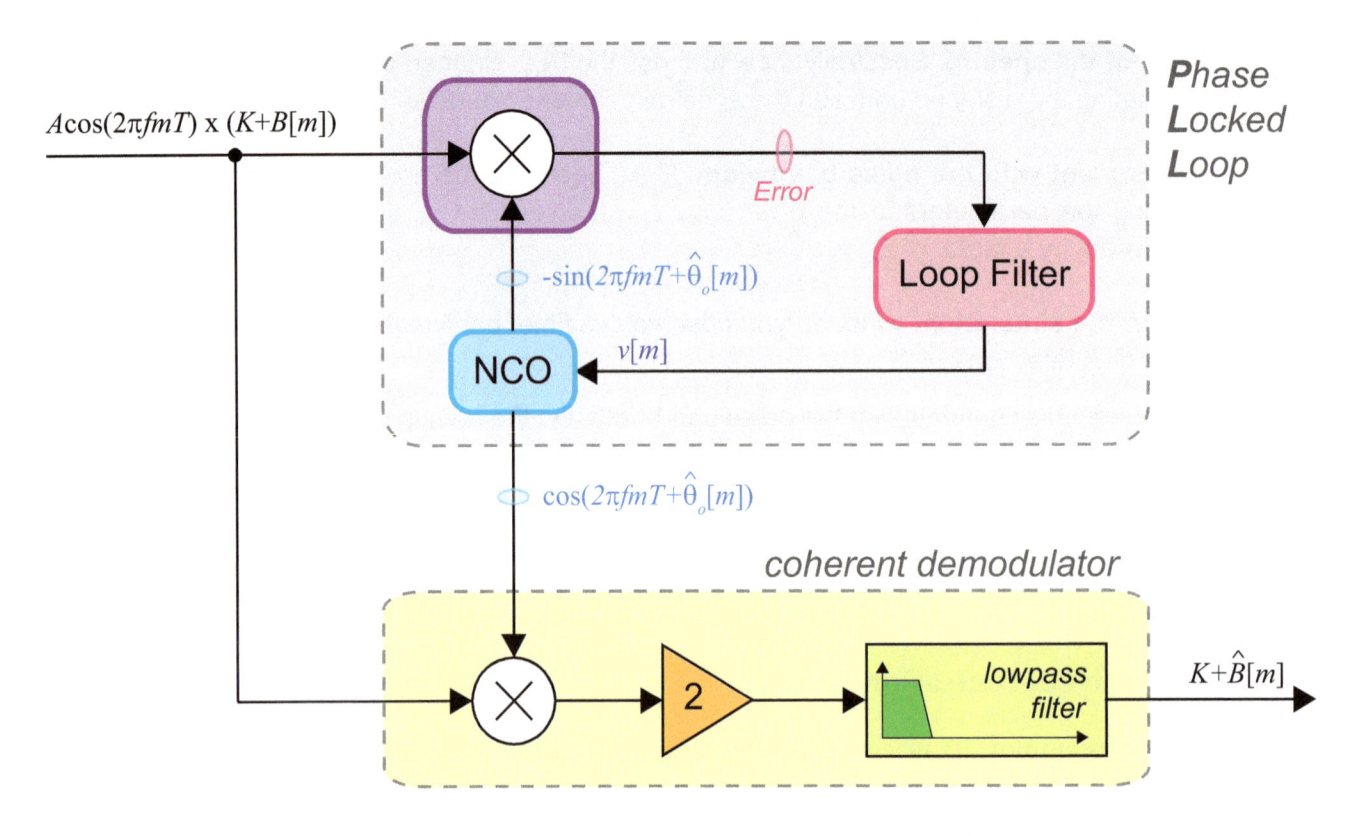

Figure 7.22: Architecture for a synchronous, PLL-based demodulator for DSB-TC AM

Exercise 7.11 Coherent Receiver for AM-DSB-TC

This example models the transmission and reception of an AM-DSB-TC signal. The receiver employs a PLL-based synchronisation circuit to regenerate the carrier and demodulate the signal to baseband.

(a) **Open MATLAB.** Set the working directory to an appropriate folder so you can open this model:

/synch/carrier/AM_DSB_TC_PLL.slx

(b) **Inspect the system.** Confirm that the receiver's local oscillator has a different initial frequency to that used to modulate the audio signal.

(c) **Run a simulation.** Observe the *Time Scope* output. You should be able to see that the transmitted and received signals look very similar. You can also try overlaying them in a single plot.

(d) Also inspect the MATLAB figure entitled *'Frequencies of Transmitted and Locally Generated Carriers'*. This shows the frequency generated by the VCO part of the PLL, as it adapts over time.

(e) Zoom into the first 0.025 seconds of the plot, and confirm that the PLL adapts within this initial period of the simulation. You should see something like the following:

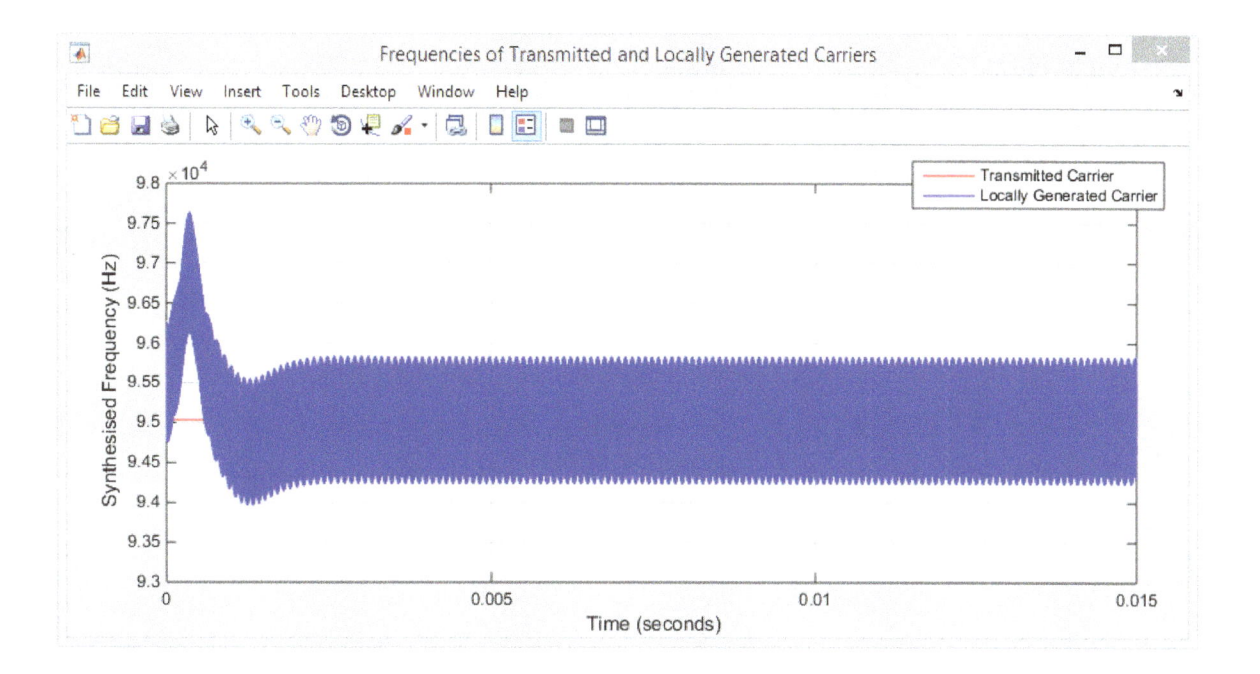

Chapter 7: Frequency Tuning and Simple Synchronisation

(f) **Listen to the audio output.** Navigate to the current directory in the main MATLAB window, then 🖱 on the file `demod_audio_pll.wav` and choosing the option *Open Outside MATLAB*. This will play the music sample using your default audio player. Listen and compare against the input audio file, `music1_mono48kHz.wav`. *Warning*, audio in this file is quite loud!

(g) Try deleting the error input to the VCO in the system (as shown below), and resimulating. This will cause the VCO to emit a fixed frequency output, rather than adjusting in response to the error signal.

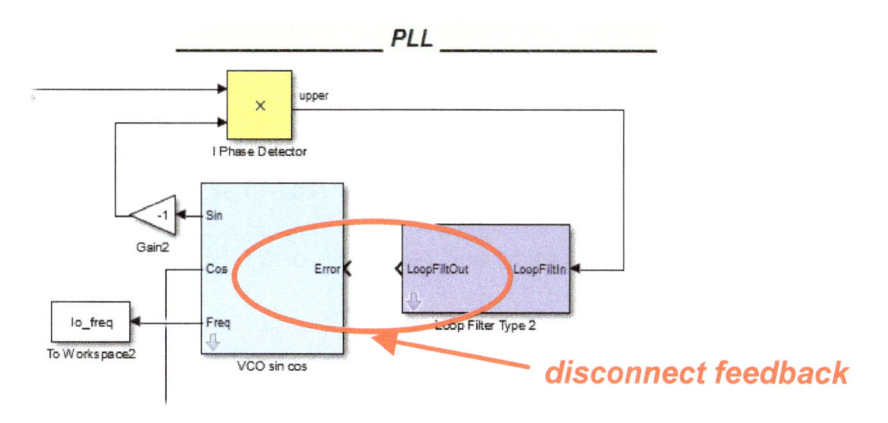

disconnect feedback

(h) Observe the output plots and listen to the resulting audio. What has happened now?

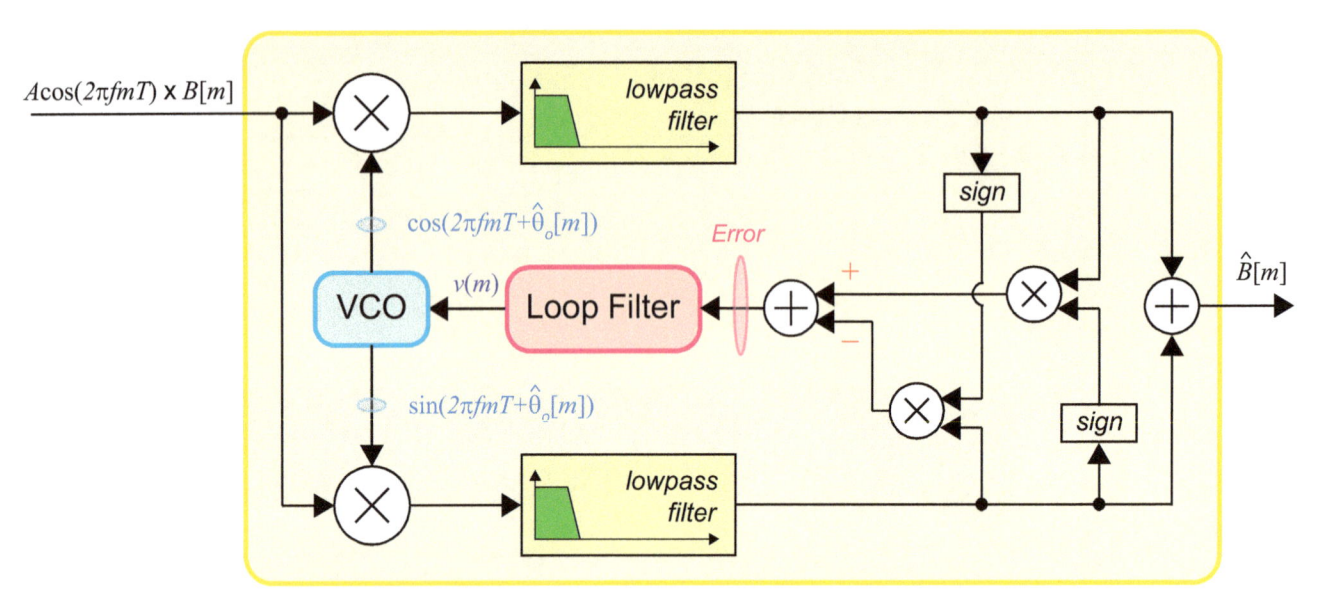

Figure 7.23: The Costas Loop for coherent demodulation of DSB-SC AM

7.10.2 Carrier Synchronisation for Suppressed Carrier AM (AM-DSB-SC)

The DSB-SC version of AM is more difficult than DSB-TC to demodulate coherently, because the carrier is not explicitly present. As a consequence, the receiver has to recover the carrier from the modulated signal that it receives, and then demodulate. The architecture commonly used for this purpose is the *Costas Loop*, which is shown in Figure 7.23.

The Costas loop has two branches, which multiply the received signal with the cosine and sine outputs of the local oscillator (VCO) and then apply a lowpass filter to each. This provides two versions of the demodulated output, which can be denoted I and Q. The Costas loop derives an error signal by combining the outputs of the two branches, then passes the error signal through the loop filter to drive the VCOs and hence adjust the local oscillator frequency and phase. The output of the VCO is a regenerated version of the received carrier.

Exercise 7.12 Coherent Receiver for AM-DSB-SC (Costas Loop)

This example models the transmission and reception of a DSB-SC AM signal. Carrier synchronisation is achieved using a Costas Loop.

(a) **Open the model:**

   `/synch/carrier/AM_DSB_SC_costas_loop.slx`

(b) **Inspect the system.** Compare the structure of the Costas Loop against the signal flow graph provided in Figure 7.23.

(c) To confirm the parameters of the Costas Loop, open *Model Properties* and view the *PreLoadFcn*. These parameters are used to configure the parameters of the loop, and entered via the masks of certain blocks, such as the Loop Filter and VCO.

(d) **Run the simulation.** This will provide 10 seconds of audio written to the output file `demod_audio_costas.wav`. Listen to this file using a media player, and compare it to the original.

(e) View the figure generated at the end of the simulation, which displays the frequency of the carrier used for modulation in the transmitter, and the frequency output by the VCO in the Costas Loop. You should observe that, after a period of adaption, the VCO produces sine and cosine outputs at exactly the same frequency as the modulating carrier signal.

(f) **Listen to the audio output.** Navigate to the current directory in the main MATLAB window, then click on the file `demod_audio_costas.wav` and choose the option *Open Outside MATLAB*. This will play the music sample using your default audio player. You can also listen and compare against the input audio file, `music1_mono48kHz.wav`. As a warning, the audio in this file is quite loud! Check that the period of frequency adaption (according to the figure) corresponds to the initial, "imperfect" section of the output audio file.

(g) If you wish, try experimenting with the parameters of the system and re-simulating to confirm the effect of your changes. You can do this in the following steps:

   Open *File > Model Properties*.

   Select and copy the contents of the *PreLoadFcn*.

   Paste the code into the MATLAB command window, change the value of `Bn` to 100, and then move the cursor to the end and press *Enter*.

   Close your media player (if open) and then re-simulate. Observe the output plot and listen to the audio to confirm the effect of your change.

(h) Try repeating the above steps for `Bn` = 1000.

7.11 Summary

Tuning and synchronisation are important aspects of implementing radio receivers, as has been explored in this chapter.

We began by considering tuning, the process of selecting a set of frequencies to demodulate to baseband for further processing. This stage is undertaken within the RTL-SDR hardware, with parameters supplied from MATLAB or Simulink.

Tuning is, however, just a first stage of adjusting the frequency of receiver based on the received signal. The need for more precise synchronisation (i.e. matching of the receiver's local oscillator to the carrier embedded within the received signal) was first of all motivated, and then examples were presented. The PLL was covered in detail, including the operating characteristics and design of PLLs, as this component forms the basis of carrier synchronisation circuits. Lastly, synchronisation circuits for the transmitted carrier and suppressed carrier variants of AM were explored via examples.

Carrier synchronisation will be further discussed and utilised in later chapters, particularly in Section 9.8 (page 358), and Exercise 10.7 (page 405), in the context of FM demodulation, and in Section 11.4 (page 448) for digital communications. When implementing digital communications systems, other forms of synchronisation are also necessary, as will be discussed elsewhere in Chapter 11.

8 Desktop AM Transmission and Reception

This chapter will focus on implementing real AM desktop SDR transmit (Tx) and receive (Rx) systems and review some of the classic AM modulation options and transmission variants. Commercially broadcast AM radio conventionally operates in the Very Low Frequency (VLF), Low Frequency (LF), and Medium Frequency (MF) radio bands, which range from 3kHz to 3MHz. As the tuners used in the RTL-SDR operate across the frequency range 24MHz to 1.75GHz (or 53MHz to 2.2GHz) [71][69], they are unable to receive any of these signals. This is an unfortunate hardware limitation, as AM signals are the simplest to demodulate, and building AM receivers would be a great place to begin your practical RTL-SDR work. There is a solution however: you can generate your own.

Various different methods can be used to create AM signals. One option is to purchase a device such as (i) the *Ham It Up* to upconvert existing AM radio signals into range; (ii) use a *USRP® software-defined radio* to create your own 'radio station'; or (iii) go down the path of designing and building a custom AM transmitter. We will investigate all three of these AM signal generation methods in depth later in this chapter. In many of the RTL-SDR receiver exercises we will receive a signal we generated from a USRP® which is a great desktop SDR signal generator also supported by MATLAB and Simulink, and excellent for deploying on the lab bench or placing at the front of a class with students in the class receiving on their RTL-SDR. If you do *not* have such a device available, it will still be possible to run and review many of the exercises using pre-stored and recorded data files in place of the *USRP®* (or visit the desktopSDR.com website and view some of the simulation videos there, to see the designs in action).

Therefore, in this chapter we first discuss methods to generate different AM signals, and move on to implement USRP® radio based modulator/ transmitter Simulink models that produce AM-DSB-SC, AM-DSB-TC, AM-SUSB and AM-SLSB (see Chapter 6). After first discussing the generation of signals, guidance on implementing some RTL-SDR AM receivers will be provided, followed by basic AM receiver examples and finally implementation of more advanced concepts such as multiplexing and demultiplexing signals to facilitate multichannel transmission using a single carrier.

8.1 Transmitting AM Signals with a USRP® Radio

The USRP® hardware family is a range of programmable FPGA based Tx/ Rx SDRs made by Ettus Research. With high speed (and high resolution) ADCs and DACs, these high quality devices take a modular approach, and allow you to use a selection of transmitter, receiver and transceiver RF daughterboards that are optimised to work at various different frequencies. By default, the FPGA on the Tx side of the USRP® radio is configured to perform Digital Upconversion (DUC) to upconvert baseband IQ samples (transferred to the device from a host computer) to an IF. The samples are converted to continuous signals using a DAC, and then mixed with a quadrature RF carrier, performing AM-DSB-SC modulation as is shown in Figure 8.1. The RF signal output by the USRP® hardware can be generalised as follows:

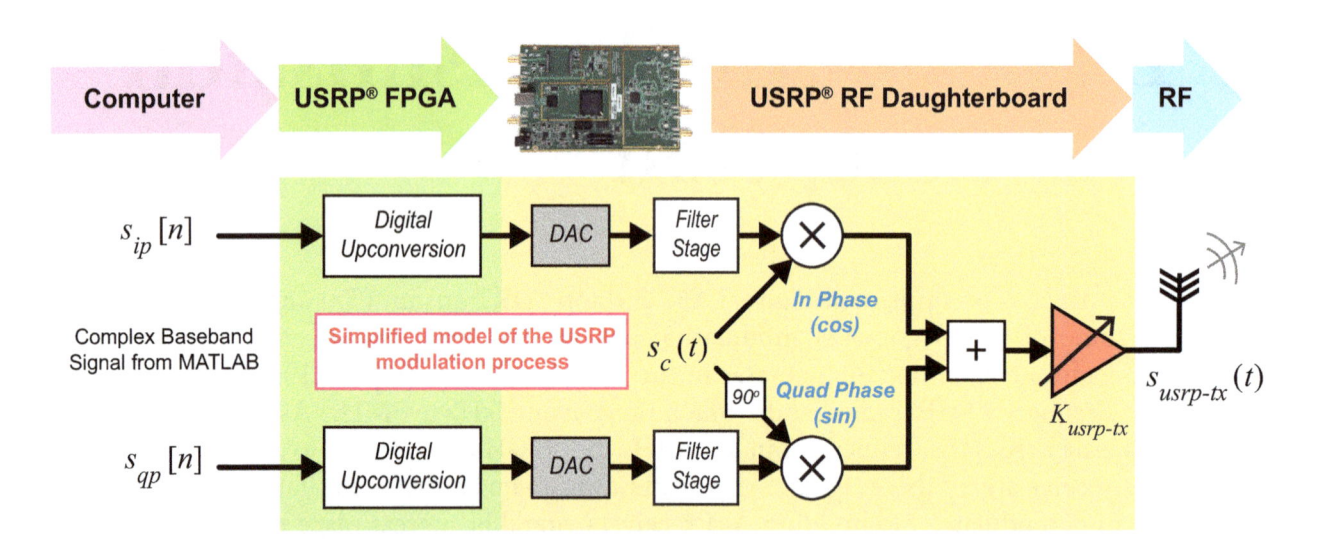

$$s_{usrp-tx}(t) = K_{usrp-tx}\left[s_{ip}(t)\cos(\omega_c t) + s_{qp}(t)\sin(\omega_c t) \right] \tag{8.1}$$

Figure 8.1: Block diagram highlighting the main processes that happen on the Tx side of the USRP® FPGA and RF daughterboard

One of the advantages of using a USRP® radio is that, like the RTL-SDR, it is fully supported by MathWorks. Downloading and installing its Hardware Support Package will install the USRP® Hardware Driver (UHD™) to your computer, and enable interfacing with the device directly from MATLAB and Simulink [66]. This add-on works with a number of the more popular USRP® hardware models, supporting USB3.0, GbE and 10GbE device connections.

The USRP® is more expensive than the RTL-SDR (however one device in a lab can transmits signals to a class of students, so it is a worthwhile acquisition). In this book, we have used the *USRP® N210 transceiver* (Figure 8.2) and the *WBX 50-2200MHz daughtercard* [87], but alternative models would be equally valid. Using this device, we have a MATLAB and Simulink controllable SDR transmitter (and receiver too).

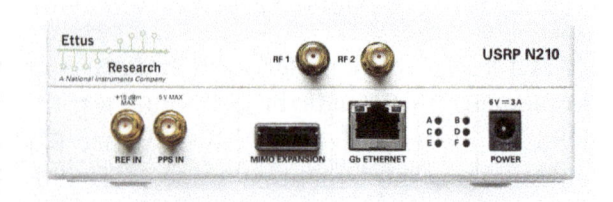

Figure 8.2: The USRP® N210

Chapter 8: Desktop AM Transmission and Reception 281

Before building any USRP® AM transmitter Simulink models, ensure you have the MathWorks USRP® Hardware Support Package installed. Details on how to do this can be found in Appendix A.2 (page 571).

8.1.1 Generating AM-DSB-SC Signals

Recall from Section 6.2 (page 203) that the AM-DSB-SC modulator is the simplest of all the AM modulators, as it only involves the mixing of an information signal with a carrier wave. As the USRP® hardware is essentially an AM-DSB-SC modulator, generation of an AM-DSB-SC signal requires only that a real (non-complex) baseband information signal is passed to it, as shown in Figure 8.3.

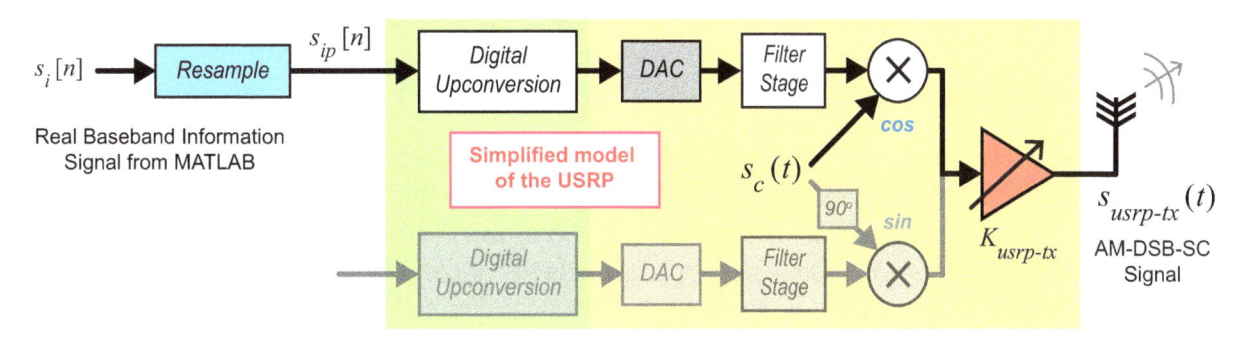

Figure 8.3: Block diagram showing Simulink/ USRP® hardware implementation of AM-DSB-SC modulator

With this in mind, constructing an AM-DSB-SC modulator is the most sensible place to start.

Exercise 8.1 USRP® Radio: AM-DSB-SC Modulator and Transmitter

In this exercise we will build a Simulink model that AM-DSB-SC modulates an audio file from a host computer onto a carrier using the USRP® hardware. In later exercises, the signal generated by this model will be received and demodulated using the RTL-SDR and Simulink.

Note: If a USRP® radio is not available, you can still complete this exercise by commenting out (or not placing) the SDRu Transmitter block.

(a) **Open MATLAB.** Set the working directory to the exercise folder,

   /my_models/transmitters/

then create a new Simulink model. Save this file with the name:

   .../usrp_tx_am_dsb_sc.slx

(b) **Place an audio source.** Navigate to ▦ > *DSP System Toolbox* > *Sources* and find the *From Multimedia File* block. Place one into the model, then ②on the block to open its parameter window. ①on the 'Browse' button, then navigate to:

   /audio_sources

and select one of the mono (..._mono48kHz) audio sources from this folder (it is preferable to use these files as they are sampled at 48kHz). Change the 'Samples per audio channel' to

'600', and on opening the 'Data Types' tab, change the 'Audio output data type' to 'single'. Apply the changes, then close the parameters window.

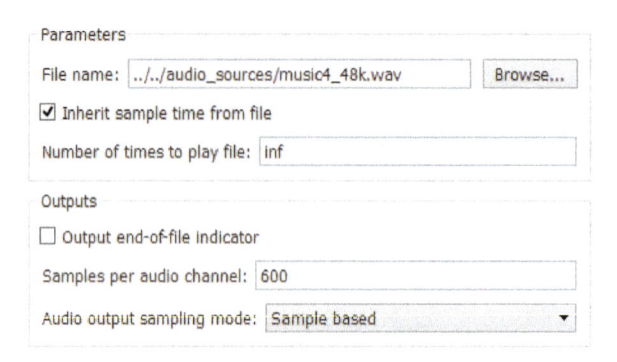

Parameters
- File name: ../../audio_sources/music4_48k.wav [Browse...]
- ☑ Inherit sample time from file
- Number of times to play file: inf

Outputs
- ☐ Output end-of-file indicator
- Samples per audio channel: 600
- Audio output sampling mode: Sample based

(c) **Implement the resampler.** Find and place two *FIR Rate Conversion* blocks from ▸ **DSP System Toolbox > Filtering > Multirate Filters**. This block is capable of decimating or interpolating data applied to its input port, which allows the sampling frequency to be changed by a non-integer factor. It also performs lowpass filtering of the data, to ensure that no aliasing or imaging occurs (see Appendix C for clarification of these terms, and background on multirate filtering in general). Opening the parameters window of the first block, change the 'Interpolation factor', 'FIR filter coefficients' and 'Decimation factor' to the following:

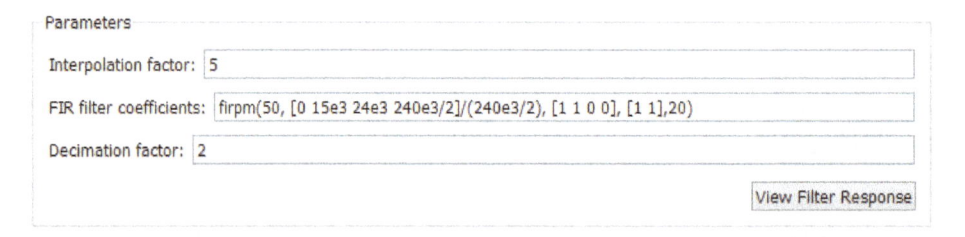

Parameters
- Interpolation factor: 5
- FIR filter coefficients: firpm(50, [0 15e3 24e3 240e3/2]/(240e3/2), [1 1 0 0], [1 1],20)
- Decimation factor: 2

[View Filter Response]

There parameters specify a desired interpolation by a factor of 5, and decimation by a factor of 2; i.e. a rate change from 48kHz to 120kHz. The 'FIR filter coefficients' are designed using the `firpm` MATLAB function (more information about this function can be found by typing 'doc firpm' into the MATLAB command window). View the filter response of the lowpass filter by [1] on the 'View Filter Response' button, and check that you understand how it as been designed. Save the changes, and then rename the block *FIR Rate Conversion o/p fs=120kHz* by [1] on the text under the block.

(d) Next, open the second *FIR Rate Conversion* block and change its parameters to the following:

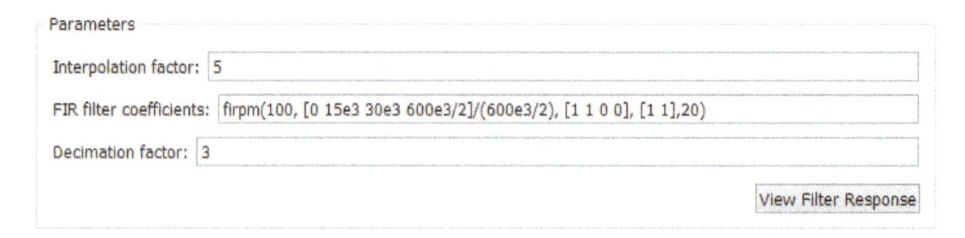

Parameters
- Interpolation factor: 5
- FIR filter coefficients: firpm(100, [0 15e3 30e3 600e3/2]/(600e3/2), [1 1 0 0], [1 1],20)
- Decimation factor: 3

[View Filter Response]

Here, we are changing the sampling rate from 120kHz to 200kHz. For clarity, rename this block *FIR Rate Conversion o/p fs=200kHz*.

Chapter 8: Desktop AM Transmission and Reception

(e) **Interface with the USRP® radio.** Fetch an *SDRu Transmitter* block from the ▦ > *Communications System Toolbox Support Package for USRP® Radio* library. Prior to use, the block needs to be configured to communicate with connected USRP® hardware. To do this, you will need to know its IP address (or USB address), and be able to 'ping' it from your computer. Instructions on how to set up this connection are given in Appendix A.2. 🖱 on the *SDRu Transmitter* block, select the appropriate 'Platform', and type the 'Address' of your USRP® radio.

(f) The RF side of the hardware also needs to be configured. Set the 'Center Frequency' parameter to the desired frequency. This value should be within the range of your RTL-SDR's tuner, e.g. in the range 25MHz–1.75GHz. If, for example, a signal is to be modulated onto a 433.9MHz carrier, you would enter '433.9e6' in this field. Set the 'LO offset' to '250e3' and the 'Interpolation' factor to '500'. By setting the local oscillator of the USRP® hardware to 250kHz, we can ensure that any harmonics it generates will not interfere with the AM signal. Choosing the value of '500' for the interpolation function means that the radio resamples the AM signal from 200kHz to 100MHz before modulating it onto the carrier.

(g) Change the 'Source' of the 'Gain' to 'Input Port', then apply the changes and close the window.

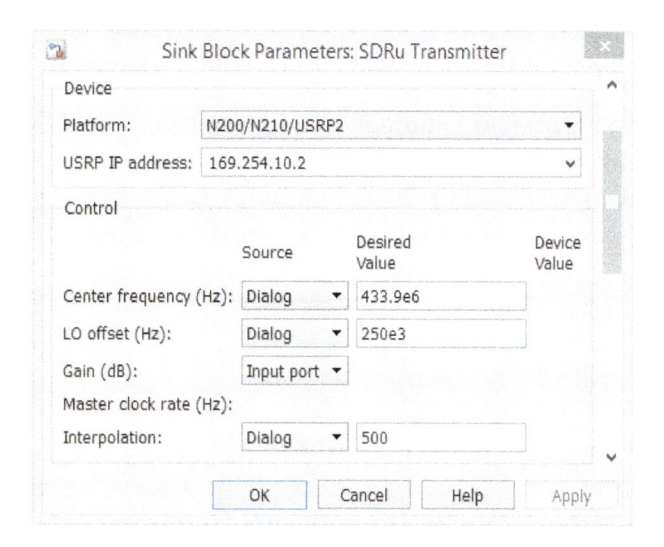

Finally, place a *Constant* from ▦ > *Simulink* > *Sources*. Set the 'Constant Value' to '10' and rename this block *Transmitter Gain (dB)*.

(h) **Connect up the blocks.** Connect the blocks up in the order shown below.

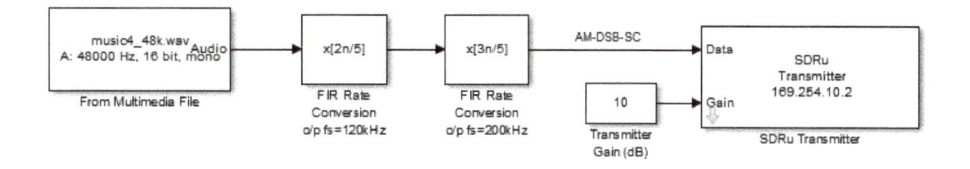

(i) **Prepare to run the simulation.** Set the *Simulation Stop Time* to 'inf' and change the *Simulation Mode* to 'Accelerator'. This will force Simulink to partially compile the model into native code for your computer, which will enable it to run faster. Finally, make sure your model is saved.

(j) **Open the reference file.** If you wish, open the following reference file to check your design (or use it as a shortcut!). Make sure that the *SDRu Transmitter* block is appropriately configured.

　　`/am/usrp_tx/usrp_am_dsb_sc.slx`

(k) Check that the USRP® hardware is turned on, and that MATLAB can communicate with it, by typing `findsdru` into the MATLAB command window. If a structure is returned as discussed in Appendix A.2, you can continue.

(l) **Run the simulation.** Begin the simulation by clicking on the 'Run' button in the Simulink toolbar. After a couple of seconds, Simulink will establish a connection with the USRP® radio and the simulation will begin. When it does, samples of the baseband information signal will be passed to the radio, modulated onto the AM carrier, and transmitted from its antenna.

Note that it is not possible to tune a consumer analogue AM radio to this signal, as its carrier frequency is well outside the normal operating range for this type of device. The RTL-SDR is however capable of receiving the signal, and we will later construct AM demodulators using the RTL-SDR that will output the audio signal transmitted in Sections 8.2 and 8.3.

8.1.2 Generating AM-DSB-TC Signals

The AM-DSB-TC modulator is very similar to the AM-DSB-SC modulator, however a DC offset is applied to the information signal before it is mixed with the carrier wave (as discussed in Section 6.3, page 210). This means we simply modify the implementation as shown in Figure 8.4.

In the next exercise we will construct this modulator by modifying the AM-DSB-SC modulator created in Exercise 8.1.

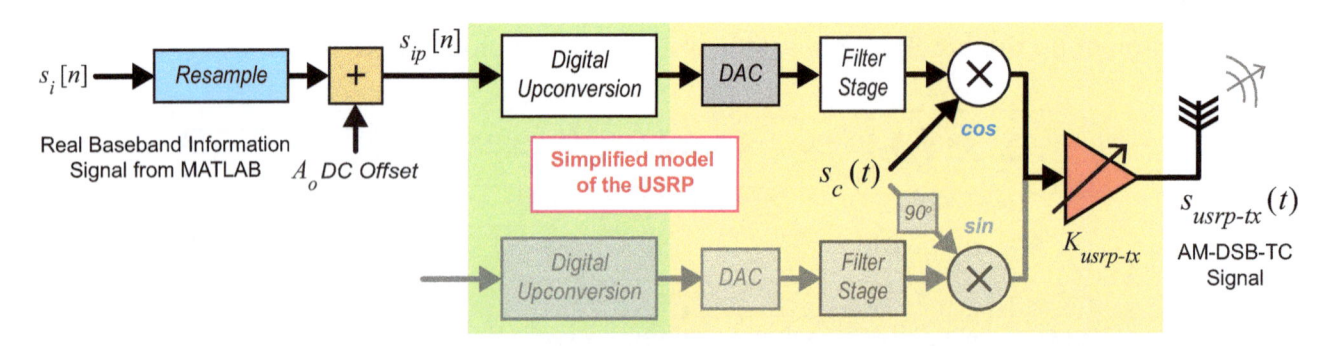

Figure 8.4: Block diagram showing Simulink/ USRP® hardware implementation of an AM-DSB-TC modulator

Chapter 8: Desktop AM Transmission and Reception

Exercise 8.2 USRP® Radio: AM-DSB-TC Modulator and Transmitter

Next we will modify the Simulink model created in Exercise 8.1 and AM-DSB-TC modulate audio output from a computer onto a carrier using the USRP® hardware. We will receive and demodulate the signal this generates in later exercises, using the RTL-SDR and Simulink.

Note: If you do not have access to a USRP® radio, this exercise can still be completed by commenting out (or not placing) the SDRu Transmitter block.

(a) **Open MATLAB.** Set the working directory to the exercise folder,

   /my_models/transmitters/

then create a copy of your AM-DSB-SC modulator model,

   .../usrp_tx_am_dsb_sc.slx

and rename it:

   .../usrp_tx_am_dsb_tc.slx

(b) **Modify the AM-DSB-SC modulator to convert it to an AM-DSB-TC modulator.** Begin by deleting the connection between the *FIR Rate Conversion o/p fs=200kHz* block and the *SDRu Transmitter* block. Move the two parts away from each other as follows:

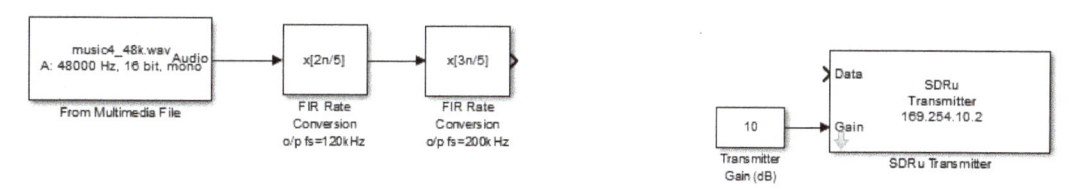

(c) Open the Simulink Library Browser and place an *Add* block from ▦ > **Simulink > Math Operations**, then a *Constant* from ▦ > **Simulink > Sources**. Set the 'Constant value' to '0.05' and rename the block *Ao DC Offset*. Connect these blocks as follows:

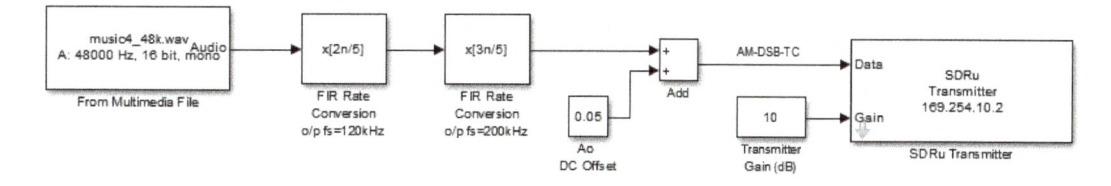

(d) **Check the USRP® radio settings.** The *SDRu Transmitter* block should be configured to communicate with your USRP® hardware (if it is not already set up). To do this, you will need to know its IP address (or USB address), and be able to 'ping' it from a computer. Guidance on setting up this connection is provided in Appendix A.2. ② on the *SDRu Transmitter* block, check that the appropriate 'Platform', and 'Address' of your USRP® hardware are entered. Make sure that the 'Center Frequency' parameter is set to the desired carrier frequency. This value should be within the range of the RTL-SDR's tuner, e.g. in the range 25MHz–1.75GHz.

(e) **Prepare to run the simulation.** Check that the *Simulation Stop Time* is set to 'inf' and that the *Simulation Mode* is set to 'Accelerator'. This will force Simulink to partially compile the model into native code for the computer, which will allow the model to run faster. Finally, make sure your model is saved.

(f) **Open the reference file.** If you do not have time to construct the transmitter yourself, open the following file. Make sure that the *SDRu Transmitter* block is appropriately configured.

/am/usrp_tx/usrp_am_dsb_tc.slx

(g) Check that the USRP® hardware is turned on, and that MATLAB can communicate with it, by typing `findsdru` into the MATLAB command window. If a structure is returned as discussed in Appendix A.2, continue to the next step.

(h) **Run the simulation.** Begin the simulation by clicking on the 'Run' button in the Simulink toolbar. After a couple of seconds, Simulink will establish a connection with the USRP® radio and the simulation will begin. When it does, samples of the baseband information signal will be passed to the radio, modulated onto the AM carrier, and transmitted from its antenna.

8.1.3 Generating AM-SSB signals

As was discussed in Section 6.4 (page 217), there are two variations of AM-SSB: AM-SUSB and AM-SLSB. Both of these signals are generated with a special type of modulator, which outputs the sum of its I and Q components. This process results in a real signal (as shown in orange in Figure 8.5) which, when plotted in the double sided spectrum, would contain the SSB component (USB, upper sideband; or LSB, lower sideband) at both the positive and negative frequencies of the complex carrier. The negative component would be a mirror image of the positive component, essentially meaning that the signal still in fact contains both lower and upper sidebands (they are just separated). If this signal was input to an *SDRu Transmitter* block, modulated by the USRP® radio's complex carrier and transmitted, a signal like the one shown in green in Figure 8.5 would be created.

Recall the equation of the AM-SSB modulator from Eq. (6.18) on page 218:

$$s_{am-ssb}(t) = s_i(t) \, \Re e\!\left[s_c(t)\right] \mp \overline{s_i(t)} \, \Im m\!\left[s_c(t)\right], \text{ where } \overline{s_i(t)} \text{ is the Hilbert transform of } s_i(t).$$

When this is input to the USRP® hardware, its output could be represented as:

$$s_{usrp-tx}(t) = K_{usrp-tx}\left[\left(s_i(t)\,\Re e\!\left[s_c(t)\right] \mp \overline{s_i(t)}\,\Im m\!\left[s_c(t)\right]\right)\cos(\omega_c t)\right]. \tag{8.2}$$

Rather than redesign the modulator to force it to only output one SSB component, this behaviour can be exploited — we can test demodulators for both AM-SUSB and AM-SLSB signals at the same time!

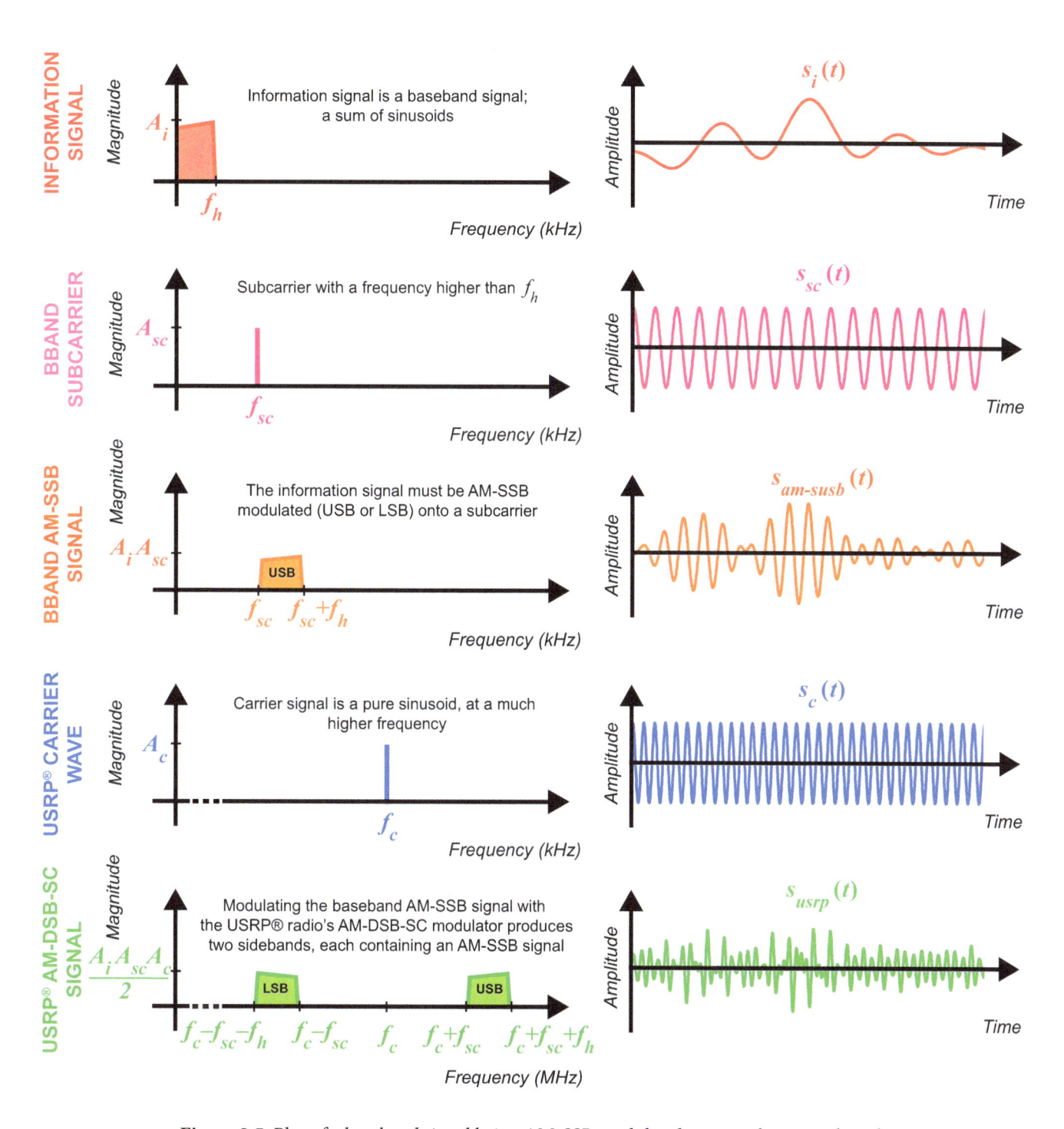

Figure 8.5: Plot of a baseband signal being AM-SSB modulated onto a subcarrier, then the AM-SSB signal being AM-DSB-SC modulated onto the USRP® radio's carrier

In the following exercise you will get the chance to generate this unusual signal by modifying the AM-DSB-SC modulator you made in Exercise 8.1. AM-SUSB modulating a baseband signal onto a subcarrier, and outputting this signal to USRP® hardware, a signal comprising of AM-SLSB and AM-SUSB components will be generated and transmitted. The block diagram in Figure 8.6 shows the order in which processes must be carried out to produce this signal.

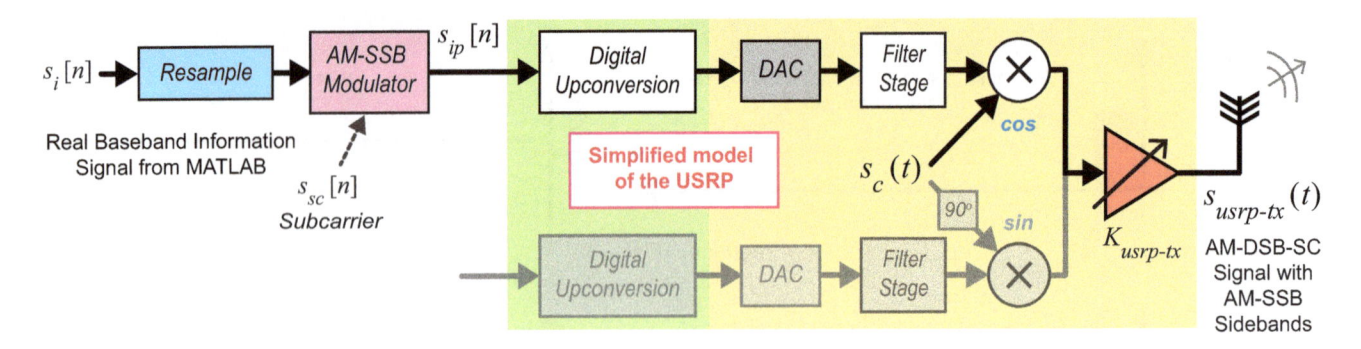

Figure 8.6: Block diagram showing Simulink/ USRP® hardware implementation of AM-SSB modulator

---

Exercise 8.3 USRP® Radio: AM-SSB Modulator and Transmitter

In this exercise you will modify the Simulink model you made in Exercise 8.1 to make it AM-SUSB modulate an audio file from your computer onto a suppressed subcarrier, then AM-DSB-SC modulate this onto an RF carrier using the USRP® radio. We will receive and demodulate the signal this generates in later exercises, using the RTL-SDR and Simulink.

Note: If you do not have a USRP® radio, you can still complete this exercise by commenting out (or not placing) the SDRu Transmitter block. Unlike Exercises 8.1 and 8.2, a modulation process is carried out in this Simulink model, so there is slightly more to see!

(a) **Open MATLAB.** Set the working directory to the exercise folder,

 `/my_models/transmitters/`

then create a copy of your AM-DSB-SC modulator model,

 `.../usrp_tx_am_dsb_sc.slx`

and rename it:

 `.../usrp_tx_am_ssb.slx`

(b) **Modify the AM-DSB-SC modulator to convert it to an AM-SSB modulator.** Begin by deleting the connection between the two *FIR Rate Conversion* blocks, and then move the two parts of model away from each other as follows:

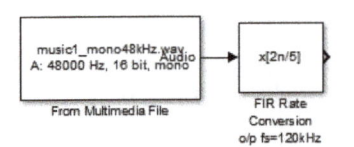

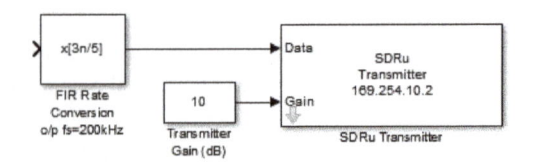

(c) Open the Simulink Library Browser and place a *Lowpass Filter* from ▦ > **DSP System Toolbox > Filtering > Filter Designs**, and an *AM-SSB Modulator* block from ▦ > **RTL-SDR Book Library > Additional Tools**. Connect these blocks up as follows:

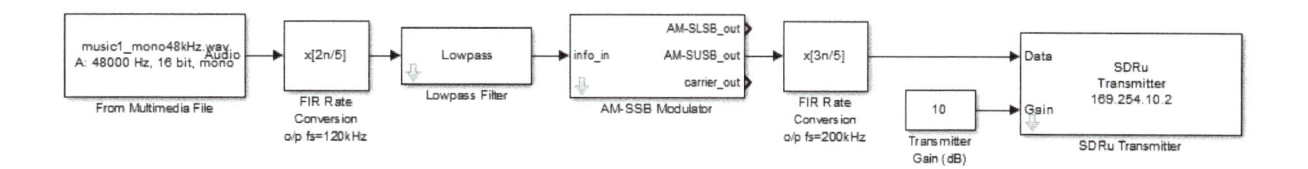

Chapter 8: Desktop AM Transmission and Reception

(d) 🖱2 on the *Lowpass Filter* to open its parameters window and configure it as shown below. Rename the block *IIR Lowpass Filter fc=15kHz*.

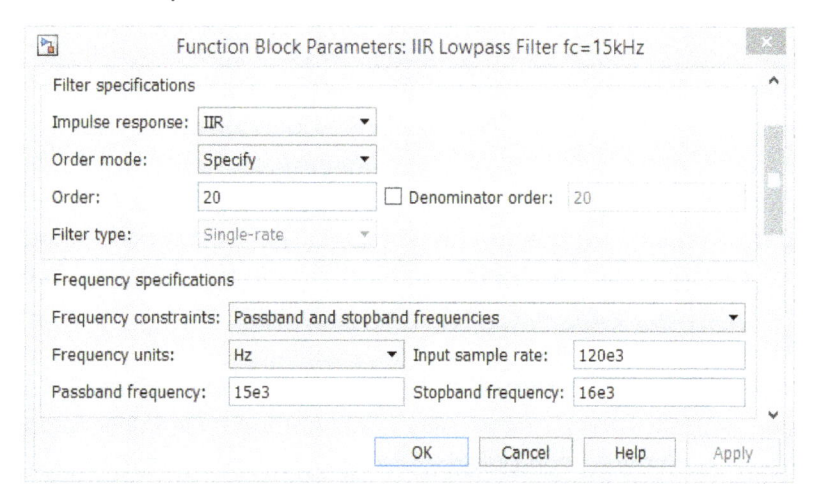

Function Block Parameters: IIR Lowpass Filter fc=15kHz

Filter specifications
- Impulse response: IIR
- Order mode: Specify
- Order: 20   ☐ Denominator order: 20
- Filter type: Single-rate

Frequency specifications
- Frequency constraints: Passband and stopband frequencies
- Frequency units: Hz   Input sample rate: 120e3
- Passband frequency: 15e3   Stopband frequency: 16e3

(e) 🖱2 on the *AM-SSB Modulator* block, and check that it is configured accordingly. If it is not, modify the values set in the mask and apply the changes.

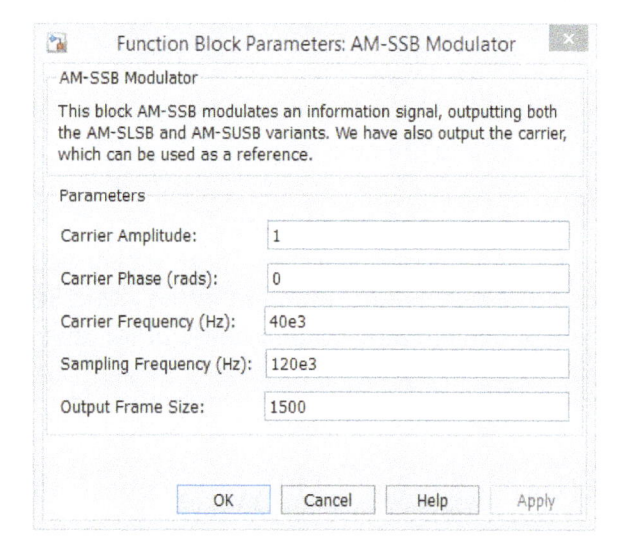

Function Block Parameters: AM-SSB Modulator

AM-SSB Modulator

This block AM-SSB modulates an information signal, outputting both the AM-SLSB and AM-SUSB variants. We have also output the carrier, which can be used as a reference.

Parameters
- Carrier Amplitude: 1
- Carrier Phase (rads): 0
- Carrier Frequency (Hz): 40e3
- Sampling Frequency (Hz): 120e3
- Output Frame Size: 1500

The AM-SSB modulator will now be configured to modulate a baseband signal onto a carrier with a frequency of 40kHz. As the modulator takes the same form as the one presented in Exercise 6.3, it will output a real signal, meaning that the negative component of its double sided spectrum will contain a mirror image of the positive component.

(f) Finally, 🖱2 on the *FIR Rate Conversion o/p fs=200kHz* block and change the 'FIR filter coefficients' to the following. (Note: the change is required because the signal input to this block contains information between 40kHz and 55kHz, which would be otherwise be filtered out!)

```
Parameters
Interpolation factor:    5
FIR filter coefficients: firpm(400, [0 55e3 65e3 600e3/2]/(600e3/2), [1 1 0 0], [1 1],20)
Decimation factor:       3
                                                                      View Filter Response
```

(g) **Check the USRP® radio settings.** The *SDRu Transmitter* block should be configured to communicate with the USRP® hardware. To do this, you will need to know its IP address (or USB address), and be able to 'ping' it from your computer. Instructions of how to set up this connection are discussed in Appendix A.2. 🖱 on the *SDRu Transmitter* block, check that the appropriate 'Platform', and 'Address' of your USRP® radio have been entered. Make sure that the 'Center Frequency' parameter is set to the desired transmit frequency, which should be within the range of your RTL-SDR's tuner, e.g. in the range 25MHz–1.75GHz.

(h) **Prepare to run the simulation.** Check that the *Simulation Stop Time* is set to 'inf' and that the *Simulation Mode* is set to 'Accelerator'. This will force Simulink to partially compile the model into native code for your computer, which allows it to run faster. Finally, save the model.

(i) **Open the reference file.** As an alternative to building the model, open the following file:

📁 `/am/usrp_tx/usrp_am_ssb.slx`

(j) Check that the USRP® hardware is turned on and that MATLAB can communicate with it by typing `findsdru` into the MATLAB command window. If a structure is returned as discussed in Appendix A.2, proceed to the next step.

(k) **Run the simulation.** Begin the simulation by 🖱 on the 'Run' ▶ button in the Simulink toolbar. After a couple of seconds, Simulink will establish a connection with the USRP® radio and the simulation will begin. When it does, samples of the AM-SSB baseband information signal will be passed to the radio, modulated onto the AM carrier, and transmitted from its antenna.

8.2 Implementing Non-Coherent AM Receivers with the RTL-SDR

Having discussed how to generate and transmit AM signals using the USRP® hardware and Simulink, we will now focus on implementing non-coherent AM receivers using the RTL-SDR. In this section there are three exercises. The first two of these will build and test envelope detectors to demodulate AM-DSB-TC signals in both MATLAB and Simulink, and in the third, we investigate what happens when you try to demodulate an AM-DSB-SC signal with an envelope detector.

Before attempting these exercises, it is worthwhile to recap on the process of receiving signals with the RTL-SDR. When the RTL-SDR is tuned to a particular frequency band, it first internally performs AM demodulation to downconvert RF signals first to an IF via the Rafael device in Figure 1.8 (page 13) and

Chapter 8: Desktop AM Transmission and Reception

then to complex baseband using the R820T. Hence, if the RTL-SDR was tuned to the exact centre frequency of an AM signal, the baseband signal output by the device would be the information signal, at baseband, and sampled at a rate of 2.4MHz (for example); you can review this in Section 6.7 (page 227).

The issue here is that, if we do not tune the RTL-SDR to exactly the centre frequency of the AM RF carrier signal, the 'baseband' signal output from the device will actually still be an AM signal, modulated but on a carrier with a very low frequency — perhaps near or at DC (0 Hz). This poses a problem, as non-coherent demodulators do not work without a high frequency carrier to create the information signal envelope (Figure 6.3, page 205). Here we consider '*high*' to be typically an order of magnitude greater than the signal bandwidth. Coherent receivers are discussed later in this chapter.

Obviously this situation is not ideal, but it can be addressed easily: we can introduce a frequency offset by deliberately tuning the RTL-SDR to a frequency *lower* than the centre of the AM signal,

$$f_{c\,(rtl-sdr\,tuner)} = f_{c\,(am\,signal)} - f_{offset} \;. \tag{8.3}$$

As long as the offset is in the range ($0 < f_{offset} < f_s/2$), the AM signal will reside at an IF (f_{offset}) inside the baseband signal output by the RTL-SDR. If the IF is a high value, e.g. 40kHz, the AM signal can then be demodulated non-coherently, as is demonstrated in Figure 8.7.

To provide an example of this in action:

You want to non-coherently demodulate an AM signal that is centred at 400MHz, and have decided on an offset of 40kHz. The first thing you do is to tune the RTL-SDR to 399.96MHz, which downconverts a band

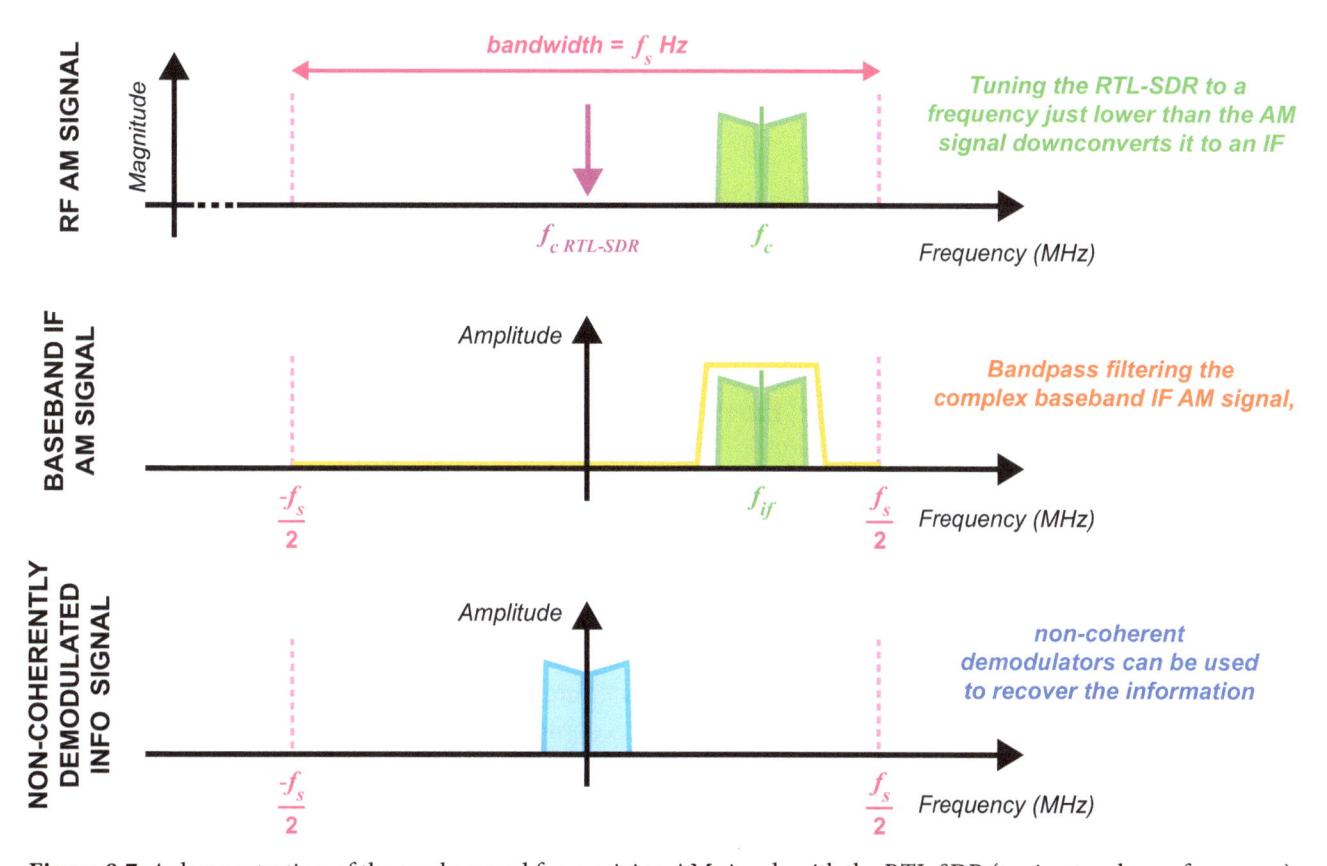

Figure 8.7: A demonstration of the workaround for receiving AM signals with the RTL-SDR (tuning to a lower frequency)

f_s Hz wide to baseband. The AM signal will be in this band, centred around 40kHz. This represents its IF. Bandpass filtering around the IF AM signal, a non-coherent AM demodulator can be used to complete the demodulation.

If this process is followed, there should be no problems using non-coherent AM demodulators to demodulate signals received by the RTL-SDR. We will test this out in Exercise 8.4, where we construct an envelope detector based non-coherent AM receiver. As discussed at the start of this chapter, it will not be possible to receive any broadcast AM signals with the RTL-SDR, so this is a situation where you will need to transmit your own.

Before starting to build any AM receiver Simulink models, make sure you have the MathWorks RTL-SDR Hardware Support Package installed. Details on how to do this can be found in Appendix A.1 (page 569).

Exercise 8.4 RTL-SDR: Envelope Detector for AM-DSB-TC Signals

In this exercise we will construct an AM receiver that uses a Complex Envelope Detector (as discussed in Section 6.8.3). Creating a new Simulink model, we will place components from the Simulink libraries that interface with the RTL-SDR and implement the demodulator. The receiver will be designed to receive an AM-DSB-TC audio signal, and will output the demodulated audio information to the computer's speakers or headphones.

Note: If you do not have an RTL-SDR, or are unable to transmit an AM-DSB-TC signal that you can receive, it is still possible to complete this exercise by substituting the RTL-SDR Receiver block with an Import RTL-SDR Data block as outlined below.

(a) **Open MATLAB.** Set the working directory to the exercise folder,

   /my_models/receivers/

   Next, create a new Simulink model. Save this file with the name:

   .../rtlsdr_am_envelope_demod.slx

(b) Open this file, and then the Simulink Library Browser.

(c) **Place an *RTL-SDR Receiver* block.** If you have an RTL-SDR available and are able to transmit an AM-DSB-TC signal, place the *RTL-SDR Receiver* block from ▦ > **Communications System Toolbox Support Package for RTL-SDR Radio**. 2️⃣ on it to open its parameter window, and change the 'Source' of 'Center frequency' and 'Tuner gain' to 'Input Port'. Enter '240e3' in the 'Sampling Rate' field to set the RTL-SDR to sample at a rate of 240kHz. Set the 'Output data type' dropdown to 'single' and enter '4096' in the 'Samples per frame' box.

If a single RTL-SDR is connected to your computer, the 'Radio address' can be left as '0'. If more than one is connected, run the `sdrinfo` command in the MATLAB command window to find the ID of the stick you desire to use. Apply these changes and then close the window. Rename this block *RTL-SDR Receiver o/p fs=240kHz* by 1️⃣ on the text under the block.

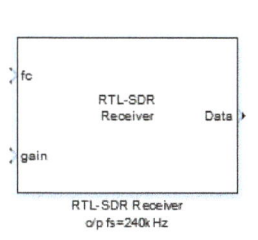

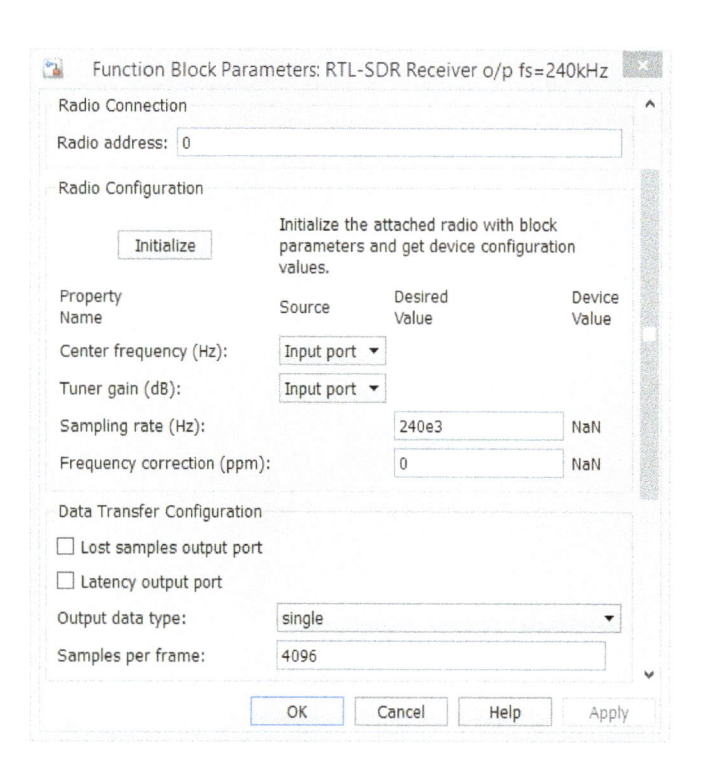

Chapter 8: Desktop AM Transmission and Reception 293

(d) Place three *Constant* blocks from ▦ > *Simulink* > *Sources*, and then an *Add* block from ▦ > *Simulink* > *Math Operations*. Modify the names of the constant blocks to *AM Signal Frequency (Hz)*, *Offset Frequency (Hz)* and *Tuner Gain (dB)*.

Set the 'Constant value' of *AM Signal Frequency* to the centre frequency of the AM signal to be received — for example, '433.9e6' for 433.9MHz. Enter '-40e3' for the value of the *Offset Frequency* to set the offset at 40kHz. Set the *Tuner Gain* block to a default value of '30' (this may need to be adjusted later on, depending on the strength of the received signal).

Connect these blocks up as follows:

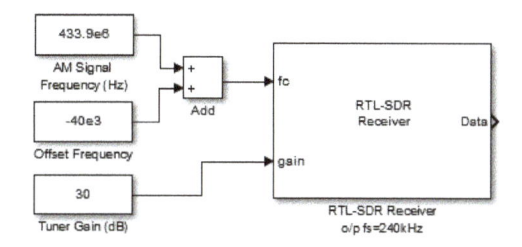

(e) **Place an *Import RTL-SDR Data* block.** If you do not have an RTL-SDR or are unable to transmit an AM-DSB-TC signal, navigate to ▦ > *RTL-SDR Book Library* > *Additional Tools* and place an *Import RTL-SDR Data* block. ② on this and change the 'File Name' parameter to reference the file:

📁 /am/rtlsdr_rx/rec_data/am_dsb_tc.mat

Leave the 'Output Frame Size' set at '4096', apply the changes and close the parameter window. If this process was successful, the block should show the sampling frequency of the recorded signal in the bottom right hand corner (in this instance, a frequency of 240kHz).

The signal output from this block should be equivalent to the signal output by the RTL-SDR configuration shown above. When this signal was recorded, the RTL-SDR was tuned to 433.96MHz, and the offset frequency was set at 40kHz.

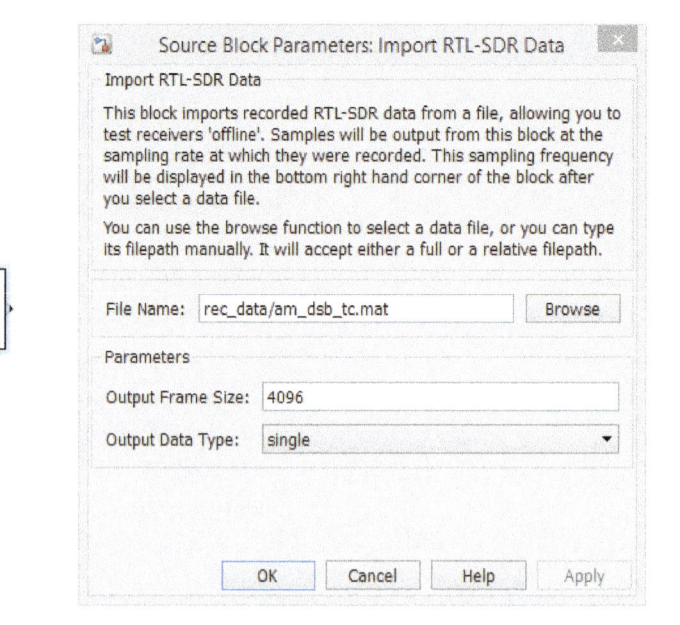

(f) **Place and configure blocks required to implement the demodulator.** Place a *Bandpass Filter* from > *DSP System Toolbox* > *Filtering* > *Filter Designs*. Open its parameters window and change the settings to the following:

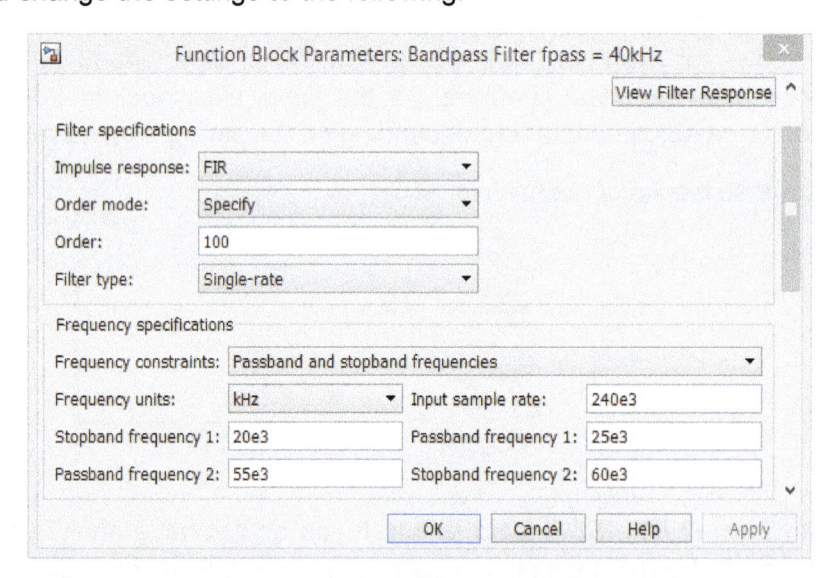

This sets the filter to pass frequencies between 25kHz and 55kHz, meaning that only the IF AM-DSB-TC signal is allowed through. Rename this *Bandpass Filter fpass = 40kHz*.

(g) Open > *Simulink* > *Math Operations* and find the *Abs* block. Place one of these in your model. This block outputs the magnitude of any signal input to it, and as discussed in Section 6.8.3, taking the magnitude of a complex AM-DSB-TC signal detects its envelope and outputs the demodulated information signal.

Chapter 8: Desktop AM Transmission and Reception

(h) Next, find the *FIR Decimation* block from ▦ > **DSP System Toolbox > Filtering > Multirate Filters**. This block decimates the data that is applied to its input port, allowing you to reduce the sampling frequency by an integer factor. It also performs lowpass filtering of the data, to ensure that no aliasing occurs. Place one of these in your model and open its parameters window. Change the 'FIR filter coefficients' and 'Decimation Factor' to the following:

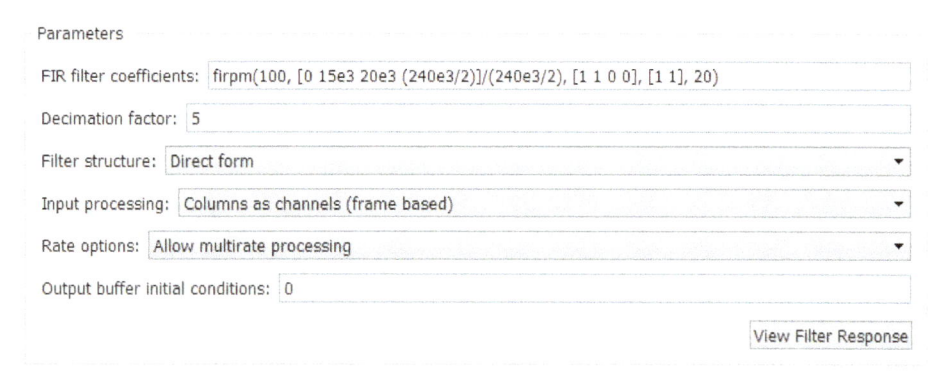

```
Parameters
FIR filter coefficients:  firpm(100, [0 15e3 20e3 (240e3/2)]/(240e3/2), [1 1 0 0], [1 1], 20)
Decimation factor: 5
Filter structure: Direct form
Input processing: Columns as channels (frame based)
Rate options: Allow multirate processing
Output buffer initial conditions: 0
                                                             View Filter Response
```

This configures the decimation factor to 5 (i.e. a rate change from 240kHz to 48kHz), and to pass frequencies up to 15kHz. Select 'Allow multirate processing' in the 'Rate options' dropdown menu, and apply the changes. Rename this block *FIR Decimation o/p fs=48kHz*.

(i) **Connect the blocks up.** Connect the blocks up as shown below.

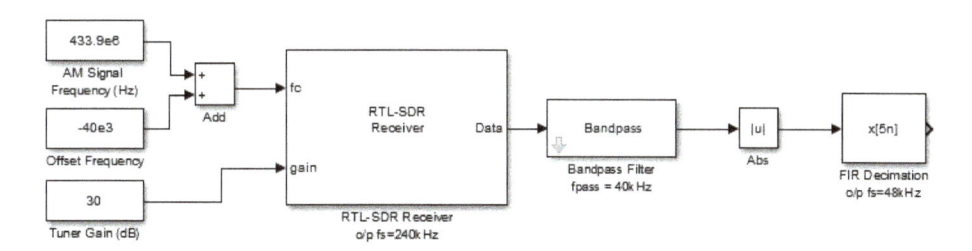

...or for an offline receiver

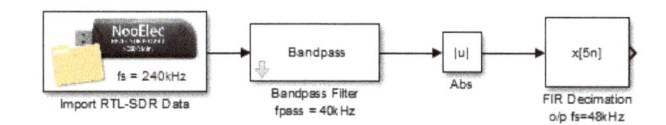

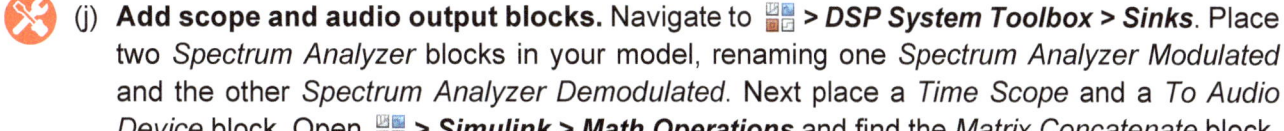

(j) **Add scope and audio output blocks.** Navigate to ▦ > **DSP System Toolbox > Sinks**. Place two *Spectrum Analyzer* blocks in your model, renaming one *Spectrum Analyzer Modulated* and the other *Spectrum Analyzer Demodulated*. Next place a *Time Scope* and a *To Audio Device* block. Open ▦ > **Simulink > Math Operations** and find the *Matrix Concatenate* block. Place this in your model, and then append all of these to the block diagram as shown in the next screenshot.

2⃣ on each of the input connections to the *Matrix Concatenate* block and give the signals meaningful names. These will remain with the signals as they are concatenated and show up in the legend of the *Spectrum Analyzer*.

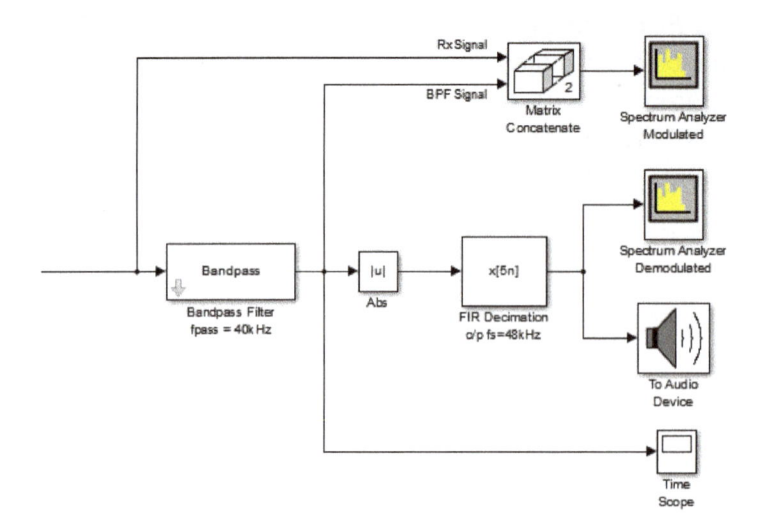

(k) **2×click** on the *Time Scope* to open it, and then navigate to 'Configuration Properties'. Open the 'Time' tab and change the 'Time span' to '512/240e3'. This will limit the scope to showing only 512 individual samples. Apply the changes then close the window.

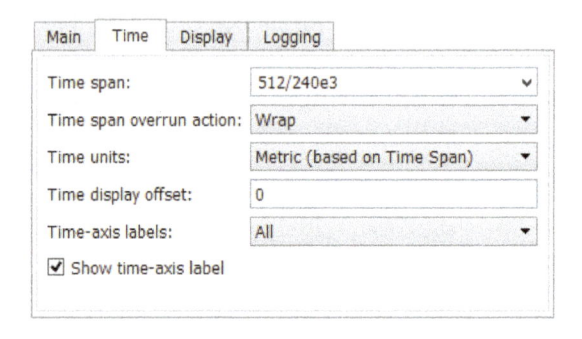

(l) **Open the reference file.** As an alternative to constructing the receiver, open the following file:

/am/rtlsdr_rx/rtlsdr_am_envelope_demod.slx

If importing data rather than connecting to an RTL-SDR, delete the *RTL-SDR Receiver* block and connect the *Import RTL-SDR Data* block in its place. Check that this block is referencing the file discussed above.

(m) **Prepare to run the simulation.** Connect speakers or headphones to your computer and test to ensure that they are working. If importing a signal, set the *Simulation Stop Time* to 60 seconds, '60'. If using an RTL-SDR, set the *Simulation Stop Time* to 'inf' by typing this into the Simulink toolbar, and the *Simulation Mode* to 'Accelerator'. This will force Simulink to partially compile the model into native code for your computer, which allows it to run faster. Check that MATLAB can communicate with it the RTL-SDR by typing `sdrinfo` into the MATLAB command window. Finally, make sure the model is saved.

(n) **Run the simulation.** Begin the simulation by **1×click** on the 'Run' ▶ button in the Simulink toolbar. After a couple of seconds the simulation will begin, the *Spectrum Analyzer* and *Time Scope* windows should appear, and the demodulated AM signal should be audible. If using an RTL-

Chapter 8: Desktop AM Transmission and Reception

SDR, adjust the frequency and gain values until the device is tuned to the desired signal Remember that the signal must be within the passband of the bandpass filter; located around 40kHz in the baseband signal.

(o) **Signal Analysis: *Spectrum Analyzers*.** The *Spectrum Analyzer* windows permit monitoring of the signal as it is demodulated in the frequency domain. You should see the spectra of two signals plotted in *Spectrum Analyzer Modulated*. Turn on the legend by navigating to *View*, *Configuration Properties* and ticking the 'Show Legend' box.

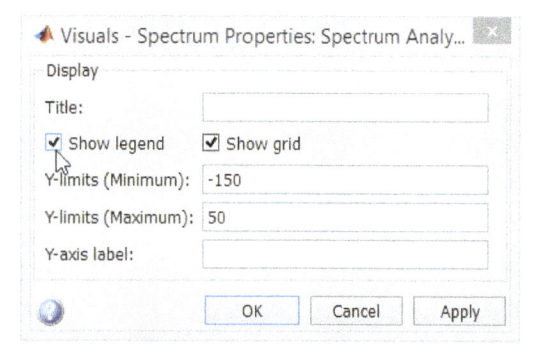

The blue line shows the spectrum of the signal received from the RTL-SDR, and the orange shows it after bandpass filtering. The AM signal is modulated at roughly 40kHz, which should be a sufficiently high frequency for the envelope detector to function correctly. *Spectrum Analyzer Demodulated* shows the spectrum of the demodulated signal, after envelope detection. Here, note that the information has been successfully shifted back to baseband.

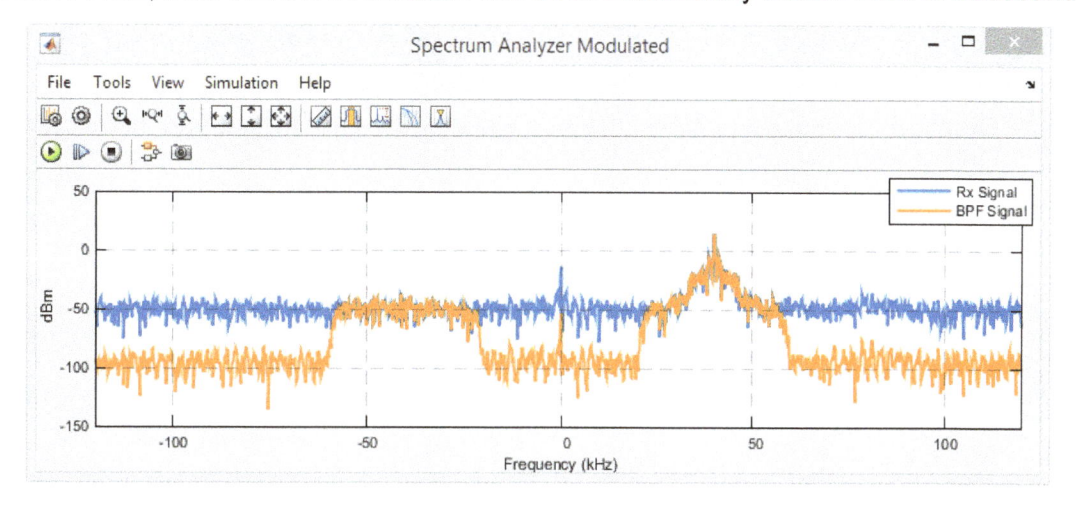

(p) **Signal Analysis: *Time Scopes*.** It should be evident from the complex signal plotted in the *Time Scope* that the amplitude of the received signal is changing. As we are only monitoring the modulated signal at the moment (its I and Q channels), we cannot confirm if the amplitude fluctuations in the music signal match this. Note that the frequency of the AM signal stays fairly constant.

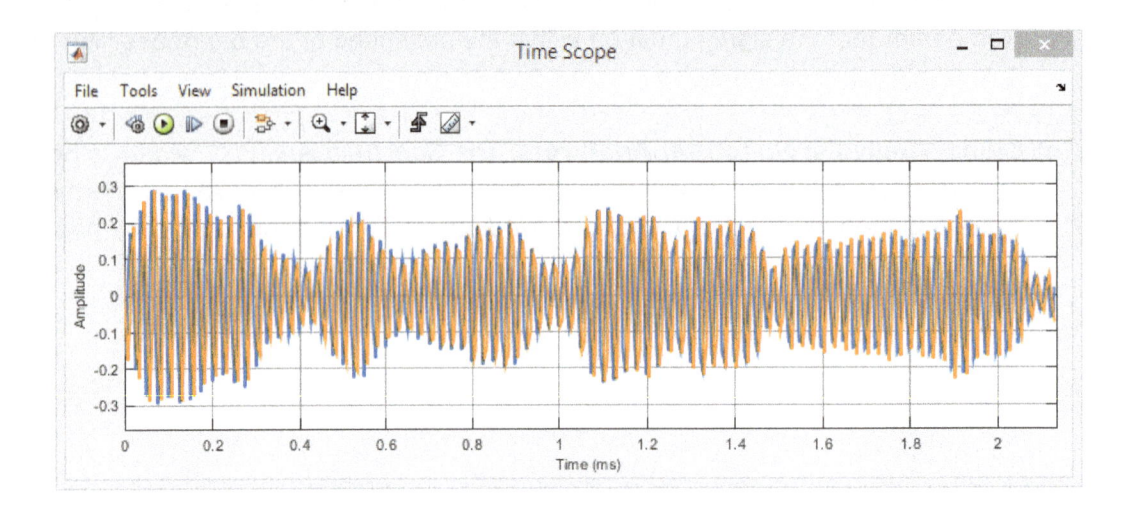

(q) Either stop the simulation or allow it to complete, 🖱 on the *File* menu of the *Time Scope*, and change the *Number of Input Ports* to '2'. 🖱 on the *View* menu and set the 'Layout' to 2x1.

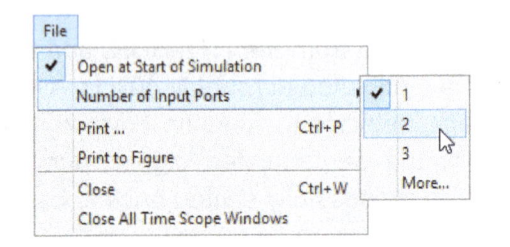

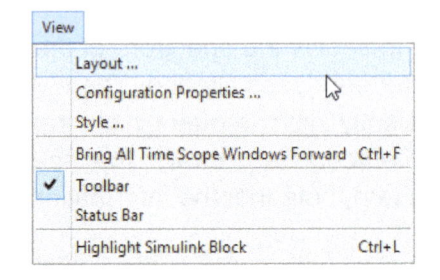

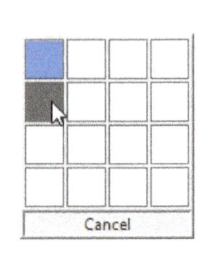

(r) Create a connection between the *Abs* block and the new input port as follows, then rerun the simulation.

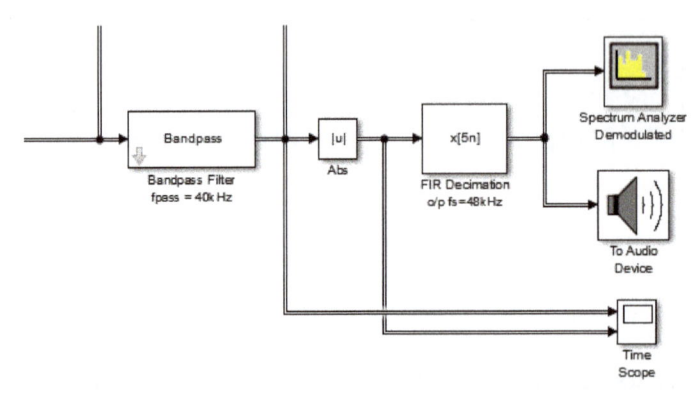

(s) The *Time Scope* should now show the IQ channels of the modulated signal in the upper plot, and the demodulated audio signal in the lower plot. This makes it much clearer that the fluctuations of the information envelope in the AM signal are caused by the amplitude changes in the music.

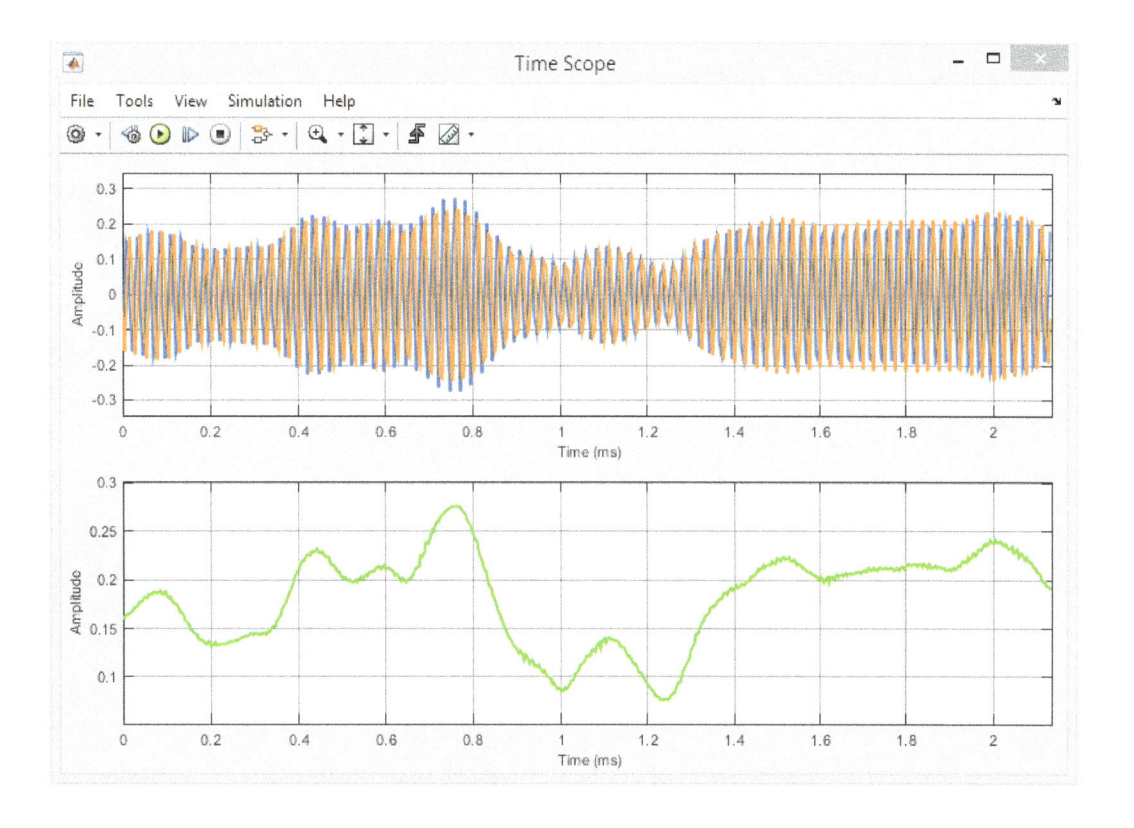

Chapter 8: Desktop AM Transmission and Reception

(t) **What is your opinion of the audio quality of the received signal?** Noise is one of the problems that plagues AM radio, and this is actually one of the reasons that led to FM being developed to replace it. It is likely that the audio signal will not sound as 'clean' as it could be, and there may be background 'hissing' that the lowpass filter in the *FIR Decimator* has not managed to remove. If you like, try increasing the number of filter coefficients in the lowpass filter by changing the '100' in the filter function string to a higher value, to see if this makes a difference.

(u) **Watch and listen to us demodulate an AM-DSB-TC signal with an envelope detector.** We have recorded a video which shows activity in the scope windows as an AM-DSB-TC signal transmitted from the USRP® hardware (with the model from Exercise 8.2) is received by the RTL-SDR and demodulated. The output audio signal has been recorded too.

desktopSDR.com/videos/#dsb_tc_envelope_demod

Exercise 8.5 RTL-SDR: MATLAB Envelope Detector for AM-DSB-TC Signals

In this exercise we will run a MATLAB function that implements the Complex Envelope Detector constructed in Exercise 8.4. The function has been designed to receive an AM-DSB-TC audio signal, and will output the demodulated audio information to the computer's speakers or headphones.

Note: If you do not have an RTL-SDR, or are unable to transmit an AM-DSB-TC signal that you can receive, you can still complete this exercise by modifying a line in the code. Guidance on this will be provided within the steps that follow.

(a) **Open MATLAB.** Set the working directory to an appropriate folder and open this file:

/am/rtlsdr_rx/rtlsdr_am_envelope_demod_matlab.m

This MATLAB function performs non-coherent AM-DSB-TC demodulation using a Complex Envelope Detector. To indicate the flow of the code, here is what an equivalent receiver would look like in Simulink:

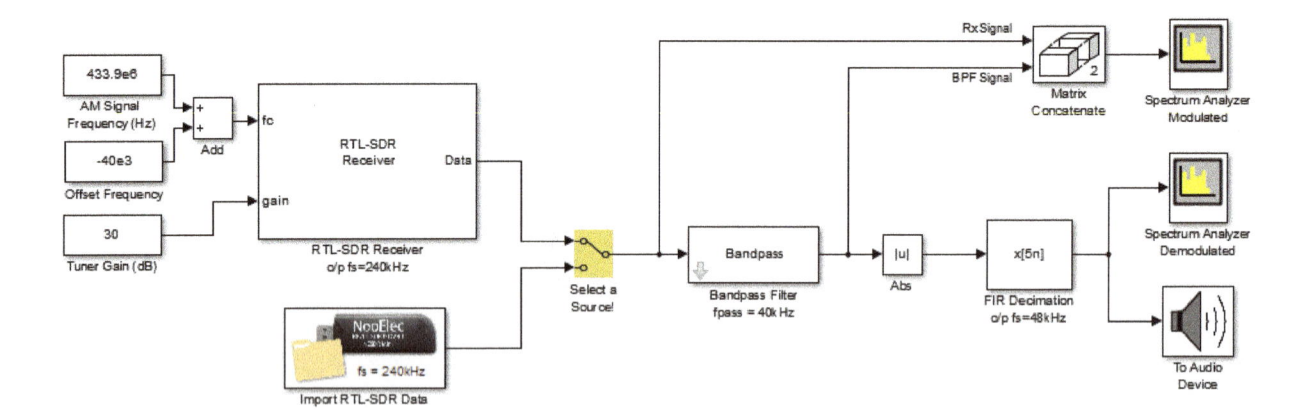

The *Manual Switch* block highlighted in yellow is used to represent the logical decision made about which data source is used.

(b) **Inspect the parameters.** There are a number of different parameters listed at the top of the `rtlsdr_am_envelope_demod_matlab` function. These are used to configure features of this receiver such as how long it should run for, and also to configure the RTL-SDR.

```
%% PARAMETERS (can change)
offline           = 0;                        % 0 = use RTL-SDR, 1 = import data
offline_filepath  = 'rec_data/am_dsb_tc.mat'; % path to AM signal
rtlsdr_id         = '0';                      % stick ID
rtlsdr_fc         = 433.9e6;                  % tuner centre frequency in Hz
rtlsdr_gain       = 30;                       % tuner gain in dB
rtlsdr_fs         = 240e3;                    % tuner sampling rate
rtlsdr_ppm        = 0;                        % tuner parts per million correction
rtlsdr_frmlen     = 256*25;                   % output data frame size (multiple of 5)
rtlsdr_datatype   = 'single';                 % output data type
audio_fs          = 48e3;                     % audio output sampling rate
sim_time          = 60;                       % simulation time in seconds
```

The value of `offline` decides whether or not a handle to a RTL-SDR system object is initialised. If you have an RTL-SDR and are able to transmit an AM-DSB-TC signal, set this to '0'. If not, you will need to import data, so set this to '1'. When importing data it is important to make sure that the `offline_filepath` string is set to reference the correct file. As with the 'Import RTL-SDR Data' block, the filepath can be specified as either a relative or full path.

Chapter 8: Desktop AM Transmission and Reception

(c) If using an RTL-SDR, modify the value of `rtlsdr_fc` to set the centre frequency of the receiver. As in Exercise 8.4, a 40kHz offset is automatically subtracted from this value to ensure that the AM signal remains modulated on a high frequency carrier.

(d) Feel free to change the `sim_time` variable to increase the run time of the simulation from the default value of 60 seconds. If you have more than one RTL-SDR attached to your computer, run the `sdrinfo` command in the MATLAB command window to find the ID of the stick you desire to use, and then configure the `rtlsdr_id` with this value. If you are changing the ID, note it needs to be entered as a `'string'` (with apostrophes).

(e) **Review the system objects.** Six different system objects are initialised in this code to create the RTL-SDR data source, implement the envelope detector, and provide spectral representations of the signals as they are demodulated. As shown below, `obj_rtlsdr` is used as the name for both the `import_rtlsdr_data` and `comm.SDRRTLReceiver` objects. This means that, regardless of which isinitialised, the rest of the code for the receiver can remain constant, calling `step(obj_rtlsdr)` for another frame of data.

```matlab
%% SYSTEM OBJECTS (do not edit)
% check if running offline
if offline == 1
    
    % link to an rtl-sdr data file
    obj_rtlsdr = import_rtlsdr_data(...
        'filepath', offline_filepath,...
        'frm_size', rtlsdr_frmlen,...
        'data_type', rtlsdr_datatype);
    
else
    
    % link to a physical rtl-sdr
    obj_rtlsdr = comm.SDRRTLReceiver(...
        rtlsdr_id,...
        'CenterFrequency', rtlsdr_fc,...
        'EnableTunerAGC', false,...
        'TunerGain', rtlsdr_gain,...
        'SampleRate', rtlsdr_fs,...
        'SamplesPerFrame', rtlsdr_frmlen,...
        'OutputDataType', rtlsdr_datatype,...
        'FrequencyCorrection', rtlsdr_ppm);
end;
```

(f) **Examine the envelope detector.** The envelope detector is implemented inside the following while loop:

```
% loop while run_time is less than sim_time
while run_time < sim_time

    % fetch a frame from obj_rtlsdr (live or offline)
    rtlsdr_data = step(obj_rtlsdr);

    % bandpass filter data to isolate AM-DSB-TC signal around 100kHz
    data_bpf = step(obj_bpf,rtlsdr_data);

    % update 'modulated' spectrum analyzer window with the new frame
    % of data, and the frame of bandpass filtered data
    step(obj_spectrummod,[rtlsdr_data,data_bpf]);

    % implement the complex envelope detector
    env_mag = abs(data_bpf);
    data_dec = step(obj_decmtr,env_mag);

    % update 'demodulated' spectrum analyzer window with new frame
    step(obj_spectrumdemod,data_dec);
    % output demodulated signal to speakers
    step(obj_audio,data_dec);

    % update run_time after processing another frame
    run_time = run_time + rtlsdr_frmtime;

end
```

Frames of AM-DSB-TC data are acquired from either the RTL-SDR or an imported data file, bandpass filtered and then passed through the envelope detector. The demodulated data is output to the computer's default audio device, and spectrum analyzers are used to display the signals in the frequency domain as they are demodulated. This code is equivalent to the receiver constructed in Exercise 8.4.

(g) **Prepare to run the function.** Connect speakers or headphones to your computer and test to ensure that they are working. If you are using an RTL-SDR, check that MATLAB can communicate with it by typing `sdrinfo` into the MATLAB command window. Finally, save the file.

(h) **Run the function.** Run the function by 🖱 on the 'Run' ▷ button in the MATLAB toolbar. After a couple of seconds the simulation will begin, the *Spectrum Analyzer* windows should appear, and it should be possible to hear the demodulated AM signal. Although the function is designed to loop for `sim_time` seconds, you can cancel its execution early using the following key combination, if desired:

CTRL + C

Chapter 8: Desktop AM Transmission and Reception

(i) **Signal Analysis: *Spectrum Analyzers*.** Very similar results should be shown in the *Spectrum Anlayzer* windows, as compared to the receiver from Exercise 8.4. *Spectrum Analyzer Modulated* will show two signals, the first being the received signal and the second being the bandpass filtered signal. When the signals are passed to the spectrum analyzer handle, unfortunately no signal names propagate to the legend; however it is still easy to identify them.

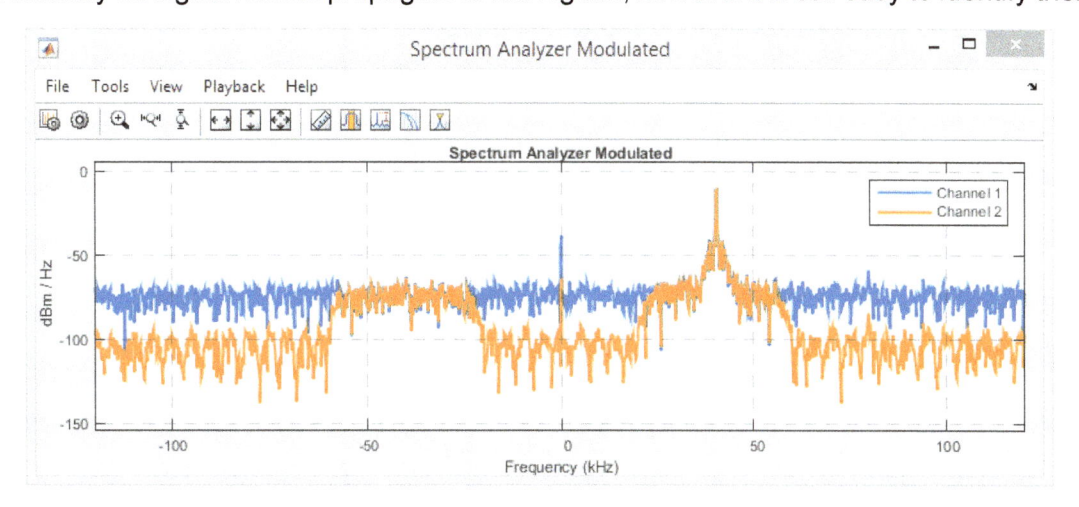

Spectrum Analyzer Demodulated shows the spectrum of the demodulated signal after envelope detection has taken place. Here, once again, it should be possible to confirm that the information has been successfully shifted back to baseband.

(j) Having now seen this receiver implemented in both MATLAB and Simulink, which form do you prefer? While Simulink is great for its graphical interface, implementing receivers in MATLAB can allow for far more flexibility to be incorporated into the design. One disadvantage of the MATLAB receiver however is that it is more difficult to tune the RTL-SDR (i.e. change its centre frequency or tuner gain) while the function is executing, because MATLAB is procedural.

AM-DSB-TC signals contain carrier components and have an information envelope. AM-DSB-SC signals on the other hand have no carrier component and no information envelope (illustrated in Figure 8.8). As discussed in Section 6.8 (page 231), envelope detectors are only able to demodulate signals that have an information envelope, as they simply smooth the gaps between the positive peaks of modulated signals. Therefore, demodulating an AM-DSB-SC signal with an envelope detector will not work!

We will explore this issue in Exercise 8.6, where we investigate what happens when attempting to demodulate an AM-DSB-SC signal with the envelope detector from Exercise 8.4, which was designed to receive AM-DSB-TC signals. Once again, this signal will need to be locally generated and then transmitted on the desktop as AM radio stations are not within the operating range of the RTL-SDR.

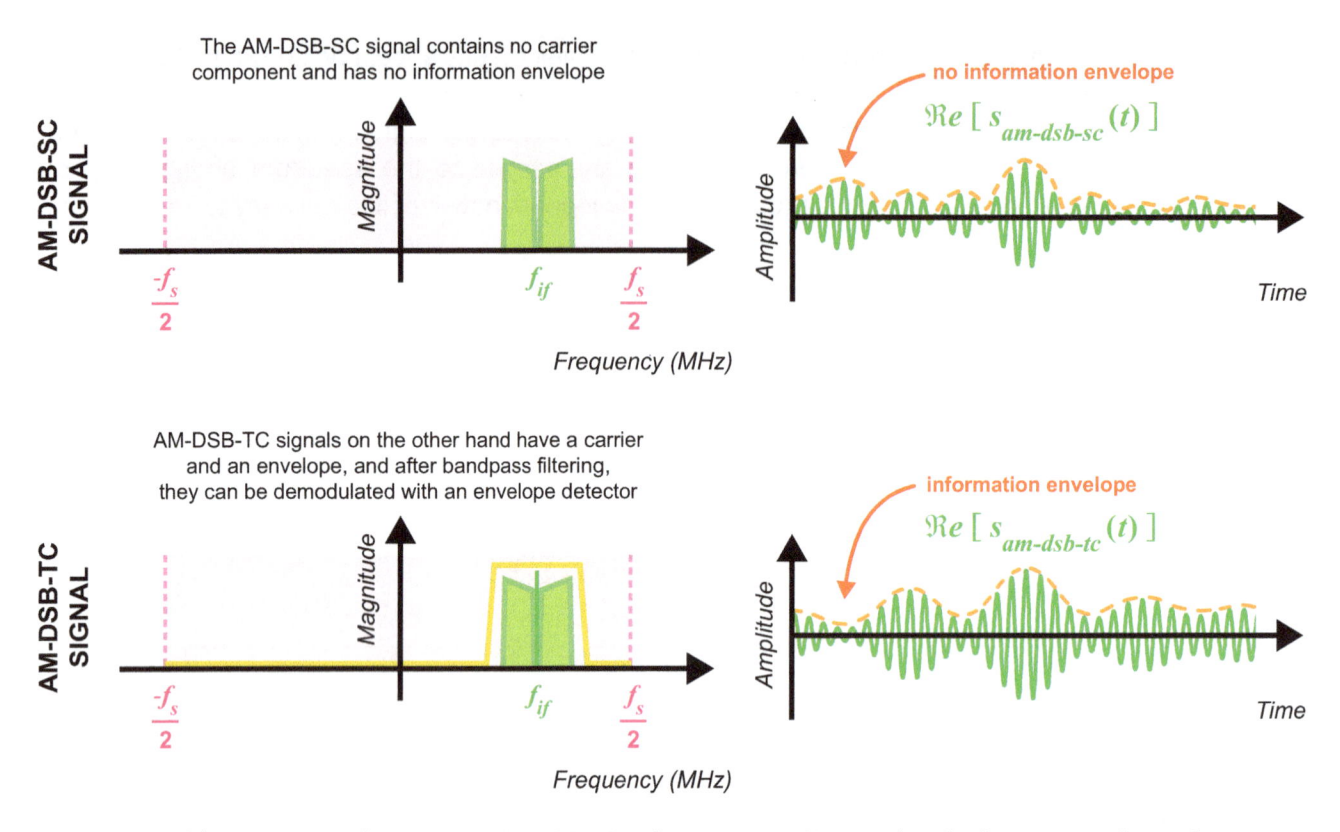

Figure 8.8: The comparison between an AM-DSB-SC and an AM-DSB-TC signal in the frequency and time domains

---

Exercise 8.6 — RTL-SDR: Envelope Detector for AM-DSB-SC Signals

We will now attempt to demodulate an AM-DSB-SC signal with the non coherent AM receiver built in Exercise 8.4. This exercise demonstrates what happens when you try to use an envelope detector to demodulate a signal with no information envelope!

Note: If you do not have an RTL-SDR, or are unable to transmit an AM-DSB-SC signal that you can receive, you can still complete this exercise by substituting the RTL-SDR Receiver block with an Import RTL-SDR Data block as discussed below.

(a) **Open MATLAB.** Set the working directory to the exercise folder,

> /my_models/receivers/

Next, open the Simulink model you created in Exercise 8.4:

> .../rtlsdr_am_envelope_demod.slx

(b) **Check the RTL-SDR data source.** If you have an RTL-SDR available and are able to transmit an AM-DSB-SC signal, you will be able to use an *RTL-SDR Receiver* block to interface with and control your RTL-SDR. If you do not have an RTL-SDR or are unable to transmit an AM-DSB-SC signal, you will need to use an *Import RTL-SDR Data* block. This will need to be configured to reference this file:

Chapter 8: Desktop AM Transmission and Reception

> 📁 `/am/rtlsdr_rx/rec_data/am_dsb_sc.mat`

Instructions of how to configure both of these blocks can be found in Exercise 8.4.

(c) **Open the reference file.** If you did not have time to construct the receiver yourself in Exercise 8.4, open the following file:

> 📁 `/am/rtlsdr_rx/rtlsdr_am_envelope_demod.slx`

If importing data rather than connecting to an RTL-SDR, delete the *RTL-SDR Receiver* block and connect the *Import RTL-SDR Data* block in its place. Check that this block is set to reference the above file.

(d) **Prepare to run the simulation.** Connect speakers or headphones to your computer. If importing a signal, set the *Simulation Stop Time* to 60 seconds, '60'. If you are using an RTL-SDR, set the *Simulation Stop Time* to 'inf' in the Simulink toolbar, and the *Simulation Mode* to 'Accelerator'. This will force Simulink to partially compile the model into native code for your computer, which allows it to run faster. Check that MATLAB can communicate with it by typing `sdrinfo` into the MATLAB command window. Finally, make sure your model is saved.

(e) **Run the simulation.** Begin the simulation by 🖱 on the 'Run' ▶ button in the Simulink toolbar. After a couple of seconds the simulation will begin, the *Spectrum Analyzer* and *Time Scope* windows should appear, and there should be an audio signal output to your speakers or headphones. If you are using an RTL-SDR, adjust the frequency and gain values until you tune the device to the signal you want to receive. Remember that the signal must be within the passband of the bandpass filter; located around 40kHz in the baseband signal.

(f) Can you hear anything that sounds like music? Does it sound horrible?! As the envelope detector smooths between the positive peaks of AM signals, any portions of the original information signal contained within the AM-DSB-SC signal that had a negative amplitude will be demodulated incorrectly. The envelope detector only outputs positive signals, meaning that the polarity of these portions inverts. Hence, roughly half of the demodulated signal is in error.

(g) **Signal Analysis: *Time Scope*.** Look at the demodulated information signal plotted in the second axis of the *Time Scope*. Can you see situations where its amplitude decreases towards zero, hits zero and inverts (like the waveforms shown overleaf)? These are the points at which the amplitude of the information signal changes from positive to negative (or negative to positive), and highlight where the envelope detector falls down when trying to demodulate AM-DSB-SC signals. Whenever the polarity changes, an error occurs.

(h) **The Human ear.** While the main point of this exercise was to highlight why envelope detectors can only be used to demodulate AM-DSB-TC signals, another interesting point is how agile the human ear is. Although the information signal has been badly distorted during the demodulation process (in that roughly half of the output signal is in error), your ear is still able to 'interpret' the music.

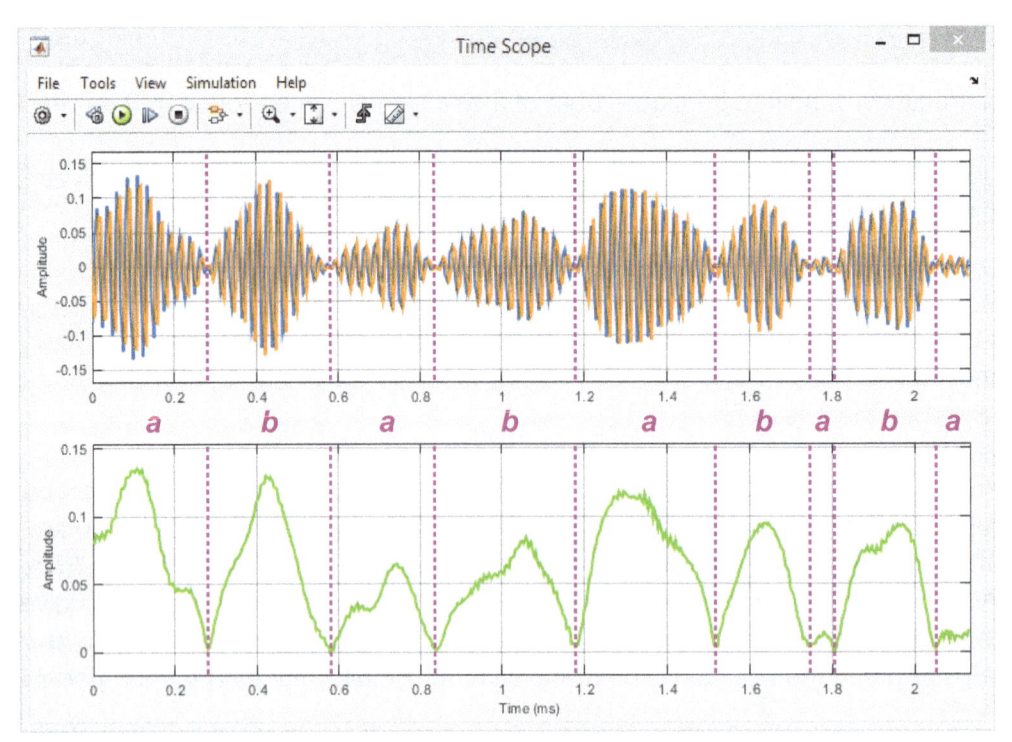

information signal polarity

- (i) **Watch and listen to us attempt to demodulate an AM-DSB-SC signal with an envelope detector.** We have recorded a video which shows activity in the scope windows as an AM-DSB-SC signal transmitted from the USRP® hardware (with the model from Exercise 8.1) is received by the RTL-SDR and demodulated. The output audio signal has been recorded too.

  desktopSDR.com/videos/#dsb_sc_envelope_demod

As AM-SSB signals have no carrier, they cannot be demodulated using an envelope detector. They are asymmetrical in the frequency domain (they have one sideband), thus they cannot be demodulated using a synchronisation loop such as a Costas Loop either. Instead, a method of synchronous demodulation is required, which simply multiplies a received AM-SSB signal with a local carrier. Used in this context the term 'synchronous' is slightly misleading, as there is no guarantee that the carrier will be of the same frequency and phase as the original modulating carrier. Whenever there is a mismatch (which is very common), a frequency shift will occur during the demodulation, resulting in somewhat distorted audio. Because the human ear is finely evolved (over a few million years) to detect speech signals, a frequency mismatch of around ±150Hz can be tolerated between the local carrier and the modulation carrier when demodulating an AM-SSB signal containing speech, however the frequency shift will cause the 'demodulated' voice signal to perhaps sound like Donald Duck or a chipmunk [96].

In Exercise 8.8, we will run an AM-SSB receiver that is pre-designed. Based on the *Phasing Method* of AM-SSB demodulation, this receiver is designed to receive only the AM-SSB signal that is generated using the USRP® radio in Exercise 8.3. The demodulation process is highlighted in Figure 8.9. The complex received signal contains two single sidebands: a USB located at f_{sc}, and an LSB at $-f_{sc}$. Mixing the complex signal with a complex sinusoid (either with frequency f_{sc} or $-f_{sc}$), the LSB or USB will be

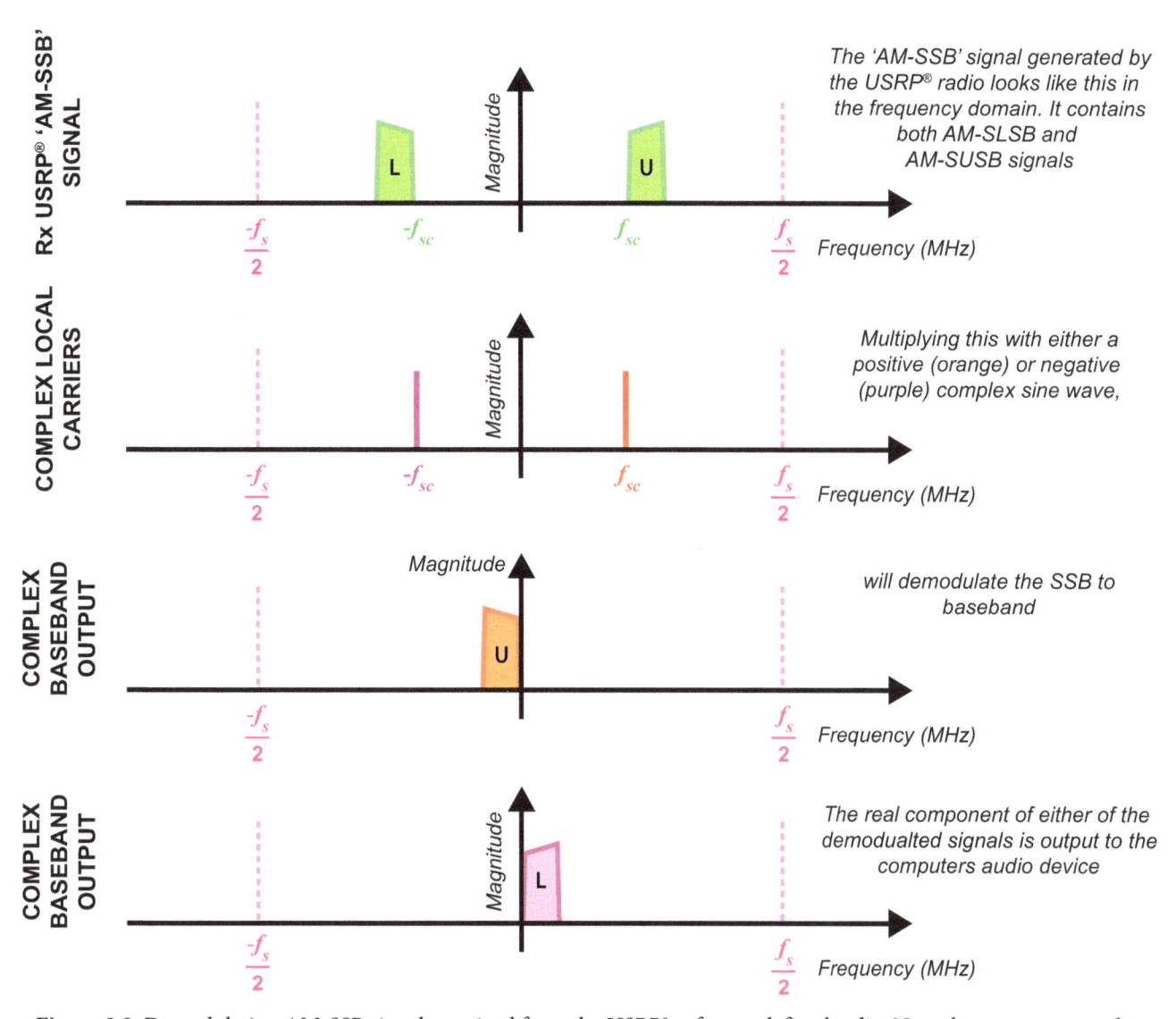

Figure 8.9: Demodulating AM-SSB signals received from the USRP® software-defined radio. Note that we represent the signals at complex baseband here (i.e. we are dealing with the output of the RTL-SDR)

demodulated to baseband. When demodulating the USB in this manner, the demodulated signal resides in the negative part of the double sided spectrum, but this can be shifted back to the positive part by dropping the imaginary component of the signal. Dropping the imaginary component has no effect on the demodulated LSB as it is already within the positive part of the complex spectrum.

If a USRP® radio is unavailable, you can still complete the exercise using the provided data file.

Exercise 8.7 RTL-SDR: Complex Demodulator for USRP® AM-SSB Signals

This exercise features a prepared AM-SSB demodulator model. Here, a complex sinusoid is used to demodulate either the AM-SUSB component of the received USRP® radio AM-SSB signal from its suppressed subcarrier at 40kHz, or the AM-SLSB component from its carrier at −40kHz. This receiver has been designed to output the demodulated audio signal to your computer's speakers or headphones.

Note: If you do not have an RTL-SDR, or are unable to transmit the AM-SSB signal from Exercise 8.4 with the USRP® hardware, you can still complete this exercise by substituting the RTL-SDR Receiver block with an Import RTL-SDR Data block as discussed below.

(a) **Open MATLAB.** Set the working directory to the exercise folder,

    `/am/rtlsdr_rx`

and then open the following file:

    `.../rtlsdr_am_ssb_demod.slx`

The block diagram of this receiver should look as follows:

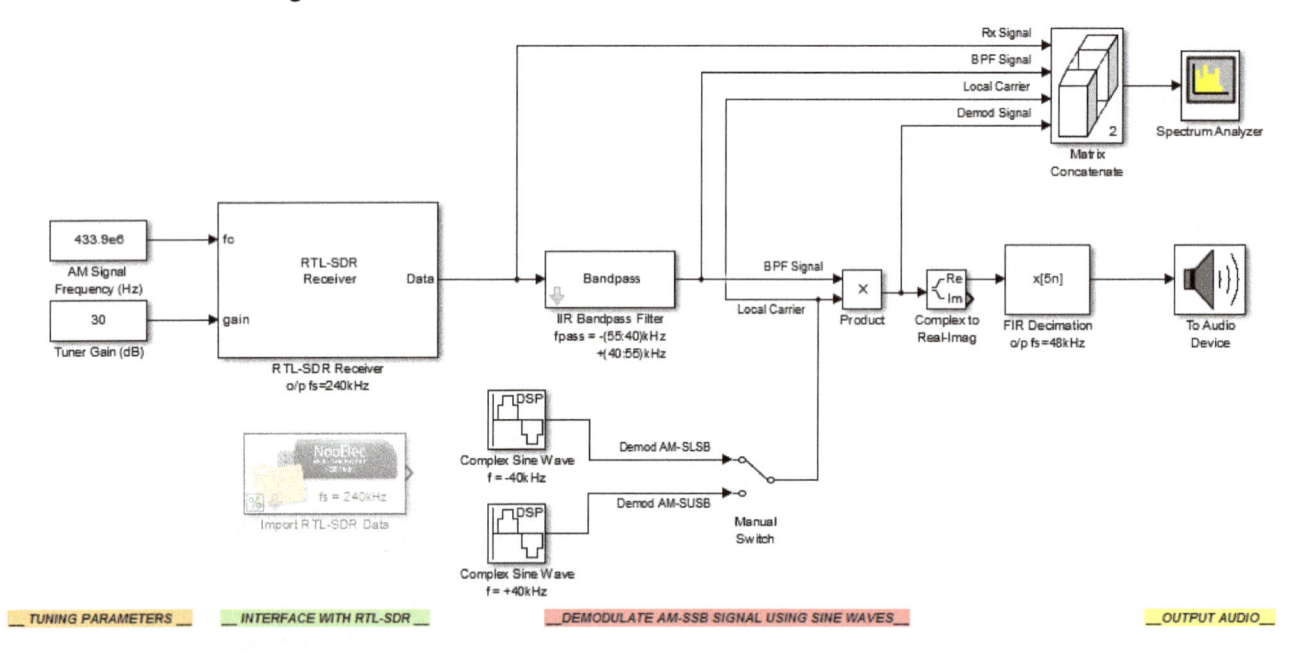

(b) **The complex sine wave.** There are two *Complex Sine Waves* in this model, one which outputs a sinusoid with a frequency of 40kHz, and the other that outputs a sinusoid with a frequency of -40kHz. Although it is not physically possible to have a negative frequency in reality, it is possible to emulate one with a complex signal. While the real component of the positive frequency sinusoid lags the imaginary component by $\pi/2$, the real component of the negative frequency sinusoid leads by $\pi/2$. This is the only difference between the two *Sine Wave* sources, but as you will see when the model runs, this makes a big difference! The *Manual Switch* block will allow you to switch between them as the simulation runs. To change between them, you will simply need to 🖱 on the block.

(c) **Check the RTL-SDR data source.** If you have an RTL-SDR available and are able to transmit an AM-SSB signal from a USRP® radio, you will be able to use an *RTL-SDR Receiver* block to interface with and control your RTL-SDR. If you do not have an RTL-SDR or are unable to transmit the signal generated in Exercise 8.3, you will need to use the *Import RTL-SDR Data* block. This should be configured to reference the following file:

    `/am/rtlsdr_rx/rec_data/am_ssb.mat`

Chapter 8: Desktop AM Transmission and Reception

If importing data rather than connecting to an RTL-SDR, delete the *RTL-SDR Receiver* block and connect the *Import RTL-SDR Data* block in its place.

(d) **Prepare to run the simulation.** Connect speakers or headphones to your computer and perform a test to ensure that they are working. If you are going to import a signal, set the *Simulation Stop Time* to 60 seconds, '60'. If you are using an RTL-SDR, set the *Simulation Stop Time* to 'inf' by typing this into the Simulink toolbar, and the *Simulation Mode* to 'Accelerator'. Check that MATLAB can communicate with the RTL-SDR by typing `sdrinfo` into the MATLAB command window.

(e) **Run the simulation.** Begin the simulation by [1] on the 'Run' ▶ button in the Simulink toolbar. After a couple of seconds the simulation will begin, the *Spectrum Analyzer* windows should appear, and an audio signal output to your speakers or headphones. If using an RTL-SDR, adjust the frequency and gain values until you tune the device to the signal you want to receive (tune the RTL-SDR to the centre frequency of the signal transmitted from the USRP® radio).

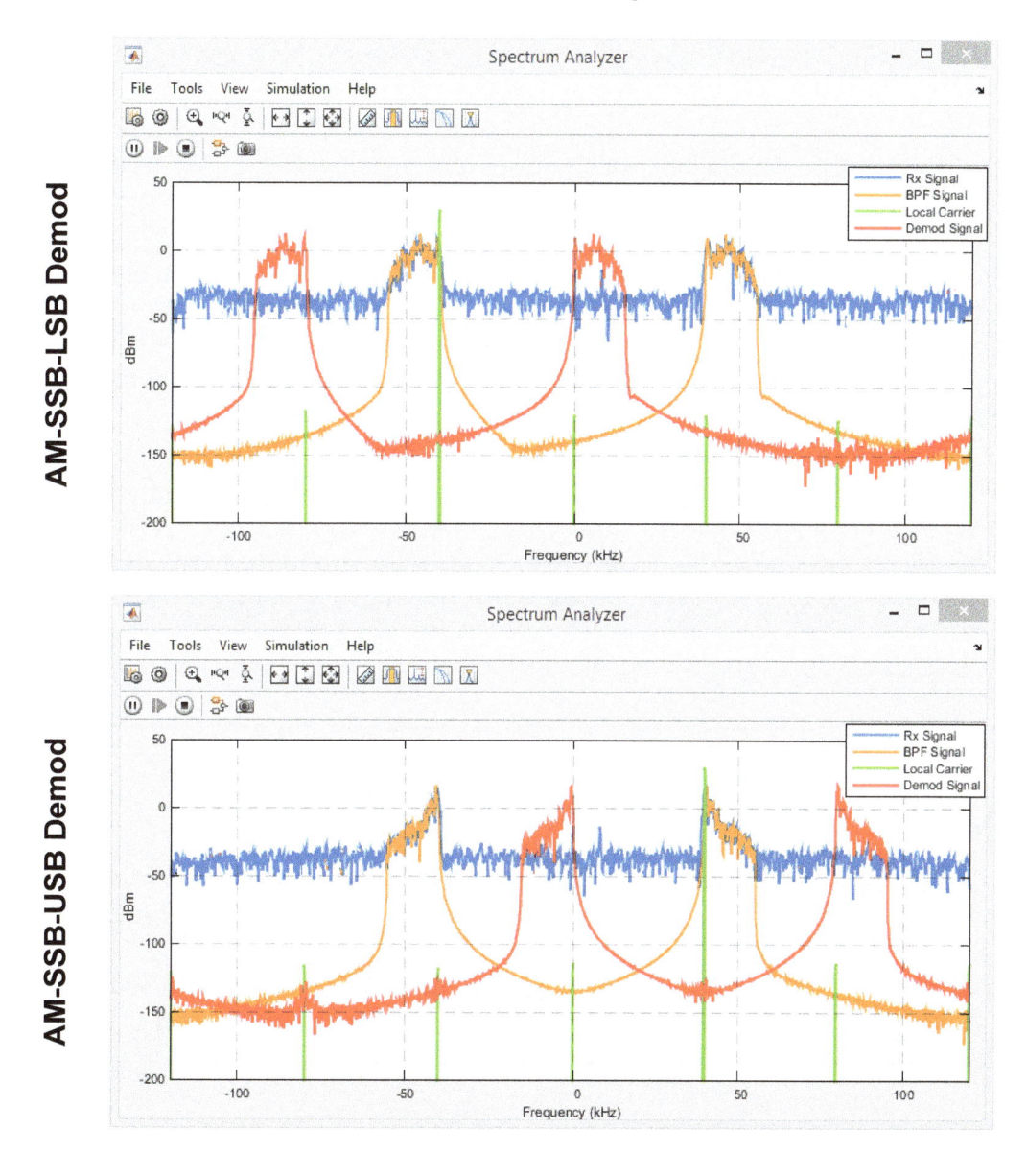

(f) **Signal Analysis.** Depending on which position the switch is in, different components in the AM-SSB signal will be demodulated. The *Spectrum Analyzer* should show something similar to the plots on the previous page when (top) set to demodulate with the negative frequency sine wave and (bottom) the positive frequency sine wave.

The blue line shows the spectrum of the received signal, and the orange shows the spectrum of the same signal after bandpass filtering. The *Complex Sine Wave* is shown in green, and the demodulated signal for each case is shown in red. Note how the *Complex Sine Waves* only have components in either the positive or negative parts of the double sided spectrum, and that demodulating the USB with the positive frequency sinusoid results in the output in the negative part of the spectrum.

(g) Regardless of the switch position, this receiver should demodulate and output a baseband audio signal to speakers or headphones. Listen to the outputs for both demodulation modes and try and identify whether a frequency shift has occurred. Tell-tale signs include voices signals sounding like they are coming from chipmunks (a positive shift) or Donald Duck (a negative shift). Neither of these is desirable, but as long as the frequency mismatch is no more than ± 150Hz, your ears should still be able to detect the full range of the voice signal.

(h) Try modifying the frequencies set in the *Complex Sine Wave* blocks to try and remove any frequency offsets. What offset is required for your signals to sounding 'right'?

(i) This exercise has hopefully highlighted one of the biggest problems that plagues the spectrally efficient AM-SSB modulation standard — a frequency offset is almost guaranteed when demodulating the signals in a receiver. Small frequency offsets can be tolerated when voice signals are demultiplexed from landline telephone systems, but in most others situations, standard AM-SSB modulation would not be a suitable option.

(j) **Watch and listen to us demodulate AM-SLSB and AM-SUSB signals.** We have recorded a video which shows activity in the *Spectrum Analyzer* window as the dual component AM-SSB signal transmitted from the USRP® hardware (with the model from Exercise 8.3) is received by the RTL-SDR and demodulated. The output audio signal has been recorded too.

desktopSDR.com/videos/#ssb_complex_demod

8.3 Implementing Coherent AM Receivers with the RTL-SDR

In Section 8.2, we confirmed why the AM envelope detector is only capable of demodulating AM-DSB-TC signals. It is only able to detect envelopes (as hinted by its name!), and these are the only AM signals where the envelope contains the information signal. Another way you can demodulate an AM-DSB-TC signal is to use a Phase Locked Loop (PLL). This will generate a sinusoid that is synchronised with the frequency and phase of the transmitted carrier, which when mixed with the received signal, will perform coherent demodulation of the information to baseband.

Chapter 8: Desktop AM Transmission and Reception 311

In Exercise 8.8, we will modify the AM-DSB-TC receiver from Exercise 8.4 to convert it to a coherent demodulator. As discussed at the start of this chapter, it is not possible receive any broadcast AM signals with the RTL-SDR, so we will need to transmit our own!

Exercise 8.8 RTL-SDR: PLL Demodulator for AM-DSB-TC Signals

Next we will modify the AM receiver constructed in Exercise 8.4 into a coherent demodulator using a *Phase Locked Loop*. The PLL will lock to the carrier in the AM-DSB-TC signal, and create a local carrier with the same frequency and phase as the modulating carrier. Multiplying this with the received signal, the audio information can be recovered. The system will be designed to receive an AM-DSB-TC audio signal, and will output the demodulated audio information to speakers or headphones.

Note: If you do not have an RTL-SDR, or are unable to transmit an AM-DSB-TC signal that you can receive, you can still complete this exercise by substituting the RTL-SDR Receiver block with an Import RTL-SDR Data block as discussed below.

(a) **Open MATLAB.** Set the working directory to the exercise folder,

   /my_models/receivers/

   Next, create a copy of the Simulink model from Exercise 8.4,

   .../rtlsdr_am_envelope_demod.slx

   and save this file with the name:

   .../rtlsdr_am_pll_demod.slx

(b) Open the new file, and then the Simulink Library Browser.

(c) **Check the RTL-SDR data source.** If an RTL-SDR is available and an AM-DSB-TC signal can be transmitted, an *RTL-SDR Receiver* block should be used to interface with and control your RTL-SDR. If you do not have an RTL-SDR or are unable to transmit an AM-DSB-TC signal, use an *Import RTL-SDR Data* block in its place. This must be configured to reference this file:

   /am/rtlsdr_rx/rec_data/am_dsb_tc.mat

   Instructions of how to configure both of these blocks can be found in Exercise 8.4.

(d) **Modify the receiver.** Because the PLL will be coherently demodulating the signal, there is no need to bandpass filter it beforehand. Delete the *Bandpass Filter*, *Abs*, *Matrix Concatenate* and *Time Scope* blocks, and their associated input and output connections. Place a *Phase Locked Loop Demodulator* block from ▦ > **RTL-SDR Book Library > Sync Tools for Analogue Comms** in the gap.

(e) ▨ on the *Phase Locked Loop Demodulator* block to open its parameters window. The 'Sampling Frequency' determines the rate at which the NCO inside it outputs samples, and the 'Quiescent Frequency' is used to set the starting value in the VCOs. The 'Noise Bandwidth' and 'Damping Ratio' parameters are used to configure the gains in the loop filter, and the

'Output Frame Size' sets the width of the output frame. Ensure that the PLL is set to 'Type 2' and is configured with the following parameters:

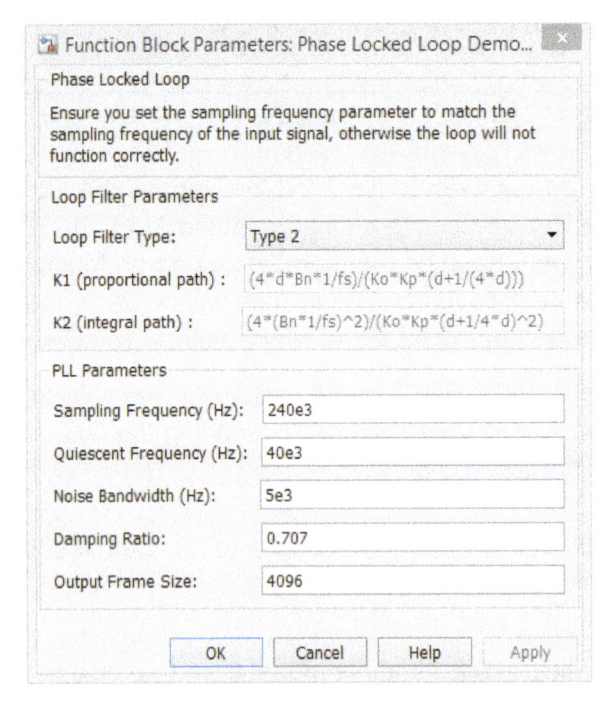

The PLL will lock to the carrier component of the input AM signal and generate a sinusoid with a matching frequency and phase. This signal is mixed with the AM signal to coherently demodulate the information from its IF carrier to baseband. The PLL will adjust constantly, ensuring that it stays locked all the time.

(f) One of the outputs ports of the PLL is titled 'freq_out'. This will output the frequency that the PLL's VCOs are locked to (and update every few seconds), to allow you to check that it they are adjusting and staying locked to the carrier. To be able to visually see the frequency changes the simplest thing to do is to place a *Gain* block from ▦ > **Simulink > Math Operations** and a *Display* block from ▦ > **Simulink > Sinks**.

Configure the *Gain* block to have a 'Gain' of '1e-3', and rename it *kHz*. This will be used to divide the PLL's 'freq_out' signal by 1000, converting it from a Hz to a kHz value. ② on the *Display* block and change its 'Format' to 'bank'. This will limit it to displaying only two decimal units. Finally, connect up the blocks as shown below.

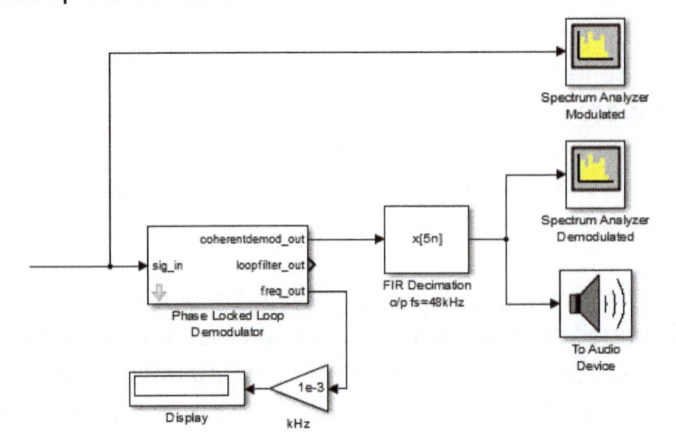

Chapter 8: Desktop AM Transmission and Reception 313

(g) **Open the reference file.** If you did not construct the receiver yourself in Exercise 8.4 (and therefore could not modify its block diagram), open the following file:

   /am/rtlsdr_rx/rtlsdr_am_pll_demod.slx

If importing data rather than connecting to an RTL-SDR, delete the *RTL-SDR Receiver* block and connect the *Import RTL-SDR Data* block in its place. Check that this block references the file discussed above.

(h) **Prepare to run the simulation.** Connect speakers or headphones to your computer and check that they are working. If importing a signal, set the *Simulation Stop Time* to 60 seconds, '60'. If using an RTL-SDR, set the *Simulation Stop Time* to 'inf' by typing this into the Simulink toolbar, and the *Simulation Mode* to 'Accelerator'. This will enable the Simulink model to run faster. Check that MATLAB can communicate with it by typing `sdrinfo` into the MATLAB command window. Finally, make sure your model is saved.

(i) **Run the simulation.** Begin the simulation by clicking on the 'Run' button in the Simulink toolbar. After a couple of seconds the simulation will begin, the *Spectrum Analyzer* windows should appear, and an audio signal should be output to your speakers or headphones. If using an RTL-SDR, adjust the frequency and gain values until the device is tuned to the signal you want to receive. The signal must be centred around the quiescent frequency of the PLL — around 40kHz in the baseband signal.

As the simulation runs, you should notice that the value shown in the *Display* block updates regularly, confirming that the PLL is adjusting.

41.33	39.86	41.96
Display	Display	Display

(j) **Signal Analysis: Audio Output.** How do you think the quality of the output signal compares to the output of the envelope detector demodulator from Exercise 8.4? In theory, the PLL should perform better than the non-coherent method; and one big advantage it has is that the signal does not need to be bandpass filtered prior to demodulation.

(k) **Change the PLL Type.** Stop the simulation, or allow it to complete, then change the *Phase Locked Loop Demodulator* 'Loop Filter Type' to 'Type 1'. Apply the changes and rerun the simulation. What happens now? Is the PLL still able to demodulate the signal? If not, why not?

(l) **Watch and listen to us demodulate an AM-DSB-TC signal with a PLL.** We have recorded a video which shows activity in the *Spectrum Analyzer* window as an AM-DSB-TC signal transmitted from the USRP® hardware (with the model from Exercise 8.2) is received by the RTL-SDR and demodulated. The output audio signal has been recorded too.

   desktopSDR.com/videos/#pll_demod

If demodulating an AM-DSB-SC signal, a coherent demodulation method that performs carrier synchronisation is required, as there is no way to non-coherently demodulate the signal. Costas Loops are suitable for this purpose. Like the PLL, these loops generate a sinusoid that is synchronised with the frequency and phase of the (suppressed) carrier, which, when mixed with the received signal, will perform coherent demodulation of the information to baseband.

In Exercise 8.9, we will modify the AM-DSB-TC receiver from Exercise 8.8 to change the synchronisation loop to a Costas Loop, enabling AM-DSB-SC demodulation. As discussed previously, we require to generate an AM signal to receive.

Exercise 8.9 RTL-SDR: Costas Demodulator for AM-DSB-SC Signals

Now we will modify the PLL based AM receiver constructed in Exercise 8.8 to enable demodulation of AM-DSB-SC signals with a Costas Loop. This loop will lock to the suppressed carrier in AM-DSB-SC audio signals to create a local carrier with the same frequency and phase as the modulation carrier. The receiver will be designed to receive an AM-DSB-SC audio signal, and will output the demodulated audio information to computer speakers or headphones.

Note: If you do not have an RTL-SDR, or are unable to transmit an AM-DSB-SC signal that you can receive, you can still complete this exercise by substituting the RTL-SDR Receiver block with an Import RTL-SDR Data block as discussed below.

(a) **Open MATLAB.** Set the working directory to the exercise folder,

> `/my_models/receivers/`

Next, create a copy of the Simulink model from Exercise 8.8,

> `.../rtlsdr_am_pll_demod.slx`

and save this file with the name:

> `.../rtlsdr_am_costas_demod.slx`

(b) Open the new file, and then the Simulink Library Browser.

(c) **Check the RTL-SDR data source.** If you have an RTL-SDR available and are able to transmit an AM-DSB-SC signal, you will be able to use an *RTL-SDR Receiver* block to interface with and control your RTL-SDR. If you do not have an RTL-SDR or are unable to transmit an AM-DSB-SC signal, you will need to use an *Import RTL-SDR Data* block. This will need to be configured to reference this file:

> `/am/rtlsdr_rx/rec_data/am_dsb_sc.mat`

Instructions of how to configure both of these blocks can be found in Exercise 8.4.

(d) **Modify the receiver.** Delete the *Phase Locked Loop Demodulator* block and substitute it with a *Costas Loop Demodulator* block from **> RTL-SDR Book Library > Sync Tools for Analogue Comms**.

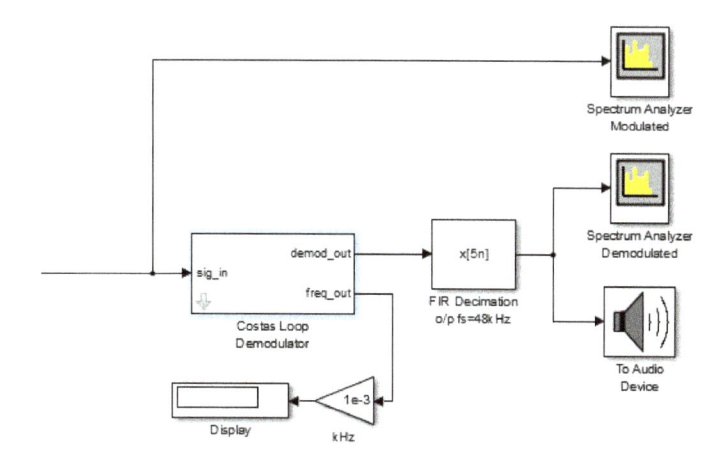

Chapter 8: Desktop AM Transmission and Reception

(e) **2-click** on the *Costas Loop Demodulator* to open its parameters window, and check that its configuration is as follows:

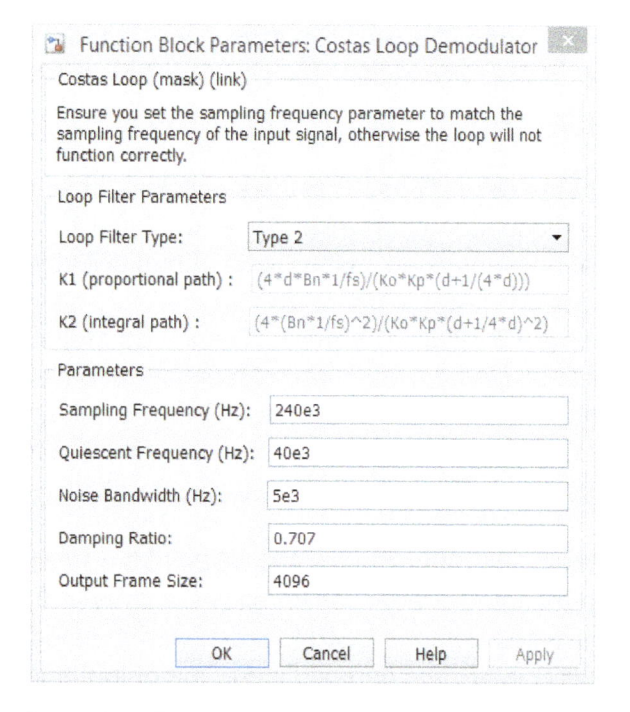

As with the PLL, the 'Sampling Frequency' determines the rate at which the NCO outputs samples, and the 'Quiescent Frequency' is used to set the step size. The 'Quiescent Frequency' is set to 40kHz as this is roughly the value of the centre frequency of the AM-DSB-SC signal. The 'Noise Bandwidth' and 'Damping Ratio' parameters are used to configure the gains in the loop filter, and the 'Output Frame Size' sets the width of the output frame.

(f) **Open the reference file.** If you did not have time to construct the receiver yourself in Exercise 8.8 (and therefore could not modify its block diagram), open the following file:

   `/am/rtlsdr_rx/rtlsdr_am_costas_demod.slx`

(g) If importing data, delete the *RTL-SDR Receiver* block and connect the *Import RTL-SDR Data* block in its place. Check that this block references the above file.

(h) **Prepare to run the simulation.** Connect speakers or headphones to the computer and perform a test. If importing a signal, set the *Simulation Stop Time* to 60 seconds, '60'. If using an RTL-SDR, set the *Simulation Stop Time* to 'inf' by typing this into the Simulink toolbar, and the *Simulation Mode* to 'Accelerator'. This will enable Simulink to run the model faster. Check that MATLAB can communicate with the RTL-SDR by typing `sdrinfo` into the MATLAB command window. Finally, make sure your model is saved.

(i) **Run the simulation.** Begin the simulation by clicking on the 'Run' button in the Simulink toolbar. After a couple of seconds the simulation will begin, the *Spectrum Analyzer* windows should appear, and an audio signal should be output to speakers or headphones. If using an RTL-SDR, adjust the frequency and gain values until the device is tuned to the desired signal. The signal must be centred around the quiescent frequency of the Costas Loop — around 40kHz in the baseband signal. Again, as the simulation runs, you should notice that the value shown in the *Display* block updates regularly, confirming that the Costas Loop is adjusting.

(j) **Signal Analysis: Audio Output.** How does the output of the *Costas Loop Demodulator* compare to the envelope detector used in Exercise 8.6? Hopefully this demodulator has actually worked(!), and recovered the information signal correctly.

(k) **Change the *Costas Loop* Type.** Stop the simulation, or allow it to complete, then change the *Costas Loop Demodulator* 'Loop Filter Type' to 'Type 1'. Apply the changes and rerun the simulation. What happens now? Is the Costas Loop still able to demodulate the signal?

(l) **Try demodulating AM-DSB-TC signals.** Have a go at demodulating an AM-DSB-TC signal with the *Costas Loop Demodulator*. You can either transmit a signal with the USRP® radio (or another transmitter) and receive it with the RTL-SDR; or import a signal using the *Import RTL-SDR Data* block (use the `am_dsb_tc.mat` file from Exercise 8.8). Does the demodulator work?

(m) **Watch and listen to us demodulate AM-DSB-SC and AM-DSB-TC signals with a Costas Loop.** We have recorded a video which shows activity in the *Spectrum Analyzer* window as AM signals transmitted from the USRP® radio (with the models from Exercises 8.1 & 8.2) are received by the RTL-SDR and demodulated. The output audio signal has been recorded too.

desktopSDR.com/videos/#costas_demod

8.4 Audio Multiplexing with the USRP® and RTL-SDR Hardware

Frequency Division Multiplexing (FDM) is one simple method for improving spectral efficiency and sharing in communications systems. FDM permits two or more information signals to be transmitted via a single cable or wireless link at the same time, by splitting the total available bandwidth of the communications channel into a number of subchannels. Each subchannel is non-overlapping, so that no interference can occur between the subchannels. The process of creating a multiplex (MPX) signal is quite straightforward. First, information signals are modulated into the subchannels using an AM

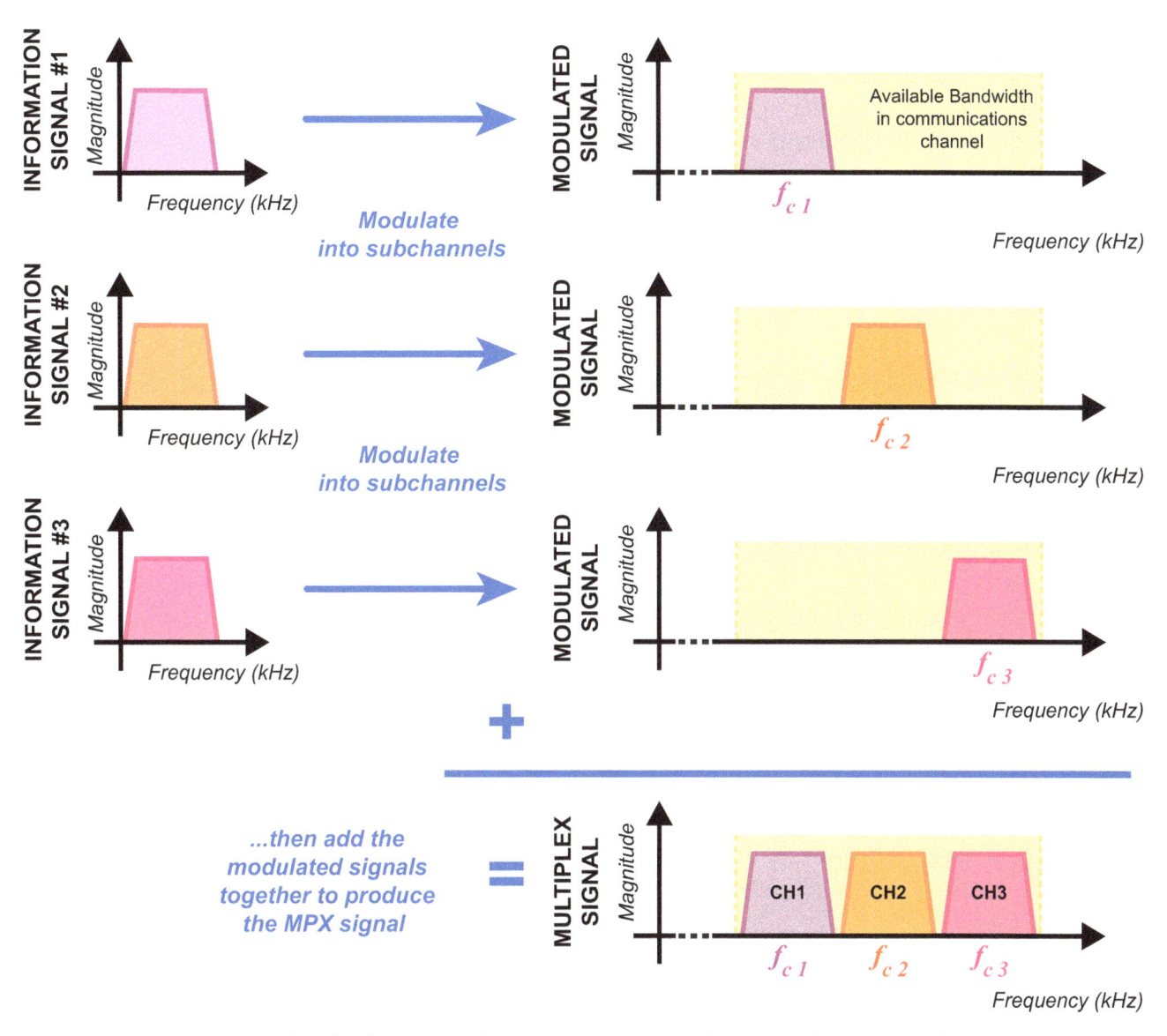

Figure 8.10: A high level overview of Frequency Division Multiplexing, showing three baseband information signals being modulated into subchannels to create a MPX signal

modulation scheme, and then the modulated signals are added together. This process is illustrated in Figure 8.10.

One of the primary applications of FDM in the 20th century was to multiplex tens of voice signals onto single, high capacity copper cables to form landline telephone networks all around the world. As most frequencies present in the human voice are in the range 100Hz–4kHz, each customer was assigned a 3.4kHz band on the trunk cable. Using FDM meant that network installation costs were greatly reduced, as individual cables did not need to be laid for each customer, and network hardware could be shared between multiple users.

To give an example, let's define a simple baseband communications channel, with a bandwidth of 15kHz. It would be possible to multiplex two real, 5kHz wide signals into this channel using AM-DSB-TC modulation. One signal would be left at baseband, in a channel spanning from 0 to 5kHz. The other would be modulated into a channel that spanned from 5kHz to 15kHz, using a real f_c = 10kHz carrier, $\cos(\omega t_c)$, where $\omega_c = 2\pi 10000t$. The second channel has double the bandwidth of the first, because the AM-DSB-TC modulation process results in a signal with two sidebands, as shown in Figure 8.11. If the carrier used to modulate the signal was real, the double sided spectrum of this MPX would look as follows. Notice how the negative part of the spectrum contains an image of channel 2.

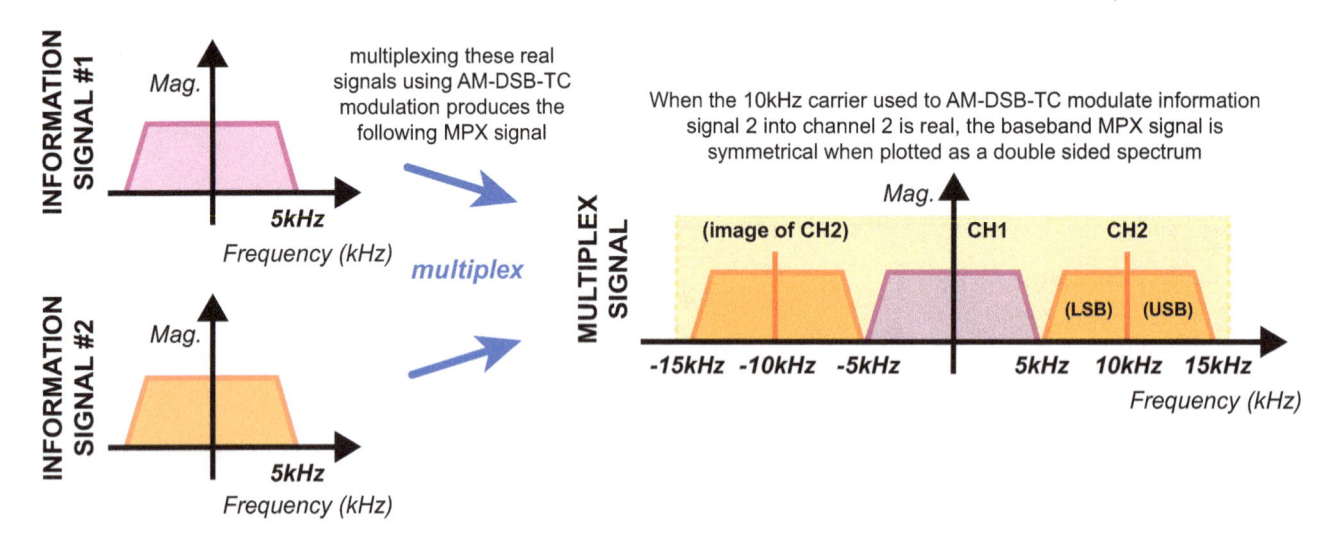

Figure 8.11: MPX design 1: two channel MPX plotted in the double sided spectrum (note that as this signal is real, the 'negative' frequency components are simply spectral images of the 'positive' frequency components)

If the 10kHz carrier was complex (i.e. a quadrature carrier) however, using $e^{j\omega t}$ rather than $\cos(\omega t)$, the double sided spectrum would be asymmetrical, as shown in Figure 8.12 (refer back to the *Complex Signals, Spectra and Quadrature Modulation* Chapter 5 (*page 171*) for a recap, if needed). No image of channel 2 is present, and this means that another 10kHz wide band has become available between -15kHz and -5kHz that can be used for another channel.

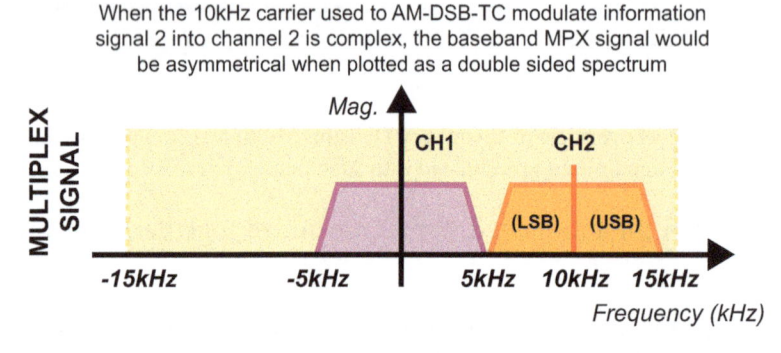

Figure 8.12: MPX design 2: two channel MPX plotted as a double sided spectrum

To AM-DSB-TC modulate a third, real, 5kHz wide signal into this free band, a complex carrier with a frequency of -10kHz is required.

As the USRP® hardware is designed to transmit quadrature signals, we are able to test out this complex modulation technique and create an RF multiplex signal containing three channels of information. The block diagram in Figure 8.14 shows the order in which processes must be carried out to produce this signal.

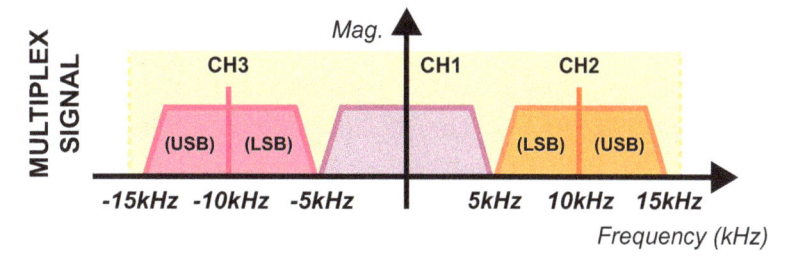

Chapter 8: Desktop AM Transmission and Reception

Figure 8.13: MPX design 3: three channel MPX plotted in the double sided spectrum

In Exercise 8.10, we will open and run a complex multiplexer/ USRP® hardware modulator and transmitter model. This will generate the MPX shown in Figure 8.13 using a number of filters, AM-DSB-TC modulators and complex sine waves, allowing three music signals to be multiplexed into a complex baseband signal. The MPX is then passed to the USRP® hardware where it is modulated onto an RF carrier and transmitted.

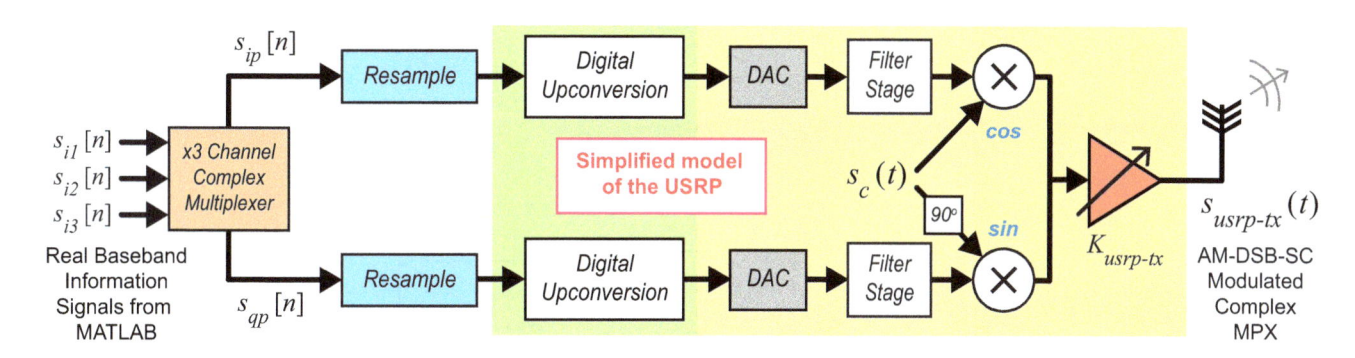

Figure 8.14: Block diagram showing the Simulink/ USRP® hardware implementation of the x3 channel complex multiplexer/ transmitter

This is a partner exercise for Exercise 8.11. If a USRP® radio is not available (or another AM transmitter — see Section 8.5), the later exercise can still be completed using a provided data file. Regardless, it is recommended to open and explore the USRP® transmitter file, to investigate how the complex MPX is generated.

Exercise 8.10 FDM AM: FDM MPX'er, AM Modulator and USRP® Transmitter

This exercise features a USRP® transmitter complex multiplexer model. Here, complex sine waves are used to modulate baseband information signals into channels, allowing for three pieces of music to be transmitted at once with the USRP® hardware.

Note: If you do not have a USRP® radio, you can still run this exercise and see the MPX signal being generated, by commenting out the SDRu Transmitter block.

(a) **Open MATLAB.** Set the working directory to the exercise folder,

```
📁 /am/usrp_tx
```

and then open the following file:

```
.../usrp_am_fdm.slx
```

The block diagram of this model should look as follows:

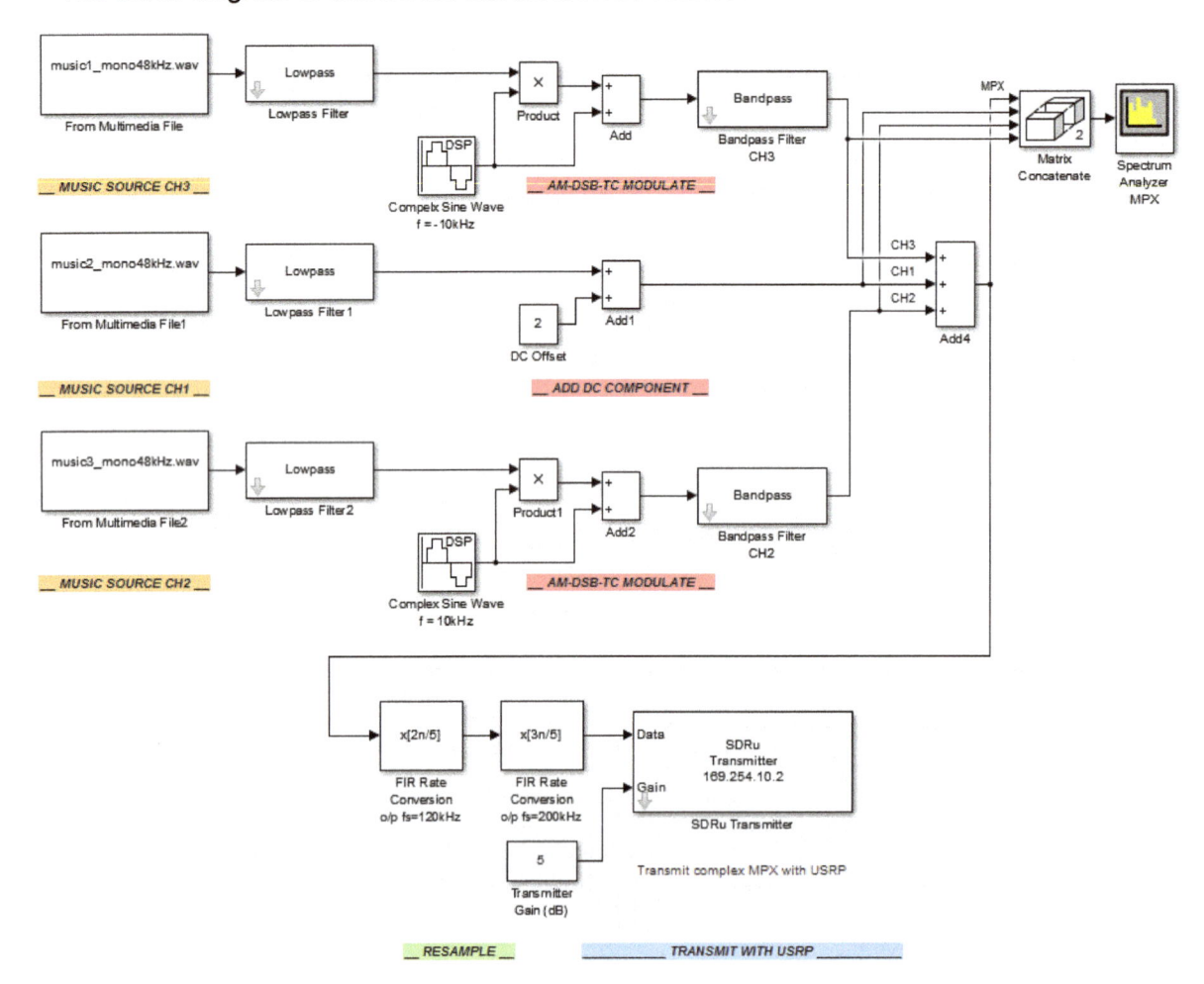

- (b) Music signals are imported to the model using *From Multimedia File* blocks, and they are then band limited to 5kHz using *Lowpass Filters*. Following this, the signals enter the modulator stage. The upper signal is AM-DSB-TC modulated by a complex sine wave with a frequency of -10kHz to create channel 3 (CH3). The middle signal remains as-is, but has a DC offset added. This ensures that channel 1 (CH1) has a carrier. The lower signal is AM-DSB-TC modulated by a complex sine wave with a frequency of 10kHz. This creates channel 2 (CH2). Finally, the MPX signal is created by adding all of these together.

    FIR Rate Conversion blocks are used to change the sampling rate of the MPX signal from 48kHz to 200kHz. The interpolated signal is then passed to the USRP® hardware via an *SDRu Transmitter* block.

- (c) **Check the USRP® radio settings.** The *SDRu Transmitter* block should be configured to communicate with your USRP® hardware. To do this, you will need to know its IP address (or

Chapter 8: Desktop AM Transmission and Reception

USB address), and be able to 'ping' it from your computer. Instructions of how to set up this connection are discussed in Appendix A.2. 2️⃣ on the *SDRu Transmitter* block, check that the appropriate 'Platform', and 'Address' for your USRP® radio have been entered. Make sure that the 'Center Frequency' parameter is set to the frequency you want to transmit on. This value should be within the range of your RTL-SDR's tuner, e.g. in the range 25MHz–1.75GHz. Check that the USRP® hardware is turned on, and that MATLAB can communicate with it by typing `findsdru` into the MATLAB command window. If a structure is returned as discussed in Appendix A.2, continue to the next step.

If you do not have a USRP® radio, comment this block out or delete it before continuing.

(d) **Prepare to run the simulation.** Check that the *Simulation Stop Time* is set to 'inf', and that the *Simulation Mode* is set to 'Accelerator'. This will force Simulink to partially compile the model into native code for your computer, which will allow it to run faster. Finally, make sure the model is saved.

(e) **Run the simulation.** Begin the simulation by 1️⃣ on the 'Run' ▶ button in the Simulink toolbar. After a couple of seconds, Simulink will establish a connection with the USRP® hardware and the simulation will begin. When it does, samples of the baseband information signal will be passed to the radio, modulated onto the AM carrier, and transmitted from its antenna.

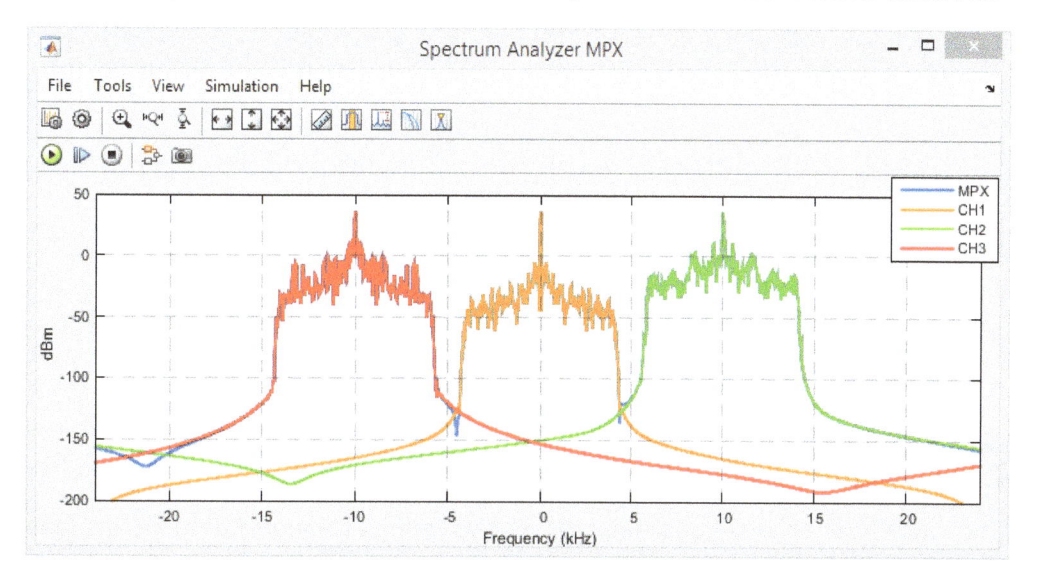

(f) **Signal Analysis.** A *Matrix Concatenate* block is used to combine four different signals so they can all be displayed in the *Spectrum Anlayzer* window. In order of input, these are the MPX, CH1, CH2 and CH3. When the simulation is running, it should be possible to confirm that each of the channels act independently of each other, and that the MPX signal is indeed the sum of all channels.

(g) It will not be possible to tune a consumer analogue AM radio to the transmitted MPX signal, as its carrier frequency is not within the AM band and would not be supported. The RTL-SDR will, however, be able to receive the signal, and we will try this out in the next exercise.

As the MPX signal is complex, the AM-DSB-SC modulation process carried out inside the USRP® transmitter does not create a signal with two sidebands — instead the complex baseband signal is simply shifted to centre around the carrier frequency specified in the *SDRu Transmitter* block, f_{c-usrp}. Tuning an RTL-SDR to a centre frequency slightly lower than the centre of the MPX, the three channels can be downconverted to IFs in the received baseband signal. Bandpass filtering around each of the channels returns the AM-DSB-TC signals contained within, and these can be demodulated using an envelope detector. This process is illustrated in Figure 8.15.

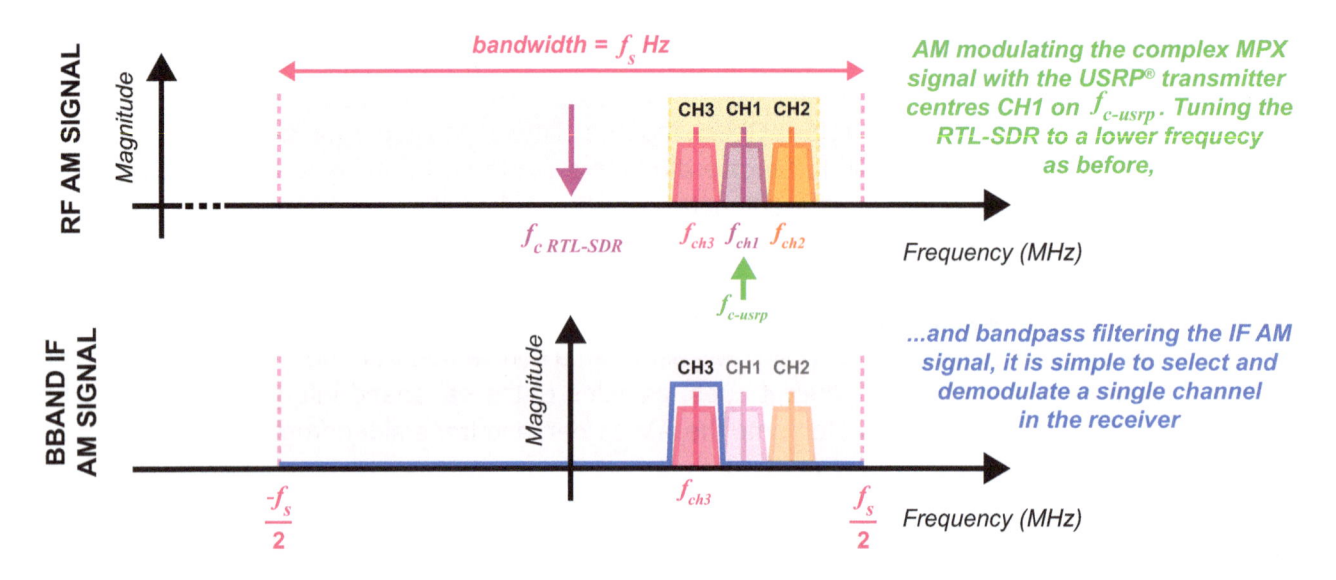

Figure 8.15: Demodulating AM-DSB-TC signals from the transmitted complex MPX signal

In Exercise 8.11, we will ask you to run a receiver we have designed for this complex MPX signal, that can demodulate and output each of the audio signals contained within the channels. If you do not have a USRP® radio and as such, are unable to transmit the signal from Exercise 8.10, you can still complete the exercise using the provided data file.

Exercise 8.11 — FDM AM: RTL-SDR AM Receiver and Demultiplexer

We will now consider an envelope detector -based receiver that will demodulate each of the channels transmitted in the complex MPX. This receiver has been designed to output the demodulated audio signal to your computer's speakers or headphones.

Note: If an RTL-SDR is not available, or you are unable to transmit the MPX signal from Exercise 8.10 with the USRP® radio, it will still be possible to complete this exercise by substituting the RTL-SDR Receiver block with an Import RTL-SDR Data block as discussed below.

(a) **Open MATLAB.** Set the working directory to the exercise folder,

> `/am/rtlsdr_rx`

and then open the following file:

> `.../rtlsdr_am_fdm_demod.slx`

Chapter 8: Desktop AM Transmission and Reception

The block diagram of this receiver should look as follows:

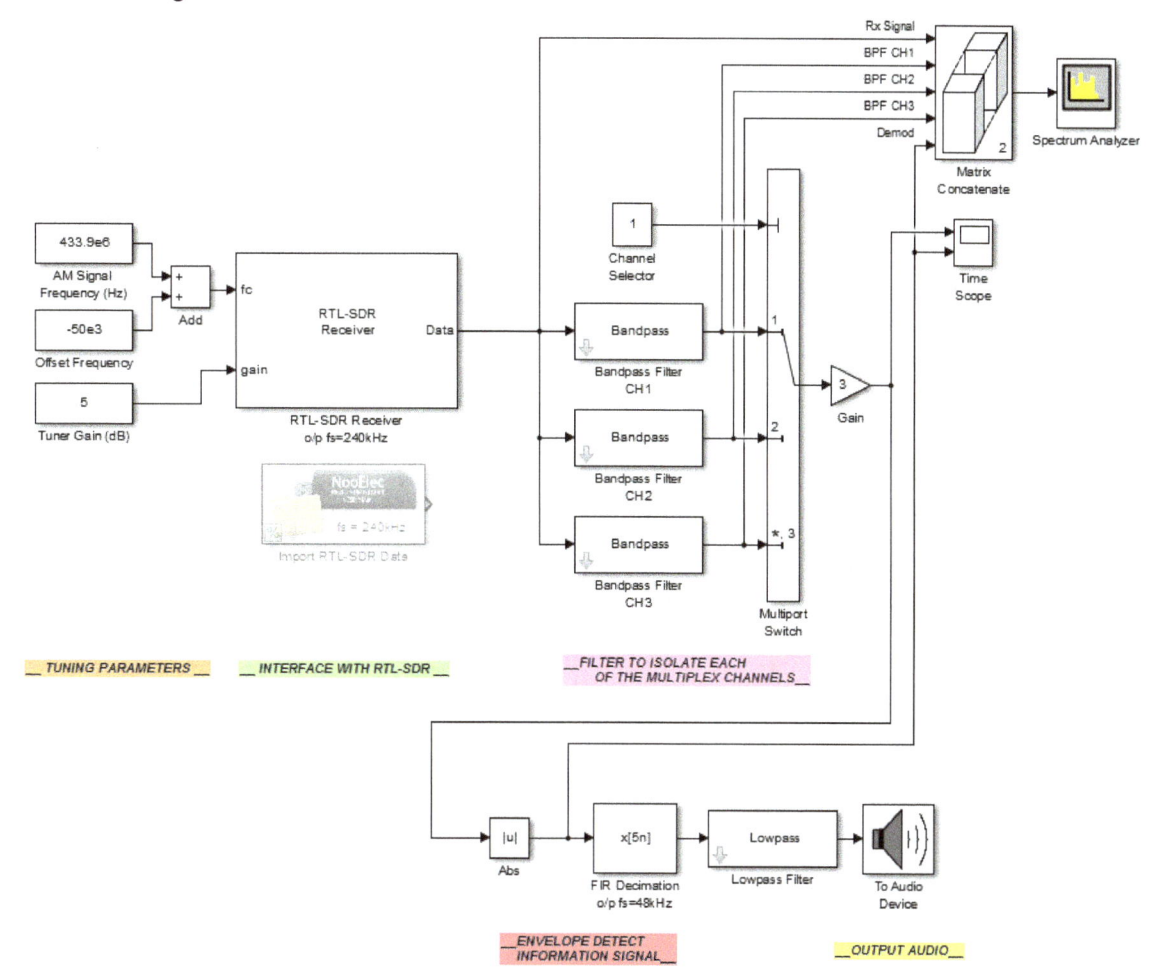

(b) If it is being used, the RTL-SDR should be tuned so that there is a frequency offset of 50kHz between the centre of CH1 and 0Hz. This ensures that all of the MPX channels remain AM-DSB-TC modulated with carriers that are of a sufficiently high frequency to be demodulated using an envelope detector. The same is true if you are importing the data file.

The IF MPX signal then enters a bank of *IIR Bandpass Filters*, where each of the individual channels are isolated, according to the following:

> CH3 = 35kHz to 45kHz, centred on a 40kHz carrier
> CH1 = 45kHz to 55kHz, centred on a 50kHz carrier
> CH2 = 55kHz to 65kHz, centred on a 60kHz carrier

A channel is selected, demodulated and output to computer speakers or headphones using the *Channel Selector* constant and the *Multiport Switch*.

(c) **Check the RTL-SDR data source.** If an RTL-SDR is available and the complex MPX signal is transmitted from a USRP® transmitter, an *RTL-SDR Receiver* block should be used to interface with and control the RTL-SDR. Otherwise, the *Import RTL-SDR Data* block should be used, configured to reference the following file:

> /am/rtlsdr_rx/rec_data/am_fdm.mat

If importing data rather than connecting to an RTL-SDR, delete the *RTL-SDR Receiver* block and connect the *Import RTL-SDR Data* block in its place.

(d) **Prepare to run the simulation.** Connect speakers or headphones to your computer and perform a test to ensure that they are working. If importing a signal, set the *Simulation Stop Time* to 60 seconds, '60'. If you are using an RTL-SDR, set the *Simulation Stop Time* to 'inf' by typing this into the Simulink toolbar, and the *Simulation Mode* to 'Accelerator'. This will allow the model to run faster. If using an RTL-SDR, check that MATLAB can communicate with it by typing `sdrinfo` into the MATLAB command window.

(e) **Run the simulation.** Begin the simulation by 1️⃣ on the 'Run' ▶ button in the Simulink toolbar. After a couple of seconds the simulation will begin, the *Spectrum Analyzer* windows should appear, and there should be an audio signal output to your speakers or headphones. If using an RTL-SDR, adjust the frequency and gain values until the device is tuned to the desired signal. Make sure that you set the centre frequency of the RTL-SDR 50kHz below the centre frequency of the signal being transmitted by the USRP® hardware.

(f) **Signal Analysis: *Spectrum Analyzer*.** The *Spectrum Analyzer* window should show a plot similar to the following:

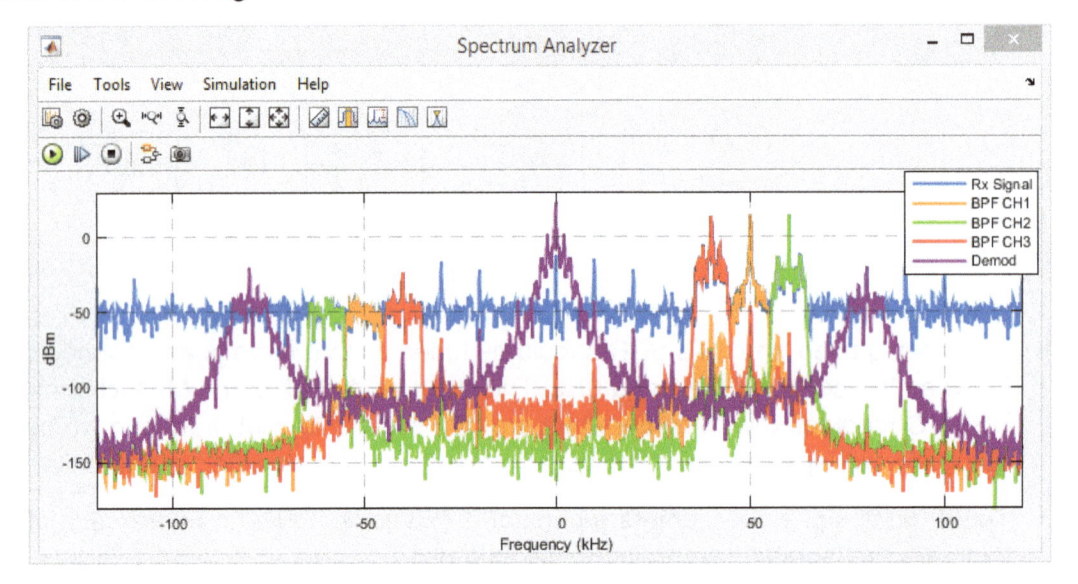

This rather colourful window shows the spectra of five different signals. The blue received signal was plotted first, so is at the back layer. If you zoom in to around the filtered channels, it should be possible to confirm that the received signal is present behind them.

CH1 is centred on 50kHz, CH2 on 60kHz, and CH3 on 40kHz, as expected. Note that each of these channels exists independently of one another, as in Exercise 8.10. Finally, the demodulated signal (which is once again a real signal, hence its symmetrical appearance in the double-sided spectrum) is shown in purple, and it is centred around 0Hz.

(g) Can you clearly hear each of the music signals when changing the value of the *Channel Selector* constant? What effect does changing the channel have on the spectrum of the demodulated output signal? Why do the spectral images change position depending on which channel is being output?

(h) **Signal Analysis: *Time Scopes*.** The *Time Scope* is configured to plot two signals in the time domain. The upper plot will show the selected channel, while it is still AM-DSB-TC modulated, and the lower will show the same signal after it has passed through the envelope detector. Any amplitude fluctuations in the demodulated signal should match fluctuations in the envelope of the modulated signal, as has been highlighted in previous exercises.

As there are three different AM-DSB-TC signals in the MPX, each with a different carrier frequency, there will be a noticeable difference in the time domain plots of the modulated signals when switching between the channels. To make a clear comparison, you can also add a second *Time Scope* and connect the three complex outputs of the bandpass filters. In the plot shown below, the signals are displayed in order of carrier frequency, rather than by channel number, which should hopefully make the differences in the carrier frequency clearer.

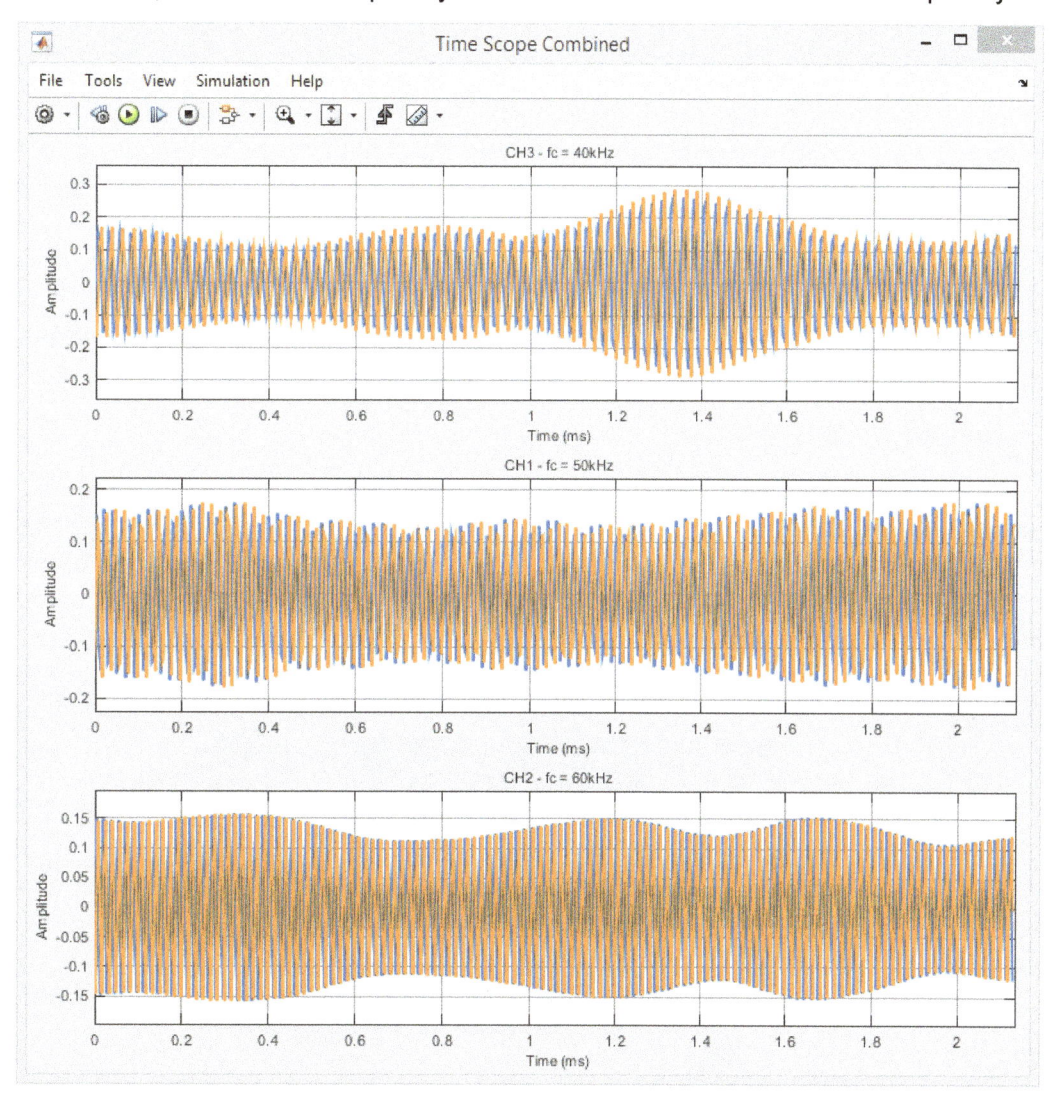

(i) This exercise (and Exercise 8.10) was designed to introduce the concept of multiplexing, which will be revisited in Chapters 9 and 10 when investigating the stereo FM MPX. It has highlighted that complex signals can carry twice the amount of information of real signals, in a given bandwidth. This idea is critical when it comes to digital communications, as will be explored in Chapters 11 and 12.

(j) **Watch and listen to us demodulate AM-DSB-TC signals from the MPX signal.** We have recorded a video which shows activity in the *Spectrum Analyzer* and *Time Scope* windows as the complex MPX signal transmitted from the USRP® hardware (with the model from Exercise 8.10) is received by the RTL-SDR and demodulated. The output audio signal has been recorded too.

desktopSDR.com/videos/#fdm_demod

8.5 Alternative Hardware for Generating Desktop AM Signals

As was discussed at the beginning of this chapter, commercially broadcast AM radio signals transmit at too low a frequency to be received with the RTL-SDR, and as a result of this, you will need to create your own. Although we initially focused on using a USRP® radio to produce AM signals, the USRP® hardware is by no means your only option. We now consider two alternatives to the USRP® hardware.

8.5.1 The Ham It Up: Upconvert MF and HF Radio Signals to 125MHz

The *Ham It Up* (HIU) is a device that upconverts (heterodynes) MF and High Frequency (HF) radio signals [300kHz–30MHz] from their respective RF frequencies to an IF, such that:

$$f_{c\,(ham\,it\,up)} = f_{c\,(am\,signal)} + 125MHz \quad . \tag{8.4}$$

The device is the result of an open source project run by Opendous [103], and it is manufactured and sold by NooElec (the retailers of the NESDR RTL-SDRs). The HIU was designed to be used with low cost SDRs such as the RTL-SDR, to enable users to receive short-wave, medium-wave and long-wave (commercial and HAM) AM radio signals. In keeping with the low cost SDR ethos, the HIU is priced at a very reasonable £25/ $45, and is available from the NooElec website [70].

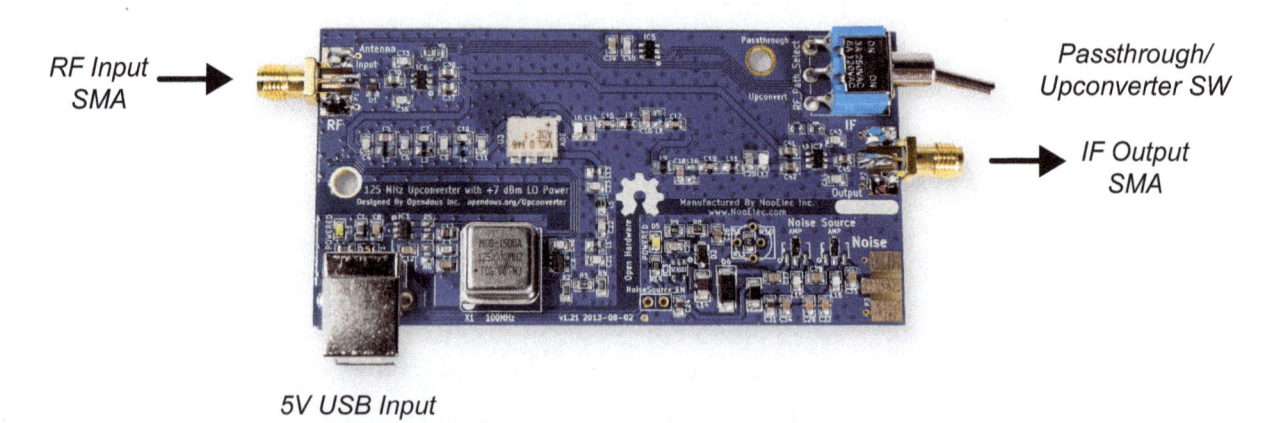

Figure 8.16: The Ham It Up, v1.2 [70]

The HIU (shown in Figure 8.16) is powered over a 5V USB connection and has in-line SMA (SubMiniature version A) connectors to which an antenna and RTL-SDR device can be attached. RF signals entering the upconverter from the *Antenna Input RF* SMA are lowpass filtered, then mixed with a sinusoidal signal from a 125MHz crystal oscillator. This process is fundamentally AM-DSB-TC modulation, and it results in the IF signal being centred around the 125MHz carrier. The IF signal is bandpass filtered so that only the carrier and upper sideband remain, and it is then output from the HIU via the *IF Output* SMA. The switch situated in the top right hand corner of the board can be used to bypass the upconverter, and connect the antenna directly to the RTL-SDR.

In Appendix E.1 (page 607) we will demonstrate how to finish construction of the HIU (you need to plug a component in!), and also test that it works. One thing to highlight about this device is that it *will not* allow you to create your own AM signals, but rather it simply upconverts existing ones. Whilst this is suitable if the sole aim is to receive signals, creation of custom signals will require a USRP® transmitter or a dedicated AM transmitter.

8.5.2 Building AM Transmitters

Designing, building, and testing AM modulator and transmitter circuits is a process the authors found to be very educational and fun! They are actually quite simple to make, and only require basic circuit building/ soldering skills.

Step-by-step instructions are included in Appendix E.2 (page 612) detailing how to construct an AM transmitter based on a commercially available modulator board called the 'RT4', which is shown in Figure 8.17. This device is an AM-DSB-TC modulator that modulates baseband information signals onto a 433.9MHz carrier.

The book would be never-ending if we presented all of the different ways that you can generate AM signals, so we have only included build instructions for this particular transmitter. Others methods can be found at desktopSDR.com.

Figure 8.17: RT4 AM Modulator (433.9MHz)

8.5.3 Transmitting AM Signals

If you have an AM transmitter available with an audio jack input, it will be possible to transmit any of the audio sources present in the

    `/audio_sources`

folder. Simply copy these to your computer audio player or smartphone, connect, and press play.

If you wish to try transmitting an AM MPX signal (which can be received and demodulated with the model from Exercise 8.11), open and run the following file:

    `/am/other_tx/amtx_am_fdm.slx`

This will generate a two channel MPX signal (saved in a single channel, mono '`.wav`' audio file format), which can be transmitted as described above.

8.6 Summary

In this chapter we have presented exercises on the theme of AM desktop SDR Tx/ Rx systems. By constructing Simulink transmitters for three of the modulation schemes presented in Chapter 6, a variety of different coherent and non-coherent AM demodulators were implemented as part of receiver designs. We introduced the concept of FDM, and demonstrated how it could be used with the USRP® radio's complex modulator to transmit multiple channels of audio signals at the same time. Moving on, the next chapter will introduce the theory behind another analogue modulation scheme — FM — which we will first work with in simulation, and later in Chapter 10, by constructing some more real-time SDR Tx/ Rx designs.

9 Frequency Modulation (FM) Theory and Simulation

Another commonly used analogue modulation standard is FM. This is a modulation scheme where a carrier wave is modulated with an information signal in a way that causes its frequency to fluctuate as the amplitude of the information signal changes. This means that, unlike AM, the amplitude of the modulated carrier signal remains constant. This modulation standard is commonly used for commercial radio stations due to its high resilience to additive noise, and is utilised by many of us to listen to our favourite stations in the car on the way to work every day.

In this chapter we will give a brief overview of the history of FM and the mathematics behind it, before moving on to discuss some of the different receivers that can be used to demodulate FM signals received by your RTL-SDR.

9.1 The History of the FM Standard

The story behind the birth of FM is one of innovation, lawsuits and tragedy. An American named Edwin Howard Armstrong first developed and demonstrated FM radio in 1933 in an effort to combat the 'static problem' that plagued AM radio transmissions. Born in 1890, Armstrong studied electrical engineering at Columbia University where he later became a Professor.

Prior to his work on frequency modulation, Armstrong spent many years making improvements to the AM radio. In 1912 he discovered that, if electromagnetic waves emitted from *Lee de Forest* glass valve radio receivers were fed back through the hardware, the signal strength increased exponentially and radio waves were generated. He called this positive feedback process *regeneration*, and it is regarded as one of the most important discoveries in the history of radio, as it meant that radio receivers could also be used as transmitters.

After designing and patenting regeneration circuits that improved the quality of AM radio receivers, Armstrong was sent to France during the First World War to serve as a Major in the Signal Corps [28].

At the time, communication methods within the Allied Forces were basic, and he took it upon himself to equip them with his inventions, waiving all patent royalty fees as he did so [89]. Before the end of the war, he also developed the *super heterodyne receiver*, a device that is still used today in almost every analogue radio, where RF signals are downconverted to an IF, allowing for cheaper hardware to be used in the radio receiver [79]. The IF downconversion concept is also used in digital and software radio architectures, including that of the RTL-SDR.

Returning to America, Armstrong became entangled in the first of his many legal battles. After a long dispute over who originally developed, and therefore owned the rights to the regenerator circuit, a US Supreme Court judge ruled against him in favour of *Lee de Forest* (the inventor of the glass valve radio), who enjoyed industrial financial backing at the time. As a result, Armstrong lost the patents he had previously been awarded.

Figure 9.1: Armstrong explaining the principles of 'super regeneration' at a meeting of the RCA, Columbia University, New York, 1922 [5]

While the court case was ongoing, Armstrong worked quietly on a different problem that existed within the radio industry: the problem of static. He discovered that, by varying the frequency of a radio carrier rather than its amplitude, transmitted signals were far more resilient to interference from noise and neighbouring transmissions. He was working on what became known as FM radio. Obtaining four patents, he acquired a license from the FCC and began work on a prototype FM radio network [28]. Armstrong had a 400ft lattice tower constructed in Alpine, New Jersey, to broadcast his FM radio signals. Although this tower was built in 1937, it is still in use today [90]!

The early trials on FM were successful, and anyone who listened to the audio quality could tell that FM was far superior to AM. Once again however, the opportunities for FM were affected by the actions of 'interested parties', with some sources suggesting that influential companies including *Radio Corporation of America* (RCA) deliberately obstructed the technology for commercial reasons [28] (another theory is that they failed to initially recognise the potential of FM [10]). The FCC changed the terms of the license Armstrong had been granted for his FM radio network, apparently in response to lobbying from companies with a commercial interest in the technology, and the frequency band he had originally been allocated (42–50MHz) was changed to 88–108MHz. This rendered all of his equipment obsolete [90]. After this incident, Armstrong challenged that RCA and NBC were infringing on the patents he held for FM technology. Soon he became caught up in legal battles, which lasted many years, affecting his family life and eventually leaving him bankrupt. Being unable to keep paying his legal team helped to drive him to depression. On the 1st February 1954, Armstrong committed suicide.

Several years went by before FM really took off, and it was not until the early 1960s that regulators started laying the ground rules for the standard. Even then, it was not until the mid 1970s before the number of FM radio stations overtook the number of AM stations. FM radio still remains popular to this day, although the 'digital switchover' looks to transition towards DAB in the near future — perhaps — but let us not write off this analogue modulation standard just yet!

Chapter 9: Frequency Modulation (FM) Theory and Simulation

9.2 The Mathematics of FM & the Modulation Index

One of the simplest forms of analogue FM modulator is the VCO. This device shown below in Figure 9.2 generates a sinusoidal signal, $c(t)$, the phase (and therefore effectively the frequency) of which changes in response to amplitude variations of an input control signal.

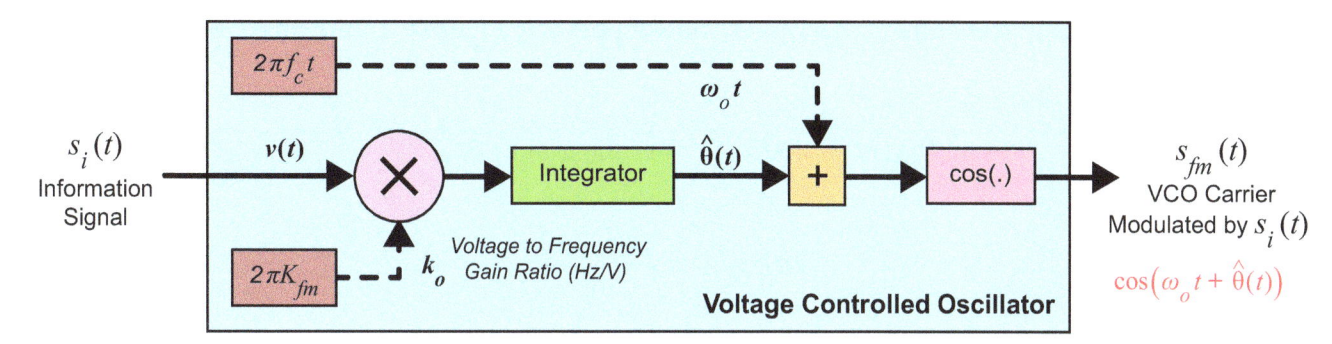

Figure 9.2: Utilising a VCO to perform FM modulation

When the control signal is input to the VCO, it is multiplied by k_o, a constant representing the 'voltage to frequency gain ratio' of the device (measured in Hz/V). The product of $v(t)$ and k_o is then integrated (changing its phase by 90 degrees). The integrated signal is denoted $\hat{\theta}(t)$:

$$\hat{\theta}(t) = k_o \int_{-\infty}^{t} v(t)\, dt. \tag{9.1}$$

The sinusoid generated by the VCO is configured to have quiescent frequency f_o and amplitude A_o. The phase of the sinusoid is determined by the instantaneous value of $\hat{\theta}(t)$. The signal output from the VCO, as introduced in Section 7.5.3 (page 248), takes the following form:

$$c(t) = A_o \cos\left(2\pi f_o t + \hat{\theta}(t)\right) \tag{9.2}$$

$$= A_o \cos\left(\omega_o t + k_o \int_{-\infty}^{t} v(t)\, dt\right).$$

When an information signal is input to the VCO's control port, and k_o is substituted with the FM modulation constant, $k_o = 2\pi K_{fm}$, Eq. (9.1) becomes:

$$\hat{\theta}(t) = \theta_{fm}(t) = 2\pi K_{fm} \times \int_{-\infty}^{t} s_i(t)\, dt. \tag{9.3}$$

Substituting Eq. (9.3) into Eq. (9.2), and replacing A_o and ω_o with the parameters of an FM carrier (A_c and ω_c) gives the FM signal [36],

$$s_{fm}(t) = A_c \cos\left(\omega_c t + \theta_{fm}(t)\right) = A_c \cos\left(\omega_c t + 2\pi K_{fm} \times \int_{-\infty}^{t} s_i(t)\, dt\right). \tag{9.4}$$

9.2.1 FM Modulation: Modulation with a Sine Wave

We shall first define an example information signal as having amplitude A_i and the frequency f_i,

$$s_i(t) = A_i \cos(2\pi f_i t) = A_i \cos(\omega_i t), \tag{9.5}$$

where $\omega_i = 2\pi f_i$. Inputting this to the VCO's control port (see Figure 9.2), the phase $\theta_{fm}(t)$ at time t becomes:

$$\theta_{fm}(t) = 2\pi K_{fm} A_i \times \int_{-\infty}^{t} \cos(\omega_i t)\, dt \tag{9.6}$$

$$= 2\pi K_{fm} A_i \times \frac{\sin(\omega_i t)}{\omega_i}$$

$$= \frac{K_{fm} A_i}{f_i} \sin(\omega_i t)$$

$$= \frac{\Delta f}{f_i} \sin(\omega_i t)$$

$$= \beta_{fm} \sin(\omega_i t)$$

where Δf is known as the Frequency Deviation, and β_{fm} is known as the Modulation Index.

Therefore the FM signal for the modulation of a single tone at frequency f_i can be expressed as:

$$s_{fm}(t) = A_c \cos\left(\omega_c t + \beta_{fm} \sin(\omega_i t)\right), \tag{9.7}$$

where: $\omega_c = 2\pi f_c$ and $\omega_i = 2\pi f_i$.

Δf represents the maximum frequency deviation of the FM carrier. We can note that the highest and lowest frequencies of Eq. (9.7) (found by differentiating its instantaneous phase) will be $f_c \pm \Delta f$.

When the amplitude of the information signal input to the VCO is 0, $s_{fm}(t)$ has frequency f_c. In all other situations, the instantaneous frequency of $s_{fm}(t)$ can be calculated with Eq. (9.8):

$$f_{fm\ inst}(t) = f_c + K_{fm} s_i(t)\ \text{Hz}. \tag{9.8}$$

Figure 9.3 shows the signal created by FM modulating a carrier with a single tone information signal. This process results in many sinusoidal terms located at frequencies either side of the carrier, and the number of these terms depends upon the value of β_{fm}. This means it is not straightforward to illustrate what an FM signal looks like in the frequency domain and it is not possible to easily and visually correlate the modulated spectrum with the baseband spectrum (recall the AM modulated spectrum for example in Figure 6.3 (page 205), where the baseband spectrum matches the upper and lower sidebands). We will review the FM spectrum later in Section 9.3.

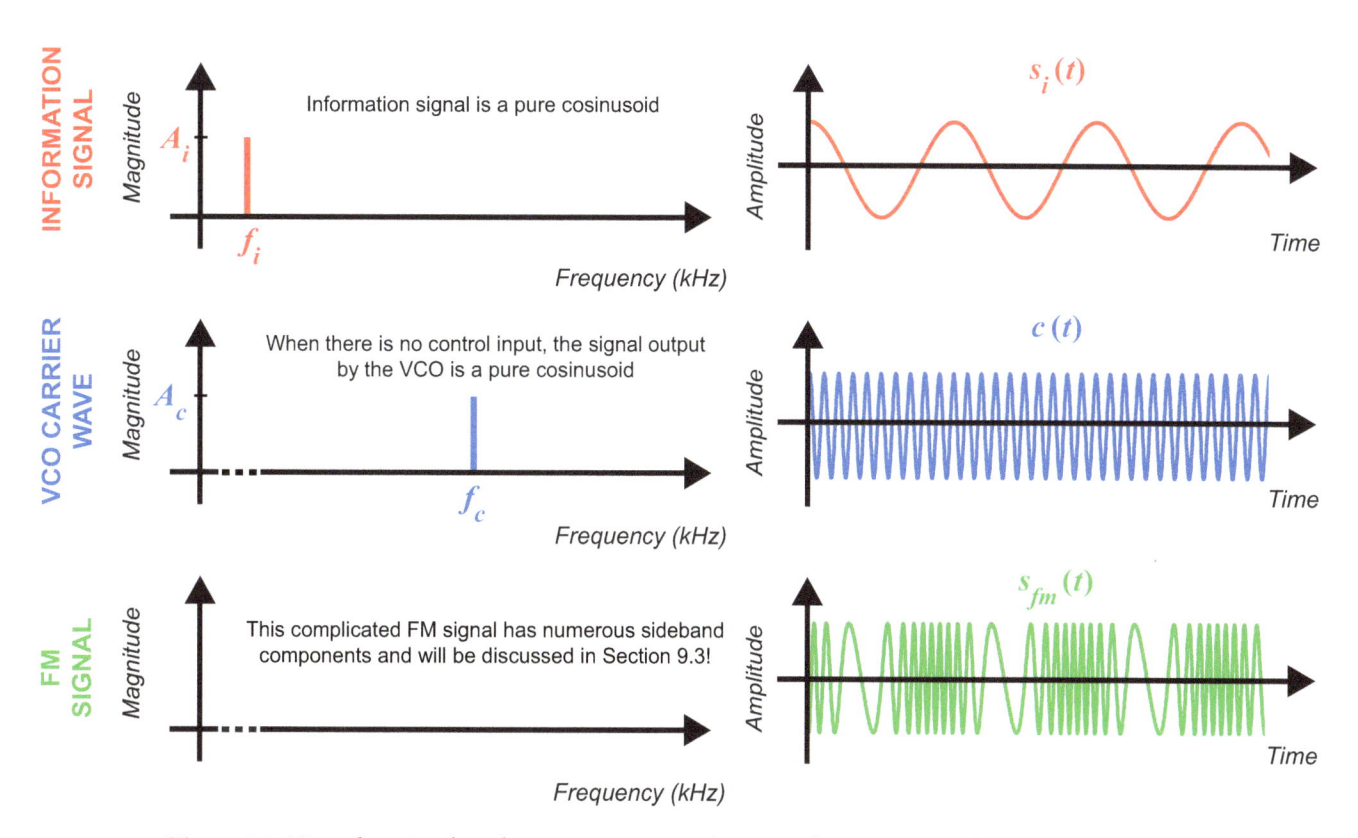

Chapter 9: Frequency Modulation (FM) Theory and Simulation 333

Figure 9.3: Time domain plots demonstrating a single tone information signal being FM modulated

It is however clear to see that when the amplitude of the information signal increases, the frequency of the modulated signal increases (and vice versa). This means that the main component of the FM signal 'wiggles' around the carrier frequency, as should have been evident in Exercise 3.4.

9.2.2 FM Modulation: Modulation with a Baseband Signal

Information signals, such as speech and music, are of course always much more spectrally rich than a single sinusoidal wave, and are composed of a set of many frequency components. Hence to more fully calculate the bandwidth of an FM signal, we need to consider more than just a single sine wave.

Based on Fourier series and transform theory, it is easy to confirm that we can decompose any signal over a period of time as a sum of sinusoids [29]. The bandwidth of an information signal $s_i(t)$ can be considered to be f_h Hz, where f_h is the highest frequency component contained within the signal. Figure 9.4 demonstrates how a carrier would look in the time domain after it was FM modulated by a baseband information signal. As for the case of single tone modulation in Figure 9.3, the frequency of the FM signal changes in sympathy with the amplitude of the information signal.

We can define an information signal that is composed of these three components:

$$s_i(t) = A_{i1}\cos(2\pi f_{i1} t) + A_{i2}\cos(2\pi f_{i2} t) + A_{i3}\cos(2\pi f_{i3} t) \tag{9.9}$$

$$= A_{i1}\cos(\omega_{i1} t) + A_{i2}\cos(\omega_{i2} t) + A_{i3}\cos(\omega_{i3} t),$$

$$\text{where: } \omega_i = 2\pi f_i.$$

Using the integral-of-a-sum rule from Eq. (B.12) from Appendix B (page 581), the phase of the FM signal $\theta_{fm}(t)$ can be found as follows:

$$\theta_{fm}(t) = 2\pi K_{fm} \times \int_{-\infty}^{t} \left[A_{i1}\cos(\omega_{i1}t) + A_{i2}\cos(\omega_{i2}t) + A_{i3}\cos(\omega_{i3}t) \right] dt \qquad (9.10)$$

$$= 2\pi K_{fm} \left[A_{i1}\int_{-\infty}^{t}\cos(\omega_{i1}t)\,dt + A_{i2}\int_{-\infty}^{t}\cos(\omega_{i2}t)\,dt + A_{i3}\int_{-\infty}^{t}\cos(\omega_{i3}t)\,dt \right]$$

$$= \frac{K_{fm}A_{i1}}{f_{i1}}\sin(\omega_{i1}t) + \frac{K_{fm}A_{i2}}{f_{i2}}\sin(\omega_{i2}t) + \frac{K_{fm}A_{i3}}{f_{i3}}\sin(\omega_{i3}t).$$

Substituting this into Eq. (9.7) results in the following (rather complicated!) FM signal:

$$s_{fm}(t) = A_c \cos\left(\omega_c t + \left[\frac{K_{fm}A_{i1}}{f_{i1}}\sin(\omega_{i1}t) + \frac{K_{fm}A_{i2}}{f_{i2}}\sin(\omega_{i2}t) + \frac{K_{fm}A_{i3}}{f_{i3}}\sin(\omega_{i3}t)\right]\right). \qquad (9.11)$$

It is less easy to tell what the modulation index or frequency deviation of this signal is, compared to the previous case of transmitting a single sine wave, as there are many parameters in the phase component. The bandwidth of the signal can still be estimated however, as will be explained in the next section.

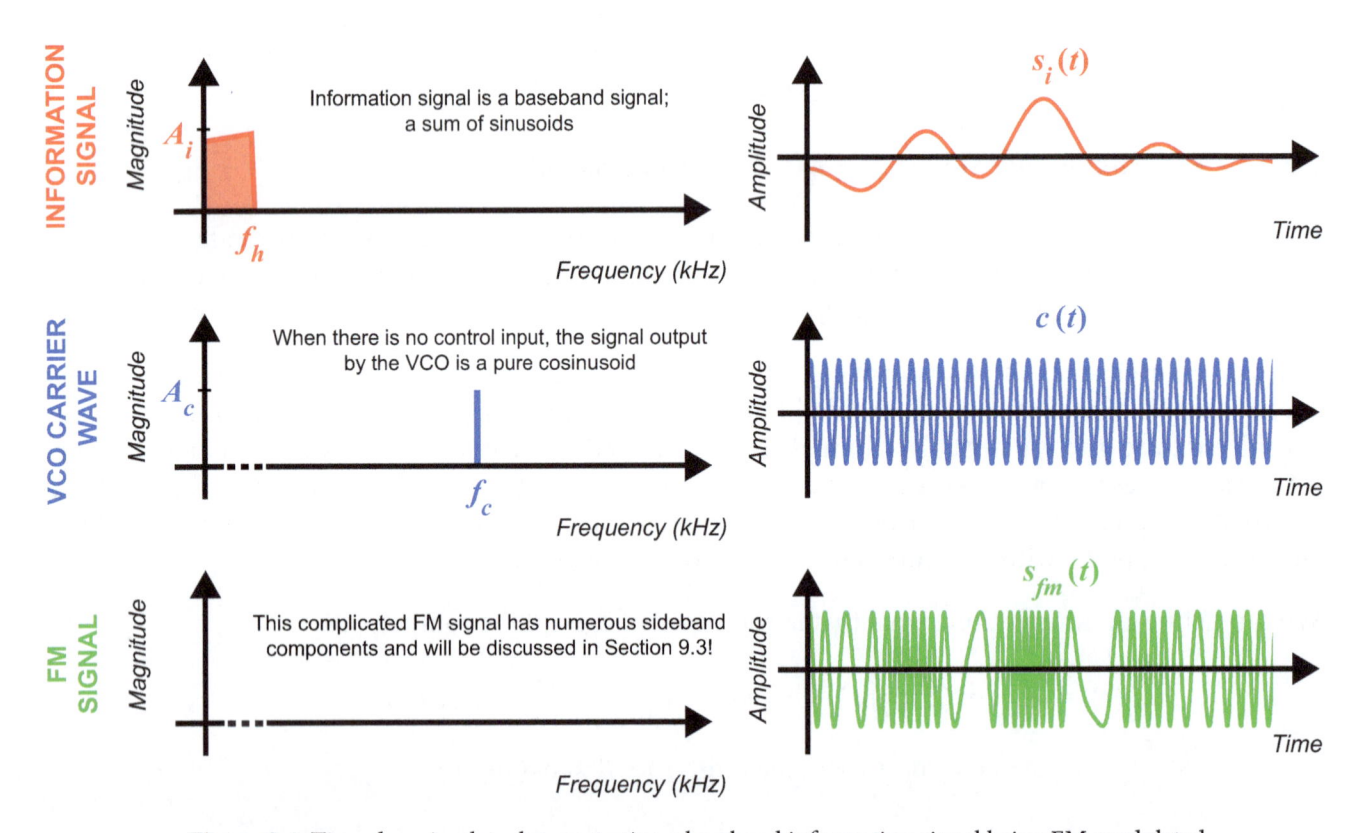

Figure 9.4: Time domain plots demonstrating a baseband information signal being FM modulated

Chapter 9: Frequency Modulation (FM) Theory and Simulation 335

9.3 FM Signal Bandwidth

Frequency modulation is either considered to be a *Narrowband* or a *Wideband* process; and the value of β_{fm} is what determines this. If the modulation index of an FM signal is $\ll 1$, it is considered to be *Narrowband FM* (NFM), while if it is $\gg 1$ it is *Wideband FM* (WFM). Next, we will discuss scenarios for NFM, before going on to cover WFM.

9.3.1 Narrowband FM

In the case of NFM, approximations can be made that help solve Eq. (9.7) because of the negligible size of β_{fm}. This is a result of the maximum frequency deviation permitted for this standard being limited to 5kHz. Assuming we are working with the single tone information signal once again, we can begin by expanding Eq. (9.7) using the sum-to-difference trigonometric identity (see Eq. (B.1) from Appendix B):

$$s_{fm}(t) = A_c \cos\left(\omega_c t + \beta_{fm} \sin(\omega_i t)\right) \qquad (9.12)$$

$$= A_c \cos(\omega_c t) \cos\left(\beta_{fm} \sin(\omega_i t)\right) - A_c \sin(\omega_c t) \sin\left(\beta_{fm} \sin(\omega_i t)\right).$$

For very small values of β_{fm}, we assume:

$$\cos\left(\beta_{fm} \sin(\omega_i t)\right) \approx 1 \quad \text{and} \quad \sin\left(\beta_{fm} \sin(\omega_i t)\right) \approx \beta_{fm} \sin(\omega_i t),$$

meaning Eq. (9.12) can be simplified using Eq. (B.5) from Appendix B to give:

$$s_{fm-nfm}(t) = A_c \cos(\omega_c t) - A_c \sin(\omega_c t) \beta_{fm} \sin(\omega_i t) \qquad (9.13)$$

$$= A_c \left[\cos(\omega_c t) + \frac{\beta_{fm}}{2} \cos(\omega_c + \omega_i)t - \frac{\beta_{fm}}{2} \cos(\omega_c - \omega_i)t \right].$$

You may recognise this equation, and this is because it closely resembles the equation of the AM-DSB-TC signal (recall Eq. (6.5)) which had the form:

$$s_{am-dsb-tc}(t) = A_c \left[A_o \cos(\omega_c t) + \frac{A_i}{2} \cos(\omega_c + \omega_i)t + \frac{A_i}{2} \cos(\omega_c - \omega_i)t \right].$$

The NFM signal is essentially the same as the AM-DSB-TC signal, with the one difference that its lower sideband (LSB) is inverted; i.e. the LSB of the NFM signal is 180 degrees out of phase with the lower sideband of the AM-DSB-TC signal, as is illustrated in Figure 9.5.

When a baseband information signal is used instead of a single tone to create an NFM signal, the same results are found, and it is simply the case that the entire lower sideband is inverted. Noting that the spectra of both NFM and AM-DSB-TC are similar means that the bandwidths of these two signals are approximately the same. Due to its efficient use of the spectrum and cheap implementation costs, NFM is commonly used for voice applications such as wireless microphones, radios in taxis and emergency service vehicles, and Ham radios.

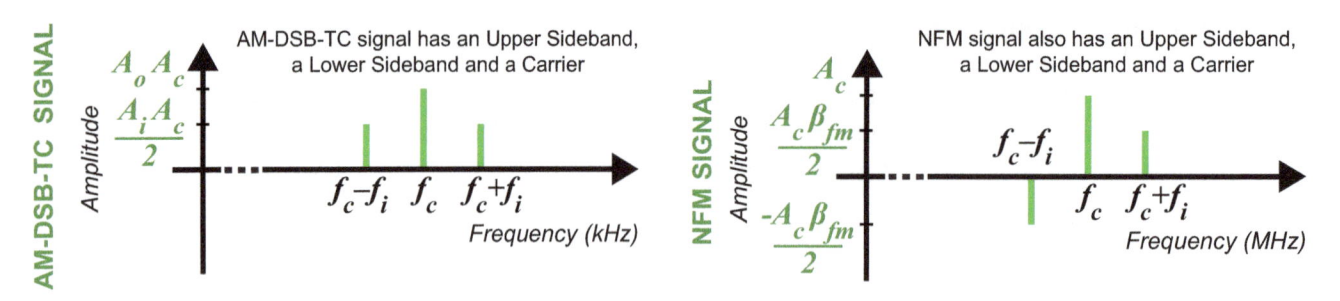

Figure 9.5: Frequency domain plots comparing the spectrums of AM-DSB-TC (left) and NFM (right) signals created with a single tone information signal

You can now complete Exercise 9.1, where we will ask you to run a simple Simulink simulation that shows what happens (in the time and frequency domains) to sinusoidal waves as they are used to modulate a carrier and produce a NFM signal.

Exercise 9.1 Narrowband FM (NFM) Simulation

In this exercise we will test out the NFM modulator. There are two different information signals included in this model, which you will be able to switch between. *Spectrum Analyzers* and a *Time Scope* are connected to various points throughout the model to allow you to monitor the signals in the time and frequency domains as they are modulated.

(a) **Open MATLAB.** Set the working directory to an appropriate folder so you can open this model:

📁 /fm/simulation/fm_nfm.slx

The block diagram should look like this:

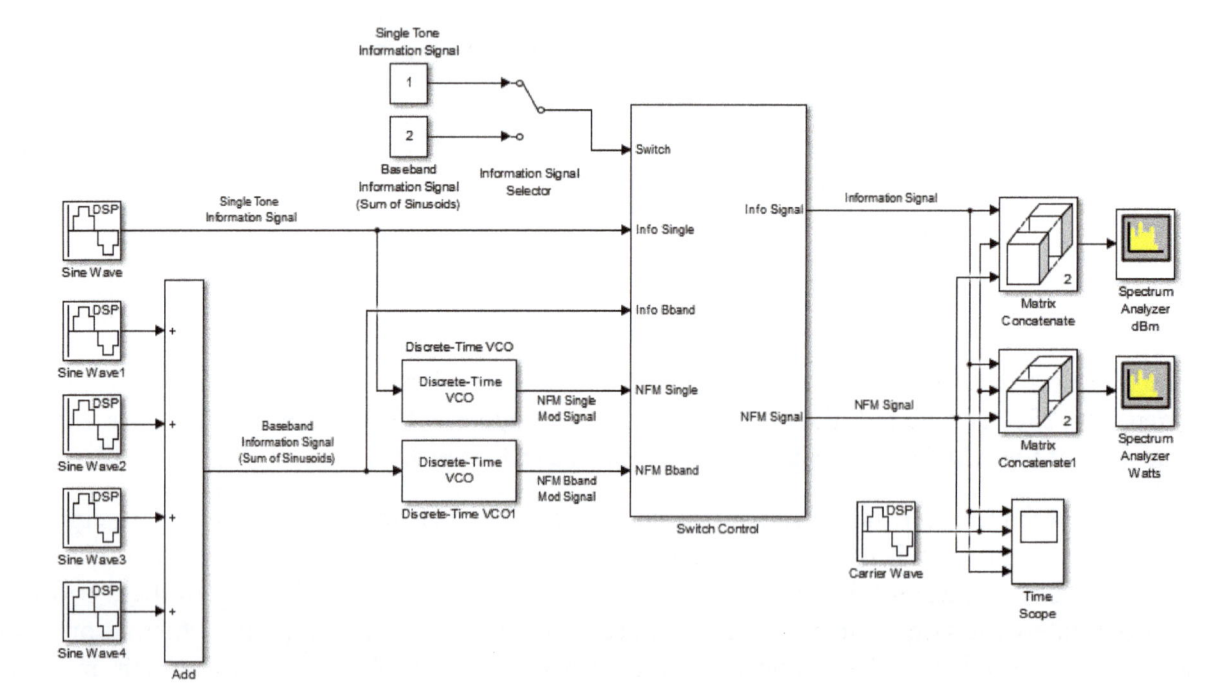

(b) **Inspect the model.** The first information signal (top) is a single tone. The other is a sum of tones. They are input to the control ports of the two *Discrete Time VCOs*, each of which output an NFM signal. The *Information Signal Selector* switch allows you to switch between them. To change the selection, simply 2 on it. Leave this set on 'Single Tone' for now.

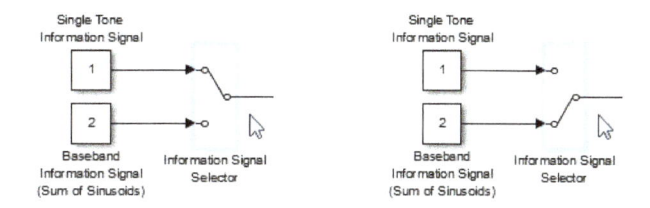

Spectrum Analyzer dBm is configured to plot the spectrums of signals input to it in dBm, while *Spectrum Analyzer Watts* is set to plot in Watts. The reason for this is to allow you to get a better feeling for how much power each sideband contains relative to the others. While the dBm scale is useful, it is easier to differentiate between signals with similar powers using Watts. For instance, a signal of 4 × greater power than a reference signal (in Watts) is equivalent to a difference of +6dBm.

(c) **Run the simulation.** Begin the simulation by 1 on the 'Run' ▶ button in the Simulink toolbar. The three scope windows should appear, and you should position these so that you can see all of them. The signals are colour coordinated: the information signal is blue, the carrier is orange, and the NFM signal is green, as shown in the legends.

(d) **Signal Analysis #1.** If you left the switch set on 'Single Tone', you should see the following in the *Spectrum Analyzer* windows:

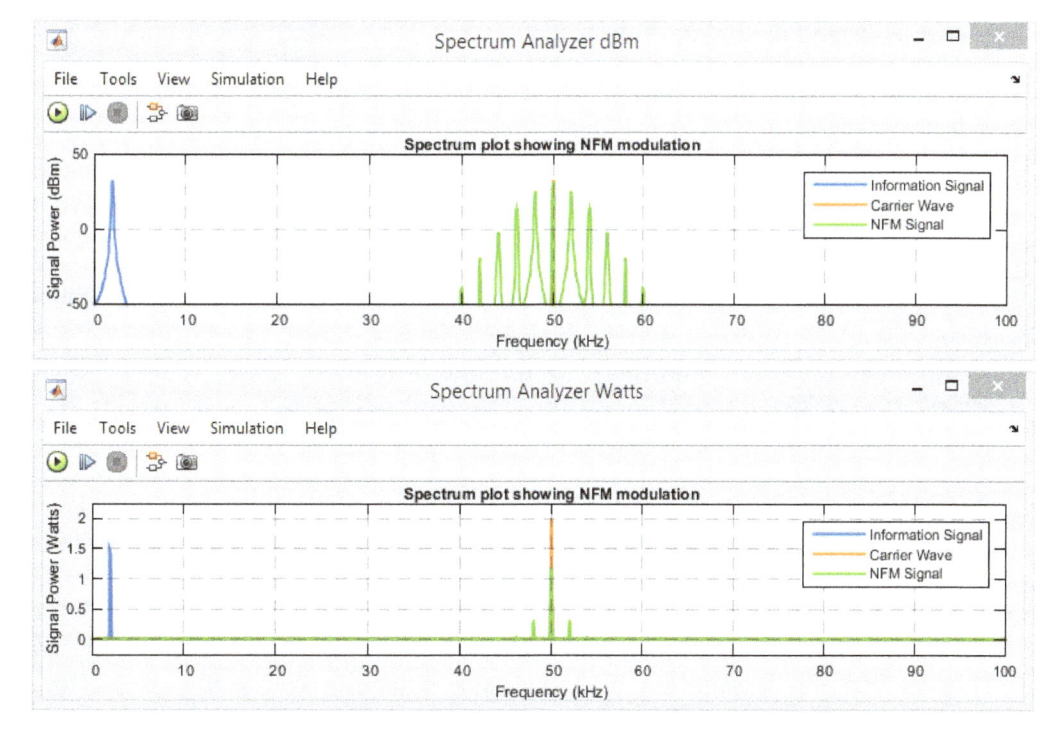

The information signal is a single tone with a frequency of 2kHz. The carrier signal generated by the VCO (also a single tone), has a quiescent frequency of 50kHz. When the information signal is used to NFM modulate the carrier, multiple sidebands are created. Only the first pair of these (located at 48kHz and 52kHz) contain a significant amount of power.

The *Time Scope* shows the information, the carrier, and the NFM signals in the time domain. It is difficult to tell if the frequency of the NFM signal changes as the amplitude of the information signal fluctuates, and this is because of the small frequency deviation associated with this type of modulator.

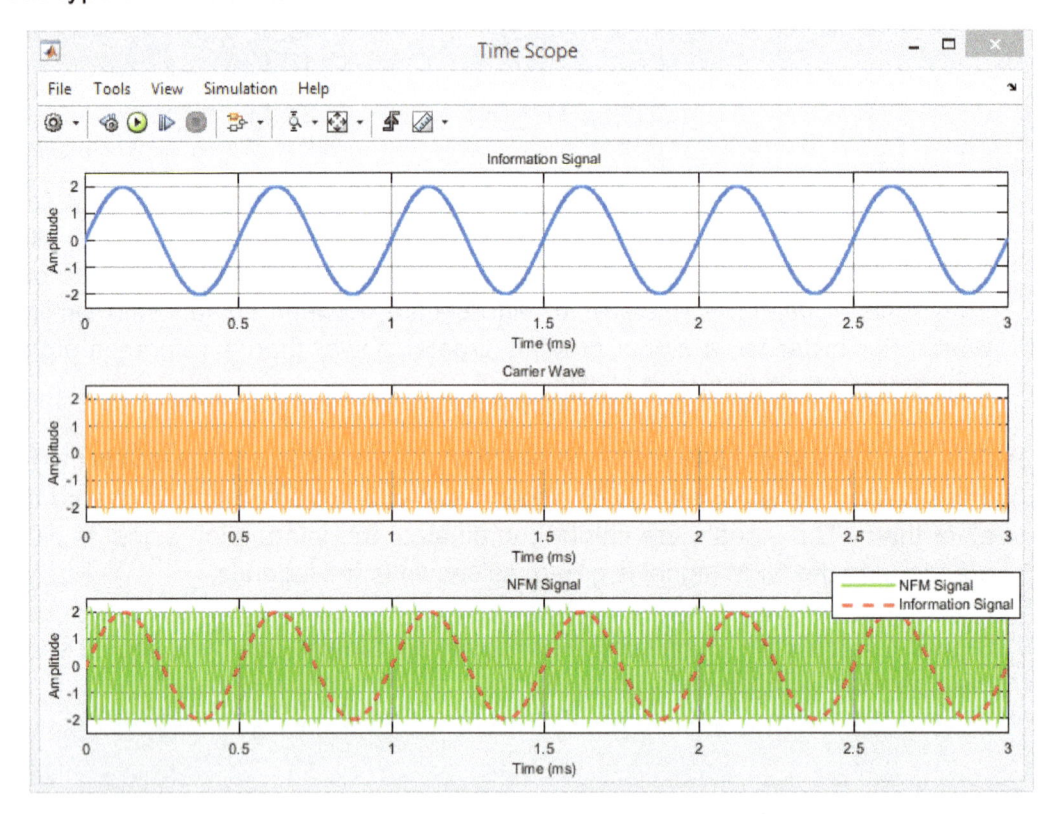

(e) **Signal Analysis #2.** Set the 'Information Signal Selector' switch to 'Baseband Information Signal', and re-run the simulation. The *Spectrum Analyzer* and *Time Scope* windows should now show views similar to those on the next page.

This time the information signal contains four frequency components, located at 1kHz, 2kHz, 3kHz and 4kHz. When it is used to modulate the 50kHz carrier, many more sidebands are created than before. It is still only the first pair of sidebands that contain significant amounts of power, giving four pairs of tones positioned at 46kHz & 54kHz, 47kHz & 53kHz, 48kHz & 52kHz, and finally 49kHz and 51kHz.

Looking at the *Time Scope* it is once again difficult to tell if the frequency of the NFM signal is changing. This is for the same reason as before, that the frequency deviation is small.

(f) **Investigate the simulation parameters.** Open the Simulink Model properties by navigating to *File > Model Properties > Model Properties*. 🖱 on the *Callbacks* tab, then on *PostLoadFcn*. This is where the parameters for the simulation are set. To find the values (for (k_o) Kvco_sum and Kvco_single, work backwards using Eq. (9.2) to find the corresponding values of K_{fm}. Make a note of the number you calculate as you will need it in Exercise 9.2.

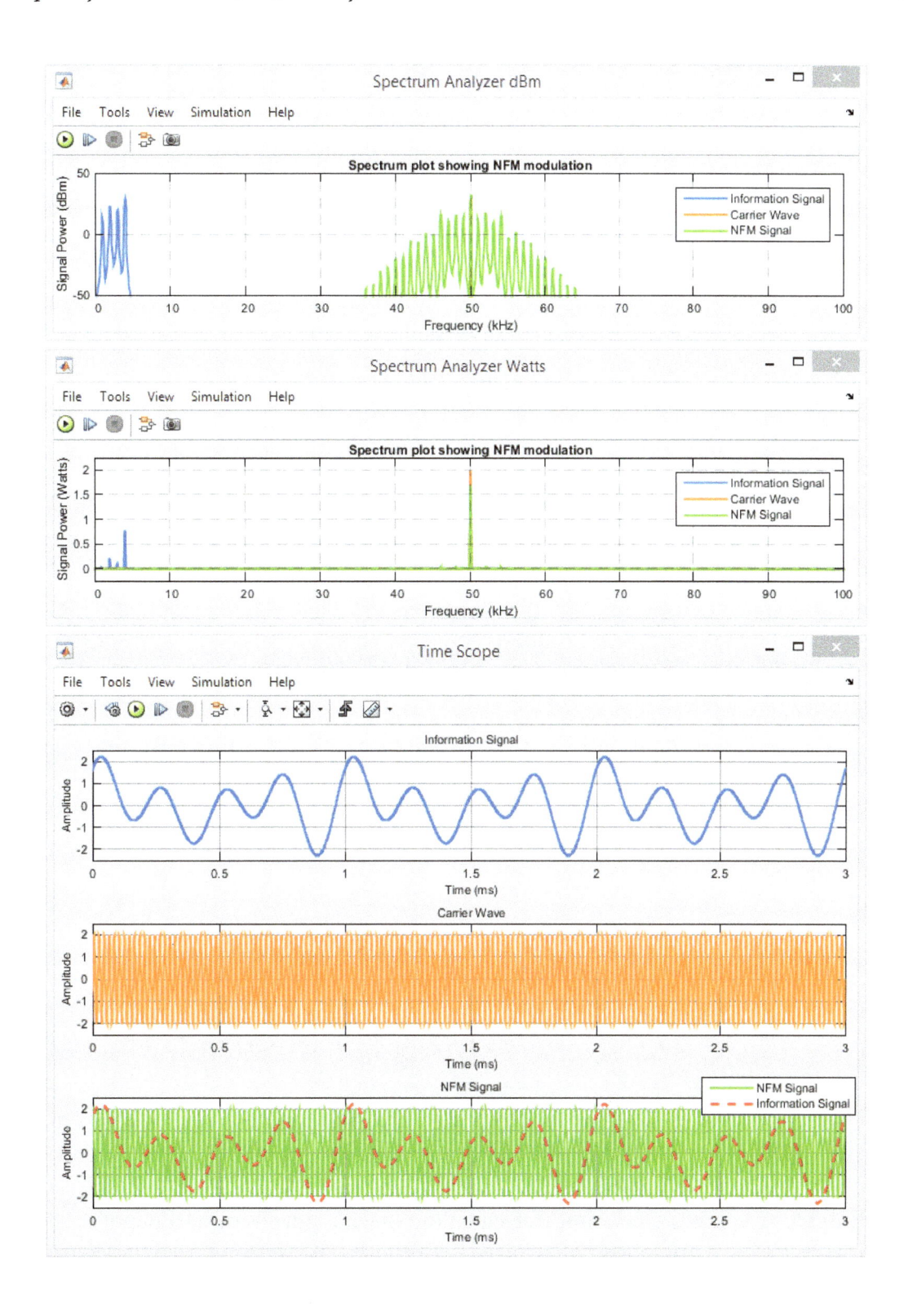

Chapter 9: Frequency Modulation (FM) Theory and Simulation

9.3.2 Wideband FM

When most people refer to 'FM radio', what they are actually talking about is *wideband* FM (WFM) radio. WFM is the standard used by commercial radio stations, and it has a frequency deviation (Δf) of 75kHz and a limited bandwidth. In the UK and many other countries, the bandwidth limit is 200kHz, which is why you will find FM radio stations positioned 0.2MHz apart on the dials of your analogue radios (102.5 FM, 102.7 FM and so on). The bandwidth has to be limited as it is in theory *infinite*, due to the creation of an *infinite* number of sidebands during the modulation process, as will be outlined. These sidebands are located at every positive and negative multiple of the information signal's frequency around the carrier ($f_c \pm n f_i$) where n is a positive integer, and the energy decreases as n increases.

To solve Eq. (9.12) for a single tone modulated WFM signal the terms,

$$\cos\!\left(\beta_{fm}\sin(\omega_i t)\right) \quad \text{and} \quad \sin\!\left(\beta_{fm}\sin(\omega_i t)\right)$$

can be expanded into their Fourier components [36]. This is typically done using Bessel functions (see below), where the sidebands are indexed using the symbol n, and J_n denotes the sideband amplitudes and is a function of the value of β_{fm}:

$$\begin{aligned}
\cos\!\left(\beta_{fm}\sin(\omega_i t)\right) &= J_0(\beta_{fm}) + 2\sum_{n=1}^{\infty} J_{2n}(\beta_{fm})\cos(2n\omega_i t) \\
&= J_0(\beta_{fm}) + 2J_2(\beta_{fm})\cos(2\omega_i t) + 2J_4(\beta_{fm})\cos(4\omega_i t) + \ldots \\
&\equiv J_0 + 2J_2\cos(2\omega_i t) + 2J_4\cos(4\omega_i t) + \ldots
\end{aligned} \quad (9.14)$$

$$\begin{aligned}
\sin\!\left(\beta_{fm}\sin(\omega_i t)\right) &= 2\sum_{n=1}^{\infty} J_{2n-1}(\beta_{fm})\sin([2n-1]\omega_i t) \\
&= 2J_1(\beta_{fm})\sin(1\omega_i t) + 2J_3(\beta_{fm})\sin(3\omega_i t) + 2J_5(\beta_{fm})\sin(5\omega_i t) + \ldots \\
&\equiv 2J_1\sin(1\omega_i t) + 2J_3\sin(3\omega_i t) + 2J_5\sin(5\omega_i t) + \ldots
\end{aligned} \quad (9.15)$$

and where we simplify the notation (for equation viewing purposes) and denote $J_n \equiv J_n(\beta_{fm})$. The actual values $J_n(\beta_{fm})$ are somewhat involved to calculate (and we don't need to calculate them here), so for our purposes we observe that it simplifies to just an amplitude value which gets successively smaller for higher frequencies as n increases, i.e. $J_{n+1} < J_n$.

Substituting Eq. (9.14) and Eq. (9.15) into Eq. (9.12) gives the Fourier equation of the WFM signal for transmission of a single tone at frequency f_i:

Chapter 9: Frequency Modulation (FM) Theory and Simulation

$$s_{fm-wfm}(t) = A_c \cos(\omega_c t)\left[J_0 + 2J_2 \cos(2\omega_i t) + 2J_4 \cos(4\omega_i t) + \ldots\right] \quad (9.16)$$
$$- A_c \sin(\omega_c t)\left[2J_1 \sin(1\omega_i t) + 2J_3 \sin(3\omega_i t) + 2J_5 \sin(5\omega_i t) + \ldots\right].$$

Finally, the $\cos() \times \cos()$ and $\sin() \times \sin()$ terms can be multiplied out using the sum-to-difference trigonometric identities from Appendix B (page 581) to give:

$$\begin{aligned}
s_{fm-wfm}(t) = &\; A_c J_0 \cos(\omega_c t) \\
&- A_c J_1 \left[\cos(\omega_c - \omega_i)t - \cos(\omega_c + \omega_i)t\right] \\
&+ A_c J_2 \left[\cos(\omega_c - 2\omega_i)t + \cos(\omega_c + 2\omega_i)t\right] \\
&- A_c J_3 \left[\cos(\omega_c - 3\omega_i)t - \cos(\omega_c + 3\omega_i)t\right] \\
&+ A_c J_4 \left[\cos(\omega_c - 4\omega_i)t + \cos(\omega_c + 4\omega_i)t\right] + \ldots
\end{aligned} \quad (9.17)$$

The $A_c J_0$ term is used to denote the power in the carrier component, and the remaining $A_c J_n$ terms are the powers of each of the sideband components. The components are shown graphically for clarification in Figure 9.6, which shows a spectral representation of a few of the sidebands that would exist for the case of a *single-tone* modulated WFM signal. Note, (recalling Eq. (9.13) and Figure 9.5) that it is only the odd numbered lower sidebands that are 180 degrees out of phase with the other sidebands.

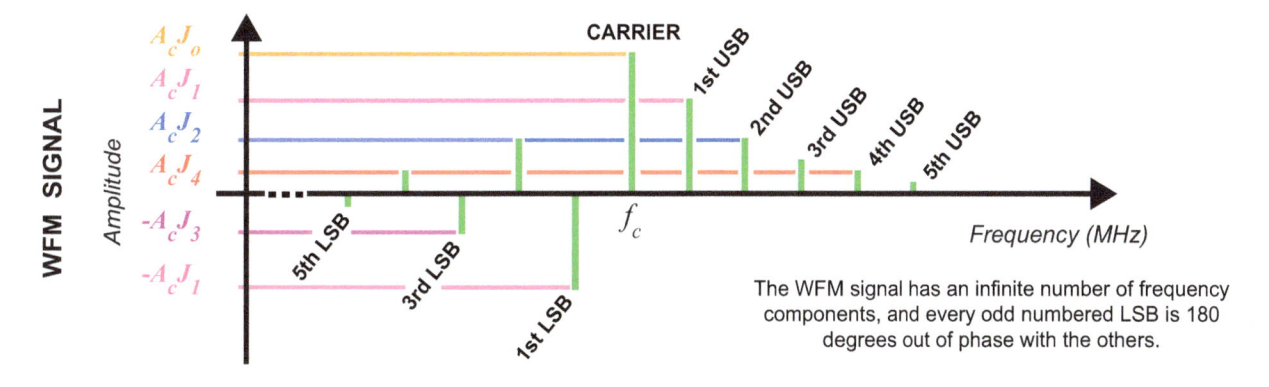

Figure 9.6: Spectral representation of a WFM signal illustrating the LSBs (lower sidebands) and USBs (upper sidebands)

The bandwidth of a WFM signal can be estimated by finding the frequencies of the highest and lowest sideband components that contain a significant amount of power. The J_n values are calculated using Bessel Functions. If, for example, a signal has a modulation index of $\beta_{fm} = 1.5$, the values in Table 9.1 could be used to show that the spectrum would only contain the carrier and four pairs of sidebands. Note that when β_{fm} is a very small value $\ll 1$, only the carrier and the first pair of sidebands contain significant power — this is the NFM case as discussed in Section 9.3.1. The values of J_n listed in Table 9.1 can be substituted into Eq. (9.17) to find the magnitude of each of the spectral components. Plotting the Bessel coefficients on a graph, as in Figure 9.7, indicates how the J_n values vary with increasing modulation index, β_{fm}.

Table 9.1: Table of 'Bessel Function of the First Kind' Coefficients

Modulation Index β_{fm}	(Carrier) J_0	J_1	J_2	J_3	J_4	J_5
0.00	1.00	-[a]	-	-	-	-
0.25	0.98	0.12	-	-	-	-
0.50	0.94	0.24	0.03	-	-	-
1.00	0.77	0.44	0.11	0.02	-	-
1.50	0.51	0.56	0.23	0.06	0.01	-
2.00	0.22	0.58	0.35	0.13	0.03	-
2.41	0.00	0.52	0.43	0.20	0.06	0.02
2.50	-0.05	0.50	0.45	0.22	0.07	0.02
3.00	-0.26	0.34	0.49	0.31	0.13	0.04

a. where '-' denotes an insignificant amount of power

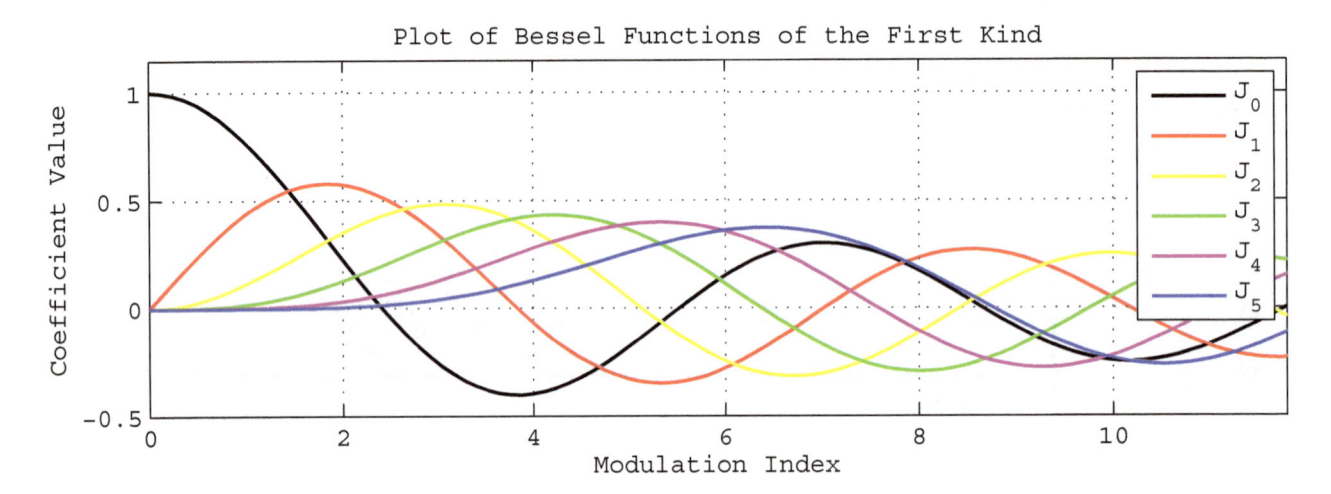

Figure 9.7: Plot of Bessel Functions

Knowing the number of sideband components that will be present in a single-tone modulated WFM signal, it is then straightforward to find an estimate of its bandwidth — simply multiply the number of sidebands, n, by $2f_i$:

$$B = 2nf_i \text{ Hz} . \tag{9.18}$$

If n is not known, it is still possible to estimate the bandwidth, as n can be found as follows:

$$n = \beta_{fm} + 1 . \tag{9.19}$$

Substituting this into Eq. (9.18), the bandwidth is found to be:

Chapter 9: Frequency Modulation (FM) Theory and Simulation

$$B = 2(\beta_{fm} + 1) f_i \quad (9.20)$$

$$= 2\left(\frac{\Delta f}{f_i} + 1\right) f_i$$

$$= 2(\Delta f + f_i) \text{ Hz}.$$

This relationship is well known as *Carson's Rule*. When the information signal used to modulate the carrier is not a pure sinusoid, f_i can be substituted with the signal's maximum frequency component, f_h.

Regulators have imposed rules on the number of sidebands that can be transmitted with WFM radio signals and, as previously mentioned, in the UK the signals must be bandlimited to 200kHz. Although this allows for roughly 98% of the total modulated energy to pass, bandpass filtering WFM signals in this way does have a small detrimental effect on the signal quality.

To illustrate how much spectral information is lost, the baseband information signal modulated and transmitted by a stereo WFM station has a bandwidth of 53kHz. The fixed frequency deviation of $\Delta f = 75\text{kHz}$ means that the theoretical bandwidth of the station according to Carson's Rule is 256kHz. The 56kHz band of modulated information past 200kHz is filtered and not transmitted, but as it only contains a fraction of the total modulated energy, the information signal can still be recovered by receivers with little distortion.

You can now try out Exercise 9.2, where we will ask you to run a simple Simulink simulation that shows what happens (in the time and frequency domains) to sinusoidal waves as they are used to modulate a carrier and produce a WFM signal.

Exercise 9.2 Wideband FM (WFM) Simulation

In this exercise we will test out the WFM modulator. As in Exercise 9.1, there are two different information signals included in this model, with the option to switch between. *Spectrum Analyzers* and a *Time Scope* are connected to various points throughout the model to allow you to monitor the signals in the time and frequency domains as they are modulated.

(a) **Open MATLAB.** Set the working directory to an appropriate folder so you can open this model:

`/fm/simulation/fm_wfm.slx`

The block diagram should look as follows, which you may notice is absolutely identical to the NFM model from Exercise 9.1! There is one major difference in this model, which is part of the model callbacks and is not immediately visible. The modulation index is >>1 due to the values set for the *voltage to frequency gain ratio* of the VCOs.

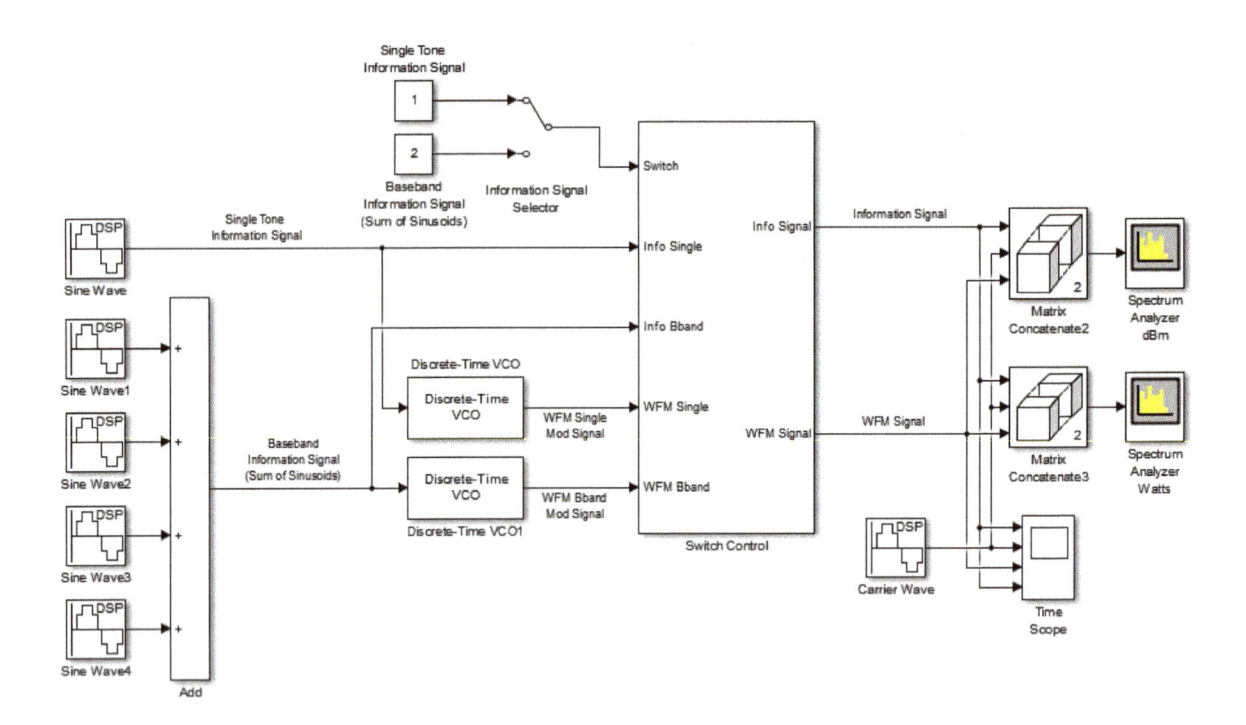

(b) **Inspect the model.** The information signals are input to the control ports of the two *Discrete Time VCOs*, which because of the higher modulation index, now output WFM signals. As before, you can switch between the information signals by simply 🖱 on the *Information Signal Selector* switch. Leave this set on 'Single Tone' for now.

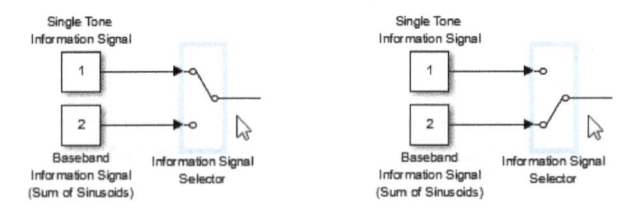

The *Spectrum Anlayzers* are configured as before, with one set to plot the frequency spectra in dBm, and the other in Watts.

(c) **Run the simulation.** Begin the simulation by 🖱 on the 'Run' ▶ button in the Simulink toolbar. The three scope windows should appear, and you should position these so that you can see all of them. The signals are colour coded for clarity: the information signal is blue, the carrier is orange, and the WFM signal is green, as shown in the legends.

(d) **Signal Analysis #1.** If you left the switch set on 'Single Tone', you should see the plots shown opposite in the *Spectrum Analyzer* windows.

The information signal is a single tone with a frequency of 2kHz. The carrier signal generated by the VCO has a frequency of 50kHz. When the information signal WFM modulates the carrier, a signal containing several sidebands is created. Each of these sidebands contains a fraction of the carrier's power, and it would appear that the most significant sidebands reside in the range ±30kHz of the carrier.

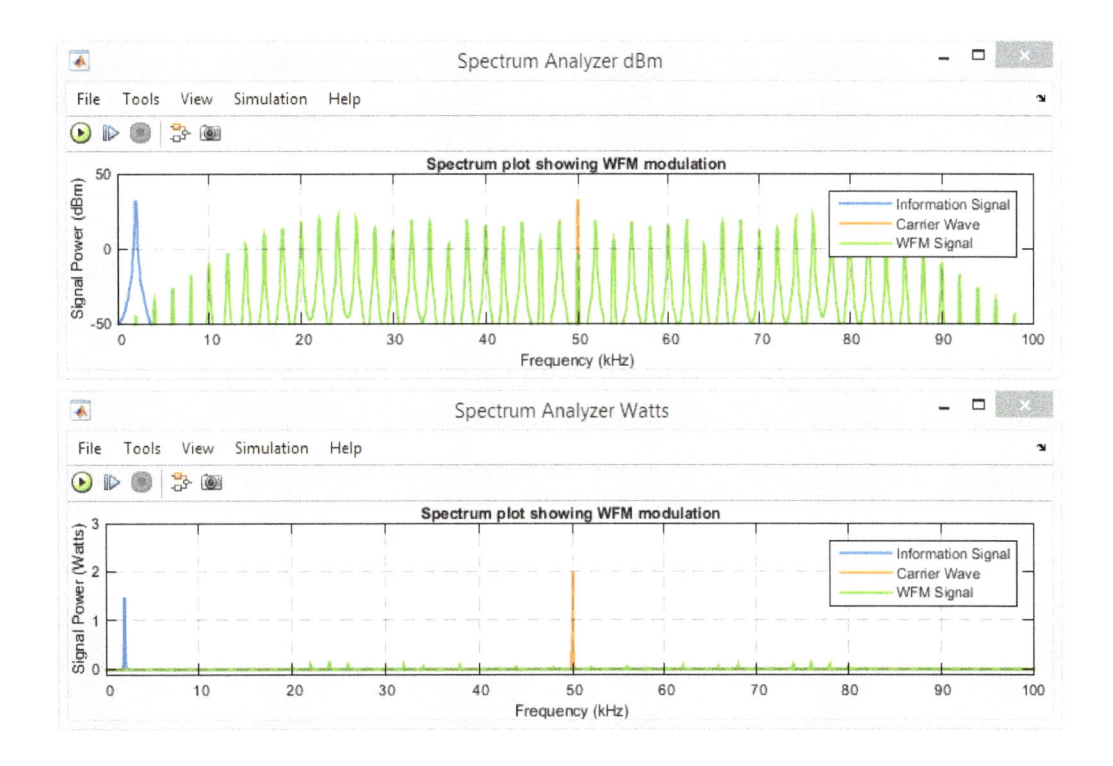

Chapter 9: Frequency Modulation (FM) Theory and Simulation

This time it is easy to see that the changing frequency of the WFM signal corresponds to the amplitude fluctuations of the information signal. Remember that the only difference between this and the model used in Exercise 9.1 is the value of the *voltage to frequency gain ratio*.

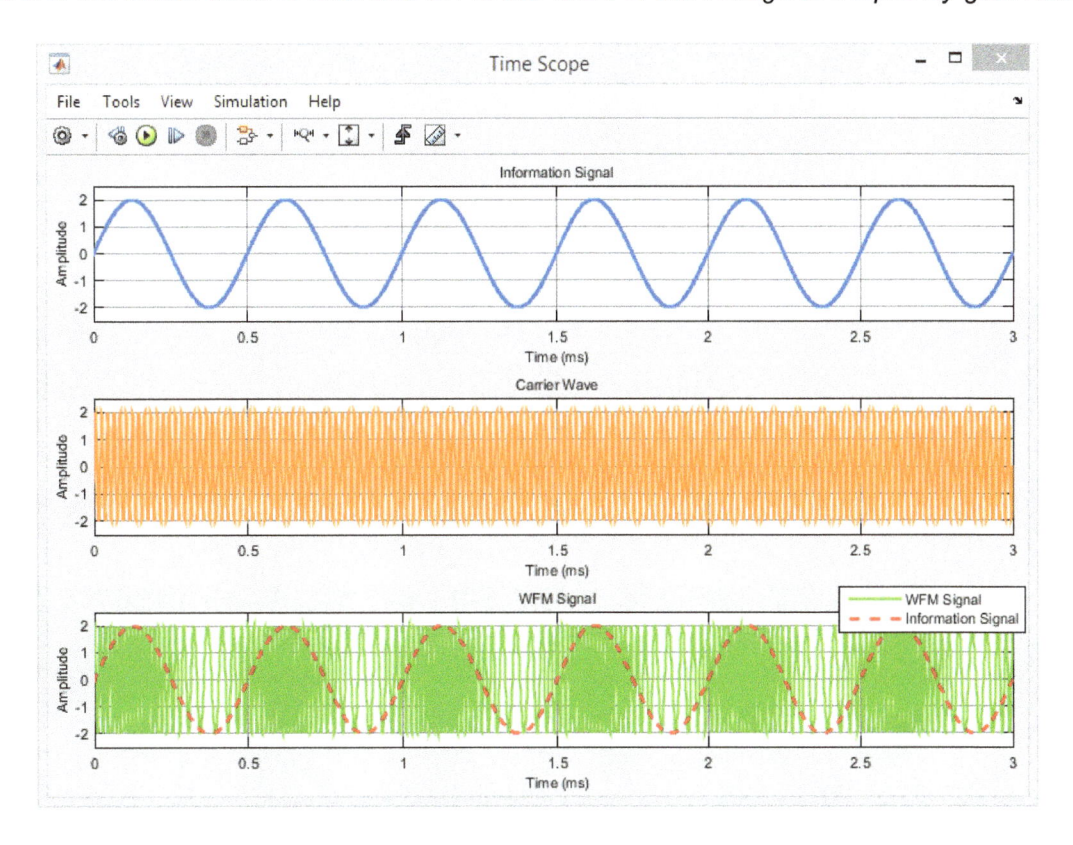

(e) **Signal Analysis #2.** Set the *Information Signal Selector* switch to 'Baseband Information Signal', and re-run the simulation. You should now see a similar view to the *Spectrum Analyzer* and *Time Scope* windows included here.

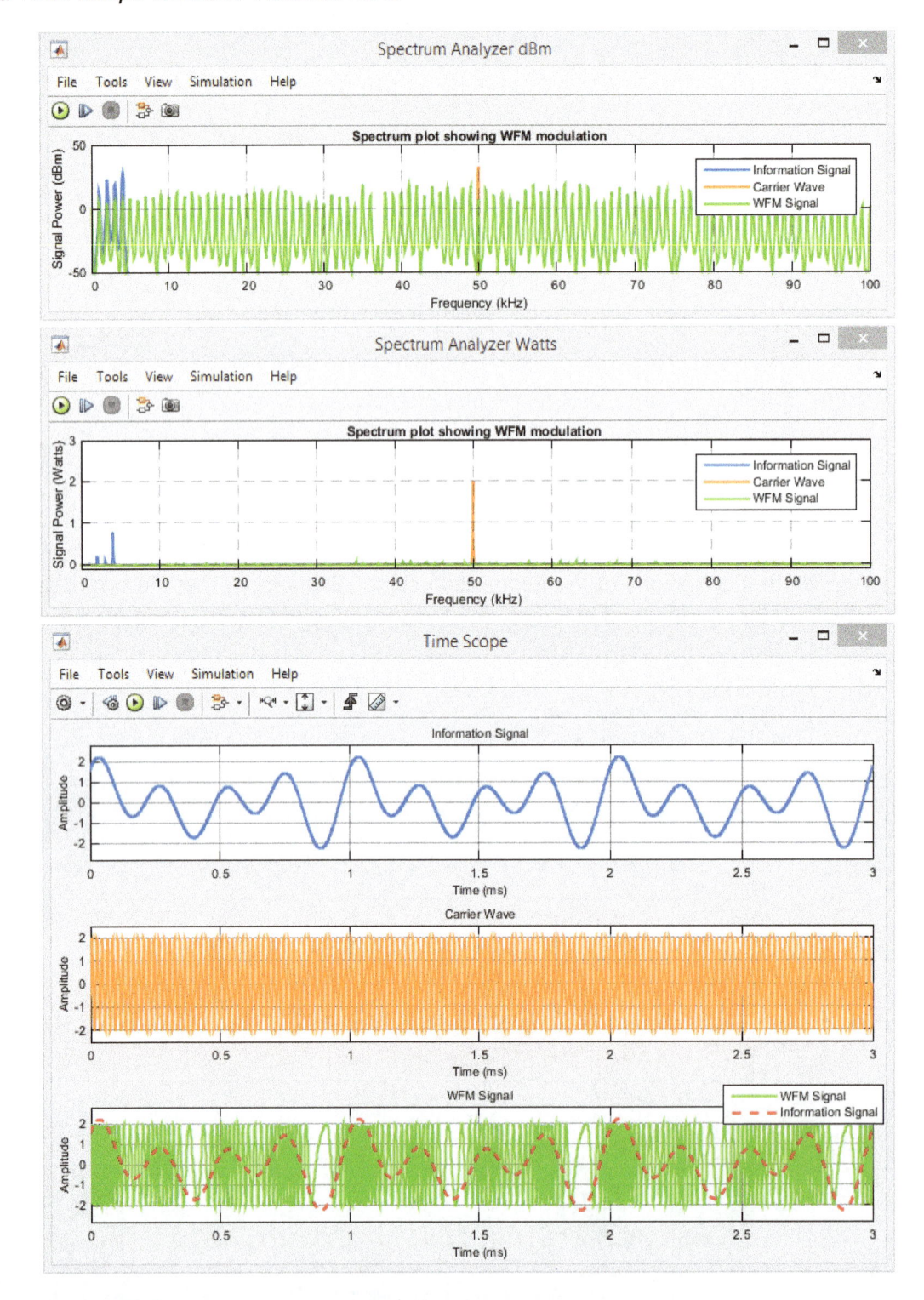

The baseband information signal contains four frequency components, located at 1kHz, 2kHz, 3kHz and 4kHz. When it is used to modulate the 50kHz carrier, many more sidebands are

Chapter 9: Frequency Modulation (FM) Theory and Simulation

created than in the 'single sine' case. Once again, these do not contain a large amount of power. As the bandwidth of the baseband signal is larger than the bandwidth of the single tone information signal, the bandwidth of the WFM signal is much wider. If the frequency of the carrier was higher, it is likely that sideband components would be seen for hundreds of kHz either side of the carrier.

As the amplitude fluctuations of this baseband signal have more variation than the sine wave example, the generated WFM signal is more representative of a 'real' WFM radio signal.

(f) **Investigate the simulation parameters.** Open the Simulink Model properties by navigating to *File > Model Properties > Model Properties*. ① on the *Callbacks* tab, then on *PostLoadFcn*. To find the values (for (k_o) Kvco_sum and Kvco_single, work backwards using Eq. (9.2) to find the values of K_{fm}. How do these values compare to the ones you calculated in Exercise 9.1?

9.4 FM Demodulation Using Differentiation

One of the standard methods used to demodulate FM signals involves taking the derivative of the received signal. For the transmitted (and perfectly received!) signal $s_{fm}(t)$, a new differentiated signal denoted as $s_{fm}'(t)$ (where the dash ' denotes the derivative) is generated by the receiver. Consider the generic FM modulated information signal from Eq. (9.4),

$$s_{fm}(t) = A_c \cos\left(\omega_c t + 2\pi K_{fm} \times \int_{-\infty}^{t} s_i(t)dt\right). \tag{9.21}$$

Noting Eq. (B.9) from Appendix B (page 581), differentiating Eq. (9.21) using the chain rule results in:

$$s_{fm}'(t) = \frac{d}{dt} s_{fm}(t)$$

$$= -A_c \frac{d}{dt}\left[\omega_c t + 2\pi K_{fm} \times \int_{-\infty}^{t} s_i(t)dt\right] \sin\left(\omega_c t + 2\pi K_{fm} \times \int_{-\infty}^{t} s_i(t)dt\right)$$

$$= -A_c \underbrace{\left[\omega_c + 2\pi K_{fm} s_i(t)\right]}_{\text{amplitude component}} \underbrace{\sin\left(\omega_c t + 2\pi K_{fm} \times \int_{-\infty}^{t} s_i(t)dt\right)}_{\text{high frequency component}}. \tag{9.22}$$

Figure 9.8 shows that when an FM signal is differentiated, there is a resulting information envelope (the amplitude component in Eq. (9.22)) multiplied by a high frequency component.

The signal $s_{fm}'(t)$ will have the appearance of an AM-DSB-TC signal in the time domain, as it has an information envelope (although unlike a standard AM envelope, the frequency of the carrier component still changes — compare Figure 9.8 with the AM envelope in Figure 6.3, page 205). The fluctuations in this envelope are directly proportional to the instantaneous frequency, $f_{fm\ inst}$, of $s_{fm}(t)$, which should itself be directly proportional to the amplitude of the original information signal $s_i(t)$.

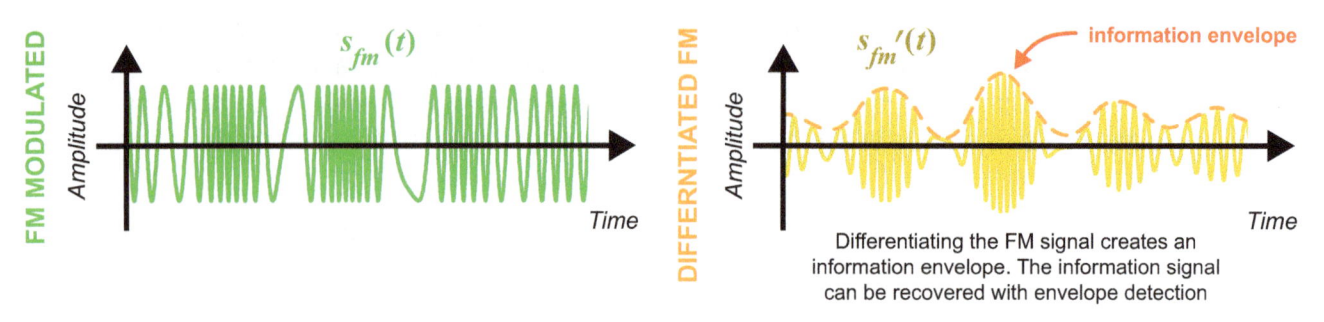

Figure 9.8: Sketch of an FM signal in the time domain before (left), and after (right) the differentiation operation

If we consider that the high frequency sinusoidal term can be removed by an envelope detector (see Section 6.7.1 on page 229), it is clear that the amplitude of the envelope is directly proportional to the amplitude of the information signal. Although it has a DC offset (ω_c, and gain of $2\pi K_{fm}$ resulting from the FM modulation constant, the amplitude fluctuations will still match those of $s_i(t)$.

A complete design of this type FM receiver will require appropriate components to select the frequency band of interest. In modern Integrated Circuit (IC) FM receivers (such as the commercial device in [100]), this process is performed by downconverting the RF signal from ~100MHz to an IF (such as 10MHz), followed by another downconversion stage to 0.5MHz or less, followed by the FM demodulator circuit. Many other methods for FM reception exist, including the simple slope overload and phase detectors, both of which will be implemented in real designs in Chapter 10.

9.5 Receiving and Downconverting FM Signals to Complex Baseband

As reviewed in Chapter 5, real RF signals received by the RTL-SDR are *quadrature demodulated* to baseband before they are sampled — this means that the baseband samples entering our MATLAB and Simulink RTL-SDR receiver designs have both I and Q components, i.e. they form a complex signal.

When an RF FM signal ($s_{fmRF}(t)$) is received by the RTL-SDR, it is mixed with a complex exponential at frequency f_{lo} (representing the overall local oscillator frequency in the RTL-SDR) to demodulate the signal to complex baseband, as illustrated in Figure 9.9. Here, we show the spectra of the various stages of the RTL-SDR downconversion process.

Assuming that the RTL-SDR receives the transmitted signal s_{fm} from Eq. (9.21), because we are working with a perfect radio channel, the signal output from the RTL-SDR can be modelled as:

$$s_{\text{RTL-SDR}}(t) = LPF\left[s_{bband}(t)\right] = LPF\left[s_{fmRF}(t)e^{-j\omega_{lo}t}\right] \tag{9.23}$$

where $e^{-j\omega_{lo}t}$ *represents the complex oscillator inside the RTL-SDR.*

Ideally the frequency of the local oscillator used to demodulate the signal and the frequency of the original modulating carrier would be the same, as this means that the RF signal will be perfectly demodulated to baseband. This is unlikely to happen however, and when it does not, meaning that the complex baseband signal is still modulated onto a low frequency 'carrier' (not a carrier in the conventional RF sense, but a low, non-zero frequency). We denote this carrier as

Chapter 9: Frequency Modulation (FM) Theory and Simulation

$$f_\Delta = f_c - f_{lo} \quad \text{or equivalently,} \quad \omega_\Delta = \omega_c - \omega_{lo}. \tag{9.24}$$

We can express the downconverted signal using Euler's Formula ($e^{j\omega t} = \cos(\omega t) + j\sin(\omega t)$):

$$\begin{aligned} s_{bband}(t) &= s_{fm\text{RF}}(t)e^{-j\omega_{lo}t} \\ &= s_{fm\text{RF}}(t) \times \Big(\cos(\omega_{lo}t) - j\sin(\omega_{lo}t)\Big) \\ &= A_c \cos\Big(\omega_c t + \theta_{fm}(t)\Big) \times \Big(\cos(\omega_{lo}t) - j\sin(\omega_{lo}t)\Big), \end{aligned} \tag{9.25}$$

$$\text{where:} \quad \theta_{fm}(t) = 2\pi K_{fm} \times \int_{-\infty}^{t} s_i(t)\,dt. \tag{9.26}$$

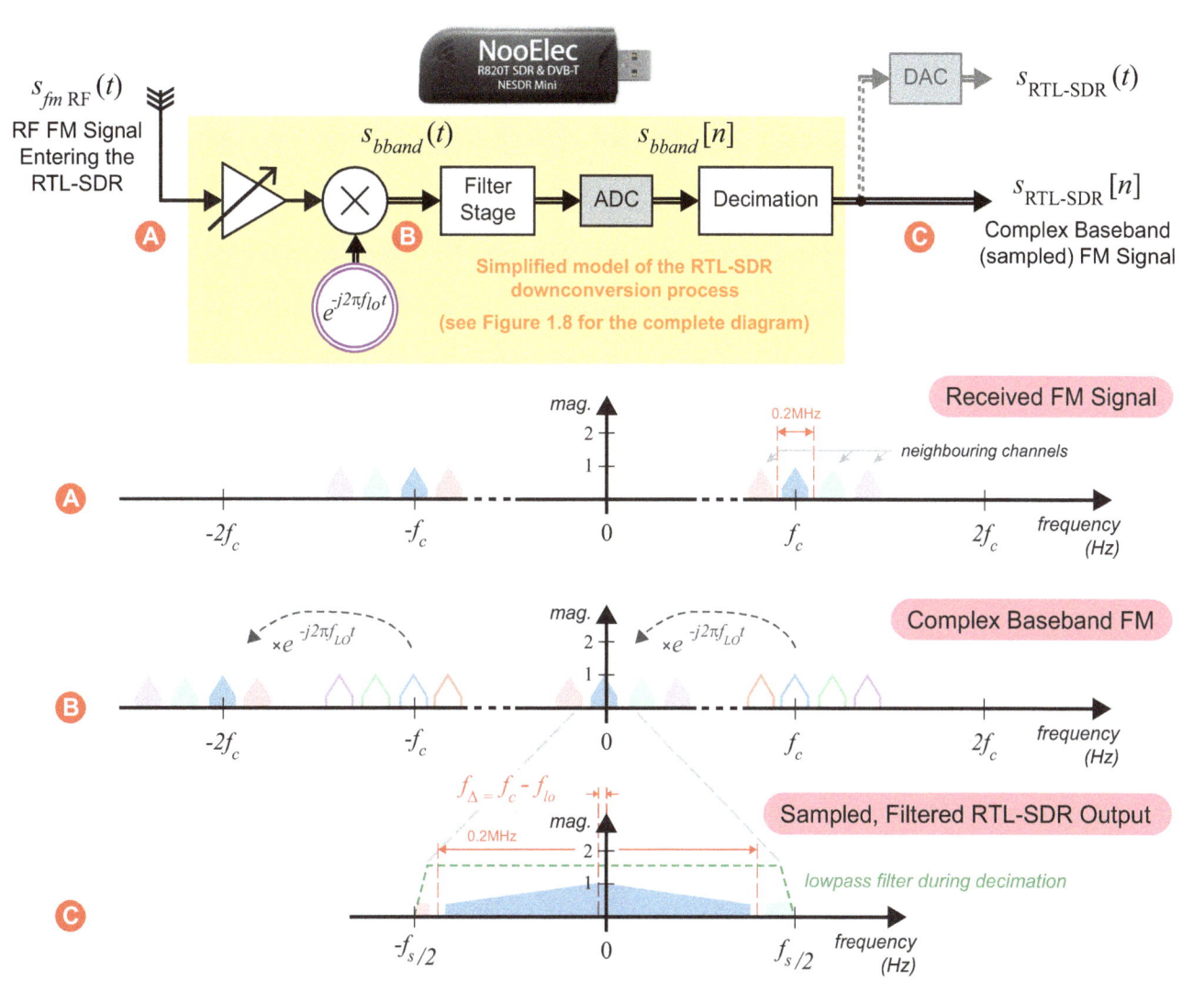

Figure 9.9: The RTL-SDR receiving an FM signal, and downconverting it to complex baseband. Note that when a frequency offset exists between the original modulating carrier f_c and the local oscillator used during the demodulation process f_{lo}, a frequency shift of f_Δ occurs in the complex baseband output

Multiplying out Eq. (9.25) gives:

$$s_{bband}(t) = A_c \cos\left(\omega_c t + \theta_{fm}(t)\right)\cos(\omega_{lo}t) - jA_c \cos\left(\omega_c t + \theta_{fm}(t)\right)\sin(\omega_{lo}t) \qquad (9.27)$$

Using Eqs. (B.4) and (B.6) from Appendix B (page 581), Eq. (9.27) becomes:

$$s_{bband}(t) = \frac{A_c}{2}\left[\cos\left(\omega_c t + \theta_{fm}(t) - \omega_{lo}t\right) + \cos\left(\omega_c t + \theta_{fm}(t) + \omega_{lo}t\right)\right] \qquad (9.28)$$

$$-j\frac{A_c}{2}\left[\sin\left(\omega_c t + \theta_{fm}(t) + \omega_{lo}t\right) - \sin\left(\omega_c t + \theta_{fm}(t) - \omega_{lo}t\right)\right]$$

high freq components   baseband components

The high frequency components are attenuated by the lowpass filters within the RTL-SDR, leaving only the complex baseband signal (which we express here in the continuous time domain for simplicity—note the signal $s_{\text{RTL-SDR}}(t)$ can be created with a DAC as shown in Figure 9.9):

$$s_{\text{RTL-SDR}}(t) = \frac{A_c}{2}\left[\cos\left(\omega_c t + \theta_{fm}(t) - \omega_{lo}t\right) + j\sin\left(\omega_c t + \theta_{fm}(t) - \omega_{lo}t\right)\right], \qquad (9.29)$$

which can be simplified by substituting $\omega_\Delta = \omega_c - \omega_{lo}$ for the '*baseband*' complex carrier:

$$s_{\text{RTL-SDR}}(t) = \frac{A_c}{2}\left[\cos\left(\omega_\Delta t + \theta_{fm}(t)\right) + j\sin\left(\omega_\Delta t + \theta_{fm}(t)\right)\right]. \qquad (9.30)$$

The baseband *carrier*, ω_Δ or f_Δ, is very close to 0Hz, or may be zero if $f_{lo} = f_c$. Substituting $\theta_{fm}(t)$ from Eq. (9.26) into Eq. (9.30) gives:

$$s_{\text{RTL-SDR}}(t) = \frac{A_c}{2}\left[\cos\left(\omega_\Delta t + 2\pi K_{fm} \times \int_{-\infty}^{t} s_i(t)dt\right) + j\sin\left(\omega_\Delta t + 2\pi K_{fm} \times \int_{-\infty}^{t} s_i(t)dt\right)\right]$$

$$= \frac{A_c}{2} e^{j\left(\omega_\Delta t + 2\pi K_{fm} \times \int_{-\infty}^{t} s_i(t)dt\right)} \qquad (9.31)$$

The RTL-SDR presents this complex FM signal as baseband samples to MATLAB and Simulink, and an example spectra of the signal is illustrated in Figure 9.9(c). This is what we must aim to demodulate in order to recover the information signal (note that the adjacent channels shown in Figure 9.9 are not part of our mathematical analysis. In the following three sections we present three different methods for demodulating the signal.

9.6 Non-Coherent FM Demodulation: The Complex Differentiation Discriminator

The first option is to implement a dual branch differentiator that deals with each of the In Phase and Quadrature Phase components separately. We shall call this the *Complex Differentiation Discriminator* (see Figure 9.16). To help explain how this works, time domain plots are included at each stage of the

Chapter 9: Frequency Modulation (FM) Theory and Simulation   351

mathematical analysis, which progressively show a WFM modulated single tone information signal being demodulated at different stages of the discriminator. This signal was recorded live from an RTL-SDR, receiving a tone at 1kHz on an FM carrier 102.5MHz. Below will show a sequence of time domain plots as we view various components of the complex differentiation discriminator.

The complex FM signal from Eq. (9.30) contains both real and imaginary components. Expressing these individually, we can denote $s_{ip}(t)$ as the I component, and $s_{qp}(t)$ as the Q component:

$$s_{ip}(t) = \Re\left[s_{\text{RTL-SDR}}(t)\right] = \frac{A_c}{2}\cos\left(\omega_\Delta t + \theta_{fm}(t)\right) \qquad (9.32)$$

$$s_{qp}(t) = \Im m\left[s_{\text{RTL-SDR}}(t)\right] = \frac{A_c}{2}\sin\left(\omega_\Delta t + \theta_{fm}(t)\right). \qquad (9.33)$$

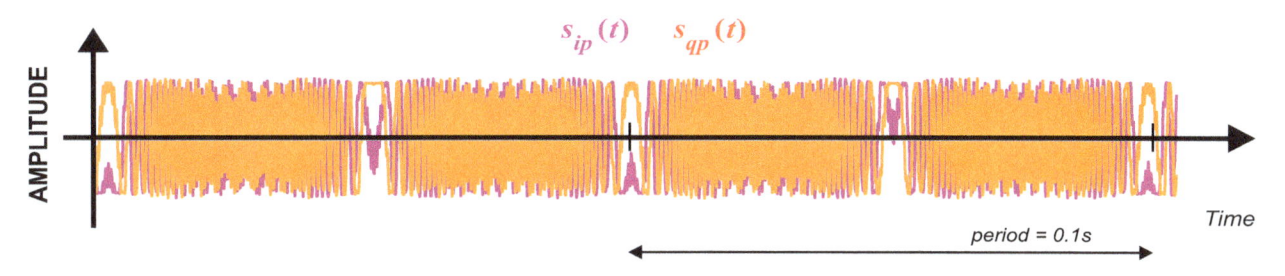

Figure 9.10: Complex Differentiation Discriminator: plot of $s_{ip}(t)$ and $s_{qp}(t)$ when transmitting a single tone. So far it looks nothing like the transmitted tone and it is difficult to view or correlate any tonal component!

Differentiating each of these components separately (using Eqs. (B.8) and (B.9)) results in the following:

$$\begin{aligned}s_{ip}'(t) &= \frac{d}{dt}s_{ip}(t) = -\frac{A_c}{2}\left[\omega_\Delta + \theta_{fm}'(t)\right]\sin\left(\omega_\Delta t + \theta_{fm}(t)\right)\\ s_{qp}'(t) &= \frac{d}{dt}s_{qp}(t) = \frac{A_c}{2}\left[\omega_\Delta + \theta_{fm}'(t)\right]\cos\left(\omega_\Delta t + \theta_{fm}(t)\right).\end{aligned} \qquad (9.34)$$

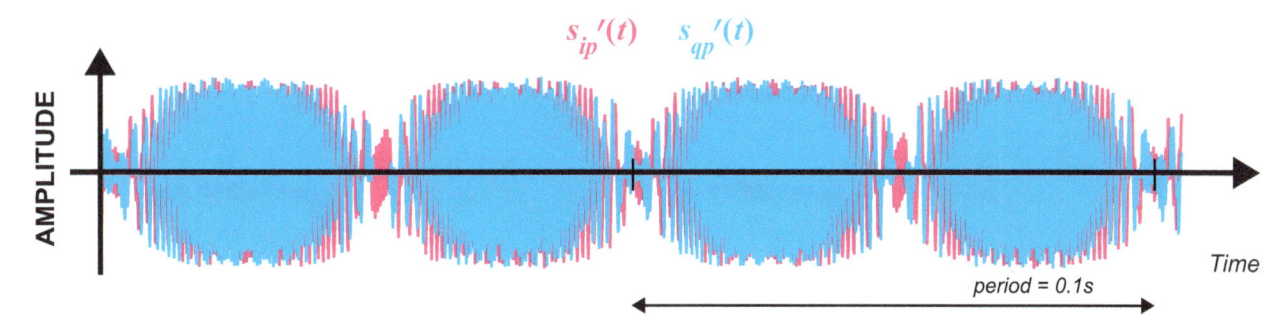

Figure 9.11: Complex Differentiation Discriminator: plot of $s_{ip}'(t)$ and $s_{qp}'(t)$ for a received signal tone.

At this stage, the waveforms of the components look a little like AM-DSB-SC signals, with an 'outline' of the modulated signal and an obvious carrier (although the frequency of the carrier still changes). There is no information envelope present however, and remembering back to what was discussed in Section 6.8 (page 231), envelope detectors can only be used to demodulate signals that have an envelope. These differentiated components from Eq. (9.33) must be manipulated before the information can be extracted.

We can begin this process by taking the differentiated equations and mixing them as follows:

$$s_{ip}'(t) \times s_{qp}(t) = -\frac{A_c^2}{4}\left[\omega_\Delta + \theta_{fm}'(t)\right]\sin^2\left(\omega_\Delta t + \theta_{fm}(t)\right)$$

$$s_{qp}'(t) \times s_{ip}(t) = \frac{A_c^2}{4}\left[\omega_\Delta + \theta_{fm}'(t)\right]\cos^2\left(\omega_\Delta t + \theta_{fm}(t)\right).$$

(9.35)

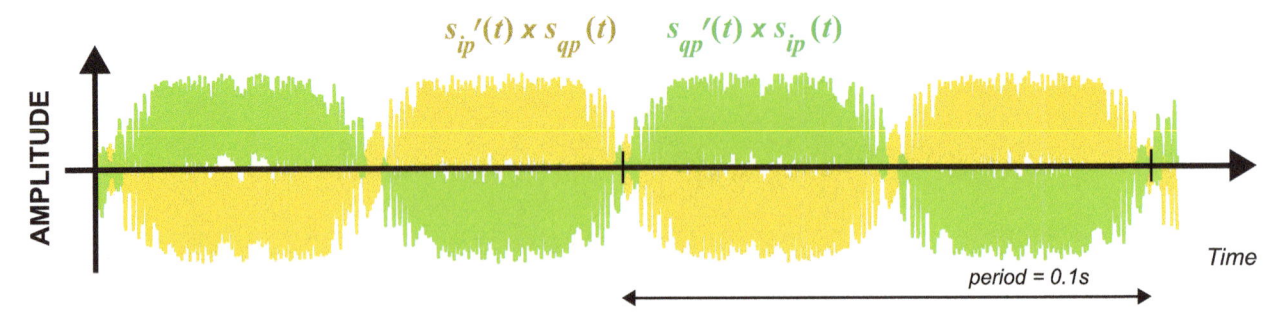

Figure 9.12: Complex Differentiation Discriminator: plot of $s_{ip}'(t) \times s_{qp}(t)$ and $s_{qp}'(t) \times s_{ip}(t)$

Combining these signals from Eq. (9.35) by subtracting $s_{ip}'(t) \times s_{qp}(t)$ from $s_{qp}'(t) \times s_{ip}(t)$ gives:

$$s_\alpha(t) = \left(s_{qp}'(t) \times s_{ip}(t)\right) - \left(s_{ip}'(t) \times s_{qp}(t)\right) \quad (9.36)$$

$$= \frac{A_c^2}{4}\left[\omega_\Delta + \theta_{fm}'(t)\right]\cos^2\left(\omega_\Delta t + \theta_{fm}(t)\right) + \frac{A_c^2}{4}\left[\omega_\Delta + \theta_{fm}'(t)\right]\sin^2\left(\omega_\Delta t + \theta_{fm}(t)\right)$$

$$= \frac{A_c^2}{4}\left[\omega_\Delta + \theta_{fm}'(t)\right]\left(\cos^2\left(\omega_\Delta t + \theta_{fm}(t)\right) + \sin^2\left(\omega_\Delta t + \theta_{fm}(t)\right)\right)$$

$$= \frac{A_c^2}{4}\left[\omega_\Delta + \theta_{fm}'(t)\right],$$

as $\cos^2(x) + \sin^2(x) = 1$ for any value of x.

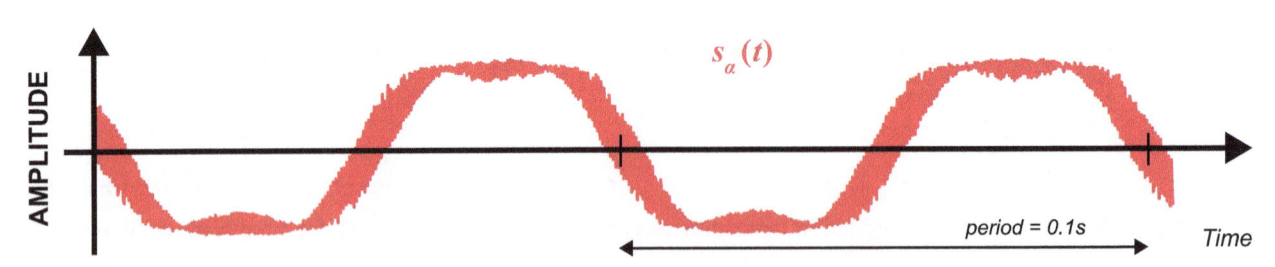

Figure 9.13: Complex Differentiation Discriminator: plot of $s_\alpha(t)$.
We can begin to see the single tone, however the peaks and troughs would appear to be clipped in some way.

This results in something now looking (vaguely!) similar to the transmitted single tone information signal. If you listen to the audio, however, it will not sound 'right' as the signal is still proportional to A_c^2.

To address this, we need to normalise the amplitude and divide through by A_c^2 [29]. We can find the instantaneous value of the received signal power $s_p(t)$ by squaring and summing the I and Q channels:

$$\begin{aligned} s_p(t) &= s_{ip}(t)^2 + s_{qp}(t)^2 \\ &= \frac{A_c^2}{4}\left(\cos^2\left(\omega_\Delta t + \theta_{fm}(t)\right) + \sin^2\left(\omega_\Delta t + \theta_{fm}(t)\right)\right) \\ &= \frac{A_c^2}{4}, \end{aligned} \qquad (9.37)$$

again noting that $\cos^2(x) + \sin^2(x) = 1$. We can now divide Eq. (9.36) by $s_p(t)$ to give:

$$\begin{aligned} s_n(t) &= \frac{s_\alpha(t)}{s_p(t)} \qquad (9.38) \\ &= \frac{\left(s_{qp}'(t) \times s_{ip}(t)\right) - \left(s_{ip}'(t) \times s_{qp}(t)\right)}{s_{ip}(t)^2 + s_{qp}(t)^2} \\ &= \frac{\frac{A_c^2}{4}\left[\omega_\Delta + \theta_{fm}'(t)\right]}{\frac{A_c^2}{4}} \\ &= \left[\omega_\Delta + \theta_{fm}'(t)\right]. \end{aligned}$$

Substituting the derivative of $\theta_{fm}(t)$ from Eq. (9.26) into Eq. (9.38), we find that we are left with a signal that is a scaled version of the original information signal, $s_i(t)$ with a DC offset, ω_Δ.

$$s_d(t) = \left[\omega_\Delta + 2\pi K_{fm} s_i(t)\right]. \qquad (9.39)$$

This completes the analysis of the demodulator [4]. Notice that Eq. (9.39) is almost identical to the amplitude component in Eq. (9.22). As highlighted in Figure 9.14, it is common to pass the demodulated signal through a lowpass filter to remove any noise at frequencies higher than the maximum frequency in the information signal. This ensures that the quality of the demodulated output is maximised.

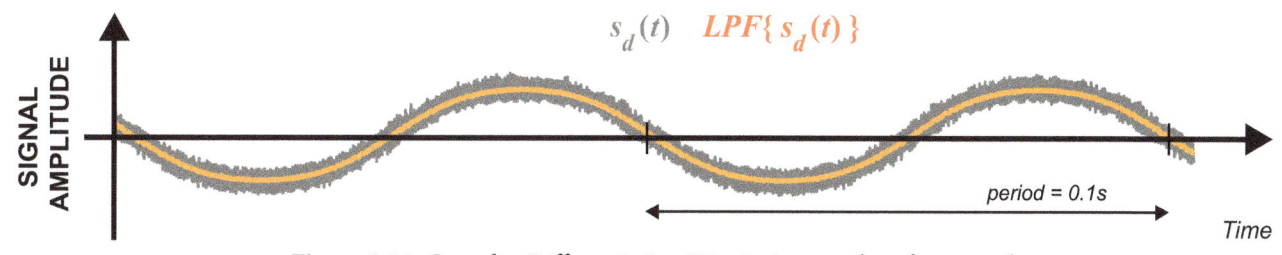

Figure 9.14: Complex Differentiation Discriminator: plot of $s_d(t)$ and $LPF\{s_d(t)\}$

9.6.1 The Digital Complex Discriminator

In order to implement the complex differentiation discriminator in the digital domain, we need to design and use digital differentiators. In the digital domain, a simple differentiator approximation can be realised with the blocks shown in Figure 9.15, where $T_s = 1/f_s$ and f_s is the sampling frequency. One simple approximation to a differentiator is given by an FIR filter, with coefficients [1, 0, -1] (or similarly coefficients of [1, -1]).

Gradient at t_0 $\quad \dfrac{dx(t)}{dt} \approx \dfrac{x[n] - x[n-2]}{2T_s}$

Approximation to differentiation of a continuous time curve by the difference between two samples.

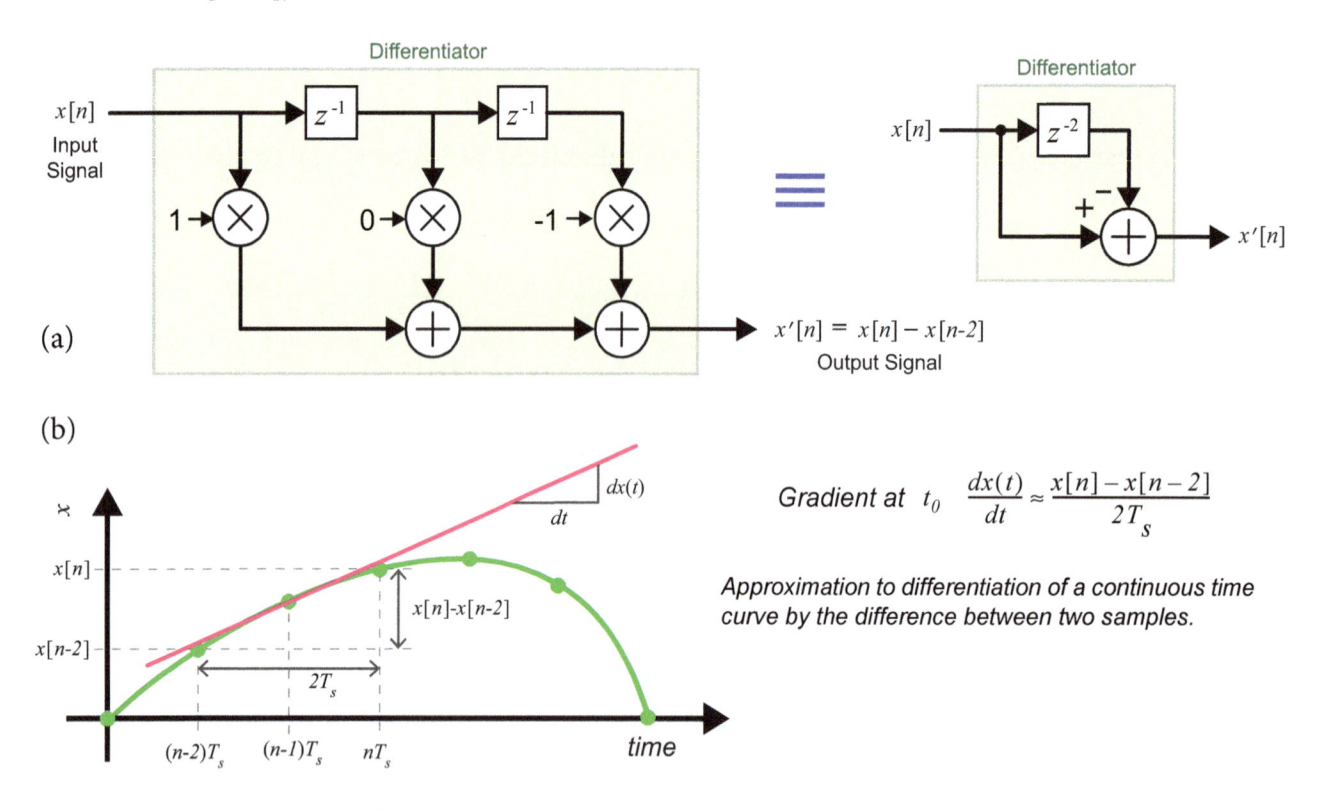

Figure 9.15: (a) DSP Differentiator, implemented with an FIR filter and (b) Approximating the tangent (gradient) at time $t_0 = (n-1)T_s$ from the difference of two samples.

Using this FIR differentiator, the digital complex differentiation discriminator can be constructed with the DSP blocks shown in Figure 9.16. The response of this discrete time demodulator in terms of I and Q samples is as given in Eq. (9.40). Notice that this is very similar to Eq. (9.38).

$$s_d[n] = \frac{\left(s_{qp}[n] - s_{qp}[n-2]\right)s_{ip}[n-1] - \left(s_{ip}[n] - s_{ip}[n-2]\right)s_{qp}[n-1]}{(s_{ip}[n-1] \times s_{ip}[n-1]) + (s_{qp}[n-1] \times s_{qp}[n-1])}. \quad (9.40)$$

The single sample delay (denoted by z^{-1}) is inserted to match the group delay of one sample period for the FIR differentiator, in order to keep the $s_{ip}[n]$ and $s_{qp}[n]$ signals synchronised. From Figure 9.15(b) we can note that the discrete differentiator should be scaled by $1/2T_s$, however this is just a constant scaling factor common to both the differentiators.

We will implement and test one of these demodulators in Section 10.2 (page 376). In the next few sections we will consider other FM receivers, starting with the *Complex Delay Line Discriminator*.

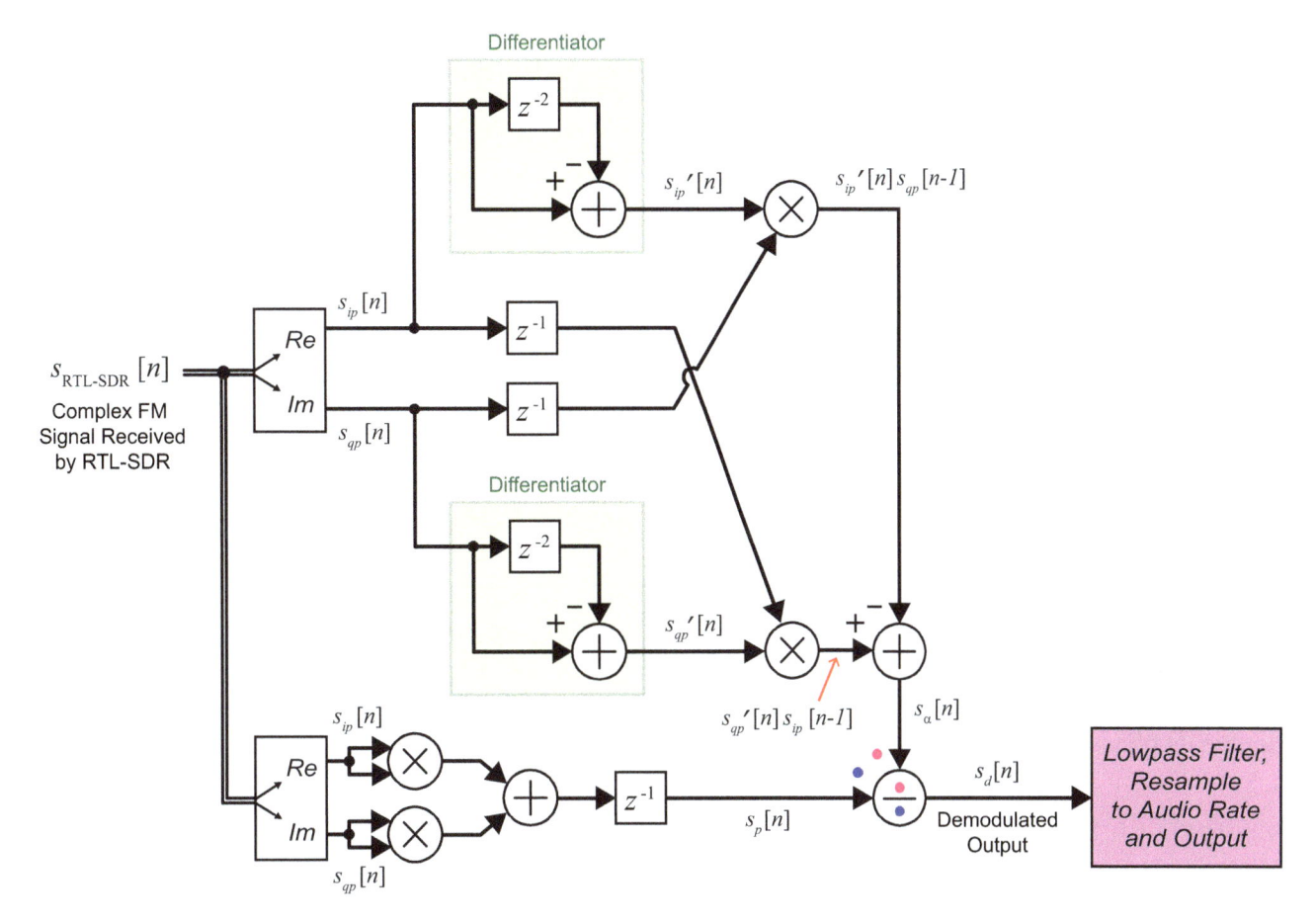

Chapter 9: Frequency Modulation (FM) Theory and Simulation

Figure 9.16: Complex Differentiation Discriminator block diagram [29]

9.7 Non-Coherent FM Demodulation: The Complex Delay Line Discriminator

Compared to the previous section's FM demodulator, a considerably simpler non-coherent FM demodulator is the *Complex Delay Line Frequency Discriminator*. This device can be implemented (based on its analogue counterpart) using the four DSP blocks shown in Figure 9.17.

To help aid understanding of how this demodulator works, time domain plots are included at each stage of the mathematical analysis, which *progressively show a WFM modulated single tone information signal being demodulated*. Like the previous example, this signal is one that we actually received with the RTL-

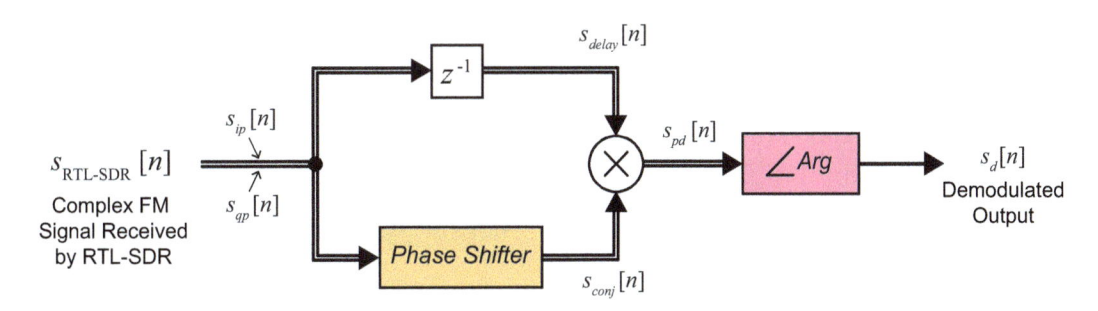

Figure 9.17: Complex Delay Line Frequency Discriminator block diagram

SDR, having generated a test signal tone at 1kHz on an FM carrier 102.5MHz, and therefore the signals shown over the next few pages are 'live' rather than simulated.

The complex FM signal from Eq. (9.31) contains both real and imaginary components. Expressing this signal in exponential form, we have:

$$s_{\text{RTL-SDR}}(t) = \frac{A_c}{2} e^{j\left(\omega_\Delta t + \theta_{fm}(t)\right)} \tag{9.41}$$

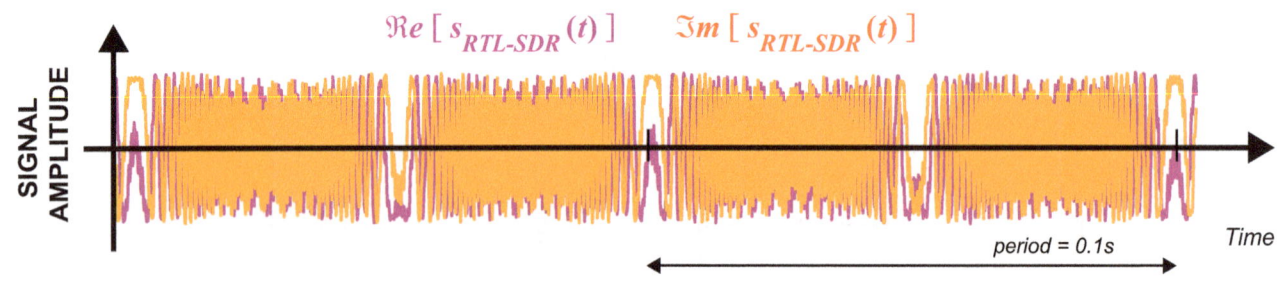

Figure 9.18: Complex Frequency Discriminator: plot of $s_{\text{RTL-SDR}}(t)$ components

This complex signal is input to two parallel blocks. One of these takes the conjugate of the signal to change its phase, and the other adds a time delay τ to the signal to retard it:

$$s_{conj}(t) = \frac{A_c}{2} e^{-j\left(\omega_\Delta t + \theta_{fm}(t)\right)} \tag{9.42}$$

$$s_{delay}(t) = \frac{A_c}{2} e^{j\left(\omega_\Delta [t-\tau] + \theta_{fm}(t-\tau)\right)}. \tag{9.43}$$

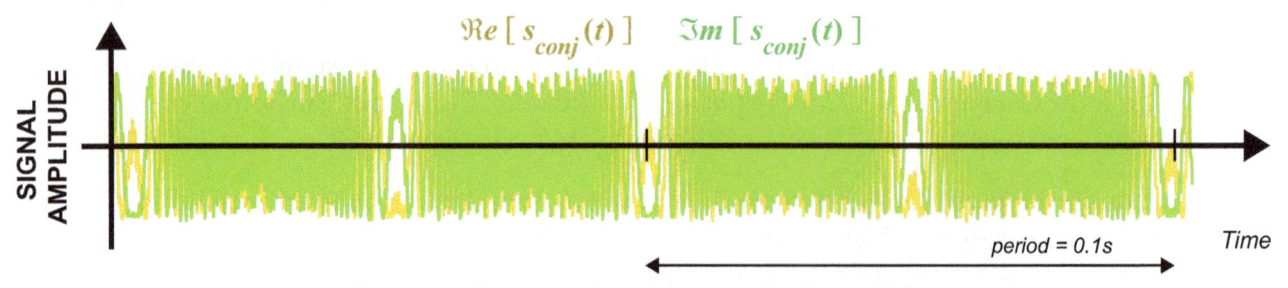

Figure 9.19: Complex Frequency Discriminator: plot of $s_{conj}(t)$ components

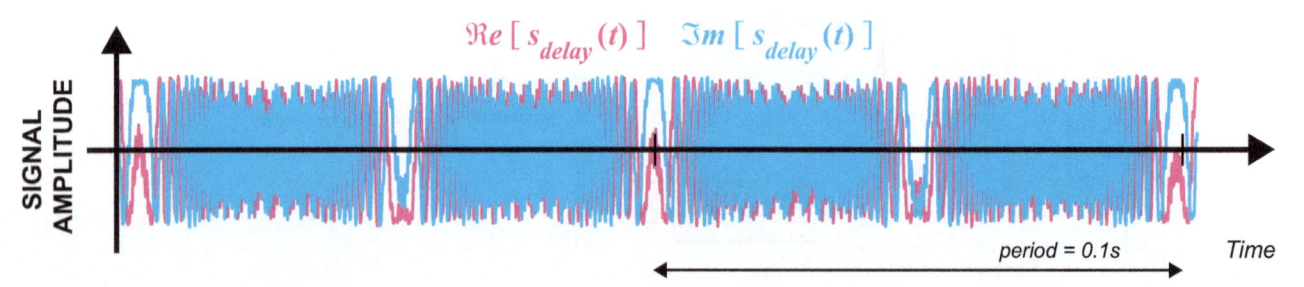

Figure 9.20: Complex Frequency Discriminator: plot of $s_{delay}(t)$ components

Chapter 9: Frequency Modulation (FM) Theory and Simulation

These signals are then mixed together in a process called *phase detection* (see Section 7.5.1 (page 246) for more information):

$$s_{pd}(t) = s_{conj}(t) \times s_{delay}(t) \tag{9.44}$$

$$= \frac{A_c^2}{4} e^{-j\left[\left(\omega_\Delta t + \theta_{fm}(t)\right) - \left(\omega_\Delta[t-\tau] + \theta_{fm}(t-\tau)\right)\right]}.$$

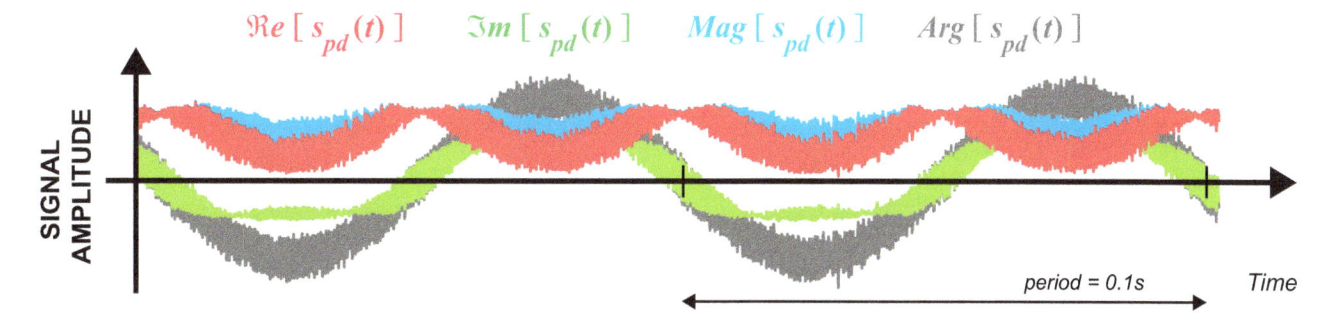

Figure 9.21: Complex Frequency Discriminator: plot of $s_{pd}(t)$ components

Note that in Figure 9.21 the real and imaginary components of the resulting signal do not resemble the sinusoidal information signal; nor does its magnitude — it is the angle of the signal that contains the information. To extract it, we simply take the argument ($\angle$):

$$s_d(t) = \angle s_{pd}(t) = -\left[\left(\omega_\Delta t + \theta_{fm}(t)\right) - \left(\omega_\Delta[t-\tau] + \theta_{fm}(t-\tau)\right)\right] \tag{9.45}$$

$$= -\left[\left(\omega_\Delta t - \omega_\Delta[t-\tau]\right) + \left(\theta_{fm}(t) - \theta_{fm}(t-\tau)\right)\right].$$

When τ is a very small value, Eq. (9.45) can be considered an approximation to a differentiation operation.

$$s_d(t) \approx -\left[\frac{d}{dt}(\omega_\Delta t) + \frac{d}{dt}\left(\theta_{fm}(t)\right)\right]. \tag{9.46}$$

Solving this, we find

$$s_d(t) = -\left[\omega_\Delta + \theta_{fm}'(t)\right] \tag{9.47}$$

$$= -\left[\omega_\Delta + 2\pi K_{fm} s_i(t)\right].$$

You should recognise this result—it is very similar to the signal that is output by the *Complex Differentiation Discriminator*! The argument of the phase detected signal varies in proportion to the original information signal, $s_i(t)$, with a DC offset.

The demodulated signal is then finally filtered to remove any noise at frequencies higher than the maximum frequency in the information signal. This improves the SNR of the demodulated output, as demonstrated in Figure 9.22.

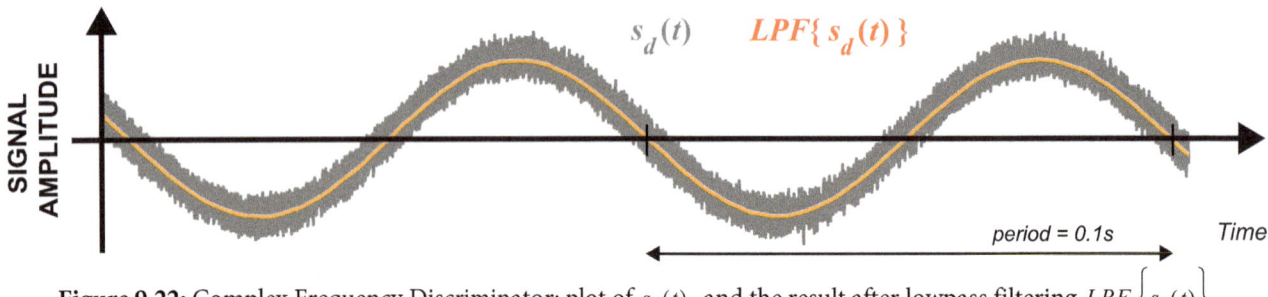

Figure 9.22: Complex Frequency Discriminator: plot of $s_d(t)$, and the result after lowpass filtering $LPF\{s_d(t)\}$

The delay line frequency discriminator is often referred to as an 'FM to AM' function, due to the fact that it converts frequency changes to amplitude changes. The output equation for the discrete time demodulator shown in Figure 9.17 in terms of I and Q samples can be expressed as follows:

$$s_d[n] = \angle\left\{\left(s_{ip}[n] - s_{qp}[n]\right) \times \left(s_{ip}[n-1] + s_{qp}[n-1]\right)\right\} \tag{9.48}$$

There will be a chance to implement and test out one of these demodulators in Section 10.2 (page 376).

9.8 Coherent FM Demodulation: The Phase Locked Loop

Phase Locked Loops can also be used as high performance coherent FM demodulators. Normally these would be positioned at the IF stage of an FM receiver, but they can also be used to demodulate baseband FM signals such as those received from the RTL-SDR. A diagram of a PLL used for this purpose is shown in Figure 9.23. We will adopt a continuous time representation for convenience, but note that the input to the PLL is actually a sampled signal (i.e. the output of the RTL-SDR hardware, and the input to a Simulink model or MATLAB script).

As introduced in Section 7.5 (page 245), PLLs have three main components: a phase detector, a loop filter, and a VCO. The phase detector mixes an input signal with the output of the VCO to produce an error signal. This error signal varies in proportion to the difference in phase between the two signals. It is input to the loop filter (a lowpass filter) to produce a control signal, which is in turn input to the VCO. The VCO integrates the control signal, and uses it to adjust the instantaneous phase of the sinusoidal output (the VCO has a quiescent frequency normally set to the expected frequency of the input signal — this can be set to 0Hz in a PLL operating at baseband). The synthesised sinusoid feeds back into the phase detector, and the process repeats. The PLL strives to make the output of the VCO match the frequency and phase of the input signal, and this is why they are commonly used in coherent demodulators for carrier synchronisation.

If we input the FM signal received using the RTL-SDR, denoted by $s_{\text{RTL-SDR}}(t)$, to the PLL, the VCO within it would aim to output a signal that has the same frequency and phase as the FM 'carrier' at frequency f_Δ. Due to the nature of the FM signal, the PLL will have to keep adjusting its frequency and phase to stay locked. There will always be a phase difference between $s_{\text{RTL-SDR}}(t)$ and $s(t)$, and the error

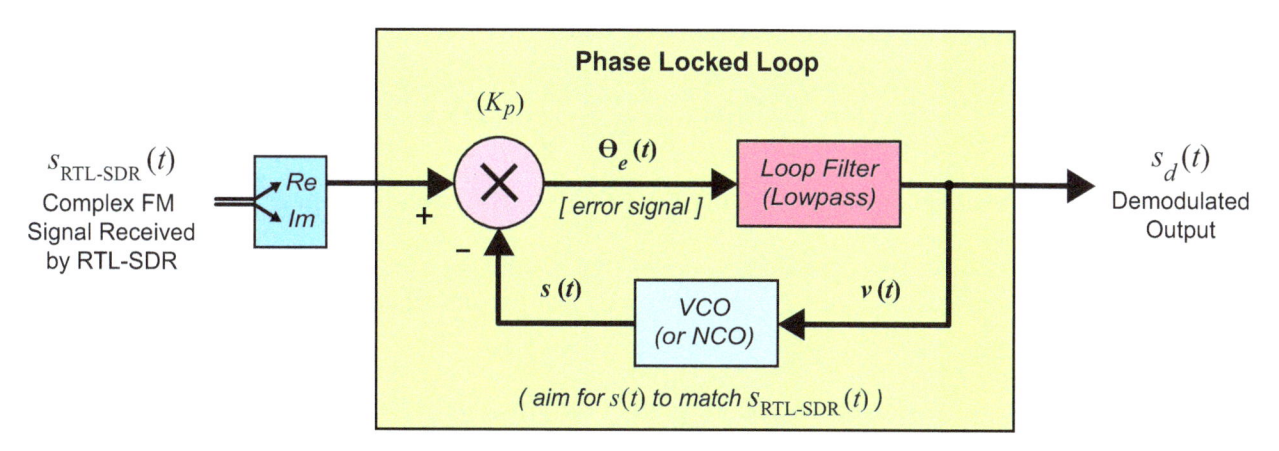

Chapter 9: Frequency Modulation (FM) Theory and Simulation

Figure 9.23: Phase Locked Loop FM Demodulator block diagram

signal resulting from phase detection will continually fluctuate. Its fluctuations will be in proportion to the amplitude changes that occur in the information signal, as information is contained within the phase component of the FM carrier. Lowpass filtering the error signal with the loop filter produces the control signal for the VCO. When the PLL is locked, the control signal $v(t)$ is directly proportional to the information signal $s_i(t)$, and can be considered as being equal to $\alpha s_i(t)$ where α denotes a constant. Referring back to Section 9.2 (where we introduced the use of a VCO as an FM modulator), inputting an information signal to the control port of a VCO makes it output an FM waveform; and when this is the case, the phase detector performs coherent demodulation! [107] [47].

To function as an FM demodulator, the internal parameters of the PLL must be chosen appropriately. The loop filter must be configured to be linear across the frequency range ($f_\Delta \pm \Delta f$) where f_Δ denotes the low 'carrier' frequency, and Δf is the frequency deviation of the FM signal. This ensures that all potential frequency variations of the FM carrier are allowed to pass.

Due to the fact that the PLL is a non-linear device, the equations required to describe the FM demodulation process are rather complicated. *NOTE: when the PLL is not locked to the FM signal it behaves as a negative feedback system that has stability problems [8], so we can only prove this process when it the PLL is locked, i.e. when the frequency of the VCO (f_o) matches the frequency of the baseband FM signal (f_Δ).*

The complex FM signal from Eq. (9.30) contains both real and imaginary components — but the PLL can only operate on one of them. In order to match Figure 9.23, we will prove this using the In Phase component:

$$\Re e\left[s_{\text{RTL-SDR}}(t)\right] = \frac{A_c}{2}\cos\left(\omega_\Delta t + \theta_{fm}(t)\right). \tag{9.49}$$

The phase detector mixes this with the inverse of the signal output from the VCO, and assuming that the PLL is 'locked', then $\omega_o = \omega_\Delta$, and

$$s(t) = A_o \sin\left(\omega_o t + \hat{\theta}(t)\right) = A_o \sin\left(\omega_\Delta t + \hat{\theta}(t)\right) \tag{9.50}$$

$$\text{where:} \quad \hat{\theta}(t) = k_o \int_{-\infty}^{t} v(t)\, dt, \tag{9.51}$$

which gives an output from the phase detector of

$$\theta_e(t) = K_p \cdot \Re\left[s_{\text{RTL-SDR}}(t)\right] \cdot -s(t). \tag{9.52}$$

This can be expanded via trigonometric identity Eq. (B.6) (from Appendix B, page 581) to form

$$\theta_e(t) = -\frac{1}{4}K_p A_c A_o\left[\sin\left(2\omega_\Delta t + \hat{\theta}(t) + \theta_{fm}(t)\right) + \sin\left(\hat{\theta}(t) - \theta_{fm}(t)\right)\right], \tag{9.53}$$

where: K_p is the gain of the phase detector.

Filtering $\theta_e(t)$ with the loop filter removes the high frequency component $\sin\left(2\omega_\Delta t + \hat{\theta}(t) + \theta_{fm}(t)\right)$, leaving the control signal:

$$v(t) = -\frac{1}{4}K_p A_c A_o \sin\left(\hat{\theta}(t) - \theta_{fm}(t)\right). \tag{9.54}$$

As the PLL is locked, $\hat{\theta}(t) \cong \theta_{fm}(t)$, so the small angle approximation leads to

$$v(t) \cong -\frac{1}{4}K_p A_c A_o\left(\hat{\theta}(t) - \theta_{fm}(t)\right). \tag{9.55}$$

Substituting Eq. (9.51) into Eq. (9.55) gives:

$$v(t) = -\frac{1}{4}K_p A_c A_o\left(\left[k_o \int_0^t v(t)\,dt\right] - \theta_{fm}(t)\right) \tag{9.56}$$

$$= -\frac{1}{4}K_p k_o A_c A_o\left(\int_0^t v(t)\,dt - \frac{\theta_{fm}(t)}{k_o}\right).$$

Differentiating this with respect to time produces the following:

$$v'(t) = -\frac{1}{4}K_p k_o A_c A_o\left(v(t) - \frac{\theta_{fm}'(t)}{k_o}\right) \tag{9.57}$$

where: $v'(t)$ is the derivative of $v(t)$
and $\theta_{fm}'(t)$ is the derivative of $\theta_{fm}(t)$,

and re-arranging and solving this for $v(t)$ yields:

$$v(t) = \frac{-4v'(t)}{K_p k_o A_c A_o} + \frac{\theta_{fm}'(t)}{k_o}. \tag{9.58}$$

Chapter 9: Frequency Modulation (FM) Theory and Simulation

As $v'(t) \ll K_p k_o$, the assumption can be made that:

$$\frac{-4v'(t)}{K_p k_o A_c A_o} \cong 0,$$

which means:

$$v(t) \cong \frac{\theta_{fm}'(t)}{k_o}. \tag{9.59}$$

Substituting the derivative of Eq. (9.26) into Eq. (9.59) completes the analysis [45] [52]:

$$s_d(t) = v(t) = \frac{2\pi K_{fm}}{k_o} s_i(t). \tag{9.60}$$

It is clear to see that the control signal resulting from PLL being locked to an FM signal is directly proportional to the original modulating information signal, $s_i(t)$, although it is different to the outputs from the other demodulators.

Because there are a number of different ways to implement a PLL, and because of the fact that the PLL is a non-linear device, describing this demodulation process in terms of samples received from the RTL-SDR is not practical. This demodulator will work well, however, and you will have the opportunity to try one out in Section 10.4 (page 396).

9.9 Demodulating Signals from Commercial FM Radio Stations

Some additional issues related to receiving signals broadcast from commercial FM radio stations must be considered, and this short section of the book guides you through these.

9.9.1 De-emphasis Filtering

As information signals are modulated for WFM radio stations, they undergo a process called *pre-emphasis*, which is where components in the signal above a certain frequency have a gain applied to them in an effort to maintain the modulation index (and therefore the bandwidth) of the station. Normally this is carried out with a filter; one that simply gives a linear gain to all frequencies past this value. In the USA this critical value was chosen to be 2122.1Hz (f_{c-usa}), and in Europe and Asia, 3183.1Hz (f_{c-eu}). These frequencies relate to analogue filter time constants of 75μs and 50μs respectively [112]. In FM receivers this must be reversed, and is done so with a process called *de-emphasis*. Again, it is common to perform this with a filter, this time one that applies an attenuation to components above the stated frequencies. These two processes are summarised by Figure 9.24.

If using a European FM radio in the USA, a gain of around 3.5dB will be applied to frequency components over f_{c-eu}, due to it having the wrong type of de-emphasis filter. Conversely, using an FM radio from the USA in Europe, you would find that an attenuation of 3.5dB is applied to these frequency components, effectively acting like a gain applied to lower frequencies. It is important therefore to use the correct de-emphasis filter for each region to avoid distorting the received signal! The effect of using an FM radio in the wrong continent are demonstrated in Figure 9.25.

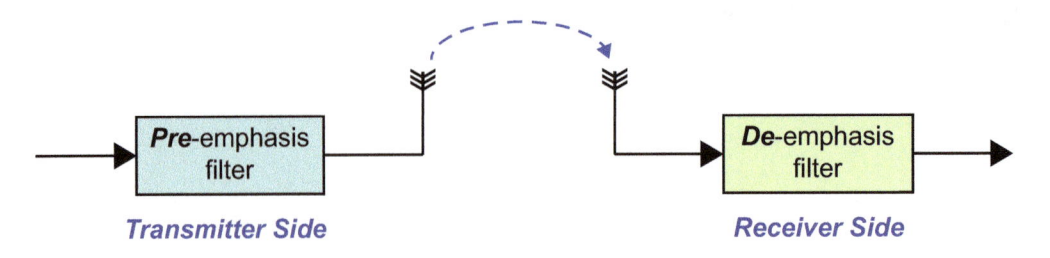

Figure 9.24: Block diagram confirming positions of pre-emphasis and de-emphasis filters

For convenience, we have constructed a de-emphasis filter Simulink block for you that can be used to implement the appropriate filter for your location. This has been packaged in the *RTL-SDR Book Library*, which you should have already added to your MATLAB path. If you have not done this yet, you should do so. Instructions can be found in Exercise 2.4.

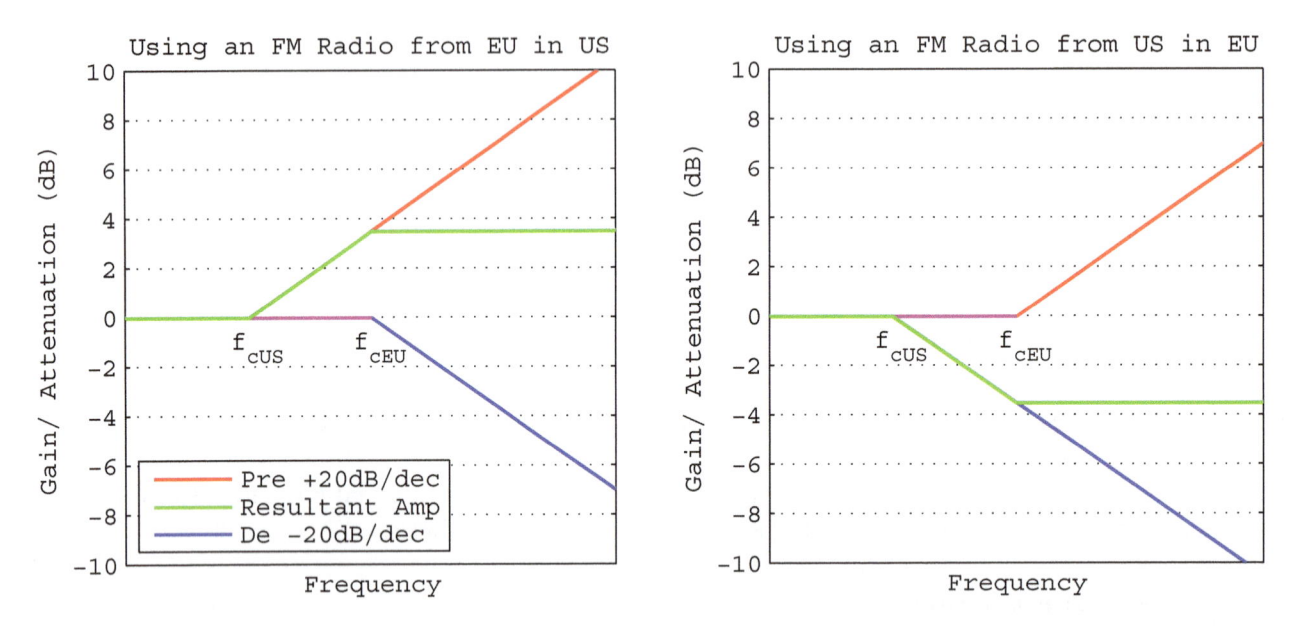

Figure 9.25: Bode plots showing the effects of using an FM radio in the wrong continent

9.9.2 The FM Radio Multiplex

It is common practice to multiplex multiple information signals together before performing modulation, as this allows for multi-channel transmission using one carrier. FDM was first used to transmit several channels of information on an FM carrier by Edwin Armstrong (pioneer of FM, and who was introduced in Section 9.1). He demonstrated that a single FM carrier was capable of simultaneously transmitting *"two radio programs, various telegraph messages, and a facsimile reproduction of the front page of The New York Times"* during a trial broadcast from the top of the Empire State Building in 1934 [28]. In the early 1960s (years after Armstrong's death), when organisations such as the FCC began putting regulations in place to prepare for the inevitable boom of FM, the multiplexing concept was revisited in an effort to facilitate the transmission of stereo audio signals.

Chapter 9: Frequency Modulation (FM) Theory and Simulation

The stereo FM MPX had to be designed in a way that allowed for both Left and Right information channels to be broadcast to new stereo FM receivers, while retaining a mono transmission for older radio hardware. It was decided that the best way to do this was to first band-limit the Left and Right information signals to 15kHz, produce separate 'Mono' and 'Stereo' components, and multiplex these together. The key to this MPX was a 19kHz sinusoid, called the 'Pilot Tone'. If the pilot tone was present in a received FM signal, the radio hardware would be informed that there was stereo information present to demodulate. A baseband spectral representation of the FM MPX is shown in Figure 9.26.

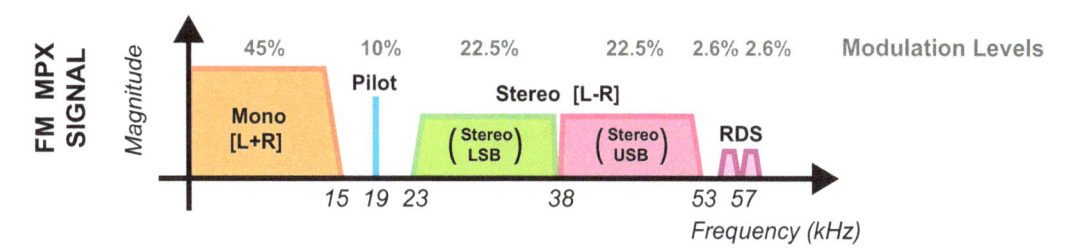

Figure 9.26: Baseband spectral representation of the FM radio multiplex

The 'Mono' component is created by adding the two information signals together [L+R], and is situated at baseband. The stereo component is created by subtracting the Right channel from the Left [L−R], and then AM-DSB-SC modulating the resulting signal onto a subcarrier with a frequency of 38kHz. This modulation process means that the bandwidth of the stereo component doubles, which is one of the reasons for the rather odd carrier frequency. Another is that the 38kHz tone can be synthesised by doubling the frequency of the pilot tone, which means that coherent AM demodulation is possible in a stereo FM receiver.

Later additions to this MPX include the Radio Data System (RDS), a service where digital information such as radio station names and current music track titles can be transmitted to radio receivers. This enables the radio in your car (for instance) to show information about radio programmes. The RDS bit stream is Binary Phase Shift Keying (BPSK) modulated, then AM-DSB-SC modulated onto a subcarrier with a carrier frequency of 57kHz. This carrier can be synthesised in a receiver by tripling the frequency of the pilot tone.

The percentages listed above each of the MPX components in Figure 9.26 relate to the permitted modulation level for each component. You may notice that, when you add up all of the numbers listed, that the result is a total modulation level of 105.2%, and this is because the original specification did not include the RDS functionality. Amplifiers are used to balance the components of the FM MPX, as demonstrated in the block diagram of a stereo FM encoder and multiplexer shown in Figure 9.27.

To generate the stereo FM signal, left and right channels of an audio signal are input to this device. It performs encoding to create the mono and stereo components, and then multiplexes them together. The MPX signal can be expressed as follows:

$$s_{fm\ mpx}(t) = A\Big[s_l(t) + s_r(t)\Big] + A\Big[s_l(t) - s_r(t)\Big]\cos\Big(2\pi f_{38k} t\Big) \qquad (9.61)$$

$$+ B\cos\Big(2\pi f_{19k} t\Big) + C\Big[s_{rds}(t)\Big]\cos\Big(2\pi f_{57k} t\Big)$$

where: $A = 45\%$, $B = 10\%$ and $C = 5.2\%$.

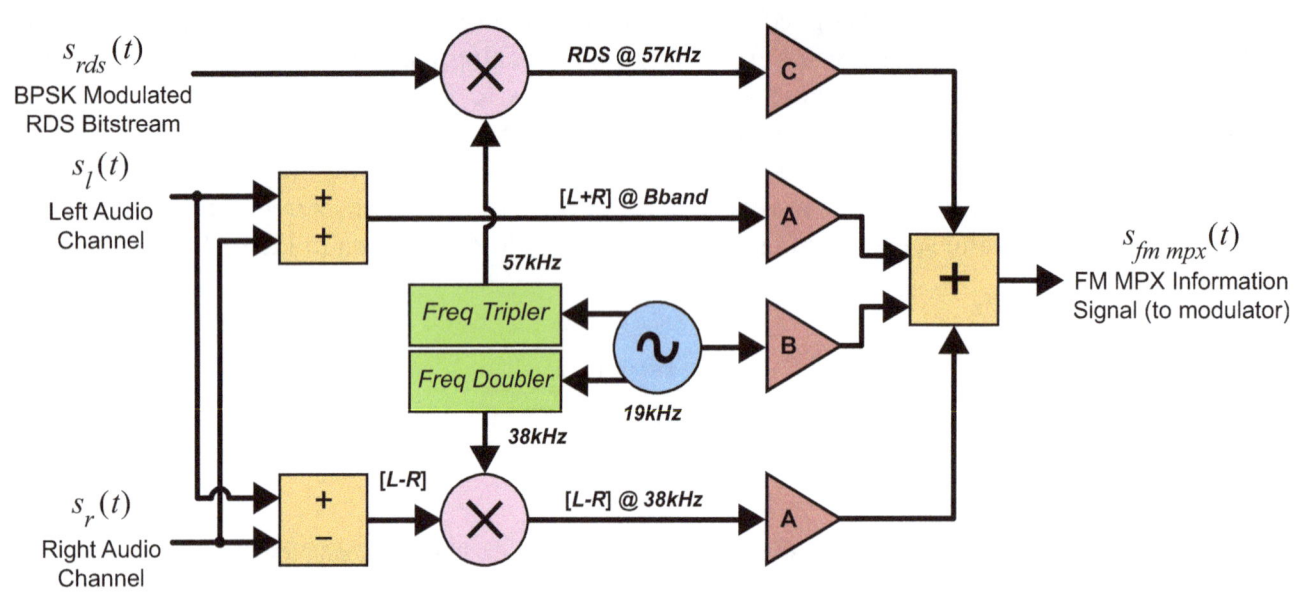

Figure 9.27: Block diagram of the stereo FM encoder and multiplexer

Next, this baseband signal (which is considered an information signal) would provide an input to an FM modulator such as the one shown in Figure 9.2, and transmitted.

After using an FM demodulator to return the FM MPX to baseband, stereo FM receivers must perform demultiplexing and decoding in order to extract the left and right channels. The first stage of this process involves using bandpass filters to isolate each of the individual MPX components. The pilot tone is passed through frequency doublers and triplers to synthesise the suppressed 38kHz and 57kHz subcarriers, and these tones are then used to coherently AM demodulate the stereo and RDS components to baseband. After lowpass filtering to ensure that no unwanted information is present, the left channel is extracted by adding the demodulated [L–R] stereo component to the [L+R] mono component, to produce *2L*. The right channel is extracted by subtracting the stereo component from the mono component, resulting in *2R*. The block diagram for this demultiplexer/ decoder is shown in Figure 9.28.

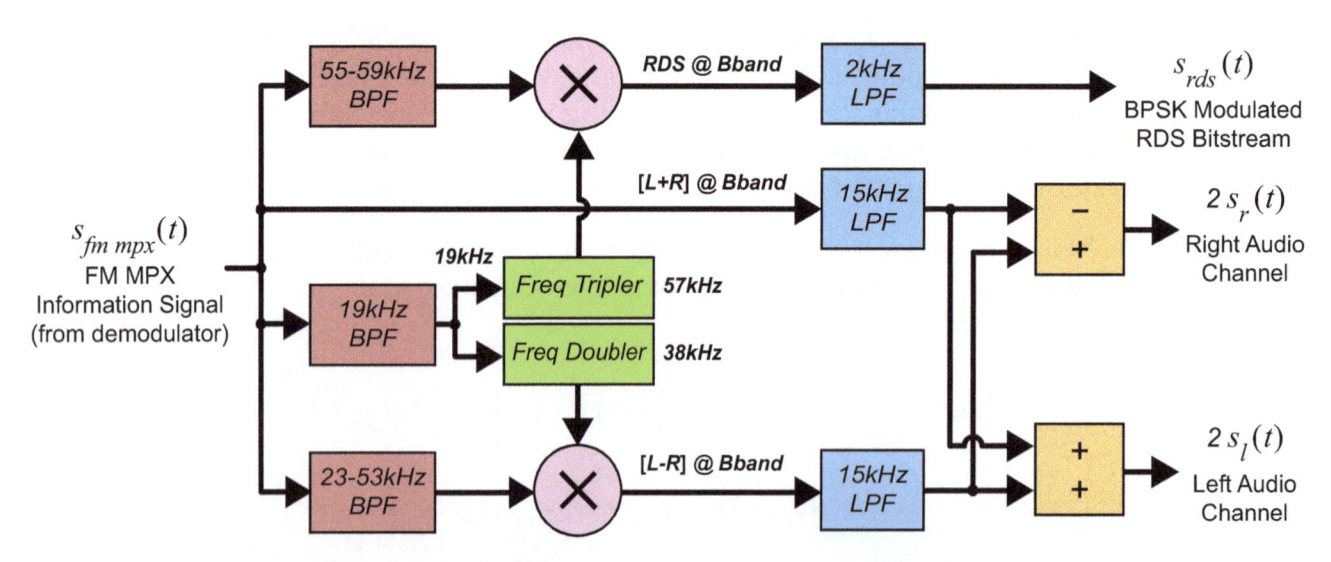

Figure 9.28: Block diagram of the stereo FM demultiplexer and decoder

Chapter 9: Frequency Modulation (FM) Theory and Simulation

Normally, radio manufacturers would utilise an IC to perform all of this processing, and due to high demand these are mass manufactured (see for example [100]). Because we are working in Simulink however, these chips are of no use to us, and FM multiplexers/ demultiplexers and encoders/ decoders must be implemented with the DSP blocks shown here. There will be an opportunity to test out an encoder/ multiplexer in Exercise 9.3.

Exercise 9.3 Stereo FM Encoder and Multiplexer Simulation

In this exercise you will run a model that creates an FM MPX signal from two baseband audio signals. *Spectrum Analyzers* are connected at various points throughout the model to allow monitoring of the signals as they undergo the multiplexing process.

(a) **Open MATLAB.** Set the working directory to an appropriate folder so you can open this model:

/fm/simulation/fm_mpx.slx

The block diagram should look like this:

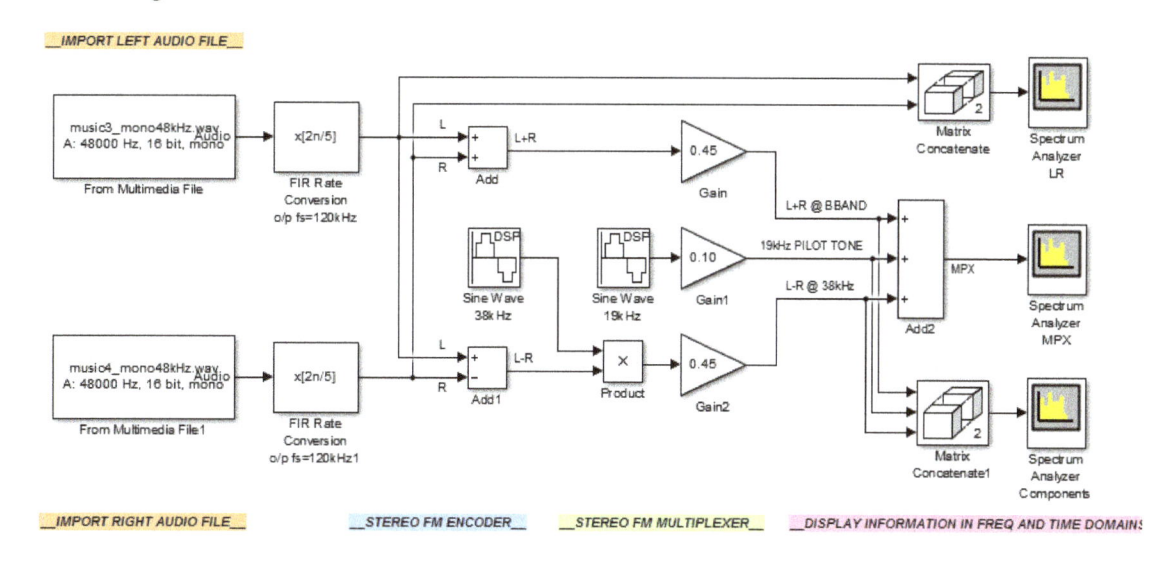

(b) **Inspect the model.** The stereo information signal comprises two imported mono audio files. Their sampling rates are increased to 120kHz using *FIR Rate Conversion* blocks (a sampling rate this high is necessary to meet the Nyquist condition), and they are then encoded to create the baseband 'mono' and 'stereo' components, [L+R] and [L−R]. The mono component is left at baseband, while the stereo component is AM-DSB-SC modulated onto a carrier with a frequency of 38kHz. The Pilot tone is created using a second *Sine Wave* block, and the three components are summed together to create the MPX signal. You should notice the similarities between this block diagram and the diagram of the FM encoder/ multiplexer in Figure 9.27.

(c) **Run the simulation.** Begin the simulation by 1 on the 'Run' ▶ button in the Simulink toolbar. The three *Spectrum Analyzer* windows should appear, and you should position these so that you can see all of them.

(d) **Signal Analysis.** *Spectrum Analyzer LR* allows you to monitor the 'left' and 'right' baseband information signals after they have been resampled to a rate of 120kHz. *Spectrum Analyzer Components* will show the three MPX components individually (mono, stereo and the pilot tone), and *Spectrum Analyzer MPX* will show them after they have been summed together to create the MPX signal. You will hopefully be able to see that the MPX signal generated in this model is very similar to the standard FM MPX spectrum shown in Figure 9.26 — albeit there is no RDS component present!

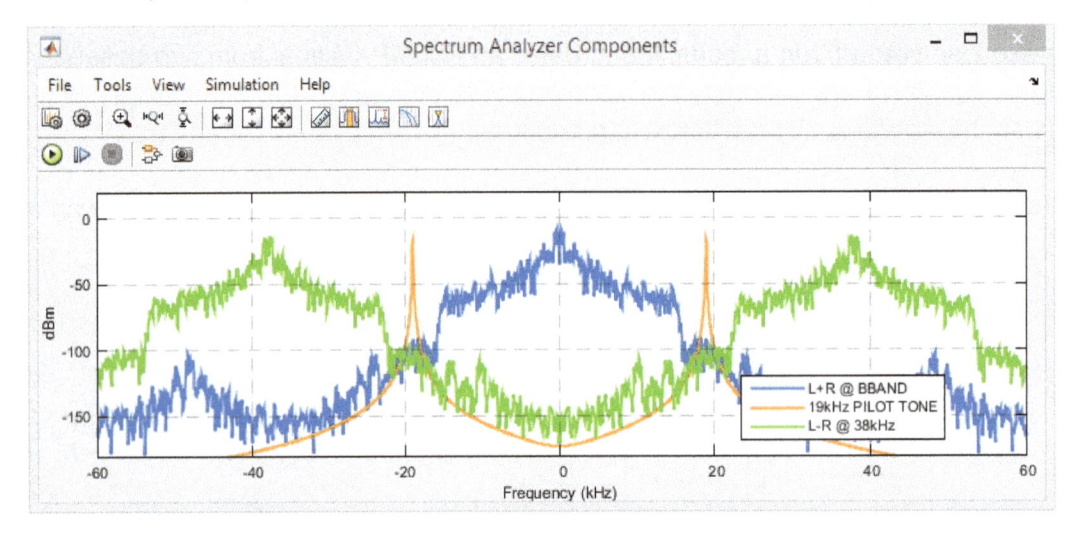

(e) We will return to FM multiplexing in the following chapter, where you will have the opportunity to construct both a stereo FM transmitter model that uses the USRP® radio, and a stereo FM receiver which uses the RTL-SDR.

9.10 Summary

In this chapter we have introduced the mathematics behind frequency modulation, and demonstrated how a VCO can be used to generate FM signals. We have emphasised the significance of the FM modulation constant, β_{fm}, and how its value determines whether an FM signal is narrow or wideband, and run simulations demonstrating this. Demodulators were presented, along with some of the additional considerations when receiving signals from commercial FM radio stations. In the next chapter, we will move on to transmit FM signals using a USRP® radio, and receive them using the RTL-SDR.

10 Desktop FM Transmission and Reception

In the last chapter we discussed how FM signals are generated and demodulated, and the additional considerations when receiving signals from commercial FM radio stations. In this chapter we will now being to implementing some real FM desktop SDR Tx/ Rx systems featuring the RTL-SDR.

Although commercial FM radio stations are broadcast in many countries around the world, the quality of the signal that can be received at any particular location may be poor. It is worthwhile trying to receive and demodulate FM radio stations in your vicinity, but having the facility to generate your own is very useful; especially if you need (or want) more control over what you receive. As with AM, a number of different types of transmitter can be used to do this. The authors found that low cost (a few $'s) *off the shelf* FM transmitters typically used with smartphones and other audio players for in-car use, were great for this purpose! We will discuss and use these and others in Section 10.7, and a number of the exercises in this chapter can also be completed with these FM transmitters outputting signals for the RTL_SDR.

Initially we will be using a USRP® radio as an FM transmitter. It is a convenient platform for transmitting custom FM signals as it can be interfaced with directly from MATLAB and Simulink. Although this requires building the modulator model yourself (and later on, stereo FM encoders and multiplexers...), a greater level of control can be exerted. It is also worth noting that the carrier frequencies of off-the-shelf FM transmitters are normally limited to the range 88MHz–108MHz, whereas the range of carrier frequencies supported on the USRP® hardware will depend on the daughterboard in use, but will typically cover a much wider range.

After the USRP® based examples, this chapter will progress through implementing other RTL-SDR based FM receivers. A number of exercises will involve building receivers using different types of FM demodulator and stereo FM decoder/ demultiplexer; and we will also run some reference models. Later in the chapter, the focus will shift to implementing some more unusual systems, for example the transmission and reception of *AM signals* using FM transmitters, and stereo multichannel transmission.

10.1 Transmitting Mono WFM Signals with the USRP® Hardware

As introduced in Section 8.1 (page 280), the USRP® radio is an SDR that can be used in conjunction with Simulink to implement a number of different radio transmitters and receivers. By default, the FPGA on the Tx side of the USRP® radio is configured to perform DUC, to interpolate baseband IQ samples (transferred to the device from a host computer) to a much higher frequency. The samples are converted to continuous time signals using a DAC, and then mixed with a complex RF carrier, performing complex AM-DSB-SC modulation. This means that to transmit an FM signal with it, the samples input to the 'Data' port of the *SDRu Transmitter* Simulink block must already be FM modulated, at baseband.

Performing baseband FM modulation is a simple process, and the associated maths is much less complicated than that of the 'standard' FM modulation reviewed in Chapter 9! Although the phase of a carrier still needs to be modulated with the integral of an information signal, the carrier is not a sinusoid; rather it is a real constant DC value, given that we are using complex modulation methods. (It might be useful to review again Figure 9.9 on page 349 — we are essentially going in reverse order: generating plot (c) and then upconverting to produce plot (a)).

The standard FM modulation process creates a real signal centred around a positive frequency carrier, the baseband FM modulation process creates an imaginary signal centred around 0Hz. While the standard FM signal at a high frequency carrier f_c takes the form

$$s_{fm}(t) = A_c \cos\left(\omega_c t + \theta_{fm}(t)\right) \quad \text{where} \quad \theta_{fm}(t) = 2\pi K_{fm} \times \int_{-\infty}^{t} s_i(t)dt , \tag{10.1}$$

we saw from Eq. (9.31) on page 350 that the complex baseband FM signal (i.e. on a low frequency "carrier" centred at f_Δ Hz, which can be 0 Hz) can be represented in complex exponential form as Me^{ϕ} or alternatively magnitude/argument form $M \angle \phi$. For a baseband FM signal centred at 0Hz, we can say:

$$s_{fm-baseband}(t) = A_c e^{-j\theta_{fm}(t)} = A_c \angle -\theta_{fm}(t) \tag{10.2}$$

$$= A_c \angle -2\pi K_{fm} \times \int_{-\infty}^{t} s_i(t)dt ,$$

where $\angle$ represents the argument of the complex baseband signal, and A_c is the DC constant (adopting a continuous time representation for convenience). A VCO is not required to generate this signal. To modulate this baseband signal onto a carrier, it simply needs to be input to the quadrature AM-DSB-SC modulator on the USRP® hardware, and upconverted to the appropriate RF centre frequency.

Substituting the complex baseband FM signal from Eq. (10.2) into Eq. (8.1) (the generalised form of the signal output by the USRP® radio, page 280), the AM modulated baseband FM signal output by the hardware can be expressed as follows:

$$s_{usrp-tx}(t) = K_{usrp-tx}\left[\Re e\left(s_{fm-baseband}(t)\right)\cos(\omega_c t) + \Im m\left(s_{fm-baseband}(t)\right)\sin(\omega_c t)\right] \tag{10.3}$$

$$= K_{usrp-tx}\left[A_c \cos\left(-\theta_{fm}(t)\right)\cos(\omega_c t) + A_c \sin\left(-\theta_{fm}(t)\right)\sin(\omega_c t)\right].$$

This can be expanded using Eqs. (B.4) and (B.5) from Appendix B (page 581) to give:

Chapter 10: Desktop FM Transmission and Reception 369

$$s_{usrp-tx}(t) = \frac{A_c K_{usrp-tx}}{2}\left[\cos\left(\omega_c t + \theta_{fm}(t)\right) + \cos\left(\omega_c t - \theta_{fm}(t)\right)\right.$$

$$\left. + \cos\left(\omega_c t + \theta_{fm}(t)\right) - \cos\left(\omega_c t - \theta_{fm}(t)\right)\right] \quad (10.4)$$

$$= \frac{A_c K_{usrp-tx}}{2}\left[2\cos\left(\omega_c t + \theta_{fm}(t)\right) + 0\right]$$

$$= A_c K_{usrp-tx} \cos\left(\omega_c t + \theta_{fm}(t)\right)$$

$$= A_c K_{usrp-tx} \cos\left(\omega_c t + 2\pi K_{fm} \times \int_{-\infty}^{t} s_i(t)dt\right).$$

Note that AM modulating the complex baseband FM signal with the USRP® radio results in the device outputting the FM signal from Eq. (10.1)!

Figure 10.1 shows how the *baseband FM modulator* can be implemented with five simple DSP blocks.

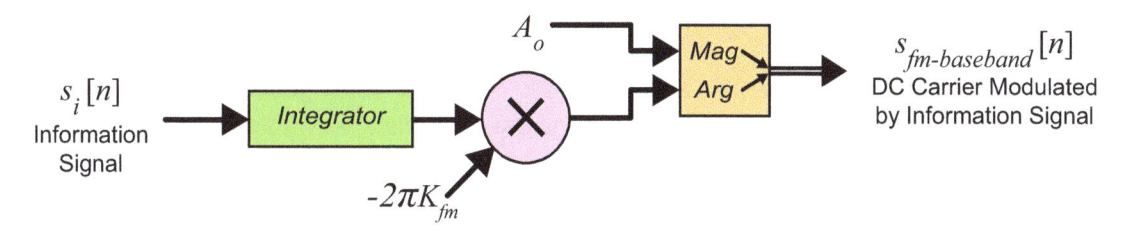

Figure 10.1: Block diagram of baseband FM modulator

To begin with, we shall focus on designing a Simulink model which will generate a mono FM signal. A high level block diagram of the processes required to implement this modulator/ transmitter is shown in Figure 10.2. In the next exercise, we will construct this modulator, and transmit the signal it creates using the USRP® radio.

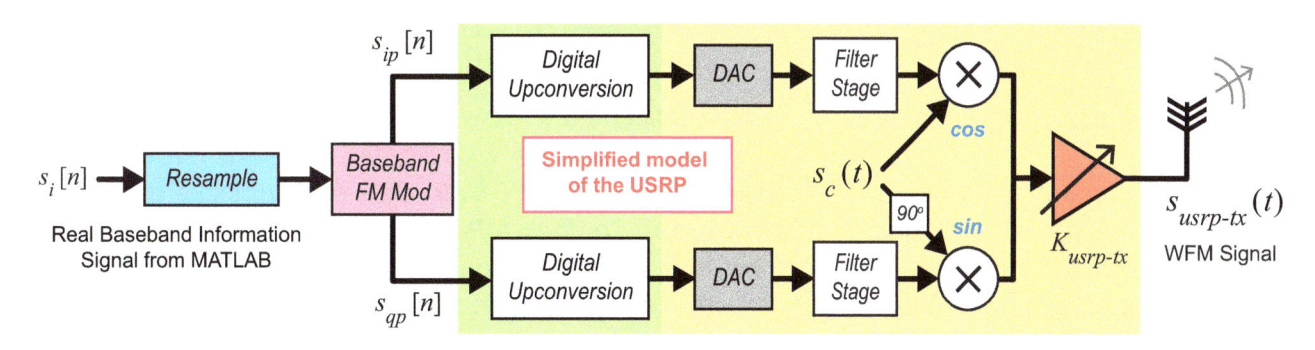

Figure 10.2: Block diagram showing a Simulink/ USRP® hardware implementation of a mono FM modulator

Before starting to build any FM transmitter Simulink models, make sure that you have the MathWorks USRP® Hardware Support Package installed. Details on this can be found in Appendix A.2 (page 571).

Exercise 10.1 USRP® Radio: Mono FM Modulator and Transmitter

In this exercise we will to build a Simulink model that will FM modulate an audio file from the computer onto a carrier using the USRP® hardware. There will be an opportunity to receive and demodulate the generated signal with the RTL-SDR and Simulink in later exercises.

Note: If you do not have a USRP® radio, it will still be possible to run this exercise and see the FM signal being generated by commenting out the SDRu Transmitter block.

(a) **Open MATLAB.** Set the working directory to the exercise folder,

   `/my_models/transmitters`

   Next, create a new Simulink model, and save this with the name:

   `.../usrp_fm_mono.slx`

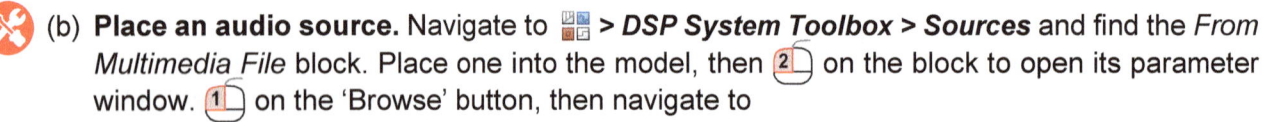

(b) **Place an audio source.** Navigate to ▦ > *DSP System Toolbox* > *Sources* and find the *From Multimedia File* block. Place one into the model, then ② on the block to open its parameter window. ① on the 'Browse' button, then navigate to

   `/audio_sources`

   and select the one of the mono audio sources from this folder (it is recommended to use these files as they are sampled at 48kHz). Change the 'Samples per audio channel' to '600', and then open the 'Data Types' tab, and change the 'Audio output data type' to 'single'.

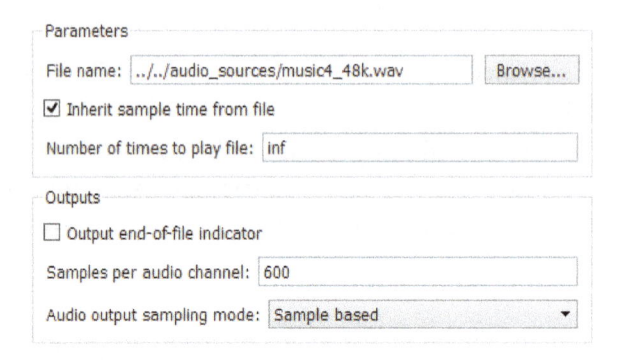

```
Parameters
File name: ../../audio_sources/music4_48k.wav    Browse...
☑ Inherit sample time from file
Number of times to play file: inf

Outputs
☐ Output end-of-file indicator
Samples per audio channel: 600
Audio output sampling mode: Sample based ▼
```

After applying the changes, close the parameters window.

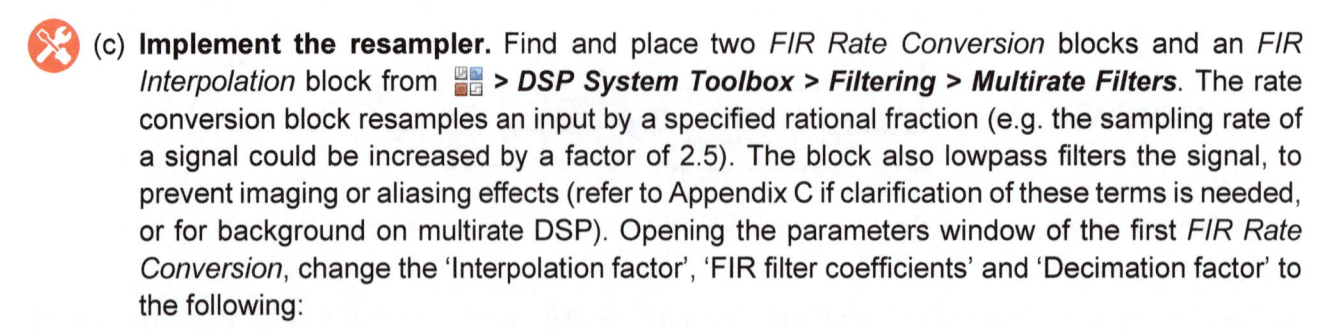

(c) **Implement the resampler.** Find and place two *FIR Rate Conversion* blocks and an *FIR Interpolation* block from ▦ > *DSP System Toolbox* > *Filtering* > *Multirate Filters*. The rate conversion block resamples an input by a specified rational fraction (e.g. the sampling rate of a signal could be increased by a factor of 2.5). The block also lowpass filters the signal, to prevent imaging or aliasing effects (refer to Appendix C if clarification of these terms is needed, or for background on multirate DSP). Opening the parameters window of the first *FIR Rate Conversion*, change the 'Interpolation factor', 'FIR filter coefficients' and 'Decimation factor' to the following:

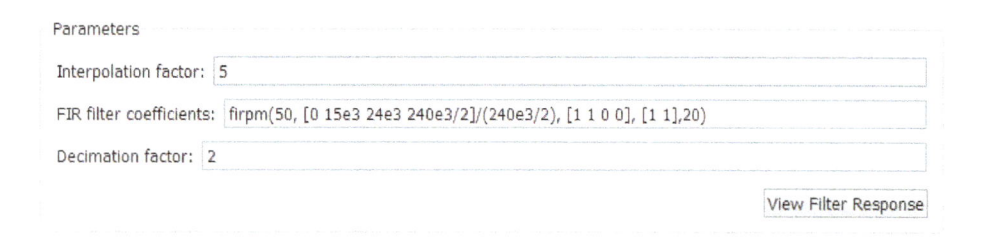

```
Parameters
Interpolation factor:    5
FIR filter coefficients: firpm(50, [0 15e3 24e3 240e3/2]/(240e3/2), [1 1 0 0], [1 1],20)
Decimation factor:       2
                                                              View Filter Response
```

This specifies that the input signal will be interpolated by a factor of 5, and then decimated by a factor of 2; i.e. a rate change from 48kHz to 120kHz. Save the changes, and then rename the block *FIR Rate Conversion o/p fs=120kHz* by 🖱 on the text under the block.

(d) Next, open the second *FIR Rate Conversion* block and change its parameters to the following:

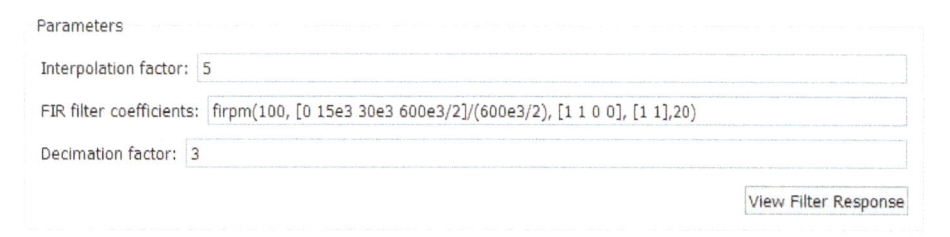

```
Parameters
Interpolation factor:    5
FIR filter coefficients: firpm(100, [0 15e3 30e3 600e3/2]/(600e3/2), [1 1 0 0], [1 1],20)
Decimation factor:       3
                                                              View Filter Response
```

Here, we are changing the sampling rate from 120kHz to 200kHz. Rename this block *FIR Rate Conversion o/p fs=200kHz*.

(e) Open the *FIR Interpolation* block and change the parameters to the following:

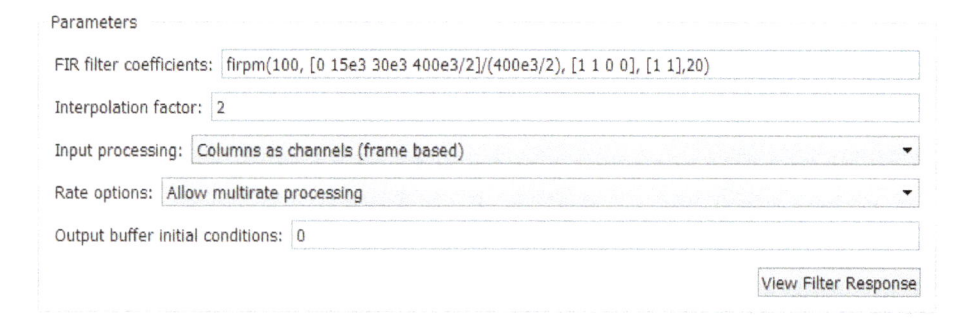

```
Parameters
FIR filter coefficients:       firpm(100, [0 15e3 30e3 400e3/2]/(400e3/2), [1 1 0 0], [1 1],20)
Interpolation factor:          2
Input processing:              Columns as channels (frame based)
Rate options:                  Allow multirate processing
Output buffer initial conditions: 0
                                                              View Filter Response
```

This interpolator is configured to change the sampling rate from 200kHz to 400kHz, which is the rate required to satisfy Nyquist conditions. Rename it *FIR Interpolation o/p fs=400kHz*.

(f) **Add filtering stages.** To ensure that spectral images in the interpolated signal are sufficiently attenuated, we will add additional filtering stages to the upconverter chain, between the *FIR Rate Conversion* and *FIR Interpolation* blocks (note that this is not presented as an optimum design — but it does run in real time on our test system!). Place two *Lowpass FIlter* blocks from ▦ **> DSP System Toolbox > Filtering > Filter Designs** into the model, and configure them as follows. Set appropriate 'Input sample rate' values (which will depend on the sampling rate of the previous block), and rename the blocks *Lowpass Filter fc=15kHz*.

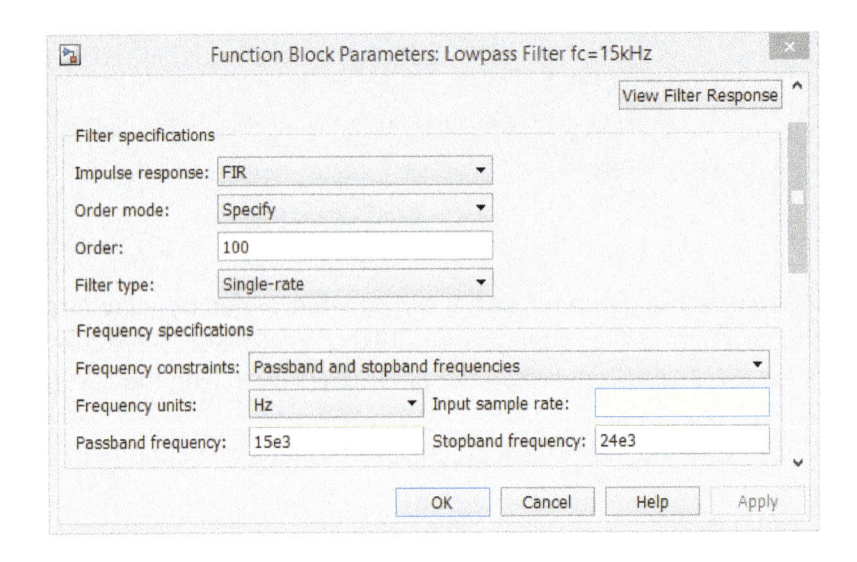

(g) Connect all of these blocks up as follows:

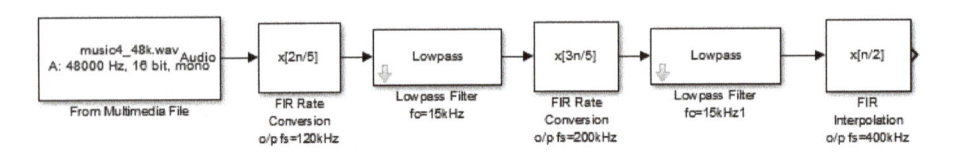

(h) **Implement the baseband FM modulator.** Open 🖱️ > *DSP System Toolbox > Filtering > Filter Implementations*, and find a *Discrete Filter* block. Place this in your model, and 🖱️ to open its parameter window. Change the 'Numerator' to '[1]' and the 'Denominator' to '[1 -1]'. Apply these changes, and rename this block *Digital Integrator*. You should see the block re-render and show the following:

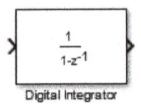

(i) Next, source a *Gain* block from the 🖱️ > *Simulink > Math Operations*. Set this to have a 'Gain' of '50', and rename the block *Frequency Sensitivity Gain 2πKfm*. Place a *Constant* from 🖱️ > *Simulink > Sources* and set its value to '0.1'. Rename this *DC Carrier Magnitude*. Complete the modulator by placing a *Magnitude-Angle to Complex* block from 🖱️ > *Simulink > Math Operations*, and connecting the four blocks as follows:

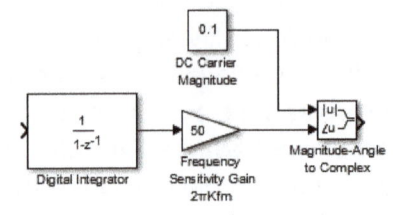

(j) **Interface with the USRP® radio.** Fetch an *SDRu Transmitter* block from the 🖱️ > *Communications System Toolbox Support Package for USRP® Radio* library. This block should now be configured to communicate with the USRP® hardware. To do this, you will need

Chapter 10: Desktop FM Transmission and Reception

to know its IP address (or USB address), and be able to 'ping' it from your computer. Instructions for setting up this connection are included in Appendix A.2 (page 571). 2 on the *SDRu Transmitter* block, select the correct 'Platform', and type the 'Address' of your hardware.

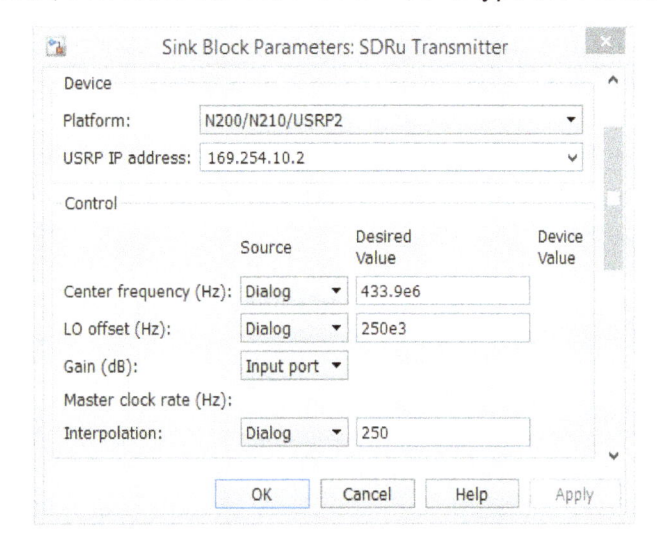

(k) The RF side of the USRP® radio should also be configured. Set the 'Center Frequency' parameter to the desired transmit frequency. This value should be within the range of your RTL-SDR's tuner, e.g. in the range 25MHz–1.75GHz. If for example, it is desired to modulate the signal onto a 433.9MHz carrier, '433.9e6' should be entered into this field. Set the 'LO offset' to '250e3' and the 'Interpolation' factor to '250'. By setting the local oscillator of the USRP® hardware to 250kHz, we can ensure that any generated harmonics will not interfere with the FM signal. Choosing the value of '250' for the interpolation function means that the USRP® radio resamples the FM signal from 400kHz to 100MHz before modulating it onto the carrier. Change the 'Source' of the 'Gain' to 'Input Port', then apply the changes and close the window.

(l) Finally, place a *Constant* from ▦ > *Simulink* > *Sources*. Set the 'Constant Value' to '1' and rename this block *Transmitter Gain (dB)*. Connect these and the other blocks as shown below.

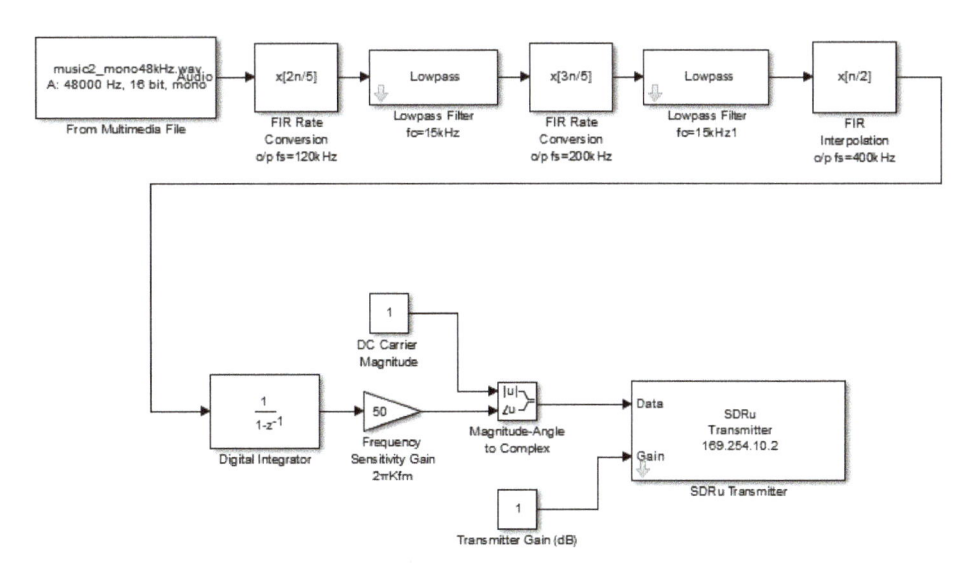

(m) **Add scopes.** When we designed the AM USRP® modulator models in Section 8.1 (page 280), there was no reason to add scopes to monitor the signal being passed to the transmitter, as it was simply the source audio signal, interpolated to a higher rate. In this situation however, we can monitor the FM modulation process as it is carried out in Simulink. Add a *Spectrum Analyzer* and a *Time Scope* from ▸ **> DSP System Toolbox > Sinks**.

(n) It would be useful to be able to monitor both the information signal and FM modulated signal in the time domain, as this will allow a comparison between the frequency fluctuations in the modulated signal, and the amplitude changes in the information signal. The *Time Scope* needs to be reconfigured to do this. **2** on the *Time Scope*, and **1** the 'File' menu, then change the 'Number of Input Ports' to '2'. **1** on the 'View' menu and set the 'Layout' to 2x1.

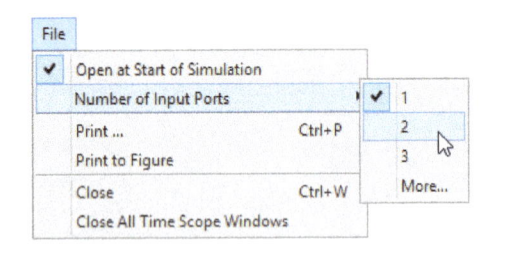

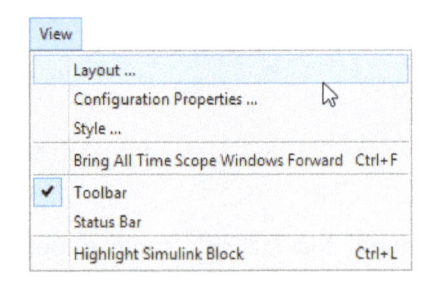

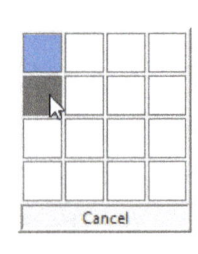

Navigate to 'Configuration Properties', and open the 'Time' tab. Change the 'Time span' to '1250/400e3'. This will limit the scope to showing only 1250 samples. Finally, connect the *Spectrum Analyzer* and *Time Scope* as follows:

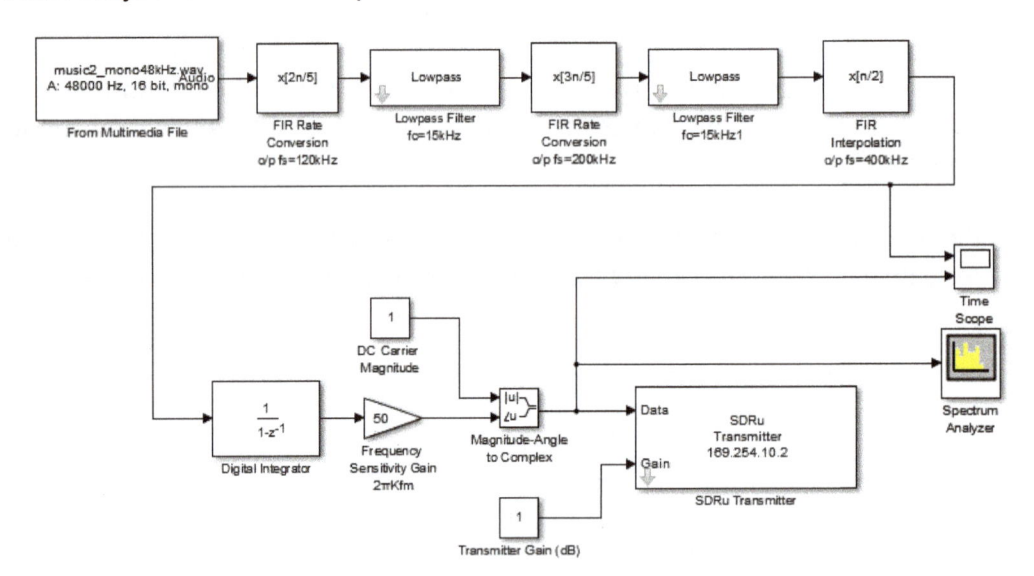

(o) **Prepare to run the simulation.** Set the *Simulation Stop Time* to 'inf' and change the *Simulation Mode* to 'Accelerator'. This will force Simulink to partially compile the model into native code for your computer, which allows it to run faster. Finally, ensure that the model is saved.

(p) **Open the reference file.** If you do not wish to construct the transmitter yourself, open the following file. Make sure that the *SDRu Transmitter* block is appropriately configured.

Chapter 10: Desktop FM Transmission and Reception 375

📁 `/fm/usrp_tx/usrp_fm_mono.slx`

Check that the USRP® hardware is turned on and that MATLAB can communicate with it by typing `findsdru` into the MATLAB command window. If a structure is returned as discussed in Appendix A.2 (page 571), continue to the next step.

(q) **Run the simulation.** Begin the simulation by 🖱 on the 'Run' ▶ button in the Simulink toolbar. After a couple of seconds, Simulink will establish a connection with the USRP® radio and the simulation will begin. When it does, samples of the FM modulated baseband information signal will be passed to the radio, modulated onto the AM carrier, and transmitted from its antenna.

If transmitting on a frequency within the range 88MHz to 108MHz, it will be possible to use a standard analogue FM radio to receive the signal. If not, the RTL-SDR can be used to receive the signal. We will construct RTL-SDR FM demodulators capable of receiving these transmitted signals in Sections 10.2 and 10.4.

(r) **Signal Analysis: *Spectrum Analyzer*.** When the simulation is running, the baseband FM modulated signal should be visible in the *Spectrum Analyzer* window. There should be a large band of energy (between 100 and 200kHz wide) shifting from side to side of the DC carrier at 0Hz. Notice that its bandwidth is constantly increasing and decreasing. This is because the *Frequency Sensitivity Gain* value is not being automatically adjusted to keep the modulation index constant. Commercially broadcast FM radio stations have to perform some kind of AGC to prevent this effect, in order to keep their signals inside their allocated bands.

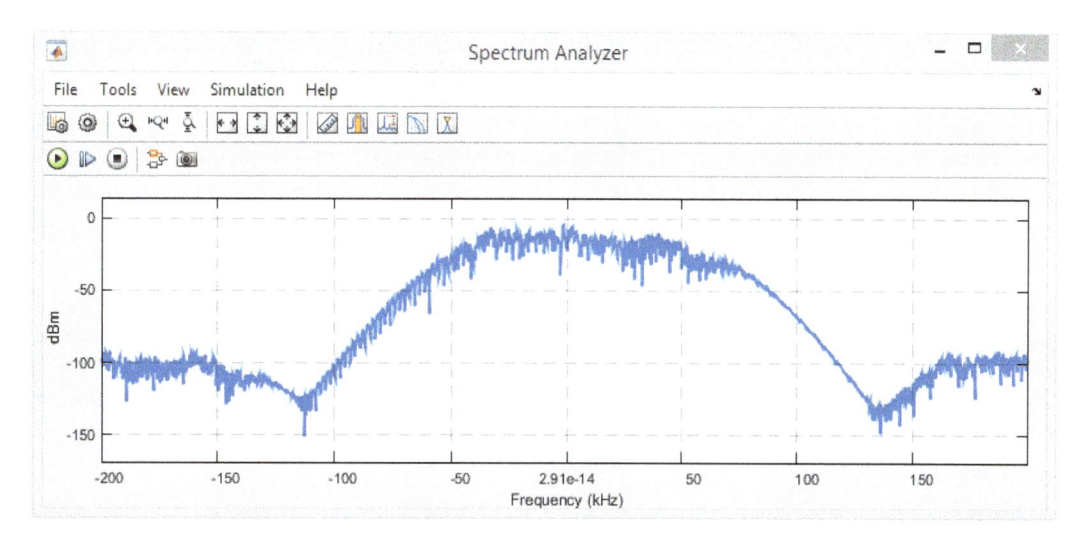

(s) **Signal Analysis: *Time Scope*.** The *Time Scope* should display two signals — the information signal and the FM modulated signal. Pause the simulation using the 'Pause' ⏸ button, and examine the plots. Can you clearly see that the frequency of the modulated signal increases as the amplitude of the information signal increases (and vice versa)? Note: you may notice there is a slight offset between the two plots; this is due to the processing time associated with the baseband modulator.

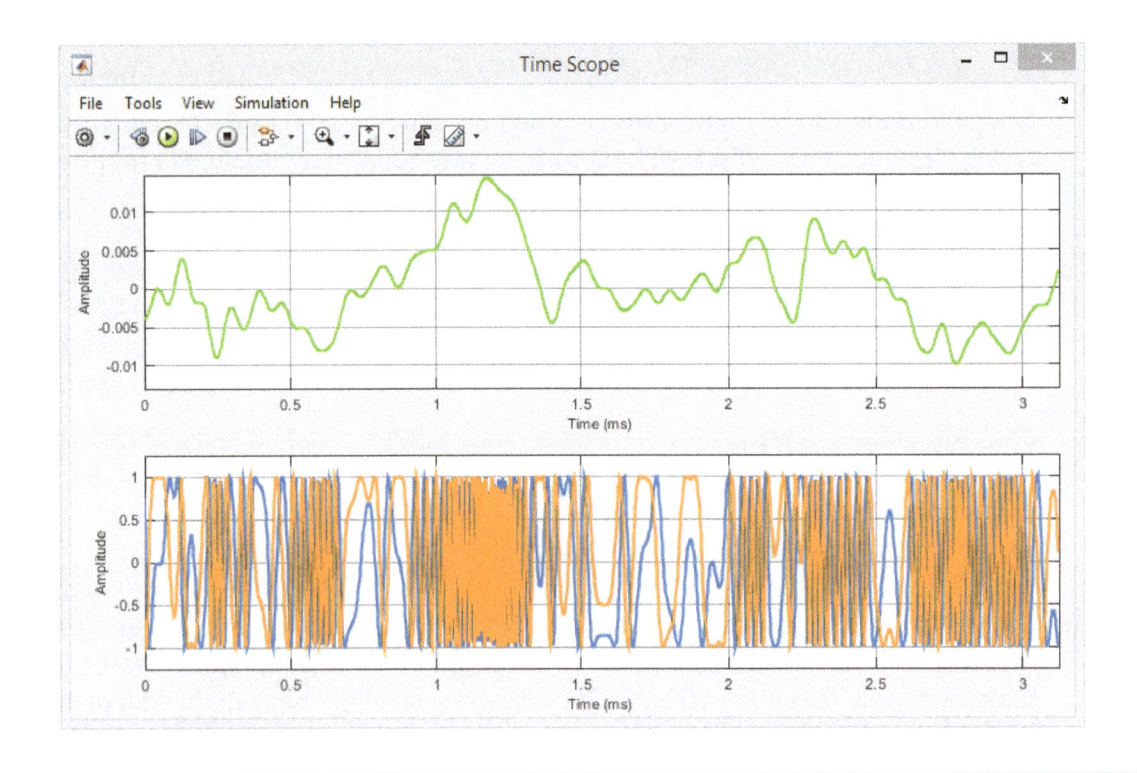

10.2 Implementing Mono FM Receivers with RTL-SDR and Simulink

Now that we have covered the generation of FM signals with the USRP® hardware, it is time to receive some with the RTL-SDR! This section contains a series of exercises, all orientated around demodulating mono FM signals. The first exercise involves building a mono FM receiver in Simulink and attempting to demodulate some of the FM radio stations that are broadcast in your area; or the signals generated with the USRP® transmitter. We provide a MATLAB script that implements a frequency discriminator, and we also set the challenge of building the Complex Differentiation Demodulator from Section 9.6 (page 332). If you do not have a USRP® radio and are unable to receive FM signals from commercial radio stations, remember that an *off the shelf* transmitter is a good alternative, or a Raspberry Pi can be used to generate an FM signal — see Section 10.7 for more information.

Before you begin building any FM receiver Simulink models, you will need to make sure you have the MathWorks RTL-SDR Hardware Support Package installed. Details on how to do this can be found in Appendix A.1 (page 569).

Exercise 10.2 RTL-SDR: Mono FM Radio Receiver (Discriminator)

In this exercise we will construct a Frequency Discriminator based FM receiver in Simulink. After creating a new Simulink model, components from the Simulink libraries that interface with the RTL-SDR will be placed, along with components to implement the demodulator. The receiver will output the demodulated audio information to computer speakers or headphones.

Note: If you do not have an RTL-SDR, or are unable to receive commercial FM radio stations and cannot transmit your own FM signal, you can still complete this exercise by substituting the RTL-SDR Receiver block with an Import RTL-SDR Data block as discussed below.

Chapter 10: Desktop FM Transmission and Reception

(a) **Open MATLAB.** Set the working directory to the exercise folder,

📁 `/my_models/receivers/`

Next, create a new Simulink model. Save this file with the name:

`.../rtlsdr_fm_discrim_demod.slx`

Open this file, and then the Simulink Library Browser.

(b) **Place an *RTL-SDR Receiver* block.** If you have an RTL-SDR available and are able to receive an FM signal (either a broadcast or a generated signal), place the *RTL-SDR Receiver* block from 🗂 > ***Communications System Toolbox Support Package for RTL-SDR Radio***. 🖱 on the block to open its parameter window, and change the 'Source' of 'Center frequency' and 'Tuner gain' to 'Input Port'. Enter '2.4e6' in the 'Sampling Rate' field to set the RTL-SDR to sample at a rate of 2.4MHz. Set the 'Output data type' dropdown to 'single' and enter '4096' in the 'Samples per frame' box.

If a single RTL-SDR is connected to the computer, the 'Radio address' can be left as '0'. If more than one is connected, run the `sdrinfo` command in the MATLAB command window to find the ID of the desired device. Apply these changes and then close the window. Rename this block to *RTL-SDR Receiver o/p fs=2.4MHz*.

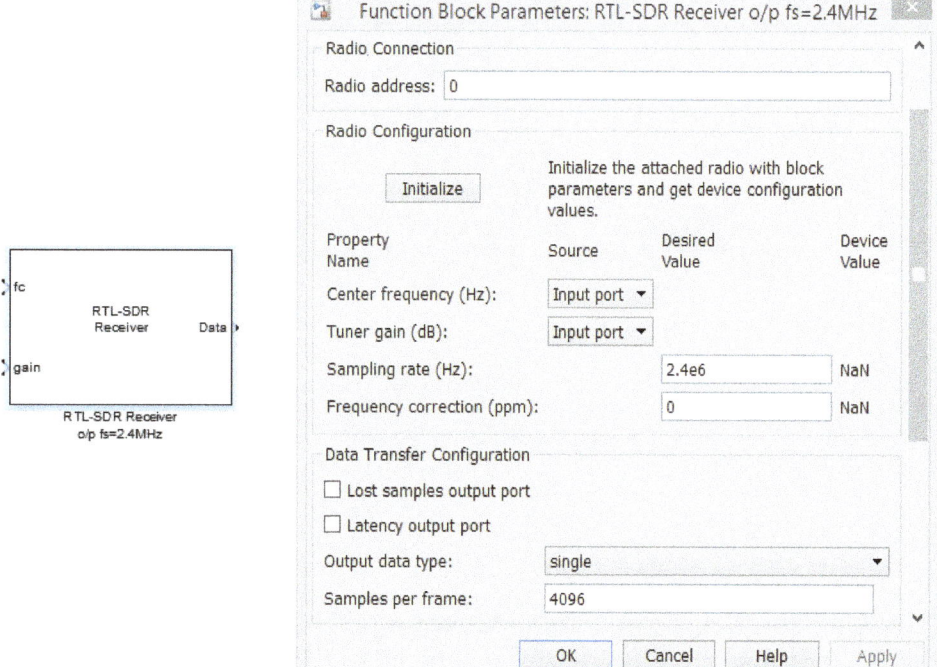

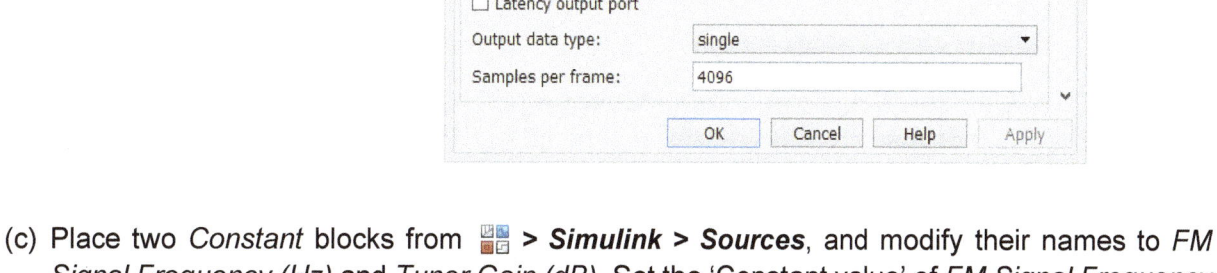

(c) Place two *Constant* blocks from 🗂 > ***Simulink > Sources***, and modify their names to *FM Signal Frequency (Hz)* and *Tuner Gain (dB)*. Set the 'Constant value' of *FM Signal Frequency* to the centre frequency of the FM signal to be received — for example, '100e6' for 100MHz. Set the *Tuner Gain* block to a default value of '10'. This may need to be adjusted later on, depending the strength of the received signal.

Connect these blocks up as follows:

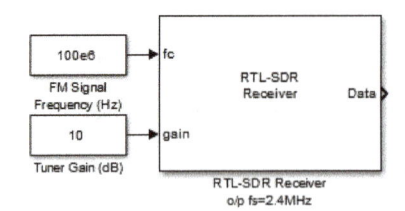

(d) **Place an *Import RTL-SDR Data* block.** If you do not have an RTL-SDR or are unable to receive an FM signal, navigate to 🎛 > **RTL-SDR Book Library > Additional Tools** and place an *Import RTL-SDR Data* block. 🖱 on this and change the 'File Name' parameter to reference the file:

📁 `/fm/rtlsdr_rx/rec_data/wfm_mono.mat`

Leave the 'Output Frame Size' set at '4096', apply the changes and close the parameter window. If this process was successful, the block should show the sampling frequency of the recorded signal in the bottom right hand corner (240kHz in this case).

The signal output by this block will be similar to the signal output by the RTL-SDR configuration shown above. When this signal was recorded, the RTL-SDR was tuned to 100MHz.

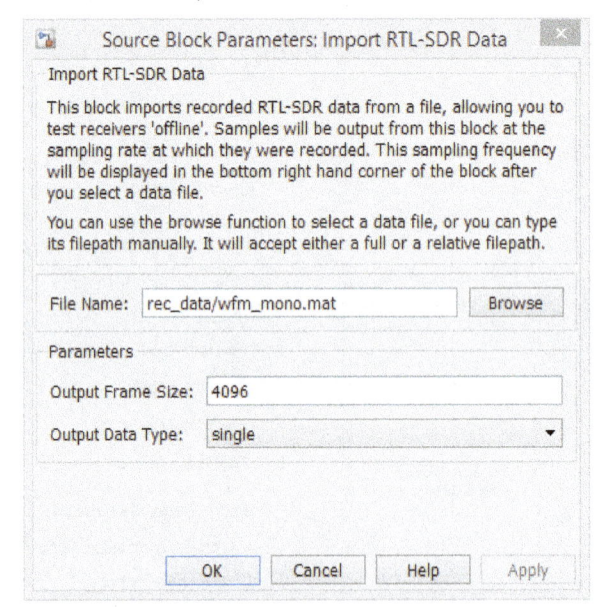

(e) **This is where things get a little interesting...** The sampling rate of the RTL-SDR is set to 2.4MHz, because it is much easier to tune to an FM station using *Spectrum Analyzers* when the receiver has a wide bandwidth. Recording 60 seconds of RTL-SDR data at 2.4MHz, however, creates a file that is around 1.1GB, and it would be impractical for us to package such large files with the book (especially considering that the first operation may be to decimate the signal!). Most of the recorded signals for the AM and FM chapters have a sampling rate of 240kHz, i.e. a tenth of the RTL-SDR rate in this model. If using an *RTL-SDR Receiver* block, you should therefore to reduce the sampling rate by a factor of 10.

Chapter 10: Desktop FM Transmission and Reception 379

(f) If using an *RTL-SDR Receiver*, place an *FIR Decimation* block from ▦ > **DSP System Toolbox > Filtering > Multirate Filters** into your model. This block decimates the data applied to its input port, allowing a reduction of the sampling frequency by an integer ratio. It also performs lowpass filtering, to ensure that no aliasing occurs. ② on it to open the parameter window, then change the 'FIR filter coefficients' and 'Decimation Factor' to the following:

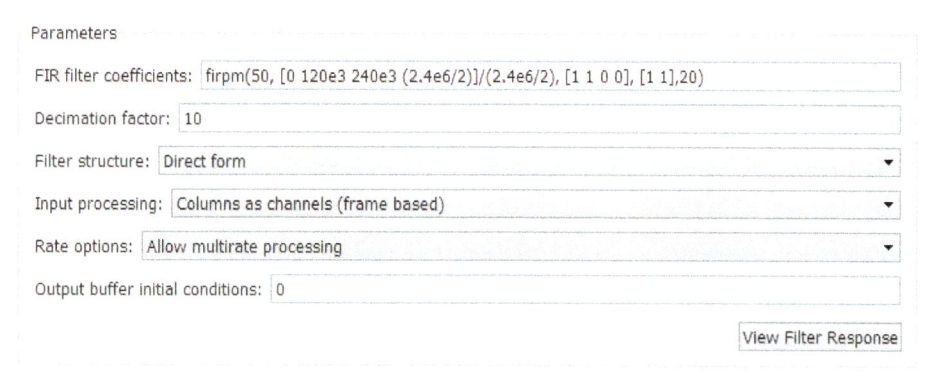

```
Parameters
FIR filter coefficients:   firpm(50, [0 120e3 240e3 (2.4e6/2)]/(2.4e6/2), [1 1 0 0], [1 1],20)
Decimation factor:         10
Filter structure:          Direct form
Input processing:          Columns as channels (frame based)
Rate options:              Allow multirate processing
Output buffer initial conditions: 0
                                                        View Filter Response
```

This configures the decimator to reduce the sampling rate by a factor of 10 (i.e. a rate change from 2.4MHz to 240kHz), and to pass frequencies up to 120kHz. Select 'Allow multirate processing' in the 'Rate options' dropdown menu, and then apply these changes. If you wish to view the filter response of the lowpass filter, ① on the 'View Filter Response' button. Rename this block *FIR Decimation o/p fs=240kHz*.

(g) **Connect the blocks.** After connecting these blocks up, you should have one of the following two configurations.

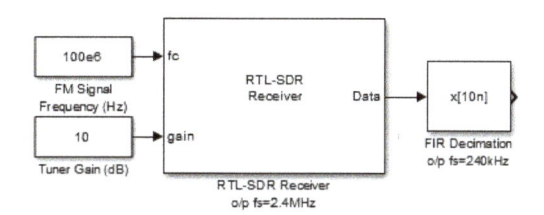

```
100e6 ──▶ fc
FM Signal         RTL-SDR
Frequency (Hz)    Receiver    Data ──▶ x[10n]
10 ──▶ gain                          FIR Decimation
Tuner Gain (dB)                      o/p fs=240kHz
                  RTL-SDR Receiver
                  o/p fs=2.4MHz
```

...or for an offline receiver

```
NooElec
fs = 240kHz
Import RTL-SDR Data
```

(h) **Implement the frequency discriminator.** Source a *Delay* block from ▦ > **Simulink > Discrete**, and *Math Function*, *Product*, and *Complex to Magnitude-Angle* blocks from ▦ > **Simulink > Math Operations**. Change the 'Function' of the *Math Function* block to 'conj', and the 'Output' of *Complex to Magnitude-Angle* to 'Angle'. Set the 'Input processing' parameter of the *Delay* block to 'Columns as channels (frame based)'.

Arrange these as shown below, and connect your RTL-SDR signal from above to the input of the *Delay* and *Math Function* blocks.

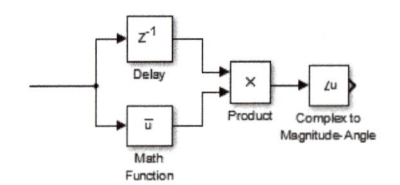

(i) **Add a decimation stage.** Place a *FIR Decimation* block from ▦ > *DSP System Toolbox* > *Filtering* > *Multirate Filters* into the model. 🖱2 on it to open its parameters window, then change the 'FIR filter coefficients' and 'Decimation Factor' to the following:

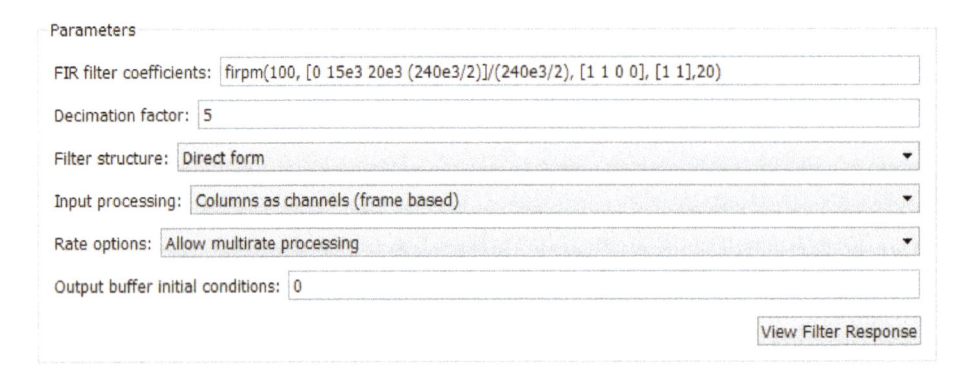

```
Parameters
FIR filter coefficients:  firpm(100, [0 15e3 20e3 (240e3/2)]/(240e3/2), [1 1 0 0], [1 1],20)
Decimation factor:  5
Filter structure:  Direct form
Input processing:  Columns as channels (frame based)
Rate options:  Allow multirate processing
Output buffer initial conditions:  0
                                               View Filter Response
```

This configures the decimator to reduce the sampling rate by a factor of 5 (i.e. a rate change from 240kHz to 48kHz), and to pass frequencies up to 15kHz. Select 'Allow multirate processing' in the 'Rate options' dropdown menu, and then apply the changes. If desired, view the filter response of the lowpass filter by 🖱1 on the 'View Filter Response' button. Rename this block *FIR Decimation o/p fs=48kHz*.

(j) The final part of the demodulator is the de-emphasis filter. Open ▦ > *RTL-SDR Book Library* > *FM Demod Tools* and add an *FM De-emphasis Filter* to the model. Open its parameter window and select the 'Filter Region' appropriate to your location. Set the sampling rate to 48kHz ('48e3') and apply the changes to reconfigure the filter.

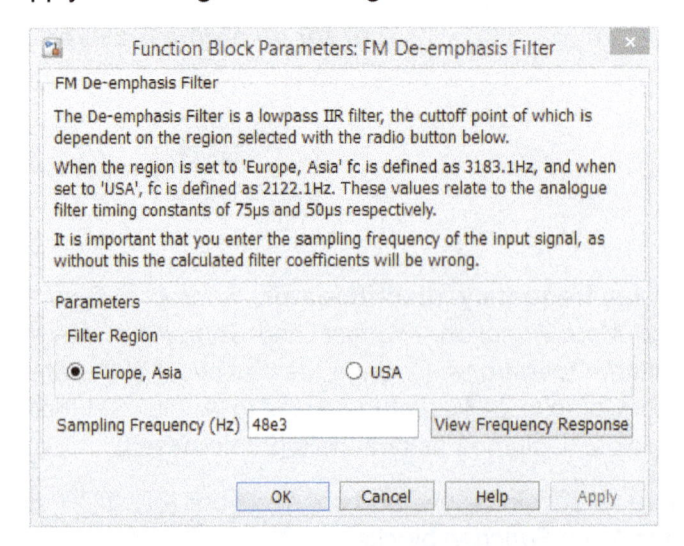

```
Function Block Parameters: FM De-emphasis Filter
FM De-emphasis Filter
The De-emphasis Filter is a lowpass IIR filter, the cuttoff point of which is
dependent on the region selected with the radio button below.
When the region is set to 'Europe, Asia' fc is defined as 3183.1Hz, and when
set to 'USA', fc is defined as 2122.1Hz. These values relate to the analogue
filter timing constants of 75µs and 50µs respectively.
It is important that you enter the sampling frequency of the input signal, as
without this the calculated filter coefficients will be wrong.

Parameters
  Filter Region
    ● Europe, Asia        ○ USA

Sampling Frequency (Hz) 48e3     View Frequency Response

            OK    Cancel    Help    Apply
```

Chapter 10: Desktop FM Transmission and Reception

(k) **Add scope and audio output blocks.** Navigate to 🔲 > *DSP System Toolbox* > *Sinks*. Place two *Spectrum Analyzer* blocks into the model, renaming one *Spectrum Analyzer Modulated* and the other *Spectrum Analyzer Demodulated*. Next, place a *Time Scope* and a *To Audio Device* block.

The *Time Scope* needs to be reconfigured to show two signals: the FM modulated signal, and the demodulated information signal. 2️⃣ on the block, 1️⃣ the 'File' menu, then change the 'Number of Input Ports' to '2'. 1️⃣ on the 'View' menu and set the 'Layout' to 2x1. Navigate to 'Configuration Properties', and open the 'Time' tab. Change the 'Time span' to '512/240e3'. This will limit the scope to showing only 512 individual samples.

(l) Finally, connect up all of the blocks as follows:

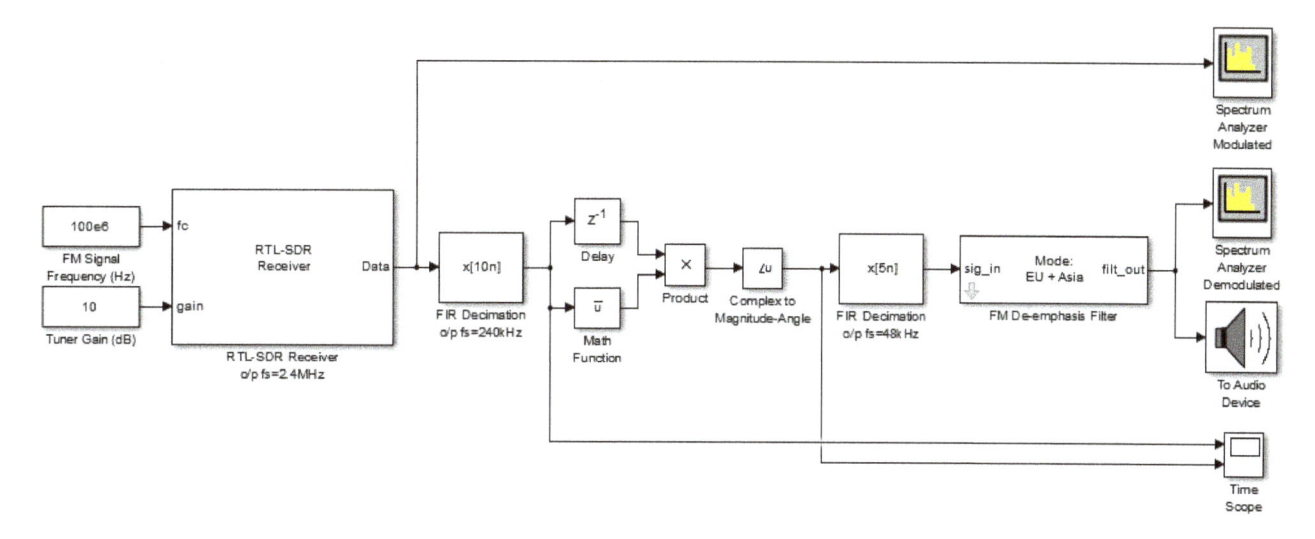

...or for an offline receiver

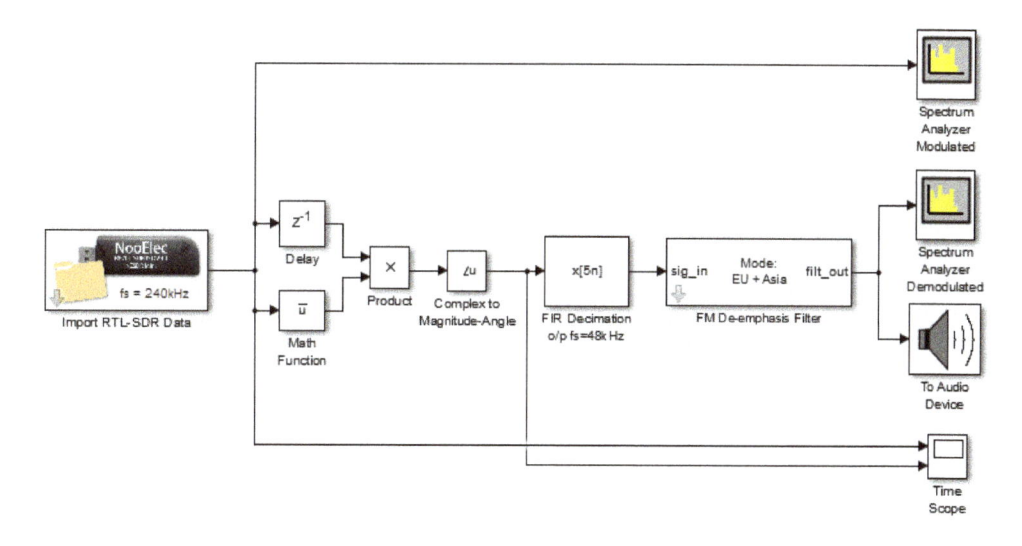

(m) **Open the reference file.** If you do not wish to construct the receiver yourself, open the following file:

📁 `/fm/rtlsdr_rx/rtlsdr_fm_discrim_demod.slx`

If importing data rather than connecting to an RTL-SDR, delete the *RTL-SDR Receiver* block and reconfigure the block diagram to the 'offline receiver' model shown above. Check that this block is set to reference the correct data file. If the *FM De-emphasis Filter* is configured for the wrong region, correct this before continuing.

(n) **Prepare to run the simulation.** Connect speakers or headphones to the computer and test to ensure that they are working. If intending to import a signal, set the *Simulation Stop Time* to 60 seconds, '60'. If using an RTL-SDR, set the *Simulation Stop Time* to 'inf' by typing this into the Simulink toolbar, and the *Simulation Mode* to 'Accelerator'. This will force Simulink to partially compile the model into native code for your computer, which allows it to run faster. Check that MATLAB can communicate with it by typing `sdrinfo` into the MATLAB command window. Finally, make sure your model is saved.

(o) **Run the simulation.** Begin the simulation by 🖱 on the 'Run' ▶ button in the Simulink toolbar. After a couple of seconds the simulation will begin, the *Spectrum Analyzer* and *Time Scope* windows should appear, and it should be possible to hear the demodulated FM signal. If using an RTL-SDR, adjust the frequency and gain values until you tune the device to the desired signal. Use *Spectrum Analyzer Modulated* to help with this.

(p) **Signal Analysis: *Spectrum Analyzers*.** The *Spectrum Anlayzer* windows allow the FM signal to be monitored in the frequency domain as it is demodulated. *Spectrum Analyzer Modulated* should show the spectrum of the signal received from the RTL-SDR, and will either have a bandwidth of 240kHz or 2.4MHz depending on the model configuration.

Spectrum Analyzer Demodulated shows the spectrum of the demodulated signal after frequency discrimination has been performed. Notice that the information signal has been successfully shifted back to baseband.

(q) **Signal Analysis: *Time Scope*.** It should be clear from the complex signal plotted in the top axes of the *Time Scope* that the frequency of the received signal is dynamic. Looking at the demodulated signal plotted in the axes below, confirm that the frequency fluctuations follow the amplitude fluctuations in the demodulated information signal.

(r) **Further Work.** Why not make your mono FM demodulator design a little more flexible by adding interactive controls for the carrier frequency and tuner gain? A control mask is available and can be found in ▦ > *RTL-SDR Book Library* > *FM Demod Tools*. The block is called *FM Tuner Freq and Gain Control*, and it can act as a direct replacement for the two constant blocks currently used to configure the RTL-SDR.

(s) **Watch and listen to us demodulate a mono FM signal with a frequency discriminator.** We have recorded a video which shows activity in the scope windows as a mono FM signal transmitted from the USRP® hardware (with the model from Exercise 10.1) is received by the RTL-SDR and demodulated. The output audio signal has been recorded too.

desktopSDR.com/videos/#discrim_demod

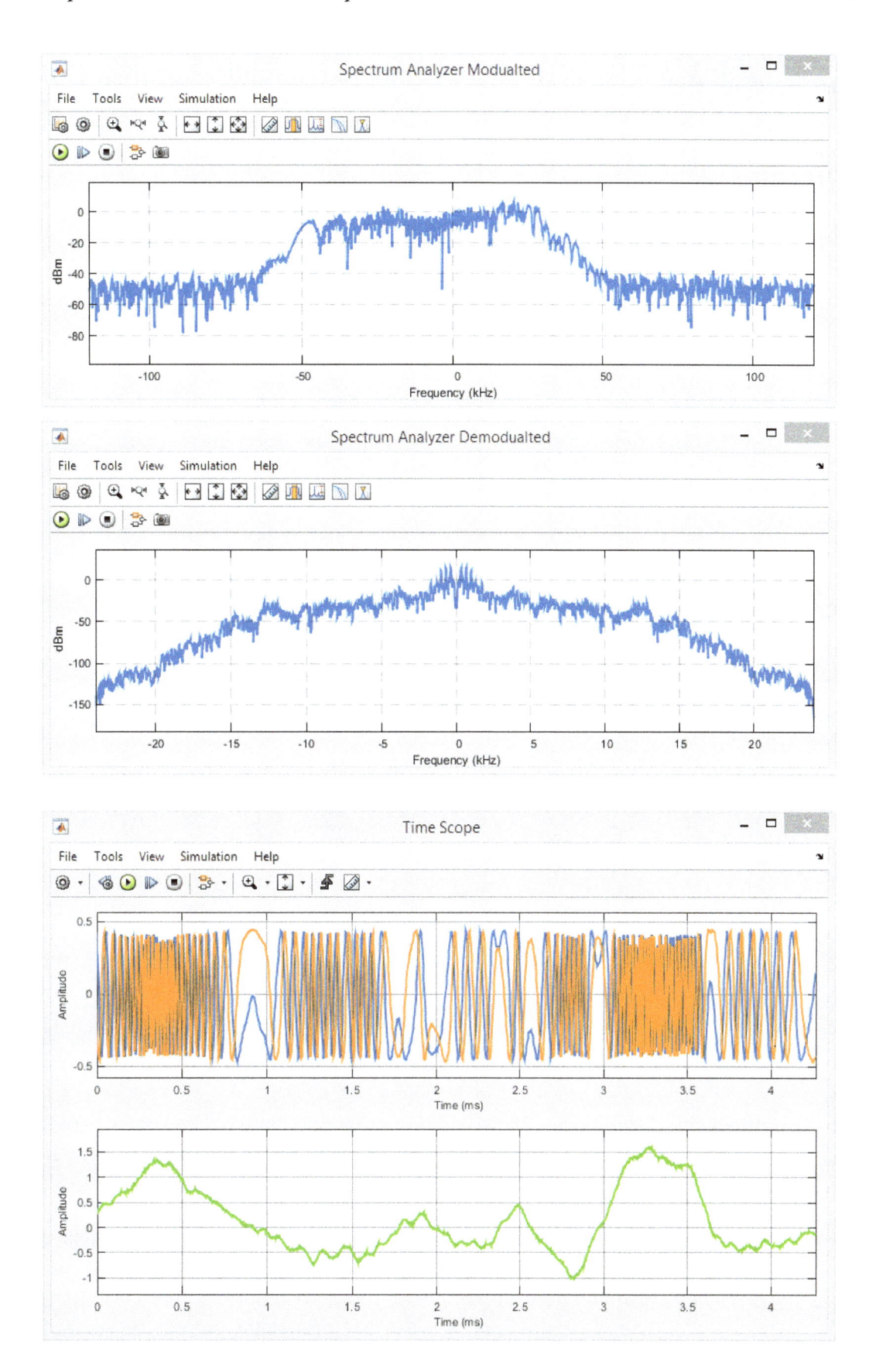

Chapter 10: Desktop FM Transmission and Reception 383

Exercise 10.3 RTL-SDR: MATLAB Mono FM Radio Receiver (Discriminator)

In this exercise we will run a MATLAB function that implements the Frequency Discriminator constructed in Exercise 10.2. The function has been designed to output the demodulated audio information to computer speakers or headphones.

Note: If you do not have an RTL-SDR, or are unable to receive a commercial FM radio station and cannot transmit your own FM signal, you can still compete this exercise by substituting the RTL-SDR Receiver system object with an Import RTL-SDR Data object as discussed below.

(a) **Open MATLAB.** Set the working directory to an appropriate folder so you can open this file:

/fm/rtlsdr_rx/rtlsdr_fm_discrim_demod_matlab.m

This MATLAB function performs FM demodulation using a Frequency Discriminator. To help explain the flow of the code, here is what an equivalent receiver would look like in Simulink:

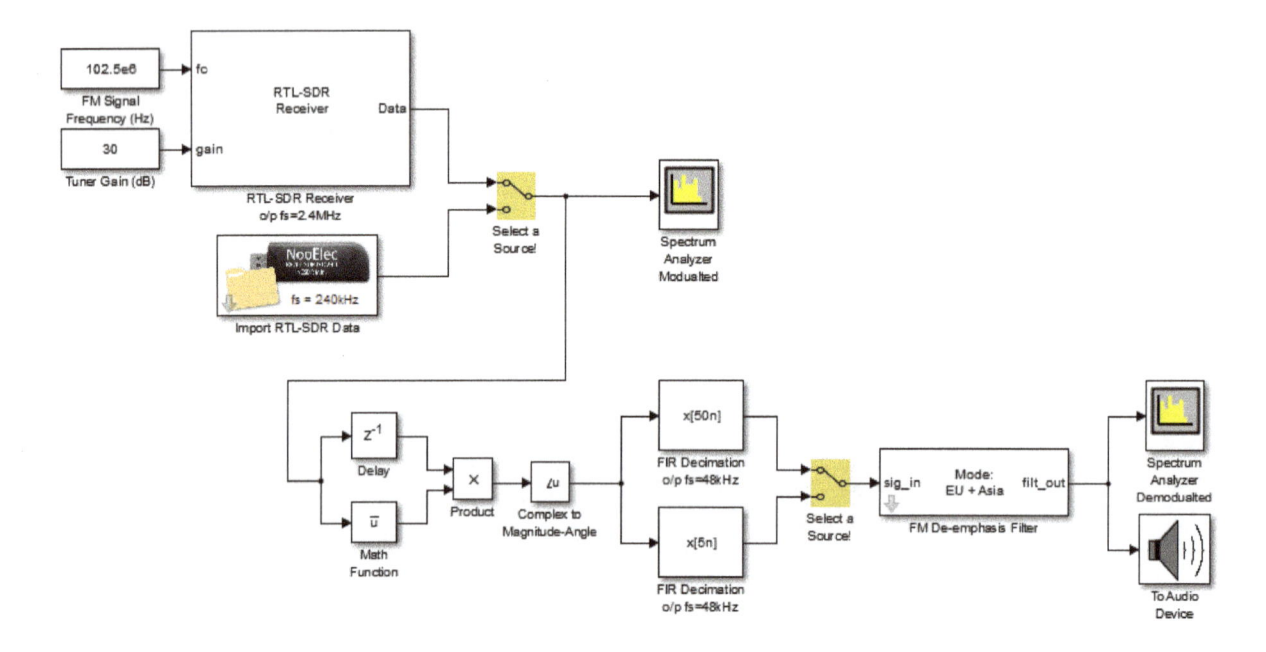

The *Manual Switch* blocks highlighted in yellow are used to represent the logical decision of which data source is used.

(b) **Inspect the parameters.** There are a number of different parameters listed at the top of the `rtlsdr_fm_discrim_demod_matlab` function. These are used to configure features of this receiver such as how long it should run for, and also to configure the RTL-SDR.

(c) The value of `offline` decides whether or not a handle to an RTL-SDR system object is initialised. If you have an RTL-SDR and are able to receive an FM signal, set this to '0'. If not, you will need to import data, so set this to '1'. When importing data it is important to make sure that the `offline_filepath` string is set to reference the correct file. As with the *Import RTL-SDR Data* block, the filepath can either be relative or full.

Chapter 10: Desktop FM Transmission and Reception

```matlab
%% PARAMETERS (can change)
offline           = 0;              % 0 = use RTL-SDR, 1 = import data
offline_filepath  = 'rec_data/wfm_mono.mat';  % path to FM signal
rtlsdr_id         = '0';            % stick ID
rtlsdr_fc         = 102.5e6;        % tuner centre frequency in Hz
rtlsdr_gain       = 30;             % tuner gain in dB
rtlsdr_fs         = 2.4e6;          % tuner sampling rate
rtlsdr_ppm        = 0;              % tuner parts per million correction
rtlsdr_frmlen     = 256*25;         % output data size (multiple of 5)
rtlsdr_datatype   = 'single';       % output data type
deemph_region     = 'eu';           % set to either eu or us
audio_fs          = 48e3;           % audio output sampling rate
sim_time          = 60;             % simulation time in seconds
```

(d) If you are using an RTL-SDR, modify the value of `rtlsdr_fc` to set the centre frequency of the receiver. Feel free to change the `sim_time` variable to increase the run time of the simulation from the default value of 60 seconds. If you have more than one RTL-SDR attached to your computer, run the `sdrinfo` command in the MATLAB command window to find the ID of the stick you desire to use, and then configure the `rtlsdr_id` with this value. If you are changing the ID, note it needs to be entered as a `'string'` (with apostrophes). Reconfigure the de-emphasis filter to your region by changing `deemph_region` to either `'eu'` or `'us'`.

(e) **Review the system objects.** Seven different system objects are initialised in this code to create the RTL-SDR data source, implement the demodulator, and provide spectral representations of the signals as they are demodulated. `obj_rtlsdr` is used as the name for both the `import_rtlsdr_data` and `comm.SDRRTLReceiver` objects. This means that regardless of which one ends up being initialised, the rest of the code for the receiver can remain constant, calling '`step(obj_rtlsdr)`' for another frame of data.

```matlab
%% SYSTEM OBJECTS (do not edit)
% check if running offline
if offline == 1
    
    % link to an rtl-sdr data file
    obj_rtlsdr = import_rtlsdr_data(...
        'filepath', offline_filepath,...
        'frm_size', rtlsdr_frmlen,...
        'data_type', rtlsdr_datatype);
    
    % reduce sampling rate
    rtlsdr_fs = 240e3;
    
    % fir decimator handle - fs = 240kHz downto 48kHz
    obj_decmtr = dsp.FIRDecimator(...
        'DecimationFactor', 5,...
        'Numerator', firpm(100,[0,15e3,20e3,(240e3/2)]/(240e3/2),...
        [1 1 0 0], [1 1], 20));
```

```matlab
        % initalise object and get sampling rate for spectrum analyzer
        step(obj_rtlsdr);
        rtlsdr_fs = obj_rtlsdr.fs;
        
    else
    
        % link to a physical rtl-sdr
        obj_rtlsdr = comm.SDRRTLReceiver(...
            rtlsdr_id,...
            'CenterFrequency', rtlsdr_fc,...
            'EnableTunerAGC', false,...
            'TunerGain', rtlsdr_gain,...
            'SampleRate', rtlsdr_fs,...
            'SamplesPerFrame', rtlsdr_frmlen,...
            'OutputDataType', rtlsdr_datatype,...
            'FrequencyCorrection', rtlsdr_ppm);
            
        % fir decimator handle - fs = 2.4MHz downto 48kHz
        obj_decmtr = dsp.FIRDecimator(...
            'DecimationFactor', 50,...
            'Numerator', firpm(350,[0,15e3,48e3,(2.4e6/2)]/(2.4e6/2),...
            [1 1 0 0], [1 1], 20));
            
    end;
```

(f) As two different sampling rates are used in this function (the RTL-SDR is set to output samples at 2.4MHz, and the imported data file will output at 240kHz), the FIR Decimation function used to reduce the sampling must be different. A `DecimationFactor` of '50' is required when obtaining samples from the RTL-SDR, while a factor of '5' is required for imported data.

For the same reason, the value of the `rtlsdr_fs` parameter is updated if a data file is being imported, as this is used during the initialisation of `obj_spectrummod`.

(g) **Examine the discriminator.** The frequency discriminator is implemented inside the following while loop. Frames of WFM data are acquired from either the RTL-SDR or an imported data file, and then passed through the frequency discriminator to demodulate them. The demodulated data is output to the computers default audio device, and spectrum analyzers are used to allow you to monitor the signals as they are demodulated in the frequency domain. This code is equivalent to the receiver you constructed in Exercise 10.2.

```matlab
% loop while run_time is less than sim_time
while run_time < sim_time
    
    % fetch a frame from obj_rtlsdr (live or offline)
    rtlsdr_data = step(obj_rtlsdr);
    
    % update 'modulated' spectrum analyzer window with new data
    step(obj_spectrummod, rtlsdr_data);
    
    % implement frequency discriminator
    discrim_delay = step(obj_delay,rtlsdr_data);
    discrim_conj  = conj(rtlsdr_data);
    discrim_pd    = discrim_delay.*discrim_conj;
    discrim_arg   = angle(discrim_pd);
    
    % decimate + de-emphasis filter data
    data_dec = step(obj_decmtr,discrim_arg);
    data_deemph = step(obj_deemph,data_dec);
    
    % update 'demodulated' spectrum analyzer window with new data
    step(obj_spectrumdemod, data_deemph);
    % output demodulated signal to speakers
    step(obj_audio,data_deemph);
    
    % update run_time after processing another frame
    run_time = run_time + rtlsdr_frmtime;
    
end
```

(h) **Prepare to run the function.** Connect speakers or headphones to the computer and test to ensure that they are working. If you are using an RTL-SDR, check that MATLAB can communicate with it by typing `sdrinfo` into the MATLAB command window. Finally, make sure the file is saved.

(i) **Run the function.** Run the function by clicking on the 'Run' button in the MATLAB toolbar. After a couple of seconds the simulation will begin, the *Spectrum Analyzer* windows should appear, and the demodulated FM signal should be audible. The function is designed to loop for `sim_time` seconds, but can be cancelled early using the following key combination:

CTRL + C

(j) **Signal Analysis: *Spectrum Analyzers*.** Very similar results should be visible in the *Spectrum Anlayzer* windows as compared to the receiver simulation in Exercise 10.2. *Spectrum Analyzer Modulated* will show the signal received from the RTL-SDR, as shown below (these screenshots were captured with importing data, hence the bandwidth of this double sided spectrum is 240kHz).

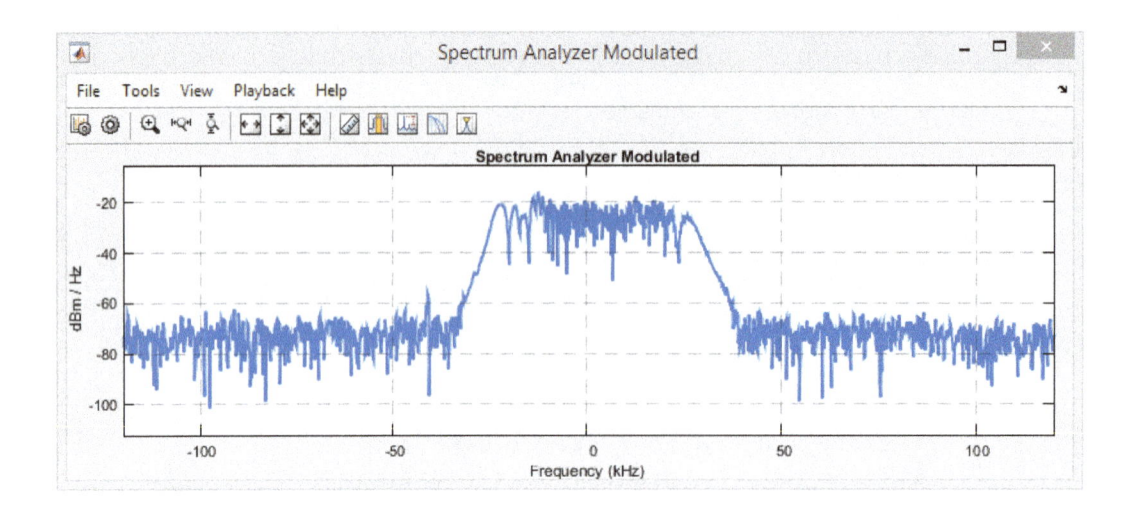

[Screenshot: Spectrum Analyzer Modulated window showing spectrum from -100 to 100 kHz, with signal peaking around -20 dBm/Hz between approximately -25 and 25 kHz]

(k) *Spectrum Analyzer Demodulated* shows the spectrum of the demodulated signal after frequency discrimination has been performed. Here, you should see that the information has been successfully shifted back to baseband.

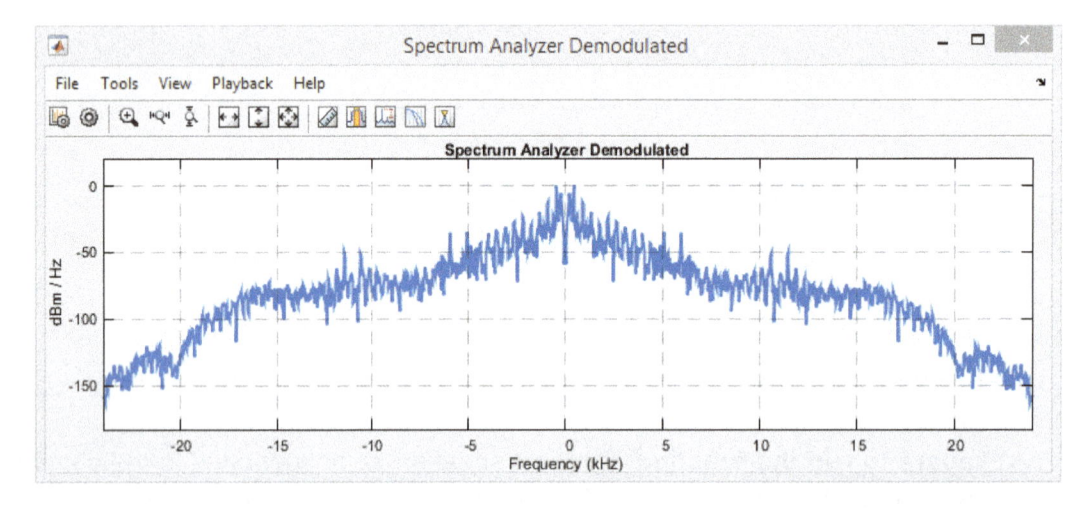

[Screenshot: Spectrum Analyzer Demodulated window showing spectrum from -20 to 20 kHz, peaking near 0 kHz]

(l) To switch between 'online' and 'offline' modes in Exercise 10.2, we had to physically modify the block diagram to change the RTL-SDR data source and reconnect *Spectrum Analyzer Modulated* to the appropriate signal. This was a little inconvenient, whereas switching source in this MATLAB function is far simpler — the only requirement is to change the value of `offline` from a '0' to a '1' (or visa versa). This highlights how receivers developed in MATLAB can have more flexible architectures than those developed in Simulink.

Exercise 10.4 RTL-SDR: Mono FM Radio Receiver (Complex Differentiation)

Now that we have implemented a Frequency Discriminator to demodulate FM signals received by the RTL-SDR, try implementing one with an Complex Differentiation Demodulator! Some hints will be provided to help get started.

Note: If you do not have an RTL-SDR, or are unable to receive a commercial FM radio station and cannot transmit your own FM signal, you can still compete this exercise by substituting the RTL-SDR Receiver system object with an Import RTL-SDR Data object as discussed below.

Chapter 10: Desktop FM Transmission and Reception 389

(a) **Create a new Simulink model.** Start by making a copy of the model created in Exercise 10.2 and save it with an appropriate file name. Strip out the discriminator parts of the model, and retain the RTL-SDR interface, the decimators, the filters, scopes and audio output.

(b) **Implement the Complex Differentiation Demodulator.** Using the diagram in Figure 9.16 (page 355), construct a Complex Differentiation Demodulator using components from the Simulink libraries. When complete, connect the *RTL-SDR Receiver* or *Import RTL-SDR Data* blocks along with the *FIR Decimator* and sink blocks, and try to run the simulation!

(c) **Open the reference file.** If you do not have time to construct the receiver yourself, or need a quick look to get started, open the following file:

/fm/rtlsdr_rx/rtlsdr_fm_complexdiff_demod.slx

The block diagram of the model should look as follows:

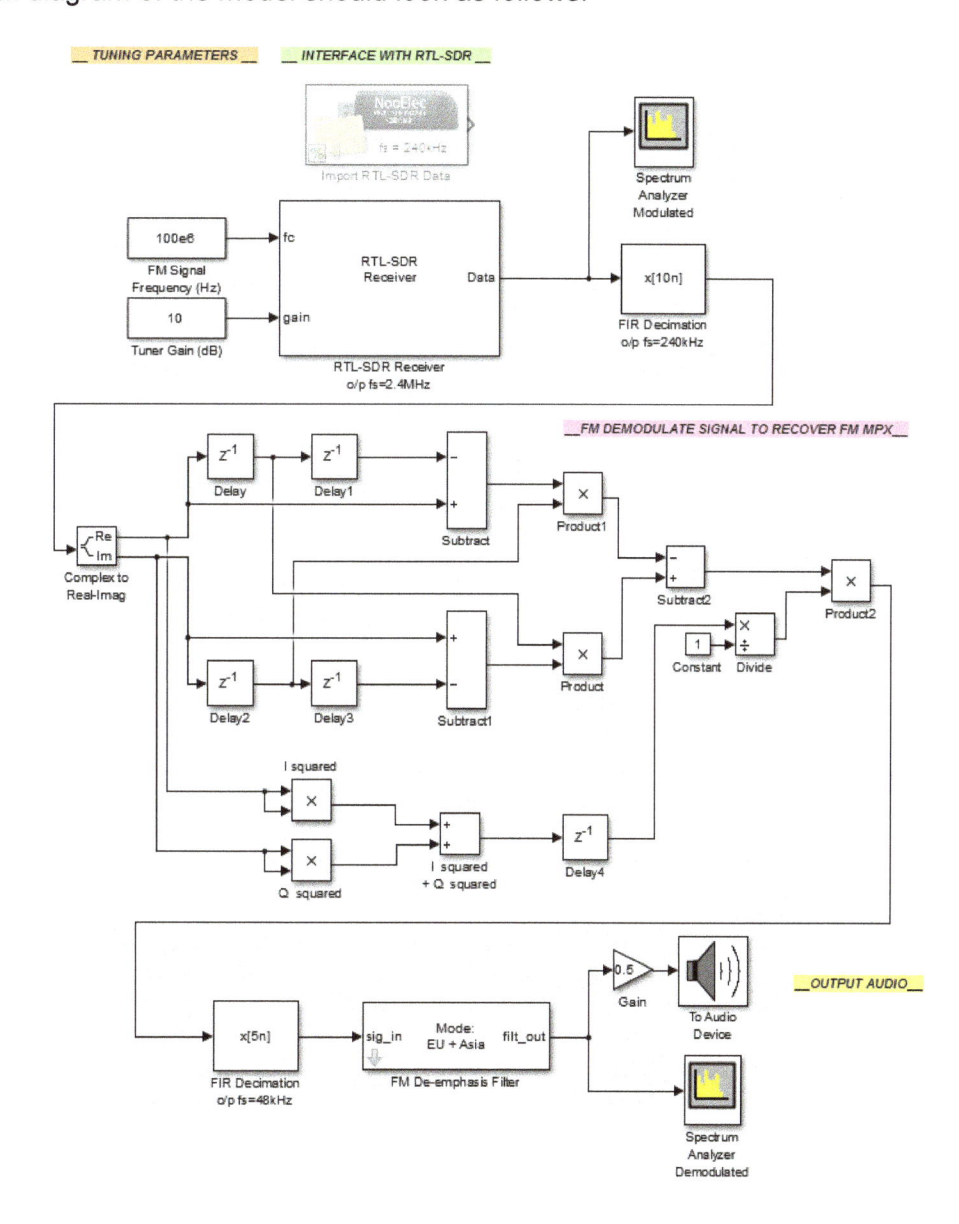

If you intend to import data rather than connect to an RTL-SDR, delete the *RTL-SDR Receiver* block and reconfigure the block diagram, taking into account the fact that the sampling rate of the data output by the *Import RTL-SDR Data* block is 240kHz. If the *FM De-emphasis Filter* is configured for the wrong region, correct this before continuing.

(d) **Prepare to run the simulation.** Connect speakers or headphones to the computer and check that they are working. If importing a signal, set the *Simulation Stop Time* to 60 seconds, '60'. If you are using an RTL-SDR, set the *Simulation Stop Time* to 'inf' by typing this into the Simulink toolbar, and the *Simulation Mode* to 'Accelerator'. This will allow Simulink to run the model faster. Check that MATLAB can communicate with the RTL-SDR by typing `sdrinfo` into the MATLAB command window. Finally, make sure that the model is saved.

(e) **Run the simulation.** Begin the simulation by 🖱 on the 'Run' ▶ button in the Simulink toolbar. After a couple of seconds the simulation will begin, the *Spectrum Analyzer* windows should appear, and the demodulated FM signal should be audible. If using an RTL-SDR, adjust the frequency and gain values until the device is tuned to the desired signal. Use *Spectrum Analyzer Modulated* to help with this.

(f) **Analysis.** Examine the signals plotted in the *Spectrum Analyzer* windows, and compare these to the signals observed in Exercise 10.2. Are they any different? Do you think this demodulation method is better than the Frequency Discriminator?

10.3 Transmitting Stereo WFM Signals with the USRP® Hardware

Now that you have created and transmitted a mono FM signal with the USRP® hardware, we can move on to something more challenging — transmitting a stereo FM signal! To do this, we will need to modify the transmitter model built in Exercise 10.1, and add a stereo encoder and multiplexer to it. To recap, a diagram for the FM stereo encoder/multiplexer is as shown in Figure 10.3. A high level block diagram of all of the processes required to generate a stereo FM signal with the USRP® hardware is shown in Figure 10.4.

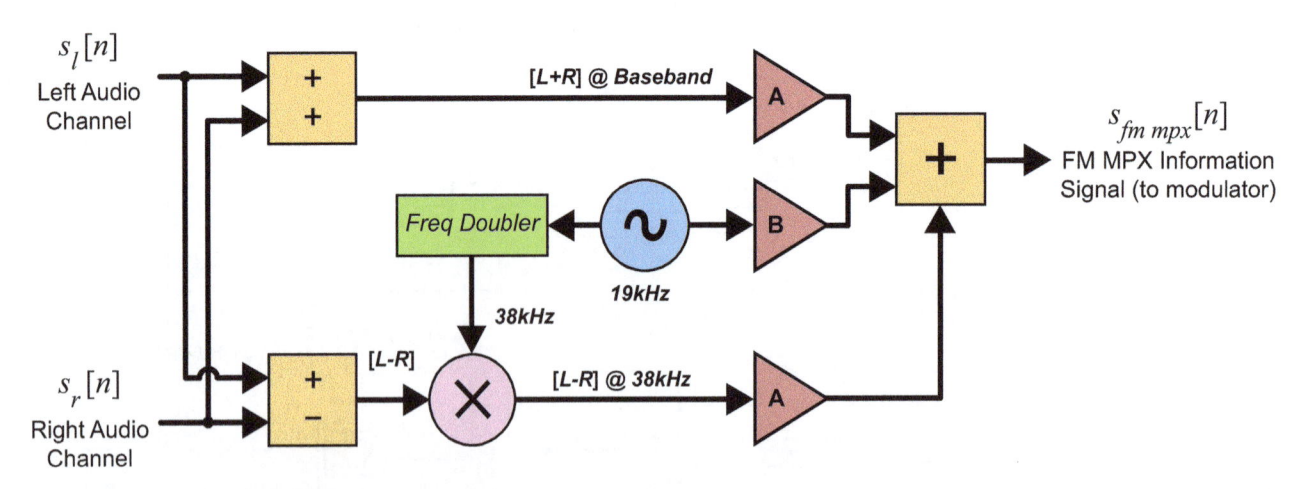

Figure 10.3: Block diagram of the stereo FM encoder and multiplexer (see Section 9.9 for more information)

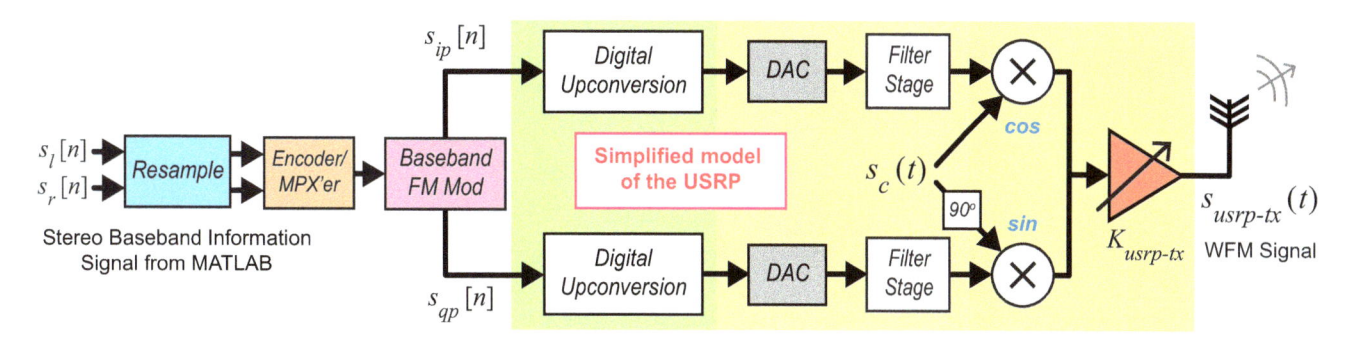

Chapter 10: Desktop FM Transmission and Reception

Figure 10.4: Block diagram showing a Simulink/ USRP® hardware implementation of a stereo FM modulator

In the following exercise you get the opportunity to construct a stereo FM encoder and multiplexer, and add this to the modulator/ transmitter from Exercise 10.1, to generate a stereo FM signal. This will then be used to transmit a stereo audio signal with the USRP® radio.

Exercise 10.5 USRP® Radio: Stereo FM Modulator and Transmitter

In this exercise we will build a stereo FM encoder and multiplexer in Simulink, that can be used in conjunction with the FM modulator from Exercise 10.1 to generate a stereo FM signal from two input audio files. This signal will be transmitted from the USRP® hardware. There will be an opportunity to receive and demodulate the signal with the RTL-SDR and Simulink in later exercises.

Note: If you do not have a USRP® radio, you can still run this exercise and see the MPX signal generated by commenting out the SDRu Transmitter block.

(a) **Open MATLAB.** Set the working directory to the exercise folder,

> /my_models/transmitters

then create a copy of the mono FM modulator model,

> .../usrp_fm_mono.slx

and rename it:

> .../usrp_fm_stereo.slx

(b) Open the new model, delete the connection between the *Lowpass Filter fc=15kHz* and *FIR Rate Conversion o/p fs = 200kHz*, and rearrange the block diagram as follows:

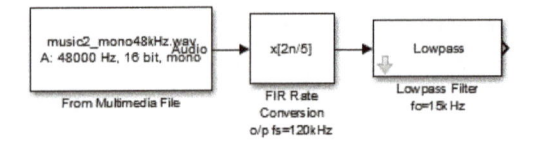

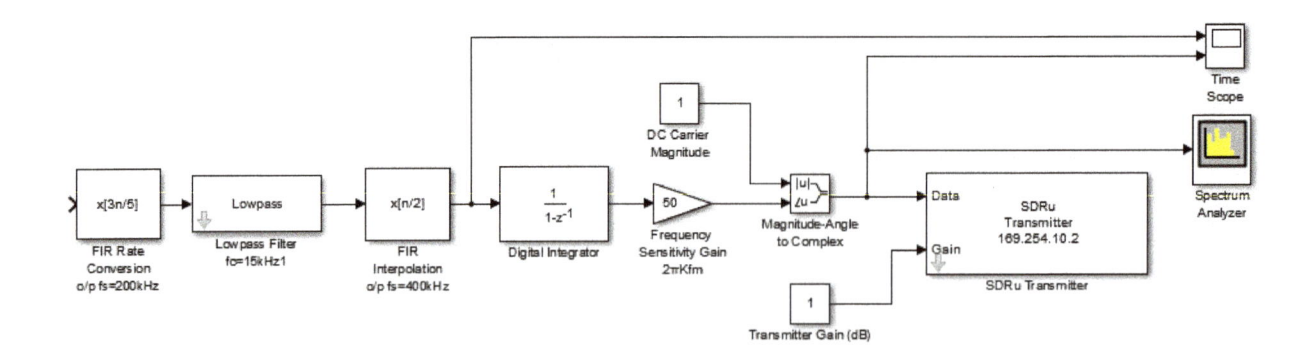

(c) **Add a second audio source.** Create copies of the first three blocks, placing these below the existing ones. [2click] on the second *From Multimedia File* and choose a different audio file from this folder:

/audio_sources

Conventionally, a stereo audio source would be used, but it is far more interesting to use two mono files, one for left and the other for right, as it should be possible to hear two distinctly different signals in the receiver.

(d) It is critical that each of the audio channels are band limited to 15kHz — if they are not, the stereo multiplexing process will not work correctly. [2click] on one of the source *Lowpass Filters* and [1click] on the 'View Filter Response' button. Note that the transition band of the filter is wide, and that the -3dB point is around 18.5kHz. This FIR filter will not sufficiently attenuate the unwanted components past 15kHz, without increasing the number of filter weights. Hence to keep the computational cost down, we redesign as an IIR filter with the following parameters:

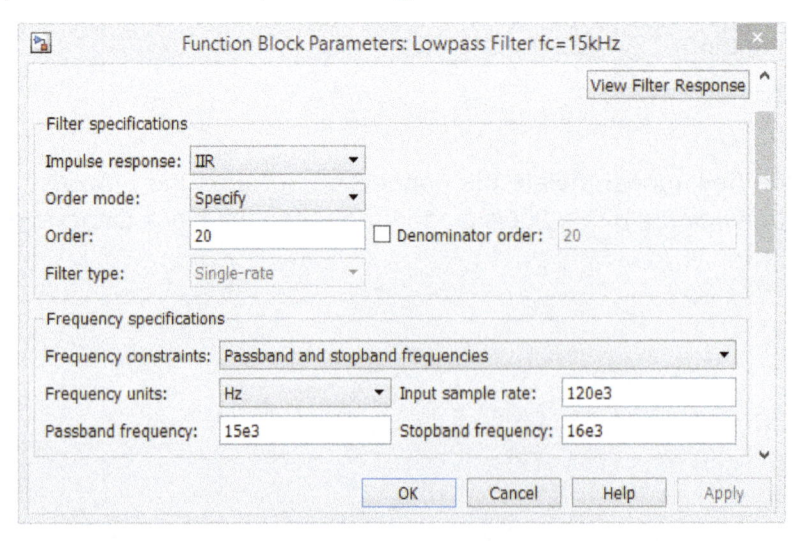

Function Block Parameters: Lowpass Filter fc=15kHz	
Filter specifications	
Impulse response:	IIR
Order mode:	Specify
Order:	20, Denominator order: 20
Filter type:	Single-rate
Frequency specifications	
Frequency constraints:	Passband and stopband frequencies
Frequency units:	Hz
Input sample rate:	120e3
Passband frequency:	15e3
Stopband frequency:	16e3

Chapter 10: Desktop FM Transmission and Reception 393

on the 'View Filter Response' button to see the difference. The -3dB point is now at around about 15.2kHz, and there is an attenuation of 50dB by 16kHz — this will work, and the filter order is reduced to 20. Change the second source *Lowpass Filter* to match.

(e) **Implement a stereo FM encoder and multiplexer.** Open > *Simulink > Math Operations* and find the *Add* block. Place three of these into the model. Leave one of them with the default settings, and change the 'List of signs' in one to '+-' and another to '+++'. From the same library, place three *Gain* blocks. Set two of these to have a 'Gain' of '0.45' and the third to have a 'Gain' of '0.10'. These gains relate to the weighting of the components as in Eq. (9.61) (page 363). Finally, fetch a *Product* block from this library and place it into the model.

(f) Open > *DSP System Toolbox > Sources* and find the *Sine Wave* block. Place two in your model. on the first of these, and change the parameters to the following:

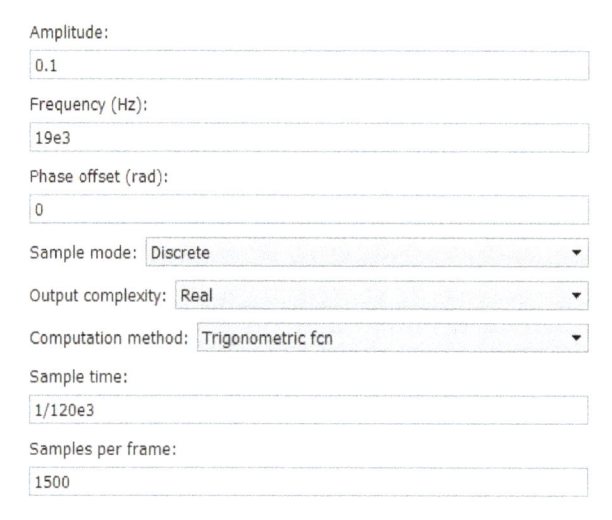

> Amplitude:
> 0.1
>
> Frequency (Hz):
> 19e3
>
> Phase offset (rad):
> 0
>
> Sample mode: Discrete
>
> Output complexity: Real
>
> Computation method: Trigonometric fcn
>
> Sample time:
> 1/120e3
>
> Samples per frame:
> 1500

Rename this block *Sine Wave 19kHz*. Repeat this process for the second block, but change the 'Amplitude' to '1', the 'Frequency' to '38e3'. Name this second block *Sine Wave 38kHz*. Connect the blocks up to form the stereo encoder/ multiplexer as shown below.

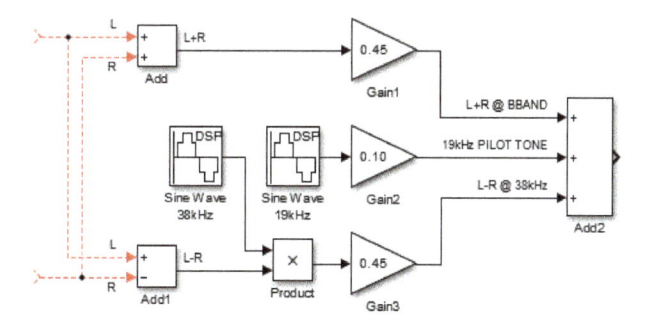

(g) Name the wires as shown in the screenshot, to help differentiate between the signals.

(h) **Connect the blocks.** Append the stereo encoder/ multiplexer to the outputs of the IIR filters, and connect the FM MPX signal (the output of *Add2* above) to the input of *FIR Rate Conversion o/p fs=200kHz*. This should result in the following system:

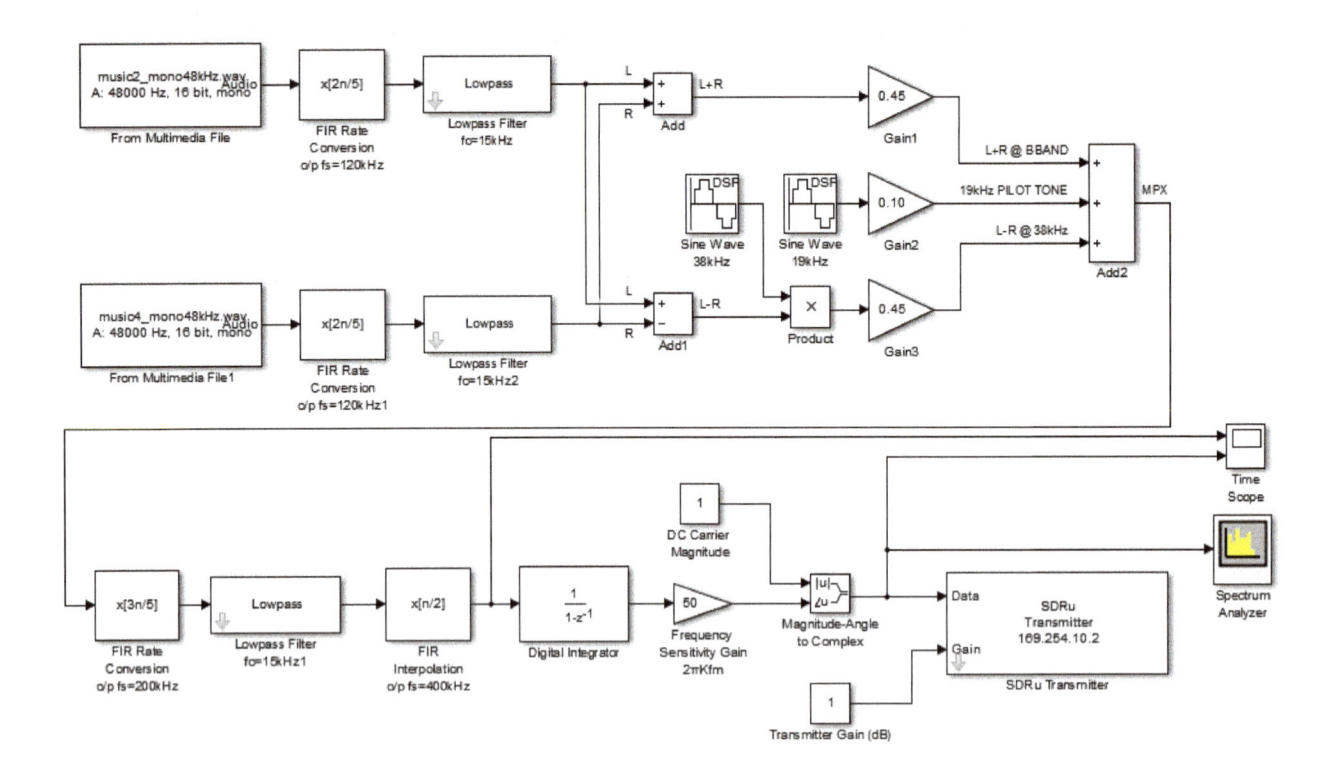

[Figure: Simulink block diagram of FM stereo transmitter]

(i) **Notice anything wrong with this system?** Look carefully at the functions of the *FIR Rate Conversion*, *Lowpass Filter* and *FIR Interpolation* blocks. The multiplexing operation results in a signal with frequency components in the range 0Hz to 53kHz. What cutoff frequencies have the filters in these blocks have been set to? They will need to be modified!

(j) Change the passband frequencies and stopband frequencies of these blocks from 15kHz and 24kHz, to 53kHz and 65kHz respectively. the 'View Filter Response' button to check that the changes have worked. It may also be worth renaming the *Lowpass Filter* block to highlight this amendment.

(k) **Shift the Spectrum Analyzer.** It would be useful to monitor the spectrum of the multiplexed signal before it is FM modulated, as this allows its conformance to the standard FM MPX to be confirmed. Leaving the *Spectrum Analyzer* connected to the FM modulated signal will show nothing new compared to Exercise 10.1, so it is worth simply connecting it to the MPX signal instead.

(l) **Prepare to run the simulation.** Set the *Simulation Stop Time* to 'inf' and change the *Simulation Mode* to 'Accelerator'. This will allow Simulink to run the model faster. Finally, make sure the model is saved.

(m) **Open the reference file.** If you are unable to construct the transmitter yourself, open the following file. Make sure that the *SDRu Transmitter* block is appropriately configured.

📁 /fm/usrp_tx/usrp_fm_stereo.slx

Chapter 10: Desktop FM Transmission and Reception   395

(n) Check that the USRP® hardware is turned on, and that MATLAB can communicate with it by typing `findsdru` into the MATLAB command window. If a structure is returned as discussed in Appendix A.2 (page 571), you can continue.

(o) **Run the simulation.** Begin the simulation by 1 on the 'Run' ▶ button in the Simulink toolbar. After a couple of seconds, Simulink will establish a connection with the USRP® radio and the simulation will begin. When it does, samples of the FM modulated baseband information signal will be passed to the radio, modulated onto the AM carrier, and transmitted from its antenna.

If a frequency within the range 88MHz to 108MHz is chosen, it will be possible to use an analogue FM radio to receive the stereo FM signal. If you can, you should find that the left speaker plays one song, and the right speaker another! If not, you will still be able to use your RTL-SDR to receive the signal, and you will have the opportunity to construct RTL-SDR FM demodulators that will output the audio signals you transmit in Section 10.4.

(p) **Signal Analysis: *Spectrum Analyzer*.** The *Spectrum Analyzer* window should allow you to see the MPX signal as it is created in the frequency domain. Does this conform to the theoretical FM MPX shown opposite?

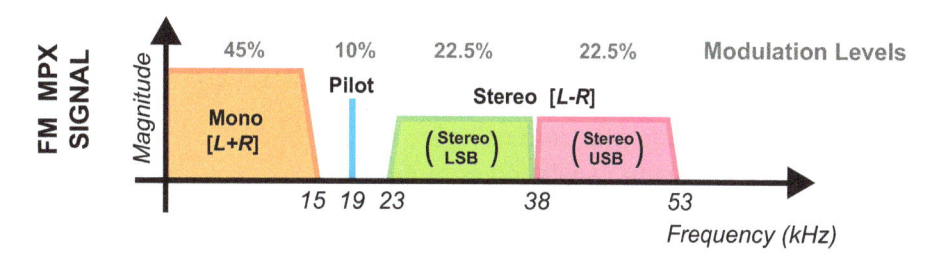

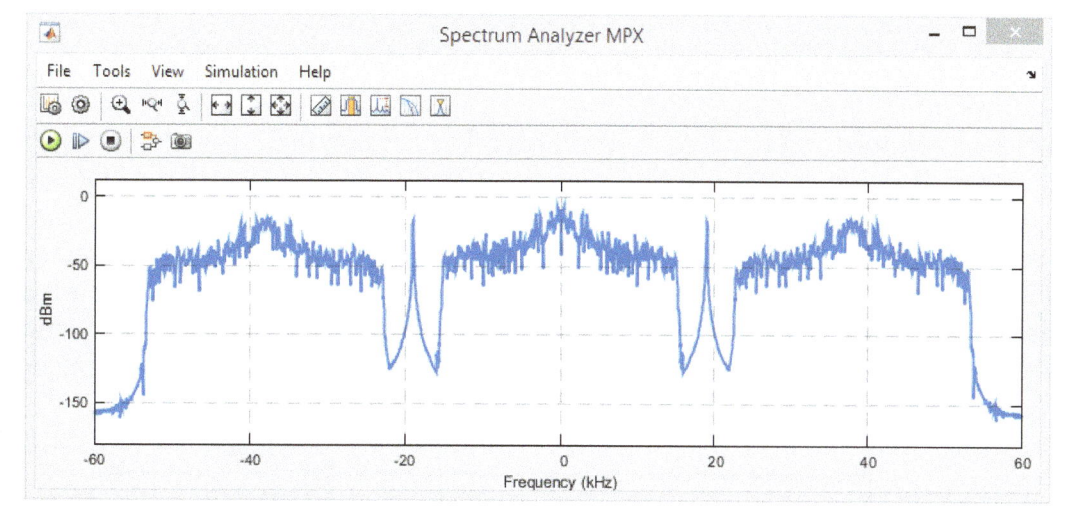

(q) **Signal Analysis: *Time Scope*.** The *Time Scope* should display two signals — the MPX signal and the FM modulated signal. Pause the simulation using the 'Pause' ⏸ button, and examine the plots. What stands out about the signals displayed here compared to the signals displayed in Exercise 10.1?

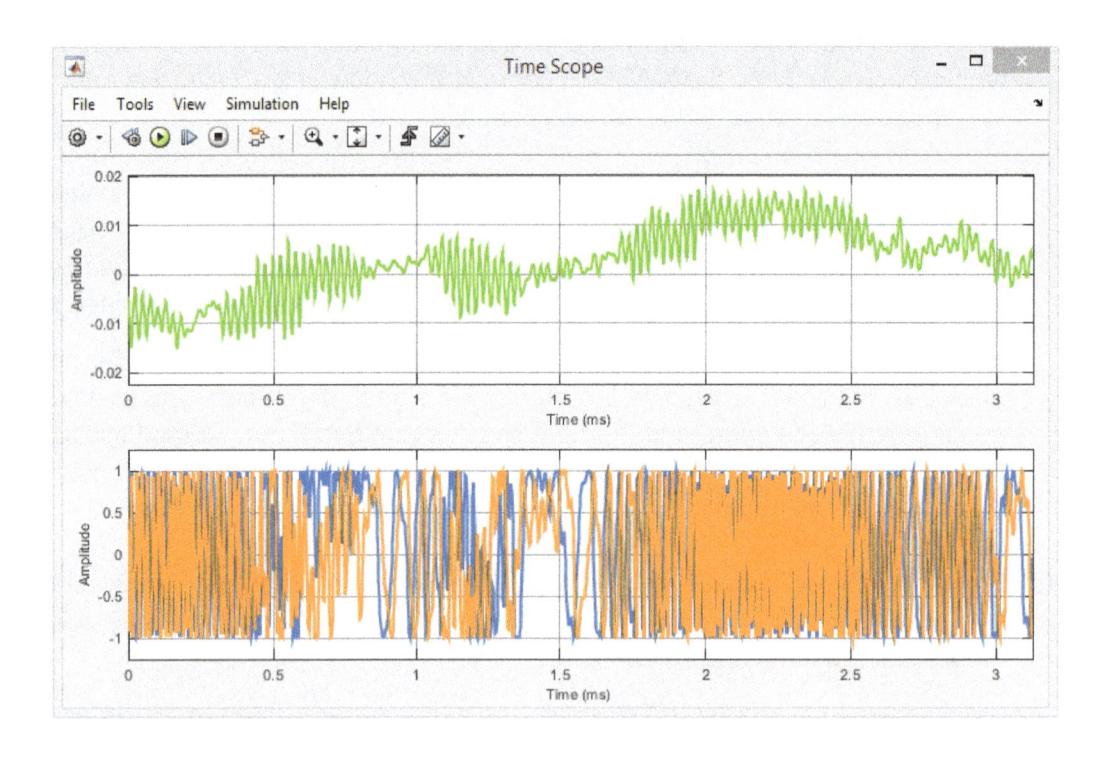

[Time Scope screenshot showing two plots: top plot with amplitude vs time (ms) showing a green waveform, bottom plot showing blue and orange waveforms]

(r) Because it is such a powerful tone, the suppressed stereo subcarrier is clearly visible in this signal. Zoom into the top plot, and navigate to *Tools -> Measurements*. You should be able to measure the period of the oscillating component to be (1/38kHz) seconds.

10.4 Implementing Stereo FM Receivers with RTL-SDR and Simulink

This section will focus on building stereo FM receivers in Simulink that should enable you to receive, demodulate, demultiplex and decode stereo FM signals. The decoded stereo signal will be output to stereo speakers or headphones. If used to receive the stereo FM signal generated and transmitted in Exercise 10.5, the left speaker will output one song, and the right speaker will output another! If you do not have a USRP® radio and are unable to receive stereo FM signals from commercial radio stations, remember that you can use an *off the shelf* transmitter or a Raspberry Pi — see Section 10.7 for more information. To recap, the diagram of a stereo demultiplexer/ decoder is shown in Figure 10.5.

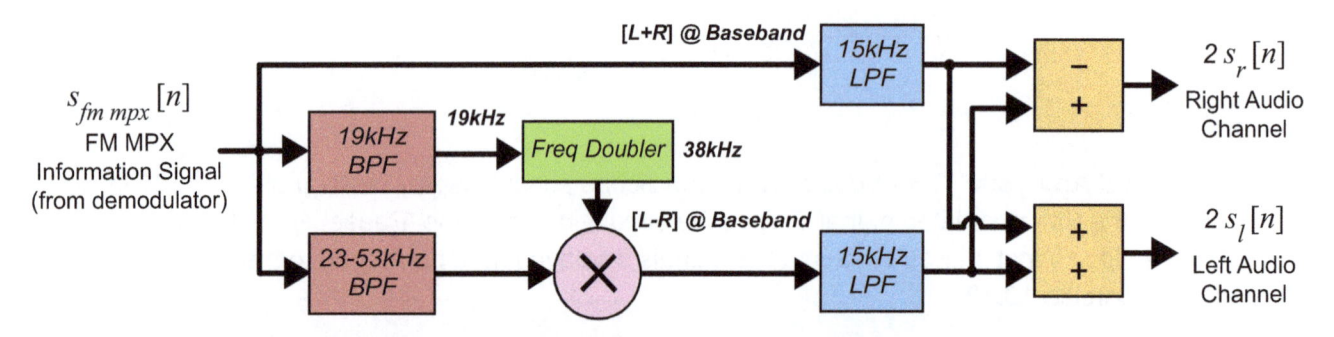

[Block diagram showing: $s_{fm\,mpx}[n]$ FM MPX Information Signal (from demodulator) feeding into 19kHz BPF → Freq Doubler → 38kHz, and 23-53kHz BPF → multiplier. Upper path [L+R] @ Baseband → 15kHz LPF. Lower path [L-R] @ Baseband → 15kHz LPF. Outputs combined to produce $2s_r[n]$ Right Audio Channel and $2s_l[n]$ Left Audio Channel.]

Figure 10.5: Block diagram of the stereo FM demultiplexer and decoder (see Section 9.9 for more information)

Chapter 10: Desktop FM Transmission and Reception

Exercise 10.6 RTL-SDR: Stereo FM Radio Receiver and Decoder (Discrim)

In this exercise we will modify the mono FM receiver model created in Exercise 10.3 to add a stereo FM demultiplexer and decoder. This will enable it to receive and demodulate stereo FM radio signals. The receiver will output a stereo audio signal to computer speakers or headphones.

Note: If you do not have an RTL-SDR, or are unable to receive commercial FM radio stations and cannot transmit your own FM signal, you can still complete this exercise by substituting the RTL-SDR Receiver block with an Import RTL-SDR Data block as discussed below.

(a) **Open MATLAB.** Set the working directory to the exercise folder,

   /my_models/receivers/

Next, make a copy of the Simulink model created in Exercise 10.2,

   .../rtlsdr_fm_discrim_demod.slx

and save this file with the name:

   .../rtlsdr_fm_discrim_demod_stereo.slx

(b) Open the new file, and then the Simulink Library Browser.

(c) **Check the RTL-SDR data source.** If an RTL-SDR is available and you are able to receive a stereo FM signal (either a broadcast or locally generated signal), it will be possible to use an *RTL-SDR Receiver* block to interface with and control the RTL-SDR. If you do not have an RTL-SDR, or are unable to receive an FM signal, an *Import RTL-SDR Data* block should be used instead. This will need to be configured to reference the file:

   /fm/rtlsdr_rx/rec_data/wfm_stereo.mat

(d) Instructions of how to configure this block and the beginning of the block diagram can be found in Exercise 10.2. Check that your design matches one of the following two configurations before continuing.

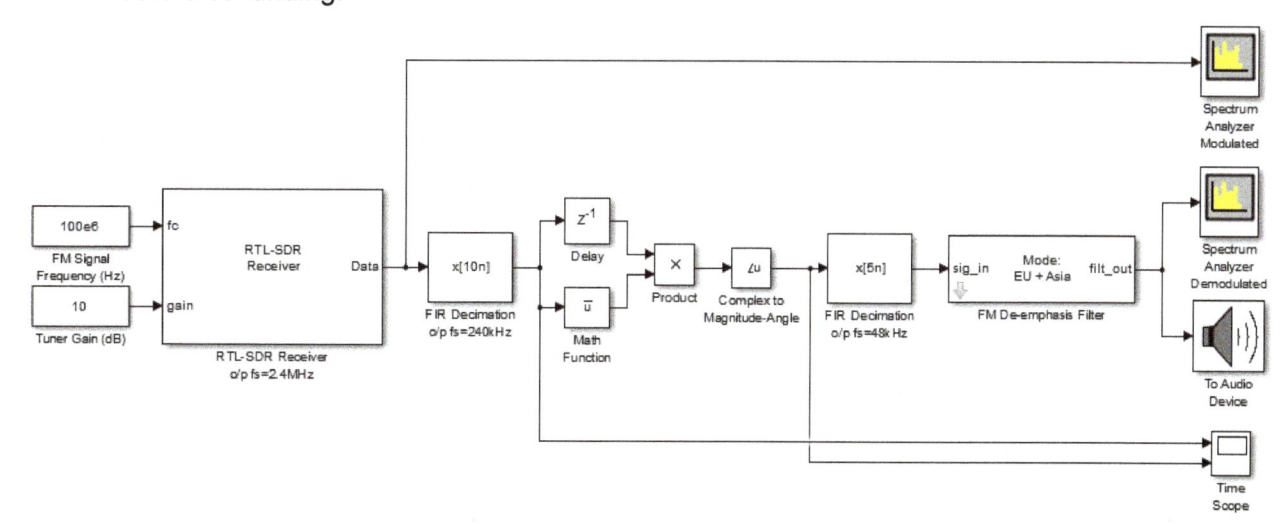

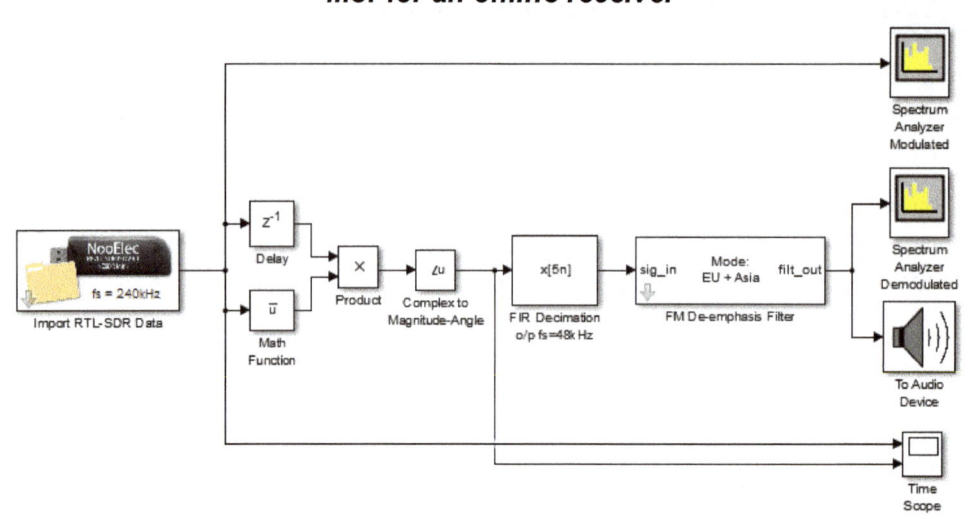

...or for an offline receiver

(e) **Change the MPX sampling rate.** Because the stereo MPX has frequency components in the range 0Hz to 53kHz, it is not possible to decimate the FM demodulated signal output from the frequency discriminator to a sampling rate of 48kHz. This must be increased to a sampling rate at least x2 53kHz to prevent aliasing from happening, and the most logical rate to choose is 120kHz. 2 on *FIR Decimation o/p fs=48kHz* and change its 'FIR filter coefficients' and 'Decimation factor' to the following.

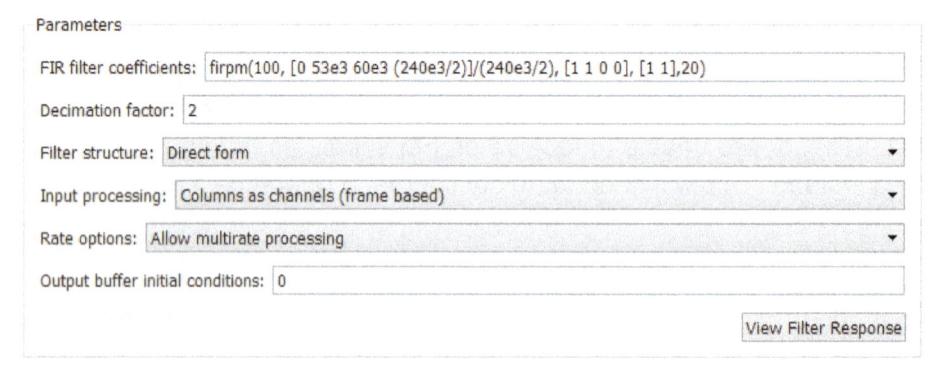

```
Parameters
FIR filter coefficients:  firpm(100, [0 53e3 60e3 (240e3/2)]/(240e3/2), [1 1 0 0], [1 1],20)
Decimation factor:  2
Filter structure:  Direct form
Input processing:  Columns as channels (frame based)
Rate options:  Allow multirate processing
Output buffer initial conditions:  0
                                                    View Filter Response
```

Rename this block *FIR Decimation o/p fs=120kHz*, and delete the connection between it and the *FM De-emphasis Filter*.

(f) **Add a stereo FM demultiplexer.** Place an *FM MPX Demultiplexer* block in an area of free space in your model. It can be found in > **RTL-SDR Book Library > FM Demod Tools**. This block has been designed to implement the stereo FM demultiplexer. A screenshot of the complete Simulink system is shown on the next page.

MPX signals entering this block are first filtered to isolate the 19kHz pilot tone. The pilot tone is input to an NCO based PLL that is configured to lock to it. A second NCO (connected to the PLL phase error signal) has identical parameters to the first, although it has double the step size. This means that phase relationship between the two NCOs is preserved, while the PLL adjusts to match the frequency and phase of the 19kHz pilot tone, but the second PLL will output a sinusoid with double the frequency of the first. This is the *frequency doubler* unit from the diagram of the stereo FM demultiplexer shown in Figure 10.5.

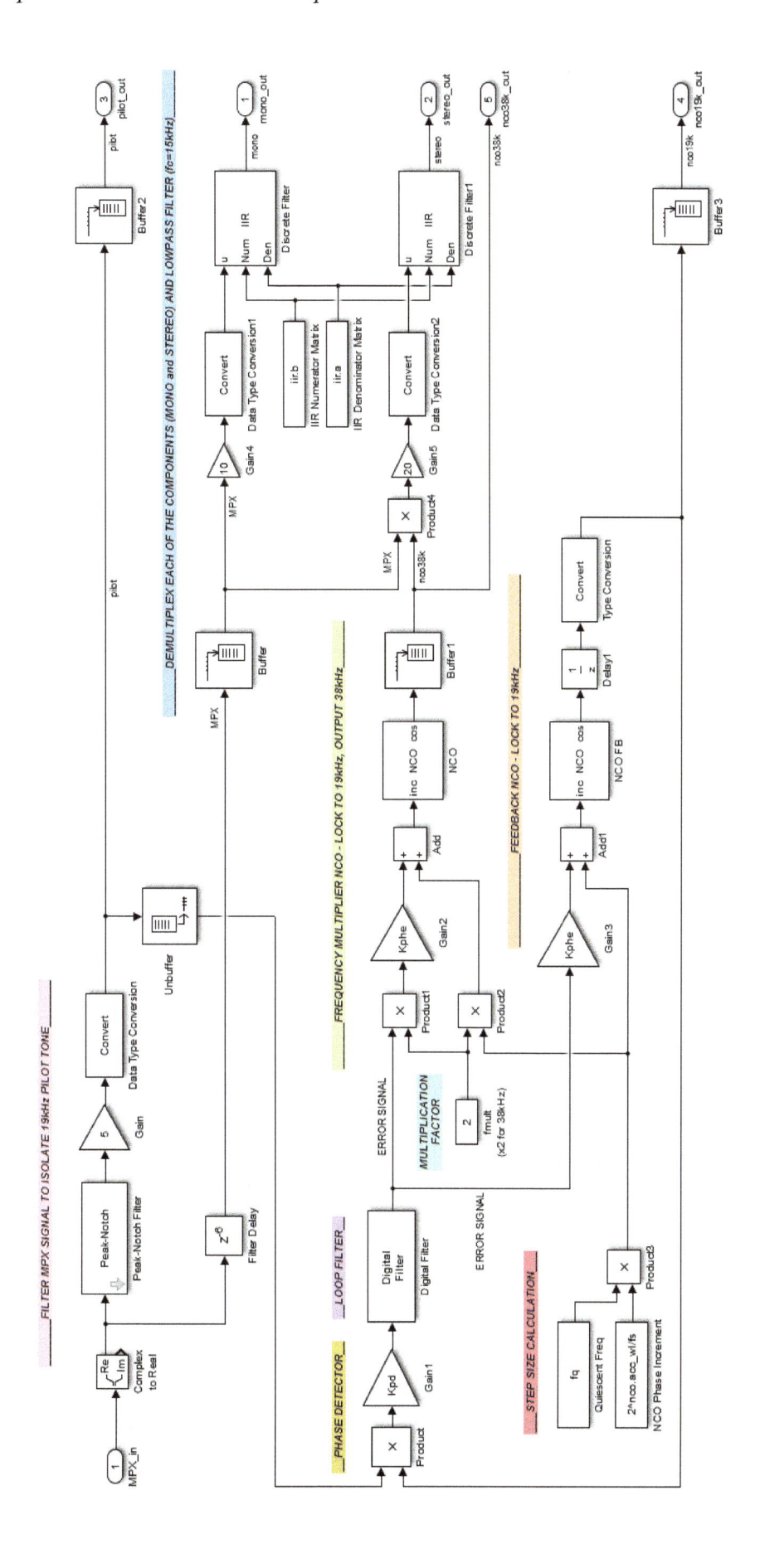

Chapter 10: Desktop FM Transmission and Reception

The 38kHz tone is used to coherently AM demodulate the stereo component of the MPX signal from its suppressed subcarrier to baseband, and then the baseband mono and stereo components are passed through an IIR filter to ensure that no frequency components over 15kHz remain. As this is only a demultiplexer block, stereo decoding still needs to be performed on the filtered mono and stereo channels output from the block. Three additional outputs are present in this system, which allow the filtered pilot tone and the two NCO signals to be monitored. These are included for testing purposes only, and are not required for any other aspect of the receiver.

(g) 2 on the block, and check that the parameters are set as follows, and connect the output of the frequency discriminator to its input:

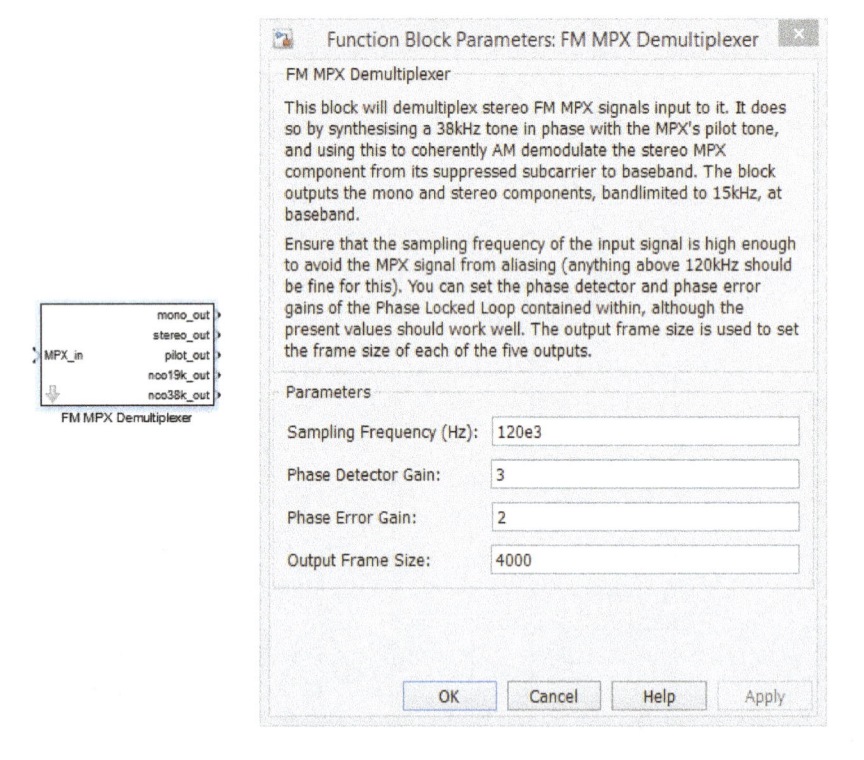

(h) The mono and stereo components output from the *FM MPX Demultiplexer* are sampled at 120kHz, which is unnecessary, as they are bandlimited to 15kHz. Reduce their sampling rates to 48kHz using *FIR Rate Conversion* blocks. These can be found in ▦ > **DSP System Toolbox > Filtering > Multirate Filters**. Parameterise these as follows:

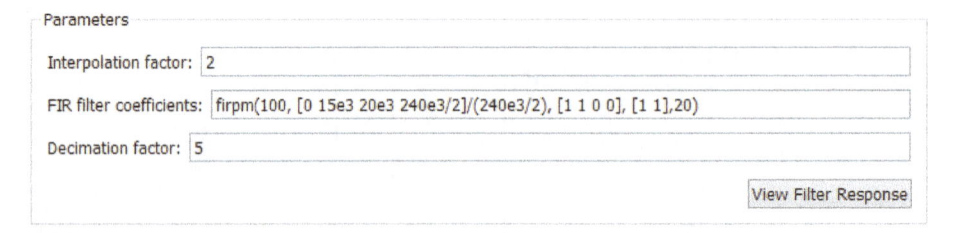

This specifies interpolation by a factor of 2 and then decimation by a factor of 5 (which implements a rate change from 120kHz to 48kHz). The lowpass filter is configured to pass frequencies up to 15kHz to prevent aliasing effects from occurring. Apply the changes, and then rename the blocks *FIR Rate Conversion o/p fs=48kHz*.

Chapter 10: Desktop FM Transmission and Reception 401

(i) **Implement the stereo FM decoder.** Place two *Add blocks* and a *Matrix Concatenate* block from ▮ > **Simulink** > **Math Operations** into the model. ② on one of the *Add* blocks and change its 'List of signs' to '+–'. Connect these blocks as shown below to implement the stereo decoder. Label the wires as shown in the screenshot to help distinguish the signals from one another.

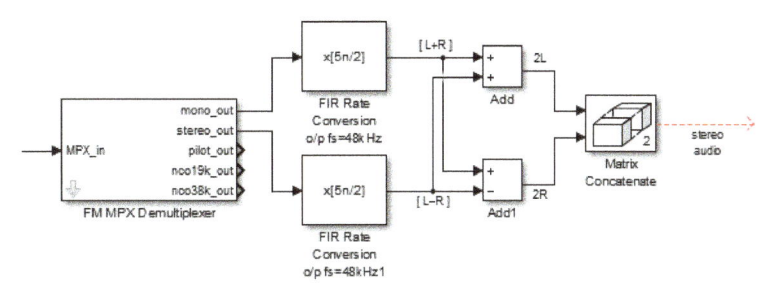

Connecting the decoder up in this way means that 2L will be recovered by adding [L+R] to [L–R], and 2R by subtracting [L–R] from [L+R]. The resulting audio channels need to be recombined into a 'stereo' signal before they can be output to speakers, which is why the 2L and 2R signals are input to the *Matrix Concatenate* block. Connect the 'stereo' signal to the input of the *FM De-emphasis Filter* to complete the receiver.

(j) **Rearrange the scopes to complete the receiver.** As this receiver performs a type of carrier synchronisation, the computer (depending in its specification) may be unable to cope with running a *Time Scope* as well as the *Spectrum Analyzers*, demodulator and the audio output all at the same time. You can try to leave it in, but we recommend that you delete it.

Currently, the first *Spectrum Analyzer* is connected to the FM modulated signal input to the frequency discriminator, and the second is connected to the demodulated audio output. It would be useful here to monitor the FM MPX before it is demultiplexed, as this allows a check that the received signal conforms to the FM standard. If importing a signal rather than receiving one live, it is not necessary to monitor the modulated signal, but if using an RTL-SDR, doing so will aid the tuning process. Either add a new *Spectrum Analyzer*, or shift *Spectrum Analyzer Modulated*, and connect it after *FIR Decimation o/p fs=120kHz*, and finally rename it to *Spectrum Analyzer MPX*.

(k) Your completed receiver should take one of the following two forms:

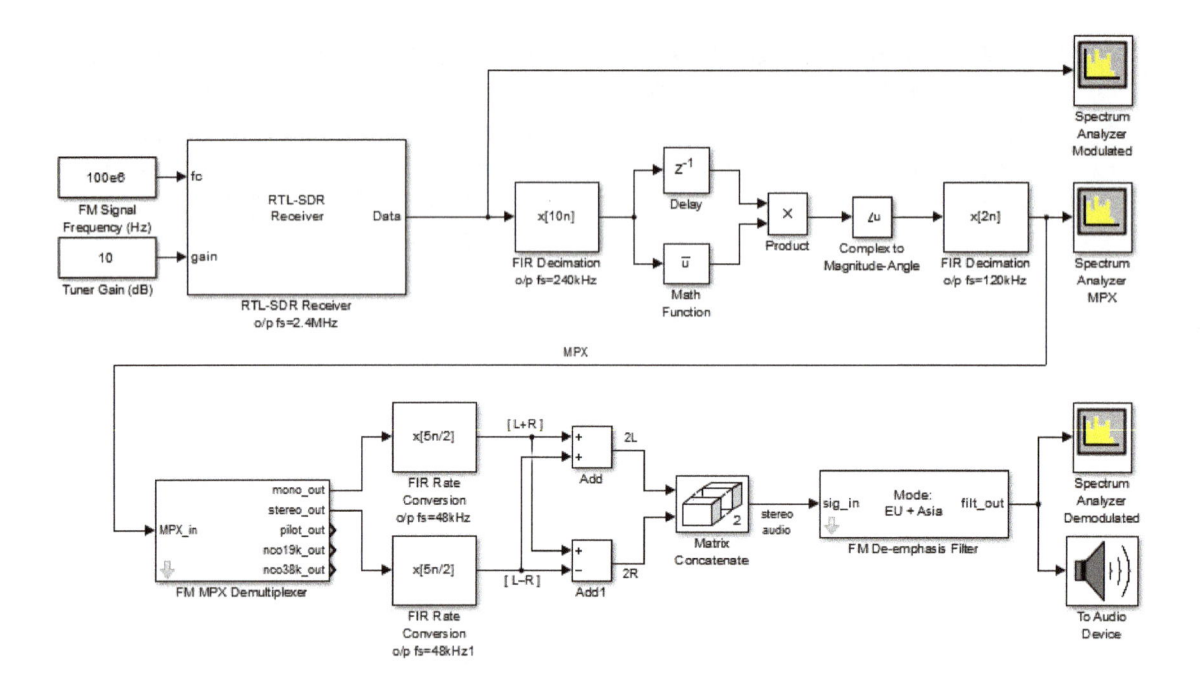

...or for an offline receiver

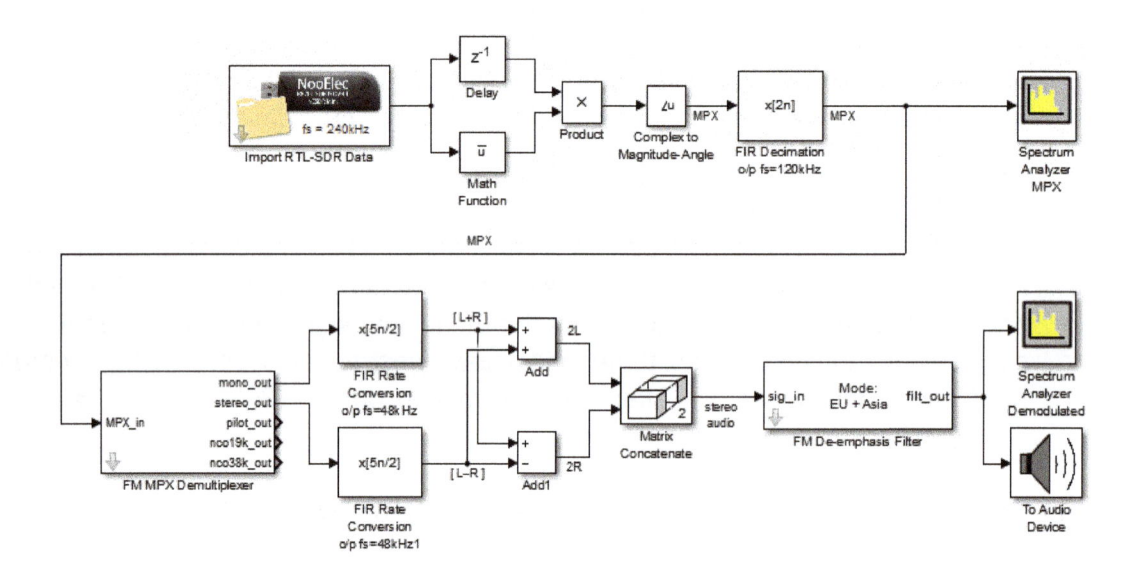

(I) **Open the reference file.** If you did not complete the prior steps, or would like to check your design against a reference model, open the following file:

/fm/rtlsdr_rx/rtlsdr_fm_discrim_stereo_demod.slx

If importing data rather than connecting to an RTL-SDR, delete the *RTL-SDR Receiver* block and reconfigure the block diagram to the 'offline receiver' model shown above. Check that this block is set to reference the correct data file. If the *FM De-emphasis Filter* is configured for the wrong region, correct this before continuing.

Chapter 10: Desktop FM Transmission and Reception 403

(m) **Prepare to run the simulation.** Connect speakers or headphones to the computer and test them. If importing a signal, set the *Simulation Stop Time* to 60 seconds, '60'. If using an RTL-SDR, set the *Simulation Stop Time* to 'inf' by typing this into the Simulink toolbar, and the *Simulation Mode* to 'Accelerator'. This will force Simulink to partially compile the model into native code for your computer, which will allow it to run faster. Check that MATLAB can communicate with the RTL-SDR by typing `sdrinfo` into the MATLAB command window. Finally, make sure that the model is saved.

(n) **Run the simulation.** Begin a simulation by 1 on the 'Run' ▶ button in the Simulink toolbar. After a couple of seconds the simulation will begin, the *Spectrum Analyzer* windows should appear, and the demodulated, demultiplexed and decoded stereo FM signal should be played. If you are using an RTL-SDR, adjust the frequency and gain values until the device is tuned to the signal you want to receive. Use *Spectrum Analyzer Modulated* to help with this.

If receiving the stereo FM signal generated in Exercise 10.5, you should be able to hear the left speaker plays one song, and the right speaker another!

(o) **Signal Analysis: *Spectrum Analyzers*.** The *Spectrum Anlayzer* windows allow you to monitor the FM signal as it is demodulated in the frequency domain. *Spectrum Analyzer MPX* should show the spectrum of the demodulated FM MPX signal, and should have a bandwidth of 240kHz. It should be possible to identify whether the signal contains a pilot tone. If it does (as for the example signal), it means you are receiving a stereo FM signal; so you should be able to clearly see activity in the 23–53kHz region, where the stereo channel is modulated onto its suppressed subcarrier. Overlaying the plot of the demodulated FM MPX from *Spectrum Analyzer Demodulated* on Figure 9.26 (page 363), you can see that the signal we received was very close to what is expected.

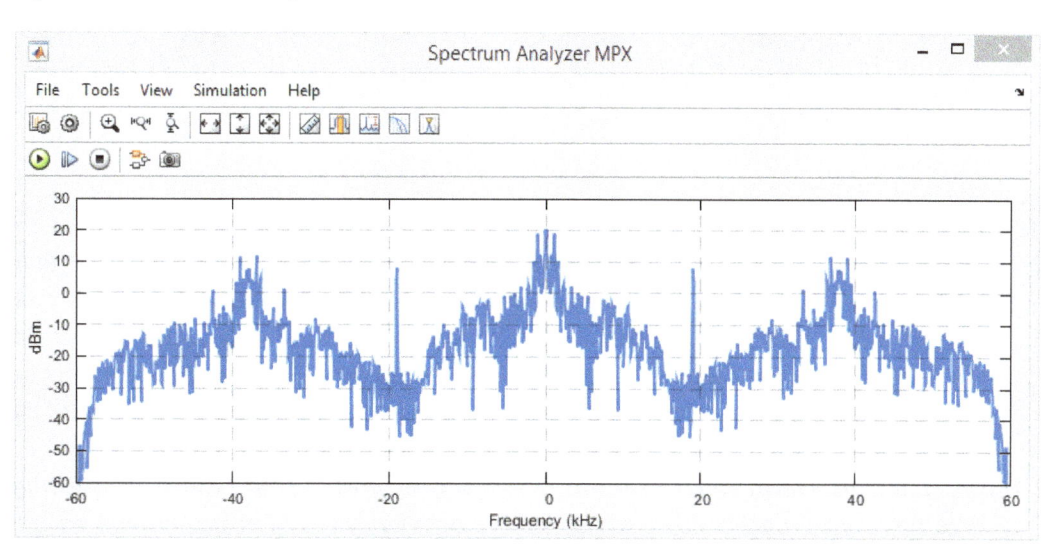

REAL FM MPX SIGNAL RECEIVED WITH THE RTL-SDR

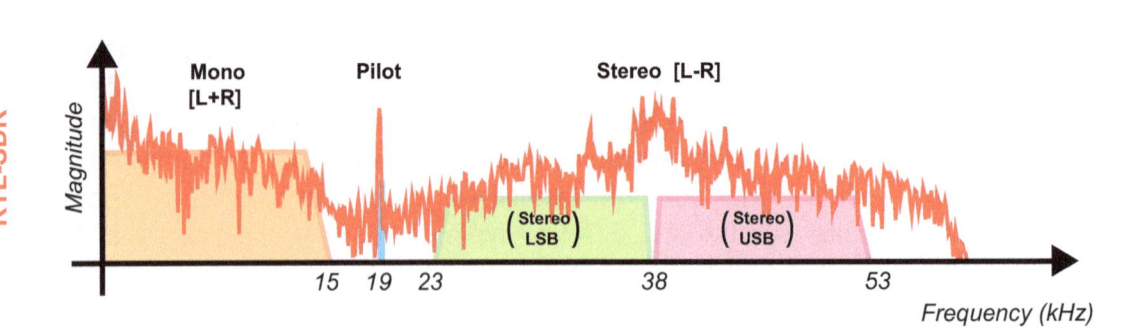

Mono [L+R] — Pilot — Stereo [L-R]
(Stereo LSB) (Stereo USB)
15 19 23   38   53   Frequency (kHz)

(p) Examine what you see. Does the signal conform to the standard FM MPX, like the example shown above?

(q) While *Spectrum Analyzer MPX* shows the MPX components in the frequency domain, *Spectrum Analyzer Demodulated* shows the spectra of the two audio channels after demultiplexing and decoding has taken place. It should be noticeable that the channels are independent of each other — especially so if receiving the custom MPX signal generated in Exercise 10.5, in which the left and right channels contain different songs.

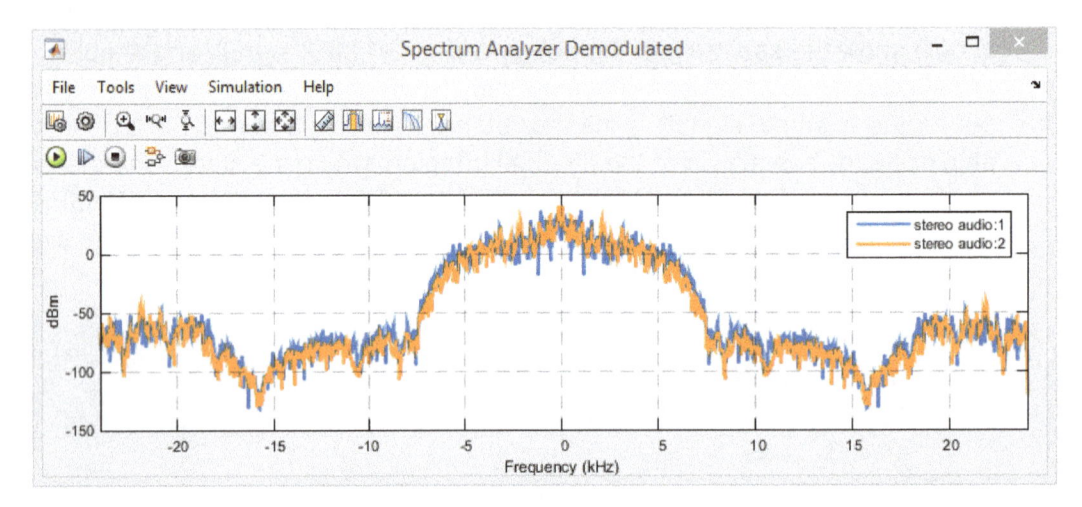

(r) **Further Work.** Why not make your stereo FM demodulator, demultiplexer and decoder design a little more flexible by adding interactive controls for the carrier frequency and tuner gain? A control mask has been created for this, and can be found in ▦ **> RTL-SDR Book Library > FM Demod Tools**. The block is called *FM Tuner Freq and Gain Control*, and this can act as a direct replacement for the two *Constant* blocks currently used to tune the RTL-SDR.

(s) **Watch and listen to us demodulate a stereo FM signal with a frequency discriminator.** We have recorded a video which shows activity in the scope windows as a stereo FM signal transmitted from the USRP® hardware (with the model from Exercise 10.5) is received by the RTL-SDR and demodulated. The output stereo audio signal has been recorded too. Stereo headphones will be needed in order to hear to the left and right channels!

desktopSDR.com/videos/#discrim_demod_stereo

Exercise 10.7 RTL-SDR: Stereo FM Radio Receiver and Decoder (PLL)

As was mentioned in Section 9.8 (page 358), Phase Locked Loops can also be used for FM demodulation. In this exercise we will run a Simulink model that uses a PLL to demodulate an FM signal. Because the PLL requires a high oversampling rate, this model cannot be run in real time, so a data file is provided and will act as the RTL-SDR data source. The stereo demultiplexing and decoding methods presented in Exercise 10.6 are utilised once again to demodulate each of the MPX components back to baseband, and extract the left and right audio signals. The receiver will output a stereo audio signal to a .wav file, which can be played back after the simulation has ended.

(a) **Open MATLAB.** Set the working directory to the exercise folder,

   /fm/rtlsdr_rx/

and then open the following file:

   .../rtlsdr_fm_pll_stereo_demod.slx

The block diagram of the model should look as follows:

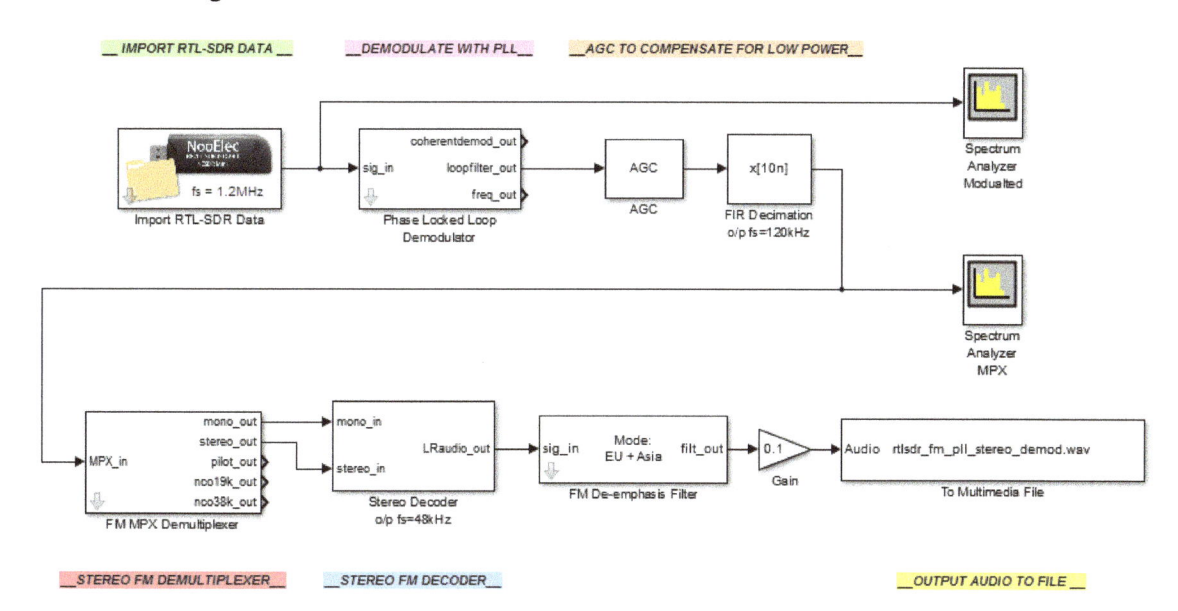

(b) The FM signal that is being imported to this receiver has been recorded at a rate of 1.2MHz, and is offset from baseband by +150kHz, as shown below.

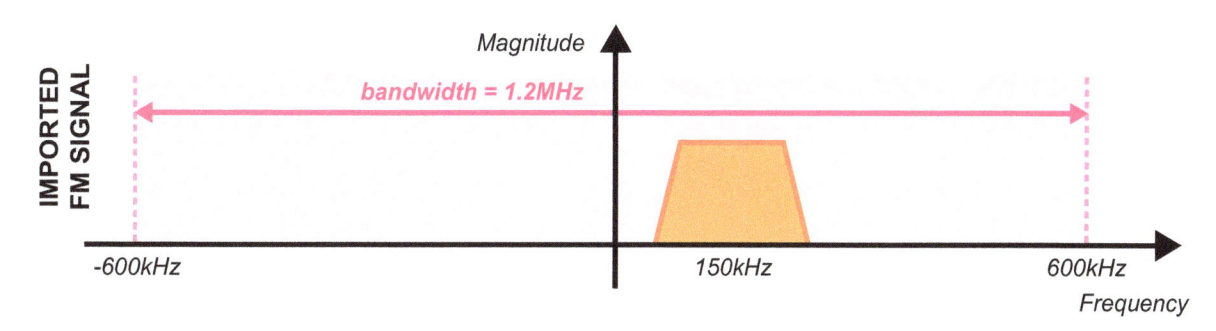

on the *Phase Locked Loop Demodulator* block to view its parameters.

The PLL has been configured to lock to this signal by setting the 'Quiescent Frequency' value to '150e3'. The 'Noise Bandwidth' has been configured to 35kHz, half the maximum frequency deviation value permitted for a WFM signal. The 'Damping Ratio' has been set to '1'. As discussed in Section 7.7.3 (page 260) of the *Frequency Tuning and Simple Synchronisation* chapter, the damping ratio of a PLL affects its ability to track changes in the frequency of the input signal. Because the frequency of the FM signal changes with response to the amplitude of the input signal, we need to have the PLL adapt quickly in order to recover all of the received information.

Recalling the discussion of Section 9.8 (page 358), note that when the PLL is locked to the FM signal, the control signal output by the loop filter (i.e. the signal that is used to configure the PLL's VCO) varies with the amplitude of the information signal. This is why the receiver uses the 'loopfilter_out' signal from the PLL, rather than 'coherentdemod_out'.

The demodulated FM MPX signal is decimated down to a sampling rate of 120kHz, and input to the *FM MPX Demultiplexer* block. After this, the mono and stereo channels are decoded to recover the left and right audio signals, the sampling rate is reduced to 48kHz, and the audio signal is output to a file.

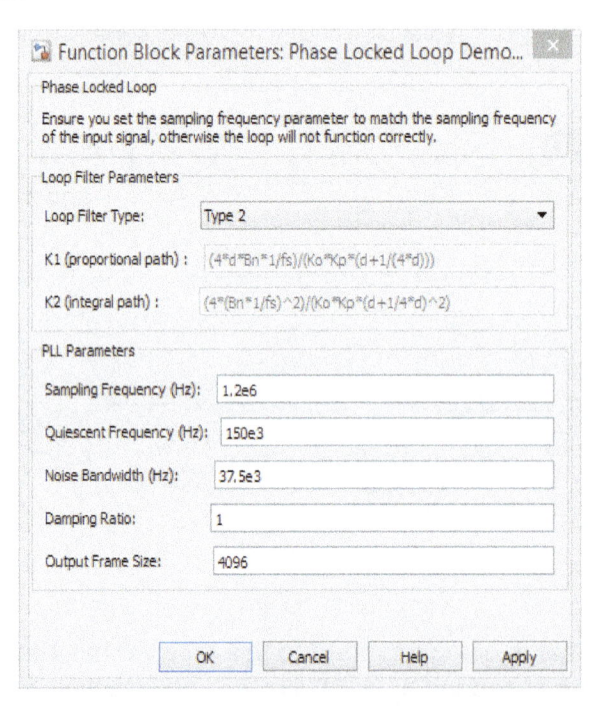

(c) **Check the RTL-SDR data source.** As this model does not run in real time, we cannot receive a live FM signal off the air (as samples will be dropped), so an *Import RTL-SDR Data* block must be used instead. Check that this is configured to reference the following file:

`/fm/rtlsdr_rx/rec_data/wfm_stereo_150kHzoffset.mat`

(d) **Prepare to run the simulation.** Check that the *Simulation Stop Time* is set to 30 seconds, '30', and the *Simulation Mode* is set to 'Accelerator'. This will force Simulink to partially compile the model into native code for your computer, which allows it to run faster.

Chapter 10: Desktop FM Transmission and Reception

(e) **Run the simulation.** Begin the simulation by 🖱 on the 'Run' ▶ button in the Simulink toolbar. After a couple of seconds the simulation will begin, the *Spectrum Analyzer* windows should appear, and the demodulated audio signal should begin being written to the file.

The *Spectrum Analyzer* windows allow the FM signal to be monitored as it is demodulated in the frequency domain. You should be able to see the FM signal residing around 150kHz in *Spectrum Analyzer Modulated*, and the demodulated FM MPX in *Spectrum Analyzer MPX*.

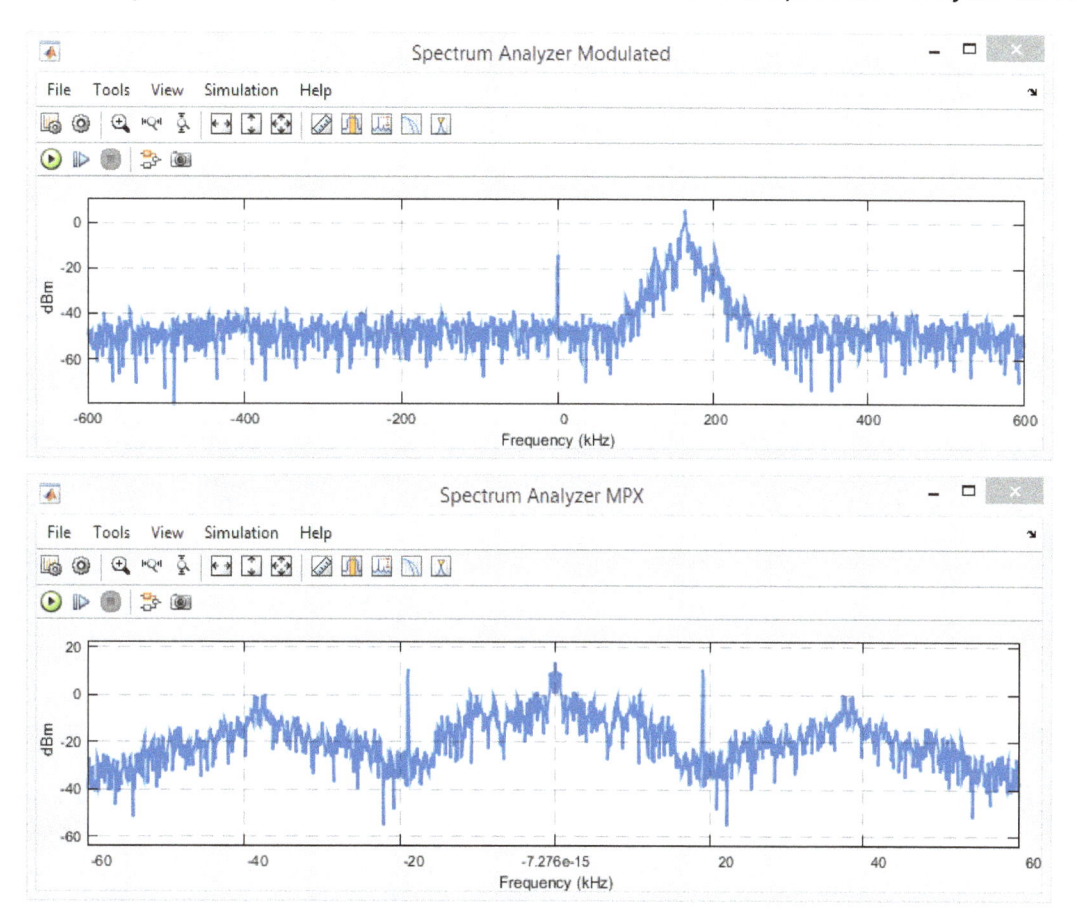

(f) **Play back the demodulated audio file.** Connect speakers or headphones to your computer and test to ensure that they are working. Once the simulation is complete, open the audio file in your default player and listen to it:

/fm/rtlsdr_rx/rtlsdr_fm_pll_demod.wav

Can you hear music? If so the PLL demodulation was successful. Hopefully the stereo demultiplexing and decoding processes have worked too, and you will be able to hear the left speaker playing one song, and the right speaker playing another!

Exercise 10.8 RTL-SDR: Stereo FM Radio Receiver (Slope Detector)

A simple and interesting FM demodulator is the *Slope Detector* method. This is essentially an AM Envelope Detector, and it demodulates the signal by extracting the energy associated with the shifting frequency. In this exercise, we will investigate how this demodulator works. More information on the demodulation method can be found here [106].

(a) **Open MATLAB.** Set the working directory to the exercise folder,

> 📁 `/fm/rtlsdr_rx`

and then open the following model:

> `.../rtlsdr_fm_slope_stereo_demod.slx`

The block diagram of the model should look as follows:

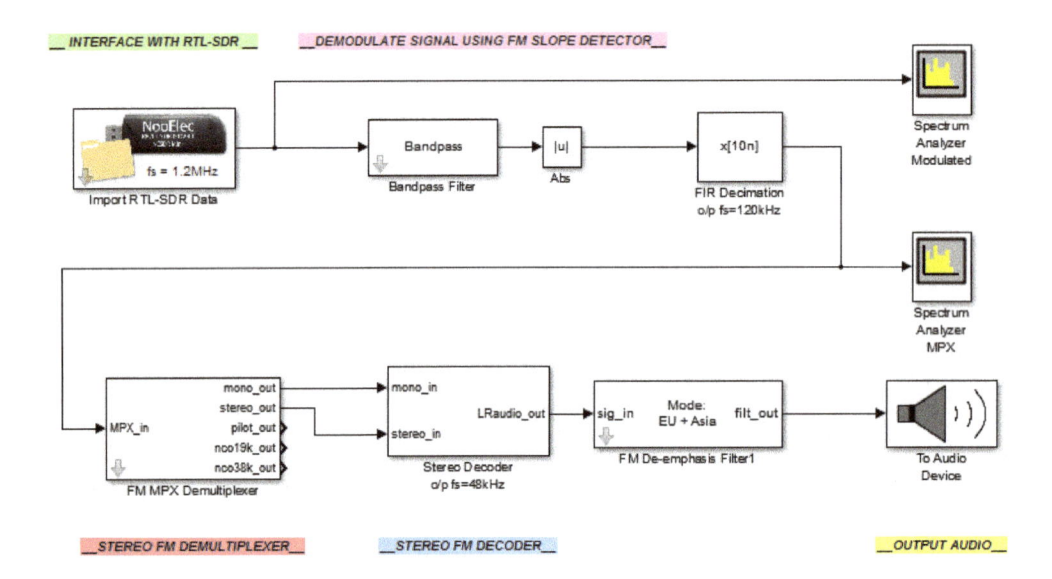

[Block diagram showing: INTERFACE WITH RTL-SDR (Import RTL-SDR Data, fs=1.2MHz) → DEMODULATE SIGNAL USING FM SLOPE DETECTOR (Bandpass Filter → Abs |u| → FIR Decimation x[10n], o/p fs=120kHz) → Spectrum Analyzer Modulated and Spectrum Analyzer MPX; then STEREO FM DEMULTIPLEXER (FM MPX Demultiplexer with MPX_in, mono_out, stereo_out, pilot_out, nco19k_out, nco38k_out) → STEREO FM DECODER (Stereo Decoder, o/p fs=48kHz, mono_in, stereo_in, LRaudio_out) → FM De-emphasis Filter1 (Mode: EU + Asia, sig_in, filt_out) → OUTPUT AUDIO (To Audio Device)]

(b) **Inspect the model.** The FM signal that is being imported to this receiver has been recorded at a rate of 1.2MHz, and is offset from baseband by +150kHz. As WFM signals have a frequency deviation of 75kHz, this means that the centre of the signal will shift between +115kHz and +185kHz. (This is the same signal that was used in Exercise 10.7.)

on the *Bandpass Filter* and note its parameters. It is configured to have a rising transition band across the frequency range 100kHz to 200kHz. As the centre frequency of the FM carrier decreases (towards 100kHz), the filter will attenuate the signal more, and when the frequency increases, it will attenuate it less. This means that, as the frequency of the FM signal fluctuates, the output of the *Bandpass Filter* will have an amplitude proportional to the frequency shifts in the input signal, similar to the waveform shown in Figure 9.8 (page 348) in the previous chapter. Passing this through the envelope detector completes the demodulation, and then the recovered audio signal can be output to speakers.

(c) **Check the RTL-SDR data source.** Check that the *Import RTL-SDR Data* block is configured to reference the following file:

> 📁 `/fm/rtlsdr_rx/rec_data/wfm_stereo_150kHzoffset.mat`

Chapter 10: Desktop FM Transmission and Reception 409

(d) **Run the simulation.** Check that the *Simulation Stop Time* is set to 30 seconds, '30', and the *Simulation Mode* is set to 'Accelerator'. This will force Simulink to run the model faster. Begin the simulation by 🖱 on the 'Run' ▶ button in the Simulink toolbar.

(e) **Can you hear the demodulated music signal?** This should be all the practical proof you need that an AM receiver as a Slope Detector can be used to receive FM signals!

10.5 Manipulating the MPX: Transmitting AM Signals with FM Transmitters

The first unconventional concept we are going to work towards is to transmit an AM signal with an FM transmitter. This can be particularly useful if you do not have any AM transmitters available but would like to complete the *Desktop AM Transmission and Reception* chapter! To do this, we will need to manipulate the FM MPX. As a recap, the spectrum of the normal FM MPX is shown in Figure 10.6 below.

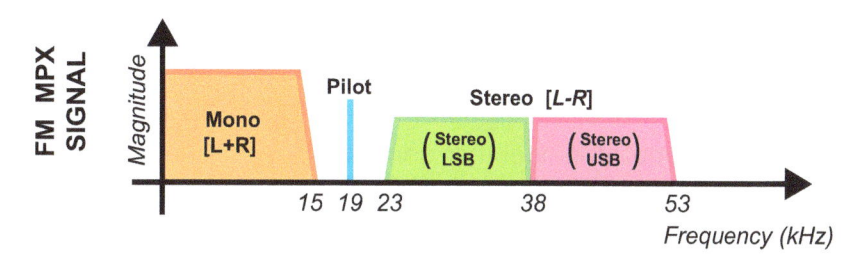

Figure 10.6: Baseband spectral representation of the standard FM radio multiplex

The mono component resides at baseband, and therefore is of no use in this situation. The stereo component is modulated onto a 38kHz subcarrier however — this is an AM-DSB-SC signal. If the MPX was manipulated in such a way that there was no mono component present, then it would look as follows:

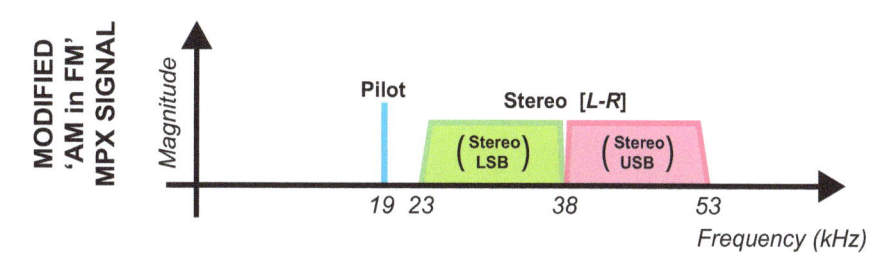

Figure 10.7: Baseband spectral representation of the *AM in FM* MPX

In this state, we would be able to test out Costas Loops in receivers to coherently demodulate the component to baseband; use PLLs and NCOs to synthesise the 38kHz carrier from the Pilot tone and use this for coherent demodulation; or investigate why envelope detectors cannot be used to demodulate AM-DSB-SC signals.

To create this *AM in FM* MPX, we will first need to generate a special baseband information signal. Recalling from Section 9.9.2 (page 362), the standard FM encoder / multiplexer generates both mono [L+R] and stereo [L−R] components; and we need to prevent it from doing this. Taking a mono information signal, α, and setting $L = \alpha$ and $R = -\alpha$, the generated MPX components will be:

$$\text{mono: } [L + R] = \Big[(\alpha) + (-\alpha) \Big] = 0 \tag{10.5}$$

$$\text{stereo: } [L - R] = \Big[(\alpha) - (-\alpha) \Big] = 2\alpha \tag{10.6}$$

In Exercise 10.9, we will open and run a signal generator/ multiplexer/ baseband FM modulator/ USRP® reference transmitter model. This will create the custom 'stereo' signal, which is then input to the same series of blocks used in Exercise 10.5 for the standard stereo FM transmitter. The baseband FM modulated signal it generates is passed to the USRP® hardware, where it is modulated onto an RF carrier and transmitted.

This is a partner exercise for Exercise 10.10. If a USRP® radio is not available, there are a number of alternatives that can be used instead, such as an *off the shelf* FM transmitter, or a Raspberry Pi. Information on using these is given in Section 10.7. Even if none of these are possible, you will still be able to complete the later exercise using the provided data file. In any case, do open and explore the USRP® transmitter file, to confirm exactly how the *AM in FM* MPX is generated.

Exercise 10.9 AM in FM: Multiplexer, Modulator and USRP® Transmitter

In this exercise we will run a Simulink model that generates a custom 'stereo' signal, then encodes, multiplexes and FM modulates this before transmitting the *AM in FM* MPX with the USRP® transmitter. The model is almost identical to the stereo FM transmitter constructed in Exercise 10.5, so we will not go into great detail discussing how the system works.

Note: If you do not have a USRP® radio, you can still run this exercise and see the MPX signal generated by commenting out the SDRu Transmitter block.

(a) **Open MATLAB.** Set the working directory to the exercise folder,

   `/fm/usrp_tx/`

   and then open the following file:

   `.../usrp_fm_amfm.slx`

   The block diagram of the model should look as follows:

(b) The mono music signal is imported using the *From Multimedia File* block, and is then resampled to 120kHz using an *FIR Rate Conversion* block. A *Gain* block is used to invert the polarity of the signal, and the 'stereo' signal is then input to the block diagram of the encoder, multiplexer and FM modulator seen previously.

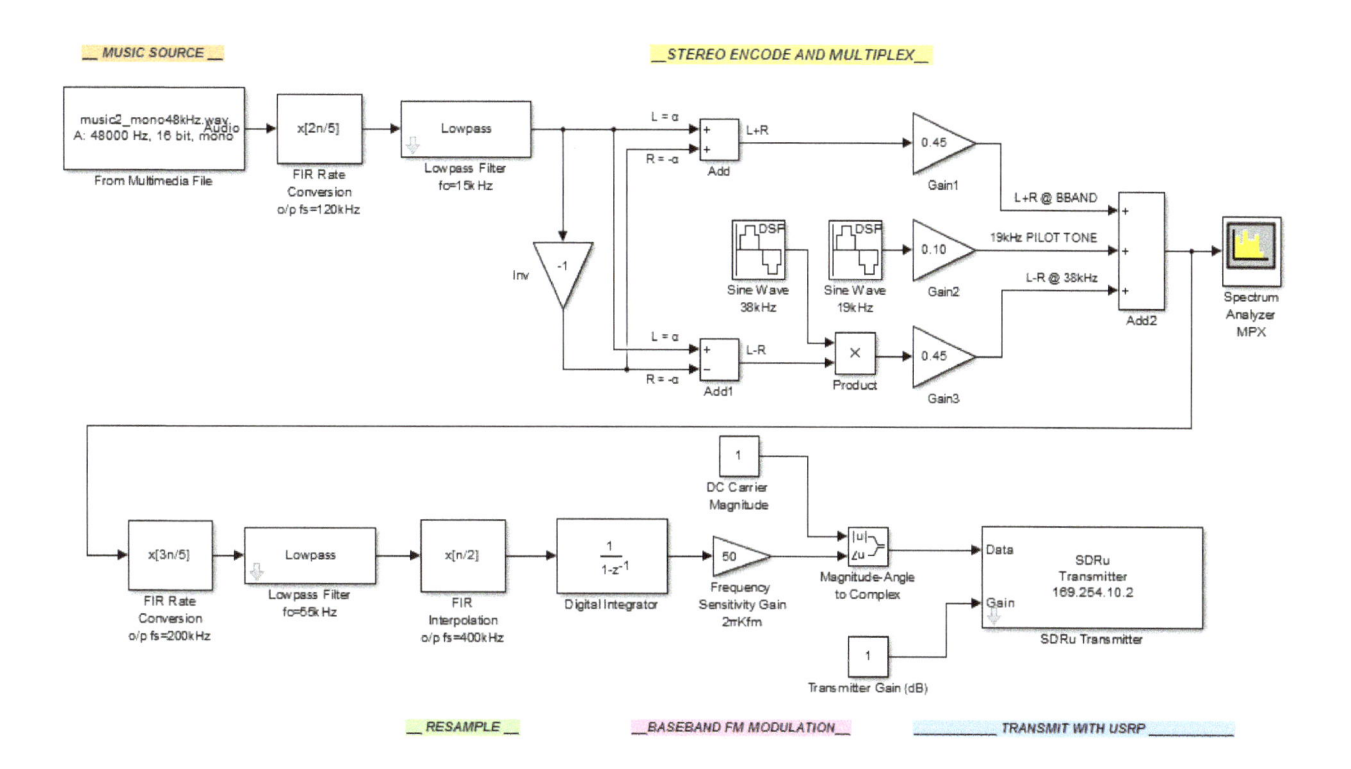

Chapter 10: Desktop FM Transmission and Reception 411

(c) **Check the USRP® radio settings.** The *SDRu Transmitter* block will need to be configured to communicate with your USRP® hardware. To do this, you will need to know its IP address (or USB address), and be able to 'ping' it from your computer. Guidance on how to set up this connection is provided in Appendix A.2 (page 571). ② on the *SDRu Transmitter* block, and check that the appropriate 'Platform', and 'Address' of your USRP® radio are entered. Make sure that the 'Center Frequency' parameter is set to the desired transmit frequency. This value should be within the range of your RTL-SDR's tuner, e.g. in the range 25MHz–1.75GHz. Check that the USRP® hardware is turned on, and that MATLAB can communicate with it, by typing `findsdru` into the MATLAB command window. If a structure is returned as discussed in Appendix A.2 (page 571), continue to the next step.

If you are not working with a USRP® radio, comment out or delete this block before continuing.

(d) **Prepare to run the simulation.** Check that the *Simulation Stop Time* is set to 'inf', and that the *Simulation Mode* is set to 'Accelerator'. This will enable Simulink to run the model faster. Finally, make sure that the model is saved.

(e) **Run the simulation.** Begin the simulation by ① on the 'Run' ▶ button in the Simulink toolbar. After a couple of seconds, Simulink will establish a connection with the USRP® radio and the simulation will begin. When it does, samples of the baseband FM signal will be passed to the radio, modulated onto an RF carrier, and transmitted from its antenna.

(f) **Signal Analysis.** The *Spectrum Analyzer* window should also appear at the start of the simulation, and this will allow the customised FM MPX to be monitored as it is generated and transmitted. It should be clear that the MPX contains the pilot tone and stereo component only.

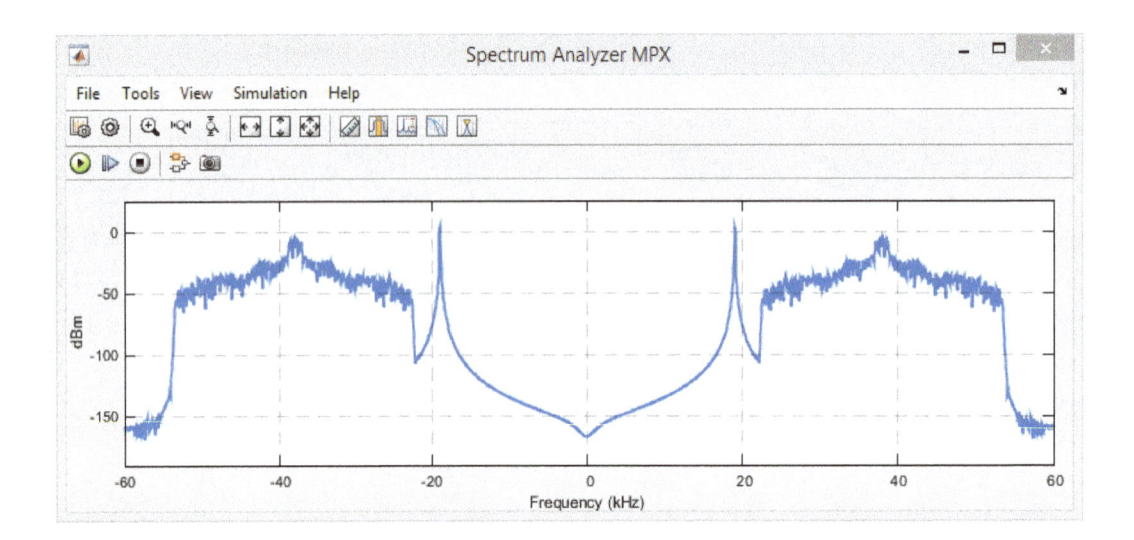

(g) When an analogue FM radio is tuned to the transmitted signal, the radio should be able to demodulate the signal correctly (provided that it is transmitted within the FM band), as the stereo decoding process should recover $L = \alpha$ and $R = -\alpha$.

The receiver for this *AM in FM* signal requires a two stage demodulation process. An FM demodulator is needed to recover the MPX, and an AM demodulator is required to demodulate the information from around its 38kHz subcarrier back to baseband. A variety of different AM demodulators can be realised for this task, and we will focus on two in particular; one coherent, and one non-coherent.

For the coherent option, we shall simply use the *FM MPX Demultiplexer* block from Exercise 10.6. This block locks to the pilot tone of the MPX signal, generates a 38kHz sinusoid that is in phase with the original suppressed subcarrier, and uses this to demodulate the stereo component to baseband. The simplest non-coherent alternative is to multiply the MPX by a generic 38kHz sinusoid. While this should work in almost the same way, the phase of the modulating carrier and the local carrier are very unlikely to match, which means that the demodulation will not be performed perfectly.

In Exercise 10.10, we will ask you to run a receiver reference model for this *AM in FM* MPX signal, that will demodulate and output the AM-DSB-SC modulated audio signal. If you are unable to transmit an *AM in FM* signal either with a USRP® radio, an *off the shelf* FM transmitter or a Raspberry Pi, you can still complete the exercise using the provided data file.

Exercise 10.10 AM in FM: RTL-SDR FM Receiver and AM Demodulator

In this exercise you will run a receiver that will demodulate the *AM in FM* signal and output the demodulated audio signal to your computer's speakers or headphones. There is the option of two demodulators—one coherent, and the other non-coherent. You will be able to switch between these in real time using the *Manual Switch*.

Note: *If you do not have an RTL-SDR, or are unable to transmit the AM in FM MPX signal from Exercise 10.9, you can still complete this exercise by substituting the RTL-SDR Receiver block with an Import RTL-SDR Data block as outlined below.*

Chapter 10: Desktop FM Transmission and Reception 413

(a) **Open MATLAB.** Set the working directory to the exercise folder,

   `/fm/rtlsdr_rx`

and then open the following file:

   `.../rtlsdr_fm_amfm_demod.slx`

The block diagram of this receiver should look as follows:

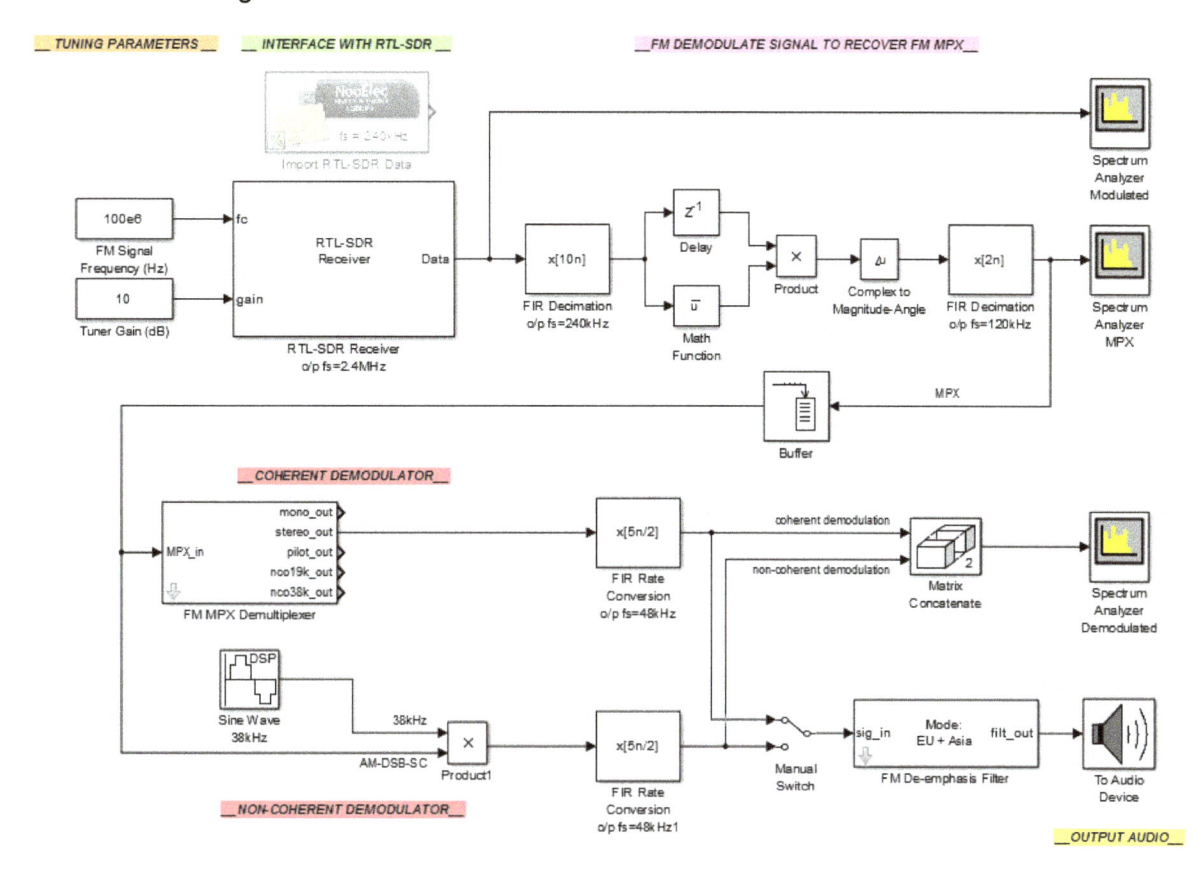

(b) If the RTL-SDR is being used, note that it should be tuned to the centre frequency of the FM modulated *AM in FM* signal. The signal is sampled at a rate of 2.4MHz (to aid with tuning), output from the *RTL-SDR Receiver* block, and then decimated to a sampling rate of 240kHz. If the *Import RTL-SDR Data* block is being used instead, it is already sampled at the required 240kHz rate, so the decimator is not required.

The FM signal is demodulated with a frequency discriminator, which recovers the *AM in FM* MPX. This is subsequently input to the two demodulators, their outputs are decimated, and then the chosen baseband signal is output to the computer speakers. You can switch between the two demodulators by 2⃣ on the *Manual Switch* block as the model runs.

(c) **Check the RTL-SDR data source.** If you have an RTL-SDR available and are able to transmit the *AM in FM* MPX signal, an *RTL-SDR Receiver* block can be used to interface with and control your RTL-SDR. As outlined previously, there are a number of different ways you can create this signal. If using a USRP® radio, run the model from Exercise 10.9. If using an *off the* *shelf* FM transmitter, plug it into the device you are going to play the special 'stereo' file from. If you are using a Raspberry Pi, copy the file across to the Pi using an SSH client.

If you do not have an RTL-SDR or are unable to transmit the signal, the *Import RTL-SDR Data* block should be used instead, and configured to reference the following file:

📁 `/fm/rtlsdr_rx/rec_data/wfm_amfm.mat`

If importing data rather than connecting to an RTL-SDR, delete the *RTL-SDR Receiver* block and connect the *Import RTL-SDR Data* block in its place. Delete the *FIR Decimation o/p fs=240kHz* block (as it is not needed), and connect the output of the *Import RTL-SDR Data* block directly to the inputs of the frequency discriminator.

(d) **Prepare to run the simulation.** Connect speakers or headphones to your computer and test to ensure that they are working. If importing a signal, set the *Simulation Stop Time* to 60 seconds, '60'. If you are using an RTL-SDR, set the *Simulation Stop Time* to 'inf' by typing this into the Simulink toolbar, and the *Simulation Mode* to 'Accelerator'. This will enable the model to run faster. If using an RTL-SDR, check that MATLAB can communicate with it by typing `sdrinfo` into the MATLAB command window.

(e) **Run the simulation.** Begin the simulation by 🖱 on the 'Run' ▶ button in the Simulink toolbar. After a couple of seconds the simulation will begin, the *Spectrum Analyzer* windows should appear, and there should be an audio signal output to speakers or headphones. If using an RTL-SDR, adjust the frequency and gain values until the device is tuned to the signal you want to receive. The *Spectrum Analyzer Modulated* can be used to help with this.

(f) **Signal Analysis: *Spectrum Analyzers*.** From *Spectrum Analyzer MPX*, it should be possible to confirm that the received signal only contains information in the stereo component, as expected. Perhaps more interesting is what *Spectrum Analyzer Demodulated* shows.

Local Carrier in Phase with Modulating Carrier

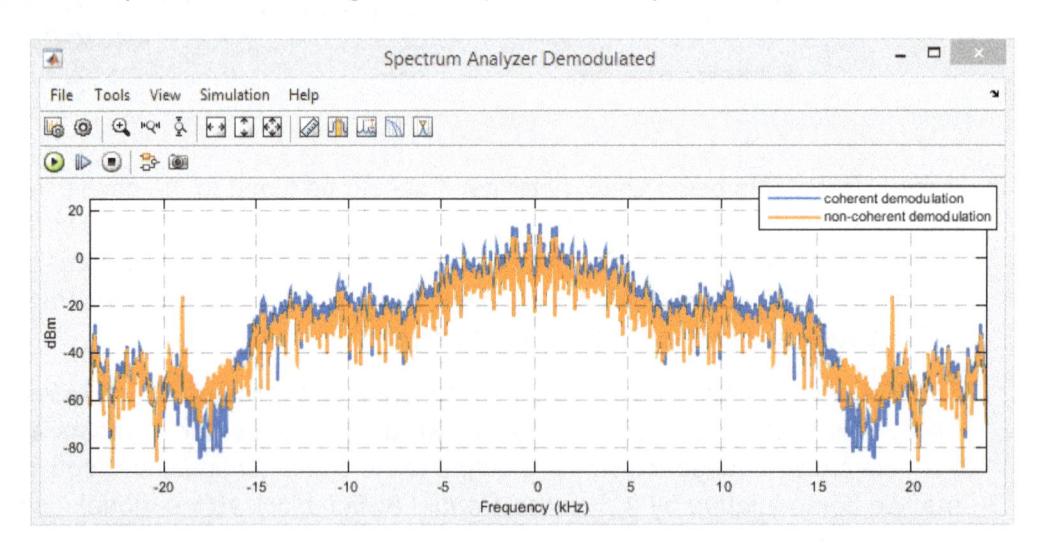

The blue line shows the coherently demodulated output in the frequency domain, while the orange line shows the non-coherently demodulated output. As the simulation runs, you may notice that the power level of the orange line drops drastically from time to time, and this is because of a phase mismatch between the modulating carrier and the local carrier used to

Chapter 10: Desktop FM Transmission and Reception 415

Local Carrier is NOT in Phase with Modulating Carrier

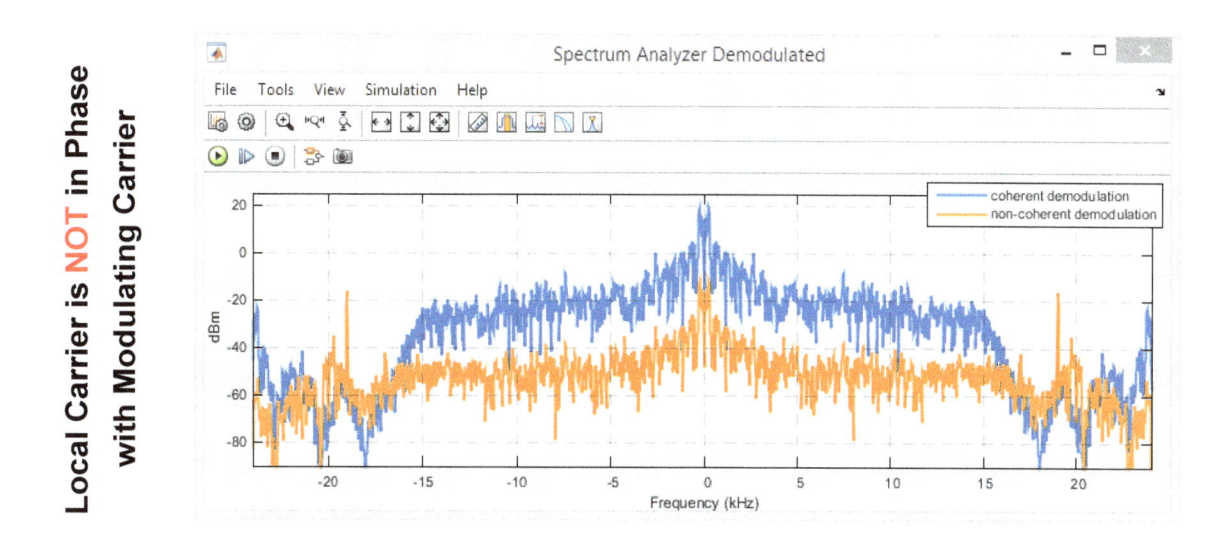

demodulate it. As the *FM MPX Demultiplexer* tracks the frequency and phase of the modulation carrier, this is not an issue with the coherent demodulation method.

(g) Switch between the demodulators using the *Manual Switch* block, and see if there is a noticeable difference in the output quality of the two signals. Do you notice a drop in volume in the non-coherent output when the phase mismatch occurs?

(h) **Test out other AM demodulation methods.** If you like, try switching out the demodulators for different ones. Try implementing any of the AM demodulators discussed in the *Desktop AM Transmission and Reception,* Chapter 8. *Phase Locked Loops* and *Costas Loops* can be found in the ▦ *> RTL-SDR Book Library > AM Sync Tools* library, and envelope detectors can be constructed with components from the ▦ *> Simulink* and ▦ *> DSP System Toolbox* libraries.

(i) **Watch and listen to us demodulate *AM in FM* MPX signals.** We have recorded a video which shows activity in the *Spectrum Analyzer* windows as the *AM in FM* MPX signal transmitted from the USRP® hardware (with the model from Exercise 10.9) is received by the RTL-SDR and demodulated. The output audio signal has been recorded too.

desktopSDR.com/videos/#amfm_demod

10.6 Manipulating the MPX: Audio Multiplexing with FM Transmitters

By nature of their design, stereo FM transmitters facilitate the transmission of two information signals simultaneously. Normally, *left* and *right* baseband signals (each band-limited to 15kHz) are encoded to create mono [L+R] and stereo [L−R] components, and these are then multiplexed together to allow both to be transmitted at the same time on a single FM carrier. The baseband stereo FM MPX is shown in Figure 10.6 on page 409.

If we were to consider the mono and stereo components as 15kHz wide communications channels, it opens the doors to the possibility of performing nested multiplexing. In Section 8.4 of *Desktop AM Transmission and Reception* (page 316), AM-DSB-TC modulation was used to multiplex 5kHz wide audio signals together; and although the focus of that section was to explore *complex* multiplexing, a similar approach could be taken here. Two 5kHz wide audio signals could be multiplexed into the mono component, and two could be multiplexed into the stereo component, as shown in Figure 10.8.

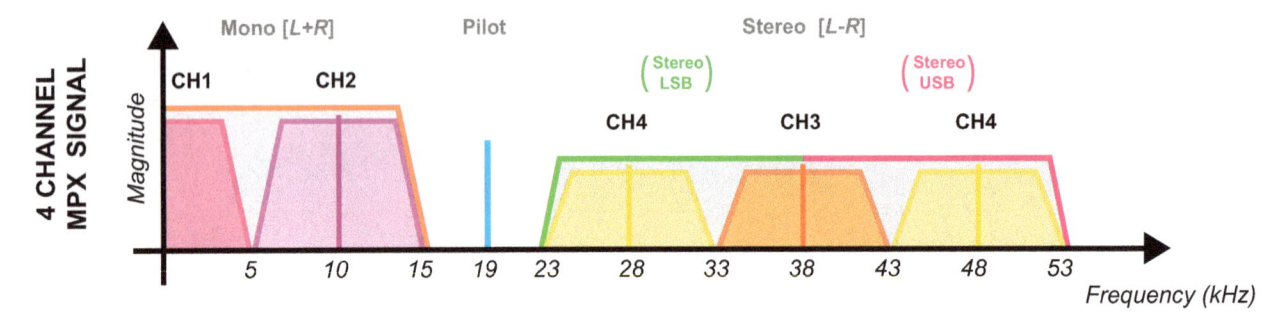

Figure 10.8: The baseband spectrum of the *FDM* stereo FM MPX

A special 'stereo' audio signal must be made for this *FDM* MPX to be generated with standard FM transmitter hardware. The four channels must be encoded and multiplexed into it as follows:

$$s_l(t) = s_{ch1}(t) + \left[1 + s_{ch2}(t)\right]\cos\left(2\pi f_{10k} t\right) \tag{10.7}$$

$$+ \left[1 + s_{ch3}(t)\right] + \left[1 + s_{ch4}(t)\right]\cos\left(2\pi f_{10k} t\right)$$

$$s_r(t) = s_{ch1}(t) + \left[1 + s_{ch2}(t)\right]\cos\left(2\pi f_{10k} t\right) \tag{10.8}$$

$$- \left[1 + s_{ch3}(t)\right] - \left[1 + s_{ch4}(t)\right]\cos\left(2\pi f_{10k} t\right)$$

When the 'stereo' signal is input to the stereo FM encoder, the following mono and stereo components are produced. These components are plotted in the frequency domain in Figure 10.9.

$$s_{mono}(t) = s_l(t) + s_r(t) = 2s_{ch1}(t) + 2\left[1 + s_{ch2}(t)\right]\cos\left(2\pi f_{10k} t\right) \tag{10.9}$$

$$s_{stereo}(t) = s_l(t) - s_r(t) = 2\left[1 + s_{ch3}(t)\right] + 2\left[1 + s_{ch4}(t)\right]\cos\left(2\pi f_{10k} t\right). \tag{10.10}$$

When these components are stereo FM multiplexed, the mono component is left at baseband, and the stereo component is AM-DSB-SC modulated onto a 38kHz subcarrier. This means that the channels will reside at the frequencies listed below. The AM-DSB-SC modulation process complicates matters slightly, as it creates an image of channel 4 between 23 and 33kHz!

channel 1	0Hz to 5kHz
channel 2	5kHz to 15kHz (dual sidebands)
channel 3	33kHz to 43kHz (dual sidebands)
channel 4	43kHz to 53kHz (dual sidebands)

Chapter 10: Desktop FM Transmission and Reception

Figure 10.9: The baseband spectra of the mono and stereo components

Substituting Eqs. (10.9) and (10.10) into the standard FM MPX equation, the complete *FDM* MPX signal can be described as:

$$s_{fdm\_fm\_mpx}(t) = 2s_{ch1}(t) + 2\Big[1 + s_{ch2}(t)\Big]\cos\Big(2\pi f_{10k}t\Big) \tag{10.11}$$

$$+ \left\{2\Big[1 + s_{ch3}(t)\Big] + 2\Big[1 + s_{ch4}(t)\Big]\cos\Big(2\pi f_{10k}t\Big)\right\}\cos\Big(2\pi f_{38k}t\Big)$$

$$+ \cos\Big(2\pi f_{19k}t\Big)$$

In Exercise 10.11 we will open and run a reference transmitter model featuring a signal generator/ multiplexer/ baseband FM modulator/ USRP®. This will create the special 'stereo' signal by encoding four different audio channels together, which is then input to the same set of blocks used in Exercise 10.5 for the standard stereo FM multiplexer and transmitter. The baseband FM modulated signal it generates is passed to the USRP® hardware, where it is modulated onto an RF carrier and transmitted.

This is a partner exercise for Exercise 10.12. If a USRP® radio is not available, there are a number of alternatives, such as an *off the shelf* stereo FM transmitter, or a Raspberry Pi. Information on these is given in Section 10.7. Even if no hardware is available with which to transmit an FM signal, it will still be possible to complete the later exercise using the data file provided. It is recommended to explore the USRP® transmitter file in any case, and confirm how the *FDM* MPX is generated.

Exercise 10.11 FDM FM: FDM & FM MPXer, Mod and USRP® Transmitter

In this exercise we will run a Simulink model that generates a custom 'stereo' signal containing four audio signals, and then encodes, multiplexes and FM modulates it, before transmitting the *FDM* MPX with the USRP® hardware. The model uses the standard stereo FM transmitter from Exercise 10.5, so we will not go into great detail discussing how the system works.

Note: If you do not have a USRP® radio, it will still be possible to run this exercise and see the FDM MPX signal being generated by commenting out the SDRu Transmitter block.

(a) **Open MATLAB.** Set the working directory to the exercise folder,

📁 `/fm/usrp_tx`

and then open the following file:

`.../usrp_fm_fdm.slx`

The block diagram of the model should look as shown below.

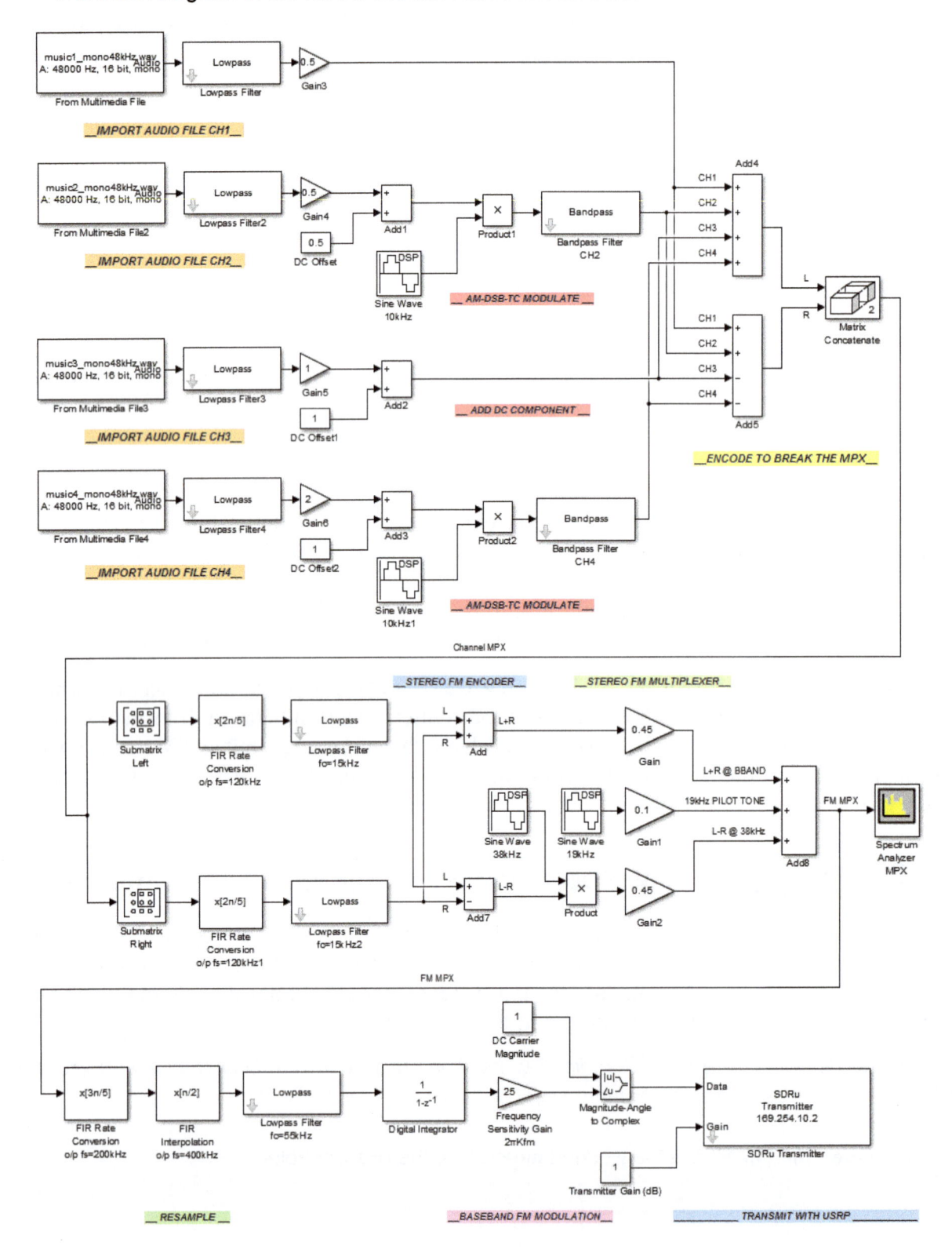

Chapter 10: Desktop FM Transmission and Reception

(b) **Four mono music signals** are imported using the *From Multimedia File* blocks, and then band-limited to 5kHz using *Lowpass Filters*. Channels 2 and 4 are AM-DSB-TC modulated onto 10kHz carriers, and then bandpass filtered to ensure that information is only present in the frequency range 5kHz to 15kHz. They, and the remaining channels, are then summed according to Eqs. (10.7) and (10.8) to create the 'left' and 'right' channels of the 'stereo' signal. The 'stereo' signal is then input to the block diagram of the encoder, multiplexer and FM modulator seen previously.

(c) **Check the USRP® radio settings.** You will need to configure the *SDRu Transmitter* block to communicate with your USRP® hardware. To do this, you will need to know its IP address (or USB address), and be able to 'ping' it from your computer. Guidance on setting up this connection is provided in Appendix A.2 (page 571). ② on the *SDRu Transmitter* block, and check that the appropriate 'Platform', and 'Address' for your USRP® radio are entered. Ensure that the 'Center Frequency' parameter is set to the desired transmit frequency (this value should be within the range of your RTL-SDR's tuner, e.g. in the range 25MHz–1.75GHz). Check that the USRP® hardware is turned on and that MATLAB can communicate with it by typing `findsdru` into the MATLAB command window. If a structure is returned as discussed in Appendix A.2 (page 571), continue to the next step.

If you are not working with a USRP® radio, comment out or delete this block before continuing.

(d) **Prepare to run the simulation.** Check that the *Simulation Stop Time* is set to 'inf', and that the *Simulation Mode* is set to 'Accelerator'. This will permit the model to run faster. Finally, make sure that the model is saved.

(e) **Run the simulation.** Begin the simulation by ① on the 'Run' ▶ button in the Simulink toolbar. After a couple of seconds, Simulink will establish a connection with the USRP® radio and the simulation will begin. When it does, samples of the baseband FM signal will be passed to the radio, modulated onto an RF carrier, and transmitted from its antenna.

(f) **Signal Analysis.** A single *Spectrum Analyzer* window will appear, which will allow you to monitor the *FDM MPX* as it is generated and transmitted. It should be possible to clearly identify each of the four channels (use Figure 10.8 to help with this).

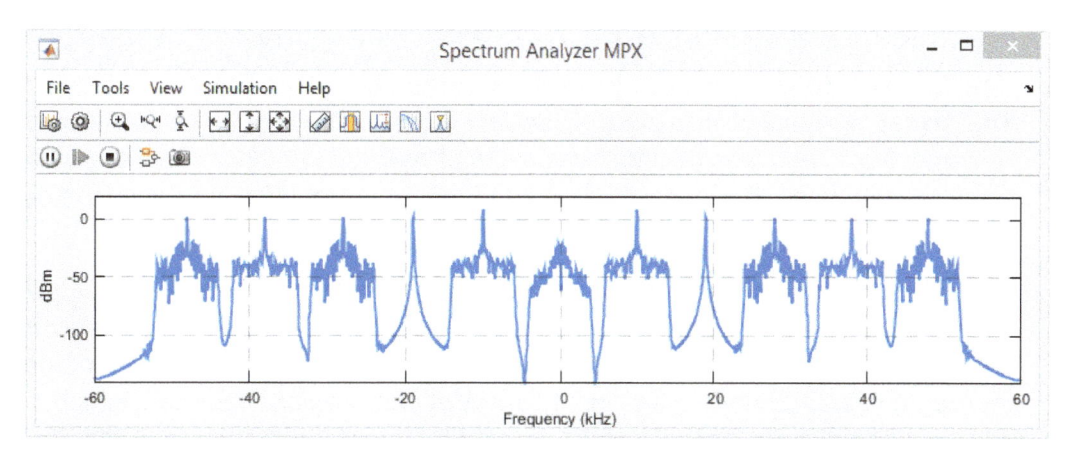

(g) If an analogue FM radio was tuned to this signal, it would output the signals from Eqs. (10.7) and (10.8), and would sound awful! Only the custom demodulator/ channel selector model that we will present in Exercise 10.12 can be used.

As we are transmitting a custom signal format, a custom receiver is obviously needed! Like the receiver for the *AM in FM* MPX, the receiver for the *FDM* MPX requires a two stage demodulation process. An FM demodulator is needed to recover the MPX, and then a series of AM demodulators are needed to demodulate each of the four audio channels from their respective carriers to baseband.

In Exercise 10.12, we will run a receiver for the *FDM* MPX signal that will demodulate each of the individual audio channels, and allow output one at a time to computer speakers. If you are unable to transmit the *FDM* MPX signal either using the USRP® transmitter, an off the shelf FM transmitter or a Raspberry Pi, you can still complete the exercise using the provided data file.

Exercise 10.12 FDM FM: RTL-SDR FM Receiver and Demultiplexer

In this final FM exercise, we will run a receiver that will demodulate each of the channels transmitted in the *FDM MPX* signal. It will be possible to switch between the channels in real time as the simulation runs. This receiver has been designed to output the demodulated audio signal to computer speakers or headphones.

Note: If you do not have an RTL-SDR, or are unable to transmit the FDM MPX signal from Exercise 10.11, you can still complete this exercise by substituting the RTL-SDR Receiver block with an Import RTL-SDR Data block as outlined below.

(a) **Open MATLAB.** Set the working directory to the exercise folder,

   `./fm/rtlsdr_rx`

   and then open the following file:

   `.../rtlsdr_fm_fdm_demod.slx`

   The block diagram of this receiver should look like the screenshot on the following page.

(b) The RTL-SDR (if used) is tuned to the centre frequency of the FM modulated *FDM MPX* signal. The signal is sampled at a rate of 2.4MHz (to aid with tuning), output from the *RTL-SDR Receiver* block, and then decimated to a sampling rate of 240kHz. If the *Import RTL-SDR Data* block is being used, it is already sampled at the required 240kHz rate, so the decimator is not required.

The FM signal is demodulated with a frequency discriminator, which recovers the *FDM MPX*. After reducing the sampling rate, the MPX enters a bank of filters, where each of the individual channels are isolated. Channels 2, 3 and 4 are AM demodulated from their carriers, and all of the demodulated channels are input to a *Multiport Switch*. You can switch between these demodulated outputs as the simulation runs by changing the value of the *Channel Selector* constant.

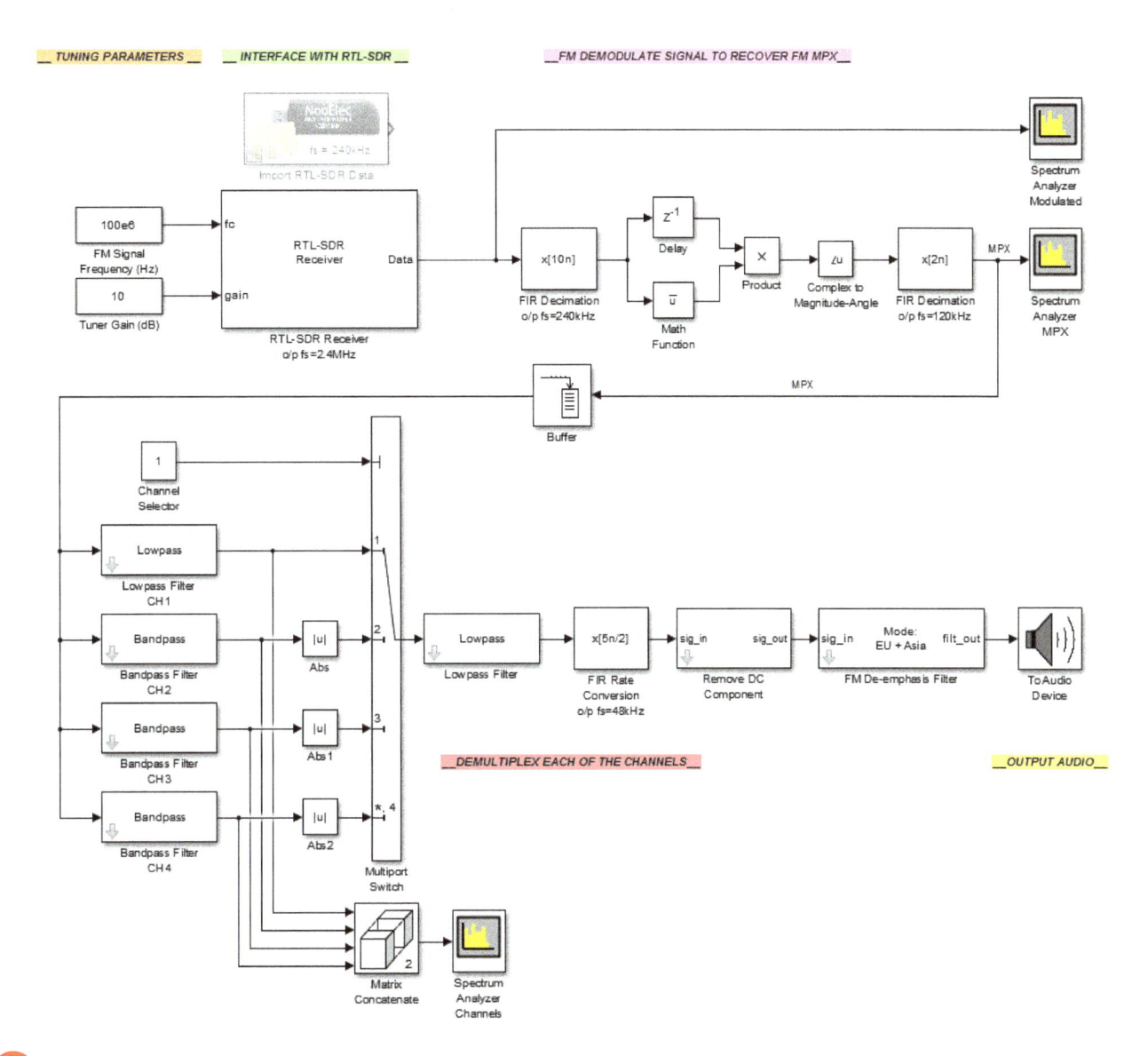

Chapter 10: Desktop FM Transmission and Reception 421

🔧 **(c) Check the RTL-SDR data source.** If you have an RTL-SDR available and are able to transmit the *FDM MPX* signal, you will be able to use an *RTL-SDR Receiver* block to interface with and control your RTL-SDR. As outlined previously, there are a number of possible ways to create this signal. If using the USRP® radio, run the model from Exercise 10.11. If an off the shelf FM transmitter is used, plug it into the device from which the special 'stereo' file will be played. If you are using the Pi, copy the file across to the Pi using an SSH client.

If it is not possible to transmit the signal using any of these methods, the *Import RTL-SDR Data* block should be used instead. This should be configured to reference the following file:

📁 `/fm/rtlsdr_rx/rec_data/wfm_fmfdm.mat`

If importing data rather than connecting to an RTL-SDR, delete the *RTL-SDR Receiver* block and connect the *Import RTL-SDR Data* block in its place. Delete the *FIR Decimation o/p fs=240kHz* block as it is not needed, and connect the output of the *Import RTL-SDR Data* block directly to the inputs of the frequency discriminator.

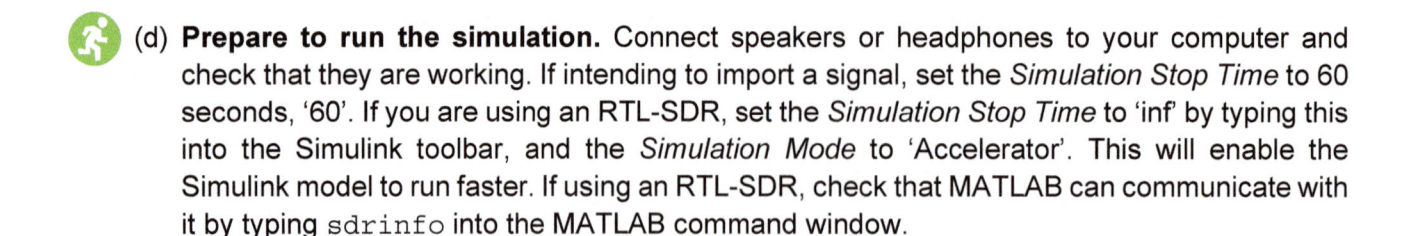

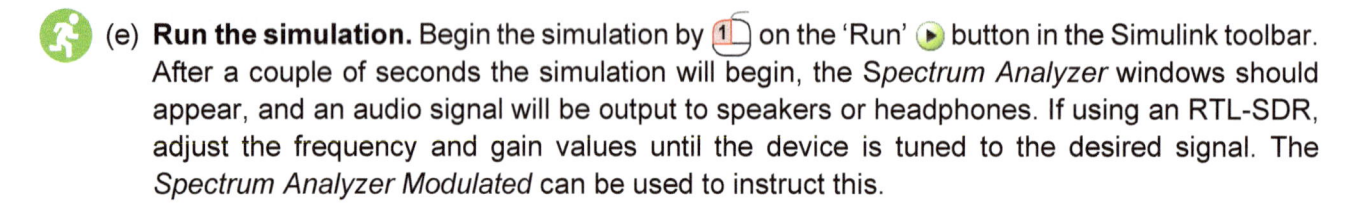

(d) **Prepare to run the simulation.** Connect speakers or headphones to your computer and check that they are working. If intending to import a signal, set the *Simulation Stop Time* to 60 seconds, '60'. If you are using an RTL-SDR, set the *Simulation Stop Time* to 'inf' by typing this into the Simulink toolbar, and the *Simulation Mode* to 'Accelerator'. This will enable the Simulink model to run faster. If using an RTL-SDR, check that MATLAB can communicate with it by typing `sdrinfo` into the MATLAB command window.

(e) **Run the simulation.** Begin the simulation by clicking on the 'Run' button in the Simulink toolbar. After a couple of seconds the simulation will begin, the *Spectrum Analyzer* windows should appear, and an audio signal will be output to speakers or headphones. If using an RTL-SDR, adjust the frequency and gain values until the device is tuned to the desired signal. The *Spectrum Analyzer Modulated* can be used to instruct this.

(f) **Signal Analysis: *Spectrum Analyzers*.** The *Spectrum Analyzer MPX* window should show something similar to the outputs from Exercise 10.11. *Spectrum Analyzer Channels* will show the four isolated channels, recovered after stereo FM demultiplexing has been performed. Channel 1 will be positioned at baseband, and channels 2, 3 and 4 will remain modulated on their 10kHz, 38kHz and 43kHz subcarriers. Zooming into the region of activity, you should be able to see the carrier peaks, as they were AM-DSB-TC modulated during the initial signal generation stage.

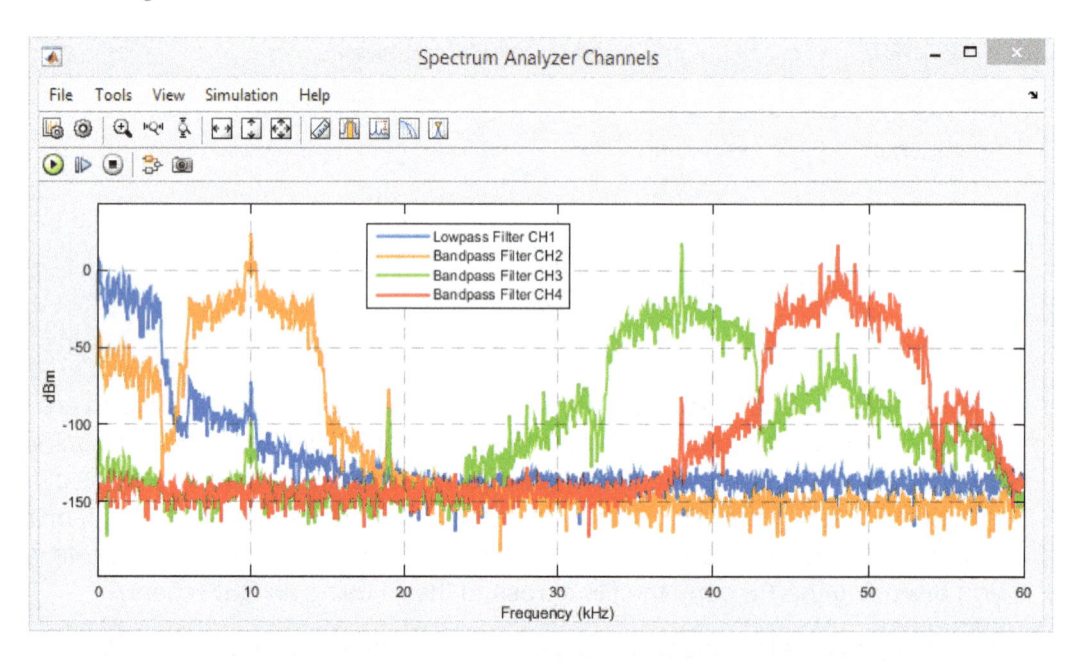

This scope is equivalent to the plot from Figure 10.8, with the image of channel 4 removed. If the signal generation and transmission processes were successful, each of the channels should be independent of the others.

(g) Can you clearly hear each of the music signals when changing the value of the *Channel Selector* constant? Do any of the channels sound of a poorer quality than the others?

Does it sound like there is more than one song being output at once? If this is the case, either the *FM MPX Demultiplexer* block is not able to lock onto the pilot tone correctly, or something

Chapter 10: Desktop FM Transmission and Reception

has gone wrong at the transmitter side. This is more likely to happen if you are using an *off the shelf* FM transmitter or a Raspberry Pi than with the USRP® hardware, as the integrity of the stereo multiplexing cannot be assured with these devices. If the results seem poor or unexpected, try importing RTL-SDR data instead, and make a comparison.

(h) **Watch and listen to us demodulate the *FDM MPX* signal.** We have recorded a video which shows activity in the *Spectrum Analyzer* windows as the *FDM MPX* signal transmitted from the USRP® hardware (with the model from Exercise 10.11) is received by the RTL-SDR and demodulated. The output audio signal has been recorded too.

desktopSDR.com/videos/#fmfdm_demod

10.7 Alternative Hardware for Generating Desktop FM Signals

As was discussed in the introduction to this section, having the facility to transmit your own FM signals can be very useful. Although the USRP® radio is most capable, there are some alternatives, and we will discuss these below. These FM transmitters can be used in place of the USRP® transmitter in Exercises 10.1, 10.5, 10.9 and 10.11.

10.7.1 'Off the shelf' FM Transmitters

One option is to purchase an *off the shelf* FM Transmitter. These devices are designed to allow modulation and transmission of information signals from an audio source such as a phone, mp3 player or laptop onto an FM carrier residing within the standard FM band (88MHz to 108MHz). They come in all shapes and sizes, some are battery powered while others draw power over USB; and they can be obtained for as little as £5/ $8 from outlets such as radioshack.com or amazon.com.

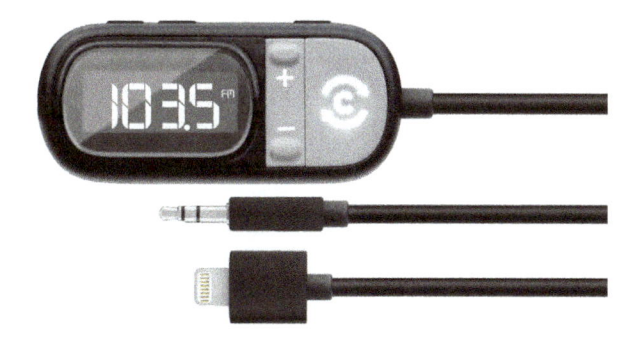

Figure 10.10: An *off the shelf* FM Transmitter [76]

For just a few dollars more, devices such as the *Belkin TuneCast* are programmable in the sense that you can set the frequency of the carrier. This can be convenient if the FM band in your area is particularly active and you are wary of transmitting on top of another station and interrupting your neighbour's favourite shows! In addition to running Exercises 10.1, 10.5, 10.9 and 10.11 with these devices, we will also develop some interesting examples in Section 12.9 (page 556) in which we transmit images using these FM transmitters, and then receive them using a complete digital communications receiver.

10.7.2 Raspberry Pi FM Transmitter: PiFM

The *PiFM* project was originally developed by Oliver Mattos and Oskar Weigl of Imperial College (London) Robotics Society back in 2012 [93]. It enables the *Raspberry Pi* to be used as a low power FM transmitter, and is possible thanks to a clever hack of the clock signals that are used to control the General Purpose Input Output (GPIO) pins. The pins are configured to perform Pulse Width Modulation

Figure 10.11: The Raspberry Pi, Model B

(PWM) to AM modulate a baseband FM MPX signal onto a carrier (similar to how the USRP® hardware creates an FM signal). The PWM signal is applied to GPIO pin 4, and after connecting an antenna to this pin, the FM radio station is transmitted from the Pi, with a range of about 75m. A configuration file allows users to set the frequency of the carrier to any value in the range 1MHz–250MHz.

The original PiFM code supported only *.wav* audio files, and required user interaction via the terminal on the Pi, both to configure and to initiate an FM transmission. In 2014 the project was revisited by an engineering intern at MAKE Labs named Wynter Woods, and rebranded as 'Pirate Radio'. He modified the source code to enable support for other audio file formats (*.mp3, .flac, .wma, .aac* and *.m4a*), and wrote the *PirateRadio.py* script that automatically runs the PiFM program as soon as the Pi has powered on. This autoplay feature meant that terminal commands were no longer required to get an FM station up and running.

MAKE Labs have published a disk image which can be downloaded from the Pirate Radio project page on their website [97]. This image is optimised in the sense that it contains boot, system and data partitions, and comes with the PiFM software pre-installed. When the authors tried downloading and using the MAKE Labs disk image, it only took about 30 minutes to get the Pi transmitting an FM signal. There were several steps in this process, which admittedly would have been simpler if we were working in a Unix computer environment, as PiFM runs in Arch Linux! A step-by-step setup guide for Microsoft Windows OS users can be found in Appendix E.3 (page 619).

Further information about this project can be found on the MAKE Labs project page [97].

10.7.3 Transmitting FM Signals

If an FM transmitter available with an audio jack input is available, you can transmit any of the audio sources provided in the

📁 /audio_sources

folder. Simply copy these to your MP3 player or smartphone, connect it up, and press play. If using a Raspberry Pi, simply transfer the files through an SSH client in Appendix E.3 (page 619).

If you wish to try transmitting an *AM in FM* signal (which can be received and demodulated with the model from Exercise 10.10), open and run the following file:

> `/fm/other_tx/fmtx_fm_amfm.slx`

To experiment with transmitting an *FDM MPX* signal (which can be received and demodulated with the model from Exercise 10.12), open and run the following file:

> `/fm/other_tx/fmtx_fm_fdm.slx`

Both of these will create stereo `.wav` audio files, which can be transmitted using the methods described earlier.

10.8 Summary

In this chapter we have presented exercises implementing various FM desktop SDR Tx/ Rx systems. We began by developing a standard mono WFM transmitter, and testing out four different receiver architectures, including one which is actually an AM demodulator! Additionally, by developing a stereo FM encoder and multiplexer, a stereo audio signal comprising of two different music files was transmitted (one for the left channel, and another for the right) with the USRP® radio and received by the RTL-SDR. The Simulink receiver was able to synchronise to the pilot tone in the FM MPX signal, and demodulate and output each of the individual channels. We also considered how a PLL could be used to demodulate an FM signal, before moving on to more abstract systems which focused on breaking the FM MPX.

With a look forward to the remaining chapters of the book, we will now switch focus from analogue to digital modulation schemes. FM will be seen again in Section 12.9 (page 556), when we develop some interesting designs to send images with a simple FM transmitter, featuring digital modulation and receiver designs.

11 Digital Communications Theory and Simulation

Wireless digital communications systems are all around us, providing voice and data connectivity via cellular phone systems, WiFi, Bluetooth, and many other standards and proprietary schemes. In this chapter, we look at some of the building blocks for digital communications, including modulation schemes, symbol and bit reception and performance in noise, differential encoding and decoding, multirate digital up- and downconverters, and more.

Digital receiver design is considered in detail, including the aspects of coarse and fine carrier frequency synchronisation, and symbol timing synchronisation. The methods developed in this chapter will later be used in Chapter 12 when we go on to develop a full baseband receiver to interface with the RTL-SDR hardware, and receive 'live' digital radio signals.

The basis for most of our discussion of digital communications is the Quaternary Phase Shift Keying (QPSK) modulation scheme, to be introduced in the next section, although others are adopted when needed.

11.1 Digital Modulation Schemes

Digital communications systems operate by mapping binary data bits to *symbols* to convey information. The method of mapping bits to symbols is defined by the chosen modulation scheme.

In this chapter, we will consider two prominent categories of digital modulation schemes: those based on *phase*, and those based on *amplitude*. Both are defined on axes with an I component (*x*-axis), and a Q component (*y*-axis), which are equivalent to the real and imaginary axes in the complex plane, respectively. Other modulation possibilities include schemes that combine aspects of both amplitude and phase; offset schemes, in which the I and Q phases change at offset time intervals; Frequency Shift Keying (FSK) schemes; and OFDM [40].

11.1.1 Digital Phase Modulation Schemes

Common examples of digital phase modulation schemes are:

- Binary Phase Shift Keying (BPSK)   1 bit per symbol
- Quaternary Phase Shift Keying (QPSK)   2 bits per symbol
- 8-Phase Shift Keying (8PSK)   3 bits per symbol

In each case, 1, 2, or 3 bits are grouped together to form a symbol, and mapped to one of the points in the constellation as shown in Figure 11.1 (BPSK), Figure 11.2 (QPSK), and Figure 11.3 (8PSK). Note from the diagrams that QPSK and 8PSK constellations may be defined in two different ways, where constellation points are situated either on or off the axes.

The scheme adopted for mapping symbols to constellation points can also vary. For example, the symbols '00', '01', '10' and '11' may be mapped to the constellation points differently to the assignments shown in Figure 11.2.

Notice that, in each of the BPSK, QPSK and 8PSK examples, the set of constellation points is spread out equally in phase. For instance, the 8PSK constellation points are each separated by an angle of 45°. Higher order PSK constellations are also possible. The constellation points are equidistant from the origin, meaning that all symbols have the same energy.

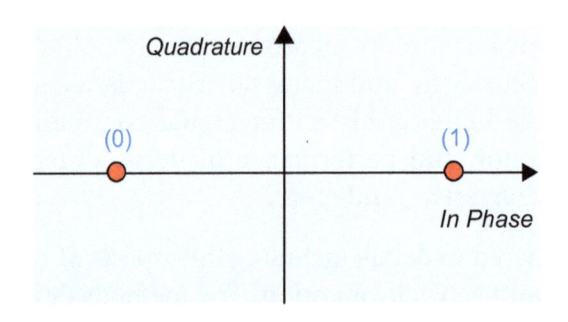

Figure 11.1: BPSK constellation

11.1.2 Digital Amplitude Modulation Schemes

The most popular amplitude-based digital modulation schemes are:

- 4-position Quadrature Amplitude Modulation (4-QAM)   2 bits per symbol
- 16-position Quadrature Amplitude Modulation (16-QAM)   4 bits per symbol
- 64-position Quadrature Amplitude Modulation (64-QAM)   6 bits per symbol

These schemes usually result in the constellation points being laid out in a square grid (variations are possible, but less common). These are shown in Figure 11.4 (a), (b) and (c) for 4-QAM, 16-QAM and 64-QAM, respectively. As with the PSK schemes, the mapping of symbols '0000', '0001' etc. to constellation points can vary from the examples shown here.

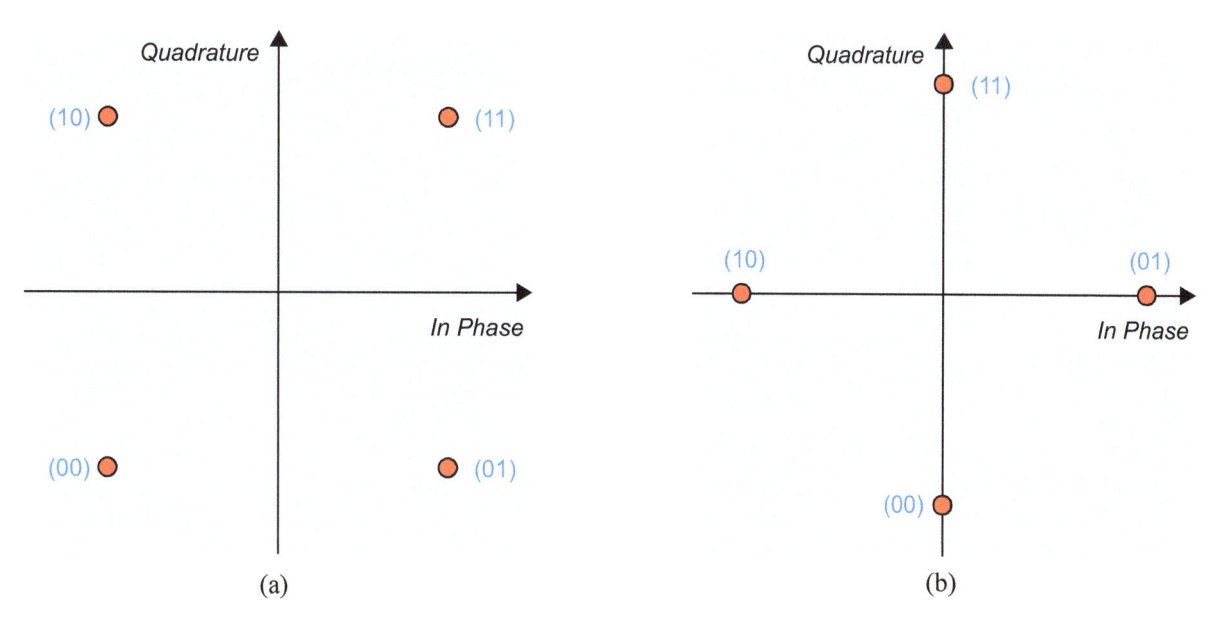

Chapter 11: Digital Communications Theory and Simulation

Figure 11.2: QPSK constellations, with symbol points: (a) at the corners of a square; (b) situated on the axes

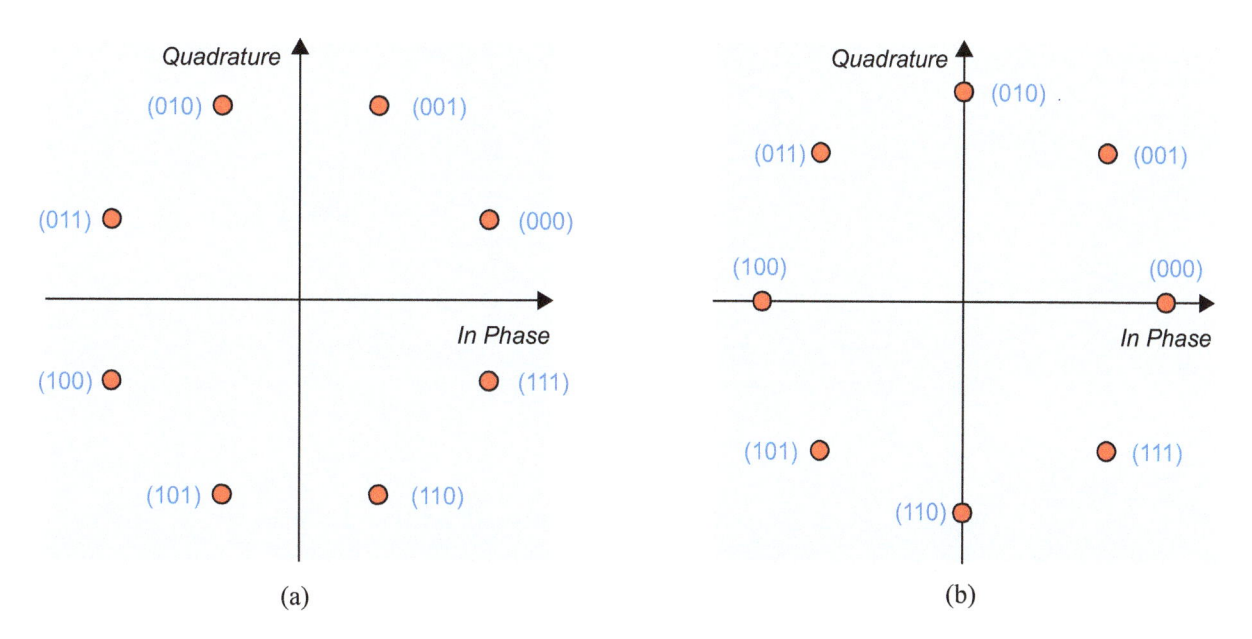

Figure 11.3: 8PSK constellations, with symbol points: (a) off the axes; (b) situated on the axes

It should be noticeable that the constellations shown in Figure 11.2(a) and Figure 11.4(b) are identical. It is true that QPSK (in the square, off-axes format) and 4-QAM are equivalent constellations, even though one is based on a phase difference and the other on amplitude.

For the QAM constellations listed earlier, symbols are formed by taking 1, 2, or 3 bits on each of the I and Q phases, to represent amplitude on that phase. For instance, 4-QAM (and equivalently QPSK) has 1 bit per phase, resulting in 2 different amplitudes on each phase, as shown in Figure 11.5. Similarly, the 64-QAM constellation has 8 different amplitudes on each of the I and Q phases (represented by 3 bits, where $2^3 = 8$). Most of our discussion will focus on QPSK, which is intuitive and widely used.

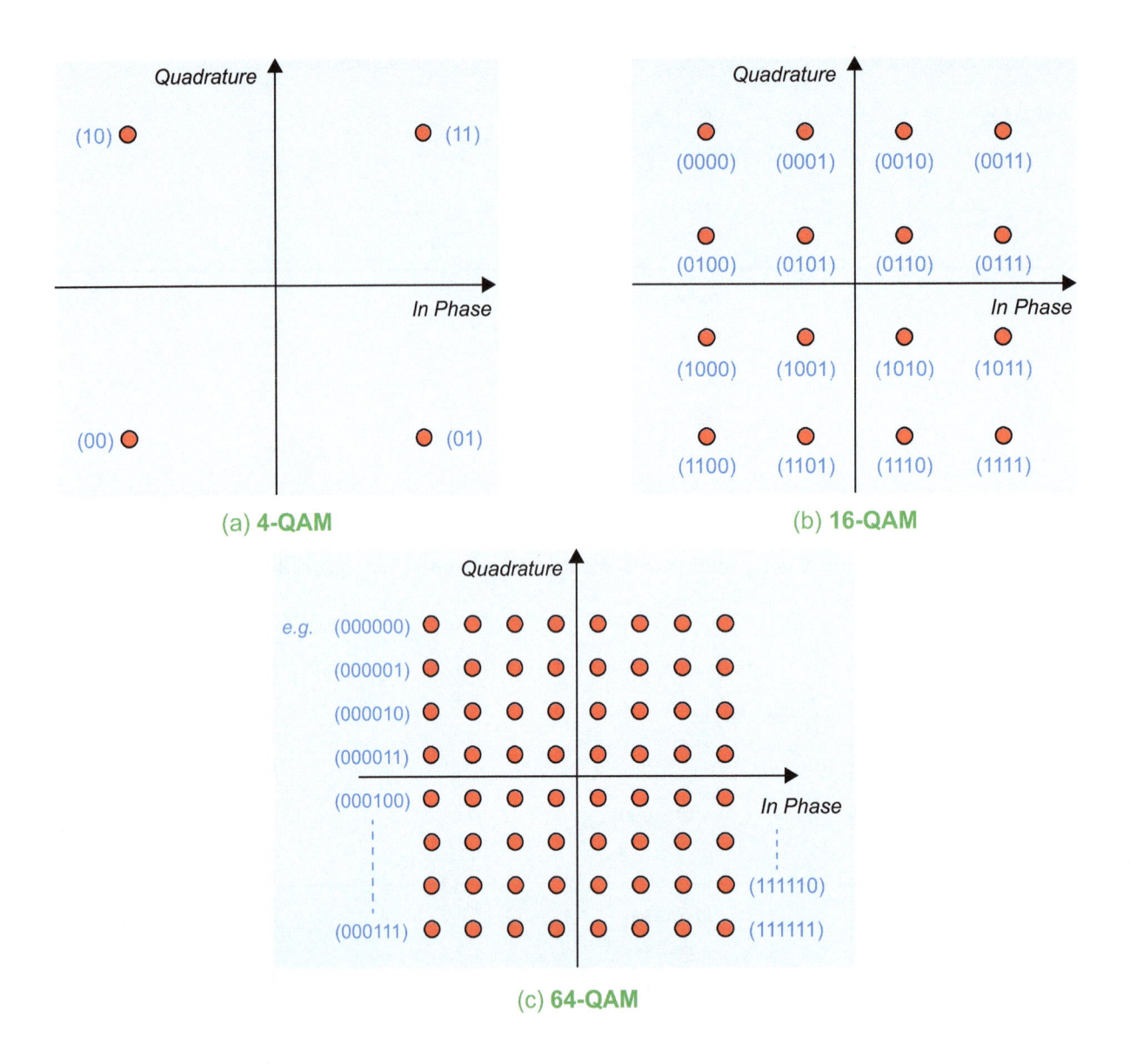

Figure 11.4: Quadrature Amplitude Modulation (QAM) constellations

11.1.3 Bit to Symbol Mapping and Demapping

Each of the schemes outlined so far is formed by mapping bits to symbols. The number of bits mapped to each symbol depends on the modulation scheme employed, ranging from 1 bit per symbol for BPSK, to 6 bits per symbol for 64-QAM.

Assuming that the symbol rate (or signalling rate) is equal, then the 64-QAM scheme can convey 6 times more information than the BPSK scheme, which is clearly an advantage. In other words, its throughput is 6 times higher. Higher order QAM modulation schemes such as 256-QAM, which has 8 bits per symbol, can provide even higher throughput.

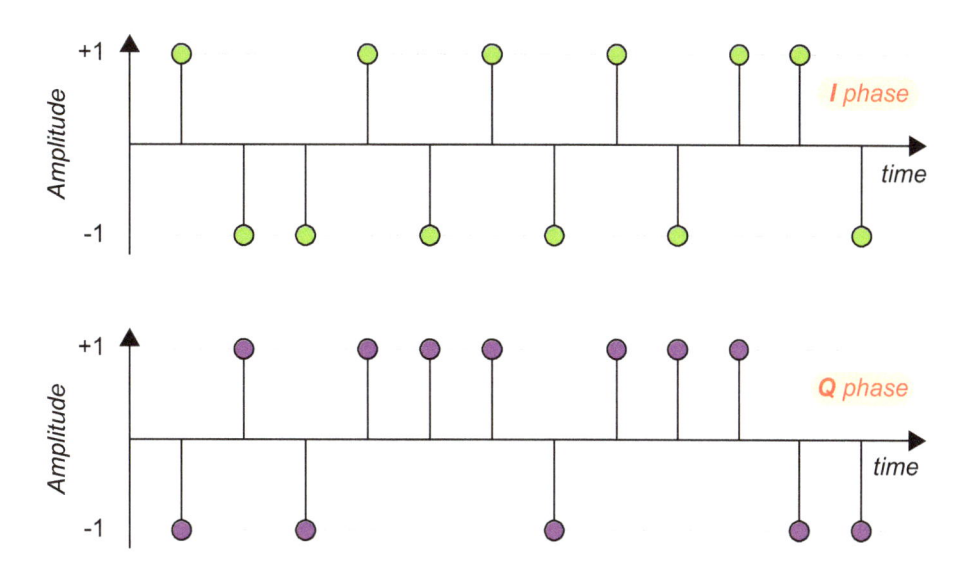

Chapter 11: Digital Communications Theory and Simulation

Figure 11.5: QPSK symbols transmitted as {+1,-1} amplitudes on the I and Q phases

The bit to symbol mapping process can be thought of as separating the binary data into groups of bits, and then transmitting the appropriate symbol. The *de*mapping process is the reverse, i.e. deciding which symbol was sent, based on a received symbol sample. Normally the received symbol samples contain some degree of error, as a result of noise, other channel effects, and symbol timing errors (the last of these to be covered when we consider timing synchronisation in Section 11.6).

Symbol demapping is performed by comparing each received symbol sample with *decision boundaries*. This is equivalent to deciding which constellation point the received sample is closest to. For simple modulation schemes such as BPSK and 4-QAM/QPSK, the decision boundaries are equivalent to the axes (the decision boundary for BPSK is the *y-axis*, while the boundaries for QPSK are the *x-* and *y-axes*, as shown in Figure 11.6(a)).

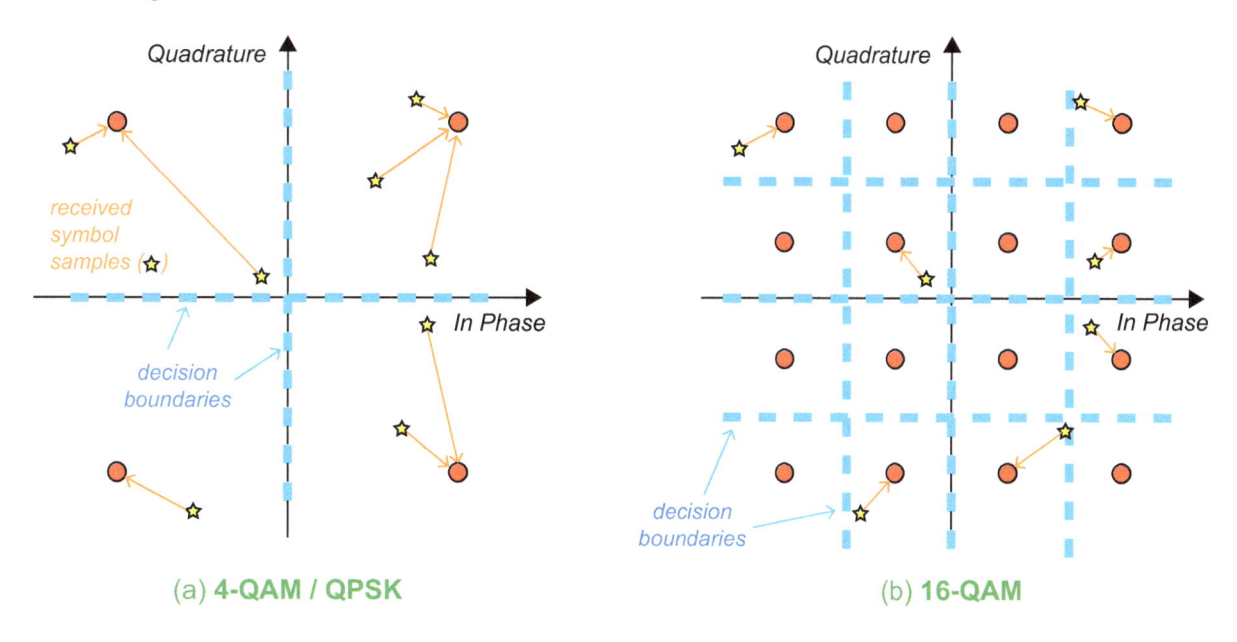

(a) **4-QAM / QPSK**

(b) **16-QAM**

Figure 11.6: Decision boundaries and symbol demapping for: (a) QPSK and (b) 16-QAM

Similarly, the decision boundaries for 16-QAM are shown in Figure 11.6(b). Here, the constellation points are more closely spaced, meaning that the demapping process is more susceptible to symbol detection errors. Thus, a given amount of AWGN can be more destructive in a higher order modulation scheme, because it more rapidly spreads the received samples beyond decision boundaries, causing the wrong symbol decisions to be made. We will explore this point further in the coming exercises.

Exercise 11.1 Bit to Symbol Mapping and Demapping (QPSK)

This exercise demonstrates bit-to-symbol mapping and demapping of QPSK (which is equivalent to 4-QAM), using the facilities of the Communications System Toolbox.

(a) **Open the system:**

   `/digital/simulation/modulation/QPSK_map_demap.slx`

(b) **View the design.** Inspect the design and note that a binary source is used to model the data bits for transmission. This model uses frame-based processing, with the source producing frames containing 2 samples (bits). This is because 2 bits correspond to 1 symbol.

(c) **Identify rates...** Can you establish what the **bit rate** and **symbol rate** are in this model?

(d) **QPSK Modulation.** Open the *QPSK Modulator Baseband* block to view its parameters, and then select the option to 'View Constellation', and inspect the output:

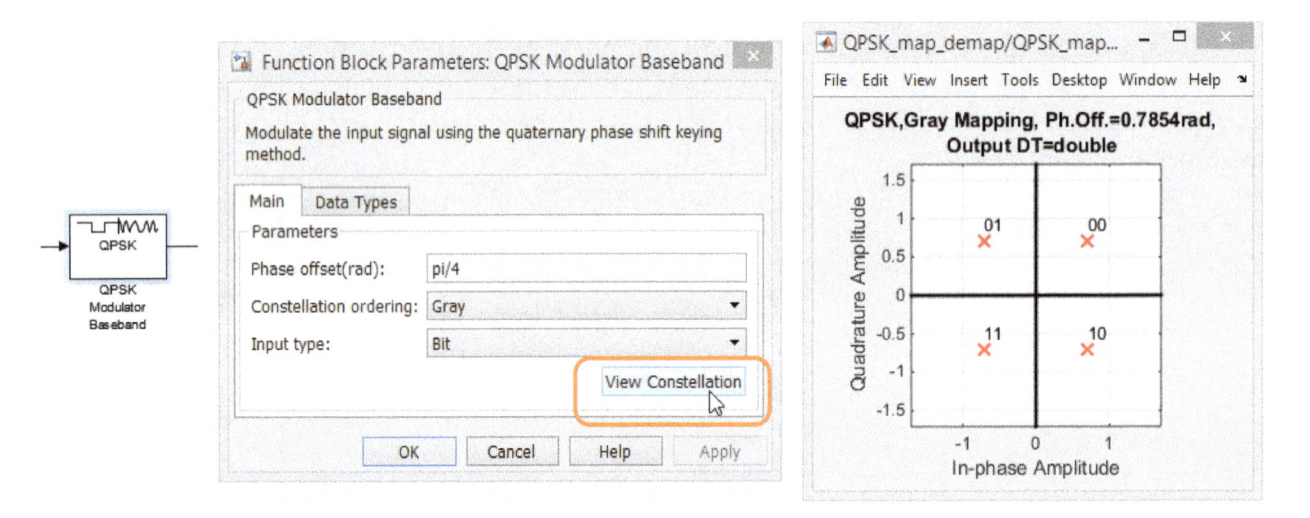

(e) You should be able to confirm that there are four constellation points, situated at approximately (±0.707, ±0.707). This means that the magnitude of each symbol is equal to 1, i.e. the symbols are all located a distance of

$$\sqrt{0.707^2 + 0.707^2} = 1$$

from the origin.

The mapping of bits to symbols should also be shown on the constellation (note that the mapping schemes used may differ from Figure 11.2 / Figure 11.4).

Chapter 11: Digital Communications Theory and Simulation

(f) **QPSK Demodulation.** Open the *QPSK Demodulator Baseband* block, and check that its parameters correspond to those of the *Modulator* block. This means that the *QPSK Demodulator* will perform the reverse process to the *QPSK Modulator*.

(g) **Simulate!** Run a simulation and view the outputs. You should be able to confirm from the *Time Scope* that the transmitted data is exactly recovered by the demapping process.

(h) Open the constellation plots and view the results. Note that the reference (expected) points are shown as red '+' signs (+), while the symbol points observed during the simulation are given by blue circles (o). They should be superimposed for all symbols, meaning that the received constellation is perfect (fully expected here, as the channel is just a wire!).

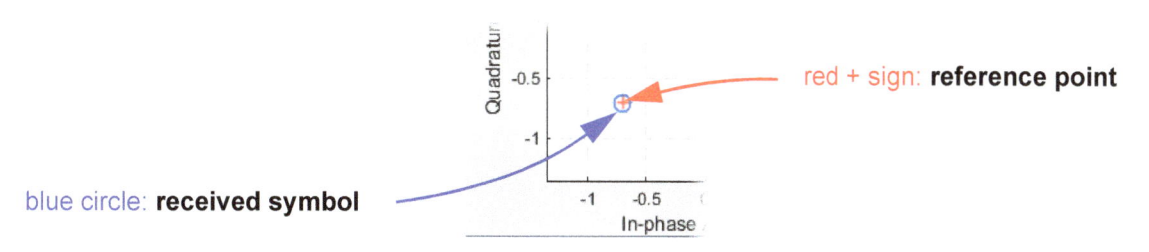

red + sign: **reference point**

blue circle: **received symbol**

(i) **The complex data type.** Finally, switch data type annotations on (*Display -> Signals & Ports -> Port Data Types*) and confirm that the symbols output by the *QPSK Modulator* are **complex**. This means that each sample comprises a *real* part (for the I component) and an *imaginary* part (for the Q component).

In the next exercise, we will configure the model differently, so that it has explicit I and Q phases, similar to Figure 11.5.

Exercise 11.2 QPSK Symbol Mapping and Demapping: Separate I and Q

So far we have considered a model with frame-based processing, that uses complex symbols. This exercise demonstrates an alternative representation in which the symbols are explicitly separated into their *In Phase* and *Quadrature* components. We will also step through the simulation slowly, to confirm how the changing symbols and IQ amplitudes are related.

(a) Open the system:

/digital/simulation/modulation/QPSK_map_demap_IQ.slx

(b) Inspect the model, and note that conversions between Complex and Real/Imaginary have been undertaken to split the signal into its In Phase and Quadrature components. Notice also the use of an *Unbuffer* block to change between frames and samples, and then a *Buffer* block to change back into frames.

(c) **Simulate!** Run a simulation and view the constellation plot, and the *Time Scope* comparing the transmitted and received bits. You should be able to confirm that the system operates as before (just with a little extra delay due to the unbuffering and buffering operations).

(d) Now view the *Time Scope IQ* block to view the individual phases. You should notice that the values transmitted on I and Q are both +0.707 and -0.707, as indicated by the constellation plot.

(e) **View Port Data Types.** Switch on port data types (*Display -> Signals & Ports -> Port Data Types*). Are the signals representing the I and Q components real or complex?

(f) 🖱 on the *Constellation Diagram*. Navigate to 'Configuration Properties', and change the 'Symbols to display' to '1'. This will limit the *Constellation Diagram* to plotting only one symbol at a time.

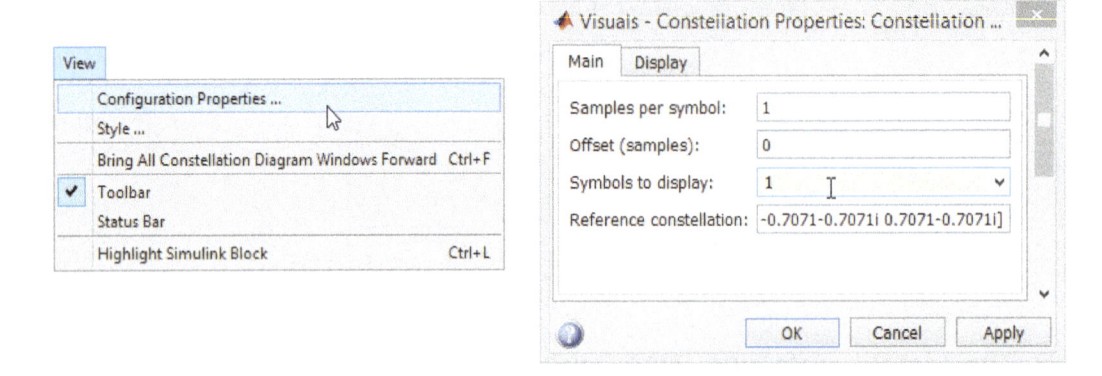

(g) **Re-simulate.** Next, run the simulation again. This time instead of 🖱 on the normal 'Run' button, 🖱 on the 'Step Forward' ▶ button instead. The simulation runs in what can be considered a 'debugging' mode, where it pauses after processing every sample. Position the *Constellation Diagram* and *Time Scope IQ* windows alongside each other, and view the changing symbols and IQ amplitudes on the constellation at the same time.

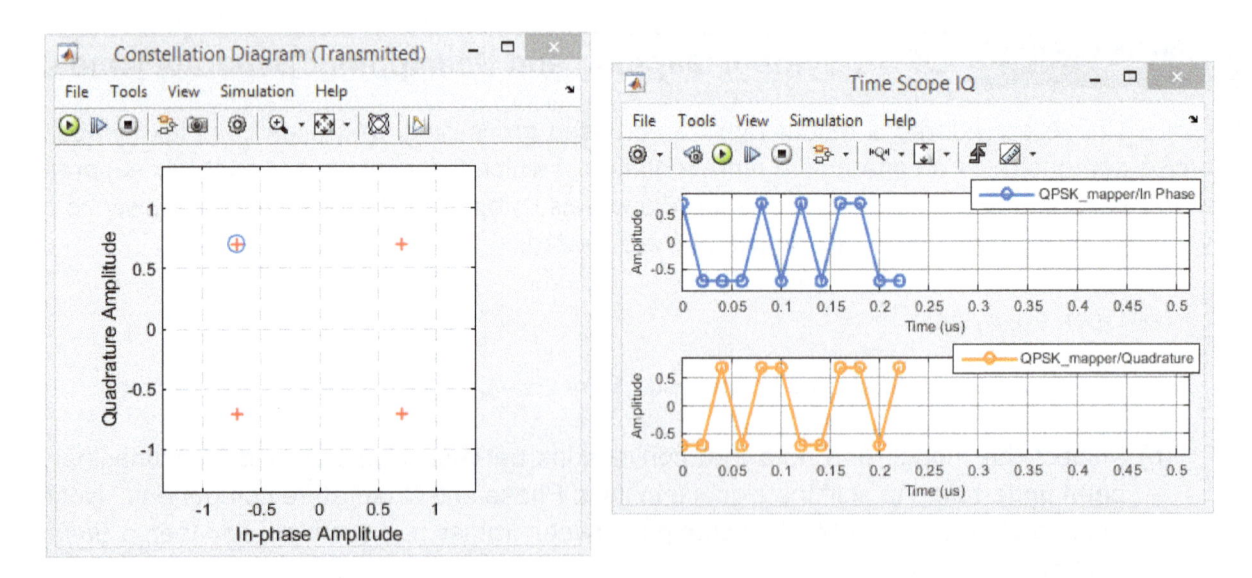

(h) You may also wish to compare the results with the mapping of bits to symbols defined in the QPSK modulator block (or refer back to the screenshot from Exercise 11.1).

Chapter 11: Digital Communications Theory and Simulation

Exercise 11.3 Bit to Symbol Mapping and Demapping in Noise (QPSK)

This exercise builds on the previous models, by adding noise in the channel. We view the impact on the received symbol constellation, and the success of recovering the transmitted bits. Note that the model returns to the complex representation (i.e. similar to Exercise 11.1), which enables a simpler block diagram.

(a) Open the system:

  /digital/simulation/modulation/QPSK_constellation_noise.slx

(b) You should notice that the channel (which was previously just a wire!) has now been made more realistic, with the addition of an *AWGN Channel*.

(c) **Set the parameters of the added noise.** Open the *AWGN Channel* block and view its parameters. The block should initially be configured to implement an E_b/N_0 figure of 10dB (or if this is not already set, please change it to 10dB). This means that a degree of noise will be injected into the transmission path.

(d) **Simulate!** Now run a simulation and view the outputs. You should notice the symbol constellation 'spreading out', something like the following:

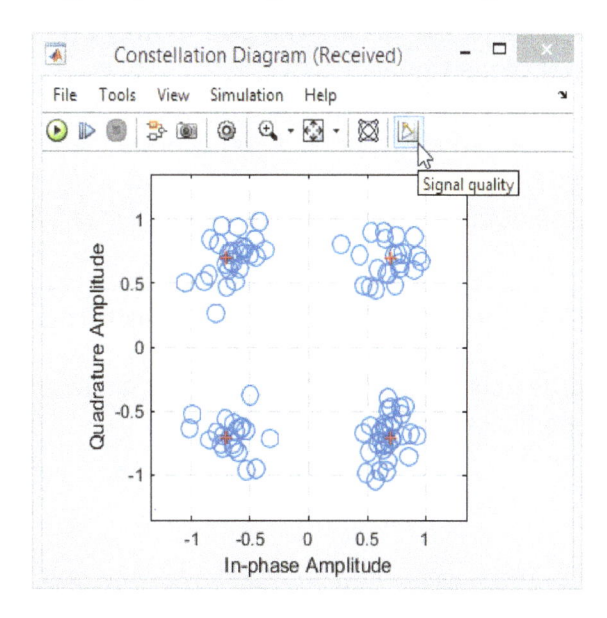

(e) You may also wish to try adding a *Time Scope* at the input and output of the channel, and resimulating to view the effect of adding noise in the time domain.

(f) **Configure Error Vector Measurements.** For a more quantitative measure of this spreading effect, you can use the 'Signal Quality' feature of the *Constellation Diagram*. This provides a figure for the Error Vector Magnitude (EVM).

To open it, 1 on the icon on the toolbar (highlighted in the last figure).

(g) A panel will open on the right hand side of the constellation plot. Expand the 'Settings' section and set the *Measurement Interval* to be 100 (if not already set). This means that an average will be calculated over a reasonable number of samples.

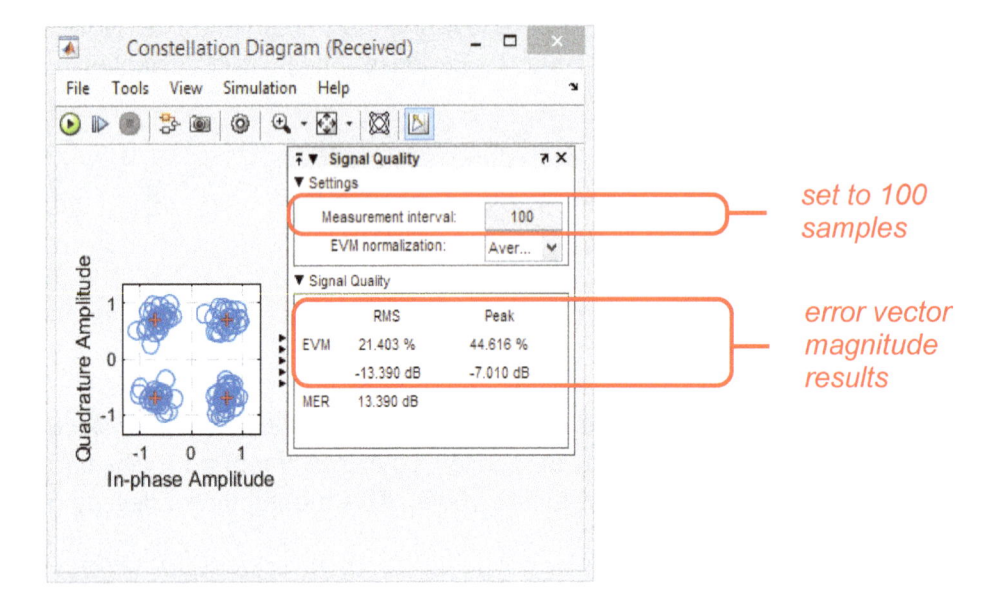

(h) **Re-run the simulation.** View the reported results for EVM in the 'Signal Quality' section. The smaller the EVM percentage, the better! The error power can also be expressed in dB form, compared to the power of the signal.

(i) **Different levels of noise.** What happens if you change the E_b/N_0 figure to 5dB, 0dB, -5dB, and -10dB? See what happens to the constellation and EVM in each case. You should also notice an impact on the number of errors shown in the *Time Scope* plots.

Exercise 11.4 16-QAM Symbol Mapping and Demapping

In this exercise, the higher order constellation, 16-QAM, is evaluated. The impact of noise on this constellation is considered in particular.

(a) Open the system:

   /digital/simulation/modulation/QAM16_constellation_IQ.slx

(b) **Inspect the model.** View the system and note that the channel has been implemented as two *AWGN Channel* blocks, because the symbols are split into the I and Q phases. Frame-based processing is used here for compatibility with the channel block.

(c) **Configure the noise.** Inspect the AWGN parameters and set both I and Q to an E_b/N_0 figure of 20dB in the first instance.

(d) **Simulate.** Run a simulation and view the outputs, in particular the constellation. Are there any errors?

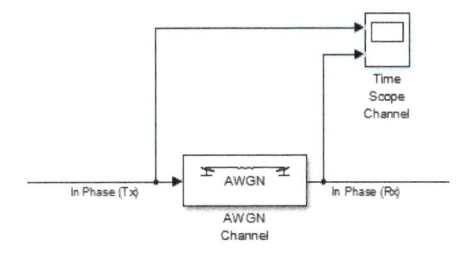

Chapter 11: Digital Communications Theory and Simulation

(e) **Add a Time Scope.** Next, add a *Time Scope* block from ▦ > *DSP System Toolbox* > *Sinks*. Configure this to have two input ports, and a layout of '1x2' (refer back to Exercise 4.11 (page 127) if needed). Connect the first input to the *In Phase (Tx)* signal, and the second to the *In Phase (Rx)* signal.

This will allow you to view the effect that the noise has on each of the I samples.

(f) **Resimulate.** Next, run a simulation to view the effect of noise on the amplitudes of the I and Q phases. The 16-QAM constellation has 4 distinct amplitudes for each of I and Q, and it is evident that the low level of AWGN corrupts the signal—but the values are still very close to the original ones.

(g) **Experiment with AWGN levels.** Try some other values for E_b/N_0 (say 15dB, 10dB, 5dB). What happens as you reduce the value of E_b/N_0?

(h) Compare the number of observed errors to the QPSK (4-QAM) constellation from earlier exercises. You should notice that there are more errors. Why do you think this is?

(i) If you wish, you can also see a more complex version of this system, in which frames of IQ data are represented with single wire connections.

📁 /digital/simulation/modulation/QAM16_constellation_noise.slx

---

Exercise 11.5 Higher Order Constellations

The exercises completed so far have looked at 4-QAM (QPSK) and 16-QAM. You might like to try investigating the higher order QAM constellations of 64-QAM (6 bits per symbol) and 256-QAM (8 bits per symbol). This can be done by modifying either of the systems presented in Exercise 11.4.

11.2 Pulse Shaping

Following conversion from bits to symbols, the next stage is transmit filtering, or 'pulse shaping'. The main purpose of the transmit filter is to appropriately limit the bandwidth of the transmission. Wireless communications are subject to spectrum regulations, which may strictly limit the bandwidth to be occupied, and the allowed emissions in adjacent bands. This specification is referred to as the *spectral mask*. The role of pulse shaping filter, in conjunction with other filters in the processing chain, is to limit the bandwidth of the signal such that it meets the applicable spectral mask. This is depicted in Figure 11.7.

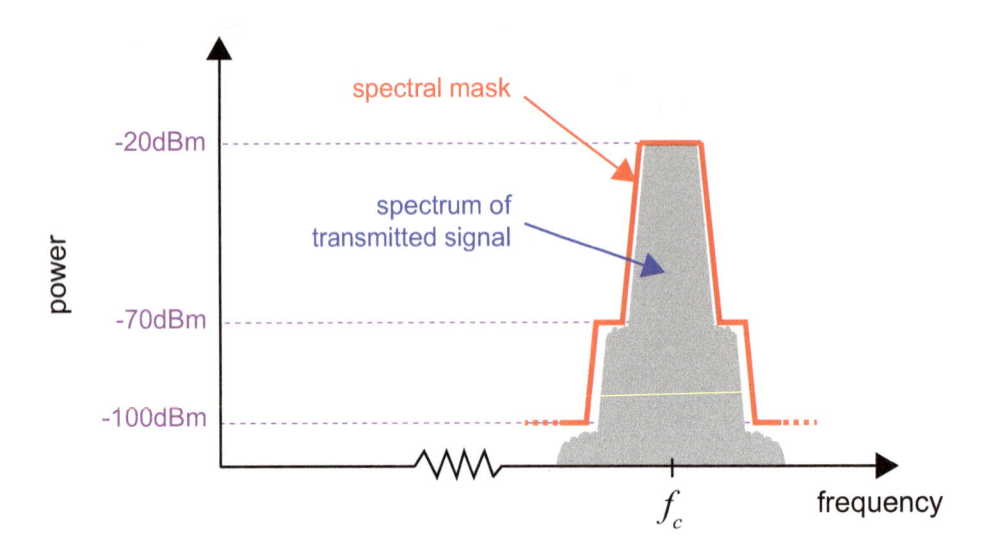

Figure 11.7: Sketch of a signal spectrum complying with a spectral mask

In doing so, the transmit filter should be designed such that it does not introduce Inter-Symbol-Interference (ISI) to the signal being transmitted. That is to say, the filter should not cause successive filtered symbols to interfere with one another.

One of the most popular pulse shapes is the *Raised Cosine* (RC). This response achieves the desired property of zero-ISI. Specifically, despite the raised cosine response usually extending over several symbol periods, the contribution from all other symbols is zero at the ideal sampling points, as illustrated by Figure 11.8.

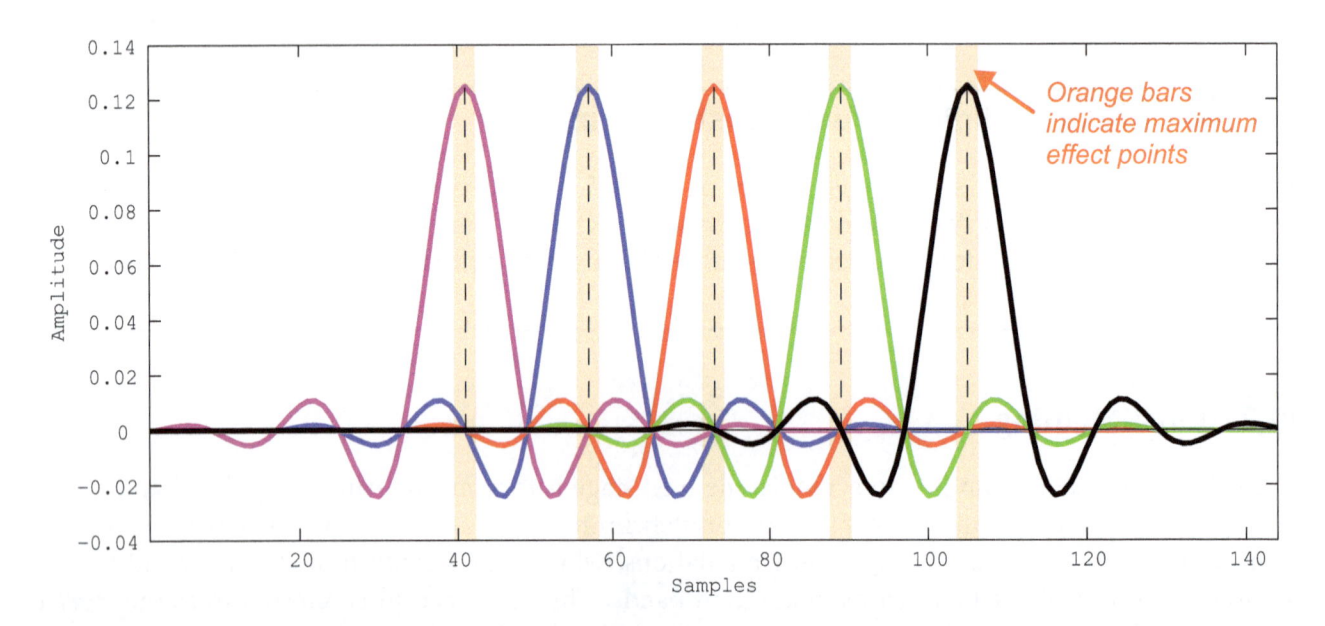

Figure 11.8: The zero-ISI property of raised cosine pulse shaping

Chapter 11: Digital Communications Theory and Simulation 439

Normally the RC response is not implemented as a single filter at the transmit side, but rather it is split across the transmitter and receiver. Hence two filters are required, and this is usually referred to as *matched filtering* [40]. Both filters are square root raised cosine filters, or simply *Root Raised Cosine* (RRC) filters, which in cascade realise the RC response. Although individually the RRC filters do not provide zero-ISI, the RC response achieved over the link as a whole—which is the important part (!)—*does* exhibit zero-ISI.

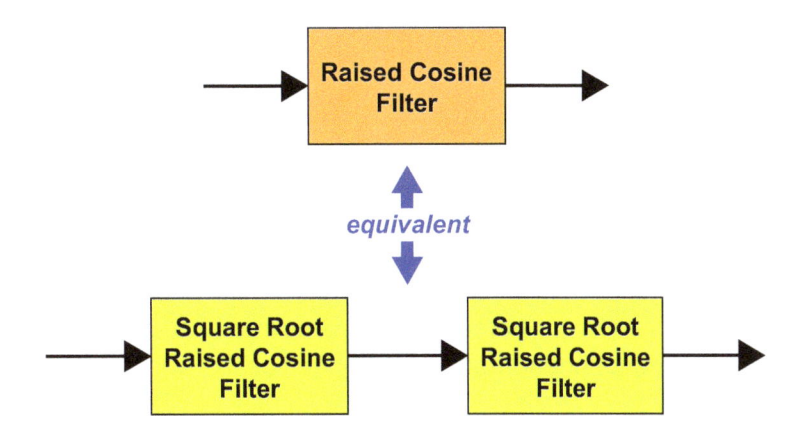

Figure 11.9: The Raised Cosine as a matched filter pair

Exercise 11.6 Pulse Shaping and Transmission Bandwidth

It is useful to compare the transmission bandwidths resulting from different types of pulse shaping filters. In this exercise, we compare rectangular with square root raised cosine pulse shaping.

(a) Open the system:

   `/digital/simulation/pulse/rect_v_RRC.slx`

(b) Notice that this system generates random QPSK data and applies pulse shaping to both I and Q phases using: (i) square root raised cosine pulse shaping; and (ii) rectangular pulse shaping.

(c) **Simulate!** Run a simulation and inspect the *Time Scope* plot to confirm that *very* different shaping is applied to the pulses!

(d) Look at the *Spectrum Analyzer* plots to see the difference between the spectra produced. Which of these would cause greater interference to users of adjacent bands? The important point is to note that the RRC shaping tightly limits the band of frequencies transmitted, unlike rectangular filtering. As shown in the annotated *Spectrum Analyzers* overleaf, rectangular filtering produces a number of significant lobes outwith the data band, and more importantly, these could caused interference to users on adjacent bands. In comparison, the RRC filtering attenuates these spectral components, to ensure the desired spectral mask is achieved.

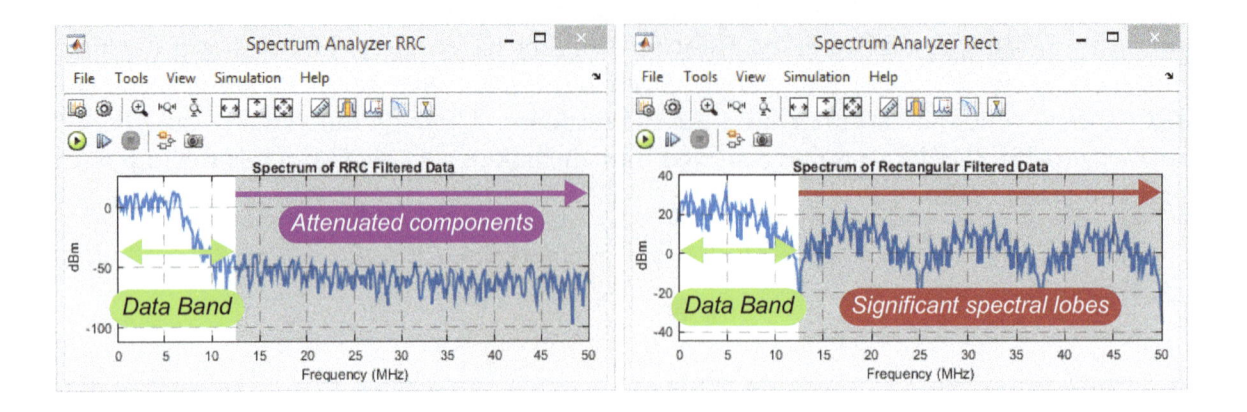

Exercise 11.7 Raised Cosine Inter-Symbol-Interference Properties

In this exercise, we observe the ISI properties of the raised cosine pulse shape, through a simple demonstration system. The effect of changing the filter specification on the magnitude response is also investigated.

(a) Open the system:

   /digital/simulation/pulse/raised_cosine_pulses.slx

(b) This system demonstrates the zero-ISI property of raised cosine pulse shaping by viewing five consecutive pulses, similar to Figure 11.8. Inspect the design and notice that delayed impulses are created and input to pulse shaping filters. The oversampling ratio (number of samples per symbol) is set when the model loads as the parameter R = 10 within the file's *Model Properties*.

(c) How many filter weights are there, and how does this relate to the parameter R? Over how many symbol periods does the response extend?

(d) **Simulate.** Run a simulation and view the results. Check that there is indeed no ISI arising from this pulse shape.

(e) **View the frequency response.** Open one of the filter blocks (they are all equivalent), and view the configuration of the block. You can also investigate the filter further, by pressing the 'View Filter Response' button at the top right hand corner of the dialogue. For the initial configuration, you should see something like this, i.e. a lowpass filter response:

(f) Zoom in along the *x*-axis (using the tool), focussing on the boxed area indicated by the red dashed lines. This indicates the transition band of the filter and it is influenced by the roll-off rate—can you see how the parameters of R = 10 and roll = 0.25 relate to this response?

(g) **Change the oversampling ratio.** The oversampling ratio can be altered very easily. Just type R = 8 (or similar) into the MATLAB command window, and press *Enter*. Then, re-run the model to see the effect of the change. You might also view the filter response and see if there are any differences!

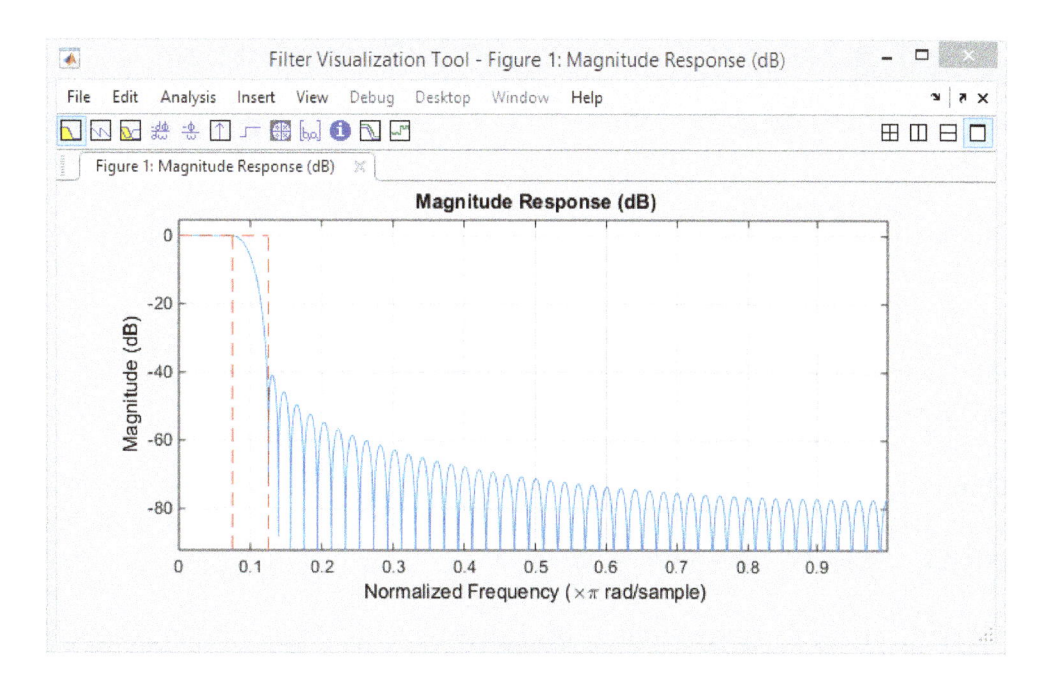

Chapter 11: Digital Communications Theory and Simulation

[Screenshot: Filter Visualization Tool showing Magnitude Response (dB) plot, with magnitude ranging from 0 to -80 dB over normalized frequency 0 to 0.9 (×π rad/sample). Red dashed lines indicate filter specification mask.]

(h) **Change the roll-off rate.** If you wish, you can also redesign the filters to use a different roll-off rate, say 0.1, or 0.7. This can be done in a similar way, by entering `roll = 0.1` at the MATLAB command prompt. After making the change, re-simulate to see the effects, and view the effect on the filter response.

Exercise 11.8 — RRC Inter-Symbol-Interference Properties

As mentioned in the introduction to this section, the RC pulse shape is commonly implemented using a pair of RRC filters. We can now confirm their equivalence via simulation.

(a) Open the system:

> `/digital/simulation/pulse/sqrt_raised_cosine_pulses.slx`

(b) **Simulate.** Try simulating this system, and confirm whether single RRC filters can provide zero-ISI. Look closely at the crossing points!

(c) **Add a second RRC filter to each branch.** Copy and paste each of the RRC filters to form a pair, similar to this:

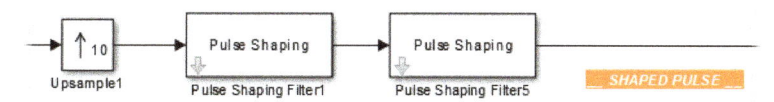

[Diagram: Upsample1 (↑10) → Pulse Shaping (Pulse Shaping Filter1) → Pulse Shaping (Pulse Shaping Filter5) → SHAPED PULSE]

(d) Now, simulate again—does the RRC *pair* provide zero-ISI? (Also, you might need to extend the simulation time to see this clearly... can you think why?)

Exercise 11.9 Matched Filtering of QPSK Modulated Data

We will now briefly confirm what the signal looks like after the second RRC filter, i.e. the matched filter in the receiver.

(a) Open the system:

   `/digital/simulation/pulse/RRC_matched.slx`

(b) Familiarise yourself with the system, and confirm the oversampling ratio (R) and raised cosine roll-off rate that have been used in this case.

(c) **Simulate!** Run a simulation and view the results. For the Q phase, the input samples have been delayed, and are shown on the same axis as the output of the matched filter. It should be easy to see the correspondence!

(d) It is worth noting the general shape of signals produced from matched filtering. Perfectly timed samples taken at the symbol rate (i.e. the green 'spikes' in the lower *Time Scope* plot) all have the same magnitude, despite the varying amplitudes of the intermediate samples. Zoom in and inspect the waveform carefully to check this is true!

The take-away point from this example is that the combined response achieved from the RRCs in the transmitter and receiver are equivalent to a single RC response, giving zero ISI over the link, when the received signal is sampled at the optimum points.

11.3 Digital Up and Downconversion

RF transmitters and receivers can have a variety of architectures, with one of the most important factors being the demarcation between digital and analogue processing. In the *Introduction* chapter of this book, the increasing scope for high-frequency processing in the digital domain was highlighted. Depending on the RF carrier frequency involved, it may be possible to undertake all processing digitally, with the conversion between analogue and digital undertaken at RF. Perhaps more commonly, baseband and IF processing occurs in the digital domain, with IF to RF modulation achieved using analogue circuitry. (These alternative architectures were shown in Figure 1.7 on page 11, and you may wish to return to that diagram for a brief recap.)

In this section, we consider methods for digital upconversion and downconversion. These methods are used to modulate and demodulate a signal, while also undertaking the multirate transitions required to move between the symbol rate, f_{symbol}, and the DAC or ADC sampling rate, f_{system}.

- **Digital Upconverter (DUC)** — In the **transmitter**, the sampling rate is raised, through a series of filtering operations, from the symbol rate (f_{symbol}) to a rate that matches that of the DAC (f_{system}) as shown in Figure 11.10.

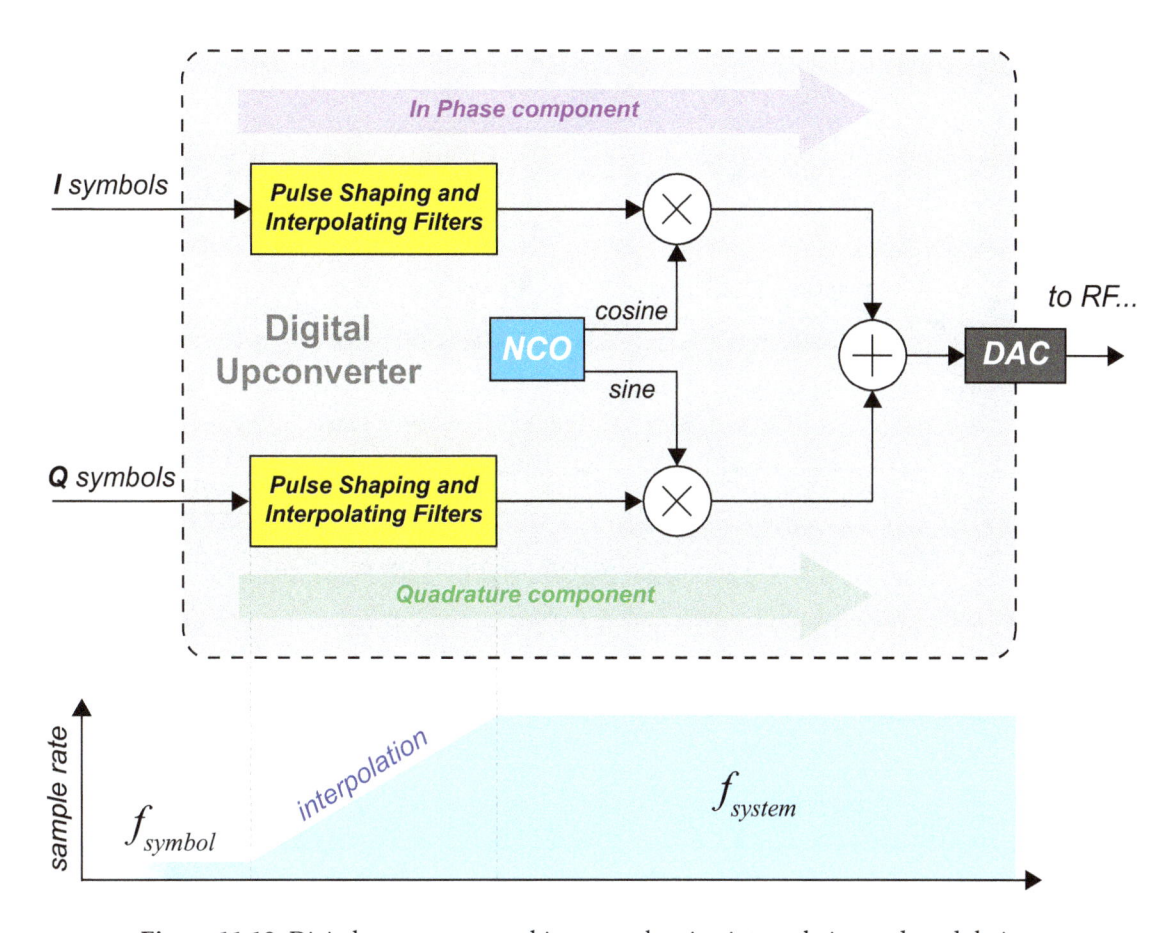

Chapter 11: Digital Communications Theory and Simulation 443

Figure 11.10: Digital upconverter architecture, showing interpolation and modulation

- **Digital Downconverter (DDC)** — In the **receiver**, a transition takes place from the ADC sampling rate, to a lower rate suitable for symbol synchronisation and further baseband processing (Figure 11.11).

Multirate filters are required for both increasing and decreasing the sampling rate. When the sampling rate is increased (i.e. *upsampling* takes place), the effect is to generate spectral images at integer multiples of the original sampling rate, and these must subsequently be removed using a lowpass filter. Together, these two operations (upsampling and lowpass filtering) are known as *interpolation* [21].

In the second case, the sampling rate is reduced using a *downsampling* operation. This must be preceded by a lowpass filter, in order to remove the frequency components that would otherwise be aliased into the baseband region. Thus, the complete process of sample rate reduction, which is referred to as *decimation*, also requires two operations: lowpass filtering and downsampling [21].

Further background information on multirate DSP, and the operations of interpolation and decimation, please refer to Appendix C (page 583).

Both interpolation and decimation are depicted in Figure 11.12, clarifying the order of the resampling and filtering operations.

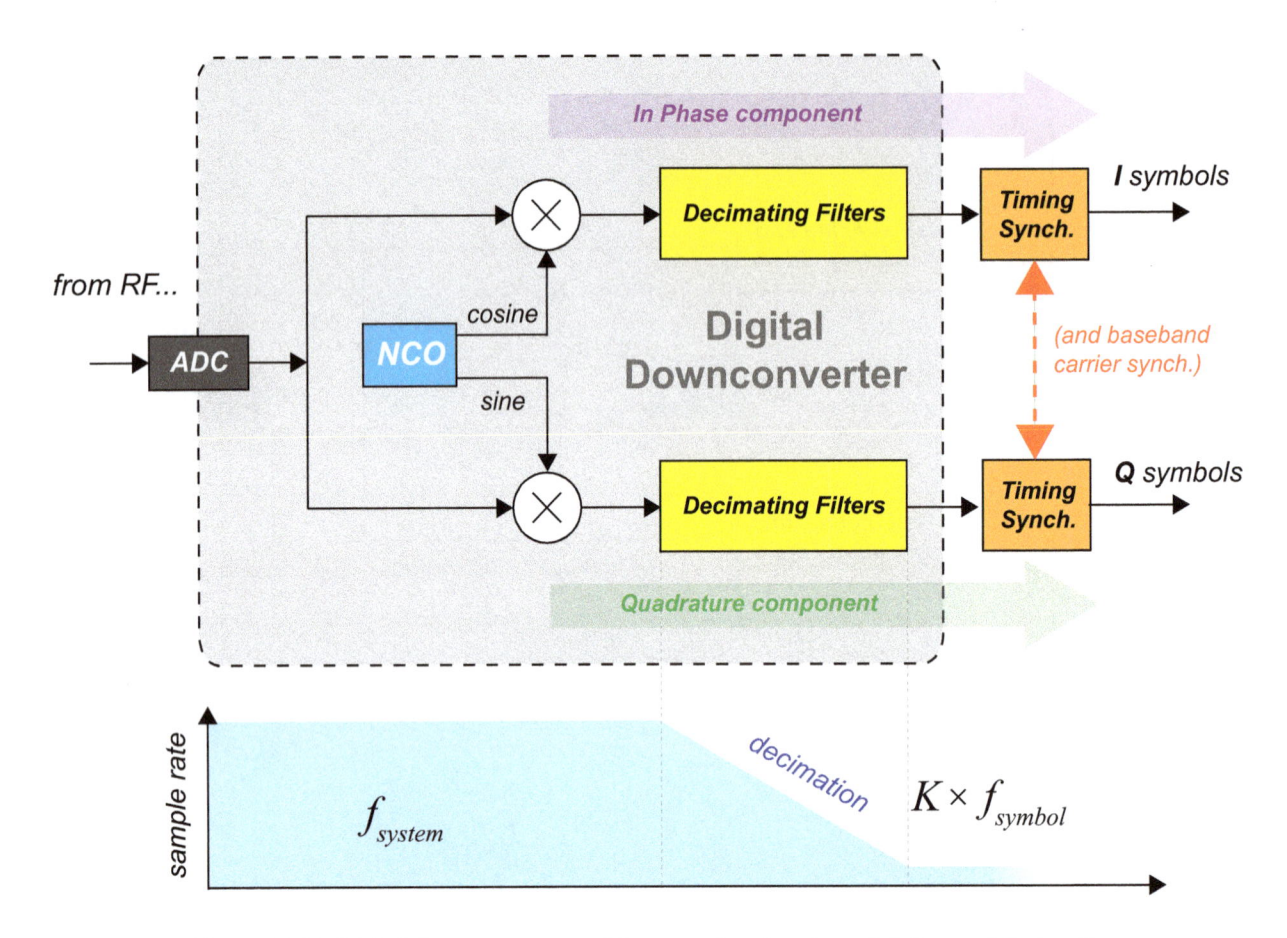

Figure 11.11: Digital downconverter architecture, showing demodulation and decimation

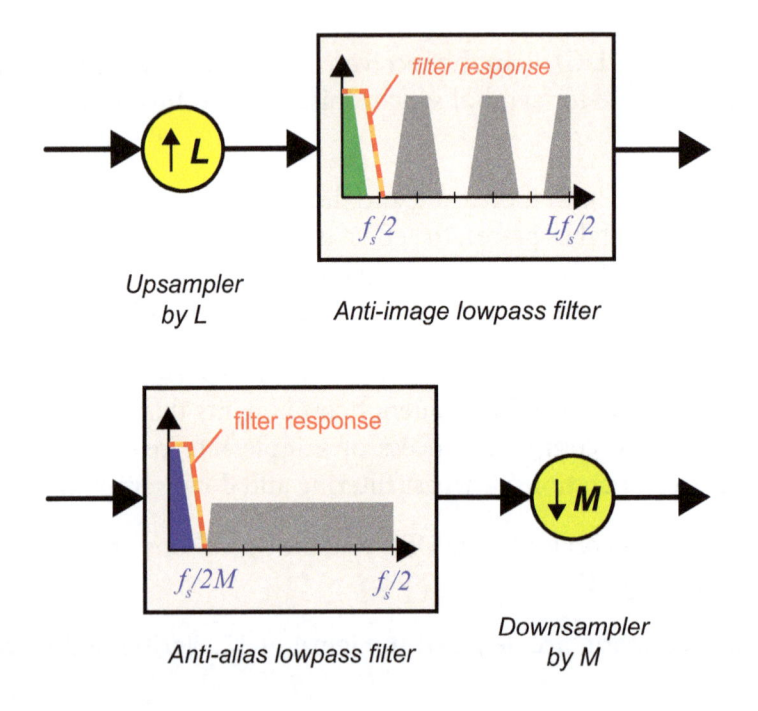

Figure 11.12: Interpolation (upper) and decimation (lower) implemented as filter-and-resampler pairs

Chapter 11: Digital Communications Theory and Simulation 445

Thus, lowpass multirate filters are required for both the DUC and DDC. The design of the filter depends on the interpolation or decimation ratio, with higher ratios demanding more exacting filter responses. Especially when the ratio

$$R_{multirate} = \frac{f_{system}}{f_{symbol}}$$

is large, it implies that a lowpass filter with a sharp cut-off is required, which in turn means that the filter is computationally expensive.

Taking the case of interpolation, an example spectrum arising from interpolating a baseband signal by a factor of 16 is provided in Figure 11.13. In this case, a lowpass filter would be designed to pass the region of the red coloured signal, while removing the spectral images (shown in blue). If the interpolation ratio was larger than 16, then there would be more spectral images to remove, and the bandwidth of the signal retained by the lowpass filter would be proportionately smaller. This would require a more expensive filter.

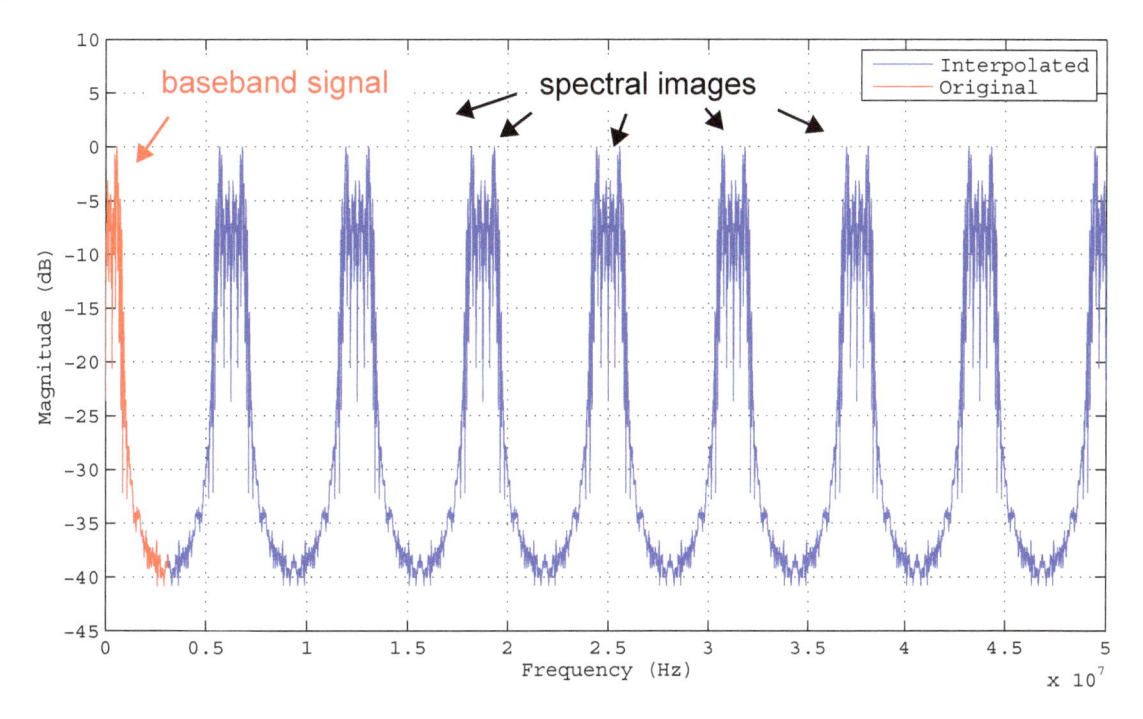

Figure 11.13: The result of upsampling a baseband signal by a factor of 16 (frequency domain)

Rather than performing the filtering with a single filter, a more efficient alternative is to partition the interpolation or decimation task into a cascade of filters, each having a more relaxed response, and undertaking a smaller rate change. For instance, if an overall decimation ratio was 200, it would be more efficiently implemented using three different filters in cascade (e.g. decimating by ratios of 20, 2, and 5 respectively—and many other combinations are possible), rather than with a single filter decimating by 200. As long as the cascaded response of the filters achieves the specification, for instance with respect to the spectral mask, then the key requirement is met. The cascaded response of a set of filters can be readily plotted and evaluated in MATLAB.

This approach of designing the DUC and DDC to reduce the computational cost is important in hardware implementations (e.g. FPGA or ASIC signal processing systems), but it is also relevant to the SDR designs considered in this book. These multirate elements are needed in order to simulate digital modulation/ demodulation between baseband and IF (or even RF), and it helps us to understand the work undertaken by the RTL-SDR (an also, for those using one, the USRP® hardware). Reducing the computational requirement of these multirate operations helps to achieve a model that is capable of running in real time.

Various types of filters can be used as part of the filter cascade, and designers may have different preferences. One notable type of multirate filter is the computationally efficient, but spectrally sub-optimal, Cascade-Integrator-Comb (CIC) filter [23]. This filter type requires very few arithmetic operations compared to other filters, but displays a characteristic 'droop' in the passband that usually requires correction. The CIC is often chosen to undertake the transitions between the highest rates, and/ or programmable rate transitions, due to its amenable computational structure and inherent flexibility. A compensation filter is commonly used in conjunction with the CIC to correct for its droop and, as this is usually implemented as an FIR, it is denoted by the acronym 'CFIR'. Other methods of mitigating or compensating for the droop also exist, e.g. [27], [34].

FIR and IIR filters can both be used within the filter cascade, with the FIR typically preferred for its linear phase response characteristics. Polyphase methods are normally applied to optimise the structure (i.e. rearrange the computation to calculate only what you need!), and this is advantageous because exactly the same functionality can be achieved with a drastic reduction in the amount of computation required.

The lowest rate filter in the chain often performs its multirate function in conjunction with pulse shaping or matched filtering (e.g. an RRC filter).

In the next two exercises, example DUC and DDC systems will be presented. These represent just one possible approach to each, with a variety of other options and configurations available. A detailed treatment of multirate DSP is not within the scope of our present book, but an overview is presented in Appendix C (page 583), and the interested reader may wish to refer to some good sources of information available on this topic, including [21] and [44].

Exercise 11.10 Digital Upconverter (DUC): Filter Cascade & Modulation

This example demonstrates a DUC that interpolates by a factor of 64, implemented using three different filters in cascade, followed by a modulation stage.

(a) Open the system:

   `/digital/simulation/DUC_DDC/DUC_filters.slx`

(b) As the file loads, it will execute a MATLAB script that designs the filters in the chain, and plots their filter responses using the Filter Visualisation Tool (`fvtool`). The most important one is the RRC-CFIR-CIC cascaded response.

(c) **Rate Transition.** What are the rate transitions undertaken by the three filters? Can you reconcile this with the overall rate transition of 64?

Chapter 11: Digital Communications Theory and Simulation 447

(d) **Run a simulation.** View the outputs in the time and frequency domains, an hence, check that the input signal has been appropriately pulse shaped, and interpolated to the rate of f_s = 100MHz.

(e) If you wish, investigate the script that designs the filter cascade `DUC1_setup.m`. This file contains comments to explain the steps involved. Note that both floating point and fixed point versions are created (the CIC filter type requires fixed point arithmetic).

(f) Next, we will add the modulation stage. Open the system:

/digital/simulation/DUC_DDC/DUC_modulation.slx

(g) Inspect the system, and compare it to Figure 11.10. Can you see the correspondence?

(h) Run a simulation and view the time and frequency domain outputs. It should be possible to confirm that the signal has been successfully modulated onto an IF carrier by the DUC. You can also try altering the parameters of the oscillator to modulate to a different frequency, if you wish.

Exercise 11.11 Digital Downconverter (DDC): Demodulation & Filter Cascade

Next, we will demodulate the signal generated in the previous exercise to baseband, and perform decimation. Like the DUC, the DDC applies lowpass filtering and decimation using several filters in cascade.

(a) Open the system:

/digital/simulation/DUC_DDC/DDC_demodulation.slx

(b) This system includes a channel model that represents noise and/or users on other frequency bands. The signal of interest is to be extracted from this signal via demodulation and filtering.

(c) Run a simulation and view the results. You should be able to identify the signal in noise (after passing through the channel), and then see that the 'wanted' signal has been correctly demodulated to baseband.

(d) Next, the signal needs to undergo decimation filtering. Open the system:

/digital/simulation/DUC_DDC/DDC_decimation.slx

This represents the DUC, followed by a channel, then the DDC and subsequent processing.

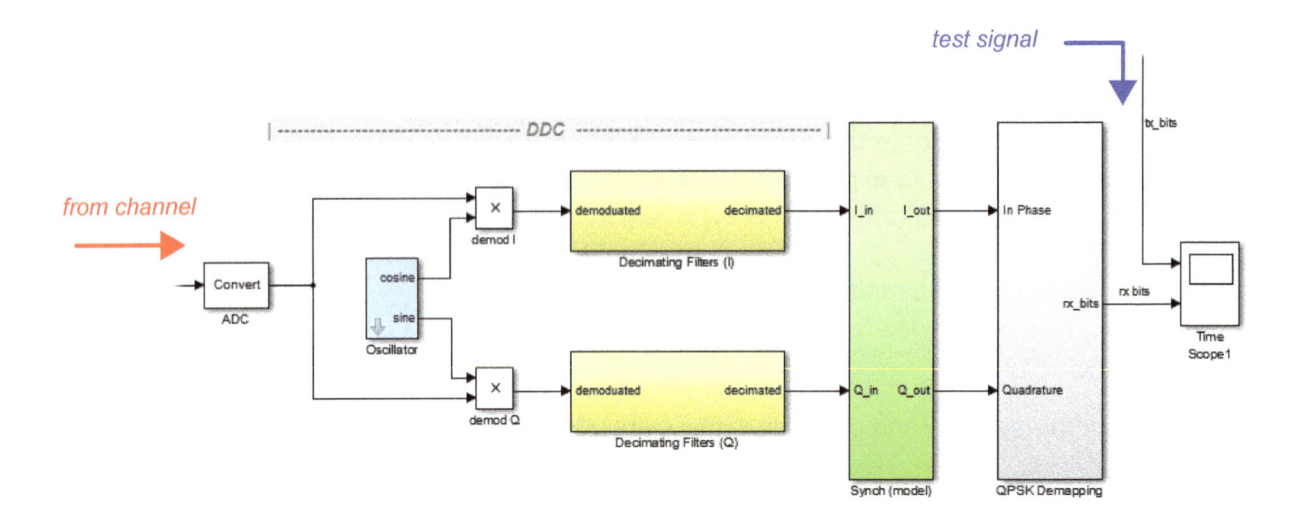

(e) Like the DUC, this system involves a cascade of filters. Take a little time to investigate the configuration (it is not the same as the DUC!). You can inspect the `DDC1_setup.m` file to see how the filters have been designed.

(f) Run a simulation and view the time domain output produced. It should be possible to confirm that the transmitted data has been correctly recovered. Thus, the DDC has successfully demodulated the received IF signal to baseband, and then reduced the sampling rate.

(g) If you wish, you can also add other *Time Scopes* and *Spectrum Analyzers* to view the signal at different stages of the DDC.

11.4 Carrier Synchronisation

When receiving a digitally modulated signal, the goal is to recover a discernible constellation from which the correct symbol decisions can be made. In the case of a simple, noisy channel, this means that the symbol samples will be in clear groups around the ideal constellation points, as shown in Figure 11.14 for the example of QPSK (we will use QPSK extensively to illustrate the points in this section).

In a realistic scenario, obtaining a constellation of this form requires both carrier synchronisation and symbol timing synchronisation to be undertaken. In this section, we will assume perfect symbol timing in order to focus on the carrier synchronisation problem.

11.4.1 Demodulation with a Fixed Frequency Carrier (Out of Carrier Synchrony)

Before going on to consider methods of carrier synchronisation, it is useful to review the effects of demodulating the received signal to baseband with a LO that is *not* frequency and phase locked to the received carrier. This motivates the requirement for carrier synchronisation.

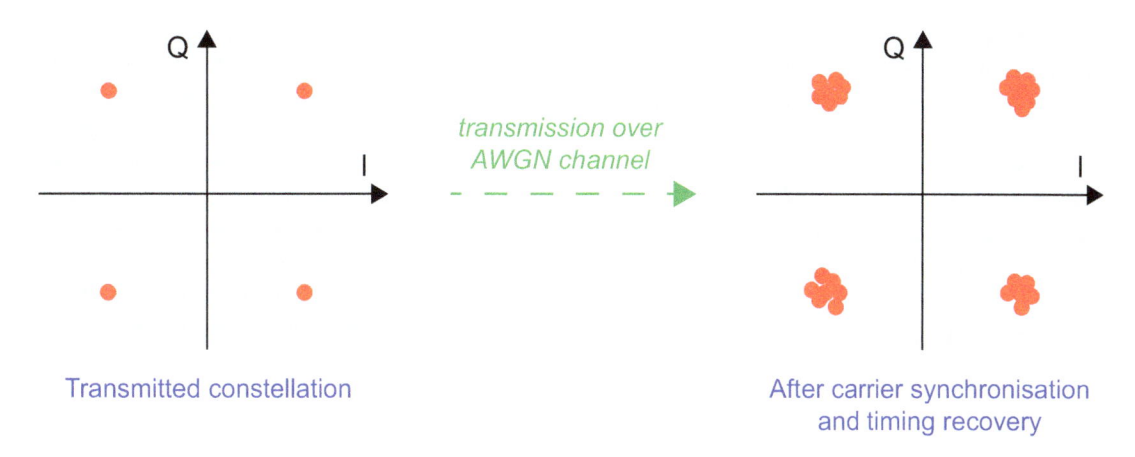

Chapter 11: Digital Communications Theory and Simulation

Figure 11.14: Transmitted and target constellations for QPSK

It may be assumed that there are three categories of problem:

1. Where the LO in the receiver has exactly the same frequency as the received carrier, but is out of phase.
2. Where the LO in the receiver has a slightly different frequency to that of the received carrier.
3. Where the frequencies of the receive LO, and received carrier signal, are changing relative to each other.

In each of these three cases, demodulating with a LO that is not synchronised to the received carrier results in an imperfect constellation. For clarity, we will neglect the impact of noise introduced in the channel.

In case (1), the result is a fixed phase offset as shown in Figure 11.15 (a). This may cause the constellation points to remain in their original quadrants (as shown here), or even to rotate into different quadrants. For case (2), the constellation rotates at a constant rate, as depicted in Figure 11.15 (b). The rate of rotation depends on the frequency error between the carrier embedded within the received signal, and the receiver's LO. For case (3), the constellation again rotates over time, but it accelerates if the frequency error is increasing, or decelerates if the frequency error is decreasing. This last possibility is represented by Figure 11.15 (c).

This lack of synchronisation means that the constellation points do not reside in the correct positions (or even the correct quadrants!), which makes it difficult or impossible to make correct symbol decisions.

As will be shown later, it is possible to compensate for these effects at baseband, as an alternative to carrier synchronising at the stage of demodulation. In both approaches, the desired outcome is to obtain a constellation similar to that transmitted, as shown in Figure 11.16 for the a fixed phase offset (case 1).

We will now go on to consider two different methods for performing carrier synchronisation:

1. Carrier synchronisation at the point of demodulation;
2. Demodulation with a fixed frequency LO, followed by carrier synchronisation at baseband.

These require different structures, as will be reviewed over the coming pages.

(a) Received constellation (*fixed phase offset*)

(b) Received constellation (*fixed frequency offset*)

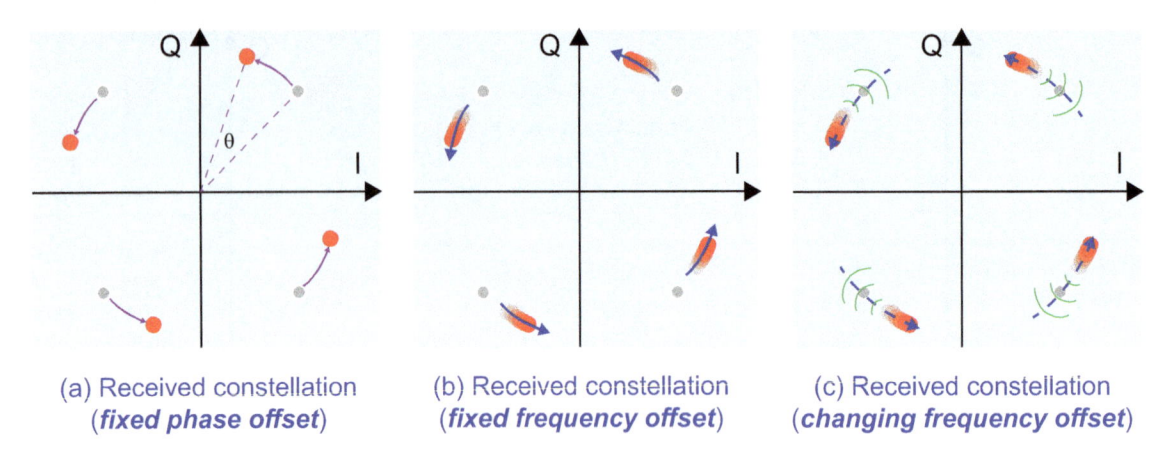

(c) Received constellation (*changing frequency offset*)

Figure 11.15: Received constellations for different phase and frequency errors (without carrier synchronisation)

carrier synchronsiation

(a) Demodulated constellation (fixed frequency oscillator)

(b) With carrier synchronisation (ideal QPSK constellation)

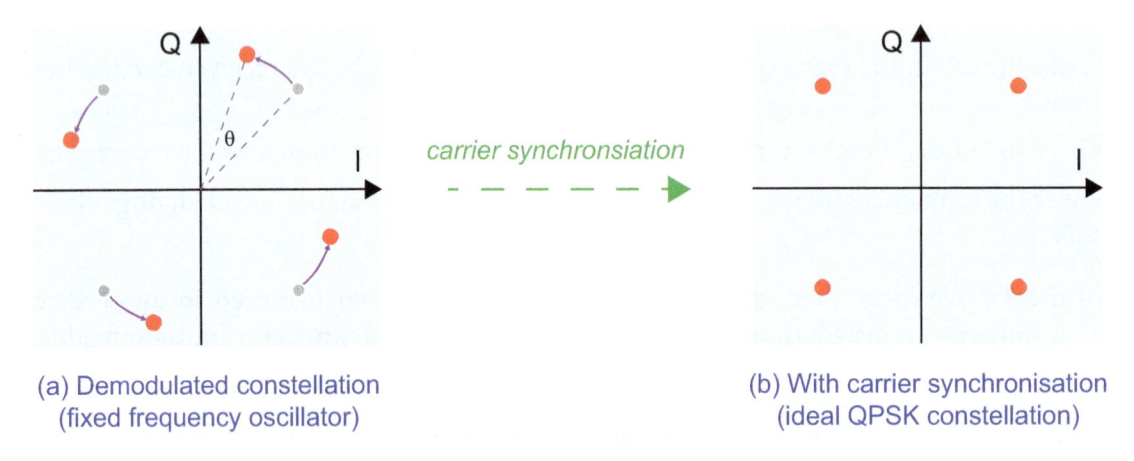

Figure 11.16: Received constellation (phase offset scenario): (a) before or without carrier synchronisation; (b) after carrier synchronisation

11.4.2 Carrier Synchronisation at the Demodulation Stage

The most intuitive method of performing carrier synchronisation is to use a synchronisation loop to generate local sine and cosine waves with the same frequency and phase as the received carrier, and multiply them with the received signal to demodulate the I and Q phases to baseband. To achieve this, a feedback loop is required that will derive a measure of phase error, and hence develop a signal to drive an NCO with quadrature (sine and cosine) outputs.

The architecture of this style of receiver is shown in Figure 11.17 [38]. The input is the received signal from the ADC, which is centred at some intermediate carrier frequency, f_c. The signal is then demodulated to baseband I and Q phases by the two outputs of the NCO, whose frequency and phase are dynamically adjusted by the feedback loop.

After demodulation, the I and Q signals pass through the matched filter, to produce the signals $x(mT)$ and $y(mT)$ respectively, where m is the sample index and T is the sample period. It is assumed that the signals are oversampled at this stage, with several samples per symbol. At the point of the sampler

Chapter 11: Digital Communications Theory and Simulation 451

indicated in red in Figure 11.17, the sampling rate is reduced to one sample per symbol, with the sampling operation controlled by the symbol timing synchronisation loop. Symbol timing is a separate synchronisation loop and it is not considered here—for now, we just assume that the samples are taken at the perfect time instants.

The symbol samples are denoted by the coordinates $(x(kT_s), y(kT_s))$ on the IQ plane, where k is the symbol sample index and T_s is the symbol period. The point given by these coordinates is compared in terms of phase with the reference constellation point, in order to derive a phase error. There are two possible approaches to this problem:

1. In a ***decision directed*** method, the transmitted data symbols are unknown. The phase error is generated based on a symbol decision, i.e. the closest symbol to the received sample.
2. In a ***data aided*** method, the receiver has knowledge of the transmitted symbols, and derives the phase error accordingly. This approach incorporates that the constellation may have rotated to any extent across the full 360° range.

In this discussion, we assume that there is no knowledge of the transmitted symbols, and therefore the *decision directed* method is used. This means that, even after carrier synchronisation, the constellation may still have a residual phase rotation (in the case of QPSK, a multiple of 90°), which will need to be compensated at a later stage in the receiver.

Using the decision directed method, a symbol estimate is made based on the each symbol sample. This is analogous to quantising the received symbol sample to the nearest available symbol.

In the case of QPSK, the samples are mapped to the nearest applicable symbols by the decision operators shown in Figure 11.17. This produces quantised symbol estimates with coordinates $(\hat{a}_o[k], \hat{a}_I[k])$. These symbol samples (before and after quantisation) have phases $\theta_r[k]$ and $\theta_s[k]$, respectively, as shown in Figure 11.18.

Each of the two phases can be calculated using an inverse tangent operation, i.e.

$$\theta_s = \tan^{-1}\left(\frac{\hat{a}_I[k]}{\hat{a}_o[k]}\right) \qquad (11.1)$$

and

$$\theta_r = \tan^{-1}\left(\frac{y(kT_s)}{x(kT_s)}\right). \qquad (11.2)$$

The phase error between these two points is simply the difference between the two phases,

$$\theta_e[k] = \theta_r[k] - \theta_s[k]. \qquad (11.3)$$

The phase error signal is subsequently upsampled and input to the loop filter. This in turn produces the feedback control signal that adjusts the frequency of the NCO.

The operation of this receiver architecture implies that the majority of the synchronisation loop operates at the sample rate of the received signal, which is normally the highest rate in the system. The only part

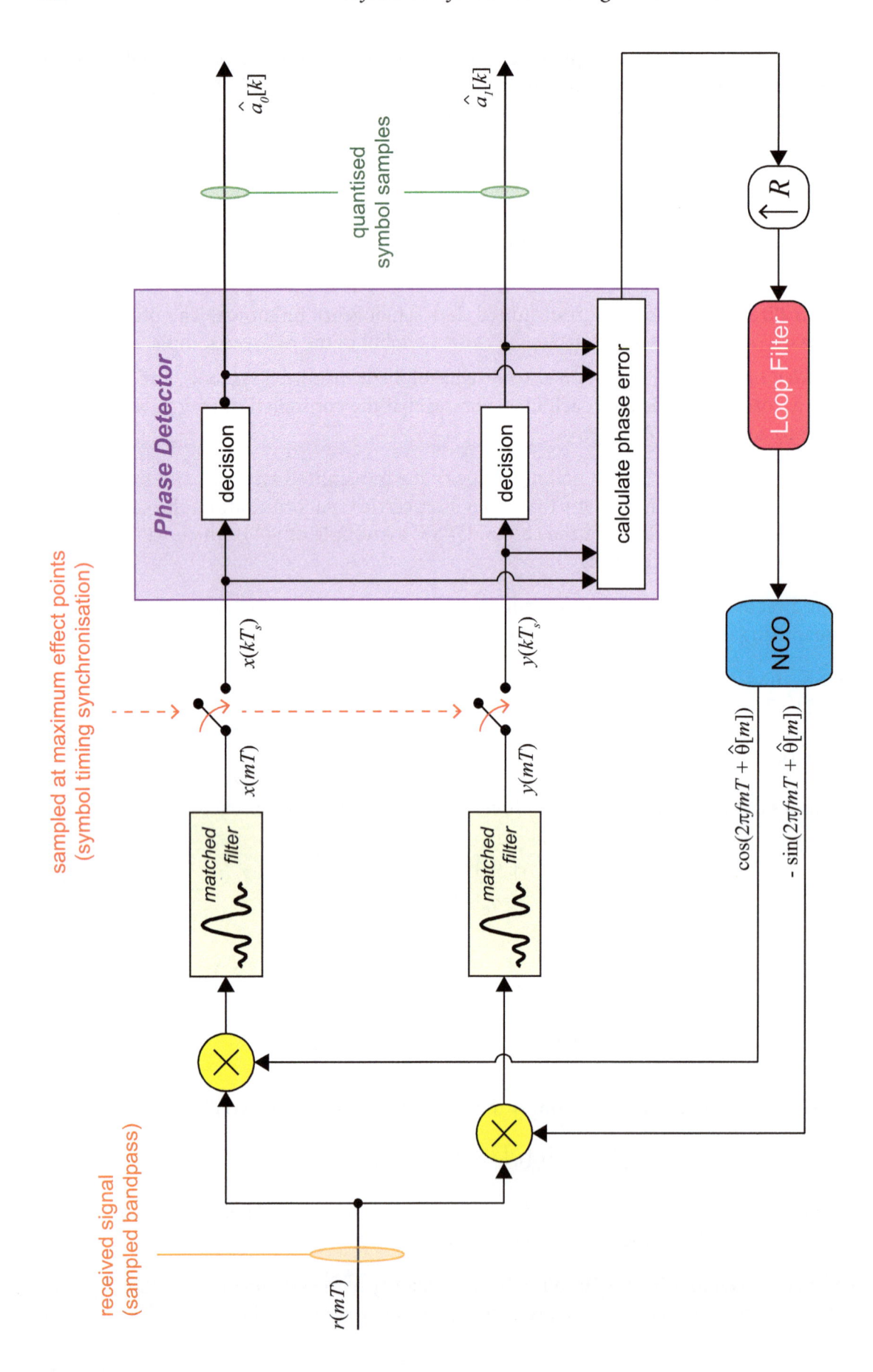

Figure 11.17: Carrier synchronisation (synchronising at demodulation stage) [38]

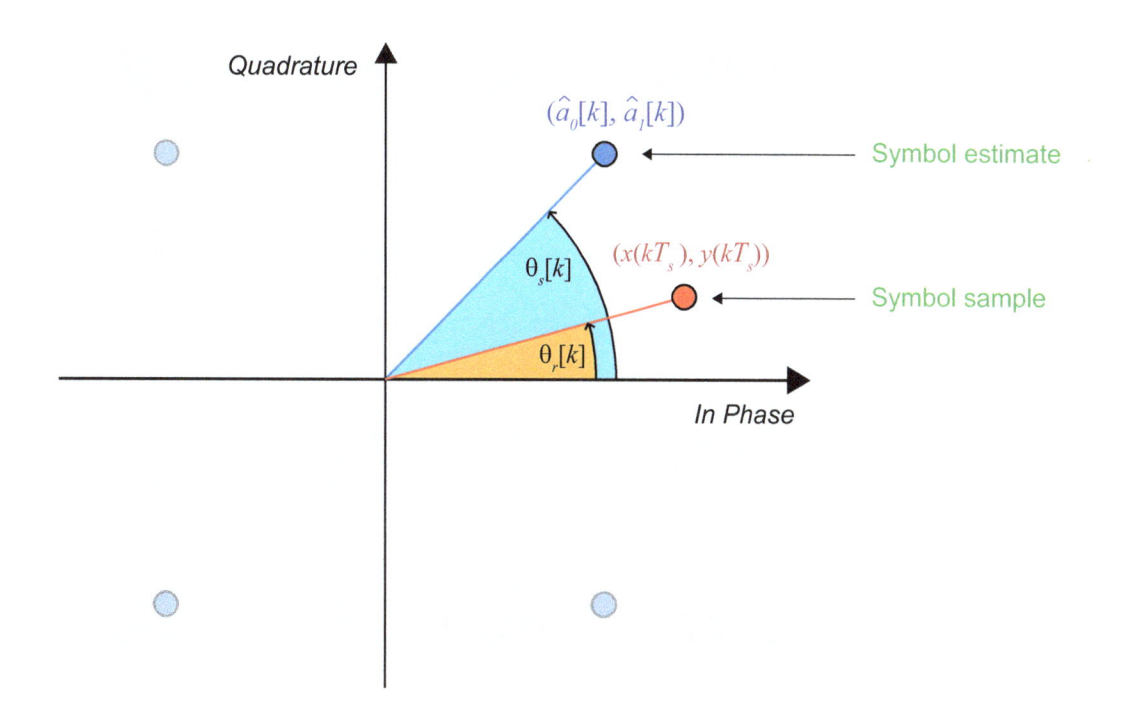

Chapter 11: Digital Communications Theory and Simulation

Figure 11.18: Calculation of the phase error [38]

running at symbol rate is the phase detector. Therefore, the level of computation is relatively high, with the loop filter and NCO running at the higher rate. Computational complexity is one practical disadvantage of this style of implementation.

We can now consider a Simulink example of this type of carrier synchronisation loop.

Exercise 11.12 Carrier Synchronisation for QPSK (Demodulation)

This example models the carrier synchronisation loop shown in Figure 11.17. The synchronisation loop controls the frequency and phase of the quadrature sine wave components that demodulate the received signal to baseband.

(a) Open the system:

   /digital/simulation/carrier_synch/QPSK_synch_demod.slx

(b) Inspect the system, and compare it with the signal flow graph provided in Figure 11.17. Notice that the phases of the constellation points are generated using the 'atan2' function. This function generates the inverse tangent with a full range across all four quadrants.

(c) **Simulate and observe the *Constellation Diagram* and *Time Scope* plots.** Open the *Constellation Diagram Rx* block and run a simulation. Watch the constellation diagram while the simulation runs. You should notice that the constellation initially appears to be moving, but after a while, the carrier synchronisation loop successfully converges and the constellation stabilises at the desired points.

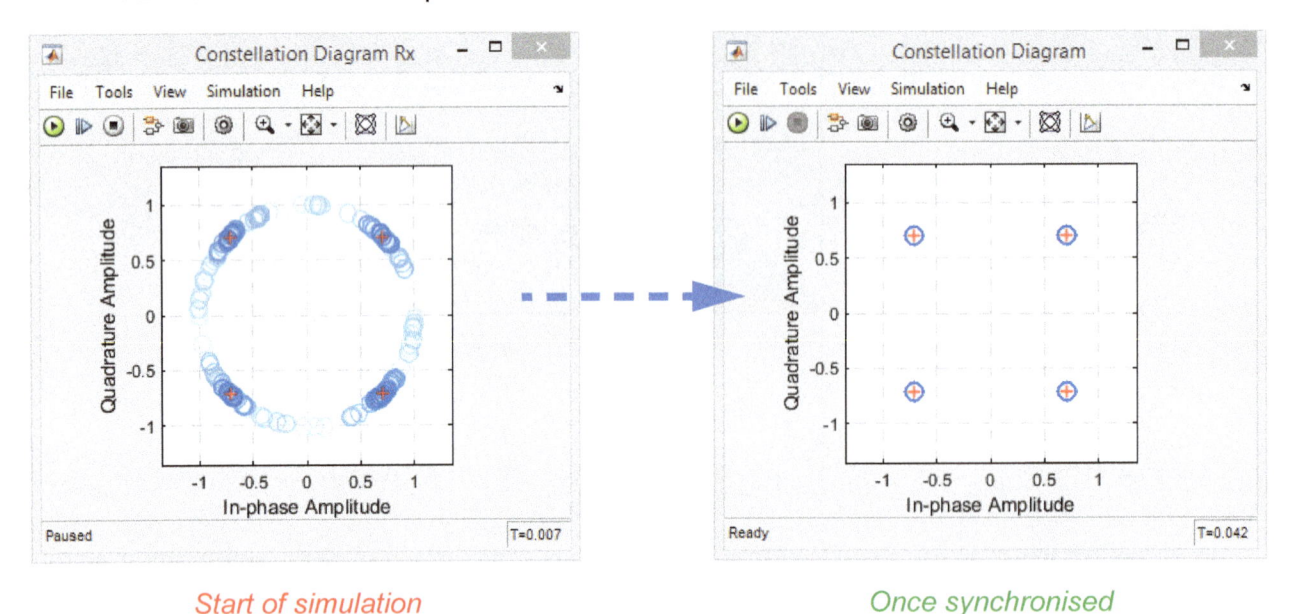

Start of simulation → *Once synchronised*

(d) Open the *RRC Samples Time Scope* and inspect the time domain waveforms to confirm that, after a little while, the samples are taken at the maximum effect points.

(e) View the *VCO Frequency Time Scope*, and notice that the frequency generated in the receiver has a transient period, before settling down to a steady value. You should see something like the following. How long does it take to adapt?

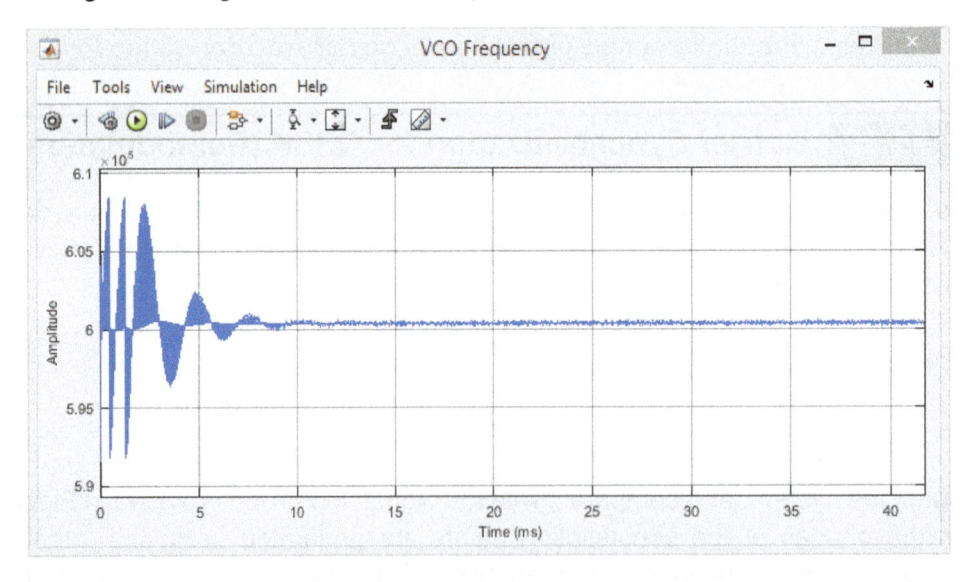

(f) **Determine the frequency once synchronised.** Still referring to the *VCO Frequency Time Scope*, find the (average) frequency synthesised after adaption. Does this correspond with the frequencies of the sine and cosine sources used to modulate in the transmitter? Note that the frequency is configured using the parameters `fc` (representing the expected carrier frequency)

Chapter 11: Digital Communications Theory and Simulation 455

and `fe` (the frequency error, which can be positive or negative). You can determine the current values of these by referring to the MATLAB Workspace.

(g) **Synchronise in noise!** Try connecting the *AWGN Channel* block and repeating the simulation to ensure that the synchronisation loop still works in the presence of noise. You can adjust the level of noise by changing the 'Variance' property in the AWGN block.

(h) Try altering the frequency error by changing the parameter in the MATLAB command window, e.g.
```
fe = 200;
fe = -550;
```
... and then resimulating.

(i) What happens if the magnitude of the error becomes too large? Earlier steps demonstrated that this style of carrier synchroniser (which synchronises at the demodulation stage) can operate successfully over a range of frequency offsets. If the deviation is too great, however, then it may take a long time or even fail to achieve lock.

11.4.3 *Fixed Frequency Demodulation with Carrier Synchronisation at Baseband*

An alternative to the architecture presented in Section 11.4.2 is to demodulate with a fixed frequency quadrature oscillator, and make a subsequent frequency correction at baseband. The likely outcome of demodulating in this way is that the constellation will have a residual frequency offset, which corresponds to a continual rotation of the constellation. Thus the synchronisation loop must 'de-rotate' the constellation back to its desired position.

This correction is applied by generating a complex exponential that exactly compensates for the low frequency rotation of the constellation. Similarly to the previous architecture, the error signal is derived from measurements of the phase difference between the sampled symbols, and their corresponding ideal constellation points (Figure 11.18). The terms of the complex exponential are generated by a component similar to the NCO, which integrates the phase input and then generates sine and cosine terms. A diagram of this architecture is shown in Figure 11.19.

Unlike the previous implementation (Figure 11.17), this system achieves carrier synchronisation by directly adjusting the symbol samples, and therefore it operates at a much lower sampling rate and requires far less computation overall.

Next, a Simulink example of this baseband carrier synchronisation loop is provided.

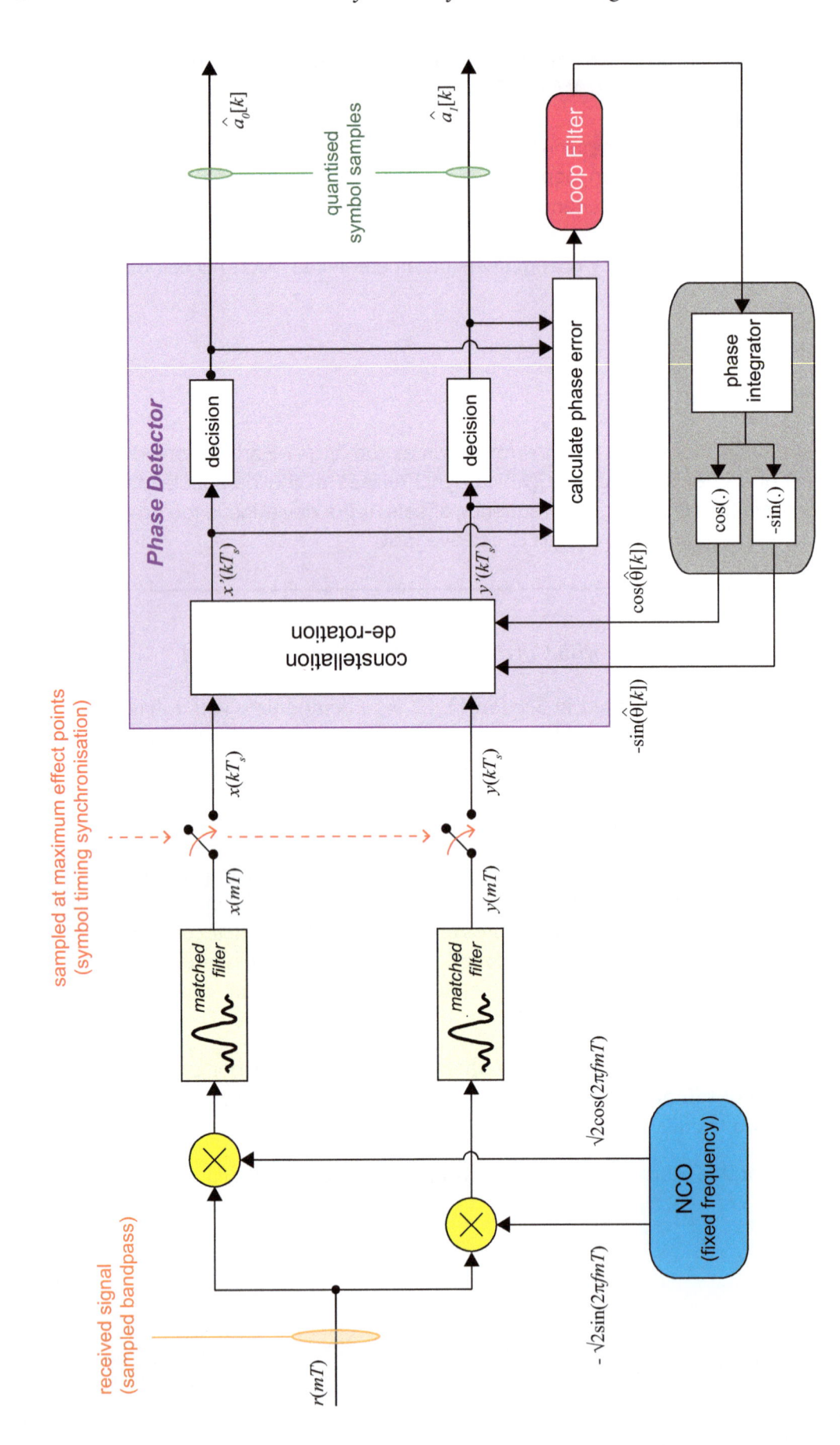

Figure 11.19: Carrier synchronisation loop operating at baseband [38]

Chapter 11: Digital Communications Theory and Simulation 457

Exercise 11.13 Carrier Synchronisation for QPSK (Baseband)

This example models the carrier synchronisation loop shown in Figure 11.19. The synchronisation loop operates at baseband, after the received signal has been demodulated using a fixed frequency oscillator. Phase error measurements are made to determine the rotation of the constellation compared to the desired position, and then a de-rotation operation is performed.

(a) Open the system:

   `/digital/simulation/carrier_synch/QPSK_synch_baseband.slx`

(b) **Inspect the system.** Take a look through this model, and compare it with the signal flow graph provided in Figure 11.19.

(c) **Confirm a carrier frequency mismatch.** Check the parameters of the sine and cosine waves in the transmitter and receiver, and confirm that they differ. Check the values of variables `fc` and `fe` in the MATLAB workspace, to find out the actual frequencies used to modulate and demodulate the signal. The carrier frequency offset will result in a residual frequency term, i.e. the constellation will rotate at a constant rate.

(d) **Simulate!** Run a simulation and inspect the various plots provided. In particular, view the *Constellation Diagram Rx* window in real time as the model runs. You should see the sample points converge towards the desired constellation during the course of the simulation. The *Time Scopes* should confirm the time taken to achieve lock.

(e) **Investigate the correction term.** Notice also that sine and cosine outputs are produced by the synchronisation loop, to de-rotate the constellation. The waveforms displayed in *Time Scope1* should include signals that adapt to a sine and cosine after the initial transient period, similar to Figure 11.20. The frequency of these terms corresponds to the rotation rate of the constellation. See if you can confirm their frequency from the time domain waveforms, or by adding a *Spectrum Analyzer* to the system.

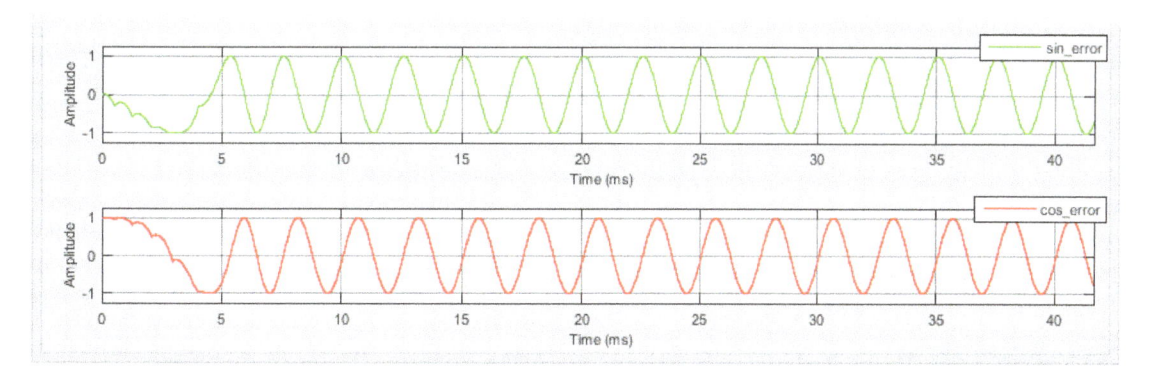

Figure 11.20: Sine and cosine correction terms generated by the baseband carrier synchronisation loop

(f) **Synchronise in noise!** Try adding noise by connecting the *AWGN Channel* block, and re-simulating. Check that carrier synchronisation is still achieved successfully. You can adjust the level of noise by changing the 'Variance' property in the AWGN block.

(g) **Experiment with different carrier frequency deviations.** If you wish, try experimenting with the frequency offset between the transmitter and receiver oscillators, by changing the value of `fe` and re-simulating. If you choose a frequency error larger than about 500Hz, then it may be necessary to extend the simulation time. For even larger frequency offsets, the loop may not (at least, as it is parameterised currently) be able to converge at all.

11.5 Timing Errors and Symbol Recovery

Information transmitted over a communications channel is subject to timing offsets that affect the sampling of the baseband signal, and hence the accurate recovery of symbols. This occurs as a result of the lack of a common time reference between the transmitter and receiver, and the delay and/or Doppler effect introduced by the channel.

Symbol timing synchronisation is performed in the receiver using a dedicated synchronisation loop, with the aim of positioning symbol samples at the best possible timing instants. The first objective is therefore to define these optimal positions. We will then examine the effects of NOT being synchronised, before going on to consider techniques and architectures for symbol timing synchronisation.

11.5.1 *Matched Filtering and Maximum Effect Points*

The recovery of symbols from a received signal requires 'demapping' of timed samples to the appropriate constellation points. We saw in Figure 11.6(a) on page 431 that, in the case of QPSK with a square constellation, the demapping process is undertaken simply according to the quadrant. Practical communications systems suffer from additive noise, which causes the symbol samples to spread out around the theoretical constellation points.

The success of the demapping process requires that samples of the received waveform are taken at the ideal positions—the *maximum effect points*—or as close as possible to these points. As was shown in Figure 11.8 on page 438, the maximum effect points are the instants where: (i) the amplitude of a single pulse is at its greatest; and (ii) the contribution of adjacent pulses is theoretically exactly zero (i.e. there is zero ISI). If the synchronisation process can time symbol samples at the maximum effect points, then the receiver will achieve best possible signal-to-noise (SNR) ratio, and the effects of ISI will be minimised.

The maximum effect points relate to the output of the matched filter. As elsewhere in this chapter, we assume that RC pulse shaping is used, hence the matched filter is a RRC.

Exercise 11.14 **Matched Filtering and Maximum Effect Points**

This exercise demonstrates the 'maximum effect points' at the output of the matched filter.

(a) Open the system:

Chapter 11: Digital Communications Theory and Simulation

> /digital/simulation/symbol_timing/max_effect_points.slx

Inspect the system, and check that the filters in the transmitter and receiver are 'matched' square root raised cosine responses.

(b) **Simulate!** Run the model and view the *Time Scope* outputs. It should be possible to confirm that the samples are consistently taken at amplitudes of +1 and -1. These are the maximum effect points (the model has been configured to take samples at the maximum effect points).

(c) **Change the sampling phase.** Try adjusting the value of the *Delay* block to 3 samples, and re-run the simulation. This means that the symbol samples are taken slightly after the maximum effect points. What effect does this have on the amplitudes of the samples?

(d) **Add noise.** Now connect the *Gaussian Noise Generator* source to the adder block, and re-simulate for both of the *Delay* values used so far (1 sample and 3 samples). You should notice that sampling at the maximum effect points produces 'good' results, even in the presence of AWGN.

(e) **Adapt to QPSK.** The example has considered BPSK modulation (or alternatively, a single phase of QPSK). Adapt and extend the model for QPSK, and introduce a *Constellation Diagram* to view the effects of sampling at the maximum effect points (or not!).

11.5.2 Types of Timing Imperfections

Sampling of the signal at the receiver is subject to timing imperfections that can result in the samples not being taken at the optimum time instants, i.e. the maximum effect points. It is useful to review these in order to motivate the need for timing synchronisation techniques.

Firstly, a ***timing phase error*** occurs if samples are taken at the correct rate (i.e. equal to the symbol rate), but at the incorrect phase. This results in the samples being taken either before or after the maximum effect points, which means that the SNR at the sampling points is not optimal. Another consequence is that ISI can also occur. This is because the symbols before and after make a non-zero contribution to the sampled amplitude, *unless* sampling takes place at the maximum effect point (refer back to Figure 11.8 on page 438 for a recap).

Figure 11.21 compares the sampling of a raised cosine shaped signal, (a) with a timing phase error, and (b) sampled at the maximum effect points. It is clear that the timing phase error causes variations in the sampled values, albeit the correct decisions (here +1 or -1) could still be made in this case. Note that these waveforms correspond to a BPSK modulated signal, or equivalently a single phase (I or Q) of QPSK.

If the timing discrepancy is restricted to a timing phase error, it implies that the *rate* of sampling is perfect. This is not a realistic assumption, because the clock oscillators in the transmitter and receiver will differ slightly, and channel effects such as Doppler can play a part too (factors similar to those affecting the carrier). A more likely scenario is a ***timing frequency error***, where samples are taken at the incorrect frequency (either faster or slower than the rate of the received symbols), meaning that the sampling phase

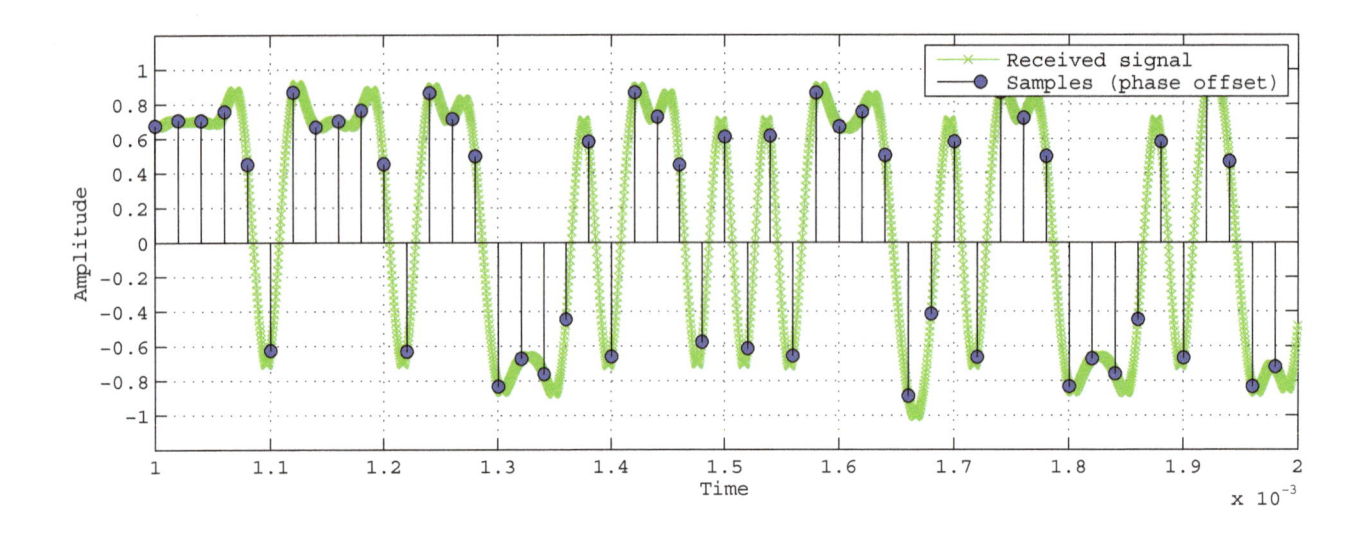

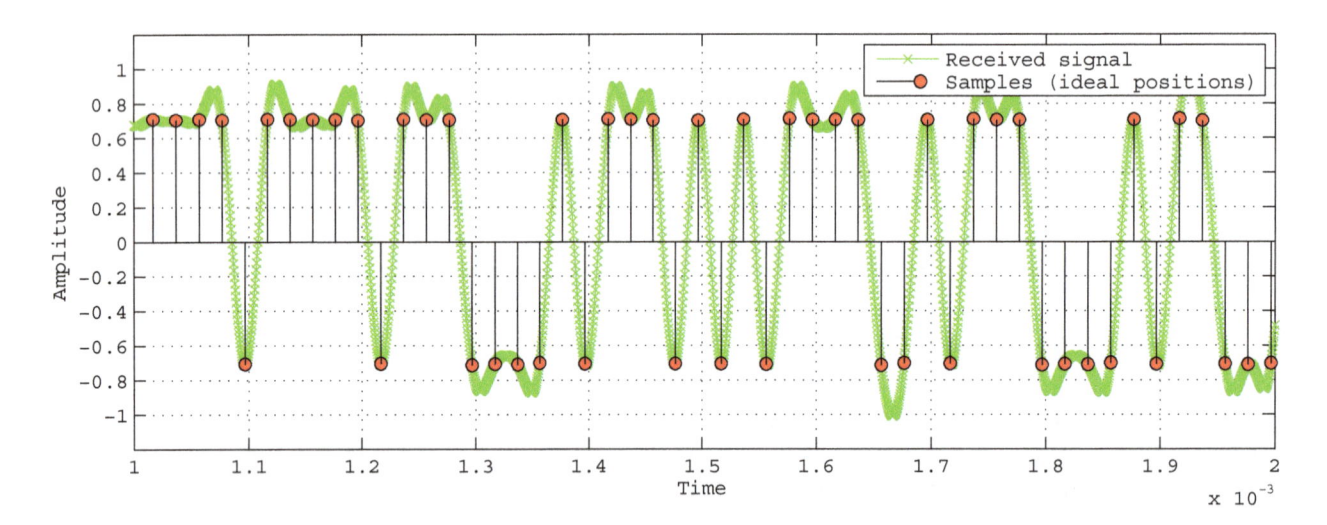

Figure 11.21: Sampling of a raised cosine shaped received BPSK waveform: (top) sampled with phase error; (bottom) with zero phase error (maximum effect points).

drifts over time. As a result, samples are taken at the maximum effect points only for a fraction of the time. We will look at the effect of timing frequency error via the exercises.

An additional defect is timing jitter, i.e. where the time between samples is subject to random variations. This effect can be attributed to the physical characteristics of the ADC, which will have some stated tolerance in terms of jitter performance. Jitter is an unavoidable phenomenon in a real, physical system (such as the RTL-SDR), but it is usually not modelled in computer simulations. We will neglect jitter effects here, because there are more significant sources of timing error that we can attempt to correct, whereas jitter is effectively a source of noise that cannot be removed.

The main problem to be tackled in the receiver is to compensate for the expected *timing frequency error*. This is accomplished using a symbol timing synchronisation method, to be covered in Section 11.6.

Chapter 11: Digital Communications Theory and Simulation

Exercise 11.15   Symbol Timing Imperfections: Sampling Phase Error

In this example, we confirm the effect of a sampling phase error (similar to Figure 11.21) with the aid of a Simulink model.

(a) Open the system:

/digital/simulation/symbol_timing/sampling_phase_error.slx

(b) Inspect the system, noting that the grey block on the left hand side models a baseband transmitter. The *Transmitter* block outputs two continuous time 'analogue' waveforms (I and Q phases) that are then sampled by the *Zero-Order Hold* blocks (representing ADCs).

(c) **Check the timing offset parameter.** 2️⃣ on the *Transmitter* block to open the configuration properties. A timing offset can be introduced to the transmitter, but this should be set to zero for this simulation (we are not considering a timing frequency offset here).

(d) Notice the yellow *Delay* block, which we will use to control the sampling phase in this simulation (the sampling *frequency* in the transmitter and receiver are exactly the same). The yellow block is initially set to a 1 sample delay.

(e) **Simulate!** Run a simulation and view the *Time Scope* and *Constellation Diagram* outputs. You should notice that the amplitudes of the symbol samples vary considerably, which causes the constellation to 'spread out' compared to the ideal positions. It should, however, be possible still to tell the symbols apart.

(f) **Check the EVM.** You can also inspect the EVM measurements using the *Signal Quality* feature of the constellation plot (refer back to Exercise 11.3 if you need a recap).

(g) **Experiment with the sampling phase.** Next, try altering the *Delay* in the yellow block, to shift the sampling phase. You should see some different results depending on the values you choose! See if you can find the 'best' delay, i.e. where the sampled symbols are closest to the constellation reference points, and the EVM is lowest. (The initial setting is not the best!).

Exercise 11.16   Symbol Timing Imperfections: Sampling Frequency Error

Following on from Exercise 11.15, we now consider a system with a sampling frequency error. This means that the sampling phase error changes over time.

(a) Open the system:

/digital/simulation/symbol_timing/sampling_frequency_error.slx

(b) View the model, and confirm that a small timing frequency offset has been applied to the transmitter block (2️⃣ to open it and view the parameter).

(c) **Simulate!** Run a simulation and inspect the plots produced. As the simulation progresses, you should notice that sampling positions drift in phase, as shown below. This results in a constellation that expands and contracts around the reference points.

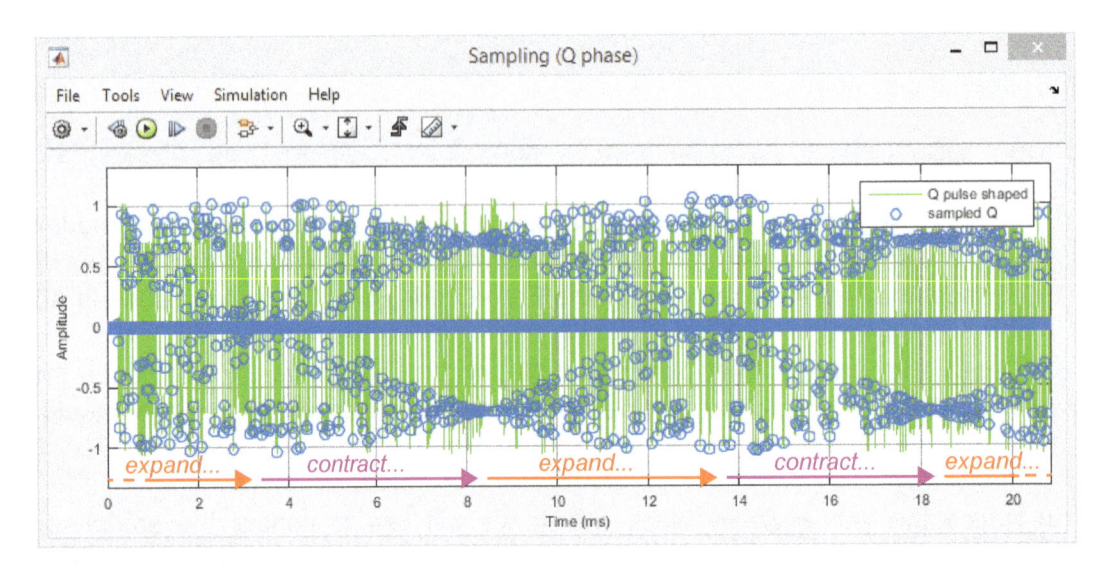

[Figure: Sampling (Q phase) plot showing Q pulse shaped signal and sampled Q points over Time (ms) from 0 to 20, with annotations indicating expand... contract... expand... contract... expand... phases.]

(d) **Alter the timing frequency.** Try changing the timing frequency offset of the *Transmitter* block to 1%, and resimulating. How does the speed of the phase drift change?

(e) Try again with a timing frequency offset of 0.1% to confirm your observations.

(f) **Experiment with timing frequency.** Finally, make the transmitter frequency lower than that of the receiver, by entering a negative value in the transmitter block. Try entering -1%... what happens this time?

11.5.3 Symbol Recovery

Having observed the effect of timing errors on the sampling process, it is useful to confirm whether this translates into symbol decision errors, and hence bit errors. We therefore return to the two models from the previous section, and add a *QPSK Demodulator* block at the output.

Exercise 11.17 Symbol Decisions: Sampling Phase Error

In this example, we extend the model from Exercise 11.15 to add a symbol decision operator, and confirm whether the error in sampling phase causes symbol errors.

(a) Open the system:

`/digital/simulation/symbol_timing/decision_phase_error.slx`

Chapter 11: Digital Communications Theory and Simulation

(b) **Inspect the section** at the far right hand side of the model. Here, a 'synchronisation delay' is added to the transmitted bit stream, so that it can be compared against the received bit stream in the *Time Scope*. If the bits are not the same, then a bit error is detected (note that any errors detected during the initial synchronisation delay phase should be ignored).

(c) **Simulate!** Run simulations for different sampling phases (set by changing the value in the yellow *Delay* block, as in Exercise 11.15) and check whether any errors result. You may need to adjust the synchronisation delay as you do so. Bear in mind that this is a 'perfect' model with no noise!

(d) **Add noise.** Try adding noise to the I and Q signals output from the *Transmitter* block, and simulating again. You might see some errors this time, depending on the level of noise you have added! It would also be useful to view the constellation while your simulations are running.

Exercise 11.18 Symbol Decisions: Sampling Frequency Error

This exercise turns to the case of a sampling frequency error. Due to the difference between the rates of sending and receiving symbols, either more or fewer symbols are sent than received, and the errors cannot be visualised in the same way as Exercise 11.17.

(a) Open the system:

   `/digital/simulation/symbol_timing/decision_frequency_error.slx`

(b) **Explore the design.** View the system, in particular the section at the right hand side. Note that the transmitted and received bits are both written to Workspace variables. MATLAB code written in the StopFcn of the model will plot these outputs at the end of a simulation (you can view this by opening *File -> Model Properties -> Callbacks -> StopFcn*).

(c) **Establish the timing frequency error.** on the *Transmitter* block and note the % timing frequency error. The offset specifies whether the transmitter produces bits faster than the receiver samples them (if there is a positive offset), or more slowly (if the offset is negative). If the setting is 0, then the two rates are the same.

(d) **Simulate!** Run a simulation and view the figure that is generated. Do the transmitted and received bits match throughout the simulation?

(e) Try altering the offset in the *Transmitter* block to a positive offset of 2%, and viewing the results. Are there bit errors this time?

(f) Zoom into the very end of the window. How do the numbers of transmitted and received bits compare? You might notice that there are either fewer received than transmitted bits, or vice versa, for instance:

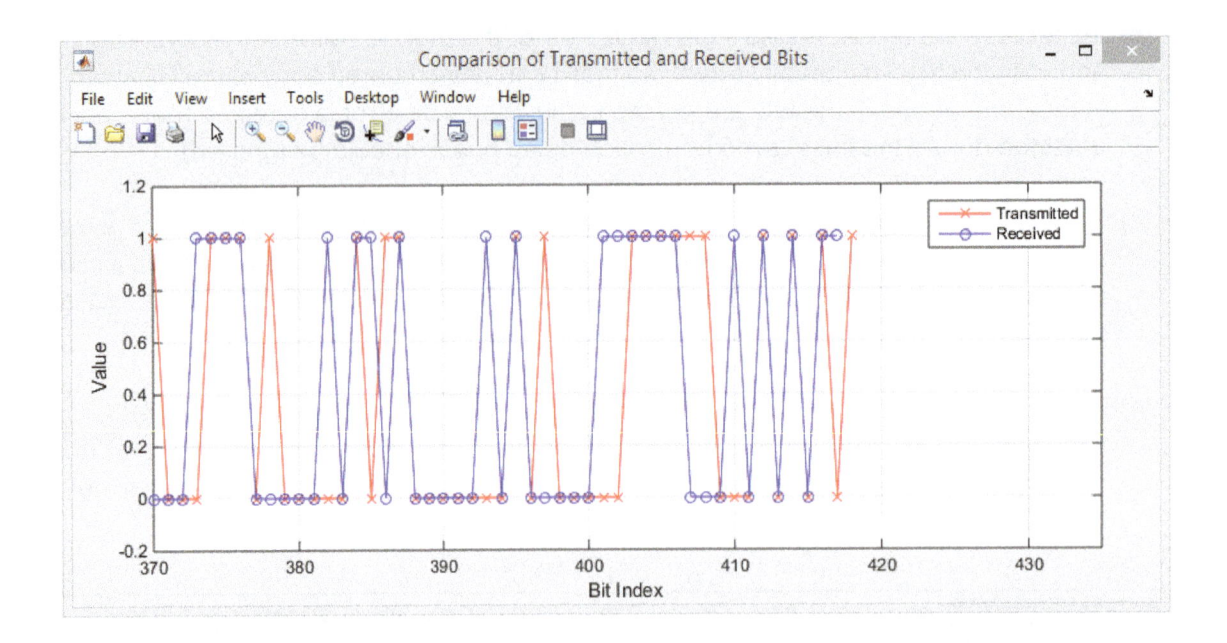

Figure: Comparison of Transmitted and Received Bits

(g) Experiment with timing frequency offsets of -2%, -0.5%, 0.5%, and any other values you would like to try, and view the results. Notice that, for very small offsets, you may need to extend the simulation time before any bit errors are evident.

Based on the observations made in the exercises in this section, it is clear that data cannot be accurately recovered in the receiver if there is an uncorrected sampling frequency offset between the transmitter and receiver. Moreover, sampling at the correct rate, but with a timing phase offset (i.e. not sampling at the maximum effect points), lowers the SNR at the point of sampling the symbols, and thus increases the likelihood of errors.

Both types of timing imperfection can be corrected in the receiver using an appropriately designed synchronisation circuit.

11.6 Symbol Timing Synchronisation

Most communications systems require symbol timing synchronisation to be performed in the receiver due to the lack of a common timing reference. Without synchronisation, phase and frequency timing offsets go uncorrected, causing errors in reception of the transmitted data. This is particularly true in the likely scenario of a timing frequency offset, where it may be impossible to recover the data.

A timing synchroniser measures the timing error and provides a feedback control signal that dynamically alters the timing parameters. The high level model of a timing synchroniser is therefore very similar to a PLL. In this case, the error detector measures timing error, rather than a phase difference between two sinusoids (as in a PLL), and the adjustable element is a Voltage Controlled Clock (VCC) or Numerically Controlled Clock (NCC) instead of the NCO or VCO needed to generate a sine wave in a PLL. The role of the NCC is to generate a sampling clock reference with a rate and phase that match those of the received signal.

A block diagram of a generic symbol timing synchroniser is shown in Figure 11.22.

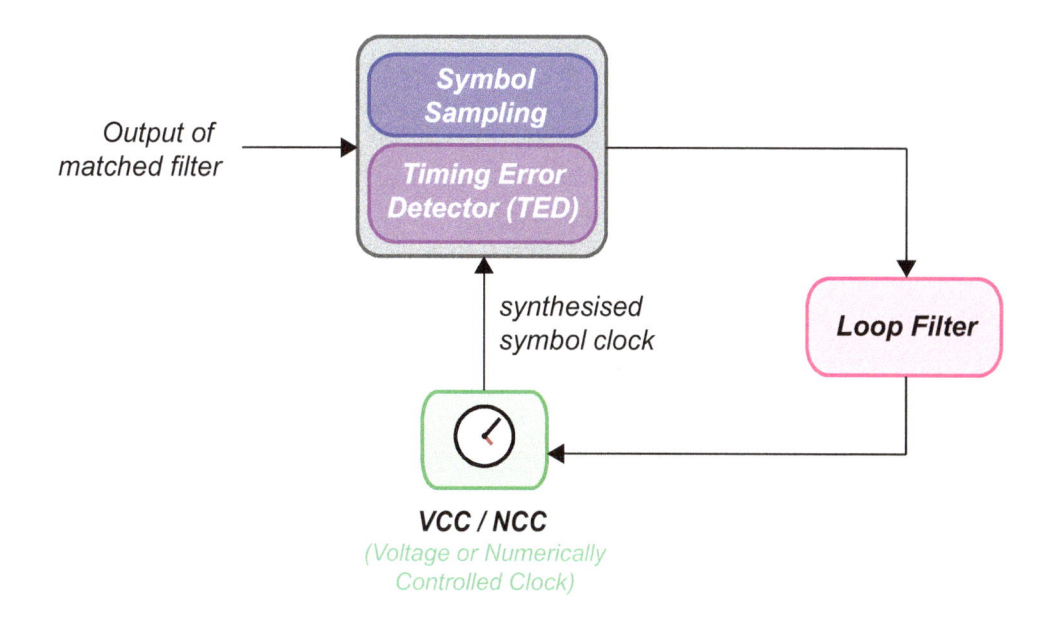

Chapter 11: Digital Communications Theory and Simulation

Figure 11.22: Model of a generic timing synchroniser

11.6.1 Timing Error Detector (TED)

The Timing Error Detector (TED) generates an error signal that instructs the timing adjustments made in the feedback loop, and it can be realised using several different algorithms. In the examples that follow, we will work with an Early-Late TED (to be covered in detail in Section 11.6.4), as this is a particularly intuitive method. Other TED algorithms include the Maximum Likelihood TED, the Mueller and Müller TED, and the zero-crossing TED (and variations) [30], [31], [38].

Allied to the TED algorithm itself, there are two general approaches to the timing synchronisation task. These are:

- **Data aided** — where the sequence of received symbols is known in advance (e.g. during a training sequence), and can be used in the generation of the timing error signal;
- **Decision directed** — where the sequence of symbols is unknown, and the timing error signal must be generated based only on the statistics of the received signal.

In the discussion and examples that follow, we will assume a *decision directed* mode of operation.

11.6.2 Timing Adjustment

The synchronisation loop has the task of adjusting the timing of symbol samples, based on measurements of the timing error. There are also alternative methods of performing the timing adjustment:

- **Oversampling** — where the input to the synchroniser is sampled at a rate several times higher than symbol rate, and the loop adjusts to select the samples closest to the maximum effect points.
- **Interpolation** — where the input signal is oversampled by a small amount (e.g. 2 x symbol rate) and then interpolated to provide the amplitude value at the desired sampling instant.

Figure 11.23 is provided to help explain the oversampling method. Here, the synchronisation loop adapts to select the samples closest to the maximum effect points. There are a limited number of samples from which to choose, with the likelihood being that the maximum effect point falls between two samples. The closer of the two available samples is therefore selected. As the 'perfect' timing instant is not available using this oversampled method, the nearest available timing instant is used instead. A small phase error is introduced by doing this, which can be denoted as $\theta_e[k]$ for sample index k. This error will vary from sample to sample, depending on the relationship between the rates involved.

For the example of an oversampling ratio of six (i.e. 6 samples per symbol), the closest available sample will be less than one twelfth of a symbol period from the maximum effect point. This timing phase error corresponds to a small error in the amplitude of the sampled pulse, as shown in Figure 11.23. A further effect is that, when a particular raised cosine pulse is not sampled at the maximum effect point, nearby symbols make a contribution to the sampled amplitude. In other words, ISI occurs (referring back to Figure 11.8 on page 438, the contribution of nearby symbols is seen to be zero *only* when samples are taken at the maximum effect points). Over time, these ISI-related errors can be considered as a source of noise. The effect is often referred to as 'self-noise' because it arises internally to the synchronisation loop [31].

For higher oversampling ratios, the maximum phase error is smaller and the resultant level of self-noise is lower. For the example of an oversampling ratio of 50, the maximum phase error between the maximum effect point and nearest available sample would be one hundredth of a symbol period, resulting in a much smaller level of self-noise. Clearly the disadvantage of increasing the oversampling ratio is that processing must take place at a higher rate, which implies a greater computational burden.

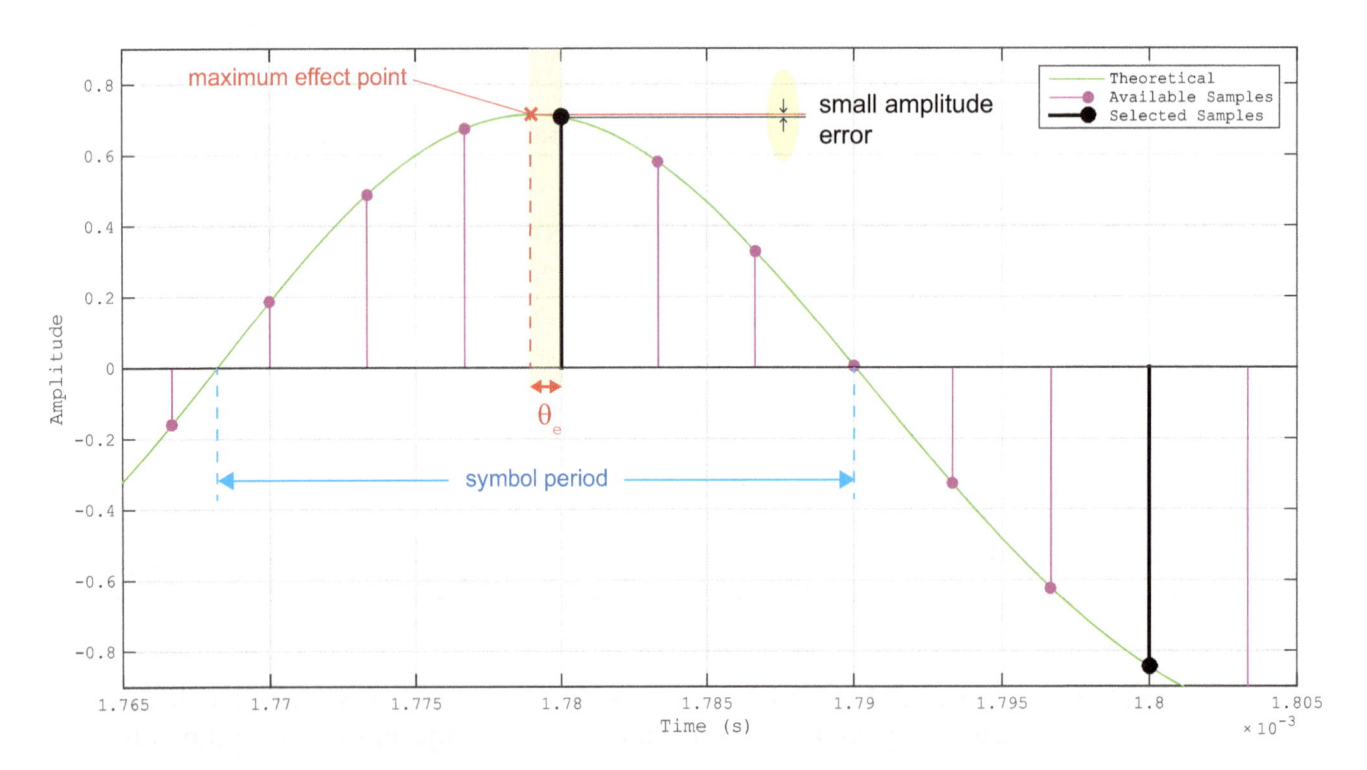

Figure 11.23: The oversampling method and symbol sampling errors

Chapter 11: Digital Communications Theory and Simulation

Later, we will look at a couple of examples comparing synchronisation loops with different degrees of oversampling.

Interpolation techniques work in a different way. Rather than 'shifting' the timing of samples to the nearest available positions, a reduced set of two samples per symbol is taken, and the amplitude at intermediate points is derived via interpolation of these available samples. Thus, the amplitude at the maximum effect point can be determined, almost as if a sample were 'created' at the desired timing instant. This approach is illustrated in Figure 11.24.

Interpolation techniques exploit the fact that the signal is sampled at a rate at least equal to Nyquist, and therefore that all information is retained via the two samples taken per symbol. Computation is necessary to generate the 'missing' information, and this can be undertaken using polyphase interpolation (which effectively generates a much larger set of samples from which to find the closest), or via piecewise polynomial interpolation techniques, implemented using a *Farrow* structure or similar [12], [38].

In this section, we will focus on oversampled techniques.

11.6.3 *Numerically Controlled Clock (NCC)*

In the oversampled synchroniser model, the role of the NCC is to emit a 1 clock-cycle pulse (or 'strobe') to control the taking of symbol samples. Once synchronised, the strobe selects the closest sample to each maximum effect point.

The NCC operates in a very similar manner to an NCO, in the sense that it uses an accumulator to control the frequency of a synthesised waveform. Rather than generating a sine wave (like an NCO), the NCC

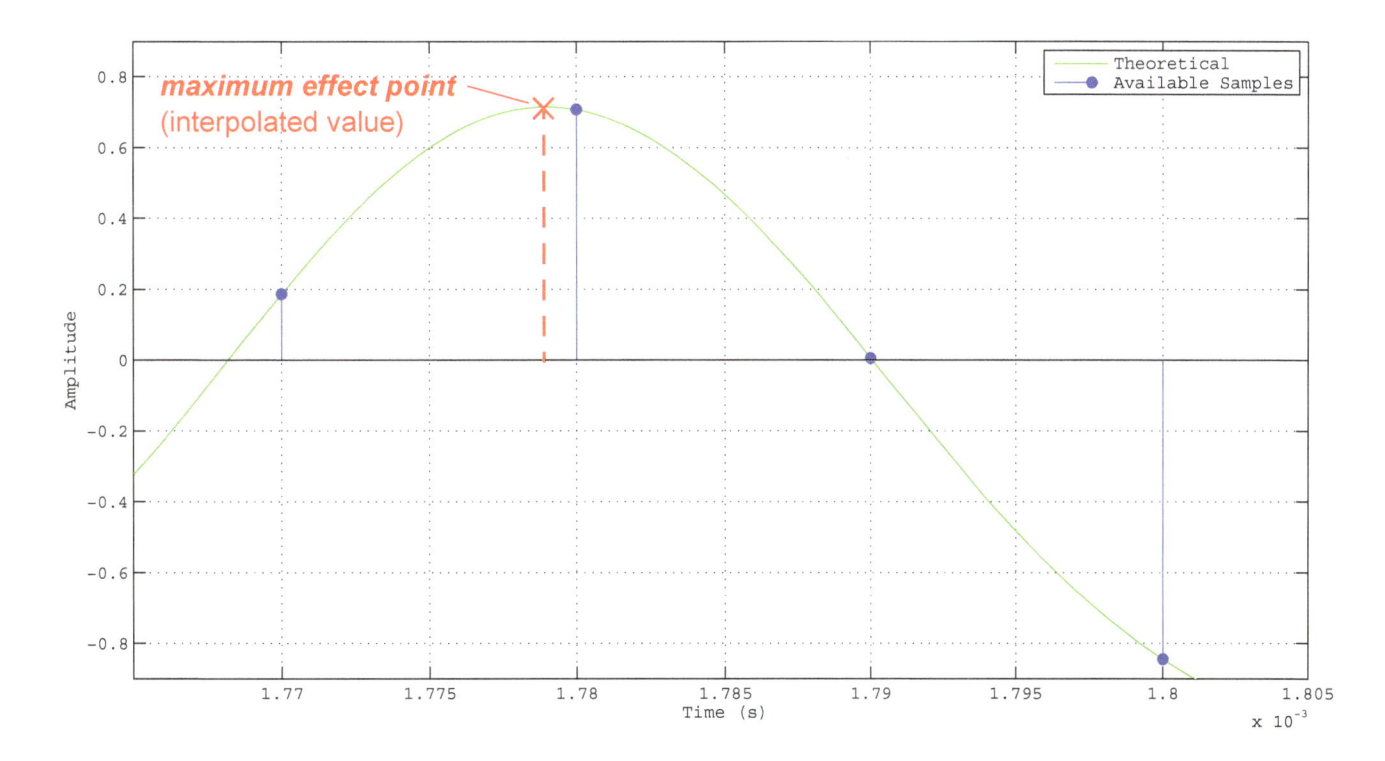

Figure 11.24: Sampling at a maximum effect point using interpolation

produces a repeating strobe signal. It achieves this by incrementing an accumulator, and detecting when the count reaches its maximum value and wraps back round to zero. Control is exerted via the step size input, with a larger step size causing the accumulator to increment more quickly and hence increase the rate of producing strobes. A sketch showing an example NCC output is provided in Figure 11.25.

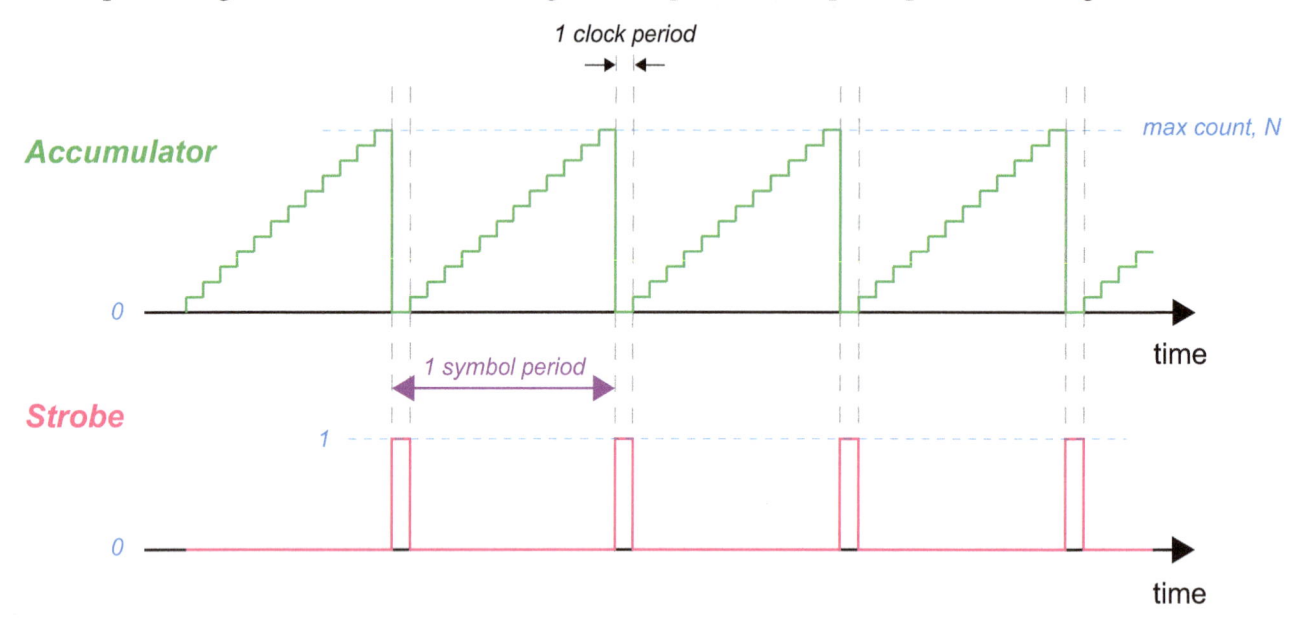

Figure 11.25: Operation of the NCC to produce strobes

In the timing synchroniser, the quiescent rate of the NCC is set corresponding to the expected symbol rate. An adjustment to the step size is applied via the filtered feedback error signal, causing the rate of producing strobes to increase or decrease accordingly.

In the systems presented in this section, the quiescent step size is 1, and the maximum count is set as N, the oversampling ratio. The count held by the NCC, $r[k]$ can therefore be expressed as

$$r[k] = r[k-1] + 1 + \mu_{fb}[k] \quad mod(N) , \qquad (11.4)$$

where $\mu_{fb}[k]$ is the feedback control signal and k is the sample index. The output of the NCC is given by

$$s[k] = \begin{cases} 1 & \text{when } r[k] < r[k-1] \\ 0 & \text{otherwise} \end{cases}, \qquad (11.5)$$

or in other words, the strobe signal is produced only when the accumulator wraps round. Thus for a feedback step size of $\mu_{fb} = 0$, the NCC would produce strobes at the expected symbol rate, i.e. once every N clock cycles. Where the adjusted step size is not an integer divisor of N, the intervals between strobes will differ over time, but the desired rate will be produced *on average*.

In the second style of implementation mentioned in Section 11.6.2, the interpolation model, the NCC accumulator output controls the interpolation process, e.g. it selects the desired component of a polyphase interpolator. In that case, the NCC would provide a numerical value rather than a strobe.

Chapter 11: Digital Communications Theory and Simulation 469

Exercise 11.19 Numerically Controlled Clock (NCC)

The NCC is an important part of the timing synchronisation circuits introduced in this chapter. Before looking at timing synchronisers in detail, it is useful to confirm via simulation how the NCC operates as an individual component.

(a) **Open the system:**

    `/digital/simulation/symbol_timing/nc_clock.slx`

(b) **View the model.** Investigate the parameters of the model, in particular the sampling rate, f_s, the oversampling ratio of the NCC, and the timing adjustment applied to the block.

(c) Evaluate the rate at which the NCC should produce strobes (based on its initial parameters). In a timing synchroniser, these strobes are used to take samples of the received, matched filtered symbol, at symbol rate. Hence calculate the expected symbol rate in this case, in kbps.

(d) **Simulate!** Run a simulation and check that the *Time Scope* shows the expected results. You should see something like Figure 11.26 (note that the observed colours may be different). The upper plot shows the accumulator signal, which is provided as an output for this example but is normally internal to the NCC, and the lower plot shows the strobe output.

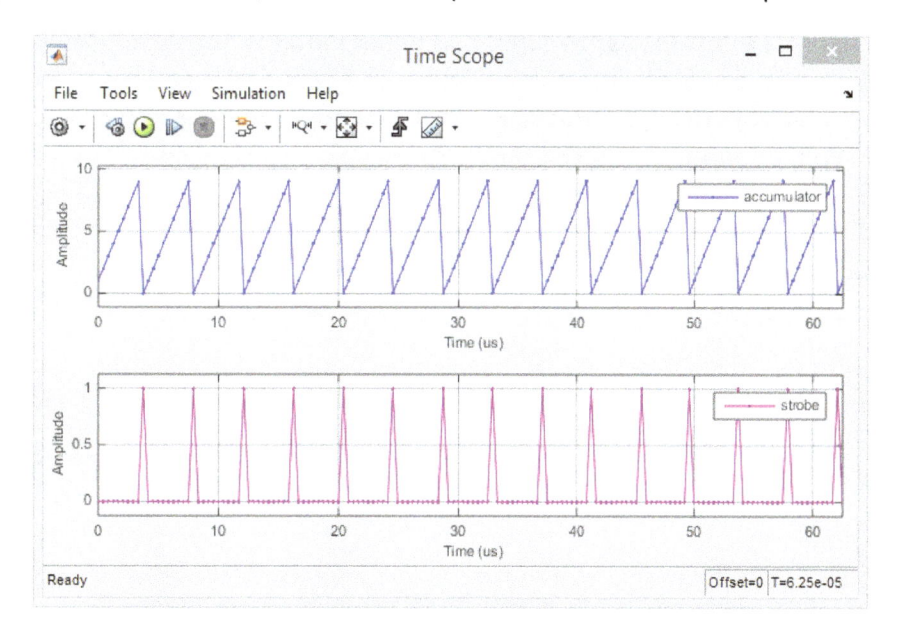

Figure 11.26: The accumulator signal and strobe output of an NCC

(e) **Adjust the input.** Try changing the value of the *Constant* block to 0.1, and then re-simulate. Do the strobes become more or less frequent? Calculate the rate of the strobes in this case.

(f) Repeat for a *Constant* input of -0.1, and view the results this time. You should notice the opposite effect.

(g) **Apply a dynamic input.** The above steps have been based on applying a constant input to the NCC. In a synchronisation loop, however, the NCC will need to adjust the rate of producing strobes based on the measured signal timing parameters. Investigate what happens when you apply a time-varying input to the NCC (you could try using a *Ramp* source block, for instance). Note that the source block may need to be followed by a *Data Type Conversion* block to set the sample time.

11.6.4 Early Late Timing Error Detector

The TED is an important part of the timing synchronisation circuit, because it generates an error signal to adjust the timing parameters of the NCC. The NCC controls the taking of symbol samples from the output of the matched filter in the receiver.

The TED computes a timing error,

$$\tau_e = \tau - \hat{\tau} \tag{11.6}$$

where τ is the ideal timing instant, and $\hat{\tau}$ is the timing estimate. The timing error is used to control the NCC, such that the timing error for the next symbol is reduced. For instance, if the timing estimate is earlier than the ideal sampling time, the control signal to the NCC will adjust the next timing estimate such that the signal is sampled slightly later, and hence closer to the ideal position. This causes the next sampling period to be extended, as shown in Figure 11.27 (noting that it may take a few iterations before the samples are placed perfectly). Conversely, if the current symbol sample is taken too late, the period until the next symbol sample is reduced.

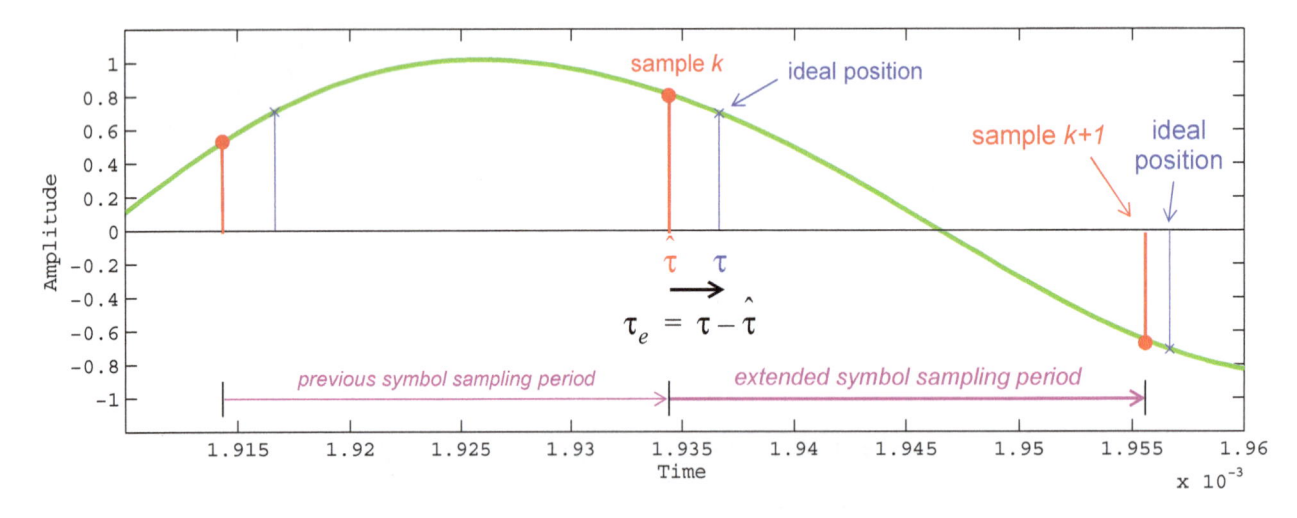

Figure 11.27: Adjusting the sampling period based on the timing error

The relationship between the symbol timing error, as defined in Eq. (11.6), and the output of the TED, is commonly described by an S-curve, similar to that used in characterising PLLs. The precise shape of the S-curve depends on the type of TED used. The general behaviour of the TED is to produce a negative

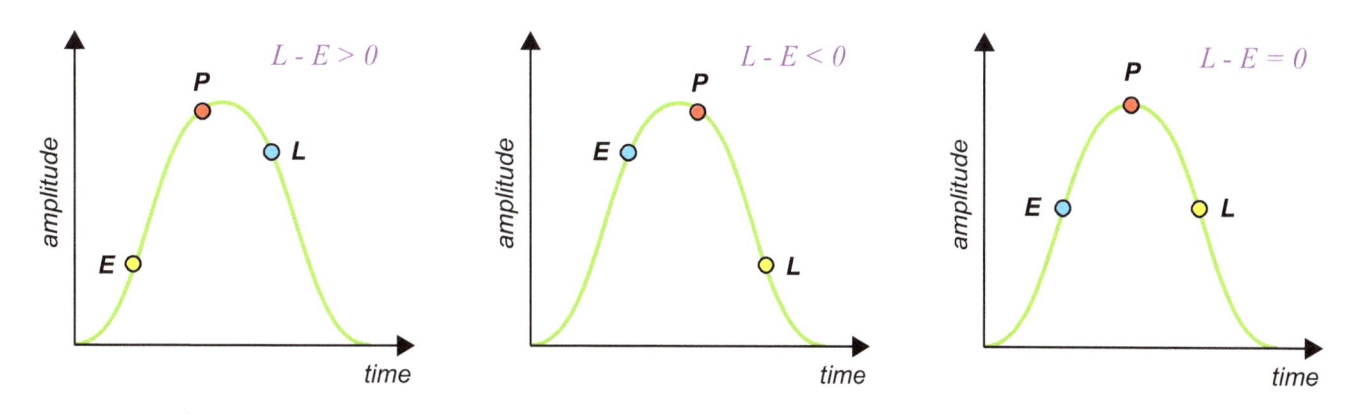

Chapter 11: Digital Communications Theory and Simulation

Figure 11.28: Examples of the error signals generated by the Early and Late branches

output when the timing error is negative, and a positive output when the timing error is positive. In both cases, the magnitude of the TED output varies with the magnitude of the timing error, and when the timing error is exactly zero, the output should also be zero, meaning that no adjustment is necessary. We will look at example S-curves shortly.

One popular method of implementing the TED is the 'Early Late' method, which operates by taking three samples of the matched filtered signal, spaced in time by an interval less than a symbol period (a spacing of half a symbol period is often chosen). These are denoted as the Early (E), Punctual (P), and Late (L) samples.

The timing error is derived from the difference between the amplitudes of the early and late samples, as shown in Figure 11.28. If the P sample is too early, then $L - E > 0$, and if the P sample is too late, then $L - E < 0$. If the P sample is perfectly placed at the maximum effect point, $L - E = 0$. The earlier or later the P sample compared to the ideal position, the greater the magnitude of $L - E$.

Note that the diagrams in Figure 11.28 illustrate the point using positive pulses only. To accommodate negative pulses, as is necessary in practice for QPSK, M-QAM and other modulation schemes, one of two approaches can be taken. Either (i) the error can be multiplied by the sign of the symbol [38], or (ii) the input signal can be rectified or squared, prior to the early-late operation. We will generally use the latter approach in the examples that follow.

More formally, and following the notation adopted in [38], the error signal produced by the Early Late TED at sample k can be expressed as

$$e[k] = \hat{a}[k]\left[x(kT_s + \Delta T_s + \hat{\tau}) - x(kT_s - \Delta T_s + \hat{\tau})\right] \qquad (11.7)$$

where the symbol period is T_s, the samples are separated by time ΔT_s, and $\hat{\tau}$ is the timing estimate. In other words, the symbol estimate $\hat{a}[k]$ is used to weight the $L - E$ quantity as described earlier. Where the sample separation is half a symbol period, Eq. (11.7) can be rewritten as

$$e[k] = \hat{a}[k]\left[x((k+0.5)T_s + \hat{\tau}) - x((k-0.5)T_s + \hat{\tau})\right]. \qquad (11.8)$$

As mentioned above, an alternative method involves squaring the signal prior to the Early Late TED, resulting in the modified expression,

$$e[k] = \left[x^2(kT_s + \Delta T_s + \hat{\tau}) - x^2(kT_s - \Delta T_s + \hat{\tau})\right]. \tag{11.9}$$

The S-curve plots the expectation of the error output generated by the TED, $g(\tau_e)$, against the normalised timing error, τ_e/T_s, where

$$g(\tau_e) = E\left\{a[k]\left[x(kT_s + \Delta T_s + \tau) - x(kT_s - \Delta T_s + \tau)\right]\right\} \tag{11.10}$$

for the case corresponding to Eq. (11.7), and

$$g(\tau_e) = E\left\{a[k]\left[x^2(kT_s + \Delta T_s + \tau) - x^2(kT_s - \Delta T_s + \tau)\right]\right\} \tag{11.11}$$

for the case corresponding to Eq. (11.9).

The decision-directed and data-aided versions produce different results, because the symbol estimates in the former case may not always be correct, especially for larger symbol timing errors.

The S-curve can be formally derived as described in [38], and it is a function of the pulse shape (incorporating the roll-off rate where the RC is used). An important point to note is that the gradient at $\tau_e = 0$ acts as the gain, K_p, of the Early Late TED. The gain is affected by the received signal amplitude and average energy, and it is undesirable for a parameter of the synchronisation loop to depend on the amplitude of the input signal. Therefore, it is common to include a prior AGC stage to ensure that the amplitude of the signal input to the TED, and hence K_p, are held constant during operation.

Figure 11.29 shows example S-curves for the raised cosine and squared raised cosine, for a roll-off rate of $\alpha = 0.35$. This represents the data-aided case, where the symbol estimates are always correct. Where decision-directed method is used, then the S-curves are likely to deviate for larger timing errors (approximately $|\tau_e| > 0.35$), due to incorrect symbol decisions being made [38]. The TED gain values extrapolated from the S-curves are $K_p = 2.69$ for the symbol-corrected version, and $K_p = 3.22$ for the squared version. The appropriate value of K_p should be used when calculating the other parameters of the synchronisation loop.

It is worth reiterating that the left hand side of the S-curve in Figure 11.29 represents negative timing errors, where the sample has been taken too late. The TED therefore produces a negative result, the magnitude of which varies with the degree of lateness. Similarly on the right hand side, the timing errors are positive, indicating that the sample has been taken too early. The output of the TED is then a positive value whose magnitude increases with the degree of timing error.

11.6.5 Early Late Timing Synchroniser

The complete Early Late synchroniser includes a Loop Filter, NCC, and other delays, in addition to the Early Late TED. A block diagram of an Early Late Synchroniser is shown in Figure 11.30.

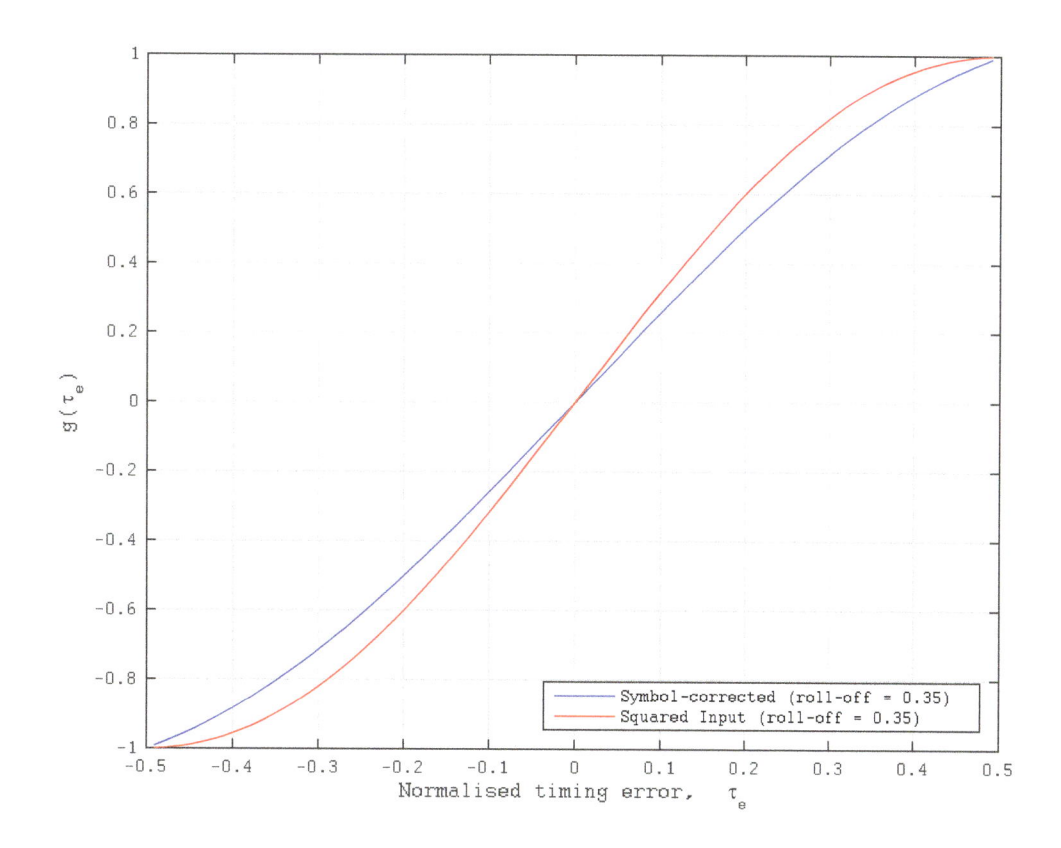

Chapter 11: Digital Communications Theory and Simulation 473

Figure 11.29: Example S-curves for the Early Late TED (data-aided)

In this timing synchroniser model, the error determined by the TED is filtered by a Loop Filter and supplied to the NCC. Thus it has a similar structure and operation to the PLLs considered earlier. It is intuitive to *add* the timing adjustment signal output by the Loop Filter to the quiescent step size of the NCC, which is the approach taken in this model. This needs to be understood in terms of the behaviour of the timing adjustment.

It was observed in Figure 11.27 that early sampling requires the subsequent symbol sampling period to be extended, which corresponds to a reduction in the frequency of the NCC. According to Figure 11.28, however, early sampling produces a positive error signal. If added to the quiescent step size, this would cause the NCC to produce strobes more frequently and thus shorten the period—opposite to the desired effect. Once the loop has converged, what actually happens is that the NCC produces strobes half a symbol period later than the maximum effect point. That is the reason why the strobes are delayed by a further half a symbol period to make a whole symbol period, prior to sampling the Punctual branch. This can be seen in the lower portion of Figure 11.30.

Alternatives exist and may be encountered in other literature, including subtracting the timing adjustment signal from the quiescent step size of the NCC, rather than adding it; or delaying the Punctual branch for a full symbol period to align with strobes output directly from the NCC.

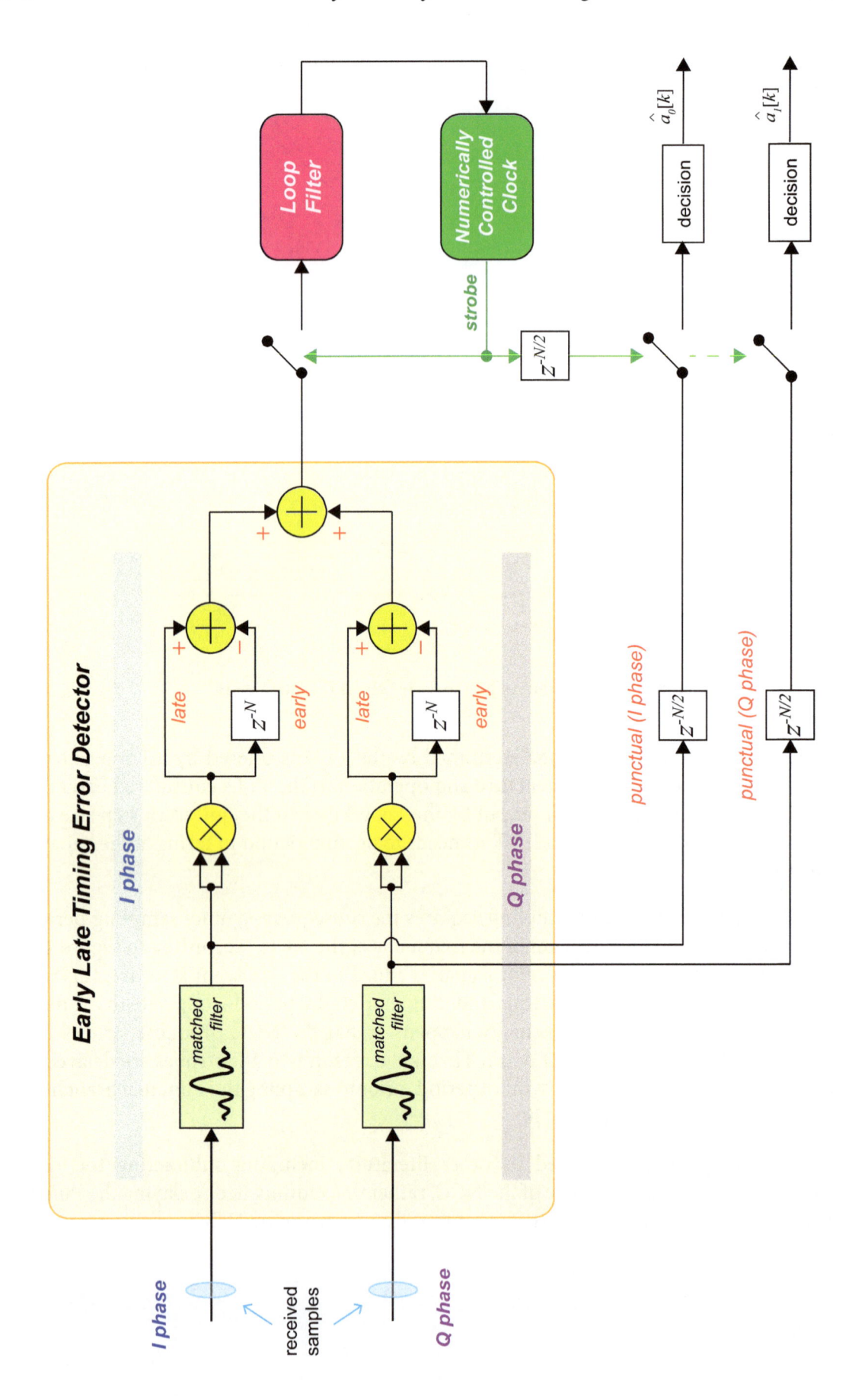

Figure 11.30: Block diagram of an Early Late Symbol Synchroniser

Chapter 11: Digital Communications Theory and Simulation 475

Exercise 11.20 — Early Late Timing Synchronisation (Raised Cosine Pulses)

In this example, the 'Early Late' method of timing error detection is used to drive a synchronisation loop. Raised cosine pulse shaping is used, implemented as a matched pair of RRC filters.

(a) **Open the system:**

   `/digital/simulation/symbol_timing/early_late_RC_48.slx`

(b) **Explore the model.** Inspect the system and review the parameters. You should be able to confirm that this system operates with an oversampling ratio of 48. This means that the input to the synchronisation loop has nominally 48 samples per symbol.

(c) **Confirm the synchronisation loop parameter design.** Check out the contents of the model's PreLoad Function (*File -> Model Properties -> Callbacks -> PreLoadFcn*). Notice that a similar set of parameters has been specified as for a PLL, including the damping ratio, bandwidth, etc. The gain value associated with the Early Late TED is set to 3.22, based on analysis of the RC pulse shape with $\alpha = 0.5$ (similar to Figure 11.29, and to be further considered in Exercise 11.21), and the gain of the NCC is 1. This information is used to compute coefficients for the loop filter parameters.

(d) **Investigate the NCC.** View the internal workings of the NCC and compare it to Figure D.6 (in Appendix D on PLLs, page 606). Note that the taking of samples is controlled by the strobe generated by the NCC, which is applied to registers (*Delays*) as an enable signal.

(e) **Simulate!** Run a simulation and view the *Constellation Diagram* as the simulation runs—you should see the samples converge towards the ideal constellation points. Inspect the other plots and ensure that the system has achieved symbol timing synchronisation. Ideally you should see the samples positioned at the maximum effect points, similar to the waveform below.

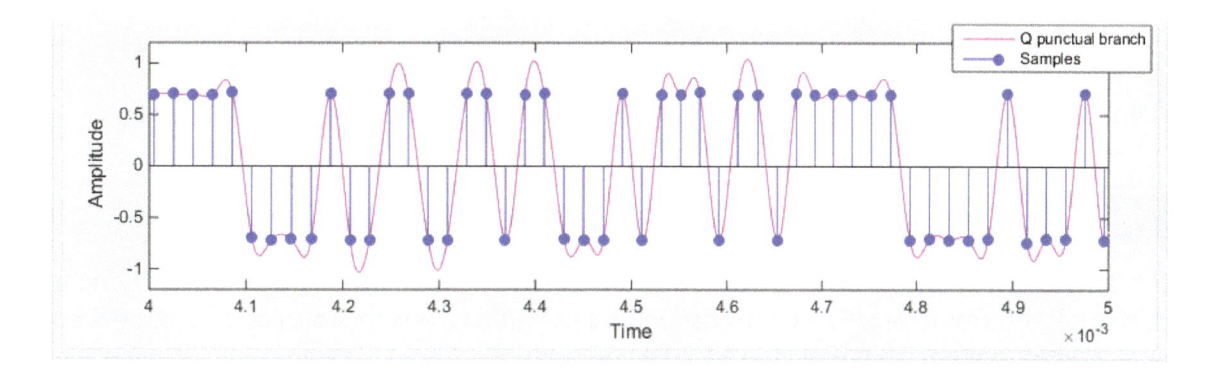

(f) Try repeating the simulation for different timing offsets (specified as percentages in the *Transmitter* block—try both positive and negative values), and see what happens as a result.

(g) **Simulate in the presence of noise.** Lastly, you might also wish to add some noise to the system to model an AWGN channel. You can do this by incorporating two *Random Source* blocks and adding their outputs to the I and Q signals, prior to the *Zero Order Hold* blocks (ADCs).

Exercise 11.21 **Early Late TED: Gain Coefficient**

The gain coefficient of the Early Late TED depends on several factors, as discussed in Section 11.6.4. The calculation of the appropriate value for the TED gain, K_s, can be undertaken during a MATLAB script as confirmed here.

(a) Open the file:

> /digital/simulation/symbol_timing/ELTED_s_curve.m

(b) Read through the code and check that you understand the steps involved.

(c) **Run the script.** Execute the script and inspect the generated plots. Alternatively, you may wish to 'Publish' the script to view the comments, code, and plots in an easy-to-read HTML format. The script can be published from the main menu, as shown in Figure 11.31.

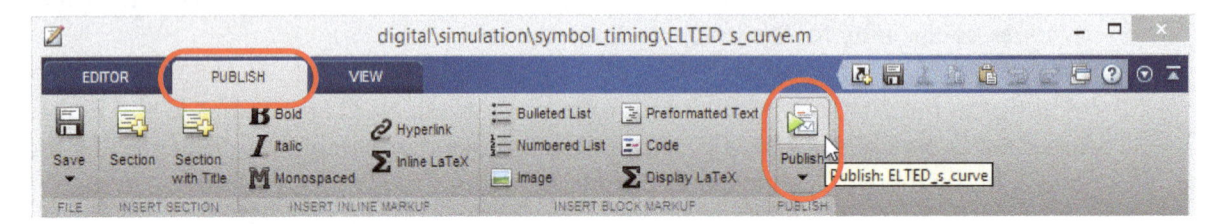

Figure 11.31: Using the 'Publish' feature to view the generation of the S-curve and TED gain parameter

(d) Note the value of K_s that has been calculated in each case.

(e) **Experiment with the parameters.** Try changing the parameters, such that the RC has a roll-off of 0.22, and is evaluated over 16 symbol periods.

(f) **Rerun.** Execute the script (or re-publish) to inspect and evaluate the results.

Exercise 11.22 **Early Late Timing Synchronisation (Design Task)**

This 'challenge' exercise calls for you to design and test your own symbol timing synchroniser for a specific set of parameters. For a real challenge, start with a blank model! You can inspect and learn from the reference system from Exercise 11.20 if you wish.

The design parameters are as follows.

Parameter	Symbol	Value
Sampling frequency	f_s	100MHz
Modulation scheme		QPSK
Data rate	R_{data}	500kbps
Pulse shape		Raised cosine ($\alpha = 0.35$)
Gain of Early Late TED	K_s	? (refer to Exercise 11.21)
Damping ratio	ζ, d	1
Normalised bandwidth		5% of symbol rate

In this section, we have looked at one particular method of performing symbol timing synchronisation. It was noted that the loop has some similarities with the PLL models considered in Chapter 7 (in particular, that the general architecture comprises an error detector, loop filter, and an *adjuster*). The design of the loop parameters to achieve the desired synchronisation characteristics has also been considered.

The exercises in this section implemented the TED using the Early Late algorithm, and an oversampled architecture was adopted. It was also assumed that the synchroniser operated in decision-directed mode. As such, the options considered here represent only a subset of possible implementations for symbol timing synchronisation. The interested reader may refer to the literature for information about other methods [31], [38].

11.7 Digital Receiver Design: Joint Carrier and Timing Synchronisation

In a 'real' digital communications system, there will be offsets in both the frequency of the carrier *and* the symbol timing parameters. Both effects must be corrected in the receiver in order to properly recover the data, and a different synchronisation circuit is required for each task. The carrier and timing synchronisation methods presented in earlier sections can be combined into a single system, as shown in Figure 11.32. Note that this diagram shows a generic symbol timing synchroniser—in the forthcoming practical exercises, we will use the Early Late design introduced in Exercise 11.20.

At this stage, we assume that the carrier and timing parameters are both subject to error, and that these errors are independent of each other. For instance, the timing error may be a few ppm (the rate of the arriving symbols could be 10ppm lower than the receiver 'expects'), while the carrier frequency offset may be 300Hz higher than the receiver's local reference.

In the next exercise, we assume that the carrier frequency offset is reasonably small, i.e. no greater than 1kHz. Large frequency offsets may be corrected with a prior stage of coarse frequency synchronisation, which will be incorporated into the model in Exercise 11.25.

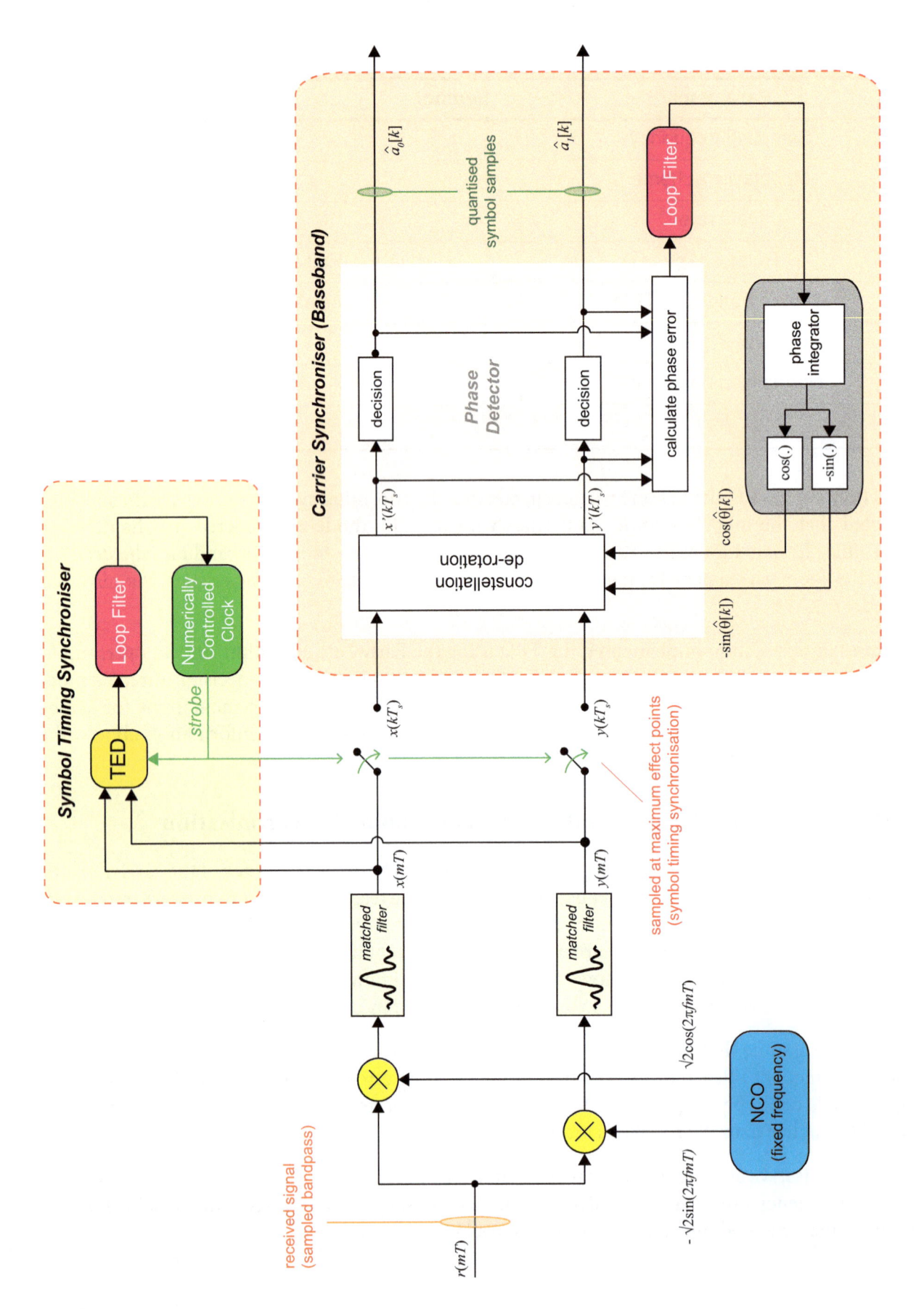

Figure 11.32: General architecture for joint symbol timing and carrier synchronisation[38]

Chapter 11: Digital Communications Theory and Simulation 479

Exercise 11.23 Joint Carrier and Timing Synchronisation for QPSK

The system presented in this exercise combines both of the carrier and timing synchronisation loops reviewed over the previous few sections. This circuit is able to simultaneously compensate for offsets in both (i) the carrier frequency, and (ii) the symbol timing parameters.

(a) Open the system:

/digital/simulation/receiver/QPSK_carrier_timing_synch.slx

(b) **Explore the model.** View the model and inspect the subsystems for carrier and symbol timing synchronisation, noting that carrier synchronisation is undertaken at baseband in this model. Compare the system with the signal flow graph provided in Figure 11.17.

(c) Check the initial parameter settings of the transmitter model, and establish the percentage deviation in frequency between the transmitter and receiver.

(d) **Simulate!** Next, open both of the *Constellation Diagram* sinks and, from there, run a simulation by 1 on the ▶ button in one of the *Constellation Diagram* windows. Watch the behaviour of the two constellations as the simulation progresses. Does the system synchronise successfully?

At the end of the simulation, a number of plot windows should appear. These display various signals that demonstrate the synchronisation process.

(e) **Analyse the simulation results.** Quite a number of different simulation plots are generated by this model, and it is useful to take a moment to consider them.

(f) Inspect the plot of the *Derotated I and Q phase* symbol samples generated by the receiver. During the initial period of the simulation, these do not show the two discrete levels expected in QPSK, but in the latter stages it is clear that the symbol samples have been consistently placed. You should see the I and Q sample points converge to (a scaled version of) +1/-1, as shown in Figure 11.33. The initial period of achieving synchronisation takes approximately 7ms in the example shown (your simulation may be different).

(g) Another plot to look at carefully is the figure entitled *Carrier Synchroniser Loop Filter Behaviour*. This helps to explain how the feedback section of the carrier synchroniser works. An example is shown in Figure 11.34. Check inside the Carrier Synchronisation block in the model to check where these signals are captured.

Note that there is some transient behaviour during the beginning of the simulation, after which the output of the loop filter settles to a steady value, and the output of the phase integrator converges to a straight line (either incrementing or decrementing, depending on the sign of the frequency error). The straight line produced by the integrator indicates a constant rate of change of phase, i.e. a fixed frequency.

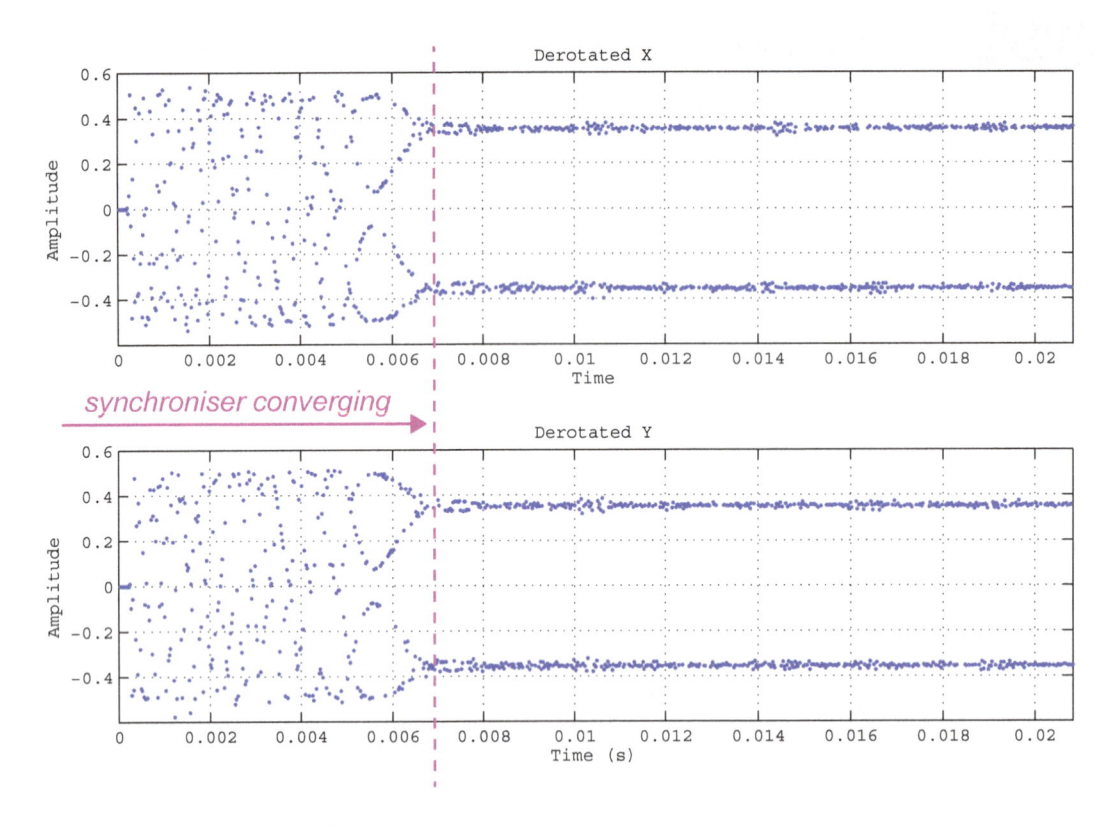

Figure 11.33: I and Q samples converging to +V/-V during the synchronisation process

- (h) **Consider the integrated phase.** Check that you can reconcile this behaviour of the loop filter and integrator, with the contents of the *Sine and Cosine Generated in the Carrier Synchroniser* plot.

- (i) **Experiment with the frequency and timing offset parameters.** Try changing the frequency offset in the transmitter model to try some different scenarios. You should see that the synchronisation behaviour changes according to the offset, and in fact it may not succeed in synchronising if the offsets are too large.

- (j) **Synchronisation in the presence of noise.** Add noise by connecting the *AWGN Channel* block, and re-simulating. Check that carrier and symbol timing synchronisation can still be achieved successfully in the presence of noise.

11.8 Coarse Frequency Synchronisation

In the previous section, it was noted that carrier synchronisation may not be successful if the initial frequency offset is too large. As an example, this scenario might be encountered if the carrier oscillators in the transmitter and receiver have wide tolerances, and there is a big difference in frequency between the outputs they generate. The problem can be mitigated by introducing a stage of 'coarse' frequency synchronisation prior to fine (PLL-based) carrier synchronisation and timing synchronisation. This first

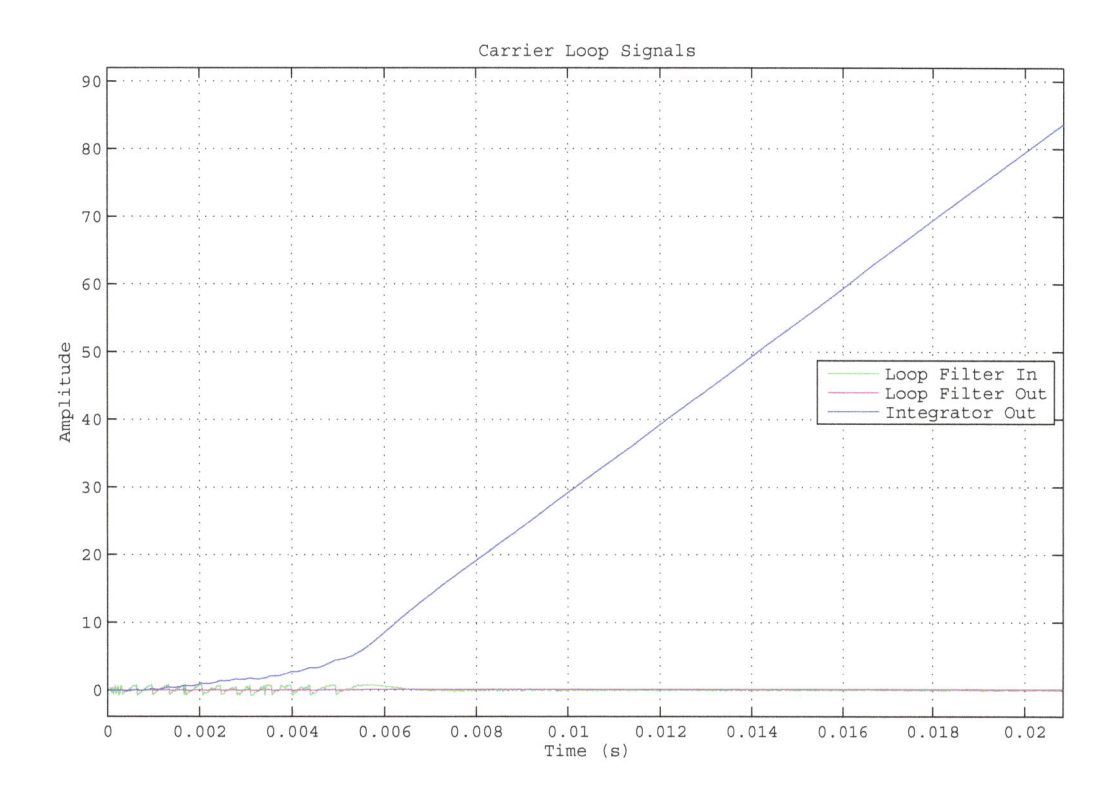

Chapter 11: Digital Communications Theory and Simulation

Figure 11.34: Behaviour of the loop filter and integrator in the carrier synchronisation loop

stage of synchronisation allows the receiver to 'tune' to roughly the carrier frequency of the desired signal, which leaves a sufficiently small offset for successful fine carrier synchronisation.

In the early sections of this book, a method of 'Eyeball Tuning' was introduced, wherein the frequency of the receiver's local oscillator was adjusted according to a visual inspection of the *Spectrum Analyzer* plot. This was effectively a form of 'coarse' synchronisation, but it involved human intervention. We will now implement coarse synchronisation in a different way, using an algorithm to automatically adjust the position of the spectrum based on observed signal characteristics.

11.8.1 Baseband Carrier Frequency Offset

Let us consider the issue of demodulation with a LO that deviates in frequency from the carrier present in the received signal. The result is that the demodulated spectrum is not centred exactly at 0Hz, but at some offset (positive or negative) corresponding to the difference between the local and received carrier frequencies. If this offset is small enough, it can be corrected using a PLL-based fine carrier synchroniser. For larger deviations, another technique is first required to centre the signal spectrum at approximately 0Hz (note that a small residual offset is likely to remain, as illustrated in Figure 11.35).

Although Figure 11.35 shows a frequency offset that is less than the signal bandwidth, it is also possible that the initial frequency offset may be more severe, and that the signal spectrum may not overlap 0Hz at all.

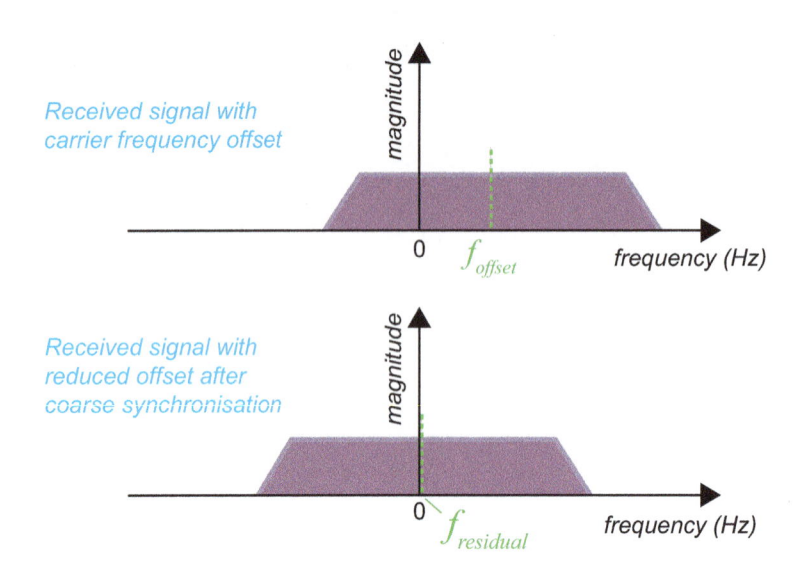

Figure 11.35: Illustration of coarse frequency synchronisation

11.8.2 Frequency Offset Estimation

Algorithms exist for performing a coarse estimate of the frequency offset, such as a frequency locked loop with band edge filtering, as described in [20]. The algorithm adopted here utilises knowledge of the received signal structure, and specifically the phase modulation index, M (equivalent to the number of PSK phases) [58]. The signal is raised to the power of M, which produces a significant tone at M times the offset frequency as a result of the signal structure. For QPSK, therefore, the signal is raised to the power of 4, and a tone is generated at 4 times the offset frequency. The actual frequency offset can then be easily computed based on the detected position of the tone—in this case a division by 4 is required.

The frequency offset estimation algorithm can be implemented by computing a FFT of the signal (raised to the power of 4), and then identifying the FFT bin with the highest magnitude. This process is depicted in Figure 11.36, for an example where f_s = 8000Hz, and the frequency offset of the received signal is 400Hz.

Given the sampling frequency, f_s, and the number of points in the FFT (N_{FFT}), the frequency spacing of FFT bins used in this algorithm is given by

$$f_{\Delta FFT} = \frac{f_s}{N_{FFT}}. \qquad (11.12)$$

The sampling frequency must be chosen according to the maximum expected frequency offset. Recalling that a tone will be produced at 4 times the offset, then the sampling frequency must be specified as at least 8 times the maximum offset.

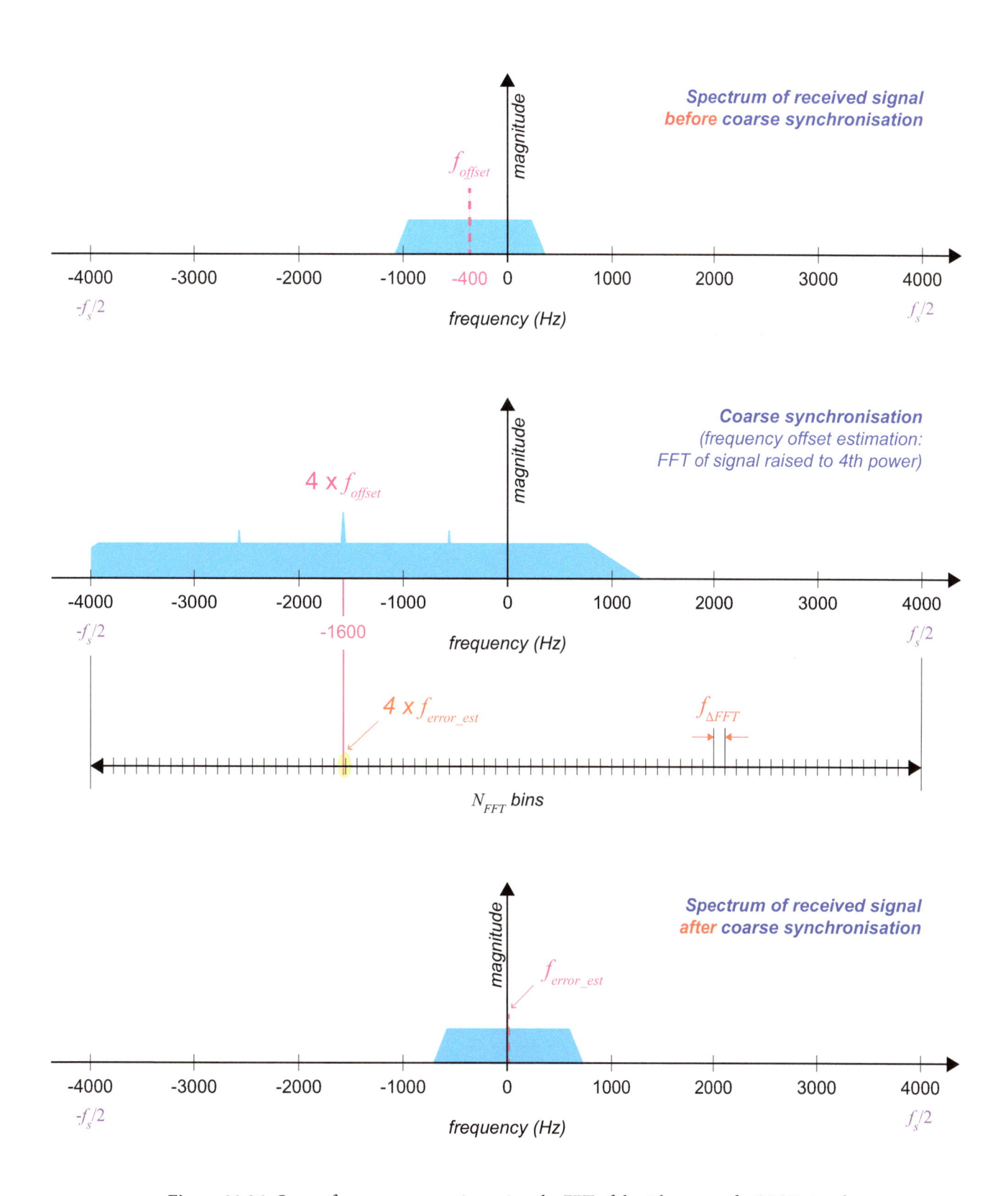

Chapter 11: Digital Communications Theory and Simulation

Figure 11.36: Coarse frequency correction using the FFT of the 4th power of a QPSK signal

There will also be an error in the frequency offset estimate, arising from the finite resolution of the FFT (i.e. the fixed number of FFT bins). The worst-case error in the frequency offset estimate is given by

$$f_{error\_est} = \frac{f_{\Delta FFT}}{2M}. \qquad (11.13)$$

This is because the error in identifying the maximum magnitude FFT bin (of the 4th power signal) is equal to half of an FFT bin-width, and this error is subsequently divided by M (here, 4) when calculating the frequency offset.

The estimation error contributes to the residual error after coarse frequency synchronisation (the other source of error is due to the frequency correction process, i.e. the error in generating the desired correction term, which will be neglected in this discussion). The number of FFT bins must therefore be chosen to provide a minimum level of frequency resolution, in particular ensuring that the residual error is less than the bandwidth of the fine carrier synchroniser.

After the computation of the FFT, the estimated frequency error is generated by: (i) finding the highest magnitude FFT bin; (ii) determining the frequency offset of that bin from 0Hz, and (iii) dividing by the modulation index, M.

11.8.3 Coarse Frequency Correction

The frequency offset generated by the estimation process described in Section 11.8.2 can subsequently be used to perform a correction, such that the signal spectrum is centred at approximately 0Hz.

The correction is implemented by generating a complex exponential term at the estimated frequency, and multiplying with the received signal. This is effectively a modulation of the signal at complex baseband. The coarse frequency offset estimation and correction stages can be summarised by the block diagram shown in Figure 11.37.

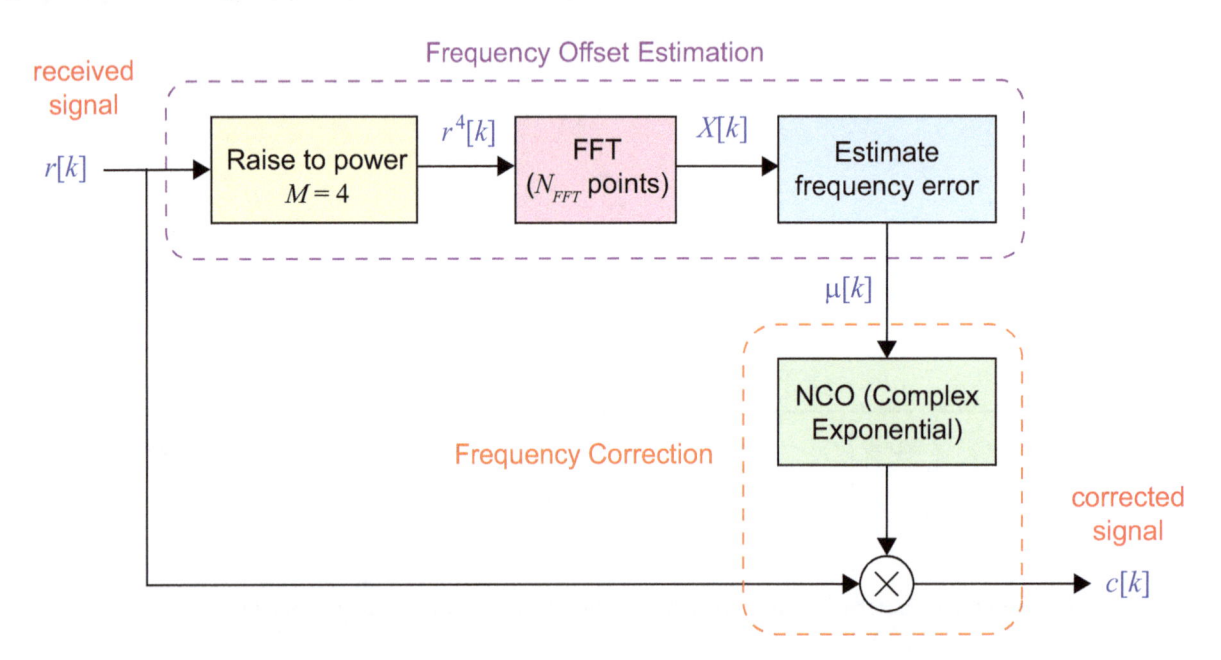

Figure 11.37: Block diagram of frequency offset estimation and correction

Chapter 11: Digital Communications Theory and Simulation

It is worth bearing in mind the interactions between sampling frequency, modulation index, maximum frequency error, FFT size, and estimation error, as discussed in the previous section. Where implementation efficiency is a concern, or the maximum expected frequency offset is particularly large, a prior stage of multirate filtering may be included to adjust the sampling frequency. For instance, if the maximum expected error was greater than 1000Hz, then the sampling rate of 8000Hz shown in Figure 11.36 would not be sufficient to accommodate the tone produced at 4 times the error frequency. In this case, an interpolating filter should be included prior to frequency offset estimation.

The examples that follow demonstrate the method of coarse frequency synchronisation, and then extend the model presented in Exercise 11.23 by incorporating a stage of coarse synchronisation prior to the fine frequency and symbol timing synchronisation loops.

Exercise 11.24 Coarse Frequency Correction

This example presents a coarse frequency synchroniser of the form shown in Figure 11.37.

(a) **Open the system**:

   /digital/simulation/receiver/QPSK_coarse_synch.slx

(b) **Explore the model.** Inspect the system, and note the frequency offset of the *Transmitter* block. Notice that the receiver model demodulates the bandpass signal using a fixed frequency oscillator, then decimates by a factor of 5.

(c) **Confirm the multirate parameters.** What are the sampling rates before and after decimation?

(d) Notice the *Buffer* block prior to the decimating filter. This implicitly sets the size of the FFT used in the coarse frequency synchronisation block.

(e) Inspect the internal configuration of the *Coarse Frequency Synchroniser*, and compare it against the block diagram shown in Figure 11.37.

(f) **Simulate!** Next, run a simulation and view the *Spectrum Analyzers* as the simulation progresses. It should be possible to spot the peak in the *Fourth Power* spectrum that allows the frequency offset to be found.

(g) Compare the actual frequency offset, as specified in the *Transmitter Model* block, with the frequency offset estimate produced by the coarse frequency synchronisation block (shown in the *Display* sink.

(h) **Experiment with the FFT size.** Try reducing the FFT size by changing the buffer size. If the FFT size is set to 256 rather than 4096, how is the accuracy of the frequency offset affected?

(i) Notice that the bandwidth of the signal (even when raised to the fourth power) is considerably smaller than the maximum supported by the sampling rate. For a more efficient implementation, you could introduce a second decimating filter after the first, to reduce the sampling rate further.

(j) **Add a second decimation stage.** Copy and paste the existing decimation filter and place it after the existing one. Open the block and set the Multirate Filter Object parameter to `Hd2`. This filter object decimates by a factor of 6, and has already been designed and created (see the *PreLoadFcn*).

(k) **Simulate the modified system.** Try re-simulating for different buffer (FFT) sizes, and frequency offsets. You should notice that, compared to the previous version (decimated by a factor of 5 rather than 5 x 6 = 30), the estimation of the frequency offset is more accurate. Why do you think this is?

Exercise 11.25 — QPSK Synchronisation (with Coarse Frequency Correction)

A limitation of the system presented in Exercise 11.23 was the extent of carrier frequency offsets that could be compensated. In this design, we add a stage of coarse frequency synchronisation prior to fine frequency synchronisation and symbol timing synchronisation. This will improve the receiver's ability to recover signals with more significant carrier frequency offsets.

(a) **Open the system**:

`/digital/simulation/receiver/QPSK_coarse_carrier_timing.slx`

NOTE: *The functionality of the AGC Simulink block was changed between releases R2015a and R2015b. We have supplied modified versions of these files, denoted by the* `_15bonwards.slx` *file names. Anyone using 15b or higher will need to use these files instead.*

(b) **Inspect the model.** Compare the system to previous examples, and notice that the model includes three types of synchronisation: coarse carrier frequency synchronisation, fine carrier frequency synchronisation, and symbol timing synchronisation.

(c) **Establish the carrier frequency offset.** Open the parameters of the *Transmitter Model* block, and note the carrier frequency offset. How significant is this offset with respect to the bandwidth of the signal?

(d) **Simulate!** Run the model, and view the *Constellation Diagrams* as the simulation progresses. During the first phase of execution, you should notice the effect of *AGC*, as the gain is adjusted in response to a moving window of input samples. The samples will move outwards from the origin as the gain is increased, as shown in Figure 11.38 for the output of the symbol timing synchroniser (in this case the constellation 'spins' because the residual carrier frequency offset has not yet been corrected).

(e) Confirm from the second *Constellation Diagram* (at the output of the fine carrier frequency synchroniser) that synchronisation has been successful, and the constellation recovered.

(f) **Find the frequency error corrected by 'fine' frequency synchronisation.** Inspect the system and establish the 'coarse estimate' of the carrier frequency to be corrected. Therefore, what was the residual frequency offset to be compensated by the fine carrier synchroniser?

Chapter 11: Digital Communications Theory and Simulation

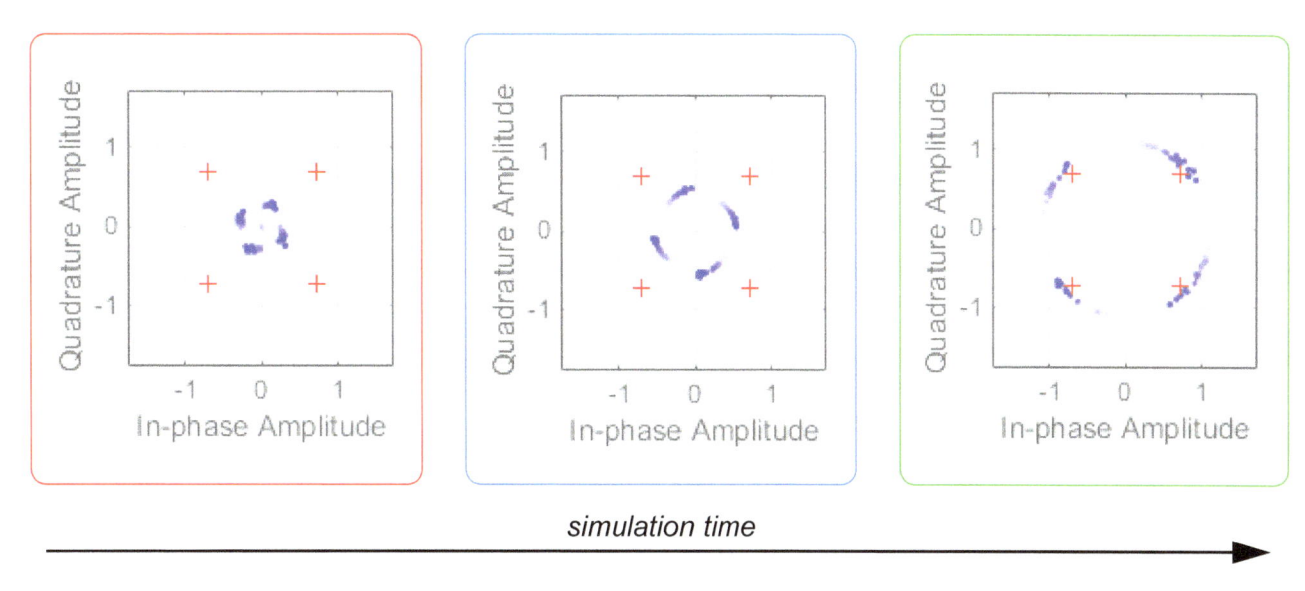

Figure 11.38: Behaviour of the constellation (output of symbol timing synchroniser) during startup

(g) Confirm your answer to the previous question, by analysing the plot window entitled *Sine and Cosine Generated in the Carrier Synchroniser*. Zoom in to 1 second on the time axis, and estimate the number of cycles within this window, to find the frequency in Hertz.

(h) Uncomment the Spectrum Analyzers and resimulate to see coarse frequency correction.

(i) **Experiment with different scenarios.** Try changing the frequency offset in the transmitter (keeping within the range -1000 to +1000Hz), and resimulating. Check if the receiver can synchronise successfully in each case.

11.8.4 RTL-SDR Practicalities

When receiving 'live' signals using the RTL-SDR, the user configures the centre frequency of the device (f_c) using either the *RTL-SDR Receiver* Simulink block, or the `comm.SDRRTLReceiver` MATLAB system object. The centre frequency is, however, subject to error due to the limited precision of the components used in the low-cost RTL-SDR. While working with various RTL-SDRs in the process of writing this book, the authors have noted that wide variations in centre frequency errors occur (up to several tens of kHz). Moreover, the frequency error of an individual device is often temperature dependent.

Frequency uncertainty of this magnitude is awkward even with the coarse frequency correction method described in this section. Therefore, the recommended procedure is to undertake the preliminary step of measuring the characteristic frequency error of the RTL-SDR, and then enter this value into the mask of the Simulink block or MATLAB system object. This will significantly reduce the initial offset that must be corrected. This procedure will be covered in more detail in Appendix A.3 (page 577), and it will be necessary to complete it prior to starting work with live reception of digital communication signals in Chapter 12.

11.9 Phase Ambiguity

In the previous section, it was mentioned that a residual constellation phase rotation may exist even after the implementation of decision-directed carrier synchronisation. This phase rotation is also called a phase ambiguity, and a further stage of processing is required to compensate for it.

For M-phase PSK signals, the carrier synchronisation process may lock to any of the M phases of the constellation. In other words, M-PSK signals can either be correctly synchronised with no phase offset, or synchronised with some non-zero phase offset that depends on the modulation scheme. In the case of BPSK, a carrier synchronisation loop could become locked correctly at a phase of 0, or with a phase offset of π, while for QPSK, the loop may lock at phase offsets of $\pi/2$, π, or $3\pi/2$, as well as at a phase of 0. In both cases, this is due to the rotational symmetry of the constellations.

Figure 11.40 illustrates a rotated QPSK constellation that has been synchronised with a phase offset of $\pi/2$; and Figure 11.39 illustrates a QPSK constellation with a phase offset of π. As can be observed, when a rotation occurs, the constellation symbol mapping changes. For instance, this means that when a '00' is sent by the transmitter, the receiver from Figure 11.39 interprets that the transmitted symbol was in fact a '11'. All detected symbols must therefore be considered erroneous in the presence of a phase offset.

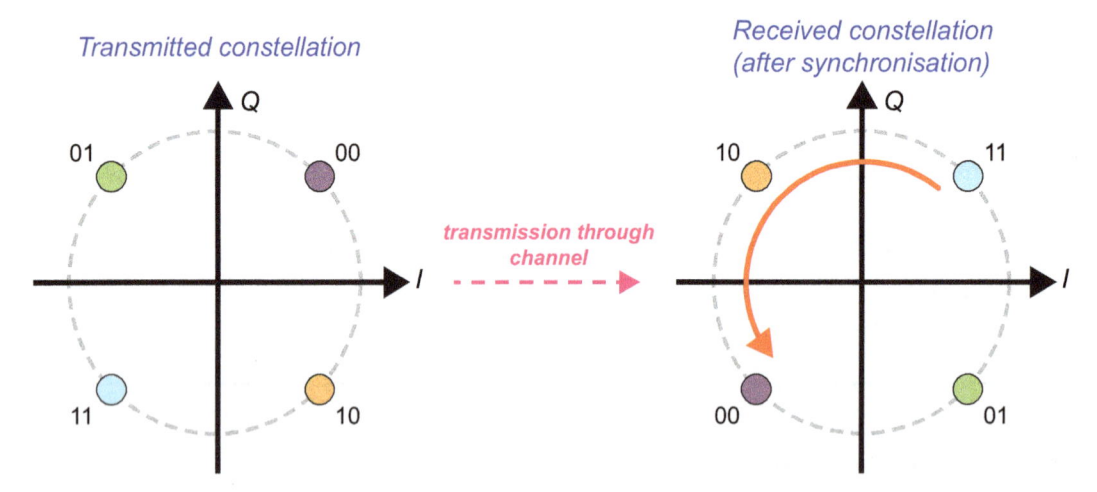

Figure 11.39: QPSK constellation with a π phase rotation

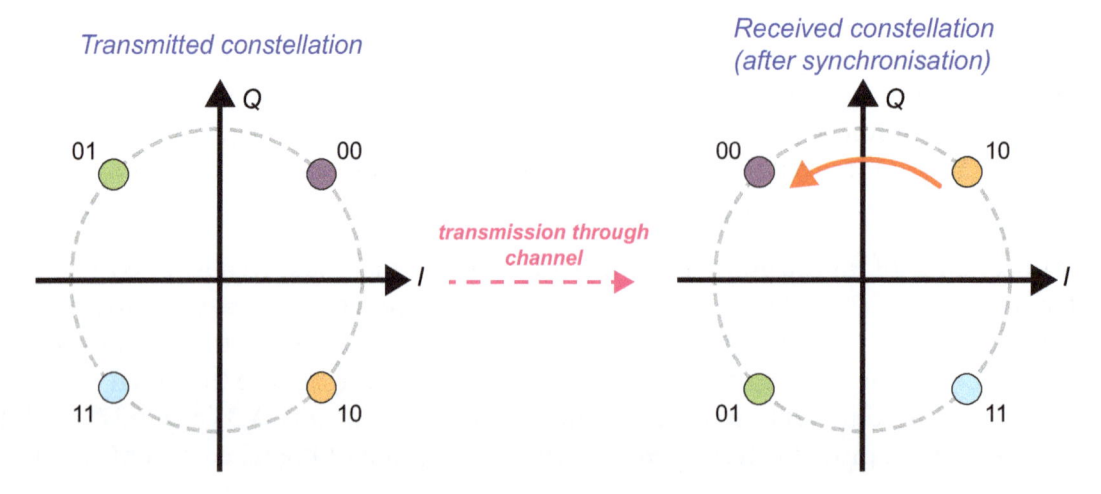

Figure 11.40: QPSK constellation with a $\pi/2$ phase rotation

Chapter 11: Digital Communications Theory and Simulation

Two main methods are normally used to resolve phase ambiguity, known as the *differential encoding* and *unique word* methods. A brief overview of both methods will be given for BPSK (as it provides simpler introductory examples) and then QPSK (to continue the construction of the receiver design) in the following two sections. More information on each of these can be found in [38].

11.10 Differential Encoding and Decoding

Conventionally, M-PSK modulation maps an input data sequence directly to a constellation comprised of symbols at different phases. This method is however not resilient to phase ambiguity in the receiver, as demonstrated by Figures 11.40 and 11.39. In comparison, differential encoding maps the data to the phase *shifts* of the modulated carrier (rather than absolute phases), by performing some simple boolean operations. Mapping the input to the phase shifts means that data can still be recovered by the receiver when there is a phase offset, because the phase shifts are preserved.

To perform differential encoding, the input data sequence is split into blocks (symbols) with length $L = \log_2 M$ bits, and the current block is encoded with the previous block to create an output based on the differences between the two.

11.10.1 BPSK differential encoding and decoding

The differential encoder for BPSK has a block length $L = 1$ (as $M = 2$), and uses the *previously encoded bit* $b_\varepsilon[n-1]$ as well as the transmitted *current input bit* $b_t[n]$ to produce the *current encoded output bit* $b_\varepsilon[n]$, where n relates to the position of the bit in the sequence. To encode the first input bit, $b_t[1]$, the initial 'previously encoded bit' is usually set to a default value of '0', as there is in fact no previously encoded bit. The BPSK differential encoding procedure is described by the logic shown in Table 11.1

This truth table is actually the same as that of an XNOR logic gate, and can be implemented using a delayed feedback loop as shown on the left hand side of Figure 11.41.

Table 11.1: BPSK differential encoder truth table

$b_t[n]$	$b_\varepsilon[n-1]$	$b_\varepsilon[n]$
0	0	1
0	1	0
1	0	0
1	1	1

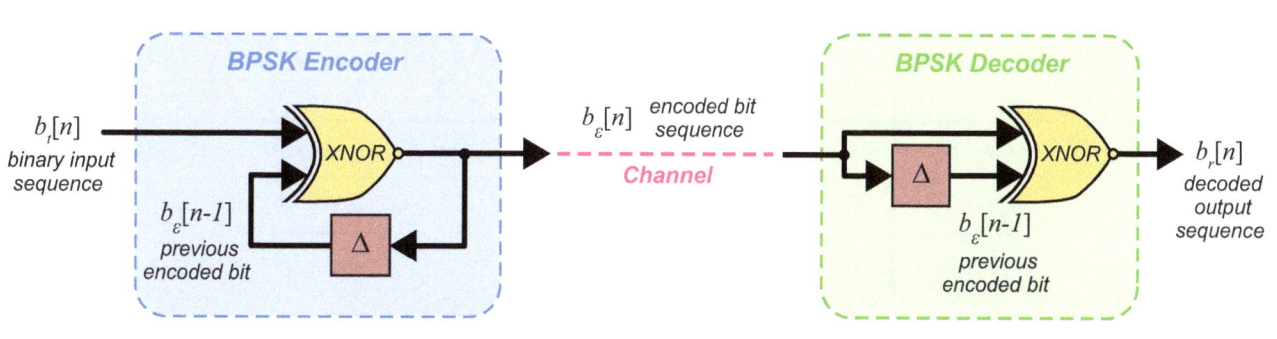

Figure 11.41: BPSK differential encoder & decoder illustration

As shown in Figure 11.41, the BSPK Encoder accepts the binary sequence $b_t[n]$ and supplies it as an input to the XNOR logic gate, along with the previously encoded bit, $b_\varepsilon[n-1]$. The encoded bit sequence $b_\varepsilon[n]$ is subsequently input to the BPSK modulator, pulse shaped, and transmitted (these steps are not depicted in the diagram).

When a receiver detects and synchronises to the signal, it is BPSK demodulated, and input to the differential decoder. Differential decoding (illustrated on the right hand side of Figure 11.41) can be modelled using the logic shown in Table 11.2. Both the current encoded bit $b_\varepsilon[n]$ and the previous encoded bit $b_\varepsilon[n-1]$ (created by the delay) are passed into the XNOR logic gate. This implements the reverse process of the encoder, and outputs the received bit sequence $b_r[n]$.

Table 11.2: BPSK differential decoder truth table

$b_\varepsilon[n-1]$	$b_\varepsilon[n]$	$b_r[n]$
0	0	1
0	1	0
1	0	0
1	1	1

Note the implicit assumption here that the channel is perfect and does not introduce errors, hence the sequence of encoded bits is equivalent at the transmitter and receiver.

To provide an example, let's consider encoding the simple input sequence $b_t = \{0, 0, 1, 0, 1, 1, 0, 1\}$. According to the encoding process described in Table 11.1, the output shown in Table 11.3 would be obtained.

Table 11.3: BPSK differentially encoded example sequence

n	0	1	2	3	4	5	6	7
$b_t[n]$	0	0	1	0	1	1	0	1
$b_\varepsilon[n-1]$	0	1	0	0	1	1	1	0
$b_\varepsilon[n]$	1	0	0	1	1	1	0	0

Decoding $b_\varepsilon[n]$ using the method from Table 11.2 gives the result shown in Table 11.4. It is clear that (in this case) the decoded sequence, $b_r[n]$, is identical to the initial input sequence, $b_t[n]$, as desired.

Table 11.4: BPSK differentially decoded output sequence

n	0	1	2	3	4	5	6	7
$b_\varepsilon[n]$	1	0	0	1	1	1	0	0
$b_\varepsilon[n-1]$	0	1	0	0	1	1	1	0
$b_r[n]$	0	0	1	0	1	1	0	1

Chapter 11: Digital Communications Theory and Simulation

To obtain the result shown above, the receiver would have to be synchronised with no phase offset. What happens when there is a phase offset of π, though? As mentioned previously, the reason for implementing differential encoding and decoding is because it makes the system resilient to phase ambiguity. Table 11.5 shows the output of the decoder when the receiver has synchronised to an incorrect phase of π.

Table 11.5: BPSK differentially decoded output sequence with receiver phase ambiguity of π

n	0	1	2	3	4	5	6	7
$b_\varepsilon[n]$	0	1	1	0	0	0	1	1
$b_\varepsilon[n-1]$	0	0	1	1	0	0	0	1
$b_r[n]$	1	0	1	0	1	1	0	1	*x1 error*

Due to the phase offset, the received encoded sequence $b_\varepsilon[n]$ in Table 11.5 is the opposite of the transmitted encoded sequence $b_\varepsilon[n]$ from Table 11.3. To a system without differential encoding and decoding, this would cause the entire output sequence to be in error; however as shown above, its inclusion means there is only a single error — the first bit. As the first bit of a transmission is usually part of a synchronisation sequence, this would not affect the transmitted data.

To observe the encoding and decoding processes in simulation, the next exercise will illustrate the practical implementation of a BPSK differential encoder and decoder pair.

Exercise 11.26 Implementation of a BPSK Differential Encoder & Decoder

We will now investigate differential encoding and decoding in a practical sense, with an input data sequence for BPSK transmission and reception. This block diagram matches the architecture shown in Figure 11.41, and it will allow a comparison to be made with the example data sequences shown earlier.

(a) **Open MATLAB**. Set the working directory to the exercise folder,

   `/digital/simulation/differential_coding`

   Next, open the file:

   `.../BPSK_diff_encode_decode.slx`

The block diagram should look as follows:

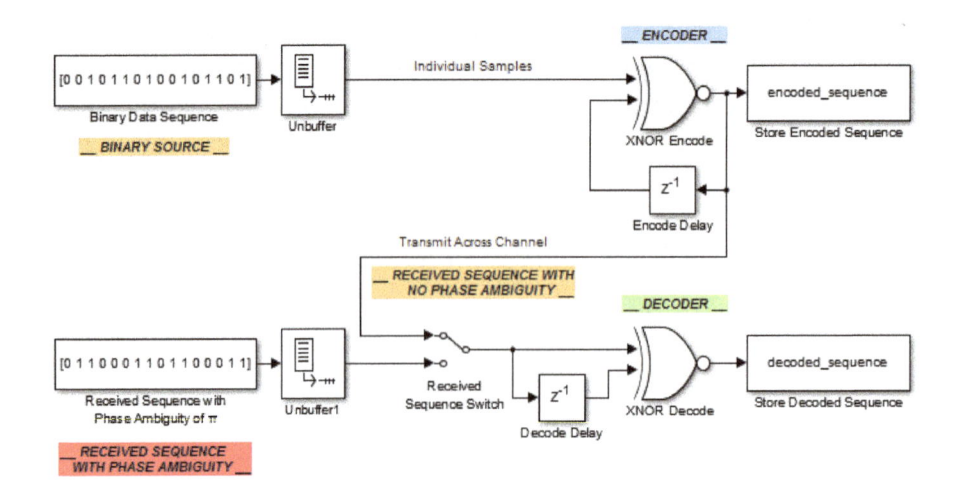

(b) **Examine the model.** The *Binary Data Sequence* block provides the input sequence for the encoder, which encodes the data exactly as illustrated earlier. The encoded output is then 'transmitted'. In the decoder, the original sequence is subsequently regenerated by the reverse process. Both the encoded and decoded sequences are saved to the MATLAB Workspace, which means that the outputs can be observed and checked against those shown in the tables presented previously. As a reminder, you should note that the input sequence is formed from two repetitions of the input sequence $b_t[n]$, which was shown in Table 11.3.

(c) There is also another input to the decoder that allows a second scenario to be tested — a receiver locking with a phase ambiguity of π. This alternate received signal emulates $b_\varepsilon[n]$ from Table 11.5, and can be chosen using the *Received Sequence Switch* in the model, as will be shown shortly.

(d) **Run the simulation.** Ensure the *Received Sequence Switch* is set to the encoded sequence as illustrated in part (a), then run the simulation by the button. Once the simulation has completed, navigate to the Workspace in the MATLAB environment.

(e) **Confirm the functionality.** To view the encoded output sequence, on the encoded_sequence variable that is stored in the Workspace. It should show you a column of bits that matches the encoded sequence, $b_\varepsilon = \{1,0,0,1,1,1,0,0\}$ that was previously observed in Table 11.3. The input sequence from Table 11.3 was repeated twice to provide the encoder input for the simulation, thus the encoder output (stored in the encoded_sequence variable) is also repeated. If this is the case, then the encoder is performing exactly as expected.

(f) Now we need to investigate the decoder output, by on the decoded_sequence variable in the Workspace. This time, you should see a column of bits that matches the decoded sequence, $b_r = \{0,0,1,0,1,1,0,1\}$ that was shown earlier in Table 11.4. Again due to the repetition in the input sequence, we observe a repeating pattern in the output sequence, which should confirm the correct functionality of the decoder. If it matches the decoded sequence shown above then the decoder is operating as expected; where this is of course the ideal case without any phase ambiguity.

Chapter 11: Digital Communications Theory and Simulation

(g) A received sequence will however often be subject to phase ambiguity issues, so let's observe how this affects the decoded output. Return to the model and change the *Received Sequence Switch* to the alternate received sequence that was introduced earlier:

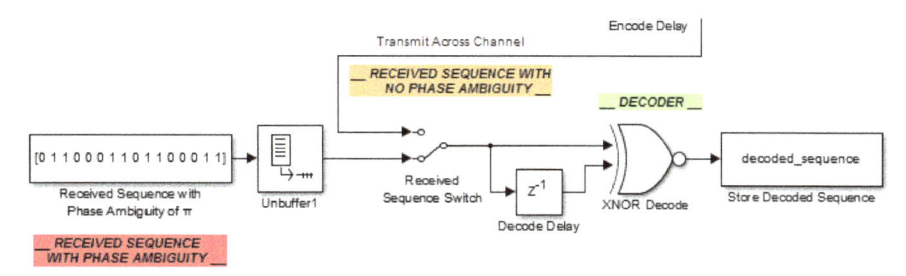

(h) Rerun the simulation by ⟨1⟩ the ▶ button, and then check the variables in the MATLAB Workspace. Although we are not concerned with the bits in the encoded sequence variable, you should note that they haven't changed from the last simulation output.

Now, look at the decoded sequence; the first bit no longer matches $b_r[n]$ in Table 11.4, as it now has an initial sequence of $\{1,0,1,0,1,1,0,1\}$. This sequence does however match $b_r[n]$ in Table 11.5, where only the first bit was in error. As expected, the implementation of differential encoding has successfully corrected the phase ambiguity of π, and ensures the remaining bits of the sequence match the original input data sequence. With this method, only the first bit is in error, rather than the whole sequence!

(i) The sequence can also be completely changed to one of your own choice if you wish to further investigate the process. You will however need to use both Table 11.1 and Table 11.2 to verify the outputs and create the received sequence that has a phase offset of π.

This exercise has demonstrated a practical implementation of differential encoding and decoding for BPSK, and also its role in successfully recovering data where a phase ambiguity exists.

11.10.2 QPSK differential encoding and decoding

The differential encoder for QPSK has a block length of $L = 2$ bits ($M = 4$) as its constellation is composed of four points at equally spaced phases. The inputs to the encoder are the *current input bit* $b_t[n]$ and the *next input bit* $b_t[n+1]$, along with the *two previously encoded bits* $b_\varepsilon[n-2]$ and $b_\varepsilon[n-1]$, which are produced using feedback loops similar to the one in the BPSK differential encoder.

To encode the first input bits, $b_t[1]$ and $b_t[2]$, the 'previously encoded bits' are usually set to default values of '0', as (once again) there are no previously encoded bits at the very first iteration. The QPSK differential encoder uses these four bits to produce the two *encoded output bits*, $b_\varepsilon[n]$ and $b_\varepsilon[n+1]$. The overall functionality of the QPSK differential encoder can be described as in Table 11.6, where the four columns on the left hand side are used to determine the next two encoded bits, given in the rightmost two columns.

Table 11.6: QPSK differential encoder truth table

$b_t[n]$	$b_t[n+1]$	$b_\varepsilon[n-2]$	$b_\varepsilon[n-1]$	$b_\varepsilon[n]$	$b_\varepsilon[n+1]$
0	0	0	0	0	0
0	0	0	1	0	1
0	0	1	0	1	0
0	0	1	1	1	1
0	1	0	0	0	1
0	1	0	1	1	1
0	1	1	0	0	0
0	1	1	1	1	0
1	0	0	0	1	0
1	0	0	1	0	0
1	0	1	0	1	1
1	0	1	1	0	1
1	1	0	0	1	1
1	1	0	1	1	0
1	1	1	0	0	1
1	1	1	1	0	0

The QPSK differential encoder is more complex than the BPSK equivalent considered previously, and its construction is shown in left hand side of Figure 11.42. The 'Comb Logic' block shown in the diagram represents the encoding process described by the truth table given in Table 11.6.

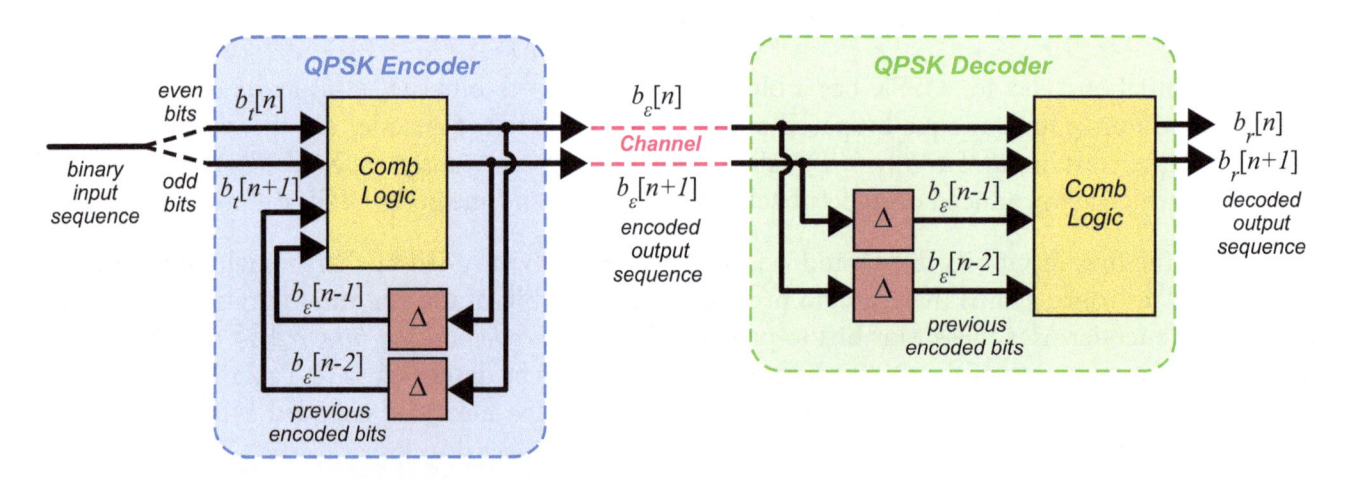

Figure 11.42: QPSK differential encoder & decoder illustration

Chapter 11: Digital Communications Theory and Simulation

A binary sequence is input to the combinational logic circuit in blocks of two bits, where $b_t[n]$ are the even-numbered bits, and $b_t[n+1]$ are the odd-numbered bits. Combined with the previously encoded bits $b_\varepsilon[n-2]$ and $b_\varepsilon[n-1]$, which are created from delaying both the even and odd encoded output bits, these are used to create the encoded output bits: $b_\varepsilon[n]$ and $b_\varepsilon[n+1]$.

The encoded bits would then be input to the QPSK modulator, pulse shaped and transmitted across the channel to the receiver.

The receiver detects and synchronises to the signal, performs demodulation, and then passes the received data to the differential decoder. The differential decoder (shown on the right hand side of Figure 11.42) can be modelled using the logic shown in Table 11.7, which is of equal complexity to the encoder. The two encoded bits $b_\varepsilon[n]$ and $b_\varepsilon[n+1]$, along with the previously encoded bits, $b_\varepsilon[n-2]$ and $b_\varepsilon[n-1]$ are input to the combinational circuit, which decodes the received sequence and outputs the original bits, $b_r[n]$ and $b_r[n+1]$.

Table 11.7: QPSK differential decoder truth table

$b_\varepsilon[n-2]$	$b_\varepsilon[n-1]$	$b_\varepsilon[n]$	$b_\varepsilon[n+1]$	$b_r[n]$	$b_r[n+1]$
0	0	0	0	0	0
0	0	0	1	0	1
0	0	1	0	1	0
0	0	1	1	1	1
0	1	0	0	1	0
0	1	0	1	0	0
0	1	1	0	1	1
0	1	1	1	0	1
1	0	0	0	0	1
1	0	0	1	1	1
1	0	1	0	0	0
1	0	1	1	1	0
1	1	0	0	1	1
1	1	0	1	1	0
1	1	1	0	0	1
1	1	1	1	0	0

As e an example, consider encoding the input sequence $b_t = \{0, 1, 0, 0, 1, 1, 0, 1, 1, 1, 1, 0, 0, 0, 0, 1\}$. Applying the logic implied by Table 11.6 would yield the result shown in Table 11.8.

Table 11.8: QPSK differentially encoded example sequence

n	0	1	2	3	4	5	6	7
$b_t[n]$	0	0	1	0	1	1	0	0
$b_t[n+1]$	1	0	1	1	1	0	0	1
$b_\varepsilon[n-2]$	0	0	0	1	0	1	0	0
$b_\varepsilon[n-1]$	0	1	1	0	0	1	1	1
$b_\varepsilon[n]$	0	0	1	0	1	0	0	1
$b_\varepsilon[n+1]$	1	1	0	0	1	1	1	1

Decoding $b_\varepsilon[n]$ using the logic from Table 11.7 provides the result shown in Table 11.9. Here, you should observe that the decoded sequence ($b_r[n]$ and $b_r[n+1]$) is identical to the initial input sequence $b_t[n]$ and $b_t[n+1]$ from Table 11.8.

Table 11.9: QPSK differentially decoded output sequence

n	0	1	2	3	4	5	6	7
$b_\varepsilon[n-2]$	0	0	0	1	0	1	0	0
$b_\varepsilon[n-1]$	0	1	1	0	0	1	1	1
$b_\varepsilon[n]$	0	0	1	0	1	0	0	1
$b_\varepsilon[n+1]$	1	1	0	0	1	1	1	1
$b_r[n]$	0	0	1	0	1	1	0	0
$b_r[n+1]$	1	0	1	1	1	0	0	1

To obtain this result, the receiver has to be synchronised with no phase ambiguity. Similar to the BPSK differential decoder, the QPSK decoding process still works when there is, for example, a phase offset of π. This is demonstrated in Table 11.10.

Table 11.10: QPSK differentially decoded output sequence with receiver phase ambiguity of π

n	0	1	2	3	4	5	6	7
$b_\varepsilon[n-2]$	0	1	1	0	1	0	1	1
$b_\varepsilon[n-1]$	0	0	0	1	1	0	0	0
$b_\varepsilon[n]$	1	1	0	1	0	1	1	0
$b_\varepsilon[n+1]$	0	0	1	1	0	0	0	0
$b_r[n]$	1	0	1	0	1	1	0	0
$b_r[n+1]$	0	0	1	1	1	0	0	1

x1 symbol error

Chapter 11: Digital Communications Theory and Simulation

The received encoded bits $b_\varepsilon[n]$ and $b_\varepsilon[n + 1]$ are the inverse of the transmitted encoded bits, due to the phase offset of π. As was demonstrated previously for BPSK, the implementation of this technique mitigates the impact of a phase ambiguity and ensures that only the first block of decoded bits is in error. We can now illustrate this with a practical example.

Exercise 11.27 Implementing a QPSK Differential Encoder & Decoder

In this exercise you will investigate differential encoding and decoding for QPSK modulation, where the encoder and decoder combinational logic will be provided by blocks from the 'RTL-SDR Exercise Components' library.

(a) **Open MATLAB**. Set the working directory to the exercise folder,

    `/digital/simulation/differential_coding`

Next, open the file:

    `.../QPSK_diff_encode_decode.slx`

The block diagram should look like this:

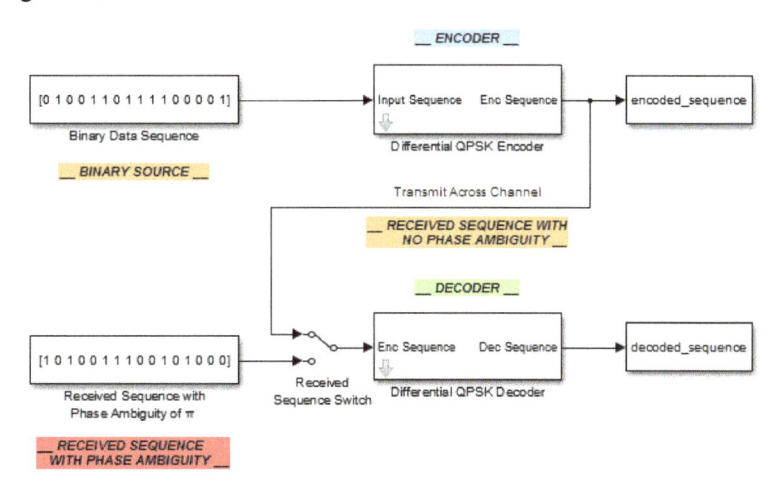

(b) **Examine the model.** The design is very similar to the previous exercise, but adapted for QPSK encoding and decoding. The *Binary Data Sequence* block provides an input sequence for the encoder. Both the encoded and decoded sequences are saved to the Workspace for observation, with the input sequence chosen to match the sequence shown in Table 11.8.

(c) As before, we also provide a second input to the decoder that allows testing of an encoded sequence with a phase ambiguity of π. This time, the alternate received signal emulates $b_\varepsilon[n]$ from Table 11.10 and will be selected using the *Received Sequence Switch*.

(d) **Investigate the encoder functionality.** To observe the combinational logic used to implement the encoder, enter the *Differential QPSK Encoder* subsystem by 1️⃣ on the grey arrow on its mask. This will show the low-level logic design that has been implemented by simplifying Table 11.6. For this design to implement the functionality of the encoder shown in Figure 11.42, the input sequence is separated into odd and even bits to form QPSK input 'blocks'. The encoded bits are then concatenated together and buffered back to the original length of the input sequence, before being output from the subsystem. These encoded bits are also used in two feedback paths that each implement a one bit delay, to turn them into the 'previously encoded bits'.

(e) **Investigate the decoder functionality.** To observe the decoder logic, enter the *Differential QPSK Decoder* subsystem via the grey arrow on the block mask. This design is also constructed from low-level logic, and represents a simplification of Table 11.7. As can be seen, the input sequence undergoes the same odd and even bit separation as before, but this time these paths have a one bit delay to generate the 'previously encoded bits' for the decoder in Figure 11.42. The decoded bits are then concatenated together and output through a buffer to create an output sequence of the desired length (this length can be set in the parameter window of the subsystem).

(f) **Simulate!** Ensure the *Received Sequence Switch* is set to the encoded sequence as illustrated in part (a), run the simulation by 1 the ▶ button, and then view the MATLAB Workspace.

(g) **Confirm the results.** View the encoded output sequence by 2 on the ☑encoded_sequence variable. It should show a column of bits that matches the encoded sequence observed earlier in Table 11.8, i.e. $b_\varepsilon - \{0,1,0,1,1,0,0,0,1,1,0,1,0,1,1\}$.

(h) Next, view the decoder output captured by the ☑decoded_sequence variable. This time, you should see a column of bits that matches the decoded sequence shown earlier in Table 11.9, i.e. $b_r - \{0,1,0,0,1,1,0,1,1,1,1,0,0,0,0,1\}$. Assuming that you can confirm a match, this confirms correct operation (for the case without any phase ambiguity).

(i) We now need to change the sequence input to the decoder, to confirm the functionality when there is a phase ambiguity in the received signal. Change the *Received Sequence Switch* to the alternate received sequence introduced earlier.

(j) Rerun the simulation, and navigate back to the Workspace in the MATLAB environment. Observe the encoded sequence variable, and determine that the output is unchanged from the last simulation.

Now, look at the decoded sequence: the first two bits (representing one data block) no longer match $b_r[n]$ & $b_r[n+1]$ in Table 11.9, as it now has an initial sequence of $\{1,0,0,0,1,1,0,1,1,1,1,0,0,0,0,1\}$. Instead, it matches the decoded sequence in Table 11.10, where this error was present. As observed previously with BPSK, differential encoding has permitted successful recovery of the data, even with a phase ambiguity.

(a) The sequence can also be changed to one of your own choice if you wish to further investigate the process. You will however need to use both Table 11.1 and Table 11.2 to verify the outputs and create a received sequence that introduces a phase offset.

Chapter 11: Digital Communications Theory and Simulation

11.11 Synchronisation with a Unique Word

Now that differential encoding and decoding has been demonstrated, this section will introduce an alternate method of mitigating a phase ambiguity — the Unique Word (UW). A UW is a defined sequence of symbols (known by both the transmitter and receiver), that are introduced to the data sequence as the signal is transmitted. Once the receiver has acquired and synchronised to the signal, the (known) UW is compared against the received sequence in an effort to detect if there is a phase offset. If an offset is detected, the received data can be corrected, removing the phase ambiguity.

11.11.1 Using the UW method with BPSK

As described in Section 11.9, synchronised BPSK signals can either have no phase offset, or an offset of π. Using the UW method with BPSK means that phase ambiguities have to be detected by comparing the received data with two versions of the UW; the UW itself, and the inverse of the UW. If the UW is detected in the received sequence, it determines that there is no phase offset, while if its inverse is found, it determines that there is an offset of π. When the outcome of the comparison stage is known, the sequence will either be output as-is (in the case of no offset), or corrected for the offset of π.

The method is depicted by the block diagram in Figure 11.43. The received sequence $b_r[n]$ is compared to the two possible versions of the UW, $b_{uw}[n]$ and $\overline{b_{uw}[n]}$, and from this the phase offset can be determined. If there is a phase offset, the bits can be corrected by either negating the bits, or using an alternative symbol-to-bit demapper.

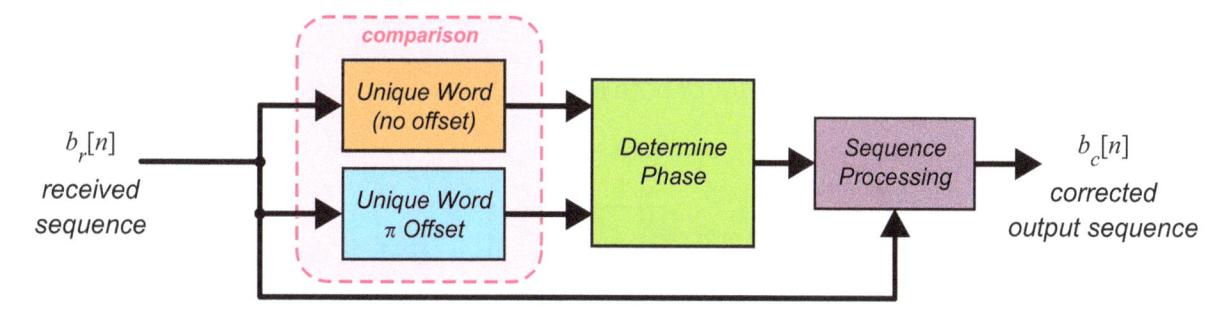

Figure 11.43: BPSK unique word receiver

We shall confirm this method with an example. A digital communications system uses a UW of $b_{uw}[n] = [01011010]$, and receives the series of bits, $b_r[n]$, given in Table 11.11. It begins by examining the sequence to find out where the start of the UW is.

Table 11.11: BPSK unique word example — no phase ambiguity

n	0	1	2	3	4	5	6	7	8	9	10	11
$b_r[n]$	0	1	0	1	1	0	1	0	0	1	1	0

The first eight bits of this sequence relate to the UW, and the remaining four bits are part of the transmitted data. In this particular example, the output of the 'Determine Phase' block would decide that the received bit stream had no phase offset, because the UW matches exactly with the received data. As a result, the receiver would output the series of bits as-is from Table 11.11 (i.e., $b_c[n] = b_r[n]$). If the receiver received the series of bits shown in Table 11.12 however, it would determine that the signal was out of phase by a value of π.

Table 11.12: BPSK unique word example — π phase ambiguity

n	0	1	2	3	4	5	6	7	8	9	10	11	
$b_r[n]$	1	0	1	0	0	1	0	1	1	0	0	1	*errors*

To correct this phase offset, the receiver would invert the sequence of bits, in order to match the transmitted sequence (i.e., $b_c[n] = \overline{b_r[n]}$). This would mean the four data bits are in fact '0110' and not '1001' as shown in Table 11.12. This method of can assess and correct any phase ambiguity for the BPSK modulation scheme.

11.11.2 Using the UW method with QPSK

Recall from Section 11.9 that synchronised QPSK signals can either have no phase offset, or an offset of $\pi/2$, π, or $3\pi/2$. Using the UW method with QPSK signals therefore means that the received sequence must be compared with four versions of the UW. The comparison allows the receiver to find the value of the phase offset, which then enables an appropriate correction to be made if needed.

The QPSK UW receiver is shown in Figure 11.44. The received sequence $b_r[n]$ is compared to the four possible versions of the UW and corrected, if required, depending upon the output of the 'Determine Phase' block. For the correction process, the sequence is better visualised in blocks of two bits, or QPSK symbols, as required for the differential decoding process. If a time offset is additionally present, either the first bit $b_r[n]$, the second bit $b_r[n+1]$, or both bits may be inverted, depending on the phase offset.

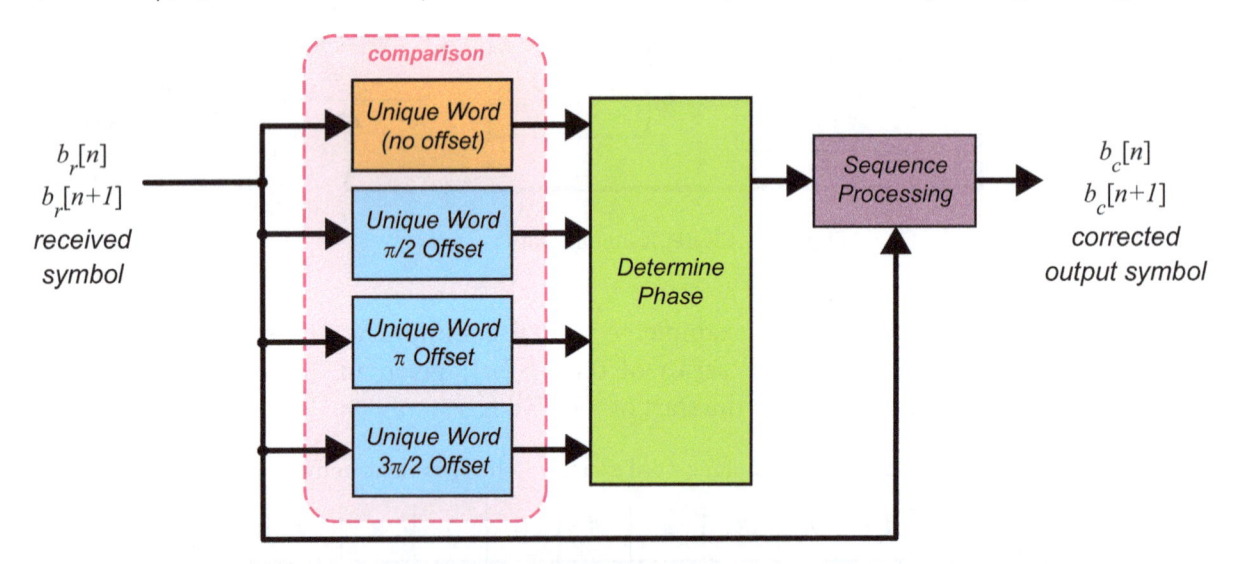

Figure 11.44: QPSK unique word receiver

Table 11.13 shows the transmitted and received symbols for each of the possible phase offsets. The first phase offset column illustrates the received codewords for no phase offset, which are also equal to offsets of any multiple of 2π. The remaining columns show the received codewords for the other incorrect phase values. The inverted bits are highlighted in red. These are the bits that must be corrected by the 'Sequence Processing' block from Figure 11.44.

Chapter 11: Digital Communications Theory and Simulation

Table 11.13: QPSK phase offsets (transmitted and received symbols)

Transmitted Symbols	Received Symbols				
	Phase Offset				
	(no offset)	$\pi/2$	π	$3\pi/2$	
00	00	10	11	01	
01	01	00	10	11	errors
10	10	11	01	00	
11	11	01	00	10	

To explain further, we shall define a scenario that uses a UW of $b_{uw}[n] = [0100101111100001]$. If the receiver detects the $b_r[n]$ sequence shown in Table 11.14, it will first attempt to identify the start of the UW.

Table 11.14: QPSK unique word example — no phase ambiguity

n	0	1	2	3	4	5	6	7	8	9	10	11
$b_r[n]$	0	0	1	1	1	1	0	0	0	1	1	0
$b_r[n+1]$	1	0	0	1	1	0	0	1	1	0	1	1

The first eight symbols in the received sequence relate to the UW, and the remaining four symbols are part of the transmitted data. In this particular example, the received sequence has no phase offset, because the UW matches exactly. This means that no correction would be required, and the receiver would output the sequence from Table 11.14 directly (i.e., $b_c[n] = b_r[n]$ and $b_c[n+1] = b_r[n+1]$).

With an offset of π, as demonstrated in Table 11.15, the received bits differ from the UW, and the receiver must perform correction.

Table 11.15: QPSK unique word example — π phase ambiguity

n	0	1	2	3	4	5	6	7	8	9	10	11
$b_r[n]$	1	1	0	0	0	0	1	1	1	0	0	1
$b_r[n+1]$	0	1	1	0	0	1	1	0	0	1	0	0

errors

This time, the decision is made by the 'Determine Phase' block, that the whole of the UW is negated, and the receiver will make the appropriate correction (i.e., $b_c[n] = \overline{b_r[n]}$ and $b_c[n+1] = \overline{b_r[n+1]}$). Correcting the sequence would convert the data bits (last four symbols) from '10010010' to their transmitted form, '01101101'.

Finally, an example is shown in Table 11.16 for a phase offset of $3\pi/2$.

Table 11.16: QPSK unique word example — $3\pi/2$ phase ambiguity

n	0	1	2	3	4	5	6	7	8	9	10	11	
$b_r[n]$	1	0	0	1	1	0	0	1	1	0	1	1	*errors*
$b_r[n+1]$	1	1	0	0	0	0	1	1	1	0	0	1	

Here, only the bits highlighted in red are in error, rather than the whole of the received sequence. Correcting them according to Table 11.13, it would be found that the data (i.e. symbols 8 to 11) is in fact '01101101' rather than '11001011'.

Using this method of assessing and correcting the received bit sequence based on a UW, any phase ambiguity that occurs in a QPSK receiver can be compensated for, and the transmitted data accurately recovered. Note that this does not consider the impact of errors caused by noise.

11.12 Summary

This chapter has covered several different aspects of digital communications. We began by reviewing digital modulation schemes, including popular variations such as BPSK, QPSK, and 16-QAM. Pulse shaping was also discussed, and shown to play an important role in defining the bandwidth occupied by transmitted signals. We focused on the RC response, although other pulse shapes are widely used.

Digital up- and down-conversion, the multirate operations necessary to translate between baseband and IF (and vice versa) were also covered and examples presented. Although the RTL-SDR hardware internally performs downconversion to baseband, the DUC and DDC are important building blocks for modelling and simulating digital communications systems.

Much of the chapter was devoted to synchronisation techniques for recovering the carrier phase and timing parameters of a received digital signal. This treatment focused on the QPSK modulation scheme, and considered also techniques for 'coarse' frequency synchronisation, which is necessary when the carrier frequency is subject to significant error (as is the case for the RTL-SDR). Two techniques for addressing the problem of carrier phase ambiguity, and its subsequent effect on symbol de-mapping, were also presented.

The next chapter will move forwards by integrating some of the concepts and receiver models covered here into real-time SDR receivers for operation with the RTL-SDR.

12 Desktop Digital Communications: QPSK Transmission and Reception

In the previous *Digital Communications Theory and Simulation* chapter, QPSK signals were generated, transmitted and received in simulation, and various offsets were purposely introduced to the Simulink models to model some of the challenges associated with real world communication channels. Although these introduced the theory behind QPSK modulation and the synchronisation systems required to receive QPSK signals, they did not fully represent the true extent of the issues — relating to interference, offsets, hardware tolerances and communication protocols — that must be considered when wirelessly transmitting and receiving signals in practice. This chapter aims to address these issues, and focuses on implementing QPSK SDR *transmit-and-receive* (Tx/ Rx) communications systems on the desktop.

In this chapter, a number of exercises will involve generating and transmitting QPSK signals with the USRP® radio, and then receiving, synchronising and demodulating them using the RTL-SDR hardware, MATLAB and Simulink. The exercises progress from simply viewing the spectrums of QPSK signals; to implementing synchronisation systems that use rudimentary communications protocols; to systems that are able to transmit text strings and images from one computer to another. Throughout this chapter, we will compare these practical digital systems with the theory from Chapter 11.

At the end of the chapter, the use of FM modulation is considered (utilising low-cost transmitter hardware), to perform the same text string and image transfer implemented with QPSK modulation and the USRP® radio hardware. Realising that not all labs and users will have access to a USRP® radio, this therefore allows readers to experience the issues of RF communications with more readily available hardware, while also introducing some further interesting concepts.

The computationally intensive models used in this chapter (especially in the later designs) demand a significant level of computer resources to perform well on a single computer. Therefore, it is strongly advised that the transmit and receive models are opened on separate computers, from this point forward. The USRP® hardware should be connected to your 'transmitter' computer, and the RTL-SDR to your 'receiver' computer. Proceeding through the exercises, file paths will be provided to the appropriate Tx and Rx models — please ensure that you open them on the correct machines!

It is important to be aware that the success of digital communications is much more dependent on the receiver being accurately tuned to the transmitted carrier frequency, than was previously experienced with analogue communications. As introduced in Chapter 7, the RTL-SDR has accuracy limitations that require compensation — particularly the frequency offset of the local oscillator. Consequently, you should take the preliminary step of determining the individual PPM correction value required for your RTL-SDR, which will be essential for all of the receiver exercises in this chapter. This value can be calculated by following the steps detailed in Exercise A.3 (page 577). The correction value should then be noted, so that it can be supplied to the RTL-SDR Receiver block in Simulink as a correction parameter.

12.1 Pulse Shaping with Real Time QPSK Transmitter and Receiver Designs

In this section, we will start by transmitting a simple QPSK signal from the USRP® transmitter to the RTL-SDR, and view the spectrum of the signal that is received. One purpose of the exercises that follow is to ensure that both your USRP® hardware and RTL-SDR receiver are set up correctly on their respective computers, before moving on to more complex designs.

To recap from Sections 8.1 (page 280) and 10.1 (page 368), the USRP® hardware is a SDR that can be used in conjunction with Simulink to implement a number of different radio transmitters and receivers. Samples of a complex baseband signal are transferred to the device and upconverted by a DUC (resident on the FPGA within the USRP® hardware). The samples are subsequently converted to analogue signals via a DAC, mixed with a complex RF carrier and then transmitted. To transmit QPSK signals with the USRP® radio, samples of a baseband, QPSK-modulated bit stream are supplied to the 'Data' port of the *SDRu Transmitter* Simulink block, as shown in Figure 12.1.

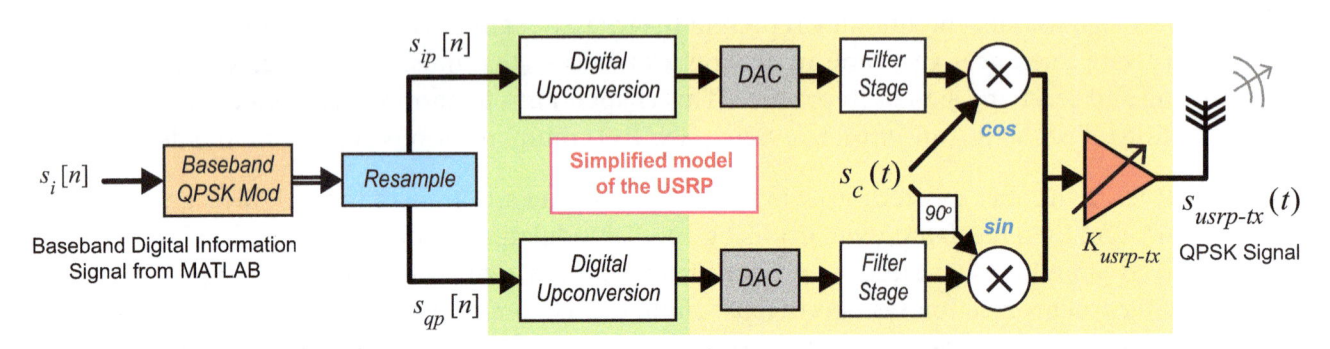

Figure 12.1: Block diagram showing a Simulink/ USRP® hardware implementation of a QPSK modulator

Additionally, this section will also feature pulse shaping; and in particular, *raised cosine* pulse shaping. Exercises 12.1 and 12.2 demonstrate how to construct transmitter and receiver designs, respectively, which together create a RC filter pair. From this, it should be possible to confirm the benefits of RC filtering with a practical communications system (i.e. constraining the transmission bandwidth and managing ISI, as previously discussed in Section 11.2).

Before building any QPSK transmitter Simulink models, you should ensure you have the MathWorks USRP® Hardware Support Package installed on the 'transmit' computer. Details of how to do this can be

Chapter 12: Desktop Digital Communications: QPSK Transmission and Reception 505

found in Appendix A.2 (page 571). If you do not have a USRP® radio, you can still run all of the USRP® Tx exercises throughout this chapter by commenting out the SDRu Transmitter block.

Exercise 12.1 **RRC Transmit Pulse Shaping with the USRP® Radio**

This exercise will involve constructing a simple block diagram that generates a random binary sequence, which will then be QPSK modulated, RRC pulse shaped and transmitted using the USRP® hardware. The model will act as an introduction to the practical transmission of digital data, while also allowing the USRP® transmitter setup to be tested.

(a) **Open MATLAB**. Set the working directory to the exercise folder,

   /my_models/transmitters/

Next, create a new Simulink model, and save it with the name,

   .../usrp_QPSK_raised_cosine.slx

(b) **Create a digital signal.** Open ▦ > *Communications System Toolbox > Comm Sources > Random Data Sources* and find the *Bernoulli Binary Generator* block. Place this block into an area of blank space in the model, preferably on the left-hand side as it will be the first block in the block diagram. 2⃞ on it, and change the 'Sample time' to '1/100e3', to set the generator to produce bits at a rate of 100kbps. Select 'Frame-based outputs' and set the 'Samples per frame' field to '2', to make it output 2-bit symbols that can be directly QPSK modulated. Change the 'Output data type' to 'boolean', apply the changes and close the parameter window.

(c) **Add a QPSK modulator.** Navigate to ▦ > *Communications System Toolbox > Modulation > Digital Baseband Modulation > PM* and find the *QPSK Modulator Baseband* block. Add it to the model, open the parameter window, and ensure 'Phase offset(rad)', 'Constellation ordering' and 'Input type' are set to 'pi/4', 'Gray', and 'Bit' respectively.

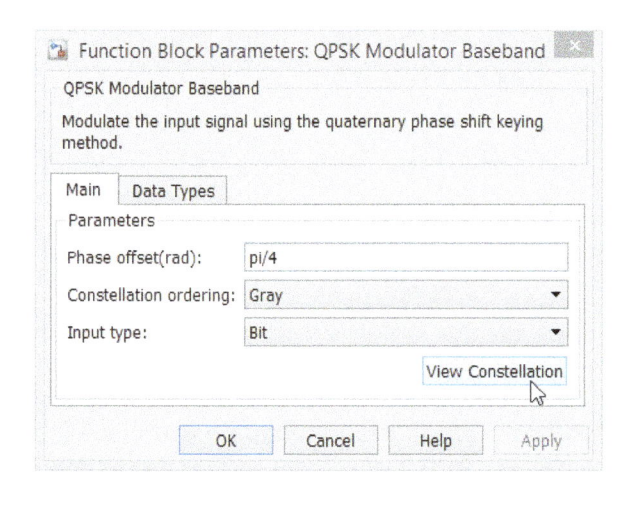

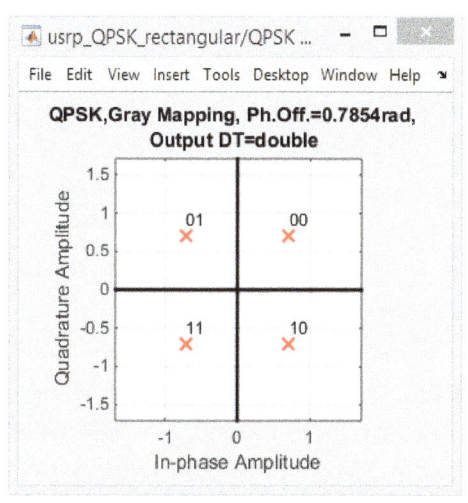

The offset value gives the QPSK constellations an offset of 45 degrees, placing the points equally in each quadrant of a Real-Imaginary axis, which can be viewed by selecting the 'View Constellation' button. Apply the changes and close the parameter window.

(d) **Add a pulse shaper.** For the reasons described in Section 11.2 (page 437), pulse shaping is required in order to transmit a digital signal with an appropriately constrained bandwidth. Place a *Raised Cosine Transmit Filter* from ▦ > *Communications System Toolbox* > *Comm Filters* in your model. 🖱️ on this block and configure it with the following parameters:

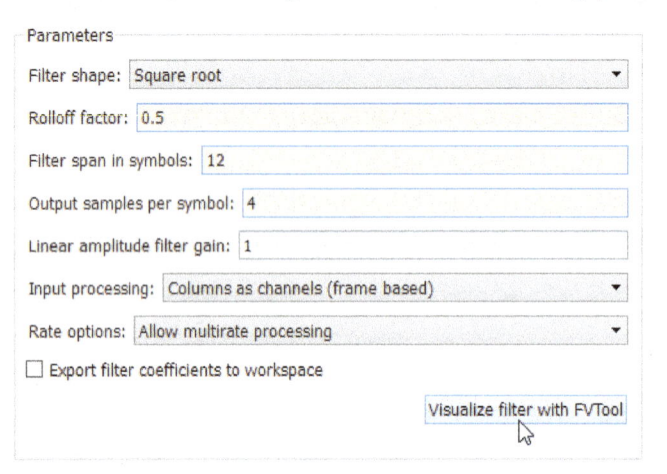

```
Parameters
Filter shape:              Square root
Rolloff factor:            0.5
Filter span in symbols:    12
Output samples per symbol: 4
Linear amplitude filter gain: 1
Input processing:          Columns as channels (frame based)
Rate options:              Allow multirate processing
☐ Export filter coefficients to workspace
                                            Visualize filter with FVTool
```

🖱️ on the 'Visualize filter with FVTool' button, and then on the 'Impulse Response' button from the toolbar. You should be able to see that the impulse response is forty-eight samples long (the 'Filter span in symbols' multiplied by 'Output samples per symbol'), and has the shape of a raised cosine pulse; as shown previously in Figure 11.8 (page 438). Passing the signal through this filter limits its bandwidth according to the desired spectral mask. Close the figure and parameter windows when you are ready to move on.

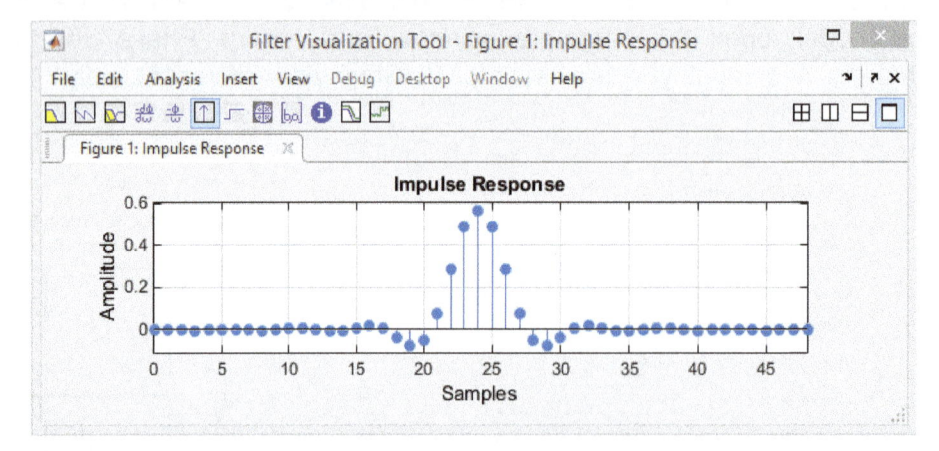

(e) To prepare the signal for output to the USRP® hardware, a *Buffer* block should be added to the model from ▦ > *DSP System Toolbox* > *Signal Management* > *Buffers*. Place it in an area of free space and rename the block *Output Buffer*. Open its parameter window, change the 'Output buffer size' to '1000', then apply the change and close the window. This block will buffer the complex samples and group them before they are output to the USRP® hardware (without this, the radio would not receive the data from the model at a sufficient rate, which would cause it to underrun, and stop the transmitter from running in real time).

Chapter 12: Desktop Digital Communications: QPSK Transmission and Reception 507

(f) **Interface with the USRP® Radio.** Fetch an *SDRu Transmitter* block from the ▦ > *Communications System Toolbox Support Package for USRP® Radio* library. This block should be configured to communicate with your USRP® hardware. To do this, you will need to know its IP address (or USB address), and be able to 'ping' it from your computer. Instructions of how to set up this connection are discussed in Appendix A.2 (page 571). 🖱 on the *SDRu Transmitter* block, select the appropriate 'Platform', and type the 'Address' of your USRP®.

(g) **Configure the USRP® Radio settings.** The RF side of the USRP® hardware also needs to be configured. Set the 'Center Frequency' parameter to the desired frequency of the transmission. This value should be within the range of the RTL-SDR tuner, e.g. in the range 25MHz–1.75GHz. If, for example, you wanted to modulate the signal onto a 602MHz carrier, '602e6' should be entered in this field. Set the 'LO offset' to '250e3' and the 'Interpolation' factor to '500'. By setting the local oscillator of the USRP® hardware to 250kHz, we can guarantee that any harmonics it generates will not interfere with the QPSK signal. Choosing the value of '500' for the interpolation function means that the radio resamples the QPSK signal from 200kHz to 100MHz before modulating it onto the carrier. Change the 'Source' of the 'Gain' to 'Input Port', then apply the changes and close the window.

(h) Finally, place a *Constant* block from ▦ > *Simulink* > *Sources*. Set the 'Constant Value' to '10' and rename this block *Transmitter Gain (dB)*.

(i) **Add a scope, and connect up the blocks.** Place a *Spectrum Analyzer* block from ▦ > *DSP System Toolbox* > *Sinks* in the model, and change its name to *Spectrum Analyzer Transmit*.

Finally, connect the blocks as shown below to complete the transmitter.

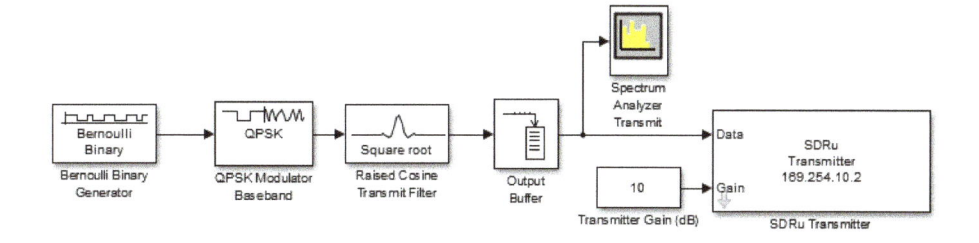

(j) **Prepare to run the simulation.** Set the *Simulation Stop Time* to 'inf', and change the *Simulation Mode* to 'Accelerator'. This will force Simulink to partially compile the model into native code for your computer, which allows it to run faster. Running the digital Tx Rx models as close to real time as is possible is very important. Finally, make sure your model is saved.

(k) **Open the reference file.** As an alternative to constructing the transmitter yourself, open the following file:

📁 /digital/usrp_tx/usrp_QPSK_raised_cosine.slx

BE CAREFUL: The *SDRu Transmitter* block is initially set to transmit on a frequency of 602MHz. Before running a simulation and transmitting radio signals, make sure that the *SDRu Transmitter* block is appropriately configured. In particular, it should be set to transmit on a frequency that you can legally use.

(l) Check that the USRP® hardware is turned on, and that MATLAB can communicate with it by typing `findsdru` into the MATLAB command window. If a structure is returned as discussed in Appendix A.2 (page 571), you can continue.

(m) **Run a simulation.** Begin the simulation by 🖱 on the 'Run' ▶ button in the Simulink toolbar. After a few seconds, Simulink will establish a connection with your USRP® radio and the simulation will begin. When it does, samples of the RRC pulse shaped QPSK signal will be passed to the radio, modulated onto the RF carrier and transmitted from its antenna.

(n) **Signal Analysis.** Examine the signal displayed in *Spectrum Analyzer Transmit*. The data shown in the scope should be similar to that shown below (without the annotations of course!).

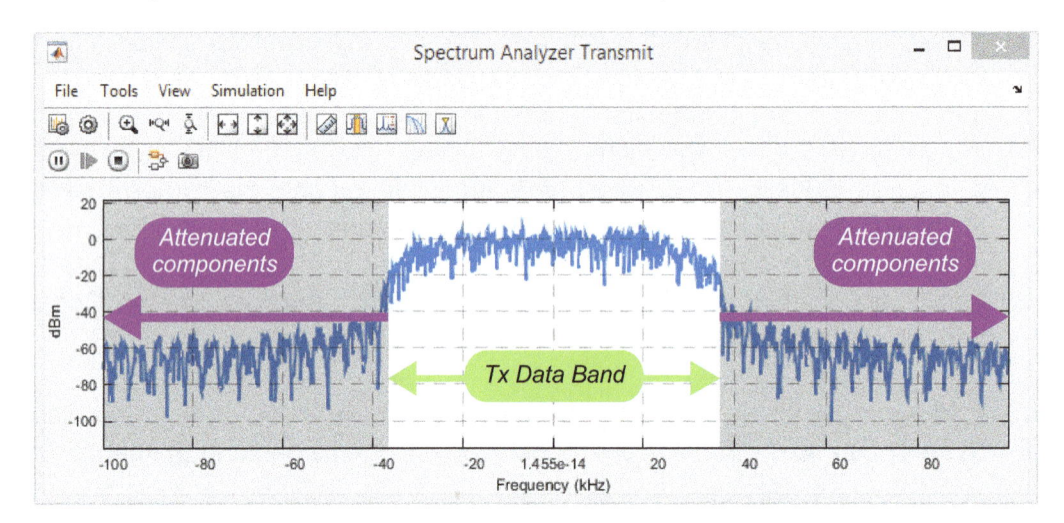

The bandwidth of the complex data signal is 75kHz (±37.5kHz either side of 0Hz), as highlighted in *Spectrum Analyzer Transmit*. We can also see from the Spectrum Analyzer that there are no high-power sidelobes present, as they have been attenuated by the pulse-shaping filter, which means that our signal conforms to the desired spectral mask. With a gain of around -60dB for these attenuated components, the transmitted signal should therefore cause minimal interference to users on adjacent frequency bands.

(o) As was discussed in Section 11.2 (page 437), the RC response is normally implemented using a pair of RRC filters, with one filter in the transmitter and another in the receiver. The next exercise will demonstrate how to design a receiver that implements a matched RRC receive filter. When you are satisfied that the transmitter model is working, you can stop it by 🖱 on the 'Stop' ■ button. If you intend to move on to the next exercise directly, then keep this model open, as it will be required later to run both the transmitter and receiver simultaneously.

Before you begin building any Simulink receiver models, make sure that the MathWorks RTL-SDR Hardware Support Package is installed. Details on how to do this can be found in Appendix A.1 (page 569). If you do not have an RTL-SDR, or are unable to transmit the signals with the USRP® radio, you can still complete all of the Rx exercises in this chapter by substituting the RTL-SDR Receiver block with an Import RTL-SDR Data block.

Chapter 12: Desktop Digital Communications: QPSK Transmission and Reception 509

Exercise 12.2 **RRC Matched Filtering in an RTL-SDR Receiver Model**

In this exercise, we will construct a simple block diagram with a matched RRC filter stage, to receive the QPSK signal generated by the transmitter constructed in Exercise 12.1. Using the RTL-SDR and this receiver model, it will be possible to view the signal after it has been transmitted through the air, and received. The purpose of this is to observe the presence of the transmitted signal on the spectrum, and also to confirm the effect of matched RC filtering.

(a) **Open MATLAB**. Set the working directory to the exercise folder,

> /my_models/receivers/

Next, create a new Simulink model, and save it with the name:

> .../rtlsdr_QPSK_raised_cosine.slx

(b) Open this new model, and the Simulink Library Browser.

(c) **Place an *RTL-SDR Receiver* block.** If you have an RTL-SDR available and are able to transmit the RRC pulse shaped QPSK signal from Exercise 12.1, place the *RTL-SDR Receiver* block from ▦ > **Communications System Toolbox Support Package for RTL-SDR Radio** into the model.

Next, ② on the block to open its parameter window, and change the 'Source' of 'Center frequency' and 'Tuner gain' to 'Input Port'. Enter '1e6' in the 'Sampling Rate' field to set the RTL-SDR to sample at a rate of 1MHz. Set the 'Output data type' dropdown to 'single' and enter '4096' in the 'Samples per frame' box.

If you have already followed the steps in Exercise A.3 (page 577), then enter the PPM correction value into the 'Frequency correction (ppm)' field. Alternatively, if you have not yet done so, follow the above exercise to obtain this correction value now, and note it for use in subsequent receiver models.

If a single RTL-SDR is connected to the computer, the 'Radio address' can be left as '0'. If more than one is connected, run the `sdrinfo` command in the MATLAB command window to find the ID of the stick you desire to use and enter the value into the 'Radio Address' field, Apply these changes and then close the window.

(d) Place two *Constant* blocks from ▦ > *Simulink* > *Sources* into the model, and modify their names to *QPSK Signal Frequency (Hz)* and *Tuner Gain (dB)*. Set the 'Constant value' of *QPSK Signal Frequency* to the centre frequency of the QPSK signal to be received — for example, '602e6' for 602MHz. Set the *Tuner Gain* block to a default value of '10' (this value may need to be adjusted later, depending on the strength of the received signal).

(e) **Place an *Import RTL-SDR Data* block.** If an RTL-SDR is not available, or you are unable to transmit the RRC pulse shaped QPSK signal, navigate to ▦ > **RTL-SDR Book Library** > **Additional Tools** and place an *Import RTL-SDR Data* block. ② on this and change the 'File Name' parameter to reference the file:

📁 `/digital/rtlsdr_rx/rec_data/qpsk_raised_cosine.mat`

Leave the 'Output Frame Size' set to '4096', then apply the changes and close the parameter window. If this process was successful, the block should show the sampling frequency of the recorded signal in the bottom right hand corner. In this instance, it should show a sampling frequency of 1MHz.

🔧 (f) **Add an RRC Matched Filter.** Add a *Raised Cosine Receive Filter* to the model from ▦ > **Communications System Toolbox > Comm Filters**, followed by a *Matrix Concatenate* block from ▦ > **Simulink > Math Operations**. Then, 🖱 on the RRC filter, and configure it with the parameters shown below.

```
Parameters
Filter shape:             Square root
Rolloff factor:           0.5
Filter span in symbols:   12
Input samples per symbol: 20
Decimation factor:        1
Decimation offset:        0
Linear amplitude filter gain: 1
Input processing:         Columns as channels (frame based)
Rate options:             Allow multirate processing
☐ Export filter coefficients to workspace
                                     [Visualize filter with FVTool]
```

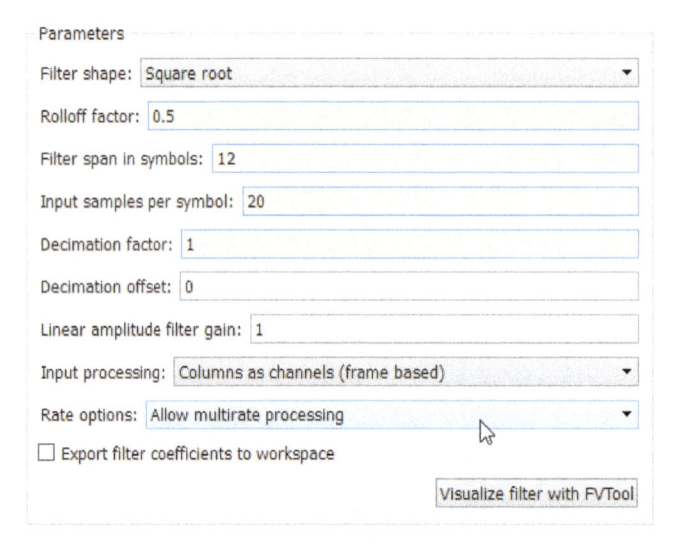

The block must be configured in this way to perform as a *matched* RRC filter. Notice that the settings do not all match the *Raised Cosine Transmit Filter* from Exercise 12.1. The 'Filter shape', 'Rolloff factor' and 'Filter span in symbols' must match, as these determine the characteristics of the filter, but the 'Input samples per symbol' and 'Decimation factor' can differ.

The 'Decimation factor' could also be matched here, but this has been set to a value of '1' (meaning that the filter does not decimate) as it allows the higher sample rate to be utilised in later receiver designs. The 'Input samples per symbol' depends upon: (i) the sample rate of the received signal (the signal output by the RTL-SDR); and (ii) the symbol rate of the transmitted signal. In this design the *RTL-SDR Receiver* block is set to sample the signal at 1MHz, and the symbol rate is set to 50kSps in the transmitter. Dividing the sample rate by the symbol rate results in the value of 20, which is entered in this 'Input samples per symbol' field.

As the pair of RRC filters used in this and the last exercise are matched, their combined response is equivalent to a single raised cosine filter. This means that the signal output from the receive filter will exhibit zero-ISI (or close to zero, taking into account the imperfections of the system) while retaining compliance with the spectral mask.

🔧 (g) **Place a scope and connect up the blocks.** Finally, introduce a *Spectrum Analyzer* block into the mode from ▦ > **DSP System Toolbox > Sinks**, and rename it as *Spectrum Analyzer Receive*. To complete the receiver design, connect the blocks in one of the two configurations

Chapter 12: Desktop Digital Communications: QPSK Transmission and Reception 511

shown below, depending on whether an RTL-SDR or recorded data is used as the source. Connecting the Spectrum Analyzer in this way will allow you to compare the received RRC filtered signal, and the signal after matched filtering — which is equivalent to raised cosine filtering. Annotate the signals as shown by 2 on the connections.

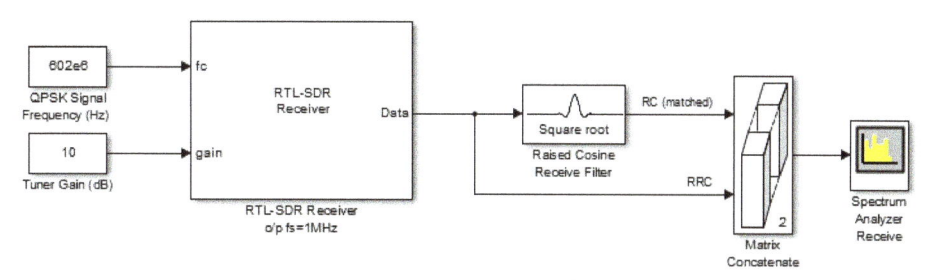

...or for an offline receiver

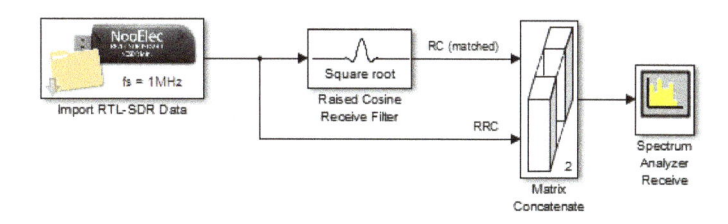

(h) **Open the reference file.** If you do not wish to construct the receiver, open the following file:

📁 /digital/rtlsdr_rx/rtlsdr_QPSK_raised_cosine.slx

If importing data rather than connecting to an RTL-SDR, delete the *RTL-SDR Receiver* block and connect the *Import RTL-SDR Data* block in its place. Check that this block references the file given above.

(i) **Prepare to run a simulation.** If importing a signal, set the *Simulation Stop Time* to 30 seconds, '30'. If using an RTL-SDR, set the *Simulation Stop Time* to 'inf' in the Simulink toolbar, and change the *Simulation Mode* to 'Accelerator'. This forces Simulink to partially compile the model into native code for your computer, which will enable the model to run faster. Check that MATLAB can communicate with it by typing `sdrinfo` into the MATLAB command window. Finally, make sure the model is saved.

(j) **Simulate!** Begin the simulation by 1 on the 'Run' ▶ button in the Simulink toolbar. After a short time the simulation will begin, and the *Spectrum Analyzer Receive* window will appear. If you are using an RTL-SDR, adjust the frequency and gain values until the device is tuned to your transmitted signal. The frequency domain scope can be used to aid this process.

(k) **Signal Analysis.** The *Spectrum Analyzer Receive* window allows the signal to be monitored in the frequency domain as it is received. It is configured to display two signals: the matched filtered RC signal, and the received RRC signal (i.e. before the matched filter). The scope should appear similar to the example shown on the following page.

The received signal was transmitted by the USRP® transmitter at a rate of 100kbps. The signal bandwidth is highlighted by the green arrow in the *Spectrum Analyzer Receive* window.

The sampling frequency of the RTL-SDR is set to 1MHz (much greater than the signal bandwidth) to enable the spectrum either side of the signal to be viewed. The spectrum obtained here shows that components in adjacent bands are significantly attenuated. (Without RRC transmit filtering, the signal spectrum would not be well constrained, and energy would leak into adjacent bands.)

Comparing the orange and blue signals, it can also be observed that the RRC filter in the receiver attenuates frequency components on adjacent bands (which might include noise and signals from other users).

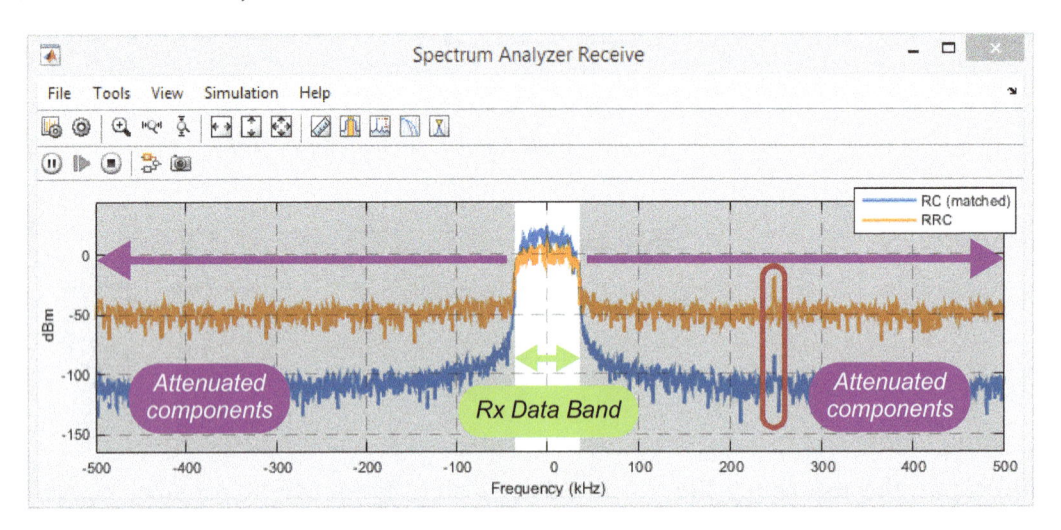

(l) The peak corresponding to the USRP® radio's LO, which was given an offset of 250kHz in the *SDRu Transmitter* block, is also visible (circled in red above). It is important to note that if the LO was configured to have an offset below about 37.5kHz, it would cause interference within the bandwidth of the desired signal.

(m) **Watch us running the receiver.** We have recorded a video which shows activity in the *Spectrum Analyzer* window, highlighting the effect that the matched RC filter has on the received signal.

desktopSDR.com/videos/#pulse_shaping

12.2 Coarse Frequency Synchronisation in a Real-time System

The next stage of processing to be added to the digital communications receiver is coarse frequency synchronisation. As explained in Section 11.8 (page 480), this first synchronisation stage is imperative to the success of the remaining synchronisation operations, as it greatly reduces any initial frequency offset present in the received signal. In practical communications systems, the offset between the *expected* and *actual* frequencies of the received carrier signal can vary. In this case the offset is assumed to be large due to the hardware tolerance issues of the RTL-SDR (recall that a preliminary step was required beforehand to enter the measured offset of the device!). The next two exercises will allow you to investigate coarse synchronisation, which is effectively an automated 'tuning' process, in a real-time system.

Exercise 12.3 Coarse Frequency Correction: Inspecting the Transmitter

In this short exercise we will open, inspect and run the transmitter reference design that will be used to generate a signal for the receiver in Exercise 12.4.

(a) **Open MATLAB.** Set the working directory to the exercise folder,

   /digital/usrp_tx/

Next, open the file:

   .../usrp_QPSK_coarse_synch.slx

The block diagram should look like this:

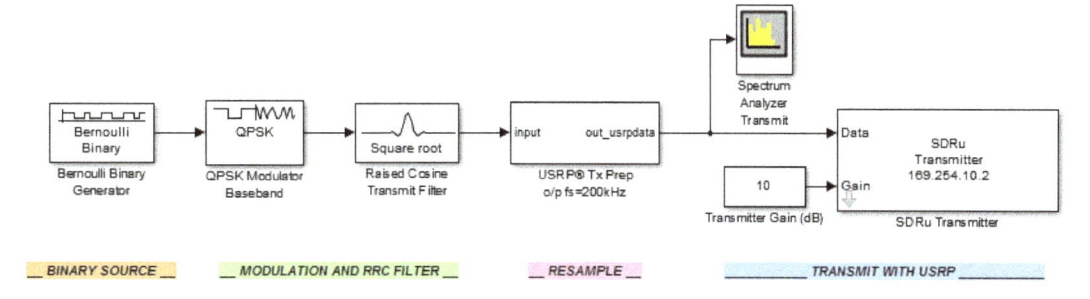

(b) **Examine the model.** Open the *Bernoulli Binary Generator* parameter window. Do you notice any differences in these parameters, compared to those from the previous transmitter model? Inspect the *USRP® Tx Prep* subsystem, and determine why it is required in this model. Notice that the subsystem contains repeated stages of the same blocks. Why do you think this is?

(c) **Check the USRP® Radio settings.** The *SDRu Transmitter* block needs to be set up to communicate with your USRP® hardware. To do this, you should follow the information provided in Exercise 12.1 to configure the block for your hardware device. Also, check that the *SDRu Transmitter* block is set to transmit on a frequency that you can legally use.

(d) **Run a simulation.** Begin the simulation by clicking on the 'Run' button in the Simulink toolbar. After a couple of seconds, Simulink will establish a connection with your USRP® radio and the simulation will begin. When it does, samples of the pulse shaped QPSK signal will be passed to the radio, modulated onto an RF carrier, and transmitted.

(e) **Signal Analysis.** Examine the signal shown in *Spectrum Analyzer Transmit*. Use the 'Zoom In X' button to inspect the signal in detail, and confirm that its bandwidth is constrained.

Exercise 12.4 Coarse Frequency Correction: Investigation with the Receiver

This exercise involves inspecting and running a receiver reference design that incorporates a *Coarse Frequency Correction* subsystem to automatically centre a received QPSK signal around 0Hz. Using this model to receive the signal generated in Exercise 12.3, it will be possible to see the theory of coarse frequency synchronisation in action.

(a) **Open MATLAB.** Set the working directory to the exercise folder,

📁 `/digital/rtlsdr_rx/`

Next, open the file:

📁 `.../rtlsdr_QPSK_coarse_synch.slx`

The block diagram should look like this:

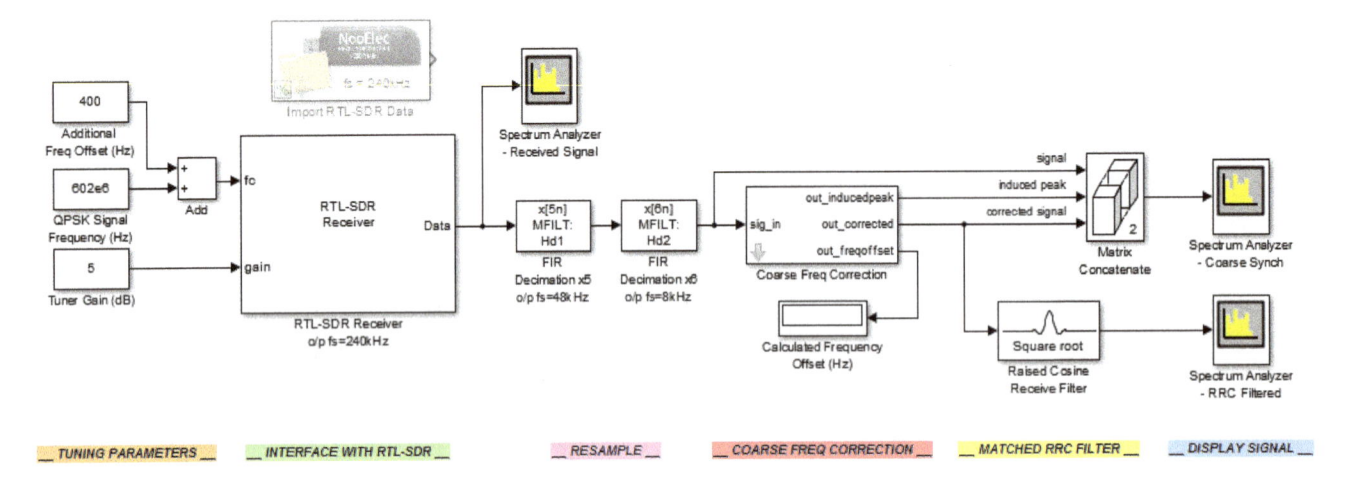

(b) **Examine the model.** Take a few moments to look at the design of the receiver, then note the sampling rate set in the *RTL-SDR Receiver* block. What benefit does this rate have over the previous rate of 1MHz for the remaining blocks in the design?

Notice that *FIR Decimation* blocks have been used (and cascaded) prior to the *Coarse Frequency Correction* stage. Thinking back to Section 11.8 (page 480), can you remember why this is important? Look at the parameters of the *Raised Cosine Receive Filter*, and note the 'Number of samples per symbol'. Can you determine why this value is used?

(c) **Check the RTL-SDR data source.** If an RTL-SDR is available and the model from Exercise 12.3 can be run to transmit the RRC pulse shaped QPSK signal, you will be able to use an *RTL-SDR Receiver* block to interface with and control your RTL-SDR. At this point, ensure that the PPM correction value for your RTL-SDR has been entered into the parameter window of the *RTL-SDR Receiver* block. Check also that the *QPSK Signal Frequency (Hz)* block is set to the appropriate value, i.e. it should match the frequency transmitted on by the USRP® hardware.

If you do not have an RTL-SDR, or are unable to transmit the signal, an *Import RTL-SDR Data* block should be used instead. This should be set to reference this file:

📁 `/digital/rtlsdr_rx/rec_data/qpsk_coarse_synch.mat`

If intending to import data rather than connecting to an RTL-SDR, delete the *RTL-SDR Receiver* block and connect the *Import RTL-SDR Data* block in its place. The duration of the recorded data is sixty seconds, so set the 'Simulation stop time' to '60'.

Chapter 12: Desktop Digital Communications: QPSK Transmission and Reception   515

(d) **Run!** Start reception by 🖱 on the 'Run' ▶ button in the Simulink toolbar. The model will soon begin to run, and the *Spectrum Analyzer* windows should appear. If using an RTL-SDR, adjust the frequency and gain values until the transmitted signal is offset by around 200Hz. Use *Spectrum Analyzer - Received Signal* to help with this.

(e) **Signal Analysis.** When the simulation is running, *Spectrum Analyzer - Coarse Synch* should display the three stages of synchronisation:

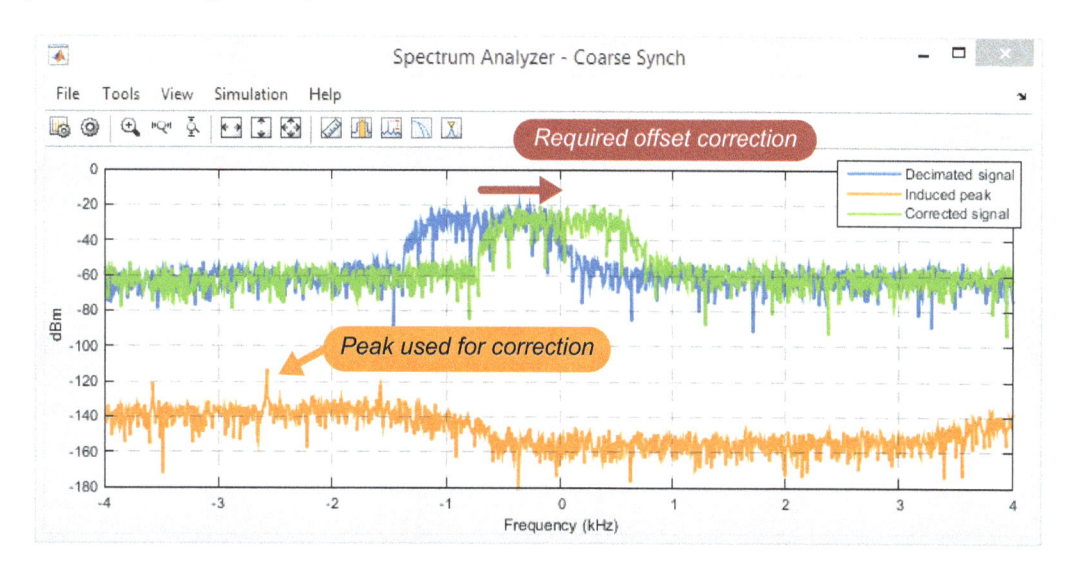

The 'Decimated signal' (blue) is the signal that was received by the RTL-SDR. Raising it to the power of 4 produces the 'Induced Peak' signal (orange) and, using the peak, the signal is shifted to create the 'Corrected signal' (green), which is centred around 0Hz. From the spectrum we can determine a rough value of frequency offset by eye, but the exact value can be observed in the *Calculated Frequency Offset (Hz)* block. The signal had an offset of 701.17Hz in the simulation shown above.

If you have any issues identifying the above features, or the spectra look much less distinct than the above, try increasing the *Tuner Gain (dB)* block until you can see a clearer correspondence with the example shown. Bear in mind that the shift may be to the left or right!

(f) Increase the value of the *Additional Freq Offset (Hz)* block, and watch the effect this has on the signals. Can you find the limitations of this synchronisation method?

(g) **Watch us running the receiver.** We have recorded a video which shows activity in the *Spectrum Analyzer* window, demonstrating coarse frequency synchronisation in action. It can be viewed at the following link.

desktopSDR.com/videos/#qpsk_coarse_synch

12.3 Carrier and Timing Synchronisation with the RTL-SDR

In Section 11.7 (page 477), a Simulink receiver design demonstrated carrier and timing synchronisation for a receiver with simulated frequency and phase offset values. Through this, we introduced the necessary subsystems for each of these synchronisation processes, and recorded some quantitative data as the system ran. Plots were produced, which revealed various characteristics about the signal and the behaviour of the receiver as it synchronised. The offset values used in Exercise 11.23 (page 479) were somewhat idealistic (i.e. they remained constant for the duration of the simulation), which is not generally the case in a realistic scenario.

In this section, we will further extend the real-time QPSK receiver; this time to include these carrier and timing synchronisation subsystems. Doing so will allow you to observe the difference in the compensation required for signals received by the RTL-SDR, and to make a comparison with the simulated receiver from Exercise 11.23. Afterwards, we will move on to a receiver that includes a *QPSK Demodulator*, as well as some other blocks necessary to output the received and demodulated bit stream.

Exercise 12.5 Carrier & Timing Synchronisation: Inspecting the Transmitter

To test the carrier and timing synchronisation in our desktop system, we require a transmitter model to produce the QPSK signal. The transmitter presented in this exercise is a modified version of the coarse synchronisation model, which omits any *Spectrum Analyzer* blocks. As we are now looking to perform real-time receiver synchronisation, it is imperative that the transmitter operates in real time too; any lag or underrun in the transmitter would greatly affect the success of reception. Removing the computationally intensive *Spectrum Analyzer* block prevents any related impairment on the processing speed of the model. Further, we no longer require this frequency domain plot, having become acquainted with the signal that the transmitter generates.

(a) **Open MATLAB**. Set the working directory to the exercise folder,

    `/digital/usrp_tx/`

Next, open the file:

    `.../usrp_QPSK_carrier_timing.slx`

The block diagram should look like this:

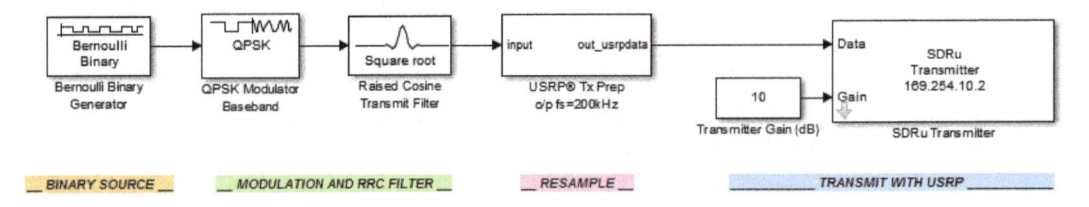

(b) **Examine the model.** When the model opens, you should notice that this block diagram possesses the same signal generation blocks as the transmitter from Exercise 12.3.

(c) **Check the USRP® Radio settings.** You will need to configure the *SDRu Transmitter* block to communicate with your USRP® hardware. To do this, follow the information provided in

Chapter 12: Desktop Digital Communications: QPSK Transmission and Reception

Exercise 12.1 to configure the settings of the block for your particular hardware. Remember to check the transmit frequency in particular, and ensure that it is set to a frequency you can legally use.

(d) **Run a simulation.** Begin the simulation by 🖱 on the 'Run' ▶ button in the Simulink toolbar. After a few moments, Simulink will establish a connection with your USRP® radio and the simulation will begin. Samples of the pulse shaped QPSK signal will be passed to the radio, modulated onto an RF carrier, and transmitted.

(e) When satisfied that the model is working, it can be stopped by 🖱 on the 'Stop' ■ button. If moving straight on to the next exercise, then keep this model open as it will be required.

Exercise 12.6 RTL-SDR and Theory Synchronisation Comparison

The reference design in this exercise is essentially the same as in Exercise 11.25 (page 486), but with the addition of the RTL-SDR interfacing block. Using this model and the transmitter from Exercise 12.5, we will investigate the effect that realistic offsets have upon the synchronisation subsystems.

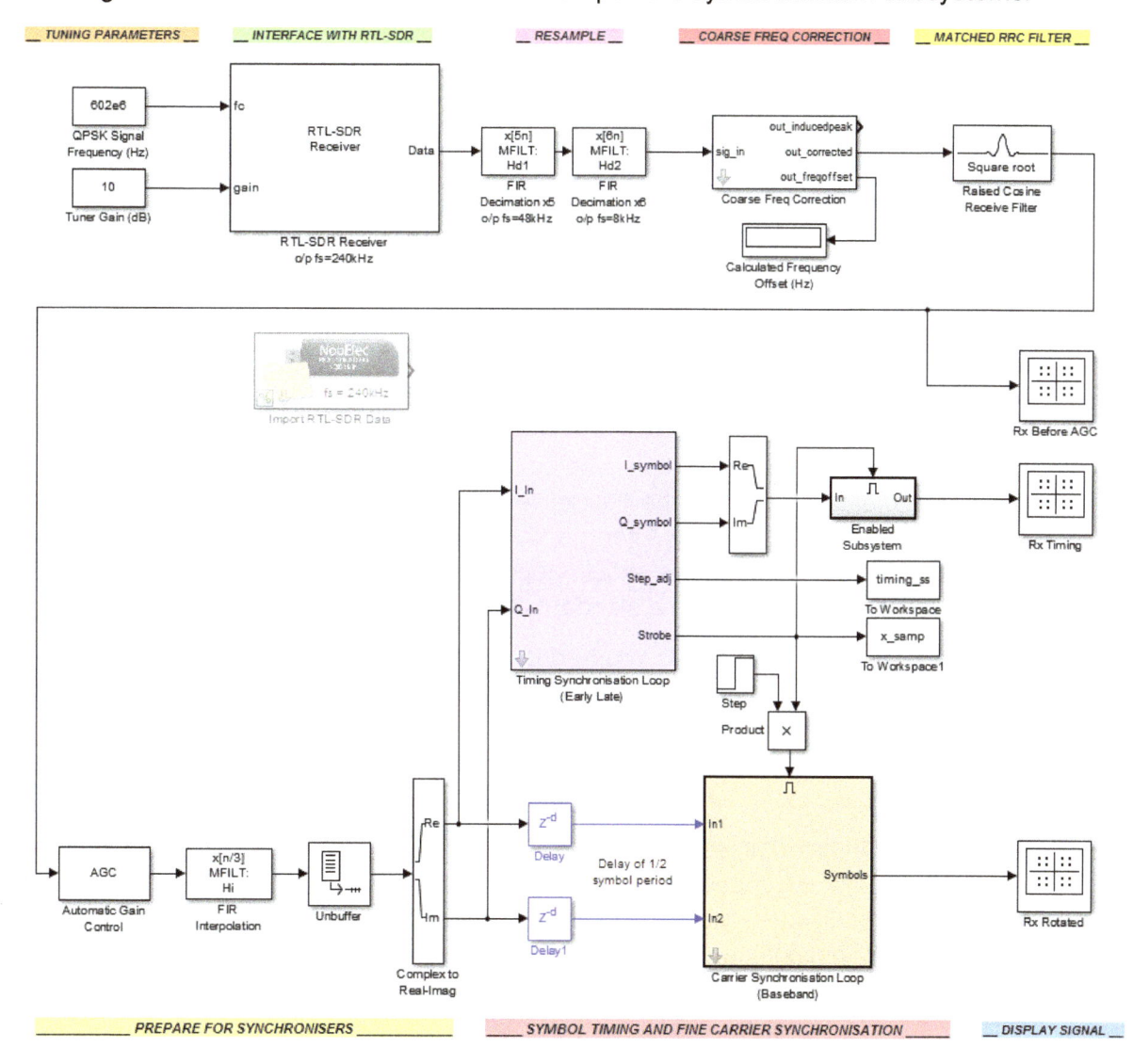

(a) **Open MATLAB.** Set the working directory to the exercise folder,

> /digital/rtlsdr_rx/

Next, open the file:

> .../rtlsdr_QPSK_carrier_timing_plots.slx

NOTE: *The functionality of the AGC Simulink block was changed between releases R2015a and R2015b. We have supplied modified versions of these files, denoted by the* `_15bonwards.slx` *file names. Anyone using 15b or higher will need to use these files instead.*

The block diagram should look similar to that shown in the earlier screenshot.

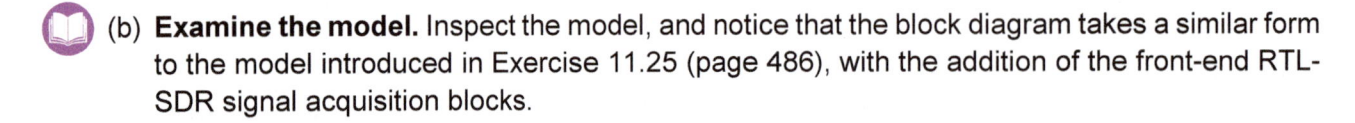

(b) **Examine the model.** Inspect the model, and notice that the block diagram takes a similar form to the model introduced in Exercise 11.25 (page 486), with the addition of the front-end RTL-SDR signal acquisition blocks.

(c) **Check the RTL-SDR data source.** If you have an RTL-SDR available, and are able to transmit the (RRC pulse shaped) QPSK signal from Exercise 12.5, an *RTL-SDR Receiver* block can be used to interface with and control your RTL-SDR. In this case, enter your PPM correction value into the parameter window. Check that the *QPSK Signal Frequency (Hz)* block is set to match the frequency transmitted on by the USRP® transmitter.

If you do not have an RTL-SDR or are unable to transmit the signal, an *Import RTL-SDR Data* block should be used instead. The block will need to be configured to reference this file:

> /digital/rtlsdr_rx/rec_data/qpsk_carrier_timing_plots.mat

If importing data rather than connecting to an RTL-SDR, delete the *RTL-SDR Receiver* block and connect the *Import RTL-SDR Data* block in its place.

(d) **Simulate!** The receiver simulation is set to run for 10 seconds, which should be long enough to gather the data required for the illustrative plots. Begin the simulation by 🖱 on the 'Run' ▶ button in the Simulink toolbar. After a couple of seconds, the simulation will begin and the *Constellation Diagram* windows should appear. Hopefully, the constellation shown in the *Rx Rotated* scope should look like a standard QPSK constellation, with 4 clusters.

The plot *Rx Before AGC* allows you to determine the strength of the received signal (before any AGC has been implemented). If the circumference of the plotted constellation is either very large or very small in the constellation diagram, then the transmitter or receiver gain may need to be adjusted to help the AGC process. For reference, the cluster shown opposite in the *Rx Before AGC* is a good example of the signal strength you should be aiming to achieve.

After the signal has passed through the AGC and timing synchronisation stages, an output similar to that shown opposite should be seen in the *Rx Timing* plot. At this point the signal should exhibit four small clusters, but it is likely that these will be spinning (either clockwise or anti-clockwise, depending on the polarity of the carrier offset). Finally, when the signal has passed through the carrier synchronisation stage, the *Rx Rotated* plot should show four clusters which are now situated statically around the reference markers.

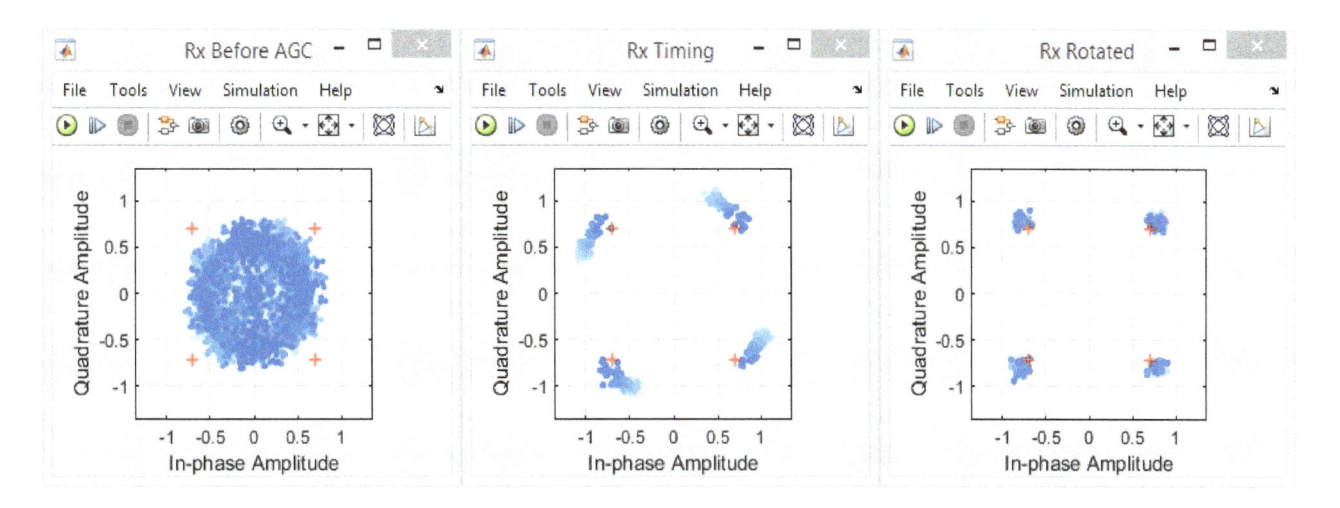

Chapter 12: Desktop Digital Communications: QPSK Transmission and Reception

If the system fails to synchronise during the period of the simulation, and particularly if the *Rx Before AGC* constellation plot looks very compact around the origin, try increasing the value set in the *Tuner Gain (dB)* block.

Further plot windows will appear at the end of the simulation, and these should show the successful synchronisation of the receiver to the signal transmitted by the USRP® transmitter. If your computer does not have enough computational power, it is likely that the receiver will not be able to synchronise using real-time received data. If this is the case, you can use imported data instead. Each time the simulation finishes, it will plot the new data in the existing figure windows.

(e) Inspect the plot titled *Derotated X and Y Components*, which shows the I and Q phases of symbols recovered by the receiver.

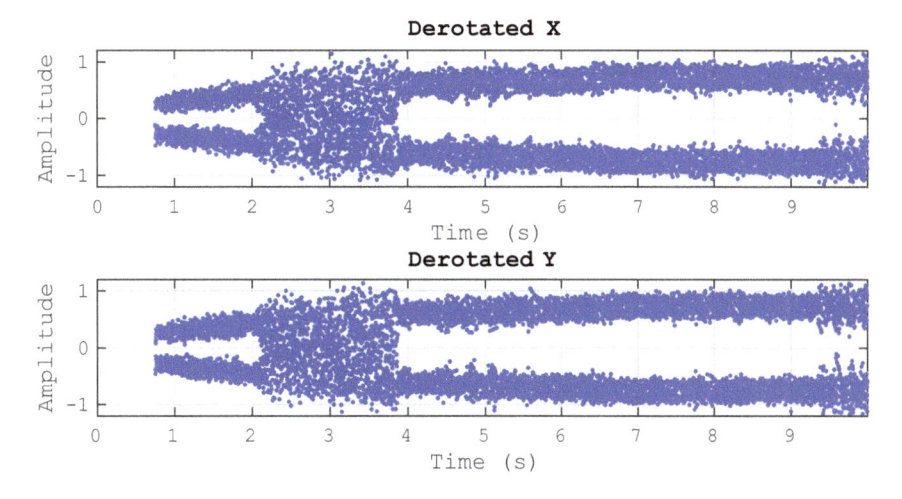

It may be noted that samples only appear from 0.75 seconds onwards, which is due to the *Step* block used to enable the *Carrier Synchronisation Loop (Baseband)* subsystem. This is implemented to prevent the subsystem becoming locked in a nul-output loop state — i.e. if, during execution, one of the computations in the feedback loop cannot compute an output resulting in a Not a Number (NaN) sample... and being a loop, it cannot recover. This would otherwise occur due to divide-by-zero errors in the carrier synchronisation subsystem during the initial stage of operation.

(f) **Data Analysis.** Compare your own figure, or indeed the one shown here, to the equivalent plot from the theory chapter (Figure 11.33 on page 480). Whereas the plot from the theory chapter showed the simulation samples converging very tightly to +V/-V, in the 'live' plot, the RTL-SDR samples converge to these approximate levels with much greater amplitude variation. This is expected, because the real signal is subject to greater noise, interference and even hardware-related drift that constantly changes the offsets affecting the signal. Although these offsets are more complex and time-varying than those considered previously with the simulated models, you should be able to recognise that the receiver still manages to synchronise successfully.

(g) Now turn your attention to the plot entitled *Carrier Synchroniser Loop Filter Behaviour*. Again you will notice the samples are plotted from 0.75 seconds onwards, because the data for this plot is also gathered from inside the carrier synchronisation subsystem.

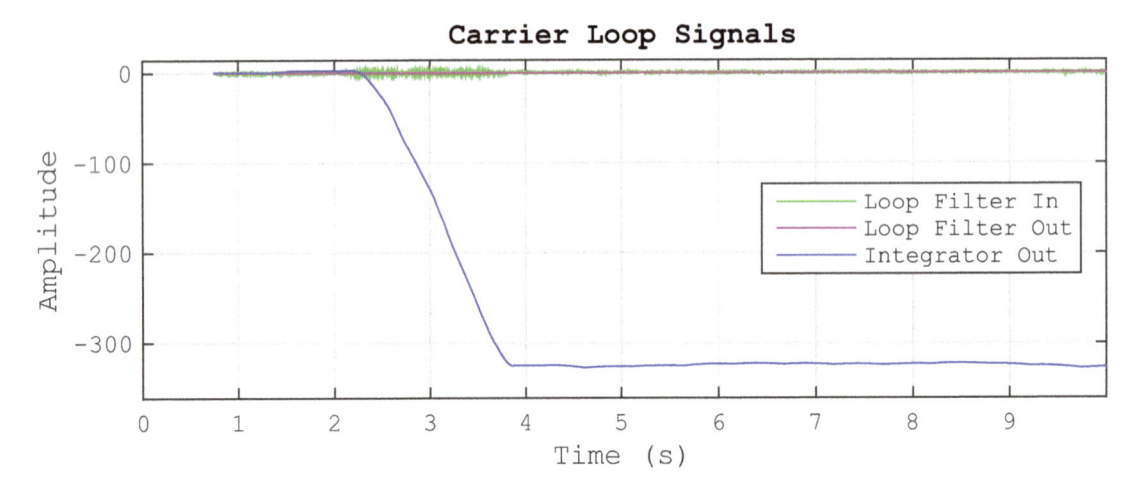

(h) Inspect the plot generated by the model, and compare it to its equivalent from the theory chapter (Figure 11.34 on page 480). The plot obtained from simulation exhibited both a steady loop filter output and a straight line phase integrator output (after a short transient period), which is rather different from what has now been plotted with the real signal data. Neither the loop filter output, nor the phase integrator output are particularly steady for quite some time, which is due to the changing offsets present in the signal. You may find that the phase integrator output will increment, decrement, or alternate during the synchronisation process, to compensate for the changing polarity of the frequency error.

It is worth noting that the receiver will synchronise more quickly when it no longer has to gather and store data in the Workspace for these plots.

(i) Before you move on, why not run the receiver model a few more times (if you haven't already done so), to see the differences in the output plots? This should allow you to notice that the offsets requiring to be compensated are rarely the same!

Exercise 12.7 Further Investigation of Real-time RTL-SDR Synchronisation

The receiver in this exercise is similar to that from Exercise 12.6, but it contains no *To Workspace* sinks, and instead introduces four new blocks that will output the demapped QPSK constellation bit pairs to

Chapter 12: Desktop Digital Communications: QPSK Transmission and Reception 521

the MATLAB command window. Using this design, and the transmitter from Exercise 12.5, we can now inspect both the *Constellation Diagrams* and the received symbols without any *Simulation Stop Time* constraints.

(a) **Open MATLAB**. Set the working directory to the exercise folder,

./digital/rtlsdr_rx/

Next, open the file:

.../rtlsdr_QPSK_carrier_timing.slx

NOTE: *The functionality of the AGC Simulink block was changed between releases R2015a and R2015b. We have supplied modified versions of these files, denoted by the* _15bonwards.slx *file names. Anyone using 15b or higher will need to use these files instead.*

The block diagram should look similar to that shown below:

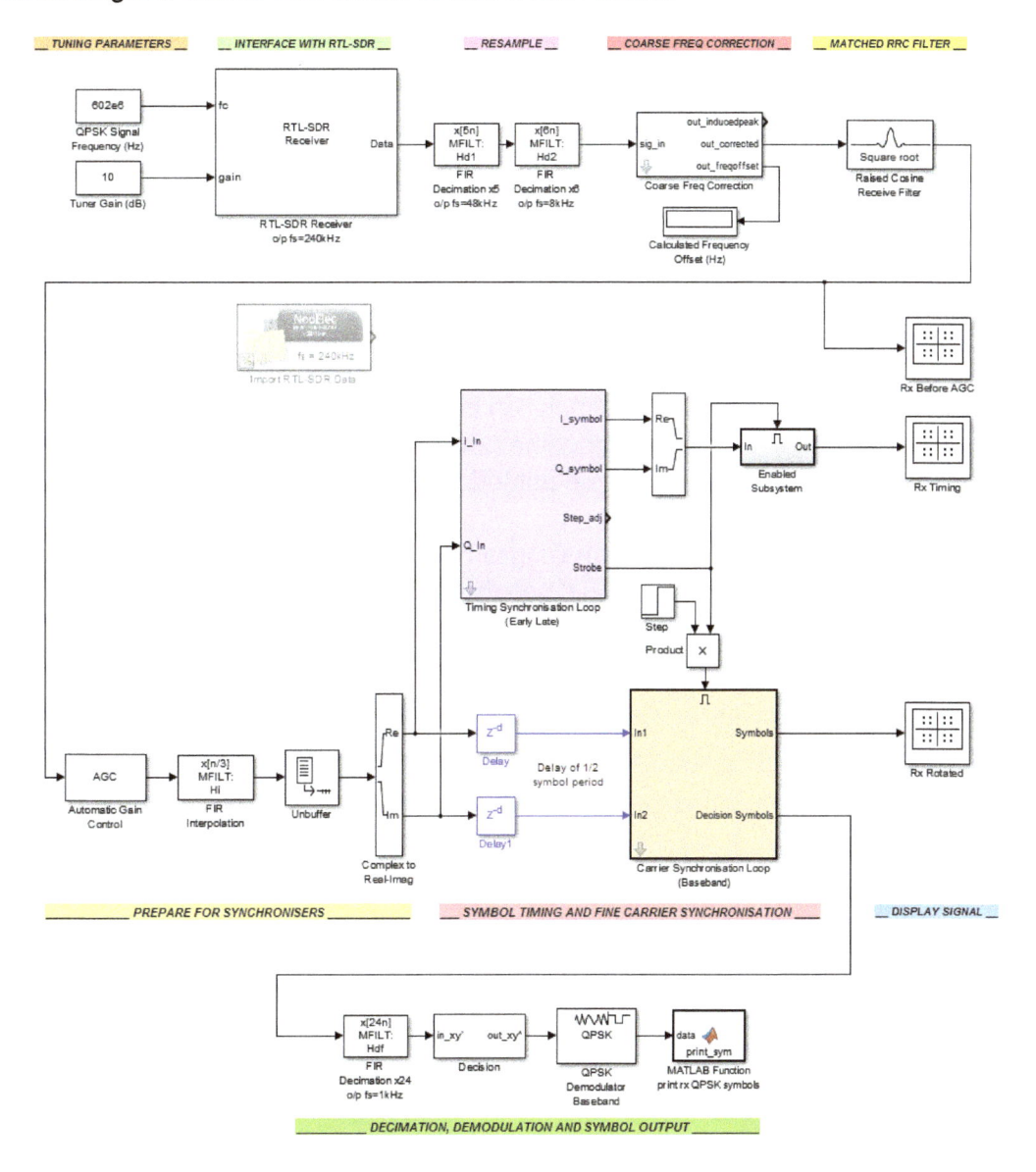

(b) **Examine the model.** Inspecting the receiver, you should notice that it is almost identical to the one considered in Exercise 12.6. Note also the final four blocks at the output of the *Carrier Synchronisation Loop* subsystem, labelled in green as *Decimation, Demodulation and Symbol Output*. These blocks implement the processing required to: (a) reduce the sampling rate back to symbol rate; (b) readjust the decimated samples back to their decided constellation points; (c) demodulate the constellations back to bits; and (d) output the received symbols to the MATLAB command window for viewing.

(c) Opening the *MATLAB Function* block will allow you to view the code that has been implemented to output the symbols. Inside this function, the data from the *QPSK Demodulator Baseband* block is converted to a string, and then printed to the MATLAB command window using the `disp()` function.

(d) **Check the RTL-SDR data source.** If an RTL-SDR is available, and you are able to transmit the RRC pulse shaped QPSK signal from Exercise 12.5, you will be able to use an *RTL-SDR Receiver* block to interface with and control your RTL-SDR. In this case, ensure that your PPM correction value has been entered into the parameter window. Also, ensure that the *QPSK Signal Frequency (Hz)* block is set to the current transmit frequency of the USRP® transmitter.

If you do not have an RTL-SDR, or are unable to transmit the signal, you should use an *Import RTL-SDR Data* block instead. This should be configured to reference the file:

/digital/rtlsdr_rx/rec_data/qpsk_carrier_timing.mat

If you intend to import data rather than connect to an RTL-SDR, delete the *RTL-SDR Receiver* block and connect the *Import RTL-SDR Data* block in its place.

(e) **Run the simulation.** Start a simulation by 🖱 on the 'Run' ▶ button in the Simulink toolbar. Once the model begins executing, the *Constellation Diagram* windows should appear, and the received symbols will begin to be printed in the MATLAB command window.

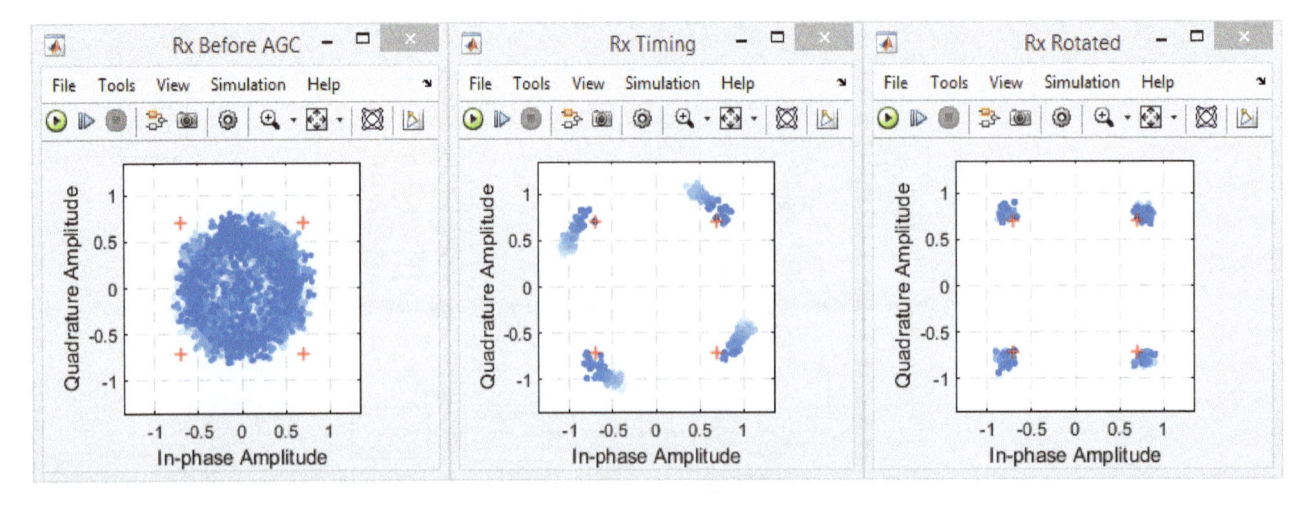

(f) **Signal Analysis: Constellation Diagrams.** Take a look at the *Constellation Diagrams*. Within a short period of time, you should see the expected views, similar to those shown below, and discussed in the previous exercise. Remember that the gain may need to be adjusted to help the AGC process i.e. the *Tuner Gain (dB)* block in the receiver model.

The constellation at the output, shown in the *Rx Rotated* plot, should show four clusters situated statically around the reference markers. If this is true, your receiver has correctly achieved carrier and timing synchronisation with the received signal. (Note that we are currently neglecting any phase ambiguity correction that may be required to correctly demap the symbols.)

(g) **Data Analysis: Printed Symbols.** As the transmitter model is configured to generate bits at a rate of 2kbps, the outputs in the command window will be very difficult to discern when the model is running. Hence, you will need to pause the model to examine them. Pause the simulation using the 'Pause' ⏸ button once you think your receiver has synchronised, and then navigate to the MATLAB command window.

Here we are simply observing that the signal has been converted back from QPSK symbols to groups of 2 bits, rather than trying to determine if these bits are correct. (Recall that the data source was random, so there is no easy mechanism of comparison here!)

```
1   1
0   1
1   0
0   0
...
```

(h) Viewing these printed symbols should consolidate the aims of the various processing stages implemented so far. The transmitter model has QPSK modulated a binary bit stream and transmitted the symbols across the RF spectrum. This signal has then been acquired by the receiver, processed and synchronised through various stages, and finally demodulated to convert the sampled symbols back to pairs of bits.

Next, we will focus on building a more complete communications system, based on this QPSK transmitter and receiver pair.

(i) **Watch us running the receiver.** We have recorded a video which shows the received signal's activity in *Constellation Diagram* windows and the output to the MATLAB command window. You are able to see the various synchronisation stages lock to the received signal, and the received codewords being printed in the command window.

desktopSDR.com/videos/#qpsk_timing_phase_synch

12.4 Developing a Simple Communications Protocol

In the chapter so far, we have reached the point of sending QPSK symbols via a wireless link. In order to send meaningful data, however, such as a string of text, we will now need to augment this system with some additional functionality to implement a basic communications protocol. This will allow, for instance, the data to be transmitted according to a particular agreed format. By imposing some standardisation on how messages are formed and formatted, and having the transmitter and receiver operate according to a '*standard*', we can ensure that messages are correctly communicated. At the most basic level, it must be ensured that the start of a message is correctly identified, so that subsequent processing of the received data can take place reliably.

To provide a little background, communications standards can use a variety of protocols and modulation techniques, and may also require information to be transmitted across different media (such as cables, fibre, and RF). In 1994, the ITU developed a model known as the *Open Systems Interconnection* (OSI) *stack* that could be used to help generalise and characterise the functionality associated with all of the different layers of digital communications systems. The model (or 'protocol stack') comprises of seven layers: Physical, Data Link, Network, Transport, Session, Presentation and Application, as highlighted in Figure 12.2 [35]. Consequently it is also commonly referred to as the 7-Layer Model.

The OSI 7-Layer Model is now relatively old in technology terms, and other models have since become popular, such as the TCP/IP model synonymous with computer networking [1]. In some systems there is a certain blurring of the layers, or even omission of layers, and hence the OSI model is not as generically applicable as it was originally. But for the purposes of this book, however, the OSI model is a useful reference and process guide to allow us to develop our very own simple proprietary protocols.

The systems presented so far have been concerned with the Physical layer (PHY), whose role is to transmit and receive electromagnetic signals, and is concerned only with the DSP operations required to pass data through a physical channel. This is where processes such as modulation, filtering, phase synchronisation and data recovery occur. According to the OSI model, the Data Link layer is where data frames are generated, and frame synchronisation is performed. Working up the stack, the Network layer normally considers issues such as addressing (allowing multiple nodes to be connected in a network), while the Transport layer deals with segmentation of data. The top three layers are concerned with the generation, manipulation and preparation of data to be transmitted and received. Each of the seven layers can be considered a separate entity, and there are numerous protocols associated with these that

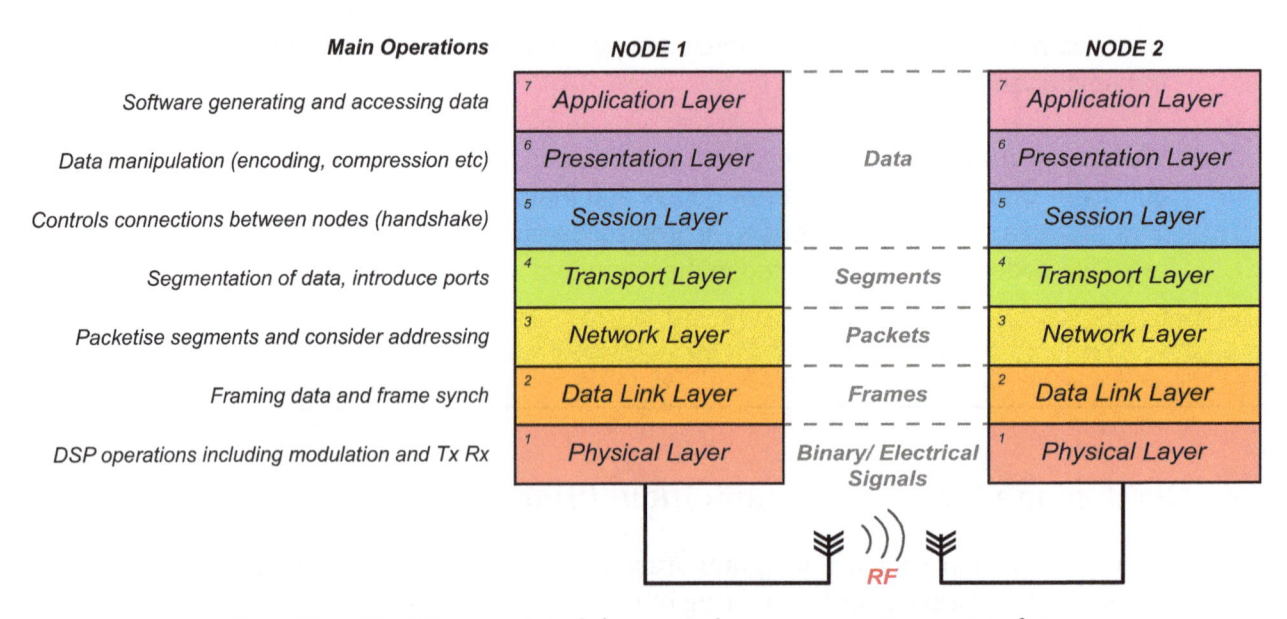

Figure 12.2: The OSI protocol stack from a wireless communications point of view

Chapter 12: Desktop Digital Communications: QPSK Transmission and Reception

you may be familiar with (e.g. HTTP, SSH, TCP, UDP, ARP, PPTP, IEEE 802.11). More information on the processes defined in the seven layers can be found in [16].

Everything we have developed to this point can be considered as based purely on the PHY layer. The QPSK transmitter has transmitted random binary sequences, and while this allowed us to test the ability of the QPSK receiver design to synchronise to the received signal, it has not been possible to verify correct reception of data, or to send more complex information like 'messages' or files. The solution is to integrate layers above the PHY in our SDR link, and to transmit data in a form that will be easily identifiable and usable at the receiver.

Note that, as we are considering point-to-point communication in this case, the task is simpler than if multiple nodes were involved. In that case, it would also be necessary to involve issues such as addressing, to ensure that messages are delivered to the correct recipients.

In the following sections, we will consider the higher-layer processes required to generate, transmit and receive character string messages over the QPSK link.

12.5 ASCII Encoding and Decoding

The American Standard Code for Information Interchange (ASCII) allows for 128 different characters — based on the English alphabet — to be encoded into 7-bit binary words [40]. This scheme falls into the Presentation layer of the OSI model. The types of characters that can be encoded with ASCII are: non-printing control characters (*NUL*, *ACK*), numbers and punctuation characters (*0-9*, *$*, ***, *space*), uppercase letters (*A-Z*), and lowercase letters (*a-z*) [16]. To give an example, uppercase 'A' is defined with the decimal value '65' in the encoding scheme, which is represented in binary as '`1000001`'.

Over the years, the ASCII encoding scheme has been used in many different applications. It permits conversion of English messages and text documents (amongst other things), between characters and ASCII binary sequences for transmission, processing, and storage on digital devices. We will eventually incorporate ASCII encoding and decoding into our transmitter and receiver designs, respectively, as this will allow useful (and intelligible) messages to be transferred.

Before we incorporate ASCII encoding and decoding in our designs, let's investigate how to implement the scheme using MATLAB code. In this short section, MATLAB scripts demonstrate the process of encoding and decoding a message of your choice. Following this, Section 12.6 will introduce the *framing* and *frame synchronisation* processes, which are required when transmitting and receiving data messages.

Exercise 12.8 ASCII Encoding using MATLAB Code

This exercise will first demonstrate a default message being converted to an ASCII binary representation, which can then be repeated using a message of your own choice. The script shows the process required to ASCII encode the message using MATLAB code, and allows you to view the outputs at different stages of the conversion.

(a) **Open MATLAB**. Set the working directory to the exercise folder,

`/digital/simulation/ASCII_coding`

Then, open the following MATLAB script:

`.../ASCII_character_encoder.m`

(b) **Examine the script.** At the top of the script there is a brief explanation of the processing that the code performs. When you have read this, turn your attention to the code itself.

```matlab
msg_string = 'Hello World!';     % enter the message string in chars
ascii_len = 7;                   % define length of ASCII codewords

% 'int8' function converts ASCII chars to equivalent decimal values
msg_dec = int8(msg_string);

% convert decimals to ASCII matrix
% 'de2bi' converts decimal representation to binary
% 'de2bi' is set to use ascii_len for number of bits per codeword
% 'left-msb' keeps left bit as the most significant bit
msg_bin = de2bi(msg_dec,ascii_len,'left-msb')';

% transform matrix to a one column vector for the output bit sequence
% binary sequence length = message string len * ASCII codeword len
bin_seq = reshape(double(msg_bin),length(msg_string)*ascii_len,1);
```

The first two lines define the message to be encoded, and the number of binary bits used to represent each character. Then, the remaining lines perform the conversion. The process can be confirmed by running the script and viewing the outputs.

(c) **Simulate!** Run the script by 🖱 on the 'Run' ▷ button in the MATLAB editor ribbon, and then navigate to the MATLAB workspace to view the `msg_dec`, `msg_bin` and `bit_seq` outputs.

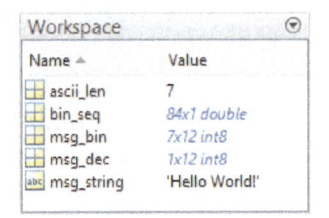

Name	Value
ascii_len	7
bin_seq	84x1 double
msg_bin	7x12 int8
msg_dec	1x12 int8
msg_string	'Hello World!'

(d) Next, 🖱 on the `msg_dec` variable, and examine the array of decimal values for each of the characters in `Hello World!`, including the space character. These values are then converted to their 7-bit binary representation, as shown in `msg_bin`, where each character is represented by a column of bits in the matrix.

(e) As a short exercise, you could either use MATLAB or some other means (e.g. an ASCII table), to perform the conversion for some of the characters of the string manually. You can compare your decimal and binary representations to those shown in these two output variables.

(f) Now look at the last variable, `bin_seq`, which shows a column vector of bits. The first seven relate the first character, the second seven to the second character and so forth. Having the

Chapter 12: Desktop Digital Communications: QPSK Transmission and Reception 527

binary data stored in column form will be useful for our later transmitter exercises, as this is the format required by our design.

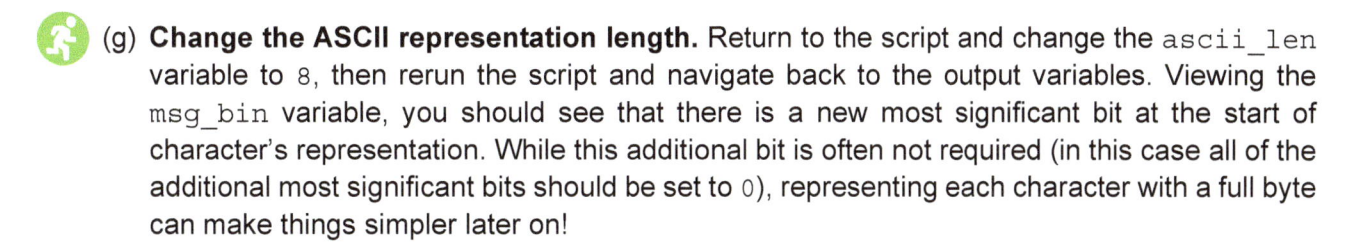

(g) **Change the ASCII representation length.** Return to the script and change the `ascii_len` variable to 8, then rerun the script and navigate back to the output variables. Viewing the `msg_bin` variable, you should see that there is a new most significant bit at the start of character's representation. While this additional bit is often not required (in this case all of the additional most significant bits should be set to 0), representing each character with a full byte can make things simpler later on!

(h) **Change the ASCII message.** Now that you have viewed the outputs for `Hello World!`, you can change the ASCII message set by the `msg_string` variable to something of your choice. Run the script again and look at the new decimal and binary values stored in the output variables. When you are satisfied with the message you have chosen and the conversion process, move on to the next exercise while keeping this script open.

Exercise 12.9 ASCII Decoding using MATLAB Code

In this exercise, we will run a MATLAB script that ASCII decodes the binary representation created in Exercise 12.8, thus reverting it back to a text string. As before, the script allows you to view the outputs at different stages of the conversion.

(a) **Open MATLAB.** Set the working directory to the exercise folder,

　　`/digital/simulation/ASCII_coding`

Then, open the following MATLAB script:

　　`.../ASCII_character_decoder.m`

(b) **Examine the script.** Once again, there is a brief overview of the code at the top of the file. Read this, and then inspect the rest of the file.

```matlab
msg_len = length(bin_seq)/ascii_len;    % calc message length in chars
msg_decoded = char(zeros(1,msg_len));   % create variable for decoded msg

% convert the binary sequence back to an English message
for count = 1:1:msg_len
    % extract binary representation of each character
    char_in_bin = bin_seq((((count-1)*ascii_len)+1):ascii_len*count)';
    % 'bi2de' converts binary representation to decimal value
    char_in_dec = bi2de(char_in_bin,'left-msb');
    % convert to character and store character
    msg_decoded(count) = char(char_in_dec);
end

% print the message to the command window
disp(msg_decoded);
```

The first few lines of code calculate the length of the message in characters, and initialise the variable `msg_decoded`, which is used to store the decoded message. The code inside the loop decodes the binary sequence back to text characters. Finally, the converted message is printed to the MATLAB command window using the `disp()` function.

(c) **Prepare to decode.** Before continuing, ensure that the `bin_seq` and `ascii_len` output variables from the `ASCII_character_encoder.m` script are stored in the MATLAB workspace, as they are required during the decoding process. If necessary, run the above script by opening it and 🖱 on the 'Run' ▷ button in the MATLAB editor, and then switch to the MATLAB Workspace to check that the required outputs are present.

(d) **Run the script.** Next, run the `ASCII_character_decoder.m` script by 🖱 on the 'Run' ▷ button, and then check that the decoded outputs have been created in the MATLAB Workspace.

Firstly, examine the `char_in_bin` variable and notice that the code has successfully extracted the binary representation for the final character of the message. Next view the `char_in_dec` variable, which holds the equivalent decimal value for this character. Finally, check that the `msg_decoded` variable holds the full character string you originally encoded.

(e) To view the outputs after the conversion of each character, insert a breakpoint at the `end` line of the `for` loop. Do this by 🖱 on the horizontal marker next to line 20 in the editor window:

Rerun the script. Each time it stops at the breakpoint, observe the variables to see the binary and decimal representation of each of the characters. To continue with the debug, 🖱 the 'Continue' ▷▷ button.

(a) 19 – / 20 / 21 / 22   (b) 19 – / 20 ● / 21 / 22

(f) Once you are familiar with the functionality of the script, remove the breakpoint by 🖱 on the red marker, and return to the encoder script to encode another message of your choice. Run the encoder script again to generate the new output variables, and then rerun the decoder script to view the decoded output.

12.6 Data and Frame Synchronisation

So far, the data used in the transmitter and receiver models in this chapter has been randomly generated. A *Bernoulli Binary Generator* source was utilised, and it was an adequate data source to allow the synchronisation circuits to be tested. Next, we will consider the transmission and reception of useful data, in the form of ASCII messages.

Unfortunately it is not as simple as adding an ASCII encoder to the transmitter, and a decoder to the receiver. The ASCII representation of the message is a concatenation of 7-bit binary codes, and correct decoding relies on knowledge of how the bits are grouped. For instance, if part of a message contained the word 'Blur', as shown in Figure 12.3, the decoding process would only be successful if the groups used to decode the characters were correctly aligned.

Chapter 12: Desktop Digital Communications: QPSK Transmission and Reception 529

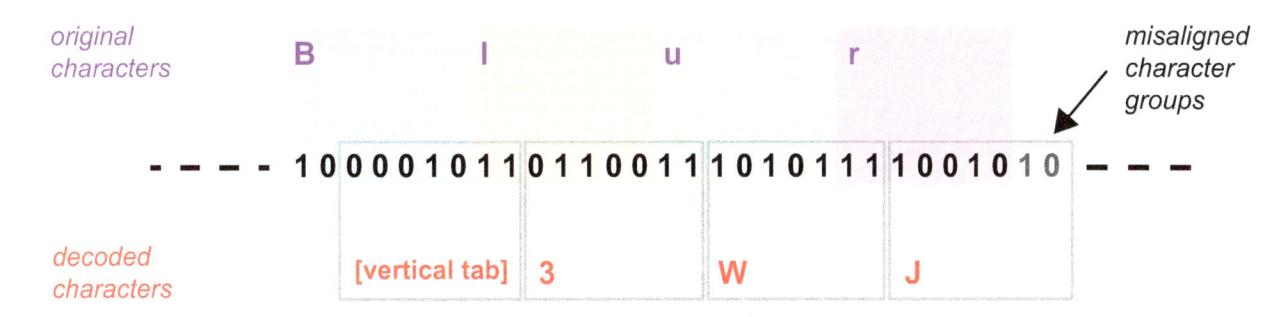

Figure 12.3: Decoding an ASCII message without correct character alignment

To transmit and receive ASCII messages successfully, therefore, it is necessary to ensure that the start of the sequence is correctly identified. Some form of rudimentary communications protocol must be implemented in order to achieve this.

12.6.1 Frame Structure

The proprietary communications protocol being developed requires a method of identifying the start of the message, in order to interpret the received data correctly. To achieve this, we will introduce a scheme of splitting the transmitted data into *frames*, which will be composed of a *payload* (containing the message), and a *header* (containing information to implement the communications protocol and aid the reception of the payload).

Although many digital communications systems use frames, their composition and structure, and the protocol implemented using frames, can differ widely. For the purposes of this book, we will simply define and use the frame structure shown in Figure 12.4. The header will be used by the *frame synchronisation* process in the receiver to achieve correct alignment of the payload, and ensure that the data is extracted and processed appropriately [40]. When a message is particularly long, it is partitioned into sections, with each section forming the payload of a frame within a sequence of frames. This is demonstrated in Figure 12.5.

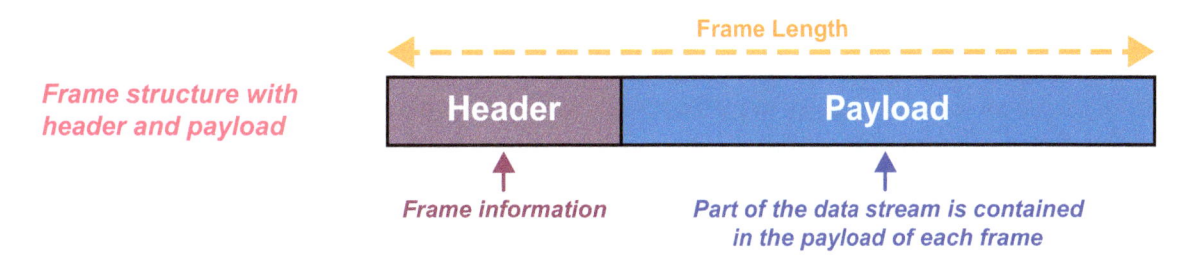

Figure 12.4: Typical structure of a frame

Payloads can be of fixed or variable length. Where the size is variable, it normally has lower and upper limits, as specified by the protocol in use. Setting a maximum payload length, and converting longer messages into a sequence of frames, is advantageous for a number of reasons. If a single frame in a series becomes corrupted, it may be possible to resend only that frame, rather than the whole message. (Adding the facility to identify lost frames and resend them is beyond the scope of this chapter, and we will not attempt to cover those aspects here.)

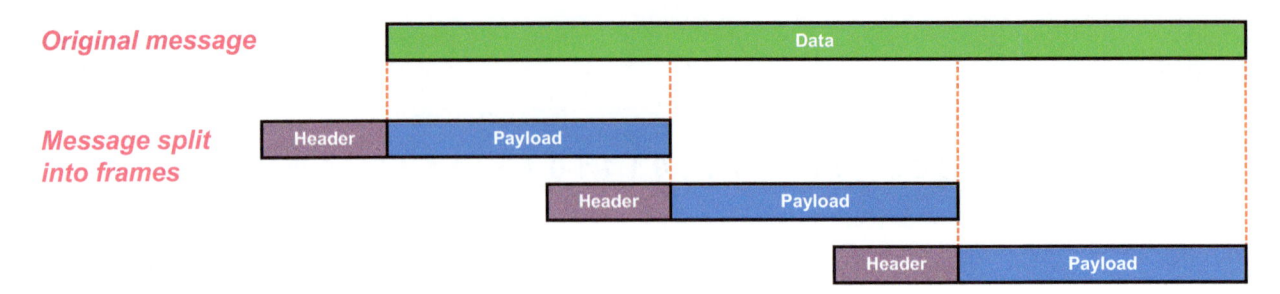

Figure 12.5: Message data segmented into frames

In both wired and wireless communications systems, *preambles* (a sequence that is normally part of the header) can be used to 'wake up' and help receivers to synchronise to the incoming data. We will go on to consider this aspect next.

12.6.2 Frame Synchronisation: Preamble and Correlation

The frame structure to be used in this communications protocol will contain a preamble within the header, to facilitate synchronisation at the frame level (in other words, to establish the position of the beginning of the payload). The design of the preamble sequence is important — the preamble must have favourable correlation properties in order to be easily identifiable.

Correlation is the process of mathematically determining the degree of similarity between two signals or sequences. For instance, the correlation between two sequences, b_1 and b_2, both of length N, may be expressed as

$$r_{12} = \sum_{n=0}^{N-1} b_1[n]b_2[n]. \tag{12.1}$$

Assuming that the sequences are formed from $\{-1, +1\}$ values, then the maximum correlation value is therefore $+N$. (It may be convenient in some circumstances to normalise the correlation operation such that the peak is $+1$.)

Correlation can be used for a variety of purposes in DSP, but in this case we consider its applicability for generating a timing measurement that will help to recover data from received frames.

Sequences suitable for a frame preamble have an *autocorrelation* function (i.e. the correlation result for the case $b_1 = b_2$) that produces a clear maximum when aligned, but a very low result when the sequences are out of alignment. The sequences are composed of ± 1 values, meaning that the product of each pair of samples is $+1$ when the sequences are aligned (as $1 \times 1 = 1$ and $-1 \times -1 = 1$), as shown in the left hand portion of Figure 12.6. Hence, the correlation output (equivalent to the summation of these individual sample products) is $r_{12} = N$, the length of the sequence. When out of alignment, the probability of $1 \times -1 = -1$ or $-1 \times 1 = -1$ occurring is about the same as either $1 \times 1 = 1$ or $-1 \times -1 = 1$, and for the type of codes we will adopt, the correlation result will be in the region $-1 \leq r_{12} \leq 1$.

Consider again the example in Figure 12.6 (note that the waveform is shown with a 'zero order hold' effect here, for the purposes of illustration).

Chapter 12: Desktop Digital Communications: QPSK Transmission and Reception

$$\sum_{n=0}^{N-1} b_1[n]b_1[n] = 13$$

$$\sum_{n=0}^{N-1} b_1[n]b_1[n-1] = -1$$

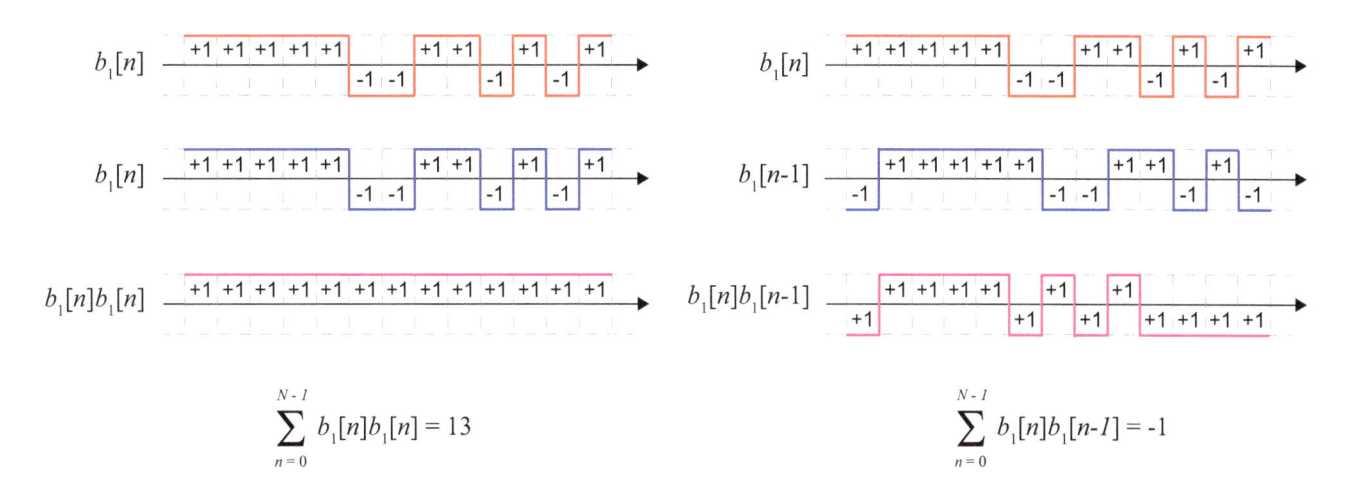

Figure 12.6: (left) autocorrelation of sequence $b_1[n]$, with zero time shift; (right) autocorrelation of sequence $b_1[n]$, with a 1 sample time shift

On the left hand side of Figure 12.6, a correlation is performed between two copies of the sequence $b_1[n]$, which are aligned in time. Notice that the result of multiplying each individual sample is +1, and thus the summation over the 13 samples of the sequence is +13.

On the right hand side of Figure 12.6, a correlation is again performed between two copies of $b_1[n]$, but in this example a shift of one sample exists. As a result, the products of individual samples are +1 approximately half of the time, while the remainder are -1. In this case, summing across the length of the sequence produces the result -1 (a low level of correlation).

Any other degree of shift could be evaluated in the same manner. Shifts of +2, +3, +4... and similarly -1, -2, -3, -4 samples produce similar results, i.e. correlation values no greater than ±1.

Taking all of these shifts into consideration, an autocorrelation function for the sequence $b_1[n]$ can be generated, as shown in Figure 12.7. As desired, this displays a peak of +13 at 0 (where the sequences are aligned) and has a low level of correlation otherwise. The example sequence chosen here is actually a *Barker sequence*, and this family of codes will be described in more detail next.

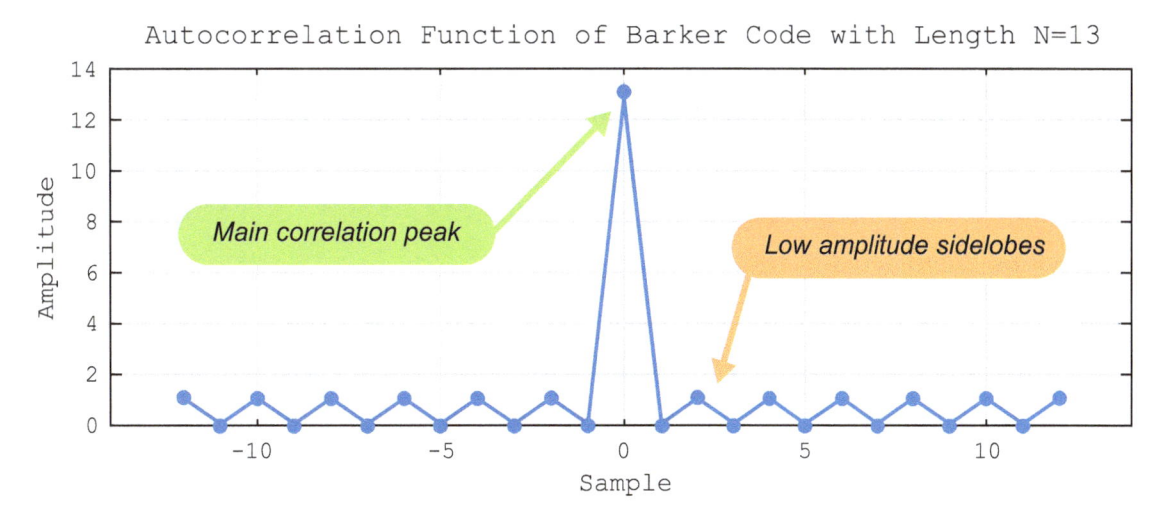

Figure 12.7: Annotated autocorrelation of a Barker-13 sequence

12.6.3 Barker Sequences

A number of different sequences have been identified that possess favourable correlation characteristics. The type we will focus on in this book is a subset known as *Barker codes* or *Barker sequences*. A Barker sequence is a finite sequence of N (+1) and (-1) values,

$$B[i] \quad \text{for} \quad i = 1, 2, \ldots, N \tag{12.2}$$

with the property that:

$$\left| \sum_{i=1}^{N-j} B[i]B[i+j] \right| \leq 1 \tag{12.3}$$

for all $1 \leq j < N$ [37].

There are several known Barker sequences, and the set has a maximum length of $N = 13$ (at least, no longer sequences have yet been found!)[3][41]. The most common of these are shown in Table 12.1. *Note:* this table omits the sequences that would be produced from inverting the sign, and separately, from reversing the time order, of the sequences shown.

Table 12.1: Barker sequences

Length	Sequences	
2	+1 -1	+1 +1
3	+1 +1 -1	
4	+1 +1 -1 +1	+1 +1 +1 -1
5	+1 +1 +1 -1 +1	
7	+1 +1 +1 -1 -1 +1 -1	
11	+1 +1 +1 -1 -1 -1 +1 -1 -1 +1 -1	
13	+1 +1 +1 +1 +1 -1 -1 +1 +1 -1 +1 -1 +1	

Barker sequences possess the key characteristic of low amplitude correlation sidelobes, as indicated in Figure 12.7; these are the outputs of the correlation process when the sequences are not aligned. The featured Barker-13 sequence will be used later, as we develop our own simple communication system.

12.6.4 Correlator Architectures

As explained in Section 12.6.2, a sequence that is known to the receiver is incorporated into the transmitted data to enable frame synchronisation. This fixed-length sequence is chosen for its desirable correlation characteristics, which allow it to be detected using a correlation process in the receiver. When the sequence is detected, the receiver is able to identify the start of a frame, and extract the data contained within the payload.

Chapter 12: Desktop Digital Communications: QPSK Transmission and Reception 533

Identification of the sequence is achieved by correlating the incoming data with a copy of the preamble stored within the receiver. A simple architecture for undertaking correlation is shown in Figure 12.8.

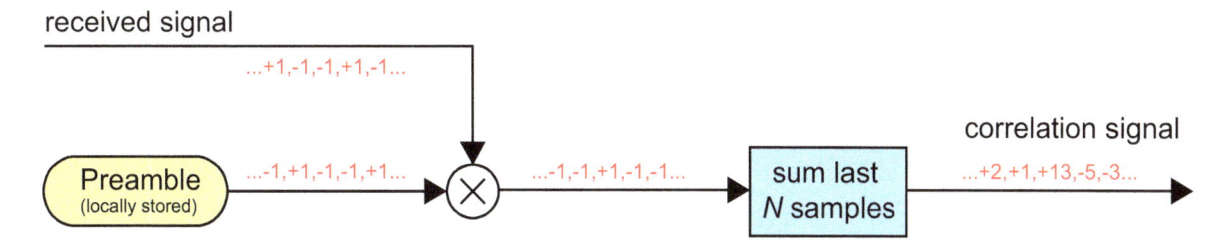

Figure 12.8: Basic serial correlator architecture

This basic, serial correlator can compute the correlation for one particular time shift only (as the stored preamble is read out sequentially in the receiver). Therefore, to generate correlation results for different degrees of time shift, multiple branches could be used to correlate the received signal with versions of the locally generated preamble that are delayed by 1, 2, 3, 4... samples. When a peak occurs in one of the branches, the preamble sequence has been detected.

A second possibility is to use a *matched filter correlator*. This approach exploits the fact that correlation is a sum of products (i.e. multiplications between input samples of the received signal, and samples of the 'known' preamble). The processing is therefore equivalent to that of an FIR filter, although note that there is no implication of filtering to change the frequency content of a signal in this case. Correlation can be achieved by time-reversing the preamble sequence, and using its samples as the filter weights. Thus all filter weight values are +1 or -1.

The reason for the time reversal can be confirmed by observing Figure 12.9, again using the example of the Barker-13 sequence. The leftmost sample of the Barker sequence, according to Table 12.1, occurs first in time, and therefore it is the first sample to enter the filter delay line. The remaining samples follow,

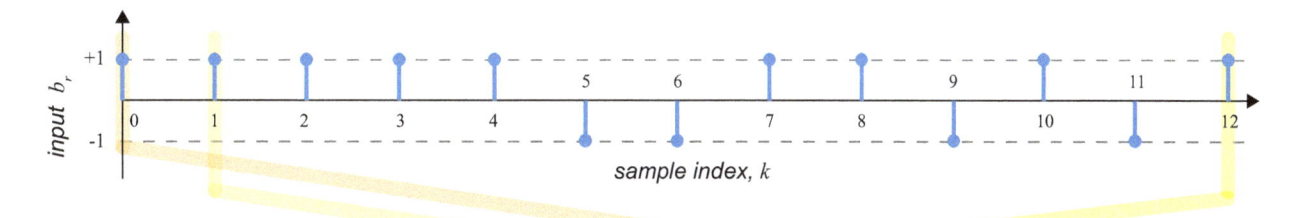

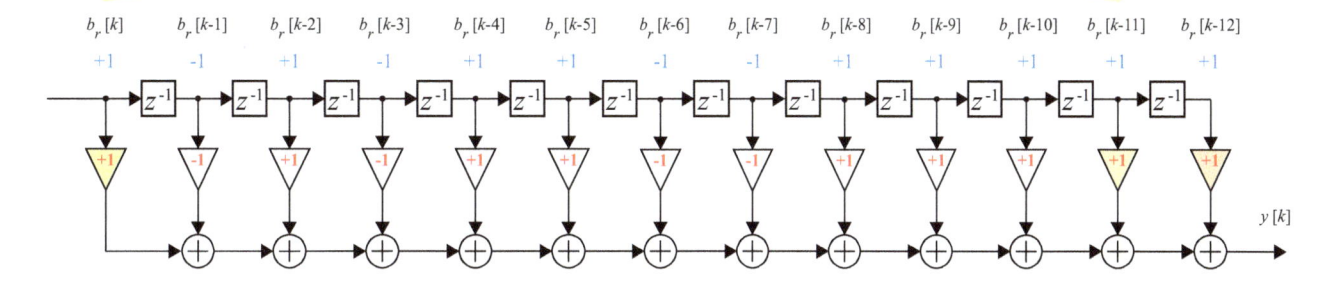

Figure 12.9: Operation of a matched filter correlator (for the example Barker-13 sequence)

and progress through the delay elements towards the right hand side. Once the delay line is fully populated, the Barker-13 sequence resides within its registers, but it is 'flipped' with respect to the sequence defined in Table 12.1.

The locally stored version of the preamble (i.e. the filter weights) must also be flipped such that the two sequences correspond. This means that, when multiplied and summed, they will produce an output equal to the length of the sequence. At this point, the computed output of the filter is equivalent to the correlation function previously defined in Eq. (12.1), and can be expressed as

$$y[k] = \sum_{n=0}^{N-1} r[k-n]w_n, \quad (12.4)$$

where $r[k]$ is the received signal to be correlated; the set of weights, W, is equivalent to the time-reversed sequence; and $y[k]$ is the output of the correlation operation at time index k. Thus, the peak output is $y[k] = 13$ for the Barker-13 sequence.

12.6.5 Frame Synchronisation System Requirements

Frame synchronisation can be accomplished using different techniques, as discussed in [40]. The structure of the transmission is significant in determining the synchronisation requirements.

For systems that are designed to transmit data continuously, frame synchronisation may be facilitated by inserting a Barker sequence (or similar) at regular defined intervals throughout the data stream. For instance, the sequence would be included in the data stream every n frames, as illustrated in Figure 12.10(a). The receiver would periodically correlate its locally stored version of the Barker sequence with the received data stream (having already performed carrier and timing synchronisation), to achieve frame synchronisation. When the computed cross-correlation crosses a defined threshold, the receiver determines that the preamble sequence has been identified within the received data.

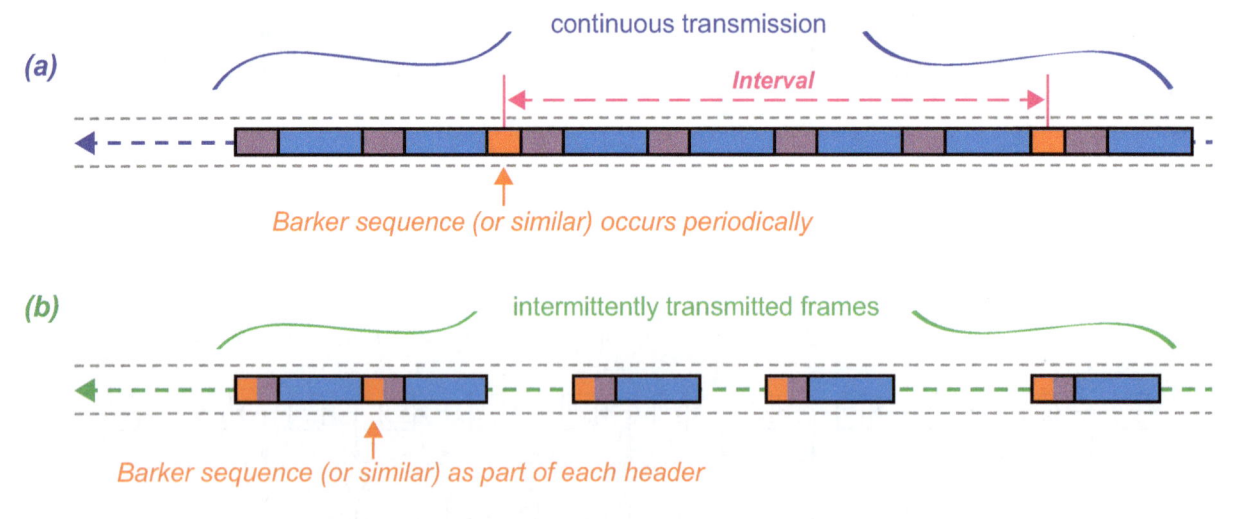

Figure 12.10: Transmission types of communication systems: (a) continuous frame transmission and (b) noncontinuous frame transmission

Chapter 12: Desktop Digital Communications: QPSK Transmission and Reception

For communication systems that transmit in a noncontinuous or intermittent manner, as illustrated in Figure 12.10(b), the Barker sequence is included in the header of each frame.

It should be noted that the inclusion of preamble data adds 'overhead' to the transmission (non-data bits that reduce the efficiency of the protocol). In some cases, the preamble may require to be long relative to the frame size (in order to prevent missed frames, or false detections [40]), and this leads to a low protocol efficiency.

12.6.6 Frame Design for Transmission of ASCII Characters

So far, all of our USRP® transmitter models have transmitted data in a continuous manner, which might suggest a periodic correlation method, i.e. the first of the two possibilities defined in the previous section. In our simple protocol, however, we will adopt a 'hybrid' method — where we will consider that individual frames are transmitted continuously (with no break in between), and we will insert a Barker-13 sequence into the header of each individual frame. Doing so means that the receiver will be able to achieve frame synchronisation quickly, and also adapt and regain synchronisation if it is lost at any point when receiving signals from the RTL-SDR in real-time (as the RTL-SDR is prone to dropping samples, adding this redundancy means that the receiver should be more reliable and successful at recovering the transmitted ASCII messages). This part of the header will be referred to as the preamble. It will initially be the *only* component of the header, but in later examples we will add further fields to extend the functionality of the protocol.

Frame synchronisation will, in this case, be achieved by continuously monitoring the cross correlation between the preamble and the received data. This means that the correlation peak associated with the Barker sequence must be clearly identifiable from any other high correlation values that arise due to the payload data, even in the presence of some reasonable level of bit errors. In designing the format of the frame, some evaluation of the cross correlation was undertaken based on this, using random data to model the payload. It was observed that some significant correlation values were produced, close in value to the peak (the payload was modelled as random bits, and therefore a string of bits has a chance of being very similar to the Barker sequence). This case is shown in Figure 12.11, for the example of fixed length frames with a total length of 200 bits. Notice that peaks occur every 200 samples, as expected. A suggested

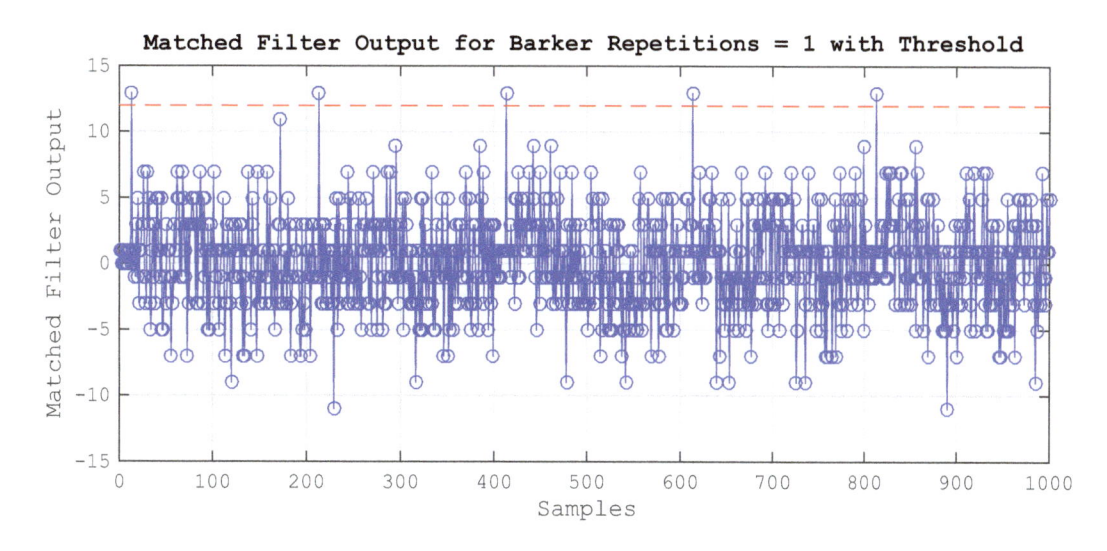

Figure 12.11: Simulated correlator output for 200-bit frames with Barker-13 sequences as frame headers

threshold value is annotated by the red dashed line (values above this line indicating detection of the preamble sequence).

Consideration was then given to using a modified Barker sequence for the preamble (specifically, 'expanding' the sequence by repeating individual bits an integer number of times). It was found that the correlation peak became more easily identifiable, with a greater amplitude difference between the peak and the next largest value, when correlating with the example frames. Of course, for a fixed frame size, increasing the length of the preamble causes the payload size to reduce, which means that the protocol becomes less efficient. The most suitable number of repetitions was found to be 3 — increasing beyond this value did not improve the prominence of the peak, while only impairing the efficiency of the protocol. In this case, the length of the preamble is 39 bits, and so a frame of length 200 bits has 161 bits of payload. This can be considered an efficiency of 161/200, or 80.5%.

The composition of the expanded Barker-13 sequence is shown in Figure 12.12 (note that it is shown in unipolar form here, i.e. 0's and 1's rather than -1's and +1's). An example correlation with frames can be seen in Figure 12.13, with a threshold marked by the red dashed line — compare this with Figure 12.11.

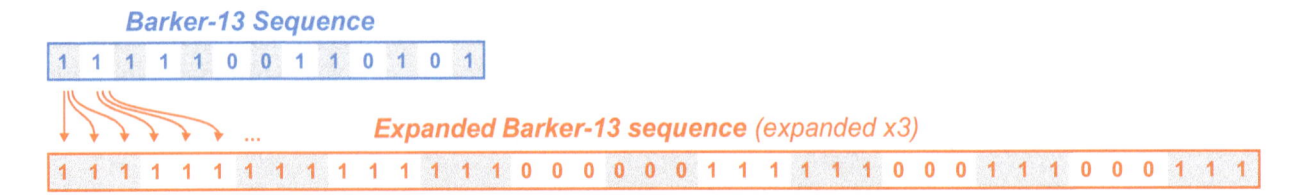

Figure 12.12: Barker sequence expansion (by a factor of 3)

As mentioned earlier, the entire header is formed from the preamble (expanded Barker sequence) at this iteration of the design. The ASCII encoded message fills the remainder of the frame, as shown in Figure 12.14.

In the next exercise we will investigate a subsystem in Simulink which incorporates the ASCII encoding process described in Section 12.5, allows the user to enter and encode their own message, and generates the frames for transmission.

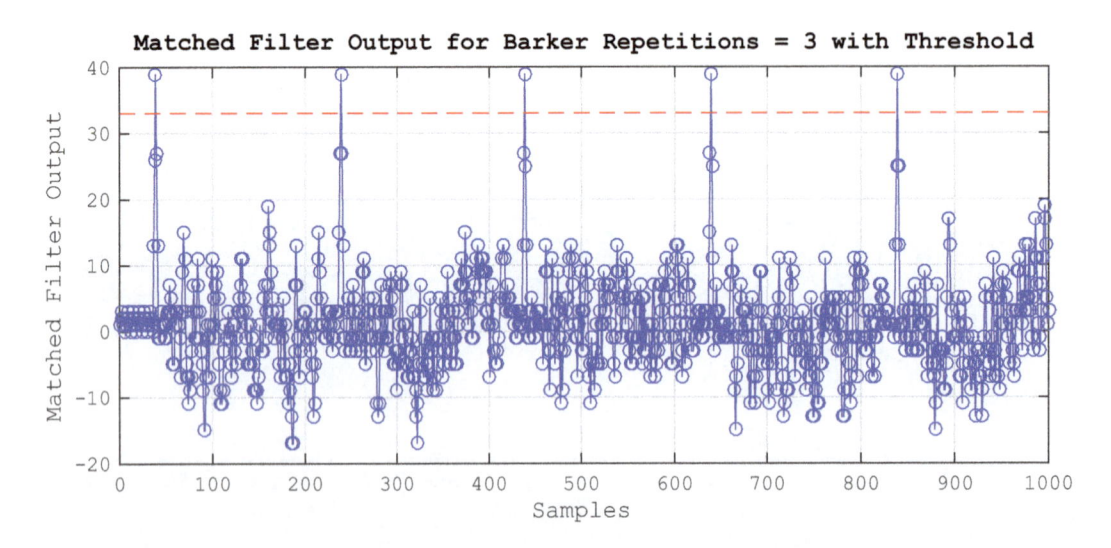

Figure 12.13: Simulated correlator output for 200-bit frames with 3x expanded Barker-13 sequences as frame headers

Chapter 12: Desktop Digital Communications: QPSK Transmission and Reception

Header	Payload
Expanded Barker Sequence	ASCII Message Data

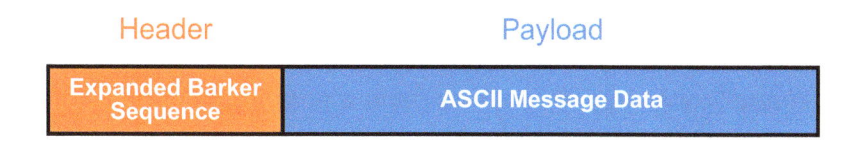

Figure 12.14: Frame structure for QPSK ASCII message transmission

---

Exercise 12.10 — Numbered ASCII Frame Generator for Transmitter Designs

The reference design presented here introduces the frame generation process, prior to inclusion in our real-time transmitter. The output from this simulation is stored in the Workspace, and shows the bit sequence of the generated frames. Later in our transmitter design, this sequence will be modulated and transmitted in the same way as the randomly generated bits that were used in previous simulations.

(a) **Open MATLAB.** Set the working directory to the exercise folder,

   /digital/simulation/ASCII_coding

Next, open the following file:

   .../ASCII_encoder_frame_gen.slx

The block diagram should look like this:

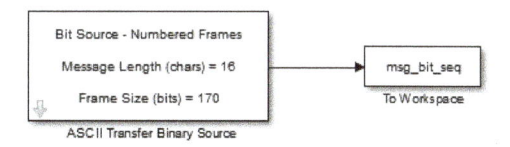

```
Bit Source - Numbered Frames
Message Length (chars) = 16        →     msg_bit_seq
Frame Size (bits) = 170                   To Workspace

ASCII Transfer Binary Source
```

(b) **Inspect the model.** In this block diagram there is an *ASCII Transfer Binary Source* subsystem that performs the message/ frame processing, and a *To Workspace* block that outputs the sequence for viewing.

The *ASCII Transfer Binary Source* block allows the user to enter a string of their choice (up to a maximum of 50 characters) in the parameter window. When simulated, the string is then encoded, and frames of data are generated ready for transmission. The frames comprise a header (the *preamble* discussed earlier) and a payload, which consists of the encoded message string, an appended frame number from 001 to 100, and padding bits to ensure the frame is of even length. Adding a frame number allows individual frames to be more easily differentiated at the receiver.

(c) You should notice that the subsystem block annotation displays two pieces of information: (i) the message length in characters (equivalent to the string length, plus the additional four characters from the appended space and the frame number), and (ii) the full frame size in bits. When a new string is entered and the change is applied in the parameter window, the block automatically updates the information on its mask.

(d) **Modify the parameters.** Open the parameter window (shown below) and replace the 'Hello World!' string with a message of your choice. Note that, if you enter more than 50 characters into the 'Message String' field, the frames generated by the block will only include characters 1 to 50, the space, and the three digits. Apply the changes to observe the effect of the changes on the block.

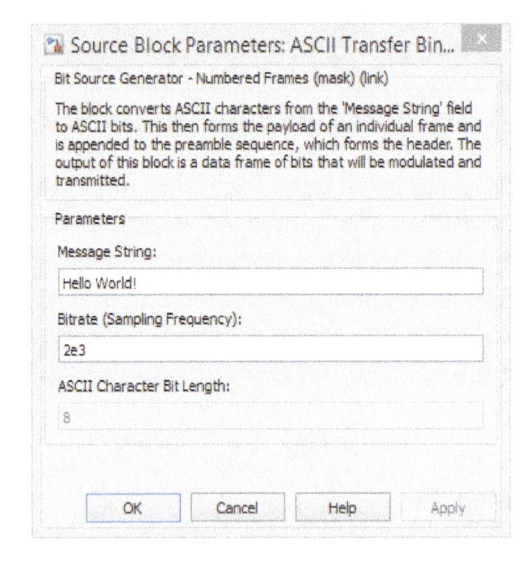

(e) Inspect the rest of the parameter window. Notice that the frames are set to be generated at the same bit rate as the transmitters considered earlier in the chapter (i.e. 2kbps), and the ASCII encoder is configured to use 8 bits for each character (this ensures that the message length is an even number of bits — useful as QPSK symbols are composed of 2 bits!).

(f) **Run the simulation.** To view the output bit sequence, run the model by 🖱 on the 'Run' ▶ button, and navigate to the Workspace in the MATLAB environment.

(g) **Examine the outputs.** 🖱 on the ⊞msg_output variable that has been stored in the Workspace during the frame generation process. It should show you something similar to the following (a different string will be displayed if you replaced 'Hello World!' in Part (d)):

```
val =

Hello World! 001
Hello World! 002
...
```

This confirms the expected result of the frame creation process. Each line corresponds to one frame.

(h) Now look at the ⊞msg_bit_seq variable. If the message was set to the default 'Hello World!' string in the parameter window, each frame will have length equal to 170 bits. At the start of this output bit sequence you should notice the preamble which, as explained earlier, has been created by expanding the Barker sequence. The preamble extends from row 1 to row 39, where row 40 contains a padding bit to make the header an even length. Rows 41–168 store

Chapter 12: Desktop Digital Communications: QPSK Transmission and Reception

the bits for each character of the `Hello World! ###` message, and the remaining two rows contain padding bits.

This frame structure repeats throughout the remainder of `msg_bit_seq`, where the only difference between frames is the iterating frame number.

(i) As a short exercise; convert the last three bytes of the first few frames back to their character representation to confirm the incrementing frame number. For the first frame, convert bits 145–168, for the second frame convert bits 315–338, for the third frame convert bits 485–508, and so forth. To remind you of how to convert back to an ASCII character, the code for the conversion of the last three bytes of the first frame is shown below:

```
char([bi2de(msg_bit_seq(145:152)','left-msb'),...
   bi2de(msg_bit_seq(153:160)','left-msb'),...
   bi2de(msg_bit_seq(161:168)','left-msb')])
```

Running the above line of code in the MATLAB Command Window should output `001` as expected for the first frame. To check the second and third frames, apply a similar manner of byte decoding by simply entering the bits from the rows described above.

(j) **View the code.** For readers who wish to view the code for the ASCII encoding and frame generation process, type 'open `ASCII_frame_gen_numbered.html`' into the MATLAB command window. A browser window will appear that contains the code used to perform both of these operations, including explanatory comments.

12.6.7 *Frame Synchronisation with a Matched Filter Correlator*

Next, we briefly introduce a receiver design based on the matched filtering approach from Section 12.6.4. This will be the focus of the next exercise.

The receiver utilises the output of a matched filter correlator to identify the beginning of each frame. Having done so, it must then reconstruct the frames from the continuous sequence of received data, and then extract the payload for further processing. These steps are depicted by Figure 12.15, and the Simulink models that follow include the appropriate coordination logic to achieve this.

It is important to note that prior stages of synchronisation in the QPSK receiver extract a symbol rate clock from the received signal (recall that the exact timing parameters must be determined from the signal using an Early Late symbol timing synchroniser, or similar, as detailed in Chapter 11). Thus, all subsequent sample-based processing should take place based on the extracted rate. Fortunately, a symbol rate clock is available in the receiver, in the form of a strobe that can be used to enable later processing. (Utilisation of this strobe will be discussed later in Exercise 12.13, when we observe this frame synchronisation incorporated into our QPSK receiver design. Beforehand, Exercise 12.11 will illustrate a practical implementation of frame synchronisation in simulation, without the timing strobe and assuming perfect timing synchronisation.)

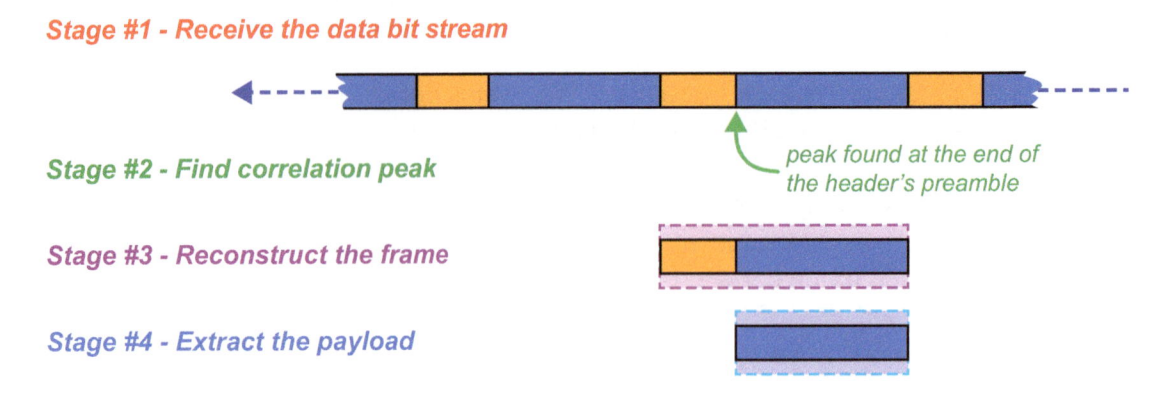

Figure 12.15: Overview of the matched filtering based frame synchronisation approach

Figure 12.16. shows a block diagram representation of the matched filter correlator, coupled with an implementation of the logic processing required to reconstruct the frames. Each value output from the filter is compared to the threshold value, and only values above the chosen threshold are considered valid correlation detections. Through this thresholding operation, we aim to reduce the possibility of false detections.

A valid detection triggers a latch, which in turn enables a data framing system to gather samples of the received data into a buffer. Doing so allows part of the bit stream to be reconstructed back into a frame with the same structure as the transmitted frames. Since the receiver explicitly knows the frame size, the framing system is designed to reset (and hence stop gathering samples) when an internal counter in the enabled system determines that a full frame has been reconstructed. As a result of the threshold trigger occurring when the last sample of the preamble has entered the filter, recovery of the full frame

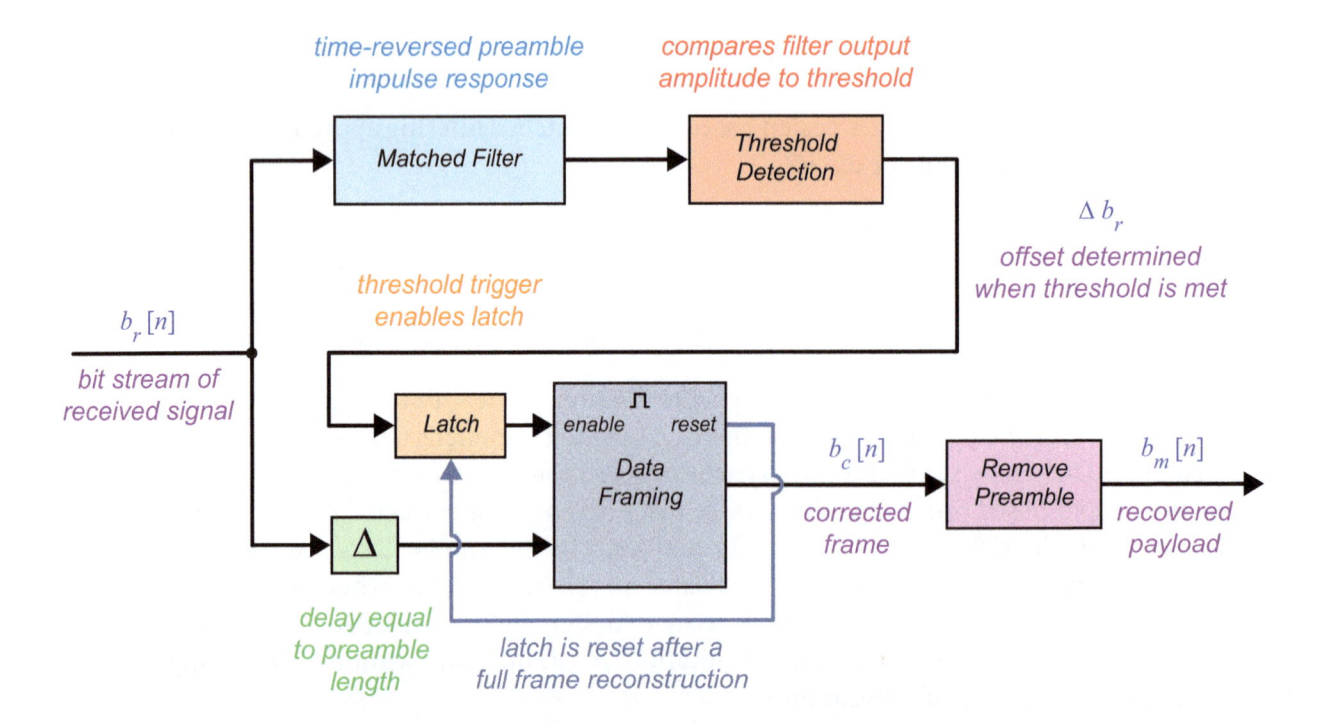

Figure 12.16: Block diagram of the matched filter based frame synchronisation

Chapter 12: Desktop Digital Communications: QPSK Transmission and Reception 541

(including the preamble) requires the bit stream entering the framing system to be delayed by the length of the preamble — re-synchronising the system to the start of the frame. The reconstructed frame is then output and the preamble is removed to obtain only the payload data.

If the received bit stream was *not* delayed before entering the framing process, it would remove the header from the data, leaving only the payload. We will opt to retain the header because, in some of the later examples, it contains information fields necessary for interpreting the frame.

Exercise 12.11 Frame Synchronisation using a Matched Filter

In this exercise we will use the *ASCII Transfer Binary Source* that was introduced in the previous exercise. It will be used for creating signals that will be passed through a time-delay channel to introduce a frame offset. Frame synchronisation is performed using a practical implementation of the matched filter correlator design discussed earlier. The purpose of this simulation is to illustrate the detrimental effect of a frame offset on the ASCII decoding process, and how frame synchronisation can correct this.

(a) **Open MATLAB**. Set the working directory to the exercise folder,

    `/digital/simulation/frame_synch`

Next, open the file:

    `.../matched_filter_offset_estimation_correction.slx`

The block diagram should look like this:

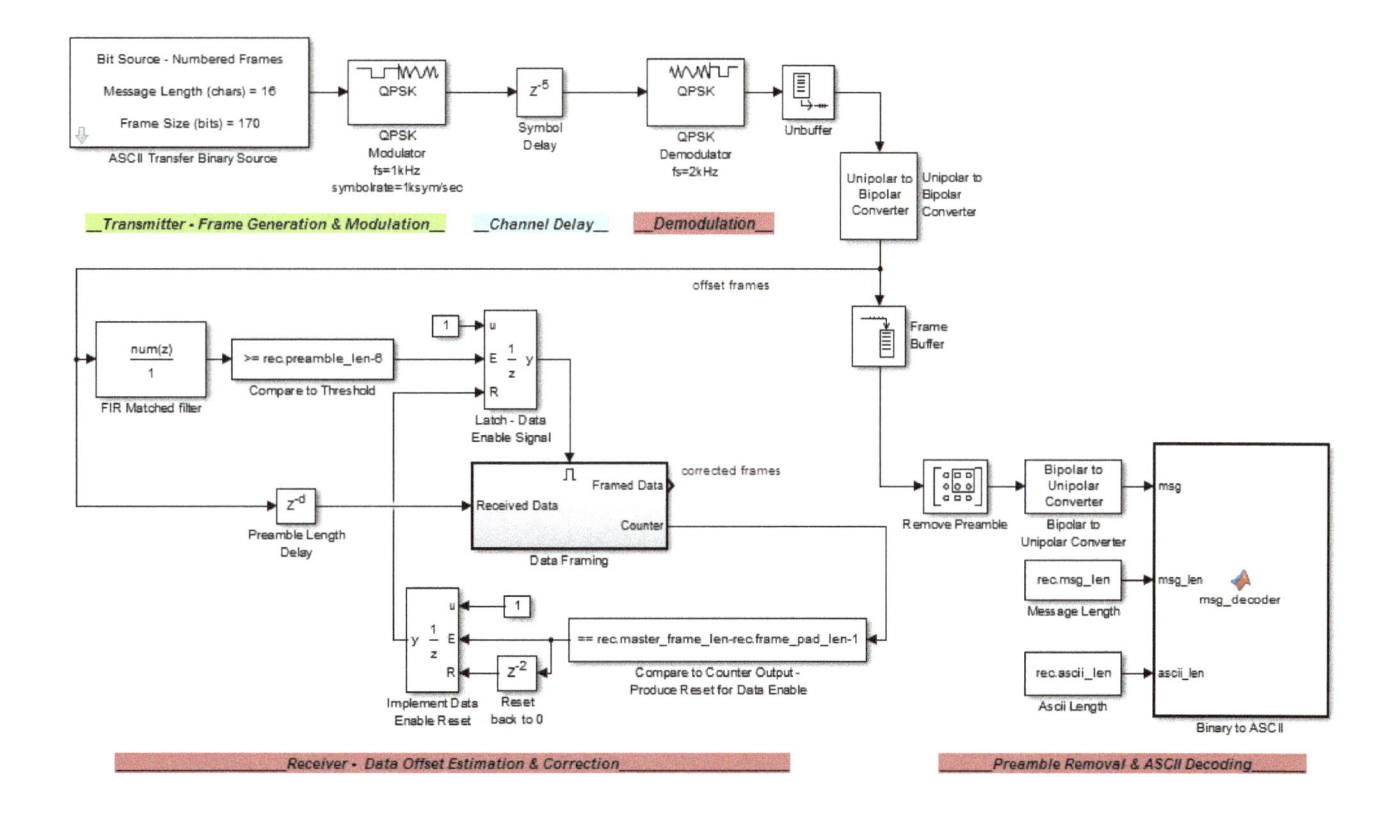

(b) **Inspect the model.** Notice that the block diagram has been gathered into groups of blocks to illustrate the task that each section performs. The area at the beginning of the block diagram shows the frame generator and the *QPSK Modulator* blocks, which emulate some of the transmitter processing.

The *Symbol Delay* block allows the user to define a delay, which introduces a symbol offset to the data as it 'travels' across the channel. The areas annotated with red labels are the demodulation; frame synchronisation processing; and preamble removal and ASCII decoding stages that will feature in our real-time receiver design.

(c) The *Preamble Removal & ASCII Decoding* section is connected to the output from the *Frame Buffer* block, which frames the 'received' data for ASCII decoding—permitting an initial observation of the output. In this initial setup, you will be able to see the effect that this offset has on the data, as no frame synchronisation will be implemented on the strings before being output to the command window.

(d) **Run the Simulation.** Start the simulation by ① the 'Run' ▶ button and view the decoded ASCII output printed in the MATLAB command window. For a symbol delay of 5, you should see a number of unintelligible messages like those shown here, which represent the converted frames of data that have been ASCII decoded in error due to this transmission delay.

As emphasised previously, even a single bit offset in a received data frame will cause every decoded character of the ASCII message to be in error.

```
ã' [    È  ÛÜ›H
ã' [    È  ÛÜ›H
ã' [    È  ÛÜ›H
...
```

(e) Return to the model and change the value set in the *Symbol Delay* block to a few different integer numbers of samples (note that the value must be positive!), and rerun the simulation each time to view the output in the command window. You should notice that changing the sample offset produces different wrongly decoded character strings.

(f) **Frame synchronisation processing.** Now ② on the *FIR Matched filter* block to open the parameter window, and note that its coefficients are set by a variable (created by the *ASCII Transfer Binary Source* block) that is stored in the Workspace. If you investigate this variable, you should notice that the values are a time-reversed version of the preamble in bipolar form.

(g) The received signal is demodulated and converted to bipolar form prior to entering the matched filter correlator, where it is processed using these coefficients. If the expected preamble is found in the received signal, the output contains a large amplitude peak that permits the data offset to be found. The filter output is then compared to a threshold value to allow the system to determine if the peak is a valid detection.

Chapter 12: Desktop Digital Communications: QPSK Transmission and Reception   543

(h) To help explain the operation of the offset correction, we will define the frame length as F, the preamble length as L, and the padding appended to the payload as P.

The *Data Framing* subsystem is enabled when the peak amplitude meets the threshold; in our case $L-6$, which sets the *Latch — Data Enable Signal* high. While the enable signal is latched high, the buffer inside the subsystem gathers samples from the received bit stream until it reaches $F-P$. When the *Counter* block inside the subsystem outputs the value of $F-P-1$, it produces a boolean result from a second comparison. The boolean output then enables another latch to create a reset signal for the *Latch — Data Enable Signal* block, setting the original enable signal low, and readying the system for the next detection.

Padding is required at the end of the payload due to the use of Simulink's *Unit Delay Enabled Resettable* block for the *Implement Data Enable Reset* in the design. Because this block introduces a delay, the frames must have padding between them to allow the system to be reset in time for the next frame.

(i) **Alter the model configuration.** To see the result of the ASCII decoding after frame synchronisation has been implemented, we need to alter the block diagram slightly. Comment out the *Frame Buffer* block and delete the connection between it and the *Remove Preamble* block. Now connect the latter to the output of the *Data Framing* subsystem, so that it receives corrected frames for decoding.

(j) The modified part of the block diagram should like that shown below:

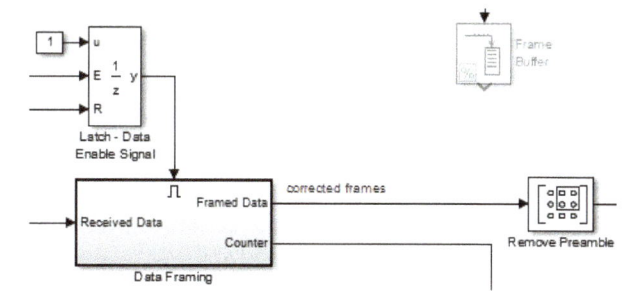

(k) **Run the Simulation.** Run the simulation again by the button and now view the corrected output printed in the MATLAB command window.

The offset of each frame has been corrected and this allows accurate conversion of the data back to the expected ASCII messages, as shown below.

```
Hello World! 001
Hello World! 002
Hello World! 003
...
```

(l) Change the sample offset a few more times to a few integer values of your choice, and rerun the simulation each time to observe the output. What do you notice?

(m) Lastly, try reducing the value of the threshold in the *Compare to Threshold* block, by increasing the amount that is subtracted from the `rec.preamble_len` variable. Rerun the simulation each time you change the value, and view the output displayed in the command window.

Doing so should show you that the threshold value is important to the success of the data recovery in this simulation, and even more so in the real-time system. If the threshold value is too low, a non-valid peak detection will enable the framing subsystem and produce unintelligible messages as the system did without frame synchronisation.

Hopefully this exercise has illustrated and reinforced the necessity of frame synchronisation in a digital communications system, to accurately recover the message payload! We will now proceed to make use of this method in our QPSK ASCII message transmission and later when sending data across the RF spectrum with an FM transmitter.

12.7 ASCII Message Transmission and Reception

Having covered both the theory and practical exercises of ASCII encoding/ decoding and frame synchronisation, we can now implement these processes in our point to point SDR link. You will be able to open and explore both the frame generator and frame synchronisation subsystems in the transmitter and receiver designs in the following two exercises.

In the *Digital Communications Theory and Simulation* chapter, Section 11.9 (page 488) introduced the concept of phase ambiguity, and explained the requirement for a phase ambiguity correction stage in digital communications systems. As we are going to be transmitting and receiving ASCII messages, this functionality must also now be considered in the transmitter and receiver designs. Phase ambiguity correction for QPSK was previously implemented using low-level logic operations during simulation, but for our real-world SDR link to operate in real time, we will substitute this with the Simulink DQPSK (Differential QPSK) block. The benefit of doing so is that the same phase ambiguity correction system can be implemented in a pre-compiled (and optimised) block, as opposed to the first principles (but computationally intensive) method we presented earlier.

Exercise 12.12 ASCII Message Tx Rx: USRP® Transmitter

Finally, we have reached the stage where we can transmit and receive ASCII messages! In this exercise we will inspect the transmitter that will be used to transfer modulated frames of data, which will each contain a string and a frame number, to the USRP® hardware.

(a) **Open MATLAB**. Set the working directory to the exercise folder,

   `./digital/usrp_tx/`

   Next, open the file:

   `.../usrp_QPSK_ascii_message.slx`

The block diagram should look like this:

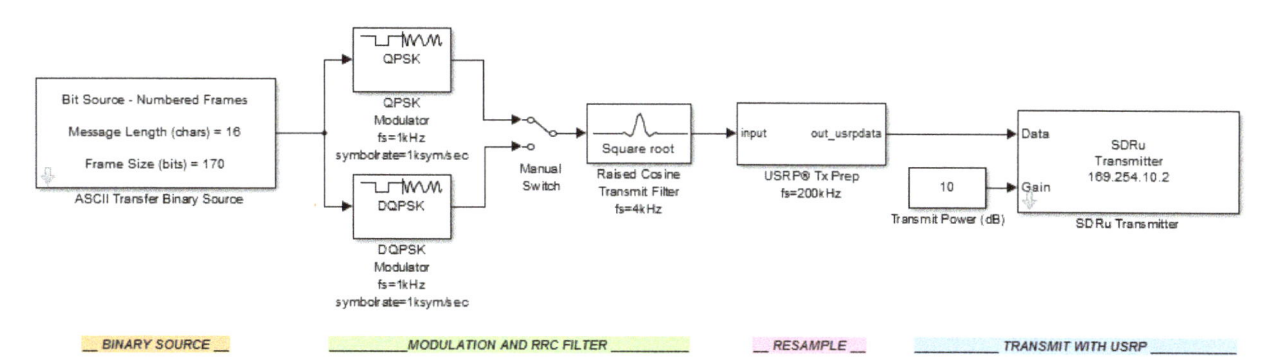

[Block diagram: Bit Source - Numbered Frames (Message Length (chars) = 16, Frame Size (bits) = 170, ASCII Transfer Binary Source) → QPSK Modulator fs=1kHz symbolrate=1ksym/sec / DQPSK Modulator fs=1kHz symbolrate=1ksym/sec → Manual Switch → Square root Raised Cosine Transmit Filter fs=4kHz → USRP® Tx Prep fs=200kHz → Gain (10, Transmit Power (dB)) → SDRu Transmitter 169.254.10.2]

__BINARY SOURCE__ ___MODULATION AND RRC FILTER___ __RESAMPLE__ _____TRANSMIT WITH USRP_____

(b) **Inspect the transmitter.** Notice that there are a few changes from the previous USRP® transmitter design. The block diagram now includes the *ASCII Transfer Binary Source* subsystem, and this is connected to both the (previously used) *QPSK Modulator* block, and a new *DQPSK Modulator* block. Including both of these modulators along with a *Manual Switch* affords us the ability to switch between the two different schemes, and see the effect that they have on the success of data recovery in the receiver.

(c) **Check the USRP® Radio settings.** You will need to configure the *SDRu Transmitter* block to communicate with your USRP® hardware. To do this, you should follow the procedure provided in Exercise 12.1 to configure the block's settings for your hardware device.

(d) **Run the model.** To check that the design is working before moving to the next exercise, ensure the USRP® hardware is connected and switched on, then start the simulation by the 'Run' button in the Simulink toolbar. After a couple of seconds, Simulink will establish a connection with your radio and the simulation will begin.

(e) When you are satisfied that the model runs without any errors, you can stop the simulation and move on to the next exercise while keeping the model open.

Exercise 12.13 **ASCII Message Tx Rx: RTL-SDR Receiver**

In this exercise, you will open and run a receiver similar to the design presented in Exercise 12.7, which now includes the additional stages required to successfully recover the transmitted frames of ASCII data. To match the transmitter, this design includes both QPSK and DQPSK demodulators and a switch to allow the full effect of phase ambiguity to be observed.

(a) **Open MATLAB.** Set the working directory to the exercise folder, and open the following file:

/digital/rtlsdr_rx/

.../rtlsdr_QPSK_ascii_message.slx

NOTE: *The functionality of the AGC Simulink block was changed between releases R2015a and R2015b. We have supplied modified versions of these files, denoted by the* _15bonwards.slx *file names. Anyone using 15b or higher will need to use these files instead.*

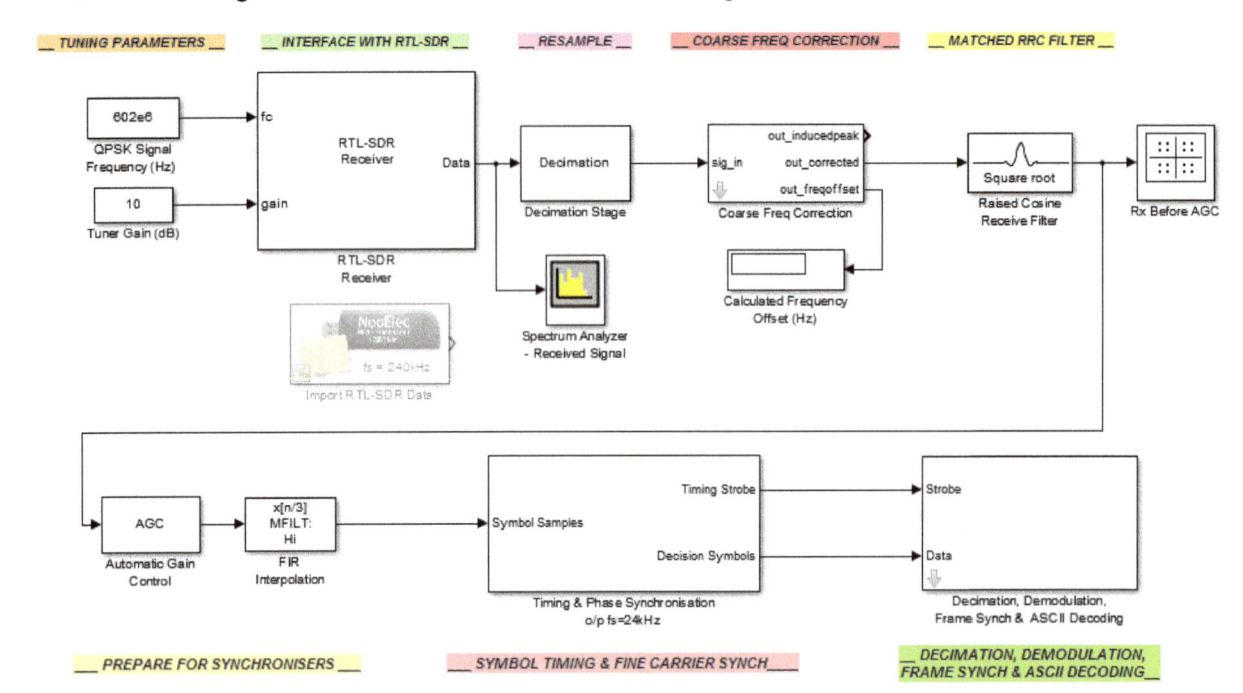

The block diagram should look similar to the following screenshot:

(b) **Inspect the receiver.** Take your time to inspect the model, as a number of things have changed since the previous receiver design. The initial decimation stages have been grouped into a subsystem, as well as the symbol timing and carrier synchronisation subsystems, as this makes for a more modular and aesthetically compact design.

The *Decimation, Demodulation, Frame Synch & ASCII Decoding* subsystem is new to the receiver design, and it performs the additional processing (reviewed earlier) to successfully recover the ASCII messages.

(c) **Receiver parameter settings.** Previously it was discussed that the receiver would require some parameters to be configured for the ASCII decoding process. If you open the *Decimation, Demodulation, Frame Synch & ASCII Decoding* subsystem's parameter window, you will see where these values are set.

The *Message Length* variable is tunable by the user and is solely dependent upon the character length of the message that has been set in the transmitter's *ASCII Transfer Binary Source*. This value can be read from the transmitter block's mask, and entered here to ensure that the transmitter and receiver match. The subsystem uses this value, along with preset information such as the preamble, to determine the length of the transmitted frames.

The *Sampling Frequency* can only be observed, as changing this value would require alterations to the parameter settings of the carrier and timing synchronisation. In doing so, there would be a high possibility that the design would become unbalanced and no longer perform as expected! As we intend this design to be plug and play, we wish to leave this value as-is.

Chapter 12: Desktop Digital Communications: QPSK Transmission and Reception   547

The *ASCII Length* is also non-tunable as its value can only realistically be greater than 7, and as discussed earlier our transmitter has this set as 8 to keep each character represented by a byte.

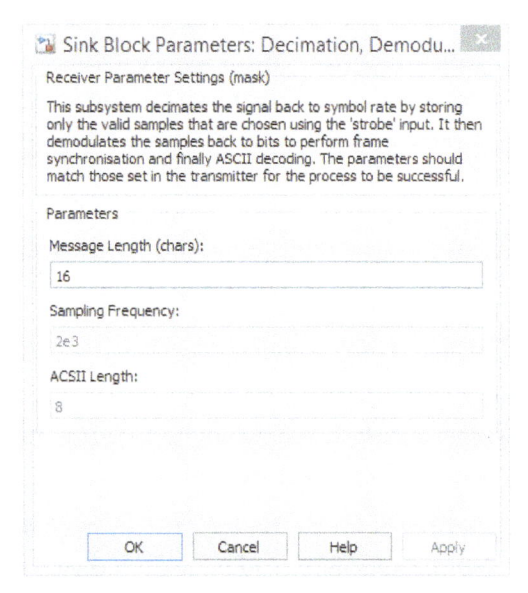

Sink Block Parameters: Decimation, Demodu...

Receiver Parameter Settings (mask)

This subsystem decimates the signal back to symbol rate by storing only the valid samples that are chosen using the 'strobe' input. It then demodulates the samples back to bits to perform frame synchronisation and finally ASCII decoding. The parameters should match those set in the transmitter for the process to be successful.

Parameters

Message Length (chars):
16

Sampling Frequency:
2e3

ACSII Length:
8

OK   Cancel   Help   Apply

📖 **(d) Decimation.** Look inside the *Decimation, Demodulation, Frame Synch & ASCII Decoding* subsystem to find the *Decimation* subsystem, highlighted below.

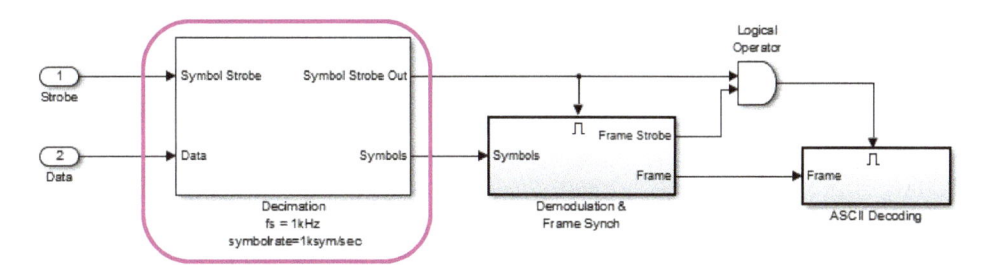

Inside the *Decimation* subsystem there is an enabled block that reduces the sampling rate back to symbol rate. If you look inside this subsystem you will see the configuration shown below, which makes use of the timing 'Symbol Strobe' to enable the process of decimation.

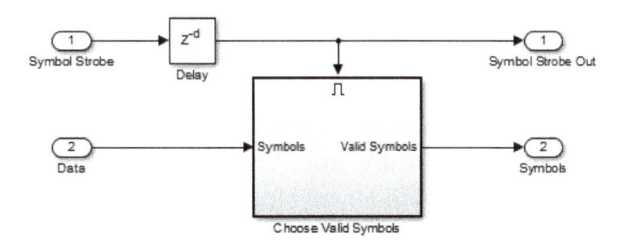

Remembering that the symbol timing and carrier phase synchronisation stages required the received data to be oversampled by a factor of 24, we must now return it to the original symbol rate in order for the data to be accurately recovered. When the timing 'Symbol Strobe' input to the *Choose Valid Symbols* block is high, it enables one valid symbol from the possible 24 (chosen by the symbol timing synchronisation stage) to be chosen and output for further processing.

(e) Returning to the *Decimation, Demodulation, Frame Synch & ASCII Decoding* subsystem of the top-level design, observe that these symbols are passed to the *Demodulation & Frame Synch* subsystem, along with the 'Symbol Strobe'. The latter is required to accurately time the remaining processes.

(f) **Demodulation & Frame Synchronisation.** Inside the enabled *Demodulation & Frame Synch* subsystem, note the *Demodulation and Phase Ambiguity Correction* subsystem. Looking inside this subsystem you will notice that — matching the transmitter setup — it has both QPSK and DQPSK demodulation blocks. When the DQPSK option is selected in both the transmitter and receiver designs, this implements phase ambiguity correction, as we will observe shortly. For now, ensure that the switch is set to the *QPSK Demodulator* option.

(g) Move back up a level to the *Demodulation & Frame Synch* subsystem, and then look inside the *Frame Synchronisation* block to see that the practical implementation of our matched filter correlator (previously introduced in Exercise 12.11), has been slightly altered here for better integration it with our real-time receiver design.

As the strobe provides symbol rate timing, the frame synchronisation section has been modified to process symbol bit pairs, rather than individual bits (meaning it runs at symbol rate). The matched filter correlation results are compared to a threshold, and values above the threshold enable valid data to be passed through the *Valid Data Selection* subsystem and framed externally by the *Delay Line* block. This modification allows the processing to remain at symbol rate while also accomplishing frame reconstruction. The trigger that is generated by the counter to reset the initial *Latch* and ready the system for the next frame is also used as a 'Frame Strobe' here; this permits the reconstructed frame of data to be output at the correct time and also provides a frame rate strobe for the final stage of *ASCII Decoding*.

(h) Moving back up to the *Decimation, Demodulation, Frame Synch & ASCII Decoding* subsystem, you will see that an AND operator synchronises the 'Frame Strobe' with the 'Symbol Strobe'. This ensures that the *ASCII Decoding* block performs its processing once per frame.

(i) Entering the *ASCII Decoding* subsystem, you will see that the preamble and its padding are removed by a *Submatrix* block, allowing the payload to be decoded back to ASCII characters. If you wish, enter into the *ASCII Decoding* subsystem (using the grey arrow), and confirm that the same ASCII decoding method is used as in previous exercises.

(j) **Check the RTL-SDR data source.** Move back to the top level of the Simulink design. If an RTL-SDR is available and you are able to transmit the ASCII message QPSK signal from Exercise 12.12, an *RTL-SDR Receiver* block can be used to interface with and control your RTL-SDR. At this point, ensure your PPM correction value is entered into the block's parameter window. Otherwise, you will need to use an *Import RTL-SDR Data* block. This should be configured to reference this file:

/digital/rtlsdr_rx/rec_data/qpsk_ascii_msg.mat

If importing data rather than interfacing with an RTL-SDR, delete the *RTL-SDR Receiver* block and connect the *Import RTL-SDR Data* block in its place.

Chapter 12: Desktop Digital Communications: QPSK Transmission and Reception 549

The length of the message that the receiver 'expects' also needs to be set in the 'Message Length (chars)' field of the *Demodulation, Frame Synch & ASCII Decoding* block. If you are using the import block, this value should be set to 16. If transmitting a signal yourself using the system from Exercise 12.16, ensure that the value matches that shown on the *ASCII Transfer Binary Source* block.

(k) **Run both the transmitter and receiver.** Ensure the transmitter is set to the *QPSK Modulator* option, and begin the simulation by 🖱 on the 'Run' ▶ button in the Simulink toolbar. Once the transmitter model has started running, return to the receiver and check that the demodulation option is set to the *QPSK Demodulator*, as described in step (f), and start the receiver simulation by 🖱 on the 'Run' ▶ button. After a couple of seconds it will begin; the *Constellation Diagram* windows should appear, and if the receiver has locked to the correct phase of the QPSK signal, the ASCII decoded strings will print to the MATLAB command window.

(l) **Receiving data without Phase Ambiguity correction.** At this point you will either be observing the receiver output decoded strings to the command window, or nothing at all (!) if it has locked to one of the three incorrect phases of the QPSK signal. As explained in Sections 11.9 & 11.10 (page 488) of the previous chapter, a phase ambiguity can cause all of the symbols to be demodulated incorrectly. Due to the construction of our frame synchroniser, the receiver will not produce an output unless it finds a valid correlation peak, which will not be found unless the symbols of the preamble are recovered accurately.

If however you have been lucky enough to lock to the correct phase, and the receiver is printing the message you set in the transmitter, then stop and rerun the simulation until you observe *nothing* being printed to the command window. Even if the receiver manages to lock to the correct phase of the QPSK signal initially, it may lose this lock after a short period of time.

As a note to those using the recorded data file, you will observe the receiver producing strings until approximately 50 seconds into the recording, at which point it locks to a different phase and the strings stop printing in the MATLAB command window.

(m) **Correcting the Phase Ambiguity.** If you are transmitting the signal using the model from Exercise 12.12, then return to it and change the modulation option to *DQPSK Modulator* by 🖱 on the *Manual Switch*. If you are using the *Import RTL-SDR Data* block instead, stop the model and configure it to reference this file:

📁 `/digital/rtlsdr_rx/rec_data/dqpsk_ascii_msg.mat`

The length of the message that the receiver expects also needs to be set in the 'Message Length (chars)' field of the *Decimation, Demodulation, Frame Synch & ASCII Decoding* block. If you are using the import block, this value should be set to 16, and if you are transmitting the signal from the design shown in Exercise 12.12, it should match the value shown on the *ASCII Transfer Binary Source* block.

Make a corresponding alteration to the receiver, by choosing *DQPSK Demodulator*. Then, start the simulation again and move to the MATLAB command window to observe the output.

(n) **Observing the output.** If using the import file, you should now see the `Hello World! ###` message being output (where the `###` stands for the changing number associated with the frames). If you are receiving the signal from the transmitter model, then hopefully the receiver will be fully synchronised and you will be observing the very same output!

A brief point to note however; as much as the authors have attempted to create a plug-and-play digital communications system, there may be small issues that affect its performance which are inherent to real world communications. Issues such as the position of the antennas, the gain that is set in each model, and importantly the RTL-SDR's PPM offset correction value, are all factors that must be considered. Hence if the output of the receiver is not currently displaying what you expect, then use a little engineering initiative to alter the above characteristics — if in the first instance stopping and restarting the model doesn't work.

(o) Recall that the design and implementation of PLLs (covered in Chapter 7, page 245), and carrier and timing synchronisation (covered in Chapter 11, page 448) requires various trade-offs to be considered. Here, in the digital communications receiver, we are subject to the trade-off of noise bandwidth and synchronisation lock time. When the transmitted QPSK signal is affected by significant interference, it can cause the receiver to lose timing and/or carrier synchronisation. In some cases, the receiver may not be able to regain lock, as the carrier frequency has been adjusted beyond the designed correction capability of our receiver. Hence, stopping and restarting the simulation is sometimes the only way to reset the system and enable synchronisation to be achieved.

(p) **Message Output Rate.** To confirm the speed at which the recovered strings are output to the command window, we should highlight that this dependent upon the transmission speed of 2kbps, and the efficiency of the frame structure.

For instance, if the frame length was 200 bits and the binary representation of the character string was only 120 bits, the receiver will print 10 frames per second where each frame contains 15 characters. If the frame length was 512 bits long and contained 54 characters, the frame print rate would drop to ~3.9 frames per second. This is because more data is being sent without modifying the (2kbps) data rate. The more characters you enter in the *ASCII Transfer Binary Source* block, the slower the output will be!

(q) **Watch us running the receiver.** We have recorded a video which shows activity in the *Constellation Diagram* windows and MATLAB command window as a QPSK signal transmitted by the USRP® hardware (with the model from Exercise 12.12) is received by the RTL-SDR, undergoes synchronisation, and then ASCII decoding. We also demonstrate the effect of phase ambiguity by switching between QPSK and DQPSK modulators and demodulators.

desktopSDR.com/videos/#qpsk_ascii

12.8 Transmitting Images Across the Desktop

Following on from the framing operations introduced for ASCII message transmission (which could be considered part of the Data Link layer if referring to the OSI model), we will next illustrate how the design can be built upon to transmit and receive greyscale images.

Previously, the constructed frames contained one discrete message in each payload, and hence there was no need to split the data across multiple frames. To transmit images however, we need to perform a *segmentation* process (associated with the OSI Transport Layer), in which large data messages are split across several frames (or even tens, hundreds or thousands of frames!). In addition, we will extend the definition of the frame header to include the fields shown in Figure 12.17.

These new fields provide simple information about the frame, which the receiver may use to recover the payload data. If a more complex communication system was being designed, to include multiple nodes, we could also consider adding address fields.

The images that will be transmitted here will be 100x100 pixel 8-bit greyscale images. Each pixel is therefore represented by one byte of data, and is equivalent to a decimal value in the range 0–255. Each column of the image (100 pixels) will be manipulated and stored in the payload of a frame, and each of these frames will be allocated a sequence number as part of the header, to aid the receiver's reconstruction of the image.

Header Information Fields:

1 – Preamble
2 – Frame Sequence Number
3 – End Flag to Show Last Frame
4 – Header Length
5 – Payload Length
6 – Payload Padding Length

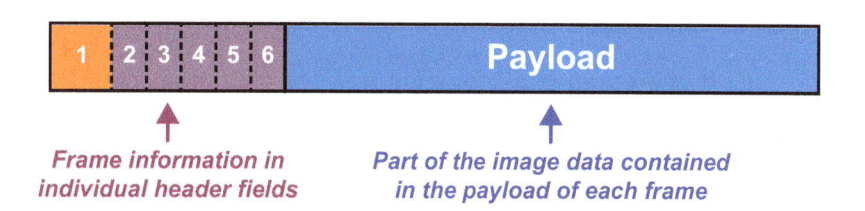

Figure 12.17: Image transfer frame structure—including header information fields

Exercise 12.14 Image Tx Rx: USRP® Transmitter

In this exercise we will observe that the *ASCII Transfer Binary Source* block has been replaced with a block titled *Image Transfer Binary Source*. This block imports a greyscale image of your choice (from a selection provided), segments the data, and then frames the segments. The frames created by this process are very similar to the ASCII frames from previous exercises, however there are now information fields contained within the header, as well as the preamble.

(a) **Open MATLAB**. Set the working directory to the exercise folder,

　　`/digital/usrp_tx/`

　Next, open the file:

　　`.../usrp_QPSK_image_transfer.slx`

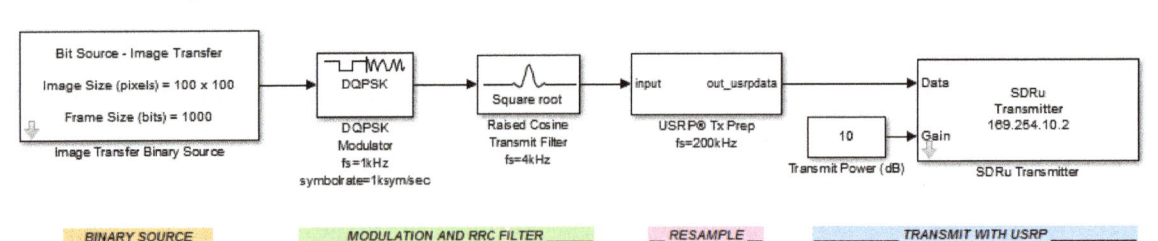

The block diagram should look like this:

[Block diagram: Image Transfer Binary Source (Image Size (pixels) = 100 x 100, Frame Size (bits) = 1000) → DQPSK Modulator (fs=1kHz, symbolrate=1ksym/sec) → Raised Cosine Transmit Filter (Square root, fs=4kHz) → USRP® Tx Prep (fs=200kHz) → Gain (Transmit Power (dB) = 10) → SDRu Transmitter (169.254.10.2)]

__BINARY SOURCE__ __MODULATION AND RRC FILTER__ __RESAMPLE__ __TRANSMIT WITH USRP__

(b) **Inspect the model.** Notice that the previous binary source has been swapped for an *Image Transfer Binary Source*, and that we will now use *DQPSK Modulation* from this point onwards.

(c) **Setting the image source.** Open the parameter window of the *Image Transfer Binary Source* block, and observe that the *Image Filename* can either be entered into the field manually, or chosen by using the *Browse* button. Several images have been provided for the image transfer exercises, matching our design's prerequisite dimensions and greyscale image depth. These images have been stored as `image#.mat`, where the `#` represents the numerical value given to each, and they can be found in the folder:

📁 `/rtlsdr_book_library`

Choosing one of these files using the *Browse* button, or entering the filename directly into the *Filename* field will allow the block to create frames for transmission. Each frame contains one column of pixels. Select an image file and close the parameter window to see that the block's mask shows values for the *Dimensions (in pixels)*, and the *Frame Size (in bits)*.

(d) If interested in discovering how the framing process is carried out, it can be viewed by typing 'open `image_frame_gen.html`' in the MATLAB command window.

The browser window that opens contains all of the code required to create the frames. At the top of this window, an illustration explains how the length of the padding in a frame's payload structure changes slightly, depending upon the amount of data that the frame contains. Along with this illustration, there is a brief explanation of the general code functionality.

The code uses the parameters set in the *Image Transfer Binary Source* mask, along with preset protocol parameters, to generate the header fields and payload data. It will then construct the frames. Once finished with the code, close it and return to the Simulink model.

(e) **Check the USRP® Radio settings.** You will need to configure the *SDRu Transmitter* block to communicate with your USRP® hardware. To do this, follow the guidance provided in Exercise 12.1 about configuring the block's settings for your hardware device.

(f) **Run the model.** To check that the design is working before moving to the next exercise, ensure the USRP® hardware is connected and switched on, then start the simulation by 🖱 the 'Run' ▶ button in the Simulink toolbar. After a couple of seconds, Simulink will establish a connection with your radio and the simulation will begin.

When you are satisfied that the model runs without any errors, you can stop the simulation and move on to the next exercise while keeping the model open.

Chapter 12: Desktop Digital Communications: QPSK Transmission and Reception

Exercise 12.15 Image Tx Rx: RTL-SDR Receiver

In this exercise we will open a receiver based on the design presented in Exercise 12.13. This time it includes the additional processing required to receive, reconstruct and display the image being transferred. The main change in this design is that the *ASCII Decoding* block used previously has been replaced with an *Image Reconstruction* block. When the receiver believes it has acquired all of the frames to display the image, the simulation will automatically pause itself, show the image, and then resume.

(a) **Open MATLAB**. Set the working directory to the exercise folder,

    `/digital/rtlsdr_rx/`

Next, open the file:

    `.../rtlsdr_QPSK_image_transfer.slx`

NOTE: *The functionality of the AGC Simulink block was changed between releases R2015a and R2015b. We have supplied modified versions of these files, denoted by the* `_15bonwards.slx` *file names. Anyone using 15b or higher will need to use these files instead.*

The block diagram should look like this:

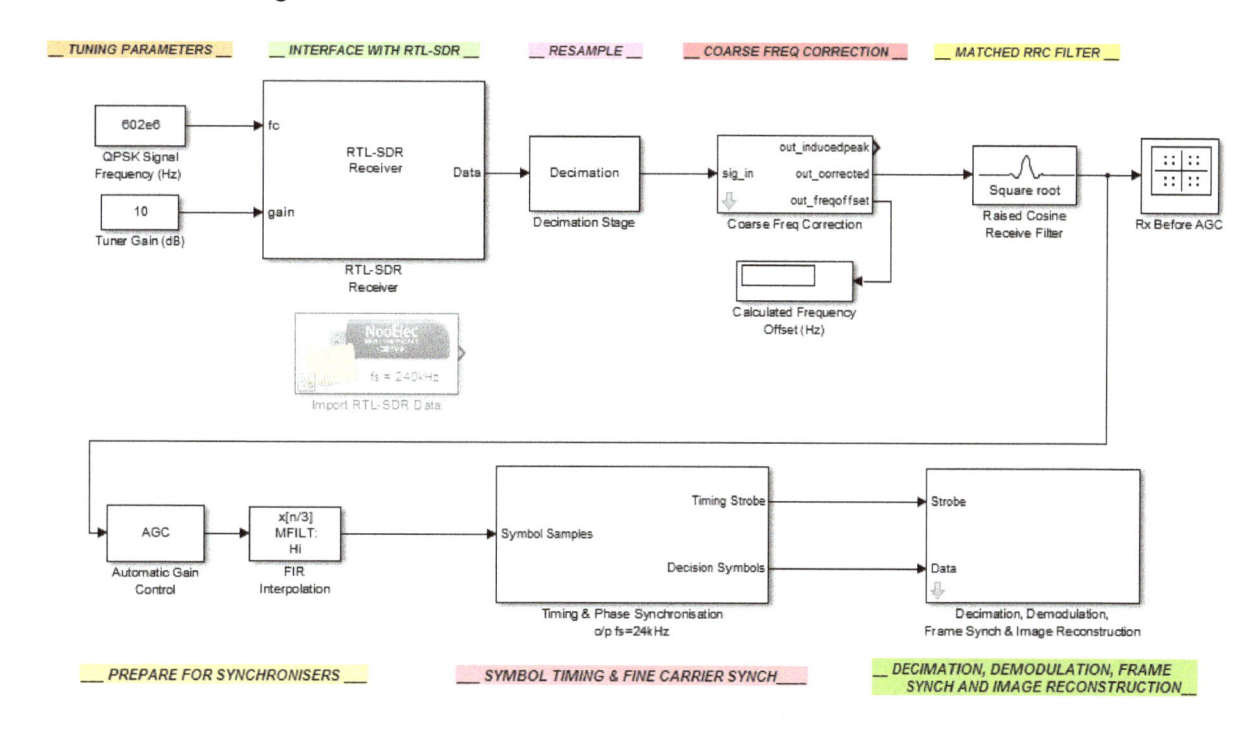

(b) **Inspect the receiver.** At first glance, the design of this model may seem no different to the previous system used for receiving ASCII messages. This is because only small changes to the design are required in order to transfer different kinds of information, as the data is always in bit or symbol form at the PHY Layer processing stage.

(c) **The final data recovery stage.** To confirm the changes that have been made, look inside the *Decimation, Demodulation, Frame Synch & Image Reconstruction* subsystem, and then inside the final processing stage, now named *Image Reconstruction*. This will show that the *ASCII Decoding* block has now been replaced by an *Image Reconstruction* block.

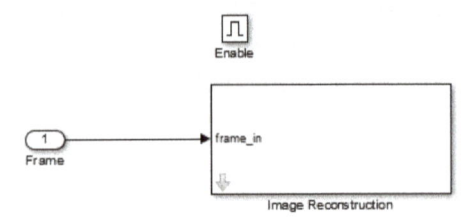

Due to the modular receiver design, we can simply swap one block for another, and reconstruct the image from the transmitted frames of data, in place of the ASCII messages.

You may also have noticed that the *Remove Preamble* block has been omitted, which means that the new *Image Reconstruction* block will receive the full transmitted frame with the additional header fields.

(d) **Image reconstruction.** Inside the *Image Reconstruction* block, the payload data and parameters from the top level mask are passed to a MATLAB function block. The code inside this block performs a number of tasks to recover and reconstruct the transmitted image. To view the code, simply 2⌘ on the *Image Reconstruction* MATLAB function block.

Firstly, it recovers information from the header fields of each frame to determine the header, payload and padding lengths, where the latter two are used to calculate the frame's data length. At this point it also recovers the sequence number from the header, and uses it during the data extraction process.

Each frame that is received until the last frame of the image, is subjected to the data extraction process, which retrieves the pixel column data from the payload and reconstructs the image by storing it as a column in a new `image` variable. Once all of the image data has been stored, it then outputs it to a figure window for viewing.

Shortly, when both the transmitter and receiver models are running, the image will be sent and received continuously and plotted in the same output figure. Note that received image quality can differ between image transmissions, which is due to the changes in interference that the signal experiences when travelling through the air.

(e) **Check the RTL-SDR data source.** Move back to the top level of the Simulink design. If you have an RTL-SDR available and are able to transmit the image/ DQPSK signal from Exercise 12.14, use the *RTL-SDR Receiver* block to interface with and control your RTL-SDR. Once again, ensure your PPM correction value has been entered into the block's parameter window. If you do not have an RTL-SDR or are unable to transmit the signal, you should use an *Import RTL-SDR Data* block instead. The block will need to be configured to reference one of the image files, where the # indicates which image is contained within the data file:

`/digital/rtlsdr_rx/rec_data/dqpsk_image#.mat`

Chapter 12: Desktop Digital Communications: QPSK Transmission and Reception 555

If you are going to import data rather than connect to an RTL-SDR, delete the *RTL-SDR Receiver* block and connect the *Import RTL-SDR Data* block in its place.

(f) **Run both the transmitter and receiver.** Start the transmitter by 🖱 on the 'Run' ▶ button in the Simulink toolbar. Once the transmitter has begun, return to the receiver and do the same. After a couple of seconds the receiver simulation will begin, the *Constellation Diagram* windows should appear, and information messages should begin to print in the MATLAB command window.

(g) **Received Images.** During the process of reconstructing the image, a message — 'Picture Column Number #' — will be displayed in the command window to give an indication of how many columns have been received. When this value reaches '100', the simulation will pause and an image will be output like the examples shown here.

Received and Plotted Images:

Young Lady Image

Wellington Statue

Two Radio Hams (W2KO) from 1980

(h) **Bit errors.** You many notice that some of the plotted images contain bit errors, similar to the example pictures. The protocol implemented for this image transfer is fairly rudimentary, and the design could be made more flexible and robust by including additional stages in both the transmitter and receiver, such as error control coding. You may wish to investigate and implement some error control coding schemes, using this design as a starting point.

As mentioned in Exercise 12.13, the signal you receive is affected by the device hardware in use, and the physical environment. Hence, some adjustments to your setup may be required to receive better quality images. Issues such as the receiver gain, the antenna position and the RTL-SDR's PPM offset correction value should be considered. If you are not receiving images like those shown here, try some simple adjustments to improve the system's performance.

(i) **Watch us running the receiver.** We have recorded a video which shows activity in the *Constellation Diagram* windows, the MATLAB command window, and the plotted image window, while a DQPSK signal transmitted by the USRP® hardware (with the model from

Exercise 12.14) is received by the RTL-SDR, and undergoes synchronisation and image reconstruction.

desktopSDR.com/videos/#qpsk_image

12.9 Transmitting Data Using FM Transmitters

While the main focus of this chapter has been on transmitting and receiving digital data using QPSK modulation, and latterly differentially encoded QPSK, we are aware that not all readers have access to a USRP® radio and are able to transmit and receive these signals "for real". While everyone should be able to run the simulation models and see the receivers working with recorded data, they will not be able to experience first hand all of the real world RF issues that have been discussed, such as the effects of interference from neighbouring transmissions in the vicinity, and the tolerances associated with the RTL-SDR. The obvious solution to this problem is to use an alternative transmitter, and here we present the transmission of data using a low cost FM Transmitter. Readers who do not have a USRP® radio may already have one of these from the *Desktop FM Transmission and Reception* exercises in Section 10.7 (page 423). If you do not, you will almost certainly want one after you have read through the rest of this section - readily available from say Amazon, and a cost of just a few quid or dollars!

As discussed in Section 10.7 (page 423), FM transmitters are designed to FM modulate and transmit baseband audio signals (band-limited to 15kHz) on a carrier that has a frequency within the FM range; 88–108MHz. Their intended use is to transmit music signals from smartphones and MP3 players, allowing the music to be received by an FM radio; or indeed an RTL-SDR with a Simulink-based FM receiver. FM signals can be non-coherently demodulated using a Complex Delay Line Frequency Discriminator (Section 9.7, page 355), and this means that the transmitted baseband information signal can be correctly recovered with relative ease when compared to the carrier phase and symbol timing synchronisation systems that must be implemented to demodulate and receive a QPSK signal. Using an FM transmitter allows you to transmit any audio signal, and as you will have seen in Section 10.7.3, it is possible to create customised baseband audio signals using Simulink. In that section, the customised signals comprised of an *AM in FM* signal, and a four channel *FDM MPX* signal.

Operating on a similar basis, we can create a baseband signal that contains digital data, as well as an audio signal. A number of different modulation schemes could be used to do this, but for simplicity we will consider the On-Off Keying (OOK) modulation scheme; the simplest technique to transmit a digital message with an analogue carrier. This is the same scheme that forms the basis of Morse Code. A binary

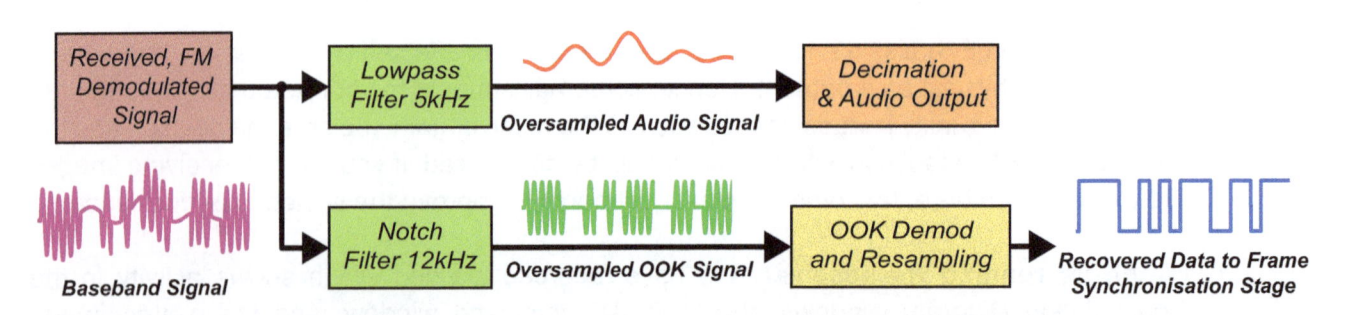

Figure 12.18: Recovering the audio and data signals from the *Data in FM* signal

Chapter 12: Desktop Digital Communications: QPSK Transmission and Reception

'1' is represented by the presence of a tone, and a '0' by its absence. Adding (or rather, multiplexing) an OOK modulated tone to a baseband music signal and FM modulating it, we are able to transmit data with a low cost FM transmitter. The architecture of the *Data in FM* signal generator is shown in Figure 12.19.

A receiver for this *Data in FM* signal is outlined in Figure 12.18. Signals received by the RTL-SDR would firstly need to be FM demodulated, in order to recover the baseband data/ audio MPX. The demodulated MPX would then be lowpass filtered to extract the audio signal, before being output to the computer's speakers. A notch filter could then be used to isolate the OOK tone, and a series of operations performed in order to recover the transmitted data.

Passing the OOK tone through an envelope detector results in a signal which closely resembles the digital bit stream, as shown in Figure 12.20. To complete the OOK demodulation, a hard decision operation has to be performed to retrieve the bit stream (the 1's and 0's). This is accomplished by centring the smoothed signal around zero, and then using the zero crossing as a decision boundary. The recovered (but oversampled) data bit stream must be downsampled to the data rate in order for the frame synchronisation system discussed in Section 12.6 to function correctly, and recover the data payload.

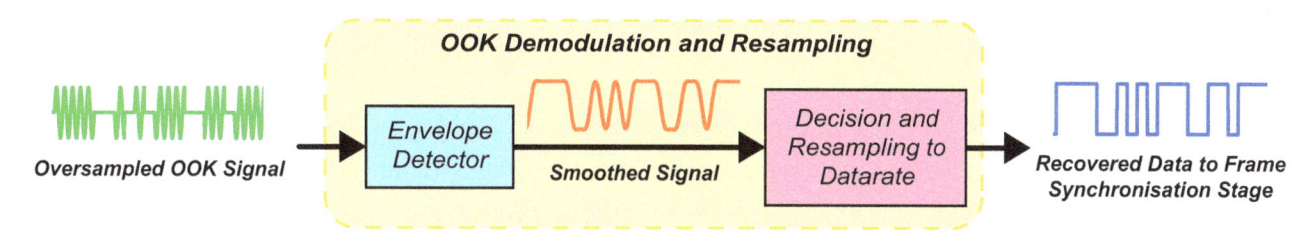

Figure 12.20: The OOK demodulation process

Considering that this signal generator and receiver pair are components of the PHY layer, we are able to directly substitute them in place of the QPSK Tx Rx PHY layer that has been developed throughout this chapter, while keeping the higher layers (framing/ frame synch, sequencing and encoding/ decoding) as-

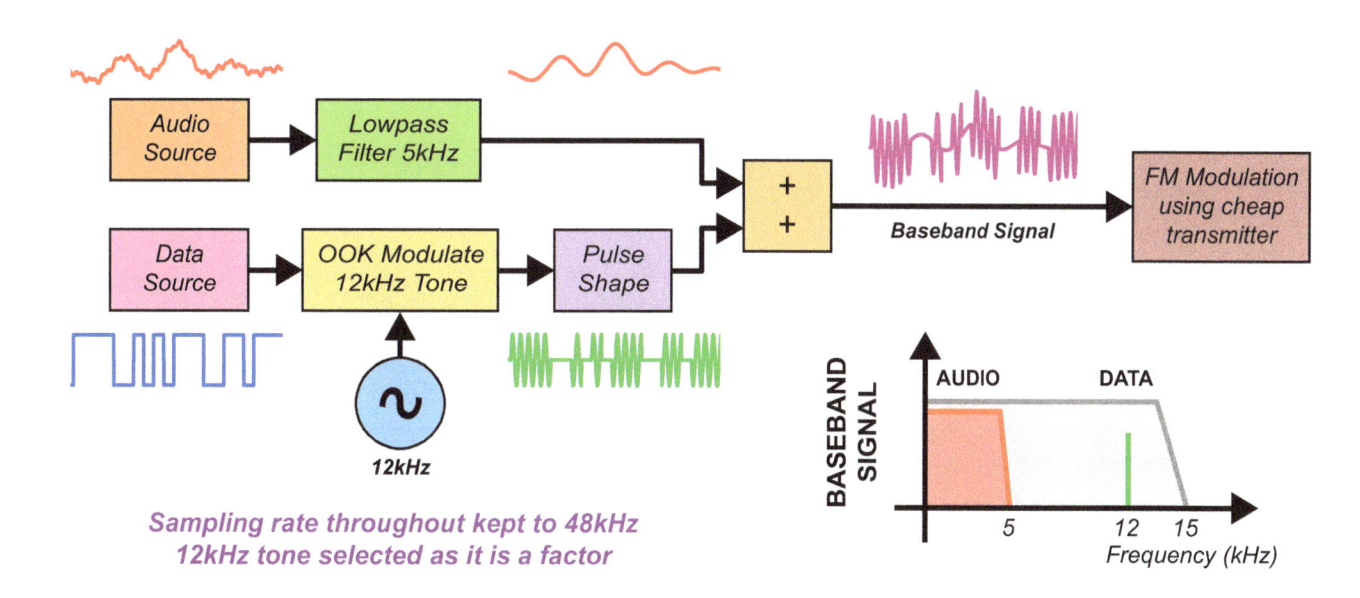

Figure 12.19: Creating the baseband *Data in FM* signal

is. In the following two sections, we will present alternative designs to the ASCII message Tx Rx pair from Section 12.7, and the greyscale image Tx Rx pair from Section 12.8, that use this *Data in FM* PHY layer.

Note: You will require an FM transmitter to complete the following exercises. The transmitter exercises create `.wav` *audio files, which can be transmitted using an off the shelf device, a Raspberry Pi running PiFM, see Exercise E.7 (page 621), or a USRP® radio set up to transmit FM see Exercise 10.1, page 370.*

12.9.1 Transmitting Audio and ASCII Strings with FM Transmitters

In this section we will present alternative designs for the ASCII transmitter and receiver pair from Section 12.7, that use the *Data in FM* PHY layer.

Exercise 12.16 **Data in FM: ASCII/ Audio Signal Generator**

In this exercise, a signal generator is used to create a baseband ASCII/ audio *Data in FM* signal. Numbered ASCII text frames are OOK modulated onto a 12kHz carrier and multiplexed with an audio signal. The result is saved to an audio file, which will need to be FM modulated and transmitted using external hardware. As previously mentioned, you may use an off the shelf FM transmitter, a Raspberry Pi, or a USRP® radio for this task.

(a) **Open MATLAB**. Set the working directory to the exercise folder,

   `/digital/other_tx/`

Next, open the file:

   `.../fmtx_datafm_ascii_audio.slx`

The block diagram should look like this:

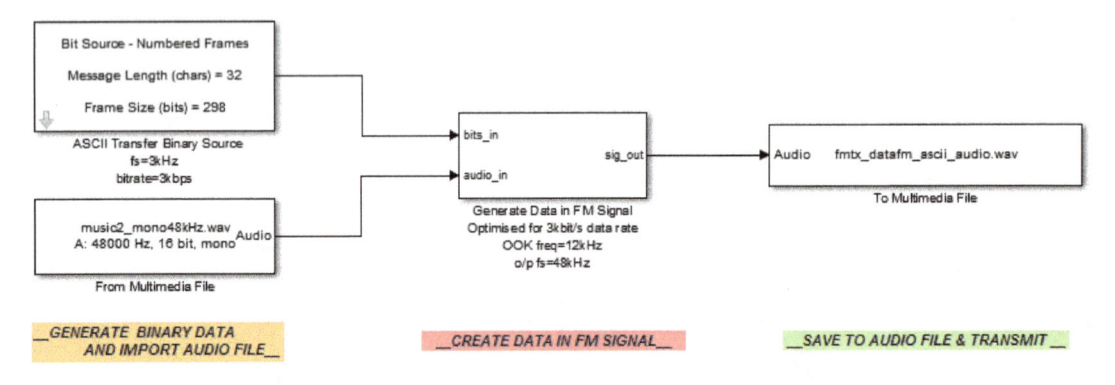

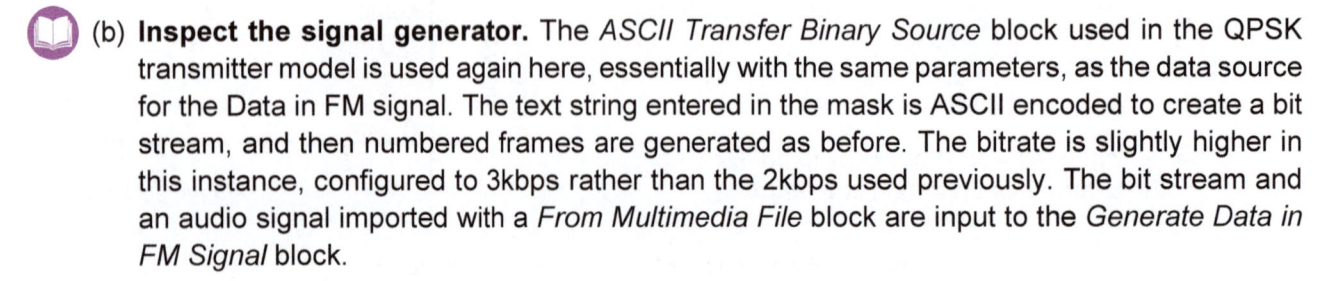

[Block diagram: Bit Source - Numbered Frames, Message Length (chars) = 32, Frame Size (bits) = 298, ASCII Transfer Binary Source fs=3kHz bitrate=3kbps → bits_in; music2_mono48kHz.wav A: 48000 Hz, 16 bit, mono, From Multimedia File → audio_in; Generate Data in FM Signal Optimised for 3kbit/s data rate OOK freq=12kHz o/p fs=48kHz → sig_out → Audio fmtx_datafm_ascii_audio.wav To Multimedia File]

__GENERATE BINARY DATA AND IMPORT AUDIO FILE__   __CREATE DATA IN FM SIGNAL__   __SAVE TO AUDIO FILE & TRANSMIT__

(b) **Inspect the signal generator.** The *ASCII Transfer Binary Source* block used in the QPSK transmitter model is used again here, essentially with the same parameters, as the data source for the Data in FM signal. The text string entered in the mask is ASCII encoded to create a bit stream, and then numbered frames are generated as before. The bitrate is slightly higher in this instance, configured to 3kbps rather than the 2kbps used previously. The bit stream and an audio signal imported with a *From Multimedia File* block are input to the *Generate Data in FM Signal* block.

Chapter 12: Desktop Digital Communications: QPSK Transmission and Reception 559

(c) **Generating the Data in FM signal.** 🖱 on the block mask for the *Generate Data in FM Signal* block and examine the system. The input bit stream is used as a switching signal to set the amplitude levels of the OOK signal to 0 and 0.1. This reduced amplitude bit stream passes through a pulse shaping filter which is designed to smooth the 0 to 0.1 transitions (and visa versa), to reduce spectral leakage. The filtered bit stream is then mixed with a 12kHz *Sine Wave* to create the OOK signal.

The input audio signal is lowpass filtered with an IIR filter to ensure that it contains no frequency components higher than 5kHz. This seeks to ensure that the audio and OOK signals do not interfere with each other. Finally, the two signals are combined together using an *Add* block, and output to the *To Multimedia File* block. Notice that a *Matrix Concatenate* block is used to group together signals of interest for viewing in a *Time Scope*.

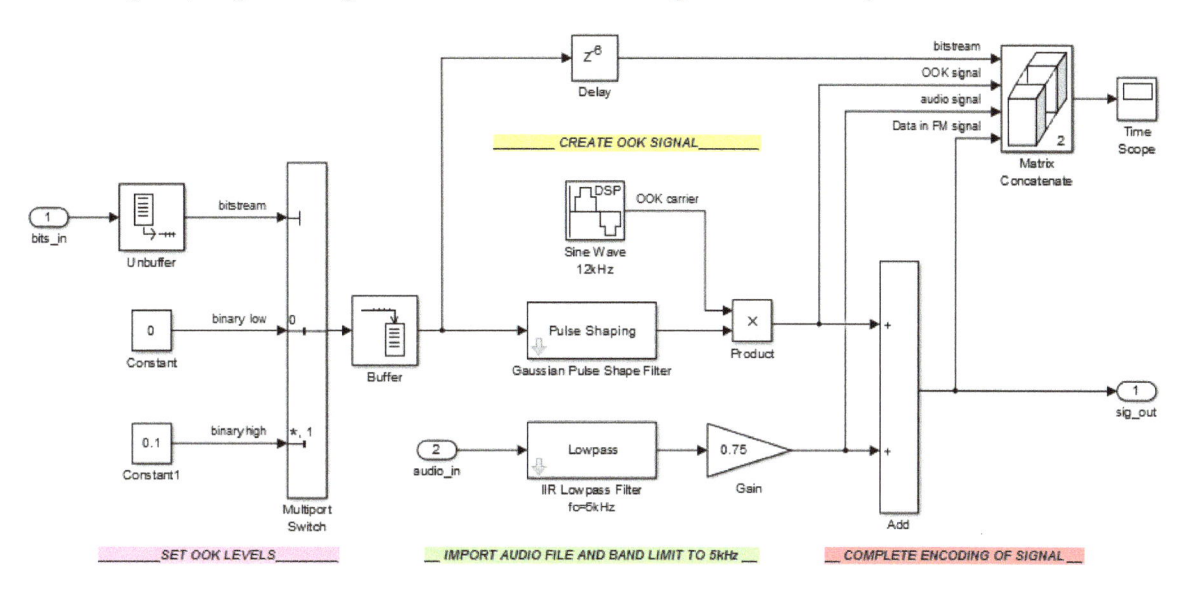

(d) **Run the simulation to generate the file.** Start the simulation by 🖱 the 'Run' ▶ button in the Simulink toolbar. Once the model initialises, the *Time Scope* window will appear, and will display the components of the Data in FM signal. Pause the simulation using the 'Pause' ⏸ button and view the scope.

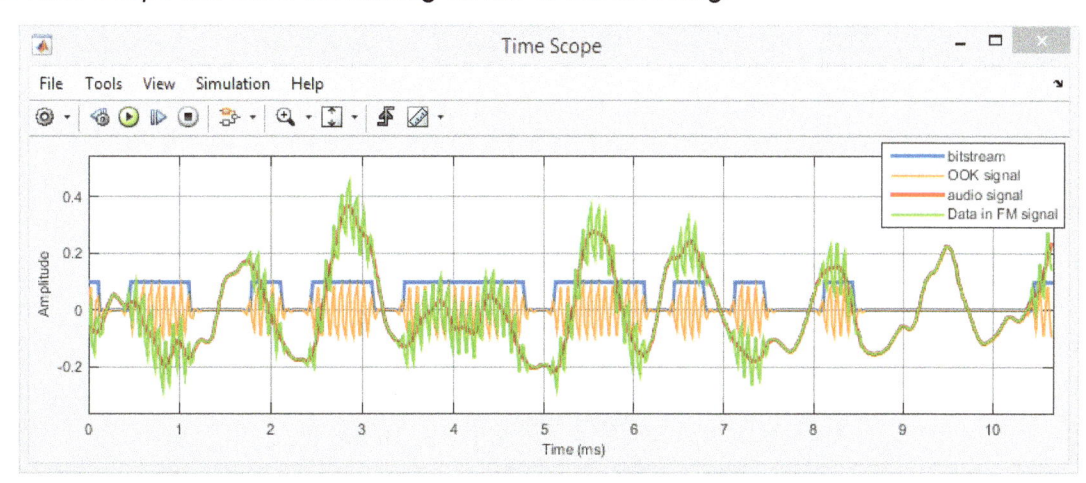

The *Time Scope* will show something similar to the following.

It is clear that the OOK signal (orange) is only 'on' when the bit stream (blue) is high. The pulses are smoothed thanks to the use of the *Gaussian Pulse Shape Filter*, which is why the 'on' pulses have rounded edges. The audio signal (red) behaves rather randomly, as expected, and the *Data in FM* signal (green) can be seen to be the sum of the OOK and audio signals.

(e) Allow the simulation to complete, and create the `.wav` audio file. This is what must be transmitted with the FM transmitter. If you are using an off-the-shelf transmitter, connect it to your computer, and play the file (note that you could also transfer the file to another device such as your smartphone, connect the transmitter, and play it from there). If you are using a Raspberry Pi and PiFM, transfer the files to the Pi using an SSH client as discussed in Appendix E.3 (page 619). If using the USRP® hardware as your transmitter, import the file to the Mono FM Transmitter model presented in Exercise 10.1 (page 370), and follow the instructions there to transmit it.

Exercise 12.17 Data in FM: RTL-SDR FM Demod & ASCII/ Audio Receiver

Next, we will run the receiver for the ASCII/ audio *Data in FM* signal. The receiver FM demodulates signals acquired by the RTL-SDR, and then recovers both the audio and ASCII information messages.

(a) **Open MATLAB**. Set the working directory to the exercise folder, and then open the file:

> `/digital/rtlsdr_rx/`
>
> `.../rtlsdr_datafm_ascii_audio.slx`

NOTE: *The functionality of the AGC Simulink block was changed between releases R2015a and R2015b. We have supplied modified versions of these files, denoted by the* `_15bonwards.slx` *file names. Anyone using 15b or higher will need to use these files instead.*

The block diagram should look similar to the screenshot:

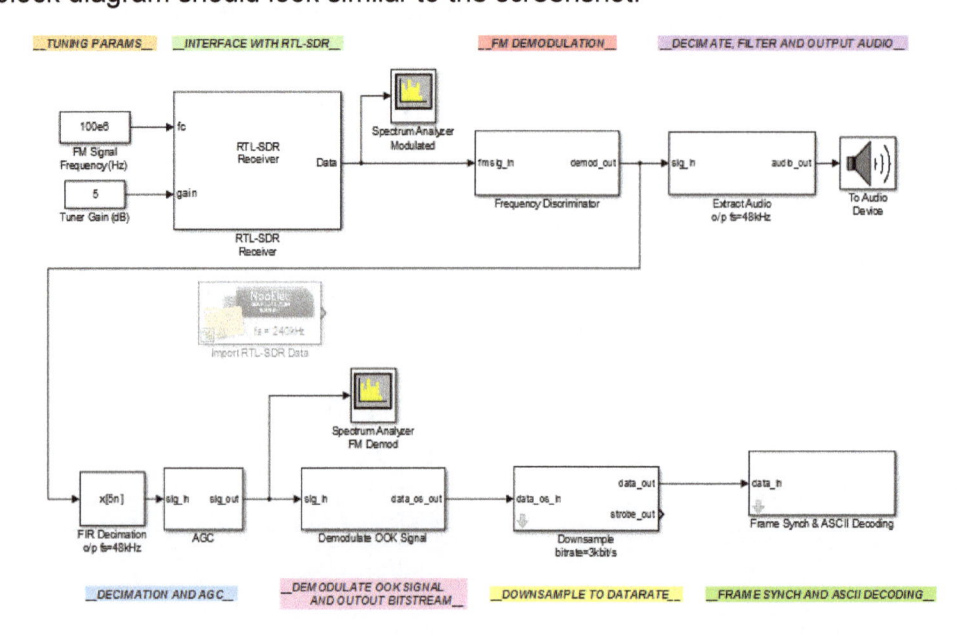

Chapter 12: Desktop Digital Communications: QPSK Transmission and Reception 561

📖 (b) **Inspect the receiver.** All that has changed between this model and the QPSK ASCII receiver is that the PHY layer processing stages have been replaced. The FM signal received by the RTL-SDR is demodulated using a frequency discriminator, and then the baseband *Data in FM* signal is filtered to isolate both of the information signals (audio and OOK). After decimation, the audio signal is output to speakers or headphones. The OOK signal is also decimated, and undergoes a stage of AGC, before being input to the *Demodulate OOK Signal* block.

📖 (c) **OOK Demodulation.** As described previously, the OOK signal is demodulated using an envelope detector and a hard decision process. 2️⃣ on the *Demodulate OOK Signal* block and explore the system contained within. As with the signal generator model, signals are grouped together using a *Matrix Concatenate* block and input to a *Time Scope*.

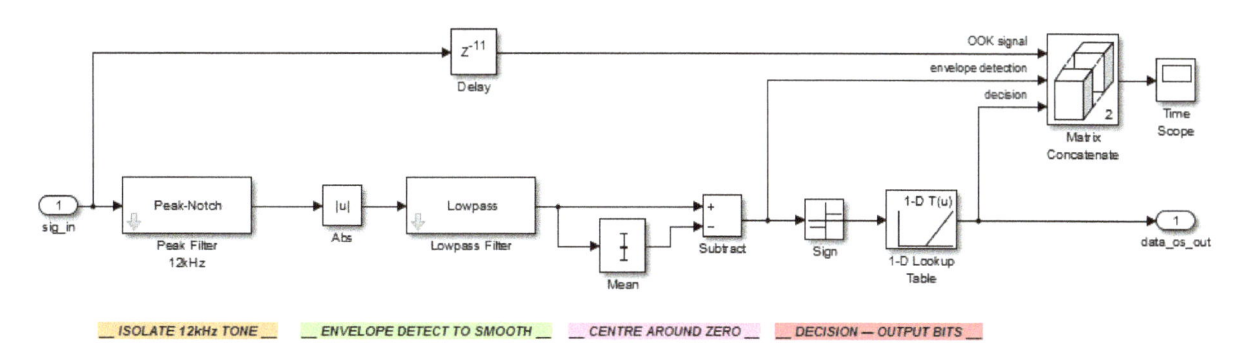

(d) After the hard decision process, the oversampled bit stream must be returned to the original data rate for the higher level processing operations (frame synch and ASCII decoding) to perform correctly. This is achieved by the *Downsample* subsystem. Return to the main block diagram and 2️⃣ on the grey arrow in the bottom left hand corner of the mask to view inside it.

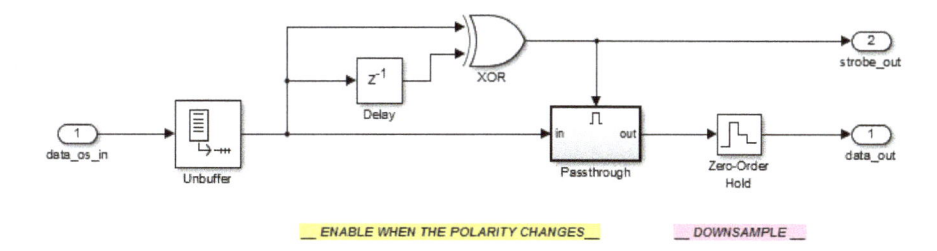

A latch is implemented using the *Delay*, *XOR* and *Passthrough* blocks, which switches polarity when the polarity of the oversampled bit stream changes. The output of the latch is input to a *ZOH* block which acts as an ADC, resampling the bit stream down to the data rate.

📖 (e) **ASCII message recovery.** The two remaining stages required to recover the ASCII message are inside the *Frame Synch & ASCII Decoding* subsystem. Frame synchronisation is performed using the matched filter approach presented in Section 12.6.7, and the ASCII decoding is undertaken with the block used throughout the latter part of this chapter.

🔧 (f) **Check the RTL-SDR data source.** If you have an RTL-SDR available and are able to transmit the ASCII/ audio *Data in FM* signal from Exercise 12.16, you will be able to use an *RTL-SDR Receiver* block to interface with and control your RTL-SDR. Ensure your PPM correction value is entered into the block's parameter window. Otherwise, you will need to use an *Import RTL-SDR Data* block. This will need to be configured to reference this file:

📁 `/digital/rtlsdr_rx/rec_data/datafm_ascii_audio.mat`

If you are going to import data rather than connect to an RTL-SDR, delete the *RTL-SDR Receiver* block and connect the *Import RTL-SDR Data* block in its place.

The length of the message that the receiver expects to recover also needs to be set in the 'Message Length (chars)' field of the *Frame Synch & ASCII Decoding* block's parameter window. If you are using the import block, this value should be set to '32'. If you are transmitting a signal yourself using the system from Exercise 12.16, you must ensure that the value matches the value shown on the *ASCII Transfer Binary Source* block.

(g) **Run the simulation.** on the 'Run' button in the Simulink toolbar to begin a simulation. After a couple of seconds the receiver model will start running, and the *Spectrum Analyzer* and *Time Scope* windows will appear. ASCII decoded strings will print to the MATLAB command window, and an audio signal will be output to your speakers or headphones. If you are using an RTL-SDR, adjust the frequency and gain values until you tune the device to your transmitted signal. You can use *Spectrum Analyzer Modulated* to help you do this.

(h) **Observing the output.** If you are using the import file you will now see the `You can transmit data in FM! ###`' message being output, where the `###` stands for the frame number associated with each particular frame. If you set a custom message prior to transmission, hopefully you should see this too!

(i) **Signal Analysis: *Spectrum Analyzer*.** *Spectrum Analyzer FM Demod* should show something similar to the following.

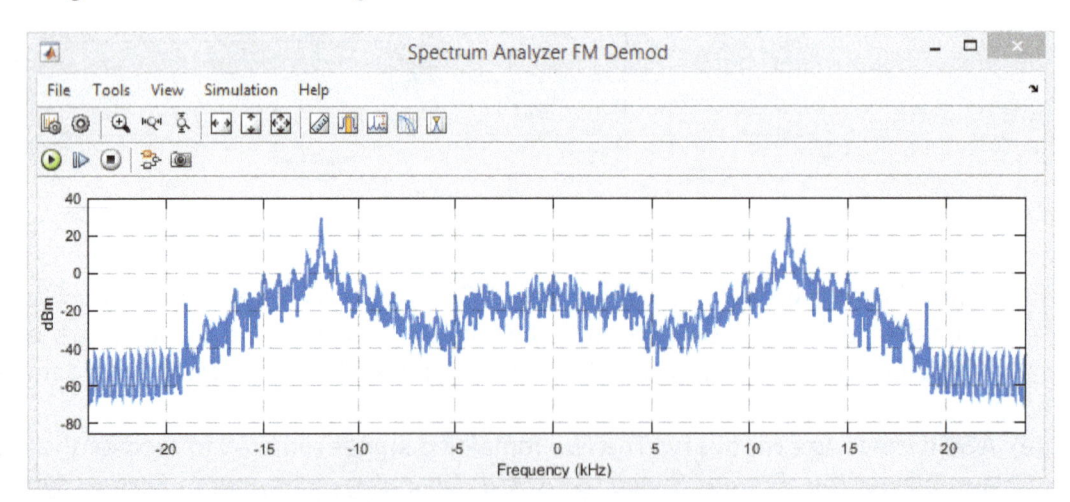

Here, it is easy to confirm that two different types of signals are present: the audio signal resides in the 0-5kHz region, and the OOK signal is centred at 12kHz. As expected, the 12kHz sine wave used to create the OOK signal is the most significant single frequency component.

(j) **Signal Analysis: *Time Scope*.** The *Time Scope* will show the OOK signal as it undergoes the demodulation process. This will look similar to the next screenshot. Here, you can see how the

Chapter 12: Desktop Digital Communications: QPSK Transmission and Reception 563

envelope detector has smoothed the OOK signal, and confirm that the hard decision process has correctly identified the 0's and 1's.

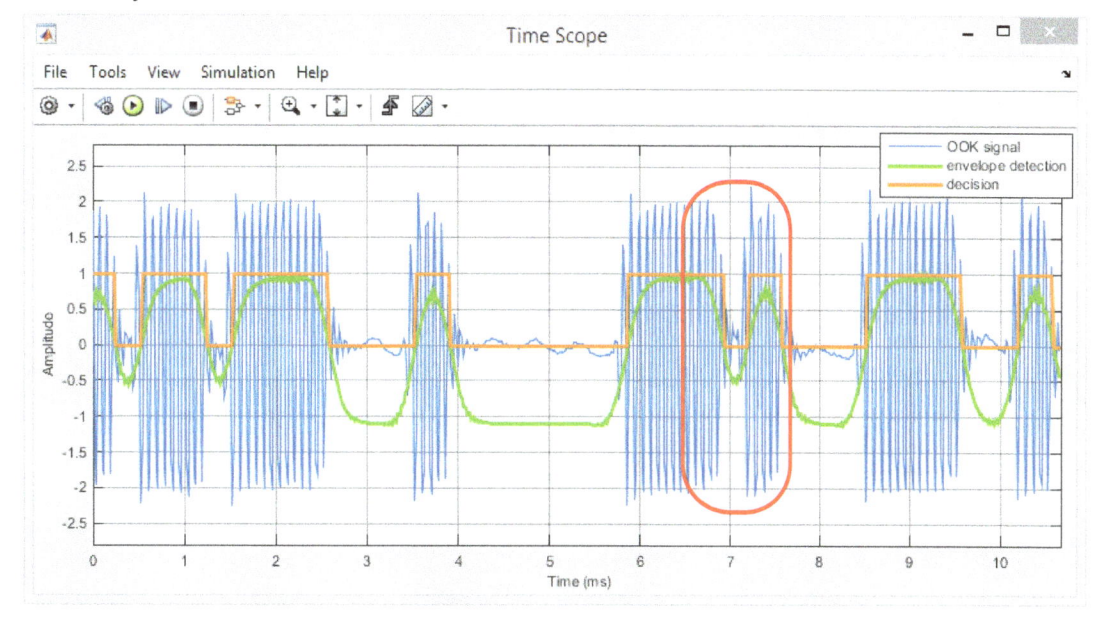

(k) **Decision errors.** Pause the simulation and examine your *Time Scope*. Do you agree with the decisions the OOK demodulator has made? Do you notice anything odd about the widths of the output pulses? When we ran the model we found that, on occasion, the zero pulse (i.e. in the ...101... highlighted above) did not last as long as it should. Why do you think this would happen? Can you think of anything you could do to mitigate this issue? *Hint: you may need to modify the parameters of the Gaussian Pulse Shape Filter in the transmitter and the Lowpass Filter in the envelope detector of the receiver.*

(l) **Watch and listen to us demodulate the ASCII/ audio *Data in FM* signal.** We have recorded a video which shows activity in the scope windows and MATLAB command window as the ASCII/ audio *Data in FM* signal is transmitted by an off-the-shelf FM transmitter and received by the RTL-SDR and demodulated. The output audio signal has been recorded too.

desktopSDR.com/videos/#datafm_ascii_audio

12.9.2 Transmitting Greyscale Images with FM Transmitters

In this final section, we provide a *Data in FM* implementation of the image transfer Tx Rx pair from Section 12.9.

Exercise 12.18 Data in FM: Image/ Audio Signal Generator

In this exercise, we will run a signal generator to create a baseband Image/ audio *Data in FM* signal. Frames of image data will be OOK modulated onto a 12kHz carrier and multiplexed with an audio signal. The result is saved to an audio file, which can be FM modulated and transmitted using external hardware. As previously mentioned, you may use an off-the-shelf FM transmitter, a Raspberry Pi, or a USRP® radio for this task.

(a) **Open MATLAB.** Set the working directory to the exercise folder,

   `/digital/other_tx/`

Next, open the file:

   `.../fmtx_datafm_image_audio.slx`

The block diagram should look like this:

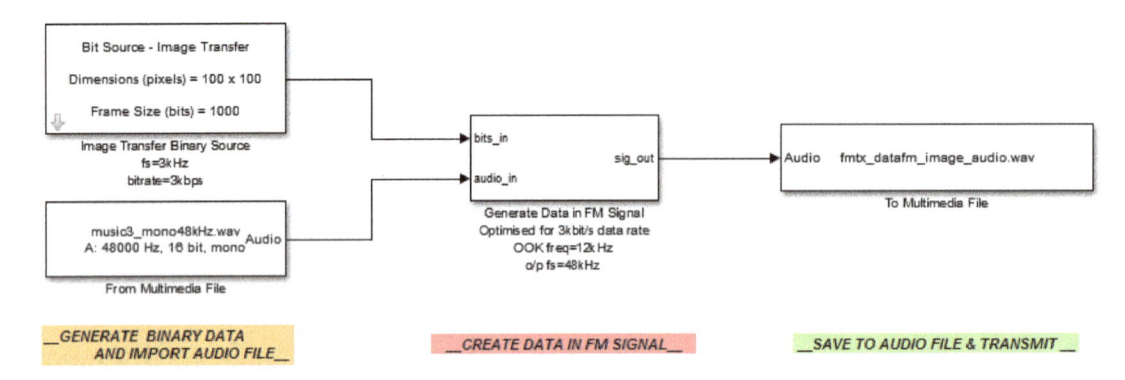

(b) **Inspect the signal generator.** The *Image Transfer Binary Source* block used in the QPSK transmitter model is used again here (essentially with the same parameters), as the data source for the Data in FM signal. The greyscale image is imported and processed to create a sequenced bit stream. The bitrate for this *Data in FM* signal generator is 3kbps. The bit stream and an audio signal imported with a *From Multimedia File* block are input to the *Generate Data in FM Signal* block, where the *Data in FM* signal is created.

(c) **Run the simulation to generate the file.** Start the simulation by the 'Run' button in the Simulink toolbar. Allow the simulation to complete, and create the `.wav` audio file. This is what must be transmitted with the FM transmitter. If you are using an off-the-shelf transmitter, connect it to your computer, and play the file (note that if you transfer the file to another device such as your smartphone, you could connect the transmitter and play it from there too). If you are using a Raspberry Pi and PiFM, transfer the files to the Pi using an SSH client as discussed in Appendix E.3 (page 619). If you are using the USRP® hardware as your transmitter, import the file to the Mono FM Transmitter model presented in Exercise 10.1 (page 370), and follow the instructions there to transmit it.

Exercise 12.19 Data in FM: RTL-SDR FM Demod & Image/ Audio Receiver

In this exercise, you will run the receiver for the Image/ audio *Data in FM* signal. The receiver FM demodulates signals acquired by the RTL-SDR, and then recovers both the audio signal and transmitted image.

(a) **Open MATLAB.** Set the working directory to the exercise folder,

Chapter 12: Desktop Digital Communications: QPSK Transmission and Reception

📁 `/digital/rtlsdr_rx/`

Next, open the file:

`.../rtlsdr_datafm_image_audio.slx`

NOTE: *The functionality of the AGC Simulink block was changed between releases R2015a and R2015b. We have supplied modified versions of these files, denoted by the* `_15bonwards.slx` *file names. Anyone using 15b or higher will need to use these files instead.*

The block diagram should look similar to the screenshot shown below.

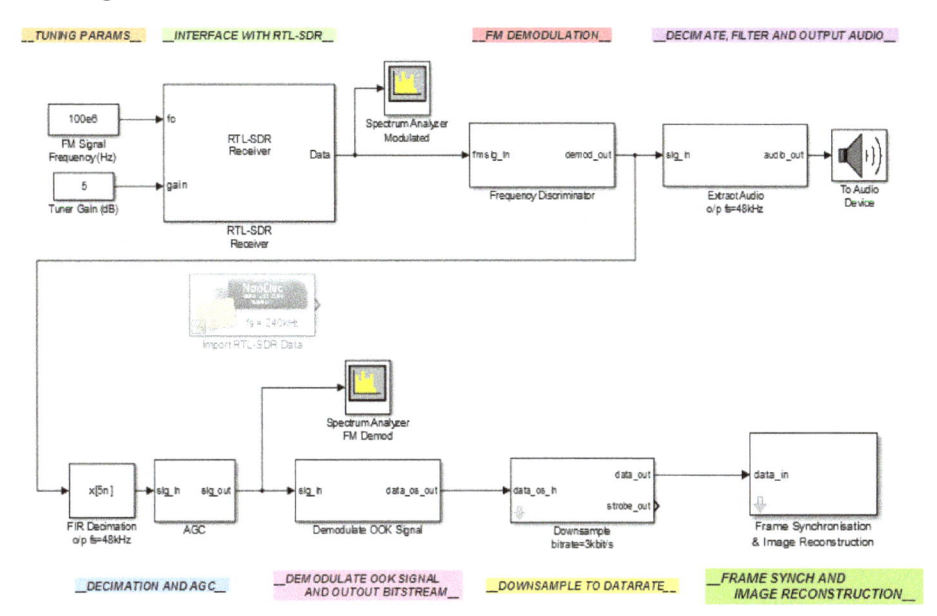

📖 **(b) Inspect the receiver.** All that has changed between this model and the QPSK Image receiver is that the PHY layer processing stages have been replaced. The FM signal received by the RTL-SDR is demodulated using a frequency discriminator, and then the baseband *Data in FM* signal is filtered to isolate both of the information signals (the audio and OOK signals). After decimation, the audio signal is output to your computer's speakers or headphones. The OOK signal is also decimated, and undergoes a stage of AGC, before being demodulated. After being downsampled to the data rate, the recovered bit stream is then passed into a frame synchronisation and image reconstruction stage.

🔧 **(c) Check the RTL-SDR data source.** If you have an RTL-SDR and are able to transmit the image/ audio *Data in FM* signal from Exercise 12.18, you will be able to use an *RTL-SDR Receiver* block to obtain the signal. Ensure your PPM correction value has been set. If you do not have an RTL-SDR or are unable to transmit the signal, you will need to use an *Import RTL-SDR Data* block. This will need to be configured to reference this file:

📁 `/digital/rtlsdr_rx/rec_data/datafm_image_audio.mat`

If you are going to import data rather than connect to an RTL-SDR, delete the *RTL-SDR Receiver* block and connect the *Import RTL-SDR Data* block in its place.

(d) **Run the simulation.** on the 'Run' button in the Simulink toolbar to begin a simulation. After a couple of seconds the receiver will begin, and the *Spectrum Analyzer* windows will appear (the *Time Scope* has been commented out to ensure the receiver processes the data fast enough). If you are using an RTL-SDR, adjust the frequency and gain values until you tune the device to the signal that you are transmitting. You can use *Spectrum Analyzer Modulated* to help you do this. During the process of reconstructing the image, a message — `Picture Column Number #` — will be displayed in the command window to give an indication of how many columns have been received. When this value reaches `100`, the simulation will pause and the image will be output to a figure window.

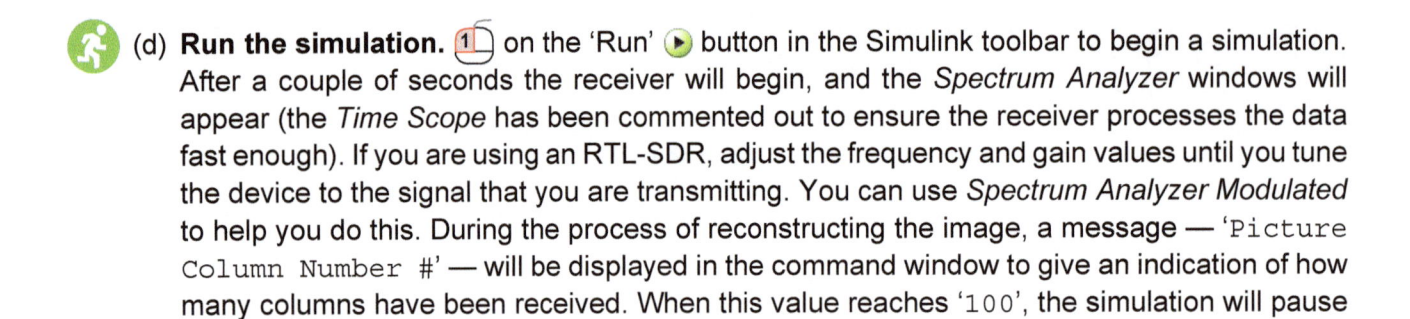

(e) **Are you able to see pictures?!** When testing this exercise out, we found that certain off-the-shelf FM transmitters worked better than others at successfully conveying the information signal in this exercise. If you are having trouble receiving a signal, switch to the imported RTL-SDR data to see what you should be observing.

(f) **Watch and listen to us demodulate the image/ audio *Data in FM* signal.** We have recorded a video which shows activity in the scope windows and MATLAB command window as the image/ audio *Data in FM* signal is transmitted by an off-the-shelf FM transmitter and received by the RTL-SDR and demodulated. The output audio signal has been recorded too.

desktopSDR.com/videos/#datafm_image_audio

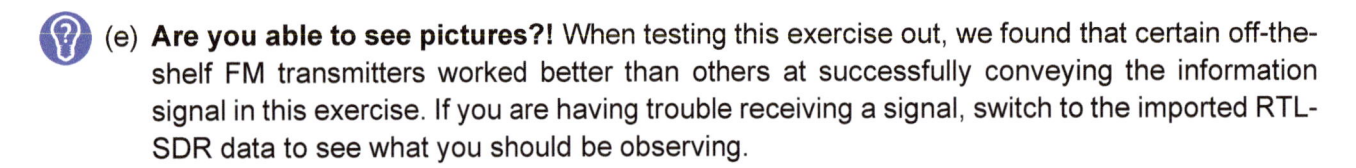

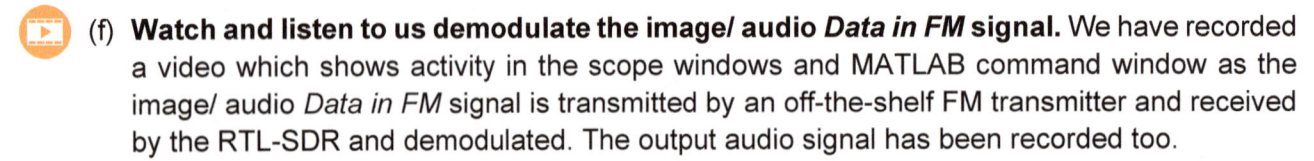

12.10 Summary

In this practical chapter we have compared real-time desktop digital communications to the theory that was introduced in Chapter 11. Through this, we have shown that the offsets and tolerances experienced by a real world communications system are much more unpredictable than theory, due to several contributing factors. We also confirmed that our design could be constructed to combat these issues and compensate for the various offsets, to achieve successful data recovery. As well as the comparison of simulation theory and practical designs, a few new topics were introduced, namely: the OSI stack, ASCII encoding, frame synchronisation and the use of protocols. By utilising these processes in our designs, we were able to not only transmit ASCII messages and greyscale images with QPSK modulation, but also with FM using inexpensive hardware.

As this is the final chapter of the book, we (the authors) hope that you can take all that you have learned from the theory and practical work, and apply it to your own intelligent and quirky SDR designs! Now that you have a firm background with RTL-SDR and using MATLAB and Simulink, we expect that more complex GPS receivers, aeroplane trackers, and maybe even your own video transfer across the desktop, will be much more manageable and achievable!

Watch out for our future and updated editions, perhaps featuring new hardware, new exercises and new ideas. You can keep track of what we are doing at desktopSDR.com. Remember – *stay tuned, the wireless revolution is just beginning!*

Appendices and Postscript

Appendix A: Hardware Setup

This Appendix contains some information you should find helpful when it comes to installing Hardware Support Packages for the RTL-SDR and USRP® hardware, and instructions on how to use the USRP® hardware to find and correct the tuner error of your RTL-SDR hardware.

Additional troubleshooting information can be found at desktopSDR.com/more/troubleshooting.

A.1 The RTL-SDR Hardware Support Package

The RTL-SDR Hardware Support Package is an add-on for MATLAB and Simulink that enables the software to interface with your RTL-SDR. Installation of this support package is a pre-requisite for all of the RTL-SDR based examples in the book.

Software Setup: Installing the Support Package

If you need to add the RTL-SDR Hardware Support Package to your installation of MATLAB and Simulink, follow the information given in reference [61] (the hyperlinked URL below)

> mathworks.com/help/supportpkg/rtlsdrradio/ug/support-package-hardware-setup.html

This page provides step by step installation instructions for users of Windows, Linux and Mac operating systems. It covers:

- Launching the *Get Hardware Support Packages* wizard,
- Selecting and installing the RTL-SDR Hardware Support Package software (which requires a free MathWorks account),
- Installing the RTL-SDR USB driver,

and also contains some basic troubleshooting information. Make sure that you choose the 'Install from Internet' option, as shown in our installation video:

> desktopSDR.com/videos/#rtlsdr_hw_supportpkg

Once you have finished the installation, please try to complete Exercises 2.2 and 2.3 (starting on page 28), to check that the installation was successful.

Hardware Setup

The NooElec NESDR RTL-SDRs are shipped with omnidirectional antennas. These can be connected to the RTL-SDR by plugging them into the device's MCX port, as shown in Figure A.1. If you wish to use a different antenna, you may need to purchase an adapter. Information about these is given in Section 3.3 (page 46). The RTL-SDR must be plugged into a USB2.0 (or higher) port in order to communicate with your computer.

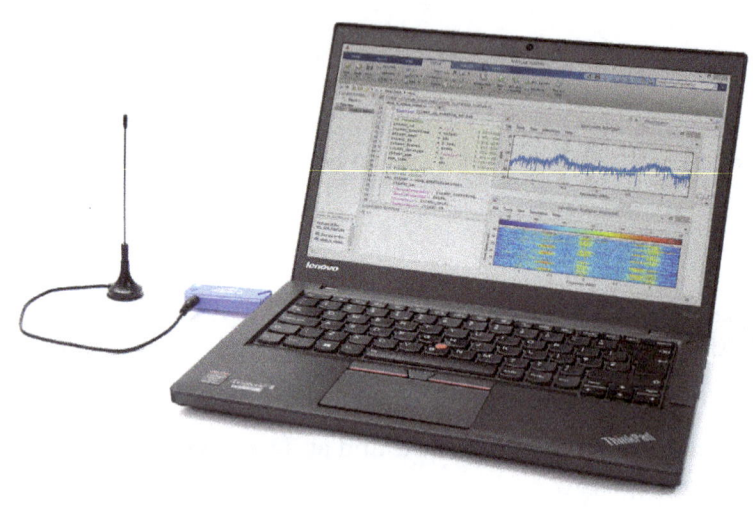

Figure A.1: The NooElec NESDR Mini 2 with antenna connected, plugged into the USB port of a laptop

Using More Than One RTL-SDR

If you have more than one RTL-SDR connected to your computer, you will need to consider radio addressing (`RadioAddress`). The first device connected will always be assigned address '0', but if there is more than one, the addresses assigned to each device will depend on the addresses of the USB port that it is connected to. The address of each device is returned by the `sdrinfo` function. *Note: to access information about the first device when a multi-dimensional container is returned, type* `rtlsdr_info{1}.details` *into the MATLAB command window*. For more information, and a diagrammatic explanation of the USB port addressing, please take a look at the 'Configure Multiple Radios' page of the Hardware Support Package documentation [62].

```
>> rtlsdr_info = sdrinfo

rtlsdr_info =

  Column 1

    [1x1 sdrr.internal.RTLSDRInfoContainer]

  Column 2

    [1x1 sdrr.internal.RTLSDRInfoContainer]
```

Figure A.2: The multi-dimensional container returned on running the `sdrinfo` function with more than one RTL-SDR connected to your computer

A.2 The USRP® Hardware Support Package

The USRP® Hardware Support Package is an add-on for MATLAB and Simulink that enables the software to interface with a USRP® radio device. The USRP® hardware is recommended for many of the examples in the book, and generally has the role of 'transmitter', creating signals that can then be received by the RTL-SDR. Installation of the USRP® support package is required in order to run these examples.

Software Setup: Installing the Support Package

If you need to add the USRP® Hardware Support Package — *Communications System Toolbox Support Package for USRP® Radio* — to your installation of MATLAB and Simulink, follow the information given in reference [68] (the hyperlinked URL below)

mathworks.com/help/supportpkg/usrpradio/ug/support-package-hardware-setup.html

This page provides step by step installation instructions for users of Windows, Linux and Mac operating systems. It covers:

- Launching the *Get Hardware Support Packages* wizard,
- Selecting and installing the USRP® Hardware Support Package software (which requires a free MathWorks account),
- Downloading the latest firmware,
- Configuring the host computers Network Interface Card (NIC) to communicate with the USRP® hardware,

and also contains some basic troubleshooting information. Make sure that you choose the 'Install from Internet' option, as shown in our installation video:

desktopSDR.com/videos/#usrp_hw_supportpkg

Note that if you are aiming to use a Bus Series USRP® radio (one with a USB3.0 connection), and are on a computer running Windows 8 or above, additional steps are required to install the USB driver. This is demonstrated in the video. Once you have finished the installation, please try to complete Exercise A.1, which checks that the installation was successful. After this, a simple way to test whether or not the device is working is to proceed onto Appendix A.3 and use the USRP® hardware to transmit a tone to the RTL-SDR.

Exercise A.1 Verify Software Setup: USRP® Hardware Support Package

This exercise will be used to confirm that the USRP® Hardware Support Package has been installed successfully.

(a) **Open MATLAB.** If you do not have MATLAB open, return to Exercise 2.1 (page 26), and repeat Part (a).

(b) **Start Simulink.** on the *Simulink Library* button in the MATLAB Home ribbon at the top of the MATLAB window. This should cause the Simulink Library Browser window to open, and it will display something similar to the screenshot shown below (your view may differ subject to your window size, and the libraries and toolboxes you have installed on your computer).

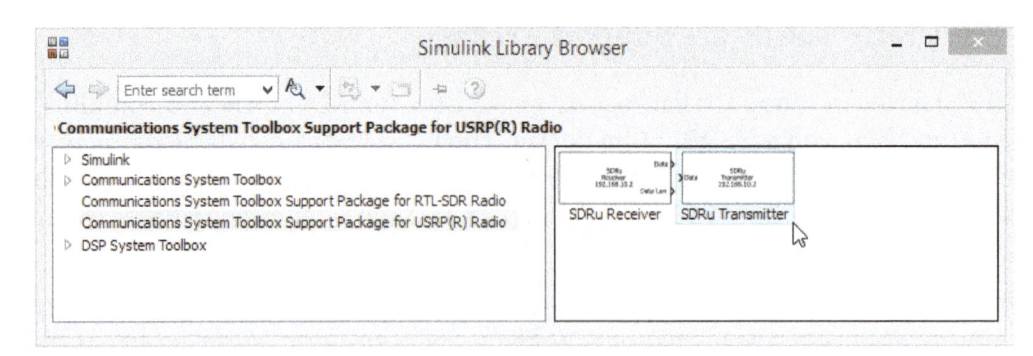

(c) **Confirm the USRP® Hardware Support is present in Simulink.** You should see that one of the libraries listed is the ▦ *> Communications System Toolbox Support Package for USRP® Radio*. This should have been installed automatically when the USRP® Hardware Support Package was added to your computer. If you 🖱 on this library, you should notice that it contains two blocks, titled *SDRu Receiver*, and *SDRu Transmitter*. These are parameterisable interfaces which can bring samples to and from the USRP® radio and Simulink in real time.

If you do *not* see the RTL-SDR library, try typing `setupsdru` into the MATLAB command line. If this command is not known, you will need to re-install the Hardware Support Package following the instructions detailed in [68].

(d) **Parameterisation of the *SDRu Transmitter* block.** When you place the *SDRu Transmitter* block in a new Simulink model, you will need to configure it with the address of your USRP® radio in order to use it. If you are using a Network or X Series device, you enter an IP address. If you are using a Bus Series USRP® radio, you enter the device's serial number, as shown below.

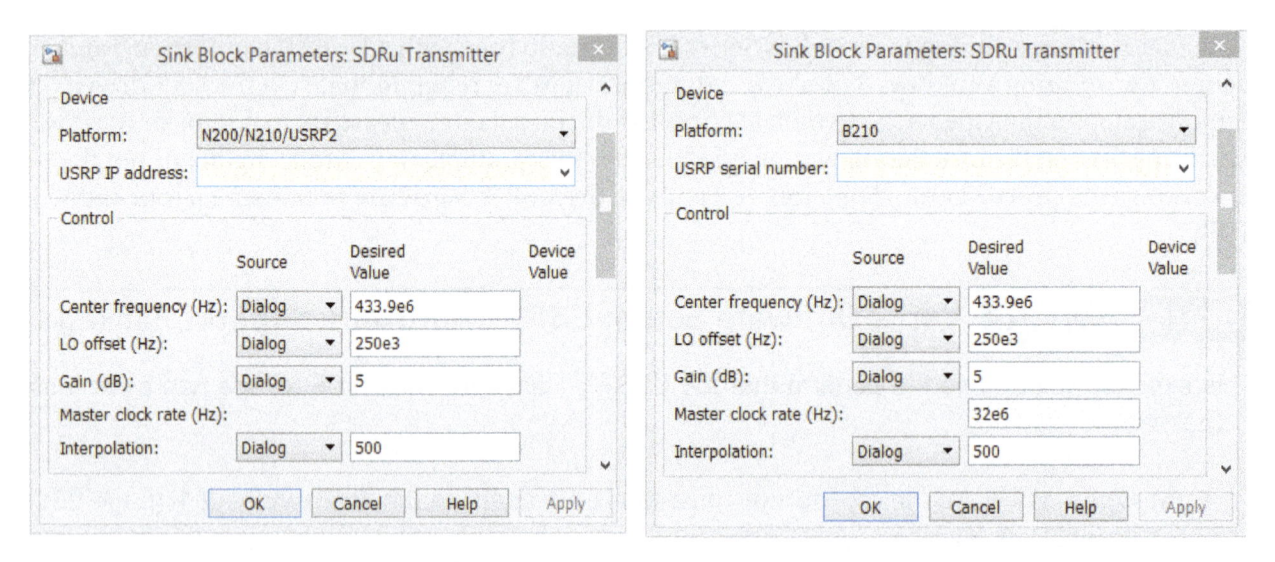

Similar to the RTL-SDR, the 'Centre frequency' parameter represents f_{rf}, the frequency of the RF carrier. The 'Local Oscillator offset' specifies the frequency of the LO. The LO commonly produces a DC spike, so setting this value wider than the bandwidth of your modulated signal is advisable. The 'Gain' specifies the overall gain of the hardware. **Note that as the device is**

Appendix A: Hardware Setup

sold uncalibrated, this is only a relative value. **To be certain of your USRP® radio's output power, you must use independent measuring equipment.**

(e) **Review the USRP® Hardware Support Package documentation.** Close the Simulink Library browser, and type `sdrudoc` into the MATLAB command window. If the Hardware Support Package has installed successfully, a help window should appear titled *Communications System Toolbox Support Package for USRP® Radio*. Perhaps take some time now to read through this documentation.

If the help window did not appear, the Hardware Support Package has not installed correctly. Repeat the installation process detailed in [68].

Exercise A.2 Verify Hardware Setup: USRP® Hardware Support Package

This exercise will be used to check that your computer can communicate with your USRP® hardware. This will either be over USB3.0, or GBE connections, depending on what particular device you have. To complete this exercise you will require a USRP® radio and the appropriate cables. If you do not currently have a USRP® radio to hand, return to this exercise once you have sourced one!

(a) **Open MATLAB.** If you do not have MATLAB open, return to Exercise 2.1 (page 26), and repeat Part (a).

(b) **BUS SERIES: Connect the USRP® hardware.** If you have not yet connected the USRP® device to your computer, plug it into a free USB3.0 port. Connecting the antenna is not necessary at this stage, but having it plugged in will do no harm.

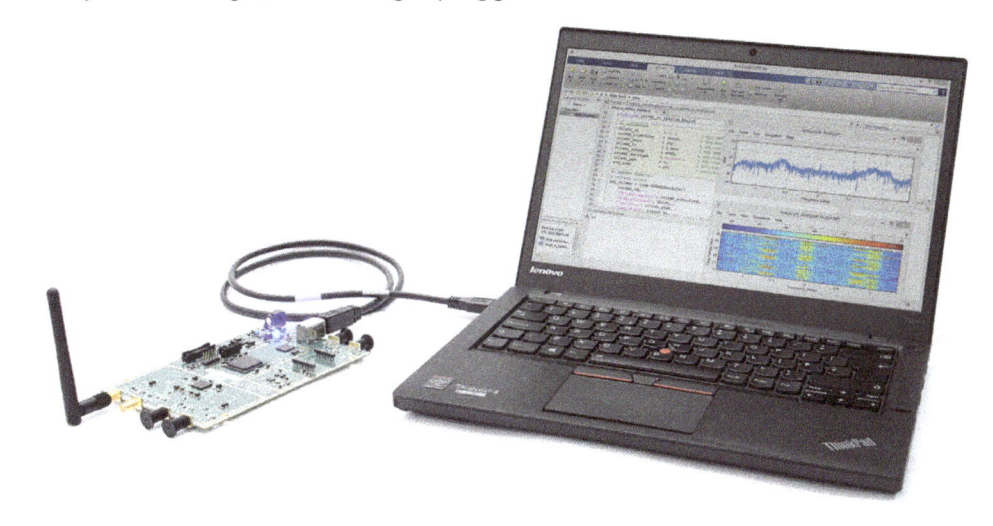

(c) **BUS SERIES: Check that the USRP® Radio is Recognised.** Enter '`my_usrp = findsdru`' into the MATLAB command window. This should prompt MATLAB to search for the USRP® hardware, download the FPGA image, and display something similar to the following:

```
>> my_usrp = findsdru

Checking radio connections...
Win32; Microsoft Visual C++ version 11.0; Boost_104900; UHD_003-vendor

---------- see libuhd version information above this line ----------
-- Loading firmware image: C:/MATLAB/SupportPackages/R2014b/usrpradio/
toolbox/shared/sdr/sdru/uhdapps/images/usrp_b200_fw.hex... done
Loading FPGA image: C:/MATLAB/SupportPackages/R2014b/usrpradio/toolbox/
shared/sdr/sdru/uhdapps/images/usrp_b210_fpga.bin...

my_usrp =

    Platform: 'B210'
   IPAddress: ''
   SerialNum: 'XXXXXXX'
      Status: 'Success'
```

If your device was *not* recognised or was *not* plugged in, then you will get a null return as shown below (you can unplug your working USRP® hardware if you wish to see this result):

```
>> my_usrp = findsdru
Checking radio connections...

my_usrp =

    Platform: ''
   IPAddress: ''
   SerialNum: ''
      Status: 'No devices found'
```

If MATLAB is unable to recognise your USRP® device (but knows what the `findsdru` command is), it is likely that the driver has not been successfully installed. Type `targetupdater` into the command window. This will open the Support Package Installer wizard you should have encountered previously. Repeat the steps detailed in [68] to reinstall the drivers. Note that if you are using a computer running Windows 8 or above, additional steps are required to install the USB driver, as detailed in our install video.

If after completing this, MATLAB is still unable to recognise your USRP® hardware, please take a look at our troubleshooting advice at desktopSDR.com, and/or in the USRP® Hardware Support Package documentation from MathWorks [67] for more information.

(d) **NETWORK or X SERIES: Connect the USRP® hardware.** If you have not yet connected the USRP® device to your computer, plug it into a free GBE port. We tend to connect it via a GBE to USB3.0 NIC, which means that the computer can remain connected to its network. Connecting the antenna is not necessary at this stage, but having it plugged in will do no harm.

Appendix A: Hardware Setup

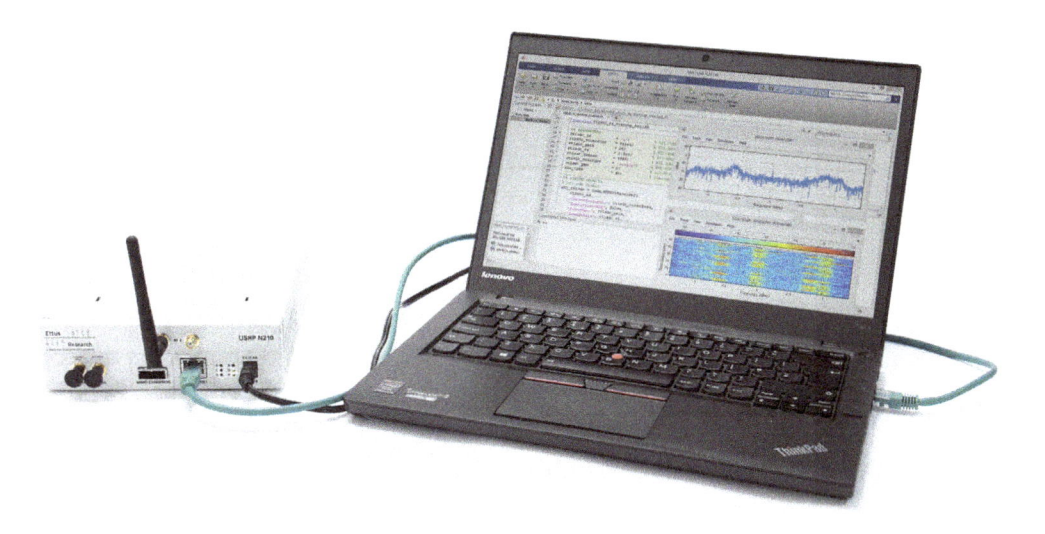

(e) **NETWORK or X SERIES: Check that the USRP® Radio is Recognised.** After ensuring that your computer's NIC is correctly configured, enter 'my_usrp = findsdru' into the MATLAB command window. This should prompt MATLAB to search for the USRP® hardware, and display something similar to the following:

```
>> my_usrp = findsdru

Checking radio connections...
Win32; Microsoft Visual C++ version 11.0; Boost_104900; UHD_003-vendor

---------- see libuhd version information above this line ----------
---------- begin libuhd warning message output ----------
The MTU (1472) is larger than the FastSendDatagramThreshold (1024)!
This will negatively affect the transmit performance.
See the transport application notes for more detail.

---------- end libuhd warning message output ----------

my_usrp =

    Platform: 'N200/N210/USRP2'
   IPAddress: '169.254.10.2'
   SerialNum: 'XXXXXX'
      Status: 'Success'
```

If your device was *not* recognised or was *not* plugged in, then you will get a null return as shown in Part (c). If MATLAB is unable to communicate your USRP® hardware (but knows what the findsdru command is), try typing dos('ping AAA.BBB.CCC.DDD') into the command window, where AAA.BBB.CCC.DDD is the IP address of your USRP® device. If the computer is able unable to communicate with it, you will receive a reply telling you that the destination host is unreachable. Check that your computer's NIC has been appropriately configured - please watch the installation video to see what to do. If everything appears to be configured correctly, it may be that your firewall is blocking connections. See what happens when you disable it.

If after completing this, MATLAB is still unable to recognise your USRP® hardware, please take a look at our troubleshooting advice at desktopSDR.com, and/or in the USRP® Hardware Support Package documentation from MathWorks [67] for more information.

Hardware Setup

A number of the more popular USRP® hardware models are able to interface with MATLAB and Simulink using the USRP® Hardware Support Package. The add-on supports devices with USB3.0, GbE and 10GbE device connections. If you purchase a Network or X Series device, you will need to install an RF daughtercard. Documentation should be provided on how to do this. The steps shown in Figure A.3 show how a WBX 50-2200MHz daughtercard would be installed into the N210 USRP® model.

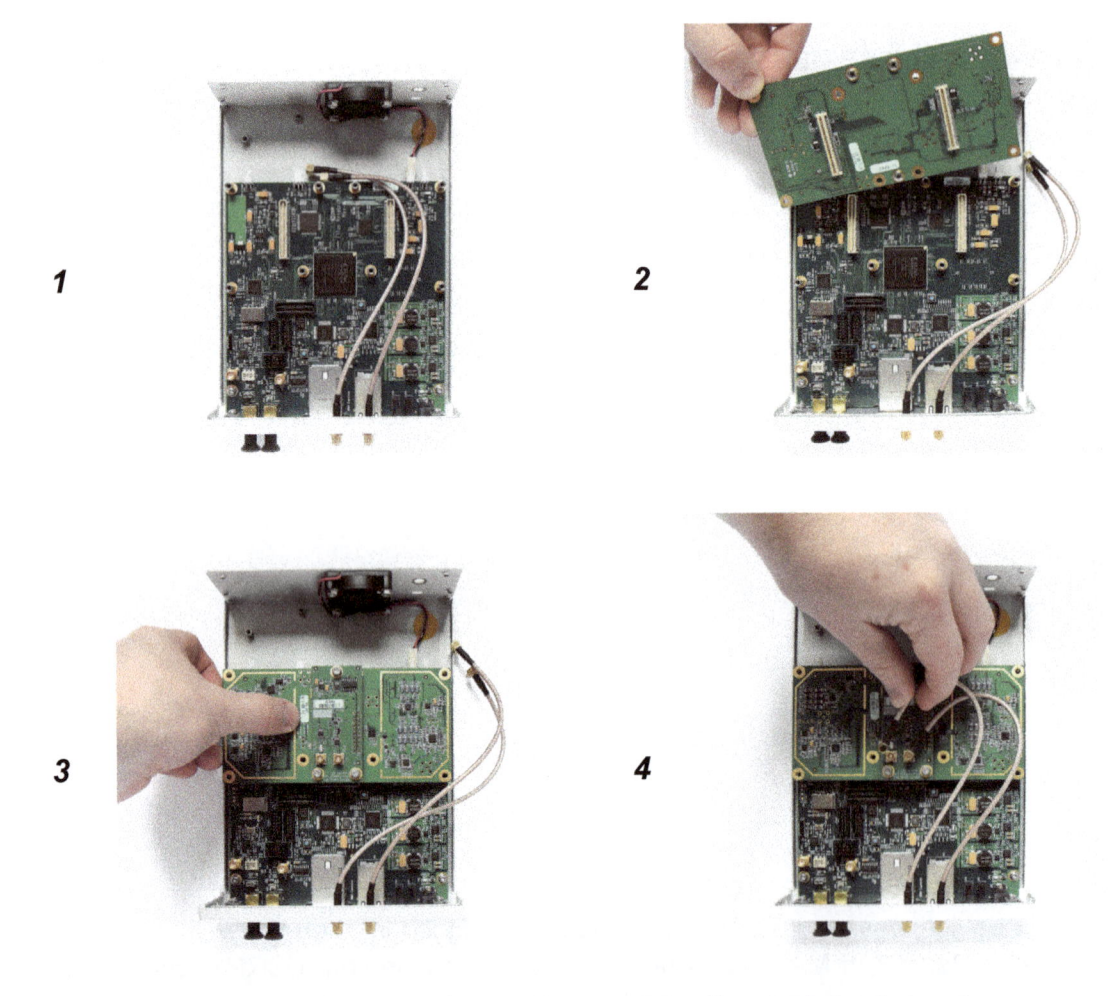

Figure A.3: Opening up the USRP® N210 and installing an RF daughtercard
(in this case, the WBX 50–2200MHz card)

Further support for setting up a USRP® radio and interfacing with MATLAB and Simulink can be found at the following link:

mathworks.com/help/supportpkg/usrpradio

Appendix A: Hardware Setup

A.3 RTL-SDR Frequency Error Correction

Parts Per Million (PPM, or ppm) is a measurement unit used to quantify very small values, in this case the LO frequency offset of the RTL-SDR. To give a simple example, an error of 10ppm is equivalent to 10/1,000,000, or 0.001%.

In this case, the frequency offset of the RTL-SDR is measured (in ppm), and the value obtained can subsequently be used as a correction parameter when interfacing with the device via MATLAB or Simulink. The RTL-SDR may have quite a significant offset, so this step can be important depending on the method of communication — it is more important for digital communication schemes; analogue communications are relatively tolerant by comparison. There is little else to explain about this, so we will simply get on and show you how to find the ppm frequency error of your RTL-SDR, and how to correct it!

Exercise A.3 **Finding the PPM Error of your RTL-SDR**

To complete this Exercise you will require an RTL-SDR and a USRP® radio. The technique that we will be using to find the ppm error of your RTL-SDR stick uses FFT matrix manipulation to calculate the offset between the centre frequency of a transmitted tone, and the centre frequency value required for your RTL-SDR downconvert the tone to 0Hz — a very similar process to coarse carrier synchronisation. You may find that your RTL-SDR has a significant offset, of the order of tens of kHz. This offset value is used to calculate the ppm error of your individual RTL-SDR, which can then be easily corrected by supplying this value to the *RTL-SDR Receiver* block as a parameter. Without performing this correction stage, it may be very difficult to use your RTL-SDR as a digital comms receiver due to the error.

Note: Every RTL-SDR will have different hardware characteristics, so you will need to complete this exercise for each device, if you have more than one. Some will need less correction that others!

(a) **Open MATLAB.** Set the working directory to:

   `/synch/rtlsdr_ppm`

There will be two files in this folder — a transmitter and a receiver. If you are working on a powerful computer, you might be able to open both of them on the same computer, in different instances of MATLAB. If not, open one of the files on each computer:

   `.../usrp_tx_ppm.slx`

   `.../rtlsdr_rx_ppm.slx`

The RTL-SDR needs to be attached to the computer running the 'receiver' model, and the USRP® transmitter to the computer running 'transmitter' model (bear in mind that the RTL-SDR and USRP® hardware will be connected to a single computer, if using only one to run both models).

(b) **The USRP® Transmitter.** The simplest way to transmit a single tone with the USRP® hardware is to input a *Sine Wave* to it. This *Sine Wave* block is configured with a frequency of 20kHz and is sampled at 200kHz. As with the previous examples, the USRP® radio is configured to resample the signal from 200kHz to 100MHz before modulating it onto the RF carrier, by interpolating by a factor of 500. The block diagram of the USRP® transmitter is as follows:

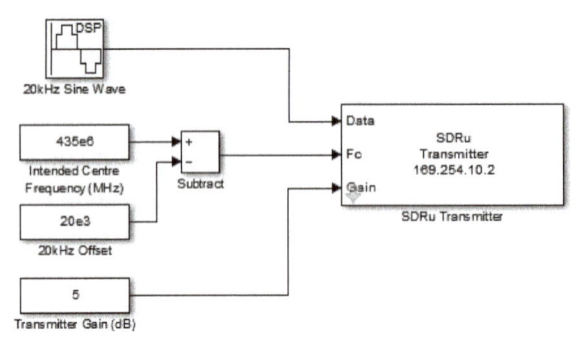

(c) **Check the USRP® Radio settings.** You will need to configure the *SDRu Transmitter* block to communicate with your USRP® hardware. To do this, you will need to know its IP address (or USB address), and be able to 'ping' it from your computer. Instructions of how to set up this connection are discussed in Appendix A.2. Then, 🖱 on the *SDRu Transmitter* block, and check that the appropriate 'Platform', and 'Address' of your USRP® radio are entered.

Make sure that the value of the *Intended Centre Frequency* block is set to the frequency you would like to transmit on — **take care** to select a frequency that you can legally use. This value should be within the range of your RTL-SDR's tuner, e.g. in the range 25MHz – 1.75GHz. Check that the USRP® hardware is turned on, and that MATLAB can communicate with it by typing `findsdru` into the MATLAB command window. If a structure is returned as discussed in Appendix A.2, you can continue.

(d) **The RTL-SDR Receiver.** RF signals received by the RTL-SDR are sampled at a rate of 2.4MHz, and then decimated to reduce the sampling rate to 400kHz. This low rate allows for a better frequency domain resolution when calculating the offset. Some matrix operations are then carried out inside the *Frequency Offset Calculation* subsystem to find the frequency offset value. Finally, a calculation is performed to find the PPM error, which is then displayed in the *PPM Correction Required* Display sink (top right corner of the model).

$$e_{ppm} = \frac{f_{offset}}{f_c \times -1^{-6}} \qquad (A.1)$$

where: f_c is the centre frequency of the RTL-SDR

The block diagram of the RTL-SDR receiver model is as follows:

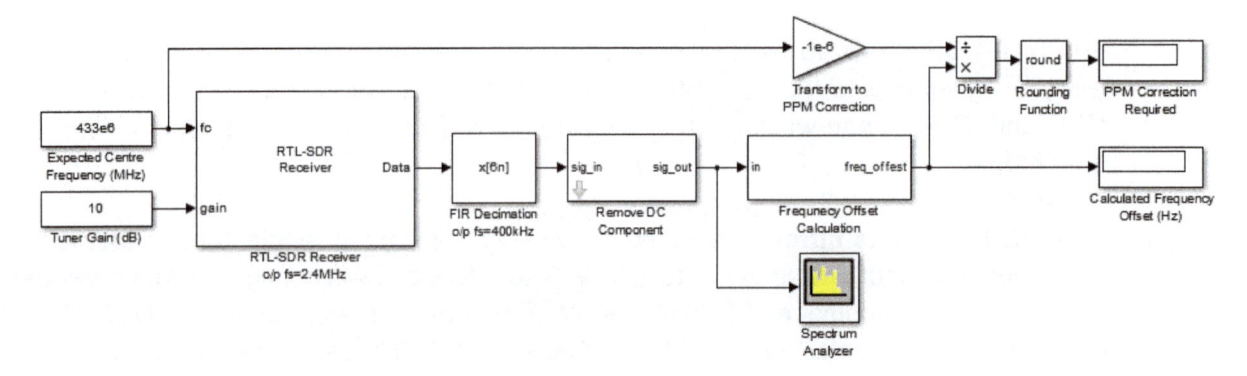

Appendix A: Hardware Setup 579

(e) **Check the RTL-SDR settings.** Check that MATLAB can communicate with your RTL-SDR by typing `sdrinfo` into the MATLAB command window. Make a note of the `RadioAddress`, and then check that this matches the value set in the *RTL-SDR Receiver* block. Check that the value set in the *Expected Centre Frequency* block is set to the same value as the *Intended Centre Frequency* block in the USRP® transmitter model.

(f) **Run the simulations.** Begin both simulations by clicking on the 'Run' buttons in the Simulink toolbar. After several seconds, the USRP® radio will begin transmitting the tone, and the RTL-SDR will begin receiving it. The *Spectrum Analyzer* window will give you a visual representation of how severe the frequency error of your RTL-SDR is — remember that the tone should be positioned at 0Hz in the baseband spectrum.

NOTE: When the authors ran these tests with all of the RTL-SDR devices we could find(!), we discovered that there were huge variations between RTL-SDR types, and even between devices of the same type! Below, some examples of the frequency offset/ PPM errors experienced are shown.

Freq Offset = -12.5kHz
PPM Error = 25

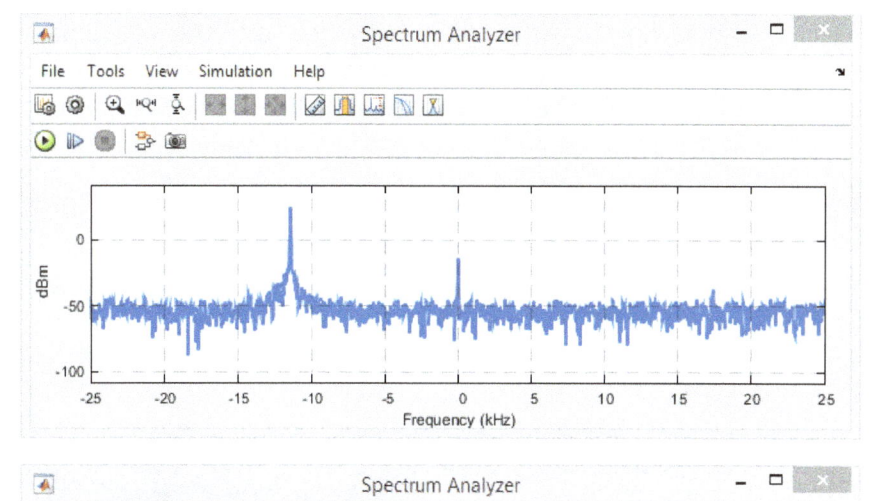

Freq Offset = 2.2kHz
PPM Error = -4

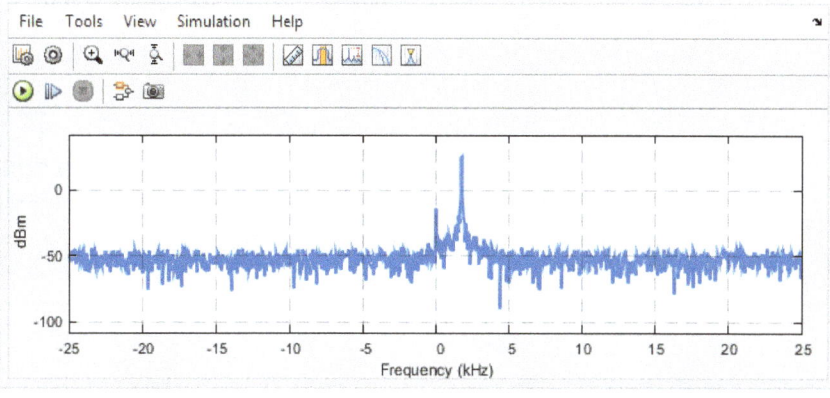

After running the simulation for around 3 minutes, the RTL-SDR should be warmed up and the offset and PPM error values should stabilise. When this happens, stop the RTL-SDR receiver model and enter the measured PPM value into the parameters of the *RTL-SDR Receiver* block. Note that the PPM value must be specified as an integer. Apply the changes, and then re-run the model.

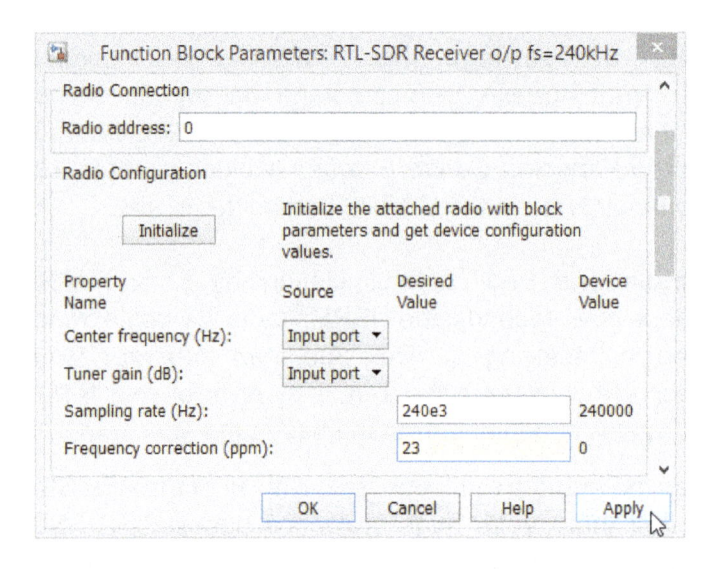

After another few minutes of use to re-heat the RTL-SDR, check that supplying the PPM correction value has had the desired effect. If the correction has worked, a PPM value of '0' should be shown in the display box, and the calculated frequency offset should be much smaller, with the centre frequency of the transmitted tone being within a hundred kHz (or so) of the RTL-SDR's centre frequency. If the PPM value is not '0', add or subtract this (depending on the sign) from the value you have already entered to the block, and try again.

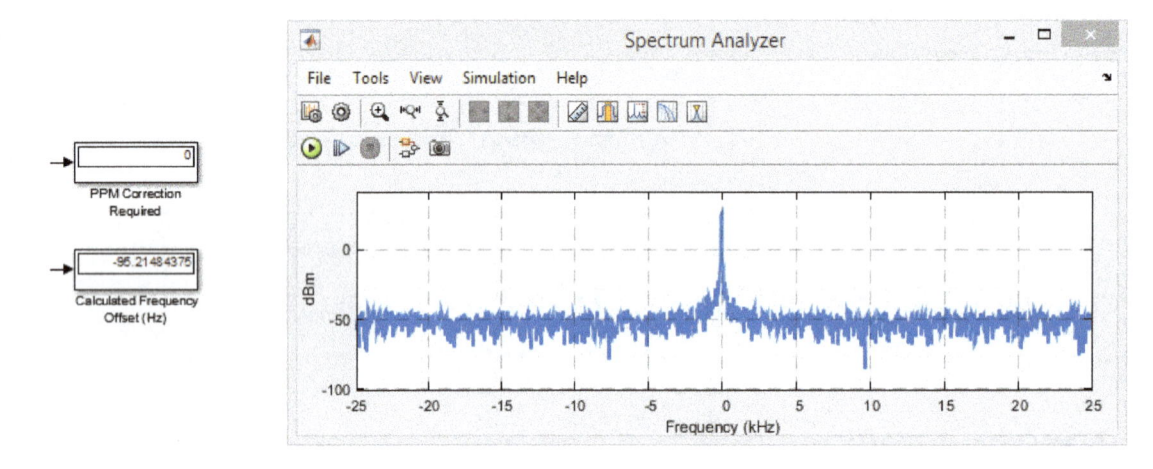

Lastly, make a note of the PPM value you determined during this Exercise — you will need it from here on! If you have more than one RTL-SDR, it might be worth labelling them with their respective correction values, for later use.

Appendix B: Common Equations

This Appendix lists some of the equations that are commonly used throughout this book.

Sum to Difference Trigonometric Rules:

$$\cos(u \pm v) = \cos(u)\cos(v) \mp \sin(u)\sin(v) \tag{B.1}$$

$$\sin(u \pm v) = \sin(u)\cos(v) \pm \cos(u)\sin(v) \tag{B.2}$$

$$\tan(u \pm v) = \frac{\tan(u) \pm \tan(v)}{1 \mp \tan(u)\tan(v)} \tag{B.3}$$

Product to Sum Trigonometric Rules:

$$\cos(u)\cos(v) = \frac{1}{2}\left[\cos(u-v) + \cos(u+v)\right] \tag{B.4}$$

$$\sin(u)\sin(v) = \frac{1}{2}\left[\cos(u-v) - \cos(u+v)\right] \tag{B.5}$$

$$\sin(u)\cos(v) = \frac{1}{2}\left[\sin(u+v) + \sin(u-v)\right] \tag{B.6}$$

$$\cos(u)\sin(v) = \frac{1}{2}\left[\sin(u+v) - \sin(u-v)\right] \tag{B.7}$$

Derivatives of Select Trigonometric Terms wrt. time

$$\frac{d}{dt}\sin\bigl(u(t)\bigr) = \cos\bigl(u(t)\bigr)\frac{d}{dt}u(t) \tag{B.8}$$

$$\frac{d}{dt}\cos\bigl(u(t)\bigr) = -\sin\bigl(u(t)\bigr)\frac{d}{dt}u(t) \tag{B.9}$$

Integrals of Select Trigonometric Terms

$$\int \sin(\omega t)\,dt = -\frac{1}{\omega}\cos(\omega t) + c \tag{B.10}$$

$$\int \cos(\omega t)\,dt = \frac{1}{\omega}\sin(\omega t) + c \tag{B.11}$$

Integral of a Sum:

$$\int u(t) + v(t)\,dt = \int u(t)\,dt + \int v(t)\,dt \tag{B.12}$$

Additional Trigonometric Rules

$$\cos^2(u) = \frac{1}{2}\big(1 + \cos(2u)\big) \tag{B.13}$$

$$\sin^2(u) = \frac{1}{2}\big(1 - \cos(2u)\big) \tag{B.14}$$

$$\sin(2u) = 2\sin(u)\cos(u) \tag{B.15}$$

Discrete Fourier Transform Pair:

$$X[k] = \sum_{n=0}^{N-1} x[n]\,W_N^{nk} \qquad k = 0, 1, 2, \ldots, N-1 \tag{B.16}$$

$$x[n] = \frac{1}{N}\sum_{k=0}^{N-1} X[k]\,W_N^{-nk} \qquad n = 0, 1, 2, \ldots, N-1 \tag{B.17}$$

Euler's Formula

$$e^{j\omega t} = \cos(\omega t) + j\sin(\omega t)\ ,\ \text{where}\ j = \sqrt{-1} \tag{B.18}$$

Complex Representations of Trigonometric Functions

$$\cos(\omega t) = \Re\!\left[e^{j\omega t}\right] = \frac{e^{j\omega t} + e^{-j\omega t}}{2} \tag{B.19}$$

$$\sin(\omega t) = \Im\!\left[e^{j\omega t}\right] = \frac{e^{j\omega t} - e^{-j\omega t}}{2j} \tag{B.20}$$

Bilinear Transform:

$$s \to \frac{2}{T}\left(\frac{1-z^{-1}}{1+z^{-1}}\right) \tag{B.21}$$

Appendix C: Digital Filtering and Multirate

Digital filters are used extensively throughout the book. This appendix briefly reviews aspects of filter design and implementation as necessary to follow this book — refer to any good DSP textbook for more!

C.1 Filter Classes and Characteristics

Filters are systems designed to change the frequency content of a signal. They are generally classified as having one of the following response types (as shown in Figure C.1):

- Lowpass
- Highpass
- Bandpass
- Bandstop

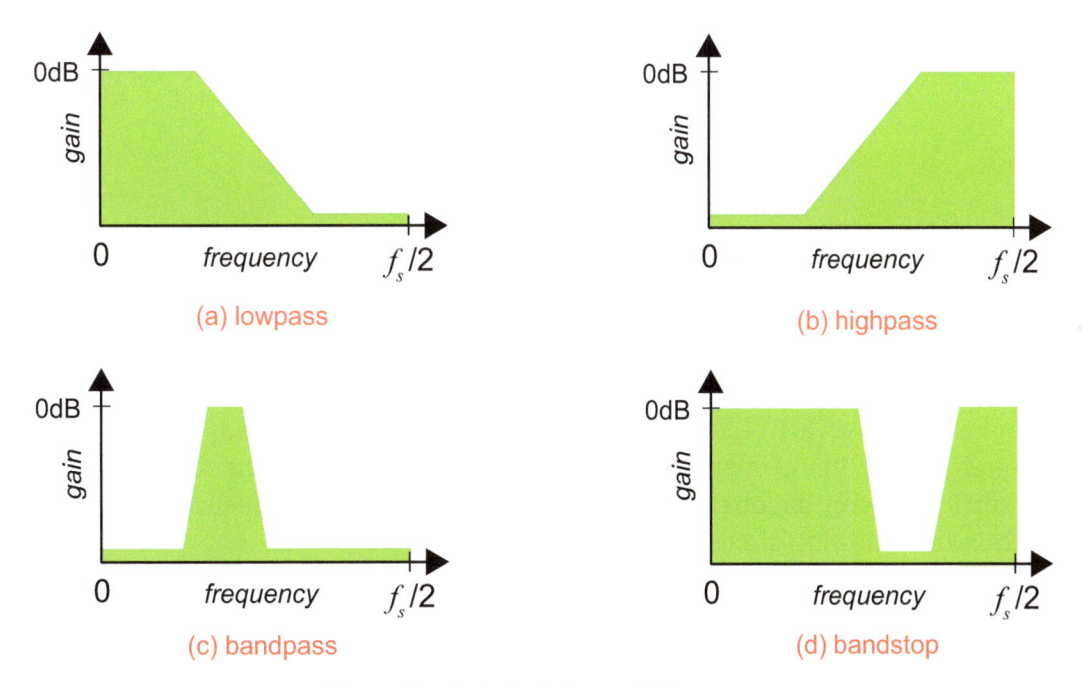

Figure C.1: Principal classes of filter response

Lowpass filters allow low frequencies to pass with approximately a 0dB gain (or a linear gain of 1), while attenuating higher frequencies by some specified degree. These regions are referred to as the 'passband' and 'stopband' respectively. **Highpass** filters are essentially the opposite, where the high frequencies are passed with ~0dB gain, and lower frequencies are attenuated. The band of frequencies between passband and stopband regions, which experiences a sliding amount of attenuation, is referred to as the transition band. Thus, **bandpass** and **bandstop** filters both have two transition bands.

The main features of a filter magnitude response are shown in Figure C.2, using the example of a lowpass filter. The engineer designing the filter can define the positions (edge frequencies) of the passband and stopband, and hence the width of the transition band. They can also specify the passband ripple and stopband attenuation parameters.

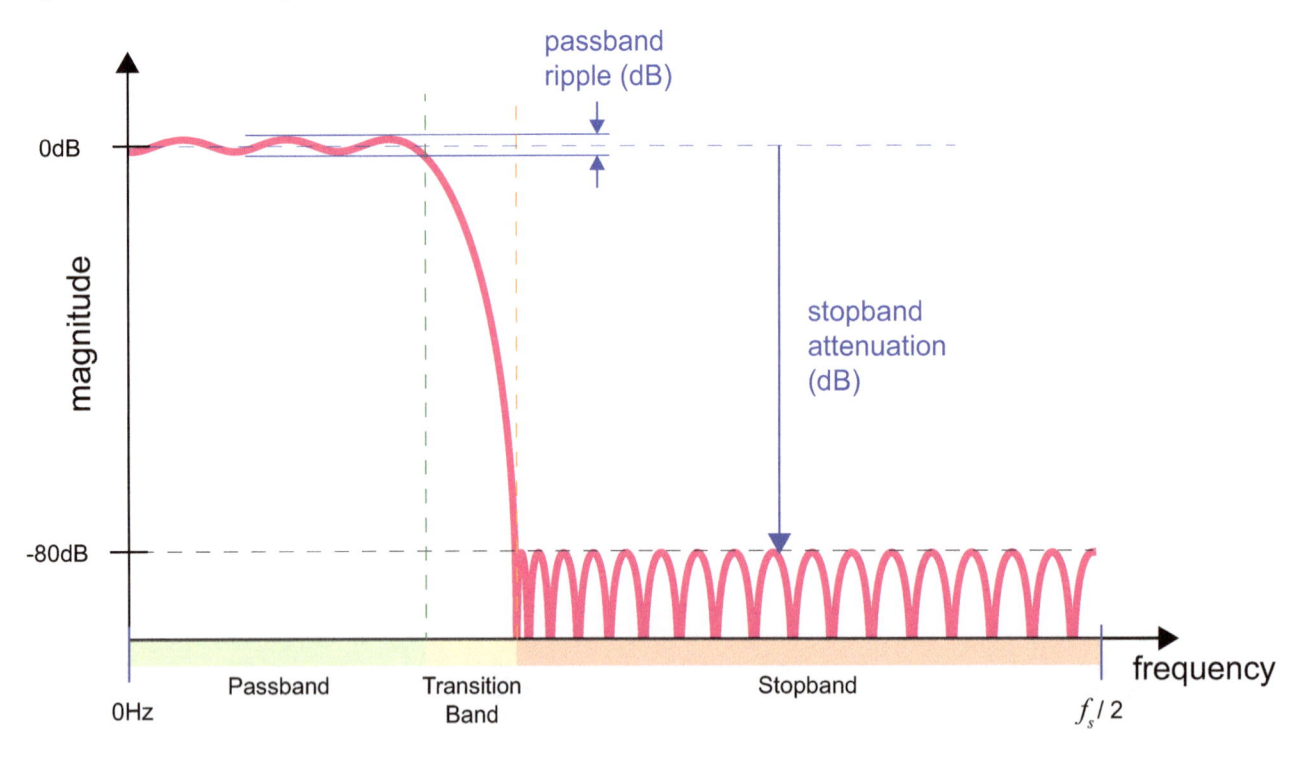

Figure C.2: Principal features of a filter magnitude response

C.2 Filter Specification and Design

To achieve any desired response, a set of filter weights are designed with the aid of a filter design algorithm. The most fundamental design choice is whether to adopt an Infinite Impulse Response (IIR) or a Finite Impulse Response (FIR) filter. In communications applications and many others, FIR filters are normally preferred as they offer *linear phase response* (provided the filter is symmetric), a characteristic which preserves the phase relationship between different frequency components present in the signal. Thus, the following discussion is based on FIRs.

MATLAB and Simulink provide tools for filter design, via functions and graphical user interfaces. It is possible to choose from a selection of design algorithms (e.g. *Window*, *Equiripple* (Parks-McLellan), *Least Squares*...) which will achieve slightly different magnitude responses in meeting the specification.

Appendix C: Digital Filtering and Multirate

The result of the design process will be to achieve a set of 'weights' (or 'coefficients') that can be used to compute a weighted sum of previous inputs,

$$y[k] = \sum_{n=0}^{N-1} w_n x[k-n] \tag{C.1}$$

where k is the sample index, w_n is the weight of index n, $x[k-n]$ is a previous input sample, and N is the total number of weights in the filter. The set of weights, W, can also be referred to as the 'impulse response' of the filter, because applying an impulse will cause the set of filter weights to be observed at the output. An example set of weight values (impulse response) is shown in Figure C.3 — notice the symmetry about the central weight.

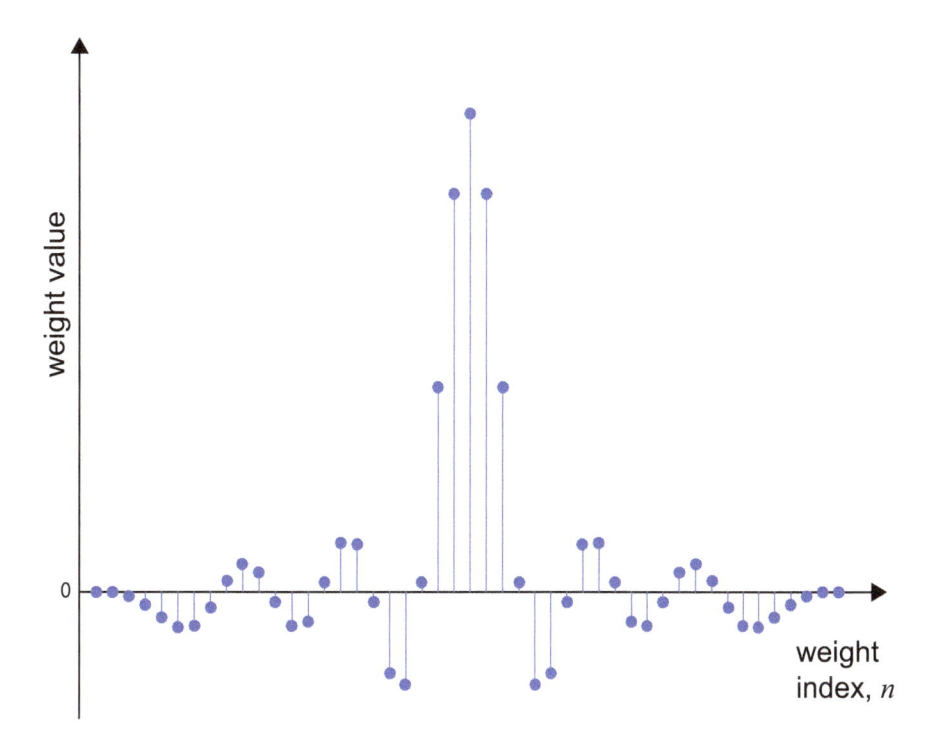

Figure C.3: An example set of filter weights

MATLAB permits filters to be designed by specifying different parameter sets, and the length of the filter can be constrained to a certain number of weights, if desired. It is also possible to specify the sampling frequency, and the edges of stopband(s) and passband(s), using a normalised scale of 0 to 1 (where 0 = 0Hz, and 1 = $f_s/2$), rather than using absolute frequencies.

The *Filter Design & Analysis Tool*, a convenient GUI for filter design, can be opened by typing `fdatool` at the MATLAB command prompt.

As a final note, it is worth highlighting that the number of filter weights is one greater than the filter 'order'. For instance, if prompted by FDATool to specify an FIR filter of fixed order, then entering an order of 40 would produce a filter with 41 (symmetric) weights.

C.3 FIR Filter Processing Architecture

The equation for an FIR filter, shown in Eq. (C.1), can be represented using a signal flow graph of the form shown in Figure C.4. Note that this shows a 5-weight filter, or equivalently, $N = 5$.

The cascade of sample delay elements, denoted by the z^{-1} components along the top of the filter, generates the required set of past inputs,

$$x[k-1], x[k-2], x[k-3]\ldots \quad \ldots x[k-(N-1)]. \tag{C.2}$$

The weight values (i.e. the set of values calculated by a filter design algorithm as described in Appendix C.2 and shown in Figure C.3), are denoted by the inputs to the multipliers, labelled $w_0, w_1, w_2 \ldots w_{N-1}$, etc. Where the weights are fixed, these multiplication operations could equivalently be represented using constant multipliers ('Gain' blocks in Simulink). The adder line at the bottom of the filter simply adds all of the *weight × delayed input sample* multiplications, and thus is equivalent to the sigma operator in Eq. (C.1).

The architecture shown in Figure C.4 is referred to as the *direct form* implementation, or *canonical* implementation, because it derives directly from the equation for an FIR, i.e. Eq. (C.1). Other forms exist (*transpose*, *systolic*, etc.) and these are often preferred for hardware implementation on ASICs and FPGAs, however the direct form FIR is entirely appropriate for the content of this book.

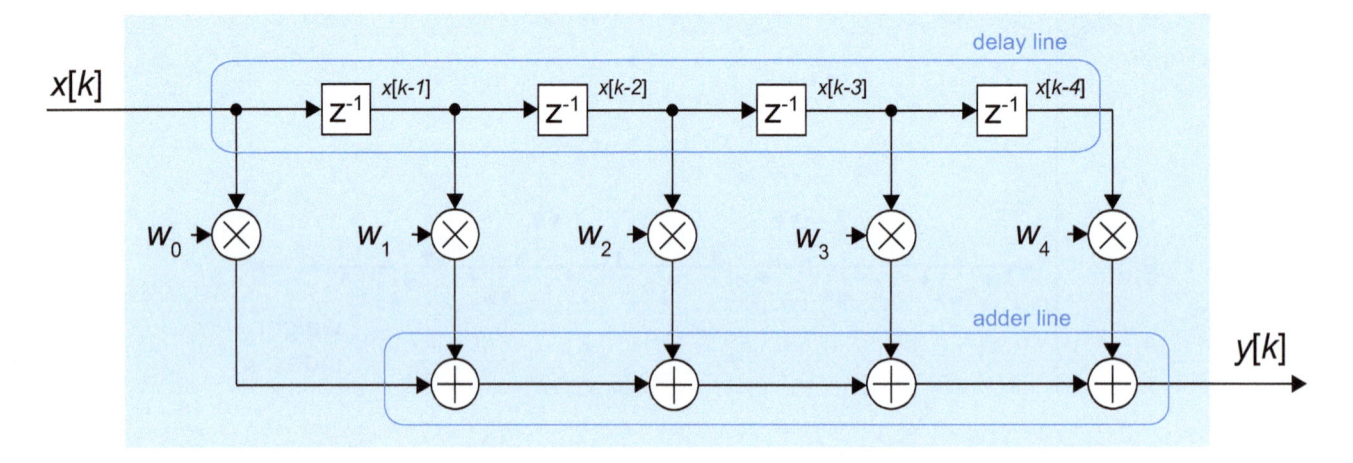

Figure C.4: Signal flow graph of the architecture of an FIR filter (direct form, 5 weights)

C.4 Computation and Trade-offs

Referring back to the FIR filter equation of Eq. (C.1) and the signal flow graph of Figure C.4, notice that this implies a total of N multiplications (weights multiplied with delayed input samples), and $N-1$ addition operations to form the sum. The larger the set of filter weights, the greater the amount of computation involved in filtering a signal.

The number of weights required to design a filter increases with:

- The number of transition bands (i.e. bandstop and bandpass filters have two transition bands, whereas lowpass and highpass filters have only one);

Appendix C: Digital Filtering and Multirate

- The level of passband ripple permitted (constraining the ripple to a very low level will require more weights, e.g. 0.1dB passband ripple with require a more expensive filter than 1dB);
- The stopband attenuation (greater attenuation will require more weights, e.g. 80dB attenuation will require more weights than an attenuation of 60dB);
- The width of the transition band, relative to the sampling rate (or the 'normalised' transition bandwidth). A narrower transition band will cause the number of required weights to increase. In fact, halving the transition bandwidth will roughly double the cost of the filter!

Certain techniques also exist to optimise the required computation, based on a given filter design, such as exploiting the symmetry of coefficients.

C.5 Multirate Filtering: The Motivation

The Nyquist sampling theorem states that a baseband signal only needs to be sampled at a rate more than twice its maximum frequency component, i.e.

$$f_s > 2f_{max}. \tag{C.3}$$

Provided this condition is met, all information present in the signal is captured.

Choosing a sampling rate in excess of that required by Nyquist will still of course be valid, but it is worth considering the potential drawbacks of such an approach. Two main points to highlight are:

- Processing samples at an unnecessarily high rate means the computational requirements are inflated. For instance, let us consider the operation of a 41-weight filter on a signal with bandwidth 5MHz. Neglecting any potential savings due to symmetry, the filter would require 41 Multiply-ACcumulate (MAC) operations to compute each result. If operated at a sampling rate of 100MHz, the filter would have to compute 41 x 100 million MACs per second. On the other hand, if the sampling rate was 10MHz (as required by Nyquist), it would instead compute only 41 x 10 million MACs per second — a reduction by a factor of 10.
- Any given response over a defined bandwidth can be achieved with fewer weights, if the sampling rate is lower. Note our last bullet point from Appendix C.4; the important point being that the transition bandwidth is *relative* to the sampling rate. Reducing the sampling rate to that defined by Nyquist permits a more relaxed filter to be used, which implies fewer weights, and thus, less computation to be performed. This can be confirmed by comparing Figure C.5, which shows two lowpass filters, each with a transition band extending from 2.5 - 3MHz, and identical specifications for passband ripple and stopband attenuation. Both are shown on a normalised frequency scale..

Putting both of these factors together, there is a clear motivation to reduce the sampling rate to match the Nyquist rate, when the initial sampling rate is much higher than the signal bandwidth. The computational requirements of performing any subsequent signal processing operations on the signal will be much lower.

On the other hand, sometimes it is necessary to increase the sampling rate of a signal, perhaps because it is going to interact with another signal at a higher sampling rate, or because a signal processing operation will cause the bandwidth of the signal to increase. An important motivating example of upwards rate transitions in SDR is when a baseband signal is modulated (digitally) to an IF frequency, prior to passing

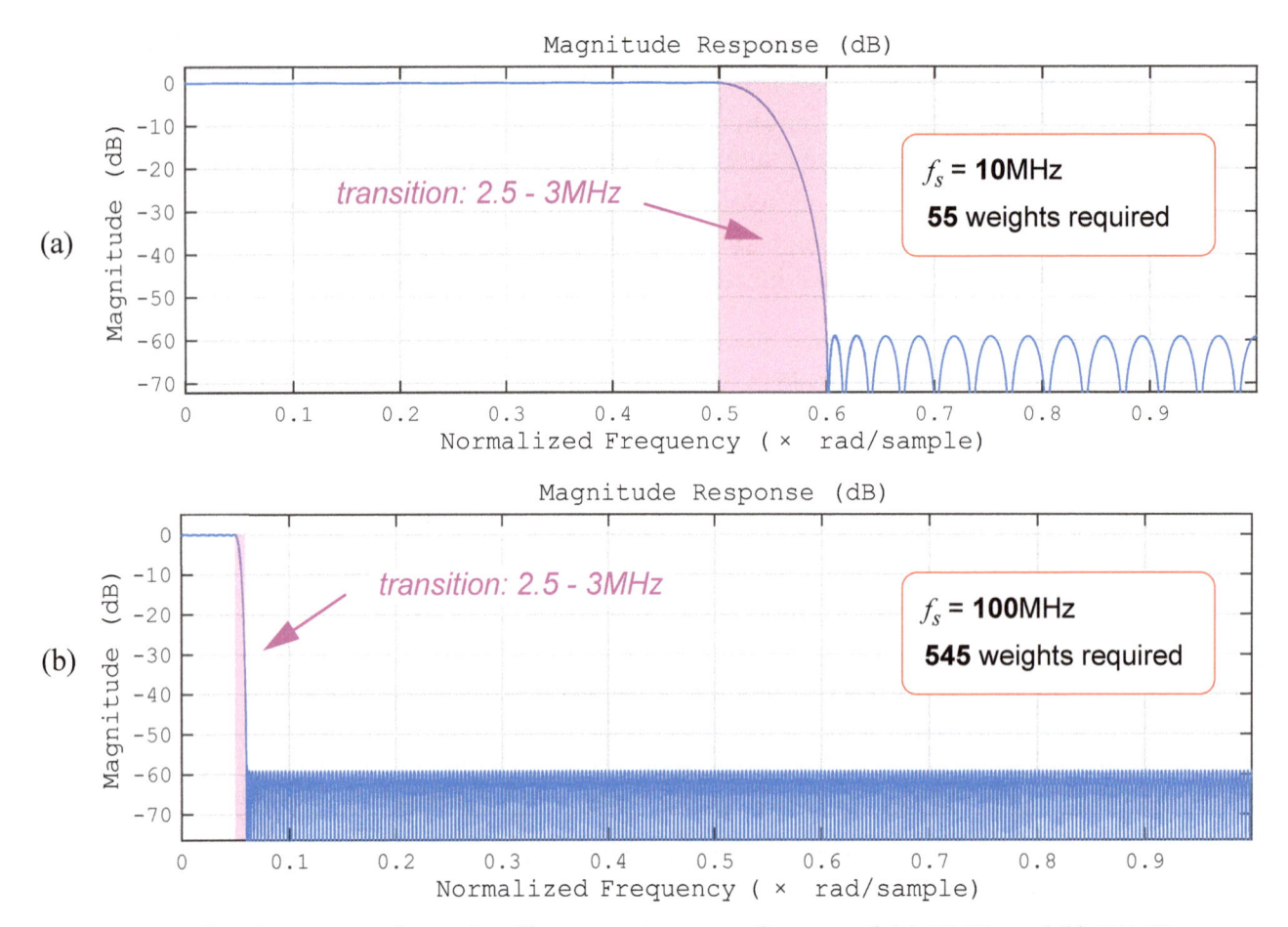

Figure C.5: Comparison of equivalent filters operating at sampling rates of: (a) 10MHz, and (b) 100MHz

the signal through a DAC. In this case, the sampling rate of the baseband signal must be increased to match the rate of the DAC, and the IF carrier signal generated at the same rate (with which the baseband signal is multiplied). A diagram illustrating this scenario can be found in Figure 11.10 on page 443.

There are many other circumstances in DSP when the bandwidth of a signal may change, and when it would be pertinent to consider adjusting the sampling rate accordingly. It is beyond the scope of our current discussion to draw out more examples, but the interested reader should be referred to [21] for a fuller discussion.

In order to actually transition between the different sampling rate required in a system, we consider the operations of *decimation* and *interpolation*. Both of these may also be referred to as *resampling* operations.

C.6 Decimation

Decimation is the process of reducing the sampling rate of a signal. In this case, we will consider only the straightforward case of decimating by an integer factor, for instance decimating a signal sampled at 80MHz by a factor of 4, to a new sampling rate of 20MHz. (Decimation by rational fractions, e.g. resampling by a ratio of 3/4 to 60MHz, is also possible but will not be considered here; neither will irrational fractions!)

Appendix C: Digital Filtering and Multirate

As will be shown next, decimation actually comprises two operations: (i) *downsampling*, the basic operation of discarding samples to change the sampling rate; and (ii) lowpass filtering, which takes place prior to downsampling, to ensure that the bandwidth of the signal is appropriately constrained.

The implication of downsampling by a factor of M is that only 1 out of every M original samples will be retained, while the other $M-1$ samples will be discarded. Thus, again for the example of $M = 4$, the result of downsampling would be that every fourth sample is retained. This example of downsampling can be seen in Figure 4.7 on page 153. In fact there are four different sets of samples that could be retained, depending on the phase of the downsampler (depicted with green, red, purple, and orange).

In the frequency domain, we must be careful that the new rate continues to fulfil the requirements of Nyquist sampling; in other words, that the new sampling rate is still in excess of twice the signal bandwidth. If this is not the case, then any signal components residing above half of the new sampling rate will be aliased (or 'folded') down into the bandwidth of interest, which will cause unwanted and irreversible distortion of the signal. (*Note: any general DSP textbook will cover sampling and aliasing, if further background information on these topics is needed.*)

To further explain the frequency domain implications, let's consider the frequency spectra in Figure C.6, where the sampling rate for two example signals (denoted A and B) has been reduced by a factor of 4 from f_s to f_d. As the signal derives from a sampled system, then spectral repetitions occur at integer multiples of the sampling rate. When sampled at the original rate, the spectra are located far apart and do not overlap. After downsampling, however, overlapping does occur for Signal B, because the signal bandwidth exceeds half of the new sampling rate, i.e. it contains frequencies above $f_d/2$. This corresponds to aliasing and is undesirable — the aliased components are superimposed onto the original signal. On the other hand, Signal A does not contain any frequencies above $f_d/2$ and so aliasing does not occur.

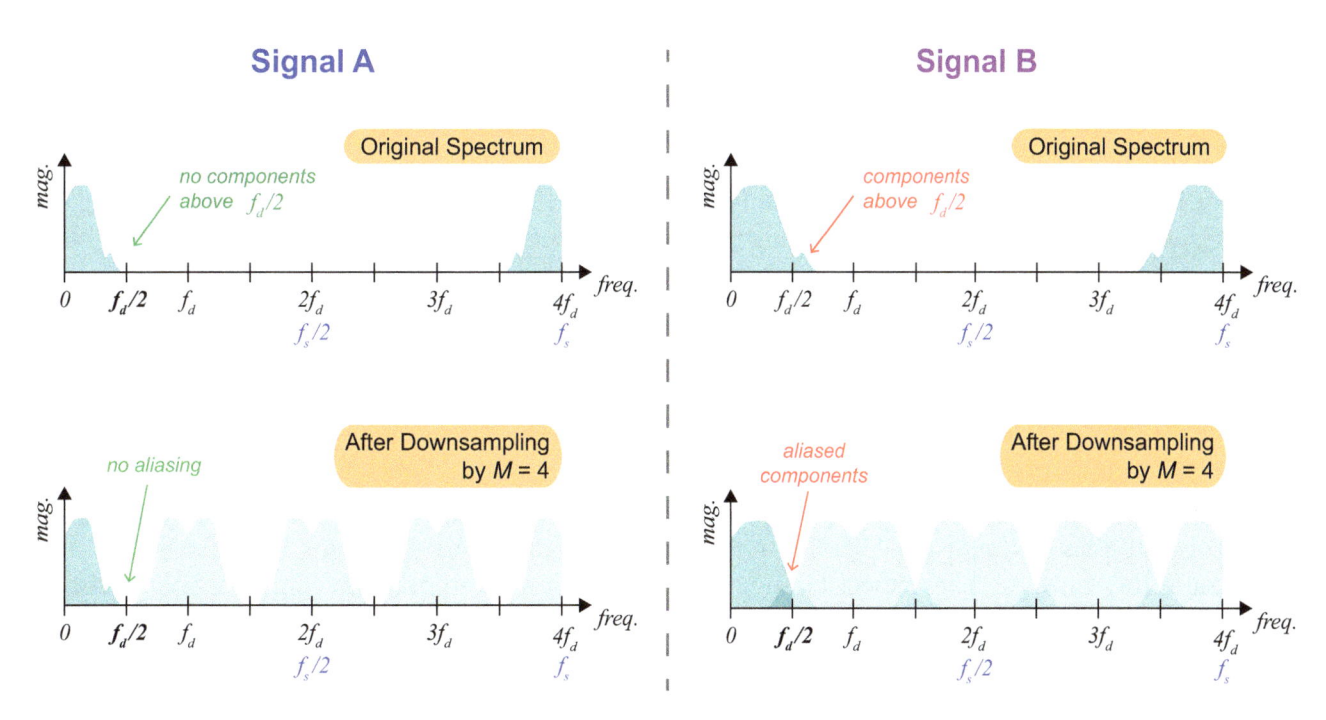

Figure C.6: Downsampling two signals (A and B) by a factor of M = 4: (left) - without aliasing; (right) - with aliasing

The simple solution to the potential problem of aliasing is to ensure that the signal is appropriately bandlimited, prior to downsampling: if no components exist at frequencies above half of the new sampling rate, then aliasing cannot happen. Bandlimiting of the signal is achieved by applying an 'anti-alias' lowpass filter, cutting off at $f_d/2$ (equivalent to $f_s/2M$). The result of this combined process is a signal at the new, lower rate, which does not suffer any aliasing effects.

The process of decimation can therefore be summarised by the block diagram shown in Figure C.7.

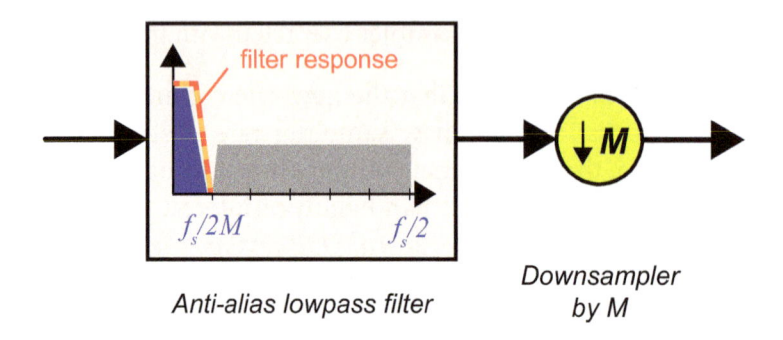

Figure C.7: Signal flow graph of a decimator: anti-alias lowpass filter, followed by downsampler

As a final comment, the reader may note from Figure C.7 that filtering takes place prior to the downsampling operation, which implies that $M-1$ out of every M output samples computed by the filter are immediately thrown away by the downsampler — a very wasteful situation! *Polyphase* techniques are normally applied when implementing decimation filters to reorder the computation, such that a mathematically identical filtering operation can be undertaken in a more efficient manner. Polyphase filters are not within the scope of this book, but are well-covered in multirate DSP books such as [21] and [44].

C.7 Interpolation

Interpolation is the opposite multirate operation to decimation, and involves raising the sampling rate of a signal. As for the decimation case, it is possible to interpolate by integer values, rational fractions and even irrational fractions, but here we will only consider the first of these, i.e. integer rate changes. For example, we might wish to raise the sampling rate of a signal with $f_s = 10\text{MHz}$, by a factor of $L = 4$, to a new sampling rate of $f_u = 40\text{MHz}$.

Increasing the sampling rate involves adding new samples between the existing ones, e.g. for every 1 sample in the original signal, 3 new samples must be inserted (in our example where $L = 4$). This is achieved by upsampling (or 'expanding') the signal by inserting zero-valued samples between the original samples. An example is shown in Figure 4.6 on page 152.

In the frequency domain, the effect of raising the sampling rate by upsampling is that the spectral images at integer multiples of the original sampling rate, f_s, now fall within the range 'visible' with the new, higher sampling rate, f_u. These spectral images must be removed, in order to achieve a signal with an equivalent spectrum to the original (in the time domain, this will be observed as a 'smoothing out' of the waveform). A suitable lowpass filter will cut off at approximately $f_s/2$ to remove all unwanted spectral images. This set of operations is depicted in Figure C.8.

Chapter :

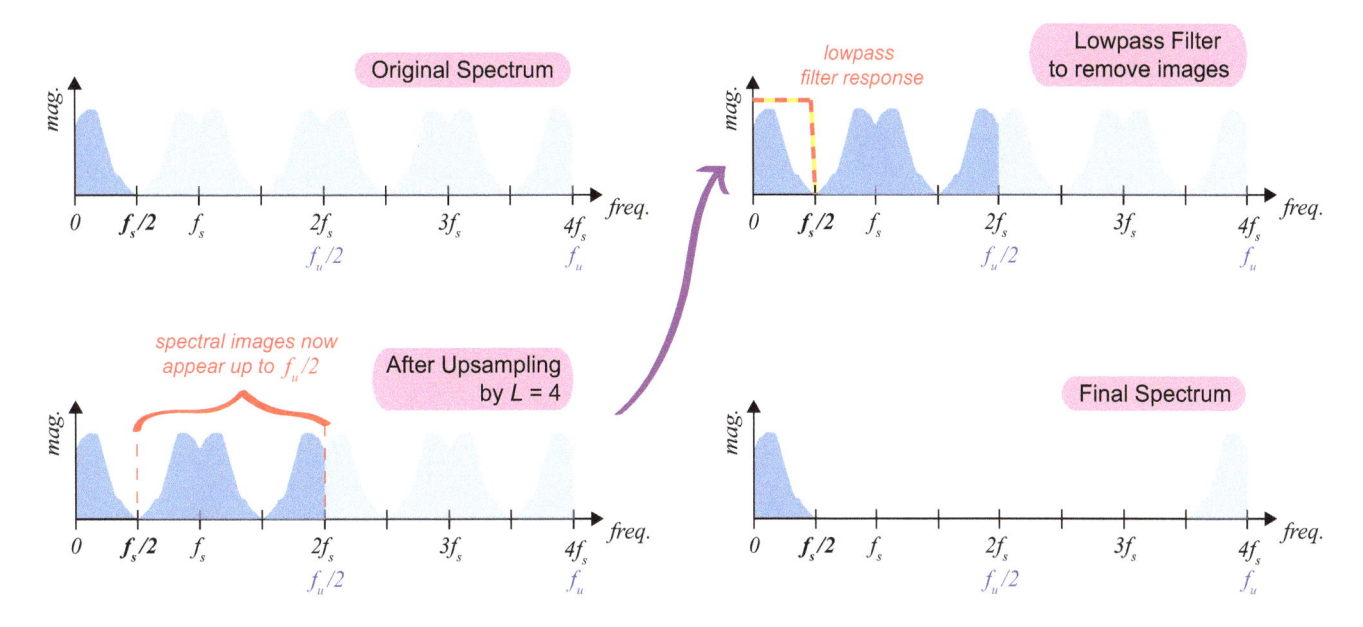

Figure C.8: Upsampling and image rejection filtering in the frequency domain

We can interpret therefore that an interpolator comprises an upsampler followed by a lowpass filter, as shown in Figure C.9.

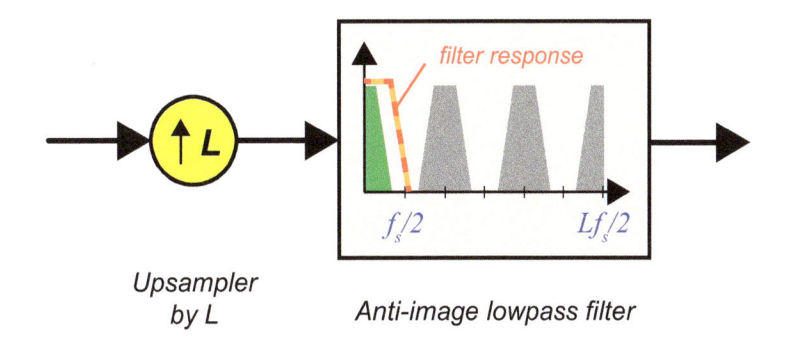

Figure C.9: Signal flow graph of an interpolator: upsampler, followed by a lowpass image rejection filter

Similar to the decimation case, it can be recognised that interpolation is an inefficient operation. In this case, it is because the upsampler inserts zero-valued samples which are then processed by the filter. A polyphase version of the interpolator can be obtained which avoids this redundant computation and is thus more efficient. For more information on polyphase techniques, the reader is again directed to one of the excellent books on the subject, e.g. [21] and [44].

Appendix D: PLL Design

This Appendix provides expanded detail on the design of Type 2 PLLs, following on from Section 7.8. The models and analysis presented in this section cover some of the important points from Appendix C of [38]. The interested reader is referred to that book for further details.

The next few subsections will:

- Derive the Z-domain transfer function for the digital Type 2 PLL.
- Derive the S-domain transfer function for the corresponding analogue Type 2 PLL.
- Through the equivalence of these two models (under the condition that the sampling rate is much higher than the signal bandwidth) form expressions linking the loop constants, bandwidth, and damping ratio.
- Review factors affecting the phase detector and NCO gains (K_p and K_o).

D.1 Digital Type 2 PLL Linear Model and Z-Domain Transfer Function

The PLL model that will be analysed is a Type 2, with a loop filter composed of a proportional and an integral path as shown in Figure D.1. The Type 2 PLL derives from the generic model presented in Figure 7.15 on page 253.

Digital PLLs are normally analysed in terms of the equivalent analogue model, and its parameters of damping ratio, ζ, and natural frequency ω_n. The aim in this section is to develop a Z-domain transfer function for the digital PLL that can be equated to the S-domain model of the analogue PLL.

The Z-domain transfer function of the PLL is given by

$$H(z) = \frac{\Theta(z)}{\Theta(z)} \qquad (D.1)$$

Noting that the loop filter transfer function is given by

$$F(z) = K_1 + K_2 \cdot \frac{1}{1 - z^{-1}} \qquad (D.2)$$

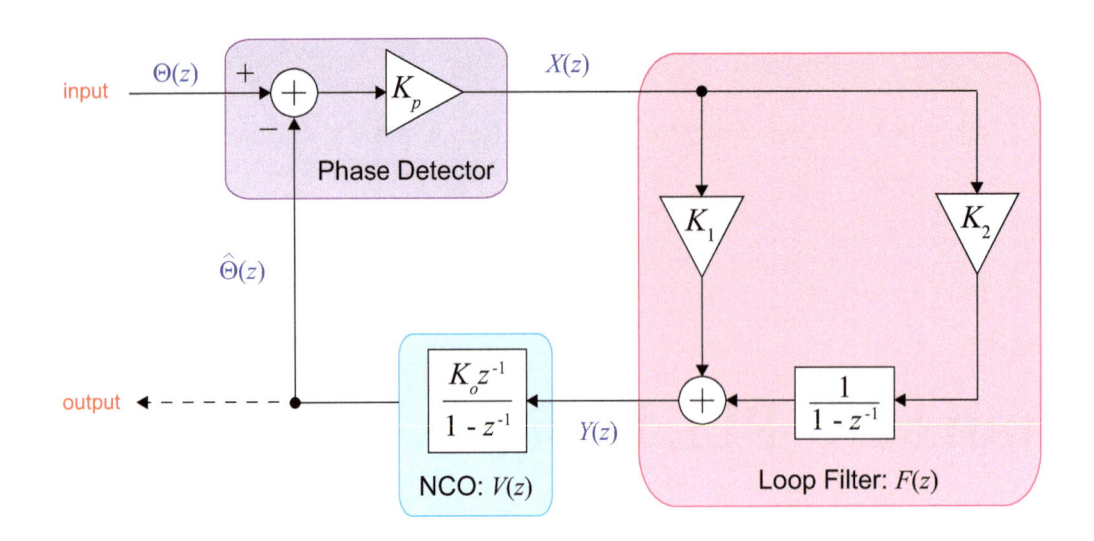

Figure D.1: Z-domain Model of the Digital Type 2 PLL

and the transfer function of the NCO is

$$V(z) = K_o \frac{z^{-1}}{1-z^{-1}}, \tag{D.3}$$

then $\hat{\Theta}(z)$ can be expressed as

$$\hat{\Theta}(z) = K_p \left[\Theta(z) - \hat{\Theta}(z) \right] \left[K_1 + K_2 \cdot \frac{1}{1-z^{-1}} \right] \left[K_o \cdot \frac{z^{-1}}{1-z^{-1}} \right]. \tag{D.4}$$

It is now possible to derive an expression for $H(z)$ in terms of the loop coefficients K_o, K_1, K_2, and K_p. First, we obtain the following by multiplying out the terms in (D.4),

$$\hat{\Theta}(z) = \left[K_p K_o K_1 \cdot \frac{z^{-1}}{1-z^{-1}} + K_p K_o K_2 \cdot \frac{z^{-1}}{1-z^{-1}} \cdot \frac{1}{1-z^{-1}} \right] \left[\Theta(z) - \hat{\Theta}(z) \right] \tag{D.5}$$

and then, making the substitution

$$\alpha = \left[K_p K_o K_1 \cdot \frac{z^{-1}}{1-z^{-1}} + K_p K_o K_2 \cdot \frac{z^{-1}}{1-z^{-1}} \cdot \frac{1}{1-z^{-1}} \right] \tag{D.6}$$

Appendix D: PLL Design

an expression for the transfer function in terms of α can be generated in a few simple steps,

$$\hat{\Theta}(z) = \alpha\left[\Theta(z) - \hat{\Theta}(z)\right] \tag{D.7}$$

$$\hat{\Theta}(z) + \alpha\hat{\Theta}(z) = \alpha\Theta(z) \tag{D.8}$$

$$\hat{\Theta}(z)(1 + \alpha) = \alpha\Theta(z) \tag{D.9}$$

giving a simple expression for $H(z)$ in terms of α,

$$H(z) = \frac{\hat{\Theta}(z)}{\Theta(z)} = \frac{\alpha}{1+\alpha}. \tag{D.10}$$

Re-substituting for α from Eq. (D.6), we now have

$$H(z) = \frac{\left[K_pK_oK_1 \cdot \frac{z^{-1}}{1-z^{-1}} + K_pK_oK_2 \cdot \frac{z^{-1}}{1-z^{-1}} \cdot \frac{1}{1-z^{-1}}\right]}{1 + \left[K_pK_oK_1 \cdot \frac{z^{-1}}{1-z^{-1}} + K_pK_oK_2 \cdot \frac{z^{-1}}{1-z^{-1}} \cdot \frac{1}{1-z^{-1}}\right]} \tag{D.11}$$

and this expression can then be simplified by multiplying the right hand side by $\dfrac{(1-z^{-1})^2}{(1-z^{-1})^2}$, to produce

$$H(z) = \frac{K_pK_oK_1z^{-1}(1-z^{-1}) + K_pK_oK_2z^{-1}}{(1-z^{-1})^2 + K_pK_oK_1z^{-1}(1-z^{-1}) + K_pK_oK_2z^{-1}}. \tag{D.12}$$

Then, after a little further rearranging...

$$H(z) = \frac{K_pK_o(K_1+K_2)z^{-1} - K_pK_oK_1z^{-2}}{1 - 2z^{-1} + z^{-2} + K_pK_o(K_1+K_2)z^{-1} - K_pK_oK_1z^{-2}} \tag{D.13}$$

$$H(z) = \frac{K_pK_o(K_1+K_2)z^{-1} - K_pK_oK_1z^{-2}}{1 - 2\left(1 - \frac{1}{2}K_pK_o(K_1+K_2)\right)z^{-1} + (1 - K_pK_oK_1)z^{-2}}. \tag{D.14}$$

Eq. (D.14) forms the final expression for the digital PLL transfer function. This expression will later be equated to the transfer function of the equivalent analogue PLL, in order to derive expressions for the loop constants in terms of damping ratio and bandwidth. First, the S-domain transfer function will be developed in the next section.

D.2 Analogue Type 2 PLL Linear Model and S-Domain Transfer Function

Next, we consider the analogue PLL model shown in Figure D.2, and its S-domain transfer function. The similarity between this analogue PLL model, and the Z-domain model of the digital PLL presented in Figure D.1, should be clear.

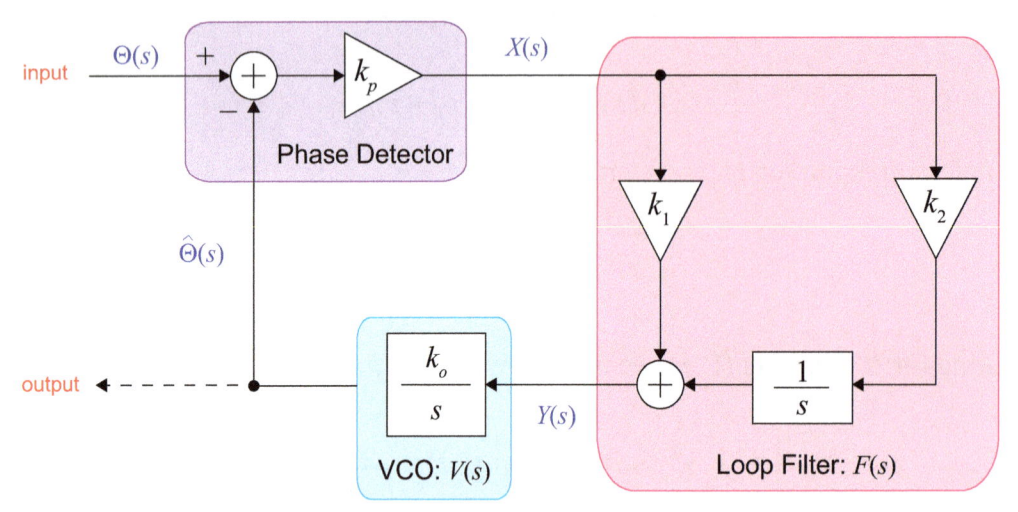

Figure D.2: S-domain Model of the Analogue Type 2 PLL

Using the same rationale, we can write the transfer function as

$$H(s) = \frac{\hat{\Theta}(s)}{\Theta(s)}, \tag{D.15}$$

and develop an expression for the output,

$$\hat{\Theta}(s) = k_p \left[\Theta(s) - \hat{\Theta}(s)\right]\left[k_1 + \frac{k_2}{s}\right]\left[\frac{k_o}{s}\right], \tag{D.16}$$

where the transfer function of the loop filter is

$$F(s) = k_1 + \frac{k_2}{s} \tag{D.17}$$

and the transfer function of the VCO is given by

$$V(s) = \frac{k_o}{s}. \tag{D.18}$$

Making the substitution

$$\beta = k_p\left[k_1 + \frac{k_2}{s}\right]\frac{k_o}{s} \tag{D.19}$$

Appendix D: PLL Design

and applying the equivalent steps to Eqs. (D.7) to (D.10), we arrive at

$$H(s) = \frac{\hat{\Theta}(s)}{\Theta(s)} = \frac{\beta}{1+\beta}. \tag{D.20}$$

It is then possible to substitute for β to form the expression

$$\frac{\hat{\Theta}(s)}{\Theta(s)} = \frac{k_p\left[k_1 + \frac{k_2}{s}\right]\frac{k_o}{s}}{1 + k_p\left[k_1 + \frac{k_2}{s}\right]\frac{k_o}{s}} \tag{D.21}$$

which can be expanded to

$$\frac{\hat{\Theta}(s)}{\Theta(s)} = \frac{\frac{k_p k_1 k_o}{s} + \frac{k_p k_2 k_o}{s^2}}{1 + \frac{k_p k_1 k_o}{s} + \frac{k_p k_2 k_o}{s^2}}. \tag{D.22}$$

Multiplying through by $\frac{s^2}{s^2}$ then produces

$$\frac{\hat{\Theta}(s)}{\Theta(s)} = \frac{k_p k_1 k_o s + k_p k_2 k_o}{s^2 + k_p k_1 k_o s + k_p k_2 k_o}, \tag{D.23}$$

following which the substitutions $2\zeta\omega_n = k_o k_p k_1$ and $\omega_n^2 = k_o k_p k_2$ can be made to produce the well known PLL transfer function,

$$\frac{\hat{\Theta}(s)}{\Theta(s)} = \frac{2\zeta\omega_n s + \omega_n^2}{s^2 + 2\zeta\omega_n s + \omega_n^2}. \tag{D.24}$$

D.3 Extraction of Digital PLL Parameters Based on Analogue PLL Equivalence

The next stage of analysis is to form a discrete time version of the analogue PLL transfer function, and to equate the two expressions (Eqs. (D.14) and D.24). This can be done by applying the bilinear transform,

$$s \to \frac{2}{T}\frac{1 - z^{-1}}{1 + z^{-1}}, \tag{D.25}$$

to the transfer function for the analogue model, given in Eq. (D.24).

The outcome of this mathematical manipulation will be that the digital loop parameters can be expressed in terms of the sampling period, T, and the 'classic' PLL parameters of damping ratio, ζ, and natural frequency, ω_n.

Applying the bilinear transform to Eq. (D.24) produces

$$H\left(\frac{2}{T}\frac{1-z^{-1}}{1+z^{-1}}\right) = \frac{2\zeta\omega_n\left(\frac{2}{T}\frac{1-z^{-1}}{1+z^{-1}}\right) + \omega_n^2}{\left(\frac{2}{T}\frac{1-z^{-1}}{1+z^{-1}}\right)^2 + 2\zeta\omega_n\left(\frac{2}{T}\frac{1-z^{-1}}{1+z^{-1}}\right) + \omega_n^2} \qquad (D.26)$$

The expression given in Eq. (D.26) can then be simplified by multiplying through by $\dfrac{(1+z^{-1})^2}{(1+z^{-1})^2}$, to produce

$$H\left(\frac{2}{T}\frac{1-z^{-1}}{1+z^{-1}}\right) = \frac{2\zeta\omega_n\left(\frac{2}{T}(1-z^{-1})(1+z^{-1})\right) + (1+z^{-1})^2\omega_n^2}{\left(\frac{4}{T^2}\cdot(1-z^{-1})^2\right) + 2\zeta\omega_n\left(\frac{2}{T}(1-z^{-1})(1+z^{-1})\right) + (1+z^{-1})^2\omega_n^2}. \qquad (D.27)$$

After expansion, the equation becomes

$$H\left(\frac{2}{T}\frac{1-z^{-1}}{1+z^{-1}}\right) = \frac{\frac{4}{T}\zeta\omega_n(1-z^{-2}) + (1+2z^{-1}+z^{-2})\omega_n^2}{\frac{4}{T^2}(1-2z^{-1}+z^{-2}) + \frac{4}{T}\zeta\omega_n(1-z^{-2}) + (1+2z^{-1}+z^{-2})\omega_n^2} \qquad (D.28)$$

and this can be rearranged to give

$$H\left(\frac{2}{T}\frac{1-z^{-1}}{1+z^{-1}}\right) = \frac{\frac{4}{T}\zeta\omega_n + \omega_n^2 + 2\omega_n^2 z^{-1} + \left(\omega_n^2 - \frac{4}{T}\zeta\omega_n\right)z^{-2}}{\frac{4}{T^2} + \frac{4}{T}\zeta\omega_n + \omega_n^2 + \left(2\omega_n^2 - \frac{8}{T^2}\right)z^{-1} + \left(\frac{4}{T^2} - \frac{4}{T}\zeta\omega_n + \omega_n^2\right)z^{-2}}. \qquad (D.29)$$

Multiplying the right hand side through by $\dfrac{T^2}{T^2}$ provides

$$H\left(\frac{2}{T}\frac{1-z^{-1}}{1+z^{-1}}\right) = \frac{4T\zeta\omega_n + T^2\omega_n^2 + 2T^2\omega_n^2 z^{-1} + (T^2\omega_n^2 - 4T\zeta\omega_n)z^{-2}}{4 + 4T\zeta\omega_n + T^2\omega_n^2 + (2T^2\omega_n^2 - 8)z^{-1} + (4 - 4T\zeta\omega_n + T^2\omega_n^2)z^{-2}}. \qquad (D.30)$$

At this point a substitution can be made where

$$\theta_n = \frac{\omega_n T}{2}, \qquad (D.31)$$

Appendix D: PLL Design

resulting in the equation

$$H\left(\frac{2}{T}\frac{1-z^{-1}}{1+z^{-1}}\right) = \frac{8\zeta\theta_n + 4\theta_n^2 + 8\theta_n^2 z^{-1} + (4\theta_n^2 - 8\zeta\theta_n)z^{-2}}{4 + 8\zeta\theta_n + 4\theta_n^2 + \left(8\theta_n^2 - 8\right)z^{-1} + (4 - 8\zeta\theta_n + 4\theta_n^2)z^{-2}} \quad \text{(D.32)}$$

which can be further simplified by removing the common factor of 4, to give

$$H\left(\frac{2}{T}\frac{1-z^{-1}}{1+z^{-1}}\right) = \frac{2\zeta\theta_n + \theta_n^2 + 2\theta_n^2 z^{-1} + (\theta_n^2 - 2\zeta\theta_n)z^{-2}}{1 + 2\zeta\theta_n + \theta_n^2 + \left(2\theta_n^2 - 2\right)z^{-1} + \left(1 - 2\zeta\theta_n + \theta_n^2\right)z^{-2}} . \quad \text{(D.33)}$$

In order to reduce the first term of the denominator to 1, we now divide through by $\frac{1 + 2\zeta\theta_n + \theta_n^2}{1 + 2\zeta\theta_n + \theta_n^2}$, which results in

$$H\left(\frac{2}{T}\frac{1-z^{-1}}{1+z^{-1}}\right) = \frac{\frac{2\zeta\theta_n + \theta_n^2}{1 + 2\zeta\theta_n + \theta_n^2} + \frac{2\theta_n^2 z^{-1}}{1 + 2\zeta\theta_n + \theta_n^2} + \frac{(\theta_n^2 - 2\zeta\theta_n)z^{-2}}{1 + 2\zeta\theta_n + \theta_n^2}}{1 - \frac{2\left(-\theta_n^2 + 1\right)z^{-1}}{1 + 2\zeta\theta_n + \theta_n^2} + \frac{(1 - 2\zeta\theta_n + \theta_n^2)z^{-2}}{1 + 2\zeta\theta_n + \theta_n^2}} \quad \text{(D.34)}$$

At this point we can equate the transfer function of the digital model from Eq. (D.14) with this discrete-time version of the analogue transfer function, Eq. (D.34), i.e.

$$H(z) = H\left(\frac{2}{T}\frac{1-z^{-1}}{1+z^{-1}}\right) \quad \text{(D.35)}$$

In particular, equating the denominators provides

$$1 - 2\left(1 - \frac{1}{2}K_p K_o(K_1 + K_2)\right)z^{-1} + (1 - K_p K_o K_1)z^{-2} = 1 - \frac{2(-\theta_n^2 + 1)z^{-1}}{1 + 2\zeta\theta_n + \theta_n^2} + \frac{(1 - 2\zeta\theta_n + \theta_n^2)z^{-2}}{1 + 2\zeta\theta_n + \theta_n^2} \quad \text{(D.36)}$$

and by comparing the z^{-1} and z^{-2} terms we can say that

$$1 - \frac{1}{2}K_p K_o(K_1 + K_2) = \frac{(-\theta_n^2 + 1)}{1 + 2\zeta\theta_n + \theta_n^2} \quad \text{(D.37)}$$

and

$$1 - K_p K_o K_1 = \frac{1 - 2\zeta\theta_n + \theta_n^2}{1 + 2\zeta\theta_n + \theta_n^2} \ . \tag{D.38}$$

It is possible to solve for $K_p K_o K_1$ from Eq. (D.38) in the following steps:

$$K_p K_o K_1 = 1 - \frac{1 - 2\zeta\theta_n + \theta_n^2}{1 + 2\zeta\theta_n + \theta_n^2} \tag{D.39}$$

$$K_p K_o K_1 = \frac{1 + 2\zeta\theta_n + \theta_n^2}{1 + 2\zeta\theta_n + \theta_n^2} - \frac{1 - 2\zeta\theta_n + \theta_n^2}{1 + 2\zeta\theta_n + \theta_n^2} \tag{D.40}$$

$$K_p K_o K_1 = \frac{4\zeta\theta_n}{1 + 2\zeta\theta_n + \theta_n^2} \ . \tag{D.41}$$

Similarly, $K_p K_o K_2$ can be obtained from Eq. (D.37), making use of the expression for $K_p K_o K_1$ derived in Eq. (D.41).

We begin by rewriting Eq. (D.37) as

$$1 - \frac{K_p K_o K_1}{2} - \frac{K_p K_o K_2}{2} = \frac{-\theta_n^2 + 1}{1 + 2\zeta\theta_n + \theta_n^2} \tag{D.42}$$

and then multiplying through by 2 to produce,

$$2 - K_p K_o K_1 - K_p K_o K_2 = \frac{-2\theta_n^2 + 2}{1 + 2\zeta\theta_n + \theta_n^2} \tag{D.43}$$

and rearranging as follows

$$K_p K_o K_2 = 2 - K_p K_o K_1 + \frac{2\theta_n^2 - 2}{1 + 2\zeta\theta_n + \theta_n^2} \ . \tag{D.44}$$

A substitution can then be made for $K_p K_o K_1$ as given in Eq. (D.41), resulting in

$$K_p K_o K_2 = 2 - \frac{4\zeta\theta_n}{1 + 2\zeta\theta_n + \theta_n^2} + \frac{2\theta_n^2 - 2}{1 + 2\zeta\theta_n + \theta_n^2} \tag{D.45}$$

Appendix D: PLL Design

which can be simplified as shown to provide the expression given in Eq. (D.47).

$$K_p K_o K_2 = \frac{2 + 4\zeta\theta_n + 2\theta_n^2}{1 + 2\zeta\theta_n + \theta_n^2} - \frac{4\zeta\theta_n}{1 + 2\zeta\theta_n + \theta_n^2} + \frac{2\theta_n^2 - 2}{1 + 2\zeta\theta_n + \theta_n^2} \tag{D.46}$$

$$K_p K_o K_2 = \frac{4\theta_n^2}{1 + 2\zeta\theta_n + \theta_n^2} \tag{D.47}$$

At this point, we have obtained expressions for the constants used within the loop. These are given in terms of the damping ratio, ζ, and θ_n, which links to the natural frequency ω_n, and sampling period, T, as given in Eq. (D.31).

For a loop filter with proportional and integral paths, the noise bandwidth of the loop, B_n, can be expressed as

$$B_n = \frac{\omega_n}{2}\left(\zeta + \frac{1}{4\zeta}\right). \tag{D.48}$$

A full derivation of loop bandwidth may be obtained from [13], [38], for this and other types of loop filter. For the purposes of this discussion, we simply note the expression of Eq. (D.48) and, recalling Eq. (D.31), rearrange to give

$$\theta_n = \frac{B_n T}{\zeta + \frac{1}{4\zeta}}. \tag{D.49}$$

Eq. (D.49) can thereafter be substituted into Eqs. (D.41) and D.47, to find $K_p K_o K_1$ and $K_p K_o K_2$ respectively. The resulting equations are:

$$K_o K_p K_1 = \frac{4\zeta\left(\frac{B_n T}{\zeta + \frac{1}{4\zeta}}\right)}{1 + 2\zeta\left(\frac{B_n T}{\zeta + \frac{1}{4\zeta}}\right) + \left(\frac{B_n T}{\zeta + \frac{1}{4\zeta}}\right)^2} \tag{D.50}$$

$$K_o K_p K_2 = \frac{4\left(\frac{B_n T}{\zeta + \frac{1}{4\zeta}}\right)^2}{1 + 2\zeta\left(\frac{B_n T}{\zeta + \frac{1}{4\zeta}}\right) + \left(\frac{B_n T}{\zeta + \frac{1}{4\zeta}}\right)^2}. \tag{D.51}$$

In digital communications, the bandwidth can also be specified relative to the symbol rate, where T_s is the symbol period and R represents the oversampling ratio, i.e.

$$R = \frac{T_s}{T}, \qquad (D.52)$$

giving an alternative expression to Eq. (D.49),

$$\theta_n = \frac{B_n T_s}{R\left(\zeta + \frac{1}{4\zeta}\right)} \qquad (D.53)$$

which may be used to produce corresponding versions of Eq. (D.50) and Eq. (D.51).

This means that we can now design the two sets of loop constants, $K_p K_o K_1$ and $K_p K_o K_2$, based on a choice of:

- Bandwidth, B_n,
- Sampling rate, T, and
- Damping ratio, ζ.

The constants K_p and K_o relate to the phase detector and oscillator, respectively, and occur as a pair. Therefore, the loop filter constants K_1 and K_2 can be calculated for specific values of K_p and K_o.

The next point to address is how the values of constants K_p and K_o are chosen or defined.

D.4 Phase Detector Gain

The previous section reviewed the mathematical expressions describing digital Type 2 PLLs. Two of the parameters included in these expressions relate to particular components of the PLL:

- K_p — the phase detector gain
- K_o — the oscillator gain

It is pertinent to understand how values for these parameters are set. We begin with the phase detector gain, K_p.

The behaviour of an ideal phase detector is to produce an output that varies in proportion to the phase difference between two input signals, where $\theta[m]$ is the phase of the input to the PLL at sample m, and $\hat{\theta}[m]$ is the phase of the corresponding locally generated sinusoid. Thus the phase error at sample m is simply given by

$$\theta_e[m] = \theta[m] - \hat{\theta}[m]. \qquad (D.54)$$

As shown in Figure D.3, the output of the ideal phase detector is

$$g(\theta_e[m]) = K_p \theta_e[m]. \qquad (D.55)$$

Appendix D: PLL Design

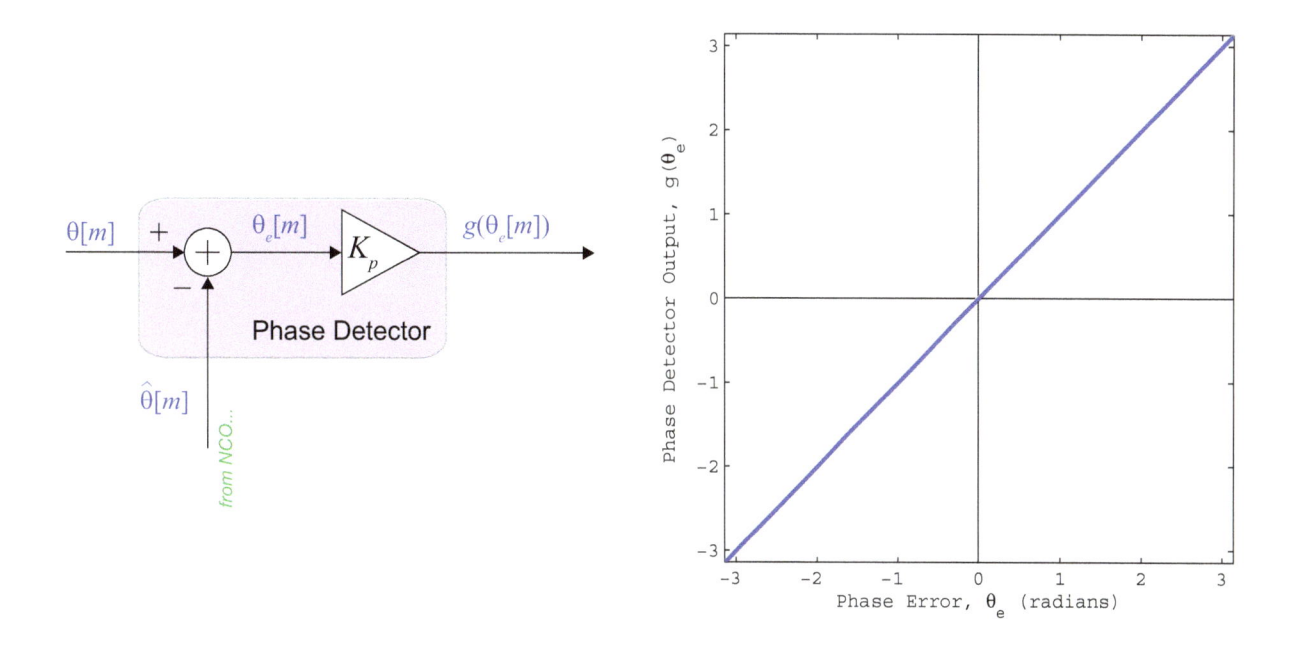

Figure D.3: An ideal Phase Detector (time domain), and its S-curve

As the output of an ideal phase detector is simply

$$g(\theta_e[m]) = \theta_e[m], \tag{D.56}$$

over the entire interval $-\pi \leq \theta_e \leq \pi$, it can be plotted as a linear function as shown in Figure D.3. Implicitly, the phase detector gain is therefore $K_p = 1$.

This graph of the phase detector output versus its input is commonly referred to as an 'S-curve'. In the general case, K_p is defined as the gradient of the S-curve at $\theta_e = 0$. When different designs of phase detector are used (for instance when the phase detector is implemented as a multiplier) the S-curve is not linear, and the value of K_p is not simply equal to 1. We will consider such an example of a non-linear phase detector next.

The other common model for a phase detector is a multiplier (as introduced in Chapter 7). In this case, the output does not vary in direct proportion to the error in the input phases, but follows a sinusoidal profile. As shown in Figure D.4, the phase detector forms the product between the PLL input with amplitude A, and a locally generated sinusoid.

The multiplier forms the product of the PLL input $A\cos(2\pi f m T + \theta_i[m])$, and the fed back component synthesised by the NCO, $-\sin(2\pi f m T + \theta_o[m])$.

The output of the multiplier is given by

$$y = \frac{A}{2}\Big[\underbrace{\sin(4\pi f m T + \theta_i[m] + \theta_o[m])}_{\text{high frequency term}} + \underbrace{\sin(\theta_i[m] - \theta_o[m])}_{\text{low frequency term}}\Big]. \tag{D.57}$$

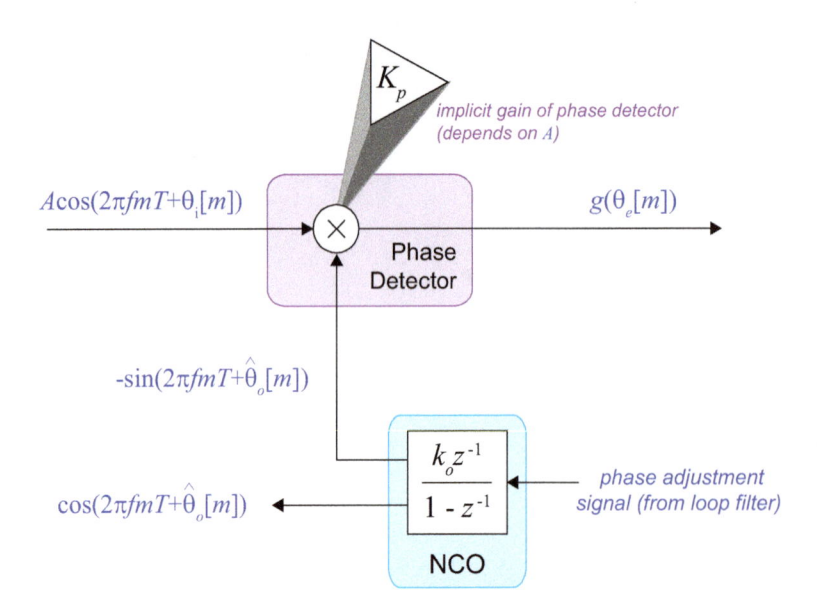

Figure D.4: Operation of a multiplier-based phase detector

The high frequency term can be neglected because it will be attenuated by the loop filter. This leaves the low frequency term, which depends only on phase. The output of the multiplier-based phase detector may therefore be expressed as

$$g(\theta_e) = \frac{A}{2}\sin(\theta_i[m] - \theta_o[m]) = \frac{A}{2}\sin(\theta_e) \ . \tag{D.58}$$

The function given in Eq. (D.58) can be plotted as an S-curve in the same way as Eq. (D.56), which results in a sinusoidal response as shown in Figure D.5 (here, for a value of $A = 1$). Notably, the response depends on A, the amplitude of the PLL input signal.

The gain of the phase detector is defined as the gradient of the function $g(\theta_e)$ at $\theta_e = 0$. In order to derive a value for $g(\theta_e)$, we apply the approximation

$$\sin(x) \approx x \tag{D.59}$$

for small values of x.

Eq. (D.58) can therefore be simplified to

$$g(\theta_e) = \frac{A}{2}\theta_e, \tag{D.60}$$

which implies that

$$K_p = \frac{A}{2}. \tag{D.61}$$

Appendix D: PLL Design

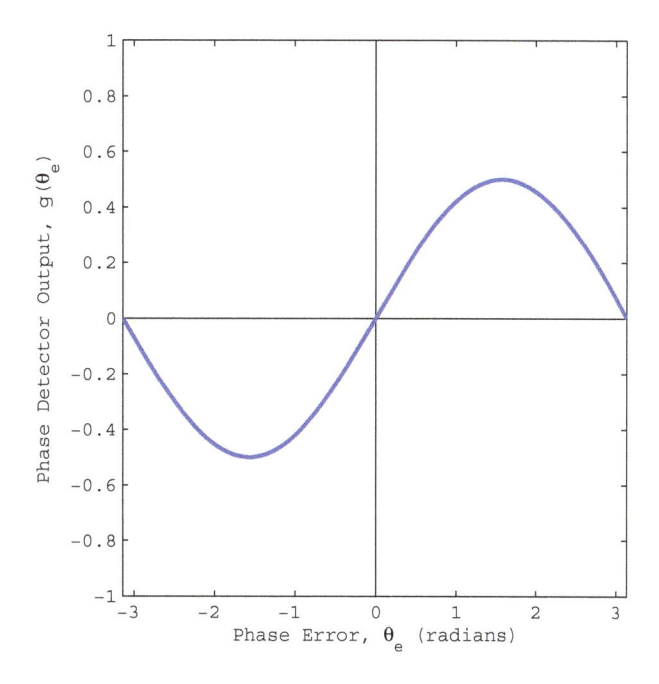

Figure D.5: S-curve for the multiplier-based phase detector (pictured for example $A = 1$)

It is therefore apparent that the phase detector gain, K_p, varies with A, the amplitude of the PLL input signal. This means that the fundamental behaviour of the PLL is linked to the amplitude of the input; for instance, if the input became weaker then the value of K_p would reduce.

The dependence of K_p on input signal amplitude is undesirable, and it is normally mitigated by employing Active Gain Control (AGC) to ensure that the value of A remains relatively constant.

In summary, then, we have considered two types of phase detectors:

- The *ideal* phase detector   (linear S-curve, $K_p = 1$)
- The *multiplier* phase detector   (sinusoidal S-curve, $K_p = A/2$)

The multiplier-based phase detector is generally adopted in the Simulink PLL examples presented in Chapter 7.

D.5 Oscillator Gain

The PLL gain coefficient for the NCO is denoted by K_o, and is shown in Figure D.6. This value is often set to a gain of $K_o = 1$, and has units radians/V. As the NCO is defined digitally, there are no physical factors affecting the choice of K_o, it is convenient to choose the trivial value of 1 (which is equivalent to a simple wire connection).

NCO operation is reviewed in Chapter 7 (Section 7.5.3, starting on page 248), where the other mathematical symbols shown in Figure D.6 are defined.

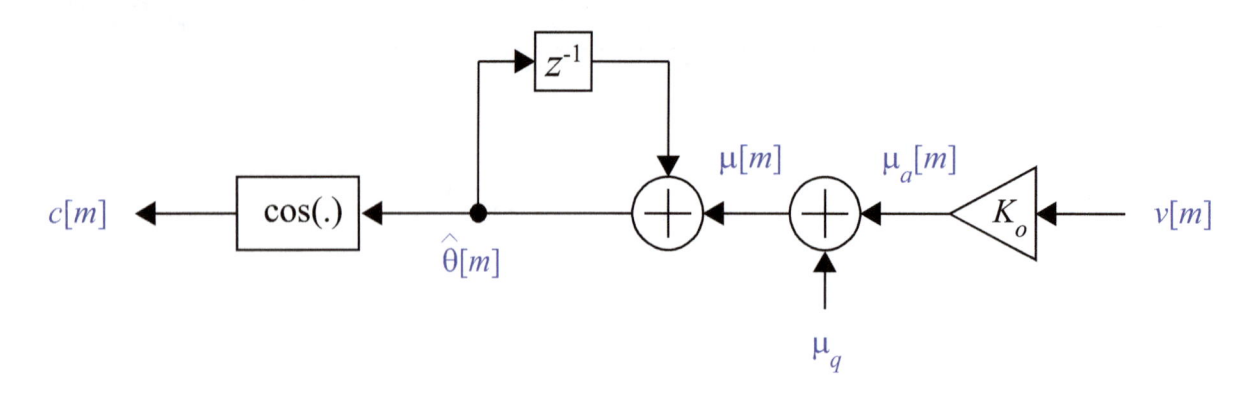

Figure D.6: The NCO structure, showing the gain coefficient, K_o

Appendix E: AM and FM Transmitters

This appendix presents a number of different ways that you can generate AM and FM signals with devices other than the USRP® hardware. We highly recommend you explore some of these, (a) because they are far cheaper options than the USRP® radios, and (b) because it is worthwhile experimenting with different RF devices.

E.1 Upconverting AM Radio Signals with the Ham It Up

If you have purchased or been given (for the lucky ones among you!) a *Ham It Up*, you probably want to know how to use it. Setting it up is very simple, however you will require a couple of extra components that are not shipped with the Ham It Up or RTL-SDR as standard. These are as follows:

> x1 AM loop antenna
> x1 SMA (male) / MCX (male) cable
> x1 USB A (male) / USB B (male) cable

Exercise E.1 will show you what to do with all of these things!

Exercise E.1 Ham It Up: Hardware Setup

This Exercise is designed to help you prepare everything you will need to get your Ham It Up (HIU) connected to your RTL-SDR. If you do not have these items lying around, you will either need to buy or construct them yourself.

(a) **Source and modify a suitable antenna.** You will need to use an AM loop antenna, to receive the AM radio signals broadcast in the air around you. These antennas are tuned to work optimally in the frequency range of 500kHz to 1.5MHz, and you will almost certainly find you have one lying around with an old HiFi unit. If you do not have one to hand, they can be purchased from various retailers, such as Amazon or RadioShack.

AM Loop Antenna

Normally these antennas simply have a bare wire attached to them, as they are designed to be connected to terminals on the back of radios, or will have a plug on the end of them (like the one pictured above). The HIU requires the antenna to be connected via a male SMA plug, and if your antenna does not have this, you will need to add one. Cut off the plug and strip the ends, then solder (or crimp) a male MCX plug onto it. This completes your antenna.

(b) **Assemble the Ham It Up.** NooElec ships the HIU in two pieces; the board (which already has the switch, filters, modulator etc. attached), and the 125MHz crystal oscillator. To complete assembly, this must be plugged into the empty 8-pin DIL socket on the board. Take the crystal oscillator, rotate it so that the black dot on it lines up with the white dot on the board, and plug it in, as shown in the photo.

Line up the dots when connecting the crystal

(c) **Source or make a cable to connect the Ham It Up to your RTL-SDR.** The cable you will need is an SMA (male) to MCX (male), and either you need to buy one, make one or cobble one together from various adaptors you have to hand. The cable should have the following ends:

Male SMA plug   *Male MCX plug*

(d) **Connect it all up.** Screw the SMA plug from the antenna into the 'Antenna Input RF' SMA socket on the HIU, and the SMA from the SMA / MCX cable into the 'IF Output' SMA. Plug the USB B to the socket on the Ham It Up, and finally connect the MCX to the antenna port on the RTL-SDR stick. You are now good to go!

Appendix E: AM and FM Transmitters

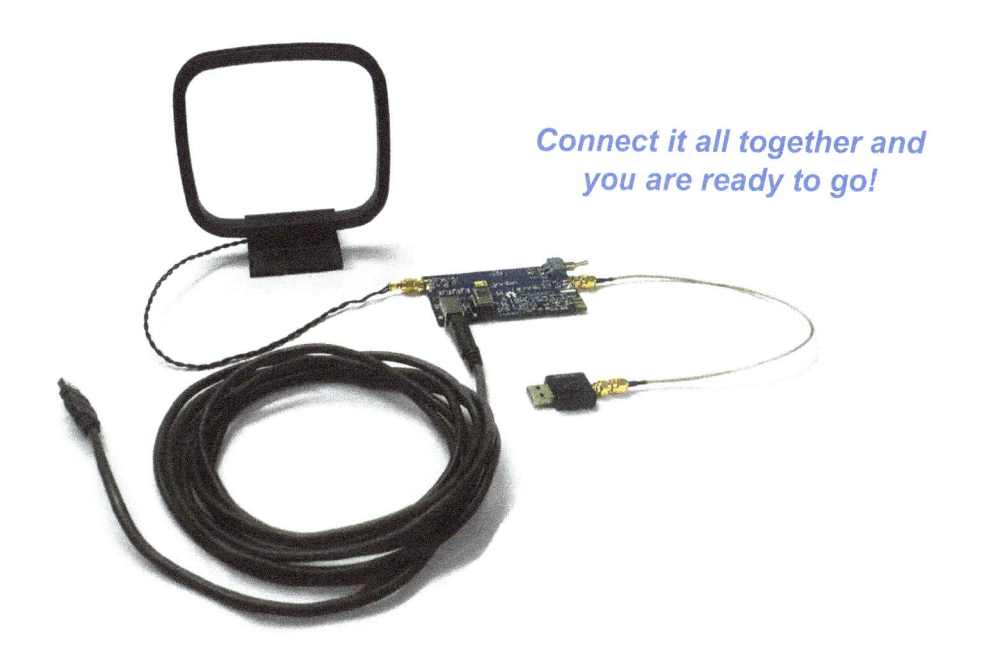

Connect it all together and you are ready to go!

Now that you have all of this bits required to use the HIU with your RTL-SDR, it is worth quickly running through what it does once again.

When switched into 'upconvert' mode, RF signals entering the HIU are fed through a lowpass filter that allows only short-wave, medium-wave and long-wave signals to pass and reach the mixer. After notch filtering to ensure no harmonics are present, the output signal from the 125MHz crystal oscillator is input to the mixer too. The IF (AM-DSB-TC) signal output from the mixer is input to a bandpass filter which suppresses the LSB, and this is then output from the board. A block diagram of the upconversion processes carried out by the HIU is shown in Figure E.1.

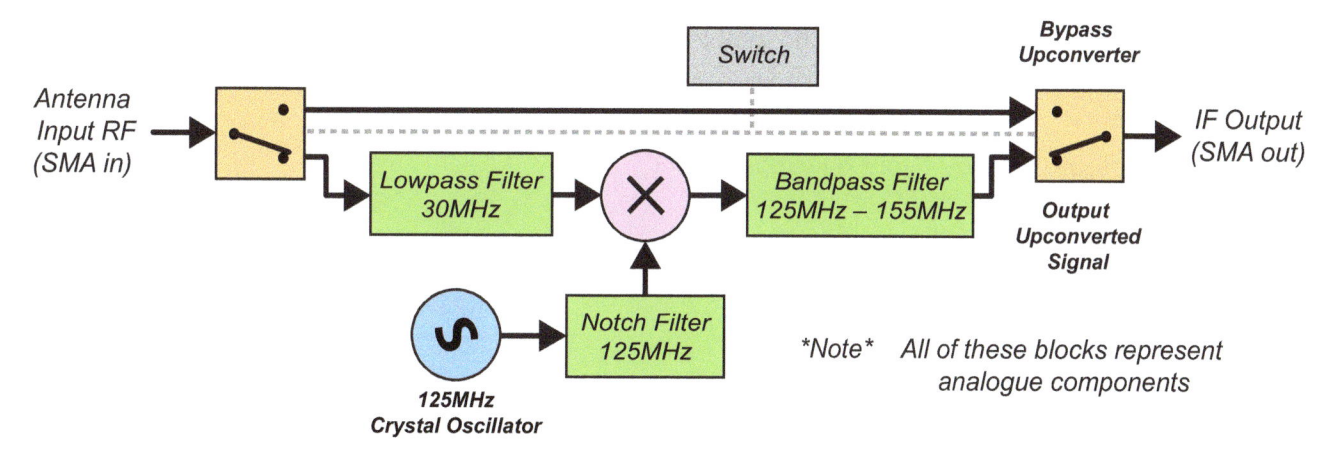

Figure E.1: Block diagram showing the main processes that take place on the Ham It Up (v1.2)

For those of you interested in an even lower level investigation of exactly what happens on the HIU, we can recommend this video [118] by YouTube user Alan Wolke. A second video uploaded on his channel covers the completion of the noise generator circuit on the board, and if you are interested in using it, this video is definitely worth a watch too [117].

When you use the HIU with your RTL-SDR, you will need to bare a couple of things in mind. As AM signals are upconverted by the HIU, their centre frequencies become:

$$f_{c\,(ham\,it\,up)} = f_{c\,(am\,signal)} + 125MHz. \qquad (E.1)$$

Envelope detectors are only able to operate when the carrier of the modulated signal input to them is of a significantly high frequency, and as was discussed in Section 8.2 (page 290), tuning the RTL-SDR to the centre frequency of an AM signal means they will not work. The workaround proposed there was to introduce a tuning frequency offset, and this method was then used in the remaining exercises in the chapter:

$$f_{c\,(rtl-sdr\,tuner)} = f_{c\,(am\,signal)} - f_{offset}. \qquad (E.2)$$

A similar process must be carried out when tuning the RTL-SDR to receive the HIU signal, as follows:

$$f_{c\,(rtl-sdr\,tuner)} = f_{c\,(am\,signal)} + 125MHz - f_{offset}. \qquad (12.5)$$

As before, the offset frequency must be within the range ($0 < f_{offset} < f_s/2$) for the AM signal of interest to reside at f_{offset} inside the baseband signal output by the RTL-SDR. Considering these points, the simplest thing to do is to modify the envelope detector receiver from Exercise 8.4 (page 292) to add the 125MHz offset to the centre frequency calculation.

Exercise E.2 — RTL-SDR: Envelope Detector for HIU AM-DSB-TC Signals

In this exercise you will modify the envelope detector you built in Exercise 8.4 to allow it to be used to tune to an AM radio signal upconverted by the Ham It Up. The receiver was designed to receive an AM-DSB-TC audio signal, and output the demodulated audio information to your computers speakers or headphones, so little needs changed in the design.

(a) **Open MATLAB.** Set the working directory to the exercise folder,

   `/my_models/receivers/`

   then create a copy of your envelope detector model,

   `.../rtlsdr_am_envelope_demod.slx`

   and rename it:

   `.../rtlsdr_am_hamitup_demod.slx`

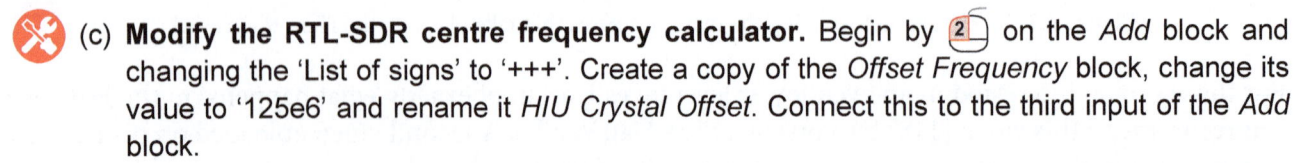

(b) You should know by now how this model works, so we will progress directly to modifying it.

(c) **Modify the RTL-SDR centre frequency calculator.** Begin by 2 on the *Add* block and changing the 'List of signs' to '+++'. Create a copy of the *Offset Frequency* block, change its value to '125e6' and rename it *HIU Crystal Offset*. Connect this to the third input of the *Add* block.

Appendix E: AM and FM Transmitters

Change the value of *AM Signal Frequency* to '800e3' as a starting point. This will now refer to the actual centre frequency of the AM radio signal. Finally, reduce the value of *Transmitter Gain* to '1'. You must do this because the HIU amplifies the signal power, and you do not want to damage your RTL-SDR.

(d) **Your modified receiver should look as follows:**

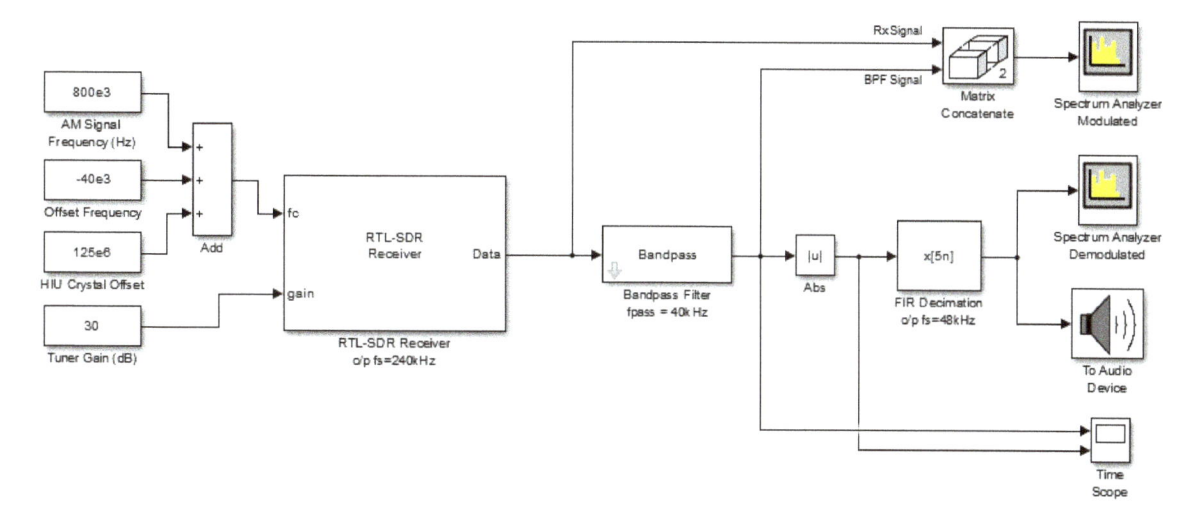

(e) **Open the reference file.** If you did not have time to construct the receiver yourself in Exercise 8.4 (and therefore could not modify its block diagram), open the following file:

📁 /am/rtlsdr_rx/rtlsdr_am_hamitup_demod.slx

(f) **Prepare to run the simulation.** Connect speakers or headphones to your computer and perform a test to ensure that they are working. Set the *Simulation Stop Time* to 'inf' by typing this into the Simulink toolbar, and the *Simulation Mode* to 'Accelerator'. This will force Simulink to partially compile the model into native code for your computer, which allows it to run faster. Check that MATLAB can communicate with it by typing `sdrinfo` into the MATLAB command window. Finally, make sure your model is saved.

(g) **Run the simulation.** Begin the simulation by 🖱 on the 'Run' ▶ button in the Simulink toolbar. After a couple of seconds the simulation will begin, the *Spectrum Analyzer* and *Time Scope* windows should appear, and you should be able to hear the demodulated AM signal. Adjust the frequency and gain values of the RTL-SDR until you tune the device to the signal you want to receive. Remember that the signal must be within the passband of the bandpass filter; located around 100kHz in the baseband signal.

(h) We found that we had to go up to the top floor of our building to receive any AM radio signals, as in the office we could barely detect signals from even the strongest locally broadcast stations. Admittedly the University of Strathclyde is in the centre of Glasgow surrounded by tall buildings, and the nearest AM transmitters are miles away, so the weak signal strength we experienced is not all that surprising! It may well be better where you are though.

As a proof of concept (even though we could barely receive anything..), the results of this exercise showed that the Ham It Up is perfectly capable of upconverting AM radio signals (which could not otherwise be received by the RTL-SDR), to bring them into its range. If you stay in an area where there are a lot of AM radio stations broadcast or there are a large number of HAM radio users, you will almost certainly experience far better results than we did.

E.2 Building an 'RT4' 433.9MHz AM Transmitter

If you are reading this it is likely you are wanting to build your own AM transmitter... if not you are in the wrong place! Building any form of analogue circuit may be a daunting concept for any DSP engineer, but the AM modulator/ transmitter circuit we are going to present to you is relatively simple, and only consists of a few components. To build it, you will only require some basic circuit building and soldering skills and the ability to follow our instructions.

The AM transmitter you are going to build is based around a 433.9MHz AM-DSB-TC modulator called the 'RT4'. These small DIL devices (around 1x2cms in size) are built by RF Solutions, and cost around £5/$8. According to the product data sheet [110], these devices are designed to transmit signals with a bandwidth up to 4kHz wide, and have a transmission range of around 70m when connected to an appropriate antenna. A photograph highlighting the functions of each of the RT4's pins is shown in Figure E.2

We have done a number of tests with this modulator, and found that it is perfectly capable of being used to transmit analogue audio signals, rather than just digital signals as the data sheet suggests. Interestingly, we found that it worked best if the information signal supplied to RT4s 'input' pin was negative, in the range $(-V_{cc} < -2V)$. Shifting an information signal into this range in real time is a simple task with the use of a differential amplifier, so this is not much of an issue.

In the following exercise we will show you how you can use the RT4 as part of an AM transmitter circuit. You will require a number of components to build this circuit, which you should be able to source from companies such as Farnell [88] or RadioShack [108]. Buying all of these components cost us less than £20.

     x1 Prototyping board
     x1 RT4 433.9MHz AM modulator
     x1 LM741 op-amp (or suitable alternative) + 8 pin DIL socket
     x2 9V PP3 batteries
     x2 PP3 battery connectors
     x1 5V regulator
     x1 3.3V regulator
     x1 3.5mm mono jack cable
     x1 UHF (400MHz +) antenna with SMA (or alternative) plug
     x1 SMA (or alternative to match antenna) solder socket
     x2 1µF capacitors
     x2 2.2µF capacitors
     x3 1kΩ resistors
     x1 5.6kΩ resistor
     + wire, side cutters, stripboard, solder (and a soldering iron)

Appendix E: AM and FM Transmitters

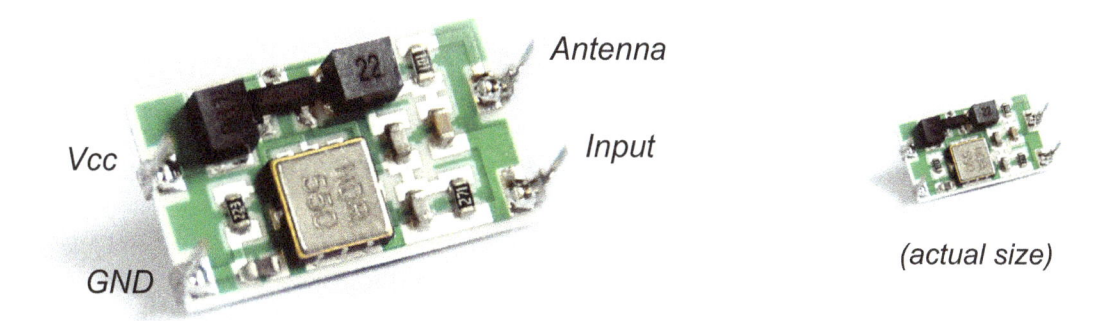

Figure E.2: A photo of the RT4, highlighting the functions of each of its pins

Additionally, items such as a voltmeter, oscilloscope and signal generator come in handy.

Exercise E.3 Build the RT4 AM Transmitter

In this exercise we will show you how to build an AM transmitter circuit using an RT4 modulator. We advise you to construct the circuit on a prototyping board and test it before beginning the process of soldering the components onto stripboard, as this means any issues can be ironed out before it is too late! A full circuit diagram of the transmitter can be found on page 616.

(a) **Construct a differential voltage source and the required power rails.** Op-amps can only output signals within the range of their power supply. As we need the op-amp in this circuit to output a negative voltage, it requires a differential power supply (-Vcc and +Vcc). This easy to make, and only requires two batteries.

(b) Link the two 9V PP3 battery connectors in series as shown in the diagram below.

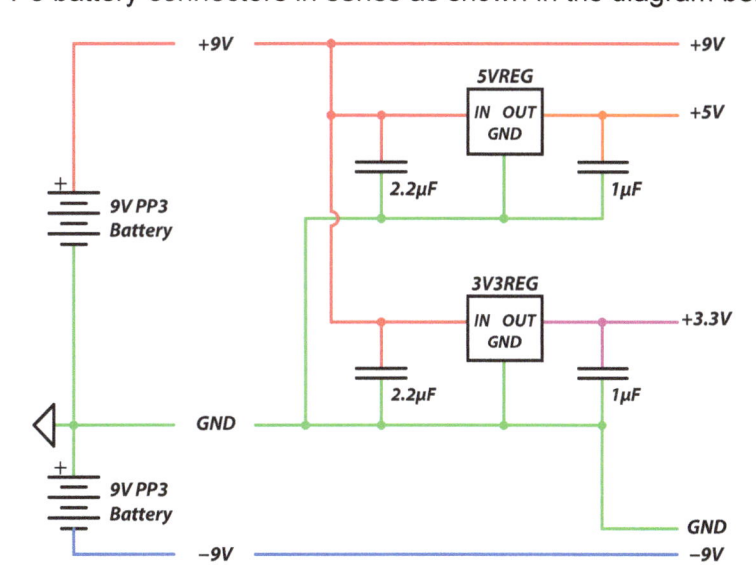

By defining the connection between them as ground, the two remaining battery contacts become +9V and -9V, relative to the ground connection. Next, take the two voltage regulators and connect their 'input' pins to the +9V source, and 'ground' pins to the ground source. Place

2.2μF capacitors between the 'input' and 'ground' pins, and 1μF capacitors between the 'output' and 'ground' pins. Although not entirely necessary, using capacitors ensures that any AC signals entering or leaving the regulators are shorted to ground.

If you are using electrolytic capacitors (it is likely you will be), ensure that you connect them the right way round, with the cathode (negative pin) tied to ground. Failure to do this might cause a small explosion!

(c) Plug the batteries into their connectors and use a voltmeter to check that all of the power rails (+9V, +5V, +3.3V, GND and -9V) are functioning correctly. If everything is working, disconnect the batteries and continue. If not, it is either the case that the batteries are not working, or that you have connected the voltage regulators incorrectly. Sort this before continuing.

(d) **Implement a differential amplifier.** You need to shift the information signal into the range −Vcc to -2V, and the easiest way to do this is with a differential amplifier. While they are normally used to amplify the difference between two voltage sources for instrumentation purposes, they can also be used as voltage subtractors. The general equation for a differential amplifier (shown in the diagram below) is as follows:

$$V_{out} = -V_1\left(\frac{R_3}{R_1}\right) + V_2\left(\frac{R_4}{R_2+R_4}\right)\left(\frac{R_1+R_3}{R_1}\right) . \quad (E.3)$$

Setting R1=R2=R3=1kΩ and R4=5.6kΩ makes Eq. (E.3) equal to:

$$V_{out} = 1.697V_2 - V_1 .$$

If we apply a constant voltage to input V1 and an information signal to V2, this means the amplifier will output 1.697*V2 volts, shifted to oscillate around -V1. If we set the constant voltage to 3.3V and use a 0.5V sine wave as the information signal, this would result in the sine wave oscillating within our desired range:

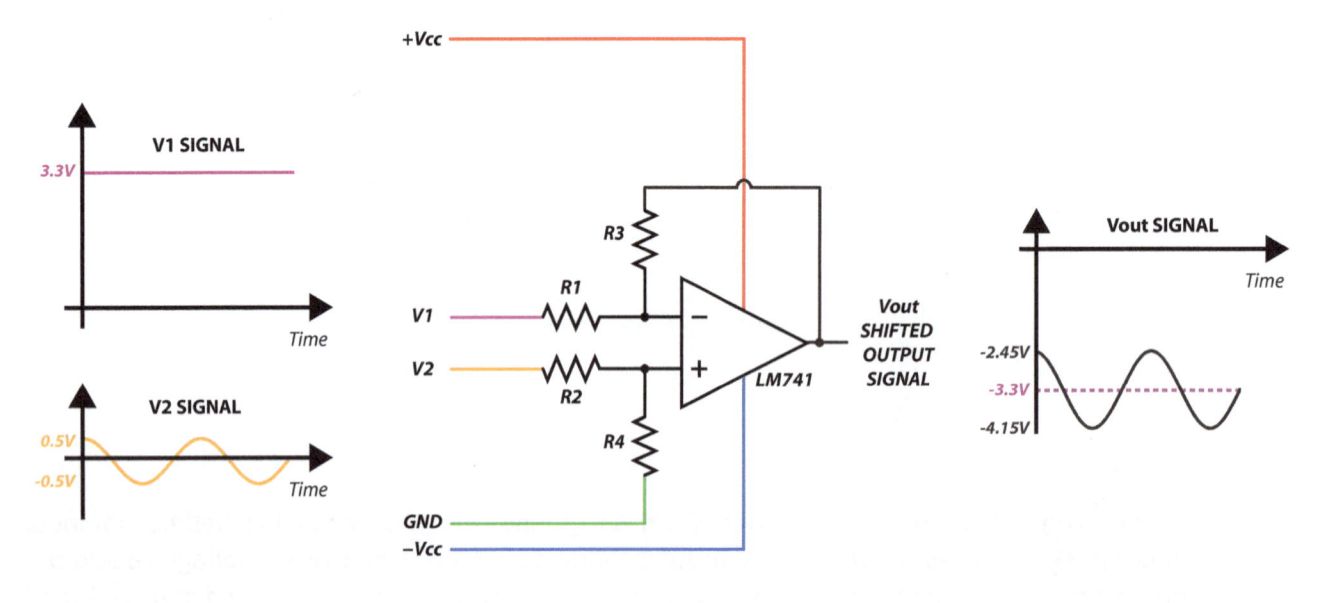

Appendix E: AM and FM Transmitters 615

(e) Take the 8 pin DIL socket and plug it into your prototyping board. Look up the datasheet for the op-amp you are using and find which pins are used for the inverting and non-inverting inputs, positive and negative power supplies, and the output. Using the diagram above, connect the four resistors to the corresponding pins of the DIL socket.

(f) Take the jack cable, cut it, and strip it so that you expose ground (outer core) and signal (inner core) contacts. Connect the signal cable to V2 and the ground cable to ground in your circuit. Connect the 3.3V source to input V1 and the differential (+9V, -9V) power supply to the positive and negative power supply rails. Finally, plug the op-amp into the DIL socket.

If you have a signal generator available, set it to generate a sine wave with a frequency of 1kHz and an amplitude of 1Vpp. Connect this to V2 and ground, and plug in your batteries. Using your voltmeter or an oscilloscope, probe the signal being input to the differential amplifier and the signal leaving it. You should find that the signal output from the op-amp matches the signal sketched in the diagram opposite; i.e. is a sine wave with an amplitude of 1.69Vpp, oscillating around -3.3V.

(g) **Complete the circuit by adding the RT4 and an antenna.** Plug the RT4 to your board, and connect the 5V power supply to its 'Vcc' pin. Ground it and then make a connection between the output of the op-amp and its 'input' pin. Connect the 'antenna' pin to the inner contact of the SMA socket, and finally, ground the outside of the SMA socket. Your completed circuit should look as follows:

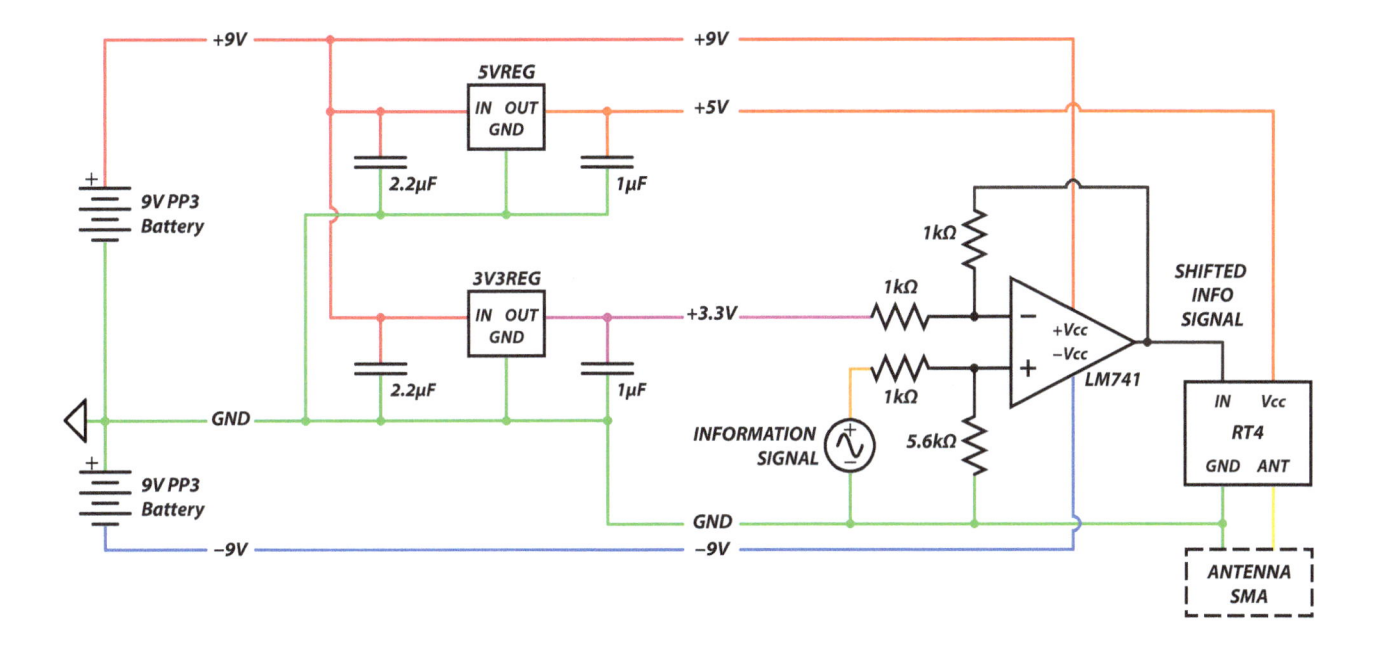

(h) **Full Circuit Diagram.** A full black and white circuit diagram is shown in Figure E.3. It (and all of the previous diagrams) were made using an online tool we highly recommend called CircuitLab [78]. It is free to register an account if you are a student and is competitively priced if you wish to register for hobbyist or professional accounts.

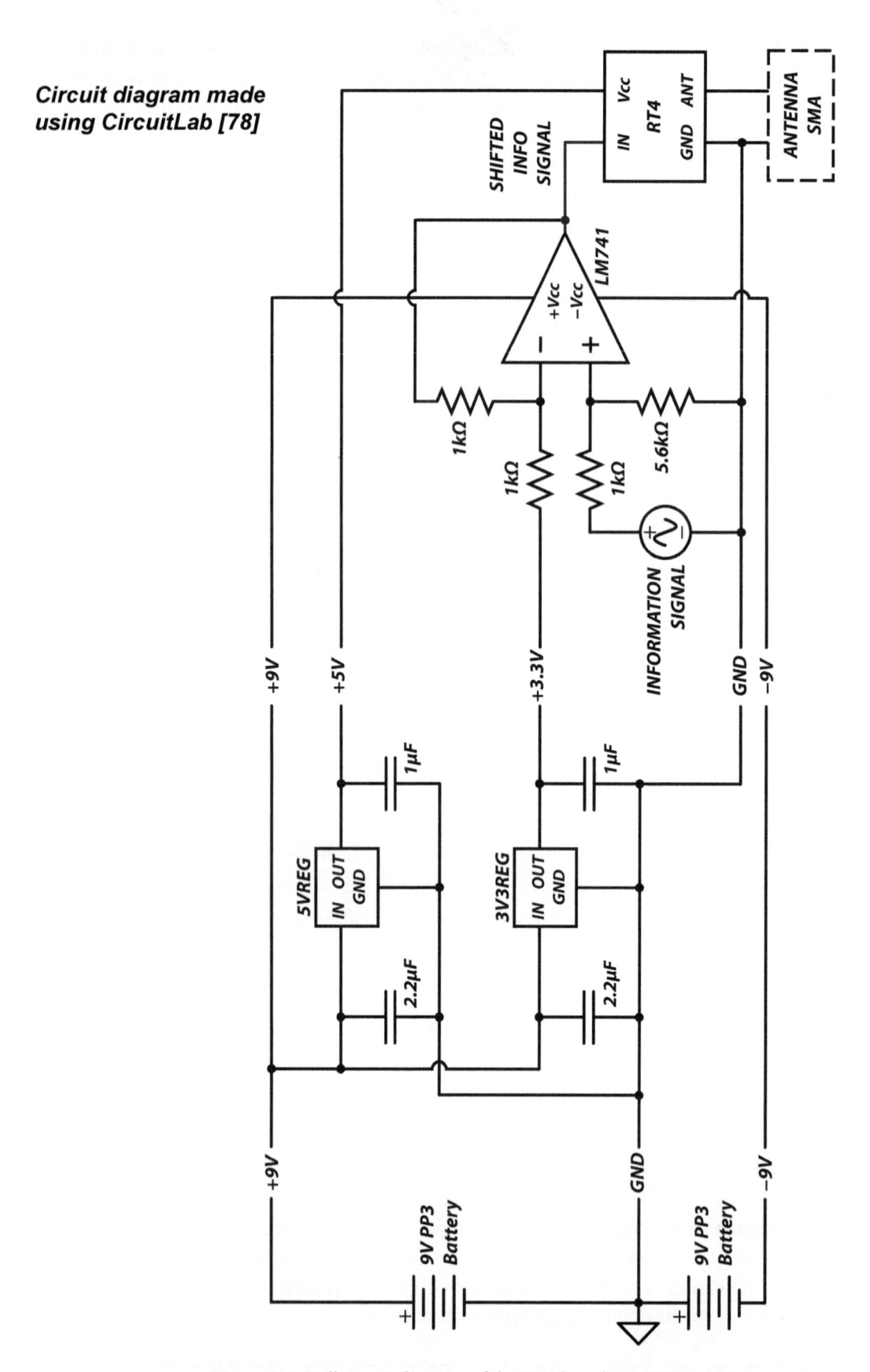

Figure E.3: Full circuit diagram of the RT4 based 433.9MHz AM transmitter

Appendix E: AM and FM Transmitters

Exercise E.4 Test the RT4 AM Transmitter

This exercise will ask you to open and run the AM envelope detector receiver model from Exercise 8.4 to test if your RT4 transmitter is working correctly. The receiver was designed to receive an AM-DSB-TC audio signal, and output the demodulated audio information to your computers speakers or headphones, so nothing needs changed in the design.

(a) **Open MATLAB.** Set the working directory to the exercise folder,

   /am/rtlsdr_rx

and then open the following file:

   .../rtlsdr_am_envelope_demod.slx

(b) You should know by now how this model works, so we will progress directly to running it.

(c) **Prepare to run the simulation.** Connect speakers or headphones to your computer and perform a test to ensure that they are working. Set the *Simulation Stop Time* to 'inf' by typing this into the Simulink toolbar, and the *Simulation Mode* to 'Accelerator'. This will force Simulink to partially compile the model into native code for your computer, which allows it to run faster. Check that MATLAB can communicate with it by typing `sdrinfo` into the MATLAB command window. Finally, make sure your model is saved.

(d) **Get transmitting!** Connect an antenna to the SMA socket, and an audio device (such as an MP3 player) to the mono jack cable sticking out of your transmitter. Plug the 9V batteries into their sockets to turn on your circuit, and being playing music from your audio source.

(e) **Run the simulation.** Begin the simulation by on the 'Run' button in the Simulink toolbar. After a couple of seconds the simulation will begin, the *Spectrum Analyzer* and *Time Scope* windows should appear, and you should be able to hear the demodulated AM signal. Adjust the frequency and gain values of the RTL-SDR until you tune the device to the signal you want to receive. Remember that the signal must be within the passband of the bandpass filter; located around 100kHz in the baseband signal.

You should be able to hear to the demodulated signal using PC speakers or headphones. If the music is not very loud, this is likely because the signal is saturating the tuner on your RTL-SDR. Try turning down the volume of the audio device or the tuner gain to improve the output signal.

Exercise E.5 Refine your RT4 AM Transmitter

If you like, you can rebuild your AM transmitter either on a stripboard or a PCB. This exercise will talk you through this.

(a) **Rebuild your transmitter on a stripboard.** While prototyping boards are great for building and testing circuits, they tend to be rather cumbersome, and are not ideal for 'final product' designs. Stripboards on the other hand tend to be lightweight and compact, and you have more control over how to lay your circuit out on them. You do not need to move your transmitter onto a stripboard if you do not want to, but apart from anything else, doing so means that when it is built, it is built and you can keep it!

Do not start by soldering components onto the board randomly; methodically plan how you will lay the circuit out, aiming to have as few strip cuts and cable patches as possible. We found a piece of stripboard roughly 10cm x 10cm was an adequate size for this. When you are sure about the positioning of the components, solder them all in place. Shown below is what we produced when we rebuilt our circuit.

(b) **Make a PCB.** For those of you who want to go one step further, you can design a compact, credit card sized PCB that contains the whole of the circuit. We designed one using Eagle, and you can find the board file in the following folder:

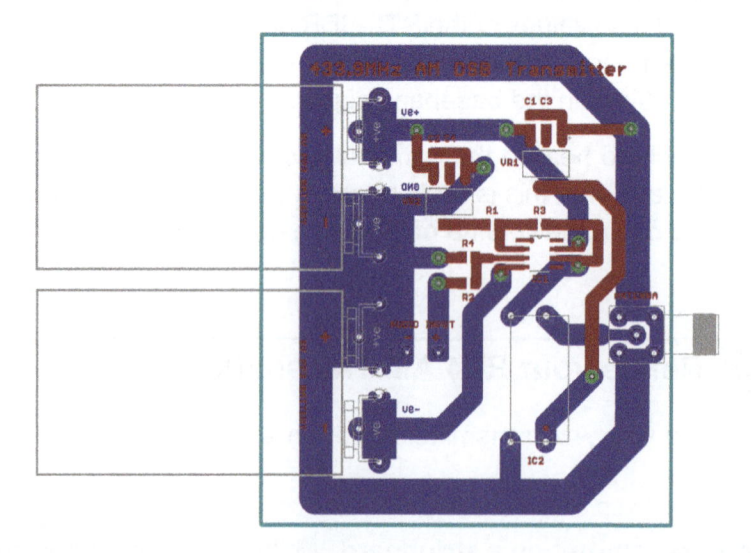

/resources/am_circuits/pcb_rt4_transmitter.brd

Appendix E: AM and FM Transmitters 619

E.3 Using the Raspberry Pi as an FM Transmitter

This section will focus on turning the Raspberry Pi (rev 1 models B and B+ only) into an FM transmitter. The following exercises are orientated for Microsoft Windows OS users, and deal only with Windows compatible software. Alternative software is available for both Linux and Mac systems, although as these are Unix based, everything you need to do can also be done through Terminal — if you so desire!

The first exercise takes your through the steps of creating a backup image of a blank SD card. Although this may sound like a pointless activity, if you do not create a backup image when the card is formatted with a single FAT32/ FAT16 partition, you will find it very difficult without help from a Linux computer to be able to use the card fully again after loading the MAKE Labs disc image onto it. This is because the image creates multiple partitions on the SD card that are associated with the Pi bootloader, the Arch Linux file system and Pirate Radio data directories, as shown in Figure E.4.

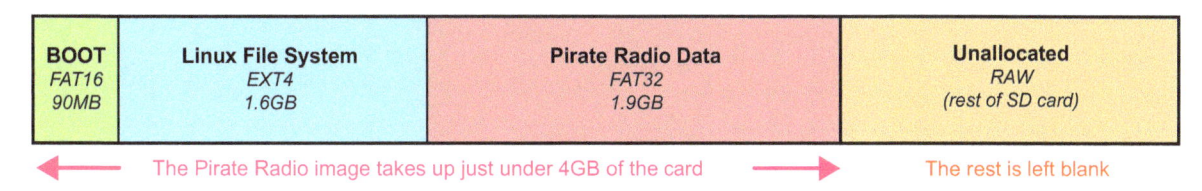

| BOOT
FAT16
90MB | Linux File System
EXT4
1.6GB | Pirate Radio Data
FAT32
1.9GB | Unallocated
RAW
(rest of SD card) |

◄──── The Pirate Radio image takes up just under 4GB of the card ────► The rest is left blank

Figure E.4: SD card partitions after loading the MAKE Labs disc image

Although the *FAT16* boot partition can be mounted in Windows, the *EXT4* file system partition cannot. When a partition fails to mount in Windows, subsequent partitions cannot be mounted either; which means that neither the *FAT32* Pirate Radio data partition or any of the remaining space on the card can be accessed. This limits the effective size of the card after the disc image has been transferred to 90MB, the size of the boot partition. By creating a backup image of the SD card before loading the image onto it, you can simply restore it to its previous state with a free piece of software called the 'USB Image Tool'. This means that it can be reverted back to a single full size partition, ready to be used by your Windows PC or digital camera again.

Exercise   E.6   PiFM: Backing Up A Blank SD Card

This exercise will guide you through the process of downloading the USB Imaging Tool, and using it to create a backup image of your blank SD card. Administrator privileges are required to run this software.

Note: All URLs last accessed August 2015

(a) **Download the USB Image Tool.** This is a free piece of software that can be downloaded from the following URL:

   alexpage.de/usb-image-tool/download

   Although it does not need to be installed, you will require administrator privileges to run this software. Extract the files from the zipped folder.

(b) Insert a blank SD card to your card reader, and make a note of what drive letter it has been assigned (e.g. 'F:').

(c) **Use the USB Image Tool to make a backup.** 2️⃣ on `USB Image Tool.exe` to open the software. Accept the UAC message that pops up. Make sure that it is set in 'Device Mode' with the dropdown in the top left hand corner of the window. Select the SD card by 1️⃣ on the icon of the device in the left hand pane. When you have checked that it is the correct one (i.e. has the correct drive letter), 1️⃣ the 'Backup' button.

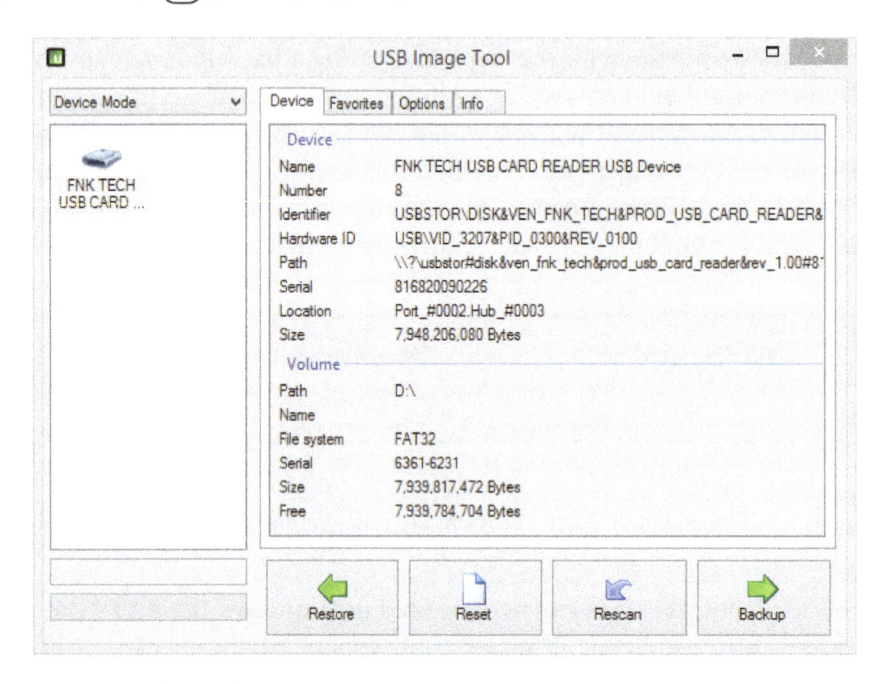

Choose to save the image in an appropriate folder. As a general note, if you backup a 4GB SD card the image will be a 4GB file. Likewise if you backup an 8GB SD card the image will be 8GB; make sure you have plenty of space to do this! When the backup is in progress, you should see the following:

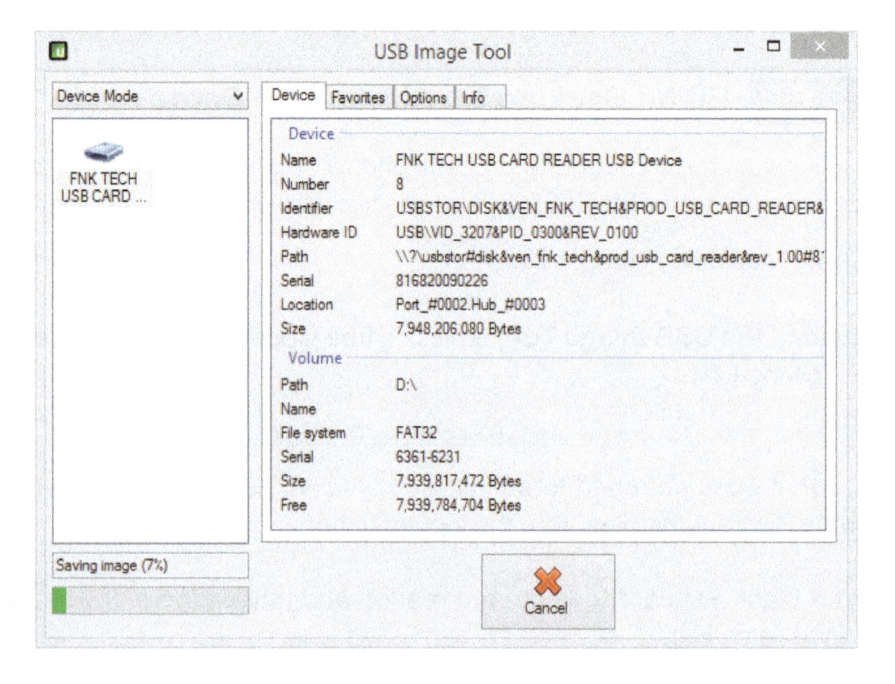

Appendix E: AM and FM Transmitters 621

Keep a hold of the file it creates. Exercise E.8 shows you how to restore your SD card to its original form using this file and the same software.

Now that you have a backup of the SD card, you are ready to start working on setting up your Raspberry Pi. The following exercise guides you through downloading the image, flashing this onto the blank SD card, editing the `pirateradio.config` file and transferring audio files to the Pi; all that is required to get a PiFM station up and running.

Exercise E.7 PiFM: Setting up a Raspberry Pi FM Radio Station

This exercise will guide you through the process of getting a PiFM station up and running. Administrator privileges are required when installing software.

DISCLAIMER: Parts of this Exercise are based on tutorials found here [97] [80].

Note: All URLs last accessed August 2015

(a) **Make the antenna.** Although designing antennas is in general a bit of a black art, the process of making an antenna for PiFM is easy. While you could sit and design a have-wave dipole antenna which would have optimal range, all you really need is around 40cm of 2.5mm^2 solid single core copper wire, some solder, and a single SIL header socket; as shown below.

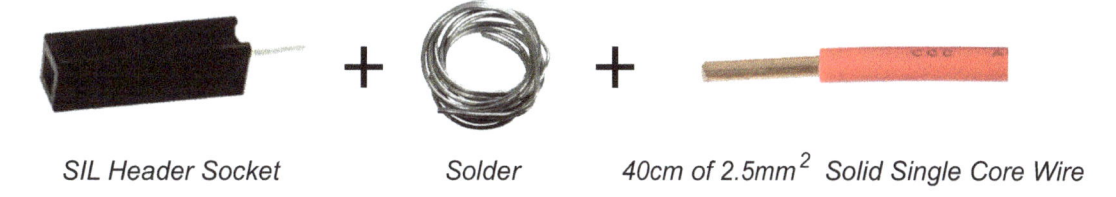

SIL Header Socket + *Solder* + *40cm of 2.5mm^2 Solid Single Core Wire*

Solder the wire onto the pin at the end of the SIL header socket, and then plug this into GPIO pin 4 of your Raspberry Pi. This completes the hardware side of this build!

(b) **Download the latest MAKE Labs PiFM image from their website.** At the time of writing this exercise, the Image file could be downloaded from the following URL:

   cdn.makezine.com/make/pifm/PiRadio.zip

Extract the image from the compressed folder. The image should have a name similar to `PiRadio_X-X.iso`.

(c) **Install software to flash the SD card.** 'Win32 Disk Imager' is a piece of free software that is able to unpackage raw disk images onto removable drives. It can be downloaded from the following website:

   sourceforge.net/projects/win32diskimager

You will require administrator privileges to install this software.

(d) Insert a blank SD card to your card reader, and make a note of what drive letter it has been assigned (e.g. 'F:').

(e) **Format the SD Card.** Open Win32 Disk Imager, and clicking on the blue folder icon, navigate to the folder containing the extracted image file. You will need to change the filetype drop down to '*.*' in order to make the `.iso` file visible. When you can find it, select it and 🖱 'Open'.

Select the drive letter of the SD card in the 'Device' dropdown menu.

SELECTING THE WRONG DEVICE HERE COULD RESULT IN YOU FORMATTING YOUR COMPUTER'S HARD DRIVE AND INCUR DATA LOSS. BE VERY CAREFUL!

(f) 🖱 on the 'Write' button to begin the data transfer. A popup message will appear asking you to confirm the overwrite process. 🖱 'Yes' to continue. The formatting process may take up to 20 minutes depending on the speed grade of the SD card, and progress will be indicated with a green bar at the bottom of the Win32 Disk Imager window.

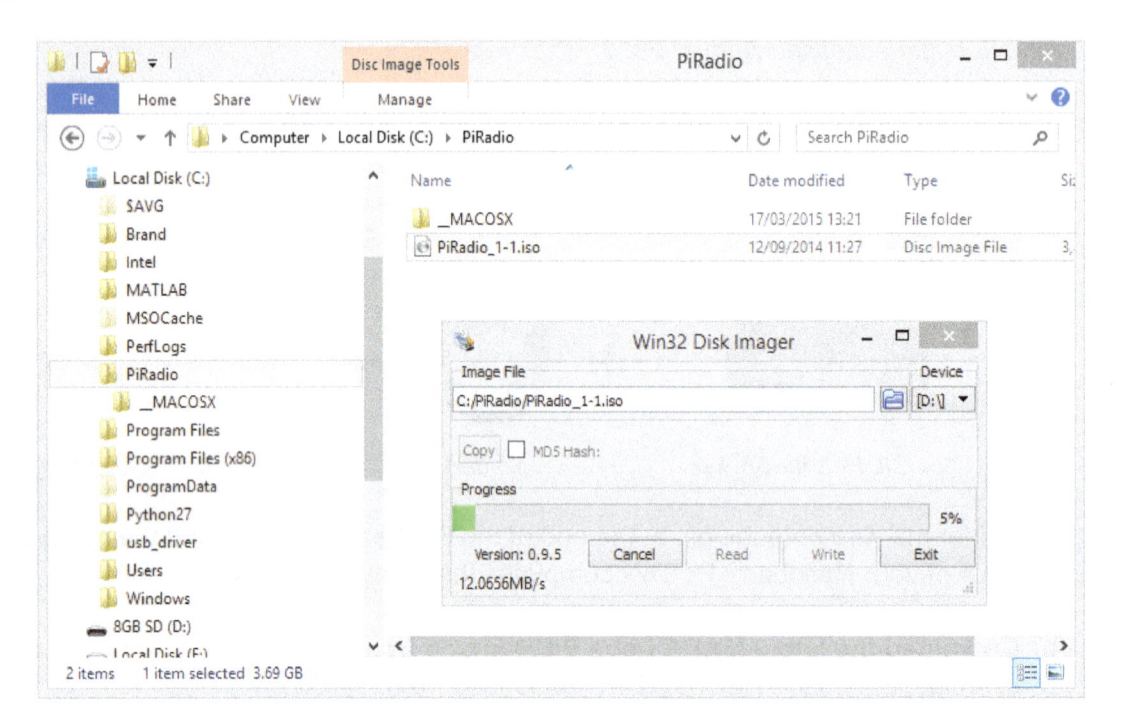

When this process is complete you will receive a confirmation message.

(g) **Install software that can establish an SSH connection.** While you could use command line software such as 'PuTTy' or 'TeraTerm' to make an SSH connection, it is easier to use 'WinSCP' for this task as it has a Windows Explorer style interface. It can be downloaded from the following website:

winscp.net/eng/download.php

You will require administrator privileges to install this software.

Appendix E: AM and FM Transmitters

(h) Plug the SD card into the slot on the bottom of the board. Connect an ethernet cable, an HDMI monitor and a USB keyboard to the Pi, and then plug in the power supply.

You are going to connect to the Pi through your network using the SSH protocol. In order to do this, you are going to need to know the IP address assigned to the Pi by your DHCP server. Turn on the Pi and look at the monitor. You should see the Arch Linux distribution starting up, and it will eventually ask for a login username. Enter the username and password,

> alarmpi login (username): *root*
> password: *root*

(i) Type the command: `ifconfig eth0` and make a note of the IP address listed next to *inet*. In the case below, the IP address is '10.0.0.2'.

```
Arch Linux 3.10.24-1-ARCH (tty1)

alarmpi login: root
Password:
Last login: Wed Dec 31 17:00:32 1969
[root@alarmpi ~]# ifconfig eth0
eth0: flags=4163<UP,BROADCAST,RUNNING,MULTICAST>  mtu 1500
        inet 10.0.0.2  netmask 255.255.255.0  broadcast 10.0.0.1
        ether b8:27:eb:65:f8:99  txqueuelen 1000  (Ethernet)
        RX packets 12352  bytes 699114 (682.7 KiB)
        RX errors 0  dropped 29  overruns 0  frame 0
        TX packets 121  bytes 18613 (18.1 KiB)
        TX errors 0  dropped 0 overruns 0  carrier 0  collisions 0
```

(j) **Connect to your Pi through WinSCP.** Open WinSCP, and enter the IP address your Pi has been assigned in the 'Host Name' field. Enter the username and password as before, and then 'Login'. After a few seconds, the connection will be established.

When the Explorer style window appears, you will see that the selected directory is 'root'. This is the not the root directory of the Linux operating system though, and to navigate to this, simply click on the '/ root' folder icon, or use the dropdown in the address bar. This is roughly what you should see when you are in the parent directory:

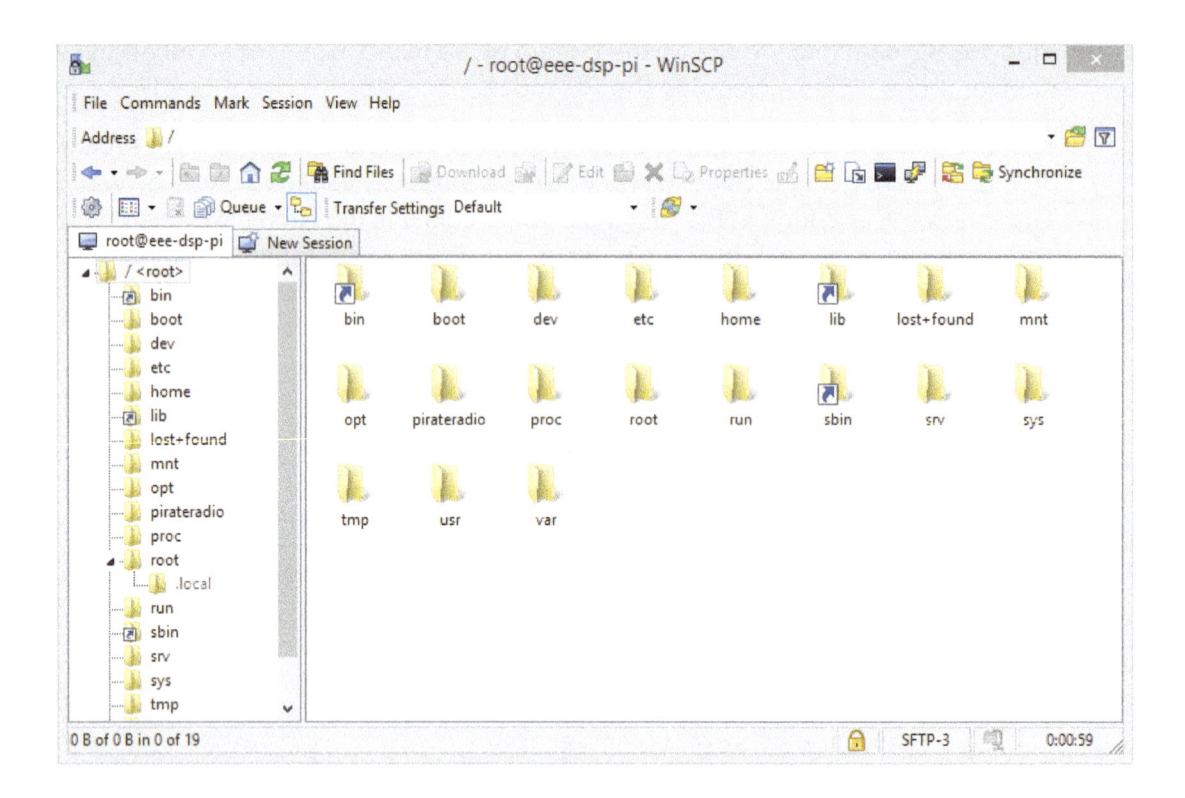

(k) 🖱️ on the 'pirateradio' folder. Inside this you should see a number of files, one of which is titled `pirateradio.config`. 🖱️ on this and select 'Edit'. The file should open in a text editor such as 'Notepad++'. Making changes to the settings in this file, you can alter the carrier frequency of the FM signal, and set music player options such as turning on 'shuffle'. The default frequency that is set in the file should be '88.9' (MHz). You can change this to whatever frequency you want in the range 1MHz–250MHz, although it would make sense to keep within the 25MHz–250MHz range so that you can receive the signal with your RTL-SDR.

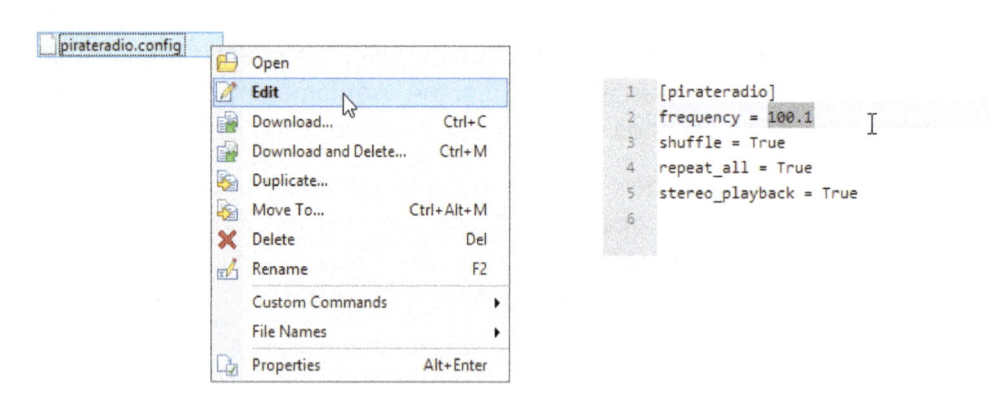

(l) When you are finished editing the file, save and close it. Back in the WinSCP window you should note that the `pirateradio.config` file is now listed in the synchronisation queue. 🖱️ on the queue listing and select 'Execute Now'.

Appendix E: AM and FM Transmitters   625

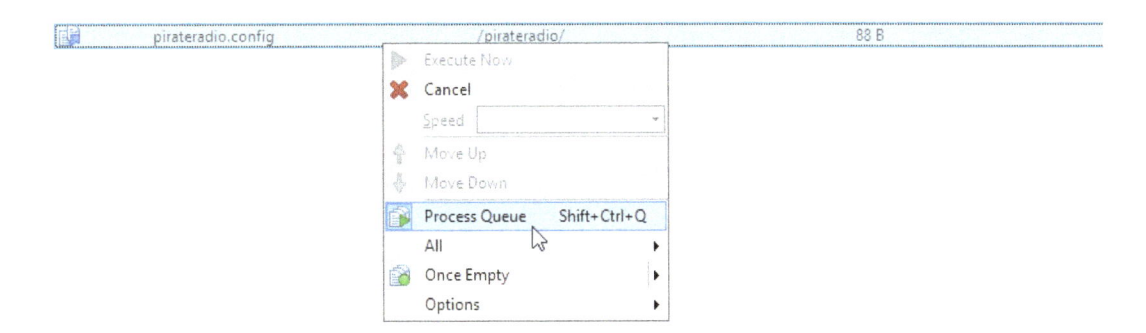

(m) **Transfer audio files to the Pi to stock your radio playlist.** The final thing you need to do in WinSCP is to transfer audio files across to the same folder that the 'pirateradio.config' file was found in. To do this, simply drag and drop files from Windows Explorer into the WinSCP window. You can use your own music here, or the files provided in:

/audio_sources

If this process is successful you will see this transfer window:

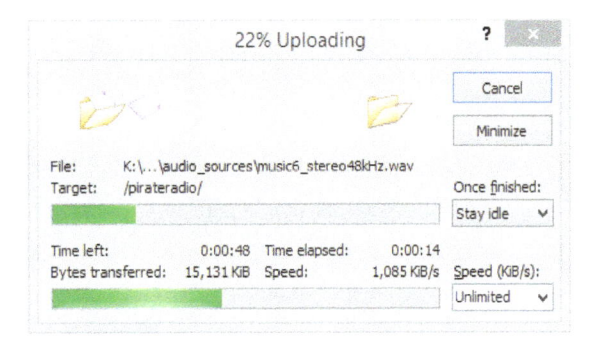

(n) **Perform a sync to ensure that all of the data has transferred successfully.** Open a terminal connection with the Pi either by [1] on the [>] 'Open terminal' button in the top toolbar or by navigating to 'Commands/ Open Terminal'. A message will appear asking if you want to open a separate shell session — you do, so [1] 'OK'.

(o) In the terminal window (titled 'Console'), enter the command '**sync**' and [1] 'Execute'. The command should be echoed back in the terminal.

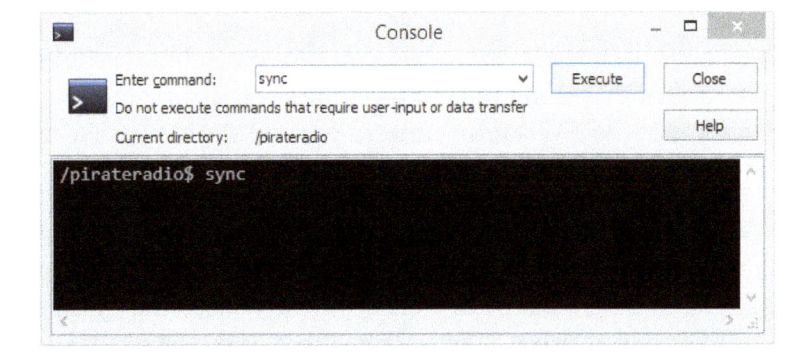

Once you have done this, close all of the WinSCP windows.

(p) **Restart your Raspberry Pi to activate your radio station.** Turn the Pi on and off again by removing and replacing the power cable. After 15 seconds or so, the OS will have fully loaded, the 'PirateRadio.py' script will run and music will start being transmitted. If you tune an analogue FM radio to the carrier you set in the configuration file, it should be able to demodulate the FM signal and allow you to listen to the audio files that are being transmitted. Likewise, if you open one of your standard stereo RTL-SDR FM receiver Simulink models, you should be able to tune to and demodulate the signals with your RTL-SDR!

Although the PiFM software apparently transmits music signals in stereo; i.e. it transmits a stereo FM MPX, we found that radio receivers (both analogue FM radios and RTL-SDR/ Simulink receivers) struggled to decode the left and right information channels successfully. We think this was due to the fact that the pilot tone was not particularly clean, and that the mono and stereo channels were not clearly defined. The NCO PLL in our stereo receiver model could not lock onto the pilot tone when trying to generate the 38kHz tone, which meant that it could not correctly coherently demodulate the stereo channel from its suppressed subcarrier. The result of this was that it sounded very noisy! The scope windows in Figure E.5 compares the MPX signal received from the Pi and the MPX received from an off-the-shelf FM transmitter.

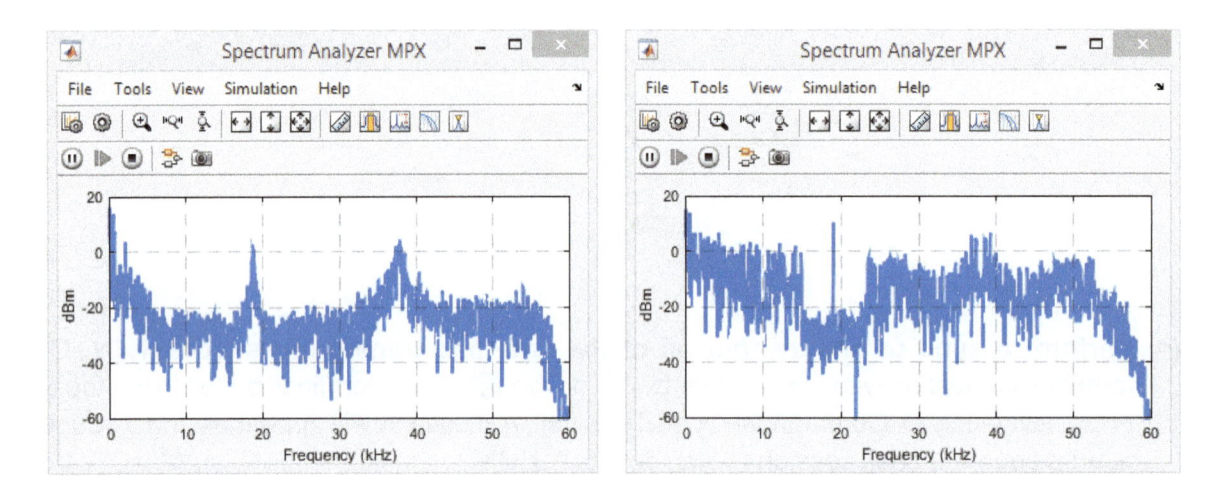

Figure E.5: Comparison: PiFM MPX signal (left), and an off-the-shelf FM transmitter MPX signal (right). Both of these signals were captured with the 'stereo FM radio receiver and decoder' model from Exercise 10.6

The problem may well have been the Pi we were using though — have a shot with your own one and let us know how you get on. Another thing that you will note is that as the signal is generated with a digital clock, there are harmonics of the FM station situated around the carrier. We found the strongest two at $f_c \pm 1.6$ MHz.

While the Raspberry Pi/ PiFM combination does create an FM radio station, we feel that you would better to use a dedicated off-the-shelf FM transmitter or a USRP® radio for the purpose of generating your own clean FM signals. That said, we are still very impressed with the PiFM software and MAKE Labs image, and it is a nice little starter project for people who have never used a Raspberry Pi before.

Appendix E: AM and FM Transmitters

While some of you may wish to leave your SD card set up for PiFM, others may want to restore it to being a standard SD card again. The final exercise in this section guides your through the restoration process.

Exercise E.8 PiFM: Restoring the SD Card

This exercise will guide you through using the USB Imaging Tool to restore the backup image to your Pirate Radio SD card. Administrator privileges are required to run the software.

(a) **Insert the Pirate Radio SD card to your card reader.** Check what drive letter the boot partition of the SD card has been assigned (e.g. 'F:'), and make a note of it.

(b) Open the USB Image Tool software again by 2️⃣ on `USB Image Tool.exe` in the folder you extracted it to. Accept the UAC message that pops up, and check you are in 'Device Mode' as before. Select the SD card by 1️⃣ on the icon of the correct device in the left hand pane. Note that the software thinks the drive is only 90MB in size, and has no name. When you have checked that it is the correct one, 1️⃣ the 'Restore' button.

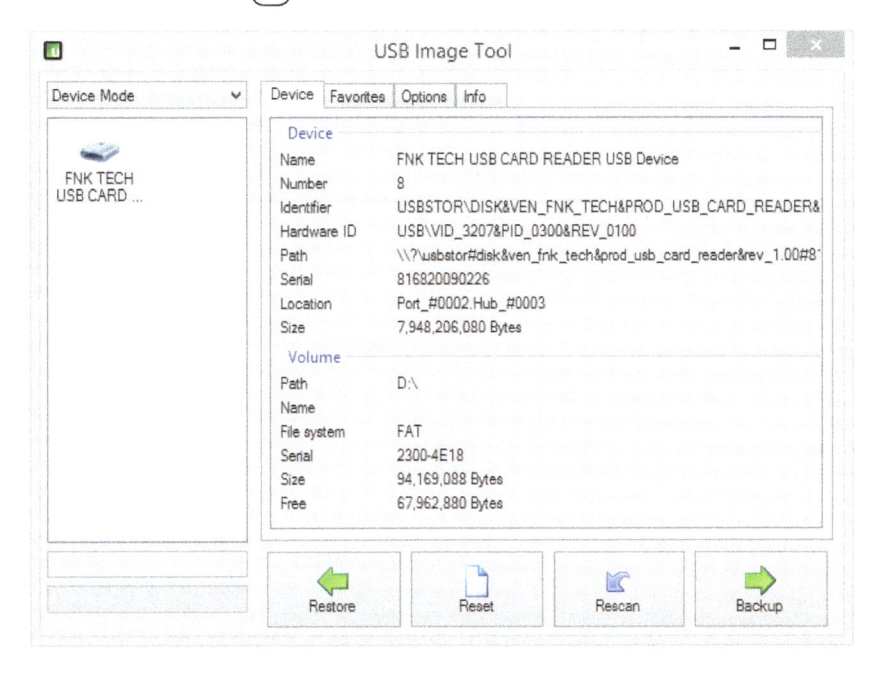

(c) Choose to load the image you created in Exercise E.6. Accept the warning message ONLY IF YOU ARE SURE YOU HAVE SELECTED THE CORRECT DEVICE! Choosing the wrong one will result in data loss.

(d) **Wait a wee while!** When the restoration is complete, the SD card should re-mount itself and re-appear in Windows Explorer. It will be restored to a single full size partition, and is now ready to be used however you want again!

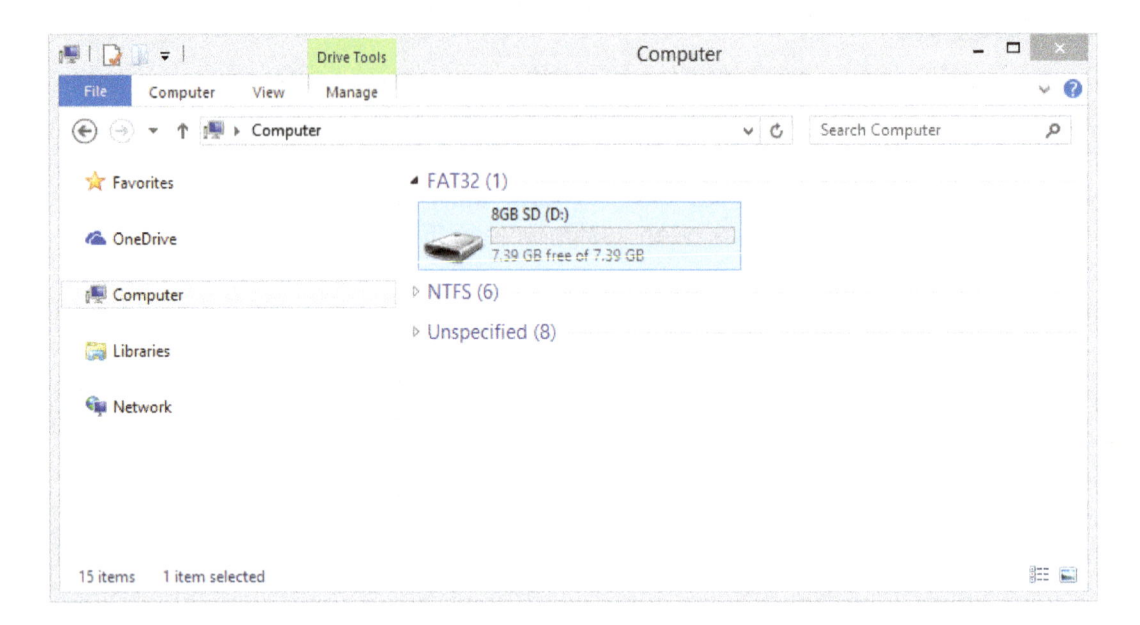

References

Books and Published Works

[1] M. M. Alani, "TCP/IP Model" in *Guide to OSI and TCP/IP Models*, Springer, 2014.

[2] C. A. Balanis, *Antenna Theory: Analysis and Design*, 2nd Edition, New York, John Wiley & Sons, 1997.

[3] R. H. Barker, "Group Synchronizing of Binary Digital Sequences," in *Communication Theory*, London: Butterworth, pp. 273-287, 1953.

[4] A. Bateman, "Quadrature Frequency Discriminator", *GlobalDSP Magazine*, vol. 1, no. 1, pp. 23, October 2002.

[5] A. L. Benson, "Armstrong of the Radio-Phone", *Hearst's International*, vol. 42, no. 5, pp. 90, November 1922.

[6] S. G. Bilen *et al*, "Software-defined radio: a new paradigm for integrated curriculum delivery", *IEEE Commun. Mag.*, vol. 52, no. 5, pp. 184-193, May 2014.

[7] J. J. Carr, *Practical Antenna Handbook*, Fifth Edition, Tab Electronics, 2011.

[8] J. Dunlop & D. G. Smith, "*Telecommunications Engineering*" (Third Edition), Van Nostrand Reinhold International, 1994.

[9] W. F. Egan, *Phase-Lock Basics*, 2nd Edition, Wiley-IEEE Press, November 2007.

[10] G. L. Frost, *Early FM Radio: Incremental Technology in Twentieth Century America*, John Hopkins University Press, 2010.

[11] M. El-Hajjar *et al*, "Demonstrating the practical challenges of wireless communications using USRP," *IEEE Commun. Mag.*, vol. 52, no. 5, pp. 194-201, May 2014.

[12] C. W. Farrow, "A Continuously Variable Fractional Delay Element," *Proc. IEEE Int. Symp. Circuits Syst.*, Vol.3, pp. 2641-2645, Espoo, Finland, Jun 7-9, 1988.

[13] F. M. Gardner, *Phaselock Techniques*, Third Edition, John Wiley & Sons, 2005.

[14] A. K. Ghosh, *Introduction to Control Systems*, Second Edition, Prentice Hall of India, 2014.

[15] J. D. Gibson, *Mobile Communications Handbook*, Third Edition, CRC Press, 2012.

[16] I. A. Glover & P. M. Grant, *Digital Communications*, Third Edition, Prentice Hall, 2009.

[17] L. Goeller, D. Tate, "A Technical Review of Software Defined Radios: Vision, Reality, and Current Status," *IEEE Military Communications Conference (MILCOM)*, pp.1466-1470, 6-8 Oct. 2014.

[18] E. Grayver, *Implementing Software Defined Radio*, Springer, 2012.

[19] A. Haghighat, "A Review on Essentials and Technical Challenges of Software Defined Radio," *IEEE Military Communications Conference*, vol. 1, pp. 377-382, 7-10 Oct, 2002.

[20] f. harris, E. Venosa, X. Chen, and C. Dick, "Band Edge Filters Perform Non Data-Aided Carrier and Timing Synchronisation of Software Defined Radio QAM Receivers", *Proceedings of the 15th International Symposium on Wireless Personal Multimedia Communications (WPMC)*, Taipei, September 2012, pp. 271 - 275.

[21] f. harris, *Multirate Signal Processing for Communication Systems*, Prentice Hall, 2004.

[22] S. Haykin, *Communication Systems*, Third Edition, New York, John Wiley & Sons, 1994.

[23] E. Hogenauer, "An Economical Class of Filters for Decimation and Interpolation," *IEEE Trans. Acoust., Speech, Signal Process*, vol. 29, pp. 155-162, April 1981.

[24] *IEEE Standard Definitions of Terms for Antennas*, IEEE Standard 145-1993, 2004.

[25] J. D. Kraus & R. J. Marhefka, *Antennas for All Applications*, Third Edition, New York, McGraw-Hill, 2003.

[26] V. F. Kroupa, *Phase Lock Loops and Frequency Synthesis*, John Wiley & Sons, 2003.

[27] A. Y. Kwentus, Z, Jiang, and A. N. Wilson, Jnr, "Application of Filter Sharpening to Cascaded Integrator-Comb Decimation Filters", IEEE Transactions on Signal Processing, vol. 45, no. 2, February 1997, pp. 457 - 467.

[28] L. Lessing, *Man of High Fidelity: Edwin Howard Armstrong*, Bantam Books Inc., 2nd Edition, 1969.

[29] R. Lyons, *Understanding Digital Signal Processing*, Prentice Hall, 3rd Edition, 2010.

[30] U. Mengali and A. N. D'Andrea, *Synchronization Techniques for Digital Receivers*, Plenum Publishers, 1997.

[31] H. Meyr, M. Moeneclaey and S. A. Fechtel, *Digital Communication Receivers: Synchronization, Channel Estimation, and Signal Processing*, John Wiley & Sons, 1998.

[32] J. Mitola, "The Software Radio Architecture," *IEEE Commun. Mag.*, vol. 33, no. 5, pp. 26-38, May 1995.

[33] J. Mitola *et al.*, "Guest Editorial on Software Radios," *IEEE J. Sel. Areas Commun.*, vol. 17, no. 4, pp. 509-512, April 1999.

[34] H. J. Oh, S. Kim, G, Choi, and Y. H. Lee, "On the Use of Interpolated Second-Order Polynomials for Efficient Filter Design in Programmable Downconversion", IEEE Journal on Selected Areas in Communications, vol. 17, no. 4, April 1999, pp 551 - 560.

[35] *Open Systems Interconnection – Model and Notation*, ITU-T Recommendation X.200, 1994.

[36] J. G. Proakis, M. Salehi, *Communication Systems Engineering*, Prentice Hall, 2002.

[37] D. Pu, A. M. Wyglinski, *Digital Communication Systems Engineering with Software-Defined Radio*, Artech House, 2013.

[38] M. Rice, *Digital Communications: A Discrete-Time Approach*, Prentice Hall, 2009.

[39] T. J. Rouphael, *RF and Digital Signal Processing for Software-Defined Radio: A Multi-Standard Multi-Mode Approach*, Newnes, 2009.

[40] B. Sklar, *Digital Communications: Fundamentals and Applications*, 2nd Edition, Prentice Hall, 2001.

[41] R. Turyn, J. Storer, "On binary sequences," *Proc. Amer. Math. Soc.*, vol. 12, pp. 394-399, June 1961.

[42] W. H. W. Tuttlebee, (Editor), *Software Defined Radio: Enabling Technologies*, John Wiley & Sons, 2002.

[43] D. R. Stephens, *Phase-Locked Loops for Wireless Communications: Digital, Analog, and Optical Implementations*, 2nd Edition, Springer, March 2013.

[44] P. P. Vaidyanathanm, *Multirate Systems and Filter Banks*, Prentice Hall, 1992.

Online Lecture Notes and Tutorials
Note: URLs last referenced: August 2015.

[45]   L. Der, "*Frequency Modulation (FM) Tutorial*", Silicon Laboratories Inc.
Available: http://www.silabs.com/Marcom%20Documents/Resources/FMTutorial.pdf

[46]   EECS Division, "*EE123 Digital Signal Processing Notes: The RTL-SDR*", University of California, Fall 2012.
Available: https://inst.eecs.berkeley.edu/~ee123/fa12/rtl_sdr.html

[47]   U. Ibrahim, "*EE5765 - FM Demodulation with the PLL*", University of Minnesota (Duluth)
Available: http://www.d.umn.edu/~ibra0130/a1-10b.pdf

[48]   R. Fitzpatrick, "*Antenna Directivity and Effective Area*", University of Texas, Austin, June 2014
Available: http://farside.ph.utexas.edu/teaching/jk1/lectures/node105.html

[49]   S. A. Tretter, "*ENEE428 Slides Chapter 5 - Amplitude Modulation*", University of Maryland, October 2014
Available: http://www.ece.umd.edu/~tretter/commlab/c6713slides/ch5.pdf

[50]   S. A. Tretter, "*ENEE428 Slides Chapter 6 - Double Sideband Suppressed Carrier Amplitude Modulation*", University of Maryland, April 2013
Available: http://www.ece.umd.edu/~tretter/commlab/c6713slides/ch6.pdf

[51]   S. A. Tretter, "*ENEE428 Slides Chapter 7 - Single Sideband Modulation and Frequency Translation*", University of Maryland, November 2014
Available: http://www.ece.umd.edu/~tretter/commlab/c6713slides/ch7.pdf

[52]   S. A. Tretter, "*ENEE428 Slides Chapter 8 - Frequency Modulation*", University of Maryland, February 2013
Available: http://www.ece.umd.edu/~tretter/commlab/c6713slides/ch8.pdf

[53]   A. Wilkinson, "*Vestigial-Sideband Modulation (VSB)*", University of Cape Town, April 2014.
Available: http://local.eleceng.uct.ac.za/courses/EEE3086F/notes/510-AM_VSB_2up.pdf

MathWorks and NooElec Website References
Note: URLs last referenced: August 2015.

[54]   MathWorks Inc., "*Communications System Toolbox*" web page
Available: http://www.mathworks.com/products/communications

[55]   MathWorks Inc., "*DSP System Toolbox*" web page
Available: http://www.mathworks.com/products/dsp-system

[56]   MathWorks Inc., "*MATLAB Home*" web page
Available: http://www.mathworks.com/products/matlab-home

[57]   MathWorks Inc., "*MATLAB for Student Use*" web page
Available: http://www.mathworks.com/academia/student_version/

[58]   MathWorks Inc., "*QPSK Transmitter and Receiver*" web page
Available: http://mathworks.com/help/comm/examples/qpsk-transmitter-and-receiver-1.html

[59]   MathWorks Inc., "*RTL-SDR Support from Communications System Toolbox*" web page
Available: http://www.mathworks.com/hardware-support/rtl-sdr.html

[60]   MathWorks Inc., "*RTL-SDR Hardware Support Package: Common Problems and Fixes*" web page
Available: http://www.mathworks.com/help/supportpkg/rtlsdrradio/ug/common-problems-and-fixes.html

[61]   MathWorks Inc., "*RTL-SDR Hardware Support Package: Configure Hardware*" web page
Available: http://www.mathworks.com/help/supportpkg/rtlsdrradio/ug/support-package-hardware-setup.html

[62] MathWorks Inc., "*RTL-SDR Hardware Support Package: Configure Multiple Radios*" web page
Available: http://www.mathworks.com/help/supportpkg/rtlsdrradio/ug/configure-multiple-radios.html

[63] MathWorks Inc., "*Signal Processing Toolbox*" web page
Available: http://www.mathworks.com/products/signal

[64] MathWorks Inc., "*Supported and Compatible Compilers (MAC) – Release R2014b*" web page
Available: http://www.mathworks.com/support/compilers/R2014b/index.html?sec=maci64

[65] MathWorks Inc., "*What Are System Objects?*" web page
Available: http://www.mathworks.com/help/comm/gs/what-are-system-objects.html

[66] MathWorks Inc., "*USRP Hardware Support Package from Communications System Toolbox*" web page
Available: http://www.mathworks.co.uk/hardware-support/usrp.html

[67] MathWorks Inc., "*USRP Hardware Support Package: Common Problems and Fixes*" web page
Available: http://www.mathworks.com/help/supportpkg/usrpradio/ug/common-problems-and-fixes.html

[68] MathWorks Inc., "*USRP Hardware Support Package: Configure Hardware*" web page
Available: http://www.mathworks.com/help/supportpkg/usrpradio/ug/support-package-hardware-setup.html

[69] NooElec Inc., "*NESDR XTR Tiny SDR & DVB-T USB Stick (RTL2832U + E4000)*" web page
Available: http://www.nooelec.com/store/sdr/sdr-receivers/nesdr-xtr-rtl2832u-e4000.html

[70] NooElec Inc., "*Ham It Up v1.2 - RF Upconverter for Software Defined Radio*" web page
Available: http://www.nooelec.com/store/ham-it-up.html

[71] NooElec Inc., "*NESDR Mini SDR & DVB-T USB Stick (RTL2832U + R820T)*" web page
Available: http://www.nooelec.com/store/sdr/sdr-receivers/nesdr-mini-rtl2832-r820t.html

[72] NooElec Inc., "*Software Defined Radio, SDR Adapters & Cables*" web page
Available: http://www.nooelec.com/store/sdr/sdr-adapters-and-cables.html

Other Website References

Note: URLs last referenced: August 2015.

[73] 3GPP (3rd Generation Partnership Project international standards organisational body), "*LTE-Advanced*"
Available: http://www.3gpp.org/technologies/keywords-acronyms/97-lte-advanced

[74] 4GUK, "*4G LTE Advanced - What you need to know about LTE-A*"
Available: http://www.4g.co.uk/4g-lte-advanced

[75] Apple Support, "*OS X Yosemite: Enable Right Click*"
Available: https://support.apple.com/kb/PH18768

[76] Amazon, "*Belkin TuneCast Auto 4*"
Available: http://www.amazon.co.uk/Belkin-TuneCast-Auto-iPod-iPhone/dp/B001ILDK6G

[77] Cellmapper, "*ARFCN Frequency Calculator*", Cellmapper.net
Available: https://www.cellmapper.net/arfcn

[78] CircuitLab (online circuit realisation software) homepage.
Available: http://www.circuitlab.com

[79] Columbia University, Living Legacies, "*Edwin Armstrong: Pioneer of the Airwaves*", 2002.
Available: http://www.columbia.edu/cu/alumni/Magazine/Spring2002/Armstrong.html

[80] ELinux.org, "*RPi Easy SD Card Setup*", 2014.
Available: http://elinux.org/RPi_Easy_SD_Card_Setup

References

[81] Elonics, "*E4000 Datasheet: Multi-Standard CMOS Terrestrial RF Tuner, Version 4*", August 2010.
Available: http://www.nooelec.com/files/e4000datasheet.pdf

[82] ETSI (European Telecommunications Standards Institute), "*Mobile Technologies GSM*".
Available: http://www.etsi.org/index.php/technologies-clusters/technologies/mobile/gsm

[83] ETSI (European Telecommunications Standards Institute), "*Long Term Evolution*".
Available: http://www.etsi.org/technologies-clusters/technologies/mobile/long-term-evolution

[84] ETSI (European Telecommunications Standards Institute), "*Short Range Devices*".
Available: http://www.etsi.org/technologies-clusters/technologies/radio/short-range-devices

[85] ETSI (European Telecommunications Standards Institute), "*UMTS*".
Available: http://www.etsi.org/technologies-clusters/technologies/mobile/umts

[86] Ettus Research, "*Ettus Research Products*".
Available: https://www.ettus.com/product

[87] Ettus Research, "*USRP N210 and Compatible Products*".
Available: https://www.ettus.com/product/details/UN210-KIT

[88] Farnell Element14 (electronic components supplies company) homepage.
Available: http://www.farnell.com

[89] The First Electronic Church of America, "*Edwin Armstrong: The creator of FM radio*".
Available: http://fecha.org/armstrong.htm

[90] S. Fybush, "*The Birthplace of FM Broadcasting, Alpine, NJ*", December 2002.
Available: http://www.fybush.com/sites/2005/site-050610.html

[91] GGToshi, "*RTL2832U and R820T Circuit Diagrams*", 閑人 Blog, June 2014.
Available: http://ggtoshi.at.webry.info/201406/article_6.html (a translated copy is available at http://translate.google.co.uk/translate?hl=&sl=ja&tl=en&u=ggtoshi.at.webry.info/201406/article_6.html)

[92] D. Hall, "*RTLSDR - RTLAMR Help*", GitHub, August 2014.
Available: https://github.com/bemasher/rtlamr/blob/master/help.md

[93] (last modified by) Hello1024, "*Turning the Raspberry Pi Into an FM Transmitter*", The Robotics Society, Imperial College London, January 2014.
Available: http://www.icrobotics.co.uk/wiki/index.php/Turning_the_Raspberry_Pi_Into_an_FM_Transmitter

[94] IEEE, "*IEEE 802.11 Wireless LANs*", The IEEE Standards Association
Available: http://standards.ieee.org/about/get/802/802.11.html

[95] W. Kadman, "*R820T Rafael Micro - High Performance Low Power Advanced Digital TV Silicon Tuner*", Radioaficion.com, March 2013.
Available: http://radioaficion.com/cms/r820t-rafael-micro

[96] R. Lyons, "*Understanding the Phasing Method of Single Sideband Demodulation*", DSP Related, August 2012.
Available: http://www.dsprelated.com/showarticle/176.php

[97] MAKE: Projects, "*Raspberry Pirate Radio*", Vol. 38, March 2014.
Available: http://makezine.com/projects/raspberry-pirate-radio

[98] S. Markgraf (email re-posted by A. Nielsen), "*R820T Tuner Support in librtlsdr*", Osmocom, September 2012.
Available: https://groups.google.com/forum/#%21msg/ultra-cheap-sdr/4oVYR34jqgg/Ybz2AVA0evoJ

[99] K. Moskvitch, "*Mars to Earth: How to send HD video between planets*", BBC News, November 2012.
Available: http://www.bbc.co.uk/news/technology-19950183

[100] NXP Semiconductor, "*TEF6901A Integrated Car Radio (single chip, FM/AM) Data Sheet*", March 2008
Available: http://www.nxp.com/documents/data_sheet/TEF6901A.pdf

[101] Ofcom, "*The United Kingdom Frequency Allocation Table*", 2014
Available: http://stakeholders.ofcom.org.uk/spectrum/information/uk-fat

[102] Ofcom, "*Laying the foundations for '5G' mobile*", January 2015
Available: http://media.ofcom.org.uk/news/2015/6ghz

[103] Opendous, "*Ham It Up Upconverter Design Details*"
Available: https://code.google.com/p/opendous/wiki/Upconverter_Design_Details

[104] A. Palosaari, "*Linux-media@vger.kernel.org — Linux video input infrastructure development discussion*", GMANE Developer Forums, February 2012
Available: http://comments.gmane.org/gmane.linux.drivers.video-input-infrastructure/44461

[105] A. Palosaari, "*Naked Hardware #6 - Ezcap USB 2.0 DVB-T Stick*", Antti's LinuxTV Blog website, November 2012
Available: http://blog.palosaari.fi/2012_11_01_archive.html

[106] I. Poole, "*FM Slope Detector*", Radio-Electronics.com
Available: http://www.radio-electronics.com/info/rf-technology-design/fm-reception/fm-slope-detector-discriminator.php

[107] I. Poole, "*PLL FM demodulator / detector*", Radio-Electronics.com
Available: http://www.radio-electronics.com/info/rf-technology-design/fm-reception/fm-pll-detector-demodulator-demodulation.php

[108] RadioShack (homepage)
Available: http://www.radioshack.com

[109] RealTek, "*RTL2832U DVB-T COFDM Demodulator + USB 2.0*".
Available: http://www.realtek.com.tw/products/productsView.aspx?Langid=1&PFid=35&Level=4&Conn=3&ProdID=257

[110] RF Solutions, "*AM Hybrid Transmitter (RT4, RT5, RTQ4, RT14)*", June 2008.
Available: http://www.rfsolutions.co.uk/acatalog/DS013-9%20AM-RTx.pdf

[111] RTL-SDR.com, "*About RTL-SDR*", RTL-SDR.com website, June 2015.
Available: http://www.rtl-sdr.com/about-rtl-sdr/

[112] D. Rudolph, "*FM Pre-Emphasis and De-Emphasis*", Radio Museum, January 2011.
Available: http://www.radiomuseum.org/forum/fm_pre_emphasis_and_de_emphasis.html

[113] (last modified by) Sylvain, "*RTL-SDR: Specifications*", Osmocom SDR website, July 2014.
Available: http://sdr.osmocom.org/trac/wiki/rtl-sdr

[114] SDR Forum, "*SDRF Cognitive Radio Definitions*", November 2007.
Available: http://www.sdrforum.org/pages/documentLibrary/documents/SDRF-06-R-0011-V1_0_0.pdf

[115] UK Free TV, "*Full service Freeview tranmitters*"
Available: http://www.ukfree.tv/maps/freeview

[116] H. Welte, "*RTL-SDR - Presentation given at FreedomHEC 2012 Taipei*", June 2012.
Available: http://sdr.osmocom.org/trac/attachment/wiki/rtl-sdr/rtl-sdr.2.pdf

[117] A. Wolke, "*Build, test, use the RF Noise Source on the Ham-It-Up RTL-SDR Upconverter*", 14th October 2014.
Available: http://youtu.be/HY1ijZbLfb8

[118] A. Wolke, "*Filter functions in an HF Upconverter used with RTL-SDR Dongle Receiver*", 1st October 2014.
Available: http://youtu.be/a4p8XInX5ZA

[119] WPS (Wireless Phone Service) Antennas, "*Mobile Phone Field Test Modes*".
Available: http://www.wpsantennas.com/pdf/testmode/FieldTestModes.pdf

List of Acronyms

A

ADC	Analogue to Digital Converter
AGC	Automatic Gain Control / Active Gain Control
AMPS	Advanced Mobile Phone System — 1G mobile phone standard for basic voice calls
AM	Amplitude Modulation
AM-DSB-SC	AM Double Sideband Suppressed Carrier
AM-DSB-TC	AM Double Sideband Transmitted Carrier
AM-SSB	AM Single Sideband
AM-VSB	AM Vestigial Sideband
ARFCN	Absolute Radio Frequency Channel Number
ARP	Address Resolution Protocol
ASCII	American Standard Code for Information Interchange
AWGN	Additive White Gaussian Noise

B

b	Bit (single data unit)
B	Byte (data unit of 8 bits)
BER	Bit Error Rate
BPF	Bandpass Filter
BPSK	Binary Phase Shift Keying

C

COFDM	Coded Orthogonal Frequency Division Multiplex

D

DAB	Digital Audio Broadcast
DAC	Digital to Analogue Converter

dB	Decibel	
dBi	Decibels over the Isotropic	
dBm	Decibels per Milliwatt	
DDC	Direct Digital Downconverter	
DFT	Discrete Fourier Transform	
DQPSK	Differential Quadrature Phase Shift Keying	
DSB	Double Sideband	
DSP	Digital Signal Processing	
DTV	Digital Television	
DVB-T	Digital Video Broadcast — Terrestrial	

E

EEPROM	Electronically Erasable Programmable Read Only Memory
ESD	ElectroStatic Discharge
EVM	Error Vector Magnitude

F

FCC	Federal Communications Commission (USA regulator)
FDM	Frequency Division Multiplexing
FFT	Fast Fourier Transform
FIR	Finite Impulse Response (discrete time filter)
FM	Frequency Modulation
FPGA	Field Programmable Gate Array
FSK	Frequency Shift Keying

G

GbE	Giga-bit Ethernet
GPIO	General Purpose Input Output
GPRS	General Packet Radio Service — Enhancement to 2G (known as 2.5G) that facilitated sending short data messages
GPS	Global Positioning System — Satellite based navigation system
GSM	Global System for Mobile Communications — 2G mobile phone standard optimised for speech
GUI	Graphical User Interface

H

HF	High Frequency radio waves (between 3 and 30MHz)
HIU	Ham It Up (upconverter that can use used with the RTL-SDR to receive AM radio signals)
HPF	Highpass Filter
HTTP	Hypertext Transfer Protocol
Hz	Hertz

List of Acronyms

I

I	In Phase component (see also *Re*)
I²C	Inter-Integrated Circuit (interface)
IC	Integrated Circuit
IDFT	Inverse Discrete Fourier Transform
IEEE	Institute of Electrical and Electronics Engineers
IF	Intermediate Frequency
IFFT	Inverse Fast Fourier Transform
IIR	Infinite Impulse Response (discrete time filter)
Im	Imaginary part of a complex signal (presented as $\Im m$ in equations, see also Q)
IoT	Internet of Things
IP	Internet Protocol
IQ	In Phase/ Quadrature complex signal
IR	Infra Red
ISI	Inter Symbol Interference
ISM	Industrial, Scientific and Medical radio frequency band
ITU	International Telecommunications Union

L

LF	Low Frequency radio waves (between 30kHz and 300kHz)
LO	Local Oscillator
LPF	Lowpass Filter
LSB	Lower Sideband
LSB	Least Significant Bit
LTE	Long Term Evolution — 4G mobile phone standard optimised for data messages
LTE-A	Long Term Evolution Advanced — '4.5G' mobile phone standard optimised for large data messages

M

MCX	Micro Coaxial (connector)
MF	Medium Frequency radio waves (between 300kHz and 3MHz)
MIMO	Multiple Input Multiple Output
MMS	Multimedia Messaging Service
mp4	Video container format (shorthand for *MPEG-4 Part 14*)
MPEG2-TS	Moving Picture Experts Group Transmission Stream
MPX	Multiplex
MSB	Most Significant Bit
MTU	Maximum Transmission Unit

N

NCC	Numerically Controlled Clock
NCO	Numerically Controlled Oscillator
NFC	Near Field Communications
NFM	Narrowband FM
NIC	Network Interface Card

O

Ofcom	Office of Communications (UK regulator)
OFDM	Orthogonal Frequency Division Multiplexing
OFDMA	Orthogonal Frequency Division Multiple Access
OOK	On-Off Keying
OSI	Open Systems Interconnection

P

-ps	Per Second, e.g. 5Mbps
PAL	Phase Alternating Line (colour TV line encoding scheme)
PCB	Printed Circuit Board
PHY	Physical Layer (cables/ RF etc.)
PLL	Phase Locked Loop
PPM	Parts Per Million
PPTP	Point to Point Tunnelling Protocol
PSK	Phase Shift Keying
PWM	Pulse Width Modulation

Q

Q	Quadrature Phase component (see also *Im*)
QAM	Quadrature Amplitude Modulation
QPSK	Quadrature Phase Shift Keying

R

RC	Raised Cosine
RDS	Radio Data Service
Re	Real part of a complex signal (presented as $\mathfrak{Re}$ in equations, see also *I*)
RF	Radio Frequency — generic term for the part of the electromagnetic spectrum in the range 3kHz to 300GHz
RRC	Root Raised Cosine
Rx	Signal Receiving (shorthand form)

List of Acronyms

S

SC	Suppressed carrier	
SDR	Software Defined Radio	
SECAM	Sequential Color with Memory (colour TV line encoding scheme)	
SMA	SubMiniature version A (connector)	
SMS	Short Message Service	
SNR	Signal to Noise Ratio	
SRD	Short Range Device(s)	
SSB	Single Sideband	
SSH	Secure Shell	

T

TACS	Total Access Communications System — 1G mobile phone standard for basic voice calls	
TC	Transmitted Carrier	
TCP	Transmission Control Protocol	
TDM	Time Division Multiplexing	
TDMA	Time Division Multiple Access	
TED	Timing Error Detector	
TV	Television	
Tx	Signal Transmission (shorthand form)	

U

UDP	User Datagram Protocol	
UHD™	USRP® Hardware Driver	
UHF	Ultra High Frequency band (300MHz to 3GHz)	
UMTS	Universal Mobile Telecommunications Service — 3G mobile phone standard optimised for data messages	
USB	Upper Sideband	
USB	Universal Serial Bus	
USRP®	Universal Software Radio Peripheral — Ettus Research SDR	
UW	Unique Word	

V

V4L	Video-4-Linux	
VCC	Voltage Controlled Clock	
VCO	Voltage Controlled Oscillator	
VHF	Very High Frequency radio waves (between 30MHz and 300MHz)	
VLF	Very Low Frequency radio waves (between 3kHz and 30kHz)	
VSB	Vestigal Sideband	

W

WAP	Wireless Application Protocol
WCDMA	Wideband Code Division Multiple Access
WFM	Wideband FM

X

| XOR | Exclusive OR binary operation |

Y

| Yagi | Yagi-Uda Antenna — directional linearly polarised antenna |

Z

| ZOH | Zero Order Hold |

#

1G	1st Generation (mobile)
2G	2nd Generation (mobile)
3G	3rd Generation (mobile)
4G	4th Generation (mobile)
5G	5th Generation (mobile)
802.11	IEEE WiFi standard

SI Prefixes

k	kilo (10^3) – e.g. kHz, 1000Hz
M	Mega (10^6) – e.g. MHz, 1000000Hz
G	Giga (10^9) – e.g. GHz, 1000000000Hz

Index

Symbols

A

Active Gain Control (AGC) 265, 268, 605
Amplitude Modulation (AM)
    demodulation
        coherent 230, 273–277
        non-coherent 231–234, 290
        theoretical 227, 241–245
        See also Demodulation
    generating signals 279, 326
        See also Transmitter
    modulation
        double sideband suppressed carrier (DSB-SC) 203–209, 242, 273, 276
        double sideband transmitted carrier (DSB-TC) 210–216, 273
        modulation index 211
        single sideband (SSB) 217–224
        vestigial sideband (VSB) 225–226
Analogue modulation schemes
    See also Amplitude Modulation (AM)
    See also Frequency Modulation (FM)
Antennas 46–49
    adaptors 48
    directivity 47
    gain 46
    polarisation 47
    tuned 47

B

Binary Phase Shift Keying (BPSK) 428
    differential encoding 489–493

C

Carrier frequency 240
Carrier frequency offset 481–487
Carrier phase 239
Carrier synchronisation 448–458
    baseband 455–458
    demodulation stage 450–455
    phase ambiguity 488–489
Cascade-Integrator-Comb (CIC) filter 446
Channel 238
    frequency shift 239
    propagation delay 239
Channelisation
    See also Multiplexing
Clock oscillator 241
Coarse frequency synchronisation 480–487, 512–515
Complex notation
    complex baseband 171
    quadrature amplitude demodulation 192
    quadrature amplitude modulation 191
    quadrature signals 182

D

Decimation 443–448, 588–590
    See also Multirate filtering
Demodulation 45, 235–238, 242–245, 449, 450
    AM coherent 230, 241–245, 310
        carrier synchroniser 227, 273–277
        Costas loop 276, 314–316
        PLL 273, 311–313
    AM non-coherent 241, 290
        complex envelope detector 233, 292–306
        complex sine waves 306–310
        envelope detector 231–233

FM coherent
 PLL 358–361, 405, 405–407
FM non-coherent
 complex differentiator 350–354, 388
 complex frequency discriminator 355–358, 376, 384, 556
 slope detector 407–409
FM stereo demultiplexer/ decoder 396
frequency offset 242–245
QPSK 428
 carrier synchronisation 450–458, 477–480
 coarse frequency synchronisation 480–487
 symbol demapping 430–436
 symbol timing synchronisation 458–477

Differential encoding 489–498
 BPSK 489–493
 QPSK 493–498
Digital Downconverter (DDC) 442–448
Digital modulation schemes 427–432
 phase modulation 428
 symbols 430, 431
 See also Binary Phase Shift Keying (BPSK)
 See also On-Off Keying (OOK)
 See also Quadrature Amplitude Modulation (QAM)
 See also Quaternary Phase Shift Keying (QPSK)
Digital Upconverter (DUC) 442–448
Doppler shift 239, 241, 258, 459

E

Early Late
 synchroniser 472–477
Early-Late
 Timing Error Detector (TED) 465, 470–472
Encoding Schemes
 ASCII 525
 differential encoding 489
Error Vector Magnitude (EVM) 435
Euler's Formula 172, 582
Exercises
 icons 22
 licence agreement ii
 mouse clicks 23
 RTL-SDR Book library 31
 structure and format 22
 support files i, 31
Eyeball tuning 54–57, 237

F

Field Programmable Gate Array (FPGA) 280, 446
Filter 583

specifications 584–585
trade-offs 586
See also Multirate filtering
Frame Synchronisation
 Barker sequences 532
 frame structure 529
 synchronisation approaches
  matched filter 539
Frequency correction
 complex exponential 455
Frequency Modulation (FM) 329
 demodulation
  coherent 358–361
  non-coherent 350–358
  theoretical 347
  See also Demodulation
 generating signals 367, 423–425
  See also Transmitter
 history 329–330
 modulation
  bandwidth 335, 340, 342
  baseband modulation 368
  Bessel function 341, 342
  frequency deviation 335, 340, 406
  modulation index 332, 335
  narrowband (NFM) 335–339
  VCO modulator 331
  wideband (WFM) 340–347
 pre-emphasis & de-emphasis filtering 361
 stereo multiplex 363–366
  demultiplexer/ decoder 364, 396
  encoder/ multiplexer 363, 365, 390–396
  manipulation 409, 415

H

Hardware support package 6
 RTL-SDR 17–19, 27–30
 USRP 280, 571–576

I

Interpolation 443–448, 590–591
 See also Multirate filtering
Inter-Symbol- Interference (ISI) 438

L

Local Oscillator (LO) 241

M

Matched filtering 439, 446
MATLAB

Index

.mat file 161, 163
arrays 116–118
command history 103
figures 111–115
functions 107–111
help browser 103, 108
introduction 100
matrices 118–120
script 105–107
set path 31
structures 120–121
system object 121–124
system objects 124
Workspace 104
See also Hardware support package
Maximum effect points 458–460
Modulation 45, 238, 242–245
See also Amplitude Modulation (AM)
See also Binary Phase Shift Keying (BPSK)
See also Frequency Modulation (FM)
See also On-Off Keying (OOK)
See also Quadrature Amplitude Modulation (QAM)
See also Quaternary Phase Shift Keying (QPSK)
Multiplexing 316–326, 363–366, 409, 415
See also Frequency Modulation (FM)
Multirate filtering 443–448, 587–591
CIC Compensation Filter (CFIR) 446
CIC filter 446
decimation 588–590
interpolation 590–591

N

Numerically Controlled Clock (NCC) 464, 467–470, 473
Numerically Controlled Oscillator (NCO) 248
frequency control 250
gain 249
operation 250
quiescent frequency 248, 249
signal flow graph 251
step size 250
use in carrier synchroniser 450
Nyquist sampling theorem 587

O

On-Off Keying (OOK) 556
Oscillator
carrier 237
clock 241
frequency offset compensation 237
tolerance 237, 240
OSI Stack 524

P

Phase ambiguity 488–489
See also Differential Encoding
Unique Word (UW) 499–502
Phase detector 246
ideal 246
multiplier 246, 247
s-curve 246
Phase Locked Loop (PLL) 235, 245
architecture 245
bandwidth 256–257, 261–263
characteristics 256
damping ratio 257, 259, 261, 263
design 256
discrete time model 252
linear model 258, 259, 261, 263, 593
loop filter 245, 248, 251, 253, 254, 257, 603
oscillator 245
oscillator gain 605
parameters 256
performance in noise 261, 272
phase detector 245, 246, 253, 259, 602
gain 604–605
ideal phase detector 265, 602, 603
multiplier phase detector 265, 603
phase error 259
PLL design 264
damping ratio 266
example 268, 270
loop filter 266
NCO gain 265
noise bandwidth 266, 272
phase detector gain 265
sample period 265
s-curve 603, 604
S-domain transfer function 593, 596
signal flow graph 253
steady state error 256–259, 263
time to achieve lock 256–257
tracking 256, 260–261
transient behaviour 260
type 248, 252, 258–260, 264
type 2 255, 257, 260, 593
Z-domain transfer function 593
Pulse shaping 437–442, 446
Inter-Symbol- Interference (ISI) 438
raised cosine 438, 504–512
root raised cosine 439
spectral mask 437

Q

Quadrature Amplitude Modulation (QAM) 428–430

Quaternary Phase Shift Keying (QPSK) 427–436, 488
    differential encoding 493–498

R

Radio Frequency (RF) 43
    bands 44, 45
    signals 43
        DAB 81–83, 93
        DTV 79–81, 93
        FM radio 60–63, 93
        GPS 239, 258
        mobile (cell) phones 64–74, 93
        satellites 93
        SRDs and car key fobs 75–79, 93
    spectrum 43
        regulation 45
        sweep 89–97
Raised Cosine (RC) 438
Root Raised Cosine (RRC) 439, 446
RTL-SDR
    a short history 12
    frequency offset 487
    hardware 10–17
        E4000 tuner 15, 96, 236
        main components 16
        R820T tuner 14, 93, 236
        R820T2 tuner 14, 236
        RTL2832U demodulator 10, 14, 236
        VCO 17
    MATLAB system object 17
    parameters
        sampling frequency 15, 17, 54, 236
        tuner gain 17, 59
    saving and re-importing signals 165–169
    Simulink block 17, 236
    sweeping the spectrum with 89–97
    tuning 14–15, 17, 54–57, 235–237
    using multiple devices 83–88
    See also Hardware support package

S

Signal
    analytic 172
    complex 172
Simulink
    commenting of blocks 141
    data types 159–161
    input and output files 161–165
    introduction 125
    library browser 125
    MATLAB path 31
    RTL-SDR block 236

    samples and frames 156–159
    spectrum analyzer 52
    USRP block 572
    See also Hardware support package
Software Defined Radio (SDR)
    architectures 8–10
    definition 1
    evolution 8
    ultimate 8
    See also RTL-SDR
Spectral mask 437
Spectrum 43
    See also Radio Frequency (RF)
Symbol constellation
    QPSK 449, 451
Symbol timing synchronisation 451, 458–460, 464–477
    Early-Late Timing Error Detector (TED) 470–477
    maximum effect points 458–460
    timing adjustment 465–467
    Timing Error Detector (TED) 465
    timing phase and frequency errors 459–462
    timing phase error 459
Synchronisation 238
    carrier 241, 273
    coarse frequency 480–487
    coarse frequency synchronisation 477
    joint carrier and timing 477–480
    See also Tuning
System object 121–124
    comm.SDRRTLReceiver 237, 487
    See MATLAB 17

T

Timing Error Detector (TED) 465
    early-late 465
Transmitter
    AM
        AM-DSB-SC 281–284
        AM-DSB-TC 284–286, 327
        AM-SSB 286–290
        Ham It Up upconverter 279, 326, 607, 609
        RT4 327, 612
    FM
        off-the-shelf 367, 423, 556
        Raspberry Pi 423, 556, 619
        WFM mono 369–376
        WFM stereo 390–396
    QPSK 504
    USRP 280, 368, 504
Trigonometric Rules 581
Troubleshooting
    See www.desktopSDR.com

Tuning 235–237
    See also Eyeball tuning
    See also RTL-SDR Tuner

U

Unique Word (UW) 499–502
USRP
    Bus series 573
    Network series 574
    overview 280
    parameters
        gain 572
        sampling frequency 572
    RF daughtercard 280, 576
    Simulink block 572
    See also Hardware support package

V

Voltage Controlled Oscillator (VCO) 248, 331
    gain 249
    quiescent frequency 248
    See also Frequency Modulation (FM)

Lightning Source UK Ltd.
Milton Keynes UK
UKHW051936050821
388292UK00002B/53

9 780992 978723